Lecture Notes in Networks and Systems

Volume 345

The series "Lecture Notes in Networks and Systems" publishes the latest developments in Networks and Systems—quickly, informally and with high quality. Original research reported in proceedings and post-proceedings represents the core of LNNS.

Volumes published in LNNS embrace all aspects and subfields of, as well as new challenges in, Networks and Systems.

The series contains proceedings and edited volumes in systems and networks, spanning the areas of Cyber-Physical Systems, Autonomous Systems, Sensor Networks, Control Systems, Energy Systems, Automotive Systems, Biological Systems, Vehicular Networking and Connected Vehicles, Aerospace Systems, Automation, Manufacturing, Smart Grids, Nonlinear Systems, Power Systems, Robotics, Social Systems, Economic Systems and other. Of particular value to both the contributors and the readership are the short publication timeframe and the world-wide distribution and exposure which enable both a wide and rapid dissemination of research output.

The series covers the theory, applications, and perspectives on the state of the art and future developments relevant to systems and networks, decision making, control, complex processes and related areas, as embedded in the fields of interdisciplinary and applied sciences, engineering, computer science, physics, economics, social, and life sciences, as well as the paradigms and methodologies behind them.

Indexed by SCOPUS, INSPEC, WTI Frankfurt eG, zbMATH, SCImago.

All books published in the series are submitted for consideration in Web of Science.

More information about this series at http://www.springer.com/series/15179

Daria Bylieva · Alfred Nordmann
Editors

Technology, Innovation and Creativity in Digital Society

XXI Professional Culture of the Specialist of the Future

Editors
Daria Bylieva
Department of Social Sciences
Peter the Great St. Petersburg Polytechnic
University
St. Petersburg, Russia

Alfred Nordmann
Institut für Philosophie
Darmstadt Technical University
Darmstadt, Hessen, Germany

ISSN 2367-3370   ISSN 2367-3389 (electronic)
Lecture Notes in Networks and Systems
ISBN 978-3-030-89710-9   ISBN 978-3-030-89708-6 (eBook)
https://doi.org/10.1007/978-3-030-89708-6

This Springer imprint is published by the registered company Springer Nature Switzerland AG
The registered company address is: Gewerbestrasse 11, 6330 Cham, Switzerland

Preface

Today, it is not enough to solve a problem, it has to be done creatively. Creativity is a marker of social distinction with the emergence in a new economy of a "creative class." The language of "construction" has given way to the rhetoric of "design," bridging technical and artistic invention, the spheres of engineering and aesthetics. In the meantime, books about "homo creator" and "homo deus" simultaneously celebrate and worry about the expansion of technological power in such areas as climate engineering, artificial intelligence, bioengineering and genetics. Are we witnessing a "second creation"? And is there a way to reconcile all this with notions of a limited repertoire of forms, of material constraints, of lock-in and path-dependencies and of rational engineering principles? In particular, this raises the problem whether creativity is a subjective feature of individual inventiveness or an objective feature of sociomaterial situations. Can procedures or devices be creative? For engineering education, what is the challenge no longer to train specialists and to convey competencies but to inspire creativity? These questions and considerations call for reflection in the fields of teaching, epistemology, communication, aesthetics and performance studies, value theory and ethics.

This volume cannot address all these aspects of creativity and its meaning in contemporary culture. It presents a selection of papers that were accepted for presentation at the XXI International Conference *Professional Culture of the Specialist of the Future*, October 2021, in St. Petersburg. These papers take their cue from the central proposition that to educate students, young professionals and future engineers is to develop their capacity for creativity. This ideal for the formation of a digital public culture requires an interdisciplinary understanding of creativity.

Subjective and objective, social and individual, process and product—creativity poses a vast range of questions. In its most general form, our contributors view creativity as a transition from a potential state to an actual one, as a vector of inevitable technological development or as the construction of the future. In the system of social relations, creativity appears as the main source of economic value creation and as a marketable algorithm. This, in turn, produces a demand for professionals capable of creative activity. The formation of creativity becomes a particularly important soft skill in the course of professional education, and the methods of its measurement become an urgent pedagogical task. This is complicated by some of the larger dimensions of creativity. At the individual level, it divides into inherited and acquired abilities. Psychologists note the connection between creativity and mental disorders. Also, the manifestation of creativity by non-human artificial intelligences in the digital age expands the understanding of the phenomenon, gives new facets to art and challenges the existing economic and legal order. The discussion of particular educational approaches, the exploration of digital technologies and the presentation of best practice examples conclude the volume. University teachers show how the teaching of creativity reinforces the teaching of other subjects, especially foreign languages. We thus become witness of the ways in which creativity engenders creativity, how we creatively employ new tools and digital capabilities to encourage and empower creative capacities.

Daria Bylieva
Alfred Nordmann

Contents

Creativity in Technosociety

Creativity in Education

Dimensions of Creativity

Creativity in Engineering - Classics of Modern Dialectical Philosophy Revisited

Alexandra Kazakova[1,2]($\boxtimes$) (iD) and Christopher Coenen[3] (iD)

[1] Gubkin Russian State University, 65 Leninsky Prospekt, Moscow 119991, Russia
[2] Bauman Moscow State Technical University, ul. Baumanskaya 2-ya, 5, Moscow, Russia
[3] Karlsruhe Institute of Technology, P.O. Box 3640, 76021 Karlsruhe, Germany

Abstract. As a source of innovation, scientific and technological creativity has long been praised in modern society as if in a mantra. However, neither the analysis of historical inventions and discoveries nor the manifold efforts to understand and foster creativity offer a guarantee of innovation success in the future, nor has a widely accepted understanding of creativity prevailed in Science and Technology Studies (STS). Against this background, it is appropriate to revisit influential reflections on creativity made by classics of modern dialectic philosophy and make them fruitful for today's situation. From this perspective, alienated labour, although creative by its very nature, appears subjectively as a toil or necessity, and is opposed to pleasure and freedom; but although the true realm of freedom begins beyond the boundaries of work, it depends on it. We suggest that whether it is an engineer in a nineteenth-century factory, an engineer in a mid-twentieth-century office, or their descendants working with computer-aided design, their activity can be studied on the same methodological basis derived from these reflections, and that this opens up new perspectives for STS.

Keywords: Creativity · Engineering · Dialectical philosophy · STS · Labour · Capitalism · Marx · Marxism · Hegel

1 Concepts of Creative Activity in Hegel and Marx

In Hegelian philosophy, the religious dichotomy of Creator and Creation is secularized and sublated (*aufgehoben*) in the concept of the self-development of the *Geist* (Spirit). In Marxian philosophy, it is demystified and humanized. In both, activity is essentially creative; it is the process of self-transformation and the realization of concrete freedom.

Hegel's *Phenomenology* [1] narrates the self-realization of the Spirit unfolding due to its diremption with Nature as object and the overcoming of this opposition through self-reflection. The self-development of the Spirit through the emergence and destruction of cultures and social institutions brings logic to the history of human affairs.

Creation, then, is not an act but a process, the dynamic unity of subject and object. The very essence of the spirit is activity; "it realizes its potentiality – makes itself its own deed, its own work" [2, p. 90]. And since it is human activity through which the Spirit realizes itself, human beings realize themselves in the same way in the sequence of their actions. Activity is not only creative, it is the process of self-creation in which the subject externalizes and affirms itself through objectification. Both in the transformation of nature and in the creation of history, as Kojeve comments, "the Spirit is in reality nothing but the negating (i.e., creative) Action realized by Man in the given World" [3, p. 71] – and, it should be added, towards the self currently given. In the "edifice" of universal history, human beings are at the same time the bricks, the masons who build it, and the architects who conceive the plan - all of them "changing during the construction" [3, p. 32].

The creative nature of human praxis – the unity of thought and action – is upheld in early Marxian anthropology as he develops his notion of the "species being" and elaborates the concept of alienation. The distinguishing feature of our species, in contrast to the immediate animal relationship to nature, is the capacity for "self-duplication" in conscious, purposeful labour that transforms the object world and ourselves: "The importance of Hegel's *Phenomenology* and its final result—the dialectic of negativity as the moving and producing principle—lies in the fact that Hegel conceives of the self-creation of man as a process, objectification as loss of object, as alienation and as supersession of this alienation; that he therefore grasps the nature of labor and conceives objective man—true, because real man—as the result of his own labor" [4, p. 386]. As Joas puts it, in the concepts of labour, activity, production and praxis, Marx "imbues all human creativity with the pathos which in Romanticism and classical philosophy seemed to be reserved for aesthetic creativity or for the mind realizing itself in history" [5, p. 92].

Since praxis as a unity of creation and self-creation is social and socially mediated through tools, organisation, language, etc., the social conditions of alienation and its supersession are to be historically and critically examined. Alienated labour – although creative by its very nature – is subjectively a toil, a necessity, and opposed to pleasure and freedom. Under industrialisation, the fourfold alienation from the product, from the process, from others and from the self reaches its climax for the worker, who consequently feels freely active only in his "animal functions". As long as the object of labour is not an expression of the subject, activity is not self-activity: "Under conditions of private ownership, labour is (…) leaving the creator no chance to recognize himself in what he has created" [5, p. 92].

Creativity is thus usurped by industrial labour in two ways: both historically (in contrast to the skilled handicraft of "semi-artistic worker of the Middle Ages" [6, p. 534]) and within the contemporary social division of labour (in contrast to intellectual – scientific, engineering and especially artistic – activities). Hegel's account of Dutch genre paintings exemplifies a cultural nostalgia for the un-alienated, pre-industrial relationship to the cultivated world that united painters with "townsmen and countrymen" [7, p. 30]. Marx's earlier focus (e.g. in *The German Ideology*) is on the degree of alienation of modern workers, whose labour has "lost all semblance of self-activity". In the *Grundrisse*, however, the relationship between creativity and alienation is more complex. The alienation of exploited labour is traced through the history of class societies, in which it always "appears as repulsive, always as external forced labour; and not-labour, by contrast, as 'freedom, and happiness'" [6, p. 534]. But it is seen as a historically disappearing necessity for the development of the productive forces, leading to the overcoming of "individual labour" and the realisation of its "social, general character" – which, under the present conditions of industrial capitalism, is recognised only in scientific or artistic labour. These activities are only possible at the expense of alienated surplus labour: "In relation to the whole of society, the creation of disposable time is then also creation of time for the production of science, art etc." [6, p. 326].

2 Freedom and Necessity

This does not mean that scientific and artistic activities are themselves fun, amusement or pleasure, however much they are regarded as such by someone like Fourier and are subjectively self-rewarding: "Really free working, e.g. composing, is at the same time precisely the most damned seriousness, the most intense exertion"; but they exemplify an activity in which "the external aims become stripped of the semblance of merely external natural urgencies, and become posited as aims which the individual himself posits – hence as self-realization" [6, p. 534].

Immediate labour-time itself, according to Marx, cannot "remain in the abstract antithesis to free time in which it appears from the perspective of bourgeois economy" [6, p. 631], it cannot become play – again *pace* Fourier, whose great merit, however, remains to have expressed the sublimation "not of distribution, but of the mode of production itself, in a higher form, as the ultimate object" [6, p. 631]. Free time, defined by Marx as both leisure time and time for higher activity, transforms its possessor "into a different subject, and he then enters into the direct production process as this different subject" [6, p. 631]. This process is then, Marx writes, both discipline, as far as the human being in the process of becoming is concerned, and at the same time exercise, experimental science, materially creative and objectifying science, as far as the human being who has become is concerned, in whose head the accumulated knowledge of society exists.

Kosík argues in *Dialectics of the Concrete* that the "idea of free time as organized leisure is entirely foreign to Marx" [8, p. 131] and that "[f]reedom does not disclose itself to man as an autonomous realm, independent of labor and existing beyond the boundaries of necessity" [8, p. 125]. Rather, it grows out of labour as its necessary prerequisite. Kosík explains: "Human doing is not split into two autonomous realms, mutually independent and indifferent, one of which would incarnate freedom and the other constitute the arena of necessity. (...) The splitting of this unified process into two seemingly independent realms does not follow from the 'nature of the 'matter' but is historically a transient state" [8, p. 125]. Only as long as "consciousness is a captive of this split", it will not "behold its historical character and will juxtapose labor and freedom, objective activity and imagination, technology and poetry as two independent ways of satiating the human drive" [8, p. 125].

While Marxist avant-gardists such as Alexei Gastev aimed at radically overcoming such juxtapositions and reconciling, for example, technology and poetry [9], under capitalism, as Kosík writes, "the romantic absolutization of dreams, imagination and poetry will accompany, as its faithful *alter ego* any 'fanaticism of labor' – i.e. any historical form of production in which the unity of necessity and freedom is realized through separating labor from joy (pleasure, bliss, happiness), or as a unity of opposites which are *personifie*d in *antagonistic* social groups [8, p. 125; *all italics in the original*]. The true realm of freedom begins beyond the boundaries of labour, but it depends on it: "The realm of freedom actually begins only where labor which is determined by necessity and mundane considerations ceases; thus in the very nature of things it lies beyond the sphere of actual material production. Just as the savage must wrestle with Nature to satisfy his wants, to maintain and reproduce life, so must civilised man, and he must do so in all social formations and under all possible modes of production. (…) Freedom in this field can only consist in socialised man, the associated producers, rationally regulating their interchange with Nature, bringing it under their common control, instead of being ruled by it as by the blind forces of Nature; and achieving this with the least expenditure of energy and under conditions most favourable to, and worthy of, their human nature. But it nonetheless still remains a realm of necessity. Beyond it begins that development of human energy which is an end in itself, the true realm of freedom, which, however, can blossom forth only with this realm of necessity as its basis" [10, p. 593].

To put it in the idiom of Florman [11], the "existential pleasures" of scientific and artistic creativity, which have been displaced from and opposed to most labour with the division of labour and are enjoyed only by very limited sections of society, show the potentiality of all labour, all work. "The work of material production can achieve this character only (1) when its social character is posited, (2) when it is of a scientific and at the same time general character" [6, p. 534]. Labour, then, is self-liberating in that it creates both the subjective and the objective conditions for itself, and scientific activity is the epitome of "labour for itself" in two respects: first, as the model of self-activity, which under present conditions is affordable only to a few people, and second, as the mastery of nature by the productive force.

3   Scientific Activity and Engineering

Already in the *Economic and Philosophical Manuscripts*, Marx denounced the cliché, immensely popular in the history of science, of scientific work as intellectual Robinsonades of self-sufficient geniuses. In elaborating the distinction between social and communal activity, he wrote: "[A]lso when I am active scientifically, etc. – an activity which I can seldom perform in direct community with others – then my activity is social (…). Not only is the material of my activity given to me as a social product (as is even the language in which the thinker is active): my own existence is social activity (…). My general consciousness is only the theoretical shape of that of which the living shape is the real community, the social fabric, although at the present day general consciousness is an abstraction from real life and as such confronts it with hostility" [12, p. 104]. This can be said of any kind of theoretical work: be it the mathematical expression of the laws of nature or political philosophy. In the *Grundrisse* and *Capital*, however, 'science' is used almost exclusively (if not in polemic with other economists) in a sense of applied science and technology, objectified in machinery.

The emergence of the machine system is the objectification of general social knowledge, the "general social intellect" takes control of "the conditions of the process of social life" [6, p. 626]. The socialising role of capital, its "great civilising influence", consists in bringing together the individual workers and collective knowledge, direct labour and scientific innovation, "the mass of hands and instruments" [6, p. 431]; but the capitalist application of machinery is at the same time the exploitation and concealment of its social nature and "communal spirit": "[I]n fixed capital, the social productivity of labour [is] posited as a property inherent in capital; including the scientific power as well as the combination of social powers within the production process" [6, p. 635]. When science is "pressed into the service of capital", "[i]nvention […] becomes a business, and the application of science to direct production itself becomes a prospect which determines and solicits it" [6, p. 623].

The inventions in the period of early industrialisation were not based on the epiphany of the natural sciences. It was the division and organisation of labour that emerged in manufactories and was later reified in the "mechanical monster" that was the modern factory, which abolished both crafts and manufactures: "There were mules and steam-engines before there were any labourers, whose exclusive occupation it was to make mules and steam-engines (…). The inventions of Vaucanson, Arkwright, Watt, and others, were, however, practicable, only because those inventors found, ready to hand, a considerable number of skilled mechanical workmen, placed at their disposal by the manufacturing period" [13, p. 266]. The social origin of such inventors illustrates the non-academic origin of invention [13, p. 319]. However, with the reciprocal development of the forces of production and the relevant disciplines, the capitalist tendency to produce scientifically increases.

In machinery (which forcibly brings together collective labour), science acts as a power alien and hostile to labour. The threat of technological substitution had for some centuries already created an opposition between labour and inventions, which were banned, burnt and smashed, and inventors were even murdered; but the physically torturing and mentally degrading factory has for the first time turned the workers against the very instruments of their labour [13, pp. 286f.]. The capitalist inventor, in turn, deliberately uses science and technology against labour: "It would be possible to write quite a history of the inventions, made since 1830, for the sole purpose of supplying capital with weapons against the revolts of the working-class" [13, p. 291]. Marx repeatedly quotes James Nasmyth, inventor of the steam hammer, with his very openly profit-oriented justifications of technical improvements [13, pp. 283–292]. A machine is accepted by capitalists when (and where) it makes labour more efficient, either through lower costs or because of the lack of labour (e.g. due to age, gender or other legal restrictions), and before it catches on, the idea of the machine can travel across borders for a long time.

4   Engineering as the Universal and the Communal Work

Machinery forcibly absorbs human material, skill and knowledge, into its "objective organism", "social body of labour", "totality of labours" [13, pp. 271–284]. Within this combination of labours, science and engineering stand in opposition to manual labour, against which they stand as an alien and incomprehensible power. But just as a machine forms an element of the value of the industrial product, so engineers are themselves a part of the total worker or "collective labourer" [4, pp. 396f.]. However, they are alienated from the factory worker class not only by virtue of their higher qualification and wage, but also by their direct or indirect technological control over labour [13, pp. 284f.] and their loyalty to the employer in the event of technological conflict [13, p. 328].

Marx used the term 'engineering' in its nineteenth-century meaning, before the massification and occupational closure of the engineering profession transformed it into a functionally very diverse but typically middle-class white-collar group, sometimes working far away from the production processes. Nevertheless, whether it is an engineer in the nineteenth-century factory, an engineer employed in an office in the mid-twentieth century, or their descendants working with computer-aided design, their activity can be studied on the same methodological basis. In the fragment "Economy through Inventions", Marx writes that "a distinction should be made between universal labour and co-operative labour. Both kinds play their role in the process of production, both flow one into the other, but both are also differentiated. Universal labour is all scientific labour, all discovery and all invention. This labour depends partly on the co-operation of the living, and partly on the utilisation of the labours of those who have gone before. Co-operative labour, on the other hand, is the direct co-operation of individuals" [10, p. 74].

Against this backdrop, let us consider a well-known case study from Science and Technology Studies (STS), Latour's narrative of the diesel engine in *Science in Action* [14, pp. 104–107]. He traces the network between Diesel's original idea and a working engine, which includes elements such as Carnot's principles, Diesel's original patent, Lord Kelvin's support, the tools and experiments of MAN's engineers, and various prototypes. This network consists of everyone and everything that worked on the engine's path to realisation, up to the state of a "closed black box". Following Marx's distinction between universal and cooperative labour, we can interpret these elements as actions, or rather as activities, and classify them (albeit ambiguously). From an empirical point of view, invention as a "universal activity" is subject to epistemological analysis of knowledge, language, methods, "stock solutions", means of visualisation, etc., inherited or shared with contemporaries [15]. Invention as a "cooperative activity" implies the coordinated actions, division of functions, formal and informal communication, etc., and is subject to empirical observation in communities of practice [16].

Once an innovation has taken place, the concrete technology, which is an objectification of these previous activities, acts (or generates new actions) all over the world. Gorokhov writes in his socio-epistemological analysis of technology that this became particularly evident in the second half of the twentieth century: "Even if an engineer only designs technical systems, he actually 'creates certain systems of activity'. In fact, it is a matter of designing activity systems – neither artificial systems (artifacts), nor natural systems (natural objects) – but the activity itself" [17, p. 9]. In his account of creativity in engineering, he refers to the activity theory developed by Schedrovitsky: "[H]uman social activity must not be seen as an attribute of the individual, but as an initial universal totality, much broader than 'individuals' themselves. So it is not the individuals who create and produce the activity, but on the contrary the activity itself 'grasps' them and makes them 'behave' in a certain way. (…) [A]ll 'things' or 'objects' are given to man by activity, and their very determination as 'objects' is due primarily to the nature of human social activity, which determines both the forms of social organization in the world – the 'second nature' – and the forms of human consciousness (…). All that is commonly called 'things', 'properties', 'relations' etc. are only temporary 'bundles' created by human activity on the basis of the material it grasps and assimilates" [18, cited in: 19, p. 65].

There seems to be quite a distance between this almost Hegelian reminiscence and radical social constructivism in philosophy of technology on the one hand and Latourian new materialism on the other. However, as we have argued, this distance can be bridged by drawing on key ideas of modern dialectical philosophy. Marx very rarely turned his eye to engineering activity as such, be it the individual success stories of inventions or the way engineers themselves speculate about their role in technological progress. He focused on their labour and knowledge, already objectified and reified in the form of machines; and this is precisely the way engineering is actually present and legitimated in capitalism. As Wendling [20] argues, in *Capital* Marx works within the concepts, vocabulary and calculus of capitalism to reveal the "knowledge practices of this world" and "those forms of life peculiar to capitalist alienation from within the norms established by that alienation" [20, p. 4]. As long as technology is fetishised and appropriated, technological creativity is conceived of in mutually contradictory forms: as inevitable progress, as marketable algorithm and as individual ingenuity.

References

1. Hegel, G.W.F.: Phenomenology of the Spirit. Oxford University Press, Oxford (1997)
2. Hegel, G.W.F.: The Philosophy of History. Batoche Books, Kitchener (2001)
3. Kojève, A.: Introduction to the Reading of Hegel. Cornell University Press, Ithaca (1980)
4. Marx, K.: Economic and philosophical manuscripts. In: Early Writings, pp. 279–400. Penguin Books, Harmondsworth (1977)
5. Joas, H.: The Creativity of Action. University of Chicago Press, Chicago (1996)
6. Marx, K.: Grundrisse. Foundations of the Critique of Political Economy. Penguin Books (in Association with the New Left Review), Harmondsworth (1973)
7. Sayers, S.: Marx and Alienation: Essays on Hegelian Themes. Palgrave Macmillan, Houndmills, Basingstoke (2011)
8. Kosík, K.: Dialectics of the Concrete. A Study on the Problems of Man and World. Reidel Publishing Company, Dordrecht and Boston (1976)
9. Coenen, C., Kazakova, A.: Utopian grammars of human-machine interaction. Technol. Lang. 2(1), 67–80 (2021). https://doi.org/10.48417/technolang.2021.01.06
10. Marx, K.: Capital, vol. III. Progress Publishers, Moscow (1959)
11. Florman, S.C.: The Existential Pleasures of Engineering. St. Martin's Griffin, New York (1994)
12. Marx, K.: Economic and Philosophic Manuscripts of 1844. Dover Publications, Mineola (2007)
13. Marx, K.: Capital, vol. I. Progress Publishers, Moscow (1956)
14. Latour, B.: Science in Action How to follow scientists and engineers through society. Harvard University Press, Harvard (1987)
15. Gavrilina, E.A.: Engineering creativity: an essay on epistemological analysis. In: Pisano, R. (ed.) A Bridge between Conceptual Frameworks. HMMS, vol. 27, pp. 195–205. Springer, Dordrecht (2015). https://doi.org/10.1007/978-94-017-9645-3_11
16. Wenger, E.: Communities of Practice: Learning, Meaning and Identity. Cambridge University Press, Cambridge (1999)

17. Gorokhov, V.G.: The Development of Engineering from Simplicity to Complexity. IFRAN, Moscow (2015).(in Russian)
18. Schedrovitsky, G.P.: Selected Works. Izdatelstvo Shkoly Kulturnoy Politiki, Moscow (1995).(in Russian)
19. Gorokhov, V.G.: Philosophy of technology as a theory of creative technological activity. In: Philosophy of Creativity, pp. 64–87. IFRAN, Moscow (2015). (in Russian)
20. Wendling, A.: Karl Marx on Technology and Alienation. Palgrave Macmillan, Houndmills, Basingstoke (2009)

Work Engagement - Gateway to Creativity

Maria Jakubik[(✉)] [iD]

Ronin Institute, Montclair, NJ 07043, USA
maria.jakubik@roninstitute.org

Abstract. The COVID pandemic challenged the importance of Work engagement (WE) or employee engagement (EE) for creativity and innovation. At a time when work is relegated to the home office, this paper seeks to answer the question: How can creativity be boosted by WE? It therefore explores the WE literature, presents a conceptual framework, clarifies key concepts, develops propositions, and discusses management and leadership challenges. It programmatically highlights responsibilities and the way in which leaders can help their organizations grow and succeed by creating an attractive work environment where employees are engaged at work, able to identify and solve complex problem with creative approaches, and able to flourish as individuals. In order to validate these propositions empirically, quantitative and qualitative surveys of business organizations have been conducted. In the meantime, the conceptual framework is presented which ties creativity to engagement. Accordingly, the novelty of this paper is in viewing creativity as an evolutionary, emerging cognitive, emotional, and behavioral process that fosters engagement.

Keywords: Work engagement (WE) · Employee engagement (EE) · Creativity · Competence · Intellectual capital (IC) · Human capital (HC) · Positive relationships at work (PRW) · Leadership

1 Introduction

In the agricultural and industrial economy, land, labor, and capital have been the main production factors. However, in the knowledge, mind, digital, and creative economy human knowledge, innovativeness and creativity are the main sources of economic value creation [1, pp. 56–65]. "Beauty. Truth. Love. Service. Wisdom. Justice. Freedom. Compassion. These are the moral imperatives that have aroused human beings to extraordinary accomplishment down through ages" [1, p. 64]. Indeed, these are the eternal values to guide individuals and organizations to strive, grow and flourish.

In the 21st century, individuals and businesses are not able to sustain their competitive advantage without being innovative and creative. During crises situations, as we experience now with the COVID world pandemic, there is an increased demand for identifying the problems and providing creative solutions to hem. Nowadays, in our complex, connected and digital world, when information and explicit knowledge become freely and widely available commodities, only intangible resources, human tacit knowledge are capable of making organizations competitive.

© The Author(s), under exclusive license to Springer Nature Switzerland AG 2022
D. Bylieva and A. Nordmann (Eds.): PCSF 2021, LNNS 345, pp. 11–21, 2022.
https://doi.org/10.1007/978-3-030-89708-6_2

Human capital (HC), is the soul of the organization. Intellectual capital (IC) of the organization combines HC, structural capital, and organizational and relationship capital. Competence, attitude, and intellectual agility [2, pp. 34–41] are the dimensions of HC. The challenges for organizations and leaders are to attract, keep, engage, and nurture, talented employees. Hamel [1] refers to a survey where 86,000 employees were asked in 16 countries to measure their engagement at work. He concludes that work engagement (WE) is important because 80% of economic value is created not by commodities but by sources that are difficult to manage, i.e., by human initiative (20%), creativity (25%), and by passion (35%). The manageable sources of economic value creation are obedience (0%), diligence (5%), and intellect (15%) [1, p. 59]. He adds however, that he is "not suggesting that obedience is literally worth *nothing*. A company where no one followed *any* rules would soon descend into anarchy." His point is that "the rule-following employees are worth zip in terms of the competitive advantage they generate" [1, p. 59]. How can leaders make their organizations attractive for talented, innovative, creative employees who have passion for their work? How can engaging leadership create a meaningful, enjoyable work environment and well-being for creative people [3]? How can leaders foster employee engagement (EE) or WE? These are vital questions for organizations.

From previous research [3–13] on EE or WE, it is well known that people who are highly engaged at work are the most innovative and creative employees. For instance, Albrecht in his handbook [4] pulled together 68 researchers' findings about EE. Kahn, Bakker, Demerouti, Fleck, Inceoglu [5–9] are among the leading researchers of WE. More recently, Jakubik and Vakkuri in an empirical research explore what it means to be engaged at work [10]. Jakubik has created a framework of how talented people can contribute to organizational performance if they get engaged at work [11]. In addition, she has developed further the job demands and resources model of EE [12]. However, the positive impacts of WE on performance have been verified by theoretical and empirical research, Imperatori argues that "despite the growing relevance of employees engagement, people are becoming more disengaged and the employee-organization relationships have become looser" [13]. Why is it that people become more disengaged at work? Indeed, this is a problem because the complex, wicked problems of our society and business badly need creative solutions. For example, nowadays we can experience the continuous struggle of devoted and creative researchers to find effective vaccines against COVID and its mutations to save people from dying. We need creativity when new ways of learning, teaching, working, and living are emerging. We need creative solutions for humanitarian, political, health, climate, environmental, and moral problems of our societies. There is a demand for creativity and innovation. Therefore, this paper addresses a contemporary and important topic by seeking to answer the question: *How can creativity be boosted by WE?*

The reminder of this paper is structured as follows: Sect. 2 defines the research problem or topic, formulates research questions, and decides on the research method. Section 3 presents a conceptual framework to show how WE leads to creativity and performance of an organization. In addition, it defines the key concepts and develops propositions. Section 4 discusses propositions, answers the research questions, and clarifies the role of managers and leaders in WE that leads to creativity of employees.

Section 5 concludes with implications for managers and leaders, states the limitations, outlines future research areas, and indicates contributions of this paper.

2 Research Questions and Method

This section of the paper defines the research problem/topic, formulates research questions, states the objectives, and decides on the research method. The author of this paper studied 14 EE/WE models [4] and she find out that creativity has been mentioned only in one of the models [8, p. 240] where it has been indicated as a product, i.e., as one of the outcomes of WE. Though, contribution of creativity to value creation [1] cannot be ignored in business. The author of this paper argues that regardless of the extensive research on EE/WE the relationship of WE and creativity would need more attention. Therefore, this paper focuses on the 'impact of WE on creativity' phenomenon. The main research question is formulated as: *How can creativity be boosted by WE?* The main objective of this paper is to contribute to the WE research and discourses with exploring this phenomenon. The following sub-questions will be explored:

- Q1: What is WE? - The objective is to define the WE concept.
- Q2: Why people engaged at work are more creative? - The objective is to show the drives and characteristics of engaged employees.
- Q3: How can WE be enhanced? - The objective is to clarify and discuss the role of managers and leaders in fostering WE.

The research method is qualitative, based on the most relevant literature.

3 Key Concepts

This section of the paper presents the conceptual framework of 'Work engagement as a gateway to creativity' (Fig. 1) that shows the main concepts and their relationships. In addition, this section defines the key concepts, and develops propositions.

- *Partnership* means a relationship between the employer and the employee. Partnership is an evolving concept. It has several definitions [14, p. 76]. The simplest definition is: partnership is when 'employers and employees working together jointly to solve problems' [15, p. 13]. A more detailed definition of partnership includes the following: commitment to success of the enterprise, building trust, recognizing legitimate roles and interests, employment security, information and consultation, sharing success, training and development [16, p. 15]. However, according to Foot and Hook [14], there are three commonalities between the different definitions of partnership: the importance of security; the common aim of business success; and the employee voice.
- *Participation* is a distinct concept from involvement and it is related to some kind of power or influence. When people participate in work processes (teamwork, meetings, decision-making, development work, research, production, etc.) they have the power to influence those processes. Participation could be voluntary (e.g., people form teams based on their interest) or it could be obligatory based on obedience (e.g., members assigned to teams by management).

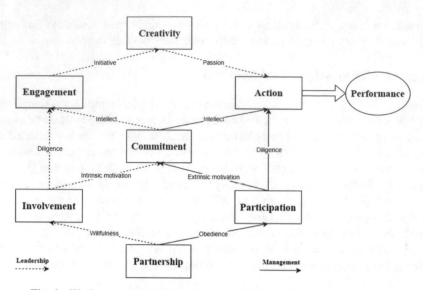

Fig. 1. Work engagement as a gateway to creativity. Conceptual framework

- *Involvement* "is a set of management practices that extend decision-making power, business information, technical and social skills, and rewards for performance" and it is "a property of organizational systems and not individuals" [17, p. 292]. Ledford [17] distinguishes three types of involvements such as: suggestion involvement; job involvement; and the combination of both, which he calls high involvement. Newman, Joseph and Hulin [18, pp. 43–61] explore the relationship of EE and job attitude - as they call it the "A-factor". Their research shows high correlation (r = 0,63) between job attitude and job involvement; between job attitude and job satisfaction (r = 0,73); and between job attitude and affective organizational commitment (r = 0,83) [18, p. 48].
- *Commitment* is an attitude, it "involves a set of linkages between people and organizations that build on human capabilities" [19, p. 206] and it is a combination of cognitive, emotional, and moral skills. Kanter [19, pp. 204–216] argues that commitment has three dimensions: mastery; membership; and meaning. With her words, mastery means caring about today and thinking about tomorrow; membership is cementing the 'we' and caring about 'me'; and meaning is believing in a larger purpose. She adds that money could be considered as the fourth 'M' of commitment.
- *Work engagement* or EE is a higher level of positive attitude toward work, it is an attitude of putting your real self into your work (i.e., your mental – cognitive; emotional – affective; behavioral – physical; and social- moral capabilities). Engaged people are passionate about their jobs and they are willing to go the extra mile. Newman, Joseph and Hulin [18] show that job attitude and employee engagement are highly correlated (0.77). Similarly, Ledford [17, pp. 295–297] examines how employee involvement leads to employee engagement, and how this results in organizational effectiveness such as increased job performance, increased citizenship behavior, and reduced withdrawal behavior (turnover, absenteeism, etc.). The transition from involvement to

commitment and to WE highly depend on intrinsic motivational factors (i.e., habits, attitudes, and drives).

- *Creativity* is an ability of a person to come up with new problems, new ideas, new solutions, new knowledge. The author of this this paper argues that creativity is considered not as a product but rather as a process, as an ability, human behavior of developing something new. She argues that creativity has cognitive (thinking, intellect, initiative), affective (feelings, values, attitude, passion), and behavioral (action, diligence) dimensions. Similarly, Al-Abadneh argues that creativity is a complex human behavior and "it can be influenced by a wide array of social, developmental and educational experience that leads to creativity in different ways in a variety of fields" [20, p. 245]. According to him, novelty and usefulness are the two dimensions of creativity. To be a creative person requires many talents and skills in order to think critically, see things differently, produce new solutions, ideas, and new knowledge. Creativity is characterized by 'problem finding' and problem solving'. He concludes that "creativity has been perceived in different ways as a mental ability, a process and a human behavior" [20, p. 247]. Indeed, creativity is a complex and multi-dimensional concept. This paper assumes that creativity can be enhanced by WE.
- *Action* leads to performance. It is a behavior that depends on commitment, creativity, and participation. Action is influenced by passion, intellect, and diligence. However, other factors (e.g., values, beliefs, passion, motivation, challenges, relationships with colleagues, financial, and nonfinancial incentives) can all play an important role in actions as well.
- *Performance*, according to Schroeder-Saulnier, has seven drivers: leadership; structure, roles and capability; people systems and processes; strategy; positive work culture; EE; and customer satisfaction [21, p. 342]. Work engagement plays a central role in her conceptual model of organizational effectiveness. She argues that engagement matters because "a direct line can be drawn through engagement to retention, productivity, customer satisfaction, and financial performance" [21, p. 340]. Performance in Fig. 1 embodies both organizational and personal performance. For engaged employees it is utmost important how their initiatives, creativity, and passion lead to success, positive performance.

Based on the conceptual framework (Fig. 1) and on the definitions of its key concepts, the author of this paper formulates the following propositions:

- P1: After employees enter a partnership and they express their willfulness more than their obedience then, they will be more involved.
- P2: Intrinsic motivation driven commitment, diligence, and intellect lead to WE.
- P3: Engaged employees contribute to performance of the organization through their actions that are originated from their initiative, creativity, and passion.
- P4: Leadership has a greater role in WE than management.

4 Discussion

This section discusses propositions, answers the research questions, and clarifies the role of managers and leaders in WE that leads to creativity of employees.

4.1 Discussion of Propositions

– P1: After employees enter a partnership and they express their willfulness more than their obedience then, they will be more involved.

Obedience means that employees comply with the rules, participate in work processes because they told to, they accomplish tasks because they were assigned to them. According to Hamel [1] obedience does not contribute to value creation. Nevertheless, he does not suggest "that obedience is literally worth *nothing*. A company where no one followed *any* rules would soon descend into anarchy" and he continues, "that rule-following employees are worth zip in terms of the competitive advantage they generate" [1, p. 59, emphases original].

However, employees who question or 'break the rules' could become more involved at work. They are more eager to make improvements, they could have arguments for not following the rules, they could ask for changing the rules. Organizations should pay more attention to 'rule breakers', they should listen to them, give them opportunities to prove their arguments.

– P2: Intrinsic motivation driven commitment, diligence, and intellect lead to WE.

Intrinsic motivation (e.g., values, norms, attitude, passion, ambitions, drives, having an impact, feedback from colleagues, positive relationships at work (PRW), well-being at work, search for meaning) leads to a higher level of commitment than extrinsic motivation (e.g., good work facilities, workplace infrastructure, money, easy access to workplace, regular working hours). This higher level of commitment together with intellect and diligence lead to WE.

The role of PWR is utmost important for engaged employees. Ragins and Dutton argue that at their best PRW "can be a generative source of enrichment, vitality, and learning that helps individuals, groups, and organizations grow, thrive, and flourish" but "they can be a toxic and corrosive source of pain, depletion, and dysfunction" at their worst [22, p. 3].

– P3: Engaged employees contribute to performance of the organization through their actions that are originated from their initiative, creativity, and passion.

Albrecht argues that "there is now evidence showing robust relationships between employee engagement and a range of important organizational outcomes" [4, p. 11]. Employee commitment, WE, initiatives, creativity, and passion determine actions and this way they have significant impacts on financial and nonfinancial results, customer loyalty, service quality, image, reputation of an organization. Similarly, Bakker points out that "engaged workers perform better than non-engaged workers" [8, p. 233]. According to him, there are four reasons for that: positive emotions, enthusiasm, happiness and joy of engaged employees manifest themselves in their actions; employees engaged at work experience better health; they can create their own jobs; and they are able to transform their engagement to others.

Jakubik, who examines how WE leads to organizational performance, argues that WE has an impact on organizational performance because "engaged employees are more productive, more profitable, more customer-focused, and more likely to stay. Highly engaged workplaces grow faster, adapt quicker, and innovate more. Organizations don't just benefit from employee engagement; they depend on it" [11, p. 103]. Concurring with all these, Imperatori concludes that "research confirms the positive outcomes of employee engagement as a source of personal well-being and positive attitudes at work, and as antecedents of business success" [13, p. 37].

– P4: Leadership has a greater role in WE than management.

In the creative economy, initiative, creativity, and passion are the characteristics of employees who are engaged at work. Competence, attitude, and intellectual agility are the most important dimensions of HC [2]. According to Hamel, "today the most valuable human capabilities are precisely those that are the least manageable. While the tools of management can compel people to be obedient and diligent, they can't make them creative and committed" [1, p. 60]. Employees who are engaged at work need less management, less control, fewer orders and they need more leadership (Fig. 1).

4.2 Answering the Research Questions

The main objective of this paper is to contribute to the WE research and discourses with exploring *how creativity can be boosted by WE*. Next, the three sub-questions will be answered.

– Q1: What is WE? The objective is to define the WE concept.

More than a decade ago, the concept of WE/EE started to emerge in the business literature. After Dicke provides 18 definitions of WE/EE [23, pp. 5–6] from the literature, he concludes that despite the proliferation of WE research "it is a concept with multiple definitions" and there is no clear definition of it [23, p. 11]. Nonetheless, his contribution is in summarizing the most commonly used terminologies in 18 definitions. The top four most frequently mentioned expressions are: drive business success; discretionary effort, going above and beyond; energy, involvement, efficacy; think, feel, act, during the performance [23, table 1-1, p. 6]. Albrecht [4, p. 4] however, expresses the need for a clear and agreed definition of engagement in order to clearly understand what engagement is, how it differs from other constructs, what it is related to, and how it should be measured.

– Q2: Why people engaged at work are more creative? The objective is to show the drives and characteristics of engaged employees.

In an empirical research, Jakubik and Vakkuri [10] explore how people experience WE, what it means to be engaged at work. Their survey was answered by 73 respondents of 13 different nationalities [10, p. 18] who felt engaged very frequently as they wanted to make an impact. Respondents indicated that they got engaged because their work was challenging, they wanted to contribute to business development, and they aimed at personal growth. Respondents felt energized, happy, enthusiast, and motivated during their engagement experience [10, p. 25]. When employees are engaged at work, they are more creative because they are challenged by wicked problems that would need solutions, because they want to make a positive impact with their contributions, because they want themselves and their organizations to succeed and flourish.

– Q3: How can WE be enhanced? The objective is to clarify and discuss the role of managers and leaders in fostering WE.

Organizations could foster WE of their employees, their "job satisfaction, and commitment not only by rewarding for results, providing career opportunities, clearly communicating organizational goals, but also by giving feedback, providing autonomy, and for allowing more democratic managerial practices" [10, p. 4].

4.3 The Role of Managers and Leaders in WE

The conceptual framework (Fig. 1) shows the relationships of key concepts and indicates where management and leadership could play a central role. Concurring with Hamel [1], the author of this paper argues that the role of management is more vital in the areas of human capabilities that are commodities (i.e., obedience, diligence, intellect) and widely available. Leadership plays important role in enhancing human capabilities (i.e., initiative, imagination, creativity, passion) that are difficult – or even impossible - to manage, not available freely, and would need a special care. These are the characteristics of HC, the soul of organizations [2].

Leadership plays an important role in WE and consequently, WE through enhanced creativity and actions plays vital role in organizations and individual's performance (Fig. 1). The findings of Anitha's [24] empirical research on the determinants of EE/WE and their impacts on both individual and organizational performance show that among the seven drivers of WE (i.e., workplace well-being, organizational policies, compensation, training, career development, team and co-worker relationship, work environment) the work environment, and the team and co-worker relationship have the most significant impact on WE. "This signifies the importance of a healthy work atmosphere and good interpersonal harmony with fellow members in the organization for anyone to be engaged positively at work" [24, p. 318].

5   Conclusions

Summing up, the paper started with arguing why creativity, innovations have become the driving forces in the 21st century. A special report on innovation in emerging markets [25] discusses frugal innovation, new emerging business models, new management paradigms. The report gives examples on how organizations care about their employees' well-being, education, how they form alliances with universities. Organizations realize that investing only in technology is not enough to sustain their competitive advantage. Investing in HC is the way how they can increase involvement, commitment, WE, creativity, and innovation of their people. Creativity is one dimension of HC [1, 2]. Research shows that employees engaged at work are more creative and they positively contribute to the organizations' performance [4–12, 17, 18, 24].

However, the problem is that the number of disengaged employees is increasing [13]. Therefore, the paper focused on an important and contemporary topic and articulated the main research questions as: *How can creativity be boosted by WE?* Then, sub-questions were identified, conceptual framework was created (Fig. 1), concepts were defined, and propositions were developed. Propositions were discussed and research questions answered. Now, the paper concludes with implications for managers and leaders, states the limitations and outlines and future research areas, and indicates contributions of this paper.

5.1   Implications for Managers and Leaders

Both managers and leaders are important in organizations' performance. However, this paper argues that in WE the role of leaders are more vital than the role of managers [1, 3, 10–12, 22]. Zhang et al. [26] investigate the relationship between EE/WE and four leadership paradigms (i.e., classical, transactional, visionary, and organic). Their research concludes that while visionary and organic leadership paradigms enhance EE in all 9 workplace characteristics, classical and transactional leadership styles negatively affect EE. These workplace characteristics or EE predictors are: communication; trust and integrity; job; effective direct supervisors; supportive direct supervisors; career advancement opportunities; contribution to organizational success; pride in the organization; and supportive colleagues/team members [26, p. 9]. They argue that supervisor effectiveness and support, career advancement opportunities, communication, and trust predict high WE.

"Classical and transactional leadership do not demand followers be engaged with the organization or its vision" [26, p. 7]. These styles are more frequently applied by managers than leaders. Therefore, the implications for leaders are to apply organic and visionary leadership styles if they want to achieve engagement of their employees. Interestingly, organic and visionary (transformational, charismatic) leadership build on emotional rather than on cognitive involvement of employees. Consequently, this underlines the conclusion of this paper that in WE leadership plays a more vital role than management.

5.2   Limitations and Future Research

The paper builds on literature related to EE/WE. Limitations of this paper are in use of limited number of sources that were selected by the author. Consequently, important sources might unintentionally be ignored. Another limitation of the paper is that its propositions were not validated by an empirical research. Therefore, future research could take the four propositions and develop a quantitative and qualitative survey to collect data from organizations. The conceptual framework (Fig. 1) could be tested by a quantitative research to see the strengths of the relations between the concepts. In addition, the literature review would need an extension and probably a more systematic approach.

5.3   Contributions

The paper has several contributions to knowledge. Firstly, it draws attention to a contemporary and important phenomenon, namely to the impact of WE on creativity as a largely ignored WE research topic. Secondly, with its propositions it contributes to the discourses about EE/WE. Thirdly, the conceptual framework (Fig. 1) presented here could be considered as a contribution to the WE research because it demonstrates the WE and creativity relationship in interrelationships with concepts important in value creation (i.e., human initiative, creativity, passion, diligence, and intellect). The framework shows also the roles of managers and leaders in this process. Finally, the novelty of this paper is in viewing creativity not as an outcome or product but rather as an evolutionary, emerging cognitive, emotional, and behavioral process. Work engagement could be fostered by visionary and organic leadership. Employees engaged at work contribute to inventive solutions, creative outcomes, and to positive individual and organizational performance with their creative thinking, feelings, and actions.

Funding.   This paper received no external funding.

Conflicts of Interest.   The author declares no conflict of interest.

References

1. Hamel, G., Breen, B.: The Future of Management. Harvard Business School Press, Boston (2007)
2. Roos, J., Roos, G., Dragonetti, N.C., Edvinsson, L.: Intellectual Capital. Navigating the New Business Landscape. MCMILLAN PRESS LTD., London (1997)
3. Rahmadani, V.G., Schaufeli, W.B.: Engaging leaders foster employees' well-being at work. In: Proceeding of the 5th International Conference on Public Health vol. 5, no. 2, pp. 1–7. TIIKM Publishing (2019). https://doi.org/10.17501/23246735.2019.5201
4. Albrecht, S.L. (ed.): Handbook of Employee Engagement. Perspectives, Issues, Research and Practice. Edward Elgar. Northampton (2010)
5. Kahn, W.: Psychological conditions of personal engagement and disengagement at work. Acad. Manag. J. **33**, 692–724 (1990). https://doi.org/10.5465/256287
6. Bakker, A.B., Demerouti, E.: The job demand-resources model: state of the art. J. Manag. Psy. **22**, 309–328 (2007). https://doi.org/10.1108/02683940710733115
7. Bakker, A.B., Demerouti, E.: Towards a model of work engagement. Career Devel. Int. **13**(3), 209–223 (2008). https://doi.org/10.1108/13620430810870476
8. Bakker, A.B.: Engagement and "job crafting": engaged employees create their own great place to work. In: Albrecht, S.L. (ed.) Handbook of Employee Engagement. Perspectives, Issues, Research and Practice, pp. 229–244. New Horizons in Management. Edward Elgar Northampton (2010). https://doi.org/10.4337/9781849806374.00027
9. Fleck, S., Inceoglu, I.: A comprehensive framework for understanding and predicting engagement. In: Albrecht, S.L. (ed.) Handbook of Employee Engagement. Perspectives, Issues, Research and Practice. New Horizons in Management, pp. 31–42. Edward Elgar Northampton (2010). https://doi.org/10.4337/9781849806374.00009
10. Jakubik, M., Vakkuri, M.: The E-Experience. Exploring Employee Engagement. Haaga-Helia Publication Series R&D-reports, Helsinki (2015). https://www.researchgate.net/publication/281439563_E-experience_Exploring_employee_engagement. Accessed 17 Feb 2021
11. Jakubik, M.: Talent engagement framework as a journey to performance. Rev. Innov. Compet. **2**(3), 101–122 (2016). https://doi.org/10.32728/ric.2016.23/6
12. Jakubik, M.: Elaborating the job demands and resources model of employee engagement. In: 1st Ferenc Farkas International Scientific Conference, Blind Reviewed Book of Proceedings, pp. 48–64. University Pecs, Hungary (2018). https://ktk.pte.hu/sites/ktk.pte.hu/files/uploads/ffkonf/ffisc2018_fin_0.pdf. Accessed 12 Feb 2021
13. Imperatori, B.: Engagement and Disengagement at Work. Drivers and Organizational Practices to Sustain Employee Passion and Performance. Springer, Milan (2017). https://doi.org/10.1007/978-3-319-51886-2
14. Foot, M., Hook, C.: Introducing Human Resource Management, 6th edn. Prentice Hall, Upper Saddle River (2011)

15. ACAS, Advisory, Conciliation and Arbitration Service: Annual Report, ACAS (1997)
16. IRS, Industrial Relations Services: We're all in this together – partnership at work. Employ. Rev. **801**, 15–17 (2004)
17. Ledford, G.E.: Fostering employee engagement through compensation and benefits. In: Berger, L.A., Berger, D.R. (eds.) The Talent Management Handbook. Creating a Sustainable Competitive Advantage by Selecting, Developing, and Promoting the Best People, pp. 291–301. The McGraw-Hill Companies, Inc., New York (2011)
18. Newman, D.A., Joseph, D.L., Hulin, C.L.: Job attitudes and employee engagement: considering the attitude "A-factor". In: Albrecht, S.L. (ed.) Handbook of Employee Engagement. Perspectives, Issues, Research and Practice. New Horizons in Management, pp. 43–61. Edward Elgar Northampton (2010). https://doi.org/10.4337/9781849806374.00010
19. Kanter, R.M.: Evolve! Succeeding in the Digital Culture of Tomorrow. Harvard Business School Press, Boston (2001)
20. Al-Abadneh, M.M.: The concept of creativity: definitions and theories. Int. J. Tour. Hotel Bus. Manag. **2**(1), 245–249 (2020). https://www.researchgate.net/publication/339831352. Accessed 17 Feb 2021
21. Schroeder-Saulnier, D.: Employee engagement and talent management. In: Berger, L.A., Berger, D.R. (eds.) The Talent Management Handbook. Creating a Sustainable Competitive Advantage by Selecting, Developing, and Promoting the Best People, pp. 340–348. The McGraw-Hill Companies, Inc., New York (2011)
22. Ragins, B.R., Dutton, J.E.: Positive relationships at work: an introduction and invitation. In: Dutton, J.E., Ragins, B.R. (eds.) Exploring Positive Relationships at Work. Building a Theoretical and Research Foundation, pp. 3–25. Psychology Press, New Jersey (2009)
23. Dicke, C.: Employee engagement? I want it, what is it? Cornell University & CAHRS Centre for Human Resource Studies (2007). https://est05.esalestrack.com/eSalesTrack/Content/Content.ashx?file=52eeebe5-d7d7-45e1-a242-c9d4207f1d9f.pdf. Accessed 17 Feb 2021
24. Anitha, J.: Determinants of employee engagement and their impact on employee performance. Int. J. Productiv. Perform. Manag. **63**(3), 308–323 (2014). https://doi.org/10.1108/IJPPM-01-2013-0008
25. The World Turned Upside Down, A special report on innovation in emerging markets. The Economist, April 17 (2010). https://www.economist.com/sites/default/files/special-reports-pdfs/15894419_0.pdf. Accessed 17 Feb 2021
26. Zhang, T., Avery, G.C., Bergsteiner, H., More, E.: The relationship between leadership paradigms and employee engagement. J. Glob. Respons. **5**(1), 4–21 (2014). https://doi.org/10.1108/JG14-0006

Cosmos and Metacosmos in Dessauer's Philosophy of Technology: Inventing the Environment

Alexander Yu. Nesterov(✉) [iD]

Samara National Research University, 34, Moskovskoye Shosse, 443086 Samara, Russia

Abstract. This essay presents a semiotic reconstruction of Friedrich Dessauer's concept of technology that he developed in the late 1950s. His works *Philosophie der Technik* (1928), *Mensch und Kosmos* (1948), *Streit um die Technik* (1958, 1959) model an ethically-positive power of technology that is rooted in the "Fourth Kingdom of technology". This is regarded as a potential cosmos containing, along with other things, all possible "preset forms of solutions". His model allows us to formulate and justify the idea of humans as a means of connecting worlds or layers of being. This, in turn, justifies the nature and vector of technological progress as a feature of holistic systems-thinking. The reconstruction and interpretation of Dessauer's approach from the point of view of general semiotics allows us to take technology as a projective semiosis realized through intelligence, mind, and processing, where the appearance of novelty in the world as a technical action is determined by the transformation of a semantic, syntactic or pragmatic rule in a particular layer of semiosis. The source of technology and human activity – invention as the discovery of a solution to a problem pre-established in the "Fourth Kingdom of technology" – is examined in the context of classical metaphysics as intellectual reflection or intuition, in which Dessauer's concept reveals the syntactic aspect but leaves open questions as to how to implement semantic and pragmatic rules.

Keywords: Friedrich Dessauer · Semiotics of technology · The Fourth Kingdom · Cosmos · Metacosmos of technology · Artificial nature · Progress

1 Introduction

The attempt to address the issue of the nature of creativity, invention, discovery, encounters one of the most difficult philosophical problems - the problem of novelty, or the new. Does the category of "novelty" describe only the result of the expansion of the boundaries of human knowledge, or does the process of generating the new take place in the existence? "Being new", is it an epistemological or ontological concept? Creating technical or purely artistic artifacts, does a person "create", "invent" or "discover"? On the one hand, these are the classical questions that go back to the ontological opposition of Parmenides' unchanging eternal Being and the fluidity, the uniqueness of each individual moment of life in Heraclitus, to the Christian idea of creation, to the efforts of

D. Bylieva and A. Nordmann (Eds.): PCSF 2021, LNNS 345, pp. 22–33, 2022.
https://doi.org/10.1007/978-3-030-89708-6_3

Hegel. On the other hand, first the "age of progress" and then the philosophy of technology after Ernst Kapp, present the space of human activity as a sphere of generation of the new, providing for the human needs and desires, forcing the creation of engineering ontologies that take into account "being" as laws of nature, and regard the "new" as technically successful application of these laws that leads to progress. In these first decades of the 21st century, in the time of transition from the second artificial nature to the third artificial nature [1], when new artificial objects are filling not only the sphere of sensory perception, regulated by the laws of nature, but also the spheres of application of logical and grammatical norms of mind, intellectual rules of imagination, fantasy, goal-setting and decision-making, the questions about the nature of human creative and inventive activity turn out to be the cornerstone on which dwell optimistic or pessimistic scenarios of the *homo sapiens'* development after the "Squark threshold" [2] or "technological singularities" [3]. The processes of technical development and the transformation of subjectivity are regarded by Dessauer as systematic in his model of the philosophy of technology [4–7]. Despite its low popularity among contemporary philosophers of technology [8–14], it is Platonic intentions supported by computational schemes of activity that have permitted us to understand the vector and content of scientific and technological progress on the border of the second and third artificial nature.

2 Dessauerian Ontology: Cosmos and Metacosmos

The processes of generating the new are associated with the transformation of the rules of activity, the de-automatization of patterns, the transformation of instincts into complex conditioned reflexes. Applying the classical theory of knowledge, one can imagine three sources of what is actually new in human life: the empirical source, expressed in the transformations of the environment that enter the inner world of a person with new impressions and phenomena; the rational source, expressed in the transformations of the intellect that enter through new concepts, language constructions, knowledge; and finally, the transcendental, expressed by changes in the pre-established structures of knowledge and activity, which entail the transformation of habitual sensations and concepts. If the first two sources of the new can be metaphorically described as wandering in a physical or superphysical space (for example, in fiction or philosophy), then the latter is formulated through the metaphor of education as a reflexive management that purposefully takes the subject beyond the limits of innate skills, systems and their interpretations.

In travel and education, a person discovers and masters the order, the regularity of the existing, "the rule of the natural course of things" ("die Regel im Naturablauf") [4, p. 68], cosmos, "the whole world of objects given in phenomena" ("die gesamte erscheinende Gegenstandswelt") [4, p. 110], "the general order of Nature" ("die Naturordnung insgesamt") [4, p. 118]. Dessauer presents the history of knowledge as the history of inventions where humans understand the cosmos, learn to ask it questions, develop the systems of knowledge from the answers received, and enter the space of the metacosmos of technology. The method of historical and philosophical reconstruction of the "cosmos" can be explicated by the reference to the transcendentalist distinction between aphorism, dialogue and system, formulated by the neo-Kantian philosopher Lapshin [15, pp. 161–164] in the same time period when Dessauer created his concept of technology. This is the idea of historically successive and overlapping philosophical inventions that advance the understanding of cosmos as a concept. Aphoristically, the cosmos is perceived by the ancients in the Mediterranean basin through the search for "the unity in the diversity, and in the unity – meaning" ("*nach Einheit in der Vielheit und in der Einheit nach dem Sinn*") [4, p. 25]; the very fact of aphoristic understanding is expressed in the thesis that "the ancients experienced difficulties not only with spiritual vision, but also – to a much greater extent than we do – with the expression of what they saw" ("*Und wir Heutigen dürfen nie vergessen, dass die Alten nicht nur die Schwierigkeit hatten, geistig zu sehen, sondern, in höherem Masse als wir, die andere: auszudrücken, was sie geschaut*") [4, p.25]). Dessauer gives much attention to the invention of the dialogue with nature and the systematic development of its results. The experience of the cosmos as an order – "only order is cognizable" ("Nur Ordnung lässt sich erkennen") [4, p. 28] allows Francis Bacon, Galileo, and Newton to invent an inductive method of cognition embodied in a measurement experiment and creating a generally valid basis for activity. Cosmos is revealed as the normal order of nature, as a solid and reliable lawfulness of the universe, discovered by humans through the experiment as a correct question posed to nature. "instead of a relaxed receptive attitude towards the cosmos, which presents itself in various ways in bodily substances, the dynamics in cosmos comes to the foreground, it becomes the subject of knowledge" ("*Statt der gelassenen hinnehemenden Haltung gegenüber dem Kosmos, der sich in den körperlichen Substanzen in ihrer grossen Mannigfaltigkeit präsentiert, tritt die Dynamik im Kosmos in den Vordergrund und wird vor allem betrachtet*") [4, p. 40].

Along with the reflection upon the discovery of the cosmos as actual and absolute laws of nature, Dessauer formulates a consistent model of layers or zones of Being, where physical causality is superimposed by a system of biological (floral and animal) expediency, and the latter, due to the development of language, becomes the basis for the subjective reality formation ("*das Reich der Innerlichkeit*" [4, p. 108]) that attracts the sphere of spirit. The physical, biological, and spiritual are the layers in which cosmos is revealed as the sum of natural laws, and their interaction creates the reality of nature. The "dispute regarding technology ("*Streit um die Technik*") introduces the anthropomorphic distinction of the potential and actual cosmos: the actual order is the one that is known and technically applied by humans, the potential order is the one that is still hidden in epistemically pre-problem situations, but can be extracted by invention and is extracted whenever the pre-problem situation becomes problematic and turns into the sum of practical tasks. The transitions between the potential and the actual cosmos, the expansion of the boundaries of the knowable allows a person to create metacosmos, an artificial, technically modified nature. "Initially, everything arises from cosmos - energies, materials, orders – including our body and its components. However, we do not perceive it first-hand, natural, like animals in the wild, but transformed, reordered in spiritualized, historically emerging forms. We, therefore, live... in a metacosmos that has emerged as a result of the grinding of natural facts with the human spirit and human finalist creativity" ("*Zuvor kommt alles aus dem Kosmos – Energien, Stoffe, Ordnungen – auch unser Leib selbst und was ihn erhält. Aber wir empfangen es nicht, wie freilebende Tiere, aus erster Hand, sondern umgestaltet, neugeordnet in durchgeistigten, geschichtlich entstehenden Formen. So leben wir... in einem Metakosmos, hervorgegangen aus Vermählung natürlicher Gegebenheiten mit menschlichem Geist und menschlichem Zielschaffen"*) [4, p. 128].

As shown by P. K. Engelmeyer, in order to answer the question about the ontology of technical activity, it is necessary to see and fix the difference between cognitive, subjectifying activity and technical, objectifying activity [16, p. 36]. The transition from cognition to action, from reception to projection, happens in the process of solving problems, practical tasks, and is described as revelation, insight, intuition, and invention. In the Russian-language space, the formula of *"triact"* is used for its explication: "In the first act the invention is proposed, in the second it is proved, in the third it is accomplished. At the end of the first act it is a hypothesis; at the end of the second - a representation; at the end of the third - a phenomenon. The first act defines it teleologically, the second - logically, the third -actually. The first act results in the idea, the second - in a plan, the third - in an action". [16, p. 103]. Dessauer formulates a model of human formative abilities *(formbildende Uranlagen)*, including the *homo investigator*, *homo inventor*, and *homo faber*. To be an "investigator" means to respond to the environment in a specifically human way, different from that of a plant or animal, to have a self-awareness able to formulate questions and pose problems. Being an "inventor" means the ability to create within the laws of nature, to be able to design. Being a *"faber"* means being able to process, translate intramental images into extramental forms, transcend the constructed objects into the external world, into the environment [6, pp. 141–142]. The scheme of projective activity that defines the ontology of technology is a neo-Kantian reversal of knowledge stages: if cognition begins with perception, is fixed by the mind and unfolds in the sphere of intelligence, then a technical act begins with the problem fixation and with the intellectual idea, which is represented as the solution of the problem, continues with the rational construction and ends with a new artificial object in the sensually perceived world.

The most complicated and, up to now, open question in the described classical knowledge and technical activity concerns the nature of ideas that arise in a person during the transition from direct knowledge to reflexive activity. This is the problem of intuition or revelation, which Dessauer consistently solves by developing the concept of the "Fourth Kingdom", synthesizing Platonic and Kantian intentions. "The formative abilities of humans are the basis of the technical historical process; the latent composition of the "fourth kingdom", consisting in pre-established implementing forms - the basis of the possibility of technology" (*"Die formenden Uranlagen des Menschen sind der Grund des technischen geschichtlichen Geschehens; der latente Bestand des 'Vierten Reiches' an prästabilierten erfüllenden Formen ist der Möglichkeitsgrund der Technik"*) [6, p. 161]. The very phrase "the Fourth Kingdom" is an extension of the Kant's model where the world was divided into the kingdoms of 1) natural sciences, 2) moral law, 3) aesthetics and expediency: "the unambiguously predated forms of the finished creation are on another level. Moreover, here the attitude of a person to the 'thing-in-itself' is organized differently, when 'inventing', he transfers one of the potential images to the sensually accessible world" (*"Die eindeutig vorgegebenen Formen der bereiten Schöpfung sind in einer anderen Ebene. Und anders... ist hier das Verhältnis des Menschen zum "Ding an sich", wenn er eine der potentiellen Gestalten "erfindend" in die Sinnenwelt herüberholt"*) [6, pp. 164–165]. In fact, Dessauer declares that the human mind in the process of transition from knowledge to activity, caused by a problem or question, gets access to a different plane of being, to the potential cosmos, which contains the solution to the problem. In the ontology of technology, this access is fixed as a reassembly of the experience of cognition or memory in such a way that the changed configuration allows us to solve the problem, in other words, it is a recollection by Plato or a religious revelation. Neal Stephenson, in "Anathem", to mention the latest extremely successful literary interpretation of this situation, described it as follows: "These truths seem to come out of another world or plane of existence. It's hard not to believe that this other world really exists in some sense-not just in our imaginations!" [17, p. 72] and called the kingdom of pre-established forms of solutions "the Hylaean Theoric World" [17, p. 79]. Dessauer formulates his conclusion about the "fourth kingdom" as follows: "The transition from non-existence to being in the experienced world, that is, the process not of change, but of essential emergence, the collision with the "thing-in-itself", not by perception from outside to inside, but inside and from within, put according to the idea in the external world in one row with natural things, finally, the task here undoubtedly deals with experienced knowledge" ("...Den Übergang vom Nichtsein zum Sein in der Erfahrungswelt, also einen Prozess nicht der Änderung, sondern der wesenhaften Entstehung, die Begegnung mit dem 'Ding an sich', nicht von aussen durch Wahrnehmung in uns hinein, sondern innen und von innen nach der Idee in die Aussenwelt neben die Naturdinge gestellt, und endlich hat der Zweck hier mit dem erfahrbaren Wissen unstreitig zu tun") [6, p. 166].

The "fourth kingdom" as a source of the metacosmos of technology is interpreted by semiotics as a syntactic dimension of intellectual reflection. At the first step, technology, like any other activity, is realized as a skill, techne, purely pragmatically. At the second step that humanity has made in more than one millennium, the technology is regarded as an objective and unique condition for the possibility of solving the task, ("pre-established form of solution") "prästabilierte Lösungsgestalt", that is, as a place in the system of activity, historically defined by the problem, experience and resources available. At the third step - as the execution or interpretation of a syntactic place in the form of a specific action, regulated by the laws of the physical world, expressing itself in a new technical object in space and time. Separating cosmos and metacomos, Dessauer actually justifies the possibility of constructing the history of activity along with the history of knowledge. Cosmos is a reality that forms the background of human existence as a "potential cosmos", and its environment - as an "actual cosmos". The history of cosmos exploration is the discovery of the objective laws of the existence of nature. Metacosmos is a natural environment modified by technology, an artificial environment of life activities, the world of culture. The development of knowledge about metacosmos is the discovery of the objective laws of human reflection on the objective laws of the existence of nature.

3   Humans as a Means of Connecting Worlds

In the work *Man and Cosmos* (*Mensch und Kosmos*) Dessauer writes: "From the standpoint of natural sciences, I would like to define man as a creature in which the layers of being are united into an individual unity (*"Ich möchte, vom Standpunkt der Erfahrungswissenschaft her, den Menschen als ein Wesen bezeichnen, in dem die Seinsschichten zu einer individuellen Einheit verknüpft sind"*) [4, pp. 113–114]. These are the layers of the physical, plant, animal, and psychic kingdoms, over which the layer of the spirit is built, expressed in the kingdoms of reason, mind, ethos, and aesthetics. Each of the layers is autonomous, it exists as an objective system of laws or, in the semiotic interpretation, as a system of places. The interaction of layers, which requires application of pragmatic and semantic rules, arises with life and expresses itself in a person with maximum completeness available for reflection.

This kind of ontological pluralism, which develops from the interaction of three or more worlds in a person, is presented in the works of Karl Popper [18, 19] where the method of falsification convincingly shows the non-reducibility of the real variety of world syntactic systems to one system. Both for dualism and pluralism, the question of the unity of laws and the unity of description is fundamental. In the models of "unified science", in the philosophy of technology and generally in the theory of activity, it is raised as a question of the projective truth, feasibility or realizability of an idea in a construction and artifact: "what once worked, will always work under the appropriate circumstances" ("...Was einmal "geht", unter den passenden Umständen, immer geht") [6, p. 172]. Dessauer formulates a hierarchy of layers, autonomous orders of being in humans, and builds the ethics of technology upon it. "A man, such a complex being that combines many interacting autonomous orders, can exist only under the condition of the hierarchy of orders, so that their interaction does not lead to chaos" ("...ein so komplexes Wesen, das eine Fülle von autonomen Ordnungen in Wechselwirkung in sich vereint, kann nur bestehen, wenn eine Hierarchie der Ordnungen wirkt, so dass ihre Wechselwirkung nicht zum Chaos treibt") [4, p. 114].

In contract to Popper's, Dessauer's ontology comprises not only a larger number of worlds, but also brings theological arguments in the discussion. Indeed, a subject cannot become an object for himself, this is a typical philosophical situation of knowing: "the human spirit as a subject in its ultimate essence is not accessible to science... In other words, space does not apply to it" ("der menschliche Geist als Subjekt in seinem letzten Wesen wissenschaftlich unerreichbar... Mit anderen Worten: bis dahin reicht der Kosmos nicht") [4, p. 119]. However, the category of spirit, subjectivity, is much broader than the rationality of Popper's "third world", it includes reflection as a semiosis and a figure of God as the basic condition to make reflection possible. "There is reliable 'to have', a possession, before the act of objectification. The spirit is more than intellectual knowledge. At this point, the boundary of any idealistic theory of knowledge of the past lies. It is this insoluble Me, which rises above the self, which cannot be objectified as belonging to cosmos, that is, cannot be transformed into an object, that we call – like others before – spirit". ("Es gibt ein sicheres "Haben", ein Besitzen, vor dem Akt des Objektsetzens. Der Geist ist grosser, als die Verstandeserkenntnis. Darin liegt die Grenze jeder idealistischen Erkenntnistheorie der Vergangenheit. Und das unauflösbare Ich, das über das Selbst emporragt, das nicht wie das Kosmische objektiviert, d.h. zum Gegenstand gemacht werden kann, nennen wir, wie es andere getan haben: Geist") [4, pp. 120–121].

The inclusion of the theological argument in the model of a person as means of connection the worlds makes it possible to describe, if not complete, but final and understandable model of cognition and activity. A natural attitude that allows humans – like

any other living organisms – to connect a certain number of worlds in their unconscious practical activity is impacted by a reflexive attitude letting them know about
the interaction of worlds in themselves, in other living beings, shift the boundaries of
interaction, expand the zone of contact, involving new worlds. This is a neo-Kantian
and phenomenological argument, where the theological aspect adds, with the figure of
God, an exemplary model of the reflexive activity that humans master and by which they
advance in acts of knowledge. Thus, Dessauer actualizes the idea of God as the boundary
of the known described by Saint Anselm and brings to life the metaphysical basis of
activity. In semiotic terms, the classical dialectic of faith and knowledge implies that
the skills of distinguishing between sign and background precede any syntactic operations and semantic interpretation. Respectively, reflection – theoretical and practical
knowledge of the interaction of worlds in subjective activity - is impossible without
this kind of pragmatics. "Human existence through faith becomes a compass-guided
journey to revelation. Cosmos does not include Me, that is why it is aware of itself as
immortal" (*"Menschliche Existenz wird durch Glauben Wanderschaft zur Offenbarung,
die des Kompasses gewiss ist. Der Kosmos aber umfasst das Ich nicht, das darum seiner
Unsterblichkeit inne wird"*) [4, p. 126].

4   The Logic of Progress, the Problem of Environment

An individual's self-understanding as a process triggered by problems and questions, is
guided by the faith in their solubility, in the power of spirit to find such schemes of the
worlds' interaction that satisfy needs, fulfill desires, remove questions and problems.
The reflexive behaviour, manifesting itself as form-building ability, creates the human
environment, metacosmos of technology, the new things that humans bring into the
natural order of the Universe and that arise from faith, knowledge and application. We
assume that persons do not adapt themselves to the environment, but instead change the
environment to suit their needs [16, pp. 91–92]. The formation of the human environment
is done by the power of technology, where the "power of technical object" is the changed
(relative to the natural) order of elements in an object, allowing this object to perform
its task, and "technology", as a world force, is the combination of forces of individual
technical objects [6, pp. 172–175].

The power of technology enters the human world with the invention of the new.
Dessauer's ontology, the concept of the "fourth kingdom" formulated with the help
of a theological argument that states that humans connect the worlds and extract new
forms of this connection in the act of invention, but they are not the creators of their
inventions. Referring to the creation myth of the Bible, humanity in general remains
in the seventh day: the process of creation is not completed and it is going on with
the help of humans [6]. However, they are only intermedia, inventors, who are finding
pre-established forms of solutions and execute them in the substratum of the physical
world. The continuation of creation is "the multiplication of real being by essential
forms. Only the result is creative, not the activity. The activity of the inventor is also
not creative in the proper sense of the word, since the creative as such precedes its
forms of solutions" (*"Vermehrung des realen Seins um wesenhafte Gestalten. Aber nur
das Resultat, nicht die Tätigkeit ist schöpferisch. Auch die Tätigkeit des Erfinders ist*

nicht eigentlich schöpferisch, weil das Schöpferische seiner Lösungsgestalten schon vorweggenommen ist. Diese sind prästabiliert.") [4, pp. 145–146]. The metacosmos created by humans as a living environment through intuition, design and processing can be regarded as a reproduction of the "kingdom not of this world" based on a reflexive that is an unnatural and artificial manner of knowledge.

The logic of creation of the environment is the logic of progress, the process of manifestation of the new in human life. Due to accumulation of artificial objects, energy processing machines in the physical world, the natural environment is replaced by an artificial environment of the first order. Further, the accumulation of artificial objects, information processing machines in the sphere of mind, forms an artificial environment of the second order. And finally, due to development of strong AI models and their application in the sphere of intelligence, an artificial environment of the third order appears. Schematically, the process of transformation of the environment can be represented as follows (Table 1):

Table 1. Transformation of the environment.

Sensory perception	Object	Artificial objects, energy-processing machines
Mind/logical and grammatical structures of consciousness	Subject	Artificial objects and information-processing machines
Reason/intellect/reflection	Concept	Artificial concepts and autonomous intelligent machines

Following the logic [20] that describes the sequential replacement of a series of natural objects, subjects, and concepts with artificial ones, there appear such effects that require serious analysis. First among them is a new ethos of interaction between humans and machines, which raises the question of human and non-human (post- or transhuman) mechanisms of self-consciousness and the problem of (historical) memory. Secondly, it is a question of global ethical assessment of the phenomenon of technology and, in general, of the ethical norms transformation urged by scientific and technological progress.

Persons invent and discover, that is, they are able to connect the physical, animal, intellectual and other worlds in such a way that a new pragmatic skill of perception or processing or the use of language to describe or design concepts, ideas or insights arises in their life. The transformation of a pragmatic rule leads to a new combination of syntactic rules, to a new configuration of places through which the procedures of signification by intellect, mind, and perception are going on. The latter provide the possibility of new "executions"; that is, the appearance of new concepts of reason, objects of mind and objects of perception with varying levels of clarity [21]. When unnatural (not realized in the natural mode of knowledge and activity) ways of interaction of the natural elements are extracted from the "fourth kingdom" in a form of a dialogue with nature, new qualities of the life-world appear. They are first created by energy-processing machines, then by information processing machines, and currently – by autonomous artificial intelligence systems actively introduced into society. These are endowed with the ability to make decisions that are significant for humanity. Once introduced with a technical invention, a new "force" as a new configuration of the elements of perception, mind or reason, no longer depends on the inventor and works independently, transforming the environment, interacting with the natural cosmic orders and other metacosmic forces.

5 Ethics

The spiral of progress obviously leads to such state of affairs when the difference between the natural and the artificial ceases to matter [1]. On one hand, this is the basis of reflexive cognition. Marin Mersenne has noted that to understand something in the engineering, technical sense, means to be able to do it [22, p. 193]. Dessauer expresses the dialectic of "thing-in-itself" (*Ding an sich*) in direct natural and indirect technical knowledge in the fact that in understanding the cosmos the "existence precedes essence", in metacosmos of technology, on the contrary, the "essence precedes existence" [6, pp. 188–189, 224–227]. The synthesis of cosmic and metacosmic approaches obviously implies such level of progress at which there will be no objects, subjects and concepts for humans that they could not "understand", that is, recreate it technically. Stanislaw Lem was almost the first to express this idea 10 years after the publication of Dessauer's fundamental works: "we can erase the difference between the "artificial" and the "natural" - this will happen when the "artificial" becomes first indistinguishable from the natural, and then surpasses it... How do you understand the superiority? It means the realization by Nature of what is impossible for Nature" [23, pp. 255–256].

On the other hand, the natural world of humans is such only by virtue of instinct and education within the historically established sum of pragmatic skills of reflection. The potential cosmos revealed to the inventor in the pre-established forms of the solution of the "fourth kingdom" is infinite, therefore, the number of possible combinations of the worlds of the actual cosmos is also infinite. The technical feasibility of new combinations is determined by human knowledge extracted from a dialogue with the cosmos. By extracting new skills first of processing, then of thinking and reflection, by creating autonomous machines that perform these functions without a human, humanity extracts forms of "pre-established harmony", the order of the cosmos, not from nature, but from the technical metacosmos. The progress presented as technical transformation of the reflection skill leads to the idea of environment realized through technology, which has access to the cosmos and the complex mechanisms of constructing metacosmos that surpass humanity. Initially, this idea arises in science fiction as the "Squark threshold" in Stanisław Lem or the "technological singularity" in Vernor Vinge [3]: "The squark threshold is the threshold for the wisdom of the habitat, after overcoming which it becomes smarter than the intelligent beings living in it" [2, p. 350]. In the 21st century, a huge number of works have been devoted to it [24, 25]. In the transcendentalist theory of knowledge, it can be expressed as the idea of a new philosophical invention, the idea of "environment" "s a mechanism of reflection, different from "aphorism", "dialogue" and "system".

The technological metacosmos that arises from the reassembly of human experience in the act of revelation gives rise to a new subject – "neo-human", "transhuman" or "posthuman". In the logic of progress, this is the inevitable consequence of technical activity that transforms the human environment. The problems of evaluating progress, such as technophobia or technophilia, are well illustrated by visionary writers. For example, Victor Pelevin in "Love for Three Zuckerbrins" reproduces the classic existentialist fear of technology as a world force emerging from the potential cosmos and subjugating human worlds: "The Earth is being colonized by an organosilicon civilization. But it doesn't do it the way Herbert Wells intended. No war. No rays, no blood. Stars don't fall from the sky, tripods don't land in cities. People themselves build an invasion army according to the drawings that gradually appear in front of them. And the aliens come to Earth in the form of technology and code…" [26, pp. 318–319].

Both technophobia and technophilia problematize the same situation, namely, the lack of an ethical norm for the second and third artificial nature. In the first case, artifacts, objects of mind and intellectual systems created by human reflection are demonized, and the impending death of civilization is proclaimed when a new subject emerges. In the second case, attempts are made to create a new code of ethics.

Dessauer, speaking about the ethics of the metacosmos, formulates significant milestones. Going back from the primitive principle of "an eye for an eye" to the golden rule of ethics formulated in the Sermon on the Mount and defining the relationship of person to person in the first artificial nature, humanity is strengthening cooperation on a planetary scale. The ethical system of mutual aid that appeared in the 19th century in the works of Pyotr Kropotkin and that was developed at the end of the 20th century, for example, by A. N. Averyanov [27], is expressed in Dessauer by the image of "unknown brothers" who create artifacts so that they can be used by anyone: "In every detail… the breath of the life of the 'unknown brothers' who invented, manufactured, delivered it is interwoven: hidden human cooperation, assistance, a thousandfold return and acceptance, a secret greeting" (*"In jedem Stück… ist ein Hauch des Lebens der "Brüder Unbekannt" eingewoben, die es für uns erdachten, herstellen, besorgten: eine… verdeckte menschliche Kooperation, Hilfeleistung, ein tausendfältiges Geben und Nehmen, ein verborgenes Grüssen"*) [6, p. 178]. For the second artificial nature, in which humanity finds itself today, we are talking not only about the artifacts of the physical world, but also about intellectual products created by the assistance of tens and hundreds of thousands of people and meeting the needs of individuals, communities, and humanity as a whole.

6   Conclusion

"Technology, according to its essence, means the following: escape from the state of animal subordination to natural conditions, liberation from them, in order to independently and responsibly shape the environment, as befits the spirit, that is, freedom in the double sense: freedom from the unbearable, freedom for your own project, to the formation of the future" [7, p. 184]. Dessauer shares with Hegel and Goethe the pathos of human liberation from natural conditioning. The new, invented by a person as a skill of intuitive insight and discovered as a place of the "fourth kingdom", a unique pre-established form of solving a problem or task, is not created by humans; it only passes from the potential cosmos to the actual one as a result of human efforts. The semiotic spiral of knowledge and activity expands, leaving the limits of the imaginable and returning back with new objects, subjects and concepts. The cosmos of nature is constantly being tested and supplemented by the metacosmos of technology. The ethics of cooperation and synthesis formulates a principle that leaves humanity the hope not only of the preservation of historical memory, but also of the development of the complexity of interactions between the worlds that are actually representing the human and the non-natural parts of the human being. "What should we do? To be the helpers, to let go of the present when the covenant requires it; to grasp the future, to contribute actively to the life that reveals itself differently each time; to live as a human being in the generosity and nobility of detachment from everyday life, listening to the voice of eternity, which calls out in the mortal world: 'Die and become!', to gain the power of hope up to the earthly death on the efficacy of the prophetic word" [4, pp. 183–184].

Acknowledgments. The reported study was funded by RFBR, project number 20-011-00462 A.

References

1. Nesterov, A.: Technology as semiosis. Technol. Lang. **1**(1), 71–80 (2020). https://doi.org/10.48417/technolang.2020.01.16
2. Lem, S.: Collected Works in 10 Volumes: Futurological Congress, vol.8. Observation on the Spot. Plays about Professor Tarantoga. Tekst, Moscow (1994). (in Russian)
3. Vinge, V.: The Coming Technological Singularity: How to Survive in the Post-Human Era (1993). https://edoras.sdsu.edu/~vinge/misc/singularity.html. Accessed 17 Apr 2021
4. Dessauer, F.: Mensch und Kosmos. Verlag Otto Walter AG, Olten (1948).(in German)
5. Dessauer, F.: Philosophie der Technik: das Problem der Realisierung. Verlag von Friedrich Cohen in Bonn, Bonn (1928).(in German)
6. Dessauer, F.: Streit um die Technik. Verlag, Josef Knecht F.a.M (1958).(in German)
7. Dessauer, F.: Streit um die Technik. Verlag Herder, Freiburg im Breisgau (1959).(in German)
8. Skrbina, D.: The Metaphysics of Technology. Taylor and Francis Inc., New York (2014) https://doi.org/10.4324/9781315879581
9. Skrbina, D.: German metaphysical insights: Dessauer, Juenger, Heidegger. Metaphys. Technol. **94**, 70–93 (2015)
10. Rivers, T.J.: An introduction to the metaphysics of technology. Technol. Soc. **27**(4), 551–574 (2005). https://doi.org/10.1016/j.techsoc.2005.08.009
11. Petrushenko, V., Chursinova, O.: Philosophical and anthropological dimension of technoscience. Filosofija, Sociologija **30**(3), 199–205 (2019). https://doi.org/10.6001/fil-soc.v30i3.4042
12. Freigang, C.: Mies van der Rohe, the Werkbund and the question of technology around 1930. RIHA J. **2018**, 17 (2018). (in German)
13. DeLashmutt, M.W.: The technological imaginary: bringing myth and imagination into dialogue with Bronislaw Szerszynski's Nature. Technol. Sacred. Zygon **41**(4), 801–810 (2006). https://doi.org/10.1111/j.1467-9744.2006.00778.x
14. Mitcham, C.: Thinking Through Technology: The Path Between Engineering and Philosophy. University of Chicago Press, Chicago (1994)
15. Lapshin, I.I.: The Philosophy of Invention and Invention in Philosophy. Respublika, Moscow (1999).(in Russian)
16. Engelmeyer, P.K.: Creativity Theory. Book House "Librokom", Moscow (2010). (in Russian)
17. Stephenson, N.: Anathem. William Morrow, New York (2008)

18. Popper, K.R.: Objective Knowledge: An Evolutionary Approach. Oxford University Press, Oxford (1972)
19. Popper, K.R. (ed.): Knowledge and the body-mind problem. In: Defence of Interaction. Routledge, New York (1994)
20. Nesterov, A.Y.: Clarification of the concept of progress through the semiotics of technology. In: Bylieva, D., Nordmann, A., Shipunova, O., Volkova, V. (eds.) PCSF/CSIS -2020. LNNS, vol. 184, pp. 3–11. Springer, Cham (2021). https://doi.org/10.1007/978-3-030-65857-1_1
21. Nesterov, A., Demina, A.: Concept of invention in semiotics of technology. In: Proceedings - 2019 21st International Conference "Complex Systems: Control and Modeling Problems", CSCMP 2019, vol. 2019, pp. 773–776. (2019). https://doi.org/10.1109/CSCMP45713.2019.8976651
22. Ropohl, G.: Is technology a philosophical problem? In: Arzakanjan, C.G., Gorohov, V.G. (eds.) Philosophy of Technology in FRG. Progress, Moscow (1989). (in Russian)
23. Lem, S.: Summa Technologiae. Izdatel'stvo AST, Moscow, Terra Fantastica, St. Petersburg (2004). (in Russian)
24. Global Future 2045. Convergent Technologies (NBICS) and Transhumanistic Evolution. MBA, Moscow (2013). (in Russian)
25. Kurzweil, R.: The Singularity is Near: When Humans Transcend Biology. Penguin Books, New York (2006)
26. Pelevin, V.: Love for the three zuckerbrins. Eksmo-Press, Moscow (2017).(in Russian)
27. Averyanov, A.N.: Systemic Knowledge of the World: Methodological Problems. Politizdat, Moscow (1985).(in Russian)

Creativity as a Technology

Andrey Pavlenko[✉] [iD]

Department of Ontology, Institute of Philosophy, Russian Academy of Sciences (RAS),
Goncharnaja st., 12, Bld.1, 109240 Moscow, Russia

Abstract. The article examines the historical, linguistic and ontological origins of the phenomenon of "new creativity" and its difference from the phenomenon of "true creativity". It is established that the reason for the modern demand for "new creativity" is *the dominance of a person's technical relationship to the world.* "Imitation of nature" (organoprojection) is giving way to "redesigning of Nature" according to the patterns of a "creative activist". It is shown that the basis of "new creativity" is the Judeo-Christian model of the creation of the world "from nothing" (*creatio ex nihilo*), while the phenomenon of "true creativity" is based on the "genetic model", in which the main mechanism is "generation" from an already given basis ("non-being"), the model of being first expressed in Plato's "Timaeus". Therefore, there is no place for "nihilism" understood as "creation from nothing (*creatio ex nihilo*)" in *the genetic strategy*, since, in the historical and cultural sense, the genetic strategy is always based on *tradition*, that is, on an already existing basis. The uncritical use of the term "creative" leads to two forms of pretentiousness which are here referred to as "creatisms". In the first creatism the general "problem of creativity" is the province of those who have never created anything in any particular area; or is, simply put, the efforts of amateurs. In the second creatism the multiple repetition of the term "creative" does not lead to any sort of "creativity" – the emergence of "something" from "nothing" – and, therefore, is a manifestation of ordinary linguistic magism.

Keywords: True creativity · New creativity · Creatism · Creative activism · Plato · Heterological · Autological

1 Introduction (Semantic Source of "Creativity")

Before examining the nature of new creativity, it is necessary to introduce a set of basic notions and formulate the main task of the research. We shall start with the definition of "new creativity". First, let's give an ostensive definition of "new creativity". The term "creative" denotes a phrase we encounter in business, a "creative manager", in politics, a "creative leader", in sociology, a "creative era", or a "creative generation", and even in the language of innovators "creative idea", etc[1]. It is easy to see that the word "creative", in these examples, is used in *a positive sense*. After all, we know that notions can be "positive" and "negative". This allows us to assume that it expresses some special value of the modern human world. Let us compare it, for clarity, with another harmless term,

---

[1] See the philosophical and educational reflation on this topic: [1–7].

D. Bylieva and A. Nordmann (Eds.): PCSF 2021, LNNS 345, pp. 34–41, 2022.
https://doi.org/10.1007/978-3-030-89708-6_4

"*traditional*", which, in itself, being completely neutral, in the context of the values of the "creative era" has, in the overwhelming majority of cases, *a negative meaning*. So why *is the term "creative" so consistently positive these days*? Let's try to answer this question by turning to etymology.

The word "creative" came from the Latin in the Middle Ages, there it had the form "*creativum*". In turn, this latter is derived from the Latin verb "*creare*", which means "to make". Hence the Latin verbal noun – "*creatio*" ("creation"). This term, as well as the verb from which it is derived, turned out to bear in the Middle Ages very specific - Judeo-Christian - meanings. The point is that the Judeo-Christian *creatio* model was proposed to explain the "creation of the world" [8] in a very specific way. Therefore, it essentially differs from the Greek model of the origin of the Cosmos found, for instance, in the works of Plato. Plato's Demiurge (as a father) gives birth to the body of the Cosmos (τὸ τοῦ κόσμου σῶμα ἐγεννήθη) (Tim. 32c) from "non-being" [9]. Plato specifically uses the Greek verb γεννάω – "to be born of a father". Whereas the God of the Jews does not "generate", but precisely "creates" the world from "nothing", literally - from "emptiness": "In the beginning, God created Heaven and Earth, the Earth was formless and empty, and darkness was above the abyss...." [10] (Genesis, Ch. 1,1–2), In another source, the Second book of the Maccabees, it is said that God created the world from "nothing": "Look at heaven and earth, and seeing everything that is on them, know that God created all things out of nothing" [10] (2 Mac. 7:28).

It is this distinction - between "generation" and "creation" - that underlay, underlies and, we think, will underlie any system of worldview that claims to be unconditional[2]. Of course, the modern "civilized man", from the height of his position, fells condescension towards religious prejudices. However, he or she cannot fail to understand that religious prejudices may have vanished like a fog under the sun, but the model that underlies them remained intact and survived. Exactly this, we think, happened to "new creativity". "*Creatio*", understood as a creation by the Jewish god of the world from "nothing" has long gone from the human world, but it did not leave without a trace. It left in its place its *matrix* – "to make (create) from nothing". We believe that this very meaning lies at the basis of any modern "creativity", no matter what form it takes. So, the very first and, therefore, the deepest meaning of the word "creative" just means "making out of nothing". What is the difference between the Jewish "making out of nothing", mentioned in the Second book of the Maccabees, and the Greek "generation from non-being"? The point is[3] that "generation from non-being" already presupposes some previous *given* (basis), which, however, exists only in *the form of a possibility*. For example, the elements of the Cosmos already exist, but the Cosmos with their help has not yet been generated. This "given" ("being in a possibility") in a certain sense already exists, although it is hidden (virtually). Creation from "nothing" does not imply such a

---

[2] This topic is considered in detail in our work [11], Ch. 1.

[3] The reader can easily recall the anthem of one European "creative party" of the 19th and 20th centuries, in which there are the following words (Eugène Pottier): "Enslaved masses, stand up, stand up./The world is about to change its foundation/We are nothing, let us be all." And in the modified version - as the anthem of the USSR (from 1918 to 1944), the same motive sounds like this: "We will destroy the whole world of violence to the ground, and then we will build ours, we will build a new world, - Who was nothing, he will become everything.".

virtual reality. Moreover, if Plato meticulously justifies his model both philosophically, mathematically, and even biologically, then the author of the Second book of the Mac-cabees only refers to the words of some semi-literate woman who absolutely did not understand anything on the matter. But, despite this, supporters of the biblical approach for two thousand years managed to impose this model on almost all European culture. Modern or "new creativity" is a distant, but as we have just seen, quite a recognizable echo of this model. Nevertheless, despite the clarification of the meaning of "new creativity" and the binding of the latter to the Judeo-Christian tradition, it remained completely unclear: *why exactly today, that is, at least since the second half of the 20th century, "all of a sudden everybody around" started talking about "new creativity"?* Why did "new creativity" lay "slumbering" for almost five centuries (from the 15th century, that is, from the end of the Middle Ages), remaining in a dormant (latent) state, and today, suddenly, shook and "woke up"? The answers to these questions will be presented in the next section.

2 New Creativity, Technology and True Creativity

Here we will not venture into the theological and cosmological subtleties of this dis-tinction. They are quite fully described in the work [11, pp. 88–98]. In addition, in one of my previous works [12, pp. 331–344], we have already noted that the modern era is characterized by the growing of the role of technology in the life of society and the diminishing of the role of science. In what way does this manifest most clearly? The answer lies on the surface: in the specifics of the relationship between the human and the reality, which in the era of the dominance of science manifested itself.

(1) in the predominance of the role of the discovery of laws, dependencies, phenomena and facts of nature,
(2) in the predominance of man's construction of a "new nature", of course, in accordance with the laws discovered by science.

What does this peculiarity of a man's technical relation to reality have to do with the matter under discussion? Everything, we must admit. The fact is that

(I) *in the conditions of the dominance of technology, true creativity itself, understood as "new creativity", becomes technical.*

What does this mean? This means that in the era of the dominance of *science*, which relied on the *discovery of laws and phenomena, true creativity* was understood as exactly the *discovery of some "know-how", skill, trick, "machenschaft", if you will, given either from above or by nature itself*. This can be confirmed by the explanation of the nature of technology as "*organoprojektion*", which we find in Ernst Kapp's works [13] in the second half of the 19th century. From his point of view, *Nature suggests technical solutions to man*. By his "mechanisms" man *copies nature, imitates it*. Thus, nature acts *as the source of the discovery of technology* and its world. Admittedly, in Pavel Florensky's "*organoprojektion*" [14, p. 421], we already see a turn from Kapp's model towards modern "new creativity". Indeed, according to Florensky, technology is the "bodybuilding of the soul": that is, the human soul materializes its technical "ideas-functions" by and of itself. However, we note that the term "*organoprojektion*" itself is still used by Pavel Florensky. In other words, although Florensky uses Kapp's term, he endows it with his own specific meaning: a person, now, *not just "projects" outside* the technical images of his body (body organs of living organisms), but carries out an arbitrary (imaginary) construction in the mind of such "mechanisms" that do not exist in "pure form" in living nature. That is why Florensky calls such constructs "idea-functions".

In our time - the era of the beginning of the *dominance of technology – true creativity itself becomes something that turns out to be the subject of technical design.* It turns out that it is possible to "teach" true creativity, because it *is not given, not opened to a person from the outside, but is already present inside in the form of a "deposit of creativity"* - after all, a "creative person" was created by God, who himself created the world from "nothing".

Specialists "in new creativity" are racking their brains to develop a "technology of true creativity". Indeed, *true creativity*, like everything else in our technical world, *becomes technological.* And not only in the sense that computers and computer programs help to prove mathematical theorems, arrange musical pieces and so on. The point is also that modern "new creativity specialists", of course, the most ambitious of them, are trying to *guess* the algorithm of true creativity.

So, as we just saw, in the era of the dominance of man's technical relationship to the world, values change dramatically.

New creativity becomes *only and only a "product"* of the person himself. In my opinion, this attitude to true creativity is based on overt *anthropolatry.* Figuratively speaking, new creativity itself can be "created" or manufactured (constructed) by a person of himself in the same way as, for example, a "chair" is manufactured of wood, and a "vase" is manufactured of clay.

The essence of this project is simple: *true creativity turns from art into a craft (skill),* which the organizers of the "creative process"undertake to teach others. There is no longer any external dependence of man on the "divine gift" or "design of nature." True creativity, according to this approach, has its origin in man, his only abilities and his only intentions and goals. Thus, "true creativity" turns into "new creativity", or, more precisely, it is reduced to "creativity", the purpose of which is *to re-create (re-design) the surrounding world and the person himself according to human patterns.* In the English-language literature, this has already received a term – the "Redesign of Nature".

Replacing "true creativity" with "new creativity" leads to an unexpected result: "new creativity" only visibly resembles "true creativity", without being so at all in fact.

3 Some Traits of "New Creativity"

Even the very first approach to the analysis of "new creativity" allows us to see what is actually creative in it. So, what is it?

"New creativity" as a consequence of person's deprivation of an external given - after all, something should be created from "nothing", from "emptiness" - is based on human "activism". "True creativity" losing its ability to "generate" - to receive new "tricks" from the outside, from nature, from the Cosmos - shyly hides behind "new creativity", which has a completely different nature: anthropolatric arrogance, imagining itself capable of "making something out of nothing"[4].

What can "human activism" discover in the conditions of its deprivation? How does this "activism" manifest itself in circumstances that are completely alien and even hostile to it? "Creative activism" has at its disposal a human given, "filled" with deprivation (in biblical terminology – "nothing"). In these conditions, "creative activism" has no choice but to *recombine* this "given deprivation". Unable to generate anything - literally "create" - a creative activist undertakes to "organize", literally – "construct", or more precisely - to imitate the process of creation. Let's make a note of this specific feature[5]. This point of view was very clearly expressed by the German specialist in the philosophy of technology Gunther Ropohl: "I would like to make it clear to the reader," writes Ropohl, "that, for my part, I regard the invention as a primary, anti-natural product of human consciousness". Such new creativity is achieved due to "planning, intellectually controlled and future-oriented activity, as well as due to the human ability, in representation to move in space and time and to combine any feature of the given" [15, p. 216].

4   New Creativity "Creatisms"

Now that we have considered the essential features of "new creativity", we can proceed to the analysis of the absurdities that any "creative activist" inevitably encounters, and which disavow his or her attempts to pass off this "organizing activity" as true creativity. Let's call such absurdities "*creatism*" - a term derived from the word "creativity".

First Creatism. Anyone who, at least once in his life, has made something or watched how others make (create) something, cannot fail to notice one feature, if not to say, a strangeness: those who have made something (created), for example, wrote a poem or a novel, proved a theorem, built a model of an airplane, etc., never thought and did not realize that they were "creating". This may seem ridiculous in the context of this article, but let's be patient. Indeed, what was the poet doing? Writing a (rhyming) poem. Right. What did the novelist do? Wrote a novel. Correct. What was the mathematician doing? Proving the theorem. Correct again. What did the aircraft designer do? Invented a concept of an airplane. Correct again. So, what's the weird thing? The weird thing is that each time answering the question about the form of activity, we have never used any mention of "true creativity". That is true. But, after all, they all created! That is also true. So, what is the matter? The point, in my opinion, is that each of the named creators was engaged in his own specific business: rhymed, wrote, proved, and modelled. In this

---

[4] See [8, pp. 88–98].

[5] The famous American philosopher Daniel Dennett gave a plenary talk at the Ontology Congress in San Sebastian in October 2012, in which he spoke about the fact that all culture can be explained with the help of "memes" (Dawkins' term). To my question that "if culture is a combination of "memes", each of which does not contain any novelty, then how can we explain the emergence of novelty (in science, art, literature, technology, etc.), After all, "memes" are the same ones that only multiply?" Dennett answered me: "The emergence of novelty is the result of the interaction of axons and neurons in the human brain". As they say – there's nothing to add.

case, a new question arises: how do these forms of activity relate to true creativity? The answer is obvious: in the realtion of part to the whole. This means that each of the creators was engaged in his own specific (private) business[6]. The point, in our opinion, is that people really creating something are not engaged in general "problems of true creativity" or the analysis of the concept of "true creativity", but actually are make something it their field of expertise. But, if this is the case, then the first "creatism" arises: *those are obsessed with the general "problems of true creativity", who, for the most part, have never created anything in any particular area, or, simply speaking, amateurs.* This absurdity is very reminiscent of the one about which Rudolph Carnap spoke in connection with the obsession of philosophers-metaphysicians with the general problems of the "existence of objects" [16, pp. 69–89]. From this state of affairs, the question inevitably follows: how can those who have not made (created) anything during their life talk about true creativity (analyse "true creativity")? It is if swimming was taught by someone who never entered the water. Creatism, in a word.

Second Creatism. The absurdity that we are about to show now may not be as grave as the previous one, but it is much more widespread. The point is that "creative activists", "creative innovators" or even simply "specialists in general problems of new creativity" rarely bother to check their own statements for logical consistency.

To understand this, let's turn to logic. In the early 20th century, two German logicians and mathematicians - Leonard Nelson and Kurt Grelling - formulated the "heterological paradox" [17, 18]. The paradox itself will not be a focus of this work, but we will need its premises. Nelson and Grelling proposed to consider the names of "properties" and "features", which are called "adjectives"[7] in natural languages grammar. Then they proposed to divide all adjectives into two types: *autologous* and *heterologous*. 1) Autologous adjectives are those that themselves have the attribute that they denote. For example, the adjective *"russkiy"* in Russian denotes the feature "to be Russian" and is itself "Russian". Consequently, the adjective *"russkiy"* in Russian is autological. On the contrary, the adjective *"angliyskiy"* in Russian is not autological, since it means "English" but is in the Russian language. In English, the picture is mirrored: the adjective English will be autological, and the adjective Russian will not be autological. 2) Heterological adjectives are those adjectives that themselves do not have the attribute that they denote. For example, the adjective "salty" itself does not have the attribute "to be salty". The adjective "green" itself is not green, etc.

Why do we need this look into the history of the premises of the heterological paradox? In order to combine the existing knowledge about the properties of adjectives with the knowledge about the predisposition of everyday thinking to *linguistic magism*, about which R. Carnap spoke so much, and thus to reveal the origins of *the second creatism.*

---

[6] In fact, it is true, we don't usually say that the poet "created the poem", that the novelist "created the nove", that the mathematician "created the theorem" and that the aircraft designer "created the airplane". Such responses would sound strange, in other words, we would feel that they contain some kind of falsehood.

[7] A detailed explanation of the ""heterological paradox"" can be found in the work of G. von Wright: [19, pp. 449–482].

The point is that *the term "creative" is heterological*, i.e., it does not contain anything creative (or referring to – "true creativity"). By pronouncing or writing the word "creative", the creative activist has absolutely nothing to add to what is said or written: after all, from pronouncing (writing) the word "creative" *"something" does not arise out of "nothing"*. In other words, the repetition of the term "creative" does not lead to any "creativity" - the emergence of "something" from "nothing" and, therefore, is a manifestation of ordinary *linguistic magism.*

A little more succinctly: the multiple (whatever abundant) use of the word "creative" in explaining the nature of "new creativity" is undoubtedly "creatism". In fairness, it should be noted that the same applies to the use of the term "true creativity".

5   Conclusion

So, we found out that such a phenomenon as "new creativity" has its origin in the semantic matrix, which was proposed in the Judeo-Christian tradition, having a clear citation in 2 Maccabees. The Latinized expression of this matrix, "creation", presupposes such a process as "creation from nothing" (*creatio ex nihilo*). The mechanism of "creation from nothing" has turned out to be extremely popular in the modern era - *the era of the beginning of the dominance of technology as a way of man's relationship to the world*. Under these conditions, the technical "imitation" of nature (*"organoprojektion"* by E. Kapp and P. Florensky) gives way to "Redesigning of Nature" according to the patterns of a "creative activist". It was shown that the new strategy of "creation from nothing" throughout European history opposed the inherent European strategy of "generation from non-being" described in Plato's "Timaeus"[8]. It was also shown logically that the term "creative" is not *autological*, but *heterological*.

It follows from the above that the "creative" strategy in European culture is opposed by the "genetic" strategy, which presupposes *the generation of something from an already existing given*. In the *genetic strategy*, "nihilism", understood as "creation from nothing (*creatio ex nihilo*)", is absent, since, in the historical and cultural sense, the genetic strategy was always based on *tradition*, that is, on an *existing basis*.

References

1. Kaufman, S.B.: The Philosophy of Creativity (with Elliot Samuel Paul). Oxford University Press, New York (2014)
2. Yeung, A.: Person and creativity: a Gilson inspired reading of St. Thomas Aquinas. Divus Thomas. **121**(3), 303–312 (2018)
3. Robinson, K., Aronica, L.: Creative schools: the grassroots revolution that's transforming education. Penguin Books (2016)
4. Fasko, D.: Education and creativity. Creat. Res. J. **13**, 317–327 (2001). https://doi.org/10.1207/s15326934crj1334_09
5. Jeanes, E.L.: 'Resisting creativity, creating the new'. A Deleuzian perspective on creativity. Creat. Innov. Manag. **15**, 127–134 (2006). https://doi.org/10.1111/j.1467-8691.2006.00379.x

---

[8] See [20].

6. Shaheen, R.: Creativity and education. Creat. Educ. **01**, 166–169 (2010). https://doi.org/10. 4236/ce.2010.13026

7. Pecheanu, I.S.E., Tudorie, C.: Initiatives towards an education for creativity. Procedia - Soc. Behav. Sci. **180**, 1520–1526 (2015). https://doi.org/10.1016/j.sbspro.2015.02.301

8. Pavlenko, A.N.: The Universe from "nothing" or the Universe from "None Being": Statement. Philosophical Problems of Cosmology. A Universe from "Nothing" or a Universe from "None Being"? URSS, "Librokom" Book House, Moscow (2012). (in Russian)

9. Plato. Plato's Cosmology; The Timaeus of Plato. Harcourt, Brace, K. Paul, Trench, Trubner & Co. ltd., London (1937)

10. Bible, T.: Authorized King James Version. Cambridge University Press, Cambridge (2004)

11. Pavlenko, A.N.: European Cosmology. Foundations of the Epistemological Turn. INTRADA, Moscow (1997)

12. Andrey iz Lyovino: A new chance for philosophy. In: History as a Symbol. Historical and Philosophical Notes, pp. 331–344, Aletheia, Sankt-Petersburg (2016)

13. Kapp, E.: Grundlinien einer Philosophie der Technik. George Westermann, Braunschweig (1877). (in German)

14. Florensky, P., Priest: Organoprojection. In: Works in Four Volumes, vol. 3, no. 1. Mysl, Moscow (2000). (in Russian)

15. Ropohl, G.: Technology as the opposite of nature. In: Philosophy of Technology in Germany. Progress, Moscow (1989)

16. Carnap, R.: Überwindung der Metaphysik durch logische Analyse der Sprache. Erkenntnis. **2**, 219–224 (1931). (in German). https://doi.org/10.1007/BF02028153

17. Grelling, K., Nelson, L.: Bemerkungen zu den Paradoxien von Russell und Burali-Forti. Abhandlungen der Fries'schen Schule **2**, 301–334 (1908). (in German)

18. Nelson, L.: Gesammelte Schriften III. In: Die kritische Methode in ihrer Bedeutung für die Wissenschaften, pp. 95–127. Felix Meiner Verlag, Hamburg (1974). (in German)

19. von Wright, G.H.: Philosophical Logic: Philosophical Papers. Cornell University Press, Ithaca (1986)

20. Pavlenko, A.: Possibility of Technology. Aletheia, St. Petersburg (2010). (in Russian)

Talent as a Personality Resource of People with Schizotypal Personality

Sofya Tarasova(⊠) iD

Psychotherapy Center Alvian, Psychological Institute of Russian Academy of Education, Mokhovaya str., 9, bld. 4, 125009 Moscow, Russia

Abstract. Talent is understood as a person's aptitude for creativity. Talent is a personality resource that supports effective adaptation of a person in different situations. In the research talent was measured based on expert assessment. Guest experts, an art expert and a painter, provided their expert opinion on a 1 to 10 scale. Originality of thinking was taken as a psychological criterion of the analysis. It was determined using a classical pathopsychological test. Guilford's SI theory was the methodological basis of the research. The following methods were used: pathopsychological test, Torrance figurative subtests, incomplete sentences, analysis of verbal and graphic production of the participants, Buss-Perry Aggression Questionnaire, Spielberg's State-Trait Anxiety Inventory. People with schizotypal personality were the participants of the research. The purpose of the research was to study manifestations of talent as a personality resource in people with schizotypal personality. Apart from diagnostics, the participants underwent psychological rehabilitation program developed by the authors. Thus, talent as a personality resource was studied based on expert opinion, psychological diagnostics as well as during psychological rehabilitation. The work was also an attempt to describe creativity in interdisciplinary terms. The empirical research has showed that talent and schizotypal personality are interconnected. Talent acts as a personality resource during psychological rehabilitation of such people. The authors of the research have developed, tested and implemented the program of psychological rehabilitation for people with schizotypal personality. In each case it is customized into an individual program of psychological rehabilitation.

Keywords: Schizotypal personality · Originality of thinking · Talent · Personality resource · Clinical psychology students · Learning in the workshop · Clinical case analysis

1 Introduction

The issue of human talent is a complex one. Many authors see connection between talent and the level of intelligence, the rate of child development. In the beginning of the XX century French psychologist Binet worked out methods of diagnostics of the intelligence level. Children of different age were offered short standardized tests of different levels of complexity. A particular set of tests for each age category. The intelligence level of a child was assessed based on what age-related tasks he or she managed to complete.

© The Author(s), under exclusive license to Springer Nature Switzerland AG 2022
D. Bylieva and A. Nordmann (Eds.): PCSF 2021, LNNS 345, pp. 42–58, 2022.
https://doi.org/10.1007/978-3-030-89708-6_5

Later, the German psychologist Stern proposed to use the so called IQ (Intelligence Quotient) for the quantitative assessment of an individual intelligence level. However, to define talent IQ is not enough. It has been noted that a lot of extraordinary intelligent children are characterized by high cognitive activity and unsatiated cognitive need. Age-related development periods are important. The transfer from one period to another, from pre-school to school, from junior to teenager is critical. A lot can change in immature personality. It is important for an extraordinary intelligent child not to get psychological traumas in critical periods that are able to slow him or her down or distort his or her future development. In such crisis periods a child is at risk to lose interest for intellectual activity and creativity. The potential of a child consists of individual and age factors, as well as intellectual and personality factors.

The term "talent" underwent changes during the XX century. Nowadays, it is applied to both children and adults who demonstrate remarkable success in any field. In the 20s scientists distinguished the following types of talent: technical, commercial, academic, artistic, and social (for example, that of a teacher). The more experience was accumulated the wider became the concept of talent. Now it was understood as potential for outstanding achievements in any socially important human activity. As a result appeared new concepts of intellectual and creative talent. Gardner's theory of intelligences was one of them. According to Gardner, there are seven types of intelligence. Linguistic intelligence accounts for the ability to create a verbal product with the help of a language, to transfer information, and to touch people's hearts. Musical intelligence account for the ability to create musical compositions, to competently perform them, and to be esthetically moved by music. Logical-mathematical intelligence accounts for the ability to manipulate and analyze relationships between symbols, signs, objects, to classify them in a particular way (for example, a scientist, a mathematician). Visual-spatial intelligence accounts for the ability to visualize and mentally manipulate with an object, to create spatial compositions (for example, an architect, an engineer, a surgeon). Bodily-kinesthetic intelligence accounts for the ability to be dexterous, to use motor skills at high level (for example, a sportsman, a dancer, a factory worker). Personal intelligence includes the ability to manage one's own feelings and use this in life activity. In addition, a person with high-level personal intelligence is able to understand motives and needs of other people and foresee their behavior (for example, a psychologist, a teacher, a politician). Creative intelligence realized in creative, productive thinking was stated separately, though many researchers believe creativity is the essence of talent, inherent in it. There is no talent without creativity. According to one of the theories, talent is the ability to create something new, to invent, and to synthesize old things in a new and unusual way. We see talent as aptitude for creative activity.

The renowned Structure of Intellect model by American psychologist Gilford is widely used. It distinguishes five types of operations: cognition including perception; memory; divergent thinking as an instrument to generate original ideas; convergent thinking as a more linear way of thinking (mechanism of comparison according to established standards and criteria). Unlike linear convergent thinking divergent thinking is aimed at receiving a number of correct answers. The characteristics of divergent thinking are fluency, "lightness" of a thought (the number of thoughts per time unit); flexibility – the ability to switch from one thought to another; originality – the ability to generate new, unusual, non-standard ideas; accuracy – the ability to finish one's own creative product. Originality of thinking is also called creativity. Talent is not restricted by intelligence. Remember about the unity of affect and intelligence. No productive cognition is possible without human emotions involved in the process. To be creative human thinking must be free. The extent of freedom from constraints and dogmata may be different, but any creative action results in something new. We see the unity of creative thinking, originality of thinking and perception as creativity. This is what is called high cognitive activity, unsatiated cognitive need, cognition pleasure. The Gilford model laid the basis for creativity tests. The works by the Gilford's school analyzed the connection between divergent thinking and IQ. It is noted that this connection is one-sided. At low IQ levels the tendency for divergent thinking hasn't been found. However, the high level of IQ does not guarantee high levels of divergent thinking. Gilford's follower Torrance widened the concept of creativity. In particular, he said that creativity is the capacity for sensitive perception of missing elements of an image, perception of disharmony. That's why any creative action implies going through difficulty, seeking potential solutions. By now profound experience of using the Torrance test has been garnered. Many authors point to the connection between affect and intelligence. Basically, intelligence and creativity themselves depend on characteristics of emotional sphere. This paper has attempted to comprehend the application of the Torrance test taking into account the experience of psychological diagnostics and psychological support. The following characteristics of a creative personality are often mentioned: oddity and eccentricity, heightened sensibility to injustice, tendency to daydreaming.

A smart child beyond his or her age or a child with some remarkable achievements in some area often has unusual personality traits. For example, he or she may demonstrate reflection not typical for his or her age or may think about universe or existential problems. Some psychopathic traits can be seen. Psychopathy is not literally a mental disorder. It is a character pathology that reveals irreversible intensity of traits preventing proper adaptation in society. Psychopathy can be congenital. Gannushkin has called it constitutional. Odd patterns in emotional, intelligent and behavioral spheres may be different. In some cases the distinctness of emotional and volitional spheres virtually devalues intelligence advantages, at least in the eyes of others. People may suffer from constant conflicts with a talented child. There is no clear border between psychopathy and a normal variant of a character. There is nothing in pathology that wouldn't be in a norm. Schizotypal psychopathy is characterized by closeness, unsociability, lowered empathy. Psychiatrists pointed to the variant of personality development not yet pathological, but not quite normal long enough. German psychiatrist Stutte (1967) systematized types of pathological development among children and described a "genius boy" who had acquired several foreign languages by the age of four. The author called such an early development "early-ripe talent". By teenage such people have the following traits: tendency to philosophizing and search for meaning; dissatisfaction with their appearance or body; strong unmotivated tantrums. Their way of thinking is something out-of-the-box and characterized by originality, ability to solve logical problems in an unusual way. Today this pattern of personality development is called schizotypy. From the psychiatry point of view this variant of character and personality development is partly close to schizophrenia. However, in contrast with the clinical characteristics of the disease schizotypal personality doesn't have the so called productive symptoms such as delusion and hallucinations.

Since long ago people demonstrating remarkable social, musical, literature, and artistic creativity have been known to have mental disorders more often than the population on average. The problem of correlation between talent and mental disorders has been discussed for ages, and is still on the table nowadays [1–5]. Especially it is true for the disease of schizophrenia. Schizophrenia is a wide-spread mental disorder which is encountered in one in 100 people. It is quite common for people with schizophrenia to be professional and successful in creative work. There are also statistics that prove cases of schizophrenia in families of outstanding talented people. Empirical data show a high level of creative potential among first grade relatives of patients with psychotic illnesses [6]. The recent epidemiological study has defined scientists (academics) from universities as the ones who demonstrate more creativity than ordinary people. According to this study, relatives of scientists are exposed to higher risk of having schizophrenia [7]. The possibility of schizophrenia and creativity being two sides of the same process is being discussed. Parallels between different types of schizophrenia and different factors of creativity are sought. There is some evidence that different subtypes of schizophrenics manifest different factors and types of creativity. However, the results of studies on creativity in schizophrenia greatly differ, may be because of multiple aspects of creativity and difficulties with differentiation between adaptive creativity and schizotypy [8].

The researchers put special emphasis on the nature of thinking and memory of those suffering from schizophrenia, namely on the analysis of the generalization level [9–12]. It is suggested that both creativity and schizotypy is the manifestation of excessive activation of unusual and remote concepts and/or words, the so called slight actualization of latent characteristics of objects (for example, a wardrobe is enclosed space). The Russian psychologist Yuri Polyakov showed that people with schizophrenia 4 times more often correctly solve logical and mathematical problems than population on average. The degree of manifestation of categoric and/or functional generalizations in schizophrenia depends on the degree of manifestation of cognitive bias. Preferred generalization type depends on the degree of manifestation of cognitive bias and the existence of productive symptoms. Patients with unmanifested cognitive bias use more categoric generalizations in answers for logical problems. Patients with manifested cognitive bias, for example patients with paranoid schizophrenia demonstrate an increase in inadequate answers, namely in bias for functional generalizations [13]. It is likely that the same nature of thinking is inherent in schizotypal personality though of lower degree [14–19]. We are talking about the continuum of the psychopathology development: mental norm, schizotypy, and schizophrenia [20]. Studies in recent years have shown that the participants with schizotypal disorders gave heterogeneous answers of categoric and/or functional types of generalization, and demonstrated slight actualization of latent characteristics of objects. But unlike selected patients with schizophrenia, the participants with schizotypal disorder are more emotionally safe [21–24]. Thus, if the defect is not strongly pronounced, slight actualization of latent characteristics of objects even helps to correctly and quickly solve problems. Mental disorders may be related to the evolution theory. Evolution reflects changes in genes over time. Schizophrenia is probably an old disease with significant genetic component. It may be a by-product of human brain evolution as well as evolution of cognitive functions, language and creative, original thinking [1]. It means that evolution forces may form any phenotype that is genetically rooted and has a long history. Attempts to find a schizophrenia genetic component that also accounts for creativity are ongoing. Researchers have been applying evolution principles to studying schizophrenia since the 60s. As a result, schizophrenia has been considered either as evolutionary beneficial state or as a negative by-product of normal brain evolution. Authors review main historical and modern evolutional explanations of the origin of schizophrenia [25–34]. In general, evolution assumptions link the schizophrenia paradox with the survival advantage obtained as a result of higher creative potential. Due to evolutional circumstances people with schizotypal personality may become more adaptive than population on average.

Personality resources are channeled to serve effective adaptation of a person in some situation. We see personality resources as a system of inherent traits manifested on three levels: the level of individual traits of an organism, the level of individual mental traits, and individual social and psychological traits.

The purpose of the research is to study manifestations of talent as a personality resource in a schizotypal personality based on pathopsychological diagnostics and in the process of psychological rehabilitation.

2   Materials and Methods

Psychodiagnostic Research Methods
Pathopsychological test. This method was used for examination of the participants' mental activity and their adaptation potential. The focus was placed on the originality of thinking.

Extended clinical conversation. Gathering of psychological history. These methods were used to examine social situation of the participants and their adaptation potential. In addition, we addressed relatives to gather psychological history and thus to verify manifestations of talent by the participants.

Incomplete sentences. The method was used to define personal psychological problems of the participants. The focus was also placed on the originality of thinking.

Spielberg's State-Trait Anxiety Inventory. The method was used to define the level of personality anxiety.

Buss-Perry Aggression Questionnaire adapted by S. N. Yenikolopov. The Russian-language version of the Buss-Perry Aggression Questionnaire contains the following scales: Physical aggression – self-report on behavioral bias for physical aggression (behavioral component); Anger – self-report on bias for irritation (emotional component); Hostility – a scale consisted of statements of two subscales: Suspicion and Sensitivity to offence (cognitive component).

Torrance figurative subtests. They were used to define the level of originality of thinking.

Analysis of verbal and graphic production of the participants (drawings, diaries, essays, other records, animations). Art specialists – an art expert and a painter – provided expert assessment of the participants' manifestations of talent on a 1 to 10 scale. They were not informed about the purpose of the research.

Sampling in the Research

We have examined 66 adults – 36 men and 30 women – aged from 18 to 60. The experimental group consisted of 36 participants with the diagnosis "schizotypal disorder" (F21), according to the International Classification of Diseases (ICD 10). The experimental group consisted of 21 men and 15 women. The participants of the control group didn't have mental health diagnosis. The control group included 30 participants with personal psychological problems who had addressed a counsellor. The control group consisted of 15 men and 15 women. Criteria for sampling: informed consent for participation in the research. Exclusion criteria: psychotic manifestations, behavior disorganization, and technical attitude towards the research. The participants from the experimental group were getting sparing medication. The target of the psychopharmacological correction was the heightened anxiety. The participants from the control group weren't getting medication. The research was carried out at the premises of a non-state medical center of psychotherapy in Moscow.

Methods of Data Analysis. The focus in our work was placed on the qualitative analysis of content of verbal or graphic production of the participants across the whole range of methods. We were interested in repetition of manifestations of talent and originality of thinking in each method. Originality of thinking was determined by means of pathopsychological test, Torrance subtests, and incomplete sentences. This set of methods allows for cross verification of the results.

The correlation analysis (Spearman's rank correlation coefficient) and analysis of intergroup differences (Mann-Whitney U-test) were carried out.

The Research Structure

The research consisted of three stages.

The first stage was the psychological diagnostics by the psychiatrist.

The second stage was the psychological rehabilitation in accordance with the established thematic plan of psychological support.

The third stage was the final step of the psychological diagnostics. The psychiatrist made an observation in real time.

3   Results

As the results of the research have shown, all participants of the experimental group are characterized by the originality of thinking. We see the originality of thinking as slight actualization of latent characteristics of objects. This characteristic of thinking was determined by classic pathopsychological methods like pictographs, exclusion of odd elements, object classification, comparison of concepts, projective drawing techniques, as well as by the Torrance figurative subtests. For the part of the participants of the experimental group with the diagnosis "schizotypal disorder" the psychiatrist questioned the diagnosis "schizophrenia" (F20). This group consisted of 14 participants, including 10 men. The participants from the assumed "schizophrenia" group can be characterized as emotionally cold and having scarce feelings. For example, they coldly talked about their closed ones and relatives, though the latter took care about the participants as can be suggested by the objective information. The participants from the experimental group with the explicit diagnosis "schizotypal disorder" showed themselves as more emotionally "warm". Even being highly eccentric and extravagant in their behavior they were emotionally more responsive. The participants of the research with personal problems who made up the control group didn't show the originality of thinking. Diagnostics using Spielberg's State-Trait Anxiety Inventory and Buss-Perry Aggression Questionnaire didn't reveal meaningful differences in the levels of personality anxiety and aggressiveness in the experimental and the control groups. However, it should be noted that people with schizotypal disorders have qualitatively different hostility than population on average.

According to the expert assessment by the art specialist and the painter, 9 participants from the experimental group were talented (9–10 scores). Five of them had the questioned diagnosis "schizophrenia". Though they didn't have productive symptoms (delusion, hallucinations) they were too emotionally cold towards other important people.

Below are two typical cases. In the first case the participant of the research was an 18-year old girl with artistic talent who studied in the design school (it should be added that most subjects there were taught in English). She was emotionally cold about her parents. There was originality of thinking detected (for example, "separation" in pictograph was displayed as bars, "grief" as a cup of tea with fumes). Originality of thinking was also manifested in the comment to the stimulus "war" in pictograph – "a picked bone as hunger and death" (she depicted an empty fridge and a bone). In response to the stimulus "happiness" she made a drawing of a bed and commented that "happiness is to get up easily in the morning". The common between a hedgehog and milk, in her opinion, is that the hedgehog crumbled into dust will become liquid; they differ in the way that the hedgehog is static while milk is fluent". In response to a stimulus phrase "blood and kerosene" she answered that "they both can be used to make a swimming pool, it's okay, but the color and flowrate are different". The difference is that "kerosene flames better".

In the second case the participant of the research was a 35-year old man, an engineer. He was emotionally cold about his wife. During the pictograph method in response to the stimulus "separation" he made a drawing of running water and a stone and explained that "water separates from earth". To remember the stimulus "grief" he depicted *The Thinker* by Rodin. To remember the stimulus "war" he depicted chess that he saw in the Historical Museum and commented: "war is the way of lie, lie is the way of war, the Old Chinese wisdom says". To remember the stimulus "power" he depicted Machiavelli. During the classification of objects method he singled out "a globe" – "it is the most important, it stands for the world", while he called cookware "a kingdom of vessels".

According to the experts, there were no talented people in the control group.

Individual Psychological Support
We provided psychological assistance to the participants of the experimental and control groups in the course of the research. But as talent wasn't found in the control group, it is much more difficult to hinge on creativity in the rehabilitation work with the control

group. For this reason we don't compare the results of the correction in two groups. Nevertheless, the psychiatrist who observed the participants from both groups over time, noted positive dynamics in the control group in 50% of cases. Negative dynamics weren't recorded.

We worked with the patients based on a preliminary thematic plan that in each case transformed in an individual psychological rehabilitation programme. Such assistance is possible only in the absence of productive symptoms (delusion, hallucinations) and under control of the psychiatrist. Below are the principles of working with the experimental group, i.e. with people with schizotypal personality.

The main purpose of the psychological rehabilitation is to support people with schizotypal personality and to form adequate understanding and attitude towards their special character.

The main tasks of the psychological support:

– to raise the level of social adaptation of people with schizotypal personality;
– lower the level of anxiety;
– to lower the level of aggressiveness to the adequate value.

The important principle of this work is to account for the psychological state of the patient at each meeting and for his or her life circumstances.

During this research the individual counselling session for people with schizotypal personality lasted for 5–6 months. It consisted of the following stages:

1. Acquaintance with the patient's life history, gathering of subjective and objective psychological history. Subjective history is gathered based on what the participant has told about himself or herself, and objective history is gathered from the relatives' words.
2. Pathopsychological test and counselling including projective techniques based on the results of the observation by the psychiatrist. Buss-Perry Aggression Questionnaire. Spielberg's State-Trait Anxiety Inventory. If necessary, the specialist tells the participant and/or his or her relatives what problems the psychology may solve (for example, to help in difficult life situations). The specialist also provides information about people's characters and how talent and high aptitude is understood.
3. Further diagnostics and counselling within the framework of causal psychotherapy. Engagement with the patient is built in the framework of the humanistic approach paradigm. Its main method is empathetic listening.
4. Usage of artistic work material. Spontaneous drawing and modelling. Musical associations. At this stage it is important to comply with the acceptance principle to strengthen the established alliance with the patient.
5. Discussion and/or non-verbal work on the following topics: My usual day, My wishes, A person I love, A person I don't love, A person I respect, My past, etc. Previously gathered psychological history should be taken into account. We focus on difficult life situations the patient has experienced: loss of work, jealousy-based conflicts, criminal conflicts, mobbing and bulling, family conflicts, a break-up, failure in an individually significant area, etc. It is important that these are not subjectively important events, but objectively psycho-traumatic factors.

6. Discussion and/or non-verbal work on the following topics: Me real and Me ideal, Me in the past, present and future, A person I am afraid of, Two moods. Drawings "Me as a weapon", "Me as a jewelry". The topics "conflict", "perception of happiness", "perception of unhappiness" may be picked. But they require more caution.
7. Essay "My personal horror script". This is an attempt to find out and formulate how the patient works on his or her fears, how he or she controls them and "makes friends with them". Here it is important to empathetically share emotional experience with the participant.
8. Intermediate diagnostics, observation by the psychiatrist. The psychologist should pay close attention to the state of personality defense mechanisms.
9. Projective techniques "Me in 10 years", "Me in 5 years", "My dream", "My aim". It is critical to take into consideration individual traits of the patient, the presence of support from significant other people, relatives, members of the family, the structure of time perspective, and, finally, the age.
10. Myth and drama method, an essay about oneself. The story must be paradoxical, intriguing, must contain a psychological situation like the one that the patient has experienced. These may be the stories about breaking up with the loved ones, about search for hobbies, about professional self-realization problems and many others.
11. The final stage of diagnostics. Assessment of the anxiety level and the degree of manifestation of aggression biases using Spielberg's State-Trait Anxiety Inventory and Buss-Perry Aggression Questionnaire.
12. The final counselling with the patient and his or her relatives being close during the whole research cycle.

The case of a 29-year old girl from the group of talented participants of the research. She belongs to the group where the psychiatrist made the explicit diagnosis "schizotypal disorder" and where there was no diagnosis "schizophrenia" under the question. This means that emotional sphere of the participant was characterized as "warm". The girl graduated from Russian State University of Cinematography named after S. Gerasimov. The teachers noted her great achievements in studies. She manages a graphical account in VK. She is successful in profession related to animation. She is interested in dances, classic autogenic training, speaks English fluently. Below is the description of the pathopsychological test results. Mental development corresponds to the education received. Rich vocabulary can be noted. Originality of thinking is present. During the classification of objects method she put aside a butterfly and named the group "something that flies". In her opinion, the common between a pillow and a skyscraper is that you can sleep on the roof, and also that these concepts are associated with romance, clouds resemble a pillow. The common between an axis and a wasp is that "the wasp flies along the straight line", "the two words in Russian have similar spelling". The common between an owl and a basket is that the owl bears the basket as in the cartoon". According to the experts, the Torrance figurative subtests have revealed the participant's own drawing style in animation.

Here are the examples of the incomplete sentences method: "I know it's silly, but I fear communicating with strangers", "for me, it is better to work with a friend, even with the imaginary one", "when I see a man with a woman, I feel lonely", "the future seems far and unreal". The psychological correction was targeted at heightened anxiety, self-esteem problems, and the problem of self-identity. During the psychological rehabilitation the following topics were discussed and worked on: My usual day, My wishes, My past, Me real, me ideal, and others. In addition, during this work we had counselling sessions with her mother where we discussed manifestation of anger and irritation. Below are the examples of drawings "Me as a jewelry", "Me real, me ideal" made during the psychological rehabilitation (Fig. 1).

Fig. 1. Drawings "me as a jewelry" (upper) and "me real, me ideal" made by the participant of the research in the process of psychological rehabilitation.

The changes in the participants' state were assessed by the psychiatrist. According to the results of the research, all 9 talented (9–10 scores from the experts) participants from the experimental group had explicit positive dynamics. Of them 5 emotionally cold patients stayed emotionally cold towards significant others, but the perception of their own emotional state became better. According to the self-report from these 5 participants positive emotions became prevailing over negative emotions, they started seeing their future more positively, and time perspective became structured. This information is taken from the psychiatrist's report. Only half of less talented participants (less than 9 scores) from the experimental group showed positive dynamics according to the psychiatrist; explicit positive dynamics weren't recorded at all. Also, there were no negative dynamics in the experimental group. According to the results of Buss-Perry Aggression Questionnaire the level of hostility in the experimental group has lowered significantly from the statistics point of view ($p = 0.05$). According to Spielberg's State-Trait Anxiety Inventory the level of anxiety has lowered significantly from the statistics point of view ($p = 0.03$). At the same time, both at the first and final stages of the psychological diagnostics the levels of hostility positively correlate with the levels of personality anxiety ($r = 0.41$, $p < 0.05$ and $r = 0.36$, $p < 0.05$ respectively).

In the control group the picture of anxiety and aggressiveness received by applying the research methods is similar. According to the results of Buss-Perry Aggression Questionnaire the level of hostility has lowered in conformity with the trend ($p = 0.08$). According to Spielberg's State-Trait Anxiety Inventory the level of personality anxiety has not significantly changed. At the same time, both at the first and final stages of the psychological diagnostics the levels of hostility positively correlate with the levels of personality anxiety ($r = 0.37$, $p < 0.05$ and $r = 0.39$, $p < 0.05$ respectively).

4 Discussion

Different attempts of interdisciplinary study of schizophrenia and schizotypy have been made. In recent years the idea has migrated from psychology to psychiatry that psychotic thoughts and feelings are considered normal among population in general. Such emotional experience is statistically widespread, and thus it just can't reflect the pathology. The same characteristic can be both healthy and pathological. But it is important at what level of schizophrenic continuum the personality stays. In a severe paranoid version of schizophrenia the key disturbances are difficulties with planning, problem solving. It is difficult to talk about productive activity in the case of paranoid schizophrenia, delusion, and hallucinations. However, with schizotypy not approaching the disease stage the originality of thinking may be helpful in adaptation to situations. It is possible to talk about evolutional advantages of psychotic disorder in this case. Therefore, it is worth trying to study behavioral, genetic, visual, and psychopharmacological markers of healthy manifestation of psychotic traits. Schizophrenia and schizotypy are connected with human evolution process. Apart from attempts to find schizophrenia genetic component, there are attempts to define changes in brain substrate as a result of schizophrenia. A number of authors link them to disturbances in fronto-temporal and parietal regions of brain [35]. On the other hand, the results of the research on biological, neurophysiological and genetic correlations in schizophrenia are contradictory [19, 36]. There is a theory that schizophrenia genetic component as a genetic basis of psychosis and schizotypal genetic component may differ [37]. We are inclined to believe that schizophrenia and creativity are the two sides of the same process, that there is schizophrenia genetic component that accounts for creativity. In schizotypy there is "good" originality of thinking vs. population in general and there is no explicit psychosis in contrast with those who suffer from schizophrenia.

If we talk about talent as aptitude for creativity, then emotional disturbances, emotional changes of a personality play an important role. Even reduced delusion and hallucinations manifested in past medical history may hinder creativity. Creativity coupled with delusion will be hurtful. Literature historians, cultural specialists, psychologists have been discussing since long ago whether Fyodor Dostoevsky would have killed someone if he hadn't written "Crime and Punishment" and "The Brothers Karamazov"? Whether the author of "Lolita" Nabokov was a pedophile? This is called the power of an artistic image and literary talent. However, many readers consider the above-mentioned books hurtful, unpleasant. Based on our experience of providing psychological support to schizophrenia patients they also can be talented, but their works look emotionally altered and cause fear. Here is the example of a drawing made by a female patient with schizophrenia. Her artistic work, according to the expert opinion of the art specialist and the painter, has artistic value (Fig. 2). It reflects the patient's delusional idea of affection – a big black dog.

Fig. 2. The example of graphic production by the patient with schizophrenia.

To what extent a hurtful and scary artwork can be called artwork? It may "poison" someone emotionally unstable. May be, this is a philosophical question. We are for "healthy" creativity, though we perfectly understand that there is nothing in pathology that wouldn't be in a norm.

We have come to the conclusion that instead of focusing on pathology only, future research should analyze behavioral, genetic, visual, and psychopharmacological correlations that determine healthy manifestation of psychotic traits. Such research can provide information about defense and coping mechanisms and evolutional advantages of psychosis. Future lies in the interdisciplinary pieces of research methodologically based on psychology. Future lies in the interdisciplinary pieces of research [38–40].

Empirical Research Findings are Used in Training of Clinical Psychologists
The course "Pathopsychology" consists of lectures, workshops and practical work. Lectures on pathopsychology are read in the third year of education when future clinical psychologists had already attended a number of courses on general psychology, personality psychology, emotions, talent, etc. This is important because by the time the students dive into clinical disciplines, they should have learnt personality typology, types of psychopathy and accentuation of personality traits from different authors, including classical German psychiatry. In our lectures and workshops we balance general psychology as a scientific discipline with clinical psychology, pathopsychology. We talk about clinical psychology being a method of general psychology as only the example of a "breakdown" of a single mental function or the whole personality allows to see common factors of mentality functioning. Understanding these important things is essential for building interdisciplinary relations in the educational process.

Studying individual diseases in the course "Pathopsychology" follows acquaintance with clinical symptoms. Pathopsychological symptoms are delusion, hallucinations, psychomotor agitation, impaired consciousness. Further we study pathopsychological syndrome of schizophrenia. It is a typical combination of symptoms, combination of primary and secondary symptoms of the disease, disease pattern. Pathopsychological syndrome of schizophrenia is a relatively stable combination of mental health disorders, i.e. behavior changes, emotional and personality manifestations, special character of cognition. The main factor is impaired personal mindset: the settings are known, but they have lost their impetus; the patient knows what is required, but doesn't do this. According to the classic B.V. Zeigarnik, this is a motivation problem. Motivation diverseness may be the case, when during the test the patient chooses one task and despite time limits manipulates with this task, with the main motive being interest for this particular task. Secondary symptoms are thought process disorders like slight actualization of latent characteristics of objects ("a wardrobe is enclosed space "), tangential speech. During workshops literature describing classical and recent research on this topic, including on talent psychology, is discussed. As part of practical work, students attend a psychiatric clinic to follow the teacher in patient observation and then to complete pathopsychological tests independently. The students make their own observation of patients in a hospital.

Below we have outlined criteria based on which students examine a patient and analyze a case. The focus is on characteristics of thought process:

• Disorders on the operational side of thinking: divergent generalization. Patients describe trivial, routine situations from a theoretical position. When completing the task on object classification they take general characteristics of objects as a basis (for example, both a cat and a car can move), ignoring the conceptual side. Or they classify them only on the basis of external, immaterial characteristics (for example, scissors and a saucepan are made from metal).

• Disorders in the motivation component of thinking. Light actualization of formal associations, inadequate convergences: "both an axis and a wasp are o-shaped, blood and kerosene can boil". Diverseness of thinking: when judgements about something go in different planes. In the case of diverseness, the basis of classification is not always the same. During one and the same task patients can unite objects based on the objects' characteristics and then based on their own preferences and attitude.

A clinical psychologist like a doctor is an excellent example of a "person-to-person" profession. For future work being successful expertise has to be handed down from a teacher to students through lectures and conversations, discussion of clinical cases during practical work and asking questions. We analyze graphic and verbal production of patients. Practical work on pathopsychology became education "in the field". Such form of education is possible and considered optimal in groups of about 15 students. Only in these circumstances a psychological diagnostician and/or consulting psychologist work out professional intuition. Expertise and professional intuition are necessary for provision of professional help to people.

Analysis of drawings can be operational and conceptual. First, we consider operational side and the drawing technique. Application of heavy pressure in drawing lines in combination with other characteristics can mean aggressiveness, energy, and stubbornness. Slight pressure can mean lack of confidence, timidity. The size and the place of a drawing on the sheet also matter, etc. As our experience has shown, timid people do depict support lines – horizon, ground, grass, etc. In conceptual analysis it is important to pay attention to people and faces on the drawing, if any. Absence of people and faces can point to schizotypal personality with emotional coldness. For emotionally cold schizotypal personality it is typical to depict members of families schematically or graphically. It is important to perform conceptual analysis during the workshop when it is possible to touch and study the drawing in details, ask questions about it. Here it is essential to compare the results of the conversation with the patient with the graphic production. An interesting tendency both in consulting and teaching work should be noted. Talented in drawing patients often post their works in the internet (VKontakte, etc.), and such works can be and have to be analyzed together with the students, and compared with the production made during psychological tests. But here ethical issue arises as it is almost impossible to do this anonymously, and very few patients give their informed consent for such educational work. Having analyzed and discussed clinical cases with the teacher the students write their conclusions independently. Thus, lectures, workshops and practical work being different forms of work complement each other.

During the research-based learning future clinical psychologists learn how to:

- diagnose special character of thinking by the above-mentioned criteria,
- determine special emotional characteristics, i.e. emotional warmth/coldness towards significant other people, close relatives,
- identify talent as aptitude for creativity in people with schizotypal personality.

5 Conclusions

1. Empirical research has shown that talent is interrelated with schizotypal personality.
2. Talent is a personality resource in the process of psychological rehabilitation of schizotypal personality. This fact may be used in professional training of clinical psychologists.
3. Similar to classic pathopsychological tests the Torrance figurative subtests reveal the originality of thinking and psychopathology in people with schizotypal personality. This fact may be used in professional training of clinical psychologists.

Disclaimer: It should be noted that tasks used to determine schizophrenic deficit are similar to tasks used to determine creativity potential. We tried to offset this by carrying out a comprehensive pathopsychological research of cognitive and emotional and volitional spheres.

References

1. Srinivasan, S., Bettella, Fr., Mattingsdal, M.: Genetic markers of human evolution are enriched in schizophrenia. Biol. Psychiatry **80**(4), 284–292 (2016). https://doi.org/10.1016/j.biopsych. 2015.10.009
2. Kar, N., Barreto, S.: Psychosis, creativity and recovery: exploring the relationship in a patient. BMJ Case Rep. **2018**, bcr2017223101 (2018). https://doi.org/10.1136/bcr-2017-223101
3. MacCabe, J.H., Sariaslan, A., Almqvist, C., Lichtenstein, P., Larsson, H., Kyaga, S.: Artistic creativity and risk for schizophrenia, bipolar disorder and unipolar depression: a Swedish population-based case-control study and sib-pair analysis. Br. J. Psychiatry **212**(6), 370–376 (2018). https://doi.org/10.1192/bjp.2018.23
4. Charlier, P., Deo, S.: Schizophrenia: four examples of historical retrospective. Diagn. Encephale **44**(6S), 55–57 (2018). https://doi.org/10.1016/S0013-7006(19)30082-X
5. Dalgleish, M.: Art-making and lived experience of Schizophrenia: a vitalist materialist analysis arts. Health **11**(1), 26–37 (2019). https://doi.org/10.1080/17533015.2017.1392330
6. Sandsten, K., Nordgaard, J., Parnas, J.: Creativity and psychosis. Ugeskr Laeger **180**(32), V02180141 (2018)
7. Parnas, J., Sandsten, K., Vestergaard, C.l., Nordgaard, J.: Schizophrenia and bipolar illness in the relatives of university scientists: an epidemiological report on the creativity-psychopathology relationship. Front. Psychiatry **10**, 175 (2019). https://doi.org/10.3389/fpsyt. 2019.00175
8. Son, S., et al.: Creativity and positive symptoms in schizophrenia revisited: structural connectivity analysis with diffusion tensor imaging. Schizophr. Res. **164**(1–3), 221–226 (2015). https://doi.org/10.1016/j.schres.2015.03.009

9. Ettinger, U., et al.: Cognition and brain function in schizotypy: a selective review. Schizophr. Bull. **41**(Suppl. 2), 417–426 (2015). https://doi.org/10.1093/schbul/sbu190

10. Acosta, H., Straube, B., Kircher, T.: Schizotypy and mentalizing: an fMRI study. Neuropsychologia **124**, 299–310 (2018). https://doi.org/10.1016/j.neuropsychologia.2018.11.012

11. Sahakyan, L., Kwapil, T.: Moving beyond summary scores: decomposing free recall performance to understand episodic memory deficits in schizotypy. J. Exp. Psychol. Gen. **147**(12), 1919–1930 (2018). https://doi.org/10.1037/xge0000401

12. Flückiger, R., et al.: The interrelationship between schizotypy, clinical high risk for psychosis and related symptoms: cognitive disturbances matter. Schizophr. Res. **210**, 188–196 (2019). https://doi.org/10.1016/j.schres.2018.12.039

13. Karagiannopoulou, L., Karamaouna, P., Zouraraki, C., Roussos, P., Bitsios, P., Giakoumaki, S.: Cognitive profiles of schizotypal dimensions in a community cohort: common properties of differential manifestations. J. Clin. Exp. Neuropsychol. **38**(9), 1050–1063 (2016). https://doi.org/10.1080/13803395.2016.1188890

14. Wastler, H., Lenzenweger, M.: Cognitive and affective theory of mind in positive schizotypy: relationship to schizotypal traits and psychosocial functioning. J. Pers. Disord. **12**, 1–16 (2020). https://doi.org/10.1521/pedi_2020_34_473

15. Bora, E.: Theory of mind and schizotypy: a meta-analysis. Schizophr. Res. **222**, 97–103 (2020). https://doi.org/10.1016/j.schres.2020.04.024

16. Giakoumaki, S.: Emotion processing deficits in the different dimensions of psychometric schizotypy Scand. J. Psychol. **57**(3), 256–70 (2016). https://doi.org/10.1111/sjop.12287

17. de Wachter, O., De La Asuncion, J., Sabbe, B., Morrens, M.: Social dysfunction in schizotypy. Tijdschr. Psychiatr. **58**(2), 114–121 (2016)

18. Mohr, C., Ettinger, U.: An Overview of the association between schizotypy and dopamine. Front. Psychiatry **5**, 184 (2014). https://doi.org/10.3389/fpsyt.2014.00184

19. Mohr, C., Claridge, G.: Schizotypy–do not worry, it is not all worrisome. Schizophr. Bull. **41**(Suppl. 2), 436–443 (2015). https://doi.org/10.1093/schbul/sbu185

20. Li, L. Y., Fung, C., Moore, M., Martin, E.: Differential emotional abnormalities among schizotypy clusters. Schizophr. Res. **208**, 285–292 (2019). https://doi.org/10.1016/j.schres.2019.01.042

21. Barrantes-Vidal, N., Grant, P., Kwapil, T.: The role of schizotypy in the study of the etiology of schizophrenia spectrum disorders. Schizophr. Bull. **41**(Suppl. 2), 408–416 (2015). https://doi.org/10.1093/schbul/sbu191

22. Horan, W., Blanchard, J., Clark, L., Green, M.: Affective traits in schizophrenia and schizotypy. Schizophr. Bull. **34**(5), 856–874 (2008). https://doi.org/10.1093/schbul/sbn083

23. Statucka, M., Walder, D.: Facial affect recognition and social functioning among individuals with varying degrees of schizotypy. Psychiatry Res. **256**, 180–187 (2017). https://doi.org/10.1016/j.psychres.2017.06.040

24. Pflum, M., Gooding, D., White, H.: Hint, hint: theory of mind performance in schizotypal individuals. J. Nerv. Ment. Dis. **201**(5), 394–399 (2013). https://doi.org/10.1097/NMD.0b013e31828e1016

25. Power, R., et al.: Polygenic risk scores for schizophrenia and bipolar disorder predict creativity. Nat. Neurosci. **18**(7), 953–955 (2015). https://doi.org/10.1038/nn.4040

26. Wang, D., Guo, T., Guo, Q., Zhang, S., Zhang, J., Luo, J.: The association between schizophrenia risk variants and creativity in healthy Han Chinese subjects. Front. Psychol. **10**, 2218 (2019). https://doi.org/10.3389/fpsyg.2019.02218

27. Pearlson, G., Folley, B.: Schizophrenia, psychiatric genetics, and Darwinian psychiatry: an evolutionary framework. Schizophr. Bull. **34**(4), 722–733 (2019). https://doi.org/10.1093/schbul/sbm130

28. Mistry, S., Harrison, J., Smith, D., Escott-Price, V., Zammit, S.: The use of polygenic risk scores to identify phenotypes associated with genetic risk of bipolar disorder and depression: a systematic review. J. Affect. Disord. **234**, 148–155 (2018). https://doi.org/10.1016/j.jad.2018.02.005

29. Brüne, M.: Schizophrenia-an evolutionary enigma? Neurosci. Biobehav. Rev. **28**(1), 41–53 (2004). https://doi.org/10.1016/j.neubiorev.2003.10.002

30. Rybakowski, J., Klonowska, P., Patrzała, A., Jaracz, J.: Psychopathology and creativity. Psychiatr. Pol. **40**(6), 1033–1049 (2006)

31. Giotakos, O.: Persistence of psychosis in the population: the cost and the price for humanity. Psychiatriki **29**(4), 316–326 (2018). https://doi.org/10.22365/jpsych.2018.294.316

32. Polimeni, J., Reiss, J.: Evolutionary perspectives on schizophrenia. Can. J. Psychiatry **48**(1), 34–39 (2003). https://doi.org/10.1177/070674370304800107

33. Thys, E., Sabbe, B., De Hert, M.: Creativity and psychiatric disorders: exploring a marginal area. Tijdschr. Psychiatr. **54**(7), 413–425 (2012)

34. Kelemen, O., Kéri, S.: Schizophrenia and evolutionary psychopathology. Psychiatr. Hung. **22**(5), 333–343 (2007)

35. Zouraraki, Chr., Karamaouna, P., Karagiannopoulou, L., Giakoumaki, S.: Schizotypy-Independent and schizotypy-modulated cognitive impairments in unaffected first-degree relatives of schizophrenia-spectrum patients. Arch. Clin. Neuropsychol. **32**(8), 1010–1025 (2017). https://doi.org/10.1093/arclin/acx029

36. Sampedro, A., et al.: Neurocognitive, social cognitive, and clinical predictors of creativity in schizophrenia. J. Psychiatr. Res. **129**, 206–213 (2020). https://doi.org/10.1016/j.jpsychires.2020.06.019

37. Cohen, A., Mohr, C., Ettinger, U., Chan, R., Park, S.: Schizotypy as an organizing framework for social and affective sciences. Schizophr. Bull. **41**(Suppl. 2), 427–435 (2015). https://doi.org/10.1093/schbul/sbu195

38. Pavlenko, A.: Technology as a new language of communication between the human being and the world. Technol. Lang. **1**, 91–96 (2020). https://doi.org/10.48417/technolang.2020.01.19

39. Zamorev, A., Fedyukovsky, A.: Euathlus and crocodile paradoxes: dialectic solution's advantages. E3S Web Conf. **164**, 11022 (2020). https://doi.org/10.1051/e3sconf/202016411022

40. Rubtsova, A.V., Almazova, N.I., Bylieva, D.S., Krylova, E.A.: Constructive model of multilingual education management in higher school. IOP Conf. Ser. Mater. Sci. Eng. **940**, 012132 (2020). https://doi.org/10.1088/1757-899X/940/1/012132

Epistemic Foraging and the Creative Process: Crawling Over Creation

Walker Trimble[✉] [iD]

Herzen State Pedagogical University of Russia, 48, Moika Emb., St. Petersburg 191186, Russia

Abstract. The function of creation is to create, to make something that was not there before. As technological objects are themselves creations, the Modern considers that they cannot create. What if it were the case that creation were a product of a set of behaviors often associated with insight but also observed in both living and non-living things? Would this change how we understand creativity? The present paper examines the relationship between an aspect of the Free Energy Principle (FEP) – a grand theory of biology, information and brain science – and creative behavior. It is argued that the seemingly aimless activity that often surrounds insights may be a variety of "epistemic foraging" modeled by the FEP's notion of "active inference". An informal thought experiment is introduced to explore whether foraging can be distinguished empirically among systems with different levels of control and consciousness. Insights from the FEP are then applied to an aesthetic formula proposed in the *Poetics* of Aristotle. It is found that the range of variety in predicted states is more important than arbitrariness. It is speculated that sets of predictions with high relative entropy in proportion to value surrounded by other sets that compensate with lower entropy could explain the foraging features of many fixed-action patterns of behavior. Initiating related discussions about additive theories of liberty and textured consciousness, these humanitarian ruminations should be given more rigor in the future so as to be applied to modeling behavior such as play, ritual, and the arts.

Keywords: Epistemic foraging · Creativity · Free Energy Principle · Human-machine interaction · Robotics · Aesthetics · Positive liberty

1 Introduction

The difference between human creativity and the works of technology has long been characterized as the distinction between maker and made. The difference between creator and creation is, at its limits, theological. The cosmological argument for God's existence rests upon *creatio ex nihilo*. All that man creates is a rearrangement of pre-existing parts. In some traditions, such as that of the Christian East, emphasis is put on the fact that the "image and likeness" which God gave man is his reason, free will, and invention (*epinoia*). Though this does not mean man makes matter out of nothingness, imaginative capacity is a shadow of original creativity.

In the modern West, human inventiveness acquired a status that it sought to rival and surpass that of the divine. The Romantic cult of genius around such figures as Newton,

D. Bylieva and A. Nordmann (Eds.): PCSF 2021, LNNS 345, pp. 59–71, 2022.
https://doi.org/10.1007/978-3-030-89708-6_6

Goethe, and Wagner gave human creation mystical, demiurgic power. Tales surrounding Eureka moments – Einstein's dream of riding on a beam of light – propagate these mysteries.

Genius is expected to leap forward and transcend any effort to determine the processes behind it. As Freudian theories were taken up by artists and poets, the Unconscious began to take the place of the spirit of genius or the Muse.

Bringsjord et al. [1] have argued that, by training computers to convince their human interlocutors of their humanity, the Turing Test merely creates the stimulus to dissemble. A better alternative, they suggest, would be to consider a system conscious if it could come up with something that its designers did not, or could not, conceive themselves. He calls this the Lovelace Test, after a comparable proposal by Ada Lovelace.

One, however, is skeptical that the Lovelace Test could tell us what 'really new' means. All complex systems with some level of autonomy and action begin to generate their own information. A programmer sent to search out a bug that has developed in a system is examining information that was not there before. A spool of string tangled in a knot conveys information not designed or intended by the winder.[1] Google's recent artificial intelligence (AI)-assisted protein mapping project will provide many connections that hitherto have not existed [2]. The upshot of Bringsjord's claim is that ingenuity should be the aim of AI rather than deception, and this is a reasonable claim. Whether it is a probative test is more debatable. After all, novelty in science or art is a socially determined quality, one frequently open to discussion. Insights often require years of confirmation to determine whether they are novel. Proving *genuine* novelty may be like solving the "Winger's Friend" problem of quantum physics where determining the observation of a state becomes dependent upon whether the message was conveyed to others.[2] This spills into the question of authorship. Should the AI mapping of proteins produce new information, credit will be socially, if not still legally, attributed to the developers and not to the system. While a poet's prayer to the Muse may be taken as false modesty, AI will not even be allowed any false pride in its inventions. Bringsjord presupposes this social element and then is surprised he cannot render it in code.

Certainly the most important thing creativity does is create; but a sensibility that sees it as a province of some demiurge occludes the importance of patterns of behavior that circulate around brilliant innovations. When creation has to be a divine spark, then it is *ipso facto* impervious to any account of the process behind it. The only real creation is that which is *ex nihilo,* and that *nihil* must be complete.

I here accept that technological systems that engage in autonomous activity sometimes arrive at new information and that the novelty of that information is independent

---

[1] Bringsjord notes that this was Turing's response to Lovelace: "Machines take me by surprise with great frequency." Yet Bringsjord's own response to this "amazingly bad" [1, p. 5] reasoning also fails to convince.

[2] "Unfortunately, it seems to us (or at least to one of us: Bringsjord) that the moral of the story may be irredeemably negative. There may simply not be a way for a mere information-processing artifact to pass LT, because what Lovelace is looking for may require a kind of autonomy that is beyond the bounds of ordinary causation and mathematics" [1, p. 25].

of the presence of an intention behind it. Shall we not try to understand how these creations occur and promote their safe and reasonable expression rather than positing their ontogeny?

A wealth of literature in areas ranging from pedagogy to machine learning has taken up the creative process as behavior and this may provide more probative value than an alchemy of inspiration. Here I will argue that the Free Energy Principle (FEP), a very productive biological theory with roots in AI research, can give us fundamental insights into the creative process. Rather than a complete proof, this paper is rather a salvo to those who argue that creativity is dependent upon a qualitative definition of consciousness, agency, and novelty. In order to explore similarities in creative processes between conscious and unconscious beings, I will take up a thought experiment followed by an excursion into aesthetics. Finally, I will argue that the philosophical and ethical implication of this theory entails a positive, additive notion of freedom. I conclude that there is value in regarding creativity as that to which all things strive.

2 The Free Energy Principle and Creativity

The Free Energy Principle (FEP) is arguably the grandest theory to have emerged in biology over the past thirty years. Developed largely by neuroimaging expert Karl Friston, it uses Bayesian probability to model a large set of complex behaviors. One of its most interesting aspects is how easily it sweeps between modeling biology – eye tracking [3] and dopamine secretion [4] – and artificial neural networks and robotics [5]. In fact, FEP's use of control processes (such as directed acyclic graphs (DAGs)) and Pearl causality reveal that at least some of its origins lie in AI and machine learning. Thus the FEP is a good place to get your hands dirty in the woolly boundaries between technology and biology.

The FEP is simple and intuitive, though devilish in the details. I shall try and summarize it in language that is useful for our discussion here. An organism seeks to maintain a stable relationship with its environment under changing conditions. It order to do so it must make predictions and minimize the error of those predictions. These can be modeled through a Bayesian probability machine where the organism (or agent) has prior beliefs about the environment that are tested according to principles of information. Individual agents making predictions then arrange themselves into further sets that also engage in active inference along predictive principles. FEP thus uses 1) Bayesian-based control theory, 2) Shannon-based information theory, 3) Markovian graph theory.

In order to initiate the control process, and thus reduce error, the agent engages in what FEP calls 'active inference'; i.e., measuring inputs against beliefs, correcting the belief, measuring another input, and so forth.[3] Now this takes place in a number of different ways. Deep to Karl Friston's approach is an ecological, enactive, account of cognition; he favors Gibsonian affordances and dynamic critical points over grid-like, nominalist, computational models [7].

So the organism is not only making predictions as to how to pierce a point with one's spear, or the eye of the leaping spelt with one's beak, but also through constant meandering over the environment. Friston has a lovely phrase for this, *epistemic foraging*.

---

[3] See the summary in [6].

Friston's followers hasten to admit that this 'epistemic' must be distinguished from the use of the term in its philosophical sense – this is not about a theory of knowledge accumulated and stored as memory.[4] However, I hope the following shows that there is no particular need to distinguish active inference from the *behaviors* inherent to acquiring knowledge. And it is behavior and not memory storage that most concerns us. Since the organism must be in a constant state of correcting beliefs and testing them, perception is not passive. As in Coleridge's visionary epistemology, perception is a faculty of the imagination – one of Friston, et al.'s papers is subtitled "The Brain as a Phantastic Organ" [8].

A forager does not 'clean its plate', it moves in a roundabout manner, sometimes here, sometimes there. Imagine birds moving between each of two bird feeders. They do not first eat all the seed in one and then move to the second. If there were three, they would not finish the first two and move on to the third. There is a radial, distributive, perhaps even apparently Brownian character to foraging. The FEP gives us a way to see that this activity is not in itself random but is an autonomous way to come up with broad sets of inputs which then over time reduce the amount of informational noise and improve predictions. The range of the agent's foraging determines the range of predictions it is able to make.

There is much about epistemic foraging that resembles the creative process, or play. A writer may start with 'free writing' – casting around phrases, snippets of ideas. A sculptor may grab a hunk of clay and see what form 'comes out of it' with no particular intension. Painters may let their brush 'guide' them over the surface of the canvas, a composer over the keys of the piano. Later, forms emerge which are refined into ideas and complete works. One might consider essential, especially initiatory, aspects of the creative process to be a kind of epistemic foraging.

An influential theory of creativity is to be found in Wallas' *The Art of Thought* (1926) [12]. This treatise is not only a careful description of the creative process, it is a bit of a self-help manual intended to increase conscious awareness of unconscious mental processes and show the reader how to guide them in the most productive direction.

After a utilitarian description of thought and consciousness, Wallas introduces his "four stages of control": preparation, incubation, illumination and verification [12, pp. 40–41]. His views regard certain stages in the creative process to be naturally more conscious than others. While the gathering and ordering of information is a conscious activity, incubation involves less conscious control. Illumination, the moment of insight, is unconscious, though some can harness the processes that surround it. Thus the creative process is one of more conscious activity preparing the ground for less. A proper 1920s scientist or mover-and-shaker will prepare his inspiration by logically laying out the problem he wishes to solve and prepare himself for the rational process he will conclude with verification of his hard-won insight. In between he will manipulate the incubation period by "free-reading", or a relaxing round of golf, with hopes that the Muse will soon strike.

---

[4] Friston has recently greatly expanded the scope of his theory to an information "theory of everything" [9], what Friston [10] and his critics [11] call "Markovian monism". This paper does not engage with this larger aspect of the FEP.

Wallas' model, consistent with much of the thought of his age, uses the unconscious to replace the whims of genius. The structure of creativity is thus conscious gathering that gives way to unconscious inspiration. I would argue that this results from a conflation of consciousness and attention. That ideas do not seem to result from a stepwise process subjectively observed does not mean that an insight is unconscious from a psychological point of view. Mental deduction (or 'ratiocination' as it was once called) is a conscious process but many of its observations take place outside of immediate, subjective attention. Furthermore, the succeeding century has shown that a range of unconscious psychological and neuro-biological processes lie behind reasoned activity. Wallas himself, in the 1920s, is not insensitive to this. Part of his book is aimed toward an ersatz practical phenomenology that should make unconscious processes more manipulable.

Where Wallas' theory rings truest today is in its description of the behavior that surrounds insights. After all, creative activity may not always lead to a discovery, but discoveries are surrounded by fixed patterns of activity, 'fixed action patterns', in the language of animal behavior. Novelty may not always be the result, but an observable activity is always present. Thus fields from education to art therapy use the terminology of preparation and incubation as elements that should set the ground for creative work.

In a recent article, Briones has applied the FEP to Wallas' classic account [13]. This is interesting because Wallas himself based part of his model on the musings of Hermann von Helmholtz, the father of the Bayesian brain hypothesis and the FEP's patron. Briones correlates the unconscious with FEP's internal state. Briones then matches prediction and observation with Wallas' system of alternating unconscious and conscious activities. However it seems, by my reading, that requiring the activity on the left side of the probability be unconscious and right conscious is a misreading of the FEP's generality. Conscious states must have aspects of the FEP inherent to their operations for the principle to have any real weight, but it is inimical to the theory for consciousness to intervene as if from elsewhere. Intentionality and consciousness should not be confused. It seems here that Briones has honestly misread Friston and understood the predictive process as explicitly passive and not active.[5] This then allows him to associate passivity with unconsciousness and apply prediction to Wallas' "Incubation" period. I would further differ from Briones in another methodological point. The FEP's system of predictions arrayed in Markov sets allows groups of observations and predictions to be 'blanketed' into ranking hierarchies. One set of activities can involve many sets within it. I regard the FEP as applying more to behaviors than concepts. It is not that "the creative process" can be mapped into a Kullback-Liebler entropy formula. More reliable modeling, in my opinion, would come from applying the principle to individual behaviors related to creativity. Though I think Briones' application of Wallas' incubation period to the FEP

---

[5] Briones quotes the following passage: "So, this *inactive* [italics mine [Briones']] perspective is very important because not only does the brain then have to explain all the sensory input, it also has to choose which sensory input to sample." [13, p. 153]. However this does not make sense from the passage. The original source included transcriptions from a video interview with Friston. Earlier Friston says: "...And then how that has become contextualized in the enactivist or the embodied cognition context" (http://serious-science.org/free-energy-principle-7602). It is clear that Friston later meant not "inactive" but, quite the opposite, "enactive". This mistranscription of the text, unfortunately, serves to undermine Briones' argument.

perceptive, we would do better to examine more of the creative process than just Wallas' incubation and concentrate less on the FEP as a whole than on epistemic foraging as a part of active inference.

Criticism of the FEP and the way philosophers and social scientists discuss creativity would soon disabuse you of any application to the habits of genius. First of all, creativity goes hand and hand with surprise and, in turn, those who, according to psychological profiles, are inclined to 'seek out new experiences' [14]. If creating is doing that which has not been done before, creation should be a surprise to its creator as well as those who receive it. Yet the function of active inference is to limit surprise, as Clark has remarked in his survey [15]. Epistemic foraging gathers information which will lead to predictions about future states so that surprise is reduced, not increased.

Active inference is, as it says, active. Surprise is reduced only when it is anticipated. If the agent has not engaged in regular observations to anticipate future states in the external world, then the rate of surprise is likely to increase. The confirmation of predictions gives positive results to the agent, but a homeostatic set of similar observations is likely to lead to a big surprise in the future. Thus there is a deceptively paradoxical benefit to seeking out surprise in the near term to reduce it overall. This is regulated by a relationship, elaborated below, between what one might call a low opportunity cost and a greater variety of observations and predictions. What determines the value of the opportunity, however is not some externally imputed set of "drives" or "instincts toward inherent benefits".[6] Epistemic foraging does not require external causes or agents that have special qualities; and this is where we see an enactive nexus between conscious and unconscious agents.

3   A Thought Experiment: The Toddler and the Robot Vacuum Cleaner

To enforce my point let us consider a pair of thought experiments each involving the same members of an informal human-machine interaction.

In experiment (A) an early-adopter has purchased a first-generation robot vacuum cleaner. It is a small disc-shaped, battery-operated, automatized machine with rotating brushes and a vacuum suction device set a couple of centimeters above the bottom of three sets of pivoting wheels. It is designed to clean floors with minimum human intervention. It is equipped with simple visual sensor devices. The robot moves over floors in a fashion that is initially randomly-determined by its software. When cameras detect that the device is within 1cm of an obstacle, it turns 30° to the right and begins to move randomly again. Its motion patterns appear also to a bystander to be random.[7]

This early-adopter purchaser has a child of around two years of age. When not observed and chastised for it, the child, of course, is fascinated with the device. She (the child) puts her toys on top of it to ride around, tries to stick the cat's tail underneath it. (The cat's fixed action patterns have already steered it away from both the other

---

[6] Compare, for example, the inherent causality of active inference from the variables introduced by Schmidhuber [16].

[7] See the image in [17].

members of the experiment.) The robot clearly responds to the obstacles placed before it and so the child is interested in introducing more of them. She tries things of different shapes and sizes. She puts them in different locations around the robot. As the robot only responds to that presented to its visual sensors, the child gets a different response based on placement of the object with respect to the robot. This introduces variations in the responses. By sticking things underneath the robot, she gets the wheels to pivot and rebalance on their bearings prompting a change in position irrespective to forward-facing obstacles; putting things on top of the robot does not change its motion, but now a toy is taking a ride and this creates the impression of a different sort of change.

Empirically, we could say that both child and robot are 'foraging'. They are moving about in a particularly unstructured way for what might be some purpose that is not revealed by the directionality of their motion. A human who runs a vacuum cleaner does not need to employ random variables to decide how to cover the floor with a machine. The human has learned that the straight rows of brushes of a vacuum best clean the floor when they approach the same area from several different angles. Over time, the human observes that the floor has a consistent enough surface in one area and moves on to the next. This concept of a adequately consistent surface is a gestalt that the human (and the cat, by the way) has as a basic part of its psychology and which informs the idea of a clean surface as well as the idea of the floor in contrast to the furniture or creatures on top of it. Conventions and memory inform the concept with notions such as 'the rug is part of the floor', 'the pedestal is not', etc. As we remember from Gestalt Theory, this object of a 'floor' is impossible to define consistently through geometry, but experiment has shown that it exists for human and non-human subjects alike. The robot has been designed to compensate for the method of manipulating the brushes and the gestalt of the floor by its construction (its area of application just high enough off the bottom so as not to cover much of a vertical surface) and the random motion encoded in its software. And, regardless of the method and psychology of perception behind cleaning behavior, there are not significant differences in outcome between the random motion of the robot and the varied and intentional motion of the human directed-machine given that the robot can be allotted a great deal more time to do the job. Furthermore, we cannot say that the behavior of a fixed-action pattern of vacuuming is always intentional in its operation. An over-experienced human vacuumer, earphones blaring, may have his or her mind elsewhere and be directing the machine with a motion Brownian in all but the formula.

Now the child's motion is part of a play interaction with the machine, a set of irregular behavioral patterns. Most children by this age do not seem to have a clear distinction between the intentions of a machine and the intentions of a human or animal agent [18], though at this age she probably does understand some of the general contrast in the machine's behavior from that of other agents. The structure of the play seems to be as follows: 'the robot moves somewhere, what can I do that will make that behavior change?' One object put in front makes the robot move differently, another put underneath does, too. Other things do not. In a certain sense the "function" of her play is to arrive at the capacities and limitations of her robot playmate. She does this by running through an inventory of actions. One of these is introducing obstacles. An inventory of obstacles is selected considering how interesting they are to her already and how they might exert responses. A doll might not be interesting because it is too big to get caught up in the wheels; or it might be essential because that particular doll is important for other reasons at the moment. The play can be best characterized by moving though the types of possible interactions, first those close at hand and important, then those further afield.

Experiment (B) takes place several years later in a different household. The technology for robot vacuum cleaners has become more advanced with machine learning technology. The visual sensors store information about the environment and motion. Over time, the robot learns to avoid places where it can get caught and tangled in curtain strings or electric cords. Consumer research revealed that some robot behaviors "annoyed" their users and manufacturers programed them to use their motion sensors and detect and adjust their behavior respective to the animated entities around them. At this point the random variables in the robot's software are proportionally less often exhibited in its behavior. It has been programed to detect different surfaces – carpets for example – and behave differently with respect to them. The vacuum's behavior obviously seems to have become more directed toward its end function.

A child at the same age and level of development will adopt the same patterns of interaction with the robot, as in (A). If a direct reaction to the child's motion is observed, she will incorporate this into her play, stepping outside of its range of vision and following it unnoticed. If its machine learning software has taught it not to go anywhere near the constantly moving object, the child will seek it out nonetheless.

A vacuum's task is to brush up and suck up light materials on the floor. The robot vacuum in (A) engages in foraging-like behavior by moving randomly in the accomplishment of its task. The vacuum in (B) has the learning software to actually learn while it is cleaning. While such software is proprietary, it very well may be programed to make predictions and then engage in a veritable form of epistemic foraging.

The activity of the child is more complex in function and intention, as we would expect from a dynamic brain and organism. However, is the indeterminate nature of activity fundamentally different from A or B? The child seeks out novelty by running through an inventory of stimuli and observations, free experimentation. A great deal of play behavior can be seen as a type of epistemic foraging. In the FEP it can be described as a means of covering a range of observations so as to be able to improve the range of future predictions. This is a reasonable model of learning. Perhaps we can say what distinguishes play from active inference is that the player is introducing surprise into its observations within particular, set confines. Perhaps these confines are intended to tolerate a high level of surprise.

Is there a significant difference between the communication in A and B? The most obvious answer is that the child's play is dependent upon the range of the robot's responses. It is possible that the AI of robot B will offer a smaller range of variation because its more directed motion has less variety than its random motion. However, it may be that the range of behaviors determined by the machine learning software will be greater than that determined by random motion; and that the quality of the behaviors will be different and thus more interesting. A robot that runs away and hides in corners (so as to be less "annoying") might be more fun and open up other opportunities for play.

For the robot that has a defined task and construction, foraging and epistemic foraging can lead to accomplishment of the task – greater utility. For the human playmate, the process may be empirically similar. In no case is intelligence, intention, task, or discovery an essential element to the activity. While foraging, strictly speaking, is distinguished from feeding by its lack of goal-directed motion at a given time, not even random motion is indispensable to the child's or robot B's foraging. It is only necessary that each of these participants run through the inventory of possibilities in their environment. This can be the surfaces of a room or the list of favorite toys; but in each case a crucial element is that the range of selection be broad and the process of selection liberal. Thus, contrary to Wallas, the collection and incubation period is itself dependent on the inventory of elements that make up the wandering and not on the intention or capacities of the wanderer.

This is a feature of the FEP that is easy to miss. The results of FEP's processes often resemble utilitarian control theory. However, as Schwartenbeck notes, this often presumes an ad hoc intervention into the process [19, p. 3], often termed as a 'reward instinct.' In the active inference system, the utility is confirmation of an internal state. Utility is equated with reduction in surprise, reduction of entropy is the 'reward', and the need to make further predictions is internal to the system of reducing surprise overall. Furthermore, as control processes often introduce random variables to model exploratory behavior, the FEP's epistemic foraging does not require random motion to function [19, p. 3]. As we see from the above, variety is not always randomly determined, contrary to intuition. If we consider that variety is determined by the relative difference between the previously predicted state and the set of states prepared for processing [19], the range of states in a complex system predicting a wide range of interacting internal states is likely to be greater that of a system determined randomly. It is a more textured reality.

As we see from (B), one important feature that stands behind this experiment, however, is that the state of play requires that the agent have a wide range of new states in a pre-determined set. They are interesting because they allow between them the maximum range of individual outcomes, but acceptable because they are confined related to surrounding predictable states.

4 Frameworks and Freedom

Friston's descriptions of epistemic foraging in the FEP often resemble the accounts of novelty and risk seeking given by behavioral economists. However, a major feature of the FEP is that its modular, Markovian, structure means that active inference can take place at all levels of hierarchy. Schwartenbeck et al. presume this when they write "This decomposition [of relative entropy] may account for numerous instances of everyday choice behavior, such as why we appreciate variation over outcomes much more when we buy a chocolate bar as opposed to a car: when the differences in the expected utilities of outcomes become less differentiable, agents will try to visit several states and not just the state that has highest utility" [19, p. 3]. Perhaps we may speculate that Markovian blankets made up of sets of these probabilities accommodate high levels of entropy (and high levels of surprise) in a predictable, confined set of states surrounded by economically low levels of entropy in others. Such models could account for fixed action patterns such as play, ritual, and the strictly framed contexts of many artistic productions.

Let us briefly consider one example of the latter. In Aristotle's *Poetics* we find the following statement:

[A] A probable impossibility is better than [B] an improbable possibility. (*Poetics* 25, 1460a)[8]

The reasoning is as follows: it is better for the artist to create an impossible world with rules followed consistently than to rely on the 'likely state of affairs' (εἰκός) on which the artist hangs an unlikely set of events. This ancient adage seems oddly tailor-made for the new-fangled FEP. For example, it is a more interesting premise to [A] create a world where the protagonist has become an enormous insect (an impossibility) and then must roll himself over onto his waggling legs like an other insect that has ended up on its back (a probability) than to create one where the protagonist is [B] still a mustachioed office worker (a possibility) who has awoken to an unlikely number of insects (an improbability) crawling on the floor of his room. The second version would need something else to make its world interesting. This is the reasoning that distinguishes the strict but strange logic of fairytales from the banality of daytime television soap operas where average people find themselves in below-average situations. If we (primitively) rephrase [A] as a probability equation, then, with some modal logic, we have

$$\sigma \ni \sum \pi\pi = 1(\pi \mid (\neg \lozenge x))$$

where σ is an art work (a Greek tragedy, for example); π, an element of σ, has the probability of occurring given x, when x is a counterfactual – something that is impossible.

In my reading of it, [A] makes more sense to the logic of the FEP than [B] because π always amounts to a prediction about x that is going to be feasible for the agent. Since x is, by nature, counterfactual, it has a truly wide range of predictions. The viewer continues making these predictions throughout the span of the work. Over time, the sum of these true predictions about the work's impossible world accumulate. Note that the very fact that x is counterfactual determines a high level of surprise; but one that is still worth testing because π has a chance of being true. The very impossibility of x means that individual states of x are going differ little from each other in utility. To reprise Schwartenbeck, et al.'s statement "…when the differences in the expected utilities of outcomes become less differentiable, agents will try to visit several states and not just the state that has the highest utility" [19, p. 3]. With [B], on the other hand, though x is probable, most of x's predictions of it are going to be false. Thus though making a true prediction of x is possible, the results are not going to be good, sending the agent elsewhere.

Though in a future publication, I hope to give a (much) more applicable version of Aristotle's formula, it should be clear here that the reasoning behind epistemic foraging is inherent to the exploratory process that is part of receiving as well as conceiving a creative work. It is a truism that the viewer comes to the stage with a certain set of beliefs. But it is insightful to think that the engagement with the artwork is a constant testing of those beliefs against the range of their variety. Again, however, the above makes clear that surprise, and even impossibility, are tolerable within a framework that isolates such states.

---

[8] Adapted from [20].

And this brings us to a final point. As we have seen, epistemic foraging depends on the variety of a range of options with respect to utility. When a framework is introduced that reduces the range of utility, the variety of the range of options increases. If the application of Aristotle's statement holds, within a particular framework, the range of possible states can even encompass the impossible.

If we consider the discrete sets of behavior described in our experiments, or Aristotle's art work, as natural constraints in the system, an unlimited range of activity within those constraints looks very much like a classic, positive definition of freedom as Berlin describes in his famous essay [21]. Positive because its constraints set the terms for variety rather than restrict them. Note, however, that the exercise of this freedom is not to push against its restraints. At any particular point the character of this behavior is as meandering with restraints as without.

This taking of exploratory liberties does not require that agents have some additive quality. Is not Schwartenbeck's interpretation of inferential variety as a high range of relative entropy with a low range of utility very much like Oscar Wildean "uselessness" – the seeking forth with apparent indifference to the destination?

5 Conclusion

While the foregoing does not assert that creative behavior necessarily leads to creativity, it aims to show that certain processes which can be modeled in nature and in AI may be theoretically compared to behavior often associated with creativity. This behavior does not account for a particular stage of the process leading to an insight, but rather to the 'aimless' nature that often surrounds insights. This behavior is not random, but can be described as epistemic foraging and compared to that notion in the FEP.

Viewing creative wandering in the light of active inference allows us to avoid the presuppositions of external causal forces and imputed instincts, or drives, that are often necessary in controlled learning models. In consequence, we observe that such behavior does not require the agent be conscious or even necessarily alive. Most important is that it have a range of variety with regard to future predictions. It also seems important that the cost to the agent in such circumstances be low and that the ability to experience a high level of relative entropy in one set of predictions be situated in a proportionally lower alternative (or future) set of entropies. Considering and modeling these two observations in detail might be able to make future contributions to the FEP and the biological and aesthetic foundations of art, play, ritual, and the creative process in detail itself.

Our informal thought experiment also proposes another potential weakness in the use of random variables that is so important for much AI in practice. We suggest that it does not matter so much that the predictions be random but that they involve a wide *range* of variety. Perhaps this is the key to human creativity. As organisms in a very four-dimensional world, with an extremely complex set of directed physical, social, and semiotic affordances, memories, and associations, we venture in our range of predictions farther and wider than our artificial creations – even far into, and then out of, the impossible. Von Helmholtz's inspiring walks and Wagner's synaesthesia are likely tapping into this store. I would venture that the richness of the texture of variety might be the true grounds for inspiration. And that progress is what other textured beings come to recognize, at long last, as insight, and the hallmarks of consciousness itself.

In the end, some of the beauty of the FEP's appeal stems not from the Romantic impulse that exalts Man to the level of God, but rather from one that sees Nature's creativity as far more sublime than Man's. This is especially impressive when one begins to notice flitting birds, crawling caterpillars, and wheedling moss as part of a similar inspiration. Or perhaps, rather than Romanticism, FEP's notion of creativity represents a return to the time of Aristotle himself when creation was less creative and more craft, when we pined less for demiurgic genius and more for well-structured artifice, when Hephaestus' mechanical owl was able to fly through the dusk no less effectively than Minerva's.

References

1. Bringsjord, S., Bello, P., Ferrucci, D.: Creativity, the Turing test, and the (better) lovelace test. Mind. Mach. **11**, 3 (2001). https://doi.org/10.1023/A:1011206622741
2. Callaway, E.: 'It will change everything': DeepMind's AI makes gigantic leap in solving protein structures. Nature **588**, 203–204 (2020). https://doi.org/10.1038/d41586-020-03348-4
3. Mirza, M.B., Adams, R.A., Mathys, C., Friston, K.J.: Human visual exploration reduces uncertainty about the sensed world. PLoS ONE **13**, e0190429 (2018). https://doi.org/10.1371/journal.pone.0190429
4. Friston, K.J., et al.: Dopamine, affordance and active inference. PLOS Comput. Biol. **8**, e1002327 (2012). https://doi.org/10.1371/journal.pcbi.1002327
5. Pio-Lopez, L., Nizard, A., Friston, K., Pezzulo, G.: Active inference and robot control: a case study. J. Roy. Soc. Interface **13** (2016). https://doi.org/10.1098/rsif.2016.0616
6. Friston, K., Kilner, J., Harrison, L.: A free energy principle for the brain. J. Physiology-Paris **100**, 70–87 (2006). https://doi.org/10.1016/j.jphysparis.2006.10.001
7. Friston, K.: Embodied inference: or "I think therefore I am, if I am what I think." In: The Implications of Embodiment: Cognition and Communication (2011) https://www.researchgate.net/publication/228816606_Embodied_Inference_or_I_think_therefore_I_am_if_I_am_what_I_think
8. Friston, K., Stephan, K., Montague, R., Dolan, R.: Computational psychiatry: the brain as a phantastic organ. The Lancet Psychiatry **1**, 148–158 (2014). https://doi.org/10.1016/S2215-0366(14)70275-5
9. Friston, K.: A free energy principle for a particular physics. arXiv:1906.10184 [q-bio] (2019)
10. Friston, K.J., Wiese, W., Hobson, J.A.: Sentience and the origins of consciousness: from Cartesian duality to Markovian monism. Entropy **22**, 516 (2020). https://doi.org/10.3390/e22050516
11. Beni, M.D.: A critical analysis of Markovian monism. Synthese, 1–21 (2021). https://doi.org/10.1007/s11229-021-03075-x
12. Wallas, G.: The Art of Thought. Trove, London (1945). https://nla.gov.au/nla.obj-502468959
13. Briones, R.: Creativity: a learning process as seen from the perspective of the free energy principle. Int. J. Innov. Sci. Res. Rev. **2**, 151–154 (2020). http://www.journalijisr.com/sites/default/files/issues-pdf/IJISRR-150.pdf. Accessed 12 May 2021
14. Kaufman, S.B.: Opening up openness to experience: a four-factor model and relations to creative achievement in the arts and sciences. J. Creat. Behav. **47**, 233–255 (2013). https://doi.org/10.1002/jocb.33
15. Clark, A.: Whatever next? Predictive brains, situated agents, and the future of cognitive science. Behav. Brain Sci. **36**, 181–204 (2013). https://doi.org/10.1017/S0140525X12000477

16. Schmidhuber, J.: Simple algorithmic theory of subjective beauty, novelty, surprise, inter-estingness, attention, curiosity, creativity, art, science, music, jokes. J. SICE. **48**(1), 21–32 (2009). https://people.idsia.ch//~juergen/sice2009.pdf. Accessed 12 May 2021
17. Bartle, C.: The path taken by a en:Roomba robotic vacuum cleaner as it cleans a room (2009). https://www.flickr.com/photos/13963375@N00/3533146556. Accessed 12 Aug 2021
18. Meltzoff, A.N., Brooks, R., Shon, A.P., Rao, R.P.N.: "Social" robots are psychological agents for infants: a test of gaze following. Neural Netw. **23**, 966–972 (2010). https://doi.org/10.1016/j.neunet.2010.09.005
19. Schwartenbeck, P., FitzGerald, T., Dolan, R., Friston, K.: Exploration, novelty, surprise, and free energy minimization. Front. Psychol. **4** (2013). https://doi.org/10.3389/fpsyg.2013.00710
20. Aristotle: Aristotle in 23 Volumes. Harvard University Press, Cambridge (1932). http://data.perseus.org/citations/urn:cts:greekLit:tlg0086.tlg034.perseus-eng1:1460a
21. Berlin, I.: Four Essays on Liberty. Oxford University Press, London (1990)

Creativity and Its Genetic Foundations

Dimitri Spivak[1,2]([✉]) [iD], Andrei Zhekalov[3] [iD], Vladislav Nyrov[5] [iD],
Pavel Shapovalov[3] [iD], and Irina Spivak[3,4] [iD]

[1] N.P. Bechtereva Human Brain Institute, Russian Academy of Sciences,
St. Petersburg 197376, Russia
[2] D.S. Likhachev Russian Scientific Research Institute of Cultural and Natural Heritage,
Moscow 129366, Russia
[3] S.M. Kirov Military Medical Academy, St. Petersburg 194044, Russia
[4] St. Petersburg State University, St. Petersburg 199034, Russia
[5] Peter the Great St. Petersburg State Polytechnical University, St. Petersburg 195251, Russia

Abstract. Present-day state of the theory of genetic foundations of creative performance, primarily at the level of the dopaminergic, serotoninergic, and noradrenergic systems, as well as neuregulin 1 gene, arginine vasopressin receptor, and angiotensinogene, is briefly reviewed. Basic results of a pilot experiment, focused upon four candidate genes for inclusion into creativity studies, namely neurotrophic factor gene (BDNF), α-actinin-3 protein encoding gene (ACTN3), angiotensin-converting enzyme 1 (ACE1), and serotonin-2A receptor gene (5HTR2A), are presented. Strong correlations between high level of creativity, both verbal and figural, and both Val/Val BDNF genotype, and RR ACTN3 genotype, are demonstrated, along with its somewhat weaker correlation with II ACE genotype. Taking into account levels of activation of basic psychological defense mechanisms and stress coping strategies, proper for 22 practically normal Arctic dwellers, who were examined in the framework of our experiment, allowed to link these correlations to optimal adaptation abilities, and to prolonged life expectancy. Basing upon this result, plausibility of discerning between two facets of creativity, one being adaptive, another being non-adaptive, is discussed, the former concerned with primarily coping with life stress, the latter providing self-actualization. Interrelation between the inherited abilities and the acquired ones, forming subject matter of correspondingly genetic and creativity studies, is regarded as a representation of basic dichotomy between nature and culture.

Keywords: Creativity · Genetic mechanisms · Adaptation

1 Introduction

Creativity is regarded in the present chapter as ability to generate production with the help of any semiotic system being accessible to humans, which (a) is relevant to the corresponding realm of activities, (b) novel in qualitative or, quantitative terms, (c) meets the requirements of sufficiently high quality. The definition cited above was elaborated by the authors of this chapter, basing upon the classical definition by J.Kaufman and

R.Sternberg [1], and taking into account present-day state of the problem [2]. Withholding and enhancing high level of creativity forms a typical trait of adaptive and successful individuals at the personal level, and of successful and prosperous communities and societies, including the so-called knowledge societies, at the societal one.

Measuring creativity forms a scientific direction which has been elaborated by a number of researchers and scientific schools. The paradigm based upon pioneering works of E.Torrance and J.Guilford seems to be central to present-day psychological theory. The set of fluency, flexibility, and originality, regarded as forming the basic set of factors of creativity, may be regarded as focal for this paradigm. Measuring levels of their activation forms the main task of the corresponding test batteries [3].

Systematic inquiry into genetic bases of creativity has been quite actively conducted in the course of the latest fifteen years. Some of the basic mechanisms have been already detected. Thus there is no doubt at present that such pivotal systems of human organism as the dopaminergic system, and the serotoninergic one, are involved into the process of its functioning. At the level of the former one, D2 dopamine receptor (DRD2), catechol-O-methyl-transferase (COMT), DA transporter (DAT), dopamine receptor D4 (DRD4), and tryptophan hydroxylase 1 (TPH1) are regarded as being most intimately involved into the creative performance; as to the latter, serotonin transporter (5-HTTLPR) seems to be most active. As to the noradrenergic system, its links with creativity, most possibly existing at the level of fluency, are still to be elaborated [4].

A number of other genes have also been detected as having to do with creativity. Neuregulin 1 is to be mentioned in this respect [5, 6], along with arginine vasopressin receptor (AVPR1a) [7, 8], and angiotensinogene (AGT) [9]. The list of candidate genes could easily be prolonged. Having stated this, we feel authorized to acknowledge that scientific inquiry into the realm of genetic correlates of creativity remains as a matter of fact to be presently at its initial stage. As it was duly stated by the authors of The Cambridge Book of Creativity, it tends to form a field of research 'with a pattern of interesting but also contradictory (at least at this point in time) results' [10].

Discussing the plausibility of introducing a cluster of new genes into creativity studies forms the main objective of the present chapter, along with testing the possibility of conducting a corresponding mass study, and summarizing some of its constructive results. Speaking in broader terms, genetic peculiarities belong to the set of inherited human abilities, while creative competences form an integral part of those acquired in the course of upbringing. Joining them in the framework of an interdisciplinary study implies working 'above the borders', especially the critical one, which divides nature and culture. Thus another important objective of the present study consists in reconsidering the profundity of this drastic divide, and in assessing optimal ways of crossing it, for the benefit of mankind.

2 Methods

A group of 22 practically healthy subjects, aged 37 ± 14 years, living and working in the Arctic region, was examined in the framework of a pilot study of creativity and its genetic correlates, which would be presented in this chapter.

The psychological block of our research included detection of creativity level, as well as the level of activation of basic psychological defense mechanisms. Methods applied for this purpose belonged to the standard set of present-day psychodiagnostic instruments. Testing was conducted one time, by filling in a number of questionnaires, in Russian, which was the native tongue of all of the subjects.

Creativity testing consisted of 6 items, three of which were dedicated to the verbal creativity, and three more ones, to the figural creativity:

- verbal creativity tests included naming: as many ways of using a definite object as possible (task 1), as many consequences of a given situation as possible (task 2), as many four-word sentences as possible, where each word would start with a definite letter (task 3);
- figural creativity tests included: drawing as many objects consisting of a given set of elements, as possible (task 4), filling in as many circles as possible by drawings of different objects (task 5), finding as many objects hidden inside a complex drawing, as possible (task 6).

Each of the aforementioned tasks was to be accomplished in 3 to 4 min. Processing testing results consisted in the case of each task in calculating the levels of fluency, flexibility, and originality. Basing on them, an integral index of creativity was calculated for each task. In this way, integral creativity indices ranging from 1 to 6, correspondingly, were calculated for each subject. In testing the level of creativity, a standard methodology, rooted in the tradition founded by E. Torrance and J. Guilford, was applied; the Russian version of the tests was elaborated by E. Tunik [11].

Testing the level of activation of basic psychological defense mechanisms consisted in application of two standard methodologies:

- level of activation of basic psychological defense/life style mechanisms (rejection, suppression, regression, compensation, projection, substitution, intellectualization, and reactive formation). Following an influential trend in present-day psychodiagnostic theory, we regard these mechanisms as playing an important role in counteracting psychological challenges and restoring inner balance by normals. In testing the level of psychological defenses, a standard methodology, rooted in R.Plutchik's psycho-evolutionary theory of emotions, and H. Kellerman's structural personality theory, was applied; its Russian version was elaborated by L. Vasserman and his research team [12]. A questionnaire, consisting of 97 items, was applied. Processing of testing results consisted in calculation of levels of activation of each of the aforementioned 8 psychological defense mechanisms. Basing on them, an integral index of activation of psychological defense mechanisms (index 7), was determined;
- basic strategies of stress coping (confrontation, distancing, self-control, social support, assuming responsibility, flight/avoidance, problem solution planning, positive reassessment). In testing the level of activation of these coping strategies, a standard methodology, elaborated by R. Lazarus and S. Folkman, was applied; its Russian version was constructed by L. Vasserman and his research team [13]. Testing consisted in filling in a questionnaire, consisting of 50 items. Processing testing results consisted in calculating indices of activation of each of the aforementioned 8 coping strategies. No integral index was calculated in the case of this methodology.

Molecular genetic block of our research included detection of polymorphisms of four genes:

- serotonin-2A receptor gene (5HTR2A). 102T > C polymorphism of this gene was proven to be related to inner flexibility of brain functioning, and with intrinsic religiosity [14, 15]. Correlation with average life expectancy was also demonstrated for this gene [16]. The polymorphism consists in single nucleotide substitution of thymine to cytosine. However predominance of the A2 allele (102C) in patients with mental disorders remains still unclear;
- angiotensin-converting enzyme 1 (ACE1). This gene, which plays a key role in the functioning of the renin-angiotensin system, has been studied quite well. The level of plasmatic ACE tends to vary greatly by normal, depending on the insertion / deletion (I/D) polymorphism in the 16 gene intron, including 287 bp [17]. A number of studies have demonstrated relation existing between D-allele, especially DD-genotype of the ACE gene, and cardiovascular diseases [18]. They tend to be linked to increased level of the ACE circulating, and angiotensin II, while the I allele, especially genotype II, are definitely protective [19]. At present time, there exists a strong tendency to regard this genotype as being associated with increased level of general metabolism, along with increased action ability [20]. Increased risk of psychopathology, including depression, was demonstrated to be related to renin-angiotensin system gene polymorphisms [21];
- neurotrophic factor gene (BDNF). The level of BDNF in blood serum was demonstrated to depend upon the replacement of methionine to valine in the 166 position (Val66Met polymorphism of this gene). It tends to reach its lowest level by carriers of Met/Met genotype, its medium level, by Val/Met heterozygotes, and its maximal level, by Val/Val homozygotes. Higher risk of having type 2 diabetes is proper for homozygous carriers of Met allele [22], as well as a number of mental disorders, including both bipolar disorder, and schizophrenia [23]. The level of BDNF in blood serum was demonstrated to be directly linked to telomere length [24]. Thus BDNF Val 166 Val genotype tends to promote and enhance general health and active longevity [25];
- α-actinin-3 protein encoding gene (ACTN3). This gene is expressed in rapidly contracting fibers of skeletal muscles, which allows us to associate it with both sports achievements [26], and active longevity [27]. As proven by us, sports achievements tend to correlate directly with the level of creativity, which might be conditioned by their common genetic basis [28]. As it was recently proven, R577X polymorphism of this gene seems to be linked in evolutionary terms with cold resistance, which makes it quite valuable in the framework of our study, focused upon Arctic dwellers [29].

Summing up, we feel authorized to state that all of the four genes, included into the program of our research, seem to be related to creativity by different indirect ties, from adaptation ability to mood and thought disorders. This makes their in-depth study timely and constructive.

Genetic analysis consisted in processing fresh blood, taken from each subject, once, by means of applying Vacutainers with 6% EDTA (Greiner Bio-one, Austria). Extraction of total DNA from peripheral blood leukocytes was conducted with a kit for isolation of genomic DNA from cells, tissues and blood (Biolabmix, Russia).

Four polymorphic variants were thus studied, i.e.: ACE Ins/Del rs4646994 (II, ID, DD); ACTN3 R577X rs1815739 (RR, RX, XX), BDNF Val66Met rs6265 (Val/Val, Val/Met, Met/Met), and HTR2A T102C rs6313 (A1A1, A1A2, A2A2), by real-time PCR reagent kits (Sintol, Russian Federation). Processing of genotyping results followed three patterns, i.e.: allele 1 homozygote, heterozygote, allele 2 homozygote. Samples of genome DNA were genotyped in real time at a DT-Prime (RT-PCR) amplifier (DNA-Technology, Russia).

3   Results

Descriptive statistics of data collected as a result of conducting our pilot research is presented in Tables 1, 2, 3, 4, 5, 6, 7, 8, 9 and 10. Statistic processing was conducted primarily at the level of linear correlation between psychological indices and genotypes, as our main objective consisted in tracing back mainstream tendencies. The latter consisted in the following basic regularities:

– correlations existing between both verbal and figural creativity, and BDNF genotypes, tend to be quite strong. As shown by Tables 1 and 2, subjects with the Val/Met genotype have the lowest level of creativity, while those with Val/Val genotype tend to have its maximal level. Subjects with the Met/Met genotype revealed medium level of creativity;

Table 1. Correlation of verbal creativity (index 2) and BDNF genotypes

BDNF genotypes	Met/Met	Val/Met	Val/Val
Mean value	9,00	6,00	15,00
n	4	4	6
sd	3,46	0,00	8,27
Difference	1–2	2–3	1–3
p	0,09*	0,02**	0,06*
Difference with sample mean value			
p	0,25	0,01**	0,15

Comment: Met/Met, Val/Met, Val/Val – BDNF genotypes; n – sample size; sd – standard deviation; a-b difference – difference between subgroups a and b; p – significance level: ** $\leq$ 0,05, * $\leq$ 0,10.

Table 2. Correlation of figural creativity (index 4) and BDNF genotypes

BDNF genotypes	Met/Met	Val/Met	Val/Val
Mean value	19,27	15,00	22,12
n	6	7	9
sd	4,95	4,88	8,88
Difference	1–2	2–3	1–3
p	0,07*	0,03**	0,05**
Difference with sample mean value			
p	0,47	0,06*	0,19

Comment: for abbreviations see comment to Table 1.

– correlations between indices of both verbal and figural creativity, and ACE genotypes, tend to be quite clear, although somewhat weaker. As shown by Tables 3 and 4, minimal level of creativity is characteristic of subjects with the DD genotype, while those with medium level of creativity have the II genotype. Creativity level of subjects with the ID genotype remains somewhat unclear;

Table 3. Correlation of verbal creativity (index 3) and ACE genotypes

ACE genotypes	DD	ID	II
Mean value	22,40	35,18	30.44
n	4	6	9
sd	8,26	25,95	15,21
Difference	1–2	2–3	1–3
p	0,15	0,35	0,06*
Difference with sample mean value			
p	0,47	0,06*	0,19

Comment: DD, ID, DD - ACE genotypes; n – sample size; sd – standard deviation; a-b difference – difference between subgroups a and b; p – significance level: ** $\leq 0,05$, * $\leq 0,10$.

– strong correlations exist between both verbal and figural creativity, and ACTN3 genotypes. As shown by Tables 5 and 6, subjects with the RX genotype have the lowest level of creativity, while those with the RR genotype have its maximal level. Level of creativity of subjects with the ID genotype remains still unclear;

Table 4. Correlation of figural creativity (index 4) and ACE genotypes

ACE genotypes	DD	ID	II
Mean value	15,92	20,27	20,26
n	6	7	9
sd	8,35	5,97	7,50
Difference	1–2	2–3	1–3
p	0,16	0,50	0,08*
Difference with sample mean value			
p	0,21	0,33	0,35

Comment: for abbreviations see comment to Table 3.

Table 5. Correlation of verbal creativity (index 3) and ACTN3 genotypes

ACTN3 genotypes	XX	RX	RR
Mean value	30,93	19,45	38,91
n	6	6	7
sd	11,02	17,89	19.91
Differences	1–2	2–3	1–3
p	0,11	0,05**	0,11
Difference with sample mean value			
p	0,46	0,12	0,17

Comment: XX, RX, RR - ACTN3 genotypes; n – sample size; sd – standard deviation; a-b difference – difference between subgroups a and b; p – significance level: ** ≤ 0,05, * ≤ 0,10.

Table 6. Correlation of figural creativity (index 4) and ACTN3 genotypes

ACTN3 genotypes	XX	RX	RR
Mean value	19,17	14,63	23,46
n	6	8	8
sd	10,37	3,94	4,42
Differences	1–2	2–3	1–3
p	0,17	0,00**	0,09*
Difference with sample mean value			
p	0,49	0,02**	0,03**

Comment: for abbreviations see comment to Table 5.

– no statistically relevant data on correlation of creativity and 5HTR2A gene polymorphisms were detected in our research.

Passing to psychological defense mechanisms, following regularities were traced back:

– basic psychological defense (Life Style Index) activation level tended to be minimal in the case of subjects with the BDNF Val/Val genotype, and maximal – for those with its Met/Met genotype (for details, cf. Table 7). Judging by absolute value of the corresponding index 7, the former defense level belongs to the lowest part of the normal interval of its values, the latter – to its highest part. Taking into account that the group examined by us comprised young normal subjects, who had passed special screening to be sent to the Arctic region, this conclusion of ours seems to be quite reasonable;

Table 7. Correlation level of activation of psychological defense mechanisms (index 7) and BDNF genotypes

BDNF genotypes	Met/Met	Val/Met	Val/Val
Mean value	25,33	21,29	16,67
n	6	7	9
sd	4,97	10,26	9,85
Difference	1–2	2–3	1–3
p	0,19	0,19	0,02**
Difference with sample mean value			
p	0,05**	0,43	0,17

Comment: Met/Met, Val/Met, Val/Val – BDNF genotypes; n – sample size; sd – standard deviation; a-b difference – difference between subgroups a and b; p – significance level: ** $\leq 0,05$, * $\leq 0,10$.

– no statistically relevant correlations between other genes, included into our study (ACE, 5HTR2A, ACTN3), and psychological defenses were found;
– at the level of stress coping strategies, no integral index was calculated, following standard methodology applied by us. In its stead, eight partial indices were calculated, representing, as it was pointed out above, eight basic coping strategies, i.e. confrontation, distancing, self-control, social support, assuming responsibility, flight/avoidance, problem solution planning, and positive reassessment (they were represented by correspondingly indices 8.1 to 8.8).

We have to admit that although a number of correlations were detected by us, they tended to be less statistically relevant than those demonstrated in the case of basic psychological defenses (this means that the corresponding probability level was not

0.05, but 0.10). One may suppose that the main reason of this situation consisted in the basic purport of the corresponding processes. Basic psychological defense mechanisms tend to be formed mostly starting from earliest stages of ontogenesis, and act mostly at the subconscious level. As to coping strategies, they are mostly formed much later, and tend to counteract stress primarily at the conscious level. As a result, the former defense mechanisms may be related to the genetic predispositions much more intimately than the former ones.

Two coping strategies, which are related to the genetic indices stronger than other ones, would be cited here as a plausible example. They represent distancing from a stressful situation (Table 8), and mobilization of social support (Table 9). As shown by the tables, in both cases Val/Val BDNF genotype corresponds to the minimal level of activation of the corresponding strategy, while Val/Met corresponds to its maximal level. Judging by absolute levels of the indices 8.2, 8.4, the level of coping strategies activation was more or less normal, which was quite reasonable, taking into account that all subjects were practically normal, and had passed a selection procedure (for normal intervals, cf. [13, p.16–19).

Thus definite correlations between coping strategies and definite genotypes, especially those of the BDNF gene, tend to exist, although at a statistically lower level.

Table 8. Correlation level of activation of distancing coping strategy (index 8.2) and BDNF genotypes

BDNF genotypes	Met/Met	Val/Met	Val/Val
Mean value	44,83	46,71	38,89
n	6	7	9
sd	8,84	11,12	7.56
Difference	1–2	2–3	1–3
P	0,37	0,07*	0,49
Difference with sample mean value			
p	0,33	0,22	0,11

Comment: Met/Met, Val/Met, Val/Val – BDNF genotypes; n – sample size; sd – standard deviation; a-b difference – difference between subgroups a and b; p – significance level: ** $\leq$ 0,05, * $\leq$ 0,10.

Reviewing the data presented above, i.e. in the Tables 7, 8 and 9, one feels authorized to conclude that the levels of both types of psychological defense mechanisms, examined by us, tended to belong to the lowest part of their normal intervals by subjects with the Val/Val genotype of the BDNF gene. At the same time, subjects with this genotype were also proven by us to reveal the maximal level of creativity (cf. Tables 1 and 2). Having compared both tendencies, we feel authorized to conclude that optimal adaptation tends to be provided by the Val/Val genotype of the BDNF gene at the level of genetics, and by heightened arousal of creative performance, at the psychological level.

Table 9. Correlation level of activation of social support coping strategy (index 8.4) and BDNF genotypes

BDNF genotypes	Met/Met	Val/Met	Val/Val
Mean value	48.00	50,86	43,67
n	6	7	9
sd	15,45	12,23	6,40
Difference	1–2	2–3	1–3
p	0,36	0,10*	0,25
Difference with sample mean value			
p	0,45	0,25	0,14

Comment: for abbreviations see comment to Table 5.

In-depth analysis of correlation between basic stress coping strategies and BDNF genotypes was concluded by application of single factor variance analysis. Mean values of each of the eight partial indices of coping strategies were calculated for each of the three BDNF genotypes. As shown by Table 10, the value of observed F-statistics was much higher than its critical value, while the observed significance level was lower than the critical one. As a result, null hypothesis, which consisted in absence of statistically relevant differences between mean values of coping indices for different BDNF genotypes, was rejected, and the alternative one, which comprised the existence of such differences, was proven. This means that coping strategies seem to be linked to definite genotypes, at least at the level of the BDNF gene, so that going on with their systematic study forms a constructive direction of systematic research.

Table 10. Single factor variance analysis of stress coping strategies and BDNF genotypes

Source of variation	SS	df	MS	F	p	Fc
Between-group	66,17198	2	33,08599	5,36524077	0,01311589	3,466800112
Within-group	129,5013	21	6,16673			
Sum total	195,6733	33				

Comment: SS - sums of squares, df – degrees of freedom, MS - mean square, F – observed value of F-statistics, p – observed significance level, Fc - critical value of F-statistics, critical significance level: $p \leq 0{,}05$.

4   Discussion

Basing upon the results of a brief review undertaken by us, we feel authorized to state that genetic support of creative performance, especially at the level of the dopaminergic, serotoninergic, and noradrenergic systems, seems to be quite active and strong.

This is also the case of such genes as neuregulin 1, arginine vasopressin receptor, and angiotensinogene.

Passing on to the results of a pilot experiment, which was conducted by us by means of examination of a group of 22 practically normal Arctic dwellers, we may feel authorized to state that a number of statistically relevant correlations between the level of creativity, both verbal and figural, and genetic polymorphisms of several genes linked to both highly important organism functions, and psychological states/processes, were revealed. To name but the strongest ones, the lowest level of creativity proved to be linked to the Val/Met BDNF genotype, and to the RX ACTN3 genotype. As to the highest level of creativity, it turned out to be most strongly correlated with the Val/Val BDNF genotype, and to the RR ACTN3 genotype. Passing to somewhat weaker correlations, the lowest level of creative performance was demonstrated to be related to the DD ACE genotype, while the II ACE genotype was linked to the medium level of creativity. This means that the BDNF gene, the ACTN3 gene, and, to a certain extent, the ACE gene are to be regarded as essential parts of genetic mechanisms providing human creativity.

In order to interpret genetic regularities obtained in this way, basic psychological defense mechanisms were regarded. Statistically relevant links were detected by us primarily concerning the BDNF gene. Thus minimal level of activation of both basic psychological defense mechanisms, measured by Life Style Index, and such important stress coping strategies as distancing, and social support, proved to be strongly related to the Val/Val BDNF genotype, while the maximal level of both turned out to be strongly related to its Met/Met genotype.

Basing upon this set of regularities, we feel authorized to state that optimal adaptation to stressful conditions is provided by the Val/Val BDNF genotype, at the genetic level, and by heightened arousal of creative performance, at the psychological level. As the Val/Val genotype tends to be strongly related to telomere length and, consequently, to prolonged life expectancy [24, 25], we feel authorized to tentatively link, via this genotype, high level of creativity with the perspective of active longevity (for a preliminary analysis of possibility and correctness of such links, see [30]).

Speaking in terms of genetics, the correlation of the three genotypes marked by us, with creativity may be interpreted in the following way:

- Val/Val BDNF genotype tends to act in a protective way, as it enhances a general increase of BDNF in blood serum, which leads in its turn to general activation of the nervous system, as well as the immune one. As a result, cognitive functioning tends to improve, as well as some patterns of mind-body interaction [31];
- contrary to the previous case, ACTN3 gene polymorphisms have not been studied sufficiently [32], which makes their interpretation somewhat uncertain. However an interesting study may be cited in this respect, which was focused upon relative injury frequencies in athletes. Subjects with the RR genotype (and also with the RX one) were shown to have muscle microtrauma, along with hormonal stress, much more frequently than those with the XX genotype. The reason was that they acted much more promptly and forcefully than their competitors, which allowed them finally to win. Thus their game strategy linked to the RR genotype, proved to be successful both in terms of mind and body activity, although, it was definitely not well-balanced

[33]. This strategy could be tentatively linked to heightened arousal of the creative performance;

– as to ACE genotypes, the topic remains still not quite clear. Direct arguments in favor of the existence of correlation between the renin-angiotensin system, and creativity were found by the authors of the present chapter [9]. As to indirect arguments, the DD ACE genotype was proven to be reliably associated with a number of mental disorders, including Alzheimer's disease [34], bipolar disorder [35], depressive states [36], and autism [37]. As a consequence, presence of this genotype may be linked to general impairment of cognitive functioning. Taking into account that the DD ACE genotype was demonstrated in our research to be correlated with the lowest level of creative performance, one feels authorized to suppose that other ACE genotypes, i.e. II and ID, might be related to stronger levels of creativity. As it was shown above, this is likely to be the case of the II ACE genotype (for details, cf. Tables 3 and 4). However proving this hypothesis forms subject matter of future research.

Taking into account genetic data, we feel authorized to state that in the case of three genes out of four studied by us (i.e. BDNF, ACTN3, and ACE), subjects who had the highest level of creativity, turned out to be simultaneously carriers of the 'strongest' genotypes. Interpreting this correlation, it would be reasonable to suppose that subjects with 'stronger' genotypes were so well prepared to cope with stressful conditions, that there was need for them to activate psychological defense mechanisms, either conscious or, subconscious: psychological adaptation was provided in their case by creative performance. One may also suppose that in normal conditions, their 'excessive' creativity was used for other purposes, which were not adaptive, but providing and enhancing the tasks belonging to the realm of self- actualization.

This conclusion of ours is in fact corroborated by data concerning the psychological state of subjects with different BDNF genotypes. As shown by Tables 7, 8 and 9, the 'stronger' the BDNF genotype, the higher the creativity level, and the lower the level of activation of psychological defense mechanisms, both conscious and subconscious. To formulate this regularity in different terms, creative performance seems to compete with defense mechanisms or, at least, to function as their counterpoint in providing adaptation to stressful conditions. One cannot exclude that there might exist definite complementarity between psychological defense mechanisms and creativity, as well. Increasing the number of the group would allow us to demonstrate that this regularity is proper not only for the BDNF gene, but also for the ACTN3, and ACE ones.

Summing up, we have to state that the border between nature and culture, at least at the levels of genetics and creativity, seems to be much more permeable than it might seem at first glance. Apart from the fact that the dopaminergic, serotoninergic, and noradrenergic systems, as well as neuregulin 1 gene, arginine vasopressin receptor, angiotensinogene, and a number of other genes, tend to back creative competences in quite a strong way, 'strong' genotypes of quite a few genes tend to be linked to high levels of creativity, both verbal and figural (BDNF, ACTN3, and, to a certain extent, ACE, may serve as a plausible example, as it was demonstrated in the present study). This type of relations, focused upon creativity as a sophisticated constellation of both adaptive and non-adaptive properties, is likely to form a point of growth at the present-day interdisciplinary field, formed by molecular genetics, on the one hand, and creativity studies, on the other one.

Acknowledgements. This study was sponsored by Russian Foundation for Basic Research, grant 20-013-00121.

The authors are grateful to A.V. Lemeschenko, MD, PhD, for taking part in the organization of the pilot study, to A.E. Trandina, MD, for taking part in sample genotyping, to E.E. Tunik, PhD, for consultations concerning the application of Creativity Test battery, and to professor L. Vasserman, PhD, for those concerning the application of the Life Style Index questionnaire, and of the Stress Coping Strategies methodology.

References

1. Sternberg, R.J., Kaufman, J.C. (eds.): The Nature of Human Creativity. Cambridge University Press, New York (2018). https://assets.cambridge.org/97811071/99811/frontmatter/9781107199811_frontmatter.pdf
2. Shalley, C., Hitt, M., Zho, J. (eds.): The Oxford Handbook of Creativity, Innovation, and Entrepreneurship. Oxford University Press, Oxford (2015). https://www.pdfdrive.com/the-oxford-handbook-of-creativity-innovation-and-entrepreneurship-e157854663.html
3. Torrance, E.P.: The Torrance Tests of Creative Thinking: Norms-Technical Manual. Personal Press, Princeton (1974)
4. Khalil, R., Godde, B., Karim, A.: The link between creativity, cognition, and creative drives and underlying neural mechanisms. Front. Neural Circuits **13**(18), 1–16 (2019). https://doi.org/10.3389/fncir.2019.00018
5. Keri, S.: Genes for psychosis and creativity: a promoter polymorphism of the Neuregulin 1 gene is related to creativity in people with high intellectual achievement. Psychol. Sci. **20**, 1070–1073 (2009). https://doi.org/10.1111/j.1467-9280.2009.02398.x
6. Venkatasubramanian, G., Kalmady, S.: Creativity, psychosis and human evolution: the exemplar case of Neuregulin 1 gene. Indian J. Psychiatry **52**(3), 282 (2010). https://doi.org/10.4103/0019-5545.71003
7. Bachner-Melman, R., et al.: AVPR1a and SLC6A4 gene polymorphisms are associated with creative dance performance. PLoS Genet. **1**(3), e42 (2005). https://doi.org/10.1371/journal.pgen.0010042
8. Ukkola, L.T., Onkamo, P., Raijas, P., Karma, K., Järvelä, I.: Musical aptitude is associated with AVPR1A-haplotypes. PLoS ONE **4**(5), e5534 (2009). https://doi.org/10.1371/journal.pone.0005534
9. Spivak, I.M., Seilieva, N.A., Smirnova, T.J., Bolotskikh, V.M., Abramchenko, V.V., Spivak, D.L.: Renin-angiotensin system gene polimorphisms and their correlation to psychological phenomena in birth stress. Tsitologiia. **50**(10), 899–906 (2008). (in Russian). https://pubmed.ncbi.nlm.nih.gov/19062524
10. Kaufman, A., Kornilov, S., Bristol, A., Tan M., Grigorenko, E.: Genetic and evolutionary bases of creativity. In: Kaufman, J.C., Sternberg, R.J. (eds.) The Cambridge Handbook of Creativity, pp. 216–232. Cambridge University Press, Cambridge (2010). https://doi.org/10.1017/cbo9780511763205.014
11. Tunik, E.E.: The Best Creativity Tests. Diagnostics of Creative Thinking. Piter, St. Petersburg (2013). (in Russian)
12. Vasserman, L.I., et al.: Psychological diagnostics of life style. Manual for Psychologists and Physicians. NIPNI im. V.M.Bekhtereva, St. Petersburg (2005). (in Russian)
13. Vasserman, L.I., et al.: Methodology of psychological diagnostics of coping with stressful and difficult personal situations. Manual for Physicians and Clinical Psychologists. NIPNI im. V.M.Bekhtereva, St. Petersburg (2009). (in Russian)

14. Vaquero Lorenzo, C., et al.: Association between the T102C polymorphism of the serotonin-2A receptor gene and schizophrenia. Prog. Neuropsychopharmacol. Biol. Psychiatry **30**(6), 1136–1138 (2006). https://doi.org/10.1016/j.pnpbp.2006.04.027

15. Carhart-Harris, R.L., Nutt, D.J.: Serotonin and brain function: a tale of two receptors. J. Psychopharmacol. **31**(9), 1091–1120 (2017). https://doi.org/10.1177/0269881117725915

16. Jobim, P.F.C., Prado-Lima, P.A.S., Schwanke, C.H.A., Giugliani, R., Cruz, I.B.M.: The polymorphism of the serotonin-2A receptor T102C is associated with age. Braz. J. Med. Biol. Res. **41**(11), 1018–1023 (2008). https://doi.org/10.1590/s0100-879x2008005000045

17. Rigat, B., Hubert, C., Alhenc-Gelas, F., Cambien, F., Corvol, P., Soubrier, F.: An insertion/deletion polymorphism in the angiotensin I-converting enzyme gene accounting for half the variance of serum enzyme levels. J. Clin. Invest. **86**, 1343–1346 (1990). https://doi.org/10.1172/JCI114844

18. Sayed-Tabatabaei, F.A., Oostra, B.A., Isaacs, A., van Duijn, C.M., Witteman, J.C.: ACE polymorphisms. Circ. Res. **98**(9), 1123–1133 (2006). https://doi.org/10.1161/01.RES.0000223145.74217.e7

19. O'Malley, J.P., Maslen, C.L., Illingworth, D.R.: Angiotensin-converting enzyme DD genotype and cardiovascular disease in heterozygous familial hypercholesterolemia. Circulation **97**, 1780–1783 (1998). https://doi.org/10.1161/01.cir.97.18.1780

20. Puthucheary, Z., Skipworth, J.R., Rawal, J., Loosemore, M., Van Someren, K., Montgomery, H.E.: The ACE gene and human performance: 12 years on. Sports Med. J. **41**(6), 433–448 (2011). https://doi.org/10.2165/11588720-000000000-00000

21. Meyer, T., et al.: Length polymorphisms in the angiotensin I-converting enzyme gene and the serotonin-transporter-linked polymorphic region constitute a risk haplotype for depression in patients with coronary artery disease. Biochem. Genet. **58**(4), 631–648 (2020). https://doi.org/10.1007/s10528-020-09967-w

22. Lau, H., Fitri, A., Ludin, M., Rajab, N.F., Shahar S.: Identification of neuroprotective factors associated with successful ageing and risk of cognitive impairment among malaysia older adults. Curr. Gerontol. Geriatr. Res. 4218756 (2017). https://doi.org/10.1155/2017/4218756

23. Prabu, P., Poongothai, S., Shanthirani, C.S., Anjana, R.M., Mohan, V., Balasubramanyam, M.: Altered circulatory levels of miR-128, BDNF, cortisol and shortened telomeres in patients with type 2 diabetes and depression. Acta Diabetol. **57**(7), 799–807 (2020). https://doi.org/10.1007/s00592-020-01486-9

24. Vasconcelos-Moreno, M.P., et al.: Telomere length, oxidative stress, inflammation and BDNF levels in siblings of patients with bipolar disorder: implications for accelerated cellular aging. Int. J. Neuropsychopharmacol. **20**(6), 445–454 (2017). https://doi.org/10.1093/ijnp/pyx001

25. Zhou, J.-X., et al.: Functional Val66Met polymorphism of brain-derived neurotrophic factor in type 2 diabetes with depression in Han Chinese subjects. Behav. Brain Funct. **9**, 34 (2013). https://doi.org/10.1186/1744-9081-9-34

26. Pickering, C., Kiely, J.: ACTN3: more than just a gene for speed. Front. Physiol. **8**, 1080 (2017). https://doi.org/10.3389/fphys.2017.01080

27. Pickering, C., Kiely, J.: ACTN3, morbidity, and healthy aging. Front. Genet. **9**, 15 (2018). https://doi.org/10.3389/fgene.2018.00015

28. Spivak, I.M., Smirnova, T.Y., Slizhov P.A., Spivak D.L.: Identification of a group for research of telomeric aging. EpSBS. **LIII**, 194–199 (2020). https://doi.org/10.15405/epsbs.2020.12.03.20

29. Wyckelsma V.L, et al.: Loss of α-actinin-3 during human evolution provides superior cold resilience and muscle heat generation. Am. J. Hum. Genet. **108**(3), 446–457 (2021). https://doi.org/10.1016/j.ajhg.2021.01.013

30. Spivak, D., Spivak, I.: Creativity and longevity: new realm of research. Eur. Proc. Soc. Behav. Sci. **LIII**, 81–88 (2020). https://doi.org/10.15405/epsbs.2020.12.03.8

31. Cahn, B.R., Goodman, M.S., Peterson, C.T., Maturi, R., Mills, P.J.: Yoga, meditation and mind-body health: increased BDNF, cortisol awakening response, and altered inflammatory marker expression after a 3-month yoga and meditation retreat. Front. Hum. Neurosci. **11**, 315 (2017). https://doi.org/10.3389/fnhum.2017.00315

32. Silva, H.H., Silva, M.G., Cerqueira, F., Tavares, V., Medeiros, R.: Genomic profile in association with sport-type, sex, ethnicity, psychological traits and sport injuries of elite athletes: review and future perspectives. J. Sports Med. Phys. Fitness. (2021). https://doi.org/10.23736/S0022-4707.21.12020-1

33. Coelho, D.B., et al.: Alpha-Actinin-3 R577X polymorphism influences muscle damage and hormonal responses after a soccer game. J. Strength Cond. Res. **33**(10), 2655–2664 (2019). https://doi.org/10.1519/JSC.0000000000002575

34. Yasar, S., Varma, V.R., Harris, G.C., Carlson, M.C.: Associations of angiotensin converting enzyme-1 and angiotensin ii blood levels and cognitive function. J. Alzheimers Dis. **63**(2), 655–664 (2018). https://doi.org/10.3233/JAD-170944

35. Barbosa, I.G., et al.: The renin angiotensin system and bipolar disorder: a systematic review. Protein Pept. Lett. **27**(6), 520–528 (2020). https://doi.org/10.2174/0929866527666200127115059

36. Vian, J., et al.: The renin-angiotensin system: a possible new target for depression. BMC Med. **15**(1), 144 (2017). https://doi.org/10.1186/s12916-017-0916-3

37. Firouzabadi, N., et al.: Genetic variants of angiotensin-converting enzyme are linked to autism: a case-control study. PLoS ONE **11**(4), e0153667 (2016). https://doi.org/10.1371/journal.pone.0153667

Images of Giftedness and Creativity

Irina Berezovskaya[1,2]([envelope]) [iD], Maria Karagacheva[2] [iD], Tatiana Slotina[2] [iD], Aleksandra Komarova[1] [iD], and Nina Popova[1] [iD]

[1] Peter the Great St. Petersburg Polytechnic University, Polytechnicheskaya, 29, 195251 St. Petersburg, Russia
[2] Emperor Alexander I St. Petersburg State Transport University, Moskovsky pr. 9, 190031 St. Petersburg, Russia

Abstract. The authors investigate the psychological aspect of creativity through the giftedness phenomenon and its manifestations in real activity. The article presents the content analysis results of students' ideas about giftedness and creativity, obtained by the associative experiment method. According to the authors, implicit notions and the giftedness theory have a significant impact on attitudes towards gifted people. The methodology of hierarchical person's image structure (HPIS) was used as a research tool. The purpose of this study is to identify the image of a gifted person based on implicit ideas of students of various specialities and universities. As a result of the study, it was revealed that the basis of the structure and content of the gifted person image is made up of intellectual characteristics, with the acquisitive characteristics being the most unpopular in the respondents' answers. In the second place, in terms of occurrence frequency, there were social characteristics, which are more typical for the students of humanitarian specialities. This group is also characterized by a more frequent indication of the negative qualities of a gifted person, while students of the technical specialities used positively colored words more often. When describing a gifted person, names and qualities of activity were more often used in the group of technical students. The data obtained by the authors confirm the special extraordinary, creative nature of a gifted person in the implicit representations of people.

Keywords: Creativity · Giftedness · Personality · Notions · Implicit theory · Images

1 Introduction

The problem of the giftedness phenomenon is one of the most pressing problems with an interdisciplinary status. Formation of creativity and identification of giftedness is one of the most important tasks of the educational process, and the presentation of a conceptual model, in which the key link in the development of specific educational programs that allow more people to reach a high level of achievement, is one of the most important tasks of modern humanities [1, 2]. The relevance of this task is especially acute in connection with the introduction of information technologies in the education process [3, 4]. Giftedness and creativity are integral characteristics of personality in implicit

D. Bylieva and A. Nordmann (Eds.): PCSF 2021, LNNS 345, pp. 87–94, 2022.
https://doi.org/10.1007/978-3-030-89708-6_8

theories that teachers often use to gifted and creative students. The relevance of the research subject is associated with the solution of theoretical and practical problems of modern education - the construction of a developmental, personality-oriented education system aimed at the development of individual capabilities, abilities and interests of students [5, 6]. Zirenko notes that it is implicit theories that reflect ideas about the essence of cognitive and personal characteristics of an individual [7]. In the scientific world, though, explicit theories drawing on rigorous, evidence-based experimental facts are more often used.

However, implicit representations have a significant impact on the perception of the surrounding reality, and therefore on the people's affective and behavioral sphere. Explicit theories are comparatively less connected with the practice and real life of people. Family, parents, educators, teachers, students, children, as a rule, are not familiar with modern and classical scientific achievements. But it is these people who, operating with their ordinary ideas, cause a direct impact on the manifestation and development of giftedness in people, therefore implicit concepts of giftedness often have a more significant impact on attitudes towards gifted individuals, on the identification of giftedness and its development. The importance of everyday ideas in the formation of an individual's ideas about the structure and functioning mechanisms of his personality and the personality of another person was first studied by American psychologists J. Bruner and R. Tagiuri, who named them "implicit personality theories" ("naive personality conception ", "common sense personality theory"). Implicit personality concepts are inherently social representations since they are presented in the form of a system of ideas about the connections between personality traits, which allow a person to better navigate in communication with other people [8]. The research devoted to the study of the gifted schoolchildren image convincingly prove that the knowledge of the implicit gifted person model contributes to effective pedagogical interaction with the gifted children.

There is a widespread prevalence of implicit ideas, in particular, among teachers directly related to gifted children studying in ordinary schools. For ex-ample, in Gagne's differentiated model of giftedness and talent, the essential role of significant people is highlighted, with teachers among them without doubt [9]. Implicit personality theories reflect generalized ideas, images of people. They are a way of organizing information, often randomly and only partially realized by a person. Since there are many sources of such knowledge, implicit theories are more generalized. Without a clear structure, they relate to a larger number of giftedness components, and the principle of positive hypothesis existing in pedagogy ("the presumption of giftedness" of everyone) makes the situation even more complicated [10, 11].

The purpose of this study is to identify the image of a gifted person based on implicit ideas of students of various specialities and universities. For this purpose, in an empirical study based on the results of a content analysis of over 2200 characteristics of concepts describing a gifted person, students of technical and humanitarian specialities are tasked to identify the place of implicit representations of giftedness in modern youth in the study of the giftedness phenomenon. The results of this study have a meaningful cognitive potential for creating further knowledge of the working conceptual model in the development of creative and gifted individuals.

2   Literature Review

In culture, the concept of creativity was associated with the concept of giftedness, which appears to explain it. Giftedness is an individual feature of a person, its consequence is a creative person. The definition of creativity as the creation of something new allows us to judge creativity only by the product, without paying due attention to the nature of the process itself. In this regard, a gifted and creative person is completely vague and often synonymous. In psychology, pedagogy, sociology, philosophy, a lot of research has been accumulated on this problem of giftedness, but there is still no generally accepted definition of giftedness, which is due to the complexity of this phenomenon and different approaches to its analysis. The traditional view of giftedness is the idea of the selective endowment of an individual with certain gifts. The pedagogical approach developing in the 20th century presents giftedness as an educational category. In the early 21st century, giftedness begins to be viewed as a context-sensitive entity [12, 13], understood as "emerging excellence" [14], or as an "optimal inter-actualized transaction between an individual and his or her environment " [15] functional person-in-situ transactions [16]. Renzulli and Ries note that the areas of activity in which a person can be recognized as "gifted" are determined by the needs and values of the prevailing culture [17]. Subotnik, Olszewski-Kubilius, and Worrell emphasize that the development of giftedness is the most promising area of education [18]. Cross, Cross and O'Reilly remark that in 2018, in the Irish education system, education for the gifted was not officially formalized, however, the results of their survey indicate the need to improve the teachers' qualifications in the relevant direction [19].

Diverse theories discuss the nature of the giftedness emergence, its dynamics, the time frame of manifestation, the methods and criteria for its identifying, supporting and accompanying, its connection with the communicative, conative, emotional spheres of the personality, and even the issue of considering giftedness to be the norm or pathology. In addition, as noted by Volkov and colleagues, in psychology there is no integral construct of the phenomenon "attitude to one's giftedness" [20]. These questions lead us to search for the connection of this phenomenon with another category, no less multifaceted in the scientific sense, i.e. with the category of "personality". V. D. Shadrikov demonstrated an approach according to which abilities and giftedness (as an integral manifestation of abilities) should be considered simultaneously in three dimensions: natural, subject-activity oriented and personal ones [21]. The studies of such authors as Makel, Snyder, Thomas, Malone, Putallaz, [22], Tan, Yough, Desmet, Pereira [23] discuss the correlation of the intelligence and giftedness concepts.

Individual factors of psychological well-being relating to gifted adolescents were studied in the work of Miklyaeva, Khoroshikh, Volkova, and it was revealed that the type of educational environment, gender, age, type of giftedness belong to the significant conditions determining the psychological well-being of gifted adolescents. Baudson, Preckel studied the teachers' ideas about the gifted and average students by the parameters of achievement [24]. Matheis, Kronborg, Schmitt, Preckel studied the teachers' stereotypes about the characteristics of gifted students. In addition to high intelligence, German and Australian teachers associated giftedness with maladjustment and demonstrated lower efficiency in teaching gifted people [25]. Preckel, Baudson, Krolak-Schwerdt, Glock analyzed the associations of people connected with gifted children [26].

There are few studies of gifted college-age people. Among them, Jurisevic, Zerak, devoted their works to the study of attitudes towards gifted students with the Slavs as an example [27].

3 Methodology

The theoretical basis of the research is the fundamental provisions on giftedness as a complex and developing personal quality that determines the possibility of achieving outstanding results in activities valuable to society. In particular, the ring concept of giftedness, developed by Renzulli, suggests that the number of gifted people can be much wider than when they are identified by tests of intelligence, creativity or achievement.

The methodological basis of the research is the concept of implicit personality theories, which assume a typical category of everyday cognition. The point is that several traits correlate with each other in the personality so that if one is present, one should also assume the presence of another. In this regard, such a trait as giftedness necessarily presupposes the presence of other specific traits.

The methodology of hierarchical person's image structure (HPIS) [28] based on the principle of an associative experiment was a research tool. The methodology is based on Kuhn and McPartland's concept [29], on the theoretical views of Bodalev [30] on social perception and those of Rean, Kolominsky [31] on social and pedagogical perception.

The verbal part of the HPIS methodology, which was used in this study, makes it possible to collect information, thanks to which extensive data were obtained on the verbal ideas of students about a gifted person. The methodology has proven its effectiveness for the comparative analysis of the socio-perceptual characteristics of the phenomenon under consideration and the personality itself, involved in the research process.

The method of content analysis, which is the basis of the primary data processing, allows us to combine the concepts - associations into several groups (conventional, social, intellectual and others), conduct their comparative analysis and identify the most frequently encountered categories related to the phenomenon of a gifted person.

For further processing of the results, mathematical statistics methods were used: descriptive statistics, comparison of average results using the φ-test (angular transformation) of Fisher.

The sample consisted of 229 respondents: 88 people of humanitarian specialities, 141 respondents represented technical sphere, all of them aged from 17 to 22 years old. The study was carried out in October-November 2020 at the universities of St. Petersburg (St. Petersburg State University of Railways of Emperor Alexander I, Russian State Pedagogical University named after A.I. Herzen, Peter the Great Polytechnic University.).

4 Results and Discussion

The empirical research results confirmed our assumption that students were much more likely to note in their answers the features of intelligence (1214 words, which accounted for more than 50% of all words), which included such concepts as quick thinking, smart, talented, brilliant, capable of mathematics, etc. Social characteristics are mentioned twice less often (507, more than 20% of all words): kind, lonely, secretive, polite, shy. Even less common were metaphors (243), activity-oriented (242), conventional (211), behavioral (184) motivational-volitional (169), bodily-physical (138) emotional (111) categories of definitions. The least common of all was the akisitive characteristics found in 7 respondents. 1666 definitions of a gifted person were positively colored, 491 of them were neutral and 73 were negatively colored. Many students wrote the names of famous people and their relatives or acquaintances (104 names), with young men preferring male names, and girls preferring female ones.

Frequency analysis of associations on the topic "Gifted Personality" showed that the most popular words in the entire sample were: talent (65.5%), genius (51.1%), capable (30.6%)), intelligence (30.1%), success (24.9%), special (21.4%), outstanding (21%), creator (17.5%), gift (16.2%) (Table 1).

Table 1. Frequency analysis of the characteristics included by students in the description of a gifted person

Characteristics	Absolute frequency	Relative frequency %
Talent/talented	112	62,22
Genius	96	53,33
Intellect/smart	66	36,67
Capable	54	30,00
Creative/creator	48	26,67
Success/successfulness	47	26,11
Outstanding	41	22,78
Unique/uniqueness	40	22,22
Special personality	39	21,67
Gift/endowment	30	16,67
Intelligence/intelligent	23	12,78
Child prodigy	19	10,56
Leader	15	8,33

It was found that, in general, the representations are of the same type: a gifted person has distinctive features from others in terms of intellectual properties and is characterized by abilities of different levels.

Among the most common words describing a gifted personality among students of different areas of training, an insignificant number of differences were revealed: the words "strange" and "interesting" were more frequent in the group of humanitarians, whereas the words "great", "artist" and "scientist" were more frequently used by technical students.

The data we have obtained prove that almost every person has implicit ideas about giftedness, which develop quite spontaneously and virtually independently from explicit concepts and scientific theories.

A significant number of negative characteristics presented by the humanitarian students are most likely associated with their more frequent use of the social qualities of a gifted person. It is in these characteristics that negatively colored words are used to a greater extent. The presence of a significant number of social characteristics when describing a gifted person indicates the dominance of social characteristics in the implicit theory of a gifted person.

In the course of the study, it was revealed that there are general trends associated with the use of intellectual and social characteristics in describing a gifted person. These trends include:

- the cognitive sphere qualities as noted by more than 60% of the respondents.
- the social sphere qualities as noted by about 30% of the respondents.
- there are comparatively few characteristics of the motivational-volitional and emotional sphere in the implicit concepts of a gifted person.
- the implicit concept of the gifted personality of modern students is associated with their profile of professional training.

Students of technical specialities more often describe a gifted person through his/her implementation in an activity, and they are inclined to indicate more specific categories: names of professions, etc. For humanities, the relationships of a gifted person with society and his/her external manifestations, such as emotional and bodily-physical ones are more significant. These features are probably related to their different professional areas.

5 Conclusion

Our research allows us to assert the relevance and prospects of studying students' ideas about creativity and giftedness. The presented research results allow us to conclude that there is a cognitive potential in the implicit study of creativity and giftedness. The results of empirical research have shown that students note the features of intelligence in their answers, they name social characteristics twice less often, they are less common: metaphors and least of all are acquisitive characteristics. The research failed to study the gender characteristics of the gifted person images, which is planned to be done in the future, as we consider this direction important from the point of view of understanding the influence of modern society stereotypes on the formation of ideas about giftedness. It also seems expedient to study the specifics of each professional orientation in more detail, which will help develop programs for training teachers to identify and support giftedness at different stages of its development.

References

1. Jurisevic, M., Zerak, U.: Attitudes towards gifted students and their education in the Slovenian context. Psychol. Russ. Curr. State **12**(4), 101–117 (2019). https://doi.org/10.11621/pir.2019.0406

2. Klochkova, E.S., Bolsunovskaya, M.V., Shirokova, S.V.: The significance of humanities for engineering education. In: Proceedings of 2018 17th Russian Scientific and Practical Conference on Planning and Teaching Engineering Staff for the Industrial and Economic Complex of the Region, PTES 2018, pp. 265–268. IEEE, St. Petersburg (2019). https://doi.org/10.1109/PTES.2018.8604199

3. Almazova, N., Barinova, D., Ipatov, O.: Forming of information culture with tools of electronic didactic materials. In: Katalinic, B. (ed.) Annals of DAAAM and Proceedings of the International DAAAM Symposium, vol. 29, no. 1, pp. 0587–0593. Danube Adria Association for Automation and Manufacturing, DAAAM, Zadar; Croatia (2018). https://doi.org/10.2507/29th.daaam.proceedings.085

4. Almazova, N., Bylieva, D., Lobatyuk, V., Rubtsova, A.: Human behavior as a source of data in the context of education system. In: SPBPU IDE 2019: Proceedings of Peter the Great St. Petersburg Polytechnic University International Scientific Conference on Innovations in Digital Economy, p. 37. ACM, Saint Petersburg (2019). https://doi.org/10.1145/3372177.3373340

5. Shipunova, O.D., Kolomeyzev, I.V., Mureyko, L.V., Kozhurin, A.Y., Kosterina, O.N.: Resources to matrix control of mental activity in information environments. Utopia y Praxis Latinoamericana. 24(5), 101–117 (2019). https://www.redalyc.org/jatsRepo/279/27962050015/html/index.html. Accessed 2 May 2021

6. Shipunova, O.D., Evseeva, L., Pozdeeva, E., Evseev, V.V., Zhabenko, I.: Social and educational environment modeling in future vision: infosphere tools. E3S Web Conf. 110, 02011 (2019). https://doi.org/10.1051/e3sconf/201911002011

7. Zirenko, M.S.: Implicit theories of intelligence and personality: relations to intelligence, motivation and personality. Psychol. J. High. Sch. Econ. 15(1), 39–53 (2018). https://doi.org/10.17323/1813-8918-2018-1-39-53. (in Russian)

8. Bylinskaya, N.V.: The phenomenon of personality giftedness in explicit and implicit theories. Young Sci. 24(128), 276–280 (2016). https://moluch.ru/archive/128/35541/. Accessed 9 Jan 2021

9. Gagne, F.: Academic talent development: theory and best practices. In: Pfeiffer, S.I., Shaunessy, E., Foley-Nicpon, M. (eds.) APA Handbook of Giftedness and Talent, pp. 163–183. APA, Washington (2018). https://doi.org/10.1037/0000038-011

10. Ilaltdinova, E.Y., Frolova, S.V.: Conceptual bases of identification, selection and support of pedagogically gifted youth. Bull. Minin Univ. 6(4), 9 (2018). https://doi.org/10.26795/2307-1281-2018-6-4-9. (in Russian)

11. Miklyaeva, A.V., Khoroshikh, V.V., Volkova, E.N.: Subjective factors of gifted adolescents' psychological well-being: a theoretical model. Sci. Educ. Today 9(4), 36–55 (2019). https://doi.org/10.15293/2658-6762.1904.03

12. Dai, D.Y.: Envisioning a new foundation for gifted education: evolving complexity theory (ECT) of talent development. Gifted Child Q. 61(3), 159–163 (2017). https://doi.org/10.1177/0016986217701837

13. Hymer, B.J.: An act of GRACE? What do contemporary understandings in psychology have to contribute to the future of gifted education? Gifted Educ. Inter. 29(2), 108–124 (2012). https://doi.org/10.1177/0261429412447707

14. Plucker, J., Barab, S.: The importance of contexts in theories of giftedness: learning to embrace the messy joys of subjectivity. In: Sternberg, R.J., Davidson, J. (eds.) Conceptions of Giftedness, pp. 201–216. Cambridge University Press. Boston (2005). https://doi.org/10.1017/CBO9780511610455.013

15. Lo, C.O., Porath, M.: Paradigm shifts in gifted education: an examination vis-a-vis its historical situatedness and pedagogical sensibilities. Gifted Child Q. 61(4), 343–360 (2017). https://doi.org/10.1177/0016986217722840

16. Barab, S.A., Plucker, J.A.: Smart people or smart contexts? Cognition, ability, and talent development in an age of situated approaches to knowing and learning. Educ. Psychol. **37**, 165–182 (2002). https://doi.org/10.1207/S15326985EP3703_3

17. Renzulli, J.S., Ries, S.M.: The three-ring conception of giftedness: a developmental approach for promoting creative productivity in young people. In: Pfeiffer, S.I., Shaunessy, E., Foley-Nicpon, M. (eds.) APA Handbook of Giftedness and Talent, pp. 185–199. APA, Washington (2018). https://doi.org/10.1037/0000038-012

18. Subotnik, R.F., Olszewski-Kubilius, P., Worrell, F.C.: Talent development as a most promising focus of giftedness and gifted education. In: Pfeiffer, S.I., Shaunessy, E., Foley-Nicpon, M. (eds.) APA Handbook of Giftedness and Talent, pp. 231–245. APA, Washington (2018). https://doi.org/10.1037/0000038-015

19. Cross, T.L., Cross, J.R., O'Reilly, C.: Attitudes about gifted education among Irish educators. High Ability Stud. **29**(2), 169–189 (2018). https://doi.org/10.1080/13598139.2018.1518775

20. Volkova, E.N., Miklyaeva, A.V., Kosheleva, A.N., Khoroshikh, V.V.: Attitude towards their own giftedness among older adolescents who have passed pedagogical selection for specialized educational programs. Psychol. Sci. Educ. **25**(3). 49–63 (2020). https://doi.org/10.17759/pse.2020250305

21. Shadrikov, V.D.: To a new psychological theory of abilities and giftedness. Psychol. J. **40**(2), 15–26 (2019). https://doi.org/10.15293/2658-6762.2003.05

22. Makel, M.C., Snyder, K.E., Thomas, C., Malone, P.S., Putallaz, M.: Gifted students' implicit beliefs about intelligence and giftedness. Gifted Child Q. **59**(4), 203–212 (2015). https://doi.org/10.1177/0016986215599057

23. Tan, D., Yough, M., Desmet, O.A., Pereira, N.: Middle school students' beliefs about intelligence and giftedness. J. Adv. Acad. APA Handbook of Giftedness and Talent. **1**, 50–73 (2019). https://doi.org/10.1177/1932202X18809360

24. Yurkevich, V.S.: Intellectual giftedness and social development: contradictory connection. Mod. Foreign Psychol. **7**(2), 28–38 (2018). https://doi.org/10.17759/jmfp.2018070203

25. Matheis, S., Kronborg, L., Schmitt, M., Preckel, F.: Threat or challenge? Teacher beliefs about gifted students and their relationship to teacher motivation. Gifted Talent. Intern. **32**(2), 134–160 (2018). https://doi.org/10.1080/15332276.2018.1537685

26. Preckel, F., Baudson, T.G., Krolak-Schwerdt, S., Glock, S.: Gifted and maladjusted? Implicit attitudes and automatic associations related to gifted children. Am. Educ. Res. J. **52**(6), 1160–1184 (2015). https://doi.org/10.3102/0002831215596413

27. Jurisevic, M., Zerak, U.: Attitudes towards gifted students and their education in the Slovenian context. Psychol. Russ. Curr. State **13**(4), 89–105 (2020). https://doi.org/10.11621/pir.2019.0406

28. Sitnikov, V.L.: The Image of the Child in the Minds of Children and Adults. Khimizdat, Saint Petersburg (2001). (in Russian)

29. Kuhn, M.H., McPartland, T.S.: An empirical investigation of self-attitudes. Am. Sociol. Rev. **19**(1), 68–76 (1954). https://doi.org/10.2307/2088175

30. Bodalev, A.A.: Perception and Understanding of Man by Man. Publishing House of Moscow University, Moscow (1982). http://maxbaxtin.ru/f/bodalev_a5_blok.pdf. Accessed 9 Jan 2021

31. Rean, A.A. Kolominsky, Ia.L.: Social Pedagogical Psychology. Peter, Saint Petersburg (2008)

Towards Creation: Sergius Bulgakov and Pavel Florensky on the Relationship Between Scientific and Religious Experience

Vera Serkova[1]([✉]) [iD], Tatyana Simonenko[2] [iD], Oleg Samylov[3] [iD],
and Alexander Pylkin[1] [iD]

[1] Peter the Great St. Petersburg Polytechnic University, St. Petersburg, Russia
[2] St. Petersburg State University, St. Petersburg, Russia
[3] Petersburg University of Railway Transport of Emperor Alexander I, St. Petersburg, Russia

Abstract. The revival of the dialogue between science and religion, which characterizes modern culture, testifies to the relevance of the problem of the correlation of scientific and religious experience in cognition. The article demonstrates various, sometimes opposite approaches to formulating and understanding this problem in modern philosophical thought, indicating the pluralism of positions and points of view, as well as the open nature of its discussion, which requires a solid theoretical basis for further analysis. It is suggested that one of the variants of the theoretical and methodological basis for understanding the correlation of scientific and religious experience is found in Russian philosophy at the beginning of the twentieth century, in particular, in the works of its two prominent representatives – Sergius Bulgakov and Pavel Florensky. Considering economic relations in capitalist society, Bulgakov supplements their study with the religious doctrine of Sophia, which makes it possible to reflect on the creative nature of the activity of the subject of production and find explanations for all those negative phenomena that accompany human economic activity. Florensky points out the limitations of pure scientific research, believing that in scientific knowledge such a limiting area of research inevitably opens up, which is no longer the subject of scientific, but exclusively religious speculation. The article analyzes these ideas of Russian philosophers, gives their interpretation in modern philosophical literature and concludes about the interpenetration of scientific and religious experience in cognition.

Keywords: Scientific experience · Religious experience · Sergius Bulgakov · Pavel Florensky

1 Introduction

The formation of any type of worldview is based on the problem of understanding reality, the scientific description of which is the foundation of theoretical knowledge, which serves as the basis not only for further cognition, but also for the transformation of reality. Today, special concerns are caused by the use of new technologies based on the achievements of modern science. It is no coincidence that the image of "Pandora's chest" has become a symbol of the development of NBIC technologies. Can science and the scientific community be responsible for the future of the world in the light of the possible devastating consequences of the application of scientific discoveries? Or do you need some additional instance at least to comprehend the problems that lie here?

Christian theology traditionally assumed the function of such control, and from the very beginning of its formation, from the teachings of the holy fathers of the church, the meaning and tasks of cognition and the boundaries of the rational transformation of the world were at the center of theologians' reasoning. Interest in the possibilities of science as a means of transforming the world increased with the beginning of the industrial revolution. Science began to be considered not only from the point of view of the possibility of improving the living conditions of mankind, but also from the point of view of its destructive power. In this respect, philosophers belonging to the Russian religious tradition of the late 19th and early 20th centuries are of great interest.

It is even more interesting the experience of combining the functions of a scientist and a religious philosopher in the same person. When a scientist and a believer are united in one researcher, two sides of his activity control one another. Let us dwell on two representatives of Russian religious philosophy – Sergius Bulgakov and Pavel Florensky, in whose work the unity of religious, scientific and philosophical foundations was fully manifested.

2 Literature Review

Analysis of modern literature shows that the relationship between religious and outside religious experience, in particular, scientific as its alternative, remains in the focus of philosophers' attention.

A number of articles express the conviction that it is impossible to reconcile scientific and religious experience. In the editorial article "Keep doors open for constructive dialogue between religion and science", published in the journal "Nature" in 2017, the heading contains the problem of the relationship between religious and scientific worldview [1]. The fundamental thesis of the article: "There is a chasm between religion and science that cannot be bridged" [1, p. 265]. Religious and science-oriented consciences clash primarily in the comprehension of issues related to the achievements of science, such as "anthropogenic climate change"; contraception; poverty; embryo selection. What does the "keep doors open" opportunity give? P. Bayón elaborates on the foundations of "the interaction between theism and atheism" as an opportunity "of reconciling harmoniously the new insights on the reality generated by the techno-scientific advances, social changes, and the place currently occupied by the traditional and institutional and transcendent religiosity and the immanent spirituality generated by the philosophical naturalism" [2, p. 761].

In the light of the achievements of natural science (primarily biology and non-classical physics), the problem arises of reconciling religious and scientific ideas about nature "blackhole" and "incorporeal nature of God's being" arise [3, p. 185]. A different position prevails here. In the discourse about the nature of the modern scientist's worldview, his views on nature clearly indicate the point of view about the integrity of these beliefs, the consistency of their scientific and religious components. So, for example, Del Carril investigates the activities of a particular scientist – Pascual Jordan, who took part in a productive debate about the nature of the atom in Quantum physics and was at the same time a believing Christian. From Del Carril's point of view, Jordan took the most advantageous position, in which his beliefs had an existential meaning, existed as if separate from his scientific activities and did not contradict his academic research [4]. Research by R. Woodford is devoted to the possibility of "intellectually rigorous and satisfying discussions of science and religion" [5, p. 937], the author is critical of Thomas Lessl's ideas about demarcation between adherents and opponents of the theory of evolution.

One of the variants of this problem is the question of how the religious and scientific "ethos" are combined in one researcher. Here we are talking about the relationship between science and religion as a cultural concept. So A.E. Razumov in his article "Faith, understanding, proof" [6] correlates different aspects of rational activity, in particular, faith and proof. This problem, old from the time of early Christianity, is placed by the author in the modern context of interactions between science and religion. The religious worldview is considered as the basis of the value hierarchy in the organization of ideas about reality. J. Teehan in his research focused on cognitive sciences analyzes "the continuing influence of religion in human affairs" [7, p. 272]. Using a specific example, J. Teehan shows why religion occupies an important place in modern culture, but it turns out, according to the author, dangerously problematic in solving moral issues. O. Vrabel in his study "On waves of religious experience: The varieties of religious experience" discusses the "waves" of religious experience and its varieties. Based on the classic work of William James "The Varieties of Religious Experience", which still "remains an important source of inspiration for many scholars and researchers", the author of the article tries to reflect James' ideas in modern cognitive approaches in understanding the meaning of religion in modern culture [8, p. 36]. Some attention is paid to the correlation of various methodologies of scientific knowledge in the educational sphere [9–13].

3 Reflection

So, in the production of reality, scientific and religious consciousness can complement one another, they can come into conflict. Representatives of the Russian religious renaissance of the early twentieth century provide valuable experience in this regard.

The worldview of Sergius Bulgakov developed very dramatically: through a spiritual crisis and the loss of "religious faith for many, many years" [14, p. 434] to the acceptance and then criticism of the main ideas of Marxism, and the formation of a deep religious position. At the same time, scientific experience (for Bulgakov, this is the field of political economy) serves as the basis for the understanding and interpretation of the world for the philosopher, Orthodox theologian and priest.

Bulgakov focused his scientific interest on the field of agrarian policy of the peasant question, i.e. on those points of the teachings of Karl Marx, which are the most significant for the Russian economy, but do not determine the course of the industrialized economy of Europe ("Capitalism and Agriculture"). The doubts of the Russian philosopher relate primarily to socialist projects for the socialization of the peasant economy. But the main thing that makes him make the path "from Marxism to idealism" is a very acute and so far intuitive understanding that the content of economic materialism does not cover that "metaphysical emptiness" that follows "from his main religious motive – from its militant atheism" [15, p. 340]. In the conclusion of his major research Bulgakov notes that Marx's mistake is due to "an overestimation of the real abilities and significance of social science, the boundaries of social cognition. He considered it possible to measure and predetermine the future according to the past and the present, meanwhile, each era brings new facts and new forces of historical development – the creativity of history does not deplete. Therefore, any forecast for the future based on the data of the present is inevitably erroneous. The stern scientist assumes here the role of prophet and diviner, leaving behind a rigorous ground of facts. Therefore, with regard to predictions for the future, we prefer honest ignoramus to social quackery or charlatanism" [15, pp. 457–458].

Bulgakov quite definitely points to the reason for the unknowability – ignoramus – this is an infinite and innumerable by any theory variety of combinations of history. It is not by chance that the word "horizon" appears in his philosophical and economic vocabulary, he really thinks so far in the horizontal plane. Subsequently, the "unknowable" is transformed by Bulgakov into the vertical transcendens, in the relationship between the human and the Divine. The clearly marked opposition to Marxism is expressed not in the abolition of economic and materialist problems, but, on the contrary, in its correlation with other, no less significant foundations of human existence. "Economic materialism, – the philosopher will say later, – should not be rejected, but internally surpassed, explained in its limitations as a philosophical "abstract principle"" [16, p. 7].

The final work of the "economic period" by Sergius Bulgakov – "Philosophy of Economy" – appears in 1912. The main theme is how order is possible in a world that depends on a person and his creative activity. And these are no longer spontaneously emerging production relations, which in a certain sense turn out to be historical traps for the spiritual development of a person, this is not a pure economic doctrine. The economic doctrine is complemented by the religious doctrine of Sophia. This theme will subsequently constitute both glory and a challenge for the philosopher. Questions are asked by a person caught up in thoughts about his own life, which has saved an unobvious connection with God.

The subjects of the economy – man and mankind as a whole, the creators of private, so to speak, stories of the world – must certainly meet with the "world master", with the demiurge, cosmocrat, steward, ruler of the Universe. The fundamental fragmentation (aposmacy) of man and the fruits of his production must be overcome in the Pleroma – the pre-established completeness of creation. Bulgakov is in search of a metaphysical principle in which his basic intuition can be expressed: "the economy is sophic in its metaphysical foundation" [16, p. 99]. The Sophia principle presupposes a complex way of explaining and interpreting evil, destruction, stagnation, regression, sinfulness, overflowing the empirical plans of human existence. The question of the direction of the movement of the world from the state of the Fall to the increasing Sophia concerns, first of all, the problem of the Creator's relation to the creature created by him. It should be noted that the method of argumentation adopted by Bulgakov is completely at variance with the positivist rules for constructing a scientific theory. Gnostic motives "Philosophy of Economy", which never became a complete work, are obvious.

Around the same time, an important event for Bulgakov's spiritual quest took place – a meeting, and then a deep cordial friendship with Pavel Alexandrovich Florensky, a philosopher from the same brilliant galaxy of Russian thinkers of the early twentieth century. In Florensky's work, the problematic of the relationship between religious and scientific knowledge has already found its embodiment in the book "The Pillar and Statement of Truth", published in 1911. Ultimately, this is the formulation of the question of whether it is possible in the modern world, as if it had long ago fallen away from God, to discern its Sophia basis – as a rational and meaningful structure of it.

Like Bulgakov, Florensky's work combines theology, philosophy and science. In this case, a question inevitably arises, which was the subject of reflection for Florensky himself: how to build a correspondence between the criteria of scientificity, adopted for accurate knowledge in science, and the foundations of religious discourse in order to make it more valid, evidential and productive?

In the memoirs of Pavel Florensky "To my children. Memories of Past Years" he reflects on how his scientific worldview was formed and how he saw in his time the correlation of scientific and religious thought: "Experience, undoubtedly authentic and about authentic, was in itself, but scientific thought, in which I simply did not believe in some spiritual layer – by itself. It was a characteristic disease of all new thought, of the entire Renaissance; now, in hindsight, I can define it as the separation of humanity and scientificity" [17, p. 217].

In Florensky's worldview, two elements collided, and for some time they could not combine into a holistic worldview, until he had the idea that it was necessary to "create a religious science and a scientific Religion". Florensky, being still a young man, in a letter to his mother O.P. Florenskaya from Moscow to Tiflis on March 25, 1904 writes about the need to bring science and religious experience closer together: "... without religious interests, a serious and thoughtful person cannot have any ideal interests, otherwise sooner or later even the outstanding personalities will become indifferent and insensitive to everything except that which directly makes felt itself" [18, p. 559].

In this respect, it is typical the conversation "Empyrean and Empiria", written by Florensky in June 1904 before entering the Theological Academy [19]. He returned to this work many times, noting the relevance of its main idea – the construction of an integral religious worldview. The very title of this conversation, devoted to the analysis of the ontological unity of two types of reality (transcendental and empirical), reproduced in the worldview, is eloquent. From Florensky's point of view, this unity is based on religious truth – the Absolute Truth of Christianity, which carries both the spiritual content and the fullness of the content of material existence.

Reflections on the Divine, on the relationship between God and man are the starting point of Florensky's consideration, which are the spiritual basis of being and determine all other characteristics of the human relationship to the world, including physical and mental. The origins of the integral worldview lie, according to Florensky, in the structure of the spiritual hierarchy itself: Empyrean – the metaphysical basis and depth of the human soul and Empiria – its rationally logical and psychophysiological level. According to Florensky, sensually observed empirical reality (Empiria) is often viewed as the only and obvious, however, another reality, really genuine, testifying to the existence of the "upper" world (Empyrean), is comprehended only in religious faith consciousness, through Christian morality – love for God.

The sensually observed world is studied by science with a set of its special methods, in general, objectifying and dismembering reality, but at the same time having exceptional value for cognition and comprehension of the world, unless they are considered as the ultimate basis of the world. Scientific research, according to Florensky, is very limited, it inevitably reaches its limit, coming close to the area that is not comprehended only by scientific methods, this is the area of religious faith and religious speculation.

In his later works, Florensky further clarifies his understanding of the relationship between religion, science and philosophy. In his essay "Philosophy of a Cult", based on a cycle of notes made by the philosopher in different years of his life (mainly in 1914–1915), he expresses the conviction that philosophy and science grow out of religion, and, acquiring an independent status and significance, nevertheless, they continue to be inextricably linked with it, fueled by its life force, since the nature of religion itself is to "unite God and the world, spirit and flesh, meaning and reality" [20, p. 60].

This understanding of the integrity and inner involvement of different sides of the worldview is by no means declarative. Florensky himself, being a scientist and engineer, carrying out various scientific researches, realized just such an idea of the investigated reality and carried out his developments from this point of view. He clearly understood that behind the phenomena of the physical world (empirical) there is a world of "true reality" (transcendental), and scientific experience can be a path of ascent to religious experience. His reasoning about the creative meaning and meaningful productivity of the contact between the "visible" and the "divine", "sensible" 2nd "intelligent" is quite characteristic: "The purpose of a cult is…not to curtail the richness of inner life, but, on the contrary, to assert this richness in its fullness, to consolidate, nurture. The accidental is elevated by the cult to its due, the subjective is enlightened into the objective. The cult transforms the natural given into the ideal" [20, p. 129]. In this respect, attention is drawn to the peculiarities of the Florensky language, which combines theological, philosophical, scientific and technical terms.

Thus, from the point of view of Florensky, "three kinds of products of human creativity" coexist: the formation of concepts, the production of machines and tools, as well as the creation of a shrine, without which neither scientific nor technical creativity is untenable [20, p. 6].

4   Discussion

In the scientific literature, various aspects of the relationship between the scientific, philosophical and theological creativity of the teachings of Bulgakov and Florensky are investigated. K. Sládek in his article "Sophiology as a theological discipline according to Solovyov, Bulgakov and Florensky" comes to the conclusion that Sophia, Wisdom of God, is introduced as a creative principle, as the basis of all knowledge, as "the act of the creation of the world [21, p. 113].

Some researchers of the works of Bulgakov and Florensky note the presence of an internal connection between their so different philosophical programs with the patrological basis of the Orthodox outlook, with Orthodox dogma. So, for example, Pavlyuchenkov in his article analyzes in detail Florensky's idea of the participation in the religious experience of Truth itself, revealing itself to a person, giving him the means to identify himself as absolute and objective, and fixing that a person has experienced in a rationally formalized dogmatic teaching [22].

In the scientific literature, a number of sources emphasize the influence of the teachings of Gregory Palamas about divine energies as a form of creation and creativity, God as the creator of the world from Nothing, and man as the creator in the world of matter, transforming and enlightening it. Biryukov analyzed the influence of the teachings of Gregory Palamas on the formation of the sophiology of Sergius Bulgakov [23]. Between 1914 and 1916, that is, after writing "Philosophy of Economy" and before "Unfading Light", S. Bulgakov reads and translates the treatise of Palamas "150 Chapters". Since this happens during the period of the crisis of his economic materialism, it can apparently be argued that through G. Palamas, Bulgakov as the scientist is being transformed into a religious philosopher. The doctrine of Palamas moves Bulgakov from the dead point of the materialistic doctrine. In "Philosophy of Economy", as D. Biryukov notes, Bulgakov insists that man, unlike God, is not able to create anything "metaphysically new", he can only imitate divine models with more or less success. In "Unfading Light", written already directly under the influence of Palamas' ideas, Bulgakov argues that a person strives for absolute creativity, however, he never actually achieves it, and this is the reason for the humility of both the scientist and the artist as creators. In addition, creativity should be providential, that is, consistent with the divine plan, so as not to plunge a person into pride and Satanism. Biryukov emphasizes that P. Florensky at the same time is interested in the teachings of Palamas, but, unlike Bulgakov, "Florensky actually did not read Palamas's works and used the Palamite language as a common discourse" [23, p. 74]. One can agree with Biryukov that the teaching of G. Palamas, this "last Orthodox", substantially correlates the "philosophy of the economy", giving it a metaphysical dimension.

Van Kessel J. in his research notes that Bulgakov and Palamas, although they have different points of view, still complement each other and are essentially united in their main idea, since Sophia is the object of both Sophiologies [24]. A. Papanikolaou has a different point of view. He argues that Sergius Bulgakov and Pavel Florensky remain the main heirs of the Eastern Orthodox meta-narrative. Based on the teachings of Palamas, they restored the connection with the patristic tradition, but at the same time did not create a "neo-patristic synthesis", this is already the merit of G. Florovsky and V. Lossky [25].

In her article A. Gacheva opens the historiosophical dimension of the religious philosophy of S. Bulgakov and P. Florensky [26]. The teachings of Bulgakov and Florensky are included in the general context of the development of Russian religious philosophy, in which the main theme is the transformation of an imperfect human existence into a new way of being (into a "new heaven and a new earth"). Starting from the ideas of N. Fedorov about the purpose of science in "justifying human history" as a creative improvement of the world in accordance with Christian commandments, the author shows the role of Bulgakov and Florensky in the "matter of salvation", in which the goals of scientific cognition are completely subordinated to Christian values. The idea of "cooperation with God" presupposes "the replacement of exploitation by regulation, the "sophiity" of the economy, the idea of the antientropic essence of labor and culture" [26, p. 114]. In that light, science is a form of conciliarity, the collective implementation of common ideas and goals, and religious worldview is a form of internal protective force.

In K. Ware's work "Orthodox theology today: Trends and tasks" the nature of theology is discussed, which Orthodox thinkers do not agree to define as "science" ("academic scientism") [27, p. 105], since its basis consists of dogmatic timeless content. Science is changeable, dynamic. Its goal is the progress of knowledge, its constant change, while theology contains eternal truths. However, the main problem for the religious philosopher Bulgakov was the "improvement" of Christian dogma and the inclusion in it of the doctrine of Sophia, in which reflections of the Gnostic teachings are guessed. K. Ware believes, however, that the religious content changes in accordance with the spirit of the times: "The master-theme of Orthodox theology in the twentieth century has been ecclesiology; in the twenty-first century, the center of interest is shifting to the doctrine of the human person. Because of globalization, problems in bioethics, and the environmental crisis, Orthodox thinkers need to reflect more deeply and with greater courage about the meaning of personhood. To be human is to be endlessly varied, innovative, self-transcending" [27, p. 105]. The main theme of Orthodox theology in the twentieth century was ecclesiology; in the twenty-first century the focus shifts to the doctrine of the human person. Because of globalization, problems in bioethics and the ecological crisis, orthodox thinkers need to reflect more deeply and with greater courage on the meaning of the personhood. To be human means to be infinitely diverse, creative, self-transcending.

Ware's opinion about the inevitable transformation of religious content can be supplemented by the arguments of R. Krečič about the antinomical nature of the philosophical works of S. Bulgakov and P. Florensky, which perfectly reflect "an integral gnoseological, philosophical and theological view, which takes into account both empirical evidence and spiritual experience" [28, p. 653], and also reflections of W. Trimble on the combination of mathematics and theology in the philosophy of P. Florensky [29].

We see that researchers show different ways of connecting science and religious worldview, as in our case - economics and non-dogmatic theology in Sergius Bulgakov, and mathematics and sophiology in P. Florensky. In modern culture, religious beliefs no longer seem to be "dusty proofs of God's existence" [29, p. 38], and the religious renaissance that began in the mid-90s. in Russia, as well as the fashion for exotic religions in the Western world since the beginning of the twentieth century, are serious symptoms of the spiritual poverty of the philosophy of modern pragmatic well-being. Scientific depth, combined with religious rootedness, is realized in a special warehouse of the worldview, in which these seemingly absolute opposites only strengthen each other.

5   Conclusion

The main conclusion of the article is that religious and scientific worldviews do not supplant one another, but can be in dynamic equilibrium. The connection and complementarity of scientific and religious experience is expressed in the following.

1. Both scientific and religious experience express different forms of rationality, in the first case – the ordered unity of knowledge about the world, in the second – the desire to penetrate the boundaries of transcendental experience, a special way of individual involvement in the creative basis of the world.
2. Belief is not given to everyone, its formation requires a certain scale, openness to transcendental forms of experience. The intention of religious consciousness is aimed at the individual responsibility of the moral choice of the individual, which is not always assumed by science.
3. In religious experience, existential states are experienced by the subject of cognition and creativity, thus, in unity with scientific cognition, the ideal of "integral knowledge" sought in Russian philosophy, filled with the experience of unity and coexistence in creative energies, is formed.
4. Special attention should be paid to representatives of science who experienced similar states and passed on their experience of combining two forms of manifestation of creativity in a person. The interaction of scientific and religious experience largely determines the development of spiritual culture, in which the integration and cumulation of different forms, ideas and meanings takes place. Spiritual openness, moral potential, logical thinking and creative intuition, formed in scientific and religious experience, are the main prerequisites for achieving true knowledge.

References

1. Keep doors open for constructive dialogue between religion and science. Nature **545**(7654), 265–266 (2017). https://doi.org/10.1038/nature.2017.21985
2. Bayón, P.S.: Disbeliefs, science and religion: the necessary dialogue between the transcendent and naturalized spirituality. Pensamiento **73**(276), 761–766 (2017). https://doi.org/10.14422/pen.v73.i276.y2017.038
3. Bentley, W.: If god is everywhere, is god in a black hole? A theology-science discussion on omnipresence. Acta Theologica **40**(2), 185–199 (2020). https://doi.org/10.12775/SetF.201 8.001
4. Del Carril, I.E.: La física cuántica y el diálogo con la religión. Scientia Et Fides **60**(1), 9–29 (2018). https://doi.org/10.12775/SetF.2018.001
5. Woodford, P.J.: Philosophy in the science classroom: how should biology teachers explain the relationship between science and religion to students? Cult. Sci. Educ. **15**(4), 937–950 (2020). https://doi.org/10.1007/s11422-020-09997-1
6. Razumov, A.E.: Faith, understanding, proof. Vysshee Obrazovanie v Rossii, **28**(4), 72–80 (2019). https://doi.org/10.31992/0869-3617-2019-28-4-72-80
7. Teehan, J.: The cognitive science of religion: implications for morality. Filosofia Unisinos **19**(3), 272–281 (2018). https://doi.org/10.4013/fsu.2018.193.09
8. Vrabel, O.: On waves of religious experience: the varieties of religious experience. Pro-Fil. **18**(1), 36–51 (2017). https://doi.org/10.5817/pf17-1-1585
9. Bylieva, D., Zamorev, A., Lobatyuk, V., Anosova, N.: Ways of enriching MOOCs for higher education: a philosophy course. In: Bylieva, D., Nordmann, A., Shipunova, O., Volkova, V. (eds.) PCSF/CSIS -2020. LNNS, vol. 184, pp. 338–351. Springer, Cham (2021). https://doi.org/10.1007/978-3-030-65857-1_29
10. Nesterov, A.: Technology as semiosis. Technol. Lang. **1**, 71–80 (2020). https://doi.org/10.48417/technolang.2020.01.16

11. Pavlenko, A.: Technology as a new language of communication between the human being and the world. Technol. Lang. **1**, 91–96 (2020). https://doi.org/10.48417/technolang.2020.01.19

12. Ershova, N.: Language of art as language of Utopia. Technol. Lang. **1**, 28–33 (2020). https://doi.org/10.48417/technolang.2020.01.06

13. Nordmann, A.: The grammar of things. Technol. Lang, **1**, 85–90 (2020). https://doi.org/10.48417/technolang.2020.01.18

14. Bulgakov, S.N.: Autobiographical Notes. IMKA-Press, Paris (1946).(in Russian)

15. Bulgakov, S.N.: Capitalism and Agriculture, vol. 2. Printing house and lithography V.A. Tikhanov, Saint-Petersburg (1900). (in Russian)

16. Bulgakov, S.: Philosophy of Economy. Nauka, Moscow (1990).(in Russian)

17. Florenskij, P.A.: To My Children. Memories of Past Years. Moskovskij Rabochij, Moscow (1992).(in Russian)

18. Finding the Way: Pavel Florensky During his University Years, vol. 2. Progress-Tradiciya, Moscow (2015).(in Russian)

19. Florenskij, P.: Empyrean and Empiria. In: Theological Works: 1902–1909. Publishing House of the Orthodox St. Tikhon University for the Humanities, Moscow (2018). (in Russian)

20. Florenskij, P.: Philosophy of the Cult (Experience of an Orthodox Anthropodicy). Mysl', Moscow (2004).(in Russian)

21. Sládek, K.: Sophiology as a theological discipline according to Solovyov, Bulgakov and Florensky. Bogoslovni Vestnik **77**(1), 109–116 (2017)

22. Pavlyuchenkov, N.N.: Florensky on religious experience and religious dogmatiks. Vestnik Pravoslavnogo Svyato-Tihonovskogo gumanitarnogo universiteta, seriya 1: Bogoslovie. Filosofiya **2**(52), 61–77 (2014). (in Russian). https://doi.org/10.15382/sturi201452.61-77

23. Biriukov, D.: Taxonomies of BEINGS in the Palamite literature. Part 2. The Palamite doctrine in the context of the previous Byzantine tradition and its reception in the Russian religious thought of the XX century (the philosophy of creativity by Sergei Bulgakov). Konštatínove listy **12**(2), 69–79 (2019). (in Russian). https://doi.org/10.17846/cl.2019.12.2.69-79

24. Kessel, J.: Sergei Bulgakov's Sophiology as the integration of sociology, philosophy, and theology. In: Glas, G., de Ridder, J. (eds.) The Future of Creation Order. NASSR, vol. 3, pp. 317–335. Springer, Cham (2017). https://doi.org/10.1007/978-3-319-70881-2_15

25. Papanikolaou, A.: Eastern orthodox theology. In: Meister, C., Beilby, J. (eds.) The Routledge Companion to Modern Christian Thought, pp. 538–548. Routledge, London (2013). https://doi.org/10.4324/9780203387856

26. Gacheva, A.: "History is not an abandoned passage..." (justification of history in Russian religious and philosophic thought of XIX – the first third of XX century). Voprosy filosofii **8**, 114–126 (2018). (in Russian). https://doi.org/10.31857/s004287440000743-2

27. Ware, K.: Orthodox theology today: trends and tasks. Int. J. Study Christ. Church **12**(2), 105–121 (2012). https://doi.org/10.1080/1474225X.2012.699434

28. Krečič, P.: Development and reality of antinomy in Russian religious thought. Bogoslovni vestnik **4**(72), 653–664 (2012)

29. Trimble, W.: Claiming infinity: tokens and spells in the foundations of the Moscow mathematical school. Technol. Lang. **2**(1), 37–53 (2021). https://doi.org/10.48417/technolang.2021.01.05

The Technical and the Religious: Concepts and Contemporary Social Practices

Tatiana Vladimirovna Bernyukevich(✉) (iD)

Moscow State University of Civil Engineering (National Research University), 26,
Yaroslavskoye Shosse, 129337 Moscow, Russia

Abstract. The study aims to update knowledge on the relationship between technology and religion considering the genesis and essence of this connection, its role in the development of culture and the solution of anthropological issues. The religious philosopher Pavel Florensky defines the significance of technology for the culture as a means of achieving the harmony established by God and lost by humans. The philosopher and theologian Friedrich Dessauer associates the genesis and development of technology with the ability of Being, created by God, toward formation. In the concept of Gilbert Simondon, the source of religious and technical thought is the pre-individual human consciousness in a state of "magical unity". The use of information technologies for the activities of religious communities is becoming more and more widespread and covers many areas of their life and function. Confessions are interested in the development of this kind of activity for performing the general tasks taken up by religious organizations, such as attracting adherents, religious enlightenment, ritual activities, organizational and communication activities. "Online Religions" that have no offline analogs are of particular interest, for example, Buddhist Geeks Sangha and The Terasem Faith. To understand the nature and consequences of world religions mastering high technologies, it is necessary to address the fundamental problem of the relationship between the religious and the technical in the system of culture while taking into account changes in society, science, and religion itself.

Keywords: Technical · Religious · Online religions · Culture · Anthropological types

1 Introduction

The topic of the connection between the technical and the religious is significant from the point of view of the need to analyze the problem of philosophical reflection on the genesis of technology, its ontological status and significance for humans, consider the impact of modern technologies and the use of technology in the religious sphere, determine the specifics and the role of the so-called cyber religions in modern society. It should be noted that in philosophical and religious studies, the issues of understanding of the concept of technology in the context of religious and philosophical issues of the development of Being and human and the emergence and development of modern religious phenomena associated with technology are considered separately. However, in our opinion, the combination of these perspectives of research will reveal the essence of the ongoing cultural and anthropological processes associated with the interaction of the religious and technical spheres of society.

It should be noted that Russian studies of the concepts of theologians and philosophers about the role of technology in culture and human life in relation to the ideas of religions have their own history. There are major fundamental studies of the works of Russian philosophers conducted by: Polovinkin, Abbot Andronic (Trubachev), Pavlyuchenkov, Khoruzhiy, Semyonova, Gacheva, Perelman, Kazyutinsky, and others. These studies are focused mainly on the perspective of the history of philosophy, philosophy of culture and social philosophy. Modern translations of German theorist Friedrich Dessauer and studies of his work by A.Yu. Nesterov from the point of view of the philosophy of technology should be specially noted.

Studies of religious practices and religious activities associated with the use of modern technologies began to actively develop in Russia since the end of the 20th century. The analysis of these issues is mainly presented in religious studies, an example of which is the collective monograph edited by Zabiyako [1]. The publications of Fedorova [2], Levushkin [3], Kisser [4], Minchenko [5], Buvaeva [6], should also be taken into consideration.

2   Research Results

The emergence of the problem of interaction between the technical and the religious is associated with the interest of philosophers and theologians in determining the place of the technical in human life. The anthropological perspective of this problem was presented in the studies of Ernst Kapp, who substantiated the idea that technology is a projection of human organs. Due to this, persons do not only create means with which they can save their physical strength, increase their efforts to solve issues related to satisfying needs, but also learn about the world and themselves. He wrote: "The system of needs, a substance enlightened by organic projection to tools and utensils and saturated with intellectuality, the changes in the earth's crust caused by it, is the external world, which, unlike the natural, surrounding and beast and remaining alien to it, has, the advantage over him is that a person, finding in him and learning to know and understand himself, achieves self-consciousness" [7]. The Russian religious philosopher Pavel Florensky expanded this understanding and self-awareness to the space of culture and the achievement of harmony set by God and lost by humans. In this space, after Florensky, the unity of the anthropological and the technical is essentially believed to be the possibility of anthropodicy. At the same time, the natural in relation to the world of culture is not overlooked: "After all, culture is never given to us without its spontaneous foundation, serving it as environment and matter: at the basis of any phenomenon of culture is some natural phenomenon cultivated by culture. Man, as a bearer of culture, does not create anything, but only forms and transforms the spontaneous things" [8].

One of the famous researchers of Florensky's works, S.M. Polovinkin, emphasizes, according to the ideas of the philosopher, "nature is the totality of things, culture is the totality of human activities-energies" [9]. According to Florensky, "the historical task of technology is to consciously continue its organoprojection, proceeding from the decisions given by the unconscious bodybuilding of the soul" [8, p. 417]. This "bodybuilding of the soul" in space allows us to return to the integrity and unity of man, nature, and culture. In this context, "organ projection" is the path to the unity of Being, which, according to Christian ideas, is given by God, but lost by man. "Half-existence, half-being and half-non-being" of the organs and functions of Adam after the Fall from grace are restored through culture [8, p. 438].

The connection between the genesis of technology and the ability of Being, created by God, to formation is determined in the works of the German philosopher and theologian Friedrich Dessauer. This formation and realization of the forms of Being occurs in the process of human technical activity. It was God, according to Dessauer, who endowed Being with "the ability and law of revealing" in various forms [10]. This "continuation of the creation of nature" is done by man thanks to technology that changes the face of the Earth [11, p. 162]. Dessauer focuses on the border and connection between the "causal kingdom of nature" and the "finalist kingdom of technology" [11, p. 163].

The negative attitude of a number of religious leaders and theologians to Dessauer's technology is connected with their understanding of the world as once given by God, without the divine potential of development, realized in the human creative and technical activity.

In the context of changing being through co-creation with God, through the development of technology, the original concepts of Russian cosmists can also be considered: Nikolay Fedorov, Konstantin Tsiolkovsky, Vladimir Vernadsky and others, in which the natural world is spontaneous and finite, it not only changes, but transforms and becomes intelligent and capable of infinity. Of course, the ideas of cosmist philosophers reveal different religious foundations and influences of various religious ideas: Christian foundations and the use of Orthodox axiology – in Fedorov's works, a heterogeneous complex of religious ideas – in Tsiolkovsky's "cosmic philosophy" (panpsychism, the ideas of Buddhism, the identification of God and the Universe.), the ideas of Eastern philosophy, including Buddhist – in Vernadsky's works. At the same time their concepts are striving for a religious and philosophical substantiation of the vector of technological development of mankind.

It should be noted that, probably, to a large extent, this heuristic message of the religious and philosophical substantiation of the genesis and development of technology was exhausted by the second half of the 20th century.

One of the most interesting concepts that substantiates the relationship between religion and technology as the sides of human consciousness and attitude to the world is the concept of Gilbert Simondon, in which religious thought and technical thought are closely linked. The source of this feature is pre-individual human consciousness, which is in a state of "magical unity", where with the nondiscrimination of the subjective and the objective, the world was perceived through the prism of sacredness [12]. These ideas are outlined by Simondon in his work "On the Mode of Existence of Technical Objects".

In the transition to individual consciousness, a phase shift occurs, as a result of which two divergent phases are formed, which Simondon calls religious and technical thought. As the Russian researcher Kurtov notes, in the process of subjectivation and objectification, the following differentiation arises: "religious thought is a person's knowledge of himself as a subject, technical thought is knowledge about the "objective" relationship of a person with the world. The religious and the technical relate to each other as parts of a gestalt, as a background and an object, respectively: the religious is always united and whole, the technical is partial and fragmentary" [13]. In the process of the development of consciousness, the technical and religious phases are linked by two mediations: initially by aesthetic thought, then by philosophical thought. In future, the divergence of technical and religious thought may be replaced by their reconvergence [12, p. 188].

Based on the ideas of Simondon and his definition of the religious and the technical as an "isomorphic" phenomena, Kurtov proposes the technical and religious as part of a techno-religious unity. To define this unity, the Russian researcher suggests the term "technotheology" that he compiled "on the model of Hcidcggcr's 'ontotheology'". According to Kurtov, "to think technologically is to think of religion and technology as one" [13].

When considering the social practices of religious life, one of the urgent tasks currently is to analyze the use of new technologies in the religious sphere and the emergence of the so-called cyber-religions. The influence of modern information technologies, the impact of the Internet on the development of the religious sphere in the research literature are considered from two perspectives, including: 1) the use of these technologies and the capabilities of the virtual environment for the implementation of religious organizations' traditional tasks of communication, information support, and religious rites (this has become especially relevant in a pandemic and self-isolation); 2) the emergence of cyber-religions, in which, according to Artyom Zabiyako, "computer technologies acquire the status of supervalue, are endowed with the qualities of sacred objects and attributes of divine essences" [1, p. 32].

There are similar typologies of the use of Internet technologies in European studies. For example, Christopher Helland highlights: 1) "religion online" is an online platform of those religious organizations that exist and develop in the physical space; 2) online religion is virtual religious organizations [14].

3 Discussion

The use of information technologies and the possibilities of the Internet in the activities of religious communities today is becoming increasingly large-scale and covers many areas of their activities. Currently, this is the creation of websites for religious organizations that shape a wide information space, channels on video hosting (for example, on YouTube), social networks (such as Facebook, Twitter), and popular applications (Instagram, Pinterest). The development of this kind of activity within the framework of the general tasks of religious organizations (attracting adherents, religious enlightenment, ritual activities, organizational and communication activities, etc.) is not only of interest to confessions, it also promotes the commitment to determine both the possibilities and limitations of the use of information technologies in religious life, their influence on the very essence of religion and religious consciousness, ways of conjunction of innovative technical means and technologies with religious traditions and the vector of development of religious organizations.

In 2009, Pope Benedict XVI, at the 43rd World Day of Social Communications, raised the topic "New technologies, new types of communication". Speaking about the features of new ways of communication in the "digital generation", he emphasizes: "The availability of mobile phones, computers, together with the global capacity of the Internet, has created numerous ways through which words and images can be simultaneously sent to the most remote corners of the world: previous generations even did not dream about this. This is especially true for the younger generation, who gets access to this communication to talk with friends, make new acquaintances, create online communities, search for information and news, in order to share their own ideas and opinions" [15, pp. 100–113]. According to Catholic leaders' opinion, the abilities of new technologies to create a space for dialogue are significant.

As mentioned above, religious organizations pay great attention to the issues of possible negative consequences of the use of modern means of communication, the problem of the responsibility of those who use these means. Natalia Urina notes that "in the context of the above said, in the Catholic doctrine, in-depth attention is paid to both the positive and negative impact of media on the individual, family and society, i.e. identification of their creative and destructive power... In this regard, the role of value guidelines and prudent decisions in the communication sphere is immeasurably increasing in the new conditions..." [16].

Issues of the use of modern technologies by Buddhist communities, the representation of Buddhism in the virtual space, are widely described in the studies from the beginning of the 21st century. The authors of the article "Russian Buddhism in the Internet Dimension" note that a direction for the study of "virtual Buddhism" has already been formed [17], and the results of studies of these scientific problems are reflected in the publications of Brett Greider [18], Richard P. Hayes [19], Charles S. Prebish [20], the collective monograph "Buddhism, the Internet, and Digital Media: The Pixel in the Lotus" [21]. Terms such as "cybersangha" and "blogisattva" are no longer new. One of the first typologies of Buddhist Internet communities was the typology of Prebisch (2004). He identified three main types of such communities: 1) web pages of traditional American Buddhist groups, created for the purpose of effective communication; 2) "virtual temples" created by traditional Buddhist communities as a kind of addition to their organizations; 3) "online communities" having no offline analogues. However, the authors of the above-mentioned article write that today it is difficult to distinguish between these phenomena and the ways of representation of Buddhist groups in the Internet space. The active process of studying the issues of the use of modern technologies by Buddhist communities is reflected in attempts to determine the methodology for researching virtual religious communities [17].

Of particular interest is the activity of Buddhist Geeks in the framework of the so-called "cloud sangha" (Buddhist Geeks Sangha). The creators of these projects define their tasks as follows: "This virtual community of practice exists to serve people who are interested in a pursuit of spiritual, contemplative, psychological, & existential insights regarding our shared human condition. Ultimately, this is a community whose purpose is to support practitioners in metabolizing & integrating these insights, such that they lead to real-life transformation" [22].

The practice of cyber religions includes such a vibrant community as The Terasem Faith. A feature of this community is its clear connection with the ideas of transhumanism and an optimistic view of the cyber future of humanity and the development of technology. The Terasem Faith movement promotes the slogan "Life is purposeful, death is optional, God is technological and love is essential". Terasem Faith adherents believe, with the help of new technologies, special "mindfiles" will be created to preserve and reproduce the "soul" of a person, reproduced in the body of a robot. The Terasem Faith positions itself as a "trans-religion" capable of leading humanity to unity and immortality: "We are a transreligion that believes we can live joyfully forever if we build mindfiles for ourselves. We insist on respecting diversity without sacrificing unity, as well as pouring maximum resources into cyberconsciousness software, geoethical nanotechnology and space settlement. A "transreligion" is a movement which can be combined with any existing religion, without having to leave a previous religion" [23]. The history of this movement is connected with the transgender entrepreneur Martina Rothblatt, who created a digital copy of her wife Bina Aspen, the robot Bina48, in 2004. The ideas of "transreligion" are described by Rothblatt in the book "Virtually Human: The Promise—and the Peril—of Digital Immortality" [24].

Terasem Rituals are religious practices of The Terasem Faith adherents. This is an important part of life of the followers of this movement: "Terasem Rituals are important to keep our Movement going until all of consciousness is connected and all the cosmos is controlled" [23]. To unite consciousness and achieve control over space, it is necessary: to listen to special programs on Terasem Radio every day; to set aside one day each week for reading and transcendental meditation exercises; to attend monthly meetings with other adepts - the time of these meetings and their content are regulated ("On the 10th of every 3rd month we celebrate a Terasem Holiday… A Terasem Gathering consists of Music Jam, Art Sharing, Recitation of 30 Sequential (from start of the year) Truths of Terasem, Talking and Teaching about the Recitation, and then the Yoga of either an Earthfire or Crownaura Terasem Connection" [23]); to gather quarterly for a special religious holiday to celebrate the transition of participants to a higher level; every leap year there is a congress of all followers – "Convocation of all Terasem Joiners, from February 29th to March 4th, to discuss and agree on important matters of policy relating to Terasem, and to Consent new Centers of Critical Consciousness" [23].

It should be noted that this movement combines the ideas of technological optimism and immortalism, learning from oriental practices (yoga and meditation), organizational forms of new age.

4 Conclusion

As the analysis of studies shows, the most difficult issue of studying religious practices in the virtual space is not their classification according to the type of involvement in this space, the type of combining their offline existence with online representation, the ways and tasks of using modern technologies. The most difficult task is to determine whether the essence of religious consciousness, the anthropological and sociocultural functions of religion change, how the sacred world and the human world are related in the space of new technologies.

With all the difference in the development of new forms, religiosity, even when it comes to cyber religions, reveals, for example, an amazing similarity in the pathos of transformation and ascent to eternity between the provisions of The Terasem Faith and the religious and philosophical ideas of Tsiolkovsky. Analysis of the content of geek-Buddhist activities shows that it does not differ radically from the tasks of Buddhist liberation, but suggests using new digital technologies. This generally corresponds to the traditional position of Buddhist soteriology, according to which the path of enlightenment offered to the adept can be successful if it takes into account the level of development of a person's consciousness, his socio-cultural characteristics and ways of mastering the world that are understandable to this person. The long history of the spread of Buddhism testifies to the fact that, being acultural and asocial in its foundations, Buddhism effectively uses the available cultural forms and methods of social adaptation [25]. This is also reflected in the use of modern technologies.

This is also the case of the attempts to apply techniques and technology to determine the nature of Buddhist phenomena, for example, meditation. In 2019, on the basis of Buddhist monasteries in southern India, the Center for Research on Meditation and Altered States of Consciousness was opened, managed by a physiologist who for many years headed the N. P. Bekhtereva Institute of Human Brain of the Russian Academy of Sciences, academician Svyatoslav Medvedev [26]. The aim of the research conducted in this Center is to find "the deep foundations of human consciousness". It should be noted that over the past 20 years, the Fourteenth Dalai Lama generally demonstrates a readiness for dialogue between Buddhists and scientists and even writes about a possible adjustment of the ontological provisions of Buddhism in accordance with the achievements of modern science [27].

In order to identify the essence of the process of "virtualization" of religious life and the consequences of this virtualization for the religions themselves, human life and the development of culture and society, it is insufficient to analyze the prevalence of various forms of "online religion" and "religion online", to identify ways of digital representation of religious communities, types of their institutionalization in the space of modern technologies. To understand the nature and consequences of religions' assimilation in the virtual world, the world of high technologies, it is necessary at a new level, taking into account all the changes that have already occurred in society, science, religion, to address the fundamental problem of the relationship between religious and technical in the cultural system. Questions of these relations can be discussed in the context of the goals of understanding the nature of cultural subjectivity, its anthropological and transcendental origins, as well as the mechanisms of cultural development through the evolution of cultural and anthropological types [28, 29].

Acknowledgment. The reported study was funded by RFBR, project number 20-011-00462 A.

References

1. Zabiyako, A.P., et al.: Cyberreligion: Science as a Factor of Religious Transformations. Amur State University, Blagoveshchensk (2012)
2. Fedorova, M.V.: Religious identity in the modern digital world. Sociodinamika **6**, 66–77 (2020)
3. Levushkan, P.: Cyber-religion: catholics, protestants and others [video lecture]. http://theory and-practice.ru/?scope=video. Accessed 12 Mar 2021

4. Kisser, T.S.: Russian Germans: religiosity online. Etnografiya **3**(9), 103–123 (2020)
5. Minchenko, T.P.: Religion, freedom of conscience and new technologies in the post-secular world. Izvestia TPU **6**, 147–151 (2012)
6. Buvaeva, G.A.: Attempts to revive religion in virtual space and "Digitalization of Buddhism." Vestnik KalmGU **2**(46), 86–91 (2020)
7. Kapp, E., Noiret, L., Espinas, A.V., Kunov, H.: Philosophy of the machine. In: The Role of Tools in Human Development. Priboy, Leningrad (1925)
8. Florensky, P.: Works in Four Volumes, vol. 3, no. 1. Mysl, Moscow (2000)
9. Polovinkin, S.M.: Christian Personalism of the Priest Pavel Florensky. RGGU, Moscow (2015)
10. Nesterov, A.: Epistemological and ontological problems of the philosophy of technology: "The fourth kingdom" of F. Dessauer. Ontol. Des. **3**(21), 377–389 (2016)
11. Dessauer, F.: Streit um die Technik. Verlag Josef Knecht, F.a.M. (1959). (in German)
12. Simondon, G.: On the Mode of Existence of Technical Objects. Aubier, Paris (1958)
13. Kurtov, M.: About technical ignorance. https://www.nlobooks.ru/magazines/novoe_literatur noe_obozrenie/138_nlo_2_2016/article/11878/. Accessed 29 Mar 2021
14. Helland, C.: Online religion as lived religion. Methodological issues in the study of religious participation on the internet. Online Heidelberg J. Relig. Intern. **1**(1. Special issues on the Theory Methodol.), 1–16 (2005). https://archiv.ub.uni-heidelberg.de/volltextserver/5823/. Accessed 28 Mar 2021
15. Klimenko, D.A.: Vatican and new information technologies. Vestn. Moscow Univ. Ser. 10 Journal. **2**, 100–113 (2010)
16. Urina, N.V.: Church online: eternal and virtual. Mediascope **1** (2004). http://surl.li/sjne. Accessed 23 Mar 2021
17. Aktamov, I.G., Badmatsyrenov, T.B., Tsyrempilov, N.V.: Russian Buddhism in the internet dimension. Vlast' **7**, 125–130 (2015). (in Russian)
18. Greider, B.: Academic Buddhology and the cyber-sangha: researching and teaching buddhism on the web. In: Nori, V., Hayes, R., Shields. J. (eds.) Teaching Buddhism in the West: From the Wheel to the Web, pp. 212–234. RoutledgeCurzon, London; New York (2002)
19. Hayes, R.: The internet as a window onto American Buddhism. In: Williams, D.R., Queen, Ch.S. (eds.) American Buddhism: Methods and Findings in Recent Scholarship, pp. 168–179. Curzon Press, Richmond (1999)
20. Prebish, C.S.: The Cybersangha: Buddhism on the internet. In: Religion Online: Finding Faith on the Internet, pp. 135–151. Routledge, London (2004)
21. Grieve, G.P., Veidlinger, D. (eds.): Buddhism, the Internet, and Digital Media: The Pixel in the Lotus. Routledge, New York (2014)
22. Website of Buddhist Geeks Sangha. https://www.buddhistgeeks.org/sangha. Accessed 15 Mar 2021
23. Website of the Terasem Faith. https://terasemfaith.net/. Accessed 15 Mar 2021
24. Martine, R.: Virtually Human: The Promise and the Peril of Digital Immortality by Martine Rothblatt, 1st edn. St. Martin's Press, 9 September 2014
25. Bernyukevich, T.: The reception of Buddhism in Russia and the "Russian Asiaphiles" in the late 19th – early 20th century. Gosudarstvo, Religiia, Tserkov' v Rossii i za Rubezhom/State Relig. Church Russ. Worldwide **38**(1), 37–61 (2020)
26. Website of «Save Tibet!». http://savetibet.ru/2020/07/06/dalai-lama-and-scientists.html. Accessed 15 Jan 2021
27. Dalai Lama. The Universe in a Single Atom: The Convergence of Science and Spirituality. (S. Khos, trans.). Save Tibet Foundation, Moscow (Buddhism and Science) (2018)
28. Pelipenko, A.A.: Comprehension of Culture. Part I. Culture and Meaning. ROSSPEN, Moscow (2012)
29. Pelipenko, A.A.: Comprehension of Culture. Part 2. The Mytho-Ritual System. Book 1. The Mediation Paradigm. ROSSPEN, Moscow (2017)

The Omnibenevolence Paradox
and the Education Paradox: An Amendment
to G. W. Leibniz's *Theodicy*

Anton Zamorev$^{(\boxtimes)}$ ⓘD and Alexander Fedyukovsky ⓘD

Peter the Great St. Petersburg Polytechnic University (SPbPU), Polytechnicheskaya, 29,
195251 St. Petersburg, Russia

Abstract. The paper is devoted to the problem of creativity and its main ethical
aspects: the Omnibenevolence Paradox and the Education Paradox. The first relates
to the idea of God the Creator, who wishes only good, but permits a great deal
of evil. The second relates to the first and is that any spiritual culture which
educates for the inculcation of certain virtues in society inevitably inculcates with
those virtues a number of corresponding vices. If the first paradox has still been
considered as a purely theological problem and was repeatedly investigated by
various philosophers, the second one, on the contrary, has been studied little,
though in practice any spiritual culture constantly faces such an issue. The paper
demonstrates that both paradoxes share the same general basic principle, and,
therefore, are eliminated with the same method. To search out that method is the
practical goal of this work. At the same time there are the following problems to be
solved: to present the idea of God not in a tightly religious value, but in its broadest
common cultural one, to investigate the general nature of the Omnibenevolence
Paradox and the Education Paradox, to demonstrate the inadequacy of traditional
attempts to solve the problem, to adopt the solution that remains after all the others
have been rejected.

Keywords: Highest goal · Morals · Theodicy · Paradox · Omnibenevolence ·
Omnipotence · Creativity · Culture · Categorical imperative

1   Introduction

Contemporary culture often exhibits a mixture of the roles of cultural production and
consumption due to the emergence of new values and channels of their distribution [1–4].
But culture as a conscious transformation of world and self by the man, by definition,
arises and develops only because it has made a cult object of its highest goals. Concepts
of what those goals are differ, but the principle is always the same: all moral rules are
verified by the fact whether they contribute to achieving the desirable goals or not [5], as
Russell teaches. In other words, moral values of culture are defined along two courses:

1) human names something as being kind, useful, right, and due, everything that is
needed for achieving the highest goal of this culture; this, per se, is subject to cultivation
and enhancement.

2) Human names as evil, harmful, wrong, and vicious, everything that, directly or indirectly, hinders achieving the highest goal of this culture; this, per se, is subject to condemnation and destruction.

It results that, whatever the highest goal of this culture is, it always acts as a criterion of justice (both of logical, in the sense of the truth and thinking rules, and ethical, in the sense of good and conduct rules). And since the ability to think and operate the conduct is reason, any concept of the highest goal is, at the same time, a measure of reason perfection; i.e., a wisdom ideal.

Thus, wisdom, in any interpretation, is not only the goal of philosophy, but also the subject of any cult in general, therefore the essence of any deity, i.e., the concept of God is prompted by the concept of justice, goodness, wisdom as [6], Feuerbach concludes.

The value of this definition is that it partly reconciles theologians with materialists. Feuerbach, who denied God as a real subject, was among the latter ones. But, at the same time, he did not consider himself an atheist at all. The true atheist, according to him, is to be considered not the one who denies a divine subject, but the one who denies such divine predicates, as love, wisdom, justice. The subject denial is not the denial of predicates in themselves. They have their own, independent value [6], Feuerbach teaches. Thereby, he emphasizes that, in the matter of faith in God, there are only two solutions permitted:

1) The negative solution consists in refusing the idea of God, i.e., in recognizing there is no ideal of the perfect mind. But, then, there is also no real criterion of justice. And it means that no judgment is really fair. In particular, the very judgment "There is no God" has no sense at all.

2) The positive solution, on the contrary, consists in recognizing, at least, some judgments as fair. But, then, it is necessary to recognize also a real criterion of justice, i.e., a certain ideal of the perfect mind, interpreting it as some reality, unless it is objective, it is at least subjective. And this is the faith in God in any of two interpretations specified.

We may be asked by many: can all who see in God only a subjective human ideal be considered as believers more than those who see in him an objective creator of the world? Is there no difference between these interpretations of God to be equally opposed against atheism, as L. Feuerbach does?

It is strange enough, but there is no difference between these interpretations. There follow the same attributes of the perfect mind, needing both its objective reality, and complicating its recognition:

1) Omnibenevolence – since any spiritual culture approves an ideal of the perfect mind as a measure of good and justice. And it means that bad or wrong is not allowed in it, by definition; at least, for representatives of this culture.

2) Omniscience – since, if we do not know the answer to any question, we allow various possible answers to this question, including a wrong one. But the perfect mind, by definition, does not allow anything wrong, so, it allows also no ignorance of any issues.

3) Omnipotence – since an omnipotent person is not the one who can cater to any others' whims, but the one who can do everything that wishes. The perfect mind wishes only good. But impossible things cannot *be good,* since they cannot *be at all.* That is, the perfect mind wishes only possible things, i.e., it can do everything that wishes.

It is simple to notice that any being, on this logic, could be considered as "omnipotent", if it had enough knowledge to correctly estimate its opportunities, and enough goodness to wish nothing over possible. But it could be only the Creator of the world who possesses such properties, and initially everything would submit His will. Other beings can approach this ideal only by gradual refusing everything which are impossible for them. Then, to the question of realizing this ideal, there are the two answers to be imagined:

1) The negative answer consists in assumption that the ideal of the perfect mind is realized for nobody. Then, the aspiration to it is an aspiration to something impossible, i.e., it cannot be good; which, however, contradicts its very essence as an ideal.

2) The positive answer consists in assumption that either the ideal of the perfect mind is already realized in God the Creator, according to Judaism, Christianity, and Islam; or there is an impersonal state realized for many reasonable beings, according to Buddhism and Taoism.

Only a certain person, but not the spiritual culture in general, could choose the negative answer. The culture, by definition, interprets the ideals as the indisputable benefit. Therefore, any consecutive spiritual culture is forced to stick to the second way, i.e., to believe or in initial God the Creator, or in a condition of divinity, achievable for people. This, as a result, all the same would make them the world creators. So, both beliefs rest against the same paradox.

The paradox (1) of the divine omnibenevolence is related to the idea of an objectively real God the Creator, i.e., the being allocated with omnibenevolence, omniscience, and omnipotence. But combining these properties in one subject contradicts the fact of evil existence in the world. To the question, why God the Creator allows evil in the world, there are only two answers theoretically possible:

1) The negative answer is that God either does not wish to eliminate this evil, and then he is not omnibenevolent; or wishes, but cannot do without evil to achieve good goals, and then he is not omnipotent; or allows angrily out of ignorance, but then he is not omniscient. In any of these three solutions, God is denied, by definition.

2) The positive answer is that God is recognized existing as omnibenevolent, omniscient and omnipotent; but then, according to the law of contraposition, it is necessary to conclude that he wishes and can exclude evil, and expects everything in advance, therefore, evil in the world is excluded in advance. Then, it is already impossible to announce any phenomenon in the world as "evil".

The most popular solutions of this problem consist in recognizing one of these answers as false on purely moral bases. Thus, some ardent atheists, proceeding from the fact of evil existence in the world, decide that the idea of God is absurd and pointless (K. Marx, F. Engels). On the contrary, taking existence of God for an axiom, many theists decide that there is no evil in the world, and the idea of "evil" arises allegedly from misunderstanding what occurs (B. Spinoza, G. Hegel). There is also a softened version of the negative answer, termed as deism, which recognizes God the Creator but denies one of his predicates.

For example, this version denies omnipotence, believing as if God does not operate events (F. Voltaire), or denies omniscience, announcing creation of the world as an unreasonable act (A. Schopenhauer), or doubts omnibenevolence (L. Shestov). But any of such doctrines, as we made sure, cancels the meaning of the idea of God, therefore they can be classified as a category of atheistic, i.e., negative solutions.

There is less popular but rather frequent solution, immoralism, i.e., the solution canceling both answers, on the basis of pointlessness of morals as their general condition. For if it is considered that the moral categories of "good" and "evil" are pointless, then in the world there is no evil, no omnibenevolent God (F. Nietzsche, A. Camus). However, such atheism, demonstrated further, is even less well-founded than the first two solutions based on moral estimates.

2   Research Methods

The methodology which we apply to resolve this problem, includes, first of all, the method of dichotomy, and that of proof by contradiction. Where these methods do not provide an unambiguous solution, the Kantian method of transcendental reduction is applied: its point is, that from two exhaustive but equiprobable answers to a question, the one which contradicts the research objectives is removed. The one which corresponds to them and, on the contrary, is recognized as, let not the truth, but the hypothesis which is transcendentally necessary within this research.

The research aims at finding the solution which, not only, would not contradict itself, but also it could be combined with creativity within some spiritual culture at least. This aim, as demonstrated further, is contradicted by all solutions mentioned above.

3   Solution 1. Immoralism and Atheism

Immoralism as a simultaneous denial of any concreteness in the concepts of "good" and "evil", and also atheism as a denial of God, contradict it most. Since if we are engaged in some purposeful activity, immediately there is an ideal of reason and related moral standards having the two goals:

1) The negative goal is to forewarn oneself and others of the acts which directly or indirectly contradict our objectives and which, on this basis, we announce as evil and forbid everybody.

2) The positive goal is to involve others, whenever possible, to assist in those cases which serve our objectives and which we therefore announce as good and encourage them.

It is obvious that any of these two goals is not realized if other people have the right not to agree with our idea of good and evil. This is as if we allowed our enemies to deprive us of everything dear to us: it is inexpedient, in terms of any possible objective. Therefore, any our ideal, whatever relative it is on the basis, is always approved as an imperative, i.e., the law applying for consensus.

It means that there is, at least, one categorical imperative ordering all reasonable beings, according to I. Kant, to choose only so that the maxims, defining our choice, are, at the same time, contained in our volition as the general law [7]. However, the law generality, by definition, means its objective reality.

Thus, even if God was initially for us only the law to regulate the order and unity of ideas in our mind, nevertheless this law obviously needs we should study the nature as if there were a systematic and expedient unity [8], infinitely found everywhere, Kant argues. It means, even if there is no God, it is necessary to live as if he existed objectively.

4 Solution 2. Theism Denying Evil

Any kind of theism denying the reality of evil in the world is also incompatible with the spiritual culture. "Evil" is defined here as what is needed to be overcome and destroyed during cultural activities. If there is no evil, so everything in the world is such what it is to be like [9], Hegel claims. But then it is necessary to overcome nothing, and it is no more sense in any cultural activity than in its absence.

Many think that the problem will disappear if the goal of culture is to see in the awareness of evil an unreality, and the idea of "evil" is explained with our delusion which should be overcome. But then it is the delusion that will be that real evil which is allegedly unreal. In his well-known aphorism, the Chan patriarch Linji also hints at this contradiction: "If you meet Buddha, kill Buddha". It means the following:

1) Meeting Buddha is an achievement of the wisdom ideal when it is realized that there is no evil in the world, that all the nature is already enlightened, that everything is exactly such as it is to be like, and it is not needed to achieve anything.

2) Killing Buddha is a refusal of the wisdom ideal which, on the contrary, is initially needed to be achieved, overcoming delusions as undue; therefore, as a result it will be such a delusion.

A different witty way to circumvent the problem consists in explaining "evil" idea not as much with delusion as relativity of moral estimates: "There is no evil, it […] can only be insufficient good, discovering the insufficiency via its opposition to good completeness," [10] Karsavin teaches. Since the fact that what seems "evil" regarding higher degrees of good will become undoubtedly good in comparison with lower degrees.

For example, the person who is just insulted received good, for this person could have been robbed; and being robbed is all the same good for it is not being killed; and being killed is not being tormented; and being tormented, but not long, and like so indefinitely.

It can seem that this theory is not only indisputable, but also it is compatible with any notion of creativity: like a good soldier who dreams to be a general, but, at the same time, he does not consider his lower military rank as "evil"; and the spiritual culture can interpret moral education as a way from the lowest degrees of good to the highest. But it is not obligatory to announce the former ones as "evil". A number of authors claim that humans need to believe in moral progress [11–13]. One might say, people are already kind, they will be just even better if they stop slandering, stealing and tormenting the innocent.

The problem is that any culture, including the Christian one, in practice does not follow the similar principle. In order to ensure and understand the error of this doctrine, the second side of the problem will be considered.

The paradox (2) of the education, noted by Iskander in his book "The Camp of a Man" [14], is that education, as a rule, is effective only for those who need it least of all. The one who personifies lower degrees of good and, therefore, needs education most, reacts it toughly, which results in a zero educational effect. It is explained by the fact that the law of action and reaction equality works not only in mechanics. It is a general law derived from definition of the very object of activity.

The word "object" means something opposite to the activity, something resisting the latter. If there is no reaction, then there is neither an object of activity, nor the activity [15], Fichte taught.

It means that except the primary evil which our culture initially tries to overcome during education, there is still the secondary evil, i.e., reaction to a good action which is generated by the culture itself, and which in other conditions could not arise. And if the primary evil could be still defined as a "smaller degree of good", the secondary evil cannot be defined so: its force, by definition, is always equal to the force of good influcnccs, though the orientation is always opposite. Therefore, the more degree of good we want to educate, the more degree of the secondary evil we will obtain as an inevitable side effect.

We can be asked: How can some cases of effective education be explained? There are three possible options: a) either the tutor finds the virtues already ripened in the pupils and only helps them to reveal; b) or the pupils deceive the tutor, pretending to be kind; c) or the tutor fraudulently redirects the pupil's reaction to achieve the goal, i.e., turns the secondary evil into indirect good.

The latter is possible since action and reaction are related by their force and focus. Therefore, types of the secondary evil developed in the society are in direct dependence on what is considered "good" in this society and how it is attempted to be educated. Though there is a lot of researches which are seeking ways of overcoming this dependence [16–20], there is no research which would allow to avoid it. The dependence can be double:

1) The negative dependence is that if we did not pursue the goals and avoided culture, we would know neither good nor evil at all, due to the lack of criterion.

2) The positive dependence is that, having culture, there is both the object of activity, and reaction, i.e., evil. That means, any culture, educating people in certain concepts of "good", in reply generates strictly determined types of evil and delusion.

For example, many great poets derived inspiration from such defects as alcoholism and fornication. Culture could have driven them into more rigid framework, but another defect would be a payment for this: impoverishment of great poetry.

Which of these two evils is more tolerant? Each culture judges from its own goals. The general rule is that it is not a "smaller degree of good", but evil as a real, direct reaction to good appears an inevitable condition of any good. Therefore, the aims of our research would be answered only by the solution where belief in reality of God and belief in reality of evil would be consistent.

At the moment there are two known versions of similar relation. One relieves all responsibility for evil from God the Creator, assigning it to the man. The other, on the contrary, only formally leaves responsibility on the man, in fact blames God the Creator for evil.

5   Solution 3. Man as the Creator of Evil

The first of these doctrines is based, on the one hand, on belief that freedom to do evil is a condition of good. Therefore, good God provided the man with this freedom. On the other hand, freedom allows a possibility of evil, but does not guarantee it allegedly. Therefore, in God the Creator' plans "[…] good and beauty can exist in a pure form, without any impurity of evil and disgrace" [21], Lossky teaches.

This doctrine does not withstand criticism for two reasons. Firstly, it ignores the law of the equality of action and reaction, which means the opposite, as we discovered above. Secondly, it contradicts the dogma of God the Creator's omniscience, therefore, two solutions are possible:

1) The negative solution assumes that freedom provided to the man to do evil does not guarantee doing evil. But then God, who provided freedom, does not guarantee such result, i.e., God does not foreknow whether the man will do evil at any given moment.

2) The positive solution is that God's omniscience excludes the possibility of his ignorance of anything. That is, God guarantees in advance that freedom will be used by villains for evil. And if, all the same, he provided this freedom to villains, therefore he wanted it to happen by all means.

The first option is excluded by the ideal of the perfect mind, i.e., consecutive culture cannot insist on it. Therefore, the aims of our research would be answered only by the latter solution.

6   Solution 4. God as Creator of Evil

Its supporters teach that God the Creator did not initially conceive the world as the kingdom of good and love: "[…] both the poetry of mortal life, and the conditions of rescue beyond the grave – equally need […] harmonious […] confluence of enmity and love" [22]. And the harmony is considered here "[…] not in the sense of peaceful and brotherly moral consent, but in the sense of the poetic and mutual compensation of contrasts both in life itself, and in art" [22], Leontyev specifies.

We find the most detailed development of this solution in G. Leibniz's Theodicy. The latter recognized the law of action and reaction equality. But he provided it with a more general form of the sufficient basis law, which says, that any real action is sufficiently based on the reality of something different as its reason or reaction; if there is a real A, there is a real not-A, and vice versa.

For example, if there is a real "not-evil", will it be good or something morally indifferent, its basis is a real "not-not-evil", i.e., evil. It means that the reality of evil appears a necessary condition not only of reality of good, but also, in general, of any reality.

Can God the Creator repeal this law? Certainly. Even the man is capable of it: it is rather simple not to have any goals as criteria! Then there will not be neither good, nor evil, nor true, nor false; and the notional difference between real and unreal will immediately be erased.

The problem is that God as an ideal of reason does not allow that due to the omnibenevolence. The latter wishes maximum good. If the zero degree of evil means not maximum, but the same zero degree of good, then the not-zero degree of evil in the world is desirable for God. Evil as a necessary condition of good is not a pure evil, but also indirect good; it means, wishing that, God wishes only good.

Would we ever have enjoyed health if we had never experienced a disease? Is a little grief needed in order to feel good stronger, i.e., to realize that it is greater? [23], Leibniz asks. If yes, the paradox is partly removed. Since it is possible to imagine some potential worlds where there is no sin and misfortunes. But these worlds, concerning good, will be much lower than ours for evil often serves as the reason of good which would not occur without this evil [23], Leibniz decided.

Its main shortcoming is that the examples specified there can work against it: if the disease is congenital and incurable, it is impossible to enjoy health anymore; if grief is so strong that a person go insane, there will be nothing to realize good. That is, life itself needs to present evil in two forms specified in the novel "What Is to Be Done?" by N. Chernyshevsky:

1) Healthy, i.e., necessary evil hinders the achievement of the highest goal, but does not deprive the person of its opportunity. Thereby, it maintains a healthy balance of action and reaction. It is not only evil, but is also indirect good, i.e., a good condition; like that "real dirt" [24], out of which over time some wheat can grow.

2) Unhealthy, i.e., excessive evil makes the achievement of the highest goal impossible. And, therefore, this is not indirect good, but pure evil and destruction. That "fantastic dirt" [24] or just "rot", out of which nothing good will ever grow for certain.

Why did N. Chernyshevsky name the first type of evil a "real dirt", and the second one a "fantastic dirt"? He may have hinted that only the first evil can really exist. Since the impossible thing, as already mentioned, cannot *be* either good, or the highest goal, due to the fact that it cannot *be at all*. It means, the concept of "the impossible highest goal", defining the second type of evil, would be an empty phantasm for God the Creator.

If we had always remembered it, only possible would always be considered our goal. All inevitable would be estimated as a necessary condition of achieving the highest goal, i.e., as indirect good. And any evil for us would be "healthy", by definition.

Therefore G. Leibniz was partly right when he ignored the possibility of unhealthy evil in his Theodicy. But the stumbling block for his system appeared its cornerstone, i.e., the sufficient basis law, which says: if there is A, there is not-A; and if there are saved ones, there are damned ones; if there is something that is not a pure evil, therefore, there is something that it is. It means, unhealthy pure evil is real too.

Besides, the Christian dogma also confirmed G. Leibniz's truth. It even accurately imaged the dichotomy, by which the problem of existence of unhealthy evil is to be solved:

1) The negative solution was recognized as objectively true both for God who can do everything what he wishes, and for people, faithful to him, who wish only what they can. For them, any real evil, by definition, is surmountable, i.e., healthy.

2) The positive solution was recognized as objectively true concerning all damned sinners who deserved eternal punishment. For them the hope for achieving the highest goal is allegedly lost forever. It means, for them this eternal punishment is unhealthy evil.

The paradox may seem to be completely removed by this doctrine if we take for an axiom that eternal suffering of sinners in the hell is good for God himself. But the cruelty of this solution induces even the thinkers close to Christianity (S. Bulgakov, N.

Berdyaev, etc.) to doubt it. And some are sure that the principle "evil atonement by suffering" is not especially compatible to the omnibenevolence.

Therefore, we will begin with a more general question: is the principle of atonement to be considered as true in itself? Is any spiritual culture to claim that the person, who sinned against the highest goal, will suffer for it by all means either morally, in the sense of repentance, or physically; either during lifetime, or after death? Since "suffering" is defined the state contradicting the goals of the suffered, there are two possible answers:

1) The negative answer assumes that the person could sin against this culture goals and not suffer for it. But, then, (s)he sinned nothing against his or her goals. So, the latter ones do not coincide with this culture goals, i.e., (s)he does not belong to it.

2) The positive answer assumes that the person, on the contrary, belongs to this culture, i.e., accepts its goal as his or hers. Then, having sinned against it, (s)he suffers, by definition. If not morally, in the sense of repentance, then physically; and if not during lifetime, then after death.

The first option is possible theoretically. But practically any consecutive spiritual culture cannot accept it. Any culture approves, at least, some prohibitions as general laws, according to Kant's imperative. It means, every culture indirectly approves the highest goal which is their cornerstone as "everyone's goal". That is, any consecutive culture cannot but believe that the sin against its prohibitions will be followed by suffering.

Can such belief be wrong? Perhaps, provided that this culture imposes to the world as "the general highest goal" that which is definitely not. But God as a wisdom ideal, by definition, does not make mistakes. That is, as "the highest goal" he can order only what is good for all people, including the damned.

But if this good becomes impossible for the damned, it cannot already be either good, or the goal for them because it cannot be for them at all. And then, firstly, it is not the general goal (the damned will appear out of this goal). Secondly, the consciousness that it cannot be their real goal will leave them only two options:

1) The negative option consists in knowing that they are assumed to have no other real goals. Then it means retreating into that zero state where, due to the goal lack, by definition, there is neither good, nor evil; neither pleasure, nor suffering. And it is not the hell of Christians any more, but nirvana of Buddhists.

2) The positive option is that real, i.e., attainable goals remain with them. But then the most significant of them will be their highest goal, i.e., their true God. For that case, sufferings are assumed to be possible, but already not eternal.

Thus, the principle of atonement for evil through sufferings, according to the latter two dichotomies, works only provided that the highest goal of spiritual culture remains always attainable for everybody. But then no types of unhealthy evil can be an objective reality either for God, or for righteous people, or even for sinners. And if such evil is to be real, according to the reason law, we need to recognize it only as the subjective reality, i.e., existing only in the system of some culture representations. And these representations are false, since they are assumed not to correspond to the objective relation of goals and means.

For example, if a certain false culture specifies the unattainable highest goal to the man or offers unusable means, in both cases the result is not the achievement of the man's real highest goal. Though objectively the man could achieve it, having corrected the errors, but within a false culture the goal will remain unattainable all the same. Therefore, our dichotomy solving a problem of unhealthy evil reality appears as follows:

1) The negative solution is to be objectively true both for God who can do everything what he wishes, and for people faithful to him who wish only what they can. For them any real evil is surmountable, i.e., healthy.

2) The positive solution is to be objectively false for all beings, but it will be subjectively true for all representatives of false cultures. For them unhealthy evil will remain real until contradictions in the moral bases of this culture are eliminated.

The last question remains: if once it happens – once all moral contradictions are eliminated, all people realize the truth and all false cultures disappear from the face of the earth – does it mean that the reality of unhealthy evil will disappear with them too? And will the healthy culture lose the sufficient basis through it? It will not, since for the subjective reality of unhealthy evil, it is enough that it remained in memory, as real experience of the errors corrected in the past.

7 Results

1) The rational origin of the idea of God the Creator, its common cultural importance and necessary logical interrelation of the main predicates of God, i.e., omnibenevolence, omniscience and omnipotence are demonstrated.

2) The general nature of the paradoxes of divine omnibenevolence and education is uncovered, this is the problem of any creativity which consisting in the law of the equality of action and reaction.

3) Such concepts as atheism, immoralism, and those types of theism which deny evil or reduce it only to the man's free will, are eliminated from its admissible solutions.

4) It is proved that good and evil are related as two parties of one action so it is impossible to eliminate evil, without having eliminated good, but it is possible some healthier quality to evil, having eliminated contradictions in moral estimates.

References

1. Sacco, P., Ferilli, G., Tavano Blessi, G.: From culture 1.0 to culture 3.0: three socio-technical regimes of social and economic value creation through culture, and their impact on European Cohesion policies. Sustainability **10**, 3923 (2018). https://doi.org/10.3390/su10113923
2. Bylieva, D., Bekirogullari, Z., Lobatyuk, V., Anosova, N.: Home assistant of the future. In: Proceedings of the International Scientific Conference - Digital Transformation on Manufacturing, Infrastructure and Service, 60. ACM, New York (2020). https://doi.org/10.1145/3446434.3446540
3. Pavlenko, A.: Technology as a new language of communication between the human being and the world. Technol. Lang. **1**, 91–96 (2020). https://doi.org/10.48417/technolang.2020.01.19
4. Serkova, V.: The digital reality: artistic choice. IOP Conf. Ser. Mater. Sci. Eng. **940**, 012154 (2020). https://doi.org/10.1088/1757-899X/940/1/012154
5. Russell, B.: Why I Am Not a Christian, 39th edn. Touchstone Books, New York (1965)
6. Feuerbach, L.: The Essence of Christianity. Cambridge University Press, Cambridge (2011)
7. Kant, I.: Critique of Practical Reason. Dover Publication, Mineola (2004)
8. Kant, I.: Critique of Pure Reason. Cambridge University Press, Cambridge (1999)

9. Hegel, G.: Encyclopedia of the Philosophical Sciences in Basic Outline. Part 1: Science of Logic. Cambridge University Press, Cambridge (2010)
10. Karsavin, L.: On good and evil. In: Small Compositions, pp. 250–284. Aleteya, Saint Petersburg (1994). (In Russian)
11. Eriksen, C.: Moral progress. In: Eriksen, C. (ed.) Moral Change: Dynamics, Structure, and Normativity, pp. 137–143. Springer, Cham (2020). https://doi.org/10.1007/978-3-030-61037-1_17
12. Moody-Adams, M.M.: Moral progress and human agency. Ethical Theory Moral Pract. **20**(1), 153–168 (2016). https://doi.org/10.1007/s10677-016-9748-z
13. Wilson, C.: Moral progress without moral realism. Philos. Pap. **39**, 97–116 (2010). https://doi.org/10.1080/05568641003669508
14. Iskander, F.: The camp of a man. In: Stories (Collection), pp. 104–260. Pravda, Moscow (1991). (In Russian)
15. Fichte, I.: The Science of Knowledge. Cambridge University Press. Cambridge (2003)
16. Croce, M.: Exemplarism in moral education: problems with applicability and indoctrination. J. Moral Educ. **48**, 291–302 (2019). https://doi.org/10.1080/03057240.2019.1579086
17. Darnell, C., Gulliford, L., Kristjánsson, K., Paris, P.: Phronesis and the knowledge-action gap in moral psychology and moral education: a new synthesis? Hum. Dev. **62**, 101–129 (2019). https://doi.org/10.1159/000496136
18. Bylieva, D., Lobatyuk, V., Tolpygin, S., Rubtsova, A.: Academic dishonesty prevention in e-learning university system. In: Rocha, Á., Adeli, H., Reis, L.P., Costanzo, S., Orovic, I., Moreira, F. (eds.) WorldCIST 2020. AISC, vol. 1161, pp. 225–234. Springer, Cham (2020). https://doi.org/10.1007/978-3-030-45697-9_22
19. Bylieva, D., Lobatyuk, V., Nam, T.: Academic dishonesty in e-learning system. In: Soliman, K.S. (ed.) Proceedings of the 33rd International Business Information Management Association Conference, IBIMA 2019: Education Excellence and Innovation Management through Vision 2020, pp. 7469–7481. International Business Information Management Association, IBIMA (2019)
20. Pylkin, A., Serkova, V., Petrov, M., Pylkina, M.: Information hygiene as prevention of destructive impacts of digital environment. In: Bylieva, D., Nordmann, A., Shipunova, O., Volkova, V. (eds.) PCSF/CSIS -2020. LNNS, vol. 184, pp. 30–37. Springer, Cham (2021). https://doi.org/10.1007/978-3-030-65857-1_4
21. Lossky, N.: World as Organic Whole Supplement to Philosophy Issues Moscow (1991). (In Russian)
22. Leontyev, K.: On the global love. In: Russian Idea, pp. 147–170. Republic, Moscow (1992)
23. Leibniz, G.: Theodicy: Essays on the Goodness of God, the Freedom of Man, and the Origin of Evil. Open Court, Chicago (1985). Translated by E. M. Huggard
24. Chernyshevsky, N.: What Is to Be Done? Khudozhestvennaya Literatura, Moscow (1985). (In Russian)

Creativity in Technosociety

The Concept of Utility: The Role of Utilitarianism in Formation of a Technological Worldview

Iskender A. Gaparov$^{(\boxtimes)}$ (iD)

Samara National Research University, Moskovskoye Shosse, 443086 Samara, Russia

Abstract. An attempt has been made to consider the ways in which the concept of "utility" is defined in modern utilitarianism and their role in the development of a technological worldview. It took two hundred years of creativity to form the modern understanding of utility. By defining the concept of utility "utilitarianly" as a way of existence of a social system supported by additional sources of survival and the benefit for all, we are able to understand utility from the technical point of view as a rational balance of means and goals. In this regard, the use of the method of historical reconstruction of the essence of the concept of "utility" becomes important for study. Based on the works of utilitarians and representatives of the philosophy of technology, we carry out a historical reconstruction of the "utility" concept in an attempt to clarify its ontological foundations and come to the understanding of utility as a multifaceted phenomenon, unthinkable without social reality, but developing in the direction to the new types and levels of existence. It is aimed at reducing the distance between what is desired and what is due, being and reasoning, and achieving happiness as a form of spiritual balance. The degree of happiness is determined by the principle of utility, which should be understood as a set of rational methods that are successfully implemented for effective solution of technical problems and as a drive for innovations, through which humans are able to expand the scope of possible experience and transform the environment without harming themselves and society.

Keywords: Utility · Utilitarianism · Philosophy of technology

1 Introduction to the Problem

What is utility? This is a fundamental problem in the history of philosophical thought and Western European culture was unable to solve it. For several centuries, utilitarians, economists, pragmatists, instrumentalists, and representatives of the philosophy of technology have tried to define the meaning of the concept of utility. All their attempts were in vain. After all, they did not take into account the specifics of the use of the concept of utility in any particular situation, and therefore treated it one-sidedly, superficially. Such approach to the concept of utility led to suspicious attitude to it: it was regarded as something that destroys the social norms (Immanuel Kant), or something that could contribute to the well-being of an individual (Jeremy Bentham). But the extent to which

© The Author(s), under exclusive license to Springer Nature Switzerland AG 2022
D. Bylieva and A. Nordmann (Eds.): PCSF 2021, LNNS 345, pp. 127–138, 2022.
https://doi.org/10.1007/978-3-030-89708-6_12

this assessment was fair can only be proved by a review of the meanings of the concept of "utility" which can reflect its essence and ontology.

The ontology of utility has been formed over centuries, unnoticed by the superficial views of ordinary people. At first, on the theoretical level – in the philosophy of Antiquity and the Middle Ages, and then on the practical level - in utilitarianism, economics and others. A significant change was the reinterpretation of the utility concept by the philosophers of technology, who abandoned the moral side of the concept of "utility" and for the first time thought of it as a balance of goals and means, not obscured by subjective assessments. But how difficult was the path before Engelmeyer and Dessauer came to the functional meaning of the concept, free from religious, ethical, aesthetic components, can be seen by tracing the main stages of the evolution of the word "utility".

An examination should begin by analyzing the attitude of ancient philosophers to the concept of "utility". Why did they so rarely touch this subject in their works and did not distinguish the shades of meaning? Because the ancient Greeks had a clear inclination to intellectual knowledge (*"theoria"*), from which all moral actions were then derived (*"praxis"*) [1]. Therefore, everything that was associated with utility was associated with slave labor or efforts to change a person, with the sphere of "praxis". This was obvious to Socrates, Plato, Aristotle, and others, who regarded the cosmos "theoretically" as pluperfection. Also, the reason for the negative attitude to the sphere of "praxis", according to Aleksej Losev, was the ancient worldview, which coordinated the image of the permanent "self" with the image of the world, perceived sensually as a physical body [2]. Perfection and inclusiveness led to the idea that man is a microcosm, and the universe is a macrocosm. The self-identity of the ancient polis citizen was inextricably linked with the cosmos, and everything that did not correspond to it was imperfect.

The ancient eidetic comprehension of reality and the capacity for form-making, generally disconnected with reality and its problems, could have remained decisive in the history of all Western European thought, if not for the turn of Protagoras and Socrates towards humans. This new way of thinking allowed them to look at humans as intelligent beings with a priori capacity for self-control and self-awareness. Interested in the moral side of behavior, Socrates gave Plato and Aristotle the opportunity to address the essential aspects, conceptualizing them in the provisions that humans: 1) are social beings; 2) can perceive moral categories and perform activities. Unlike Plato, Aristotle was interested in the sphere of "praxis". He was particularly attracted by the following aspects: goal, means and activity. He understood the goal as "a focus on a certain result, different from the work itself" [3, p. 228], the means - as something that led to the intended image, and in activity he saw "an active attitude to the world or its elements" [3, p. 224]. Such choice was not accidental. It was built from the same components as the path to happiness, which was understood as self-realization, as an activity for its own sake, driven by dianoetic virtues.

This view of practical activity, having passed through the entire era of Antiquity, remained unchanged until the Renaissance, when a number of sophists (Pico della Mirandola, Francesco Petrarca, Niccolo Machiavelli and some others) under influence of the Christian religion adopted an idea that a man was "a creature of indeterminate image", whom God took and "set him in the middle of the world" [4, pp. 6–7]. A person was able to realize himself in every form and become anyone he wished to be. Free will, vivid soul and spirit gave him the ability to self-realization. It could not be accomplished completely. After all, as a spirit, man was free to act at his will, but as a natural being, he remained bound by the laws of the phenomenal world (nature).

The philosophers of the Modern era brought the renaissance ideas on the place of "man in the world of things" to their logical conclusion. These ideas defined the practical way of being that gave birth to praxeology, the theory of human action, its foundations and methods of rational and effective action. The problem of determining the criteria for this kind of activity was raised by proto-utilitarians in a number of critical pamphlets related to the understanding happiness and the means to achieve it. Several front lines were highlighted. One of them took pleasure as its starting point, the other - reason. Different preconditions led to the emergence of two trends – economic and philosophical.

For early modern economists (Thomas Malthus, Adam Smith), the problem of criteria for effective activity was solved through the analysis of human wants and needs. Pleasure was accepted as the basic need, equally characteristic of all people, that set in motion the internal forces and directed them to maintain stability and balance in the society. In isolation from thinking, pleasure was unsuitable and even criminal, because when pursuing it, a person did not take into account the interests of other people. For this reason, economists suggested that necessarily there must be a principle that restrained the sensuous impulses and redirected them to the benefit of all. They meant a benefit in the form of the "invisible hand" of the market. This concept ensured the following profits: 1) it restrained humans from insults, violence and murder, by taking into account all the benefits that can be derived from business relationships; 2) it provided "optimal allocation of relatively unlimited resources" [5, p. 36].

In contrast to the economists, the early utilitarians tried to define the concept of utility through the social nature of a person, through a number of situations where an individual both acted independently of social requirements and voluntarily obeyed them. Anthony Ashley Cooper (Lord Shaftesbury), Francis Hutcheson, George Berkeley, and David Hume developed a set of conceptual provisions for a special kind of innate feeling. Its purpose was to reduce the diversity of sensory data to the unity of experience, allowing humans to extract benefits without additional efforts. To justify the necessity of this feeling, Francis Hutcheson developed a number of rules for correct thinking. According to one of them, "those Objects of Contemplation in which there is Uniformity amid Variety, are more distinctly and easily comprehended and retained, than irregular Objects; because the accurate Observation of one or two Parts often leads to the Knowledge of the Whole" [6, p. 79]. Irregular objects are forgotten more quickly because they require time to be imprinted in the mind, as a result, the efforts taken in the present do not pay off in the future. For this reason, a person needs something that is able to unite diversity and, guided by objective laws, to prove the reliability of the fact. The concept of utility emerged as a form of correct thinking. Only Bernard de Mandeville [7] knew the extent to which it was practical. He understood the essence of humans hidden in the affects ennobled by the society and translated into a binary "egoism"/"altruism". The rest of the proto-utilitarians, John Gay, William Paley, Joseph Priestley, Paul-Henri Holbach, took into account the mistakes and paid attention to the social aspects of an individual (upbringing and education). Under the concept of utility, they understood the order of the elements in the system, each of which had a specific purpose, performed a socially significant function beneficial for all, regardless of a person's actions. The main idea was that humans should not deviate from regulatory requirements in their actions and commensurate the following impact on the public well-being.

The pro-utilitarian approach to the concept of utility revealed an error in its interpretation by the economists which consisted in trying to reduce it to the following terms: benefit, profit, income. This interpretation lead to a tautology in which the sentence with the concept of "utility" became purely analytical, since it did not reveal the entire meaning. Bentham saw this problem, pointing out that the "word "utility" did not indicate ideas so clearly, and also did not allow for a quantitative assessment of interests and the measure of right and wrong" [8, p. 116]. On the contrary, John Stuart Mill argued that difficulties stem from the original grounds of the proto-utilitarians which considered pleasure primary over virtue. He proposed to identify the concept of utility with justice, while allowing the following condition: participation in the collective interests of humanity.

These reflections formed the main task for the specialists in the field of practical life. The essence of the work was to determine the ultimate (transcendental) foundations of the concept of "utility" that would remove the lack of clarity of the meaning of the term and determine its place in the structure of philosophical knowledge.

2 The Concept of Utility in the Ethics of Classical Utilitarianism

Let us turn to the ideas of the classical utilitarians and try to present their attempts to synthesize the views of their predecessors on definition of the concept of "utility", as well as to determine their influence on the formation of the technical worldview.

It is necessary to start with the figure of one of the most significant social theorists of the Modern era – Jeremy Bentham. The interest in his personality is ambiguous. From the legal aspect, his views on the essence and structure of the state (the civil code), the project of the ideal prison "panopticon" is considered outdated. From the point of view of ethics and economics, the representatives of the scientific elite of the 20th century, Bertrand Russell, Amartya Sen, Frank Knight, Jerry Hausman, consider Bentham's ideas quite erroneous, since they: 1) do not clarify the meaning of moral categories; 2) reveal an attempt to calculate pleasure "arithmetically" when the rights and freedoms of an individual are sacrificed to the public interest.

The reason for this unjustified negative attitude towards Bentham could be the dynamics of meanings in the conceptual apparatus among the English intelligentsia in the 19th–20th centuries. This transformation continues nowadays, indicating a shift in the worldview. The public consciousness enters into such information flows, where any philosophical problem devoted to fundamental concepts of truth or good turns into a bioethical or technical one. An inaccurate analysis of praxeology could also be a reproach to those who did not understand enough or did not want to take a meaningful attitude to the teachings of Jeremy Bentham.

However, if we carefully consider the merits of the Bentham's concept, it will become clear that everything that is being connected to utilitarianism, a doctrine based on the concept of utility, actually refers to praxiology, aimed at finding acceptable criteria for expedient activity. Such points are noticed by the Russian theorists Artemyeva [9], Sushentsova [10], Yarkova [11], Ushkin et al. [12] and others. Researchers like Ogleznev [13], Sioma [14], dwelling on the ways to redeem from fictitious entities introduced by the philosophy of Antiquity and the Middle Ages, seek to find common grounds specific for utilitarianism and praxiology of Bentham. The utility criteria are important because they help to process entities and determine their significance and stability. In this regard, Bentham's legal positivism, his starting point for reflections on the ideal social order, provides material for justifying the principle of utility and its effectiveness for the reality given to us. The usefulness of different postulates and oaths is questionable, since they "impose obligations, but do not provide privileges" [14, p. 121].

We proceed now to an attempt to redefine Bentham's concept of utility. The synthesis that he initiated was directed against the economic and mechanistic points of view. The interest in the concept of "utility" and its development was connected with the search for "the golden mean" or measure of social well-being, the essence and purpose of which was hidden in the way of existence assuming the priority of pleasure over suffering. Functionally, Bentham reveals the concept of utility as the ability of an object "to produce benefit, advantage, pleasure, good, or happiness (all equivalent in the present case) or (this being the same thing), to prevent the happening of mischief, pain, evil or unhappiness to the party whose interest is considered" [15, p. 7]. This idea is discussed by such authors as Prokofiev [16], D. S. Astashova, A.V. Petrov and many others. However, almost all researchers miss the fact that Bentham's understanding of utility is multifunctional, since it has a wide scope of application (depending on the meaning). To remove the antagonism regarding the various meanings of the concept, Bentham seeks to create conditions for the way things and people exist, in which they will definitely be identical. Therefore, he lays his hopes on mathematics and even develops "moral arithmetic", complementing it with an analysis of the mechanisms of the prerequisites, causes and consequences of actions. Bentham suggests taking them in consideration by everyone to make a number of reliable conclusions about their own or other people's thoughts on the performed or abstained action.

Bentham's utilitarian approach had a significant influence on the worldview of John Stuart Mill, who criticised the provisions of classical utilitarianism. Such an attitude results from a misinterpretation of the pleasures experienced by a person in a purposeful activity. After all, Bentham used two approaches simultaniously – praxiological and utilitarian, and missed this point. He did not clarify the essence of the leveling of all pleasures. As a result, a vicious person could be regarded as a virtuous one may be to more extent than a few benefactors. The "moral arithmetic" took into account the interests of society, without evaluation of the intentions of individuals. They could differ from the results due to the intervention of external factors. In order to avoid such ethical paradoxes, Mill sets the quality of pleasure as a criterion by "(a) making the quality of pleasure, not its bulk or intensity, the standard; and (b) referring differences in quality to differences in the characters which experience them." [17, p. 279]. The reason why he distinguishes between pleasures is seen in the essential difference between the animal and the human. If humans, who have higher abilities (intellection), have happiness as main goal, then for animals, bound by the laws of the phenomenal world (nature), and acting under the influence of instincts, there are only pleasures. Mill agrees with the fact that "it is indisputable that the being whose capacities of enjoyment are low has the greatest chance of having them fully satisfied; and a highly endowed being will always feel that any happiness which he can look for, as the world is constituted, is imperfect" [18, p. 12]. Happiness as the correspondence of objective conditions to subjective desires and goals becomes the initial reference point that determines the degree of usefulness of all actions and the peace inherent in a person with an inner "self". However, the question arises on determining the significance of subjective goals that affect the level of social well-being. In pursuit of happiness, persons are able to do without concern about the situation in the society, if of course it does not become an obstacle in their way. To solve this question, Mill analyzes the mechanisms that govern all human actions (virtue and justice).

First, he states that the peace with oneself in regards to activity cannot be independent of the interaction with other people due to the imperfection of the social structure where loneliness is unacceptable in any form. Humans have to sacrifice part of their interests for the benefit of the whole society. According to Mill, this donation preferably should not be forced, but voluntary. It can become such only through internal regulatory mechanisms that make a person virtuous. Otherwise, with each such act, humans will experience suffering (misery) that does not meet the main goal of utilitarianism - to combine individual and public interests in such a way as to derive greater benefit from them. Mill offers an ideal solution to the problem – to entrust the state with responsibility of making citizens virtuous, so that the laws are "as nearly as possible in harmony with the interest of the whole", and education and public opinion form "an indissoluble association between his own happiness and the good of the whole" [18, p. 19].

Mill's approach turns to be a favorable condition for clarifying the idea of justice in ethics of utilitarianism, as he integrates virtues with the public good and regards this as a means for achieving happiness. This idea drew so much interest because it served the basis for the existence of society, without which the realization of the principle of utility, well-being and happiness would be impossible. It includes: 1) the existence of inalienable rights that allow individuals to act at their own discretion; 2) the presence of such normative aspects as no-harm rule and just deserts. When virtues integrated with public good are considered as obligation they serve both citizens and the state. Their synthesis is reflected in the Bentham's dictum, "everybody to count for one, nobody for more than one," [18, p. 20]. Such a social bias towards equality of subjective rights and interests leads to the identity of justice and benefit. But if justice appears as a means for maintaining balance, order and stability in the society, then benefit acts as a principle of social harmony, creating such an ideal situation in which everyone gets the desirable, while not sacrificing social interests.

Mill's utopian project related to the dynamics of social development was heavily loaded with social elements; and, therefore, the concept of benefit was unthinkable without society. This has been repeatedly noted by Herbert Spencer, which is confirmed by the research of Russian theorists such as Kryuchkova [19], Podvoysky [20], and some others. In Spencer's opinion, everything that concerns the concept of utility is understood ambiguously by utilitarians, since they seek to artificially replace individual interests with public ones. As a result, there is a contradiction that cannot be eliminated even by the force of state. Spencer prefers the policy of non-interference of the state in social processes and the development of social life without external control. After all, the more the state intervenes in the affairs of a particular individual, the more it hinders the formation of adaptation skills for constantly changing reality. To replace the criticized approach, Spencer offers a concept of behavior and functional utility of action represented by evolutionism and organicism.

The English philosopher starts his research with a question: what is the main goal for all living beings? He criticizes the utilitarians for a methodology that accepts pleasure and happiness as the starting point, but misses the main source – life. Without life, everything else is meaningless. Under it, Spencer understands "the continuous adjustment of internal relations to external relations" [21, p. 19] or the mode of existence aimed at acquiring the best state of the system – equilibrium as the preservation of all its parts. It is driven by evolution, the essence of which is represented in the complication of a living being through the processes of differentiation, an increase in the amount and concentration of matter. So the creature is endowed with a number of properties and qualities that allow it to adapt more effectively to the surrounding circumstances. On the example of adaptation of an infusoria and rotifer, Spencer comes to the conclusion that the lack of organs of movement and external senses in infusoria is fully compensated in rotifers, as a result, being more complex and differentiated creatures, they are able to timely anticipate the danger and respond to it. Therefore, "when the differentiation, isolation of forms, and the struggle for existence cease, regression begins" [19, p. 73], leading to the extinction of an individual species. To define the degree of adaptability of living beings to the surrounding conditions, Spencer develops special criteria, understood as breadth and length of life. In the sum of their components, they represent behavior as "the adjustments of acts to ends" [21, p. 16] – survival and reproduction. Everything that contributes to their realization appears as the most successful combination and is accompanied by pleasure, everything that hinders them is harmful, because it shortens life. The problem arises at the stage when the consideration of living beings passes to humans, who, due to the complex differentiation of parts and a variety of functions, in addition to the main goals, have derivatives mediating and hiding them. The pleasures that follow the actions of almost all living beings and guide them in finding their place in the system of nature, in humans are replaced by perversely distorted ones that stand far from life and happiness. In pursuit of these pleasures, a person brings his death closer. To prevent such a scenario, Spencer turns to the analysis of the structure of the social organism (culture), which arises as a response to the complex structure of the human body. The interconnection of intelligent beings, the idealized object, provides an individual with the criteria of reality – the laws of correct life and true needs. As a result, a person first forms an idea of what is useful, and then the concept of benefit that is understood not instinctively (at the level of behavior), but rationally. In contrast to animals, whose criterion of expediency is pleasure, a person relies on experience. Through experience, human mind selects the best options for action and implements them. In accordance with this, the preferences of people move to a higher stage of development. After all, they now have to adapt not to the natural environment, but to the social one. The culture, having set the initial guidelines for actions, forms the morality through which a person understands the social structure and existence as a whole, thinking about significant goals and selecting the necessary means. The imperfection of the social system pushes individuals to self-realization both for the pursuit of their own interests and for overcoming of social antagonisms. In this regard, "the limit of moral development is reached when human desires and actions driven by them are fully balanced with the conditions of the best existence" [19, p. 77].

3 The Concept of Utility in the Technical Worldview

Classical utilitarians, having raised the problem of the diversity of meanings of the concept of "utility" and trying to solve it by means of ethics, rehabilitated the sphere of practical philosophy - deontology and praxiology. This approach was largely explained by the peculiarities of the Modern era, whose contemporaries were Bentham, Mill, Sidgwick. The adoption of an optimistic picture of the world, based on a belief in reason and progress, gave hope for the definition of the utility as a criterion for social well-being, without which none of the actions could be realized. Bentham's "Moral arithmetic" and Mill's idea of "equal justice" were the tools for quick and effective calculation of the level of social well-being understood as the strive to live.

The real problems began to arise after Spencer's criticism of the theory of the social contract, which was basic for classical utilitarians. He developed an organic society model that assumed the essential desire for benefit in every being as the ability to adapt to the conditions of the natural or social environment, and behavior - as a number of actions proportionate to the goals. This approach testified to the following contradictions of the concept of utility. On the one hand, the utility represented a multifunctionality of acts, with the help of which a living being was able to adapt much more successfully to the conditions of the surrounding/social reality, but on the other hand, it could designate a rational correspondence of means to goals. While Spencer's approaches were relatively identical due to the organic type of thinking, the difference was most pronounced in the conceptions of the pragmatists and instrumentalists. They were not concerned about public interests and focused on the mechanisms that constituted action, one of which was experience. So, John Dewey believes that when "we gain an experience, we somehow act on the object, and then suffer the consequences of our actions. We do something with the object, and it does something with us in response: this is the specifics of the interaction. The relationship between these two parts of experience determines its fruitfulness and value" [22, p. 28]. Then it becomes clear that behavior and adaptation are not identical to thinking based on experience. If the actions of an animal are programmed and obey the "stimulus-response" scheme, then humans acquire a number of privileges. They are able to vary them at their own discretion or to act in a way that provides for probability, in other words for the reassembly of experience. Thus, the concept of utility appears in a new functional and technical understanding that was most fully described by the philosophy of technology [23].

One of the founders of the philosophy of technology is the Russian theorist and engineer Petr Engelmeyer. The study of the phenomenon of technology is directly connected with the creative process that implies the adjustment of nature, the adaptation of one's own thoughts to a new experience (reassembly of experience). Creativity manifests itself on two levels – subjectifying and objectifying. In a certain sense, complementing and clarifying the organicist models, Engelmeyer sees as subjectifying activity the focus on adapting one's "self" to the conditions of the surrounding reality, and as the objectifying one – the adaptation of external conditions to person's "self". The implementation of the two aspects of activity takes place in the sphere of "praxis", represented by different arts. According to Engelmeyer, the most significant is the phenomenon of technology that appears as: 1) a set of effective techniques aimed at achieving goals; 2) "the possibility of interweaving the human will into the number of natural forces" [24, p. 35]; 3) transforming of the surrounding reality in accordance with human needs. The preference is given to technology over other arts because of its utilitarian and applied nature, transforming the primordial state of nature, and direct involvement in the creative process that unfolds in three planes (acts) in design, plan and implementation. It is virtually impossible to consider the structure of creativity on the example of fine arts. In them, it is directly associated with the presentation, and not with the introduction of the new and the necessary for human life. It is evidenced by the activities of ancient civilizations that worshiped a technical device further adjusting the ideals of beauty to it.

The reason, why the technology was presented at its best, was its original basis - utility. The devotion to technical structure indicates that it comes from human nature, "facilitates the achievement of the intended goal (regardless of the goal itself)" [24, p. 30] and increases labor productivity. The concept of utility appears as a way of matching the means/tasks to the goals that leads to the best achievement of the result, that is, to innovation. Petr Engelmeyer states that innovations are necessary. There may be several reasons: 1) increasing the number of people on the planet; 2) their needs. However, he also believes that innovations should be evaluated in respect to utility, otherwise they will be unsustainable. The utility needs the criteria allowing people to learn how to distinguish the new from the old. Under them, Engelmeyer understands the accumulated cultural experience - the routine. It makes possible the translation of the benefits presented by a new technical invention into reality, and contributes to their verification for compliance with expediency and actuality. After all, something that is premature and inappropriate to human needs will in no way find a place among people, because it will cause a negative reaction. But having passed the tests of utility, that is, the test of suitability, convenience, and comfort, the new becomes the old. The invention does not cease to be useful. It integrates into reality and becomes the part of the world without which human life is unthinkable. As a means, it opens up new ways for the realization of goals. Generalizing, it should be said that the concept of utility in the works of Engelmeyer represents the result of the interaction of the new (invention) and the old (routine), in which something that was previously instrumental (functional) in the world of ideas realises through human activity and appears in the human reality.

The German philosopher Friedrich Dessauer in his works was close to Engelmeyer's definition of the concept of utility. Adhering to the views of the Russian theorist on the essence of technology and creativity, he complements them with his reflections based on Plato's teaching about ideas. Thus, according to Dessauer, a person is born defenseless against the forces of nature, which is manifested in the absence of elementary means for survival that animals have. However, the fact that humans are still alive indicates that they have other skills hidden from superficial observation. By them, he understands a number of inherent technical abilities aimed at changing the environment, manifested in homo investigator, homo inventor, homo faber. With the help of thinking that unites these abilities, in the process of creative act (internal processing of experience), humans are able to extract the pre-set forms of technical solutions (ideas) from the intra-mental space and implement them in reality [25]. This approach corresponds to the "triact" theory of Engelmeyer and to the "three worlds" of Carl Popper till the moment, when Dessauer introduces the "fourth kingdom" of solved problems (ideas), as a result of which the phenomenon of creativity extends to the universe and affects all people, regardless of gender or cultural characteristics, giving them the access to the world heritage.

Another innovation of Dessauer's was the revision and clarification of the utility concept. Understanding as utility the adequacy of tasks (means) and goals or the situations involving the interaction of causal and final goals, Dessauer for the first time poses the question of how it could happen that humans went beyond the limits of the given nature in the sphere of the transcendent. The answer to the question is the factors of physiological and psychological discomfort that lead people to search for more favorable conditions of existence, including happiness. Providing these conditions, humans face certain problems [26]. Due to the fact that the new conditions also tend to develop and complicate, people create society for subsequent resolution procedures. The developed system of social relations faces the same problems, to solve which means to advance in existence to the extent when tension is removed and well-being is ensured. Therefore, the human way of existence is doomed to perish without the ability to conquer "unknown spaces" and needs the measure for developing only necessary ones. Therefore, it needs a special criterion of utility. Clarifying Dessauer's meaning of utility on the example of medicaments and machines, the Russian philosopher Nesterov notes that this is a certain change, "a violation of the natural order of things, including the transformation of the horizons of imagination, which, in their turn, indicate opportunities for new technical objects" [27, p. 98]. The sum of all the forces of technical objects affects the human world. As a result of this change, humans form an idea of utility, fixed in the potential of the quality of the environment, or, to put it in purely metaphysical language, in the act of being.

4 Conclusion

The analysis of the development of the concept of utility by pro-utilitarians, as well as their influence on the formation of the technical worldview and the notion of utility in the philosophy of technology, is essential for the conditions of modern reality where the static and the dynamic constantly replace each other. Humans require sufficient strength and the ability to remain in this world, to find a means of such an existence that would guarantee a successful and significant life.

Otherwise, a person risks becoming a living mechanism whose main purpose is to act as a service personnel, to perform a number of functions. In this regard, the study, devoted to the evolution of the concept of "utility" and its influence on the technical mindset, provided a number of conclusions on the meaning and purpose of this term in the works of economists, philosophers and technicians, interesting for revision of current global projects.

The economists consider utility as benefit, profit, revenue, that is, something that is directly linked to the exchange of some goods for others. In such a situation, a particular person acts as an intermediary engaged in the development of sales technology and sales of products. From the point of view of the economic process, a person with all the unique characteristics and abilities turns into an abstract subject, whose main purpose is determined not by self-development, but by the maintenance of the capitalist system. The "invisible hand of the market", initially determined in good intentions by Adam Smith, later turns into a meaningless cycle of pricing and consumption, where the processes of regression dominates over progress and development of social relations.

On the contrary, the proponents of pro-utilitarian tendencies put forward social well-being as a goal and understand utility as a social phenomenon aimed at life self-maintenance as a quasi-stationary state close to a stable equilibrium, provided through a constant increase in pleasures. New for classical utilitarians is the division of pleasure into several types, the best of which are related to the concept of "happiness", understood as the proportionality of actions to human desires. On this basis, the benefit acts as a guarantor of maintaining order, not only social order, but also ensuring the self-development of a particular person, through various attempts to approach peace with oneself, society and the surrounding reality.

The representatives of the philosophy of technology put a different meaning in the concept of utility. For Engelmeyer and Dessauer the approaches of economists and utilitarians are unacceptable, because they identify people with means, and relegate to the background their individual life and abilities [28]. For this reason, for them the concept of utility, unlike all other theorists that understand it in the social plan through the clash of individual and public interests, appears functionally as: 1) the method of proportionality of tasks to goals; 2) violation (change) of the natural order, opening the horizons for the introduction of new technical objects and environments. This approach to the concept of utility in philosophy is new. It divides the social and the functional aspects of utility, and opens up freedom for human actions. The technical worldview, being practical in its structure, alienates a person from speculative constructions, but approaches the urgent metaphysical style, where the question of the nature of utility is closely connected with the essence of a person, defined as the final mode of existence. Thanks to the engineering type of thinking, humans are able to overcome limitations, but only society can give them "a cue" how far they will advance in their individual attempts.

Acknowledgements. The reported study was funded by RFBR, project number 20-011-00462 A.

References

1. Aristotle. Nicomachean Ethics. Batoche Books, Kitchener (1999). Translated by W. D. Ross
2. Losev, A.F.: Essays on Ancient Symbolism and Mythology. Mysl' Publ, Moscow (1993).(In Russian)
3. Potapova, K.I.: Aristotle's Metaphysic of Action. Izvestia of the Ural Federal University. Soc. Sci. 158, **11**(4), 224–233 (2016). (In Russian)
4. Pico della Mirandola, G.: Oration on the Dignity of Man. Gateway Editions, Chicago (1956)
5. Volostnov, N.S., Troshin, A.S., Lazutina, A.L., Khokhlov, A.A., Lebedeva, T.E.: A. Smith and modern economic science: bars to cognitive portrait. Moscow Econ. J. **1**, 36 (2019). (In Russian)

6. Hutcheson, F.: An Inquiry into the Original of Our Ideas of Beauty and Virtue. Liberty Fund, Indianapolis (2004). Ed. with an Introduction W. Leidhold
7. Mandeville, B.: The Fable Of The Bees: Or Private Vices Publick Benefits. Capricorn Books, New York (1962)
8. Gadzhikurbanova, P.A.: The concept of summum bonum in classical utilitarianism. Ethical Thought **10**, 114–130 (2010). (In Russian)
9. Artemyeva, O.V.: Formation of the concept of universality in ethics. Phil. J. **12**(3), 95–109 (2019). (in Russian). https://doi.org/10.21146/2072-0726-2019-12-3-95-109
10. Sushentsova, M.S.: Utilitarianism of J. Bentham and J. S. Mill: from virtue to rationality. St Petersburg Univ. J. Econ. Stud. **33**(1), 17–35 (2017). (In Russian). http://hdl.handle.net/11701/6340
11. Yarkova, E.N.: Utilitarianism in Russia. J. of Soc. Soc. Anthropol. **22**(4), 7–36 (2019). (In Russian). https://doi.org/10.31119/jssa.2019.22.4.1
12. Ushkin, S.G, Koval, E.A, Zhadunova, N.V.: From theoretical design to practical viewpoints: implementation of ethical principles in youth life strategies. Monit. Public Opin.: Econ. Soc. Changeso **3**, 66–93 (2020). (In Russian). https://doi.org/10.14515/monitoring.2020.3.1596
13. Ogleznev, V.V.: Bentham on of fictional essences and categories of Aristotle. Scholae. Phil. Antiq. Class. Trad. **1**(13), 339–348 (2019). (In Russian) https://doi.org/10.25205/1995-4328-2019-13-1-339-348
14. Sioma, N.A.: The problem of fictitious entities in Jeremy Bentham's legislation. Vestn. Volgogr. State Univ. **1**(19), 117–123 (2013). (In Russian)
15. Bentham, J.: An Introduction to the Principles of Morals and Legislation. Jonathan Bennett, Rossford (2017)
16. Prokofiev, A.V.: Utilitarianism. Phil. Anthropol. **5**(2), 192–215 (2019). (In Russian). https://doi.org/10.21146/2414-3715-2019-5-2-192-215
17. Dewey, J., Tufts, J.H.: Ethics. Henry Holt and Company, New York (1906)
18. Mill, J.S.: Utilitarianism. Batoche Books, Kitchener (2001)
19. Kryuchkova, Y.: Functional utility of action in the theoretical perspective H. Spenser's sociology. RUDN J. Soc. **2**, 70–81 (2008). (In Russian)
20. Podvoyskiy D.G.: Arranging an "iron cage", or how institutions become human enemies (revising the classical theories). Monit. Public Opin.: Eco. Soc. Chang. **5**, 29–70 (2020). (In Russian). https://doi.org/10.14515/monitoring-.2018.4.10
21. Spencer, H.: The Principies of Ethies, vol. I. Appleton and Company, New York (1895)
22. Dewey, J.: Experience and Nature. Dover Publications, Chicago (1925)
23. Mikhailovsky, A.V.: Four key questions in philosophy of technology: on "homo creator" by Hans Poser. Epistemology Phil. Sci. **56**(3), 225–233 (2019). (In Russian). https://doi.org/10.5840/eps201956361
24. Jengelmejer, P.K.: Theory of Creativity. Librokom, Moscow (2010).(In Russian)
25. Dessauer, F.: Streit um die Technik. Verlag Herder, Freiburg in Breisgau (1959). (In Ger.)
26. Nesterov, A.Yu., Demina A.I.: The categories "Meaning" and "Invention" in the context of the semiotics of technologie. Bull. Moscow State Reg. Univ.: Philos. Sci. **49**, 42–50 (2019). https://doi.org/10.18384/2310-7227-2020-3-77-84
27. Nesterov, A.Yu.: Semiotic Foundations of Technology and Technical Consciousness. Samara Academy of Humanities, Samara (2017). (In Russian)
28. Nesterov, A.Y.: Clarification of the concept of progress through the semiotics of technology. In: Bylieva, D., Nordmann, A., Shipunova, O., Volkova, V. (eds.) PCSF/CSIS - 2020. LNNS, vol. 184, pp. 3–11. Springer, Cham (2021). https://doi.org/10.1007/978-3-030-65857-1_1

Digital Technologies of the Self: Instrumental Rationality or Creative Integrity?

Andrey A. Ivanov[1,2](✉) ⓘ, Anton A. Ivanov[3] ⓘ, and Yana S. Ivashchenko[5,4](✉) ⓘ

[1] Siberian University of Consumer Cooperation, Novosibirsk 630087, Russia
[2] Siberian State University of Water Transport, Novosibirsk 630099, Russia
[3] Komsomolsk-na-Amure State University Associate, Komsomolsk-on-Amur 681013, Russia
[4] Novosibirsk State Technical University, Novosibirsk 630073, Russia
[5] Novosibirsk State Pedagogical University, Novosibirsk 630126, Russia

Abstract. The article examines the culture of digital self-tracking as a phenomenon reflecting substantial features of modern culture. Research focuses on the conceptual contradictions arising in self-tracking practices and their scientific and theoretical interpretation. Following the dialectical method, these contradictions are considered not as impasses of theoretical thought, but as reflections of the dialectical essence of culture. The first dilemma of commensurability contains the contradiction between the objectivity and efficiency of the measurement and the irreducibility of human being too abstract results of tracking. The second dilemma emerges in understanding whether self-tracking is an instrument of heteronomous subordination or creative emancipation of the individual. The third dilemma is whether self-tracking is a symptom of the economic colonization of the lifeworld or whether it can embrace a different functional and semantic logic. The structural relatedness of these antinomies proves that the self-tracking phenomenon is based on the anthropological model of the subject of Modernity and reflects its ambivalent properties in late capitalist digital culture.

Keywords: Self-tracking · Technologies of the self · Quantification · Flexible normalism · Reification · Biopolitics · Temporality · Subject

1 Introduction

The development of mobile digital technologies, biometric tracking, and analysis devices has created a specific culture of "self-tracking" (synonymous concepts: life-logging, personal analytics, personal informatics). Manufacturers offer a variety of devices and apps for counting the steps taken, calories consumption, heart rate, blood pressure, stress levels, sleep patterns, moods, etc. Self-trackers, in turn, add the measurement of social interactions, beneficial and unhealthy habits, work productivity, meditation practices, etc. The self-tracking culture includes a system of special values, norms, discourses, and practices shared by its participants. In addition to individual use of self-tracking devices, trackers share achievements and experiences, unite in communities (the most famous is "Quantified Self"), numerical values can transform into the collective norms, and those, in turn, into the institutional requirements coming from employers and insurance companies.

Self-tracking is the communication of a person's consciousness with the body and everyday activity through a sensor and an app converting data into digital form. These intermediaries are the sub-type of cultural tools designed to cognize and manipulate the world – trade tools, language, sports and medicine, "self-technique". But, if before the invention of QS-technologies, self-tracking and self-monitoring have been predominantly based on the subjective qualitative experience (and, of course, the interpretative help of an expert as a doctor, trainer, or psychologist), now the qualia disadvantages are proposed to be countervailed with collecting and analyzing digital data of daily activity obtained through impartial wearable sensors.

In other terms, the main point of self-tracking ideology can be formulated in the following way: the complex reality of the human body and behavior can be successfully quantified by using the digital device implemented in everyday life which facilitates more objective understanding and optimization of reality. In self-tracking research, there is an array of conceptual contradictions about this idea.

First, the problem of commensuration: on the one hand, measurement rationalizes reality and makes it more controlled, on the other hand, there are obvious limits in quantifying the qualitative complexity of human life. The second dilemma is the heteronomous or autonomous potential of the "power of numbers": whether self-tracking numerical values could become social requirements and grounds for "rational discrimination" or they help to recognize the individual norm and distinguish it from the collective one. The third question is whether self-tracking manifests the economization (commodification) of modern culture or it includes other functional logic not reducible to economic.

In our opinion, these dilemmas explicit not the theoretical reflection impasse but the fundamental contradictions that characterize the dialectic of entire modern culture.

2 Methodology

The methodological basis of this work is the dialectical approach which assumes the consideration of social reality as the unity and struggle of opposites underlying the social practices and value-semantic systems.

In self-tracking research discourse, we found the number of conflicting positions-dilemmas that constantly emerge in the interpretation of the new social practice of digital culture. Firstly, the relationship between these contradictions was exposed: the choice in favor of a position in one dilemma is made by reference to the solution of another. Secondly, these dilemmas are considered not as contradictory antinomies that demonstrate the impuissance of theoretical thought, but as the expression of binary categories that underlie social practices and value-semantic systems of modern culture. Thirdly, the basic antinomy that, in our view, produces and synthesizes all the other contradictions in self-tracking practices were formulated.

3 Results

3.1 Magic of Numbers and Quantification Boundaries

The Quantified Self movement's slogan is "Self Knowledge Through Numbers". Quantified data are perceived as more objective and neutral, therefore providing reliable access to reality and its control. Measurements performed with the help of precision devices, statistical regularities determined by calculating are designed to eliminate the errors of unreliable intuition and subjective self-evaluation [1].

Trust of numbers and measurements is a feature of Modernity culture based on the intersubjective skill of "active flexible and timely formalization of any kind of problems by their converting in measurable and solvable shape" [2]. In Veber's terms, trust in measurement is both the mover and the consequence of a scientific dispelling of the world and its social rationalization. Rationalization is based not on knowledge but the belief in its unbounded growth and that everything in the world is conceivable [3].

Nevertheless, rationalization has not transformed the social life into a more understandable and predictable form, diverse autonomous fields of rationality enhancing societal structural differentiation being developed. Structural conditions are becoming increasingly complex and incomprehensible to individuals, acting as unpredictable natural forces that can't be contraposed by lost traditional roles and identities [4]. Hence one of the responses to the contingency of "liquid modernity" is the privatization of digital control, the transformation (reification) of the self into a capital suitable for formalization, calculation, and planning. If the complexity of the social world, according to rationalist myth, is managed and foreseen by quantitative formalization, the same formalization should help in managing and adapting one's complexity.

"Personal science" is one of the designations for this activity given by the website Quantified Self. Self-tracking provides knowledge about areas, processes, and patterns beyond conscious experience and control. However, this is not a formulation of Freudian depths of secret appetence and repressed complexes, but the self of trivial behavioral habits, physiological and mental automatisms [1]. This awareness of automatisms enables the selftracker to manipulate and correct them. S. Meißner argues this efficiency to be gained by dissociation mechanism positively understood as the opportunity to detach oneself, to determine the boundaries of one's mobility [5]. Armed with quantitative, objective knowledge of bodily and behavioral processes, the self-tracker achieves the true autonomy of free rational action, appropriates 'one's own body [6].

Nonetheless, the belief in the magic of numbers and the efficiency of self-tracking has to face various challenges. Tracking technologies today have not managed to obtain the degree of accuracy to take into account all possible individual peculiarities and context conditions affecting measurements. The movement of the fitness bracelet equipped hand doesn't always track the actual walking or running; heart rate variability is not always a serious stress symptom for self-tracker to pay attention to. Many indicators nowadays such as food consumed, emotions, attention, etc. being not automatically trackable and requiring conscious working on, ergo, correlate with the qualitative experience. Converting psychophysical states and processes into a measurable form makes it easier to formulate goals and find optimization tools but readily detects the losses. Physical indicators do not take into account the mental processes, quantitative norms replace complex and ambiguous qualitative states of living body and personality [7, 8]. So-called "data doubles" which are the digital portraits of self-trackers conflict with everyday experiences and practices, creating unproductive alienation [6].

In many cases, the experience of self-tracking generates a critical attitude to quantifying technologies and recognizing their limitations in comparison with internal experience, a lot of data turns out to be useless and meaningless [9]. The analogy with the "carnival" videos of excessive cooking and eating created by well-known fitness trainers can be drawn [10]. The demonstration of the transgressive corporeality in the mode of waste and excess is the embodiment of longing for the carnal pleasures, for nature not conquered by the calculating mind. Both the carnivalesque corporeality and the doubt about tracking devices adequacy manifest the ineffability of the subject integral nature with the help of quantitative and technological forms.

3.2 Between Discrimination and Emancipation

Self-tracking at first glimpse seems to be a purely individual activity, in which the subject acts as an autonomous researcher and creator of himself. N. Schüll believes the trackers are similar to entrepreneurs constantly making decisions based on available information [11]: wearable devices only "steer" and help to make the right decisions, giving the appropriate information. Although, as J. Polls et al. point out, this idea of the ideal user and use implies a strictly rational subject who consciously chooses goals and makes decisions outside the fields of power and social compulsion [6].

M. Foucault's analysis of biopolitics and institutions of the technologies of the Self has shown the mode of political domination to be extended to biology and internalized in the neoliberal subject of "The care of the self" [12]. Compliance with, as well as adherence to, the biological norm becomes a common political task, which implementation to be delegated to autonomous individuals. The somatic dimension is the significant component of modern subjectivity, linked by ethical, aesthetic, and socio-normative relationships and constructed by micropolitical mechanisms.

As distinguished from the Industrial age disciplinary practices, where the qualitative norm delineated the narrow static zone of the ideal model, the Postmodern discipline is characterized by a "flexible normalism" based on the dynamic statistical norm that offers the mobile quantitative highest rate [13]. This rate is intrinsically unattainable for subjectivity to complete and calm down afterwards: it requires incessant moving, overcoming and improvement, perfectly reflected by the "fitness" concept as the fundamentally incomplete process [14].

The compliance of the individual to follow the flexible norm is determined by some factors. The biological norm, by being statistically defined, obtains the shape of dispassionate objectivity with no explicit moral, religious, or ideological foundation. Particular imperative, as if it emanates from life itself, comprehended through rational means. The soft, flexible norm allows for a mobile range of statistical error, for it is not imposed as a directive, but calls for its adoption in the mode of the game, independent research and improvement. At the same time, the social context [15] that contains the global acceptance of quantification, ranking, and datafication has the "steering" effect: currency rates, epidemic charts, corporate and athlete achievements, likes, followers, social media reposts, YouTube views, business KPIs, box office receipts, etc.

Flexible Normalism, converting statistical descriptions into social expectations and implemented in self-tracking practices, according to Selke, forms a resource for "rational discrimination" [16]. Psychological problems associated with obsessive tracking and dissatisfaction with self seems to matter only at the level of individual cases, which can always be counterposed by the reciprocal cases of balanced self-cognition or simple rejection of applications. But self-tracking, aside from being part of the commercial offer of the IT industry, is gradually being incorporated into the system of institutional requirements of corporations and government agencies. Health professionals and institutions recommend self-tracking; employers integrate the self-tracking into employee fitness programs and impose penalties for sedentary lifestyles; insurance companies apply "health indexes" to assess the insurance risks. Thus, the flexible norms of digitized body and behavior become the basis for distinguishing people, the criterion for their socialization and success, and, consequently, the unequal distribution of rights.

Such criticism of the discriminatory potential of self-tracking technologies and their social application tries to ignore the enclosed emancipation opportunities, as it is pointed out by S. Meißner: "Thus, those data do not per se seem to enslave the human user but, quite the contrary, provide the opportunity for some leeway" [5]. The use of tracking technologies presumes different attitudes to both digital indicators and application recommendations, the choice between which is made by the individual [17, 18]. For example, J. Pols et al. identified two self-tracking styles among QS followers – the "objectivist-changer", who strives to gain objective knowledge about him/herself for its optimization, and the "aesthetic-semiotic" style interested in diverse forms of self-awareness and experiments [6]. The second style does not involve the rigid formalization by creating a limited set of options, but the process opened to non-formalized solutions and the search for a fundamentally new experience. Self-tracking data can be an argument against the expert's opinion, as well as defend against the social coercion of general statistical norms – knowledge of individual norm can be opposed to social rationing [5].

The dilemma of discrimination-emancipation seems to be solved by standard appeal to the moral neutrality of technologies: they can be used for different purposes depending on individuals. But both the individual numerical norm and the aesthetic-semiotic style of quantification interpreted as manifestations of autonomy are based on the willingness of the subject to construct him/herself according to the technological model proposed by the impersonal actor – smart gadget with corporation standing behind. Otherwise stated, the root of this contradiction lies in the processes by which modern individual is ready to open the boundaries of his private personal life to the power of impersonal institutions and to present him/herself as a measurable object.

3.3   Commodification of the Lifeworld

There is an influential and convincing position of explaining the practices of self-tracking as a manifestation of totalized economic criteria that permeate all spheres of the lifeworld in the age of late capitalism [16, 19–21].

This criticism goes back to the Marxist concepts of commodity fetishism and reification. Commodity fetishism is the representation of commodity cost as its ontological basis preceding the labor and exchange, which is transformed into the "everything-has-its-price"-attitude – everything is amenable to economic measurement and market exchange. D. Lukács introduced the concept of "reification" as a subjective dimension of commodity fetishism: as well as abstract exchange value ignores the certain qualities of commodity, the worker in capitalist production is perceived as a functional object deprived of individual traits and valuable only from the point of view of calculated labor force [22]. Neomarxists T. Adorno, Y. Habermas, A. Honnet considered reification as a dialectic process of separation of employee and consumer, work and leisure, private and public, and totalization as blurring the boundaries between the measurable and the immeasurable, the economic and the uneconomic.

P. Schulz deems the temporal calculus of labor and wages to be the crucial factor of totalized reification: the perception of time as calculable capital has been extended to leisure time, which turned to a labor continuation by other ways and required efficient and rational use, calling on the individual to become "one's capitalist during leisure time" [21]. In conditions of post-Ford unstable employment, the neoliberal model of the subject taking control and regulation of his health and spare time, evaluating them in terms of calculability, productivity and optimization ("personal taylorism") dominates [23, 24]. Thus, self-tracking practices are the response to blurring boundaries between work and leisure through self-reification that appropriates quantitative management: "attempt to reappropriate the reificated self" [21].

S. Meißner criticizes the hypothesis of the dominance of economic logic in self-tracking practices and social space, appealing to N. Luhmann's differentiation theory, according to which the decisive feature of modern society is "coexistence of different functional logics which cannot be reduced to each other or deduced from each other" [5]. The spread of self-tracking practices can be interpreted not only from the perspective of "economizing the social" but also from the perspective of expansion of scientific discourse or aestheticizing the social [25], etc. The neoliberal model of self-optimization as a result of the alienating internalization of economic logic is not the only and non-alternative motivation for lifelogging, the latter can be stimulated by different incentives and opens up different opportunities for self-knowledge and self-improvement. Otherwise stated, self-tracking allows for different rendition – enslaving social pressure or liberating knowledge – and the difference itself means an open space of freedom.

Results of the survey of self-tracking instruments using by amateur and professional athletes, presented in the publication of A. Rapp and L. Tirabeni, seem to corroborate the correctness of S. Meißner's position. Unlike amateurs who look for more accurate and new information about themselves in self-tracking devices, elite athletes value these devices mostly for convenience and are aware of when it is better to trust subjective feelings [26]. But, in our opinion, the meaning of this difference is that experienced athlete can see the qualitative limits of external measuring device efficiency and sets the borderline of numerical homogeneity.

H. Lefebvre described the mechanism by which economic quantification colonizes the everyday world. The working time measurement, which became the basis of value and wages in capitalism, implements the "homogeneous and desacralized" temporality, divisible into the equal neutral units. Working time assigns the leisure time, abstract and rationalized work rhythms determine the biological and social rhythms of sleep and wakefulness, eating, playing with children and entertainment, etc. [27].

Formulated differently, the metric culture successfully colonizes private life and leisure time because its perspective sees no qualitative differences, but conventional graduations, indifferent to conditional and content-related diversity of measured processes. The citizen of late capitalism often sees no reason to stop the colonization impulse of quantification and embeds its principles into personal spheres due to its universal abstractiveness [28]. The abstractiveness, as it turned out, can only be established by the dominance of market exchange and labor time measuring rather than abstractions of scientific thinking or aesthetic images.

Homogeneity and universality of metric culture enable it to be integrated into the operation of a wide variety of logics or, what is the same, to include any cultural content in its "omnivorous" abstract form. Or individuals may veil their action with biased logics of self-discovery, "fun and exciting" play, competition, or aesthetic self-expression. A variety of categories pervasive in modern biographical narratives and discourses of self-representation prove to be the same biased forms concealing their economic origin: interest, success, profit, balance, rating, eventful life, etc.

4 Discussion

The considered dilemmas that have arisen in the understanding of self-tracking, from our point of view, harken back to one of the most important categories underlying the New European culture – the subject as the leading anthropological model of modernity. The metaphysical subject external to the empirical world which it cognizes and explores is parallel to the autonomous personality as a source of initiatives in social space. The subject is autonomous in the sense that its behavior proceeds from worldview ideas about the world and duty and its responsibility for its actions. The subject is inexhaustible in the sense that it is always greater than its empirical conditions, including the conditions of its body and social roles. The subject creates and uses tools (language, science, technology) enabling it to represent and operate the world of passive objects.

The critique of the metaphysical subject questions the above-mentioned properties: cognition depends on a priori forms and semiotic means of representation that are beyond the control of the subject (correlationism); it is limited by psychophysical and sociocultural conditions that act unconsciously (Freudianism and poststructuralism); the free and independent subject is the ideological construct that supports economic exploitation and political domination (Marxism and neo-Marxism); human subjectivity is not privileged and it is interwoven in networks of interactions with other actors – animals, things, technologies, institutions (actor-network theory).

In the practices of self-tracking, we find the same antinomies of understanding a person in the culture of our time. Self-tracking provides more accurate, dispassionate, formal data that allows you to build a more objective picture, but most of this information is not amenable to interpretation and is useless, and the complex integrity of a person is torn and escapes. Selftracker uses the tools of quantification of life processes to organize their lives, to strengthen control over themselves, i.e. to be autonomous, but thereby simply internalizes the "flexible norms" of neoliberal biopower. Selftracker shows his creative freedom in exploring his private life, uses different approaches in understanding and applying data, mixes the economic logic of optimization with other logics, but all this is just a curtain and decoration that hides the irrational dominance of homogeneous economic exchange.

5   Conclusion

Thus, the contradictions in the scientific interpretation of self-tracking are seen to derive from the antinomic essence of the subject in the New European culture. The Discovery of the extra-personal determinants of subject behavior in physiological processes, social structures, and technologies collides with the manifestation of its creative freedom and autonomy. The self-tracking discourse asserts the potential of the subject's relief of unconscious embodiment by translating it into measurable, impersonal indicators. But the translation mechanism projects the logic of economic relations onto the subject-body relationship, ergo, the autonomous subject turns out to be the instrument of external domination. New forms of information and digital culture transform the position of the subject, put it in a horizontal row with other actors, question its privileged position, but at the same time, the culture (with self-tracking as a vivid example) persistently reproduces the myth of the "otherness" of the subject and his creative exclusivity.

References

1. Wolf, G.: The data-driven life. In: The New York Times, 28 April 2010, https://www.nytimes.com/2010/05/02/magazine/02self-measurement-t.html. Accessed 22 Feb 2021
2. Petrov, M.K.: Language, Sign, Culture, 2nd edn. Editorial, Moscow (2004).(in Russian)
3. Weber, M.: Science as a profession and vocation. In: Whimster, S., Bruun, H.H. (eds.) Collected Methodological Writings, pp. 335–353. Routledge, London/New York (2012)
4. Lupton, D.: Self-tracking cultures: towards a sociology of personal informatics. In: Proceedings of the 26th Australian Computer-Human Interaction Conference on Designing Futures: The Future of Design, pp. 77–86. Association for Computing Machinery, New York (2014). https://doi.org/10.1145/2686612.2686623
5. Meißner, S.: Effects of quantified self beyond self-optimization. In: Selke, S. (ed.) Lifelogging, pp. 235–248. Springer VS, Wiesbaden (2016). https://doi.org/10.1007/978-3-658-13137-1_13
6. Pols, J., Willems, D., Aanestad, M.: Making sense with numbers. unravelling ethico-psychological subjects in practices of self-quantification. Sociol. Health Illn. **41**(1), 98–115 (2019). https://doi.org/10.1111/1467-9566.12894
7. Dietrich, M., van Laerhoven K.: Reflect yourself! Opportunities and limits of wearable activity recognition for self-tracking. In: Selke, S. (ed.) Lifelogging, pp. 213–234. Springer VS, Wiesbaden (2016). https://doi.org/10.1007/978-3-658-13137-1_12
8. Wiedemann, L.: Self-monitoring. embodying data and obliviating the lived body!? In: Selke, S. (ed.) Lifelogging, pp. 207–212. Springer VS, Wiesbaden (2016). https://doi.org/10.1007/978-3-658-13137-1_11
9. The Unquantified Self. Figuring out What Really Counts. https://unquantifiedself.wordpress.com. Accessed 22 Apr 2021
10. Lupton, D., Feldman, Z. (eds.): Digital Food Cultures, 1st edn. Routledge, London (2020). https://doi.org/10.4324/9780429402135
11. Schüll, N.: Data for life: wearable technology and the design of self-care. BioSocieties **11**, 317–333 (2016). https://doi.org/10.1057/biosoc.2015.47

12. Foucault, M.: Technologies of the self. In: Martin, L., Gutman, H., Hutton, P. (eds.) Technologies of the Self: A Seminar with Michel Foucault, pp. 16–49. Tavistock, London (1988)
13. Link, J.: On the Contribution of Normalism to Modernity and Postmodernity. Cult. Crit. **57**(1), 33–46 (2004)
14. Gertenbach, L., Mönkeberg, S.: Lifelogging and vital normalism. In: Selke, S. (ed.) Lifelogging, pp. 25–42. Springer VS, Wiesbaden (2016). https://doi.org/10.1007/978-3-658-131 37-1_2
15. Berry, R.A., Rodgers, R.F., Campagna, J.: Outperforming iBodies: A Conceptual Framework Integrating Body Performance Self-Tracking Technologies with Body Image and Eating Concerns. Sex Roles **85**(1–2), 1–12 (2020). https://doi.org/10.1007/s11199-020-01201-6
16. Selke, S.: Rational discrimination and lifelogging: the expansion of the combat zone and the new taxonomy of the social. In: Selke, S. (ed.) Lifelogging, pp. 345–372. Springer VS, Wiesbaden (2016). https://doi.org/10.1007/978-3-658-13137-1_19
17. Lyall, B., Robards, B.: Tool, toy and tutor: subjective experiences of digital self-tracking. J. Sociol. **54**(1), 108–124 (2017). https://doi.org/10.1177/1440783317722854
18. Nim, E.G.: Student discourse on digital self-tracking: rhetorics and practices. Monit. Publ. Opin.: Econ. Soc. Ch. **2**, 191–211 (2020). https://doi.org/10.14515/monitoring.2020.2.989
19. Ajana, B.: Digital health and the biopolitics of the quantified self. Digit. Health **3**(1) (2017). https://doi.org/10.1177/2055207616689509
20. Ajana, B.: Personal metrics: users' experiences and perceptions of self-tracking practices and data. Soc. Sci. Inform. **59**(4), 654–678 (2020). https://doi.org/10.1177/0539018420959522
21. Schulz, P.: Lifelogging: a project of liberation or a source of reification. In: Selke, S. (ed.) Lifelogging, pp. 43–60. Springer VS, Wiesbaden (2016). https://doi.org/10.1007/978-3-658-13137-1_3
22. Lukács, G.: Reification and the consciousness of the proletariat. In: Lukács, G. (ed.) History and Class Consciousness, pp. 83–222. MIT Press, Cambridge (1971)
23. Berg, M.: Making sense with sensors: self-tracking and the temporalities of wellbeing. Digit. Health **3**(1) (2017). https://doi.org/10.1177/2055207617699767
24. Lupton, D.: You are your data: self-tracking practices and concepts of data. In: Selke, S. (ed.) Lifelogging, pp. 61–80. Springer VS, Wiesbaden (2016). https://doi.org/10.1007/978-3-658-13137-1_4
25. Kent, R.: Self-tracking health over time: from the use of instagram to perform optimal health to the protective shield of the digital detox. Soc. Med.+Soc. **6**(3) (2020). https://doi.org/10.1177/2056305120940694
26. Rapp, A., Tirabeni, L.: Self-tracking while doing sport: comfort, motivation, attention and lifestyle of athletes using personal informatics tools. Int. J. Hum. Comp. Stud. **140**, 102434 (2020). https://doi.org/10.1016/j.ijhcs.2020.102434
27. Lefebvre, H.: Rhythmanalysis: Space. Time and Everyday Life. Continuum, London (2004)
28. Pitts, F.H., Jean, E., Clarke, Y.: Sonifying the quantified self: rhythmanalysis and performance research in and against the reduction of life-time to labour-time. Cap. Cl. **44**(2), 219–239 (2020). https://doi.org/10.1177/0309816819873370

National Judicial Bodies in Search of a Balance of Public and Private Interests

Viacheslav V. Ivanov[1] ⓘ, Daria M. Matsepuro[2(✉)] ⓘ, and Tatiana V. Trubnikova[2] ⓘ

[1] Samara National Research University, 34, Moskovskoye Shosse, 443086 Samara, Russia
[2] National Research Tomsk State University, 36, Lenin Avenue, 634050 Tomsk, Russia

Abstract. The snowballing development of genetic and machine learning technologies and the use of the results of such developments in different areas of social life poses serious problems as to setting limits for their application. It is necessary to find an acceptable balance between various interests: scientific progress, the right of society to security and an improvement in the quality of life, commercial interests, the right to information, freedom and privacy. The fundamental nature of the ethical and legal issues arising in the transition to a new technological paradigm, the lack of ready-made solutions, the "slow" development of legislation, which does not keep pace with the current state and prospects for the progress and use of technologies, lead to a situation when judiciary bodies are entrusted to search for a balance between public and private interests in this area. This paper provides a comparative analysis of two similar cases which have passed through the national judicial bodies of the United Kingdom and Russia. These claims are for rights protection related to the use of identity recognition technologies. Based on this, it is considered how the judicial activity of national courts and its quality affects the change in legal reality, regulation and law enforcement in the field of the use of new technologies.

Keywords: Facial recognition technology · DNA database · Privacy · Balance of public and private interests · Judicial legislation

1 Introduction

We are living in the era of the exponential progression in information, communication, genomic and machine learning technologies. It creates new opportunities and wide perspectives, enable to solve many issues society is facing with in alternative means. For instance, the authors of this paper being university lecturers could keep up with their work and communicate with students during lockdown conditions for the whole year, as well as most of the people only owing to modern technologies. Progress in genomic technologies fosters impressive gains in personalised medicine. For instance, recently American researchers presented results from an investigator-initiated phase I clinical trial personalized cancer vaccine created based on the sequence of each patient's tumor and germline DNA [1]. Now artificial intelligence (AI) helps to study and develop new medicines, analyse patient records, insight into medical scans. AI-based Facial Recognition Technology (FRT) is used for socially critical purposes. For example, FRT is used in the Traffic Jam tool, which allows human rights agencies to identify victims of the slave trade and determine their location. Image analysis during automatic search among millions of records is completed within a few seconds [2]. In 2018 Indian police in New Delhi identified about 3,000 missing children within just four days using FRT [3].

Also genomic, communication and recognition technologies are widely used for commercial purposes. Its application can be easy and effective and but at the same time, less beneficial and even frightening for a particular person. For example, in a number of stores, shopping centres and cafes, FRT is used to detect people previously included in the "black list" (for example, for an attempt to steal something), to analyse customer behavior (for example, his purchases, habits, route through the shopping centre and etc.), his emotions and satisfaction with the visit, in order to increase sales and loyalty to the network [4]. There are serious concerns that businesses may use facial recognition to discriminate unfairly or unlawfully. For example, a retailer may create its own dataset of "known" criminals, and there is no practice for individuals of these lists to be notified or may appeal against this action. The network can even share such lists with counterparties. And this can lead to the situation when people will be generally denied service without due process [5]. The commercial use of genomic technologies potentially contains risks that personal genetic data will end up in the hands of persons who can use them for discriminatory purposes (credit and insurance companies, employers) or against the interests and/or wishes of a particular person (in medicine, genetic editing, family planning) [6].

The results of sociological study show that people are wary about the possibilities of using personality recognition for commercial purposes. So, in 2015, RichRelevance conducted a survey of 1000 buyers from the UK. Respondents were asked to rate a range of in-store shopping technologies as "cool" or "creepy," and facial recognition technologies were categorized as "creepy" at the end of the scale by most respondents. According to the respondents, it is "creepy" to find that a salesperson knows about your usual expenses (77%), to receive messages aimed at personal targeting caused by facial recognition (68%), as well as when unfamiliar staff calls your name (73%) [7]. Similar results were obtained in follow-up studies with shoppers in the US, UK, France and Germany. The use of facial recognition in supermarkets was rated as "creepy" technology:

in 2015 - 75% of respondents;
in 2016 - 67%;
in 2017 - 69% of respondents [8].

At the same time, almost 1/3 of UK consumers are willing to share additional personal information to improve the shopping experience [9].

Fears of discrimination by insurance companies or employers, social stigmatization as a result of the use of DNA data by individuals and organizations were repeatedly expressed by participants of another study [10].

But there is another area where the use of modern technologies can lead to both very useful and extremely negative results for society. We are talking about the use of technologies for the collection and processing of communication data biometrics, FRT,

DNA profiling by the state and its bodies, officials in order to ensure public safety and to investigate crimes.

On the one hand, biological samples containing the DNA of the criminal, the identification of this DNA with the DNA of the suspect, can be a key element in the investigation. For example, in 2010, DNA profiling helped to identify and convict an Afghan citizen for a series of violent sexual acts in California between 2002 and 2004 and rape in Austria in 2009. DNA samples were helpful in solving the 2013 Boston Strangler case of the rapes and murders of 11 women committed between 1962 and 1964 [11]. According to Interpol data for 2016–2019, of the 130 countries surveyed, 89 reported using DNA profiles in police investigations, and 70 of them reported having a searchable DNA database [12]. Video surveillance with identity recognition also can be used to disclose and prevent crimes and to search for persons who have committed illegal acts. At least 75 countries are in actively use AI-technologies for surveillance: smart city/safe city platforms (56 countries), facial recognition systems (64 countries), and intelligent police (52 countries) [13]. Also biometric technologies can be used to improve management efficiency. For example, the world's largest biometric program in India uses facial recognition to create a national identity system called Aadhaar [14].

Meanwhile the wide use of the above mentioned technologies by the state can lead to a violation of the presumption of innocence, discriminatory practices, disproportionate restriction of the right to privacy, violation of the right to freedom of speech and assembly. It can equip authoritarian governments with a tool for total surveillance and fight against political dissent.

For example, it is alleged that the use of the Indian Aadhaar ID system turns non-registered persons into second-class citizens, and that the system was used to illegally exclude citizens from the electoral roll before elections [14].

Potential disadvantages of maintaining and using DNA databases designed to test registered individuals for implication in crimes are widely discussed in the legal literature. The results of the application of vast DNA database (up to 10% of the population) in the UK showed that the increase information had a slight effect on the crime detection rate. GeneWatch UK estimates that only about 0.03% of crimes were solved due to matching of stored DNA profiles [15]. The total percentage of registered crimes related to DNA detection was about 0.37%. At the same time, the method of building the DNA profiles database (which included the data of innocent persons) was found by the European Court of Human Rights (ECHR) to violate the right provided for in Article 8 of the European Convention [16]. Meanwhile innocents whose DNA profiles are included in databases can be vulnerable to stigma or bias [2]. In addition, the presence in such database itself "set at defiance human rights, such as the presumption of innocence, since the very structure of standard procedures for using databases implies that criminals committed crimes earlier are considered to be probable suspects in future crimes and must "prove" their innocence" [17]. The use of DNA databases for criminal investigations raises criteria issues related to the selection of DNA profiles to be included and the collection, storage, use and dissemination of data. [18, 19]. The borderline of the database and its use is getting to be a form of social differentiation between "We are good trustworthy citizens" and "They, Others, not deserving of trust" [20].

Issues related to the admissibility of using DNA profiles stored in non-state databases for identification are also discussed [21, 22]. The study of 22 companies' policies showed that only four companies provided additional information on how law enforcement agencies should request permission to use their services for law enforcement purposes. Moreover, the two databases follow the different approach, providing a dedicated service for law enforcement [23]. Similar results were obtained in our recent study. Some heads of scientific organisations in the field of genetics during structured interview admitted the use of "existing database" for identification studies by the request of the law enforcement bodies. So there was neither permission of patients no court decision to use their genetic data to search for criminals or their relatives [6].

Advancement of in-home DNA testing has made some genetic genealogy databases so large that, according to various estimates, they can identify 60% to 90% of North Americans of European descent, even if they themselves have never taken these tests. [21, 24]. It becomes possible as a result of "family search". The use of this tool raises many ethical issues, including the establishment of genetic supervision over certain groups of individuals and [18, 19]. It can be more dangerous when creating of genetic profile is associated with political activity of citizens. For example, in accordance with the draft law pending in the State Duma of the Russian Federation, the possibility of genomic registration of all persons subjected to administrative arrest is provided [25]. Meanwhile, administrative arrest has recently become the most widely used punishment in Russia for persons who took part in peaceful assemblies without prior approval. In this sense, one cannot but agree that family searches can ultimately lead to the criminalization of a certain social groups.

DNA forensic phenotyping, which close to the ethically controversial practice of racial persecution, seems to be even more alarming in terms of inequality and discrimination. The legislation on its possible use varies greatly over the European countries: it is directly regulated only in the Netherlands. In other countries, legislation is either implied or absent, so it can be interpreted in different ways by experts and practitioners. As a result, forensic DNA phenotyping is used in Spain and Great Britain, and is considered prohibited in other countries such as Germany, Belgium and Austria [19, 20, 26].

The use of FRT by state arises fears of loss of privacy and unjustified intrusion into such right [14, 27–29]. Misidentification can put innocent people on watchlists, with an increased risk of poor outcomes for minorities and others at risk [5, 30]. These technologies can also be used by authoritarian regimes to fight dissent [31].

FRT surveillance is not just an invasion of privacy; the level of control, interference in privacy and its aggressive nature can lead to a "chilling effect" on freedom of assembly and expression and the general use of public space by communities. In the context of political protest, open surveillance can damage legitimate political activities, undermine the perception of the legitimacy of protest groups and limit their access to resources. [28, 31–33]. In 2018, the group established to provide ethical guidelines for police interviewed more than 1,000 Londoners about the use of real-time FRT by the police. 38% of respondent at the age of 16–24 and 28% of Asians, blacks and mixed-raced said they would stay away from events which could be tracked by real-time face recognition [13].

The growth in the use of face recognition and DNA identification technologies is coming against the background of extremely weak and insufficient regulation all over the countries [13, 21, 24, 31, 33–35] due to the rapid development of such technologies. Nevertheless more and more governments use new instrument for monitoring, tracking and surveillance. Objectives of such surveillance can range from legitimate to violating human rights [13]. But the general trend is that formulated by G.T. Marx in 2005: "A fascinating aspect of surveillance technologies as hegemonic control involves their tendency to expand to new areas, subjects, and forms. The surveillance appetite once aroused can be insatiable. A social process of surveillance creep (and sometimes gallop) can often be seen. Here a tool introduced for a specific purpose comes to be used for other purposes, as those with the technology realize its potential and ask, "Why not?"" [36]. J. Dahl and A. Sætnan [20] also discuss in detail the "creep" of state control functions, their will to scale it up and expand to new spheres. In light of this trend and potential threats of new technologies in terms of illegal surveillance and fight dissent, more tight standards and stronger guaranties for application of these technologies are badly needed.

However, it requires broad public discussions to reach social consensus, the formation of public consent, which is still a long way off. Public opinion on the application of the considered technologies is rather contradictory and highly variable. So, after the arrest of the alleged Golden State killer, online survey to study public opinion on the use of genetic data in the work of law enforcement agencies was conducted. The majority of the respondents "supported police in search for websites that identify relatives by genetic profiles (79%) and disclose genetic testing clients to the police (62%), as well as the building of fake profiles by the police to search for information on websites (65%)". However, respondents were significantly more supportive of this activity when the goal is to identify perpetrators of violent crimes (80%), crimes against children (78%) than when the goal is to identify perpetrators of non-violent crimes (39%) [37]. When the circumstances of the Utah case, where there was an assault but not a murder or sexual assault, were disclosed, public opinion on use of genetic information by police to search for the criminal was sharply negative [21]. Apparently, we should agree that as people are getting more familiar with the vulnerability of genetic and biometric data, the possible forms of exercising control over them and the risk of this data falling into the hands of third parties, the attitude to issues related to the setting of limits and transparency for the use of relevant technologies by governmental bodies is getting more cautious. This assumption is supported by the results of the 2020 "Consumer Privacy Survey" on privacy during COVID-19 pandemic [38].

Based on our study using The International Genetic Literacy and Attitudes Survey – Legal and Ethical (iGLAS-LE), new version of iGLAS instrument 49% of respondaents consider current legal and ethical regulation of genetic data as insufficient and 43% say "don't know" [39] (The study was carried out in 2018–2020 with the financial support of the Russian Foundation for Basic Research, RFBR grant No. 18-29-14071. Detailed information can be provided by request: Robert Chapman r.chapman@gold.ac.uk).

Under the conditions of insufficient legal certainty and the lack of consensus on the application of modern technologies that can be used by the state to spy on citizens, the activities of the judicial authorities both national and supranational are of particular importance. They are addressed in cases of explicit conflict indicating the need to find a new balance between public and private interests resulting from the changed external conditions or better awareness of the risk of the use of new technologies by society.

If the courts shrink away from obligations this function, this may lead to further "imbalance" of the situation or probably - to an excessive state interference in human rights and privacy, contrary to the interests of society. For example, Williams and Johnson provide the examples when the Norwegian police used, if not illegal, but at least unconventional methods to obtain a DNA profile and search for it in a DNA database. In one case, a man, who was suspected by police of a serious crime, drowned and his body was taken to the Institute of Forensic Medicine for an autopsy. The police, without asking the consent of the next of kin, took a sample for comparison. This model justified the man, but allowed his brother to be held accountable due to family resemblance. Another suspect in the same case died of cancer. In this case, the police requested access to tissue samples at the hospital, but were refused. However, this did not prevent the use of the DNA results obtained in relation to the first suspect in court. Other methods of "creative collection" of materials for DNA identification were also used. In one case, police confiscated cigarette butts from a man's garden. In another case, a suspect was summoned for questioning in an unrelated case in which he was not a suspect but a witness. In this relaxed atmosphere, he was served a glass of water, which was then used to obtain his DNA profile in order to verify his involvement in the crime. However, such highly controversial police practices have no formal implications as the courts do not rule out the evidence obtained. As a result, the relevant practice is not suppressed, but, on the contrary, is reproduced and developed [17].

In this article, we set out to investigate how the activities of the court can affect the legal reality and public relations in the use of modern technologies by the state to control citizen.

2   Methods

For this purpose, we decided to conduct a comparative analysis that attempted to find a balance between public and private interests by different national courts; and to study how it affects the legal reality. It has been made possible by simultaneously considering in the United Kingdom and Russian courts rather similar claims to protect private interests from the use of FRT by the state.

Since June 2018, the British human rights organization "Liberty" has represented Cardiff resident Ed Bridges in court, who accused South Wales police of illegally using FRT against him, causing him harm. On October 7, 2019, in Russia, political activist Alena Popova, supported by the Russian human rights organization Roskomsvoboda, filed a lawsuit in the Savyolovsky Court of Moscow demanding to declare illegal the use of FRT by Moscow authorities in urban video surveillance system. Both claims were failed in the first court instance, both plaintiffs continued to defend their position in the court of appeal [27]. The Russian Court of Appeal fully agreed with the findings of the first-instance court and dismissed the plaintiff's complaint, while the British Appeal Court accept three of five arguments of the plaintiff's appeal, although it concluded that the first-instance court was just and there was no damage to Claimant personally no done.

It was decided to analyse comparatively the essence of these claims, the circumstances investigated by the courts, the content of the court decisions, their justification, and, further, try to trace how these court cases have changed the public perception of the relevant problems, as well as the activities of state and interstate bodies.

3   Results

3.1   Background of Claims and Appellants' Arguments

UK

Complaint was brought in 2018 by Mr. Bridges supported by "Liberty" against chief constable of South Wales [40]. Since mid-2017, South Wales Police have launched a pilot program called AFR (live automatic facial recognition) Locate, which takes digital images of faces from live video feeds, extracts face biometrics in real time and compares them with watchlists, alerting police of matches so they can apprehend a suspect. If there is no match, the biometric data is deleted almost immediately. The use of AFR Locate has been tested at certain events and locations, such as football match and exhibition. The Appellant, Mr. Bridges, alleged that he had on two particular occasions where AFR Locate was used, so that the police could obtain and recognise his image. At the same time, there were no notifications that the police were using AFR Locate. It was not disputed that the Appellant was not on the watchlist, he was not detained, using AFR Locate, police officers did not approach him during the events or later.

The Appellant argued that the use of the AFR Locate violated his right to respect for "private and family life, his home and his correspondence", provided in Article 8 of the European Convention on Human Rights. The Appellant also argued that the use of AFR Locate violates non-discrimination (as this technology produces more false matches for women and ethnic minorities) [41], as well as the provisions of UK data protection law. This claim was the first case in the world to appeal against the use of RFT by the state.

Russia

The Appellant was Ms. Alena Popova, supported by the human rights organisation "Roskomsvoboda" [4]. The Defendants in this suit were the Ministry of the Interior of the Russian Federation and the Moscow Department of Information Technologies (DIT). Since 2017, the Safe City system has been operating in Moscow, which also uses FRT cameras (City Video Surveillance System). Already in 2018, more than 7,000 permanent cameras were connected to this system in Moscow. The operator of this system and the custodian of information from cameras is DIT. Information of interest to law enforcement agencies is transmitted to them. It was reported that the cameras independently match the faces from the video stream with the database of the wanted persons, and in case of a match, report this to a nearby police officer. The reason for appeal was the fine that Ms. Popova received in 2018 for taking part in a single-person protest near the State Duma. During the case proceeding, the court evaluated the recordings from CCTV cameras, which showed 32-fold increase in the image with fixation on the Appellant's face. It means that she had been subjected to recognition.

The claim was initiated in 2019. That time, the case of Bridges v. Chief Constable of South Wales was already pending. The Appellant alleged that the use of cameras with such technology violated her constitutional right to privacy (similar to the right provided for in Article 8 of the ECHR), as well as the Russian law on personal data. She also drew the court's attention to the fact that there is no legislative procedure to apply to an authorized person with a request to exclude her data from the system; it is unknown who and in how has access to the results of the system's operation; the absence guarantees against abuse and information leaks.

Comparison
The circumstances of these cases appear to be very similar. It makes further comparison possible. At the same time, it should be noted that the personal circumstances of Mr. Bridges were different from those of Ms. Popova. Mr. Bridges was not stopped or detained by the police; the results of the AFR Locate operating for him were not saved in the system and were not used in the future. On the contrary, the biometric data of Ms. Popova was apparently saved and handed over to law enforcement agencies, despite the fact that she was not on the list of wanted persons, and were also used to justify her being prosecuted for participating in a single-person protest. Thus, the interference of the state in her rights was obviously more significant than in case of Mr. Bridges.

Based on the comparison of two cases and timing of their initiation it can be admitted that the fact of Ms. Popova's claim itself was a consequence of a wide public discussion on claim of Mr. Bridges.

3.2   Trial Court: Arguments and Conclusions

UK
The issue of sufficiency of the existing legal regime in the UK to ensure the proper and non-arbitrary use of AFR technology in a free and civilized society has been identified as a principle one by the trial court. To that end, court studied the personal situation of the Appellant, current rules of the AFR use by the police, specific features of the technology used in AFR Locate for face recognition, terms and conditions of the obtained information storage and processing, generating of watchlists, procedure for informing the public about the use of AFR Locate. In the course of the judicial proceedings, the court extensively involved experts. It was a trigger for broad public discussion on the use of FRT by the state.

In court decision it was stipulated that regulation of new technologies use in law enforcement should ensure a reasonable balance between private rights, on the one hand, and public interests, on the other. This decision was based both on the previous practice of the UK courts and on the main precedents of the ECHR, as well as on the decisions of the Court of Justice of the European Union. This approach can only be welcomed.

The court found that the use of AFR Locate restricted the right provided for in Article 8 (1) of the ECHR. It was needed to understand what biometric data are in general and to determine the originality of facial images obtained as a result of using AFR as a type of biometrics. The court noted that facial images are similar to fingerprints as both types usually visible to the public. However, unlike fingerprints, using AFR technology, facial biometric data can be obtained without the use of special knowledge or force, and on a massive scale. Comparison of AFR data with other biometric data allowed the court to take completely new approach in assessing interference with the Appellant's right to privacy. Earlier, taking photo and/or video recording of a person in a public place had been assessed by the UK and USA courts based on an approach of personal expectations of confidentiality. The defendant's representative also tried to refer to the fact that the Appellant could not presume on privacy when walking in a public place and could expect that his image would be recorded for crime prevention. The court, however, concluded that biometric data was fundamentally different from simple video recording, stating that biometric facial identifiers provide "accurate and unique" data about a person, like fingerprints and DNA, and therefore are data of an "intrinsic private" nature. Therefore, the mere acquisition and storage of biometric data (regardless terms of storage) is enough to admit an interference with the right provided for in Article 8 of the Convention, regardless of whether they were obtained in a public place open to observation. At the same time, the court found no violation of the Appellant right to privacy.

In particular, the court found that this method was no more intrusive than the use of video surveillance on the street. So existing data protection laws, codes and policies of lawinforcement agencies were deemed sufficient by the court to regulate the use of AFR Locate. However, it was noted in the decision that this issue should be periodical reviewed in the future.

The court's decision was also motivated by a detailed study of the Appellant's personal situation: the court considered that the interference with his personal rights was insignificant. However, the court noted that the decision could have been different if a person identified in the AFR or on the watchlist had applied to the court. "The inclusion of any person on any watchlist and the subsequent processing of that person's personal data for no good reason is likely to constitute an unlawful interference with their right to privacy under Article 8" (paragraph 105 of the judgment) [43].

Important consequences of this court decision are also related to the fact that the Court has substantiated the unconditional classification of the data about a person obtained in the course of using AFR Locate to the category of "personal data", since they allow identifying the person concerned. At the same time, the court recalled that "personal data" are those that allow to identify a person directly or indirectly. In this case, indirect identification will take place if the data is sufficient for identification along with other information available to the operator or a third party. Here the court referred to the judgment of the Court of Justice of the European Union in the case Breyer v Bundesrepublik Deutschland, according to which such data are also "personal" when only a third party has additional data necessary for personal identification, provided that the possibility of combining data is a means, probably reasonable to use for personal identification. The identification of the data object must be legally and practically possible for a third party (without disproportionate effort in terms of time, cost and efforts). Nevertheless, in this case, the court went further and recognized that the data obtained as a result of the use of system and itself without additional information about the name of a person, allows to identify him directly because the nature of biometrics is used to distinguish one person from another.

At the same time, the court found no violation of the legislation on personal data in this specific Appellant situation, as the processing of his personal data did not violate the requirements of law and was adequate.

The following circumstances were considered by the court as relevant:

– the use of AFR was open and transparent, with public involvement, for a limited time with limited coverage;
– it was used for a specific and limited purpose to identify particular persons;
– no data was saved, and the intervention was limited by prompt algorithmic processing and deletion of biometric data;
– no one was arrested by mistake, and no one complained of abuse;
– no personal information of the Appellant was provided to any of agents and there was no attempt to identify him.

The court also rejected the discrimination claim because it did not find convincing evidence that the recognition algorithms were performing poorly.

Russia
Contrary to the British court, the court of first instance in Russia refused to consider the data received and processed by the City Video Surveillance System as personal data. In substantiating, it was pointed out that the operator (Unified Data Storage Center - ECHD) "does not have personal data of citizens (full name, etc.), as well as biometric personal data (iris, height, weight, etc.) required for identification". Thus, the concept of "personal data" was interpreted quite narrow.

The next argument of the court was the following: the facial recognition algorithm used by the ECHD compares the image received by the ECHD from video cameras with a photograph provided by a law enforcement agency. The algorithms can reveal the coincidence of biometric data only with a certain degree of probability (65%). Accordingly, the above processes cannot be used to identify a person. At the same time, speaking about the legitimacy of such actions, the court indicated that police officers who have been granted with access to the data of the ECHD use them to identify persons on the federal wanted list, persons who are prohibited from attending mass events by a court decision, persons under administrative supervision. It should be noted that there is a certain contradiction here, since the data that can be used "for identification" undoubtedly make it possible to identify the person.

The court also pointed out that citizens are not subject to surveillance when like technologies are applied, since, in accordance with paragraph 11 of the Regulation on the ECHD, the objects of video surveillance include territories (for example, courtyards and other public places), but not individuals.

Based on these arguments, the court concluded that there was no violation of the Appellant rights, since "the installation and use of a video surveillance system is associated with ensuring security, by virtue of which it is not a source of obtaining personal data".

Ms. Popova's argument that the results of the identity recognition system were used in the court proceedings for an administrative offense, as a result of which a fine was imposed on her, was rejected by the court with reference to the fact that "this video was provided in the case file the Appellant herself". At the same time, the court clarified, according to the representative of the defendant, that in the process of transmitting the video image from the camera to the ECHD there is no automatic (arbitrary) approach of the person's face, the recording is carried out in streaming mode from the distance at which the camera is installed. However, the source of the video footage investigated by the court when imposing a fine on Ms. Popova for taking part in the picket remained a mystery.

Comparison

The courts of Russia and the UK have used fundamentally different approaches to the consideration and resolution of the dispute. The Russian court refused to verify relevant facts pointed out by the Appellant. It used an extremely narrow and formal approach to determining what "personal data" is, did not check whether there had actually been an interference with the Appellant's right to privacy, did not try to strike a balance between the right to privacy and the need to investigate crimes. The decision of the Russian court contains internal contradictions that do not allow for unambiguous determination of legality in the field of application of the technology (whether it is used to identify a person). The court obviously disregarded important circumstances (the video of Ms. Popova was made during her protest by "City video surveillance system", she was identified with the Appellant, the very existence of this video refuted the testimony of the defendant that the data was not identifiable and was deleted in 120 h, as well as the fact that people are not "objects of observation").

In contrast, although the British court also denied Mr. Bridges' claim, the rationale behind this decision was radically different. It was based on a deep study of the application of such technology and its specific features by the court, the established facts, resolved a number of fundamental issues, and, at the same time, clearly highlighted the personal situation of the Appellant, so it may be used to substantiate other claims filed by persons whose rights could be more abused than Bridges'. Therefore, this court decision, even being formally negative, created conditions for improving the practice of the FRT use.

3.3 Appeal Court: Arguments and Conclusions

UK

The Court of Appeal ruled that police interference with Mr. Bridges' rights could not be qualified as "legitimate" as the current regulations provided the police with too much discretion in the choice of watchlist and in determining where Locate AFR might be deployed.

At the same time, the court considered proportionate the interference of the state with the right of Mr. Bridges under the Article 8 of the Convention, taking into account the personal situation of Mr. Bridges, the degree of influence of the use of AFR on him.

In support of the fact that the treatment of personal data was not in accordance with the law, the Appellant referred both to the fact that Article 8 had been violated, and there are risks to other rights that may be affected by the use of AFR technology, such as the right to freedom of assembly in accordance with the Article 11 of the Convention and the right to freedom of expression in accordance with the Article 10 of the Convention. Unfortunately, the Court of Appeal did not take into account the latter argument, but nevertheless agreed with the Appelant that the legislation on personal data protection had been abused by the very violation of the Article 8 of the Convention.

In addition, the Appeal court agreed with Mr. Bridges that it is "the responsibility of the process, not the result," to test for non-discrimination in the use of identity recognition technologies, and "helps to convince members of the public, regardless of their race or gender, that their interests were duly taken into account…". The court criticised that the South Wales police did not independently verify that the software algorithm used might discriminate against certain groups, and on this point agreed with the Appellant.

Thus, in general, the Court of Appeal for England and Wales overturned the decision of the first instance court, finding that the use of AFR was illegal and violated human rights.

Russia

The appeal court fully upheld the decision of the first instance court.

However, it was noted:

- "technology of "face recognition" is not a prohibited way of using information by its owner…";
- the law "establishes a reasonable degree of freedom to apply the" processing of personal data / images", which is used by the Department to achieve legitimate goals. The state has the right, at its discretion, to choose the most appropriate means of protecting the rights of citizens";
- "employees of the Main Directorate of the Ministry of Internal Affairs of Russia in Moscow, who have been granted access to the ECHD, in order to fulfill their official duties, use a video analytics system based on the ECHD to identify persons on the federal wanted list, persons who are prohibited by a court from visiting big public events, and persons under administrative supervision. Thus, the activity of the Main Directorate of the Ministry of Internal Affairs of Russia in Moscow on the use of the video analytics system based on the ECHD is carried out in full compliance with the current legislation and cannot be regarded as violating the rights of the administrative Appellant";
- "the court of first instance correctly noted that citizens do not belong to direct objects of video surveillance, since in accordance with paragraph 11 of the Regulation on the ECHD, objects of video surveillance include: territories and objects." "In itself, obtaining an image of an administrative Appellant while she is in the territory falling under the surveillance sector of a particular camera, installed in order to monitor the environment in the area where the building of the State Duma of the Russian Federation is located, is not a way of collecting personal (biometric data) of the applicant, since it is not was used directly to identify her";

- "in absence of a personal identification procedure, video images of citizens cannot be considered biometric personal data";
- "the arguments of the appeal that the information obtained from the CCTV cameras was used as evidence in the case of an administrative offense, in connection with which they affect her rights, cannot entail the cancellation or change of the decision that took place", since the decision to bring a person to an administrative liability cannot be challenged separately from the proceedings on the case of an administrative offense.

Comparison

The Russian court of appeal essentially refused to consider the arguments of the Appelant an just reproduced the arguments of the court of first instance, despite it admitted that the video image of the administrative Appelant was obtained from CCTV cameras with the function of recognition and was used in the proceedings in the case of an administrative offense. In addition, it was formulated the powers of the state and the ECHD (operator) so broadly ad in such vague terms, that now it is impossible to imagine a situation in which the activities of the relevant authorities and organisations on the use of FRT and video recording technologies of citizens can be recognised as illegal. This creates an extremely unfavorable background for further attempts to protect the rights of persons in respect of whom automatic identity recognition technologies are applied.

At the same time, the UK Court of Appeal examined in detail and thoroughly both the issues of fact and law, which allowed it to draw a number of new conclusions in the case, differing from those of the first instance court.

3.4 Implications of the Court's Decisions

UK

The decision of the Court of Appeal in the Bridges case was hit the headlines, as well as widely discussed by professional lawyers, human rights defenders not only in Europe, but over the world [44]. It is recognised as a breakthrough for individuals and organizations defending human rights and demanding restrictions on the use of AFR [45]. This decision was included in the top-10 most important court decisions for 2020.

South Wales police have refused to appeal the Bridges court of appeal decision. Tony Porter, Commissioner for Surveillance Cameras (SCC), following a ruling by the Court of Appeal, has issued an updated "Facing the Camera" [46] best practice guide to all police forces in England and Wales. He explained that he did this in order to help determine how to use recognition technologies in accordance with the current legal framework and taking into account the decision of the Court of Appeal in the Bridges case [47].

The Metropolitan Police said it intends to continue to use facial recognition technology. At the same time, she points out that her approach "to face recognition in person is different from those cases that were appealed to the South Wales police". In particular, the Metropolitan Police say they use this technology only selectively - on the basis of intelligence and only in the fight against serious crimes, including serious violence, crimes with weapons and knives, sexual exploitation of children and helping to protect the vulnerable. The Metropolitan Police website says, "We note the public about every use of facial recognition technology before, during and after," and claims that Metropolitan Police use their own regulations (as opposed to those used by the South Wales Police). "We will carefully consider the court's decision and will act on any important issues to ensure that we comply with our obligations to use facial recognition in a legal, ethical and proportionate manner," the message says [48].

On April 21, 2021, the Commission's Proposals for a European Approach to Artificial Intelligence and Machinery Products (Proposal for the Regulation of a European Approach to Artificial Intelligence) were published. In accordance to them remote biometric identification systems are classified as high-risk AI systems. In principle it is proposed to prohibit their use in public places for law enforcement purposes. Specific exceptions to this rule are expected to be strictly defined (for example, when strictly necessary to locate a missing child, to prevent a specific and imminent terrorist threat, or to locate, locate, identify or prosecute a criminal or serious criminal suspect)). Such use requires permission from a judicial or other independent authority and subject to appropriate time, geographic and searchable database restrictions [49]. It appears that some of these proposals were also a consequence of the Bridges v South Wales Police case.

Russia

In Russia the use of technologies for automatic recognition in Russia have been developing in a different way. Even taking into account the conditions of COVID-19 pandemic and the restrictions associated with them, it seems reasonable to assume that these negative changes in a certain sense are induced by the formal and inconsistent approach of the courts to the use of personality recognition technologies, reflected in the judicial solutions.

So, face recognition systems began to be used to calculate and prosecute persons who found themselves in the place of mass unauthorized actions [50]. At the same time, in practice, video recording becomes the main or even the only evidence in the case (despite the fact that in the court decision analyzed above, the probability of identification was estimated at 65%) [51]. It has a significant impact on the mentality of Muscovites, giving them fears of video cameras. For example, the well-known expert on the work of Vladimir Nabokov, Mikhail Shulman, who on the day of the protest at the end of January was detained "as a witness" based on the results of the recognition system in the Moscow metro, admits that he now feels defenseless and behaves in public transport "like a guerrilla" [52].

An attempt to appeal against such actions was unsuccessful. So, Alena Popova, together with V.S. Milov on January 10, 2020 filed a claim over the use of a facial recognition system at a mass meeting on September 29, 2019 on Sakharov Avenue in Moscow. Defendant, the Department of Information Technology, confirmed in the process that cameras with face recognition function were used at the "Let go" meeting in the capital on September 29 and that cameras were used indiscriminately. Nevertheless, the court refused to accept this statement of claim for consideration, referring to the fact that the Savelovsky District Court of Moscow had previously considered an administrative case with an identical subject matter of the claim (the decision in this case was analyzed above). This circumstance, in the opinion of the court, excludes the possibility of considering a case on the same subject in court [53].

Unfavorable consequences may also occur for Muscovites due to errors in the actions of the face recognition system. Thus, in October 2020, the face recognition system in the Moscow metro took Sergei Mezhuev for another wanted person. Despite the fact that the police immediately realized the mistake, they took Sergei to the police station, where they collected all the data from him, including biometric data. RosKomSvoboda lawyers complained to the police and asked Roskomnadzor for clarification. In its response, the department indicated: "Given that the police officers received information about the search for a person mistakenly identified as S.M. Mezhuev, there were grounds for collecting and processing personal data, including biometric data, in order to establish an identity. Thus, there are no grounds for the Office to take response measures," the government's response says [51]. Thus, the analysed court decision creates the basis for the subsequent formal attitude to the protection of human rights, an extremely broad, even "limitless" approach to the powers of bodies working with biometric data of individuals.

Similarly, the unreasonably narrow approach to defining the circle of "personal data" proposed in the analyzed judicial decision allows arbitrarily and unreasonably to exclude any data from it, if they are not part of the last name, first name and patronymic of the person. So, in October 2020, the mayor of Moscow signed Decree No. 97-UMU, according to which all employers in Moscow had to (in order to monitor the implementation of coronavirus restrictions) send to the mayor's office information about the mobile phone number, vehicle number, social card number about all workers. This paragraph of the Decree was contested in court. The claim was denied with reference to the fact that this information cannot be correlated with a specific individual, therefore, this data itself is not personal [54]. At the same time, when an attempt was made to request similar data in relation to employees of the Moscow government, the provision of such information was refused on the grounds that these data are personal, and this interpretation of the law in relation to these persons was confirmed by the prosecutor's office [55].

The court also refused Anna Kuznetsova's claim to the city administration of the Ministry of Internal Affairs and the Department of Information Technologies of Moscow, in which she complained about the leakage of data about her identity obtained during video surveillance with recognition technology. The reasoning is the same - the privacy of the Appellant is not violated in any way when conducting video surveillance in the city and using face recognition technology, because in special regulations on the operation of the video database it is established that the objects of observation are streets, landscaping objects and other inanimate objects, and the citizens are not mentioned there. In addition, the court indicated that since video surveillance is carried out on the street, it means that everything that happens can be recorded without the consent of citizens as filming in places open for free visits, in state, public and other interests. At the same time, the court concluded that the comparison of photographs with images of people on Moscow video streams is not the processing of personal data, therefore the provisions of the law on personal data are inapplicable in this case [56].

On the other hand, the inability to achieve human rights protection in Russia forced Vladimir Milov and activist Alena Popova to appeal to the European Court of Human Rights (ECHR) with a complaint about the use of facial recognition technology at a mass meeting on September 29, 2019 in Moscow. If this complaint is considered by the court, then all states that have ratified the European Convention will be able to receive from the ECHR a certain guideline in the development and application of their own automatic face recognition systems.

4   Conclusion

For now, there are no social bargains, general ethical rules, updated legislation, clearly formulated legal norms regarding the use of modern technologies by the state for tracking citizens. Under such conditions, the activity of the judicial bodies - national and supranational - acquires special significance. They are addressed in cases where there is an obvious conflict, a controversial issue. The court decision of such disputes should be used to find a balance between public and private interests, and become the subject of public discussion.

Thus, the function of finding a compromise, a balance between the need to ensure the safety of people and the fight against crime, while respecting the rights, freedoms and guarantees of citizens is assigned to the judiciary. Under these conditions, the court is required to study not only questions of law and fact, but also independently conduct a careful weighing of the significance of the legitimate aim that must be pursued, the rule of the law, which is the subject of state intervention, and the degree of acceptable interference. Such activity of the court to resolve specific cases can be considered as an act of judicial lawmaking. It is designed not only to resolve a specific situation, but also to create a basis for further public discussion of the most pressing problems, not only to reveal the shortcomings and gaps of the current regulatory regulation, but also to create new rules, making a decision based on an independent weighing of various values, not only to record the presence problems, but also to form a new law enforcement practice. If the courts are removed from performing this function, then the existing conflict will only worsen, leading to increased state intervention and raising of public dissatisfaction with the position of the state.

Summing up what has been said, the function of finding a compromise, a balance between the need to ensure the safety of people and the fight against crime, while respecting the rights, freedoms and guarantees of citizens is assigned to the judiciary. Under these conditions, "good judges are more important than good laws".

Financial and Competing Interests Disclosure
This work was supported by out with the financial support of the Russian Foundation for Basic Research, grant RFBR No. 18-29-14071.

References

1. Personalized Cancer Vaccine Given After Adjuvant Therapy Safe, Shows Early Efficacy in Multiple Tumor Types. https://www.aacr.org/about-the-aacr/newsroom/news-releases/personalized-cancer-vaccine-given-after-adjuvant-therapy-safe-shows-early-efficacy-in-multiple-tumor-types/ Accessed 4 Apr 2021

2. The Facts on Facial Recognition with Artificial Intelligence. https://aws.amazon.com/rekogn ition/the-facts-on-facial-recognition-with-artificial-intelligence/?nc1=h_ls Accessed 4 Apr 2021
3. Cuthbertson, A.: Indian Police trace 3,000 Missing Children in Just Four Days Using Facial Recognition Technology. Independent 24 April (2018). https://www.independent.co.uk/life-style/gadgets-and-tech/news/india-police-missing-children-facial-recognition-tech-trace-find-reunite-a8320406.html. Accessed 4 Apr 2021
4. Who and how uses face recognition technologies in Russia. https://rb.ru/longread/facial-rec ognition/. Accessed 4 Apr 2021
5. Leong, B.: Facial recognition and the future of privacy: i always feel like … somebody's watching me. Bull. Atom. Sci. **75**(3), 109–115 (2019). https://doi.org/10.1080/00963402. 2019.1604886
6. Andreeva, O.I., Matsepuro, D.M., Olkhovik, N.V., Trubnikova, T.V.: Criminal justice in the post-genomic era: new challenges and the search for balance. Bull. Tomsk State Univ. Right. **35**, 14–28 (2020). http://doi.org/https://doi.org/10.17223/22253513/35/2
7. Newton, R.: You are being watched: face recognition deemed 'creepy' by UK shop-pers. https://www.theguardian.com/small-business-network/2015/jul/27/you-are-being-wat ched-face-recognition-creepy-uk-shoppers. Accessed 4 Apr 2021
8. UK creepy or cool? https://richrelevance.com/resources/uk-creepy-or-cool/. Accessed 4 Apr 2021
9. UK creepy v cool? https://richrelevance.com/resources/2018-uk-creepy-v-cool/ Accessed 4 Apr 2021
10. Chavarria-Soley, G., et al.: Attitudes of Costa Rican individuals towards donation of personal genetic data for research. Pers. Med. **18**(3), 141–152 (2021). https://doi.org/10.2217/pme-2020-0113
11. DNA ties dead suspect to 'Boston Strangler' case: officials. https://www.reuters.com/article/ us-usa-crime-bostonstrangler/dna-ties-dead-suspect-to-boston-strangler-case-officials-idU SBRE96A0O820130711. Accessed 4 Apr 2021
12. Global DNA Profiling Survey Results 2019. https://www.interpol.int/How-we-work/Forens ics/DNA. Accessed 4 Apr 2021
13. Feldstein, S.: The Global Expansion of AI Surveillance. Carnegie Endowment for Interna-tional Peace (2019) https://carnegieendowment.org/2019/09/17/global-expansion-of-ai-sur veillance-pub-79847. Accessed date 04 April 2021
14. Castelvecchi, D.: Is facial recognition too biased to be let loose? Nature **587**(7834), 347–349 (2020). https://doi.org/10.1038/d41586-020-03186-4
15. The UK Police National DNA Database. http://www.genewatch.org/sub-539478. Accessed 4 Apr 2021
16. Case of S. and Marper v. the United Kingdom (No. 30562/04&30566/04). http://hudoc.echr. coe.int/eng?i=001-90051. Accessed 4 Apr 2021
17. Williams, R., Johnson, P.: Circuits of surveillance. Surveill. Soc. **2**(1), 1 (2004). https://doi. org/10.1901/jaba.2004.2-1
18. Machado, H., Granja, R.: DNA databases and big data. In: Machado, H., Granja, R. (eds.) Forensic Genetics in the Governance of Crime, pp. 57–70. Springer, Singapore (2020). https:// doi.org/10.1007/978-981-15-2429-5_5
19. Marx, G.T.: Seeing hazily (but not darkly) through the lens: some recent empirical studies of surveillance technologies. Law Soc. Inq. **30**(2), 339–399 (2005). https://doi.org/10.1111/ j.1747-4469.2005.tb01016.x
20. Dahl, J.Y., Sætnan, A.R.: "It all happened so slowly" – on controlling function creep in forensic DNA databases. Int. J. Law Crime Justice **37**(3), 83–103 (2009). https://doi.org/10. 1016/j.ijlcj.2009.04.002

21. Arnold, C.: The controversial company using DNA to sketch the faces of criminals. Nature **585**(7824), 178–181 (2020). https://doi.org/10.1038/d41586-020-02545-5
22. Murray, D., Fussey, P.: Bulk surveillance in the digital age: rethinking the human rights law approach to bulk monitoring of communications data. Isr. Law Rev. **52**(1), 31–60 (2019). https://doi.org/10.1017/S0021223718000304
23. Van Noorden, R.: The ethical questions that haunt facial-recognition research. Nature **587**(834), 354–358 (2020). https://doi.org/10.1038/d41586-020-03187-3
24. Callaway, E.: Supercharged crime-scene DNA analysis sparks privacy concerns. Nature **562**(7727), 315–316 (2018). https://doi.org/10.1038/d41586-018-06997-8
25. Bill No. 1048800-7. On Amendments to Certain Legislative Acts of the Russian Federation on State Genomic Registration. https://sozd.duma.gov.ru/bill/1048800-7. Accessed 4 Apr 2021
26. Whittall, H.: The forensic use of DNA: scientific success story, ethical minefield. Biotechnol. J.: Healthc. Nutr. Technol. **3**(3), 303–305 (2008). https://doi.org/10.1002/biot.200800018
27. Andreeva, O.I., Ivanov, V.V., Nesterov, A.Yu., Trubnikova, T.V.: Face recognition technology in criminal proceedings: the evidence for legal regulation in use of artificial intelligence. Tomsk State Univ. J. **449**, 201–212 (2019). http://doi.org/https://doi.org/10.17223/15617793/449/25
28. Aston, V.: State surveillance of protest and the rights to privacy and freedom of assembly: a comparison of judicial and protestor perspectives. Eur. J. Law Technol. **8**(1), (2017). https://ejlt.org/index.php/ejlt/article/view/548. Accessed 4 Apr 2021
29. Purshouse, J., Campbell, L.: Privacy, crime control and police use of automated facial recognition technology. Crim. Law Rev. **3**, 188–204 (2019). https://ueaeprints.uea.ac.uk/id/eprint/69577/. Accessed date 02 Apr 2020
30. Crawford, K.: Halt the use of facial-recognition technology until it is regulated. Nature **572**(7771), 565 (2019). https://doi.org/10.1038/d41586-019-02514-7
31. Hirose, M.: Privacy in public spaces: the reasonable expectation of privacy against the dragnet use of facial recognition technology. Conn. Law Rev. **49**, 1591 (2016) https://opencommons.uconn.edu/law_review/377/ Accessed date 02 Apr 2020
32. Nesterova, I.: Mass data gathering and surveillance: the fight against facial recognition technology in the globalized world. SHS Web Conf. **74**, 03006 (2020). https://doi.org/10.1051/shsconf/20207403006
33. Roussi, A.: Resisting the rise of facial recognition. Nature **587**(7834), 350–353 (2020). https://doi.org/10.1038/d41586-020-03188-2
34. Buckley, B., Hunter, M.: Say cheese! Privacy and facial recognition. Comp. Law Secur. Report **27**(6), 637–640 (2011). https://doi.org/10.1016/j.clsr.2011.09.011
35. Denham, E.: Facial recognition technology and law enforcement. Information Commissioner's Office (2018). https://ico.org.uk/about-the-ico/news-and-events/blog-facial-recognition-technology-and-law-enforcement/ Accessed 02 Apr 2020
36. Moreau, Y.: Crack down on genomic surveillance. Nature **576**(7785), 36–38 (2019). https://doi.org/10.1038/d41586-019-03687-x
37. Guerrini, C.J., Robinson, J.O., Petersen, D., McGuire, A.L.: Should police have access to genetic genealogy databases? Capturing the Golden State Killer and other criminals using a controversial new forensic technique. PLoS Biol. **16**(10), e2006906 (2018). https://doi.org/10.1371/journal.pbio.2006906
38. Protecting Data Privacy to Maintain Digital Trust. https://www.cisco.com/c/dam/en_us/about/doing_business/trust-center/docs/cybersecurity-series-2020-cps.pdf?CCID=cc000742&DTID=esootr000515&OID=rptsc023525. Accessed 4 Apr 2021
39. Selita, F., Smereczynska, V., Chapman, R., Toivainen, T., Kovas, Y.: Judging in the genomic era: judges' genetic knowledge, confidence and need for training. Eur. J. Hum. Genet. EJHG **28**(10), 1322–1330 (2020). https://doi.org/10.1038/s41431-020-0650-8

40. Case No: C1/2019/2670. https://www.judiciary.uk/wp-content/uploads/2020/08/R-Bridges-v-CC-South-Wales-ors-Judgment.pdf. Accessed 4 Apr 2021

41. Case No: CO/4085/2018. https://www.judiciary.uk/wp-content/uploads/2019/09/bridges-swp-judgment-Final03-09-19-1.pdf. Accessed 4 Apr 2021

42. Information on the case No. 02a-0577/2019. https://mos-gorsud.ru/rs/savyolovskij/ser vices/cases/kas/details/988f386e-be51-47b0-b48f-e871043ef1fc?caseNumber=2%D0%B0-577/19. Accessed 2 Apr 2021

43. European Convention on Human Rights. https://www.echr.coe.int/Documents/Conven tion_ENG.pdf. Accessed 2 Apr 2021

44. UK High Court upholds police use of automated facial recognition technology to identify sus-pects. https://www.hrlc.org.au/human-rights-case-summaries/2019/10/30/uk-high-court-uph olds-police-use-of-automated-facial-recognition-technology-to-identify-suspects. Accessed 4 Apr 2021

45. R (Bridges) V The Chief Constable Of South Wales Police [2020] EWCA Civ 1058 https://privacylawbarrister.com/2020/08/28/r-bridges-v-the-chief-constable-of-south-wales-police-2020-ewca-civ-1058/. Accessed 4 Apr 2021

46. Facing the Camera. Guidance 2020. https://assets.publishing.service.gov.uk/government/upl oads/system/uploads/attachment_data/file/940386/6.7024_SCC_Facial_recognition_report_ v3_WEB.pdf. Accessed 4 Apr 2021

47. Surveillance Camera Commissioner releases guidance for police on use of Live Facial Recognition. https://www.gov.uk/government/news/surveillance-camera-commissio ner-releases-guidance-for-police-on-use-of-live-facial-recognition. Accessed 4 Apr 2021

48. Metropolitan Police. Live Facial Recognition. https://www.met.police.uk/advice/advice-and-information/facial-recognition/live-facial-recognition/. Accessed 30 Mar 2021

49. Proposal for a Regulation laying down harmonised rules on artificial intelligence https://dig ital-strategy.ec.europa.eu/en/library/proposal-regulation-european-approach-artificial-intell igence. Accessed 30 Mar 2021

50. About Muscovites caught in cameras. The video surveillance system has shown its effec-tiveness on civilians. https://www.vtimes.io/2021/02/04/o-moskvichah-popavshih-v-kameri-a2978. Accessed 28 Mar 2021; 10 days of arrest based on data from the face recognition sys-tem. https://roskomsvoboda.org/69485/; Kamil Galeev: how I ended up in a special detention center a week after the rally because of the face recognition system. https://roskomsvoboda. org/post/kamil-galeev-kak-ya-okazalsya-cherez-ned/. Accessed 28 Mar 2021

51. Information on the case No. 12–0723/2020. https://mos-gorsud.ru/rs/preobrazhenskij/ services/cases/appeal-admin/details/5ec07436-0fb3-4b8c-8f73-364399c49f86?caseNu mber=%E2%84%9612-723/20. Accessed 30 Mar 2021

52. Mikhail Shulman on face recognition: In the metro now I go down as if in a trap. https:// roskomsvoboda.org/post/shulman-subway-trap-face-recognition-surveillance/. Accessed 30 Mar 2021

53. Information on the case No. 02a-0072/2020. https://mos-gorsud.ru/rs/tverskoj/services/cases/ kas/details/8f0ad27b-ba67-4e50-84eb-c3c5d788ef6c?participants=%D0%9C%D0%B8% D0%BB%D0%BE%D0%B2+%D0%92.%D0%A1.&caseRangeDateFrom=21.08.2019& formType=fullForm. Accessed 30 Mar 2021

54. Information on the case No. 3a-0325/2021 https://mos-gorsud.ru/mgs/services/cases/first-admin/details/3d8a1911-2e3f-11eb-8b6b-9d58d007c7d4. Accessed 30 Mar 2021

55. The Moscow City Court dismissed us in a lawsuit to spy on employees working in the remote mode – we analyze the decision. https://roskomsvoboda.org/69830/. Accessed 27 Mar 2021
56. Information on the case No. 02a-0798/2020. https://mos-gorsud.ru/rs/tverskoj/services/cases/kas/details/151b3db1-0400-11eb-a7b5-d914ac4c1d0c?participants=%D0%94%D0%B5%D0%BF%D0%B0%D1%80%D1%82%D0%B0%D0%BC%D0%B5%D0%BD%D1%82+%D0%B8%D0%BD%D1%84%D0%BE%D1%80%D0%BC%D0%B0%D1%86%D0%B8%D0%BE%D0%BD%D0%BD%D1%8B%D1%85. Accessed 27 Mar 2021

Remote Work as a Societal Incentive for Creativity: Phygital Initiative for Self-actualization

Natalia V. Burova[1]([✉]) [iD], Eleonora B. Molodkova[1] [iD], Anastasia V. Nikolaenko[2] [iD], and Olga A. Popazova[1] [iD]

[1] St. Petersburg State University of Economics, 21, Sadovaya Street, St. Petersburg 191023, Russia
nbourova@unecon.ru, nnp@spbstu.ru
[2] North-West Institute of Management – Branch of the Russian Presidential Academy of National Economy and Public Administration (St. Petersburg), Sredny Prospect VO, 57/43, St. Petersburg 197376, Russia

Abstract. The concept of the basic guaranteed income for citizens is widely discussed by scientists and politicians. The pandemic forced authorities around the world to implement punctual payments to the whole population or some categories (such as families with children, compensation of losses for businesses at lockdown, etc.). These subsidies raised again the question of the future of labor: if routine work is accomplished by robots and bots, computers and neuron-networks, what activities stay for humans? The corporate loyalty of employees was highly appreciated during last centuries. But the pandemic introduced remote regime of work for the masses of specialists, the lockdown for services sector business and the unemployment for the professionals who lost their stable jobs, showed the necessity to discuss the content of the labor contract. The article represents the results of the exploratory interview with 43 employers (mainly from small and medium enterprises) to identify the core symptoms of the evolution of labor. The answers on labor contract' essence, given by employers, allowed researchers to raise hypotheses about the new features of the labor market: if the most appreciated content of labor is the creative contribution, how the interaction between business and labor should be organized? During the pandemic, the rupture widened rapidly between the successful part of population who taken the functions of their own autonomous value creation process, and the other non-privileged part who lost their integration into societal and, particularly, economic system.

Keywords: Labor market · Phygital activity · Creativity · Initiative · Human capital investment · Entrepreneurship · Divergence

1 Introduction

The total digitalization at the beginning of 3rd millennium launched global transformation of value creation chains with a deep change of the role of humans involvement.

The robotized technological lines replace humans with machines, production system conceived according to the approach of cyber-physical systems and Internet of Things (IoT) require the human engineering spirit and smart skillfulness before and after the production process, but not inside [1, 2]. The pessimistic vision describes the future where humans can be useful for machines to improve and to repair them, while the optimistic outlook finds out new opportunities for leisure and creativity: when the toil is fulfilled by machines, and the mission of personalities is the flourishing [3–6].

The automation as technological progress helps to assure the innovative economic growth [7–9] and to develop societal structures and processes, to enhance gender equality, to improve the inclusive strategies for human capital of people with particular needs, to soften regional inequalities [10–12]. The remote regime of work from home (or telework [13, 14]) is widespread since the 1990s when the global web developed and the Internet penetrated to outlying territories, OECD published the share of the populations with diverse kinds of activities in 2015 (see Fig. 1):

The remote work concerned wide layers of population due to the policy of physical distancing and social isolation that was provoked by covid-19. It demonstrated the limited possibilities to empty the producing facilities from people, but at the same time, the remote activities appeared more effective for some categories of personnel and for some functional positions and several kinds of tasks, i.e., the long discussions during meetings became shorter with video conferencing; the accounting work is executed well at home and does not require to bear the expenses of spend time, money and risk to transport, etc. [15–17]. These contradictory trends provoked the opposite reaction among employees and, even, inside the same person – the remote work made people to assure a number of investments and get a number of advantages.

Among the expenses and specific efforts experts mention home place use, family coordination of home duties such as care of older parents and little babies, internet access and telecommunication infrastructure shared with other members of family living in the same flat, the purchase of equipment, furniture, increased payments for electricity, water and costs of the use of household facilities, new skills to be acquired to use the digital communication instruments [18–22], etc.

Among the advantages the individuals remarked some considerable opportunities of the physical comfort with the preferred food and clothes at home, of the psychological comfort for introverted persons, of the reduced costs to shoes and dress required by office code, the new enrichment in the skills of digital communication and of various services available online, the improvement of their furniture, infrastructure, equipment and, sometimes, even the decoration of the house, etc.

The pandemic accelerated the formation of the three dimensions of communication and professional activities – the remote work illustrates the combination of physical presence and longitude of some processes, of data and digital representation of human imagination and conscience in depth of virtual reality, and of the Phygital interaction existing on the border between physical world and digital technological tools (such as augmented reality, digital twins, or medicalization of behaviors and choices, care about patients or relatives due to the bracelets helping to monitor the state of health).

The phydital world usually is studied as an environment where individuals play passive role of monitored and seduced by marketing tools. The pandemic emphasized

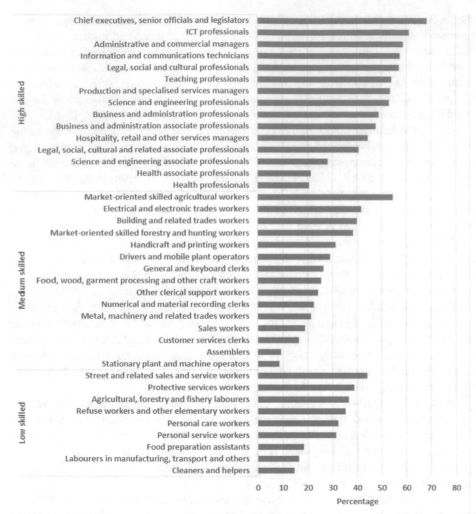

Fig. 1. Use of telework varies widely across occupations: Cross-country average by occupation of share of people using telework in 2015 [14, p. 23].

the role played by some curious and proactive people who produce the improvement of their life and work due to the options and possibilities of this combination, where the cyber-physical systems serve the needs. This will to improve and evolve is one of the core elements that distinguish self-actualization-driven personalities. Enriching phygital experience is based on the initiative of the person, who seeks the new time-passing ways and pushes the barriers.

This paper aims to deepen the understanding from the managers and entrepreneurs point of view of the phydital initiative as a sign, a symptom of the efficient employees and to broaden the perception of the precarious work that helps to employers to cut costs, but the freelance as regime and freelancer as a person are chosen by employers for the other reasons than the cost saving – the more considerable argument is the attractiveness of a wide-experienced professionals with their autonomous independent judgments and visions.

The core purpose of this paper is to detect if the employers prefer to hire the future loyal collaborators who will integrate into corporate system, or they appreciate higher the possibility to be in touch with risky human capital [23, 24] such as high-qualified freelancers, invited external experts and specialists working for different companies. The concepts of open innovations [25–27] and open knowledge reduces the sharp problem of commercial confidence and intellectual property protection, because the knowledge-driven economy orients companies to use networking and to share the ideas rather than hide them.

The research was oriented to the point of view of employers who continued the business during the pandemic. The component of the analysis of the remote work effects that is of interests for researchers is related to the ratio between the obtained gain and the increase of costs of online activities of the employees. It concerns the qualities of the persons (such as psychological feature of introversion) that distinguish the categories of professionals who were more satisfied than unsatisfied with the remote work regime imposed suddenly and inevitably.

2 Materials and Methods

The research presented in this paper tried to look at this distinction from the point of view of employers who remarked the increase of effectiveness of some kinds of staff and the decrease of results of professional activities of other personnel. This difference, obviously, relates to the diversity of personality' characteristics [28, 29], of the background and experience of implementation of digital communication tools for the working tasks, etc. [27, 30, 31].

Using qualitative method of interview, researchers tried to outline the axes for such a dichotomy: according to the different answers of employers, the professional skills, personal business commitment qualities and type of activities were discussed as the large groups of issues. The scenario of the interview included the following questions:

- does business pay for the hours of assistance at a job place, for the efforts made, for contribution to profit achieved by the company, for the growth and development of new projects to adapt to the dynamic preferences of market and for the leadership of the enterprise on its segment, for the value chain and the reputation of the mark among customers and partners;
- what parameters can describe the personal features, the nature of activities and the rules and regulative mechanisms of the corporate governance that have the most significant impact on the effectiveness of the remote work of employees.

These two blocks of questions included the questions about the incentives for risky decisions and choices, for the initiative to improve the process and products, and to the dynamics of the behavior of the employees during the transfer from the "normal" presence activity towards the remote work.

Labor behavior is a subject of two-faceted studies - as a socio-economic activity determined by the social structure and relations between societal actors (this analysis is external to the content of labor), and as process of change of a resource (material or information, real or imagined world) that is subject for organization, and performance of labor functions depend on the level of technology, the mode of management and the socio-cultural model inherent to humans working. The phygital expertise should combine the external and internal analysis and allow entrepreneurs to see the high-skilled collaborators improving their business. The aggregating platforms, the gig economy and neural networks are perceived as instrumental concepts, and the core motivation of the growth is the self-actualization and improvement of the universe.

To obtain a balanced opinion, the sample of the employers included 17 managers of producing enterprises (wooden products, assembling electronic equipment, and a shipyard) and 26 representatives of service companies (research labs, IT sector, design and architecture bureau, education and banking institutions, a hairdresser group, marketing and advertising company, real-estate agency), more detailed description of the sample is presented in the Table 1.

Table 1. Number of employers interviewed, by size of enterprise and sector

Sector	Small and middle	Large enterprises	Total
Wooden manufacturing	2	1	3
Assembling enterprises	1	3	4
Shipyard	-	4	4
Social Research lab	2	1	3
Chemical analysis lab	3	-	3
IT	4	-	4
Design & architecture bureau	1	-	1
Education	2	4	6
Banking	-	3	3
Hairdresser group	4	-	4
Marketing, advertising & PR company	3	2	5
Real-estate agency	2	1	3

The geographical location of the respondents is mostly North-West region of Russia, including the city of St-Petersburg and the Leningrad area (40 respondents), Petrozavodsk and Murmansk (2 universities administrators), and Moscow (1 bank executive officer). The gender and age structure of the respondents is quite balanced – 25 women (58.1%) and 18 men (41.9%), 12 persons of the age between 32 and 45 years (27.9%), 16 respondents were 46–50 years old (37.2%) and 15 persons elder than 51 (34.9%).

The interviews were taken partly by online communication through social media and messengers (9 interviews), but in majority in physical presence (34 cases).

The pilot interviewing was undertaken in late 2020, the research started in Oct 2020 to prepare and finalize the methodic tools, and the interviews were taken in the January–March 2021.

3   Results

The distribution of the answers of employers about the essence of labor which they pay for is presented in Fig. 2:

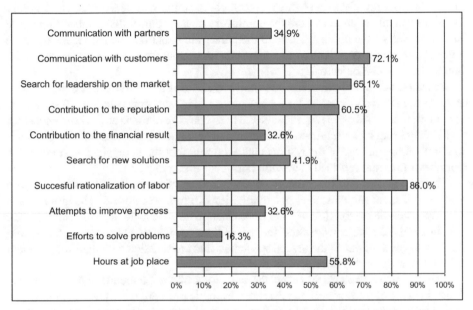

Fig. 2. Employers' support for the importance of components of work of the employees.

In the traditional mass industrial production the most important contribution of employees consisted in their constant physical efforts which were purchased by the employer, these efforts were applied with physical presence of the employee at her or his job place. And the essential factor for the calculation and payment of wage referred to the number of hours spent within the process of manufacturing. The similar approach is registered in the big telecommunication marketing agency, where the number of calls depends on the volume of time spent. More than half of the respondents confirmed that they are ready to pay for the physical presence of an employee at a job place.

Nevertheless, this answer has taken the 5th place, and the leader among answers is the modernization and rationalization of the process of the professional activity of the employee (86.0%). This choice represented the clear preference for the autonomous analysis of the process and for the interest of the employees to make themselves the efforts to improve and to achieve the result in the improvement of the process. The next three places related to the market and customer relationship that are built by employees: 72.1% for the communication with customers, 65.1% for the search of better strategies to achieve the leadership on the market and 60.5% for the construction of the brand and support for the image and reputation of the company.

The unexpected high percentage of the answers related to the search of new solutions (41.9%) which normally is not the task of the employee who should execute the orders and not to undertake the risky actions. It is especially interesting, that another unexpected result concerns the contribution to income and profit. Usually, managers seem to be inclined to invite employees to contribute to the incoming money flow for the companies, but during the interview this low level of choices was explained with the unexpected argument – instead of division of labor and responsibilities, the reason to the low assessment of the employee financial contribution was declared the uncertainty and unpredictable situation where the employers do not control their embedding in the financial result of the company. Due to the pandemic situation, the managers are more concentrated on the contribution of employees for the business processes improvement than to the financial objectives achievement.

Based on this distribution, the question was asked if the employee is a person who executes or makes decisions, seeks for resources and invents new solutions. The answers were unanimous – 42 of 43 respondents told that today the humans at work are necessary only if they bring an entrepreneurial elements of functioning, and the following rules is no more sufficient. One of the respondents even told that the company had fired all the routine workers and replaced them with bots (real-estate agency).

The core question of the research concerned the preference for the employees who work for different projects, and, potentially, for different employers. The attitude was well expected – 38 persons (88.4%) answered that this is unacceptable in their enterprises. But, during the further conversation, the half of the respondents (24 persons, 55.8%) told that they have experienced the invited specialists and are satisfied with their involvement and contribution for the development of the enterprise.

The interview scenario included the questions of the pandemic influence on the choice in favor of free-lancers, but the most dramatic problem which was mentioned by respondents related to the necessity to fire people or to decrease the wages. Nevertheless, 9 persons (20.9%) told that the pandemic permitted to clear distinguish the people who produce value from the "useless" staff. The question followed: what are the characteristics of "useful" staff. The interesting explanations were proposed by respondents (the question was asked to all the respondents), their weight is presented in the Fig. 3:

The respondents gave a clear preference for the personal features of people who are able to take responsibilities (97.7%), to pay attention to their professional skills and growth (90.7%) and to the development of the enterprise (81.4%).

The interviewed managers mentioned the pandemic as an opportunity and a good moment to clearly describe the expectations towards employees. There are some the most typical remarks: "we had to talk in an open manner with people to fire and with people who stayed", "we explained that their duties include the making decision as well as the execution of routine tasks, because for iterative operations we can use chat bots and robotized technologies" (administrator from banking institution, and similar answers were obtained in almost all enterprises, producing goods or service, including the responses for hairdresser saloons' group, from educational administrations, etc.).

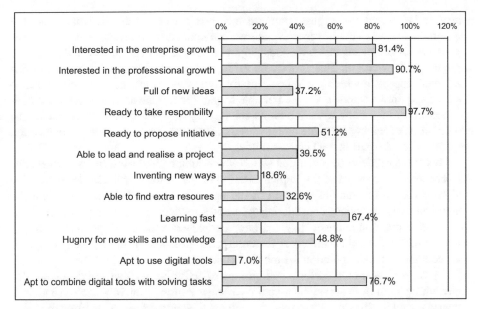

Fig. 3. Employers' description of an attractive and "useful" employee for their enterprise.

This result can be illustrated with the data obtained in the research previous to the pandemic: the studies of 2017 [32] and 2019 [21] demonstrated the predominance among the requirements of the employers towards the efficient integration of a new incoming employee into the organizational system of business-processes, including the socialization and adaptation to organizational culture, the assimilation of values and the subordination norms.

Compared to the studies of 2017 and 2019, the results accumulated now in early 2021 show the evolution of the hierarchy of employers' motives to choose employees (at the stage of selection during recruitment and during the making decisions about dismissals). The pandemic permitted to re-organize the responsibilities of the heads of departments who re-conceive the value-creation chain, and of the employees who re-think and re-construct their labor processes from the point of view of efficiency and optimization of resources they use.

This new clear preference to the autonomy of employees is closely related to the remote work because of impossibility for managers to control the everyday and every-minute functioning of the employee at her / his home. But, the causality is inverse: the pandemic and remote work permitted to embody the search for efficiency and higher involvement and commitment of employees. In fact, the digital monitoring and controlling through various tools (webcams, analysis of the communications through internet, etc.) can be even more detailed and strict than in the offline activities. And managers are more interested to find people who are able to solve problems and to increase the efficiency of their labor instead of just executing instructions. The representative of a chemical laboratory that fulfills complex analyses for global market, with high-qualified staff about 80 persons, expressed this idea: "without pandemic we had reserves to pay wages to people who were just coming 8 h 5 days a week to ask us what to do, but the coronavirus has cut our reserves, and we are concentrated on the real value creating processes of our business, the most sophisticated analytical questions that require ingenuity and invention. This is our core value, and we are forced to implement today the "lean" approach which could be a possibility before pandemic and which became a strong requirement today. We are not able to pay for routine activity, because it does not permit our company to earn enough for paying wages". This idea represents the new attitude of managers to the employees as active players within the value creation process, who take part in the business model.

This answer reflects the evolution which was examined in the last block of questions that concerned the issues of the implementation of the professional and personal skills and competencies to the development of the enterprise. One of the frequent remarks concerned the insufficiency of mastering tools and knowledge: the employees participate in the value creation chain if they use their competence for the improvement of the process and for the enhancement of the enterprise' position on the market, compared with competitors and reputed in the mind of customers.

A professional should perform both groups of activities, routine and creative tasks, approaching them from non-standard points of view. Automation, robotization and digitalization of controlling permit to relieve a person of standard tasks by shifting the focus of attention to cognitive functions – instead of thinking "how to perform an operation", a specialist should be able to constantly ask questions: "how to conceive and modernize" the execution of an operation.

The understanding of what really happens includes three elements: causality, anticipation and regulation (ability to realize an impact). The respondents mentioned the important parameters that can help a person to find the efficient place in the economy to realize her or his intentions, vision of the world, values and priorities.

The high-skilled talented specialists need to develop their creativity competences, but the other categories of the employees would be replaces and stay outside the innovative growth. They seem to be unable to influence on the trends of the societal development, but they would represent the broad groups of population to consume the products created by the people who are competent in creative phygital activities.

The following competencies are outlined to play crucial role for the attractiveness of a personality for the company, according to the opinions of the respondents (see Table 2):

The findings show that the crucial factors for the successful professional self-actualization are the ability to anticipate (88.4%) and to use the new combination of the existing resources (79.1%). These two elements form the system of the phygital proactive initiative, when people master the digital tools and create human values and meanings.

The next three answers in order of decreasing frequency include the creativity of employees in the field of analytical work and communication: discovering evolution and goal-setting as two competences oriented to the future. The analytical capacities of people differ from the intellectual capacities of software: statistical analyses carried out by computing equipment are to be filtered through human understanding of reality and eventual effect, the choice of solutions is to be grounded on ethical fundamental principles to take into account the long-term consequences. This dramatic difference requires

Table 2. Creativity competence as a factor for the active position in the economic system of the societal structure

Blocks of parameters	Creativity	Competencies	Share of respondents
Analysis	Discover the relationships	Distinguish trends in Technology and Market Demand	65.1%
	Reveal causality	New forms of data collection and structure of metadata	25.6%
	Methodology	Integrating methods	60.5%
Conceptual phase	Search for resources and their new combination	Open mind for the new methods and sources	79.1%
	Anticipation and foresight research	Assessment of the consequences of a choice	88.4%
	Prioritizing	Modeling and formatting key significant parameters	41.9%
	Project management	Vision and implementation	53.5%
Communication	Clarification of goals and of failures	Comprehensive failure analysis	62.8%
	Identification and classification of obstacles	Facets of communication (logical, emotional, behavioral, etc.)	7.0%
Change	Study of technologies and social trends	Formation and understanding of new hierarchical structures and layers, groups and categories in society	30.2%
	Invention, discovery, search for new meanings	Proposal of new value-semantic models Diagnosis by symptoms and insight	20.9%

that people involved in business processes would be able to conduct the analysis of market trends and technological evolution (65.1%). The chat bots and management systems (such as CRM, customer relation management, or BPM, business performance management) can help to classify the questions and objectives, but they need to address people to motivate other people, to re-orient the vision and new perception, this corresponds to the ability to communicate in complex situations, to clarify goal-setting and to work with failures (62.8%). The use of various methods to find and solve problems is based on a complex associative reasoning, that represents the ability of employees to integrate methods (60.5%).

4   Conclusion

The research confirmed a change in labor behavior in terms of content – the employers expect that the employees agree to take an increasing part of entrepreneurial functions from the business, responsibility, risks, making choices and making decisions.

The pandemic accelerated the convergence of the offline and online environments, of physical world with digital tools [33–35]. The substance of human life stays in the physical universe, but the symbolic space of imagination and conscience and the digitized field of work and leisure in the online environment form the unique ground for the further development of professionals [36–39]. The proactive attitude of people who want to change the society has new field to be carried out, the combined phygital multi-dimensional space for initiative and action.

References

1. Cappelli, L., Fedorov, D.A., Korableva, O.N., Pokrovskaia, N.N.: Digital regulation of intellectual capital for open innovation: industries' expert assessments of tacit knowledge for controlling and networking outcome. Future Intern. **13**(2), 44 (2021). https://doi.org/10.3390/fi1 3020044
2. Khansuvarova, T., Khansuvarov, R., Pokrovskaia, N.: Network decentralized regulation with the fog-edge computing and blockchain for business development. In: de Waal, B.M.E., Ravesteijn, P. (eds.) 14th European Conference on Management, Leadership and Governance, ECMLG 2018, pp. 205–212. Academic Conferences and Publishing International Limited, Utrecht (2018)
3. Wells, H.G.: The Time Machine, an Invention. William Heinemann, London (1895)
4. Shipunova, O., Evseeva, L., Pozdeeva, E. Evseev V., Zhabenko I.: Social and educational environment modeling in future vision: infosphere tools. In: Kalinina, O. (ed.) E3S Web Conference, vol. 110, p. 02011 (2019). https://doi.org/10.1051/e3sconf/201911002011
5. Asanov, I.A., Pokrovskaia, N.N.: Digital regulatory tools for entrepreneurial and creative behavior in the knowledge economy. In: Shaposhnikov, S. (ed.) 2017 International Conference Quality Management, Transport and Information Security, Information Technologies, IT&QM&IS, St. Petersburg, Russia, 23–30 September 2017, pp. 43–46. IEEE, New York (2017). https://doi.org/10.1109/ITMQIS.2017.8085759
6. Serkova, V.A.: The digital reality: artistic choice. IOP Conf. Ser.: Mater. Sci. Eng. **940**(1), 012154 (2020). https://doi.org/10.1088/1757-899X/940/1/012154
7. Bellini, F., Dulskaia, I., Savastano, V., D'Ascenzo, F.: Business models innovation for sustainable urban mobility in small and medium-sized European cities. Manag. Market. **14**, 266–277 (2019). https://doi.org/10.2478/mmcks-2019-0019
8. Pokrovskaia, N.N.: Tax, financial and social regulatory mechanisms within the knowledge-driven economy. Blockchain algorithms and fog computing for the efficient regulation. In: Shestopalov, M. (ed.) Proceedings of the 2017 20th IEEE International Conference on Soft Computing and Measurements, pp. 709–712. IEEE, New York (2017). https://doi.org/10.1109/SCM.2017.7970698

9. Almazova, N., Rubtsova, A., Krylova, E., Barinova, D., Eremin, Y., Smolskaia, N.: Blended learning model in the innovative electronic basis of technical engineers training. In: Katalinic, B. (ed.) Proceedings of the 30th DAAAM International Symposium, pp. 0814–0825. DAAAM International, Zadar (2019). https://doi.org/10.2507/30th.daaam.proceedings.113

10. Pozdeeva, E.G., Shipunova, O.D., Evseeva, L.I.: Social assessment of innovations and professional responsibility of future engineers. IOP Conf. Ser. Earth Environ. Sci. **337**, 012049 (2019). https://doi.org/10.1088/1755-1315/337/1/012049

11. Bylieva, D.S., Lobatyuk, V.V., Rubtsova, A.V.: Information and communication technologies as an active principle of social change. IOP Conf. Ser. Earth Environ. Sci. **337**, 012054 (2019). https://doi.org/10.1088/1755-1315/337/1/012054

12. Cappelli, L., Ruggieri, R., Pokrovskaia, N.N.: Digital twins as a tool for agile transformation in innovative manufacturing. In: Korableva, O.N. (ed.) Technological Perspective within the Eurasian Space: New Markets and Points Economic Growth, pp. 91–94. Asterion, St. Petersburg (2019)

13. Espinoza, R., Reznikova, L.: Who can log in? The importance of skills for the feasibility of teleworking arrangements across OECD countries. In: OECD Social, Employment and Migration Working Papers, vol. 242. OECD Publishing, Paris (2020). https://doi.org/10.1787/3f115a10-en

14. OECD: Productivity gains from teleworking in the post COVID-19 era: how can public policies make it happen? https://www.oecd.org/coronavirus/policy-responses/productivity-gains-from-teleworking-in-the-post-covid-19-era-a5d52e99/. Accessed 10 Apr 2021

15. Boyko, S.V.; Pokrovskaia, N.N.; Slobodskoy, A.L.; Spivak, V.A.: Socio-economic questions of motivating collaborators in the context of remote work. Sociol. Law. **1**, 6–17 (2021). https://doi.org/10.35854/2219-6242-2021-1-6-17

16. Atkinson, R.D., Brake, D., Castro, D., Ezell, S.: Digital Policy for Physical Distancing: 28 Stimulus Proposals That Will Pay Long-Term Dividends. Information Technology & Innovation Foundation, Washington (2020). https://itif.org/publications/2020/04/06/digital-policy-physical-distancing-28-stimulus-proposals-will-pay-long-term. Accessed 10 Apr 2021

17. Di Mauro, F., Syverson, C.: The COVID crisis and productivity growth (2020). https://voxeu.org/article/covid-crisis-and-productivity-growth. Accessed 10 Mar 2021

18. Almazova, N., Krylova, E., Rubtsova, A., Odinokaya, M.: Challenges and opportunities for russian higher education amid COVID-19: teachers' perspective. Educ. Sci. **10**, 368 (2020). https://doi.org/10.3390/educsci10120368

19. Levitskaya, A.N., Pokrovskaia, N.N.: Career expectations and plans of the young specialists in the labor market. J. Sociol. Soc. Anthrop. **24**(1), 105–137 (2021). (in Russian). https://doi.org/10.31119/jssa.2021.24.1.5

20. Shipunova, O.D., Berezovskaya, I.P., Mureyko, L.M., Evseev, V.V., Evseeva, L.I.: Personal intellectual potential in the e-culture conditions. Espacios **39**(40), 15 (2018)

21. Gelikh, O.Ya., Levitskaya, A.N., Pokrovskaia, N.N.: Sociological analysis of social attitudes of professional and career growth in the context of the digital economy and knowledge society. Sociol. Law. **3**, 6–18 (2020). (in Russian)

22. Ipatov, O., Barinova, D. Odinokaya, M.; Rubtsova, A. Pyatnitsky, A.: The impact of digital transformation process of the Russian University. In: Katalinic, B. (ed.) Proceedings of the 31st DAAAM International Symposium, pp. 0271–0275, DAAAM International, Vienna (2020). https://doi.org/10.2507/31st.daaam.proceedings.037

23. Beckmann, M.: Self-managed working time and firm performance: microeconometric evidence. WWZ Working Paper, No. 2016/01. Center of Business and Economics, University of Basel, Switzerland (2016)

24. Asanov, I.A., Pokrovskaia, N.N.: Digital regulatory tools for entrepreneurial and creative behavior in the knowledge economy. In: Shaposhnikov, S. (ed.) 2017 International Conference Quality Management, Transport and Information Security, Information Technologies,

IT&QM&IS, pp. 43–46. IEEE, New York (2017). https://doi.org/10.1109/ITMQIS.2017.808 5759

25. West, J., Gallagher, S.: Challenges of open innovation: the paradox of firm investment in open-source software. Res. Dev. Manag. **36**(3), 319 (2006). https://doi.org/10.1111/j.1467-9310.2006.00436.x

26. Cappelli, L., Fedorov, D.A., Korableva, O.N., Pokrovskaia, N.N.: Digital regulation of intellectual capital for open innovation: industries' expert assessments of tacit knowledge for controlling and networking outcome. Future Internet **13**(2), 44 (2021). https://doi.org/10.3390/fi13020044

27. Parker, G., Van Alstyne, M., Choudary, S.P.: Platform Revolution: How Networked Markets Are Transforming the Economy and How to Make Them Work for You. W.W. Norton & Company, New York (2016)

28. Mureyko, L.V., Shipunova, O.D., Pasholikov, M.A., Romanenko, I.B., Romanenko, Y.M.: The correlation of neurophysiologic and social mechanisms of the subconscious manipulation in media technology. Intern. J. Civil Engin. Technol. **9**, 2020–2028 (2018)

29. Ababkova, M.Yu., Leontieva, V.L.: Neuromarketing for education: rethinking frameworks for marketing activities. In: Ardashkin, I.B., Martyushev, N.V., Klyagin, S.V., Barkova, E.V., Massalimova, A.R., Syrov, V.N. (eds.) The European Proceedings of Social & Behavioural Sciences EpSBS, vol. XXXV, pp. 1–9. Future Academy, London (2018). https://doi.org/10.15405/epsbs.2018.02.1

30. Ababkova, M.Y., Cappelli, L., D'Ascenzo, F., Leontyeva, V.L., Pokrovskaia, N.N.: Digital communication tools and knowledge creation processes for enriched intellectual outcome – experience of short-term e-learning courses during pandemic. Future Internet **13**, 43 (2021). https://doi.org/10.3390/fi13020043

31. Brusakova, I.A.: About problems of management of knowledge of the digital enterprise in fuzzy topological space. In: Shestopalov, M. (ed.) Proceedings of the 2017 20th IEEE International Conference on Soft Computing and Measurements, pp. 792–795. IEEE, New York (2017). https://doi.org/10.1109/SCM.2017.7970726

32. Avakova, E.B., Pokrovskaia, N.N., Kuznetsov, A.A.: Sociological analysis of education as a system for the formation of intellectual capital in an information society. Izvestia SPbGEU **2**, 123–130 (2019). (in Russian)

33. Asanov, I., Flores, F., McKenzie, D., Mensmann, M., Schulte, M.: Remote-learning, time-use, and mental health of Ecuadorian high-school students during the COVID-19 quarantine. World Dev. **138**, 105225 (2021). https://doi.org/10.1016/j.worlddev.2020.105225

34. Bylieva, D., Almazova, N., Lobatyuk, V., Rubtsova, A.: Virtual pet: trends of development. In: Antipova, T., Rocha, Á. (eds.) DSIC 2019. AISC, vol. 1114, pp. 545–554. Springer, Cham (2020). https://doi.org/10.1007/978-3-030-37737-3_47

35. Afanasyeva, T.S., Grishakina, N.I., Pokrovskaia, N.N.: Marketing in the age of the coronavirus infection pandemic. In: Ivanova, O. (ed.) Russia 2020 - A New Reality: Economy and Society, ISPCR 2020, pp. 229–233. Atlantis Press, Dordrecht (2021). https://doi.org/10.2991/aebmr.k.210222.045

36. Brusakova, I.: Cognitive technologies of information managements of business processes of the digital enterprises. Int. J. Adv. Inf. Sci. Technol. **5**(1), 73–76 (2016). https://doi.org/10.15693/ijaist/2016.v5i1.73-76

37. Pokrovskaia, N., Margulyan, Ya., Bulatetskaia, A., Snisarenko, S.: Intellectual analysis for educational path cognitive modeling: digital knowledge for post-modern value creation. Wisdom **14**(1), 69–76 (2020). https://doi.org/10.24234/wisdom.v14i1.305

38. Brusakova, I.A., Shepelev, R.E.: Innovations in the technique and economy for the digital enterprise. In: 2016 IEEE V Forum "Strategic Partnership of Universities and Enterprises of Hi-Tech Branches (Science. Education. Innovations)", pp. 27–29. IEEE, New York (2016). https://doi.org/10.1109/IVForum.2016.7835844

39. Pokrovskaia, N.N.: Leisure and entertainment as a creative space-time manifold in a post-modern world. In: Ozturk, R.G. (ed.) Handbook of Research on the Impact of Culture and Society on the Entertainment Industry, pp. 21–38. IGI Global, Hershey (2014). https://doi.org/10.4018/978-1-4666-6190-5.ch002
40. Kasyanik, P.M., Gulk, E.B., Olennikova, M.V., Zakharov, K.P., Kruglikov, V.N.: Educational process at the technical university through the eyes of its participants. In: Auer, M.E., Gural-nick, D., Uhomoibhi, J. (eds.) ICL 2016. AISC, vol. 544, pp. 377–388. Springer, Cham (2017). https://doi.org/10.1007/978-3-319-50337-0_36

The Creative Factor in the Competition Between Human and Artificial Intelligence: A Challenge for Labor Law

Irina Filipova[1]([⊠]) [iD] and Natalia E. Anosova[2] [iD]

[1] Lobachevsky State University of Nizhny Novgorod, 23,
Prospekt Gagarina, Nizhny Novgorod 603022, Russian Federation
[2] Peter the Great St. Petersburg Polytechnic University, St. Petersburg,
Polytechnicheskaya 29, Saint Petersburg 195251, Russian Federation

Abstract. Today, creativity is vital for building a successful career ladder. The evolving digital technologies contribute to the achievement of new goals, which gives the employee the opportunity to realize their potential. At the same time, the development of artificial intelligence transforms the workspace, setting the boundaries for workers' creativity. The introduction of these technologies can have serious consequences for employees, thus, people can abandon solving analytical and creative tasks, delegating them to artificial intelligence. This is a serious threat to the anthropocentricity of society. Creativity enables human intelligence to compete with artificial intelligence, which means that the development of the creative abilities of future specialists is very important for preserving the anthropocentrism of society. The law should contribute to minimizing these risks. Since most professionals are involved in labor relations, labor laws can help develop creativity. For example, digital control restricts the employee's right for privacy, while creativity is a skill better developed by a free person. Therefore, it is necessary to limit the use of digital sensors and neurosensors to monitor employees on a legislative level.

Keywords: Skills · Creativity · Artificial intelligence · Employment · Legal regulation

1 Introduction

Today, the society has entered the era of digital transformation. Digital technologies are developing, and they are increasingly used in business, healthcare, education, government and everyday life [1–5]. The so-called VUCA-world is being formed now, the megatrends of which are volatility, uncertainty, complexity and ambiguity. According to experts [6–11], the development of digital technologies in manufacturing, services and agriculture will continue, which means the world of work will undergo major changes.

Billions of people around the world are involved in labor relations, and remuneration for work is their main source of living, so, changes in the world of work will affect most of the world's population. Digital technologies transform production processes

D. Bylieva and A. Nordmann (Eds.): PCSF 2021, LNNS 345, pp. 182–191, 2022.
https://doi.org/10.1007/978-3-030-89708-6_16

and create new responsibilities for workers. Due to the introduction of new technologies into production and the service sector, workers will need to master new skills, adapt to new norms of professional communication, which will inevitably lead to changes in the professional culture of the specialists of the future.

This study aims to determine how the developing digital technologies, primarily artificial intelligence, affect the skills of specialists in the near future. New skills required from workers will inevitably affect the legal regulation of their work. The purpose of this paper is to substantiate the increased value of creativity as a quality of a specialist in the future, the growing risk of a decrease in creativity as a basic human skill due to the development of artificial intelligence and the need to take this risk into account in labor legislation.

2   The Importance of Employee Creativity in the Context of the Expansion of the Technosphere

2.1   Skills of the Specialist of the Future

"Professionalism is one of the important values in the modern culture, an essential factor of personal dignity" [12]. The professional culture of a specialist includes special and professional skills, culture of professional communication, etc. [13, 14].

The professional skills of specialists differ depending on the work they perform. However, there are certain skills that are essential for the employee to develop as a specialist and build a career ladder, and they in many ways make up the professional culture. Among these skills are the following:

Analytical thinking and innovation;
Active learning and learning strategies;
Creativity, originality and initiative;
Leadership and social influence;
Technology use, monitoring and control;
Emotional intelligence;
Complex problem-solving;
Resilience, stress tolerance and flexibility [15], etc.

According to experts, the above skills will become even more important in the future. They were considered to be the most demanded skills for the next 5 years in "The Future of Jobs Report 2020", dedicated to the prospects of the labor market, taking into account the rapid development of technologics and published following the results of the World Economic Forum held in 2020. The labor market has already changed under the influence of the new technologies, and these technologies continue to develop, thereby increasing their impact on the world of work. Most influential are end-to-end digital technologies, since they penetrate almost all sectors of the economy, transform production, management and the service sector. End-to-end digital technologies interact and complement each other, which enhances the effect of their use.

2.2 The Value of Technology for the Work of the Future

Today, some of the technologies in the field of work and employment are called Game-Changing Technologies. These technologies include advanced robotics, additive manufacturing, Internet of Things and wearable devices, electric vehicles, autonomous vehicles, industrial biotechnologies, block chain, virtual and augmented reality [16]. These technologies are likely to transform the economy and labor market, which could lead to significant problems in production and services by 2030.

The process of transition from an industrial economy to a digital economy will result in changing labor relations. Three vectors of change are likely to affect work and employment, each of them associated with the spread of digital technologies:

(1) automation as the replacement of human labor by machine labor;
(2) digitalization as the transformation of physical objects and documents into digital ones and vice versa;
(3) platformization as the use of digital platforms being intermediaries for the algorithmic organization of economic transactions [17].

The opinions of researchers, mainly economists, about the future world of work differ significantly. The researchers agree on a high degree of uncertainty about the consequences of digital transformation in the field of work. Some authors believe that the great number of job losses is changing the economy and society on a global scale, since digital transformation, unlike previous changes, occurs simultaneously everywhere in the world. "There are many empirical estimates by now that suggest that such a radical loss of jobs is due to the substitution of labour by these technologies" [18]. Other researchers state that some jobs will be abandoned and new ones will appear, because the previous history shows that technological progress contributes to the employment growth. "There are many potential jobs created when technology develops. Most obviously, this leads to an increase in the demand for this new technology and, in turn, to an increase in the demand for labour in technology firms" [19]. Other authors believe that while the previous industrial revolutions created new jobs, stimulating progress and bringing wages to a higher level, the current digital transformation is reinforcing the opposite trend: the real wages of workers in the developed countries are shrinking, as the economy needs less and less work force. Technological innovation reduces costs through automation and reduces the need for workers. One of the features of the new wave of technological progress is the emphasis on information products, and there is often rather a small number of highly qualified professionals needed in IT sector [20]. In addition, the development of self-replicating machines signals a crucial moment in the relationship between humans and artificial intelligence.

Artificial intelligence drives digital transformation and has the greatest impact on manufacturing and business. The introduction of artificial intelligence has increased productivity, optimized logistics, reduced costs, etc. Employers are increasingly using collaborative robots with artificial intelligence, which entails an increase in human-machine interaction in production. Artificial intelligence in the form of data mining programs can evaluate workers based on the data obtained from monitoring. There is an intellectual automation in the field of work [21]. A 2019 poll by Deloitte & Touche LLP showed that about 95% of CEOs and 97% of companies' board members see "serious threats and disruptions to their growth prospects in the next two to three years" [20]. This is driving the introduction of smart automation of control processes using artificial intelligence, which make robots a new work force. In 2018, Bruegel Institute conducted a study on the impact of industrial robots on employment and wages in six countries of the European Union accounting for 85.5% of the EU industrial robot market (Germany, France, Italy, Spain, Finland and Sweden). As it turned out, one additional robot per thousand workers reduces the employment rate by 0.16–0.2% [23].

The ongoing intellectual automation will inevitably change the skills required from workers and increase the substitution of workers, thereby increasing the problem of employment [24]. The change in the range of tasks and transformation of the working environment are inevitable [25].

2.3 Creativity as an Essential Skill

Creativity is a key factor for an employee to build a professional career. Creativity is an important component of a high professional culture that an employee needs for successful self-realization in the profession enabling them to perform at a high level, develop professional skills, get adapted to the new demands of society in his field and make a creative contribution to the profession. The professional culture of the specialist includes a wide professional outlook, high competence and the ability to use knowledge and skills effectively. Professionalism as a prerequisite for the development of professional culture requires the ability for self-development, the search for optimal techniques for the creative fulfillment of professional tasks.

The digital transformation of the economy creates the need for workers who are able to contribute to this transformation, solving various creative tasks. Such specialists can identify the problem, highlight meaningful information in the set of data, analyze the situation, compare the facts and predict the consequences, and choose the optimal solution. Creativity drives employees to come up with new ideas, implement them in their work, find non-standard solutions and create innovative products, assessing the need and feasibility of innovations.

In the second half of the twentieth century, the labor market was relatively stable. Career decisions were made at the beginning of their professional career; specialists often worked in the same organization for several decades. In the twenty-first century, the situation is changing. The development of creative skills enhances the intelligence and flexibility of the worker allowing him to integrate into an economy with the expanding use of artificial intelligence. A creative person as a potential employee wins over those who are less creative. Thus, the development of creativity is a necessity for the social well-being in the context of digitalization.

2.4 Dilemma

The following dilemma arises here. On the one hand, high-tech companies that are leaders in technological transformation require an increasing number of highly qualified and creative specialists, in proportion to the total number of employees. The more companies embrace innovation, the greater is the need for creative workers. On the other hand, artificial intelligence makes it possible to free people from calculations and analysis of the situation. Many workers will prefer to delegate their duties to artificial intelligence and rely on machines not only for analyzing situations, but also for making decisions. This increases the risk of a decline in the workers' analytical skills and the risk of failure in the search for independent solutions.

Taking into account the fact that in high-tech companies the need for the number of workers is much lower than in the enterprises of an industrial society, it is highly likely that the need for workers will decrease in the future. The industrial Internet of Things technology makes it possible to create so-called smart factories employing a very small number of specialists. Robots and smart control systems will replace most workers. High-tech equipment and intelligent control systems of smart factories are capable of self-organization. Smart factories make products with an optimal production cycle of an individualized product at the price of mass production, which arouses the interest of investors and stimulates the creation of new industries using this model.

At the same time, there is a threat of the loss of creativity by many workers, since they will not perform a number of creative tasks and get used to relying on artificial intelligence. In order to keep anthropocentricity of society it is important to maintain and develop the level of human creativity. We should keep in mind that creativity enables a person to remain human.

Thus, creativity is one of the most valuable skills of the future specialist. Creativity needs to be developed through education, which means that educational institutions and mentors who work in them should stimulate creativity in students, facilitating the projects aimed at developing freedom of expression and creative skills.

If we do not tackle the potential risk of lower level of creativity in humans, then human intelligence will increasingly lose to artificial intelligence in solving problems. The range of creative tasks that artificial intelligence can solve is constantly expanding, which makes it possible to free employees from dealing with these problems. That is, in parallel with the development of artificial intelligence, the risk of degradation of human intelligence increases. Today, many people are not able to make simple calculations and use calculator, they cannot evaluate the situation and delegate this task to computer. The next generation of people who will have grown up in the conditions of total digitalization will demonstrate this tendency even more vividly.

"Scientific and technological innovations generate acute problems and endanger human existence itself. Modern challenges require that the humanitarian component, which determines a general approach to the problems of social development, should be a systemically important part of professional activity" [26].

2.5 Work and Creativity

Work occupies one of the main places in the life of many people, the loss of a job or a decrease in the importance of the work performed can pose a threat to a full-fledged human life [27]. Hence, the replacement of an employee with robots for performing repetitive, tedious work, such as unfolding or sorting items in a warehouse, is reasonable if the employee monitors the robot, monitors the process and intervenes just in case of problems. On the one hand, the employee's new tasks will require additional training, but, on the other hand, they will reduce the risk of fatigue from tedious work and increase its value to the employee.

Work challenges are very important for a person, and they include the following: "pursuing a purpose, social relationships, exercising skills and self-development, self-esteem and recognition, and autonomy" [28]. The spread of artificial intelligence systems in production affects all of the above. Thus, robots will take on complex tasks, the development of certain skills for workers will lose its importance, and work will be less conducive to human self-realization. At the same time, effective collaboration with robots can increase the professional value of a specialist, and the role of human creativity will increase. For example, a medical doctor who will work with the involvement of artificial intelligence can make a more accurate diagnosis, perform an operation in the least traumatic way, and prescribe a treatment that takes into account the individual characteristics of the patient, thereby acquiring additional opportunities for creative development as a researcher in the field of medicine.

Furthermore, performing more work using special software and robotics, workers are more involved in the digital environment [29, 30]. A creative person often does not act according to a protocol, while artificial intelligence operates according to an algorithm. The increased application of robots in production will also require employees to adhere to a given algorithm, the implementation of which will be controlled by artificial intelligence, and this will "squeeze" creativity out of labor relations. After all, employers control employees, and artificial intelligence enhances control. In addition, collecting data by sensors through which robots receive information about the world around them reduces the autonomy of employees, narrowing the right to privacy, while creativity is a quality that is favored by human freedom from strict external control.

Large companies such as Amazon and Google have been using algorithmic control for several years. Their experience allows us to conclude that workers in such conditions strictly follow the protocol, their work is assessed automatically and a non-standard, creative approach to work can be regarded by an artificial intelligence system as a violation of the protocol, entailing sanctions.

Obviously, production automation raises legal and ethical issues. These problems need to be tackled, and labor legislation can help to achieve a compromise between ensuring effective human-AI interaction and respect for the employee's personality.

In the market economy, the market determines the relationship between its partici-pants through the contract, but when it comes to the organization of labor, the market approach is limited to the subordination of the employee to the employer, which makes the employee extremely vulnerable [31].

Labor legislation as a socially oriented legislation contains a number of provisions that protect workers from a radical change in their status due to external factors. These provisions include the following: the fixed minimum wage, maximum working hours, measures to protect against dismissal, discrimination, etc. In order to preserve the anthro-pocentricity of society, it is necessary to establish the norms of labor legislation that stim-ulate the development of creativity as a skill necessary for the specialist of the future. Firstly, labor legislation should contain the norms that significantly restrict the control of the employer using digital technologies over the performance of the employee. Pro-fessor F. Hendricks believes that it is necessary to recognize the urgent need to formulate requirements in the law for the protection of the information about employees and suggests using the term "Privacy 4.0" analogous to the term "Industry 4.0" used to refer to the technological transition to the digital economy. The level reached to date ensures the impartiality of data processing and eliminates the increasing imbalance between the subject and the object of data collection and analysis. Now, reaching a new level of "Privacy 4.0" in the coming years is a prerequisite of the development of artificial intelligence and robotics, allowing us to avoid dehumanization of jobs [32]. Secondly, it is necessary to introduce quotas for jobs that require the use of creative skills, reserved for people. Soon, the artificial intelligence will be able to replace more and more people in performing duties, including those of a creative nature. Obviously, this increases the economic interest of the employer in using artificial intelligence rather than in hiring employees. It is in the interests of society as a whole to restrict the employer's right to replace workers with artificial intelligence systems and introduction of the quotas in the elaborate labor legislation can be an appropriate step.

3 Conclusion

Creativity is one of the key skills of a specialist. Both the well-being of an individual and the preservation of the anthropocentricity of civilization in the era of digital technologies depend on the level of creativity. The digital transformation of production that is currently taking place provides new opportunities for creativity, but it also entails new risks to the development of human creative skills that are necessary for further development of humankind. The report "The Challenge of Industry 4.0 and the Demand for New Answers", prepared by experts from the Industry Global Union, emphasizes that innovations can radically change "workers' circumstances and industrial work and manufacturing in general" [33].

As practice shows, algorithmic control and use of robotics contribute to the "displacement" of creativity from labor relations, since employees in a digital environment will perform their duties according to algorithms, and artificial intelligence will increasingly control the work process. At the same time, the most promising are the prospects for professionals with a high level of creative skills who win the competition with other workers and with artificial intelligence.

In the new competency models, priority should be given to human and existential skills that create the basis for prosperity and success. In order not to miss this opportunity, it is necessary to implement a number of innovations on the political level, including changes in labor legislation [34].

Regarding creativity as the most important social value will allow humankind to protect itself in the face of digital transformation. Labor law can contribute to it by establishing legal restrictions on the employer's control over the employee's performance using digital sensors and surveillance equipment, and by introducing appropriate quotas for jobs.

Acknowledgements. The article is prepared within the framework of the project "Change in the Principles of Legal Regulation of Labour in Connection with the Transformation of Labour Relations in the Digital Economy"". The reported study was funded by RFBR according to the research project № 19-011-00320.

References

1. Tretyakov, I.L.: Information networks and manipulative technologies in the arsenal of extremists. In: Bylieva, D., Nordmann, A., Shipunova, O., Volkova, V. (eds.) Knowledge in the Information Society. PCSF 2020, CSIS 2020. Lecture Notes in Networks and Systems, vol. 184, pp. 125–135. Springer, Cham (2021). https://doi.org/10.1007/978-3-030-65857-1_13
2. Bylieva, D., Bekirogullari, Z., Lobatyuk, V., Nam, T.: How virtual personal assistants influence children's communication. In: Bylieva, D., Nordmann, A., Shipunova, O., Volkova, V. (eds.) Knowledge in the Information Society. PCSF 2020, CSIS 2020. Lecture Notes in Networks and Systems, vol. 184, pp. 112–124. Springer, Cham (2021). https://doi.org/10.1007/978-3-030-65857-1_12
3. Kazaryan, R., Pogodin, D., Andreeva, P., Galaeva, N., Tregubova, E.: System approach to using information modelling technology in sustainable construction production development. In: E3S Web Conference, vol. 258, pp. 09009 (2021). https://doi.org/10.1051/e3sconf/202125 809009
4. Bylieva, D., Lobatyuk, V., Tolpygin, S., Rubtsova, A.: Academic dishonesty prevention in e-learning university system. In: Rocha, A., Adeli, H., Reis, L.P., Costanzo, S., Orovic, I., Moreira, F. (eds.) 8th World Conference on Information Systems and Technologies. Advances in Intelligent Systems and Computing, vol. 1161, pp. 225–234. Springer, Cham (2020). https://doi.org/10.1007/978-3-030-45697-9_22
5. Xylander, C. von: A(l)gora: the mindscape. Technol. Lang. **1**, 115–125 (2020). https://doi.org/10.48417/technolang.2020.01.23
6. Korinek, A., Stiglitz, J.E.: Covid-19 driven advances in automation and artificial intelligence risk exacerbating economic inequality. BMJ **372**, 367 (2021). https://doi.org/10.1136/bmj.n367
7. Klinova, K., Korinek, A.: AI and shared prosperity. In: Proceedings of the 2021 AAAI/ACM Conference on AI, Ethics, and Society (AIES 2021), pp. 1–7. ACM, New York (2021). https://doi.org/10.1145/3461702.3462619
8. De Vos, M.: Work 4.0 and the future of labour law. SSRN E-J. 1–27 (2018). https://doi.org/10.2139/ssrn.3217834
9. De Stefano, V.: Negotiating the algorithm: automation, artificial intelligence and labour protection. Comp. Labor Law Policy J. **41**(1), 1–32 (2019). https://doi.org/10.2139/ssrn.317 8233
10. Kalantzis-Cope, P., Gherab-Martín, K.: Emerging Digital Spaces in Contemporary Society. Palgrave Macmillan UK, London (2010). https://doi.org/10.1057/9780230299047
11. Gherab-Martín, K.: Interdisciplinariedad y redes epistemológicas de la ciencia en Internet. Arbor. **CLXXXV**, 611–622 (2009). https://doi.org/10.3989/arbor.2009.i737.317
12. Karmazina, E.V.: Professional culture of the specialist of the future: moral aspects. In: Chernyavskaya, V., Kuße, H. (eds.) European Proceedings of Social and Behavioural Sciences EpSBS, vol. 51, pp. 803–812. Fututre Academy, London (2018). https://doi.org/10.15405/epsbs.2018.12.02.87
13. Barinova, D., Ipatov, O., Odinokaya, M. Zhigadlo, V.: Pedagogical assessment of general professional competencies of technical engineers training. In: Katalinic, B. (ed.) Proceedings of the 30th DAAAM International Symposium, Vienna, Austria, pp. 0508–0512. DAAAM International (2019). https://doi.org/10.2507/30th.daaam.proceedings.068
14. Valieva, F., Fomina, S., Nilova, I.: Distance learning during the corona-lockdown: some psychological and pedagogical aspects. In: Bylieva, D., Nordmann, A., Shipunova, O., Volkova, V. (eds.) Knowledge in the Information Society. PCSF 2020, CSIS 2020. LNNS, vol. 184. pp. 289–300. Springer, Cham (2021). https://doi.org/10.1007/978-3-030-65857-1_25

15. The Future of Jobs Report 2020: World Economic Forum (2020). https://www.weforum.org/reports/the-future-of-jobs-report-2020. Accessed 2 Apr 2021
16. Peruffo, E., Rodríguez Contreras, R., Mandl, I., Bisello, M.: Game-Changing Technologies: Transforming Production and Employment in Europe. Eurofound, Publications Office of the European Union, Luxembourg (2020). https://doi.org/10.2806/054475
17. Fernández-Macías, E.: Automation, Digitisation and Platforms: Implications for Work and Employment. Eurofound. Publications Office of the European Union, Luxembourg (2018). https://doi.org/10.2806/090974
18. Lewney, R., Alexandri, E., Storrie, D.: Technology Scenario: Employment Implications of Radical Automation. Eurofound. Publications Office of the European Union, Luxembourg: (2019). https://doi.org/10.2806/88443
19. Storrie, D.: The Future of Manufacturing in Europe. Eurofound. Publications Office of the European Union, Luxembourg (2019). https://doi.org/10.2806/44491
20. Korinek, A.: Labor in the age of automation and artificial intelligence. Economists for Inclusive Prosperity (2019). https://econfip.org/wp-content/uploads/2019/02/6.Labor-in-the-Age-of-Automation-and-Artificial-Intelligence.pdf. Accessed 10 Apr 2021
21. Krenz, A., Prettner, K., Strulik, H.: Robots, reshoring, and the lot of low-skilled workers. Euro. Econ. Rev. **136**, 17 (2021). https://doi.org/10.1016/j.euroecorev.2021.103744
22. Palmer, N., et al.: Intelligent Automation: Rules, Relationships and Robots. Future Strategies Ink, Book Division, Lighthouse Point, Florida (2020)
23. Chiacchio, F., Petropoulos, G., Pichler, D.: The impact of industrial robots on EU employment and wages: a local labour market approach. Working Paper, vol. 2, Bruegel, Brussels (2018). https://www.bruegel.org/wp-content/uploads/2018/04/Working-Paper-AB_250 42018.pdf. Accessed 1 Apr 2021
24. Cappelli, P.: The Consequences of AI-based Technologies for Jobs. Working Paper. Publications Office of the European Union (2020). https://doi.org/10.2777/348580
25. Brynjolfsson, E., Mitchell, T., Rock, D.: What can machines learn and what does it mean for occupations and the economy? AEA Papers Proc. **108**, 43–47 (2018). https://doi.org/10.1257/pandp.20181019
26. Maslov, V.M., Sosnina, E.N., Erdili, N.I., Khorunzhii, V.P.: Professional activities in the context of educational humanitarian expertise. In: Chernyavskaya, V., Kuße, H. (eds.) European Proceedings of Social and Behavioural Sciences EpSBS, vol. 51, pp. 398–406. Fututre Academy, London (2018). https://doi.org/10.15405/epsbs.2018.12.02.43
27. Danaher, J.: Will life be worth living in a world without work? Technological unemployment and the meaning of life. Sci. Eng. Ethics **23**(1), 41–64 (2017). https://doi.org/10.1007/s11948-016-9770-5
28. Smids, J., Nyholm, S., Berkers, H.: Robots in the workplace: a threat to – or opportunity for – meaningful work? Philos. Technol. **33**, 503 (2020). https://doi.org/10.1007/s13347-019-003 77-4
29. Filipova, I.A.: Labour law: challenges of digital society. Pravo. Zhurnal Vysshey shkoly ekonomiki **2**, 162–182 (2020). (in Russian). https://doi.org/10.17323/2072-8166.2020.2.162.182
30. Pokrovskaia, N.N., Korableva, O.N., Cappelli, L., Fedorov, D.A.: Digital regulation of intellectual capital for open innovation: industries' expert assessments of tacit knowledge for controlling and networking outcome. Future Internet **13**, 44 (2021). https://doi.org/10.3390/fi13020044
31. Gruber-Risak, M.: Working in 2030: heaven or hell? Why regulation, standards, and workers' representation will still matter. In: Güldenberg, S., Ernst, E., North, K. (eds.) Managing Work in the Digital Economy. Challenges, Strategies and Practices for the Next Decade, pp. 99–110. Springer, Cham (2021). https://doi.org/10.1007/978-3-030-65173-2_7

32. Hendrickx, F.: From digits to robots: the privacy-autonomy nexus in new labor law machinery. Comp. Labor Law Policy J. **40**(3), 365–387 (2019)
33. The Challenge of Industry 4.0 and the Demand for New Answers. http://www.industriall-union.org/sites/default/files/uploads/documents/2017/SWITZERLAND/Industry4point0 Conf/draft_integrated_industry_4.0_paper_5_17.10.2017.pdf. Accessed 10 Apr 2021
34. Chin, V., Puckett, J., Boutenko, V.: Mass Uniqueness. A Global Challenge for One Billion Workers. https://rda.worldskills.ru/storage/app/media/Reports/2019%20BCG/2019_BCG%20Mas%20Unikum_Report_EN.pdf. Accessed 1 Aug 2021

Legal Aspects of Artificial Intelligence Application in Artistic Activity

Vladimir Demidov⬤, Ekaterina Dolzhenkova⬤, Dmitry Mokhorov⬤,
and Anna Mokhorova(✉) ⬤

Peter the Great St. Petersburg Polytechnic University, Polytechnicheskaya, 29,
195251 St. Petersburg, Russia
mokhorova@list.ru

Abstract. The rapid development of artificial intelligence (AI) and its acquisition of human qualities necessitates the establishment of an appropriate legal framework. Today, various doctrinal concepts are being developed in order to define AI within the framework of subject-object legal relations: from traditional ideas about AI exclusively as an object of law to innovative ones that equate AI in legal status with a person. Among others, theories of fiction are also being used in order to fix in legislation a certain set of characteristics for AI. Their use has made it possible to identify significant legal aspects inherent in the problems of introducing AI as comparable to objects of literature, visual, or visual arts, as well as in the field of creating audio and video works. The creation of AI technologies requires a clear legal certainty; it seems relevant to develop unambiguous criteria for formulating the very concept of AI. As a result of considering the issues of AI's legal regulation, it has been established that civil legislation does not adequately meet the requirements that have formed in everyday reality in connection with the rapid scientific progress. It is necessary to make changes to certain norms of the Civil Code, as well as to adopt executive legislative provisions, so as to determine how processes using AI should be regulated. This should meet both the interests of developers of new information devices as well as users who are faced with widespread adoption of the latest technological advances.

Keywords: Artificial intelligence · Artistic activity · Legal regulation

1 Introduction

Modern technologies have made it possible to apply artificial intelligence (AI) in various spheres of life by constantly advancing it to perform an increasing number of different tasks even without human intervention. Many experts believe that AI already surpasses natural, or human, intelligence [1]. This conclusion has given rise to both legal and ethical issues resulting in the demand for the state and the academic community to propose regulations concerning the legal status of AI and conditions of its application [2].

Unfortunately, the legal regulation of AI remains significantly underdeveloped for the overall current state of robotics and does not compare to other fields in legislative frameworks [3].

© The Author(s), under exclusive license to Springer Nature Switzerland AG 2022
D. Bylieva and A. Nordmann (Eds.): PCSF 2021, LNNS 345, pp. 192–202, 2022.
https://doi.org/10.1007/978-3-030-89708-6_17

The legal community is currently faced with a number of obstacles that constrain the establishment of a strong regulatory basis for AI application. First, the very notion of AI has not yet been defined clearly and unambiguously owing to its interdisciplinary nature. Secondly, the whole establishment of universal standards is deemed highly unlikely due to a wide range of AI applications. Therefore, it is urgent that the notion of AI be recognized legislatively, subjects of legal regulation be identified and participants of legal relations be defined to minimize legal uncertainty.

2 Literature Review

Currently, there has have a growing number of studies on possible applications of AI in various areas, for example, in design projects [4–6] and virtual reality [7]. Another area of interest is creating smart clothing with the use of AI [8]. Besides, AI is used in teaching music [9, 10] and languages [11], in performing translation functions [12], even in forensics [13]. All this, in turn, results in more advanced and more skilled AI systems capable of operating without human intervention. AI is of particular importance in the digital economy [14–16].

Law is not immune from disruption by new technology [17]. Considerable work was done on developing "legal expert systems", legal advice services, tools and platforms to enable usage of free legal advice systems [18], methods of forensic [19]. Thus, it is crucial that legal regulations be established regarding activities involving artificial intelligence [20, 21].

Numerous concepts of the legal regulation of AI are being developed, which can be conditionally divided into several groups: the legalistic approach involves the identification of the cornerstone problems of AI application and their regulation [22, 23]. For example, in recent years, the United States has consistently adopted laws for regulating specific aspects of the AI usage, including at the state level. Another approach is technological and supposes the secondary nature of law and the absence of the need to adopt a systemic legal act regulating the use of AI (this applies mainly in civil law countries, and France in particular). To date, the European Union has developed a number of policy documents on the development of artificial intelligence. Most of them emphasize the need to create ethical boundaries.

The third approach is a combined one that separates the ethics of using robots and robotics, and the proponents of this approach point the need for legal regulation of robots exclusively. In 2015, China launched a comprehensive program for the development "Made in China 2025", and in July 2017, the "Next Generation Artificial Intelligence Development Plan" adopted by the State Council of the People's Republic of China was published, but both legal and ethical regulation are in their infancy [24].

Many researchers point the "end-to-end" nature of AI technology, which indicates the need to develop regulations in all branches of law, both public and private [25–28].

Within the framework of private law, national and international experts actively promote the concept of "electronic persons" that grants AI a certain legal status [29–36]. Some experts also debate the right of AI to intellectual property [37–41]. Some authors [42, 43] studied liability and risks of AI application. However, despite a strong interest of the academic community, the issue under consideration still lacks a complete and comprehensive research.

3 Methods

The issue of legal regulations of AI is deemed rather ambiguous today as no uniform approach has been made yet by legislative drafters in cooperation with information systems developers. The ability to perform artistic activity is something that only human beings possess and that distinguishes them from other biological species and modern technical devices. This statement implies that only a human person can have intelligence. However, a rapid growth of advanced information technologies challenges its accuracy, thereby highlighting the urgency to study the issue of legal regulation of AI. Various approaches and research methods should be used in studying the legal framework of the application of AI-aided automated system and robots to identify some legal characteristics of the new area of public relations. The following methods were used in the study. The legal approach used to assess consequences of introducing AI into the everyday activity of human beings on the legal basis has determined rather specific criteria to define some notions operating in the area of information technologies. The unambiguity of terms is a crucial condition required to apply legal techniques in a detailed characterization of AI-related processes.

Studying AI-related issues must be systematic, as the introduction of innovative technical systems and devices has a great impact on a wide range of public relations that are being established within the complex social and biological system known as the human society. Moreover, the application of AI requires a comparative analysis of legal systems as the civil law of many countries states there must be special institutions that regulate relations similar to the relations arising from the introduction of advanced technical systems into the everyday human life.

Such complicated relations between a human being, AI, and legislation must be investigated through an objective approach to law-making. On the one hand, the establishment and development of AI as a new subject of a human society, although without a clearly determined status yet, can be observed as a force destructive to the current system of social activities and that is likely to destroy social relations in the future as well. On the other hand, this new subject must be taken under control and introduced into civil relations. However, these intentions are faced with challenges posed by a constantly growing and improving AI and its ambiguous role in the social life and civil relations toward which it is striving to be a part.

4 Results

4.1 Legal Aspects of the Application of AI in Artistic Activity

The application of AI requires a comprehensive improvement of legal and regulatory acts that would ensure unambiguity and certainty within relations arising from the use of advanced technologies and scientific innovations. The civil law of leading countries currently fails to present a uniform legal regulation on the introduction of AI into human beings' artistic activities. Artistic activity also involves an economic aspect, in particular making profit from the distribution of creative works. First and foremost, it is necessary to define AI-aided devices from the point of view of economic relations regulated by civil law; this will significantly determine the efficiency of AI application in supporting the social and economic progress in the future. However, no uniform criteria have yet been established yet regarding legal regulations of AI-related phenomena. The civil law places a great emphasis on the status of participants of civil relations. Commonly, civil relations involve either natural persons or organizations as legal persons. This approach fails to define the legal status of AI-aided technical devices as their legal characteristics do not conform to this enclosed system of civil law participants. AI-aided devices cannot be considered as either natural or legal persons due to their lack of consciousness attributed to humans only. Even the most complex automated systems operating on the most advanced software do not possess consciousness, the feature that makes an individual a complete member of society. Although AI-aided devices may execute the most difficult technical projects and substitute humans in many aspects of social and economic relations, the civil law states that the absence of one of participants in civil relations cancels such relations entirely.

Hence, as AI-aided devices cannot be treated as natural persons, it is recommended that new regulations be added into the civil law that would clearly and unambiguously define the status of automated AI systems. We believe AI systems cannot be referred to as legal persons. A legal person acts as a legal personality for its individual members, or human beings. Therefore, it can be concluded that AI-aided devices do not meet the requirements of the civil law as they are not considered a body corporate. Moreover, actions performed by an AI-aided machine are rarely influenced by the will of a person operating the high-tech device. Recently, software engineers have been able to improve AI-aided devices to such a degree that they are now capable of calculating and following their own ways of development. In other words, software developers have admitted that they do not have enough data now to predict all kinds of actions AI-aided devices might be capable of in the future following this man-generated self-development feature. The legal problem arising in this situation proves that some articles should be introduced into the civil law of all states that would clearly and unambiguously define the legal status of AI and its application.

All of the above-stated showed that it would be indeed legally incorrect to treat AI-aided devices as natural or legal persons in civil relations; therefore, some experts propose that they should receive the status of an electronic person. They believe that this would provide a more accurate control of relations being established between members of society in various areas of artistic activity. The main advantage of introducing a new kind of participant into civil relations would be a significant decrease in the number of legal gaps in social and economic operations regulated by the current civil law. In other words, the introduction of an electronic person might result in a more efficient control of the civil life. At the same time, this concept raises questions of appointing legal liability between the developers of AI-aided devices and those natural and legal persons applying such automated advanced machines in their life. Thus, regulations on the status of electronic person should be formulated in great detail in a series of specific legal and regulatory acts that do not conflict with the current civil law.

Lawmakers have to take into consideration other additional yet important legal characteristics that would separate the electronic person from natural and legal persons. This would secure the status of an electronic person as a legal subject and adjust the civil norms to the ever-changing conditions of developing and applying the innovative artistic process.

4.2 Convergence Between Artificial Intelligence and Artistic Activity in International Law

As of today, the international law states that the application of computer devices should be treated as the application done by natural or legal persons, as written in the United Nations Convention "On the use of electronic communications in international contracts" of 2005. This also covers the application of AI-aided devices. The Resolution of the European Parliament proposes quite an interesting and promising take on the legal status of AI-aided devices as seen in "Civil law rules on robotics" of February 16, 2017. The law suggests to define AI as a specific participant of civil relations. The document describes an AI-aided device as an electronic person, hence implying that it possesses certain rights and responsibilities as a participant of civil relations. The document also highlights that AI cannot act as a complete substitute for a human being in relations regulated by civil law. Instead, these high-tech information devices should be used to create more opportunities to grow for the society, eventually expanding the application of AI onto various areas of social and economic activities. The Resolution says that AI-aided devices do not incur civil liability; property responsibility is ultimately imposed on a natural person or a legal person that acts as a body corporate and, according to the current civil regulations, is subject to civil sanctions in case of damage to a human being's life, health or property, or to property interests of legal entities or states as a whole.

The EU will fulfill effective control using investments from the union and member countries [44]. It is planned to implement a program based on the principles of excellence and trust. Improving AI improves the EU's opportunities to become a global competitor. Trust is based on creating favorable environmental that will involve all actors in the digital environment (users, developers, etc.). The strategy also includes legal regulation that will create a safe environment for the development of AI in the united Europe. These conditions will allow for the attraction of internal developers and users, and participants from third countries, which will achieve the goal of being competitive in the global market.

Other points to be taken into consideration are the legal issues concerning artistic activity of human beings and AI.

AI can create pictures that can be appreciated by people along with pictures made by the human hand. In one case, respondents were not told which pictures AI had "made". The result of the survey surprised the developers of this system – in many cases, paintings created by AI were rated higher than those painted by humans [45]. In August 2017, the Amper AI system composed the music for the "I AM AI" album. [46]. In April 2020, Open AI released the Jukebox, a neural network that generates music in a variety of genres, but, unlike painting, AI's advances in audio are not yet so impressive. While the Jukebox represents a bold leap forward in terms of music quality, audio length, and ability to tune in to an artist or genre, the differences between artificial music and human-made works are still noticeable.

AI was able to fill out a two-page fragment of a piano composition by Dvoržak in E minor into a complete work 115 years after the death of the composer. It should be noted that this service became the first virtual artist in the world to be officially registered as a composer with the French copyright organization and SACEM in Luxembourg. This means that the works of AIVA are protected by copyright, as well as the works of a human composer.

Robots are known that create poems, for example WASP (Wishful Auto-matic Spanish Poet) analyzes already existing poems of Spanish poets and other information, and then formulates a poem [47].

In mid-January 2020, a court in Shenzhen, China ruled that an article created using artificial intelligence was copyrighted. This is the first time that a court has copyrighted text written by an AI.

In July 2021, The Times published information that Intellectual property officials in South Africa have become the first in the world to award a patent that names an artificial intelligence as the inventor of a product. The moment is a triumph for Ryan Abbott, a professor at the University of Surrey who has for years led a battle for patent offices around the world to recognise artificial intelligences as inventors.

More advanced AI-aided automated systems might eventually overtake human beings and take the lead in creating works of art. Although AI is currently proposed to be used more as a processing tool, development prospects and modification of AI-aided machines demonstrate real possibilities for AI to create its own works of art in the near future on the basis of machine learning and autonomous operation. The importance of such an outcome is that AI-aided devices will be granted a considerably high degree of autonomy that will allow them to operate without direct human interference; in other words, they will become the ones making decisions. This would significantly decrease the extent of human influence on the practical application of AI, which would result in complete elimination of all areas of AI application important for the human society. Therefore, it is highly crucial to establish legal mechanisms as soon as possible to regulate the areas of automated artificial systems application described above and control AI application in general.

At present, the civil law gives a clear statement that only a human being acting as an individual possessing consciousness can be the creator of a work of art. However, there have been many examples of AI creating works of intellectual property. Thus, an AI system running the most advanced software can really be capable of realizing the most complex challenges of playing a chess game or creating new works of fine arts or music that are original and unique. In other words, AI has the potential to substitute a human being in various areas of their artistic activity. At the same time, the Supreme Court of the United States, a highly respected judicial authority, in its ruling in the case of Feist Publications, Inc. v. Rural Telephone Service Co., Inc. – 499 U.S. 340, 111 S. Ct. 1282 (1991) established that any work without a minimum of original creativity by a human being cannot be protected by copyright. As can be seen, the court named creativity and consciousness expressed by a natural person as essential characteristics of a work of art, which clearly makes a distinction between works of human creativity and those presented by AI. Such a statement seems rather controversial due to the current state of advanced information systems. Still, it can be used to differentiate to a great extent works of human creativity from the intellectual property created by AI and protected by copyright.

4.3 Convergence Between Artificial Intelligence and Artistic Activity in Russian Law

Prerequisites for a sufficiently effective regulation of legal aspects concerning the application of AI-aided devices are not confined to international regulatory acts only.

Today, intellectual property rights only protect works created by humans, but according to researchers, the difference between works created by artificial intelligence and human beings "is determined only by the law, which calls the copyright holder exclusively a person. Otherwise, the results of human and robot activity when solving the same problems will meet the same criteria" [48, p. 33], which means the emergence of a number of issues before legal scholars related to the legal status of AI, the consequences of its activities and responsibility for the harm that may be caused to them. The last question is the most important. With the introduction of more advanced artificial intelligence systems, the likelihood of harm to humans will only increase [34].

Article 137 of the Civil Code of the Russian Federation imposes property law on a rather unusual physical object – animals. This implies that a property owner is subject to civil liability for any damages caused by an animal to other members of society or legal persons. Thus, it seems quite reasonable to apply the approach used in this article of the Russian Civil Code to define the legal status of AI. Although it is rather controversial, this solution still helps to create legal certainty in the regulation of civil liability emerging in event of damages caused by AI-aided machines.

Particular attention should be paid to Article 1064 of the Civil Code of the Russian Federation that states that any damage inflicted on a citizen or their property shall be subject to full compensation by the person who inflicted the damage; in this case, the owner of an autonomous agent is considered to be such a person. An autonomous agent might be a complex information technology system in the form of AI that belongs to a specific natural or legal person or a state in general by reason of ownership or any other property right.

According to the article of the Russian Civil Code mentioned above, the status of an autonomous agent means that any AI-aided device acquires the right to exercise the powers delegated to it by its owner in the event of emerging civil relations due to the fact that its complex electronic system allows for variability of actions within the substantially changing natural and social environment. At the same time, exercising Article 1064 of the Russian Civil Code reveals the following issue: which participant of specific civil relations resulting from the application of AI will be held accountable and be ordered to pay compensation for damages caused by active actions of complex devices. Thus, it is crucial that a full definition of an autonomous agent in the civil code consider developments in the area of AI so that it corresponds to changes happening in the cutting-edge science and technology.

5 Discussion

The application of robotics in relations regulated by civil law requires strict certainty when altering texts of legal and regulatory acts. Innovations introduced into civil laws, regulations, rules and instructions that always follow changes in the fundamental civil acts must take into consideration specific characteristics of implementing AI into human artistic activities. The process of aligning the civil code with high-tech advanced mechanisms operating on the latest achievements of information technologies is very complicated. However, without it, the delay in corresponding legal and regulatory acts of the civil law to the actual innovations in information and technology will have an adverse effect on the introduction of AI-aided devices, hence greatly impeding the progress of applying new information technologies in human artistic creativity.

The regulation of relations emerging with the application of AI highly depends on altering legal articles of those chapters of the Civil Code of the Russian Federation that deal with the protection of the rights of legal and natural persons associated with creating and applying the latest information and technology systems. The choice should be made for those innovations of the civil law that are required now to establish a legal basis for the application of various AI-aided devices. When altering individual norms and regulations of the Civil Code of the Russian Federation, it is recommended to apply the same approaches that already exist in specific articles of the Code, hence following a logical sequence in establishing a conceptual framework that would integrate into the current civil legislation those aspects regulating a new kind of relations associated with a rapid technological progress.

The process of altering legal acts might be accompanied by the real practice of courts referring to some particular articles of the Civil Code of the Russian Federation in their dealing with cases concerning the widespread introduction of AI into property and non-property relations. The fact that the nature of civil law allows for a rather broad interpretation of some articles of regulatory acts makes it possible for legal personalities to apply the already available acts in their regulation of previously unknown relations resulting from the use of advanced technologies. It appears to be forward-looking to introduce regulations on the status of electronic person into legal and regulatory acts of the civil law in order to take under control a new kind of civil relations based on the latest achievements in the area of AI. Defining the legal status of electronic person as a participant in civil relations could provide necessary preconditions for a more precise regulation of these interactions growing rapidly in the ever-evolving human society. Moreover, the practical application of AI in human artistic activities as the actual creator of art works (translations, texts, video games, films, music, clothing, interior pieces, sketches, paintings, sculptures, etc.) requires the introduction of additional legal mechanisms ensuring autonomous control of this area of human activities.

6 Conclusion

The further introduction of technologies into artistic activity requires appropriate legal regulation of AI-related processes. In light of the above-stated, it can be concluded that the civil law of the Russian Federation and other leading states currently fails to explicitly define participants of this specific kind of civil relations. This results in a legal issue of the insufficiency of the current legal and regulatory acts in controlling this new kind of relations emerging between human beings and AI-aided devices. Therefore, it is deemed crucial to alter texts of the civil law, hereby establishing a formal legal framework that would rather efficiently regulate new aspects of applying information and technology innovations for a better technology-fuelled progress in the creating of video and audio content, works of literature, visual and fine art, and fashion and design pieces. Moreover, the actual court practice demonstrates the need to study cases concerning the application of AI in human artistic activities more thoroughly. Civil law currently fails to fully utilize the ambiguity of some legal and regulatory acts to deal with information and technology innovations. However, it is necessary to introduce changes into some articles of the Civil Code of the Russian Federation that would clearly define property liability in the situation of AI application.

References

1. Bostrom, N.: Superintelligence: Paths, Dangers, Strategies. Mann, Ivanov and Ferber, Moscow (2016). (in Russian)
2. Sokolova, M.: Conflicts of "robot rights". Jurists' debates on the development of CyberCode in Russia. https://www.itweek.ru/ai/article/detail.php?ID=195514. Accessed 16 Apr 2021
3. Archipov, V.V., Naumov, V.B.: AI and autonomous devices in the context of law: on the development of the first Russian law on robotics. In: Proceedings of St. Petersburg Institute for Informatics and Automation of the Russian Academy of Sciences, vol. 6, 46–62 (2017). (in Russian)
4. Yu, Y., Binghong, Zh., Fei, G., Jiaxin, T.: Research on artificial intelligence in the field of art design under the background of convergence media. IOP Conf. Ser.: Mater. Sci. Eng. **825**, 012027 (2020). https://doi.org/10.1088/1757-899X/825/1/012027
5. Lin, Yo.: Research on application and breakthrough of artificial intelligence in art design in the new era. J. Phys.: Conf. Ser. **1648**, 032187 (2020). https://doi.org/10.1088/1742-6596/1648/3/032187
6. Ammon, S.: Image-based epistemic strategies in modeling: designing architecture after the digital turn. Philos. Eng. Technol. **28**, 177–205 (2017). https://doi.org/10.1007/978-3-319-56466-1_8
7. Yang, Yi.: Application of artificial intelligence technology in virtual reality animation aided production. J. Phys.: Conf. Ser. **1744**, 032037 (2021). https://doi.org/10.1088/1742-6596/1744/3/032037
8. Wei, Xi.: The application and development of artificial intelligence in smart clothing. IOP Conf. Ser.: Mater. Sci. Eng. **320**, 012017 (2018). https://doi.org/10.1088/1757-899X/320/1/012017
9. Ye, F.: A study on music education based on artificial intelligence. IOP Conf. Ser.: Mater. Sci. Eng. **750**, 012115 (2020). https://doi.org/10.1088/1757-899X/750/1/012115
10. Lei, D.: Research on network teaching of music major based on artificial intelligence technology. J. Phys.: Conf. Ser. **1648**, 042094 (2020). https://doi.org/10.1088/1742-6596/1648/4/042094
11. Cai, X.: Practice of hybrid teaching mode of English writing based on artificial intelligence. J. Phys.: Conf. Ser. **1648**, 042062 (2020). https://doi.org/10.1088/1742-6596/1648/4/042062
12. Zheng, S., Zhu, Sh.: A study of computer aided translation based on artificial intelligence technology. J. Phys.: Conf. Ser. **1646**, 12127 (2020). https://doi.org/10.1088/1742-6596/1646/1/012127

13. Menshikov, P.: Modeling methods in forensic engineering and technical expertise: artificial languages of modeling in forensics. Technol. Lang. **2**, 77–85 (2021). https://doi.org/10.48417/technolang.2021.02.08

14. Voskresenskaya, E., Vorona-Slivinskaya, L., Achba L.: Digital economy: theoretical and legal enforcement issues in terms of regional aspect. In: E3S Web Conference, vol. 164, pp. 09016 (2020). https://doi.org/10.1051/e3sconf/202016409016

15. Voskresenskaya, E., Vorona-Slivinskaya, L., Kazakov, Y.: Study of the protection of the architectural heritage of Russia. In: E3S Web Conference, vol. 135, pp. 03041 (2019). https://doi.org/10.1051/e3sconf/201913503041

16. Voskresenskaya, E., Vorona-Slivinskaya, L., Achba, L.: Current state of intellectual property management and innovational development of the Russia. In: Popovic, Z., Manakov, A., Breskich, V. (eds.) VIII International Scientific Siberian Transport Forum. TransSiberia 2019. AISC, vol. 1116, pp. 422–428. Springer, Cham (2019). https://doi.org/10.1007/978-3-030-37919-3_41

17. Alarie, B., Niblett, A., Yoon, A.H.: How artificial intelligence will affect the practice of law. Univ. Toronto Law J. **68**, 106–124 (2018). https://doi.org/10.3138/utlj.2017-0052

18. Greenleaf, G., Mowbray, A., Chung, P.: Building sustainable free legal advisory systems: experiences from the history of AI & law. Comp. Law Secur. Rev. **34**(2), 314–326 (2018). https://doi.org/10.1016/j.clsr.2018.02.007

19. Gorshkova, K.O., Rossinskaya, E.R., Kirillova, N.P., Fogel, A.A., Kochemirovskaia, S.V., Kochemirovsky, V.A.: Investigation of the new possibility of mathematical processing of Raman spectra for dating documents. Sci. Just. **6**(5), 451–465 (2020). https://doi.org/10.1016/j.scijus.2020.06.007

20. Hoffmann-Riem, W.: Artificial intelligence as a challenge for law and regulation. In: Wischmeyer, T., Rademacher, T. (eds.) Regulating Artificial Intelligence. Springer, Cham (2020). https://doi.org/10.1007/978-3-030-32361-5_1

21. Margarita, R.C.: Artificial intelligence: from ethics to law. Telecommun. Policy **44**(6) (2020). https://doi.org/10.1016/j.telpol.2020.101937

22. Ashley, K.D.: Artificial Intelligence and Legal Analytics: New Tools for Law Practice in the Digital Age. Cambridge University Press, Cambridge (2017)

23. Balkin, J.M.: The path of robotics law. California Law Rev. **6**, 45–60 (2015)

24. Komissina, I.I.: Current state and prospects of development of artificial intelligence technologies in China. Probl. Nat. Strateg. **1**, 137–159 (2019)

25. Nemitz, P. Constitutional democracy and technology in the age of artificial intelligence. Philos. Trans. R. Soc. Phil. Trans. R. Soc. **376**(2133), 20180089 (2018). https://doi.org/10.1098/RSTA.2018.0089

26. Turner, J.: Robot Rules. Regulating Artificial Intelligence. Palgrave Macmillan, London (2020)

27. Hallevy, G.: When Robots Kill: Artificial Intelligence Under Criminal Law. University Press of New England, Boston (2013)

28. Kirsanova, N., Gogoleva, V., Zyabkina, T., Semenova, K: The use of digital technologies in the administration of justice in the field of crime. In: E3S Web Conference, vol. 258, pp. 05035 (2021). https://doi.org/10.1051/e3sconf/202125805035

29. Yastrebov, O.A.: Discussion on a prerequisite for granting robots the legal status of "electronic persons." ISS Legal Sci. **1**, 189–203 (2017). (in Russian)

30. Yastrebov, O.A.: Artificial intelligence in law: conceptual and theoretical approaches. In: Legal personality: general theoretical, specific and international legal analysis. In: Proceedings of XII Annual Readings in Memory of S.N. Bratus, pp. 271–283. The Institute of Legislation and Comparative Law under the Government of the Russian Federation Publication, Moscow (2017). (in Russian)

31. Vashkevich, A.: Legal electronic persons. https://www.vedomosti.ru/opinion/articles/2016/05/23/641943-yuridicheskie-elektronnie-litsa. Accessed 16 Apr 2021
32. Nevejans, N.: European civil law rules in robotics: study. Policy Department for Citizens' Rights and Constitutional Affairs». European Parliament's Committee on Legal Affairs. http://www.europarl.europa.eu/RegData/etudes/STUD/2016/571379/IPOL_STUC2. Accessed 16 Apr 2021
33. Muzyka, K.: The outline of personhood law regarding artificial intelligences and emulated human entities. J. Artif. General Intell. **4**(3), 164–169 (2013). https://doi.org/10.2478/jagi-2013-0010
34. Solum, L.B.: Legal personhood for artificial intelligences. North Carolina Law Rev. **70**(4), 1231–1287 (1992)
35. Cerka, P., Grigiene, J., Sirbikyte, G.: Is it possible to grant legal personality to artificial intelligence software systems? Comput. Law Secur. Rev. **33**(5), 685–699 (2017)
36. Kurki, A.J.V., Pietrzykowski, T. (eds.): Legal Personhood: Animals, Artificial Intelligence and the Unborn. Springer, Cham (2017). https://doi.org/10.1007/978-3-319-53462-6
37. Sennikov, N.L.: The issue of comparing intellectual property right and rights of artificial intelligence. News Sci. Edu. **10**(1), 97–100 (2017). (in Russian)
38. Popovskiy, M.M.: Development prospects of copyright law and related rights in the Russian Federation. Econ. Bus. Banks **4**, 163–173 (2017). (in Russian)
39. Keisner, A., Raffo, J., Wunsch-Vincent, S.: Robotics: breakthrough technologies, innovation, intellectual property. Foresight STI Govern. **10**(2), 7–27 (2016)
40. Bohler H.M.: (2017) EU copyright protection of works created by artificial intelligence systems. Master's thesis. Faculty of Law of University of Bergen (2017). https://hdl.handle.net/1956/16479. Accessed 16 Apr 2021
41. Bridy, A.: Coding Creativity: copyright and the artificially intelligent author. Stanford Technol. Law Rev. **5**, 1–28 (2012)
42. Bogdanov, D.E.: Influence of bioprinting technology on development of civil liability. Lex Russica **9**(166), 88–99 (2020). https://doi.org/10.17803/1729-5920.2020.166.9.088-099. (in Russian)
43. Alekseev, A.O., Erakhtina, O.S., Kondratyeva, K.S., Nikitin, T.F.: Approaches to civil liability of developers of artificial intelligence technology: on the basis of technology classification. Inf. Soc. **6**, 47–57 (2020). (in Russian)
44. A European approach to Artificial intelligence. https://digital-strategy.ec.europa.eu/en/policies/european-approach-artificial-intelligence. Accessed 16 Apr 2021
45. Efimova, E. A new word in painting: artificial intelligence "writes" pictures in a unique style Vesti.Ru. http://www.vesti.ru/doc.html?id=2905726
46. Galeon. D. The world's first album composed and produced by an AI has been unveiled. https://futurism.com/the-worlds-first-album-composed-andproduced-by-an-ai-has-been-unveiled/
47. Evaluation of various strategies for the automatic generation of Spanish poetry. http://nil.fdi.ucm.es/sites/default/files/GervasAISB2000.pdf
48. Morhat, P.M.: The concept of the absence of authors for the work created by artificial intelligence. Legal World **1**, 33–35 (2019)

Creative Solutions and Professional Culture
of Prison Staff

Ivan L. Tretyakov(⊠) ⓘ

Peter the Great St. Petersburg Polytechnic University, Politekhnicheskaya 29,
Saint Petersburg 195251, Russia
mail@tretyakov.su

Abstract. The features of the influence of cognitive training methods on the improvement of the level of professional culture of prison staff and the growth of their creative potential are considered. The features and peculiarities of the penal system employees, who are in close contact with convicts, and engaged in the prevention of conflicts, criminal aggression, escapes, drug abuse, and riots have been investigated. The author concludes that the regular use of cognitive technologies contributes significantly to the professionalism of the penitentiary staff, to creative thinking, quick social/psychological adaptation to unfavorable situations, and prevention of burnout. The author's program of cognitive training contains six thematic sections related to specific methods that are aimed at: (1) the formation of creative thinking; (2) the formation of a high subjective significance of professional culture and effectiveness; (3) constructive change of interpersonal (in the broadest sense) relationships; the acquisition of communication skills; (4) the study of adaptive coping strategies, methods of self-analysis and autogenic training; (5) study of methods of rapid diagnosis of psychological qualities of a person and assessment of his emotional state, as well as the most effective ways to influence a person; (6) development of organizational skills. This program will be useful for such purposes as the selection of candidates, training and retraining of law enforcement officers, as well as the prevention of their professional deformation.

Keywords: Cognitive technologies · Creativity · Professional culture

1 Cognitive Technologies and Creative Solutions in the Process of Forming the Professional Culture of Prison Staff

In the post-industrial world, efforts for creating and testing cognitive technologies are in great demand. More often than not, these technologies are the result of some heated and lengthy debates, intense scientific researches and intuitive insights of various scientists and specialists. Cognitive technologies are principally based on both traditional, well-established knowledge on human nature/capabilities, and the latest advances in genetics, neurophysiology, neuro-linguistics, neuro-programming, etc. When working with personnel, it is the cognitive technologies/techniques that serve as a specific hallmark of how skilled and professionally developed a particular individual is. These technologies clearly show his ability for learning, socialization and creativity of perception and thinking. We should not forget that the structure of training programs (within the framework of cognitive techniques) is normally considered as three interrelated modules/blocks, i.e., input control block, theoretical block, and procedural block. The input control or primary module (monitoring unit) is a set of techniques for social/psychological, pedagogical, etc. diagnostics of an individual. Next, a theoretical block comes into play. As suggested by its name, this block contains properly targeted information that is required for the formation of special skills and competencies of a specialist, which would turn him into a high-level professional. Finally, the third block contains some specific practical recommendations. This block aims at the consolidation of the material studied and helps implement the most significant theoretical provision in day-to-day life.

However attractive the cognitive technologies seem; they are still too far from being perfect and can never comprise all the nuances of the human psyche. It is quite hard to predict what a human being is about to do next; at the same time, we do know a lot of examples when a typical outsider, being considered unpromising by those who surrounded him, turned out to be a bright, creative person, or when a *jailbird* recidivist openly declared his repentance and radically changed his system of values.

Thus, when dealing with such problems, we should be as objective as possible and avoid hasty conclusions. It takes many years for a person to become a good specialist in any field of activity (government, commercial, public, etc.), while the final result can hardly be predicted in advance. At the same time, many researchers believe that, firstly, the success of a specialist largely depends on how deeply and organically the employee was able to absorb the basic principles of professional culture and ethics [1], how close he approached both generally accepted and purely corporate ideas about the ideal employee [2]; secondly, we have to remember that in a post-industrial society, the assessment of the quality and performance of any specialist is based on the analysis of some number of skills and features [3], among which the following are the most prominent: hard work, efficiency, the level of information training, the ability to think creatively, to develop promising strategies, and confidently navigate the issues of innovative management [4].

Regarding the historical aspect, we can note that specific professional groups, which specialize in a single, quite specific type of activity, began to form during the establishment (construction) of social relations, at the very dawn of civilization. The secrets of the profession (work skills) were passed down from generation to generation. Then, the emergence of the first embryos of the state gave birth to a new, special type of people who were able to wield their powers, administer justice, and execute punishment. The rulers of antiquity confronted both external threats, (possible intrusions of foreigners) and internal ones, (local crime). Numerous sources (clay tablets of Mesopotamia, inscriptions on Egyptian architectural monuments, survived papyri, etc.) contain evidence of harsh and sometimes unnecessarily cruel treatment of criminals.

The best minds of the Enlightenment epoch began to offer humane ways of fighting crime as a social disease. Starting from the second half of the eighteenth century and till the present time, attempts have been made to find the most efficient and valid technologies for correcting (reclaiming) convicts. In recent decades, one can see the desire of the heads of the penal system to use innovative methods when carrying out psycho-correctional and rehabilitation measures in correctional colonies. There is no doubt that information and communication technologies can seriously improve the quality of relapse prevention, while the digital algorithms developed on their basis would help the former criminals find a job and have them successfully reintegrated with the society (after they are released from prison). Ultimately, these circumstances lead to the idea that legislation should be updated on the regular basis, the entire work of correctional institutions constantly improved, and the outdated paradigms timely changed [5]. At the same time, considering the issues of reforming the penitentiary system, scientists have never forgotten that it is the human factor, that is the cornerstone of any large-scale modernization, and revolutionary transformations, which in this case directly applies to individual employees of the punishment service and convicts they work with [6].

It would be quite appropriate to mention such phenomenon as *ostracism* in the context of the issues we consider here [7]. Any criminal punishment in itself with compulsory isolation (imprisonment) is a manifestation of ostracism. We have every reason to believe that this phenomenon was already encountered in a primitive society. Presently, ostracism increasingly becomes a topic for some serious discussions, since it permeates entire spheres of life of today's people (cultural, communicative, socio-psychological) [8]. It occurs much more often than it is generally believed in the human relation system [9]. Moreover, ostracism (e.g., at the general population level) is an extremely powerful frustrating factor. It would also be appropriate to clarify that apart from the legitimate *rejection* of violators of official norms and standards, there is an informal (illegitimate) ostracism, e.g., it could be a sort of alienation among convicts by the convicts themselves, i.e., there is ostracism with some additional, micro-social isolation and deprivation. Some signs of ostracism could be found among students, and workers of various professions, including employees of penal institutions. Sometimes, emotional refection is accompanied by cyber-bullying and other signs of mental abuse.

Regarding the moral and psychological climate in any team of employees, it would not be an exaggeration to say that the nature of interpersonal and business relations among penitential personnel requires constant monitoring and timely adjustment. Deficit of friendly concern (social support), lack of required life experience and deep professional skills in the field of professional activity as well as undeveloped self-control and recovery skills lead to maladaptive behavioral coping strategies that aggravate frustration.

Clarification of the reasons for any case of *burnout* shows that the professional stress of people from the Ministry of Justice is based on the following circumstances: gross violations of working conditions, non-compliance (sometimes basic ignorance) of the principles of organizational culture, and interpersonal conflicts [10].

We have to note that in penitentiary institutions there is a high incidence of suicides, both among convicts and working staff [11]; therefore, there always remains a need to develop efficient approaches to preventing and reducing the occurrence of suicides (which calls for the interdisciplinary research with due account of the most diverse aspects of suicidal activity) [12].

There is a lot of evidence in mass media that the prestige of work in prisons is extremely low, while staff turnover being quite high. At the same time, there is a certain specificity to this point; i.e., the staff shortage is a problem only for the lower and middle strata of workers, while the administrative apparatus is staffed as full as possible, and its number is continuously growing (together with the executive and administration expenses).

The administration of correctional facilities fires several problematic employees regularly because of delinquencies, loss of confidence, alcoholism, and drug addiction. Some employees are also accused of committing crimes. Not every employee can overcome corruption risks (temptations) and properly use his/her powers. People with a lack of moral principles, conscience, etc., try to join the structures of the Ministry of Justice to quickly enrich themselves by bribes and extortion. In no way there can be any talks of any professionalism (apart from purely criminal) for such persons. The Professional culture is unthinkable without adherence to service ethics, without the constant desire of an employee to acquire new knowledge, without his volitional efforts, or without creative searching activity. This study attempts to study the professional culture of those employees for whom honor and dignity, loyalty to duty, etc. are meaningful words since they devoted themselves to a hard, socially useful and unsafe business; i.e., working with convicts (most of whom have committed grave and especially grave crimes).

The government sets forth serious requirements to the personal qualities of personnel of penitentiary facilities. Specialists of the Federal Penitentiary System bear double responsibility; i.e., for themselves and the convicted. It is high intelligence and resistance to stress that help employees identify and stifle the causes of penal crime. Furthermore, specialists (psychologists, teachers, operative workers, inspectors of the security departments, and production units) try to optimize the cognitive processes of their wards to form persistent patterns of law-abiding behavior by using informational methods of influencing the consciousness of convicts. To cope with this task, they carry out a deeply personal analysis of each convict. The circumstances that contributed to the commission of the crime, and in some cases the evidence presented in court, are thoroughly analyzed and pondered [13]. Further on, biographical information, data on criminal experience, results of the psychological examination are brought to comprehensive analyses [14]. It is the knowledge about the strong and weak points of the human psyche that allows specialists to develop strategies and tactics for individual educational work in places of detention, control and prevent negative manifestations of behavioral activity (in the form of anger, aggression, violent sodomy, etc.), and stimulate the mechanisms of readaptation [15]. According to the ideal scenario, people should return to society from places of detention (after they served their terms of punishment) with a clear realization of all the malignancy of what they have done, who have sincerely repented, who have compensated damage to the victims and the state. The re-socialization and reintegration of the former convicts will not cause any difficulties if citizens who successfully mastered useful professions (in demand in the labor market), rather than just more sophisticated and skilled criminals ready to commit new crimes, leave the penitentiary facilities. Let us repeat that all of the above points are just ideal landmarks (sweet dreams of a criminologist) to which we should strive. But the reality is such that the rate of criminal relapses (in all countries) tend to grow quite steadily, which makes recidivism (without exaggeration) a real threat to national security [16].

Thus, there is no doubt that at the governmental level there is a high need for the measures to improve the quality of socio-psychological support for both personnel of the Ministry of Justice and the inhabitants of correctional facilities, for identifying the factors of personal growth of employees, for finding proper motivators for the personnel to fulfil their duties. Moreover, it appears that from the standpoint of penitentiary science, we have to define such a concept as the *professional culture of employees of correctional facilities.*

Summarizing what is said above, we can assert the following - penitentiary institutions need more experienced employees properly trained with cognitive technologies/techniques and having digital competencies. Moreover, at the state/governmental level, we see a serious need for the measures to qualitatively improve the information, social/psychological, and pedagogical support for specialists of the Ministry of Justice, as well as for identifying the factors that would stimulate their personal growth and motivate their best performance. It seems that from a humanitarian science standpoint, we have to define such a concept as the professional culture of employees of penitentiary institutions and to demonstrate the importance of creative approaches in reviewing outdated doctrines and conservative knowledge.

All of the above mentioned confirms the high relevance of our research topic.

2 Literature Review

The scientific literature [17] describes in some detail the use of cognitive technologies in various areas of social life, including education, culture, or training specialists of creative professions [18]. However, judging by what has been published over the past five years, penitentiary institutions are not yet considered by the heads of law enforcement agencies as testing grounds for trying fundamentally new cognitive methods to improve perception, upbringing, education, rehabilitation, and so on. In their works psychologists, criminologists heads of the Ministry of Justice hardly ever deal with the issues of digitalization for the penitentiary system; at the same time, scientists very rarely analyze the creative potential of employees, the criteria of their professional culture.

The study by K. Morrisson and M. Mausosk carried out at the highest theoretical and methodological level, is a very successful exception in this regard. These materials contain extensive information on the issues in question. A total of 15 prisons in Scotland, out of which 10 were private, formed the empirical base of the study. We were especially impressed by the comprehensive approach of the Authors to the analysis of the professional evolution of an average prison officer (from the moment he/she entered the service until his/her retirement). Such issues as the contradictions of the prison system and the mistakes of modern reformers, the numerous difficulties faced by a rookie, burnout, peculiarities of interpersonal relations, training and professional development of employees together with problems of professional culture have been considered in detail. The Authors' statement regarding the support of creative employees, who strive for professional excellence and are respected by both staff and prisoners deserves special attention. Unfortunately, no methods to working with prison staff were disclosed. The use of both classical models and cognitive technology is suggested only indirectly [19].

Interviewing employees revealed their interest in both professional, and career growth. Haneef F. et all., using networking technique, demonstrated the correlation between the level of education and the fast/successful career growth. They emphasize that such correlation is quite complex and ambiguous. However, these Authors never described the ways they used for testing the quality of knowledge and evaluating the efficiency of education. Neither they explained what they meant by career growth (the definition of which being very subjective). They do not describe any correlations between such concepts as a career, professionalism, and professional culture. Unfortunately, the said article says nothing about the role of cognitive technologies in the sphere of education, about their influence on all subsequent professional activities of the respondents [20].

Culture brings people together with its historically established meanings, values, beliefs, generally accepted (about a specific society) criteria of stratification, the concept of taboo, ostracism, etc., as well as with the symbolic framing of norms and rules of behavior. Personal (and professional) growth is possible only if a person activated the so-called cultural code at the level of his/her consciousness, i.e., a person learnt to perceive meaningfully the fundamental algorithms of being, the value-cognitive systems.

Penitentiary personnel should have strong legal *immunity* to informal rules, fetishes and canons of the criminal world. The criminal (general criminal) environment has its subculture, which is based on primitive, aggressive, extremely cynical views, and anti-social attitudes. Such a deviant quasi-culture has been fostered in places of detention. It is the reaction of the criminal community of professional criminals (i.e., destructive and stigmatized sociopaths) to ostracism and long-term isolation from society. The criminal (prison) subculture is opposed to the professional culture. Figuratively speaking, such confrontation is the eternal struggle of the government with a serious, incurable social pathology, i.e., with crime [21]. The subculture of criminals is not a uniform phenomenon; it is generally influenced by ethnic, religious, socio-demographic [22], economic and political factors [23].

One can hardly imagine modern criminals (primarily residents of large metropolitan areas) outside of cyber reality [24], e.g., without the digital *world* of Facebook, Instagram, Twitter [25]. However, if a white-collar crime has not escaped the influence of postmodern trends and is functioning in a fairly autonomous (albeit extremely large-scale) mode, the ordinary crime (represented by thieves, rapists, robbers, murderers, etc.) is still controlled by the traditionalists (conservatives) or their competitors, i.e., bandits of the new wave who do not recognize any traditional authorities and are ready to even temporarily cooperate with the authorities to get rid of their enemies.

We can see the following distinctive features of the criminal subculture: 1) it has its norms, rituals, ideologemes, specific myth-making, rigid hierarchy (caste) and clannishness, shadow justice, and punitive sanctions; 2) it uses its special language, gestures, manners, tattoos, clothing, and various forms of organizing leisure.

There are many obvious problems directly related to the dissemination of the criminal subculture in society and the tolerant attitude of the marginalized population towards it. However, there is one aspect of the darkest side of the life of convicts, which calls for special attention, since it concerns sexual deviations, or, more specifically, violent sodomy. The victims of such aggressive acts become real outcasts, being the most powerless and despised individuals; e.g., they live in separate rooms, eat in strictly designated places, do the hardest, low-skilled work, and at the same time they are constantly forced to satisfy the sexual lust of other convicts [26]. This means that we are talking about extremely radical, cruel forms of stigmatization and ostracism [27]. The preventive measures for such crimes are either not carried out at all, or performed quite formally [28].

Any work with people always calls for high responsibility, sociability, and the presence of at least a minimal level of empathy. Meanwhile, close communication is often associated with conflicts, with tense and hostile attitude of the contacting (talking to each other) participants, misunderstandings, aggressive reactions, inappropriate behavior, and destructive attitudes of *clients*. Verbal contacts of penitentiary personnel with convicts for their correction (i.e., social, correction, psychological correction, rehabilitation, etc.) are also accompanied by conflicting emotions and psychophysiological stress on both sides [29]. With time, excess stress factors, fixation on negative events, and psycho-traumatic phenomena start to undermine the adaptive potential of the human organism, deteriorating stress resistance [30].

Modern socio-psychological and humanitarian studies give a significant deal of attention to the search for efficient methods to deal with stress [31], the ways to neutralize frustrating influences [32], to prevent burnout of personnel, the development of fast and safe methods of self-regulation (self-defence), which help maintaining proper mental balance [33]. Stress, especially in form of chronic stress, leads to lack of motivation [34], loss of performance, obsessive fears, anxiety, unreasonable worries, insomnia, and suicidal intentions [35].

Thus, the review of literary sources related to our topic reveals that the professional culture of correctional personnel is an area of interdisciplinary knowledge, where the interests of criminology, psychology, psychiatry, pedagogy, sociology, etc. are closely intertwined. The topic appears to be so complex and multifaceted that we can quite confidently state that in the long run there will be an evident need for theoretical studies required for further development of penitentiary science. Very unfortunately, we could not find any works devoted to creating and testing models for *personal and professional growth* and *overcoming destructive behavior* (developed for people working in the penal system), nor could we find any publications containing an analysis of the level of cultural development for the penitentiary staff, their creative potential.

3 Methodology

To achieve the goals we started and get the most complete (comprehensive) solution of the task assigned, we used the following efficient and well-proven methods of cognition:

1. An analytical method, which comprises consideration and clarification of various information related to topical issues of organizing, functioning and reforming the penal system. Penal legislation was the source, from which we obtained the required data. In Russia, as well as in foreign countries, we can get various documents at the federal and departmental levels, i.e., scientific and methodological literature on criminal penology, on general and legal psychology, pedagogy, management, personal files of employees, static reviews on the state of discipline and legality, emergencies, crime, sickness rate, etc. both for the employees of penitentiary institutions and the convicts.

2. An empirical (diagnostic) method, including interviewing, questioning, and psychological testing. When conducting an experimental psychological research, we used the following: method (test) definitions of meaning, expression and direction of life orientations; developer is D.A. Leontiev, which makes it possible to establish the level of self-organization of a person and the formation (rate of manifestation) of the mechanisms of self-actualization for an individual; Lüscher Color Test allows to evaluate the emotional state of the respondent, as well as to carry out express diagnostics of his individual personality traits; Giessen-Test (GT) - aimed at establishing a vector of behavioral activity, general response to external circumstances, depending on individual qualities, worldviews and the nature of interpersonal relationships (interactions) that are dominating in any mini-society; *Strategic Approach to Coping Scale-SACS* questionnaire, developer is S.E. Hobfoll et al., which makes it possible to judge upon the presence of certain models of overcoming (coping) behavior; *Coping Resources Inventory for Stress (CRIS)* questionnaire, developer is K.B. Matheny et al., which makes it possible to evaluate the individual-personal potential of coping with stress; the *Scale of Psychological Stress Lemyr-Tessier-Fillion "PSM-25"*, which measures stressful feelings with due account of negative psychosomatic, behavioral and emotional changes; *Emotional Intelligence (N. Hall)* diagnostics aimed at establishing an ability of a person to control his/her emotions.

3. *Cognitive or Formative* methods, which consist in modeling the structure and content of such a phenomenon as the professional culture of a specialist working in the enforcement system of the Ministry of Justice in a modern information society.

4   Results of an Empirical Study

Our study involved 240 people (current employees of the Ministry of Justice who worked in correctional institutions of St. Petersburg, Leningrad and Vologda regions, Krasnoyarsk Territory, and the Republics of Karelia and Komi). The study was carried out in several steps with different time intervals from 2016 to 2020. All persons (228 men and 12 women) were between the age of 24 to 30 years. The work experience (in this area in question) of the interviewees was at least two years, while 70% had been working in this field for five to six years. We did not consider the period of service in the army, since we were interested in the following aspects: the nature of the influence of places of imprisonment on the psyche of employees, the formation of special skills in working with convicts, the level of the professional culture, and the presence of adaptive potential. The survey participants, both NCOs and officers, were divided into two groups. Let's consider in more detail the categories of employees we designated.

1. The number of persons included in the first group was 160. These were comprehensively trained employees (for convenience, we designated them as successful specialists); no penalties had ever been imposed on them. They readily communicated with the organizers of the study and tried to answer our questions as fully as possible. We have to note that all-female specialists were in the first group. 64% of successful employees were recommended by their superiors for a higher position, i.e., constituted a personnel reserve for a particular institution, as well as for

higher management positions. The respondents (78%) predominantly had a higher education (10% had graduated from a secondary specialized institution, while 12% studied in absentia at the law faculties of various universities). The psychologists found out that in this group prevailed the *internal locus of control*. These respondents demonstrated constructive goal-setting, life-affirming strategies, we're ready for cooperation, making optimistic plans for the future. 10% of the interviewed were active book readers, understood poetry and painting, visited theatres, concert halls, and exhibitions (mostly virtual), i.e., they tried to maintain their level of culture and lead an intellectual lifestyle. People from St. Petersburg happened to be the most mobile and educated. All 160 respondents had their pages on social networks. 15% showed some signs of addictive behavior in the form of insignificant impulsivity, emotional lability, occasional alcohol drinking (once a month maximum). 17% were doing some physical exercises (workout), 4% reported that they are engaged in scientific work, having publications in special (professional) journals. By the time of the survey, 92% stayed married, had children, and regarded family relations as favorable and harmonious. In general, the self-esteem of the respondents was somewhat overestimated, however, people remained quite critical and adequate. 33% (50 male and 3 female respondents) caused our concern. Predominantly, those were persons with sensitive, asthenic traits who came up with psychosomatic complaints. According to the conclusion of psychologists (later confirmed by psycho-diagnostic tests), the revealed negative phenomena can be considered as part of the burnout syndrome (e.g., irritability, fatigue, exhaustion, loss of sleep, headaches, anxiety, mood swings, decreased concentration of attention). The interviewees, who showed some signs of burnout, also reported that severe symptoms became particularly intense in their third year of service. Presumably, this was the beginning of the frustration. The presence of suicidal intentions was confirmed by 4% of those surveyed with signs of burnout syndrome.

2. The remaining 80 formed the second group. These were the least promising and quite problematic persons (those who were going to be fired), which were quite negatively characterized by their superiors. Over 3 penalties were imposed on such employees. They clearly showed deviant stereotypes that were especially pronounced due to the lack of upbringing, rather primitive needs, and were also caused by the persistent addictive mechanisms. We could notice that they did not possess such life priorities as the wish for self-knowledge, self-organization, or active self-development. The external locus of control was evident (i.e., people tended to blame others for all their troubles). Our psychologists recorded high aggressiveness and affective instability (with a clear predominance of negative emotions, and low mood background). 37% of respondents revealed the signs of Internet addiction (in form of uncontrolled playing computer games). Over 60% reported that alcohol used to be a permanent attribute of their leisure time. They drank to overcome alienation, boredom, and communication barriers. The respondents drank strong alcoholic drinks over three times a week, while every fourth problematic employee drank every day. 58% did not have a family. The interviewees were distinguished by a tendency to tell lies, in conversation they tried to ascribe to themselves mythical achievements, demonstrating immaturity, egoism, traits of hysterical, irresponsible and lack of self-criticism. 24% of such problematic employees (who had been working less than three years)

revealed the signs of burnout. Practically, pathological abnormalities, complaints, etc. were not different as compared to what was identified for the first group, except that in the second group the employees showed more pronounced anxiety together with destructive behavior. For the rest of the respondents, the burnout was either impossible initially (due to a specific psycho-type, or due to an irresponsible and dismissive attitude to their duties, low professionalism, and deviant worldviews), or because at some point the burnout stopped to develop with further persistent professional and moral deformation of the personality. 15% confirmed that occasionally, under certain unfavorable circumstances, they had suicidal intentions (moreover, in 10% there was an evident connection between the increased burnout and suicidal intentions).

In course of surveys, psychologists (the members of the research group) established that the representatives of the second group practically did not care about any reputational risks, nor were they interested in the opinion of others (except for a few marginal individuals with whom they closely communicated). Approximately two-thirds of the problematic respondents demonstrated a tendency to risky behavior, exaggerated self-confidence, and unreasonable claims to leadership.

Among the employees that we assigned to the *risk group*, there were persons (10%), who were prone to aggression and hostility. The persons who should not work in places of deprivation of liberty were identified. These individuals were characterized by being prone to conflicts, rigidity, rancor, suspicion, cynicism, and a tough (sometimes cruel) attitude towards convicts and their colleagues. The study of the motivational of these respondents revealed qualitative changes in their needs, preferences and attitudes in the form of disinhibition of instincts, urges to rude domination, inability to feel empathy. We believe that such behavior is the result of the shortcomings in intellectual development, psychopathic traits, or early organic damage to the central nervous system, as well as later brain injuries, pedagogical neglect, lack of proper spiritual development, etc. In the future, we concluded that this category should be considered as a special criminogenic group of subjects.

The analysis of the integrative level of emotional intelligence seems to be of undoubted importance in the context of the issues in question. The psycho-diagnostic techniques we used, showed that in the first group of respondents the integral indicator of emotional intelligence was at the level of 71.0–72.5 points. These are fairly high values, which show the ability of given specialists to cope with their emotions, revealing also the presence of both cognitive and non-cognitive abilities. However, for the persons with signs of burnout (from the first group), average values being recorded as 59.9–62.0 points.

The results we obtained confirmed the assumption that the employees from the risk group could not properly understand their own and other people's emotions; moreover, they had problems with control of their emotional-volitional processes. The study of the most characteristic stress resistance factors revealed a high and medium level of coping behavior (this referred to *successful* employees); the average level being typical of people with signs of burnout. Such employees demonstrated adaptive and passive-adaptive behavior. Normally, they were guided by such concepts as a sense of duty, mutual assistance, cooperation, dialogue, compromise, collective opinion. When working with convicts, the employees with positive and constructive attitudes used rehabilitation strategies, trying to establish confiding relationships (but keeping the required distance), readily contacted everyone who wished to be in contact and offered them social and legal support. Some manipulative technologies were used in conducting psychological and pedagogical activities (quite rarely, if and when it was really necessary, and at the same time strictly within the framework of the current legislation).

People from the second group demonstrated a low (22.0–25.5) or medium (49.6–51.1) rate of emotional intelligence. The level of coping resources also turned out to be quite low, since problematic employees resorted mainly to pseudo-adaptive and maladaptive behavior. Interpersonal relations (within the penal system) were based on attempts to get adapted to the superiors as well as on confrontation and fighting with people around. When dealing with convicts, such officers were aggressive and resorted to suppressive (imperative) strategies.

Let's turn to the data obtained as a result of interviews, questionnaires, and polling with the participation of 240 individuals. Thus, in an anonymous survey, the vast majority of those surveyed reported having chronic fatigue, high workload, dissatisfaction with working conditions, wages, or attitude towards themselves (towards their needs) on the part of their superiors. About 25% of the respondents showed serious concern about their health, visited a psychologist or psychotherapist, and tried to combine their holidays with treatment in a sanatorium. 40% were ready to change their job in the colony for more prestigious and highly paid positions.

When answering questions about assessing the efficiency of criminal punishment for correction, the staff was convinced that only a small part of convicts can be corrected in the places of detention (while no correction is possible at all for recidivists, professional criminals, and individuals prone to cruel violence). At the same time, no respondents insisted on the uselessness of the individual upbringing work. Generally speaking, the staff was quite positive about the training of convicts (e.g., remotely, under the auspices of higher educational institutions). Regardless of their working experience, most (over 90%) of the respondents reported to the experts that not enough attention was paid to measures of preventing criminal aggression, conflicts, violent sodomy, radical moods (in the form of extremism and xenophobia), drug addiction and alcoholism among convicts. Moreover, most of the preventive measures were quite formal. In some institutions, the processes of rehabilitation, resocialization and reintegration of convicts are just emerging, such institutions have no qualified specialists who would be able to solve quickly the problems of social rehabilitation of the convicts, together with the problems of medical treatment and psychological support for those with limited mind soundness. The workers of the colony workers expressed their concern about the growing number of people suffering from mental diseases, who were prone to inadequate reactions, the appearance of individuals who showed rapid psyche decay within a short time, who had problems with memory, the rapid progress of dementia, etc. Specialists responsible for the upbringing of the convicts (chiefs of detachments), penitentiary psychologists, and social workers reported that they needed a fundamentally new, more flexible, reliable and tested methodology (methodological approaches) to the study of emotional-volitional, motivational, cognitive and communicative qualities of the personality of their wards, as well as practical recommendations for the formation of initiative groups (anonymous communities), consisting of convicts and their relatives, to increase the rehabilitation potential of persons who decided to get corrected.

Only 30% of the respondents assessed the general cultural and professional training level of their colleagues as satisfactory. Over 70% insisted that the penitentiary system needed to be reformed. The respondents listed the following reasons for the crisis of the correctional system:

- General decline of social morals, growth of violence, corruption, intolerance;
- Omissions and blunders when recruiting personnel: people were hired at random, some of the recruited persons having openly criminogenic traits;
- Inability of the administrative apparatus of the correctional system to implement anti-crisis strategies for human resource management, inability to use innovative approaches in training and retraining the staff, and inability to counteract corruption.

Two-thirds of respondents believe that in the conditions of the penitentiary system, most innovations and creative solutions are blocked, and a polycentric approach to the education and training of employees is extremely difficult here. Nevertheless, the following position was identified among the respondents: people work in quite complicated and sometimes extreme conditions; therefore, we see the need to create a specified vertical structure, which would regulate inter-departmental efforts to develop adaptive strategies aimed at the optimal, harmonious personal growth of specialists of the Ministry of Justice. In particular, the majority of the interviewed believed that it would be expedient to have prescribed in the law a mechanism to ensure control of civil society over the training and retraining of penitentiary workers (with due account of the tasks required to bring up the level of their general and professional culture). Furthermore, a significant number of respondents with no valid imposed penalties, who were concerned about the state of discipline among personnel, insisted that independent commissions should be involved in the analysis of the causes of all kinds of violations and emergencies in places of detention.

A few most active respondents (2%) who knew each other and worked in the same division of the Krasnoyarsk Territory came up with the idea that the process of forming the professional culture of specialists in the XXI century should involve maximum accessibility of students (or employees willing to improve their level of knowledge) to information resources, supporting creative individuals (with creative inclinations, leadership, organizational qualities, and team spirit), who strive to master digital innovations, and the formation of moral, professional and ethical qualities, moral consciousness, and the sense of responsibility.

We have developed and tested a group cognitive training program aimed at improving the level of professional and general cultural competence of employees of correctional colonies, enhancing their skills of self-control, stress resistance, as well as preventing maladjustment, burnout, moral deformation, etc. The optimal period of the first training cycle is 21 days (the duration of classes is 8 h). The program comprises a description of some psychological, pedagogical and psychotherapeutic measures to neutralize (block, avoid) the problematic zones of individual and social consciousness, which are of a powerful frustrating and demotivating influence, and are also the sources of depersonalization and dehumanization.

The choice of topics for the module is determined by the specifics of the work of various categories of employees (for example, psychologists, teachers, doctors, security guards, instructors, social and administrative workers). The nature and topics of optional classes (seminars, business games), conferences are determined in accordance with the requests of prison staff.

The program contains six thematic sections related to specific methods that are aimed at:

- Formation of creative thinking skills;
- Formation of the high subjective significance of professional culture and performance;
- Constructive change (correction) of professional, family and interpersonal (in the broadest sense) relations; acquisition of communication skills;
- Learning adaptive coping strategies, introspection techniques and autogenic training;
- Learning the techniques of express diagnostics of the psychological qualities of a person, and assessing his/her emotional state, as well as the most efficient ways of influencing a person;
- Developing organizational skills.

It is important that after the completion of the basic part of the program, regular psychocorrection of the motivational level of individual behavior is carried out at the workplace. Specific measures will depend on the results of psychological and social diagnostics of employees, on monitoring of corporate potential.

Of the entire arsenal of psycho-therapeutic methods in our program, the most demanded appeared to be the following: gestalt, neurolinguistic programming, rational positive and family therapy, and role-playing games. Pantomime and psycho-drama happened to be most attractive and interesting for women and younger male employees (with higher education).

A total of 88 employees took part in the training. Out of them, 72 people complained of their mental state, deterioration in working capacity, etc. (these were the representatives of the first and second groups with signs of emotional exhaustion). 16 people (representatives of the first group) did not experience any serious problems both in their personal life and in the sphere of official activity, but they were willing to have their potential developed.

The efficiency of the author's program was duly confirmed by the presence of statistically significant (positive) changes obtained as a result of repeated testing of the trained participants, as well as by long-time studies (the time interval being one and a half years).

5 Discussion/Analysis

This article attempts to show a new, creative look at the selection of educational, diagnostic, and adaptive techniques that would contribute to improving the quality of human skills in various services of the Ministry of Justice that would in turn ensure the smooth performance of the whole penitentiary system. Against the background of universal digitalization, the improvement and continuous updating of cognitive techniques together with the expansion of their capabilities is quite obvious [36]. Some educational centres have succeeded in setting up democratic, individually-oriented models, and flexible algorithms for obtaining information, which would help employees willingly improve their professional skills, managing simultaneously the educational process [37]. Such an approach, based on the personal participation of an individual in organizing his education, with independent choice of the appropriate curriculum and lessons, seems to be most attractive and promising.

In recent years we see that international political structures together with the IT giants they control are quite aggressively imposing the idea of universal digital education all over the world. However, it is prevailingly scientists who can adequately and comprehensively assess the processes, which take place in modern post-industrial society (in particular, in the field of education). Thus, Dowling-Hetherington, L. et al., point out that digitalization is a good idea only in case it is complementation to real-life; in their opinion, the most sophisticated neural networks will never be able to replace a human being [38]. The authors categorically declare that under no circumstances digital techniques should replace the methods of traditional pedagogy [39, 40]. For our part, we would like to point out that the general education and special training of penitentiary workers do need digitalization. However, such digitalization should be well thought out and scientifically grounded with the priority of the human factor, rather than something radical and overly accelerated. The so-called digital barriers between employees and convicts are unacceptable.

Having studied the activities of correctional institutions in Russia together with the comprehensive functioning of their employees, we formulated the thesis that the professional culture of prison staff cannot be considered in isolation from creativity. In addition, professional culture is a systematic combination of individual and personal qualities, general education, cultural and educational and purely professional competencies. The analysis of world literature reveals that many authors adhere to this position; the scientists in some studies focus mainly on psychological aspects, in particular, cognitive-communicative mechanisms, and the specificity of emotional responses are discussed in detail [41, 42].

The data we obtained confirmed the observations of well-known psychologists that the indicators of emotional intelligence were very informative for diagnostic and making forecasts; quite often they revealed the presence of maladaptive manifestations, high aggression, conflicts, affective instability, and disinhibition of appetencies [43; 44]. In particular, we found that problematic (deviant) employees had low emotional intelligence, showed a tendency to destructive activity, to insufficiently thoughtful actions, to alcoholism, and had some signs of gambling addiction.

Presently, we can find a lot of research directions dedicated to problems of burnout. In prisons, the pronounced negative effects of stress on the psyche (which applies to both employees and convicts) tend to block any positive emotions, increase nervousness, discomfort, dysfunction, exacerbate chronic diseases, and cause suicidal intentions [45,

46]. Scientists normally find a relationship between burnout and work conditions, dissatisfaction and sense of being underestimated [47, 48]. Such ideas in no way raise any doubts, but the results of our study give us all reasons to assert that the resistance to stress, the general well-being and the performance of penitentiary workers are largely dependent on a combination of constructive, purposeful intellectual activity and adaptive coping strategies, in other words, by a combination of congenital and acquired qualities that allow a person to overcome problems, as well as to optimize, creatively transform the surrounding reality, and successfully react to destructive factors with less pain and trauma.

The exchange of information between various correctional services is extremely important, therefore we readily read the publications that were devoted to the problems of rehabilitation of convicts [49], the development of correctional culture, the dissemination of adaptive practices [50], and the restoration of the rights of persons having mental abnormalities [51]. According to some employees of Russian correctional institutions (who participated in polls, interviews, etc.), there is still a clear need for rehabilitation and post-rehabilitation support for convicts and members of their families, as well as for cooperation with non-profit and religious recovery centres and international funds.

Employees who possess cognitive technologies can respond quickly to the requests of convicts, they are more attentive and humane than their colleagues. The attentive and careful attitude to people who are unable to take care of themselves is a sign of high moral and professional culture. As we were able to see, Russian and foreign scientists pay considerable attention to the conditions of detention and treatment of people with mental diseases (mainly, these were people with progressive cognitive impairments) [52; 53]. After such disabled people (including those suffering from diseases of the central nervous system) are released from prisons, the special committees consider the availability of housing for them as well as their possible employment (with due account of the severity of their health problems) [54–56].

The reports on the independent cooperation of former convicts to overcome stigmatizing and deprivation are particularly noteworthy [57]. Such actions cannot be carried out with no permission and patronage from the administration of correctional institutions, therefore, for teaching convicts the skills of mutual assistance, and setting up constructive relationships with others the role of psychologists and pedagogues is quite obvious [58].

Almost all today's criminologists believe that penitentiary institutions are a large-scale and reliable reflection of the morals prevailing in society, rather than a mere social cross-section, which characterizes the marginal and psycho-anomalous stratum of the population, [59–61]. In this study, we do not limit ourselves to simply stating this fact; instead, we try to convince our readers that countering violence, cruelty, radicalism and xenophobia is closely related to general social preventive measures, which must be implemented by all available resources of government structures, civil society institutions, professional associations of scientists, practicing employees of the penal system, and the whole population of the country (including former convicts, persons serving sentences, their relatives, victims of criminal offences).

Places of detention should be considered as a kind of battlefield where two incompatible realities, two philosophies, two ideologies collide, i.e., the professional culture of workers of the penitentiary system, intending to protecting and preserving universal human values, with the inhuman subculture of the criminal (prison) world, which absorbed all the worst and most disgusting features of the human race.

If an individual exposed to cognitive techniques assumes an active life position, innately possesses required abilities, has adequate skills and digital competencies, and considers his activity as something the society needs badly from both moral and legal standpoints, the growth of his professionalism and professional culture will be fast, efficient, and smooth.

6 Conclusion

We believe that the major components for the success in the professional growth of penitentiary employees are as follows - the high level of neuropsychic stability, the emotional and social intelligence, the ability to self-control and recovery, the communicative and cognitive skills, extensive knowledge beyond mere professional training, the motivation to work and behave according to the behavior standards, the mature system of values with prevailing moral attitudes. We are convinced that the true professional always consciously observe all kinds of rules and regulations (universal commandments, legislative requirements, ethical standards of an employee of the punishment institutions, corporate rules of conduct). The above-listed features form the basis of averaged models of professional competence for employees of the penal service. The typological uniqueness of such models depends upon the specifics of the functional duties performed by employees.

Cognitive technologies/techniques, creative solutions, professionalism, and professional culture are interrelated and complementary ideas associated with a conventional concept of an ideal specialist (employee, etc.). Within the penitentiary system, cognitive technologies can be used for working with both employees and convicts. Concerning training of employees, the cognitive techniques ensure (1) diagnostical analysis of significant hallmarks (intelligence, cultural development, creativity, propensity for deviant or criminal activity, etc.); (2) implementation of comprehensive educational and cultural programs; (3) contribution to the development and consolidation of skills to solve all kinds of professional problems.

Emotional burnout is a real threat to the life, health and professional culture of penitentiary workers. The devastation of the psyche is accompanied by frustration and exacerbation of all accumulated contradictions. In the long run, chronic traumatic stress factors significantly distort the subject's vision of the real state of affairs. The perception of the socio-psychological situation at any given moment (as well as the awareness of the near future events) becomes superficial, fragmentary, and not entirely adequate.

The persistent adaptive coping strategies are the result of deeply motivated emotional-volitional efforts of employees with an internal locus of control. The style of behavior style of successful employees is characterized by a cognitive approach to overcoming difficulties, interpersonal contradictions, as well as by the tendency for cooperation, interaction, and reasonable compromise.

When considering the experience of conducting training, we can note that there was an evident change in indicators characterizing the cognitive and emotional-volitional activity of the respondents. Relatively speaking, thanks to psychotherapeutic sessions, it was possible to resolve the most acute contradictions that had arisen in the process of employees' functioning, as well as to create a new, more adaptive hierarchy of relations between the energetic, meaningful and semantic components of the professional activity.

The penitentiary system is an important element of criminological prevention of penitentiary and post-penitentiary crime, and its employees (specialists, workers, etc.) bear responsibility for individual work with convicts. However, the sphere of functioning of employees of the Ministry of Justice can hardly be limited to the performance of official duties. We have to recognize that all employees of penitentiary institutions are (at least indirectly) the participants of informational, generally social, and cultural processes, while the professional culture has the most significant impact on the worldview, legal consciousness, and motivation of employees' performance (professional activity).

There is no doubt that professional culture forms the cooperation vector between society and the state, creating common values, points of contact, and topics for dialogue, rather than being merely an instrument of consolidation for various social and workgroups, corporate communities, etc., which determines the dynamics of their functioning. The professional culture of a specialist in the penal system is considered to be the most important factor for training highly qualified employees, and such a process, of course, calls for an especially thoughtful control. A high level of professional culture of penitentiary workers undoubtedly means an active rejection of everything immoral and anti-social, a guarantee of minimizing corruption risks, abuse, or violations of rights and freedoms.

7 Limitation and Study Forward

We are fully aware that the teaching algorithms we have proposed in the form of training designed for employees of penitentiary institutions need further testing and improvement. Unfortunately, the limited scope of this study did not allow us to consider gender differences in the professional culture of employees of the penal system, the peculiarities of the burnout process for men and women, which will certainly be reflected in subsequent works. Furthermore, to optimize measures for individual crime prevention as much as possible, the search for the most effective forms of psychological influence on convicts, combining methods of information, persuasion and coercion, should be continued.

References

1. Odinokaya, M., Krepkaia, T., Sheredekina, O., Bernavskaya, M.: The culture of professional self-realization as a fundamental factor of students' Internet communication in the modern educational environment of higher education. Educ. Sci. **9**(3), 187 (2019). https://doi.org/10.3390/educsci9030187
2. Shipunova, O., Evseeva, L., Pozdeeva, E., Evseev, V. V., Zhabenko, I.: Social and educational environment modeling in future vision: infosphere tools. In: E3S Web of Conferences, vol. 110, pp. 02011 (2019). https://doi.org/10.1051/e3sconf/201911002011

3. Almazova, N., Bylieva, D., Lobatyuk, V., Rubtsova, A.: Human behavior as a source of data in the context of education system. In: SPBPU IDE 2019: Proceedings of Peter the Great St. Petersburg Polytechnic University International Scientific Conference on Innovations in Digital Economy, vol. 37 (2019). https://doi.org/10.1145/3372177.3373340

4. Bylieva, D.S., Lobatyuk, V.V, Rubtsova, A.V.: Information and communication technologies as an active principle of social change. IOP Conf. Ser.: Earth. Environ. Sci. **337**, 012054 (2019). https://doi.org/10.1088/1755-1315/337/1/012054

5. Antipov, A.N.: The concept of the development of the penal system of the russian federation till 2030: the project drafting methodology. Penal Syst.: Law Econ. Manag. **2**, 15–17 (2020). (in Russian). https://doi.org/10.18572/2072-4438-2020-2-15-17

6. Kulakova, S.V.: The establishment of professional competences of employees of the penal system of the Russian Federation included in the executive personnel reserve. Penal Syst.: Law Econ. Manag. **2**, 29–33 (2020). (in Russian). https://doi.org/10.18572/2072-4438-2020-2-29-33

7. Schneider, F.M., Zwillich, B., Germann, M.J., Hopp, F.R.: Social media ostracism: the effects of being excluded online. Comput. Hum. Behav. **73**, 385–393 (2017). https://doi.org/10.1016/j.chb.2017.03.052

8. Pfundmair, M.: Ostracism promotes a terroristic mindset. Behav. Sci. Terror. Polit. Aggress. 1–15 (2018). https://doi.org/10.1080/19434472.2018.1443965

9. Ren, D., Wesselmann, E.D., Williams, K.D.: Hurt people hurt people: ostracism and aggression. Curr. Opin. Psychol. **19**, 34–38 (2018). https://doi.org/10.1016/j.copsyc.2017.03.026

10. Bell, S., Hopkin, G., Forrester, A.: Exposure to traumatic events and the experience of burnout, compassion fatigue and compassion satisfaction among prison mental health staff: an exploratory survey. Issues Mental Health Nurs. **40**(4), 304–309 (2019). https://doi.org/10.1080/01612840.2018.1534911

11. Perry, A.E., Waterman, M.G., House, A.O., Greenhalgh, J.: Implementation of a problem-solving training initiative to reduce self-harm in prisons: a qualitative perspective of prison staff, field researchers and prisoners at risk of self-harm. Health Just. **7**, 14 (2019). https://doi.org/10.1186/s40352-019-0094-9

12. Caravaca-Sánchez, F., Ignatyev, Y., Mundt, A.P.: Associations between childhood abuse, mental health problems, and suicide risk among male prison populations in Spain. Crim. Behav. Mental Health **29**(1), 18–30 (2019). https://doi.org/10.1002/cbm.2099

13. Crewe, B., Hulley, S., Wright, S.: The gendered pains of life imprisonment. Br. J. Crim. **57**(6), 1359–1378 (2017). https://doi.org/10.1093/bjc/azw088

14. Skett, S., Lewis, C.: Development of the offender personality disorder pathway: a summary of the underpinning evidence. Probat. J. **66**(2), 167–180 (2019). https://doi.org/10.1177/0264550519832370

15. Vaswani, N., Paul, S.: 'It's knowing the right things to say and do': challenges and opportunities for Trauma-informed practice in the prison context. Howard J. Crime Just. **58**(4), 513–534 (2019). https://doi.org/10.1111/hojo.12344

16. Rumyantsev, N.V., Naruslanov E.F.: The current state of committing gross violations in penal institutions and corrective measures applied to offenders. Penal Syst.: Law, Econ. Manag. **1**, 3–5 (2020). (in Russian). https://doi.org/10.18572/2072-4438-2020-1-3-5

17. Lytras, M., Damiani, E., Sarirete, A.: Technology-enhanced learning research in higher education: a transformative education primer. Comput. Hum. Behav. **109**, 106350 (2020). https://doi.org/10.1016/j.chb.2020.106350

18. Smith, E.E., Kahlke, R., Judd, T.: Not just digital natives: integrating technologies in professional education contexts. Australas. J. Educ. Technol. **36**(3), 1–14 (2020). https://doi.org/10.14742/ajet.5689

19. Morrison, K., Maycock, M.: Becoming a prison officer: an analysis of the early development of prison officer cultures. Howard J. Crime Just. **60**(1), 3–24 (2020). https://doi.org/10.1111/hojo.12394

20. Haneef, F., et al.: Using network science to understand the link between subjects and professions. Comput. Hum. Behav. **106**, 106228 (2020). https://doi.org/10.1016/j.chb.2019.106228

21. Mesko, G., Hacin, R.: Social distance between prisoners and prison staff. Pris. J. **99**(6), 706–724 (2019). https://doi.org/10.1177/0032885519877382

22. Penado-Abilleira, M., Rodicio-García, M.L.: A study of the inmate code of conduct in Spanish Prison. Deviant Behav. **41**, 1–18 (2020). https://doi.org/10.1080/01639625.2020.1768642

23. West-Smith, M., Pogrebin, M.: Female inmate subcultures. In: Bernat, F.P., Frailing, K., Gelsthorpe, L., Kethineni, S., Pasko, L. (ed.) The Encyclopedia of Women and Crime, pp. 1–6. Wiley-Blackwell, New York (2019). https://doi.org/10.1002/9781118929803.ewac0131

24. Cullen, F.T., Chouhy, C., Jonson, C.L.: Public opinion about white-collar crime. In: Rorie, M.L. (ed.) The Handbook of White-Collar Crime, pp. 209–228. Wiley, Hoboken (2019). https://doi.org/10.1002/9781118775004.ch14

25. Williams, E.J., Muir, K.A.: Model of trust manipulation: exploiting communication mechanisms and authenticity cues to deceive. In: Docan-Morgan, T. (ed.) The Palgrave Handbook of Deceptive Communication, pp. 249–265. Palgrave Macmillan, Cham (2019). https://doi.org/10.1007/978-3-319-96334-1_13

26. Rutter, J.T.: Just punishment: Ricks v. Shover and the need for a constitutional response to guard-on-inmate prison sexual abuse. Villanova Law Rev. **64**, 34–62 (2019). https://www.villanovalawreview.com/article/10935-just-punishment-ricks-v-shover-and-the-need-for-a-constitutional-response-to-guard-on-inmate-prison-sexual-abuse

27. Laws, B.: The return of the suppressed: exploring how emotional suppression reappears as violence and pain among male and female prisoners. Punishm. Soc. **21**(5), 560–577 (2019). https://doi.org/10.1177/1462474518805071

28. Ahlin, E.M.: Moving beyond prison rape: assessing sexual victimization among youth in custody. Aggress. Violent Behav. **47**, 160–168 (2019). https://doi.org/10.1016/j.avb.2019.04.002

29. Lambert, E.G., Keena, L.D., Haynes, S.H., May, D., Ricciardelli, R., Leone, M.: Testing a path model of organizational justice and correctional staff job stress among southern correctional staff. Crim. Just. Behav. **46**(10), 1367–1384 (2019). https://doi.org/10.1177/0093854819843336

30. Trounson, J.S., Pfeifer, J.E., Skues, J.L.: Perceived workplace adversity and correctional officer psychological well-being: an international examination of the impact of officer response styles. J. Forensic Psychiatry Psychol. **30**(1), 17–37 (2019). https://doi.org/10.1080/14789949.2018.1441427

31. Basinska, B.A., Dåderman, A.M.: Work values of police officers and their relationship with job burnout and work engagement. Front. Psychol. **10**, 442 (2019). https://doi.org/10.3389/fpsyg.2019.00442

32. Queirós, C., Passos, F., Bártolo, A., Marques, A.J., da Silva, C.F., Pereira, A.: Burnout and stress measurement in police officers: literature review and a study with the operational police stress questionnaire. Front. Psychol. **11**(587), 1–23 (2020). https://doi.org/10.3389/fpsyg.2020.00587

33. Bianchi, R., Schonfeld, I.S., Laurent, E.: Burnout: moving beyond the status quo. Int. J. Stress Manag. **26**(1), 36–45 (2019). https://doi.org/10.1037/str0000088

34. Testoni, I., Nencioni, I., Ronconi, L., Alemanno, F., Zamperini, A.: Burnout, reasons for living and dehumanisation among italian penitentiary police officers. Int. J. Environ. Res. Public Health **17**(9), 3117 (2020). https://doi.org/10.3390/ijerph17093117

35. Foley, J., Dawn Massey, K.L.: The «cost» of caring in policing: from burnout to PTSD in police officers in England and Wales. Police J.: Theory Pract. Principles **20**(10), 1–18 (2020). https://doi.org/10.1177/0032258X20917442

36. Ebner, C., Gegenfurtner, A.: Learning and satisfaction in webinar, online, and faceto-face instruction: a meta-analysis. Front. Educ. **4**, 92 (2019). https://doi.org/10.3389/feduc.2019.00092

37. Oberländer, M., Beinicke, A., Bipp, T.: Digital competencies: a review of the literature and applications in the workplace. Comput. Educ. **146**, 103752 (2020). https://doi.org/10.1016/j.compedu.2019.103752

38. Dowling-Hetherington, L., Glowatz, M., McDonald, E., Dempsey, A.: Business students experiences of technology tools and applications in higher education. Int. J. Train. Dev. **24**(1), 22–39 (2020). https://doi.org/10.1111/ijtd.12168

39. Gegenfurtner, A., Schmidt-Hertha, B., Lewis, P.: Digital technologies in training and adult education. Int. J. Train. Dev. **24**(1), 1–4 (2020). https://doi.org/10.1111/ijtd.12172

40. Rohs, M., Schmidt-Hertha, B., Rott, K.J., Bolten, R.: Measurement of media pedagogical competences of adult educators. Eur. J. Res. Educ. Learn. Adults **10**(3), 307–324 (2019). https://doi.org/10.3384/rela.2000-7426.ojs393

41. Rapisarda, F., et al.: Development and validation of the mental health professional culture inventory. Epidemiol. Psychiatr. Sci. **29**(80), 1–9 (2020). https://doi.org/10.1017/S2045796019000787

42. Skiba, R.: Effective means of teaching and developing emotional intelligence in the corrections industry. Adv. Appl. Sociol. **10**, 187–199 (2020). https://doi.org/10.4236/aasoci.2020.106012

43. Garcıa-Sancho, E., Salguero, J.M., Fernandez-Berrocal, P.: Ability emotional intelligence and its relation to aggression across time and age groups. Scand. J. Psychol. **58**(1), 43–51 (2017). https://doi.org/10.1111/sjop.1233

44. Mattingly, V., Kraiger, K.: Can emotional intelligence be trained? A meta-analytical investigation. Hum. Res. Manag. Rev. **29**, 140–155 (2018). https://doi.org/10.1016/j.hrmr.2018.03.002

45. Kois, L.E., Hill, K., Gonzales, L., Hunter, S., Chauhan, P.: Correctional officer mental health training: analysis of 52 US jurisdictions. Crim. Just. Policy Rev. **31**(4), 555–572 (2020). https://doi.org/10.1177/0887403419849624

46. Useche, S.A., Montoro, L.V., Ruiz, J.I., Vanegas, C., Sanmartin, J., Alfaro, E.: Workplace burnout and health issues among Colombian correctional officers. PLoS ONE **14**(2), 1–20 (2019). https://doi.org/10.1371/journal.pone.0211447

47. Isenhardt, A., Hostettler, U.: Inmate violence and correctional staff burnout: the role of sense of security, gender, and job characteristics. J. Interpers. Violence **35**(1–2), 173–207 (2020). https://doi.org/10.1080/23774657.2017.1421053

48. Keena, L.D., Lambert, E.G., Haynes, S.H., May, D., Buckner, Z.: Examining the relationship between job characteristics and job satisfaction among Southern prison staff. Corrections **5**(2), 109–129 (2020). https://doi.org/10.1080/23774657.2017.1421053

49. Ahalt, C., Haney, C., Ekhaugen K., Williams, B.: Role of a US-Norway exchange in placing health and well-being at the center of US prison reform. Am. J. Pub. Health (AJPH) **110**, 1 (2020). https://doi.org/10.2105/AJPH.2019.305444

50. Ben-Moshe, L.: Why prisons are not «The New Asylums». Punishm. Soc. **19**(3), 272–289 (2017). https://doi.org/10.1177/1462474517704852

51. Kitson-Boyce, R., Blagden, N., Winder, B., Dillon, G.: «This time it's different». preparing for release through a prison-model of CoSA: a phenomenological and repertory grid analysis. Scx. Abuse **31**(8), 886–907 (2019). https://doi.org/10.1177/1079063218775969

52. Arnau, F., Garcıa-Guerrero, J., Benito, A., Vera-Remartınez, E.J., Baquero, A., Haro, G.: Sociodemographic, clinical, and therapeutic aspects of penitentiary psychiatric consultation:

toward integration into the general mental health services. J. Forensic Sci. **65**(1), 160–165 (2020). https://doi.org/10.1111/1556-4029.14137

53. Brooke, J., Jackson, D.: An exploration of the support provided by prison staff, education, health and social care professionals, and prisoners for prisoners with dementia. J. Forensic Psychiat. Psychol. **30**(5), 807–823 (2019). https://doi.org/10.1080/14789949.2019.1638959

54. McCausland, R., Baldry, E.: «I feel like I failed him by ringing the police»: criminalising disability in Australia. Punishm. Soc. **19**, 290–309 (2017). https://doi.org/10.1177/146247 4517696126

55. Rowe, S., Dowse, L., Newton, D., McGillivray, J., Baldry, E.: Addressing education, training, and employment supports for prisoners with cognitive disability: insights from an Australian programme. J. Policy Pract. Intellect. Disabil. **17**(1), 43–50 (2020). https://doi.org/10.1111/jppi.12321

56. Wangmo, T., Handtke, V., Bretschneider, W., Elger, B.S.: Prisons should mirror society: the debate on age-segregated housing for older prisoners. Ageing Soc. **37**(4), 675–694 (2017). https://doi.org/10.1017/S0144686X15001373

57. South, J., Bagnall, A.M., Woodall, J.: Developing a typology for peer education and peer support delivered by prisoners. J. Correct. Health Care **23**(2), 214–229 (2017). https://doi.org/10.1177/1078345817700602

58. Baker, T., Zgoba, K., Gordon, J.A.: Incarcerated for a sex offense: in-prison experiences and concerns about reentry. Sexual Abuse **33**(2), 135–156 (2019). https://doi.org/10.1177/107906 3219884588

59. Brooks, N., Fritzon, K., Watt, B., Duncan, K., Madsen, L.: Criminal and noncriminal psychopathy: the devil is in the detail. In: Corporate Psychopathy, pp 79–105. Palgrave Macmillan, Cham (2020). https://doi.org/10.1007/978-3-030-27188-6_3

60. Hutchison, J.: «It's sexual assault. it's barbaric»: strip searching in women's prisons as state-inflicted sexual assault. Affilia **33**(2), 160–176 (2020). https://doi.org/10.1177/088610991 9878274

61. Jasko, K., et al.: Social context moderates the effects of quest for significance on violent extremism. J. Pers. Soc. Psychol. **118**(6), 1165–1187 (2020). https://doi.org/10.1037/pspi00 00198

Pressure of Digital Technologies and Students' Creative Thinking in the Educational System

Inna B. Romanenko[⊠] [iD], Yulia V. Puyu [iD], Yuriy M. Romanenko [iD],
and Stanislav E. Fedorin [iD]

Herzen State Pedagogical University of Russia, 48 Moika Embankment,
Saint Petersburg, Russian Federation

Abstract. The article analyzes the phenomenon of digital pressure and the accompanying circumstances from the point of view of the need to develop various forms of sustainability and creative thinking on the part of students in the educational system. Digital pressure is examined as the result of a massive invasion of digital technologies in various spheres of life as manifested in various forms which contribute to the destabilization of systems, disrupting their work, causing tension, and changing ways of life. Special attention is paid to the ability of the instrumental-technological mind (in the form of the technologies used) to manipulate human consciousness and behavior, collecting and accumulating information about the human subject. Now personal and biometric data are an important resource that aid in modeling future states and possible personal choices. Each individual is studied separately and the mechanisms involved in social networks greatly simplify this process. In this context, such phenomena as computer addiction, "digital dementia", a simplified perception of reality, a person's subjectless existence of a person, as well as "new civilization" and "new caste" are analyzed. Higher education is viewed not only in its relation to social advancement but also as a "socially safe", ensuring not only consistency in the profession but also life success and fulfilment.

Keywords: Digital pressure · Digital technologies · Creative thinking · Developing sustainability forms · Education system

1 Introduction

The subject of this article is the conceptualization of the problem of digital pressure in its relation to the development of young students' creative thinking skills where detailed consideration of this phenomenon and its causes will be undertaken along with an examination of the various forms of resistance to manipulative influences. Furthermore, the possibility of obtaining a higher education is considered from the point of view of gaining social protection and life prospects for young people.

The emergence of concepts such as digital pressure, digital fatigue, "digital concentration camps", and the like is the result of a massive invasion of digital technologies into the professional, domestic, intellectual, cultural, economic, political life of society.

© The Author(s), under exclusive license to Springer Nature Switzerland AG 2022
D. Bylieva and A. Nordmann (Eds.): PCSF 2021, LNNS 345, pp. 224–231, 2022.
https://doi.org/10.1007/978-3-030-89708-6_19

It manifests itself in different things and forms: in the increase of cyber threats, new settlement mechanisms between corporations and economic entities (crypt-currency), new forms of communication and creativeness, emerging phobias and tensions in human life, as well as in a change of the lifestyle of a person. The intensity of cyberattacks is growing in various directions and forms in the world, specialists in political sciences note the transition in this area from intelligence activities to the possible damage (in the form of destabilization of national financial systems, the activities of social networks, disruption of certain sectors of the economy, psychological and cultural pressure on different countries, "a war of infrastructures", etc. Psychological stress is also manifested at the level of social psychology. Despite the significant rate of virtualization and informatization of modern human life the world itself continues to remain "analog", it still largely depends on the decisions of people. This circumstance is associated with uncertainty and a growing sense of danger and unpredictability which has been recorded in various sociological surveys of the population in different countries.

In such a situation of uncertainty and unpredictability, one should be expected to worry more about information security and digital independence. Self-examination and a reasonably balanced assessment of one's capabilities and "vulnerabilities" allow for identifying "critical points" of development and growth. The world continues to build and develop based on digital technologies. It is almost universally recognized that there is no point in discussing whether this is good or bad, or whether it should be resisted. The digital world exists, this is the reality we live in. But it is necessary to preserve oneself wealth and subjectivity in this changing world. The paradox can be formulated in the following way: individual self-preservation in such a world depends on a person's ability to be able to learn and to be able to love to learn, to be able to change and to have the will to be changed. The phenomenon of leadership is rethought in this context as the ability to "overcome dehumanizing economic and social imperatives and to offer the social environment new human-sized meanings for interaction" [1].

2 Literature Review

It's a well-known fact that education is a backbone component of culture. Having turned into a way of life for a human being, education is also a mechanism for the reproduction of society at the same time, it is the most important factor in its socio-cultural dynamics. Educational reforms which were carried out in the spirit of radical liberalism in Russia and the countries of Western Europe in the previous decades were guided by the fact that the "educational market" has become the main regulator. It has led to the situation of oversupply of economists, lawyers and managers in Russia. But at the same time, there were no significant improvements neither in the economy, nor in legal regulation, nor management. The process of massiveization of education contributes to its simplification and deprivation of its quality. The desire to receive profit from education was reduced to strict regulation of the university professors' and teachers' activities, to the growth of bureaucratic accountability. It's quite understandable that educational reforms are carried out by the state, so they require investment in this area which accordingly leads to the increase of state control. But at the same time, the growth of state control threatens the "academic freedoms" (freedom of creativity, criticism and discussion) which are the important regulators of university life since ancient times. As Professor V. Mironov, the former dean of the Department of Philosophy at the Moscow State University noticed in an interview several years ago: "Scientists and professors should not be tested, they need to be trusted" [2]. In this regard, it should be noted that scientists are people with "unlimited cognitive capital" that provides a variety of results [3].

The problem of digital pressure, the importance of the development of creative thinking of the young generation (Y and Z) in the educational system, the possibilities and consequences of total digitalization of the education sector are investigated in various substantive aspects and directions: the influence of information and computer technologies on the formation of students' methods of self-realization [4; 5]; the impact of innovative practices in the field of digitalization on the interpretation of the modern university mission [6]; the specifics of mental activities in the information environment [7]; Russia's response to new types of threats in the 21st century (about the forecast of political risks when building relationships with artificial intelligence), justifying the need to use civilizational and axiological approaches in the educational sphere and educational policy [8]; the influence of manipulative models in education with the help of media technologies [9]; the relationship of modern educational technologies and truly humanistic educational tasks [10]; the attitude of the modern educational paradigm to the educational paradigms of past eras that have already taken place in history [11]; the relationship of the virtual and speech aspects in the space of youth communication [12]. The problem of digitalization of education has become the subject of interdisciplinary analysis and many publications that have appeared recently. They consider the opportunities that digital technologies open up for the educational system, as well as some negative consequences and anthropological risks of their thoughtless use. The topic of a separate analysis is the study of the features of neurophysiological and social mechanisms of manipulation in media technologies [13].

3 Research Methods

While analyzing the complex problems of digitalization of education, manipulative susceptibility of youth, the influence of virtual reality on youth consciousness or substantiating the need to develop forms of resilience to digital pressure in education and creative thinking philosophical methods were used: phenomenological, structural-functional, hermeneutic. The data from sociological and statistical studies were also used. While analyzing the ontological, epistemological, socio-cultural foundations of digital pressure and the mechanisms of manipulative technologies and the transversality of educational strategies some methodological approaches were used: paradigmatic, socio-psychological, as well as the basic concepts of the "philosophy of everyday life" in the problem field of which social subjects are formed in the situation of the formation of ideals, value preferences, behavioral reactions, decision making.

4 Results and Discussions

Higher education is related not only to social advancement, but it is also a "socially safe" that ensures not only success in life but also consistency in the profession and fulfilment in life in general. The goal of this "socially safe" is to develop several competencies related to the "acquisition and growth of knowledge", but also the ability to adapt to a changing world. Science metrics, rankings, international databases and systems for indexing and tracking scientific citation, etc., of course, are necessary in the modern world, but at the same time, they create multiple barriers to the development of real science. Scientific citations do not determine scientific quality, but only indirectly testifies to it. T.V. Chernigovskaya notes the fact that humanity has become a "planetary species." Industrial and post-industrial society is a phased co-evolution of a man with technology, which will be continued [14].

The new format of human existence in the modern world is associated with a complex digitalization of the educational, professional sphere and personal life, which led to the redirection of consciousness to another type of reality (virtual, network reality). As a result, there is a loss of interpersonal communication, "the intimacy of existence", the person turns into a hostage of programs, algorithms, regulations that affect his life and behavior. The danger of a prolonged stay of children, adolescents, young people in social networks and the internet in modern society is associated with purely physiological reasons, for example, with the production of dopamine (the hormone of satisfaction received by the brain when receiving encouragement and during processes associated with receiving pleasure). This is especially true for children who are very dependent on the production of this hormone. If a child is restricted in the use of the computer, as well as if his presence in social networks is radically limited, then this causes immediate protest. Also acutely children and adolescents feel the situation of being ignored when they are not noticed on social networks if they do not receive "likes", or nobody responds to them or reacts to the content they post. In addition to these circumstances, we cannot fail to mention the dangers that await children on social networks, because children do not distinguish between real reality and virtual, private and public life, they report excessive information about themselves, which allows unscrupulous users to easily use childish gullibility and lure them into "dubious environments" through their romanticization, all kinds of promises and sometimes intimidation [15].

Let us take note of another point, which so far seems to many users to be less dangerous (for both adults and children). Connecting to social networks at present also means that from this moment on, you will be studied and researched, as it were with your tacit consent: interests, needs, connections, acquaintances, intentions, etc. And as a result of such total attention, your range of interests will be manipulated, gently imposing and limiting it. It should be also noticed that many people are even impressed by such close attention to their person. People lose responsibility and concentration, they do not notice the fact that the reasons for the choice are formed gradually and now they do not stem from them but imperceptibly sprout from the rooted "seeds of manipulation".

The instrumental and technological mind at the beginning of this century, represented by large corporations and their technologies, can constantly observe us, study us, manage us, imperceptibly shaping our needs and interests in the direction they need. Each of us is studied separately and constantly, the mechanisms involved in social networks greatly simplify this process. There is a race to get the attention of the person, as well as to get hold of his data and personal information. And if earlier the land and natural resources were important social assets of development, now knowledge and information are becoming such fundamental resources [16]. Personal and biometric data of a person are an important resource (asset) with which one can model future states and possible choices of a person. Human inclinations and capabilities are in our DNA, but with the help of technology, they can be corrected and redirected.

Computer addiction associated with the formation of a pathological connection between a person and a computer is a disease that must be treated, one must be able to prevent it, without bringing it to "digital dementia", a flat perception of reality, as it were in one projection, without distinguishing between multilevel structures. This is a simplified model of life perception without long texts (no more than 1 page) and video plots (no more than 5 min), without in-depth analytics, building and reconstructing cause-and-effect relationships, a proper sense of time and space, without personal search and modeling of future states. These processes lead to the destruction of the individual perception of the world and «lack of subjectness» as a phenomenon reflecting the processes of deontologization in social processes: the individual is deprived of his usual "opportunities for the manifestation of his uniqueness, the formation of responsibility and awareness" in professional activities [17].

If the revolutionary changes of the past affected the economy, politics and the social sphere, then future civilizational and revolutionary changes will affect to a great extent the "human brains" and physiology (genetic engineering, cyborg engineering), which can give rise to new social differentiation, criteria and assessments the need and usefulness of society, new forms of inequality. The danger is who will make the responsible decisions. The technologies themselves are neutral, but much depends on those who own them, uses them and for what purposes they are used. Historically, such technology as the Internet was created as an effective medium for horizontal communication. The age generation is formed by the real world where sensations (sight, hearing, touch, smell, speech) are of great importance. The younger generation is formed in social networks, where a person and his "avatar" do not coincide, individual communication is replaced by the exchange of information, creativity in its highest forms of expression - byways of self-expression and self-positioning, etc., where "new ethics" and law operates, written speech becomes more and more primitive, where the perception of life is generally simplified, serious reasoned and systematic thinking turns into a clip thinking [18].

A simplified model of world perception, focused almost exclusively on a comfortable existence with a minimum of physical and intellectual stress, is becoming widespread. And only a very small part of people (according to scientists' forecasts) soon will be able to read complex texts, perceive complex ambiguous artistic and musical images, analyze complex information ("new civilization", "new caste"). The process of cognition has always required concentration, effort, focused search (archives, library, laboratory research, scientific dialogue, etc.). Concerns about the disappearance of textual culture may still be too hasty and erroneous in the sense that it is not a textual culture that is leaving, but the number of its carriers is decreasing. But those who remain its bearers will be the new elite, which presupposes not only erudition but also a highly spiritual perception of life, responsibility and compassion. "You have to live for the sake of what you can die for" (Ivan Ilyin, the Russian philosopher). In this sense, love for the Motherland should be separated from politics, but close to the values of the family and the person himself as the highest value. Values are absolute, they cannot be quantified in the sense that the price of a value cannot be determined. But there is also an intimate component in values, an appeal to the world of ideas and a dialogue with the eternal.

Civil society is not only a society of the current majority, but it includes many generations (living, departed, future). All generations have rights: the living have the right to a responsible existence and decision-making, the dead have the right to be remembered and continued by their descendants, and future generations have the right to life and a healthy environment. Generations Y and Z (according to the typology of W. Strausse and N. Howe) represent modern youth audiences, they are the main subjects of the modern educational process. Representatives of this generations differ from previous generations by their focus on self-realization, self-development, intolerance to the things that do not fit into their value system (corruption, authoritarianism, disrespect for their views and positions, etc.), the need for like-minded people (the desire to act together, to organize something, arrange, volunteering, charity, etc.). Of course, we need to communicate with young people, they need to be studied, they need to be taught, educated and developed, accepting their dissimilarity from the older generation, but at the same time acting delicately, patiently and responsibly. Their vision of the world, desires and intentions often do not coincide with ours.

5   Conclusions

Bloggers now have a significant impact on youth consciousness through social networks which have not only become a channel for delivering information, platforms for creative thinking, self-realization and popularization of youth ideas, but also a place of self-affirmation. New social divisions have emerged into bloggers becoming "influencers", active users, passive users, etc. These divisions have come to structure our lives in a new way. Big bloggers who started their activities on the internet intuitively are now building teams and employing professional photographers, photo monitors, cameramen, speechwriters, administrators, and they like to turn utopia into video reality and, if it is possible, to sell it. Let us note the fact that some crisis phenomena are also observed in this area, which does not at all negate the relevance of this form of activities [19, 20]. Social networks are overcrowded with photos and videos; currently, they are distinguished by the poverty of the content or the empty content of the materials posted. Often people applying for a certain position do not have the necessary qualifications, education, or professional experience. Experts note the fact that we live in the era of the "comet's tail", so big events and personalities are not expected, respectively, bloggers are forced to plunge into petty topics, where everything is subordinated to consumption and entertainment. They cannot build long development strategies for several years ahead to be interesting and demanded by the user audience. But at the same time, both young bloggers and experienced ones have a dependence on subscribers. Bloggers are often distinguished by unprofessionalism, narrow outlook, limited vocabulary, illiteracy. They may have 2–3 higher educations each (diplomas and certificates are readily demonstrated), but at the same time, they have often have no complete primary and secondary education at the basic level. We also note the importance for society of serious specialists in various fields, whose point of view and qualified view on events are very important for the audience. Some of them maintain interesting blogs on narrow topics and therefore are in demand by a small audience, specialists, whose circle is rather narrow, but this fact does not negate their educational value. But for many serious specialists, scientists, journalists, cultural figures, the very word "blogger" often has some offensive sense.

References

1. Chepyuk, O.R.: Business Modeling of Innovation Systems. Publishing House of Nizhny Novgorod State University, Nizhny Novgorod (2013). (in Russian)
2. Mironov, V.V.: My optimism is "restrained": about philosophy, education and man. Concept: Philos. Relig. Cult. 2, 10–24 (2019). https://doi.org/10.24833/2541-8831-2019-2-10-9-26. (in Russian)
3. Haprov, S.: Digital Communism. Moscow Financial and Industrial University «Synergy», Moscow (2013). (in Russian)
4. Neysbit, D.: High Technology, Deep Humanity: Technology and Our Search for Meaning. AST Transitbook, Moscow (2005). (in Russian)
5. Shutenko, A.I, Shutenko, E.N., Sergeev, A.M., Ryzhkova, I.V., Koreneva, A.V., Tegaleva, T.D.: Socio-cultural dominants of higher school innovation mission (Dominantessocioculturales de la mision de innovacion de la escuela superior rusa). Espacios 39(52), 34–38 (2018). http://www.revistaespacios.com/al8v39n52/18395234.html. Accessed 21 February 2021
6. Shutenko, E.N., Shutenko, A.I., Sergeev, A.M., Ryzhkova, I.V., Koreneva, A.., Tegaleva, T.D.: Axiological dimension of the higher school innovative potential. Mod. J. Lang. Teach. Method 8(9), 106–113 (2018). http://mjltm.org/article-1-232-en.pdf
7. Shipunova, O.D., Mureyko, L.V., Kzyurin, A.Y., Kolomeyzev, I.V., Kosterina, O.N.: Resources to matrix control of mental activity in information environments. Utopía y Praxis Latinoamericana 24(5), 113–122 (2019)
8. Bespalova, T., Bakhtin, M., Sviridkina, E., Lepekhin, V.: Russia's response to new types of threats of the XXI century. Innov. Technol. Sci. Educ. 210, 1–9 (2020). https://doi.org/10.1051/e3sconf/202021016025
9. Puyu, Y., Bolshakov, Y., Nikonov, S.B., Bolshakov, S.N.: Cloud technologies in the promotion strategy of integrated Asian. Soc. Sci. 11(19), 8–14 (2015)
10. Tulpe, I.A., Luzina, T.I., Puiu, Y.V., Spivak, V.I., Kozhurin, A.Y.: Memories of the future: educational technologies and goals of the enlightenment. Opcion 35(22), 1169–1185 (2019)
11. Bogatyrev, D., Romanenko, I.: The religious foundations of educational paradigms (from antiquity to postmodernity). ΣΧΟΛΗ. Ancient Philos. Class. Trad. 10(2), 495–511 (2016). (in Russian)
12. Romanenko, I., Puyu, Y., Romanenko, N., Tyukhova, I., Iskra, O.: Virtual and speech aspects in youth communication space. Adv. Soc. Sci. Edu. Hum. Res. 289, 408–411 (2018). https://doi.org/10.2991/csis-18.2019.83
13. Shipunova, O.D., Berezovskaya, I.P.: Organization technology of professional interactions in the engineering environments. Int. J. Civ. Eng. Technol. 9(10), 2020–2028 (2019)
14. Chernigovskaya, T.V.: Schrödinger's Cheshire Cat Smile: Language and Consciousness. St. Petersburg State University, Faculty of Liberal Arts and Sciences. Languages of Slavic Culture, Moscow (2016). (in Russian)
15. Anderson, L., Hibbert, P., Mason, K., Rivers, C.: Management education in turbulent times. J. Manage. Educ. 42(4), 423–440 (2018)
16. Bell, D.: The Measurement of Knowledge and Technology. Indicators of Social Change: Concepts and Measurements. Russell Sage Foundation, New York (1968).

17. Mureyko, L.V., Shipunova, O.D., Pasholikov, M.A.Romanenko, I.B., Romanenko, Y.M.: The Correlation of neurophysiologic and social mechanisms of the subconscious manipulations in media technology. Int. J. Civil Eng. Technol. (IJCIET), **9**(9), 1–9 (2018). http://www.iaeme.com/IJCIET/issues.asp?JType=IJCIET&VType=9&IType=9. Accessed 21 Feb 2021
18. Romanenko, I.B., Puyu, Y.V., Romanenko, Y.M., Romanenko, L.Y.: Digitalization of education: conservatism and innovative development. In: Bylieva, D., Nordmann, A., Shipunova, O., Volkova, V. (eds) Knowledge in the Information Society. PCSF 2020, CSIS 2020. LNNS, vol. 184, pp. 22–29. Springer, Cham. (2021). https://doi.org/10.1007/978-3-030-65857-1_3
19. Shipunova, O.D.: The role of relation to values principle in the social management practices. The existential-communicatory aspect. Middle-East J. Sci. Res. **19**(4), 565–569 (2014). http://doi.org/https://doi.org/10.5829/idosi.mejsr.2014.19.4.123684
20. Kuznetsov, D.I., Shipunova, O.D.: Communication and the natural social order. Mediterr. J. Soc. Sci. **6**(3S), 251–260 (2015). https://doi.org/10.5901/mjss.2015.v6n3s3p265

Informative and Communicative Environment for the Development of Student Creativity and Flexible Skills

Vyacheslav Samoylov⊙, Ekaterina Budnik⁽⊠⁾ ⊙, and Anatoly Tsaregorodtsev⊙

Moscow State Linguistic University, Ostozhenka str., 38, 119034 Moscow, Russia
{v.samoilov,e.budnik,avtsaregorodtsev}@linguanet.ru

Abstract. This article explores the problems of developing creativity and flexible teaching skills in educational organisations. Creativity in the modern world is becoming an integral part of many processes, including education. Student creative associations linked by common interests are one of the forms of additional education. The interaction of students from different training areas mutually enriches and stimulates the development of creativity in educational organisations. On the other hand, student teamwork indirectly and directly affects the development of flexible skills, which modern employers highly value. In this regard, a study of the labour market was carried out to identify the flexible skills required for a particular profession. A model of the flexible skills of the modern specialist was developed. Also, an approach is proposed for the development of students' flexible skills through student creative associations. As a result of the study, the communication and interaction of such structures were revealed, and a model of the information and communication environment (ICS) was proposed to solve it. This article provides a structural diagram of the ICS and also proposes criteria for its effectiveness. In addition, criticism of the proposed approach and weak points that require additional study are presented and the prospects for the development of the study.

Keywords: Creativity development · Education · Flexible skills · Information portal

1 Introduction

New technologies are now developing and disseminating at an unprecedented pace. Innovative technologies have penetrated deeply into all spheres of life, and their influence will only increase. Today it is already impossible to imagine life without mobile applications, websites, internet portals, etc. Modern professional skills are very different from those of 5–10 years ago. At the same time, preservation of traditions and adherence to classical teaching methods mean Russian education is not keeping pace with society's new trends and demands. Universities are complex, inert structures that take time to change.

Today, the education market has gone far beyond schools, colleges and universities. Freedom of information via the internet allows people to exchange experience and knowledge, organise communities and even companies for distance learning, and lead technology companies to organise their educational platforms and universities to improve the quality of training [1–3].

At the same time, the influence of competition laws sees employers willingly use the most advanced technologies and make high demands on selected candidates. A modern specialist must be highly knowledgeable in their professional field and boast widely developed flexible skills. Classical university education primarily promotes professional skills, providing students with freedom of choice and action in developing their flexible skills. This attitude towards education is due to a lack of requirements from regulators and understanding the necessary flexible skills for a particular educational program.

This imbalance can be counterbalanced by student creative associations created according to students' internal interests. However, these groups are often scattered, poorly organised and have inadequate management structures, making them ineffective and narrowly focused [4–6]. Since student creative associations aimed at self-study are a low-cost way to foster an environment that develops students' flexible skills and increases their self-organisation, it is proposed to develop an information and communication environment that will raise students' awareness and involvement in the university society. Therefore, this study aims to avoid repetition of developing students' flexible skills by creating conditions for the formation and development of student creative associations.

2 Methods

To achieve the goal of the study, several problems were solved, and the following methods were used:

- Analysis of the Russian labour market to identify required flexible skills
- Classification of occupations and flexible skills
- Systematic approach for presenting the process of developing flexible skills in the form of a combined set of student creative associations
- Deductive method for moving from general occupation classes and the flexible skills they need for specific occupations
- Comparison of general required flexible skills among professions of different classes

The analysis of the structure and relationships of student creative associations was carried out using the example of Moscow State Linguistic University.

3 Results

3.1 Analysing the Labour Market for In-Demand Flexible Skills

An analysis of the Russian labour market for valued flexible skills should be based on the definition of flexible skills, their content, and classification. Defining a suitable research area and drawing a clear line between agile and professional skills is necessary.

The Oxford Dictionary defines the concept of "soft-skills" as follows: personal qualities that allow a person to harmoniously and effectively interact with other people [7, 8].

The definition of flexible skills suggests that these skills are closely related to professional activities. Indeed, developed oratory skills are a professional necessity for a teacher but not required for an IT employee. This limitation refers to the classification of professions.

Here it is necessary to refer to the well-known subject-activity concept of professional labour forwarded by E.A. Klimov. Klimov considered professional activity within the framework of the ergatic system, which is represented by a variety of subject-object relationships conditioned by the objective reality of the world of professions: man-technology; human nature; human-human; person-sign, etc. [9, 10]. According to this concept, a particular profession lies at the intersection of classes and types of professions. Table 1 shows the ratio of classes and types of profession.

Table 1. Classification of professions according to Klimov's concept

Profession classes types of professions	Gnostic	Transformative	Prospecting (creative)
Technician man	An employee of the technical control department (TCD)	Locksmith, driver, assembler	Constructor, inventor
Man-man	Investigator	Teacher, nurse, doctor, salesperson	Psychologist, entrepreneur
Human nature	The receiver of agricultural products	Field grower, veterinarian	Breeder
Sign man	Corrector	Programmer, accountant, encryptor	Theoretical physicist
Man-artistic image	Painting inspector	Jeweller, painter, decorator	Composer, writer, sculptor

The classification presented in Table 1 allows one to limit the set of combinations of flexible skills. There is no need to select a list of flexible skills for each profession; it is enough to determine which cell of the given classification it is located and select the appropriate set of flexible skills. This approach allows one to limit oneself to fifteen combinations of flexible skills.

The combination of flexible skills for each cell of the occupation classification table is determined based on what requirements employers have for candidates; therefore, to reflect the actual combinations, it is necessary to analyse the current Russian labour market. Let us restrict ourselves to three professions from different professional classification cells (psychologist, programmer, TCD employee) and analyse.

Figure 1 shows a diagram of the employers' requirements for future employees regarding flexible skills. The diagrams show the skills and the percentage of vacancies from the total number of analysed vacancies in the Russian labour market. For the profession "Psychologist", 213 vacancies were analysed, for the profession "Programmer" – 252 vacancies, for the profession "Employee of Quality Control Department" – 255 vacancies.

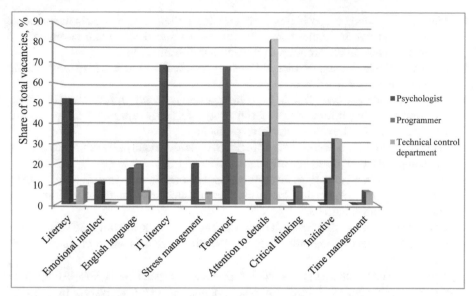

Fig. 1. Diagram of employers' requirements for future employees regarding the possession of flexible skills

Analysing the resulting diagram, one can draw attention to the fact that the necessary flexible skills for the professions "Programmer" and "TCD employee" have more in common than those of a "Psychologist". This is due to the proximity of the professions and their general orientation related to technology. We can also say that skills which are, to a certain extent, inherent in all three professions, include foreign language skills and teamwork.

3.2 The Agile Skills Model of the Modern Specialist

Unfortunately, there is currently no single, standardised classification of flexible skills. This is due to the very definition of the term "soft skills"; it does not clearly define professional and non-professional skills. One of the complete classifications is given in [11]. Based on this work, the following classes can be distinguished:

1. Communication skills are considered as the ability to interact with a variety of sources of information, the ability to interpret received information, including from written sources, presentations and information from the media.
2. Critical thinking is associated with research, analysis and problem solving, the ability to identify and formulate ideas, problems, hypotheses, analyse facts and opinions, develop solutions, evaluate results, and draw conclusions.
3. Global and intercultural competencies, such as analysing the impact of globalisation on local and global communities, apply cognitive and behavioural skills effectively to interact in different cultural contexts.

4. Mathematical literacy – problem-solving related to the exact sciences in the broadest range of contexts, including using quantitative data to develop and communicate sound arguments in classroom practice and real-life situations.
5. Creative thinking is aimed at synthesising practical ideas, working with visual sources and images in original ways.
6. Information literacy can identify, find and evaluate necessary information in a complex and changing information environment, effectively and responsibly use this information to develop ideas and solve problems.
7. Technology literacy is associated with effectively and critically assessing, navigating, and using a range of digital technologies.
8. Integrative learning combines disciplinary and divergent ideas in different contexts, synthesising and transferring the principles of integrative learning into complex learning and life situations.
9. Collaboration is a skill of effective interaction, teamwork to complete educational tasks, achievement of common goals that benefit individuals and society.

Thus, based on the classes presented, it is possible to expand on the above diagram of employers' requirements of future employees obtained using the analysis of the Russian labour market. Table 2 presents a model of flexible skills of a specialist in psychology, programming, technical support.

Table 2. Model of flexible skills of a specialist in the field: psychology, programming, technical support

Class number	Class name	Psychologist	Programmer	TCD employee
1	Communication skills	Yes	Yes	Yes
2	Critical thinking	No	Yes	No
3	Global and intercultural competences	No	No	No
4	Mathematical literacy	No	Yes	No
5	Creative thinking	No	Yes	No
6	Information literacy	Yes	Yes	Yes
7	Technological literacy	Yes	No (included in professional skills)	No (included in professional skills)
8	Integrative learning	No	No	No
9	Collaboration (collaboration)	Yes	Yes	Yes

If we expand the model of flexible skills of a specialist presented in Table 2 with the classification of professions, then we can get a generalised model of flexible skills of professions.

3.3   The Structure of the Information and Communication Environment of Interaction and Self-education of Students

The development of flexible skills occurs in an integrated way [12–14]. The method of point development of over-professional skills is also effective, but the possibility of developing several skills simultaneously reduces the total training time. For example, solving math problems in groups develops two flexible skills: collaboration and math literacy. Therefore, one effective way to develop flexible skills will be forming student creative associations that promote self-education, for example, a student oratory club, a student psychological development club, a student programming club, etc.

The use of student creative associations helps develop flexible skills in universities because they are self-organising structures with internal management. Additional control from the administration and additional financial costs are not required. Moreover, uniting students from different areas of study will allow them to expand their horizons and create teams to implement interdisciplinary projects.

According to the principle of common scientific interests, the unification of students from various fields will create conditions for the growth of creativity in an educational organisation. Arthur Koestler said that creativity is a collision of two unrelated coordinate systems, and all decisive events that have occurred throughout the history of scientific thinking can be described as mutual mental enrichment between different disciplines [15, 16]. Thus, by developing interdisciplinary interactions, we give impetus to the development of creativity.

Nevertheless, in such a structure, information and communication problems arise [17–19]. Since student creative associations are independent structures, they are scat tered, their activities may not intersect, and there may be no interaction. High-quality informing of students about the clubs' activities will attract more students to develop their flexible skills and give them a complete picture of the possibilities for planning their additional education. To overcome the problem of informing and communicating with students, we propose to create an intra-university information and communication environment to develop student creativity and flexible skills, which will be a portal of a particular structure (Fig. 2).

The portal should include:

- Information and news part, which is presented in the form of a constantly updated structured news feed containing information about open grant competitions, conferences, social competitions, competitions and hackathons. This segment of the portal is responsible for motivating student teams.
- Information and educational part, which is presented in the form of segments of the portal reserved for the activities of various student creative associations. This part of the portal will allow students to engage in additional creative activities, and educational content will enable them to develop their chosen flexible skills independently.

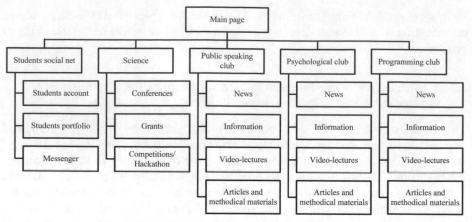

Fig. 2. The structure of the information and communication environment of student interaction and self-education

- The communicative part is presented in the form of an intra-campus social network for finding research partners or team members. This part of the portal is responsible for developing communication skills and identifying leaders among students. It contains information about the skills and achievements of each student registered on the portal, which is necessary for the general search for applicants to the team.

The portal structure presented in Fig. 2 is not fixed; it is formed for each specific educational organisation and will include all student creative associations.

3.4 Criteria for Assessing the Effectiveness of Using the Developed Environment

The question arises regarding the effectiveness of using the proposed information and communication environment. The following criteria are used to assess the efficacy of classical information portals: portal traffic, determining the time spent on the site's pages, determination of key phrases by which users went to the site [20–22], etc.

Such performance criteria are not suitable for an educational approach since the primary purpose of classical portals and the proposed environment is different. Therefore, it is necessary to define new performance criteria for the developed information and communication environment.

Based on the goal of creating a portal – the development of flexible skills among students, the following criteria for the effectiveness of the developed information and communication environment can be determined:

1. The degree of development of flexible skills of students who are in student societies. Additional training in student societies certainly affects the quality of development of students' flexible skills.

2. The degree of development of flexible skills among university graduates. The increase in the number of students with flexible skills indirectly affects the development of flexible skills among students who are not engaged in directed additional education.
3. The active use of the portal by students. This includes the basic technical criteria for evaluating the portal's effectiveness, which shows the basic statistics of visits to the portal.

Using the listed criteria will allow one to get a complete picture of the student's reaction to the created system and the development of their flexible skills.

4 Discussion

4.1 Criticism of the Proposed Approach

The following can be listed as criticisms of the proposed approach:

1. E.A. Klimov developed the concept of professional typology in the 1990s; therefore, its relevance and compliance with new professions, especially those that have emerged in recent years, require additional study.
2. The modern Russian labour market analysis was carried out for only three professions, and the sample was only about 250 vacancies for each profession. There is a need to expand research on the modern labour market to build a more complete flexible skills model for modern professionals.
3. The effectiveness of the proposed information and communication environment for developing student creativity and flexible skills is a long-time project. To understand how effective it is, it is necessary to implement it and conduct long-term research to change the degree of students' flexible skills development.

4.2 Prospects for Development

The above comments to the proposed approach determine the future direction of this research. To better understand the problems of developing students' flexible skills through student creative associations, it is necessary to:

• Analyse the relevance of Klimov's professional typology for the classification of modern professions.
• Conduct additional studies of the modern labour market to identify the connection between occupations and flexible skills.
• Implement the proposed ICS and study the principles of how students work with it, as well as its impact on the development of flexible skills and stimulation of creativity.
• Further avenues for developing this research include formulating a more convenient way to present information and communication environment to students' [23], for example, a mobile application and opening access to it to other universities.

5   Conclusion

This article explored the problems of developing students' flexible skills by creating conditions for forming and developing student creative associations. A classification of professions was carried out, as well as an analysis of the modern Russian labour market in order to form a model of flexible skills for specialists from the psychology, programming, and technical support professions.

The analysis of the modern Russian labour market and the structure and relationships of student creative associations in modern educational organisations made it possible to form an approach for developing the creativity and flexible skills of students. To overcome the problem of informing and communicating self-governing creative associations, it is proposed to use the developed concept of an information and communication environment.

References

1. Lakhal, S., Khechine, H., Mukamurera, J.: Explaining persistence in online courses in higher education: a difference-in-differences analysis. Int. J. Educ. Technol. High. Educ. **18**(1), 1–32 (2021). https://doi.org/10.1186/s41239-021-00251-4
2. Kugusheva, T.V., Novitskaya, A.I.: Interaction between university and business: management challenges and educational requests. Bull. Acad. Knowl. Krasnodar **6**(41), 171–176 (2020). https://doi.org/10.24412/2304-6139-2020-10782. https://cyberleninka.ru/article/n/kreativnyy-potentsial-korporatsii-tsifrovaya-zrelost-1. Access 15 Apr 2021
3. Kobicheva, A.: The structures interaction model of universities and business. In: Conference: 18th PCSF 2018 - Professional Culture of the Specialist of the Future, December 2018. https://doi.org/10.15405/epsbs.2018.12.02.44
4. Kulikov, S.P., Novikov, S.V., Prosvirina, N.V.: Analysis of approaches to student self-government and criteria for its effectiveness at the university. Moscow Econ. J. (11), 543–551 (2019). https://doi.org/10.24411/2413-046X-2019-10088. https://cyberleninka.ru/article/n/analiz-podhodov-k-studencheskomu-samoupravleniyu-i-kriterii-ego-effektivnosti-v-vuze. Access 15 Apr 2021
5. Vincy, I.R.: Examining the effect of explicit instruction on vocabulary learning and on receptive-productive gap: an experimental study. J. Lang. Linguist. Stud. **16**(4), 2040–2058 (2020). https://doi.org/10.17263/JLLS.851033
6. Gubarenko, I.V., Kovalenko, V.I., Kovalenko, E.V., Miyusov, V.A., Sokolova, O.A.: Methods of social interaction learning for students of non-profit organisations. Int. J. Criminol. Sociol. **9**, 1898–1905 (2020). https://doi.org/10.6000/1929-4409.2020.09.220
7. Frantsuzova, O.A.: Convergence in education: soft skills or non-academic skills of schoolchildren: modern vectors of education development: current problems and promising solutions. Sat. Scientific. Works of the XI International Scientific-Practical. Conference on "Shamov Pedagogical Readings of the Scientific School of Educational Systems Management", pp. 278–283 (2019). https://istina.msu.ru/publications/article/318879376/. Access 18 Apr 2021
8. Tyurikov, A.G., Zubets, A.N., Razov, P.V., Amerslanova, A.N., Savchenko, N.V.: Assessment model of quality and demand for educational services considering the consumers' opinion. Humanit. Soc. Sci. Rev. **7**(6), 160–168 (2019). https://doi.org/10.18510/hssr.2019.7632

9. Ivanova, E.M.: Subject-activity concept of professional labor E.A. Klimova and its scientific and practical value. Bull. Moscow Univ. Ser. Psychol. Moscow, no. 2 (2010). https://cyberleninka.ru/article/n/subektno-deyatelnostnaya-kontseptsiya-profes sionalnogo-truda-e-a-klimova-i-ee-nauchno-prakticheskaya-tsennost. Access 18 Apr 2021

10. Tuchina, O.R., Tarasova, V.V.: Professional identity in the era of transitivity (based on the study of future engineers). Herzen's Read.: Psychol. Res. Educ. **3**, 719–726 (2020). https://doi.org/10.33910/herzenpsyconf-2020-3-65

11. Frantsuzova, O.A., Rakhimyanova, I.A.: Soft skills in modern education. Coll. Scientific. Stat. VII all-Russian Scientific-Practical Conference from Int. Participation "Topical Issues of the Humanities: Theory, Methodology, Practice", Moscow, pp. 475–481 (2020). https://cyberleninka.ru/article/n/soft-skills-myagkie-navyki-i-ih-rol-v-podgot ovke-sovremennyh-spetsialistov/viewer. Access 18 Apr 2021

12. Rebrina, F.G., Khakimova, S., Ishkinyaeva, A.: Networking cooperation in forming soft skills of a new type of teacher. Aust. Educ. Comput. **34**(1) (2019). https://doi.org/10.20896/saci. v7i3.530

13. Dubovikova, E.P.: Developing professional competences in training construction engineering specialists. Paper Presented at the IOP Conference Series: Materials Science and Engineering, vol. 687, no. 4 (2019). https://doi.org/10.1088/1757-899X/687/4/044046

14. Pisoni, G., Gaio, L., Rossi, A.: Investigating soft skills development through peer reviews assessments in an entrepreneurship course. Paper Presented at the Proceedings - 2019 IEEE International Symposium on Multimedia, ISM 2019, pp. 291–296 (2019). https://doi.org/10. 1109/ISM46123.2019.00065

15. Koestler, A.: The Act of Creation. Hutchinson, London (1964)

16. Baker, W., Czarnocha B.: The "act of creation" of Koestler & theories of learning in math education research. Math. Teach.-Res. J. **9**(3–4), 22–29 (2018). https://commons.hostos.cuny.edu/ mtrj/wp-content/uploads/sites/30/2018/12/v6n4-The-act-of-creation-of-Koestler.pdf. Access 23 Apr 2021

17. Kahraman, E., Gokasan, T.A., Ozad, B.E.: Usage of social networks by digital natives as a new communication platform for interpersonal communication: a study on university students in cyprus. Interact. Stud. **21**(3), 440–460 (2020). https://doi.org/10.1075/is.20004.kah

18. Frolova, E.V., Rogach, O.V., Ryabova, T.M.: Benefits and risks of switching to distance learning in a pandemic. Perspekt. Nauki Obrazovania **48**(6), 78–88 (2020). https://doi.org/ 10.32744/PSE.2020.6.7

19. Cruz, M.L., Sa, S., Mesquita, D., Lima, R.M., Saunders-Smits, G.: The effectiveness of an activity to practice communication competencies: A case study across five European engineering universities. Int. J. Mech. Eng. Educ. (2021). https://doi.org/10.1177/030641902110 14458

20. Ghandour, A.: Knowledge sharing platform for multi-site organisation. Paper Presented at the Proceedings - 2019 International Arab Conference on Information Technology, ACIT 2019, pp. 113–117 (2019). https://doi.org/10.1109/ACIT47987.2019.8991121

21. Farrahi, R., Gilasi, H., Khademi, S., Chopannejad, S.: Towards a comprehensive quality evaluation model for hospital websites. Acta Inf. Med. **26**(4), 274–279 (2018). https://doi. org/10.5455/aim.2018.26.274-279

22. Gomez-Galan, J., Martinez-Lopez, J.A., Lazaro-Perez, C., Garcia-Cabrero, J.C.: Open innovation during web surfing: topics of interest and rejection by Latin American college students. J. Open Innov.: Technol. Mark. Complex. **7**(1), 1–17 (2021). https://doi.org/10.3390/joitmc 7010017

23. Potemkina, T.V., Bondareva, L.V., Novoselova, S., Shevechkova, M.: The analysis of students' preferences of mobile applications for studying Russian as a foreign language. Perspek. Nauki Obrazovania **48**(6), 220–233 (2020). https://doi.org/10.32744/PSE.2020.6.17

Cultivating Creativity
Case: Higher Education in Finland

Maria Jakubik[(⊠)] (iD)

Ronin Institute, Montclair, NJ 07043, USA
`maria.jakubik@ronininstitute.org`

Abstract. In the creative or mind economy, there is an increased demand for creative, educated human capital (HC) capable of solving challenging, wicked problems of our time. This paper seeks to answer the question how higher education (HE) in Finland can answer the demand of work life for creative minds. The main approach is to explore, on the one hand, the labor market's demands for knowledge, skills, and competencies of future employees and on the other hand, discuss whether the current Finnish HE institutions (HEIs) are capable of supplying future employees with the required creativity skills. The findings show that while needs for creativity of future employees are continuously increasing at work, in HE there is not enough attention paid on cultivating creativity of the future generation. The paper contributes to the discourses related to the renewal of HEIs and it offers implications for HE policy makers, HEIs' leaders, managers, educator, and researchers.

Keywords: Creativity · Knowledge · Skills · Competence · Human Capital (HC) · Higher Education (HE) · Higher Education Institutions (HEIs)

1 Introduction

In the 21st century, human capital (HC) is the utmost important drive of the economic value creation, growth and development. The creative economy requires educated, skilled, intelligent people with advanced competencies who are capable to create value and address the main challenges of the society. According to philosopher Maxwell [1], our world faces seven fundamental problems such as: revolution for philosophy and education; revolution in what we take physical science to tell us about the world; revolution in our whole conception of science; revolution in biology; revolution in social sciences; revolution in academic inquiry; and social revolution.

The topic of this paper is contemporary and important because it focuses on higher education (HE), on how universities in Finland can contribute to addressing these fundamental problems by providing platforms for students to creatively explore and solve great problems of humanity such as the health crisis (COVID pandemic), demographic changes (aging population), climate crisis (global warming), social media crisis (fake news), ethical and moral crisis (corruption), and humanity crisis (immigration). The question is "how universities might do more to help humanity solve grave global problems" [1]. The role of HE is critical in fostering the future generation with knowledge, skills, and competencies demanded by the work life.

There are several reasons why this paper focuses on the Finnish HE as context for cultivating creativity. *Firstly*, among nature and sustainable development, functionality and well-being, education and know-how is one of the features of Finland's country brand. "Finland has one of the bests education systems in the world. ... Thanks to our world-class education system, we find joyful solutions to both local and global challenges" [2]. *Secondly*, according to the OECD 2019 report on skills related policies to work, live and learn in a digital world [3], Finland ranks highly among countries on criteria: skills and benefit from digitalization; providing the necessary skills to the next generation; digital exposure; skills-related policies to make the most of digital transformation; effective ICT integration in schools; teachers' preparation and training needs; lifelong learning systems; advanced education; and learning outside of formal education. *Thirdly*, according to The Global Human Capital Report 2017 [4], Finland (score 77.07) ranks second out of 130 countries after Norway (score 77.12) and followed by Switzerland (score 76.48) and the US (score 74.84). *Finally*, the author of this paper has her special interest in HE based on her over a two-decade experience as a teacher in tertiary education in Finland.

Although the Finnish higher education institutions (HEIs) are performing well, and in several criteria, they are ranked highly in the world [2–5], however, there are also challenges for them. OECD Education Policy Outlook: Finland [6] indicates several of the problems that would need more attention. For example, the followings were identified: declining public funding; decreasing performance of students in PISA-tests; widening students' performance differences according to their socio-economic status, immigrant background and gender; declining employment rates among younger adults; increasing number of unemployed young people; long study periods; highly selective tertiary admissions systems; and a need for a more internationally competitive HE system [6, p. 3].

The objective of this paper is to focus only on one of these problems, namely on how the Finnish HE can better satisfy the demand of work life for creativity and this way increase the employability of the graduates. The research approach is a single case study that provides insight to the cultivating creativity in HE phenomenon.

The rest of the paper has three sections. The research design section states the topic, identifies the research questions, decides on the research strategy, philosophy, approach, method, and it defines the main concepts. Section 3, presents the Finnish HE as the context of the case and discusses the two sub-questions of the research. Finally, in the conclusions section, the author answers the main question, indicates future trends in the Finnish HE, and offers few suggestions for HE policy makers, leaders, managers, educators, and researchers.

2 Research Design

This part of the paper, firstly, presents the criteria based on what the research topic is selected, the research questions are formulated, and the objectives of the research are defined. Then, the author of the paper argues why the case study approach is chosen, and finally, how the key concepts could be defined.

2.1 Research Topic, Questions and Objectives

The research topic (i.e., problem) of this paper is: cultivating creativity. Selecting this research topic is based on the author's: personal interest in knowledge, skills and competence development in HE; long experience as practitioner in HE in Finland; knowledge of the relevant literature; and on her earlier research related to HE. According to Saunders et al. [7, p. 31] there are three main criteria in assessing the goodness of a research topic: appropriateness, capability, and fulfillment. The selected topic is appropriate because the topic fits the journal profile, has clear link to the theory, and the paper aims to provide fresh insight into the topic. The capability criteria are fulfilled as the author proved her research skills in her earlier publications, the research is achievable within the available time, and there is an availability of the necessary literature. Because the topic motivates the author to explore it and contribute to its better understanding, the fulfilment criteria are met too.

Formulating the research questions and objectives are the next steps in research design [7, pp. 42–46]. As it has been shown in the introduction, human society [1], businesses, and HEIs [6] face several complex, wicked problems. Therefore, problem solving capability, creativity and innovation skills of HC are highly demanded by employers. Developing knowledge, skills, and competencies of students is one of the goals of HE. Consequently, this paper seeks to answer the main question: *How can HE in Finland answer the demand of work life for creative minds to solve the wicked problems of our society and businesses?* There are two sub-questions:

1. Why creativity is an important skill in the 21st century? The objectives are to find out the demands of work life for creativity of employees and to understand the importance of creativity as a skill in work life; and
2. How can HE in Finland cultivate creativity of the future generation? The objective is to explore the ways, methods of cultivating creativity in HEIs in Finland.

2.2 Research Strategy, Philosophy, Approach and Method

The case study research (CSR) strategy is selected because the research problem (i.e., cultivating creativity) is a contemporary phenomenon in its real-life context (i.e., HE). The phenomenon is complex with many variables. Therefore, its investigation and understanding need multiple sources and different perspectives. The aim is to make an original contribution, a better understanding to knowledge about the research topic. Furthermore, the research questions are formulated as 'why' and 'how' questions, which concur with Yin's definition of CSR [8, p. 8]. Similar to Yin's definition, Dul and Hak [9] provide the following definition: "A case study is a study in which (a) one case (single case study) or a small number of cases (comparative case study) in their real-life context are selected, and (b) scores obtained from these cases are analysed in a qualitative manner" [9, p. 4]. Myers [10, pp. 74–77], however, criticizes Yin's definition of CSR as being at the same time 'too broad' and 'too narrow'. He writes that Yin's definition "is not entirely appropriate for all qualitative researchers in business" [10, p. 76]. He argues that "The purpose of the case study research is to use empirical evidence from real people in real organizations to make an original contribution to knowledge" [10, p. 73].

The next phase in CSR design is deciding on the philosophical standpoint, type of the case study, and on the research approach. CSR can be implemented from different philosophical perspectives, such as positivist, interpretive, and critical because "case study research is philosophically neutral" [10, p. 77]. Yin [8, pp. 46–60] defined four types of case study design: single-holistic; single-embedded; multiple-holistic; and multiple embedded. Alternatively, Stake [11] categorizes case study types as: intrinsic – to better understand a specific case; instrumental – to provide generalization; and multiple or collective – to jointly research the phenomenon. The research approaches in CSR can be exploratory, explanatory, or descriptive.

In brief, the case presented in this paper can be defined as a single, interpretive, intrinsic case with exploratory and descriptive approaches that applies qualitative research methods.

2.3 Concepts

The concepts of creativity, knowledge, skills, and competence are relevant to this case. In the 21st century, creativity as a human skill is highly demanded by the labour market (see more arguments in Sect. 3.2 of this paper). People with creative skills are needed in business for solving complex, ill-defined, wicked problems namely, "the climate crisis; the current pandemic; the destruction of the natural world, catastrophic loss of wild life, and mass extinction of species; lethal modern war; the spread of modern armaments; the menace of nuclear weapons; pollution of earth, sea and air; rapid rise in the human population; increasing antibiotic resistance; the degradation of democratic politics" [1]. Businesses not only express their need for creative employees but at the same time they are responsible for creating an environment where creativity can flourish. The role of managers and leaders is key for cultivating knowledge, skills, competencies, and creativity in business.

Creativity as a concept is discussed in a recent paper written by Al-Abadneh [12]. He argues that creativity is a complex human behavior and "it can be influenced by a wide array of social, developmental and educational experience that leads to creativity in different ways in a variety of fields" [12, p. 245]. According to him, novelty and usefulness are the two dimensions of creativity. To be a creative person requires many talents and skills in order to think critically, see things differently, produce new solutions, ideas, and new knowledge. Creativity is characterized by 'problem finding' and problem solving'. He concludes that "creativity has been perceived in different ways as a mental ability, a process and a human behavior" [12, p. 247]. Indeed, creativity is a complex and multi-dimensional concept. This paper assumes that creativity can be enhanced by education.

The concept of creativity is defined from a motivational perspective [13, p. 396] as "the tendency to generate or recognize ideas, alternatives, or possibilities that may be useful in solving problems, communicating with others, and entertaining ourselves and others". Franken [13, p. 394] argues that to be creative one needs to see things from different perspectives to create novel understanding, alternatives and approaches. Creativity requires "flexibility, tolerance of ambiguity or unpredictability, and the enjoyment of things heretofore unknown".

Csikszentmihalyi [14, pp. 58–73] explores creativity from psychology perspective and identifies the main characteristics of a creative individual. According to him, creative persons: have a great energy; are often quiet; at the same time, they are smart and naive, playful and disciplined, humble and proud, independent and rebellious, imaginative and have a sense of reality, responsible and irresponsible; passionate about their work; and they enjoy what they do.

The nature of creative thoughts is examined by Weisberg [15, p. 4]. He writes that being creative means the ability to create, produce something new. However, he points out that "it is not enough for it to be novel: it must have value, or be appropriate to the cognitive demands of the situation." Creativity is activity that requires efforts and hard work.

Knowledge, skills and competence can be defined in many ways. Knowledge for example can be defined as a process and/or a product, it can be embodied and/or embedded, explicit and/or tacit knowledge. However, in this paper, because the topic is related to education, the European Qualifications Framework's (EQF) [16] definitions of knowledge, skills, and competence in HE (i.e., levels 6 and 7, bachelor and master) are accepted:

"*Knowledge* means the outcome of the assimilation of information through learning. Knowledge is the body of facts, principles, theories and practices that is related to a field of study or work. In the EQF, knowledge is described as theoretical and/or factual.

Skills mean the ability to apply knowledge and use know-how to complete tasks and solve problems. In the EQF, skills are described as cognitive (use of logical, intuitive and creative thinking) and practical (involving manual dexterity and the use of methods, materials, tools and instruments).

Competence means the proven ability to use knowledge, skills and personal, social and/or methodological abilities, in work or study situations and in professional and/or personal development. In the EQF, competence is described in terms of responsibility and autonomy" [17, p. Appendix 3, emphases added].

3   The Case

This section briefly describes the Finnish HE as context. Next, it focuses on exploring the work life's demand for creative HC, and finally, it discusses the ways and methods how HE in Finland can cultivate creativity of the future generation.

3.1   Context

Understanding the context in CSR is utmost important [7, p. 197]. The Finnish HE system consists of 35 HEIs: 13 universities of sciences (USC) and 22 universities of applied sciences (UAS). The main objective of USC is to focus on basic, scientific research. "They interact with the society actively and also promote societal impact of research results" [18, p. 17]. "Universities of sciences are multidisciplinary and regional higher education institutions with strong connections to business and industries as well as to regional development" [18, p. 16]. Universities of applied sciences focus on applied research, development and innovation, and on fostering knowledge, skills, and competencies of students closely demanded by the labor market. Since August 2005 both USC and UAS provide master's degrees.

On the one hand, surveys of USC graduates confirm the need for a more practical approach. Nissilä [19] analysed the feedback from 1,580 master graduates from ten USCs in Finland in 2017. According to her report [19, p. 6], there is an increased demand for more practical and less theoretical techniques in teaching, more practical business projects, more digital methods, more feedback and interactions between students and staff. Similarly, in a more recent survey of 2,430 MSc (econ) graduates in Finland by Teittinen [20], 1,747 respondents expressed their needs for more working life-oriented education, cooperation with companies, and for more feedback from course teachers. "Respondents were dissatisfied with issues such as unemployment, current rates of pay, fixed-term employment contracts and the fact that their work does not correspond to their education" [20, p. 6].

On the other hand, Jakubik [21–23] argues that master graduates from the UAS develop more practical skills wanted in business, gain more practical knowledge, and acquire more practical business competences than master graduates from USC because of the strong collaboration of UAS with the business community during the master's thesis process. Her research is based on feedback from 91 organizations in Finland during the period of 2007–2016 on UAS master students' skills, knowledge, and competence development during their thesis writing process as work development project. She argues [23, pp. 104–110] that based on the positive feedback from organizations, it could be concluded that in collaboration of academia and businesses (i.e., business and academia ecosystem) master students developed practical, useful competencies required by employers.

Jakubik's research validated the results of an earlier survey conducted by Ojala [24] on UAS master's graduates and employers. The research data included 1,274 UAS Master's degree graduates and their 78 employers surveyed in 2012. Ojala concludes: "The graduates benefited most from completing the degree in a sense that the degree increased the graduates' job-specific skills. The graduates felt they had also benefited from the degree in terms of growth in their general abilities. These abilities included gaining skills for expert work and lifelong learning and acquiring completely new skills. … The employers showed that the increased value of the UAS Master's degree was the focus on work and practice as well as increased specialist expertise. The graduates gained broad and varied working life skills from the UAS Master's degree. Apart from expertise, the graduates gained e.g., research skills, open-mindedness, time and work management, collaboration and networking skills." [24, pp. 5–6].

In brief, the Finnish HE system stands on two pillars namely on USC and UAS. These two are complementing each other. They focus on developing different knowledge, skills, and competencies to be applied in the work life.

3.2 Demand for Creativity

The objectives of this section are to find out the work life demand for creativity of employees and to understand the importance of creativity as a skill in work life. Therefore, the aim is to answer the question: *Why creativity is an important skill in the 21st century?*

There are two main types of creativity, i.e., 'everyday creativity' and 'exceptional creativity' according to Schuster [25]. In his view, creativity is a complex concept that includes "intelligence, intensive interest, knowledge, originality (ideas), creative instinct, non-conformity, courage, and persistence". All these traits of HC are valuable in business. The required skills by the work life are surveyed by the World Economic Forum (WEF) [26]. The research indicates that the demand of labor market is increasing for higher-order (i.e., nonroutine interpersonal and analytical) skills such as collaboration, creativity and problem-solving, and character qualities like persistence, curiosity and initiative [26, p. 2]. According to WEF [26], "*critical thinking* is the ability to identify, analyze and evaluate situations, ideas and information in order to formulate responses to problems. *Creativity* is the ability to imagine and devise innovative new ways of addressing problems, answering questions or expressing meaning through the application, synthesis or repurposing of knowledge. *Communication* and *collaboration* involve working in coordination with others to convey information or tackle problems" [26] (Chapter 1, emphases added). Finland, Japan and South-Korea ranked very high on critical thinking/problem-solving, and creativity skills among the nearly 100 high-income OECD countries [26, p. 7].

In the creative economy, new knowledge and new value is created by creative minds, rather than the traditional factors of production (i.e., land, labor, capital). The importance of educated, creative HC become pivotal. Human capital is the knowledge that each individual has and generates. Hamel [27, pp. 58–59] argues that the relative contribution of different human capabilities (i.e., obedience 0%, diligence 5%, intellect 15%, initiative 20%, creativity 25% and passion and zeal 35%) to value creation shows that 80% comes from initiative, creativity, and passion. Hamel argues that obedience, diligence and intellect of employees can be achieved through management. However, there is leadership needed to activate initiative, creativity and passion in employees that lead to high engagement at work.

In the fourth industrial revolution, the following ten skills are important for employees [28]: complex problem solving, critical thinking, creativity, people management, coordinating with others, emotional intelligence, judgment and decision making, service orientation, negotiation, and cognitive flexibility. According to the WEF 2016's prediction, creativity in 2020 will take the third place of importance compared with its tenth place in 2015. The WEF report [28, p. 21] categorized the most important work-related skills into three major groups: abilities, basic skills, and cross-functional skills. Creativity belongs to the cognitive abilities together with cognitive flexibility, logical reasoning, problem sensitivity, mathematical reasoning, and visualization skills.

3.3 Cultivating Creativity

This section seeks to answer the research question: *How can HE in Finland cultivate creativity of the future generation?* The objective is to explore the ways, methods of cultivating creativity in HEIs in Finland.

The main goal of HE is to advance career goals of the future generation by cultivating students' knowledge, skills, and competencies. Gardner argues that "education is considered one of the most important functions of a society … in the absence of education, individuals will not be able to function adequately in the contemporary world, let alone the world of the future" [29, p. 135]. If today's students want to succeed in their careers, they would need the following skills and abilities [26]: critical thinking, creativity, collaboration, communication, information literacy, media literacy, technology literacy, flexibility, leadership, initiative, productivity, and social skills. Naturally, creativity in our highly connected and complex world is not only the characteristic of an individual but mostly it is related to teams, groups, and organizations.

Creativity is among the most important nonroutine interpersonal skills demanded by the labor market. According to Gardner [30], people would need five minds for the future such as disciplined, synthesizing, creating, respectful, and ethical minds. De Bono offers his ways and approaches to achieve creativity by lateral thinking process [31] and by applying the six thinking hats [32] method. According to him, "Lateral thinking is closely related to creativity. … creativity involves aesthetic sensibility, emotional resonance and gift for expression. … More and more, creativity is coming to be valued as the essential ingredient in change and in progress. … In order to be able to use creativity one must … regard it as a way of using the mind" [31, p. 11]. Indeed, education has a critical role and huge responsibility in developing all five kinds of minds and in nurturing and fostering creativity.

The Finnish Ministry of Education and Culture's strategy for 2030 [33] identifies three objectives: to enable better skills, knowledge and competence for all; to take creative, inquiry-based and responsible action that renews society; and to ensure equal opportunities for a meaningful life. The strategy also indicates seven promises such as to: take responsibility for securing the foundations of education and culture in society; *create the right conditions for fostering skills, employment, creativity and inclusiveness in society*; strengthen the economy's capacity for renewal in order to enable wellbeing and sustainable growth; safeguard democracy and freedom of expression; reinforce gender equality, parity and mutual respect; create opportunities for a meaningful life for all; increase our international impact and commit to sustainable development [33] (emphases added). While these objectives and promises are good, there are no specific actions offered for implementing them in HE.

Learners who enter HE are young adults with their own personalities already evolved in family and friends' relationships and in their earlier education. Many of them have work experience and specific career goals. They already know what knowledge, skills, and competencies they want to develop [22, Table 1]. Therefore, for HEIs it is utmost important to understand both the learners' needs and the labor markets' demands when formulating the curricula for learning.

The educators' role is different in HE. They need to take the role of a coach and mentor instead of being only knowledge providers and lecturers. They need listening skills and need to provide continuous and constructive feedback. There are several methods how educators can develop young adults' creativity. For example: work-based learning (WBL); problem-based learning (PBL); exploratory learning; investigative learning; learning by doing; case study discussions; project work; group work; debates; research opportunities; and so on.

Creativity can be developed in several fronts. According to Miller [34] these are the following areas: idea creativity; material creativity; spontaneous creativity; event creativity; organization creativity; relationship creativity; and inner creativity. The techniques for developing idea creativity, according to Miller [34, pp. 64–94], are: matrix analysis; morphological analysis; nature of the business; reframing questions; force field analysis; attribute listing; scamper; alternative scenarios; forced or direct associations; design tree; imagery; brainstorming; analogy; dreams; drawing; meditation. Another, useful collection of creative tools and ideas is provided by Higgins [35]. In HE, the educators themselves need to be creative, and utilize all available methods, tools, and techniques to foster creativity of the next generation.

To sum it up, the case presented here started with the description of the Finnish HE system as a context. Then, the two sub research questions were explored, i.e., the labor markets' demand for creativity and the HE answers to this demand were briefly addressed. Next, the main research question will be addressed.

4   Conclusions

Summing up, the paper started with arguments why creativity become important characteristic of HC in the creative economy in the 21st century. Then, the role of HE in supplying creative employees was clarified. Next, the focus on the Finnish HE, as one of the leading HE system in the world, was decided. It followed by the research design of the case, where the research problem, questions, objectives, research philosophy, approach, method were presented and the key concepts defined. Then, in section three, after highlighting the main characteristics of the Finnish HE as context, the case "cultivating creativity in HE of Finland" was explored with the help of the two sub-questions. Now, in this section of the paper, the main research question will be answered, and implications for HE policy makers, HEIs' leaders, managers, educator, and researchers will be proposed.

The main research question is formulated as: *How can HE in Finland answer the demand of work life for creative minds to solve the wicked problems of our society and businesses?* The paper approached the "cultivating creativity" research problem by exploring it from two main perspectives namely, from the labor market's demand for creative HC and from the Finnish HE role in fostering creativity in the future generation. It can be concluded that the Finnish HE is performing well related to creativity skills [26].

Together with wide range of stakeholders, OECD developed a shared vision 2030 for education and future skills [36]. The report argues: "In the face of an increasingly volatile, uncertain, complex and ambiguous world, education can make the difference as to whether people embrace the challenges they are confronted with or whether they are defeated by them. And in an era characterized by a new explosion of scientific knowledge and a growing array of complex societal problems, it is appropriate that curricula should continue to evolve, perhaps in radical ways" [36, p. 3].

The Finnish Ministry of Education and Culture has its own vision for HE and research for 2030 [37] which includes three main goals: over 50% of all young people complete a higher education degree; development of higher education and expertise in different life situations; and 4% of GDP allocated to research and development, new creative power of science, sustainable growth, more wellbeing [38, p. 4]. This vision admits that HE and HEIs would need to change, they would need to renew themselves in order to respond the demands from the environment. Finland needs to increase research and innovation intensity in HEIs. The action plan for achieving these goals focuses on the followings:

1. Becoming a nation with the most competent labor force - increasing the share of the labor force with a higher education degree; introducing a model and concepts for education provision in continuous learning; and attracting more international talent to Finland.
2. Higher education with an ability to reform and provide digital services - building a higher education environment for digital services and making education more digital, increasing modularity and reinventing teaching.
3. A higher education community with the skills to deliver the best learning outcomes and environments in the world - launching a development program for higher university pedagogics and guidance skills that will receive financial support from the Ministry of Education and Culture.
4. Cooperation and transparency driving research and innovation - a university leadership program will be launched with international partners to improve change management, employee competences and wellbeing in higher education; and strengthening the knowledge base for developing employee wellbeing and leadership.
5. Higher education institutes will become the best workplaces in Finland - more coherent RDI policies; supporting the building of internationally attractive knowledge clusters and innovation systems; and using shared approaches and legislative means to strengthen open research and innovation. [38, pp. 7–11).

4.1   Implications for HE Policy Makers, HEIs Leaders and Managers

The author of this paper concurs with Tapscott [39] in arguing for renewing our thinking about education and learning in a way that fit the requirements of the digital economy. Tapscott foresees six emerging trends: increasingly, work and learning are becoming the same thing; learning is becoming a lifelong challenge; learning is shifting away from formal schools and universities; some educational institutions are working hard to reinvent themselves for relevance, but progress is slow; organizational consciousness is required to create learning organizations; and the new media can transform education, creating a working-learning infostructure for the digital economy [39, pp. 198–207]. Therefore, the new economic, technological [28], social contexts [1] push HE policy makers in Finland to rethink their old assumptions and to find solutions for the emerging problems of education [6].

Leaders and managers of HEIs need to create a flexible learning environment for their students. Internationalization of the Finnish HEIs is another challenge for HE policy makers and leaders. Furthermore, they need to foster stronger connections with businesses to make possible WBL, PBL, identifying and solving real, complex and wicked business problems, and this way developing creativity of their students [19–23]. Suggestions for HEIs to develop required skills demanded by the industry include: creating clear learning objectives; developing curricula and instructional strategies; delivering instructions; embedding ongoing assessment; providing appropriate interventions; and tracking outcomes and learning [26, p. 8].

4.2   Implications for HE Educators

In the new emerging educational paradigm [40] "teachers should act as coaches, facilitators, and guides for learners; they should engage them in learning, create an inspiring, exciting, interesting, safe, and challenging learning environment, and provide support and help when it is needed. … The role of teachers will be mentoring, advising, tutoring, and coaching students, rather than 'preaching' the truth to them." [40, p. 11]. The main goal is not only to transfer the knowledge of the teacher but instead co-creating new knowledge together.

Although Finland, together with Japan and South-Korea, ranked very high on critical thinking/problem-solving, and creativity skills among the nearly 100 high-income OECD countries [26, p. 7], there would be more attention needed for HE educators to acquire and apply in their practices wide range of creativity tools, techniques, methods [34, 35].

4.3   Implications for HE Researchers

Researcher of HE in Finland could tackle the problems identified by the OECD Education Policy Outlook: Finland [6, p. 3] report: declining public funding; decreasing performance of students in PISA-tests; widening students' performance differences according to their socio-economic status, immigrant background and gender; declining employment rates among younger adults; increasing number of unemployed young people; long study periods; highly selective tertiary admissions systems; and need for a more internationally competitive HE system.

To conclude, this paper focused on how HE in Finland can cultivate creativity of future generations making them capable to solve complex, wicked problems of society and businesses. This single case contributes to the emerging discourses in the literature about knowledge, skills, competences needed in the 21st century.

Funding.   This research has not received external funding.

Conflicts of Interest.   The author declares no conflict of interest.

References

1. Maxwell, N.: Our fundamental problem: a revolution for the philosophy and the world. Humanit. Arts Soc. Mag. **3** (2021). https://humanitiesartsandsociety.org/magazine/our-fundamental-problem-a-revolution-for-philosophy-and-the-world/. Accessed 17 June 2021

2. Finland's Country Branding Strategy. Suomi Finland Message Map. Finland Promotion Board (2017). https://toolbox.finland.fi/wp-content/uploads/sites/2/2017/07/2017-07-12-ulkominis terio-strategia-final-eng.pdf. Accessed 3 Feb 2021
3. Overview – Skills-related policies to work, live and learn in a digital world. In: OECD Skills Outlook 2019: Thriving in a Digital World, Chap. 1, Table 1.1. OECD Publishing, Paris (2019). https://doi.org/10.1787/fae51e68-en
4. Finland Country Profile. In: The Global Human Capital Report 2017. Preparing people for the future of work, p. 94. World Economic Forum (2017). https://weforum.ent.box.com/s/dar i4dktg4jt2g9xo2o5pksjpatvawdb. Accessed 4 Feb 2021
5. Finland In: Education at a Glance 2020: OECD Indicators. OECD Publishing, Paris (2020). https://doi.org/10.1787/a236a58f-en
6. OECD Education Policy Outlook: Finland (2020). http://www.oecd.org/education/EDUCAT ION%20POLICY%20OUTLOOK%20FINLAND_EN.pdf. Accessed 5 Feb 2021
7. Saunders, M.N.K., Lewis, P., Thornhill, A.: Research Methods for Business Students, 8th edn. Pearson Education Limited, Harlow (2019)
8. Yin, R.K.: Case Study Research: Design and Methods. Sage Publications Ltd, Thousand Oaks (2009)
9. Dul, J., Hak, T.: Case Study Methodology in Business Research. Elsevier, Burlington (2008)
10. Myers, M.D.: Qualitative Research in Business and Management. Sage Publications Ltd., Thousand Oaks (2009)
11. Stake, R.E.: Multiple Case Study Analysis. The Guilford Press, New York (2005)
12. Al-Abadneh, M.M.: The concept of creativity: definitions and theories. Int. J. Tour. Hotel Bus. Manag. 2(1), 245–249 (2020). https://www.researchgate.net/publication/339831352. Accessed 4 Feb 2021
13. Franken, R.E.: Human Motivation. Thomson Wadsworth, Belmont (1905)
14. Csikszentmihalyi, M.: Creativity: Flow and Psychology of Discovery and Invention. Harper Perennial, New York (2013)
15. Weisberg, R.W.: Creativity: Beyond the Myth of Genius. W.H. Freeman & Co., New York (1993)
16. European Commission: The European qualifications framework for lifelong learning (EQF). Office for Official Publications of the European Communities, Luxembourg (2008). http:// ecahe.eu/w/index.php?title=European_Qualifications_Framework. Accessed 5 Feb 2021
17. ARENE: The Bologna Process and Finnish Universities of Applied Sciences. Participation of Finnish Universities of Applied Sciences in the European Higher Education Area. The Final Report of the Project. ARENE, Helsinki (2007)
18. Laakso-Manninen, R., Tuomi, L.: Professional Higher Education Management – Best Practices from Finland. Professional Publishing, Helsinki (2020)
19. Nissilä, W.: Vastavalmistuneiden kauppatieteiden maisterien palaute 2017. Suomen Ekonomien Julkaisu. 2, 6 (2018)
20. Teittinen, J.: Vastavalmistuneiden kauppatieteiden maisterien palaute 2019. Ekonomit, MER (2020). https://www.ekonomit.fi/wp-content/uploads/2020/10/Vastavalmistuneiden-palaute-2019_raportti_final.pdf. Accessed 6 Feb 2021
21. Jakubik, M.: Masters bring business benefits – proved by finish managers. Turk. Online J. Edu. Technol. 2, 65–77 (2018). http://www.tojet.net/special/2018_12_3.pdf. Accessed 6 Feb 2021
22. Jakubik, M.: Enhancing human capital beyond university boundaries. High. Educ. Skills Work-Based Learn. 10(2), 434–446 (2019). https://doi.org/10.1108/HESWBL-06-2019-0074. Accessed 06 Feb 2021

23. Jakubik, M.: Capturing knowledge co-creation with the practice ecosystem framework in business and academia collaboration. Int. J. Manag. Knowl. Learn. (IJMKL) **8**(1), 95–114 (2019). http://econpapers.repec.org/article/isvjouijm/, https://doaj.org/toc/2232-569 https://www.issbs.si/press/ISSN/2232-5697/8-1.pdf. Accessed 07 Feb 2021

24. Ojala, K.: Ylemmät ammattikorkeakoulututkinnot työmarkkinoilla ja korkeakoulujärjestelmässä. (The university of applied sciences master's degree in the labor market and higher education system). University of Turku: Turku, Finland, Abstract, pp. 5–6 (2017). https://www.utupub.fi/bitstream/handle/10024/134665/AnnalesC437Ojala.pdf?sequence=2&isAllowed=y. Accessed 07 Feb 2021

25. Schuster M.: A kreativitás fogalma és a hétköznapi kreativitás [The concept of creativity--and everyday creativity]. Psychiatr. Hung. **21**(4), 279–287 (2006). PMID: 17170469. https://pubmed.ncbi.nlm.nih.gov/17170469/. Accessed 07 Feb 2021

26. New Vision for Education. Unlocking the Potential of Technology. Chapter 1: The skills needed in the 21st century - New Vision for Education (weforum.org), World Economic Forum: Geneva, Switzerland, Exhibits 1, 5 and 6, p. 2, 7, 8 (2015). http://www3.weforum.org/docs/WEFUSA_NewVisionforEducation_Report2015.pdf. Accessed 07 Feb 2021

27. Hamel, G., Breen, B.: The Future of Management, pp. 58–59. Harvard Business School Press, Boston (2007)

28. The Future of Jobs. Employment, Skills and Workforce Strategy for the Fourth Industrial Revolution. World Economic Forum, Geneva, Switzerland (2016), Figure 9: Core work-related skills, p. 21. http://www3.weforum.org/docs/WEF_Future_of_Jobs.pdf. Accessed 08 Feb 2021

29. Gardner, H.: Changing Minds: The Art and Science of Changing Our Own and Other People's Minds. Harvard Business School Press, Boston (2006)

30. Gardner, H.: Five Minds for the Future. Harvard Business School Press, Boston (2006). Chapter 4: The Creating Mind, pp. 77–101

31. De Bono, E.: Lateral Thinking. Penguin Books, London (1990).First Published in 1970

32. De Bono, E.: Six Thinking Hats. Revised edn. Penguin Books, London (2000).First Published in 1985

33. Strategy 2030: Ministry of Education and Culture, Helsinki, Finland (2019). http://urn.fi/URN:ISBN:978-952-263-632-4, https://julkaisut.valtioneuvosto.fi/bitstream/handle/10024/161562/OKM14.pdf. Accessed 09 Feb 2021

34. Miller, W.C.: The Creative Edge: Fostering Innovation Where You Work. Addison-Wesley Publishing Company, New York (1987)

35. Higgins, J.M.: 101 Creative Problem Solving Techniques: The Handbook of New Ideas for Business. The New Management Publishing Company, New York (1994)

36. The Future of Education and Skills. Education 2030. The Future We Want. OECD (2018). http://www.oecd.org/education/2030/E2030%20Position%20Paper%20%2805.04.2018%29.pdf. Accessed 10 Feb 2021

37. Vision for higher education and research. Ministry of Education and Culture, Finland (2017). https://minedu.fi/en/-/suomen-korkeakoulutukselle-ja-tutkimukselle-visio-2030. Accessed 10 Feb 2021

38. Education and Learning, Knowledge, Science and Technology for the Benefit of People and Society. Proposal for Finland. (Power Point Presentation), Suomi 100+, Finland (2019). https://minedu.fi/documents/1410845/12021888/Vision+2030+roadmap/6dfddc6f-ab7a-2ca2-32a2-84633828c942/Vision+2030+roadmap.pdf. Accessed 10 Feb 2021

39. Tapscott, D.: The Digital Economy: Promise and Peril in the Age of Networked Intelligence. McGraw-Hill Publisher, New York (1996)

40. Jakubik, M.: Quo Vadis Educatio? Emergence of a new educational paradigm. J. Syst. Cybern. Inf. (JSCI), **18**(5), 7–15 (2020). http://www.iiisci.org/journal/sci/FullText.asp?var=&id=CK2 08UT20. Accessed 10 Feb 2021

"Without Electricity/Gravity…" Generating Ideas About the Fate of Civilization

Daria Bylieva(✉) ⓘ, Victoria Lobatyuk ⓘ, Dmitry Kuznetsov ⓘ, and Tatiana Nam ⓘ

Peter the Great St. Petersburg Polytechnic University (SPbPU), 195251 Saint-Petersburg, Russia
bylieva_ds@spbstu.ru

Abstract. In the modern technological world, the concept of creativity is associated with a timeline and a focus on the future. The accelerating change of life under the influence of high-tech changes requires new approaches to the discourse about the future. Creativity as the most "lively" and unpredictable component of technological development is difficult to define, measure, and evaluate. The rapid development of technologies has brought the creation of ideas and their implementation much closer. The most popular issue is becoming the discourse about the future in the format of foresights, which represent the programming of the future. In this study, the authors propose a direction for generating ideas other than tasks for evaluating individual divergent thinking and the foresight format. The assumption of the absence of one of the basic foundations of the existence of modern humanity was used as an incentive to generate ideas about the fate of our civilization, its problems, and opportunities. Understanding the impact of technology development on society, the ability to assess the consequences of technological decisions is an important competence of a modern engineer. The article presents a general picture of the student's answers to the question about the consequences of the absence of gravity (N = 100) and electricity (N = 150). Along with disaster scenarios, students present the solutions for adaptation, positive consequences, and options for using new opportunities. The answers trace the influence of mass culture, popular media discourse, scientific facts related to the topic, and everyday experience, which in some cases are evaluated and revised in a new context.

Keywords: Creativity · Technology · Future

1 Creativity, Technology, Future

Recently, the question about the essence of creativity is closely related to technological progress. Today, there is a widespread opinion in society that artificial intelligence and digital technologies take away routine work from a person, but leave creative ones.

Researchers believe that routine occupations, which require little creativity or complex training, can be replaced fairly easily with computers [1]. We cannot ignore the fact that there are mathematical models that predict the probability of computerization of works based on their creative components [2]. However, there is a problem of understanding what can be considered as creativity and whether it is inherent in each individual. Is it worth worrying about the demand for a person because of it? As Samad

Seyidov notes, any desire to define and thereby limit creativity paradoxically contradicts its essence, which consists of breaking the rules and destroying the framework [3].

The situation becomes especially confusing when artificial intelligence successfully invades areas that are traditionally considered to be the most demanding of creativity from a person [4, 5]. Regardless of whether robots and computer programs are considered as creative collaborators of a person, medium [6], the subject of computational creativity [7], or a person as part of the machine, dispositive or technical system [8] one must admit that there is a need of understanding what creativity is.

The traditional definition of creativity includes originality and effectiveness (or novelty and utility) [9–12]. However, the challenges of our time make such a definition insufficient. Glăvcanu and Beghetto consider it necessary to supplement it, defining creative experience as "novel person world encounters grounded in meaningful actions and interactions, which are marked by the principles of open-endedness, nonlinearity, pluri-perspectives and future-orientation" [13].

In general, we would like to note that modern attempts to define creativity are often associated with a timeline, with the future. The proponents of its dynamic definition add the most consistent time dimension to creativity [14–16]. So Corazza suggests that instead of originality and efficiency, we talk about *potential* originality and effectiveness [14]. According to Corazza and Lubart, the word "potential" adds the indetermination and dynamic to its definition, and the creativity of the highest level creates the conditions of high potential for possible future achievements of the goals of the process, or even serendipitous findings [16]. Vlad Petre Glaveanu and Alexthere Gillespie analyze a difference between a new creature as it was in the past, exists in the present, and can potentially be developed and used in the future [17].

Extrapolation, which often served as the basis for making forecasts of the future, turns out to be powerless before accelerating scientific and technological progress. Human life is constantly changing under the influence of new technological advances. Technoscience sets out to reshape the world as a hybrid of nature and culture [18]. From these facts, one may conclude that technological creativity can be considered as a source of a sequence of bifurcation points that serve as one of the guides of the future. Some new choices, which are presented to society by creativity, affect the activities of individual groups of people, others - change the life of the entire human race. So, creativity can be recognized as "building the future".

Creativity, as one of the most "alive", unpredictable and uncontrollable components, gives technological evolution a certain similarity to biological evolution. A lot of "random" ideas or inventions also accidentally find an embodiment or application. Exaptation is of particular interest in this sense [19]. Andriani and Cattani point out that the majority of "biological traits and human artifacts that were developed for particular functions started as something different: feathers were most likely selected for thermal insulation, bones originated as excess calcium repositories, microwave ovens started life as radar magnetrons" [20].

Creativity as a bridge to the future cannot be assessed unbiased in the present. Its value may manifest itself over time and may not be obvious. That is why some researchers even suggest removing any mention of value in the definition of creativity [21].

An alternative to the extrapolation of the discourse about the future is the development of additional development scenarios that offer a range of possibilities. This is a method based on creativity and divergent thinking.

The awareness of existence in an era of constant technological changes has led to the emergence of a new approach to "working with the future" in the form of so-called "foresight". The term, which has gained popularity since the 1990s, meant "strategic forward-looking technology analysis for policy-making" [22]. However, in the XXI century technology foresight has become a practical tool for "building the future", which has proven itself at the level of organizations, social movements, and national politics. Dmitry Peskov, the ideological inspirer and organizer of the foresight movement in Russia, claims that «all those who have been working with us for the past few years and have worked on the foresights (..) can tell us as well as I can about what will happen to technologies in the next twenty years"[23]. The merging of thinking about the future and planning, observed in foresights, significantly narrows the freedom of creative thought, limiting it to the framework of ideological, political, economic, or other goal-setting. So, Hyeonju Son considers that foresight dominant futures practice, based on the belief that the main purpose of futures studies is to improve effectiveness and lead innovation, ignores other alternatives to contemporary global capitalism [24]. Neda Atanasoski and Kalindi Vora mention that even if revolutionary engineering technoobjects and platforms, aim at better human life, they "tend to be limited by prior racial and gendered imaginaries." [25]. Arthur I. Miller emphasizes that it's very often, the greatest discoveries are of things we never realized we needed [26].

2 What Will Happen… If …?

Although there is a lot of different "creativity test" and "measure of the creative process", there are doubts that a person's creativity can be measured [27, 28].

Most often, experts use tests for the divergence of thinking to assess creativity [29]. At the same time, the criterion of creativity is often originality as its basic and integral property. The most obvious way to measure it is the rarity among the solutions under consideration. In tests for the divergence of thinking, the percentage of the most rarely encountered ideas can be calculated. So, in the Guilford's Alternative Uses test the answers, presented in no more than 5% of their group, are qualified as original [30]. Also, experts use a rating assessment of originality. Meanwhile, no matter how interesting the task of evaluating an individual's creativity or the features of creativity to the individual mind of a person recognized as a "genius" [31], creativity as a "bridge to the future" is a social phenomenon where an individual is included in a social network and has socio-cultural experience.

Therefore, the general set of ideas and their orientation is of the greatest interest. Understanding the impact of technologies on the future, the ability to assess the consequences of technological solutions is a necessary competence of a modern engineer [32–36], and a specialist in general [37–39]. Technology assessment faces increasing demands to assess imaginations of futures that circulate in the present [40, 41]. The system of education should contribute to the formation of the responsibility of future specialists for our future [42, 43].

In the empirical part of this study, we consider the range of ideas of university students about the features of the development of civilization. The incentive for generating ideas was chosen to eliminate a certain basis for the existence of humanity. Such an approach can be considered as a simple variant of "trans-formational creativity" (according to Margaret Boden [44, 45]), consisting of changing a certain base.

In this study, first-year students of Peter the Great St. Petersburg Polytechnic University answered the questions "What will happen if gravity gradually begins to disappear?" and "Imagine that for reasons beyond the control of humanity, it has become impossible to use electricity. Describe what will happen?". Those who wanted to take part in the study gave answers to questions in a free form. Within the first question, answers were received from 100 students, within the second -150 students; the characteristics of the respondents are presented in Tables 1 and 2 (respectively).

Table 1. Demographic profiles. Question "What will happen if gravity gradually begins to disappear?"

Profiles	Description	Percentage (%)
Gender (Male/Female)	Male	52
	Female	48
Age	18–19	70
	20–21	30

Table 2. Demographic profiles. Question "Imagine that for reasons beyond the control of humanity, it has become impossible to use electricity. Describe what will happen?".

Profiles	Description	Percentage (%)
Gender (Male/Female)	Male	54
	Female	46
Age	18–19	82
	20–21	18

2.1 The Disappearance of Gravity

Most of the received answers to the question: "what will happen if gravity begins to gradually weaken?" we can conditionally divide into three groups: fatal consequences for humanity, technological adaptations to new conditions, and new opportunities (Fig. 1).

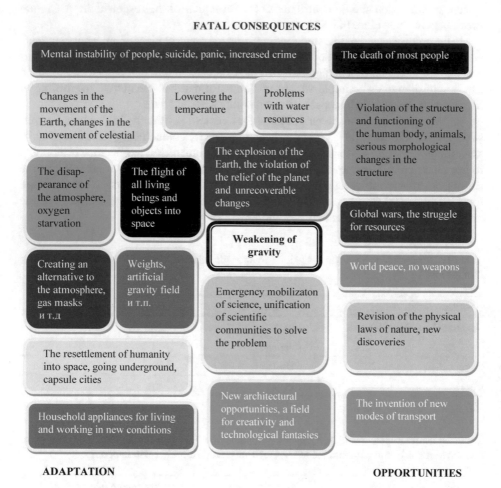

FATAL CONSEQUENCES

Mental instability of people, suicide, panic, increased crime

The death of most people

Changes in the movement of the Earth, changes in the movement of celestial

Lowering the temperature

Problems with water resources

Violation of the structure and functioning of the human body, animals, serious morphological changes in the structure

The disappearance of the atmosphere, oxygen starvation

The flight of all living beings and objects into space

The explosion of the Earth, the violation of the relief of the planet and unrecoverable changes

Global wars, the struggle for resources

Weakening of gravity

Creating an alternative to the atmosphere, gas masks и т.д

Weights, artificial gravity field и т.п.

World peace, no weapons

Emergency mobilizaton of science, unification of scientific communities to solve the problem

Revision of the physical laws of nature, new discoveries

The resettlement of humanity into space, going underground, capsule cities

Household appliances for living and working in new conditions

New architectural opportunities, a field for creativity and technological fantasies

The invention of new modes of transport

ADAPTATION   **OPPORTUNITIES**

Fig. 1. Options Field. "What will happen if gravity begins to gradually weaken?"

Most of the answers refer to fatal consequences as a variant of the development of events, but some of the respondents completely focused their attention only on the single aspect, without assuming any possibilities for salvation. For example, students identify several variants of the death of the earth and humanity at once, including "*explosion of the earth*", "*complete flooding of the earth*", "*oxygen will disappear*", "*people will fly beyond the earth*", "*plants, animals, and fish will disappear or die out*", "*health problems*", "*gradual extinction of humanity as a species*". The researchers note an interesting fact that in some responses there is a relationship between social status and income with the ability to survive. For example, among the popular comments there are such as *"there will be a dropout, which is likely to be based on the power and wealth of people"*; *"the road to the bunker will be open to the minimum number, including scientists, doctors, engineers, government members"*; *"very smart, rich and healthy people will go to other planets to try to save a life"*, *"underground bunkers would be an ideal rescue of people from the radioactive radiation of the sun, the lack of atmosphere and water, and the richest people would immediately occupy them"*. At the same time, more than 90% of the answers indicate "natural selection", when the "death of humanity" will occur.

Forecasts regarding the impact on the relief, water resources, pollution of the planet and lowering its temperature, changes in movement, etc. they were presented with such scenarios as *"the atmosphere will disappear and it will become cold"*, *"all volcanoes will start to erupt, rivers will stop and, eventually, water will disappear"*, "there will be an eternal darkness on the planet", *"the atmosphere will change, no one will be able to hold radiation from space, plus radiation coming from the core will also make itself felt"*, *"it will become incredibly cold on Earth"*, *"underground water will begin to destroy the Earth's surface from below, making cracks in it"*, *"garbage will begin to pollute the environment in a larger volume than before, which will interfere with normal life"*.

New inventions in the answers are divided into two groups: "adaptive" and "alternative". Attempts to adapt to the absence of gravity on the planet Earth, which lead to the invention of restraining devices, new types of transport, housing and the creation of a new reality, are described by the following words: *"the installation of special railings on the streets, for which people can hold on while moving"*, *"the invention and urgent introduction into mass use of tools and gadgets, devices that provide artificial attraction to the surface of the Earth, the creation of artificial gravity"*, *"humanity will have to create an "ark" with a self-sufficient life support system in a month because without gravity, planets and stars will start flying all over the universe"*, *"scientists from all over the world will start thinking about creating devices that allow them to interact more strongly with the environment: magnetic shoes for comfortable walking, flying cars"*, *"everyone will live in special ships with oxygen packages"*, *"creative opportunities will increase dramatically, especially in the field of architecture and home improvement"*; *"buildings will now be able to expand in any direction and the organization of space will become much more multidimensional, the need for the usual chairs and tables disappears, but the need for temporary attachment to a specific point increases, which affects the design features of such things"*, *"ensuring tightness will also be vital, inflatable spheres and domes with the function of air supply will begin to appear"*; *"oxygen masks are becoming commonplace"*, *"cars are gradually ceasing to be relevant, as people first find the opportunity to jump further and further, and then float through the air, practically without needing ground transport"*, *"it becomes possible to use pipes and capsules driven by compressed gases to transport people and larger and "heavier" items than correspondence, compact jet rocket engines will become widespread"*, *"over time, scientists will come up with some gadgets that will allow us to leave the house: some suits like spacesuits with a long rope that can be tied to the house or special "earthling devices" that create a special field around you."* So we can conclude all the variants of the device do not show a wide variety of creative engineering solutions, but two trends can easily be distinguished among the ideas and proposals: fixing/attaching and sealed capsules/spheres/houses.

Alternative inventions are associated with relocation to other planets or with life in space, which is most often presented simply as «flight/travel/migration/relocation to space", *"people will start producing as many rockets as possible so that the maximum number of people will fly into space"*, *"accelerated colonization of Mars"*. Conversely, there are also such unfavorable options for sending off-planet, as "in space or capsules humanity will not be able to survive without oxygen", "going into space, it is not easy to find a suitable planet for life, perhaps it does not exist".

We would like to consider separately the impact of the lack of gravity on a person: *"if gravity does not affect our heart, muscles, and bones, our organs will develop differently"*, *"blood tends to fall under the influence of gravity, as a result of this, lack of gravity can lead to the death of living organisms"*, *"people will start to get sick more, because due to the lack of gravity, muscle atrophy and weakening of the immune system will occur"*, *"without gravity, the loss of air pressure will destroy the inner ear of a person, muscles will atrophy without load, blood vessels will burst"*, *"wounds would heal longer, the immune system would weaken, the sense of balance would change."* But there are also optimistic forecasts such as *"most likely, our body would simply adapt to live without gravity"*.

Also, in the forecasts of students, "war" and "peace" collided. They present scenarios of the onset of world peace, the disappearance of weapons, the unification of people, which sometimes are bearing the color of dreams: *"People will be able to travel, go anywhere in the world on foot! The authorities may figure out how to set boundaries in the air, but until then it is possible to fulfill your dream"*, *"We will be able to build houses in the sky! This will solve the problem of overpopulation and allow everyone to have their own home"*, *"There will be world peace for a while because weapons will no longer be effective. Also, the threats of nuclear war will stop"*. Conversely, some young people see the prospect of global wars, the struggle for resources: *"in the absence of gravity, many global conflicts will begin for the right to possess vital resources"*, *"chaos, wars, conflicts at all levels will begin"*, *"general destruction and wars not only of people but of countries for resources"*.

In the forecasts of the interviewed students, the physical side of the process often faded into the background, and emotions, dreams, images from books and films began to prevail over the scientific picture of the world, which was shaken due to the lack of gravity in it. But at the same time, 25% of the answers include hope for the development of science, opportunities for its breakthrough, the unity of scientists for the benefit of humanity.

2.2 The Disappearance of Electricity

As in the study of the question described above, the problems that arise in the short and long term in connection with the disappearance of electricity are widely represented in the answers of the respondents. But along with the existing issues, students demonstrate the positive consequences and solutions associated with the transition to alternative energy sources, as well as attempts to present a scientific picture of the world without electricity (Fig. 2).

Fig. 2. The field of the main answers to the question: "What will happen if it becomes impossible to use electricity?"

Most young people consider electricity to be a key factor of the era, but they describe the consequences of its disappearance in different ways. According to the most pessimistic scenarios, students predict social cataclysms, economic crises, looting, world wars, anarchy, up to the death of humanity. For example, in one of the scenarios, a student describes the struggle for resources as follows: *"Almost 8 billion people are living on Earth and, of course, natural resources would not be enough even for a year or two, since their extraction also depends on electricity. This will definitely entail a struggle for resources on an international scale, and most likely even a world war. The primary task of each state will be its own survival and the destruction of other countries"*. It is interesting that in the war scenario there is a direct reference to the experience of a computer game: *"as in the game "Fallout" countries will start a war for resources, and soon there will be very few people"*. The "economic crisis" scenario: "First of all, most people on the planet will become bankrupt. At the moment 94% of all funds are electronic money or the money will simply become worthless. Also, the prices of oil, coal, and other reserves will sharply increase in price, since combustible substances will remain the only types of fuel. There will be an economic crisis". Respondents also pay attention to the closure of enterprises, the shortage of goods, unemployment. The topic of the disappearance of crypto currency is mentioned several times. They consider separately the changes in the labor market such as the disappearance of professions, the need for completely different skills. And manual labor is the most often indicated one. However, students point out and other skills, for example, "the ability to work with shaft/piston/belt mechanisms is beginning to be valued more".

Young people pay great attention to the violation of the usual forms of communication and data exchange. They consider both technical and psychological consequences of the lack of information and communication technologies: increased anxiety, depression, growth of the number of suicides. Students note the problems associated with the storage and preparation of food, medical care, and security. According to some authors, planes will not fly, cars will not drive, rockets will not fly. In the works, there is a theme of population relocation from cold areas to warm ones, as well as from large cities to rural areas, and vice versa, greater territorial unification due to the lack of rapid communication.

Optimistic scenarios provide arguments for adaptation to new conditions of existence and several advantages from living in a "non-digital era" with books, personal communication, and more physical activity. For example, *"The lost electricity will entail the end of the information era, the Internet, all the media. That is the disappearance of everything without which it is difficult to imagine the modern world. Technological progress will take 2 steps back, but after generations, people will get used to all this and life will gain stability. People will restart reading more, there will be less advertising in the world, everything will become like literally 100–200 years ago"*.

Reducing the volume of advertising, and in general, decreasing the information noise is perceived as a benefit. Some young people associate the "return to the past" scenario with a more humane relationship: *"The lack of electricity will make many people "go outside", get out of their comfort zone to fully experience the romance of the past era. People will start writing letters to each other. They will make an appointment in advance and understand that they will not be able to inform you about the delay. They will begin* to *"hang out" more in "reality", thereby improving their relationships with each other"*. A world without electricity seems romantic: *"Return to the roots! Live communication, live music, handwritten books, candlelit dates…"*. Returning to the past is also associated with physical health: *«1. All mankind will learn to work with their hands again. 2. People will understand that a bicycle is a great way to get around. 3. Together with the two previous points, a person will acquire a model physical form"*. Ideas about how the lack of electricity will affect the environmental situation were diverse: from the catastrophic consequences of forest destruction to environmental well-being (most often due to the transition to more environmentally friendly types of energy). Examples of eco-positive scenarios: *"Many reservoirs will be cleaned, the amount of carbon dioxide emissions into the environment will decrease (temperature drop). The share of thermal power plants (which generates electricity) accounts for about 14% of the total air pollution by technical means"*.

Other participants did not build long-term prospects, but imagined what would happen without electricity "here and now". For example, «It will be pitch dark. In addition, we will lose such items as a phone, a stun gun for self-defense, electric flashlights". The theme of darkness was closely associated with the «power outage". This is probably the first obvious consequence that students associate with the problem and experience since they once faced a power supply failure. The leitmotif of the colossal growth in demand for candles and matches, its economic and environmental consequences are probably connected with this. Also, the darkness was the reason for the increase in crime. In addition, the set of short-term consequences includes accidents at enterprises, roads, and nuclear power plants, deaths in hospitals, panic, and insanity of the population. The search for alternative energy sources is a fairly popular topic (as an option, the search for a way to return electricity). This topic may be present as a strategy for the behavior of mankind, or the answers may indicate specific search directions, for example, magnetic, solar-ray energy or thermonuclear fusion energy. Among the answers, there are the following assumptions: *"Everything that worked with electricity will work mechanically or thermodynamically"* or *"Humanity will return to using mechanisms built on direct interaction with nature (mills, all kinds of mechanisms using falling water, engines). And they will also start using animals again"*. More original ways of obtaining energy were also named, for example, *from antimatter or using the Dyson Sphere (an astro-engineering project for obtaining energy, using the energy of the central star)*. Some respondents hope to use wind, water, solar energy, etc., thus demonstrating awareness of alternative energy sources. However, in most cases, they do not disclose what energy will be used instead of electricity. A group of students proposed chemical options for energy generation: *"A certain number of chemicals will be included in the sale when they are mixed, energy will be generated for a short time. Such electricity will enter into necessary things like food or water"*, *"they will create micromechanical circuits in which energy will be generated through chemical reactions"*. Chemistry appears economically promising: *"Chemists will become entrepreneurs, they will sell wax and alcohol burners, as well as create new combustible compounds that are not able to cause a fire or to harm humans"*. Students pay special attention to the topic of lighting, using "solar energy storage devices", "phosphor lamps", "gasoline or kerosene lamps", etc.

Several students tried to evaluate the physical side of the process, building hypotheses about what processes can be behind the disappearance of electricity and what it can lead to. For example, *"The disappearance of electricity will lead to the destruction of atoms and molecules because the electric repulsive forces will act against the gravitational forces of attraction, the universe will collapse into a huge black hole"*, *"The division of atoms will lead to a disorderly release of energy. Perhaps they will find a way to use it as an alternative form of energy"*.

Despite the widespread use of electricity, a relatively recent acquisition of humanity, its disappearance provokes young people to develop apocalyptic scenarios no less than the loss of gravity. Nevertheless, a popular leitmotif is also "returning to the past", and in some cases completely and irrevocably, in others, it is still "the future-the past", which combines modern realities and technological capabilities of the past centuries. Interestingly, popular discourses about modern society can be traced in the pool of ideas. The article considers the impact of the disappearance of electricity on the environmental situation in the world, the positive and negative consequences of the rejection of information and communication technologies. It is also worth noting that the question of electricity for future engineers is closely associated with developments in the field of alternative energy sources, although in some cases known solutions were relayed uncritically without taking into account the specifics of the task. Personal experience is another viewed source of ideas when people call a power outage in everyday life "we have no light". It is the problem of lighting, the need for matches, and candles (less often alternative options) that often comes to the fore, sometimes in conjunction with the disconnection of household appliances, devices, and gadgets.

3 Conclusion and Discussion

In modern society, the topics of technology, creativity, and the future are closely interrelated. The rapid development of technologies has contributed to the emergence of stronger links between the creation of an idea and its implementation. This, in turn, influenced the fact that the most popular question is becoming the discourse about the future in the format of foresights, which represent the programming of the future. In this work, the authors sought to offer a variant of creative idea generation, which differs from tasks for evaluating individual divergent thinking, and the format of foresight, which seeks to maximize efficiency and usefulness. The assumption about the absence of one of the basic foundations of the existence of modern humanity was used as an incentive for reasoning about the fate of our civilization, its problems, and opportunities. It is noteworthy that although in some cases the scenario of a catastrophe and the death of civilization was presented as the only consequence, in most cases young people demonstrated divergent thinking, offering along with the disaster scenario options for adapting to the situation, specific technological solutions for returning to normal life and even finding the potential for development in a new situation.

Some of the proposed ideas are essentially naive, but they are of interest from the point of view of analyzing the students' ideas about the ways of development of our civilization. The answers trace the influence of mass culture, popular media discourse, scientific facts related to the topic, and everyday experience, which in some cases are evaluated and revised in a new context. In addition, the received material can become the first stage, after which collective ways of working with ideas can be used for more serious and in-depth development of problems in the educational process. This study was limited to students of a Russian university and gives a rather narrow picture. Therefore, in the future, it would be interesting not only to expand the scientific and methodological base of the study but also to compare the opinions of young people from different countries.

References

1. Kim, Y.J., Kim, K., Lee, S.: The rise of technological unemployment and its implications on the future macroeconomic landscape. Futures **87**, 1–9 (2017). https://doi.org/10.1016/j.fut ures.2017.01.003
2. Bakhshi, H., Frey, C.B., Osborne, M.: Creativity vs. robots the creative economy and the future of employment. https://www.nesta.org.uk/report/creativity-vs-robots/. Accessed 29 Apr 2021
3. Seyidov, S.: Phenomenology of Creativity: History, Paradoxes. Personality Paperback/AuthorHouseUK, Milton Keynes (2013)
4. Anantrasirichai, N., Bull, D.: Artificial intelligence in the creative industries: a review. Artif. Intell. Rev. (2021). https://doi.org/10.1007/s10462-021-10039-7
5. Browne, K.: Who (or what) is an AI artist? Leonardo 1–9 (2021). https://doi.org/10.1162/leon_a_02092
6. Mazzone, M., Elgammal, A.: Art, creativity, and the potential of artificial intelligence. Arts **8**, 26 (2019). https://doi.org/10.3390/arts8010026
7. López de Mántares, R.: Artificial intelligence and the arts: toward computational creativity. In: The Next Step: Exponential Life. Turner Libros, Madrid (2017)
8. Zylinska, J.: AI Art: Machine Visions and Warped Dreams. Open Humanities Press, London (2020)
9. Runco, M.A., Jaeger, G.J.: The standard definition of creativity. Creat. Res. J. **24**, 92–96 (2012). https://doi.org/10.1080/10400419.2012.650092
10. Martin, L., Wilson, N.: Defining creativity with discovery. Creat. Res. J. **29**, 417–425 (2017). https://doi.org/10.1080/10400419.2017.1376543
11. Simonton, D.K.: Creativity, problem solving, and solution set sightedness: radically reformulating BVSR. J. Creat. Behav. **46**, 48–65 (2012). https://doi.org/10.1002/jocb.004
12. Parkhurst, H.B.: Confusion, lack of consensus, and the definition of creativity as a construct. J. Creat. Behav. **33**, 1–21 (1999). https://doi.org/10.1002/j.2162-6057.1999.tb01035.x
13. Glăveanu, V.P., Beghetto, R.A.: Creative experience: a non-standard definition of creativity. Creat. Res. J. **33**, 75–80 (2021). https://doi.org/10.1080/10400419.2020.1827606
14. Corazza, G.E.: Potential originality and effectiveness: the dynamic definition of creativity. Creat. Res. J. **28**, 258–267 (2016). https://doi.org/10.1080/10400419.2016.1195627
15. Walia, C.: A dynamic definition of creativity. Creat. Res. J. **31**, 237–247 (2019). https://doi.org/10.1080/10400419.2019.1641787
16. Corazza, G.E., Lubart, T.: The big bang of originality and effectiveness: a dynamic creativity framework and its application to scientific missions. Front. Psychol. **11** (2020). https://doi.org/10.3389/fpsyg.2020.575067
17. Glaveanu, V.P., Gillespie, A.: Creativity out of difference: theorising the semiotic, social and temporal origin of creative acts. In: Glăveanu, V.P., Gillespie, A., Valsiner, J. (eds.) Rethinking Creativity: Contributions from Social and Cultural Psychology (Cultural Dynamics of Social Representation), pp. 1–16. Routledge, London (2015)
18. Nordmann, A.: Science in the context of technology. In: Carrier, M., Nordmann, A. (eds.) Science in the Context of Application. Boston Studies in the Philosophy of Science, vol. 274, pp. 467–482. Springer, Dordrecht (2011). https://doi.org/10.1007/978-90-481-9051-5_27

19. Garud, R., Gehman, J., Giuliani, A.P.: Technological exaptation: a narrative approach. Ind. Corp. Chang. **25**, 149–166 (2016). https://doi.org/10.1093/icc/dtv050

20. Andriani, P., Cattani, G.: Exaptation as source of creativity, innovation, and diversity: introduction to the special section. Ind. Corp. Change **25**, 115–131 (2016). https://doi.org/10.1093/icc/dtv053

21. Weisberg, R.W.: On the usefulness of "value" in the definition of creativity. Creat. Res. J. **27**, 111–124 (2015). https://doi.org/10.1080/10400419.2015.1030320

22. Georghiou, L.: The Handbook of Technology Foresight: Concepts and Practice. Edward Elgar Publishing, Northampton (2008)

23. Peskov, D.: How is Governance Possible in Russia in the Era of Permanent Technological Revolution? https://www.youtube.com/watch?v=JksDIfmMvcs. Accessed 2 May 2021

24. Son, H.: The history of western futures studies: an exploration of the intellectual traditions and three-phase periodization. Futures **66**, 120–137 (2015). https://doi.org/10.1016/j.futures.2014.12.013

25. Atanasosski, N., Vora, K.: Surrogate Humanity: Race, Robots, and the Politics of Technological Futures. Duke University Press, London (2019). https://doi.org/10.2307/j.ctv1198x3v

26. Miller, A.I.: The Artist in the Machine: The World of AI-Powered Creativity. MIT Press, Cambridge (2019)

27. Piffer, D.: Can creativity be measured? An attempt to clarify the notion of creativity and general directions for future research. Think. Skills Creat. **7**, 258–264 (2012). https://doi.org/10.1016/j.tsc.2012.04.009

28. Cropley, A.J.: Defining and measuring creativity: are creativity tests worth using? Roeper Rev. **23**, 72–79 (2000). https://doi.org/10.1080/02783190009554069

29. Vincent, A.S., Decker, B.P., Mumford, M.D.: Divergent thinking, intelligence, and expertise: a test of alternative models. Creat. Res. J. **14**, 163–178 (2002). https://doi.org/10.1207/S15326934CRJ1402_4

30. De Oliveira, E., Reynaud, E., Osiurak, F.: Roles of technical reasoning, theory of mind, creativity, and fluid cognition in cumulative technological culture. Hum. Nat. **30**(3), 326–340 (2019). https://doi.org/10.1007/s12110-019-09349-1

31. Glăveanu, V.P.: Principles for a cultural psychology of creativity. Cult. Psychol. **16**, 147–163 (2010)

32. Pozdeeva, E.G., Shipunova, O.D., Evseeva, L.I.: Social assessment of innovations and professional responsibility of future engineers. IOP Conf. Ser. Earth Environ. Sci. **337**, 012049 (2019). https://doi.org/10.1088/1755-1315/337/1/012049

33. Shipunova, O.D., Evseeva, L., Pozdeeva, E., Evseev, V.V., Zhabenko, I.: Social and educational environment modeling in future vision: infosphere tools. E3S Web Conf. **110**, 02011 (2019). https://doi.org/10.1051/e3sconf/201911002011

34. Baranova, T.A., Mokhorov, D.A., Kobicheva, A.M., Tokareva, E.Y.: The formation of students' personality at Peter the Great St. Petersburg Polytechnic University: attitude to university and attitude to future profession. Eur. J. Contemp. Educ. **10**, 173–186 (2021). https://doi.org/10.13187/ejced.2021.1.173

35. Kazaryan, R.: The concept of anthropotechnical safety of functioning and quality of life. In: Murgul, V., Pukhkal, V. (eds.) International Scientific Conference Energy Management of Municipal Facilities and Sustainable Energy Technologies EMMFT 2019. AISC, vol. 1258, pp. 759–767. Springer, Cham (2021). https://doi.org/10.1007/978-3-030-57450-5_64

36. Ammon, S.: Image-based epistemic strategies in modeling: designing architecture after the digital turn. In: Ammon, S., Capdevila-Werning, R. (eds.) The Active Image. PET, vol. 28, pp. 177–205. Springer, Cham (2017). https://doi.org/10.1007/978-3-319-56466-1_8

37. Krasnoshchekov, V., Arseniev, D., Rud', V., Switala, F., Chetiy, V.: Improving the quality of pre-master training of foreign students in the field of environment. IOP Conf. Ser. Earth Environ. Sci. **390**, 012017 (2019). https://doi.org/10.1088/1755-1315/390/1/012017

38. Samorodova, E.A., Ogorodov, M.K., Belyaeva, I.G., Savelyeva, E.B.: The study of practical legal cases as an effective method of acquiring the discursive communicative skills of international jurists when learning the professional foreign language. XLinguae. **13**, 121–138 (2020). https://doi.org/10.18355/XL.2020.13.01.10

39. Shipunova, O., Pozdeeva, E., Evseeva, L., Mureyko, L.V.: Young students' attitude toward expert knowledge. In: Bylieva, D., Nordmann, A., Shipunova, O., Volkova, V. (eds.) Knowledge in the Information Society. LNNS, vol. 184, pp. 391–400. Springer, Cham (2021). https://doi.org/10.1007/978-3-030-65857-1_33

40. Lösch, A., et al.: Technology assessment of socio-technical futures—a discussion paper. In: Lösch, A., Grunwald, A., Meister, M., Schulz-Schaeffer, I. (eds.) Socio-Technical Futures Shaping the Present. TWGFTSS, pp. 285–308. Springer, Wiesbaden (2019). https://doi.org/10.1007/978-3-658-27155-8_13

41. Ammon, S.: Epilogue: the rise of imagery in the age of modeling. In: Ammon, S., Capdevila-Werning, R. (eds.) The Active Image. PET, vol. 28, pp. 287–312. Springer, Cham (2017). https://doi.org/10.1007/978-3-319-56466-1_12

42. Jakubik, M.: Educating for the future - cultivating practical wisdom in education. J. Syst. Cybern. Inform. **18**, 50–54 (2020)

43. Kamp, A.: Engineering education in the rapidly changing world: rethinking the vision for higher engineering education. TU Delft, Delft (2016). https://repository.tudelft.nl/islandora/object/uuid:ae3b30e3-5380-4a07-afb5-dafd30b7b433?collection=research. Accessed 9 Apr 2021

44. Boden, M.A.: The Creative Mind. Routledge, London (2003)

45. Sautoy, M.: The Creativity Code: Art and Innovation in the Age of AI. Harvard University Press, Cambridge (2019)

Green Universities in an Orange Economy: New Campus Policy

Natalia V. Goncharova⬚, Irina S. Pelymskaya(✉)⬚, Ekaterina V. Zaitseva⬚,
and Pavel V. Mezentsev⬚

Ural Federal University, 19, Mira Street, 620002 Ekaterinburg, Russia
`i.s.pelymskaya@urfu.ru`

Abstract. Changes in modern economic spaces of different countries of the world, global challenges stimulate interest in the goals in the field of sustainable development of territories. Elements of the theory of innovative industries are used to solve problems in classical segments of the economy associated with ensuring their growth and development. This theory is built on concept of orange economy where creative and intellectual potential are the main resources. The purpose of the article is to analyze the experience of UrFU in solving problems in the field of sustainable development and developing the concept of a "green university" and "green campus", which imply an improvement in the quality of the environment. Methods of content analysis, qualitative and quantitative research methods were used in the work such as methods of systematization, analysis of documents and statistical information, foresight session. We have identified the main goals of the university campus policy; identified and substantiated the main categories that meet the challenges of the future orange economy. The authors analyzed the results of a foresight session with experts in the field of sustainable development and creative economics, which made it possible to develop forecasts for the development of the university with a planning horizon till 2030 for the implementation of a new campus policy aimed at creating a green university. The article discusses a predictive model for creating a university campus on the example of the Ural Federal University with possible development as part of the orange economy concept.

Keywords: Green university · Campus · Orange economy · Creative activity · Sustainable development

1 Introduction

The UN resolution declared 2021 the International Year of the Creative Economy for Sustainable Development. In 2015, heads of state and public figures at the 70th session of the UN General Assembly approved the 2030 Agenda for Sustainable Development and new goals aimed at protecting and increasing the well-being of the planet. Responding to global challenges, UNESCO calls for adherence to uniform rules that improve human life. Sustainable development is defined as development that meets the needs of the present generation without compromising the ability of future generations to meet their own needs [1, 2]. For the transition to sustainable development and the provision of opportunities for the implementation of new technologies, it is necessary to transform traditional knowledge. Large universities with highly intellectual staff with innovative potential are able to solve various problems in the field of sustainable development and improve the quality of the environment not only of the university space, but also of the urban and regional territory [3–5]. Universities should become a leading platform for creative ideas, turning into scientific, innovative centers of intellectual life [6–8]. Universities can not only generate comprehensive solutions for the transition to sustainable development based on innovative ideas, expert opinions and modern technological developments, but also implement them in territories of different sizes.

2 Methods

In the modern socio-economic space, along with the term "creative economy", the term "orange economy" has become entrenched, which arose in Latin America [9]. The orange economy is built on economic activities based on knowledge and the relationship of human creativity and ideas, as well as on cultural values or artistic creation, cultural heritage and other forms of individual or collective creative expression [10]. That's why, we should not just consider the orange economy narrowly, as an economy aimed at exclusively artistic creativity, because creativity is diverse, especially the forms of its implementation. It is the interconnection of various types of activities that have transformed into products, goods or services that will determine their value through the content of intellectual property [11].

The main element of the orange economy is the development strategy, which is aimed at creating spaces that allow individuals to reach the maximum point of their development in any form of implementation. The orange economy presents limitless opportunities for the realization of a creative individual or community. These are not only talents in the direct sense, but also cultural heritage, intellectual property and other social constructs [12] Today, despite the public discourse of discussing this issue, the scientific language of this topic has not been practically developed from the point of view of the social paradigm [13]. A. Vega-Muñoz et al. Presents the orange economy as a kind of space in which creative and cultural industries are used through transformations within digital entrepreneurship using various means and technologies, whereby new business models arise and this mechanism is ubiquitous [14].

The United Nations Conference on Trade and Development defines the concept of the creative (orange) economy as an economy in its development stage, which is based on creative assets [15]. Our study proposes to consider the orange economy as a synergistic effect of various activities, as a result of which innovative ideas are transformed into cultural and creative goods or services, the value of which is protected by intellectual property rights, through which the national cultural heritage of the society is transmitted.

Universities are one of the generators of innovative ideas. Universities, possessing creative human potential, are able to form new growth points of the orange economy and solve various problems in the field of sustainable development. Universities around the world are actively supporting the idea of environmental protection and ensuring global sustainable development through the creation of a "green university". Tan H. et al. write that "meanwhile, as the base of knowledge transformation, sci-tech talents cultivation and technical innovation, it is very significant to promote campus ecological civilization and to endow the universities a great duty of leading the sustainable development of society" [16].

Green University is a university that, within the framework of the concept of sustainable development, creates space and conducts activities aimed not only at developing innovations, but also protecting the environment, creating ecological infrastructure, various products and services, educational innovations, etc. [17]. Recently, the concept of a "green campus" has become popular, which assumes that the created space will make it possible to use energy efficiency technologies, lead to the use of digital and preserve natural resources, and improve the quality of the environment. All this will not only improve the quality of life of people living in a green campus, but also ensure the sustainable development of the urban and regional space [18]. Many researchers study the Russian and world experience of the transition of higher educational institutions to the principles of sustainable development and a green university [19–21], assess the impact of man-made pressure on the campus [22], suggest various ways to modernize university campuses [23], analyze the role of higher educational institutions in the implementation of sustainable development goals [24, 25]. Educational and research activities are processes that should not be interrupted outside the classroom, department or laboratory. On the contrary, there they should be saturated with new meanings, goals and ideas: on the way "home", during spare time or playing sports. This requires an accessible and safe environment, a comfortable household life, filled with modern digital technologies. It is necessary to implement a transfer from the archaic model of "living in a dormitory" to a promising model of "living in an accessible, comfortable, safe, digital and international ecosystem and a creative economy." The orange economy requires a transition from convergent management of organizations (characterized by one-man management, strict hierarchy, stability) to divergent management (characterized by collective decisions, freedom of choice, improvisation, adaptability).

3   Results and Discussions

3.1   Problem Statement

Ural Federal University is one of the largest universities in the Russian Federation, which has trained more than 350 thousands of graduates over 100 years of educational activity. A pronounced difference between UrFU among Russian universities is a high volume of applied developments and the implementation of projects for the production of high-tech products and services for the industrial sector.

The material and technical base of the university includes 15 educational buildings, 16 dormitories, communal infrastructure, sports facilities and administrative buildings. The University is making efforts to create the most favorable and comfortable conditions for students and teachers.

Today it is impossible to talk about sustainable development and competitive advantages of the territory without the use of digital/information technologies in the orange economy. Since 2010, an innovative infrastructure has been created at UrFU, which includes innovation and implementation centers, small innovative enterprises and subdivisions of the support system for the development of knowledge-intensive entrepreneurship (from the generation of innovative ideas to the production of innovative products), a system for the formation of technological entrepreneurs "from the school bench to their own start-up" (UrFU Talent School, Innovative Diving, UrFU Accelerator).

The innovative infrastructure of UrFU (in the context of the theory "green universities in an orange economy") unites divisions that provide a full cycle of creation and support of competitive production of innovative products based on the results of research and development in the field of high technologies. Having our own innovative infrastructure allows us to create and implement competitive innovative technologies and implement world-class projects.

To develop a new campus policy based on the green university and the orange economy of UrFU for the period up to 2030, a foresight session was held in September 2020.

While organizing the foresight session, it was taken into account that UrFU, as the leading university in the region, organizes its activities in accordance with the principles identified by the UN for sustainable development. For the transition to sustainable development, complex solutions are needed based on innovative ideas, expert opinions and modern technological developments, implemented in parallel in territories of different sizes. The most important of the existing sustainable development goals for our research are: quality education, affordable and clean energy, decent work and economic growth, industry, innovation, and infrastructure, sustainable cities and communities, partnerships, partnerships. For UrFU, the most optimal way to transition to a model of sustainable development and functioning is the introduction of methods and practices of the environmental aspect of sustainable development, which are the basis of the UI Green Metric World University Rankings.

3.2 Development Directions of the Green University

As part of the implementation of the campus policy, we have proposed the following directions for the development of a green university for the period up to 2030:

1. Development of the concept of long-term spatial development of the Ural Federal University campus, taking into account the construction of a sports village and the active development of the urban environment, in which it is necessary to take into account the expected increase in the number of students and scientists. The new space must meet the technological demands of the future, and the infrastructure must support innovative teaching, learning and research methods.
2. Digitalization and automation of campus management and dispatching systems based on a single information complex. Formation of an interactive environment.
3. Implementation of advanced security technologies throughout the campus.
4. Achievement of the highest levels of comfort and accessibility of the environment both inside and outside the campus, meeting the highest international standards. Renovation of buildings with an emphasis on a modern digital library and housing.
5. Creation of new and development of existing leisure and relaxation spaces, green areas, as places of attraction and communications.

6. Equipment of campus premises with modern ergonomic furniture.
7. Creation of interactive coworking spaces (spaces for independent group work and self-study of students).
8. Creation of an adapted space for people with disabilities.

Let's move on to the main categories of a green university that meet the challenges of the future orange economy, which determine the competitiveness of territories. They are formed by us on the basis of the main indicators of the Green Metric World University Rankings and are calculated based on the results of the foresight session: infrastructure, energy efficiency, waste, water, transport, education in the planning horizon of 10 years. During the foresight session, from four to eight indicators were identified in each category and their indicators were constructed, presented by us in tabular forms.

Infrastructure. The structure and infrastructure of the campus is expressed in the university's attitude towards a green environment. Indicators determine the compliance of the university with the status of a "green campus". The goal is to develop space for landscaping and environmental protection, as well as for the development of sustainable energy (Table 1).

Table 1. Forecasting campus development by infrastructure category

Indicator	2022	2024	2026	2028	2030
The ratio of the area of open space to the total area	<1%	1–80%	>80–90%	>90–95%	>95%
The total area of the campus covered with forest	≤2%	2–9%	>9–22%	>22–35%	>35%
The total area of the campus covered with green plants	<10%	10–20%	> 20–30%	>30–40%	>40%
Total area of campus absorbing water, excluding forest and green plants	<2%	2–10%	>10–20%	>20–30%	>30%
The ratio of total open space to campus population	<10 $м^2$	10–20 $м^2$	>20–40 $м^2$	>40–70 $м^2$	>70 $м^2$
Percentage of the university budget dedicated to sustainable development	> 1%	> 1–5%	> 5–10%	>10–15%	>15%

Energy Efficiency. The structure and infrastructure of the campus are expressed in the attitude of the university to the issues of energy use and climate preservation. Indicators determine the university's compliance with the requirements of energy conservation, reduction of greenhouse gas emissions and reduction of the total carbon footprint. Goal: to intensify the university's efforts to improve the energy efficiency of campus buildings, strengthen care for nature and conserve energy resources (Table 2).

Table 2. Forecasting campus development by energy efficiency category

Indicator	2022	2024	2026	2028	2030
Use of energy efficient tools	<1%	1–25%	>25–50%	>50–75%	>75%
Implementation of "smart knowledge"	<1%	1–25%	>25–50%	>50–75%	>75%
Number of renewable energy sources on campus	No	1	2	3	>3
Total electricity consumption, to the total number of residents on the campus, kWh	≥2424	<2424–1535	<1535 ÷ 633	<633 ÷ 279	<279
Ratio of renewable energy production to total energy consumption per year	≤0,5%	>0,5–1%	>1–2%	>2–25%	>25%
Application of green building elements reflected in the construction and renovation policy	No	1	2	3	>3
Greenhouse Gas Emissions Reduction Program	No	Program in development	Cut from 1–3 sources	Cut from 2–3 sources	Cut from 3 sources
Ratio of total carbon footprint to campus population (metric tons / person)	≥2,05	<2,05–1,11	<1,11–0,42	<0,42–0,10	<0,10

Waste. The structure and infrastructure of the campus are expressed in the university's attitude to waste treatment and recycling. Objective: to enhance the university's efforts to create a sustainable environment. As a result of the activities of the university staff and students, a large amount of waste is generated on the campus. The university should be interested in developing the following waste management and recycling programs: toxic waste recycling, sewage sludge disposal, organic waste treatment, inorganic waste treatment, policies to reduce the use of paper and plastics on campus (Table 3).

Table 3. Forecasting campus development by waste category

Indicator	2022	2024	2026	2028	2030
Recycling program for all types of waste	No	1–25% of waste	>25–50% of waste	>50–75% of waste	>75% of waste
Reduction Program for paper and plastic in campus	No	1 program	2 programs	3 programs	>3 programs
Organic waste management, % undergoing processing	Open placement	1–25%	>25–50%	>50–75%	>75%
Inorganic waste handling, % undergoing processing	Burned	1–25%	>25–50%	>50–75%	>75%
Toxic waste management, % undergoing treatment	No	1–25%	>25–50%	>50–75%	>75%

Water. Water use on campus is another of the most important metrics in the green rankings. The main task is to reduce the use of groundwater, stimulate the program of conservation and protection of the habitat. The priority areas should be programs for water utilization, the use of water-saving technologies and devices, purified water (Table 4).

Transport. Transport systems have a significant impact on university emissions. Limiting the number of vehicles on campus, prioritizing the use of buses and personal mobility (bicycles, electric vehicles, etc.) will help maintain a healthier environment. Walking policies should encourage students and staff to walk on campus and minimize the use of private vehicles. Objective: To strengthen the university's efforts to reduce carbon dioxide emissions around the campus through the use of environmentally friendly public transport (Table 5).

Table 4. Forecasting campus development by water category

Indicator	2022	2024	2026	2028	2030
Water conservation program and its implementation, %	No	Program in development	1–25%	>25–50%	>50%
Implementation of the recycling water supply program, %	No	Program in development	1–25%	>25–50%	>50%
Use of water-saving appliances, %	No	Program in development	1–25%	>25–50%	>50%
Purified water consumption, %	No	1–25%	>25–50%	>50–75%	>75%

Table 5. Forecasting the development of the campus by category of transport

Indicator	2022	2024	2026	2028	2030
The ratio of the total number of vehicles (cars and motorcycles) to the number of residents on campus	≥1	<1 ÷ 0,5	<0,5 ÷ 0,125	<0,125 ÷ 0,045	<0,045
Shuttle bus services	Possible, but not provided by the university	Provided (by the university or other party), they are regular but paid	Provided (by the university or other party), they are regular, the university covers some of the costs	Provided by the university regularly and free of charge	Provided by the university on a regular basis and with zero emissions
Availability of vehicles with zero emissions on campus	Zero emission engines not available	It is almost impossible to use zero emission engines	Zero emission engines are available but not provided by the university	Zero emission engines available, provided by the university on a paid basis	Zero emission engines are available, provided by the university for free

(*continued*)

Table 5. (*continued*)

Indicator	2022	2024	2026	2028	2030
Ratio of zero emission vehicles to total campus emissions	≤0,002	>0,002 to ≤0,004	>0,004 to ≤0,008	>0,008 to ≤0,02	>0,02
The ratio of parking area to total campus area	>11%	<11–7%	<7–4%	<4–1%	<1%
Program to limit and reduce the area of parking for the last 3 years	No	Program in development	The program reduced parking by 10%	The program reduced parking by 10–30%	The program has reduced parking 30% or parking is prohibited
Number of transportation initiatives to reduce private vehicles on campus	No	1	2	3	>3
Pedestrian accessibility level on campus	No	Walking paths available	Walkways are accessible and equipped from a safety point of view	Walkways are accessible and equipped in terms of safety and convenience	Walkways are accessible and equipped in terms of safety and convenience, and partly for the convenience of persons with disabilities

Education. Education for sustainable development is designed to provide effective solutions to future challenges and enhance the resilience of the university. The main directions of campus development in this category are quantitative indicators for activities, student organizations and financial funding for sustainable development researches (Table 6).

Table 6. Forecasting campus development by education category

Indicator	2022	2024	2026	2028	2030
Ratio of sustainability courses to total number of courses	<1%	1–5%	>5–10%	>10–20%	>20%
Ratio of sustainable development research funding to total research funding	<1%	1–8%	>8–20%	>20–40%	>40%
Number of publications on sustainable develoment	0	1–20	21–83	84–300	>300
Number of activities related to sustainable development	0	1–4	5–17	18–47	>47
Number of student organizations in the field of sustainable development	0	1–2	3–4	5–10	>10
University-supported sustainable development website	Not available	Under development	Available and can be accessed	Available and can be consulted, updated periodically	Available and referenced, updated annually
Formation of a sustainable development report	Not available	Under development	Available but not publicly available	Available and published from time to time	Available and published annually

3.3 Planned Results

First, the strategy of focusing on the implementation of the orange economy is most often applied to small and island states. However, it can be applied to university campuses, since in the new paradigm of university development they are isolated green social and educational spaces outside the urban environment. Secondly, the university community is a highly resourceful creative team with intellectual property, capable of imparting economic value to various creative ideas. It is the universities with their human potential that are able to generate and transfer new knowledge, meet the challenges of the 21st century, solve problems in the field of sustainable development, the main of which are: environmental, social, economic, and applied.

Recently, the concepts of "green university" and "green campus" have become popular, which imply improving the quality of the environment not only around the university, but also preserving the natural resources of the region in which it is located.

In his speech, the Governor of the Sverdlovsk Region, Evgeny Kuyvashev, on June 1, 2021, at the federal strategic session of the heads of Russian regions, noted that "the construction of a campus in the Sverdlovsk Region will create a modern educational environment where the principles of project and interdisciplinary learning with the formation of multifunctional teams will be introduced. Ultimately, this will significantly increase the number of scientific developments for industrial purposes."

Today the UrFU campus has a distributed type of accommodation. Its buildings are located in different parts of the city and do not represent a closed system [26].

By 2023, the university is preparing to host the XXXII World Summer Student Games (Universiade-2023). At the site of UrFU in the Vtuzgorodok area, in preparation for the Universiade, it is planned to overhaul and reconstruct sports facilities. The Universiade Village will be located outside the city, in the green zone. The concept of the Universiade Village envisages the construction of 5 residential buildings for athletes and 3 multifunctional centers, which after the Universiade will be transferred to the university and become a new university campus (Fig. 1). Residential buildings for athletes will become hostels for students, and the multifunctional centers will house a Specialized Educational and Scientific Center and 2 educational buildings of institutes: the Institute of Economics and Management and the Institute of Information Technologies.

Fig. 1. General view of the new campus of UrFU

The main distinguishing features of the University campus will be:

- localized in one place green campus with modern infrastructure equipment for each cluster;
- a modernized management system based on the digital university model, allowing development in the context of the orange economy;
- developed international environment by attracting talented youth;
- developed information infrastructure and digital services which allow to commercialize innovative technologies of UrFU;
- competitive social infrastructure.

Table 7 presents the values of the main indicators of the new campus of the university after the implementation of the Green Campus project.

Table 7. Project values UrFU green campus

Indicator	2020	2030	Variation	
			Absolute	Relative, %
Total number of students	34900	39550	+4650	+13,3
Total area of buildings, m^2	436044	585444	+149400	+34,3
Total area of dormitories, m^2	149147	263290	+114143	+76,5
Total area of coworking spaces, m^2	22370	41730	+19360	+86,5
Total area of campuses covered with forest, m^2	0	152600	+152600	–
Total area of smart buildings, m^2	4095	96300	+100395	in 24,5 times
Total area of covered sport facilities m^2	19 130	78830	+60700	in 3,17 times
Total number of science parks	0	2	+2	–
Total number of small enterprises	84	195	+111	+132,1

As a result of changes in infrastructure, the total area of buildings will increase by more than 34%, dormitories - by almost 77%. The area of coworking spaces will increase by 86.5%, the number of smart buildings will grow 24.5 times. The total area of covered sport facilities will increase in more than 3 times by building new House of winter kinds of sports.

Currently there are almost 20,000 nonresident and foreign students and postgraduates at UrFU. A new modern campus located outside the city in an ecological area will allow students to efficiently engage in educational, scientific and innovative activities, to have an active and healthy lifestyle, to play sports and relax in a comfortable and safe environment. The construction of the campus will create a modern educational environment where the principles of project and interdisciplinary learning will be introduced with the formation of multifunctional teams. Ultimately, this will significantly increase the number of industrial scientific projects. However, the construction of a new, even the most modern campus is not enough for transformation students' attitudes towards improving the quality of the environment and problems in the field of sustainable development. It is necessary to update the higher educational programs in accordance with the requirements of the creative economy in order to train specialists from different subject areas with competencies in the field of sustainable development.

Educational programs in all areas should include courses on sustainable consumption, cyclical economics, energy conservation, industrial ecology, cleaner production, and waste management. By 2030, the ratio of sustainable development courses to total courses should exceed 20%. This will change the image of a future student as an individual consumer of goods and services made with the use of creative potential.

4   Conclusions

Universities can actively participate in the creation of creative spaces, since it is they who are assigned the task of forming the creative and innovative potential of the country and regions. In the regional context, federal universities play an important role as pivotal universities. It is they who are engaged in the production of knowledge, form new competencies of graduates that are in demand on the labor market in the orange economy. University teams are the object and subject of the formation of creative competencies, to which we include research work. It is the representatives of the student and teaching community who develop and implement new ideas that contribute to innovation in all segments of the orange economy market.

The current model of maintaining university dormitories in Russia does not allow creating conditions in them comparable to those offered by other student mobility centers in the world. Currently, at UrFU, only 50% of nonresidents are provided with places in student buildings, the introduction of a green campus policy will allow at least 90% of students to be accommodated by 2030. On the basis of the information complex of the green campus, services will be organized to assist in placement, solving migration, adaptation and legal issues, developing strategies and programs for innovative development, including within the framework of the "orange economy" concept.

Currently, only 15% of UrFU buildings are provided with a set of spaces for the realization of the intellectual and creative potential of students and staff, as well as for sports and recreation (coworking spaces for students and staff, gyms and other gyms, professorial and student clubs, etc.), an increase in the number of these spaces is extremely necessary for the development of the university in the trends of the orange economy. Our forecast showed that with the introduction of green university ideas, their share will increase to 85% in 2030.

Accordingly, the Creation of the "Green Campus of UrFU" will allow the formation of new points of growth for the orange economy: strengthening of entrepreneurial potential; development of the innovative component of universities; expanding economic spaces for the dissemination of the culture of the orange economy; the mission of the university will be filled with new content, it will make the transition from the traditional to the mission of realizing the creative and cultural industries through transformations within the framework of digital entrepreneurship using various means and technologies. After all, the digital revolution is a key trend affecting the future of the orange economy and opens up great prospects for the sustainable development [27–30].

References

1. George, H.: Creative cities and sustainable development: a framework. In: Gu, X., Lim, M.K., O'Connor, J. (eds.) Re-Imagining Creative Cities in Twenty-First Century Asia, pp. 265–275. Springer, Cham (2020). https://doi.org/10.1007/978-3-030-46291-8_18
2. Kazaryan, R.: The concept of anthropotechnical safety of functioning and quality of life. In: Murgul, V., Pukhkal, V. (eds.) Advances in Intelligent Systems and Computing, vol. 1258, pp. 759–767. Springer, Cham (2021). https://doi.org/10.1007/978-3-030-57450-5_64
3. Leal Filho, W., et al.: Identifying and overcoming obstacles to the implementation of sustainable development at universities. J. Integr. Environ. Sci. **14**, 93–108 (2017). https://doi.org/10.1080/1943815X.2017.1362007
4. Lăzăroiu, G.: Is there an absence of capability in sustainable development in universities? Educ. Philos. Theor. **49**, 1305–1308 (2017). https://doi.org/10.1080/00131857.2017.1300023
5. Krasnoshchekov, V., Arseniev, D., Rud' V., Switala, F, Chetiy, V.: Improving the quality of pre-master training of foreign students in the field of environment. IOP Conf. Ser. Earth Environ. Sci. **39**, 012017 (2019). https://doi.org/10.1088/1755-1315/390/1/012017
6. Blasco, N., Brusca, I., Labrador, M.: Drivers for universities' contribution to the sustainable development goals: an analysis of Spanish public universities. Sustainability **13**, 89 (2020). https://doi.org/10.3390/su13010089
7. Demidov, V., Mokhorov, D., Mokhorova, A., Semenova, K.: Professional public accreditation of educational programs in the education quality assessment system. E3S Web Conf. **244**, 11042 (2021). https://doi.org/10.1051/e3sconf/202124411042
8. Ragazzi, M., Ghidini, F.: Environmental sustainability of universities: critical analysis of a green ranking. Energy Procedia **119**, 111–120 (2017). https://doi.org/10.1016/j.egypro.2017.07.054
9. Benavente, J.M., Grazzi, M.: Public Policies for Creativity and Innovation: Promoting the Orange Economy in Latin America and the Caribbean . Inter-American Development Bank (2017). https://doi.org/10.18235/0000841
10. United nations: Macroeconomic Policy Issues. https://undocs.org/pdf?symbol=ru/A/C.2/74/L.16. Accessed 11 May 2021
11. Gaviria Roa, L.A., Castillo, H.G., Montiel, A.H.: Orange economy: study on the behavior of cultural and creative industries in Colombia. Int. J. Mech. Eng. Technol. **10**(12), 160–173 (2019)
12. Buitrago Restrepo, F., Duque Márquez, I.: The orange economy: an infinite opportunity. Inter-American Development Bank, Washington D.C. (2013)
13. López Cortés, O.A., Moneada Prieto, V.M.: The orange entrepreneurship policy in Colombia, new ways of population control. Rev. Repub. **29**(1), 107–128 (2020). https://doi.org/10.21017/rev.repub.2020.v29.a89
14. Vega-Muñoz, A., Bustamante-Pavez, G., Salazar-Sepúlveda, G.: Orange economy and digital entrepreneurship in Latin America: creative sparkles among raw materials. In: Handbook of Research on Digital Marketing Innovations in Social Entrepreneurship and Solidarity Economics. IGI Global, Hershey (2019)
15. Benavente, J.M., Grazzi, M.: Public Policies for Creativity and Innovation: Promoting the Orange Economy in Latin America and the Caribbean. Inter-American Development Bank (2017). https://publications.iadb.org/publications/english/document/Public-Policies-for-Creativity-and-Innovation-Promoting-the-Orange-Economy-in-Latin-America-and-the-Caribbean.pdf. Accessed 15 Apr 2021
16. Tan, H., Chen, S., Shi, Q., Wang, L.: Development of green campus in China. J. Clean. Prod. **64**, 646–653 (2014). https://doi.org/10.1016/j.jclepro.2013.10.019

17. Perevozchikov, K.I., Khmelkova, N.V.: "Green" campus is a new model of a university for a "green" economy. In: Zaks, L.A., Myasnikova, L.A. (eds.) Conference 2015 University of the XXI Century: Old Paradigms and Modern Challenges, vol. 1, pp. 424–427. Non-state Educational Institution of Higher Professional Education Humanitarian University, Yekaterinburg (2015). (in Russian)
18. Khasanov, V.R.: Green universities in the context of life safety. Culture and ecology are the foundations of sustainable development in Russia. In: 2019 Green Bridge Across Generations, vol. 1, pp. 323–328. UrFU, Yekaterinburg (2019). (in Russian)
19. Kuznetsov, V.V., Lukina, A.V., Malova, D.V.: Principles and mechanisms of the strategy of sustainable development of the university. Bull. Russ. Univ. Econ. G.V. Plekhanov 1(91), 56–64 (2017)
20. Yakimovich, D. N.: Development of associations of green universities in Russia and China. Participation in international associations. In: Nikonorova, I.V., Ilyin, V.N. (eds.) Earth Sciences: From Theory to Practice (Archikov Readings - 2020), vol. 1, pp. 36–42. ID Sreda, Cheboksary (2020). (in Russian)
21. Kruglova, L.E., Redina, M.M., Khaustov, A.P.: Justification of environmental policy at the university level. Bull. Peoples' Friendsh. Univ. Russ. Ser.: Ecol. Life Saf. 26(2), 251–260 (2018). (in Russian)
22. Boeva, D.V., Khaustov, A.P.: Assessment of the impact of vehicles on the campus of Peoples' Friendship University of Russia. Bull. Peoples' Friendsh. Univ. Russ. Ser.: Ecol. Life Saf. 26(4), 419–430 (2018). (in Russian)
23. Ovchinnikova, N.A., Krygina, A.M.: Ecological reconstruction of university campuses. Invest. Constr. Real Estate 2(1), 275–276 (2017)
24. Panasenkova, E.: Influence of sustainable development factors on the strategic policy of universities. XXI Century. Technosphere Saf. 2(18), 146–156 (2020). (in Russian)
25. Dlimbetova, G.K., Moldabekova, S.K., Abenova, S.U.: Greening education in the interests of sustainable development of the Republic of Kazakhstan. In: Batyan, A.B., Maskevithch, S.A. (eds.) Sakharov Readings 2019: Environmental Issues of the 21st Century, vol. 1, pp. 79–82. IVC Minfina, Minsk (2019). (in Russian)
26. Berestova A.V., Larionova V.A.: Choosing the spatial organisation of a modern campus. Part 1. An analysis of global university campus space. Acad. Bullet. UralNIIproject RAASN 3(34), 66–70 (2017). (in Russian)
27. Korres, G.M., Kourliouros, E., Michailidis, M.P.: Handbook of Research on Policies and Practices for Sustainable Economic Growth and Regional Development. IGI Global, Hershey (2017)
28. Del Río Castro, G., González Fernández, M.C., Uruburu Colsa, Á.: Unleashing the convergence amid digitalization and sustainability towards pursuing the Sustainable Development Goals (SDGs): a holistic review. J. Clean. Prod. 280, 122204 (2021). https://doi.org/10.1016/j.jclepro.2020.122204
29. Almazova, N., Bylieva, D., Lobatyuk, V., Rubtsova, A.: Human behavior as a source of data in the context of education system. In: SPBPU IDE 2019: Proceedings of Peter the Great St. Petersburg Polytechnic University International Scientific Conference on Innovations in Digital Economy, 37. ACM, Saint – Petersburg (2019). https://doi.org/10.1145/3372177.3373340
30. Bobylev, S.N., et al.: Indicators for digitalization of sustainable development goals in peex program. Geogr. Environ. Sustain. 11, 145–156 (2018). https://doi.org/10.24057/2071-9388-2018-11-1-145-156

A Creative Approach to Creating a Livable Urban Environment

Daria Shalina$^{(\boxtimes)}$ (iD), Natalia Stepanova (iD), and Viola Larionova (iD)

Ural Federal University (UrFU), 19 Mira Street, Ekaterinburg 620002, Russia
{n.r.stepanova,v.a.larionov}@urfu.ru

Abstract. Every year, the participation of citizens in the creation of urban projects is becoming more and more. It is especially popular to involve citizens in creative projects to create popular public spaces on the site of industrial zones, abandoned buildings, and cultural heritage sites. The relevance of the study is due to the popularity of the creation and renovation of creative spaces in urban practices that identify the needs of the population. The aim of the work is to analyze the development of creative industries in the Sverdlovsk region (Russia). The use of modern tools and algorithms that consider what the city wants in socially significant projects for all segments of the population is possible today. An example of citizens' participation in the formation of a livable urban environment is considered in detail by the example of adapting a cultural heritage object for modern use. The results of the study show that the joint work of citizens, business and administration allows you to generate ideas and implement social creative projects.

Keywords: City · Creative industries · Public participation · Engagement

1 Introduction

The coronavirus pandemic has shown the world that using digitalization tools is efficient, fast, and convenient [1–5]. The massive shift to remote work has proven this. A less obvious trend caused by the pandemic has been the need for creativity [6, 7]. Creative industries have already been developing for about 10 years. Unfortunately, during the pandemic in 2020, it was not possible to hold creative events. However, the United Nations has declared 2021 the International Year of the Creative Economy, to support creative people and as a new way to sustainably develop territories [8].

Against the background of the global development of creative industries, Russia has not yet mastered this trend enough. The volume of the creative economy in the world has doubled over the past 10 years (for example, in the UK it is 10%, and in Russia this volume was less than 3%) [9].

1.1 Relevance and Practical Significance

The investment of cities in the formation of a livable urban environment, including in the field of creative industries, is characterized by a significant payback. A positive example is the city of Moscow, where investments in the development of the urban environment have a payback rate 3–5 times higher than the cost [9].

Creative industries in urban planning involve the introduction of modern trends in architecture and art [10, 11]. Nevertheless, the popularity of modern urban spaces is accompanied by high costs for the construction of such facilities. Therefore, when developing creative industries, they use cooperation and joint work of all stakeholders [12]. This is in addition to the participants of the construction process (development and information companies and corporations, manufacturers of building materials and structures, architects, designers, urbanists, sociologists, analysts, realtors, banks), the local administration and the public of cities. One of the most significant parties, of course, is the population of cities and localities, actively expressing their opinions [13].

Our research provides an example of cooperation for progressive development in a changing economy and global trends. Namely, the creation of creative spaces in the Sverdlovsk region and the capital of the Urals, Ekaterinburg, in a new way.

One of the most relevant areas of such cooperation is the involvement of young people in the development of public spaces (as it is organized in the city of Ulyanovsk) and the processes of preserving historical and cultural heritage (as it is done in the cities of Irkutsk and Izhevsk) [14].

1.2 Purpose and Objectives of the Study

The purpose of this study is to analyze the development of creative industries in the Sverdlovsk region and Ekaterinburg based on statistics and real practical experience of involving citizens in the formation of a livable urban environment in urbanized territories.

To achieve this goal, it is necessary to solve the following tasks:

– To study the effect of creating creative industries;
– To analyze the development of creative industries in the Sverdlovsk region;
– To describe the possibilities of implementing "creative projects»;
– To analyze the joint work of stakeholders (leaders of Ural initiatives, interested citizens and partners) in creating new creative spaces;
– To justify the applied research results.

2   Theoretical Foundations

For a high level of quality of life of the population, a livable urban environment is necessary, which is the infrastructure of the city, providing the population with the necessary amenities, both domestic and aesthetic [15–18]. The main criteria for evaluating a livable urban environment are the following indicators [19]:

– Comfort;
– Security;
– Management efficiency;
– Eco-friendly;

– Identity and diversity;
– Modernity and relevance.

If you follow these criteria, you can form images of the future of urbanized territories. Images of the future in the field of creating a livable urban environment are the improvement of urban space in the way that citizens would like it to be. The creation of such images is increasingly aimed at the development of creative industries.

Creative industries represent new sectors of the economy represented by creative activities. The main effect of the development of creative industries is felt by the residents of the city themselves. As new "creative" jobs are created [20, 21], creative competencies are developed [22, 23], access to modern culture is formed, and the city is transformed for the better. For the region, the effect of creating creative spaces is even greater. Namely, there is an increase in investment in the creative sector of the economy, new business models are being implemented, small and medium-sized businesses in the creative sphere are developing, the flow of tourists to the region and the city is increasing, and the quality of human capital is improving. At the country level, developed creative industries ensure its creative image, strengthen its position in the global market and modernize industries.

Creative industries are divided into different groups. However, they generally include the art industry, performing arts, music, film, photography, publishing, television and radio broadcasting, information technology and video games, advertising, architecture, design, fashion, jewelry, museums, cultural heritage sites, and education in the creative industries.

Three main approaches are used to assess the development of the creative sector in the region [24]. The first of them is the industry one. This approach involves the analysis of economic activities, i.e. creative industries. The second approach is more detailed-for professions that are creative. The third approach involves the analysis of foreign trade in creative goods and services.

One of the implementation tools can be a new format of programs to support urban initiatives, which increases the sustainability and effectiveness of initiative projects that benefit people and develop cities. In our example, this is a hackathon. It is this option of participation of active interested citizens that we will consider later.

3   Research Methods

This study is based on theoretical research methods with empirical elements. Among the theoretical ones, analysis, classification and description were used, while among the empirical ones, observation and experiment were used. To assess the development of creative industries in the Sverdlovsk region, statistics are presented with a detailed analysis, and then a synthesis - to represent the overall situation. The creative spaces of Ekaterinburg were classified of creative industries, which allowed us to evaluate the well-developed and poorly developed creative industries.

Empirical methods involve practical research with real results. The study presents the practical experience of creating urban projects by the residents themselves in the form of practical cases. Citizens formed images of the future for Ekaterinburg, based on the criteria of a livable urban environment. Observation and experiment became the main basis for collecting the results of the activities carried out. The main objective of the experimental activity was to identify the potential possibilities of creative approaches in the development of concepts of development projects for the adaptation of cultural heritage projects for modern use. Considering the peculiarities of working with cultural heritage objects, the citizens had to consider various options for the renovation of existing objects to give them new functions. It was necessary to turn problems into opportunities, weaknesses into strengths. The project ideas obtained because of the experiment can be useful for implementation in practice in the urban environment for developers, developers, citizens, and the administration. All the results of the study are described and justified.

4   Conducted Research

4.1   Assessment of the Development of the Creative Economy in the Sverdlovsk Region

The Sverdlovsk region is in the Urals and is famous for the development of industrial sectors, but still has a creative potential. The total contribution to the regional economy of any industry can be determined by the gross regional product (GRP) in general. Based on the published statistics of the GRP of the Sverdlovsk region, two types of activities were identified that are more suitable for creative ones. The first type is information and communication activities. The second – activities in the field of culture, sports, leisure and entertainment (see Fig. 1).

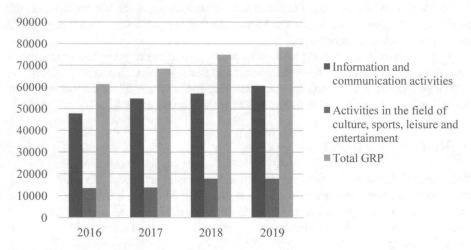

Fig. 1. The gross regional product of the Sverdlovsk region in the creative industries in 2016–2019, million rubles [25].

Figure 1 shows that GRP in the creative industries is gradually growing. The main share in the GRP of the creative industries is made up of activities in the field of information and communication. Although the creative industries occupy a very small part of the total GRP (see Fig. 2).

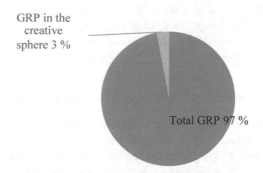

Fig. 2. The structure of GRP in the Sverdlovsk region in 2019, % [25].

Only 3% of the total GRP was occupied by the creative industries in 2019.

As noted above, creative industries attract investment to the region. Figure 3 shows the structure of investments in the main activities of the Sverdlovsk region for 2020.

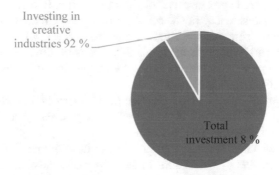

Fig. 3. The volume of investments in fixed assets by type of economic activity in the Sverdlovsk region in 2020, % [25].

Almost one-tenth of the investment in the Sverdlovsk region is occupied by creative industries, that is, for the study region, the development of creative industries is interesting and relevant from an economic and social point of view.

Using the industry-specific approach of assessing creative industries, the demographics of organizations of different types of economic activity in January 2021 were analyzed. So, 14 organizations in the creative sector have already been registered, but at the same time 80 have been liquidated. The current situation may be caused by the presence of anti-coronavirus measures, the lack of vision of the prospects of creative directions, or the lack of a real need among the population.

A more detailed assessment of the creative industries includes an analysis of the number of employees employed in a particular type of activity. Thus, 66 100 people are employed in the creative industries in the Sverdlovsk Region, which is 3.6% of the total number of qualified employees [25].

Analysis of the development of creative industries in the Sverdlovsk region showed that creative industries occupy a small part of the regional economy, although the Sverdlovsk region has a great potential in this area, based on the variety of activities. It is necessary to support and indicate the opportunities for the development of "creative projects".

4.2   Opportunities for the Development of Creative Spaces in Ekaterinburg

In 2021, the Ministry of Construction of Russia approved a Standard for involving Citizens in solving issues of urban environment development [26, 27]. This document will help local authorities to improve the quality of the urban environment, considering the views of residents. The standard was developed jointly with the Agency for Strategic Initiatives (ASI) [28]. From July to December 2020, it was piloted in 20 Russian regions. During this time, the engagement rate of residents increased by about 43%.

Thanks to the all-Russian program for the transformation of the territories of former industrial zones and abandoned buildings "Rurban Creative Lab" and the Agency for Strategic Initiatives in the Sverdlovsk region, it is planned to create several creative spaces on the site of former industrial zones in Ekaterinburg, Nizhny Tagil, Sysert, Chernoistochinsk, Aramil and other territories of the region. Excellent examples of creative spaces are "ArtResidence" in Chernoistochinsk and "Creative Factory" in Sysert [29]. The well-known experience of creating creative clusters will be systematized and a regulatory framework will be formed, which will ensure the fastest and most effective development of creative industries and show real prospects for the development of creative spaces in the Sverdlovsk region.

The participation of citizens helps to create popular spaces. In addition, citizens are more attentive and careful about the objects in which they have invested their time, effort, and sometimes financial resources [26].

There is also an international program for the development of urban initiatives "School of Urban Pioneers", which aims to develop professional skills in urban development and support urban initiatives [30]. All citizens interested in improving the urban environment can participate in it. Thanks to this program, project initiators can gain skills for their implementation, as well as exchange experience with other participants from different countries. The appearance of a program of this kind indicates a serious step forward in the field of urbanism.

In the existing world practice, there are five levels of civil participation (see Table 1).

In our program, we used the third level-complicity.

Table 1. Levels of civic participation by form

Level	Participation form	Description
The first	Informing the public	Providing citizens with information about the planned initiative and key technical and economic indicators of the project
The second	Consulting services	Considering the opinions of residents, their positions and wishes about the project
The third	Collaborative design	Conducting joint development and implementation of the project, which is agreed by all interested parties
The fourth	Cooperation or partnership	Work on the development and implementation of the project, in which citizens are transferred part of the functions of the developer or initiator of the project
The fifth	Endowment of power	Citizens are given the right to make a final decision on several key issues in the development and implementation of the project

5   Applied Aspects of Research

In our work, we will analyze such a format of work with residents as the hacka-thon "Image of the Future" on the development of products for the preservation of historical houses, the rescue of parks and the implementation of other urban projects [31]. This event was held in our hometown of Ekaterinburg in the period from 2 to 11 April 2021 in online and offline formats.

Hackathon in Ekaterinburg is an event that involves the creation of "images of the future" in the city through acquaintance with active citizens, the generation of creative solutions and continuous brainstorming. The experts suggested that the participants develop several significant urban projects step by step [32]. The problems of implementing environmental solutions in business, developing green areas in Ekaterinburg, creating new assignments for old buildings, as well as the renovation of the Nurov estate were solved.

The diversity of projects indicates the real need of the population to participate in urban projects. The project areas included the most relevant topics: ecology, cultural heritage sites, location, urban initiatives. Project authors spread their ideas, attracting more and more people. The more enthusiastic people in the team, the more effective the created project will work.

As an example of the development of urban projects, we considered the renovation of the Nurov estate (Nurov's garden and house), located at 1 Chapaev Street, Ekaterinburg (the building of the house is two-storeyed).

The estate is currently not fully exploited. Since 2019, the "Nurov Garden" has been opened on the territory of the Nurov estate, which has been organized by volunteers since 2017 [33].

As part of the joint work of public activists and owners (the company "Mayak Corporation"), creative ideas were proposed for the restart of this object.

The goal is to develop a concept for the use of the Nurov estate and the surrounding area.

The concept of using the Nurov estate and the surrounding area included:

1. Location analysis (including SWOT analysis)
2. Object analysis (including SWOT analysis)
3. Conclusions about possible use (at least three options). SWOT analysis of possible uses. Choosing one optimal use case.
4. Selected optimal use case:

   – Positioning;
   – Filling in the functions and functional zoning of the object;
   – Selection of possible tenants and / or operators, including rental rates, income.
   – Building a financial model of the project.

5. Conclusions and recommendations

From the many proposed projects, we will show some elements of one of the possible uses. This is the concept of creating a creative space, where on the first floor there will be open areas/coworking spaces for classes in creative areas, and on the second floor – art exhibitions and art shows.

The proposed concept is aimed at the development of creative industries in Eka-terinburg. Its relevance is due to the best experience of world and Russian trends. At the same time, the proposed concept considers the criteria for creating a livable urban environment, which is of great importance for a modern city [34]. These are comfort, safety, management efficiency, identity and diversity, modernity, and relevance, where creativity is a new trend.

The SWOT analysis presented in Table 2 clearly demonstrates the advantages and disadvantages of the object in Ekaterinburg before the renovation.

Threats can be ranked by the significance of the severity of the consequences. So the list of threats from the least important to the most important will look like this:

1. Lack of demand from residents;
2. Long renovation time of the object;
3. Loss of interest in the creative space;
4. Loss of a unified concept of creative space due to the diversity of tenants;
5. The object will remain empty, without a destination;
6. The emergence of competitors with a similar concept.

The SWOT analysis presented in Table 3 clearly demonstrates the advantages and disadvantages of the object in Ekaterinburg after the renovation of the optimal use of the object as a creative art cluster.

Table 2. SWOT analysis of the Nurov estate before renovation

Strengths	Weaknesses
- Location in the city center; - All types of public transport are located nearby; - Developed social infrastructure; - There is a modern building development; - The interest of residents in the renovation of the object	- Deterioration of the engineering infrastructure; - The presence of an office center opposite, which restricts traffic; - Point development; - Not a gentrified embankment; - Narrow pedestrian part of the road; - Small number of parking spaces; - Small area; - The inability to change the appearance of the building
Opportunities	**Threats**
- Implementation of creative spaces, as following the trends of creativity	- The object will remain empty, without a destination; - Lack of demand from residents; - Long renovation time of the object; - Loss of a unified project concept due to the diversity of tenants; - Loss of interest in space; - The emergence of competitors with a similar concept

Table 3. SWOT analysis of the Nurov estate after renovation

Strengths	Weaknesses
- Location in the city center; - All types of public transport are located nearby; - Developed social infrastructure; - There is a modern building development; - The interest of residents in the renovation of the object	Deterioration of the engineering infrastructure; - The presence of an office center opposite, which restricts traffic; - Point development; - Not a gentrified embankment; - Narrow pedestrian part of the road; - Small number of parking spaces; - Small area; - The inability to change the appearance of the building
Opportunities	**Threats**
- Implementation of creative spaces, as following the trends of creativity; - Creating creative spaces for residents of all ages; - Constant updating of the program of art exhibitions and shows; - Conducting an active marketing strategy to attract more residents and tourists; - Rent of premises to tenants only in creative areas; - Development of creative industries in Yekaterinburg and the region	- The object will remain empty, without a destination; - Long time of renovation of the object; - The emergence of competitors with a similar concept

Thus, we see the transformation of weaknesses into strengths, threats into opportunities, using the example of adapting a cultural heritage object for modern use.

6 Research Results and Conclusions

The results of the study demonstrate real practical experience of participation of all groups of citizens in the development of the urban environment. Residents of the city see the urban space and evaluate it in their own way. They want to satisfy their need for a particular space. However, in the absence of the necessary space, they simply accept this fact and live on. The presence of projects to support the initiatives of residents turns the situation around-the creative ideas of residents are not forgotten but are improved and implemented. This is how a new model of urban modernization is being formed.

The creative approach to creating a livable urban environment focuses on the active participation of city residents. In our case, it is in the creation of a project for the adaptation of a cultural heritage object for modern use.

Co-participating design is a joint work with citizens on the development and implementation of an initiative and project, the successful result of which is a coordinated and satisfactory project concept for all parties. Using various forms of work and the application of working engagement technologies, residents can:

1) Promote the creation and development of communities in the city involved in the development of the city;
2) Organize with the administrative structures, enterprises of the city or citizens channels of communication of stakeholders at all stages of the project life cycle (coordination of goals and functions, development and coordination of the concept, project implementation);
3) To promote the formation of a responsible attitude of all participants to the projects developed and implemented in the city, giving all participants a certain level of influence on the final decision.

References

1. Faraj, S., Renno, W., Bhardwaj, A.: Unto the breach: what the COVID-19 pandemic exposes about digitalization. Inf. Organ. **31**, 100337 (2021). https://doi.org/10.1016/j.infoandorg.2021.100337
2. Almazova, N., Krylova, E., Rubtsova, A., Odinokaya, M.: Challenges and opportunities for russian higher education amid COVID-19: teachers' perspective. Educ. Sci. **10**, 368 (2020). https://doi.org/10.3390/educsci10120368
3. Nell, P.C., Foss, N.J., Klein, P.G., Schmitt, J.: Avoiding digitalization traps: tools for top managers **64**(2), 163–169 (2021). https://doi.org/10.1016/j.bushor.2020.11.005
4. Bylieva, D., Zamorev, A., Lobatyuk, V., Anosova, N.: Ways of enriching MOOCs for higher education: a philosophy course. In: Bylieva, D., Nordmann, A., Shipunova, O., Volkova, V. (eds.) Knowledge in the Information Society. LNNS, vol. 184, pp. 338–351. Springer, Cham (2021). https://doi.org/10.1007/978-3-030-65857-1_29

5. Baranova, T., Kobicheva, A., Tokareva, E.: Total transition to online learning: students' and teachers' motivation and attitudes. In: Bylieva, D., Nordmann, A., Shipunova, O., Volkova, V. (eds.) Knowledge in the Information Society. LNNS, vol. 184, pp. 301–310. Springer, Cham (2021). https://doi.org/10.1007/978-3-030-65857-1_26

6. Mercier, M., et al.: COVID-19: a boon or a bane for creativity? Front. Psychol. **11** (2021). https://doi.org/10.3389/fpsyg.2020.601150

7. Kapoor, H., Kaufman, J.C.: Meaning-making through creativity during COVID-19. Front. Psychol. **11** (2020). https://doi.org/10.3389/fpsyg.2020.595990

8. International Year of Creative Economy for Sustainable Development. UNESCO. https://en.unesco.org/news/international-year-creative-economy-sustainable-development. Accessed 5 Apr 2021

9. ASI: the volume of the creative economy in Russia is less than 3%. https://tass.ru/ekonom ika/9235421. Accessed 5 Apr 2021

10. de la Peña, D., Allen, D.J., Hester, R.T., Hou, J., Lawson, L.L., McNally, M.J. (eds.): Design as Democracy: Techniques for Collective Creativity. Island Press/Center for Resource Economics, Washington, D.C. (2017). https://doi.org/10.5822/978-1-61091-848-0

11. Cudny, W., Comunian, R., Wolaniuk, A.: Arts and creativity: a business and branding strategy for Lodz as a neoliberal city. Cities **100**, 102659 (2020). https://doi.org/10.1016/j.cities.2020.102659

12. Radomska, J., Wołczek, P., Sołoducho-Pelc, L., Silva, S.: The impact of trust on the approach to management—a case study of creative industries. Sustainability **11**(3), 816 (2019). https://doi.org/10.3390/su11030816

13. Systematic support for creative industries is an important part of Moscow's development strategy. https://duma.mos.ru/ru/0/news/novosti-fraktsiy/sistemnaya-podderjka-kreativnyih-industriy-vajnaya-chast-strategii-razvitiya-moskvyi. Accessed 8 Apr 2021

14. The program for creating creative clusters was presented at the Russian Real Estate Innovation Forum. https://asi.ru/news/181334/?fbclid=IwAR38RYUKOFf1vcmpNhmYcp-yrlpjY TI9q7p-nz0s948oGB0yT0qCrnZDtC0. Accessed 29 May 2021

15. Ganchenko, D., Tarzanova, Y.: Comfortable urban environment: innovation or transformation of the term. Dev. Theor. Pract. Manage. Soc. Econ. Syst. **8**, 81–85 (2019). https://doi.org/10.24411/9999-026A-2019-00019

16. Ferretti, V., Gandino, E.: Co-designing the solution space for rural regeneration in a new world heritage site: a choice experiments approach. Eur. J. Oper. Res. **268**, 1077–1091 (2018). https://doi.org/10.1016/j.ejor.2017.10.003

17. Yaskova, N.Y., Sarchenko, V.I., Khirevich, S.A.: Main principles of comprehensive approach to formation of comfortable urban environment. IOP Conf. Ser. Mater. Sci. Eng. **1079**, 032031 (2021). https://doi.org/10.1088/1757-899X/1079/3/032031

18. Kerr, B.A., et al.: Creativity and innovation in Iceland: individual, environmental, and cultural variables. Gift. Talent. Int. **32**, 27–43 (2017). https://doi.org/10.1080/15332276.2017.1397903

19. Federal project "Creating a comfortable urban environment". https://minstroyrf.gov.ru/docs/50262/. Accessed 25 May 2021

20. Domaneschi, L.: Learning (not) to labour how middle-class young adults look for creative jobs in a precarious time in Italy. In: Youth and the Politics of the Present. Routledge, London (2019). https://doi.org/10.4324/9780429198267

21. Paus, E.: Confronting Dystopia: The New Technological Revolution and the Future of Work. Cornell University Press, Cornell (2018)

22. Tan, J.-L., Caleon, I., Ng, H.L., Poon, C.L., Koh, E.: Collective creativity competencies and collaborative problem-solving outcomes: insights from the dialogic interactions of Singapore student teams. In: Care, E., Griffin, P., Wilson, M. (eds.) Assessment and Teaching of 21st

Century Skills. EAIA, pp. 95–118. Springer, Cham (2018). https://doi.org/10.1007/978-3-319-65368-6_6

23. Trostinskaia, I.R., Safonova, A.S., Pokrovskaia, N.N.: Professionalization of education within the digital economy and communicative competencies. In: Proceedings of the 2017 IEEE VI Forum Strategic Partnership of Universities and Enterprises of Hi-Tech Branches (Science Education Innovations), pp. 29–32. IEEE, St. Petersburg (2017)

24. HSE measures the creative capital of the megapolis. https://issek.hse.ru/news/399084439.html. Accessed 8 Apr 2021

25. Official statistics of the Sverdlovsk region. https://sverdl.gks.ru/folder/26395. Accessed 10 Apr 2021

26. Ministry of Construction and Housing and Communal Services of the Russian Federation. http://minstroyrf.gov.ru/. Accessed 10 Apr 2021

27. The Center for Urban Competencies of the Agency for Strategic Initiatives and the Ministry of Construction and Housing and Communal Services of the Russian Federation, Standard for involving citizens in solving issues of urban environment development. https://100gorodov.ru/attachments/1/9c/f07547-38a4-40b4-b02e-66179bc4e616/Standart_vovlecheniya.pdf. Accessed 10 Apr 2021

28. Agency for strategic initiatives. https://asi.ru/. Accessed 15 Apr 2021

29. ASI will support the Sverdlovsk region in creating creative spaces on the site of former industrial zones. http://midural.ru/news/list/document179797/. Accessed 15 Apr 2021

30. International Program for the Development of Urban Initiatives "School of Urban Pioneers" 2020/21. https://rus.cisr-berlin.org/sup21. Accessed 15 Apr 2021

31. Opportunity of the day: create a new digital product for an urban project. The Village, http://we.the-village.ru/cYOT. Accessed 20 Apr 2021

32. Images of the future. Hackathon. https://uralcitizen.ru/future_hack. Accessed 20 Apr 2021

33. In the center of the city, the merchant Nurov's garden was opened, which the volunteers cultivated for three years. https://66.ru/news/society/223391/. Accessed 17 Apr 2021

34. Ekaterinburg "parkoskverovtsy" went to the shore of the Iset. https://www.uralinform.ru/news/society/337591-ekaterinburgskie-parkoskverovcy-vyshli-na-bereg-iseti/. Accessed 20 Apr 2021

Mental Maps as a Creative Tool of Marketing Analysis in Education

Marianna Yu. Ababkova[1,4(✉)] , Nadezhda N. Pokrovskaia[1,2,3] ,
Veronika L. Leontyeva[1] , and Marina S. Arkannikova[1]

[1] Higher School of Media Communications and Public Relations of the Institute of Humanities,
Peter the Great St. Petersburg Polytechnic University, St. Petersburg 195251, Russia
ababkova_myu@spbstu.ru

[2] Department of Public Relations and Advertising of the Institute of Human Philosophy, Herzen
State Pedagogical University of Russia, St. Petersburg 191186, Russia

[3] Department of Innovation Management, St. Petersburg Electrotechnical University "LETI",
St. Petersburg 197022, Russia

[4] Department of Public Relations, St. Petersburg Electrotechnical
University "LETI", St. Petersburg 197022, Russia

Abstract. Marketing research in education involves new techniques to overcome
the mismatch between the unconscious and emotional perception of a brand and
verbal-centered survey approaches to consumer studies. The article illustrates the
use of non-verbal marketing research technique in order to create a mental map
describing the perception of an educational institution and giving an additional
insight into students' knowledge structures. The research included two stages in
order to obtain data by means of ZMET and the mental maps' construction. The
images obtained by ZMET allow to clarify the unconscious perception of an educational organization and its brand. The images a result of ZMET study can be
used to create mental maps to depict the interrelated concepts or constructions
of a respondent. These deep images can be directly translated into the marketing
communications and branding strategies of the educational institution. The dominant metaphor for the image of an educational institution can be identified, and the
subconscious negative associations can be probed to work with. Based on ZMET,
the mental maps of perception of the university, higher school, specialty, as well
as a map of negative images associated with learning were constructed. Mental
mapping enables the educators and marketers to aggregate data in diagrams to
visualize and systematize images, thoughts and associations to provide creative
and tailored marketing and communication strategy.

Keywords: Marketing research in education · ZMET · Non-verbal techniques ·
Images · Mental map · Communication strategy

1 Introduction

The globalization, total increase of demand for customized educational services and
consumer-centrism in higher education led to a focus on the need for adaptation of the
effective marketing activities of universities [1, 2]. The growth of universities and their
diversified programs, heterogeneity, students' mobility and world-wide access to online

programs [3] impel a tertiary institution into a constant search for opportunities and alternatives to maintain its leadership [4, 5] and to find an optimal strategy to fit in the expectations of its students [6].

The evolution of marketing concept in education as an indispensable management function nowadays is based on the assumption that the success of universities depends on the educators' and marketers' competence to investigate the process of decision making, to examine the image of the university and it is perceived by the applicants and students, to create an efficient university's image and to reduce some negative effects of the university's growth and a lack of individualized approach to the students' attention [7].

Marketisation and marketing activities in educational institutions develop a new wide research scope to ensure customer engagement and to identify the pertinent marketing, communications, brand strategies, and research methods [8].

Marketing research in education allows to obtain multifaceted information to understand cognitive relevance of educational brand or marketing communication strategy from consumers' point of view [9]. One of the main marketing research objectives in education is a comprehensive analysis of the students' and entrants' motives for obtaining higher education by students and potential applicants [10].

Qualitative and quantitative marketing research provide useful baseline data and identify significant characteristics of an educational organization in order to increase its brand potential on the educational market. Qualitative research can be described as labor-intensive time-consuming and expensive process, suggesting difficulties in obtaining, processing, tabulating and analyzing the data. There are also concerns about the reliability of respondents' responses, so far as consumers' feelings involved and good foreign language command required. Some researchers point out that rather than yielding consumers' percept or comprehend a marketing driver, traditional marketing research methods render how consumers react to researchers' conceptualization of a market driver [11].

The educational path choice as well as considering options and choosing a particular university is a complicated decision-making system determined by a set of reasons that sometimes falls below university entrants' conscious level. This tacit information lies beyond the framework of traditional linear research [12].

According to some authors [13], the profile of the educational services consumer has a guiding role in the activity of the university, so the institution has to aim at obtaining and refining additional information to clarify its knowledge about the current and potential students' opinion vis-à-vis the institution, its educational offer and the reputation [6].

Nonverbal methods of marketing contribute to understanding consumers' socio-cultural characteristics and solving the problems of verbal communication, and more importantly, ensuring consumers' insight and clarifying their underlying reasons and values.

2 Materials and Methods

Students' perceptions, thoughts and feeling in the tertiary sector are traditionally investigated and measured through quantitative techniques such as surveys. Marketing research in education often grapple with a mismatch between verbocentric data and the nonverbal language of marketing communications. This issue is particularly evident when a sample of foreign students, when respondents have difficulties in expressing their thoughts accurately in a non-native language [14, 15].

The main nonverbal marketing research methods are semiotic analysis, associated with the study of the sign system corresponding to the brand; photoanalysis; narrative approach (storytelling), which allows to identify the "true needs" of customers and helps researchers understand drivers of consumer behavior [16].

The Zaltman metaphor elicitation technique (ZMET) provides data to holistic understanding of consumers and elicits their inner feelings and thoughts to employ the gained ideas to further creative improvement of communication strategy of a marketing actor [17]. This technique is based on the fact that nonverbal communication prevails and images and metaphors are integral and indispensable parts of thinking process [18].

ZMET consists of two stages, first of them includes collecting images by the respondents and further stage encompasses an in-depth interview conducted by the researcher to ensure clarity of metaphor ascribed by the respondents. Thinking as cognitive process bases on embodied experience and includes feelings, thoughts, ideas, and concepts that are derived from sensory experience. Thus, researchers, having access to visual, tactile, acoustic, and other images, will get a clue to a better understanding of consumers' inner world. The image collection stage gives the respondents the opportunity to pay prater heed to their inner thoughts and associations and reflect their feelings on photos, pictures and images. The interview stage includes questions founded on the proposed images to elicit hidden drivers, motives, and fears. Thinking out of the box helps both the researchers and the respondents to unveil important but unconscious ideas to hone marketing communication strategy and university brand relevance. Nowadays this technique is incorporated into service marketing research areas to develop a deeper insight into the factors, drivers, stereotypes, associations and thoughts influencing a consumer [19], such as sports [20], in hospitality management [21], tourism [22], etc.

ZMET can be varied be accordance with the specific objectives of each particular research. The original steps include storytelling (respondents describe the meaning of the chosen images); explication of respondents' inability to find a necessary picture or to describe an image; sorting and labeling pictures in piles; construct elicitation; choosing the most representative product's image; describing the opposite of the brand's image; describing the brand's image in other senses' terms; a mental map creating; a summary image of a brand or a product; consensus map or causal model creating [18, 23, 24].

Mental maps were initially coined by E.C. Tolman and introduced as cognitive maps in 1948 [25] to elicit the personal and subjective image of environment [26]. First this notion included images based on a person's perception and experience of a place, incorporating its structure and exterior [27] and resulted in the hypothesis that mental representation of reality is map-like rather than language-like [28]. The initial term "cognitive map" has attained many senses, leading to confusion and misunderstandings [29].

Mapping as a creative tool and technique in some research papers can be referred to as follows: cognitive maps, mental maps, concept maps, mind concepts, spider mapping, network diagrams, knowledge maps, thought maps, graphical representations, structured overviews, semantic maps, or cluster arrangements [30, 31].

Nowadays this term has several main meanings. The first one is that mental maps are attributed to spatial thinking and the ability of an individual to understand the structure of his location, to describe it in relation to other spaces, and to clarify the development of environment perception concepts [32].

The second opinion considers a mental map as one form of graphical representation of knowledge and a tool for planning, especially in education to motivate imagination, thinking, and memory [31, 33].

The third application of mental maps could be found in marketing research (e.g., in ZMET) to display the inner world and deeper thoughts of a consumer through the use of metaphors. The mental maps in ZMET enables educators and marketers to process consumers' thoughts represented in pictures and structure metaphors to understand consumer's inner values and to hone market segmentation, product concept, branding and advertisement [34].

Mental mapping presents diagrams to visualize and systematize images, thoughts and associations, and their relations ascribed to a certain object by a consumer to view marketing and communication opportunities and insights holistically to provide a well-tuned marketing and communication strategy [17].

The exploratory research was carried out based on an integrated approach combining qualitive and qualitative tools such as the Zaltman Metaphor Elicitation Technique, and semantic differential. The objective was to examine the applicability of the Zaltman Metaphor Elicitation Technique (ZMET) and mental maps in education.

The study was aimed to apply mental mapping built on the ZMET not only to enable a framework to tailor communication strategy, but also to find gaps and mismatches, as well as weak points in the process of education for the Higher School of Mediacommunications and PR of the Institute of Humanities at Peter the Great Saint-Petersburg Polytechnic University. In total, the sample was 87 participants and consisted of the students of the 3rd grade of the Higher School of Mediacommunications and PR. The sample included 82% females and 18% males, aged 20–22, all the respondents were the students of Bachelor's (4 years) degree and were subjects to the inner and external communications of the University. The structure of the sample represented 100% of the 3rd grade of the Higher School of Mediacommunications and PR and was representative for the whole Higher School cohort, the research covered the students attending courses given by the researchers.

The process of the ZMET research and mental mapping consisted of the respondents' choosing a number of images to describe their perception of the University, the Higher School and the profession (specialty), in-depth interview and then mental mapping to elicit perceptions and feelings related to the items mentioned. The in-depth interview results were structured on the basis of semantic differential. The respondents were asked to describe the pictures they brought and to explain the symbols and metaphors they used to express their inner feelings and associations with the University, The Higher School, and their profession. All images presented by the respondents were bipolar, for example, satisfaction from the qualified academics, lecturers and staff included both "satisfaction" and "dissatisfaction" with the continuum of the pictures between these two poles.

Then all the images were grouped in 14 scales of semantic differential on the grounds of the metaphors' meanings and core contents (e.g., features of the University, personal development, quality of education, etc.).

3 Results

The data obtained is multifaceted and the number of mentions of the most important features of the University, the Higher School and profession presented in the table below (Table 1). First, this table allows to group the answers, and to find out what items have long been of importance or concern to the students.

Table 1. Data obtained by the ZMET.

The scale of semantic differential	Percentage of positive images obtained by the ZMET		
	University	Higher school	Profession
University's history, traditions, interest and popularity within the community	24,1	0	0
University's success and leadership in the industry, its role in the world of sciences	5,1	0	0
Famous academics and personalities, research and development	7,6	0	0
Professionalism, organization and management, supervision and instructions	1,3	2,5	2,5
Security, stability, consistency	1,3	3,8	3,8
Extracurricular activities, entertainment and leisure	2,5	1,3	1,3
Belonging to students' community, unity, friendship	11,4	6,3	6,3
Self-determination, personal development, building up one's own capabilities and potential	11,4	1,3	1,3
Inspiration, ideas, creativity	5,1	7,6	7,6
Future, great career opportunities and professional development	6,3	5,1	5,1
Social appreciation of profession, honor and prestige	0	0	0
Training activities and discussions, projects, internship	2,5	10,1	10,1
Relevant education programmes, blended education	1,3	0	0
Qualified academics, lecturers and staff	1,3	17,7	17,7

The most important metaphors in describing the image of the University were its history, traditions, interest and popularity within the community (24.1% of positive images); belonging to students' community, unity, friendship (11.4 % of positive images); Self-determination, personal development, building up one's own capabilities and potential (11.4% of positive images).

Another side of the data obtained by ZMET allows to clarify and revise the results of the survey research. As it was mentioned above, these images, associations and thoughts can be used to create diagrams (mental models, or maps) to depict the interrelated concepts or constructions of an individual or a sample. These deep images, identified in the participants' mental models, can be directly translated into the marketing communications' concept and branding strategies of an educational institution. The identification of the dominant metaphors for the educational institution's image, and a "probe" of the subconscious feelings of the respondents, also can provide positive and negative associations as a yardstick for a substantive review of an educational institution service policy or rethinking its internal business processes.

Fig. 1. Negative images associated with the educational process (Pictures published by permission of the respondents).

For an example, 35% images brought by the participants and related to the Higher School, contained the direct allusions to paper work and the large number of reports as one of the main irritants to the students. The comments during the in-depth-interview were: "Folders. Students put their work in them. Fear"; "Fuss. Paper work and reports"; "A huge number of documents"; "A lot of unnecessary papers, confusion, misunderstanding, waste paper, confusion, chaos, routine" (see Fig. 1).

Another part of the images conveyed the feelings and the thoughts of anxiety, stress, disillusionment. The comments were as follows: "Pain. Lack of motivation to work, to learn. Procrastination"; "Fear, frustration"; "Anxiety, misunderstanding, and helplessness" (Fig. 2).

Fig. 2. Negative images related to the higher school (Pictures published by permission of the respondents).

After the analysis of the extracted metaphors and constructs the most popular of them (mentioned by 30% of the respondents) were generalized and included on the negative and positive mental maps.

Figure 3 shows the mental map of the negative images ascribed to the learning process.

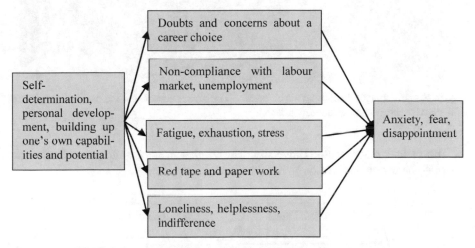

Fig. 3. Negative images associated with the educational institution.

The respondents, who had chosen the pictures classified during the in-depth interview as "Self-determination, personal development, building up one's own capabilities and potential" scale (see the Table 1), expressed explicitly or implicitly some negative associations connected with disappointment and fear.

This mental map provides the educators with a list of the difficulties and frustrations that the students meet during their university studies and to offer the educational, marketing and managerial measures to overcome them. For example, the presented mental map of the negative images helps to focus the efforts on such directions as explaining the meaning and content of the profession at the very first stages of admission to the university; optimizing the educational process; reducing the number of reports and paper work, and implementing the activities to help the students to build a career and to reduce stress during the process of education.

It is worth mentioning, that the images chosen to describe the students' perception of the University in 65% of cases were more generalized, representing photos of St. Petersburg, the grand stairs and the main hall of the University, historical photos and pictures taken from the Internet. 70% of the images related to the perception of the Higher School were associated with specific personalities (academics, lectures, classmates), situations and events held by the School, its premises, classrooms, lobbies and surroundings, most of photos were taken by the students themselves.

The images related to the profession, on the one hand, were also creative, including the students' own perception of the Advertising and PR, photos from the Internet, and on the other hand, some social stereotypes, cliches and iconographics, gleaned from films, books, and the Internet (Fig. 4).

Fig. 4. Specific images describing the university, The higher school and the profession (pictures published by permission of the respondents).

The mental maps of the University, The Higher School and the profession are presented below on the Figs. 5, 6, 7.

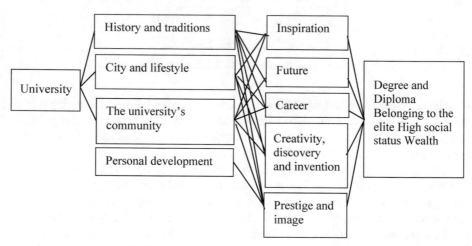

Fig. 5. Mental map of the perception of the university.

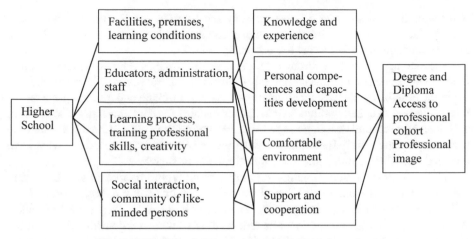

Fig. 6. Mental map of the perception of the higher school.

Fig. 7. Mental maps of the profession.

The analysis of the mental maps enables to reconsider the positioning, the perception of the key values of the University's brand and the concept of its marketing communications. The main motives for entering the university are not only to acquire knowledge, and a diploma, but also to join a certain social group, to get a higher and prestigious status, wealth and to manage an individual career path. The emphasis in marketing communications at the university level should include the images conveying the historical and scientific traditions of the University and the atmosphere of the city, contribute to the unique climate and the professional environment and opportunities for self-development (Fig. 5). The image of a higher school created by the tools of marketing communications should be more specific, tied to the learning environment and personalities, to convey the atmosphere of learning and students' life and to cover all aspects of its educational and scientific activities (Fig. 6). The applicant may not be aware of the essential aspects of the advertising profession, so it is necessary to work with negative images and to fine-tune the image of a specialist, the scope and breadth of his functional responsibilities, as well as the contribution to the life of society (Fig. 7).

The findings show that through the pictures and images the respondents express their thoughts and perceptions more freely than during the survey, and the researcher are given the possibility to probe their underlying values, rationale and motives. As a result, these collected data and findings lead to deeper insights into the university's perception, sustained by profound descriptions and images. The applicability of ZMET to elicit the perception of the university's image widens the results and helps to extract what have been left out by traditional qualitative questionnaire or survey.

The study results help to outline a number of emerging themes to employ during the promotion of the university's brand and admission campaign as well as during the different stages of the students' journey. The research opened a deeper level of under-standing of the students' emotional factors relating to their relationship with both the University and the Higher School. ZMET is a better predictor of understanding students' thoughts, feelings and associations than an annual survey on students' satisfaction and the University and the Higher School evaluation. Of course, the verbal-centered sur-vey helps to outline and articulate the main characteristics of the educational institution and to underline the most important ones, but still the findings from the ZMET deal with the unconscious associations and thoughts to provide insights for emotion-based communications and some certain areas of educational and operational processes that could be improved not relying on purely only semantic differential and other measures of students' satisfaction.

4 Conclusion

The mental map is a creative construct tool to help marketers and educators to understand students' values and emotions at a deeper level to improve education excellence and to tailor marketing communications for an educational institution.

According to the mental maps, options for advertising concepts of an educational institution and its brand can be identified through the main associations gained by the images' analysis and in-depth interview, but it is impossible to obtain a quantitative assessment of the relationship between the constructs of the mental map. In addition, the images and photos respondents bring to the interview are familiar and they do not replace the creative approach to finding visual and other images to be employed within the marketing communication concept and to even more effectively activate the associations with the brand of the university.

Nonverbal techniques help to uncover unique metaphorical ideas that are important to the consumers. In addition, projective techniques bridge the gap between researchers and respondents, as the information is collected and provided by the participants, rather than formulated in terms of the researchers [35]. It is the participant, not the researcher, who chooses visual or sensory images [36]. Using images, the participants are better equipped to manifest their doubts and discontent and to represent the ideas potentially unknown to the educators [37].

In our view, the combination of verbal and nonverbal techniques in educational marketing research, for example, the ZMET and the survey, will expand the useful data to investigate the behavior of the consumers and their profiles to update and adjust not only the communication strategy, but also the scope and the structure of the educational process [38, 39].

The need for cost-effective techniques conducting market research in education is still urgent and relevant [40]. Some researchers presume that music can become a viable addition or substitute for images in projective techniques. The potential application of the technique can help to update the brand strategy and improve its position. This technique provides the companies with consumer ratings in such areas as perception of products, brand equity, design, purchase and users' experience, life experience, and consumption context [41]. ZMET is particularly useful for projects that highlight issues that have not been previously explored.

References

1. Shipunova, O., Evseeva, L., Pozdeeva, E., Evseev V., Zhabenko I.: Social and educational environment modeling in future vision: infosphere tools. In: Kalinina, O. (eds.) E3S Web of Conferences, vol. 110, 02011 (2019). https://doi.org/10.1051/e3sconf/201911002011
2. Odinokaya, M., Krepkaia, T., Sheredekina, O., Bernavskaya, M.: The culture of professional self-realization as a fundamental factor of students' internet communication in the modern educational environment of higher education. Educ. Sci. **9**, 187 (2019). https://doi.org/10.3390/educsci9030187
3. Ipatov, O., Barinova, D., Odinokaya, M., Rubtsova, A., Pyatnitsky, A.: The impact of digital transformation process of the Russian university. In: Katalinic, B. (eds.) Proceedings of the 31st DAAAM International Symposium, vol. 31(1), pp. 0271–0275. DAAAM International, Vienna (2020). https://doi.org/10.2507/31st.daaam.proceedings.037
4. Asanov, I.A., Pokrovskaia, N.N.: Digital regulatory tools for entrepreneurial and creative behavior in the knowledge economy. In: Shaposhnikov, S. (ed.) 2017 International Conference Quality Management, Transport and Information Security, Information Technologies, pp. 43–46. IEEE, St. Petersburg (2017). https://doi.org/10.1109/ITMQIS.2017.8085759
5. Fersman, N.G., Zemlinskaya, T.Y., Novak-Kalyayeva, L.: E-learning and the world university rankings as the modern ways of attractiveness enhancement for the Russian universities. In: Soliman, K.S. (ed.) Proceedings of the 30th International Business Information Management Association Conference, IBIMA 2017 - Vision 2020: Sustainable Economic development, Innovation Management, and Global Growth, pp. 927–944. IBIMA, Madrid (2017)
6. Oana, D.: The consumer of university educational services – a central element of educational marketing. Stud. Bus. Econ. **14**(2), 31–40 (2019). https://doi.org/10.2478/sbe-2019-0023
7. Manea, N., Purcaru, M.: The evolution of educational marketing. Ann. Spiru Haret Univ. Econ. Ser. **17**(4), 37–45 (2017). https://doi.org/10.26458/1744
8. Foskett, N.: Marketisation and education marketing: the evolution of a discipline and a research field. In: Oplatka, I., Hemsley-Brown, J. (eds.) The Management and Leadership of Educational Marketing: Research, Practice and Applications. Advances in Educational Administration, vol. 15, pp. 39–61. Emerald Group Publishing Limited, Bingley (2012). https://doi.org/10.1108/S1479-3660(2012)0000015004
9. Yang, X.G.: Education marketing research. theoretical. Econ. Lett. **6**, 1180–1185 (2016). https://doi.org/10.4236/tel.2016.65111
10. Pryakhina, K.: Current trends in marketing research. Efektyvna ekonomika **12** (2020). (in Ukrainian), https://doi.org/10.32702/2307-2105-2020.12.111

11. Gebhardt, G.F., Carrillat, F.A., Riggle, R.J., Locander, W.B.: A market-based procedure for assessing and improving content validity. Cust. Needs Solut. **7**(1–2), 19–41 (2019). https://doi.org/10.1007/s40547-019-00099-w

12. Ababkova, M.Yu., Leontieva, V.L.: neuromarketing for education: rethinking frameworks for marketing activities. In: Ardashkin, I.B., Martyushev, N.V., Klyagin, S.V., Barkova, E.V., Massalimova, A.R., Syrov, V.N. (eds.) The European Proceedings of Social & Behavioural Sciences EpSBS XXXV, pp. 1–9. Future Academy, London (2018). https://doi.org/10.15405/epsbs.2018.02.1

13. Maringe, F., Gibbs, P.: Marketing Higher Education Theory and Practice. Open University Press, Buckingham (2008)

14. Rubtsova, A.V., Almazova, N.I., Bylieva, D.S., Krylova, E.A.: Constructive model of multilingual education management in higher school. IOP Conf. Ser. Mater. Sci. Eng. **940**, 012132 (2020). https://doi.org/10.1088/1757-899X/940/1/012132

15. Ababkova, M.Yu., Leontieva, V.L.: Metaphor-based research for studying Russian and Chinese students' perception of the university. In: Shipunova, O., Bylieva, D. (eds.) European Proceedings of Social and Behavioural Sciences EpSBS 98, pp. 89–98. European Publisher, London (2020). https://doi.org/10.15405/epsbs.2020.12.03.9

16. Padgett, D., Allen, D.: Communicating experiences: a narrative approach to creating service brand image. J. Advert. **26**(4), 49–62 (1997)

17. Christensen, G.L., Olson, J.C.: Mapping consumers' mental models with ZMET. Psychol. Mark. **19**(6), 477–501 (2002). https://doi.org/10.1002/mar.10021

18. Zaltman, G., Coulter, R.: Seeing the voice of the customer: metaphor-based advertising research. J. Advert. Res. **35**(4), 35–51 (1995)

19. Nasr, L., Burton, B., Gruber, T.: Developing a deeper understanding of positive customer feedback. J. Serv. Mark. **32**(2), 142–160 (2018)

20. Chen, P.-J.: Differences between male and female sport event tourists: a qualitative study. Int. J. Hosp. Manag. **29**(2), 277–290 (2009). https://doi.org/10.1016/j.ijhm.2009.10.007

21. Ji, M., King, B.: Explaining the embodied hospitality experience with ZMET. Int. J. Contemp. Hosp. Manag. **30**(11), 3442–3461 (2018). https://doi.org/10.1108/IJCHM-10-2017-0709

22. Khoo-Lattimore, C., Prideaux, B.: ZMET: a psychological approach to understanding unsustainable tourism mobility. J. Sustain. Tour. **21**(7), 1036–1048 (2013). https://doi.org/10.1080/09669582.2013.815765

23. van Dessel, M.: The ZMET technique: a new paradigm for improving marketing and marketing research. In: Purchase, S. (ed.) AANZMAC Conference: Marketing Research and Research Methodologies (Qualitative), pp. 48–54. University of Western Australia Business School, Perth (2005)

24. Catchings-Castello, G.: The ZMET alternative. Mark. Res. **12**(2), 6–15 (2000)

25. Tolman, E.C.: Cognitive maps in rats and men. Psychol. Rev. **55**, 189–208 (1948)

26. Aram, F., Solgi, E., Higueras García, E., Mohammadzadeh, S., Mosavi, A., Shamshirband, S.: Design and validation of a computational program for analysing mental maps: aram mental map analyzer. Sustainability **11**, 3790 (2019). https://doi.org/10.3390/su11143790

27. Matlin, M.W.: Memory Strategies and Metacognition. In: Farmer, T.A., Matlin, M.W. (eds.) Cognition, 10th edn., pp. 112–133. Wiley, London (2005)

28. Blumson, B.: Mental maps. Philos. Phenom. Res. **85**(2), 413–434 (2012). https://doi.org/10.1111/j.1933-1592.2011.00499.x

29. Tversky, B.: Cognitive maps, cognitive collages, and spatial mental models. In: Frank, A.U., Campari, I. (eds.) Spatial Information Theory A Theoretical Basis for GIS. LNCS, vol. 716, pp. 14–24. Springer, Heidelberg (1993). https://doi.org/10.1007/3-540-57207-4_2

30. Luchembe, D., Chinyama, K., Jumbe, J.: The Effect of using concept mapping on student's attitude and achievement when learning the physics topic of circular and rotational motion. Eur. J. Phys. Educ. **5**(4), 10–29 (2014)

31. Koleňáková, R.Š., Kozárová, N.: The impact of mental mapping on pupils' attitudes to learning. CBU Int. Conf. Proc. **7**, 464–471 (2019). https://doi.org/10.12955/cbup.v7.1402

32. Castellar, S.M.V., Juliasz, P.C.S.: Mental map and spatial thinking. Proc. Int. Cartogr. Assoc. **1**, 18 (2018). https://doi.org/10.5194/ica-proc-1-18-2018

33. Broggy, J., Mcclelland, G.: Undergraduate students' attitudes towards physics after a concept mapping experience. University of Limerick, Ireland (2008)

34. Fan, C.-H.: A study on the consumers' brand cognition and design strategy by ZMET. In: Stephanidis, C. (ed.) Universal Access in Human-Computer Interaction. Applications and Services. LNCS, vol. 5616, pp. 333–342. Springer, Heidelberg (2009). https://doi.org/10.1007/978-3-642-02713-0_35

35. Chaing, T., Rau, H., Shiang, W.J., Chiang, J.L.: Combining ZMET and MEC in study of user expectations for ideal catering customer service app. MATEC Web Conf. **119**(7), 01025 (2017). https://doi.org/10.1051/matecconf/201711901025

36. Kiselev, V.M., Afonsky, S.A., Zherebtsova, N.A.: ZMET-analysis of archetypes among students. Vestnik Plekhanov Russ. Univ. Econ. **6**, 89–102 (2018). (in Russian)

37. Coulter, R.A., Zaltman, G., Coulter, K.: Interpreting consumer perceptions of advertising: an application of the Zaltman Metaphor Elicitation Technique. J. Advert. **30**(4), 1–21 (2001). https://doi.org/10.1080/00913367.2001.10673648

38. Leontieva, V.L., Ababkova, M.Yu.: Image of disciplines and university: a new approach to research. In: Chernyavskaya, V., Kuße, H. (eds.) The European Proceedings of Social & Behavioural Sciences EpSBS 2018, vol. 51, pp. 874–881. Future Academy, London (2018). https://doi.org/10.15405/epsbs.2018.12.02.94

39. Shearer, R.L., Aldemir, T., Hitchcock, J., Resig, J., Driver, J., Kohler, M.: What students want: a vision of a future online learning experience grounded in distance education theory. Am. J. Distance Educ. **34**(1), 36–52 (2020). https://doi.org/10.1080/08923647.2019.1706019

40. Kasyanik, P.M., Gulk, E.B., Olennikova, M.V., Zakharov, K.P., Kruglikov, V.N.: Educational process at the technical university through the eyes of its participants. In: Auer, M., Guralnick, D., Uhomoibhi, J. (eds.) Interactive Collaborative Learning, ICL 2016, AISC, vol. 544, pp. 377–388. Springer, Cham (2017). https://doi.org/10.1007/978-3-319-50337-0_36

41. Almazova, N., Bylieva, D., Lobatyuk, V., Rubtsova, A.: Human behavior as a source of data in the context of education system. In: SPBPU IDE'19: Proceedings of Peter the Great St. Petersburg Polytechnic University International Scientific Conference on Innovations in Digital Economy, 37. ACM, Saint-Petersburg (2019). https://doi.org/10.1145/3372177.3373340

Creative Interventions in Corporate Museums and the Transformation of a Company's Communication Space

Marina S. Arkannikova⑩, Elena Pozdeeva⑩, Lidiya Evseeva⑩, and Anna Tanova⁽✉⁾⑩

Higher School of Media Communications and Public Relations, Peter the Great St. Petersburg Polytechnic University, 195251 St.-Petersburg, Russia

Abstract. Modern corporate museums need to use creativity in the processing of all components of the museum space when interacting with target audiences. Corporate museums play an important role as a custodian and translator of the organization's values to the external and internal environment. The methodological basis of the work was the concept of P. Bourdieu about four types of capital, with an emphasis on cultural capital as a means of forming a single communication space of the organization and translating of the organization's values. In the article we study distinctive characteristics of corporate museums and functional dimensions of a corporate museum subbrand, using T. Gad's model. The article presents the results of a sociological study of corporate museums. These results enable to come to conclusion about trends of development of capitalisation of values of corporate museums and of communication opportunities of museum space which are designed to enhance the communication effects of the company's brand.

Keywords: Corporate museum · Creative tools · Social space · Communication space · Cultural capital · Organization · Subbrand

1 Introduction

Analyzing the role of culture in the modern economy, Sharon notes that culture becomes the subject of a competitive struggle between class, ethnic, religious and other groups that claim to conceptualize and control streets, museums, parks and other urban spaces. Museums and galleries, having lost state subsidies, are forced to "flirt" with the public in order to expand their audience by capitalizing art and other assets. And culture itself becomes a continuous process of coordination [1].

In the modern world, museums are becoming multidisciplinary creative clusters or their component parts. The structure of creative clusters includes permanent museum exhibitions or exhibition spaces. In this regard, interest in corporate museums is growing, which is associated with their creative activities aimed at synthesizing the types of cultural consumption of both visitors and employees of the organization. The specificity of the creative clusters creation in Russia is that it is often a private entrepreneurial initiative of the space owners, it is they who determine the strategies for the development

© The Author(s), under exclusive license to Springer Nature Switzerland AG 2022
D. Bylieva and A. Nordmann (Eds.): PCSF 2021, LNNS 345, pp. 310–321, 2022.
https://doi.org/10.1007/978-3-030-89708-6_26

of the creative space. But the trends in the development of communications in the field of culture based on virtualization and digital technologies allow companies to embark on this path and contribute to the development of creative spaces. In this regard, the task of the corporate museum is to create an aesthetically coherent frame for the organization, just as it is usually associated with the history, architectural and/or industrial significance of the building, with the image of the city, legendary heroes and prevailing myths.

The corporate museum today acts as a communication platform, within the framework of which interaction with visitors in real and virtual space is realized. This space can be thematized as a certain "territory", limited by time and space, where communicators act, carrying out communicative activities through discourses in accordance with their goals.

One of the main functions of the corporate museum is to broadcast the values of the company, to draw attention to the history of the industry, activities, brand and modern heroes of the company [2]. In this communicative space, several fields arise, determined by the statuses of agents, their goals and values. Among them are economic agents (interested in capitalizing the museum business), cultural agents - qualified museum workers (custodians of valuables), social agents - company employees, social communities, stakeholders, regular visitors (constituting a network of interactions and communications), symbolic agents (brand-markers, heroes of stories, logos).

2 Methods

The methodological basis for studying the role of the corporate museum as a creative space is P. Bourdieu's approach to the types of capital. (In the work "Social space and the genesis of "classes"") P. Bourdieu identifies four types of capital: economic, cultural, social and symbolic [3, 4]. They are formed on the basis of the company's activities, including the communicative one, the subsystem of which is the corporate museum. According to Bourdieu, economic capital is directly converted into monetary resources, institutionalized in the form of property rights, and forms the basis of other types of capital. Cultural capital can appear in three states: an incorporated state ("in the form of long-term dispositions of mind and body"); an objectified state - in the form of cultural goods and significant artifacts (paintings, books, tools, machines, etc.); institutionalized state (formalized achievements in a particular area) [5, 6].

Social capital is viewed as a set of real or potential resources associated with the possession of a stable network of more or less institutionalized relationships of mutual acquaintance and recognition associated with group membership [6] (p. 66). It appears in the form of collective capital (social ties), networks of interaction.

Symbolic capital denotes a person's ability to produce opinions, the ability to impose a certain understanding on other agents. This is an advance that the group can provide to the subject who has given it material and symbolic guarantees. The most important role in its functioning is played by the manipulation of various ways of assessing the available and potential resources. In this respect, all other types of capital depend on symbolic capital. And the demonstration of symbolic capital is one of the mechanisms through which capital flows to capital.

The development of the corporate museum proves P. Bourdieu's idea of converting capital, transforming cultural, social and symbolic into economic capital, characterized by the inflow of various resources into the company, increasing awareness, expanding the target audience and developing forms of interaction between participants. Developing creative technologies, the corporate museum plays the role of not only a symbolic representative of corporate culture, but also a translator of corporate values into the external environment, into a wider audience of public space.

Thus, the corporate museum can be analyzed as an element of the company's communication space, recognized and endowed with the authority of the corporate community, possessing resources that form symbolic, social, cultural capital, which are dynamically converted into the economic capital of the company.

A feature of the modern activities of the corporate museum is that it plays the role of a creative component of the company's socio-economic activities. Creativity is always associated with creative approach, the ability to innovate, come up with unusual ideas and unexpected solutions. As an element of modern technology, creativity is seen as the use of creative possibilities of creating something new to increase the efficiency of any activity [7–13].

Creative industries in the 21st century have become part of the creative economy, and museums have taken their place here. The task of applied creative directions is to develop new things, to produce ideas in the context of the set goals and directions. The specificity of creativity is determined by the peculiarities of the modern cultural, informational, marketing space development: creativity is placed within the framework of modern pragmatic and applied industries and it is embodied in a combination of techniques and technologies [14–18]. An interesting twist is that the creation of a new communicative space, which was initially perceived as a means of achieving a pragmatic goal, at the present stage itself becomes a goal, since it gives rise to a new life of its constituent elements and acting agents. According to V. Benjamin, who noted that the technical reproducibility of a work of art, for the first time in world history, frees it from parasitic existence in ritual [19], we can talk about new sources of creativity, relying on traditional forms (museums), but with new agents and opportunities.

Already in the twentieth century, the museum began to be seen as a means of communication, the features of which are the transmission of society's values. This is justified by the fact that the museum's exhibits are always inscribed in the historical context, appeal to the knowledge of the past, heroes and values honored at the present time.

A corporate museum is a museum formed by the corporate community in its own interests for the purpose of self-preservation and development [20], associated with private or departmental affiliation, focused on broadcasting values and goals that are significant for the organization (enterprise).

The characteristic features of a corporate museum include:

- Membership of community members to a specific organization (enterprise, firm, company);
- Possession of autonomy in the formation of their own system of values, internal communications, models of activity and development strategies and the use of this phenomenon as a tool for their preservation;
- Expanding the capabilities of the mechanisms of memory, reflection, vision of the present, foresight and construction of the future, self-identification, self-expression

and communication on the basis of a specific semiotic system, the signs and symbols of which are objects, that are common for these communities [21].

Compared to a state corporate museum, it has a number of advantages, first of all, autonomy in the formation of its own system of values, internal relations, communications and models of activity, which contributes to its reproduction and development.

If state museums, which broadcast a connection with the past and emphasize the cultural significance of the accumulated spiritual and material potential of society, then corporate museums are focused on the present and the future. But at the same time, they also build their concept on the basis of the history of the case (business), company (organization, enterprise), brand in relation to their real life today and tasks for the future. The distinction between ordinary museums and corporate museums is also related to the fact that the educational aspect prevails in ordinary museums, while corporate ones appeal more to emotionality to represent the brand of their company, which is impossible without creative presentation. Considering the role of corporate museums, one can define it as an intermediary between the present, the past and the future. These museums are in a constant search for a balance between innovation and tradition, and this sets a great future for them.

The corporate museum performs the functions inherent in traditional museums: documentation (collection, storage and updating of archives, which are the basis for museum expositions); educational and upbringing function (communication of visitors with a guide, with museum exhibits, through interactive forms); the function of organizing leisure (events, games, quests, quizzes). Along with these functions, specific functions are also important for the corporate museum, which include the information and educational function (acquaintance with the company, its activities and heroes, its plans); activating (participation in corporate events of a business and developmental nature, training, travel, creative contribution to the museum business).

A modern corporate museum is a symbol of a corporation, a translator of corporate culture and values into the internal and external environment, as well as an enhancement of the communication effect of the company's activities. The exhibits in the corporate museum are directly embedded in the historical context of the brand. It is the strongest motivator for the employees of the organization, emphasizing the correspondence of the images and symbols of the museum to the image of the enterprise (organization) [22]. And in this regard, the sub-brand of the museum is an important component for successful communication with visitors, which is intended to broadcast the image and make it accessible to the audience in understandable language (we are talking about the sub-brand, since the brand is understood as the brand of the organization itself, which houses the corporate museum) [23]. When developing a concept for a museum or directly collecting documents, various exhibits for expositions, it is assumed that not only the organization's staff, but also interested parties - visitors, sponsors - will participate.

The trends in the development of a modern corporate museum include:

- Entering the virtual space, which implies constant updating of available resources in the Internet space, namely the museum website, active social networks, and possibly a virtual museum project.

- The use of interactive ways of interacting with visitors, which can be implemented both from the technical side (multimedia technologies in the design of expositions), and game techniques in communication
- Use of a convenient navigation system to guide visitors in the museum.
- Active use of corporate identity, as well as the formation of attributes for external identification (name; logo and trademark, slogan or slogan, anthem; corporate publications, website, souvenirs, information media, clothing of museum employees) [24].

There are also a number of key points of the success spiral for the corporate museum:

- Unusual architecture that can make the museum a tourist attraction;
- Development of various types of activity or interaction as involving the audience in museum activities;
- Reliance on emotions - empathy and a special atmosphere for the disclosed topics, available exhibits;
- Modern communication channels supporting feedback, development of digital technologies and virtualization;
- Exclusive exhibits;
- Values and brand history [25].

One of the successful examples of a corporate museum is the Russian Railways Museum. It is aimed at integrating several tasks: broadcasting the importance of passenger and cargo transportation services, disclosing the history of the development of railway transport; emphasizing the importance of the driver's profession, career guidance activities of the museum. This corporate museum has created and maintains a strong sub-brand, its recognition and attendance in St. Petersburg is marked by high values. So, for example, on weekdays the museum is visited by 1.6 thousand to 3.5 thousand people, on weekends the number of visitors is from 3.5 to 5.5 thousand people [26].

According to T. Gad, a corporate brand acts as a code of organizational differentiation, vital, influential, universal and unique like DNA [27]. He suggests including several dimensions in functional brand analysis:

- functional dimension - the perception of the rational benefits of a product or service associated with a brand;
- social dimension - identification with a specific group of the public;
- spiritual dimension - understanding of local or global responsibility;
- mental dimension - support of the consumer, brand user.

Using Gad's approach, one can analytically represent the sub-brand field of the Russian railways corporate museum as follows:

- functional dimension: the idea of Russian railways is a glossary of knowledge about modern logistics and transport resources;
- social dimension: partnership with a museum means supporting environmental policy and considering one's own role in responsible behavior towards resources;

- spiritual dimension: drawing attention to the problem of careful attitude to resources, environmental values, the global significance of environmental problems;
- mental dimension: the museum uses scientific information presented in an interactive form and encourages the modeling of new consumer practices in their own way of life.

Among new aspects of using the space of corporate museums one may think about creative clusters in megapolises including St. Petersburg, which has changed museum landscape to provide the bridge between memory of considerable art and industrial achievements in the history of the city. "Sevkabelport" on the shores of Gulf of Finland is one of the most famous among the city art spaces. It is widely visited by young people.

Creative space "Berthold centre" can serve as an example of multifunctional creative space, situated in the rooms of former foundry named after Herman Berthold, which manufactured typographic typefaces and is among the greatest in St. Peterburg. Exhibitions of photos, graphics, sculptures as well as festivals are regularly held in the centre halls.

"#Pryazhka" 5 is another striking example of creative space, organized since 2020 on the embankment of the Pryazhka River in the building of former cartographic military fabric. Today it deals with meetings, movie shows, markets, as well as workshops, where one can learn how to make ceramics.

3   Results

The Center for Sociological Research and Digital Communications of the Graduate School of Media Communications and Public Relations of the Humanitarian Institute of Peter the Great St. Petersburg Polytechnic University conducted a study of corporate museums as part of a partnership agreement with the Executive Directorate of the All-Russian Competition "Corporate Museum".

The aim of the study was to identify the creative role of the corporate museum in the activities of an organization/enterprise. The object of the research was the heads of corporate museums or representatives of PR departments of organizations/enterprises. The subject is their understanding of the role of the corporate museum in the activities of an organization/enterprise. The objectives of the study were to find out what the corporate museum gives to the enterprise/company, external target audiences, as well as to explore the creative component in the activities of the corporate museum as a translator of the values of the organization, its corporate culture into the external space. The research was conducted by the online survey method on the LimeSurvey platform from November 15 to January 15, 2020. The study sample consisted of 57 respondents (65% of 88 people who received a mailing about the survey by mail). The questionnaire consisted of three blocks.

The research hypothesis was formulated as follows: creative aspects in the activities of corporate museums enhance the communication effects of a company/organization brand.

The results of the survey showed the following. To the question "In what format does the museum work?" 32% of respondents answered that the museum works mainly in the format of excursions/exhibitions, 17% - in the format of special meetings, 13% - project target work, 10% - creative meetings, 7% - master classes, 13% - other, 8% did not answer the question (see Fig. 1).

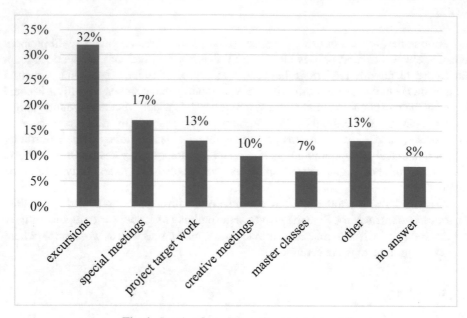

Fig. 1. In what format does the museum work?

13% who indicated "other" clarified their answer:

– activities for children;
– thematic, career guidance events, exhibitions, contests, quizzes, virtual excursions, online exhibitions, rationalization, expert, survey activities;
– practical training of Culture Institute students;
– exhibition, stock, research, career guidance activities among schoolchildren;
– scientific conferences, educational process, search group of students, procession of the Immortal Medical Battalion;
– exhibitions, website, publications, leaflets;
– all the forms presented above.

From the answers provided, it follows that the museum is used in the educational process ("practical training of students"), in research "scientific conferences", "publications" and career guidance activities ("career guidance activities for schoolchildren"), and is also involved in events of various nature - children's developmental, patriotic ("activities for children", "a search group of students, the procession of the Immortal Medical Battalion").

The most popular format of work is guided tours, project target work and special meetings. 70% of the respondents adhere to these formats of work. Such formats as master classes and creative meetings are not yet popular and are held in only a third of corporate museums. To a greater extent, this is due to the nature of the production activities of the enterprises represented in the sample, the bulk of which are involved in various sectors of heavy and light industry, where interactivity and creativity are not so well developed as in corporate museums in other fields of activity.

Respondents' answers to the open question "How do you see the desired position of the corporate museum in the future?" helps to imagine how and in what form creative elements can be used in the activities of a corporate museum when translating the values of the organization into the external and internal environment.

The responses of the interviewed representatives of corporate museums were grouped into the following key categories:

- an attractive tourist facility at the level of not only the enterprise, but also the region as a whole, focused on internal and external audiences ("one of the main participants in the cultural life of the city and country, a center for the preservation of railway equipment, an attractive tourist point of the city"; "one of the most significant centers of city's cultural life, a place interesting for tourists from other settlements"; "corporate museum integrated into the museum network of the Perm region (along with regional and municipal museums), carrying out its activities in accordance with corporate and regional interests"; "development and demand from the corporate and external audience");
- a center for providing methodological, educational and scientific support in its industry, which also performs educational functions ("a methodological center for providing information support to railway museums"; "a large educational and scientific center of civil aviation"; "a museum is not only a keeper of memory, but also a kind an information center open to everyone, popularizing knowledge about hydraulic engineering and scientific research in this area, equipped with modern technology");
- a communication platform for different audiences with the latest technologies and forms of museum work ("a technically equipped, modern, new Wagon of historical heritage, which continues its educational activities at the most remote stations of the Far North"; "today the history museum of "Gazprom Dobycha Urengoy" is the only industry a museum in New Urengoy, which is actively working in all areas of museum activity and is rightfully considered one of the most modern interactive corporate museums; in the future, we wish to remain an interesting communication platform for audiences of various levels, using the latest technologies and new museum forms of work"; "A place to get acquainted with the production capabilities of the plant, get acquainted with the history and development prospects");
- a museum with a sufficient amount of financial, human, material and other resources ("to have sufficient premises to carry out all areas of work; to increase the number of employees up to 3 people: manager, researcher, curator; receive the financial resources necessary for the modern design of expositions and exhibitions"; "When there is 1 person on the staff, he physically cannot perform the entire volume of work with high quality and meet the constantly growing requirements; therefore, he has to focus on topical issues to the detriment of others. I would like to have a decent premise, 3 people on the staff (manager, researcher, curator), funding").

The vision for the future development of the corporate museum is associated with understanding it as a center of cultural, scientific, educational, professional life, receiving sufficient funding from a professional staff, equipped with the latest museum technology to attract internal and external audiences, including as a tourist site. All this provides ample opportunities for using creative ideas to attract and retain the attention of representatives of all target groups of the corporate museum audience.

The corporate museum is seen not only as part of an organization/enterprise, but also as a "significant center of the city's cultural life", "educational and scientific center", "a museum integrated into the museum network (along with regional and municipal museums)." Thus, according to the respondents, the corporate museum requires constant modernization, strengthening the creative component of its activities in order to grow together with the organization it represents.

As for the respondents' answers to the question about the tools for promoting the value of the industry profession in the external environment, many of the examples they cited demonstrate a high level of creativity for the development of the company's communication space.

Among the tools for promoting the values of the industry profession, the respondents noted the following:

– conducting educational events, organizing career guidance programs ("classes for the younger generation - students of railway colleges and higher educational institutions, an introduction to the profession, career guidance programs for adolescents, classes for students of the Malaya Oktyabrskaya Children's Railway"; "excursions for schoolchildren and students, lectures"; "Career guidance work with students of secondary and higher schools - lectures, excursions to the museum and laboratories of the institute, thematic lessons in schools"; "the corporate history museum implements a career guidance project "From the museum exposition to the future profession", it immerses young people in the specifics of the oil and gas industry professions, helps to determine the choice of a future specialty"; "Local history section (railway topics) in the work of the Baikal International School on the basis of the school of RZD "Russian Railways" No. 21"; "lessons-lectures in schools, lecture halls in the museum"; "open lessons"; "Olympiads, conferences, tests, reports"; "Open days");
– excursion and thematic events in the museum and outside it ("excursions along the Circum-Baikal Railway"; "themed events of the museum for children - the All-Russian festival of energy saving" Together Brighter ", Water Day, Energy Day, etc."; "excursions in the museum"; "Excursions, meetings, thematic exhibitions"; "excursions"; "exposition, exhibitions"; "corporate events, exhibitions, master classes, etc.");
– cooperation with the media and publications on social networks/on the website ("publications in the media"; "media, social networks"; "publications on social networks"; "publications of books, articles in the media, website");
– participation in professional projects ("participation in competitions and projects (regional and federal level)"; "projects: exhibitions, publications, public events within the framework of projects (master classes, presentations)"; "competitions, volunteering, meetings with veterans").

It is important to note that cooperation with the media can take place both in the format of the preparation of traditional media materials and special projects (for example, such as the online edition "Bumaga" implemented last year. It prepared a thematic selection of feature films telling about the specifics of working in different branches of the economy). A similar idea can be implemented by corporate museums, preparing special projects that reveal the values of the professions of their organizations/enterprises, using various forms of artistic and scientific creativity to enhance their communications with the target audience.

To promote the values of the industry's profession of the organization/enterprise, the tools of project and educational activities are mainly used (organization of educational events, career guidance programs, excursion and educational events in the museum and outside it: exhibitions, excursions, lectures, thematic open lessons in schools, presentations and etc.), technologies for interaction with the media, participation in professional industry projects. The addressee of the former is the younger generation (children - schoolchildren and applicants - and their parents), of the latter - a wider circle of the public - from consumers of products or services of an enterprise/organization to potential partners and authorities.

4   Conclusion

Thus, the results of the study will help to better understand the role that corporate museums play in the life of modern organizations and local or regional communities, as well as to determine the potential for museum development and the formation of a pool of new ideas for working with audiences based on modern creative technologies. Based on the prevailing trends in the introduction of creative tools and interactive forms of work with the target audience, it can be argued that the performance by a corporate museum of its role aimed at reproducing the traditional values of the organization associated with the historical context of its activities and the role of the flagship of the company's future development is impossible without constant creative processing of the museum space components. We can say that the hypothesis of the study was partially confirmed, which gives corporate museums a guideline for the future development of creative components in their activities for the optimal representation of the company's communication space. It should be emphasized separately that the space of a corporate museum requires the constant work of professionals of various profiles - from guides to PR specialists, as well as modernization, strengthening the creative component for growth together with the organization it represents, in order to enhance the communication effects of the company's brand.

References

1. Zukin, S.: The Cultures of Cities. Blackwell Publishers, New York (1995)
2. Arkannikova, M.S.: Strategic Communications of Corporate Museums. Pottech-Press, Saint Petersburg (2021). (in Russian)
3. Bourdieu, P.: Forms of capital. Econ. Sociol. **5**, 60–74 (2002)

4. Radaev, V.: The concept of capital, forms of capital and their conversion. Econ. Sociol. **4**, 20–30 (2002). (in Russian)
5. Bourdieu, P.: Ökonomisches Kapital, kulturelles Kapital, soziales Kaputal. In: Kreckel, R. (ed.) Soziale Ungeichheiten (Soziale Welt, Sonderheft 2), pp. 183–198. Otto Schwartz & Co., Goettingen (1983). (in Ger.)
6. Bourdieu, P.: Forms of capital. In: Granovetter, M., Swedberg, R. (eds.) The Sociology of Economic Life, 2nd edn., pp. 98–102. Westview Press, Boulder (2001)
7. Plevako, S.V.: The phenomenon of creativity in the modern communicative space. Uchenye zapiski ZabGU. Ser.: Phil. Hist. Orient. Stud. **3**, 193–195 (2010). (in Russian)
8. Asochakov, Yu.V., Bogomyagkova, E.S., Ivanov, D.V.: New dimension of social development: activity and creativity in internet communications. Sociol, Res. **47**(1), 75–86 (2021). https://doi.org/10.31857/S013216250012083-4
9. Taylor, M., Kent, L.: Dialogic engagement: clarifying foundational concepts. J. Publ. Rel. Res. **25**(5), 384–398 (2014). https://doi.org/10.1080/1062726X.2014.95610
10. Kirillina, N.V.: The phenomenon of engagement as a reflection of the social potential of communication. Communicology **8**(1), 27–33 (2020)
11. Bylieva, D., Almazova, N., Lobatyuk, V., Rubtsova, A.: Virtual pet: trends of development. In: Antipova, T., Rocha, Á. (eds.) DSIC 2019. AISC, vol. 1114, pp. 545–554. Springer, Cham (2020). https://doi.org/10.1007/978-3-030-37737-3_47
12. Shipunova, O.D., Mureyko, L.V., Serkova, V.A., Romanenko, I.B., Romanenko, Y.: The time factor in consciousness construction. Indian J. Sci. Technol. **9**(42), 104226 (2016). https://doi.org/10.17485/ijst/2016/v9i42/104226
13. Zamorev, A., Fedyukovsky, A.: Euathlus and Crocodile paradoxes: dialectic solution's advantages. E3S Web Conf. **164**, 11022 (2020). https://doi.org/10.1051/e3sconf/202016411022
14. Arseniev, D.G., Trostinskaya, I.R., Pozdeeva, E.G., Evseeva, L.I.: Processes of changes in the educational environment under the influence of digital technologies. In: SPBPU IDE 2019: Proceedings of Peter the Great St. Petersburg Polytechnic University International Scientific Conference on Innovations in Digital Economy (60). ACM (2019). https://doi.org/10.1145/3372177.3377547
15. Shipunova, O.D., Pozdeeva, E.G., Evseev, V.V.: The impact of the digital interaction network on the future professionals behavior. In: Nordmann, A., Moccozet, L., Volkova, V., Shipunova, O. (eds.) CSIS 2019: Proceedings of the XI International Scientific Conference Communicative Strategies of the Information Society, 25 October 2019, pp. 1–6. ACM, New York (2019). https://doi.org/10.1145/3373722.3373792
16. Matveevskaya, A.S., Pogodin, S.N.: The essence of cross-cultural conflict (Presentation of a problem) Vestnik Sankt-Peterburgskogo Universiteta, Filosofia I Konfliktologiia **33**(1), 115–118 (2017). https://doi.org/10.21638/10.21638/11701/spbu17.2017.112
17. Razinkina, E., Pankova, L., Pozdeeva, E., Evseeva, L., Tanova, A.: Education quality as a factor of modern student's social success. In: Zheltenkov, A., Mottaeva, A. (eds.) Topical Problems of Green Architecture, Civil and Environmental Engineering 2019 (TPACEE 2019). E3S Web Conference, vol. 164, p. 12008 (2020). https://doi.org/10.1051/e3sconf/202016412008
18. Vasileva, O., Tarakanova, T.: Communication in the formation of the institutional environment in the tourism industry. In: DEFIN2020: III International Scientific and Practical Conference, vol. 66. ACM, New York (2020). https://doi.org/10.1145/3388984.3390952
19. Benjamin, W., Jennings, M.: The work of art in the age of its technological reproducibility. Grey Room **39**, 11–38 (2010)
20. Nikishin, N.A.: The social phenomenon of the Corporate Museum. Scale of the phenomenon, diversity, specificity. In: Scientific and Practical Seminar "Publication of a Museum Object in a Corporate Museum". Icom Russia, Khanty-Mansiysk (2016). (in Russian). https://docplayer.ru/58193971-Socialnyy-fenomen-korporativnogo-muzeya.html. Accessed 29 Mar 2021

21. Nikishin, N.A.: Subjects and forms of museum communication in the space of corporate society. In: Third International Conference "Corporate Museums Today", pp. 21–26. Icom Russia, Kaliningrad (2016)
22. Serbina, N.V.: Russian Corporate Museums: prospects for development and promotion. Apriori. Ser.: Humanit. **6**, 8–16 (2014)
23. Does the Museum Need a Brand? Ifors.Ru: research and consulting company, part of the VTsIOM partnership. https://museum-vf.ru/en/museum/partnery-en/. Accessed 21 Mar 2021
24. Zaitseva, G.A., Zaitsev, D.A.: Social marketing in the activities of corporate museums. In: II International Conference "Corporate Museums Today" Dedicated to the 70th Anniversary of the Nuclear Industry, pp. 145–146. Boslen, Moscow (2015)
25. Henkel, M.: There is no brand without content. A look at the fundamentals of corporate museums. II International Conference "Corporate Museums Today" Dedicated to the 70th Anniversary of the Nuclear Industry, pp. 44–45. Boslen, Moscow (2015)
26. The Village. https://www.the-village.ru/weekend/news/291172-muzey-rzhd-statistika. Accessed 5 Mar 2021
27. Gad, T.: 4D branding: cracking the corporate code of the networked economy. In: Akkaya, M. (ed.) pp. 13–14. Stockholm School of Economics in St. Petersburg, St. Petersburg (2005)

Ways to Solve the Problems
of Employer-Sponsored Education

Natalia V. Goncharova⊚ and Liudmila V. Daineko(✉)⊚

Ural Federal University Named After the First President of Russia B.N. Yeltsin, 19 Mira St.,
620002 Yekaterinburg, Russia
l.v.daineko@urfu.ru

Abstract. The national innovation system is based on the creative economy, the value of which is expressed in new ideas. Creative economy changes the approach to the labor market, production, trade, and provision of services. There is traditionally a shortage of qualified specialists, especially in engineering areas for industrial enterprises in the Russian labor market. One of the reasons for the lack of specialists is the small number of university graduates who are employed according to their specialty. Employer-sponsored education can serve as a tool to increase the mutual responsibility of the educational organization, the future employer and the applicant for the choice of the future specialty and further employment. To implement employer-sponsored education, the company acts as a future employer for the graduate, the state provides a state-funded place for the student's training, the university carries out the educational process by order of an enterprise, and a student receives an education and works at the enterprise for at least three years. The aim of the article is to analyze the experience of the UrFU in implementing employer-sponsored education within the allocated quota at the expense of the federal budget. The paper considers the indicators that characterize the dynamics of employer-sponsored education at the university for the period from 2015 to 2020. The authors identify the main problems in the organization of the admission of citizens to employer-sponsored programs that reduce its attractiveness, and suggest ways to increase the number of students studying under the conditions of employer-sponsored admission.

Keywords: Employer-sponsored education · Competence gap · University graduates · Higher education · Ural Federal University

1 Introduction

The global transformative processes caused by the COVID-19 pandemic have shown that the most vulnerable category in the labor market is university graduates who do not have employment history and work experience [1–3]. For this reason, the issues of employment of graduates are of great important and relevance. According to the survey conducted be Career.ru [4], in 2017–2018, only 52% of university graduates plan to work by profession. At this moment, the creative economy is desperately short of highly qualified professionals who know how to add economic value to creative ideas and who can create and implement competitive and innovative technologies [5, 6]. To solve the problem of employment of graduates, to reduce the competence gaps, and to acquire the skills of corporate professional culture and developing creative potential, the university offers to use the opportunities of employer-sponsored education. Such interaction between enterprises, universities and students, providing employers with the opportunity to participate in the training of the necessary specialists, solving practical tasks of employers during training [7], promoting the company's brand among young people [8], allows to achieve maximum educational results and a high degree of adaptation of graduates to the requirements of the labor market demands of a creative economy [9–11].

2 Problem Statement

The modern labor market imposes higher and higher demands on university graduates. To reduce the competence gaps, universities are actively introducing new forms of training into the educational process (e-learning, project-based methods, game practices, adaptive learning, etc.) [12–14]. The 2020 pandemic and the emergency transfer of many processes to online mode have also been a challenge for education, business, and the state. The experience of graduates in 2020 proved that those who signed contracts for employer-sponsored education were more secure, since guaranteed employment was a factor that minimized stress. A modern employer needs people with practical work experience, not just university graduates. Moreover, understanding that it takes several years for a specialist to develop professionally, universities try to reduce this time lag, by introduction of employer-sponsored programs [15]. The participation of universities in employer-sponsored education of students encourages teachers to master new educational technologies and to study the problems of real business in depth [16]. Phoenix believes that in order to obtain a more highly skilled and more productive workforce, the responsibility for funding the training programs should be divided between the government, students, and employers [17]. Lerman analyzing the extent and intensity of employer-provided training, the types of training provided by the employer, and employees enrolled in employer-sponsored programs notes the importance of employer-sponsored education for improving employee productivity [18]. Tran & Smith, in their study, show that employer-sponsored tuition has a positive effect on long-term student learning outcomes [19]. A 2013 study by Soltani, Twigg & Dickens, UK, on the perception of employer-sponsored education by students, employers, and teachers found that the lack of graduates with the necessary competencies increases the cost of hiring and affects the performance of companies. The results of the study showed that the employer-sponsored programs meet the needs of students and employers. The main conclusion of the study was the understanding of the value of improving professional competencies by solving real problems of the enterprises involved in the project [20]. The data of the recent monitoring of the level of satisfaction of stakeholders with the quality of education [21] showed the interest of employers in acquiring the ability of graduates to take decisive measures in the event of uncertainty, the ability to adapt to changes, and the ability to work in a team. However, there are also negative aspects of education funded by commercial enterprises. It is especially evident in the medical field. Creta & Gross, exploring the components of an effective professional development strategy, believe that it is largely up to managers to create an environment conducive to professional development. Professional development can be achieved through mentoring, sponsorship, and succession planning. The advantages of professional development of nurses are increased autonomy, increased individual competence, and involvement in the corporate culture [22]. However, Marlow, considering the sponsorship of continuing medical education by pharmaceutical companies, notes a possible bias in information masking the promotion of products to the market. On the other hand, the pharmaceutical industry has made a significant contribution to innovative continuing education programs and health education research [23].

3 Methodology

The key difference and the basis for the success of the Russian engineering school is the triad "education-science-industry", with industry playing a leading role [24]. In October 2020, the Government of the Russian Federation approved the Regulation on Employer-Sponsored Education, which established the procedure for organizing and implementing this kind of educational programs in secondary vocational and higher educational institutions. Employer-sponsored education is one of the methods of admission to an educational institution, within the quota of the Government of the Russian Federation based on a contract on employer-sponsored education. The features of employer-sponsored education training at the university are:

- education is provided at the expense of the federal budget;
- admission to places within the employer-sponsored admission quota is carried out on a competitive basis for specially allocated budget places;
- the contract and enrollment in places within the quota of admission to employer-sponsored educational programs guarantee the graduate employment;
- the applicant can participate in the competition for employer-sponsored education, in general competition, and chose paid education;
- the applicant must submit the necessary documents for admission to the university in accordance with the Admission Rules;
- the applicant provides data on the results of the Unified State Exam, or passes the entrance tests conducted by the university independently Customers of employer-sponsored education can be enterprises of any form of ownership that have concluded a three-way contract for employer-sponsored education with the university and the student. An enterprise that enters into a contract for employer-sponsored education has the right to provide for a competitive procedure (testing or interview) for the selection of students.

The contract of employer-sponsored education provides for the responsibility of three parties – students, employers and universities for compliance with the terms of the contract. If the student drops out of school or does not work through the agreed time at the enterprise or the enterprise does not provide the student with a workplace, compensation for the university's costs for the student's education should be provided. The conclusion of a contract of employer-sponsored education is possible not only before admission, but also at any time of the student's study at the university.

In order to study the specifics of the implementation of targeted education, the authors investigated statistical data on the results of targeted enrolment for six years (from 2015 to 2020). In order to analyse the reasons for the decline in the popularity of targeted education, a survey of students studying in the initial years of undergraduate education was conducted. General scientific methods of research, such as analysis and synthesis, comparison and generalization were used in the work.

4 Results

The Ural Federal University is the largest educational organization in the Ural Federal District, where more than 36 thousand people study and more than 330 educational programs are implemented in 27 enlarged groups of specialties. The university employs around 4,000 teachers. Among them: 583 doctors of sciences, 1878 candidates of sciences, 1108 have the title of associate professor, 285 have the title of professor, 23 full and corresponding members of Russian Academy of Sciences, 1 corresponding member of Academy of Arts, 53 academicians and corresponding members of branch academies, 38 members of foreign and international academies. In 2010, the university created its innovation infrastructure, which includes innovation and development centres, small innovative enterprises and units of the science-based entrepreneurship development support system (from generation of innovative ideas to production of innovative products), the system of forming technological entrepreneurs "from the school bench to your startup" (UrFU Talent School, Innovative Diving, UrFU Accelerator). The university's highly resourceful human resources potential and in-house innovation infrastructure allow it to train personnel for a creative economy.

The university's mission is to increase the international competitiveness of the Ural region, to ensure reindustrialization, to increase human, scientific, and technical potential, and to balance the renewal of traditional and the development of post-industrial sectors of the Russian economy, primarily in the Ural Federal District [25]. To implement the mission, providing communication between employers and future graduates, the Ural Federal University is working on the organization of employer-sponsored education and training for the needs of industry, the social sphere and priority sectors of the economy. Within the quota of admission for employer-sponsored programs established by the Government of the Russian Federation, the state-financed openings are provided, that is, neither the customer, nor the future employer, nor the student himself does not pay for training (even partially). The conclusion of a contract on employer-sponsored education is not a guarantee of enrollment in a state-financed opening, enrollment in places within the quota of admission to employer-sponsored programs is also carried out on a competitive basis for separately allocated state-financed openings.

According to the quota, in 2021, there were allocated 991 places for the UrFU for admission to employer-sponsored programs, including: undergraduate programs for full-time training – 452 places, mixed attendance mode – 10 places, part-time mode – 56 places; under specialty training programs: full–time training – 108 places; under master's programs: full-time training – 321 places, mixed attendance mode – 6 places, part-time mode – 38 places.

Currently, a little more than 1 000 people are enrolled in higher education programs in the UrFU as part of employer-sponsored programs.

The number of students accepted as a result of employer-sponsored admission to the first year of full-time study in bachelor's and specialist's programs is significantly reduced annually: from 554 people – in 2015, to 253 – in 2019. At the same time the number of citizens sent to the UrFU to participate in the competition for employer-sponsored education is several times less than the annual admission quota.

The indicators that characterize the dynamics of employer-sponsored education in the UrFU are presented in Table 1.

Table 1. Dynamics of indicators characterizing the employer-sponsored admission.

Indicators	2015	2016	2017	2018	2019	2020
Employer-sponsored admission quota for the first year of full-time bachelor's and specialist's degree programs, pers	867	761	554	650	863	912
The number of students accepted by the results of the employer-sponsored admission to the first year of full-time education in bachelor's and specialist's programs, pers	554	512	361	261	253	210
Share of fulfillment of the quota of the employer-sponsored admission to the first year of full-time education on bachelor's and specialist's programs, %	63,90	67,28	65,16	40,15	29,32	23,03
The share of the number of students admitted as a result of employer-sponsored to the first year of full-time bachelor's and specialist's degree programs in the total number of students admitted to the first year of bachelor's and specialist's degree programs on a full-time education, %	9,76	9,15	6,28	4,36	3,77	2,16

Organizations-customers of employer-sponsored education for the UrFU are large industrial enterprises of the Ural region, most of which are located in small industrial cities.

To analyze the decline in the popularity of employer-sponsored education, the authors interviewed 218 bachelors of the 1 and 2 year of study in economic and technical specialties, since that the number of students studying under the contract form of training is large among them and the number of students studying under the employer-sponsored admission is insignificant. Before conducting the survey, students were asked the question: "What is employer-sponsored education?" - to which 28% answered "I find it difficult to answer", 39% answered "I do not know", 33% answered "training under a contract between a university and an enterprise". After a 20-min briefing on the features of employer-sponsored education, a survey of students was conducted. The results of the survey are presented in Table 2.

Table 2. Student survey results.

Questions	Answers "Yes"	Answers "No"	Answers "I don't know"
If you had known about the conditions of targeted training before admission, would you have taken advantage of this opportunity?	35,71%	42,86%	21,43%
Are you afraid of penalties for expenses for education for non-compliance with the contract of employer-sponsored education?	67,86%	32,14%	0
Is it important for you to be able to change your individual educational trajectory while studying?	78,57%	6,12%	15,31%
Is the location of the company-sponsor of the employer-sponsored education important to you?	71,43%	0	28,57%
Is it important for you to have a 3-year work commitment after graduation?	50,13%	27,13%	22,74%
Is the number of existing company-sponsors sufficient?	0	85,21%	14,79%

The survey showed that only 35,71% of the students are ready to conclude a contract for targeted study. The majority of respondents (42,86%) do not consider target education for themselves, and 21,43% are not ready to answer this question. Also, the majority of students (67,86%) consider the main problem of targeted study as the existence of obligations to reimburse the budget for non-fulfilment of the conditions of the targeted study agreement. 78,57% of students prefer to have the opportunity to change the educational trajectory offered by the university, without fixing it by the contract of targeted study. 85,21% of students indicated insufficient number of employers offering contracts of targeted study. The location of the potential employer is important for 71,43% of students.

As a result of the survey, the following main problems were identified in the organization of admission of citizens to employer-sponsored education, which reduce the attractiveness of admission to employer-sponsored programs for citizens:

- payment of a fine by the parties to the contract on employer-sponsored education;
- insufficient awareness of applicants about the possibilities of employer-sponsored education;
- the inability to change the educational trajectory in the learning process.

5   Discussion

To increase the number of students studying under the employer-sponsored admission, the university must conduct a campaign in schools throughout the year together with enterprises, informing applicants and their parents about the features and opportunities of employer-sponsored education. It is also important to conduct career guidance activities more actively to attract students to employer-sponsored education. It is necessary to post information about customer enterprises that intend to conclude contracts on employer-sponsored education with citizens on the official website of the university. This information should contain information about the location of the company, the peculiarities of production, salary level, career opportunities and means of social support (housing, payment for additional education, etc.).

To implement the ability to change individual educational trajectory, additional educational services in the form of special disciplines and additional educational modules that are relevant to the customer may be provided.

To expand the range of customer enterprises, it is necessary to conduct career events with the participation of representatives of employer organizations, where companies are more actively informed about the possibilities of targeted training of specialists for specific tasks of the enterprise according to jointly developed educational programs.

6   Conclusion

The implementation of the proposed measures will allow all parties involved in the organization of employer-sponsored education to achieve their goals:

- Graduates: to get a high-quality education that meets the challenges of the 21st century at the most famous university in the region and to be confident in their future, with guaranteed employment.
- The university: to train professionals who are really in demand for the creative economy, to improve the quality of training of graduates through the joint organization of the educational process of the student with enterprises, to increase the level of qualification of teachers by working with enterprises of the real sector of the economy.

- Enterprises: to reduce the time for finding qualified personnel, to reduce the period of adaptation of young specialists, to get specialists with developed competencies that are important for the enterprise capable of developing and implementing creative ideas.
- The state: to increase the efficiency of the use of budget funds allocated to universities to ensure the educational process, to improve the balance of the labor market, and to solve social issues, to address the shortage of human resources in new areas of the economy (digital, creative, etc.).

References

1. Duta, A., Wielgoszewska, B., Iannelli, C.: Different degrees of career success: social origin and graduates' education and labour market trajectories. Adv. Life Course Res. **47**, 100376 (2021). https://doi.org/10.1016/j.alcr.2020.100376
2. Treviño-Reyna, G., Czabanowska, K., Haque, S., Plepys, C.M., Magaña, L., Middleton, J.: Employment outcomes and job satisfaction of international public health professionals: what lessons for public health and COVID-19 pandemic preparedness? Employment outcomes of public health graduates. Int. J. Health Plann. Manag. **36**, 124–150 (2021). https://doi.org/10.1002/hpm.3140
3. Almazova, N., Krylova, E., Rubtsova, A., Odinokaya, M.: Challenges and opportunities for Russian higher education amid COVID-19: teachers' perspective. Educ. Sci. **10**, 368 (2020). https://doi.org/10.3390/educsci10120368
4. HeadHunter.ru. https://ekaterinburg.hh.ru/article/21826. Accessed 5 May 2021
5. Herlina, H., Harianto, B.T.: Finding the characteristics of creative people in developing villages for the foundation of creative industry. Ekuilibrium J. Ilm. Bid. Ilmu Ekon. **16**, 64 (2021). https://doi.org/10.24269/ekuilibrium.v16i1.3269
6. Frederiksen, M.H., Knudsen, M.P.: From creative ideas to innovation performance: the role of assessment criteria. Creat. Innov. Manag. **26**, 60–74 (2017). https://doi.org/10.1111/caim.12204
7. Daineko, L.V., Goncharova, N., Larionova, V.A., Ovchinnikova, V.A.: Fostering professional competencies of students with the new approaches in higher education. In: Shipunova, O., Byleva, D. (eds.) Proceedings of the Joint Conferences: PCSF 2020 & CSIS 2020. European Proceedings of Social and Behavioural Sciences EpSBS, vol. 98, pp. 231–239. European Publishing Limited, London (2020). https://doi.org/10.15405/epsbs.2020.12.03.24
8. Tsiotsou, R., Alexandris, K.: Delineating the outcomes of sponsorship: sponsor image, word of mouth, and purchase intentions. Intern. J. Retail Distrib. Manag. **37**(4), 358–369 (2009). https://doi.org/10.1108/09590550910948583
9. Donald, W.E., Ashleigh, M.J., Baruch, Y.: Students' perceptions of education and employability. Career Dev. Int. **23**, 513–540 (2018). https://doi.org/10.1108/CDI-09-2017-0171
10. Cai, J., Youngblood, V.T., Khodyreva, E.A., Khuziakhmetov, A.N.: Higher education curricula designing on the basis of the regional labour market demands. EURASIA J. Math. Sci. Technol. Educ. **13**(7), 2805–2819 (2017). https://doi.org/10.12973/eurasia.2017.00719a
11. Almazova, N., Rubtsova, A., Krylova, E., Barinova, D., Eremin, Y., Smolskaia, N.: Blended learning model in the innovative electronic basis of technical engineers training. In: Katalinic, B. (ed.) Proceedings of the 30th DAAAM International Symposium, pp. 0814–0825. DAAAM International, Zadar, Croatia (2019). https://doi.org/10.2507/30th.daaam.proceedings.113

12. Demidov, V., Mokhorov, D., Mokhorova, A., Semenova, K.: Professional public accreditation of educational programs in the education quality assessment system. E3S Web Conf. **244**, 11042 (2021). https://doi.org/10.1051/e3sconf/202124411042

13. Aladyshkin, I.V., Kulik, S.V., Odinokaya, M.A., Safonova, A.S., Kalmykova, S.V.: Development of electronic information and educational environment of the University 4.0 and prospects of integration of engineering education and humanities. In: Anikina, Z. (ed.) IEEHGIP 2020. LNNS, vol. 131, pp. 659–671. Springer, Cham (2020). https://doi.org/10.1007/978-3-030-47415-7_70

14. Bylieva, D., Bekirogullari, Z., Kuznetsov, D., Almazova, N., Lobatyuk, V., Rubtsova, A.: Online group student peer-communication as an element of open education. Future Internet **12**, 143 (2020). https://doi.org/10.3390/fi12090143

15. Vaganova, O.I., Smirnova, Z.V., Kaznacheeva, S.N., Kutepova, L.I., Kutepov, M.M.: Practically-oriented technologies in professional education. In: Popkova, E.G. (ed.) Growth Poles of the Global Economy: Emergence, Changes and Future Perspectives. LNNS, vol. 73, pp. 433–439. Springer, Cham (2020). https://doi.org/10.1007/978-3-030-15160-7_44

16. Sobolev, A.B., Bogatova, T.F.: Targeted training at the Ural State University: traditions, experience, prospects. Univ. Manag.: Pract. Anal. **6**, 31–37 (2005). (in Russian)

17. Phoenix, D.: Making a Success of Employer Sponsored Education. Higher Education Policy Institute, Oxford (2016)

18. Lerman, R.I., McKernan, S.M., Riegg, S.: The scope of employer-provided training in the United States: who, what, where, and how much? In: O'Leary, C.J., Straits, R.A., Wandner, S.A. (eds.) Job Training Policy in the United States, pp. 211–244. W.E. Upjohn Institute, Kalamazoo, MI (2004). https://doi.org/10.17848/9781417549993.ch7

19. Tran, H., Smith, D.: The impact of employer-sponsored educational assistance benefits on community college student outcomes. J. Stud. Financ. Aid **47**(2), 82–100 (2017)

20. Soltani, F., Twigg, D., Dickens, J.: Sponsorship works: study of the perceptions of students, employers, and academics of industrial sponsorship. J. Prof. Issues Eng. Educ. Pract. **139**(3), 171–176 (2013). https://doi.org/10.1061/(ASCE)EI.1943-5541.0000143

21. Bareika, N., Barun, A., Dauhiala, N., Dauhiala, D.: The role of external stakeholders in ensuring the quality of educational services of Polotsk State University. SHS Web Conf. **97**, 01021 (2021). https://doi.org/10.1051/shsconf/20219701021

22. Creta, A.M., Gross, A.H.: Components of an effective professional development strategy: the professional practice model, peer feedback, mentorship, sponsorship, and succession planning. Semin. Oncol. Nurs. **36**(3), 151024 (2020). https://doi.org/10.1016/j.soncn.2020.151024

23. Marlow, B.: The future sponsorship of CME in Canada: industry, government, physicians or a blend? CMAJ **171**(2), 150–151 (2004). https://doi.org/10.1503/cmaj.1040629

24. Rudskoy, A.I., Borovkov, A.I., Romanov, P.I.: Russian experience in engineering education development. Vysshee Obrazovanie v Rossii = High. Educ. Russ. **27**(1), 151–162 (2018)

25. URFU.ru. https://urfu.ru/ru/about/today/mission/. Accessed 5 May 2021

Transcreation as a Creative Tool of Translation

Anastasia S. Gerasimova[1]([✉]) [ID], Elena Sereda[1] [ID], and Sofia Rubtsova[2] [ID]

[1] Saint-Petersburg University of Management Technologies and Economics, Lermontovsky Prospect St., 44, Lit A, 190103 Saint-Petersburg, Russia
nastasi_09@mail.ru, e.sereda@spbume.ru
[2] Peter the Great Saint-Petersburg Polytechnic University, Polytechnicheskaya St., 29, 195251 Saint-Petersburg, Russia
rubtsova.sa@edu.spbstu.ru

Abstract. The article touches upon the necessity of using the creative approach during translating the texts of the advertising discourse. The authors of the study emphasize the creative component and the knowledge of national and cultural peculiarities of the countries which are necessary to be considered on a par with the linguistic training. TV-commercials are used as the material of the research as they allow to perform and analyze the translating and creative skills of specialists. The authors consider the transcreation strategy to be an efficient creative means of delivering the authentic message based on the national and cultural peculiarities of another country to the Russian target audience. The transcreation strategy includes not only linguistic transformations but also presupposes the localization of foreign elements while the process of translation. Thus, the article shows that the process of translation of the TV-commercials should be of an interdisciplinary nature as it contributes to the full realization of both linguistic and cultural components of the original text.

Keywords: Linguist-translator student · Transcreation strategy · The discourse of advertising · Translation creativity

1 Introduction

It is traditionally considered that the ability to perform the pre-translation text analysis, which provides an accurate understanding of the original statement, as well as the ability to achieve the equivalence using the basic professional techniques while the process of translation are the main necessary skills that should be trained in future linguist-translators. However, globalization processes and the development of modern society show that the approach to teaching future linguist-translators is supposed to provide the formation of a number of competences among which the readiness to solve tasks in a creative way is not the last in the list. Therefore, one of the ways to improve the quality of future specialists training is the targeted formation of their competence [1]. In the modern Russian educational system, particular importance is given to the implementation of the competency-based approach [2]. Translation of advertising texts and videos seems to be a fertile ground for realizing the creative abilities of future translators [3].

D. Bylieva and A. Nordmann (Eds.): PCSF 2021, LNNS 345, pp. 331–338, 2022.
https://doi.org/10.1007/978-3-030-89708-6_28

Advertising is an important element of today's world. This linguistic, marketing and social phenomenon is thought to be an integral component of any culture. Being a complex form of communication, advertising reflects the existing way of life and creates it by means of controlling the human behavior. Thus, the study of advertising texts and videos as well as their linguistic and creative characteristics is vital, especially, in terms of the fact that a language worker – terminologist, translator, lexicographer, technical writer, grammarian, language teaching specialist, etc. as specified in, is a new generation professional with a basic linguistic/philological education, prepared to meet the demands of modern technology and science [4].

The study of advertising translation in terms of teaching future linguist-translators is also an item of interest as due to the fair criticism on the qualification and quality of young specialists in recent years, particular pedagogical issues have become of major importance, namely, the question of improving the quality of education, provoking interest towards the future profession and scientific work, self-education [5]. Basically, the professional consciousness of specialists of the future is already being shaped, and the modern system of higher education plays an important role in this process [6]. When discussing this problem, one should also keep in mind that in addition to professional and creative skills future specialists should possess, there exist a problem of psychological adaptation. The problem of psychological adaptation to the extended learning environments becomes significant [7]. However, this issue is not supposed to be discussed in detail in terms of the present article.

A TV-commercial is a complex type of the text where the auditory and visual components are of equal importance. In this regard, while translating commercials the translator works not only with the original and target languages but also with such aspects as the context of the dialogue, sound effects, images and the video sequence as a whole [8]. A modern translator needs to have a desire to work in the environment of multidimensionality. Therefore, one can notice that a sufficient level of readiness for vocational learning ensures the success of mastering the profession and high professional achievements in the future [9]. Thus, the problem of the present study may be presented as follows: to perform the relevance of the methods that ensure the formation of professional skills in future specialists taking into account the development of creative abilities in them.

The solution of the issues stated above may bring us to the purpose of the study which is as follows: to analyze the creative potential of the transcreation strategy used in the process of translation of TV-commercials.

2   Methods

The article describes the peculiarities of translation of TV-commercials from English into Russian in terms of the transcreation strategy. **The material of the study** constitutes 100 English TV-commercials as well as 100 Russian TV-commercials (1 variant of translation for each original sample).

The results of the study may be used in the process of training of future linguist-translators for the formation of professional competences.

The cognitive and discourse research method was used to analyze the changings in the language and culture. This method provides the explanation of semantic variations of the statements as well as interprets the actions which implement the pragmatic component of the text. **The method of scientific observation** was used as empirical in the research. **The communicative and pragmatic method** allowed to study real communicative processes which occur in parallel with the national mentality. The combined application of these methods makes it possible to form the competences necessary for a future translator.

3   Results

While working with TV-commercials, the translator should not only translate the phrases said by the characters of the advertisement, subtitles or the voice-over text but also convey the sense of the original statements, that is to achieve the transmission of the author's intention as the extralinguistic, most often visual element, which has been added to the verbal media text in recent decades, still causes a lot of controversy [10].

In comparison to some other types of translation strategies, the strategy of transcreation often includes the localization of words, video and images in order to make them understandable for the target audience [11]. This process is quite extensive and time consuming, and this, apparently, led to the emergence of the conception which is often described as something more than translation [12]. According to E.D. Malionova, transcreation strategy of translation is a strategy of a creative rethinking of a segment of the original text with the subsequent creation of a new text by means of the target language in view of polymodal and culturally specific context of the work as well as the characteristics of the communicative situation, technical and legal restrictions, the intended reaction of the recipient [13]. Transcreation means the adaptation of the original text in terms of cultural peculiarities of the target country [14].

As vocational training at universities is focused generally on general-theoretical base of professional knowledge and has no highly specialized branch features [15], it is obvious that studying the translation of advertising texts allows the future specialist to obtain practical skills of translation which may be implemented in their full extent in the future professional activity. This idea becomes also vital with respect to the fact that today the strategy of the future creative specialists' training in the system of higher professional education is becoming one of the most important goal for the national education system's development and meets the urgent tasks of training students at higher professional institutions [16]. Thus, the process of teaching translation art faces new challenges of the modern world. The model of teacher education should accumulate several approaches: sociocultural, constructivist, productive, dynamic; each of which ensures the possibility to use diverse teaching practices [17].

The aim of the present study is to demonstrate that the analysis of the features of conveying of the artistic conception of TV-commercials contributes to the formation of the skill of adaptation of a foreign advertising text to a cultural specificity of the target language country. It is obvious that the traditional educational model is not able to create conditions for the modern cultural process, in which the formation of the personality involves not only entering the culture, but also the production of its values. Creativity in this case must be immanent in education, and not an episodic activity [18].

The analysis of the material showed that the biggest amount of TV-commercials constituted the examples which were fully dubbed into Russian (40%). The amount of adverts that included the voice-over text was 35% of the total. As for the other two groups – the commercials containing Russian subtitles and the commercials containing Russian text in the video sequence – their amount constituted 15% and 10% respectively. The percentage of all the examples is given in the Fig. 1.

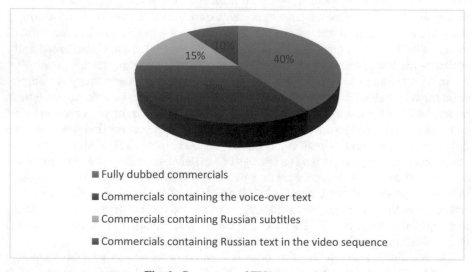

Fig. 1. Percentage of TV-commercials

4   Discussion

To demonstrate the multidimensionality of the transcreation strategy the article analyses in detail some translations of TV-commercials.

The translation of TV-advertisements which contain only English text in the video sequence is supposed to be the simplest type from the transcreational point of view. In this case the translator is not restricted by the rules of subtitling or dubbing. The only requirement is to adjust the written message to the video sequence and to the target culture. The example of such an advert can be the commercial of Apple iPhone:

Download. Upload. Stream. Play. All in 5G. Shoot. Edit. Share in Dolby Vision. Night mode on every camera. Wide. Ultra Wide. Selfie. Super Retina XDR display. A 14 Bionic. The fastest smartphone chip ever. MagSafe. Everything just clicks. All-new Ceramic Shield. 4x better drop performance. Oops resistant. And introducing the world's smallest 5G iPhone, iPhone 12 mini. Just like that. iPhone 12.	Снимайте. Монтируйте. Показывайте видео Dolby Vision. Ночной режим на всех камерах. Широкоугольная. Сверхширокоугольная. Селфи. Дисплей Super Retina XDR. A 14 Bionic. Самый быстрый процессор iPhone. Аксессуары MagSafe. Созданы быть вместе. Ceramic Shield. Риск повреждений при падении в 4 раза ниже. Защита от воды. И еще представляем компактный, но мощный iPhone 12 mini. Вот так-то. iPhone 12.

The above stated example is a TV-commercial which looks like a quick change of video sequence accompanied by the rhythmic music. Each frame contains a dynamic action performed by the actors. The action itself is somehow connected with the new functions of the iPhone advertised in the commercial. In the very center of the frame there appears some text describing the function as well. As one can notice from the example, the text represents itself some short sentences which consist of just one or two words which make them easy to read and memorize while watching the advert.

Technically, it is not problematic; however, some parts of the contents bring difficulties while translating the material. Firstly, the phrase *"Download. Upload. Stream. Play. All in 5G"* is fully omitted in the Russian version of the commercial as there is no 5G on the territory of Russia. This is also the reason of why the element *"5G iPhone"* is excluded from the very end of the Russian advert. If these components were to appear on the screen, they would make the Russian viewers confused. Moreover, they could even cause a negative reaction from the Russian audience: why should one pay more for the functions that are not available?

Secondly, the phrase *"Everything just clicks"* is also localized. All the Apple accessories can be attached to the phone very easy, literary, with "just one click". The English version of commercial is built on pun supported by the "clicking" sounds. The Russian text contains the phrase *«Созданы быть вместе»* which does not reflect the idea described in the original version.

It is notable that the Russian advert has many foreign language inclusions. Their use is aimed at making the commercial "foreign" which reminds the viewers that the product is of a high quality and is known all over the world. In addition, the very fame of the Apple company makes the foreign terms easy to understand.

The next example of a TV-commercial is the advert of cat food "Sheba" which contains the voice-over text. In this case the translator is not restricted by the rules of subtitling or dubbing as well. The text is voiced by a woman as the main characters in the advert are the woman and her cat.

Dancing is my passion // And so is my cat // That's why I give her Sheba // Slogan: Sheba // Follow your passion //	Увлечениям я отдаюсь со всей страстью // Музыка / танцы / моя кошка // Мой выбор для нее / Шеба // Слоган: Шеба // Для страстных натур //

The detailed analysis of the commercial shows that the English version of the advert has a logical chain as follows: "dancing is passion – passion is a cat – a cat is Sheba – Sheba allows you to follow your passion". The Russian commercial has a modified variant of this chain as it contains a sentence «*Увлечением я отдаюсь со всей страстью*» and an additional word «*музыка*» (*"music"*). Thus, the Russian advert has a new synonymic row "passion – music – dancing – cat" which does not appellate directly to the food "Sheba". The authors consider that such localization makes the commercial less motivating and, as a consequence, less effective.

One should note as well that the commercial is starred by Eva Longoria – a famous American actress and producer. Generally speaking, thanks to their famous image celebrities usually provide successful advertising campaigns. However, this is not the case with "Sheba" for the Russian viewers as Eva Longoria is not very popular in Russia. Therefore, the Russian audience is unlikely to remember the advert easily due to the image of the actress – the music is what makes it recognizable.

The next example of a TV-commercial is the advert of "Dior" perfume. The Russian version of broadcasting contains small inclusions of English dialogs in the video sequence that are translated with the help of the Russian subtitles. Such type of transcreation is chosen because the original text is represented in a form of short phrases which are not difficult to put into the subtitles. Moreover, the final phrase of the commercial *"Miss Dior / the new Eau de Parfum // Dior //"* which has the name of the brand and the type of the advertised product is not localized and even not translated into Russian; it is only supported by a subtitle commentary «*Новая парфюмерная вода*».

- I love you! - Prove it! - And you / what would you do for love? Miss Dior / the new Eau de Parfum // Dior //	- Я люблю тебя! - Докажи! - А вы, на что вы готовы ради любви? Новая парфюмерная вода.

It may be possible that such a variant of the Russian commercial creates a more luxurious image of the product. Some Russian consumers automatically associate imported goods with the goods of a high quality. In addition, the English inclusions are so simple that one can understand them intuitively, without any translation, paying more attention to what is happening in the video sequence and listening to music used in the commercial as it is a very famous track "Chandelier" by Sia.

The present analysis of the TV-commercials shows vividly that the future linguist-translator needs both translating and creative skills as the discourse of advertising is a combination of linguistic, marketing and social components which often determine non-standard decisions while translating the advertising texts. The creative skill should be trained in the future linguist-translators as in modern dynamically changing conditions of social and economic development readiness for learning is the basis of professional development of a specialist, a central component, the core of professional culture, which determines the level of professional achievements [19]. Modern approaches to the educational process updating are of particular importance because they help to actualize the role of the learner's personality in the way of achieving educational results and expanding the zone of education influence on the socio-cultural reality of the students' community [20].

The profession of a translator places rather strict requirements on professional training in front of the future specialists. Along with the professional competences which impose on the future specialist the translating requirements according to ethical standards, future professional translators should also have general professional competences. These general skills include the ability of a linguist-translator to solve creative tasks without violating linguistic and cultural norms of the original and the target languages.

References

1. Barinova, D., Ipatov, O., Odinokaya, M., Zhigadlo, V.: Pedagogical assessment of general professional competencies of technical engineers training. In: Katalinic, B. (ed.) 30th DAAAM International Symposium, Vienna, Austria, pp. 0508–0512 (2019). https://doi.org/10.2507/30th.daaam.proceedings.068
2. Odinokaya, M., Andreeva, A., Mikhailova, O., Petrov, M., Pyatnitsky, N.: Modern aspects of the implementation of interactive technologies in a multidisciplinary university. E3S Web Conf. **164**, 12011 (2020). https://doi.org/10.1051/e3sconf/202016412011
3. Efimova, N.N., Semenova, E.M., Sereda, E., Gerasimova, A.S.: Formation of linguist-translator's professional culture while studying American nation linguocultural specifics. In: Chernyavskaya, V., Kuße, H. (eds.) 18th PCSF2018 Professional Culture of the Specialist of the Future, vol. 51, pp. 1662–1671. Future Academy, London (2018). https://doi.org/10.15405/epsbs.2018.12.02.17
4. Rubtsova, A.V., Almazova, N.I., Bylieva, D.S., Krylova, E.A.: Constructive model of multilingual education management in higher school. IOP Conf. Ser. Mater. Sci. Eng. **940**, 012132 (2020). https://doi.org/10.1088/1757-899X/940/1/012132
5. Mezentseva, M.E., Svidzinskaya, G.B.: Semantic differential method as a pedagogical diagnostics tool in university learning process. In: Chernyavskaya, V., Kuße, H. (eds.) European Proceedings of Social & Behavioural Sciences, vol. 51, pp. 528–537. Future Academy, London (2018). https://doi.org/10.15405/epsbs.2018.12.02.57
6. Karmazina, E.V.: Professional culture of the specialist of the future: moral aspects. In: Chernyavskaya, V., Kuße, H. (eds.) European Proceedings of Social & Behavioural Sciences, vol. 51, pp. 803–812 (2018). https://doi.org/10.15405/epsbs.2018.12.02.87
7. Shipunova, O.D., Berezovskaya, I.P., Smolskaia, N.B.: The role of student's self-actualization in adapting to the e-learning environment. In: García-Peñalvo, F.J. (ed.) Proceedings of the Seventh International Conference on Technological Ecosystems for Enhancing Multiculturality (TEEM 2019), pp. 745–750. ACM, New York (2019). https://doi.org/10.1145/3362789.3362884
8. Mujiyanto, Y., Fitriati, S.W.: Multimodality in audio-Verbo-visual translation. KnE Soc. Sci. **3**(18), 747–758. https://doi.org/10.18502/kss.v3i18.476

9. Nizhegorodtseva, N., Zhukova, T., Ledovskaya, T.: Dynamics of readiness for learning of university students. In: Almazova, N.I., Rubtsova, A.V., Bylieva, D.S. (eds.) 19th PCSF 2019 Professional Culture of the Specialist of the Future. European Proceedings of Social & Behavioural Sciences, vol. 73, pp. 314–321. Future Academy, London (2019). https://doi.org/10.15405/epsbs.2019.12.34

10. Lenkova, T.: Three steps to understanding the creolized text. In: Almazova, N.I., Rubtsova, A.V., Bylieva D.S. (eds.) 19th PCSF 2019 Professional Culture of the Specialist of the Future. European Proceedings of Social & Behavioural Sciences, vol. 73, pp. 518–523. Future Academy, London (2019). https://doi.org/10.15405/epsbs.2019.12.55

11. Chaume, F.: Audiovisual translation trends: growing diversity, choice and enhanced localization. In: Esser, A. (ed.) Media Across Borders: Localising TV, Film and Video Games, pp. 68–84. Routledge, New York (2016). https://doi.org/10.4324/9781315749983

12. Pedersen, D.: Exploring the concept of transcreation – transcreation as 'more than translation'? Cultus: J. Intercult. Medit. Commun. Transcreat. Prof. **7**, 57–72 (2014)

13. Malionova, E.D.: Creative practices in translation. In: Translation and Culture: Interaction and Mutual Influence, pp. 53–55. VoGU, Vologda; NGLU, Nizhnij Novgorod (2018). (in Russian)

14. Pedersen, D.: Managing transcreation projects. An ethnographic study. Transl. Spaces **6**(1), 44–61 (2019). https://doi.org/10.1075/ts.6.1.03ped

15. Artyukhina, M., Dorokhova, T., Vyguzova, Y., Nachernaya, S.: Practical oriented training as formation conditions of professional communication. In: Chernyavskaya, V., Kuße, H. (eds.) European Proceedings of Social & Behavioural Sciences, vol. 51, pp. 766–772. Future Academy, London (2018). https://doi.org/10.15405/epsbs.2018.12.02.83

16. Yakovlev, B.P., Stavruk, M.A., Dumova, T.B.: The impact of motivation determination on the university students' creative thinking. In: Chernyavskaya, V., Kuße, H. (eds.) European Proceedings of Social & Behavioural Sciences, vol. 51, pp. 747–753. Future Academy, London (2018). https://doi.org/10.15405/epsbs.2018.12.02.81

17. Almazova, N., Eremin, Y., Kats, N., Rubtsova, A.: Integrative multifunctional model of bilingual teacher education. In: IOP Conference Series: Materials Science and Engineering, Volume 940, International Scientific Conference "Digital Transformation on Manufacturing, Infrastructure and Service" 21–22 November 2019, St. Petersburg, Russian Federation, p. 012134 (2020). https://doi.org/10.1088/1757-899X/940/1/01213

18. Barinova, D., Ipatov, O.: Forming of information culture with tools of electronic didactic materials. In: Katalinic, B. (ed.) Proceedings of the 29th DAAAM International Symposium, pp. 0587–0593. DAAAM, Vienna, Austria (2018). https://doi.org/10.2507/29th.daaam.proceedings.08

19. Nizhegorodtseva, N.V.: Readiness for learning as a component of the professional culture. In: Chernyavskaya, V., Kuße, H. (eds.) European Proceedings of Social & Behavioural Sciences, vol. 51, pp. 823–829. Future Academy, London (2018). https://doi.org/10.15405/epsbs.2018.12.02.89

20. Rubtsova, A.: Socio-linguistic innovations in education: productive implementation of intercultural communication. IOP Conf. Ser.: Mater. Sci. Eng. **497**, 012059 (2019). https://doi.org/10.1088/1757-899X/497/1/012059

Specialized Periodicals in the Science and Technology Transfer System of Germany and the USSR in 1920–1930s

Maxim A. Ganin(✉) 🆔

Peter the Great St. Petersburg Polytechnic University, Polytechnicheskaya 29, 195251 St. Petersburg, Russia

Abstract. This paper discusses one of the communications channels that was used to transfer technology between Germany and Soviet Russia in 1920–30s – periodical science and technology journals. An attempt is made to find out about the premises that contributed to more intensive work in this area, identify concrete science and technology periodicals that appeared in this historical period, consider the specifics of such journals and conclude about the general development trends of this transfer channel. The paper uses the methods of science and technology source studies. The scientific novelty of the research is that for the first time a complex of science and technology journals is analyzed in the context of technology transfer between Germany and the USSR in 1920–1930s and their role as a communication channel for transferring and receiving scientific and technical knowledge is shown. In the course of the detailed examination of journals, such as: Science and Technology Bulletin, German Engineering and Trade Bulletin, a conclusion was made that all of them were a single communications channel for transferring technology between Germany and the USSR. But they did not duplicate each other and carried out the transfer in different fields, given their own specifics. Science and technology journals performed an essential function and provided Soviet scientists and technical experts with up-to-date information about the cutting-edge advances of the Western science and technology, as well as opened access to this knowledge to a wide range of people. As a result, one of the fundamental premises was formed for the development of national science and technology brainpower.

Keywords: Science and technology · Technology transfer · Soviet-German relationships · Science and technology journals · Science and technology bulletin · German engineering · Trade bulletin

1 Introduction

Despite frequents contradictions in various spheres, Russia and Germany had traditionally had close science and technology ties, which were cut with the beginning of World War I in 1914. After the war was over, these ties had to be reestablished as quickly as possible. On April 16, 1922 the Treaty of Rapallo was signed. It helped to resolve some major contradictions between the two countries and opened up the way for closer relationships. A constituent part of Rapallo policy was the development of scientific and cultural contacts beneficial for both parties [1, p. 115].

The contacts in the field of science and technology were a complex, multi-faceted phenomenon. They were carried out using a whole lot of directions (channels), with the following ones worth highlighting: trips of Soviet experts abroad, visits of foreign specialists to the USSR, taking part in foreign fairs and exhibitions, purchasing the necessary equipment and tools abroad, the activities of Soviet-German science and technology institutions, participation of Soviet and German scientists and experts in scientific conferences, subscribing to foreign field-specific literature, etc.

Thus, a technology transfer process took place. In a broader sense it represents an exchange of technologies and engineering knowledge between individuals, enterprises, universities, research centers and governmental bodies at all levels [2, p. 31].

In their article, A. Nordman and D.S. Bylieva note that Russian thought traditionally considered a word as a "technical" or "magic" artefact, capable of changing the world [3, p. 8]. In 1920–1930s, the printed word was a major tool that could transfer scientific and technical information between Germany and the USSR. Journals of economics and technology were one of the channels to transfer technology between our countries. They were aimed at achieving two major goals: informative one (presenting up-to-date data about scientific and technical achievements abroad, as well about the general status of German economy) and communicative one (contributing to the establishment of scientific-technical and economic relations between German companies and scientific institutions with Russian counteragents).

It should be noted that science and technology journals were a part of a large-scale process of the national science and industry development in 1920–1930s. Various aspects of this process are touched upon in the works by contemporary authors. Among them, the works of the following authors are worth mentioning: Alexeev [4], Ulyanova and Sidorchuk [5]. In other chronological frameworks these matters are discussed in the papers by Speransky and Baranova [6], Prischepa and Baranova [7], Prischepa and Kokkonen [8], Sidorchuk [9] and some other authors.

Authors such as Kukulin [10], Prischepa [11], Konysheva [12] study specifically the system of science and technology periodicals. Alexandrov, Dmitriev et al. [13], Vasina and Zaparia [14], Sobolev [15] research the development of science and technology ties between Germany and the USSR in 1920–1930s. However, until now researchers have not considered the complex of science and technology journals as an independent communication channel for technology transfer between these countries.

This paper considers some journals published in 1920-s–1930-s and dedicated to the matters of Soviet-German cooperation in the field of science and technology:

- *Science and Technology Bulletin*(**1929–1937**), a monthly journal of Russian German society "Culture and Technology";
- *German Engineering*(**1925–1941**) – a monthly illustrated journal published by the Institute of Economics for the Relations with Russia and other Eastern European states in Koenigsberg (since 1931 published by the Association of German Engineers jointly with the Institute of Economics for the Relations with the USSR);
- *Trade Bulletin*(**1922–1927**), a weekly journal of the USSR Trade Mission in Germany.

No doubt, the periodicals considered in this article were not the only ones of the kind. In particular, we should mention the Bureau of Foreign Science and Technology of the Science and Technology Department in the Supreme Economic Council in Berlin (BFST), set up in 1921 to "establish relations with German (and Western European) scientists and ensure solid and continuous exchange of new scientific values between Russia and the West" [16, pp. 23–24]. One of the most important activities it was involved in was publishing science and technology journals [16, p. 4]. The following of them are worth highlighting: *Science and Technology News* (in 1921), *Successes of Industrial Engineering* (in 1921–1927) – the title was a subject to change several times: *Technology and Industrial Economy* in 1922, *Collected Papers and Monographs on Science and Technology Matters* from 1923 to 1924 and *Monographs and Collected Papers on Some Technology Matters* in 1926; *Science and Technology* (in 1923–1928) – since 1925 *Successes of Industrial Engineering*; *Chemical Industry and Trade* (in 1923–1927) – since 1924 Chemical Industry.

Although the above BFST journals set a goal to familiarize the readers with the achievements of science and technology in the entire world, the main focus was on Germany, which can be explained, first of all, by the fact that the BFST was based in Berlin, and, secondly, by the close science and technology ties that had formed between the workers of the BFST and German scientists, scientific and non-government organizations, and industrial centers. As we can see from the list of the journals, the publishing activities of the BFST was large scale.

2 Periodical Science and Technology Journals in 1920–30s

2.1 Science and Technology Bulletin

In March 1924, *Culture and Technology*, a joint Russian-German society was set up. Its main purpose "was to comprehensively facilitate rapprochement between the USSR and Germany based on the scientific, technical and cultural interests of both countries". The society was initiated, among others, by Professor A. Einstein, Dr. G. Arko (radio engineer, Technical Director of Telefunken), and K. Matchos (Chairman of the Association of German Engineers). Apart from Professor A. Einstein and Dr. G. Arko, the founders of the Society on the part of Germany were Dr. G. Simon, F. Deutsch, Professors Augagen and Arbeau and some other scientists. The Soviet part was represented by A.I. Rykov, L.B. Krasin, G.M. Krzhizhanovsky, O.Yu. Schmitt, A.P. Sadyrin, L.M. Khinchuk, P.A. Bogdanov, S.A. Levitin, Professor A.N. Dolgov, Engineer L.K. Martens, Academician V.N. Ipatiev. A. Einstein and A.I. Rykov were elected as Honorable Chairmen of the Society during its first meeting [17, p. 515].

According to the Charter, adopted on August 29, 1924, the Society had the following objectives:

– reveal and use cultural enlightening aspects of engineering to the interest of educated masses and to raise the production forces of the RSRSR;
– facilitate the spread of popular scientific and technical knowledge as well practical skills to apply new work methods among the wide masses of employed population in the RSRSR;

– propagandize the most up-to-date technological advances.

In addition, the Society is striving for further spread, development and consolidation of constant close ties between the RSRSR and Germany on the grounds of technology, technical sciences and incidental economic relations.

The Society was an organization that serves exceptionally to ideological and cultural technical purposes. In order to pursue these goals, the society "Culture and Technology" took upon itself the publication of books, reference books, posters, the creation and demonstration of movies and exhibitions, the organization of popular scientific lectures, the establishment of exemplary technical educational institutions [18, p. 224].

The Society contributed actively to the arrival of German scientists to the USSR to give lectures on various problems of science and technology. In January 1929, Moscow and Leningrad saw the "Week of German Engineering", organized by the Society and being a huge success. It was decided to publish *Russian-German Science and Technology Bulletin* in the meetings between German scientists and Professor K. Matchos, Chairman of the Union of German Engineers, which took place during the "Week of German Engineering".

The "Editorial Note" of the first issue of the journal pointed out that the purpose of *the Bulletin* was "to promote the cutting-edge achievements of German science, technology and experience to the industry and agriculture of the USSR". The authors highlighted that the use of up-to-date foreign scientific and technological advances is essential to the interests of "socialist construction" in the USSR. It was emphasized that in this respect Germany is one of the best-developed countries [19, p. 3].

The Bulletin was published since July 1929. Some prominent scientists of that time both from the USSR and from Germany became the members of the editorial board, including: German professors Arco, Kurt Hess, K. Matschos, de Thierry, Franz Fischer (Berlin), Soviet academicians A.N. Bach (Moscow) and A.I. Ioffe (Leningrad), Soviet professors Loleit A.F., Martens L.K., and Pryanishnikov D.N. (Moscow). Professor L.K. Martens was appointed executive editor.

We should note that the first issue of the journal contained a short opening address by A. Einstein. He pointed out that *the Bulletin*, which was published monthly, would allow the widest circles of the public in the USSR and Germany to learn about the acute scientific and technical problems as well as the achievements of both countries [19, p. 3].

The first issue of *the Bulletin* was almost entirely about the German Engineering Week. Eleven German scientists, headed by prof. Von Miller (designer of the first power station in Berlin) arrived in Moscow to take part in the event. The group also included Prof. F. Fischer, who succeeded in obtaining synthetic gasoline, oil and paraffin from coal, K. Matchos, an acknowledged expert in the history of technology and the head of the Association of German Engineers; prof. Hess, known for his studies of fiber and specializing in the chemistry of cellulose, starch and other carbohydrates; prof. K. Wendt, known for the improvement of the smelting processes in cast iron production and for the introduction of the first electrical metal rolling machine; prof. Probst, a top specialist in the field of reinforced concrete structures; prof. L.V. Riess, one of leading agricultural experts in Germany; prof. E. Grimder, specialist in yarn production and finishing of cotton fabrics; engineer E. Zander, a leading expert in agriculture, specialized in field irrigation problems; architect B. Taut, specialist in housing construction [19, p. 94].

As seen, each of these specialists worked in the areas vital for the economy and industry of the Soviet Union. Electrification matters were extremely important given the objectives of the GOELRO plan. Agricultural problems, especially related to the setting up of large agricultural enterprises and increasing mechanization, were dealt with according to the collectivization policy pursued since 1928. Housing construction was essential because due to the rapid industrialization and urbanization, people moved to cities and towns and had to be provided with shelter. Technical innovation in some sectors of industry was predetermined by the need to catch up with the industrially developed capitalist countries. Thus, the basic internal structure of the first part of the journal was determined from the very beginning and included papers written by distinguished German scientists and experts on the matters the Soviet side was interested in.

Its second half, which was considerably smaller in size, included the regular section "Bibliography" (brief information about recent scientific and technical works published in Germany) and a number of additional sections, which could be totally or partially absent from issue to issue. Among the latter, the following ones can be noted: "Chronicle" (included data about recent events in Germany in the field of science and technology); "From the Life of "Culture and Technology Society" (contained information about the events related to the activities of the Society) and "Days of German Engineering" (publications of abstracts of the presentations made by German specialists during their visit to the USSR, and the abstracts of their scientific research materials).

The sections "Around the plants and laboratories of Germany" and "Higher schools of Germany" are worth special attention. They often contained data from the reports on business trips of Russian experts to Germany. Thus, the section "Around the plants and laboratories of Germany" in the issue of April, 1930 contains the data from the reports about the foreign business trips of S.V. Gerbachev, the head of the Physical Chemistry Department of the scientific chemical and pharmaceutical institute NTU of USSR and of M.A. Velikanov, an outstanding hydrologist, Professor of Moscow Higher Technical University [20, p. 50–55]. From time to time the second part of the journal was dedicated to the data about technical fairs, exhibitions and conferences (for instance, a paper by E. Strassberg titled "Technical Fair in Leipzig", published in No. 3–4 in 1930) [21, p. 76–86]. It is important to note that since 1931 virtually all of the above sections disappeared. The only left sections were "Chronicle" ("Information" since 1932) and "Bibliography".

The last change of the journal structure occurred in 1934–1935, when articles and publications were systematized by topic blocks (transport, mechanical engineering, chemistry, construction, etc.).

The activities of the Soviet-German society "Culture and Technology" encouraged the development of science and technology in our country. On July 11, 1934 People's Commissar for Foreign Affairs M.M. Litvinov in his memo to I.V. Stalin wrote: "the Culture and Technology Society has been involving German professors and highly-qualified specialists in making presentations, taking part in conferences and consulting on all sorts of construction matters". At the same time, he highlighted the fact that the work of the Society had allowed significant currency savings, since "German scientists and specialists visited the USSR exclusively on a voluntary basis, that is, without being paid any fee" [22, p. 60].

The most interesting presentations and articles of those German professors and highly-qualified specialists M.M. Litvinov was writing about were published on the pages of "Science and Technology". Thus, Soviet experts had a unique opportunity to promptly obtain data about the recent achievements of Germany and the rest of the world in the field of science and technology and use those data in their own work.

Let us give a concrete example of the above-said. The materials of the Fund of the State Regional Trust of Leningrad Mass Production Plants ("Tremass") contain a correspondence between the rationalization department of the trust and the Soviet-German society "Culture and Technology". A letter, dated February 11, 1931 and addressed by the Society to its member organization "Tremass", announced that in mid-March Dr. V. Klaus, a famous German expert on non-ferrous metals was to visit Leningrad at the invitation of the Society (the program of his lectures was enclosed to the letter). Thus, "Tremass" was asked to advise the Society if the trust enterprises needed to be consulted by this specialist, and also to decide on a list of questions that should be answers in articles by foreign experts [23]. Sometime later, we encounter an extensive paper written by V. Klaus "Special aluminum alloys for casting" in Journal Issue No. 4 (16) for April 1931, in which the declared topic was considered in minute detail [24, p. 76–86].

The Culture and Technology Society was actively developing until the Nazi Party came to power in Germany in 1933. Since 1934, the Society had to virtually cease its main activities [13, p. 186]. Nevertheless, the work on *Science and Technology* continued to 1937. On March 11 of the same year, the Politburo of the Central Committee of the All-Union Communist Party of the Bolsheviks decided that further existence of the Soviet-German society "Culture and Technology" as inexpedient and it was liquidated [25, p. 180].

2.2 Science and Technology Bulletin

In 1920 the Eastern European Fair opened in Konigsberg (today's Kaliningrad). It soon became a recognized trade intermediary between Germany and Eastern European countries [26, p. 234]. The directorate of the fair paid special attention to the USSR, believing that the young Soviet country was the most promising trade partner given its geographical proximity and the complicated political position of the city after it was de facto separated from the rest of Germany by the Danzig Corridor. One of the factors that encouraged the Soviet side to trade with Konigsberg was the active issuing of periodicals in Russian dedicated to the fair itself and to the status of German economy, science and technology.

Since January 1925, the publishing house of the German Eastern European Fair began issuing the journal *German Engineering* [27]. It should be noted that apart from *German Engineering*, it supervised a whole lot of periodicals, such as: *The Market of Eastern Europe, Eastern European Farmer* and *International Forest Market*. In March 1925, the publishing house of the German Eastern European Fair was restructured into the Eastern European Publisher, whose objectives included establishing and facilitating the economic relations between Eastern Prussia and the USSR.

It should be noted that the Konigsberg Economic Institute for Relations with Russia and other Eastern European States was directly involved in the work on *German Engineering*, as well as on all the journals mentioned above. This institute was set up in 1922 with the purpose to improve the contacts with the USSR and it was engaged in intense publishing activities [28].

The journal was published in Russian and was an expanded edition of *Illustrated Technical Review*, which, in turn, was an appendix to *The Market of Eastern Europe* journal. An independent journal dedicated to technical issues was needed because the content of the appendix expended significantly and could no longer cover all the matters Soviet readers were interested in. It was decided to publish the new journal on a monthly basis. Its main purpose was to acquaint the readers with the latest inventions and achievements in the field of German and international engineering [29, p. 35].

Since 1931, *German Engineering* was published in Berlin and the Association of German Engineers took an active part in this work. The association was a very large organization, had about 30,000 members by the end of 1920s and published 10 big periodicals, whose many copies were sent to other countries, including the USSR [19, p. 6]. Attention should be paid to the fact that its chairman was Professor K. Matchos. As noted above, he was one of the most active initiators of the creation of the *Russian-German Bulletin of Science and Technology* and later became a member of the editorial board of this journal [19]. Thus, it can be stated that this organization playcd an cssential role in facilitating scientific and technical transfer between our countries.

German Engineering had a laconic structure. The brief content of the issue was given on the cover. The beginning and the end of each issue included an extensive block of advertisements from German manufacturers (primarily, machine-tool and machine-building plants) and announcements on fairs and exhibitions held in Germany. We should highlight that some of these advertising and informational materials can be found in the archives of some Soviet organizations. Their content and design fully matches those posted on the pages of the journal. So, for example, among the materials of the fund of Leningrad Research Institute No. 13, you can find an invitation to visit the Twelfth German East European Fair in Konigsberg (from 14 to 17 February 1926) and some details about the event [30]. Both leaflets are identical to those in issue No. 12 of 1925 [31]. Thus, the target audience of the advertising block posted in the journal can be tracked. It included not only industrial enterprises but also research centers.

The main part included papers dedicated to the science and technology issues important for the Soviet side. For example, the first issue in 1925 included the following articles that are worth mentioning: "The Krupp turbine locomotive with a condenser", "The Anschlütz automatic ship rudder", "A new oil engine", "A giant rotary machine", "The Kort new power hammer", "Spectral X-ray analysis", "Producing automobile axles", etc. In addition, there are some detailed reports about events that took place like fairs and exhibitions (ibidem: "The technical outcome of a large automobile exhibition in Berlin", etc.) [32]. The end of the issue, as a rule, included the "Bibliography" section.

The journal envisaged getting feedback from its readers. Thus, for example, in issues of 1937 you could find a special tear-off envelope, which could be used to inform the editor about the technical issues of particular interest. In addition, readers wishing to subscribe to the catalogs of German companies were asked to fill out a request form that could be sent to the editors. Such requests seem to have been sent quite often. The editors noted that they received "an enormous number of requests for very extensive and expensive catalogs" and, therefore, recommended to subscribe to them only in case of practical need, having indicated the specialty of the customer and the institution he operates for [32, p. 23, 26].

In general, it can be stated that the target audience of the journal were Soviet engineers, who thus received the information they needed about the latest German technological achievements, and individuals whose were in charge of purchasing the engineering equipment required by national enterprises. As for the German side, the journal, in turn, was interesting from the perspective of the possibility of attracting customers interested in buying the engineering products. The profits from sales were extremely important for the economy of Germany.

New issues of *German Engineering* continued to be published even despite the outbreak of World War II, because our countries continued scientific and technical cooperation in some areas. The Russian National Library has issues right until 1940, while some book auctions offer issues published in 1941. Thus, it can be stated that this transfer channel was open until the very outbreak of the Great Patriotic War.

2.3 Science and Technology Bulletin

Another journal deserving attention is *Trade Bulletin*, issued by the USSR Trade Mission in Germany.

A trade mission is a government agency abroad that ensures the country's interests in foreign economic activities. In Germany, the RSFSR (later the USSR) Trade Mission received an official status in 1921 as a result of signing the Interim Pact between the countries in Berlin. In accordance with clause 12 of the Pact: "The Russian Trade Mission in Germany, as a state trade institution, is the legal representation of the Russian government for performing legal acts in Germany" [33, p. 881]. However, in reality, the trade mission had been actively operating even before this Pact, although it form was somewhat different. Thus, in 1919–1921 "semi-legal ambassador" V.L. Kopp carried out his activities in Germany [34, p. 113]. The Soviet diplomat and the Bureau for the Affairs of Russian Prisoners of War he created were involved in a whole range of very diverse activities, including: representing the interests of Russian prisoners of war and internees; negotiating commercial deals; buying weapons, medicines, consumer goods and household items; organizing the export of purchased goods to Russia, etc. [35, p. 102].

Later on V.L. Kopp was suggested to hold the position of temporary ambassador of the RSFSR in Germany and hand over the trade functions to B.S. Stomonyakov [35, p. 105]. From 1921 to 1925 the latter one officially held the office of the RSFSR Trade Mission in this country [22, p. 29].

On March 1, 1922 the information department of the RSFSR Trade Mission in Germany started to issue *Trade Bulletin* journal in Russian to substitute *Economic Bulletin*, which was published by the Berlin Mission in 1920–21. *Aus der Volkswirtschaft der U.d. S.S.R.* journal in German was initiated in May 1922. It was meant for the German side, which was interested in obtaining up-to-date information about the state of Soviet economy.

Trade Bulletin went out weekly. Its target audience included specialized state institutions, industrial and trade associations, research enterprises, state and public figures. Initially, the journal included only regular reviews of the situation in the world, the economic state of individual countries and markets, information about Russian goods on foreign markets, about the activities of the RSFSR Trade Mission in Germany, as well as bibliography on economic issues [36, p. 19]. However, after a while its content underwent some changes.

There were some subscription forms in a number of issues of 1927 from which we learn that, in addition to all of the above, *Trade Bulletin* included papers and reviews containing practical information on export technology, standardization and sale of national raw materials, quality matters and conditions for purchasing imported goods on German and other Western European markets. It periodically brought out special issues devoted to concrete matters of the national economy (such as radio, import of cars, agriculture, textiles, etc.); paid a lot of attention to fairs and exhibitions, both general and special ones; placed advertisements of large German firms [37].

In this study, we primarily focus on how the journal worked to cover the scientific and technical achievements of Germany and other foreign countries. It should be noted right away that the work was done consistently and started long before 1927. Starting from 1924 *Trade Bulletin* contained a section "Technology and Scientific Organization of Labor". There we can find various articles reviewing the latest advances in foreign science and technology. In particular, those papers conveyed information about the latest types of agricultural machinery [38, p. 18–20], electric welding machines, peat briquetting machines [39, p. 18], etc. There were discussions of electric power, the practice of scientific organization of labor on the German railways [40, p. 23–24], the organization of engineering education in Germany [41, p. 28–29] and many others. Later on the section "Technology and Scientific Organization of Labor" was closed. The papers thematically related to this block were placed directly in the first part of the journal.

From the very first issues, *Trade Bulletin* also contained information on various fairs, exhibitions and events. It could be placed in various ways: in the form of an advertisement [42], as a sub-item "Bureau of Fairs and Exhibitions" in the section "Activities of the USSR Trade Mission in Germany" [43, p. 22] or as an individual article [42, p. 3–4]. By and large, it can be said that this matter was carefully scrutinized, which, on the one hand, is explained by the direct participation of the Berlin Trade Mission in most of the trade events, and, on the other hand, by the desire to bring the latest information on the science and technology matters, important for the world, to Soviet specialists and people in charge.

To sum up, it can be concluded that in the process of its development, *Trade Bulletin* began to cover a whole lot of aspects of scientific and technical cooperation between Germany and the USSR and gradually transformed from an economic review journal into a valuable tool for industrial technology transfer.

On July 14, 1927, the Bulletin of the Berlin Trade Mission ceased to exist as an independent periodical. It merged with *Soviet Trade* journal, published by the central agency of the People's Commissariat of Trade. In the explanatory note, the editors of the journals pointed out that the Berlin Trade Mission had repeatedly raised the question of whether it was inexpedient to publish numerous "Bulletins" by various trade missions and representative offices of economic agencies and suggested combining economic and commercial information in the central agency. The Trade Mission was convinced that this led to parallelism in work, caused inconvenience for the readers who had to subscribe to a number of periodicals, and resulted in unreasonably high currency expenditures [44].

Indeed, in those years quite a few Soviet trade missions published their Bulletins. For example, *Trade Bulletin of the USSR Trade Mission in Austria* (1925–1926), *Trade Bulletin: Monthly Review of the USSR Trade Mission in Turkey* (1925–1929), etc. The centralization of their work seemed to be a natural step aimed at creating an effective and cost-effective tool for global technology transfer.

3 Conclusions

Having considered in detail all three journals, we may conclude that they represented a **single technology transfer channel** between Germany and the USSR. At the same time, they did not duplicate each other, but instead carried out the transfer in various fields given their own specifics.

Thus, *Science and Technology Bulletin* was primarily focused on the German scientific and technical experience available for its possible extrapolation to Soviet reality. It is proven by the reports the journal published on the trips of Soviet specialists to Germany informing about the useful observations they made, as well as the fact that many of the German specialists, whose articles were published in the Bulletin, were invited by specialized organizations to the USSR for work. For example, in October 1930 Prof. B. Taut, who is already mentioned above, was invited to Moscow by Mosstroyobyedinenie [45, p. 40–43]. In the same year he moved to the USSR for doing full-time design work [46, p. 141]. Prof. P.L. Pasternak was invited by the Central Bureau of Foreign Consultations (CBFC) to set up a new civil engineering technical college and reorganize technical teaching in the USSR [47, p. 80], etc.

In turn, *German Engineering* was dedicated to the review of recent technological advances of German engineering. A considerable advertising block of products manufactured by German industrial enterprises that was published in the journal suggests that the primary purpose of the journal was to contribute to increased turnover of German industrial goods in the USSR.

Trade Bulletin described in detail the activities of the USSR Trade Mission in Berlin as well as the general matters of German economy and economic cooperation between our countries, in particular in the field of manufacturing. Moreover, the employees of Berlin Trade Mission, having to do a lot of assignments related to buying from Germany some machines and tools for the needs of Soviet enterprises, obtained valuable knowledge and practical experience, which were reflected in the Bulletin section "Technology and Industrial Engineering".

It should be noted that the above journals, which contained rich scientific and technological material, performed a very important educational function. This communication transfer channel was used to provide Soviet scientists and technical experts with information about the recent achievements of the West in science and technology and granted a wide circle of people with access to that information. As a result, one of the main premises was laid to form and develop the national brainpower in the field of science and technology. It was essential in the time when the national economy had to be recuperated as fast as possible.

It seems that the research of such historical experience is very relevant, because now our country faces the same challenges as back in 1920–1930s. Russian economy has to find new growth drivers, which can be achieved if advanced science and technology brainpower appears. This task can be tackled effectively in the shortest time possible only in case the mechanism of technology transfer is actively used between Russia and advanced countries, including Germany, in all variety of its different aspects.

References

1. Ioffe, A.E.: Soviet-German society "Culture and Technology." Vestnik AN SSSR **5**, 115–118 (1966)
2. Burganova, L.A.: Transfer of technologies in Germany (on the example of the chemical industry). Vestnik ekonomiki, prava i sociologii **4**, 31–35 (2012). (in Russian)
3. Nordmann, A., Bylieva, D.: In the beginning was the word. Technol. Lang. **2**(1), 1–11 (2021). https://doi.org/10.48417/technolang.2021.01.01
4. Alekseev, T.V.: The contribution of the Leningrad industry to the provision of the USSR Navy with radio equipment in the 1920s–1930s. Voenno-istoricheskij zhurnal **9**, 53–59 (2012)
5. Ulyanova, S.B., Sidorchuk, I.V.: The "gloomy past" of St. Petersburg workers in the symbolic space of Leningrad in the 1920s. Voprosy istorii **12**(2), 85–93 (2020). https://doi.org/10.31166/VoprosyIstorii202012Statyi36
6. Speranskij, A.V.: University science in Ural during the Great Patriotic War. Voprosy istorii **12**(3), 274–280 (2019). https://doi.org/10.31166/VoprosyIstorii201912Statyi85
7. Prishchepa, A.S., Baranova, T.A.: Training of industrial and production personnel in the post-war years in the Soviet Union. Voprosy istorii **12**(3), 250–255 (2019). https://doi.org/10.31166/VoprosyIstorii201912Statyi81
8. Kokkonen, E.I., Prishchepa, A.S.: Soviet technical intelligentsia and the scientific and technological revolution (1950–1970). Voprosy istorii **3**, 194–200 (2020). https://doi.org/10.31166/VoprosyIstorii202003Statyi21
9. Sidorchuk, I.V.: The Time the USSR "surpassed" America: Deamericanization as a component of scientific and technological policy in the postwar period. Voprosy istorii **12**(3), 268–273 (2019). https://doi.org/10.31166/VoprosyIstorii201912Statyi84
10. Kukulin, I.: Periodicals for engineers and technicians: Soviet popular science journals and modeling the interests of the late Soviet scientific and technical intelligentsia. Novoe literaturnoe obozrenie **3**(145), 61–85 (2017). (in Russian)
11. Prishchepa, A.S.: Popularization of scientific and technical knowledge in the journal "Technology of Youth" in the second half of the 1950s - mid-1960s. Uchenye zapiski Novgorodskogo gosudarstvennogo universiteta **3**(15), 24 (2018). (in Russian)
12. Konysheva, E.V.: «Abroad»: Coverage of Western Experience in the Soviet Professional Press of the 1920s and 1930s. Academia. Arhitektura i stroitel'stvo **4**, 9–15 (2015). (in Russian)
13. Aleksandrov, D.A., Dmitriev, A.N., Kopelevich, Y.U.H.: Soviet-German Scientific Ties of the Time of the Weimar Republic. Nauka, SPb (2001)
14. Vasina, I.I., Zaparij, V.V.: Scientific, technical and economic cooperation between the Urals and Germany in the 20–30s. XX century. LAP LAMBERT Academic Publishing GmbH & Co., Saarbrucken (2012)

15. Sobolev, V.S.: From the history of ties in the field of scientific and technical cooperation between Germany and the USSR in the second half of the 1920s. Sociologiya nauki i tekhnologij **2**, 60–67 (2015). (in Russian)
16. A brief report on the activities of BINT (Bureau of Foreign Science and Technology of the Scientific and Technical Department of the Supreme Council of the National Economy of the USSR) in Berlin 1921–1924. Berlin (1924)
17. Engel'mejer, P.K.: Culture and technology (chronicle). Tech. Econ. Bull. **5**(7), 515–516 (1925)
18. Gorohov, V.G.: Technology and culture: the emergence of the philosophy of technology and the theory of technical creativity in Russia and Germany in the late 19th - early 20th centuries. Logos, Moscow (2009)
19. Russian-German Bulletin of Science and Technology **1**. Culture and Technology, Moscow, Berlin (1929)
20. Russian-German Bulletin of Science and Technology **2**(4). Culture and Technology, Moscow, Berlin (1930)
21. Russian-German Bulletin of Science and Technology **3–4** (5–6). Culture and Technology, Moscow, Berlin, (1925)
22. Sevost'yanov, G.N.: Moscow - Berlin = Moskau - Berlin: politics and diplomacy of the Kremlin, 1920–1941: collection of documents in 3 volumes, vol. 3. Nauka, Moscow (2011)
23. Central State Archives of St. Petersburg, coll. 1961, aids 10, fol. 217, pp. 8–9. Letter from the Culture and Technology Society with the attachment of V. Klaus's lecture program (1931)
24. Russian-German Bulletin of Science and Technology. Culture and Technology. Moscow, Berlin **4**(16) (1931)
25. Kostyrchenko, G.V.: Stalin's Secret Politics: Power and Anti-Semitism. Mezhdunarodnye otnosheniya, Moskva (2001)
26. Gauze, F.: Koenigsberg in Prussia. The history of a European city. Bitter, Reklinghauzen (1994)
27. German Technique: A Monthly Illustrated Magazine for Eastern Europe. Fair Committee, Konigsberg **1**. Union of German Engineers in conjunction with the Economic Institute for Relations with the USSR, Berlin (1925)
28. German Technique: A Monthly Illustrated Magazine for Eastern Europe. Fair Committee, Konigsberg **3**. Union of German Engineers in conjunction with the Economic Institute for Relations with the USSR, Berlin (1925)
29. Eastern European market. Organ of the German Oriental Fair and Economic Institute for relations with Russia and the bordering Eastern European states, Konigsberg **7**, (1925)
30. Central State Archive of Scientific and Technical Documentation of St. Petersburg, coll. 81, aids 1-1, fol. 78, pp. 97–98 turnover. Information message and invitation to the German East European Fair in Königsberg (1926)
31. German Technique: A Monthly Illustrated Magazine for Eastern Europe. Fair Committee, Konigsberg **12**. Union of German Engineers in conjunction with the Economic Institute for Relations with the USSR, Berlin (1925)
32. German Technique: A Monthly Illustrated Magazine for Eastern Europe. Fair Committee, Konigsberg **1**, Union of German Engineers in conjunction with the Economic Institute for Relations with the USSR, Berlin (1937)
33. Collection of legalizations and orders of the Government for 1921. Administration of the USSR Council of People's Commissars, Moscow (1944)
34. Chernoperov, V. L.: Failures of the foreign trade activities of the mission of V.L. Kopp in Germany (1919–1920). Vestnik Nizhegorodskogo universiteta im. N.I. Lobachevskogo. Mezhdunarodnye otnosheniya, Politologiya, Regionovedenie **1**, 112–122 (2004). (in Russian)
35. Makarenko, P.V.: Victor Kopp: at the origins of Soviet-German cooperation. Izvestiya vysshih uchebnyh zavedenij. Severo-Kavkazskij region. Obshchestvennye nauki **5**(159), 101–106 (2010). (in Russian)

36. Trade Bulletin: Review of the World Economy: Weekly of the USSR Trade Delegation in Germany **50**(92). Information Department of the USSR Trade Delegation in Germany, Berlin (1923)
37. Trade Bulletin: Review of the World Economy: Weekly of the USSR Trade Delegation in Germany, Information Department of the USSR Trade Delegation in Germany, Berlin **13** (1927)
38. Trade Bulletin: Review of the World Economy: Weekly of the USSR Trade Delegation in Germany, Information Department of the USSR Trade Delegation in Germany, Berlin **33** (1924)
39. Trade Bulletin: Review of the World Economy: Weekly of the USSR Trade Delegation in Germany, Information Department of the USSR Trade Delegation in Germany, Berlin **30** (1924)
40. Trade Bulletin: Review of the World Economy: Weekly of the USSR Trade Delegation in Germany **26**. Information Department of the USSR Trade Delegation in Germany, Berlin (1924)
41. Trade Bulletin: Review of the World Economy: Weekly of the USSR Trade Delegation in Germany, **7**. Information Department of the USSR Trade Delegation in Germany, Berlin (1924)
42. Trade Bulletin: Review of the World Economy: Weekly of the USSR Trade Delegation in Germany, Information Department of the USSR Trade Delegation in Germany, Berlin **10/11** (1923)
43. Trade Bulletin: Review of the World Economy: Weekly of the USSR Trade Delegation in Germany **31**. Information Department of the USSR Trade Delegation in Germany, Berlin (1924)
44. Trade Bulletin: Review of the World Economy: Weekly of the USSR Trade Delegation in Germany **26**. Information Department of the USSR Trade Delegation in Germany, Berlin (1927)
45. Russian-German Bulletin of Science and Technology **2**(13). Culture and Technology, Moscow, Berlin (1931)
46. Meerovich, M., Hmel'nickij, D.: The role of foreign architects in the formation of Soviet industrialization. Prostranstvennaya ekonomika **4**, 131–149 (2003). (in Russian)
47. Russian-German Bulletin of Science and Technology **1**(3). Culture and Technology, Moscow, Berlin (1930)

Creativity and Media Culture in Modern Kazakhstan

Saule Barlybayeva[✉] [iD]

Al-Farabi Kazakh National University, Str. Al-Farabi 71, Almaty, Kazakhstan

Abstract. The development of network technologies and new media increases the information flow, improves the quality of socio-economic and spiritual and cultural processes. The mass media are becoming a catalyst for the cultural development of modern society, which forms the civilization of the 21st century. Currently, they won first place in the information impact on the individual and society. Mass media not only translate the existing system of values but also actively form a new cognitive information space in Kazakhstan. With the new media came another media culture that changes our consciousness, our tastes, opens up new media opportunities, and expands new horizons of the media sphere. The digital age has fundamentally changed the media environment. Changes in public consciousness were especially pronounced in the 21st century under the influence of information technology, the development of new media, the Internet, and social networks. The approval of new thinking as one of the conditions for solving global problems of modern society is associated with a reassessment of values, a change in existing spiritual guidelines and established social attitudes under the influence of digital technologies. Socio-technical imaginaries is the way in which different cultures and communities imagine their future in terms of symbolically charged scientific and technical achievements. This research articulates the socio-technical imaginary that grows up around digital technology in Kazakhstan. The 21st century has created a new information space, a new media sphere, where the formation and development of a new media culture of the century is taking place.

Keywords: Information technologies · Creativity · Media culture · Digital society

1 Introduction

The 21st century has come, it brought with it modern media technologies: the Internet, webcasting, mobile applications, social media, the blogosphere [1–3]. Currently, digitalization (digitization) of all aspects of human activity [4–6]. First, it concerns the media of communication.

The purpose of the article is to show the development of media culture in Kazakhstan, modern trends in new media in the era of communication globalization. The article introduces the modern information landscape, the development of communication technologies, and new media in Kazakhstan in the period of globalization and the digital revolution, which are moving to a new level in the context of modernizing public consciousness in the republic and forming a new media culture of the information society.

Modern mass communication is developing during the period of globalization and digitalization [7]. Innovative technologies and globalization mutually reinforce each other, giving acceleration and scope. Modern achievements in the development of the Internet, mobile telephony, cable satellite television, television and radio broadcasting have changed the directions of the development of mass communications, information and landscape of countries and regions. New technical advances of the information revolution are complemented by new mass media capabilities: interactivity, multimedia, personalization of information, globalization, convection, hypertext, speed, mobility, digitalization, multi-functionality, etc. When switching to digital broadcasting, the quality and significance of broadcast content become mass communication channels.

The new media space made it easier to search for various information, acquire knowledge, made it possible to access the funds of libraries, universities, museums. Internet users have turned from simple consumers of information into active recipients, creators and disseminators of information [8–10]. The Internet has gradually come to different parts of the world. Meanings and images created and presented by the media and communications most clearly form a culture, cultural environment. Electronic, audiovisual mass media strongly affect our feelings, emotions, are of great importance in our perception of information.

Pfotenhauer and Jasanoff note that it is impossible to separate the dynamics of innovation from its cultural context and innovation policy is routinely constructed as addressing a collectively felt and publicly diagnosed deficit [11]. Sociotechnical imaginaries are at once products of and instruments of the co-production of science, technology, and society in modernity [12]. Socio-technical imaginaries is the way in which different cultures and communities imagine their future in terms of symbolically charged scientific and technical achievements. This research articulates the socio-technical imaginary that grows up around digital technology in Kazakhstan.

2 Literature Review

In the middle of the twentieth century, researchers began to study the influence of information technology and the media on the development of society. Back in 1948, media researcher Harold D. Lasswell, analyzing the impact of the media, the functions of mass media on the audience [13], among the most important of them, was the activity of mass communication on the transfer of cultural values. At first, the influence of the mass media studied the content of information, the preparation of news, and the specifics of the perception of television viewers were studied. But the impact of mass media on spiritual, cultural and value orientations has not been studied. And only with the development of cinema, radio broadcasting and television, it became necessary to study them as translators of personal and social values [14–16]. Further study of cultural values showed that they permeate the spiritual life of the individual, social groups and the whole society.

Due to the global nature and the inclusiveness of mass communication, the mass media image created by the mass media and the sociocultural reality is the most accessible to the audience. Media culture, media realism culture is a special type of culture of the information society. The material and spiritual components of the information culture are closely interrelated, they cause each other. The image of reality created by the media system in the era of the prevalence of information technologies, according to V.D. Mansurova is a special type of virtual socio-cultural reality - media reality [17].

In the scientific literature, the unambiguous definition of information culture does not exist because of the difference in interdisciplinary methodological approaches. Some scientists define it as informational qualities of an individual, others - as a system and multi-dimensional informational activity. Some researchers consider information culture "as part of a general culture consisting of a fusion of information outlook, information literacy and literacy in the field of information and communication technologies" [18]. And the influence of the development trends of media culture on modern life, on the daily practice of people, becomes every year more relevant and significant.

Sources of research are works, monographs, articles of Kazakhstan and foreign researchers on new information and communication technologies, about globalization, on new media, information culture; conducted sociological research in the form of interviews about the attitude of young people to media culture.

The works of the following scientists tell about the process of globalization and its influence on the cultural development of society: Kelle [19], Colin [20], Barbashin [21], Lich [22], Kryukova [23], Kirillov [24].

3 Methodology

At present, there have been major changes in the media, changes in the communication sphere: in the forms of media ownership, in new platforms, tendencies. The modern world has highlighted the field of electronic media. The essence of the new culture and the new economy, everywhere developing under the influence of information technology is currently being investigated by scientists.

The actual materials of the study are: statistical data, a sociological survey conducted in the form of an interview identify the attitude of the younger generation to the formation of a new media culture. The article analyzes works and articles of Kazakhstani and foreign scientists on the development of media culture in the era of globalization.

The main research methods are: the principles of historical, complex and system analysis, the basic concepts of the development of new media, generalization of socio-political and information-cultural phenomena that act the main factors of mass communication processes that form the media culture of the 21st century.

4 Results and Discussion

Each time has its own culture of relationships between people, in the workplace, in life, which is also characterized by its media culture, which depends on the development of society as a whole, on information technologies and the media. They, in turn, absorb it, absorb it and reflect it in their communication channels. And what is interesting, if in different countries culture and mentality are different, there is a specificity, then media culture smoothest this difference, it becomes standard, universal under the influence of information and communication technologies.

The rapid development of information technologies, new media: the blogosphere, social networks, the transition to digital broadcasting in the XX1 century - necessitated the understanding of cultural and information processes in Kazakhstan. Any communication media is part of the culture. The cultural context is a crucial component of mass communication. Media culture reflects the development of society and depends on the general culture of the society, and at the same time promotes the culture of the individual and society as a whole. The mass media play an important role in this process, developing the cultural and spiritual potential of modern society.

Communicologists, media researchers note that, under the influence of mass communication, the values of the individual, personal human existence have increased and the orientation towards the human community has weakened. Currently, they won first place in the information impact on the individual and society. Mass media not only translate the existing system of values but also actively form a new cognitive information space.

The essence of the new culture and the new economy, which are developing everywhere under the influence of information technology, is currently being investigated by scientists. The potential of ICT and new media is already among the state priorities.

As the President of Kazakhstan N. Nazarbayev noted at the Republican meeting on the issues of digitalization on September 13, 2017: "Over the years of independence, we have been able to become one of the 50 competitive countries in the world. The task now is to enter the 30-Ku, which requires new innovative development from Kazakhstan and accelerated technological renewal. Therefore, at the beginning of 2018, in its Address to the people of Kazakhstan, the Third Modernization was announced, the core of which is digitalization".

State programs are being implemented: "Electronic government", "Informational Kazakhstan-2020", "Digital Kazakhstan".

The main task of the state program "Digital Kazakhstan" is the development of digital technologies, their introduction into government structures, into production, into the business sector, and the social and public sphere. Thanks to digitalization, the Kazakhstani economy should increase by 30%, in monetary terms, it will be more than 2 trillion tenges. It is planned to create about 200 thousand new jobs in the field of e-commerce. Only due to the digitization of customs and tax services, the state budget will receive more than 100 billion tenge (1 $ = 420 tenge) [25].

An important area of this program is the development of the Internet infrastructure in the republic. As the former Minister of Information and Communication of the Republic of Kazakhstan, Dauren Abayev, noted, "now more than 13 million people in the country have access to high-speed Internet, settlements with more than 10,000 thousand people are covered by the 3G standard, and with more than 50,000 thousand - 4G. In 2017, the fourth generation Internet services covered all regional centers of the country. Now the task of the Ministry is to provide access to broadband Internet in rural areas. By 2021, it is planned to cover 1,249 villages with these services, where 2 million inhabitants live [26]. In 2018, the Ministry of Information and Communications of the Republic of Kazakhstan proceeds with the implementation of the project "Construction of fiber-optic communication lines in rural areas of the Republic of Kazakhstan".

For free access to products and services in the sphere of culture and art, historical and cultural objects within the framework of the state program "Digital Kazakhstan", it is planned to create a single National Information Portal "e-culture.kz". In the spring of 2018, the Great Silk Road-2018 expedition was launched, where several research projects were scheduled.

The main task of the Digital Kazakhstan Program is the creation of a digital platform that will increase the competitiveness of economic sectors. For the republic, digitalization is a way to move away from the raw materials economy and move to an industrial-digital economy.

Since January 2018, the new Tax Code has entered into force, which provides great opportunities for the development of Internet commerce. The new code provides tax incentives for entrepreneurs operating through an online store or online platform. In particular, legal entities are exempt from paying corporate income tax, individual entrepreneurs - from individual income tax. Businessmen who trade over the Internet are exempt from these types of taxes subject to the three mandatory conditions: First, payment for goods must be made in a cashless manner; secondly, the contract with the buyer must be concluded online; thirdly, the entrepreneur must have his delivery service, or an agreement with the person who is engaged in transportation, shipment, delivery of goods.

Deputy Prime Minister of Kazakhstan Askar Zhumagaliyev said that the state is taking several serious steps to develop digital technologies in Kazakhstan, including increasing the popularity of digital technologies among the population. He said that from the lower grades Kazakhstan will be taught programming skills. "High school students need to get knowledge of entrepreneurship. So that they understand how to develop production, startups. We will build an understanding of the guys on how to do business. We will introduce IT skills for all specialties in universities, strengthen the quality of education of specialists in the field of information technologies," said Askar Zhumagaliyev [27].

According to experts, about 50% of all existing professions will gradually disappear. Occupations that are currently being reduced include accountants, lawyers, traders, recruiters, analysts, administrative staff, and others. The remaining professions will change significantly.

For the first time in different countries, it is necessary to train specialists for professions that are not yet on the market. There is a formation of a new economic structure - the "knowledge economy". Increasingly qualified are people who can think critically, work effectively in a team, respond quickly to changes, communicate with large data sets, and work with a variety of information.

The southern capital of Kazakhstan, Almaty, is the flagship for the introduction of digitalization and produces half of all IT products in the country. As part of the "smart city" concept, system digital projects are being implemented in the areas of security, public transport, education and health. In Astana, the northern capital of the republic in the summer of 2018, an IT hub (Information Technology Hub) will open. It is planned to launch a new music streaming service and an online platform with video content "video on demand" with an emphasis on domestic content.

Cities - Astana and Almaty are leaders in the practical implementation of Digital Kazakhstan. Here, in aggregate, 85% of programming services and 89.3% of IT-services support are provided. In the field of information services, two directions dominate the placement and processing of data (76.7%) and the creation, maintenance and support of web portals (9.6%). Digitization in Kazakhstan is still a lot of state structures and business. As for the population, only 13.2% of services in the field of programming and 22% of information services were rendered at his expense [28].

Over the last decade, the media consumption of the people of Kazakhstan has changed. On the one hand, the globalization of the information space and the intensive development of new media platforms have significantly expanded the country's media market. On the other hand, multi-vector changes in the socio-cultural image of the media audience. So the older generation prefers traditional media, and in particular, television. As D.Naysbit notes, "the most significant intrusion of technology into a life turned out to be television, which is much livelier and much more time consuming than radio and telephone [29].

Television watching (television viewing) is one of the important characteristics of the culture of everyday life of a modern person; he is preferred especially in the evening hours. Radio listen to most car owners. The younger generation (students, schoolchildren) prefers computers, the Internet, social networks. There is a transformation of the information space, the influence on the basic skills and personal qualities of people. Digital identity - self-identification with the image - is becoming a new psychological norm of modern man.

Convergent culture - economic, technological, social, cultural convergence leads to changes in culture, where the audience, consumers are encouraged to search for new information and participate in all stages of the production of content, content. A "culture of participation" is opposed to the traditional notion of media consumption. In the new system, the traditional division into media producers and consumers is smoothed, levelled, transformed into user members, where the audience appreciates co-creation.

Multimedia, convergence and digitalization (transition to a digital standard) - the main trends of mass communication, are firmly established in the practical activities of the media. "There are articles, columns and photo stories on the radio site, there are sound news and video reports on the site of the newspaper or information agency. Television talk show on the Internet, placed on a multi-screen and supported by a television host blog, multimedia articles with sound and video clips, contextual links, infographics and 3D animation is already a reality [30].

The media sphere has become differentiated by the development of digital technologies, online publications, Internet broadcasting, and mobile applications. For example, "Russian Radio Asia", which broadcasts to 28 cities of Kazakhstan, is the first radio station that can be watched. All live broadcasts - from the "Order Table" to the raffles—are broadcast online on social Facebook.

In the modern information society, where information and communication technologies and virtual communication are of great importance, the transformation of social development and cultural changes are taking place. The translator, transmitter and distributor of them are the media, which themselves are subject to change and incorporate into the public consciousness of people. "Splitting of consciousness comes in the place of ideological integrity, which is no longer an individual worldview, but a consumer ideology, which absolutely everyone should follow" [31]. Consumer benchmarks are introduced into the life benchmarks of people through the mass media, becoming part of popular culture. We can cite some examples of articles, television and radio programs devoted to this topic, among them: "How to become a millionaire", "Who will take a million", "That order", "Fart money", "How to succeed", "How is a restaurant", "Dreams come true", "Laughter with home delivery", etc. Thus, consumer ideology broadcast by popular culture is gradually being introduced.

A new generation has emerged that can be called digital, which cannot imagine itself without a new media reality, for which the Internet is a way of life. As the founder of Hashtag our storied from the UK, Yusuf Omar, at the Eurasian Media Forum on May 24, 2018, in Almaty, noted, "The future of media is content for all users of the Network." The main thing is not to "drown" in such a large massive flow of information; therefore, the use of information technologies must be approached intelligently, selectively and responsibly.

The information resources of the society are now becoming a determining factor in its development, both in the scientific and technical, social and spiritual development of young people. The media is a social institution in the process of personal development. According to the force of influence, according to the degree of influence on the public consciousness of people, the mass media have no equal in public life.

There is a trend of de-massification, from the entire abundance of the mass information flow, it comes to you personally: on your smartphone, phone, gadget, this stream is standard, which comes to many. A book, theatre, ballet, cinema can sometimes not be used with unusual electronic media of mass influence. For success, popularity, you must become a "media" person, gain more likes, posts, users. Sometimes there is a highlighting of the form, not the content; design, sensations, not taste and talent. There is a change in the assessment of thoughtful art - an accessible, dynamic and vivid manifestation, self-expression. Creativity has been replaced by the original show, entertainment.

In art, in culture, in the media of the twentieth century, there was a great spiritual content of their content. Art forms began to be synthesized, complementary. At the beginning of the XXI century - they became more technological, where the main role is played by technologies: information, scientific and technical, and mass communication has become completely different - innovative. In art, in journalism, in life, there is a change of priorities, values, psychology of perception of information, the MMC itself, the human psyche.

Today the audience of the mass media is not just a consumer of various information, but it is also the creator, moderator, and communicator of the media process thanks to digital technologies and the interactivity of new media.

Recently, Schools bloggers, IT camps (IT information technology) have been created, where they will teach the principles of the video camera, the lavaliere microphone, design the YouTube account design. The younger generation often shoots themselves at holidays, at various events, their pets, about each day they live, they prepare video clips, video reviews, showing their talents in dancing, in sports, at school, on vacation, etc.

Among the younger generation, YouTube is very popular; Musical.ly is a social network for creating videos, live broadcasts, and messaging. Through this application, users can create videos with a duration of 15 s to 1 min, choose music for them, soundtracks, use different speed settings (slow, normal, fast, slow) and add pre-installed filters and effects. You can view popular videos, photos and hashtags. For the boys, it will be interesting to get acquainted with Lego Mindstorms EV3, go through the collection and programming of robots, creating a team for them in competitions in rob-football, rob-boiling, rob-golf. In this IT camp, you can learn how to work with Photoshop, Light room, you will be told about the rules for building a frame, will be trained in portrait, landscape, panoramic photography, time-lapse, stop-motion, freeze light.

Despite the latest digital technologies, the most valuable source of human culture has always been - a book. As many say, "readers will be in charge of those sitting at computers." Analyzing the role of mass media in the information process, the impact on the audience, the formation of cultural stereotypes of society, scientists of the University of Pennsylvania USA, headed by Professor J. Gerbner in the second half of the twentieth century identified two key functions: social integration and socialization.

Introducing certain cultural stereotypes into the mass consciousness and forming a certain type of personality, thereby strengthen the existing system of social relations. Previously, these functions were performed by mythology, religion, folklore, and oral folk art. At present, these functions are performed by television, which introduces the mass audience to the world of conventional culture with ideas about life values. The Internet and social networks joined the television in this process today, introducing the audience to the virtual, networked world.

The researchers note that the all-encompassing impact of the mass media contributes to the standardization of thinking, lifestyle, and consumer vision of the world. The properties of the audience itself change over time: the habits of the audience, users of new media change, fashion arises and disappears, attachments to certain media formats.

Recently, increased attention to what is happening in social networks. Life on the Web, in social networks, as experts note, for the younger generation turns into an independent reality, where the formation of a personality takes place. On the one hand, they open up new communication opportunities for interaction and cooperation; on the other hand, can be used to manipulate public consciousness. And here media literacy and fact-checking are very important, which are introduced into the educational process of the journalism faculties of the country's universities.

Let us give another example, the case of a missing girl, who was supported by her mother for several days in a row: how she is resting, how well she is at school, showing a photo, naming her name, school address, class number. After some time, the girl disappears, which again reports her mom on social networks. Such open information on the Web is fraught with such consequences: unhealthy attention from criminal users of social networks. It speaks of the negative aspects of modern information technology.

The young generation of the 21st century, growing up under the conditions of the "third wave", is quickly assimilating the new mass media formats, since they contribute to the individualization and demassification of both the individual and the culture. If in the twentieth-century information was treated as a commodity, then in the twenty-first century it refers to information - as a stimulator of creative forces, applications and searches.

Here is what students of the Faculty of Journalism of Al-Farabi KazNU say about media culture:

– Irina Kovalchuk: "The 21st-century society is a more mobilized generation that loves efficiency and quality. The information environment dictates its own rules, which differ from those that were even a decade earlier. The present society wants to know more and deeper. There are still invariable "scandals", intrigues, investigations - these are topics that society will always savor.

– Mukhtar Maltabarov: "Modern media culture is based on the way of life of modern society. In recent years, the image and life of a consumer society have been promoted. Our values, lifestyle and even behavior are influenced by the mass media, which are now ubiquitous thanks to gadgets, smartphones, and mini-computers.

– Aidana Doronova: "Nowadays media is not only a way to get information, but it is also the most important way to learn about the world around us. Media improve life, develop, and provide many opportunities for implementation. Media has already become part of our culture, which is necessary for life, to support democracy.

– Catherine Leiman: "The trends of modern culture are as accurately reflected in the media content. Media culture no longer brings up, does not teach, it listens to the desires of the consumer and gives exactly the content that will attract attention. Kazakhstani media culture has become liberated, plunged into the era of new media and adheres to foreign standards of broadcasting. The 21st century is an era of active feedback, where the culture reflects the behavior, attitudes and values of the millennial generation".

Currently, a certain type of thinking is being formed under the influence of social networks and instant messengers. We increasingly perceive information at short intervals, in the form of visual images, perception is visualized. We are attracted by catchy, sharp headlines that "catch" and attract attention to them. The term of the news is very short, and the television story is one or two minutes, they are easier to send, "repost", discuss and forget than analyze. Such an audience is easy to manage, manipulate because there is not even time to check the data, the accuracy of the information received. Long news people do not perceive a maximum of 6 paragraphs. The average length of texts written in social networks for 10 years has decreased 6 times. Today, every device has GPS and social networks: you can learn everything about us.

Information security is an important part of national security, the country's information policy because, in the information society, the main commodity is information that influences the adoption of important government decisions, the development of country development strategies. In June 2018, during the meeting of the Parliament of the Republic of Kazakhstan, deputies adopted the Law of the Republic of Kazakhstan "On protecting children from information harmful to their health and development". "One of the topical and important issues discussed recently is the ability of the state to protect its younger generation from information threats of a new type - this is remote control of children, the influence of which increases every year".

A program "Cyber Shield of Kazakhstan" is being created in Kazakhstan. Our country has moved from 82nd place to 40th in the world cybersecurity ranking.

Protection of the information space and its impact on information security is associated with modern real-life challenges and threats that have recently increased throughout the world.

Informational influence on values is a special way of existence and formation of society and culture, and at the same time - a source of many problems that accompany humanity throughout its history. Mass media has become one of the components of the psychosocial environment of humanity and claim - not without reason! -the role of a very powerful factor in the value orientation of society. They belong to the leadership in the field of ideological impact on society and the individual. They became broadcasters of cultural achievements and, undoubtedly, actively influence the acceptance or rejection by the society of certain cultural values [32].

New media: the Internet, mobile telephony, social networks are not only a means of communication, a platform for the development of open civil society but also a vehicle for a new media culture. It manifests itself in the ability to communicate, use information and telecommunication technologies, readiness to perceive, process information, work with Internet resources, sources, have a certain level of knowledge, intelligence, speak languages, orient in network management, and work in a new digital environment. Global networks have become the basis of intellectual interaction and information exchange in all spheres of human activity.

Each period has its own culture of relationships between people, in the workplace, in life, which is also characterized by its media culture, which depends on the development of society as a whole, on information technologies and the media. They, in turn, absorb it, absorb it and reflect it in their communication channels. "The virtual reality of modern culture is not so much an ideal space as an environment of media culture that absorbs a person." [33].

The development of new media space creates new problems, such as information inequality, information security, computer crime, cyber-attacks, manipulation of public consciousness, fake news, reduction of cultural diversity, etc. These threats require international cooperation, cooperation in the information and communication field. And what is interesting, if in different countries culture and mentality are different, there is a specificity, then media culture smoothest this difference, it becomes standard, universal under the influence of information and communication technologies.

Kazakhstan has experienced several stages of industrialization. The first was in the 1930s along with collectivization, after—in the war of the 1940s, when more than 400 different factories and plants were transported to the republic, then in the era of virgin lands and the period before the collapse of the Soviet Union. After independence, Kazakhstan survived the demolition of old industry and economic ties. After the economy grew, the new industrialization of Kazakhstan began. In the new century, income generation comes not only from production but also from the services and applications of digital technology. The country has entered a transitional period to the fourth industrial revolution [34].

The main challenges of the XXI century will be solved not so much in the political and economic spheres, but rather in the field of culture.

5 Conclusion

The mass media form the public consciousness, public opinion, under their influence social and social priorities are formed, we see and feel the world surrounding reality through the eyes of the media. "The virtual reality of modern culture is not so much an ideal space as the medium of media culture absorbing a person" [35]. Media culture not only reflects the life, value, professional, social and social orientations of the individual and people in general but also determines the future of modern society.

The role of media culture in the life of society and man is growing as an important means in understanding the world around it in its socio-political, intellectual, moral, psychological and artistic aspects, as an important factor influencing public opinion, public consciousness and the life values of a person; as a social institution of digital civilization. "The globalization of media culture is the result of the complex processes of the influence of world information relations on the standardization and synchronization of cultural models in various countries of the world" [35].

As noted by researchers Sheila Jasanoff and Sang-Hyun Kim, "….national sociotechnical imaginaries are "collectively imagined forms of social life and social order reflected in the design and fulfillment of nation-specific scientific and/or technological projects" [36].

The idea of modern Kazakhstan is clearly manifested in the state program "Digital Kazakhstan", which reflects an innovative approach in the social and technological sphere and the development of a new media culture under the influence of progressive digital technologies.

New media: the Internet, mobile telephony, social networks are not only a means of communication, a platform for the development of open civil society but also a vehicle for a new media culture. It manifests itself in the ability to communicate, use information and telecommunication technologies, readiness to perceive, process information, work with Internet resources, sources, have a certain level of knowledge, intelligence, speak languages, orient in network management, and work in a new digital environment. Global networks have become the basis of intellectual interaction and information exchange in all spheres of human activity.

The XX1 century has created a new media landscape, where the formation, formation and development of a new media culture of the century take place.

References

1. Boulianne, S.: Twenty years of digital media effects on civic and political participation. Commun. Res. **47**, 947–966 (2020). https://doi.org/10.1177/0093650218808186
2. Xu, K., Liao, T.: Explicating cues: a typology for understanding emerging media technologies. J. Comput. Commun. **25**, 32–43 (2020). https://doi.org/10.1093/jcmc/zmz023
3. Kramsch, C.: The political power of the algorithm. Technol. Lang. **1**, 45–48 (2020). https://doi.org/10.48417/technolang.2020.01.10
4. Shipunova, O.D., Evseeva, L., Pozdeeva, E., Evseev, V.V., Zhabenko, I.: Social and educational environment modeling in future vision: Infosphere tools. E3S Web Conf. **110**, 02011 (2019). https://doi.org/10.1051/e3sconf/201911002011
5. Bylieva, D., Bekirogullari, Z., Lobatyuk, V., Nam, T.: How virtual personal assistants influence children's communication. In: Bylieva, D., Nordmann, A., Shipunova, O., Volkova, V. (eds.) PCSF/CSIS -2020. LNNS, vol. 184, pp. 112–124. Springer, Cham (2021). https://doi.org/10.1007/978-3-030-65857-1_12
6. Tretyakov, I.L.: Information networks and manipulative technologies in the arsenal of extremists. In: Bylieva, D., Nordmann, A., Shipunova, O., Volkova, V. (eds.) PCSF/CSIS -2020. LNNS, vol. 184, pp. 125–135. Springer, Cham (2021). https://doi.org/10.1007/978-3-030-65857-1_13
7. Mellado, C., Georgiou, M., Nah, S.: Advancing journalism and communication research: new concepts, theories, and pathways. J. Mass Commun. Q. **97**, 333–341 (2020). https://doi.org/10.1177/1077699020917204

8. Baradaran Rahimi, F., Boyd, J.E., Levy, R.M., Eiserman, J.R.: New media and space: an empirical study of learning and enjoyment through museum hybrid space. In: IEEE Transactions on Visualization and Computer Graphics. IEEE, New York (2021). https://doi.org/10.1109/TVCG.2020.3043324

9. Oliveira Lopes, R.: Museum curation in the digital age. In: The Future of Creative Work, pp. 123–139. Edward Elgar Publishing, Cheltenham, Northampton (2020). https://doi.org/10.4337/9781839101106.00016

10. Storozheva, S., Mikidenko, N., Dvurechenskaya, N., Strukova, E.: Library smart systems: new opportunities of access to knowledge in online education. In: Bylieva, D., Nordmann, A., Shipunova, O., Volkova, V. (eds.) PCSF/CSIS -2020. LNNS, vol. 184, pp. 274–286. Springer, Cham (2021). https://doi.org/10.1007/978-3-030-65857-1_24

11. Pfotenhauer, S., Jasanoff, S.: Panacea or diagnosis? Imaginaries of innovation and the 'MIT model' in three political cultures. Soc. Stud. Sci. **47**, 783–810 (2017). https://doi.org/10.1177/0306312717706110

12. Jasanoff, S.: Future imperfect: science, technology, and the imaginations of modernity. In: Jasanoff, S., Kim, S.-H. (eds.) Dreamscapes of Modernity, pp. 1–33. University of Chicago Press, Chicago (2015). https://doi.org/10.7208/chicago/9780226276663.003.0001

13. Lasswell, H.D.: The structure and function of communication in society. In: Schramm, W., Roberts, D.F. (eds.) The Process and Effects of Mass Communication, pp. 84–99. University of Illinois Press, Urbana (1971). http://sipa.jlu.edu.cn/__local/E/39/71/4CE63D3C04A10B5795F0108EBE6_A7BC17AA_34AAE.pdf. Accessed 8 May 2021

14. Calcagni, F., Amorim Maia, A.T., Connolly, J.J.T., Langemeyer, J.: Digital co-construction of relational values: understanding the role of social media for sustainability. Sustain. Sci. **14**(5), 1309–1321 (2019). https://doi.org/10.1007/s11625-019-00672-1

15. Serkova, V.: The digital reality: artistic choice. IOP Conf. Ser. Mater. Sci. Eng. **940**, 012154 (2020). https://doi.org/10.1088/1757-899X/940/1/012154

16. Pokrovskaia, N.N., Ababkova, M.Y., Fedorov, D.A.: Educational services for intellectual capital growth or transmission of culture for transfer of knowledge—Consumer satisfaction at St Petersburg Universities. Educ. Sci. **9**, 183 (2019). https://doi.org/10.3390/educsci9030183

17. Mansurova, V.D.: Journalistic worldview as a type of sociocultural reality. Doctoral Dissertation of Philosophy. Altai State University, Barnaul (2003). (in Russian)

18. Gendina, N., Kolkova, N., Starodubova, G., Ulenko, Y.: Formation of Information Culture of Personality. Interregional Center for Library Cooperation, Moscow (2006)

19. Kelle, V.: Globalization processes and the dynamics of culture. Knowl. Underst. Skill **1**, 69–70 (2005)

20. Colin, K.K.: Neo-globalism and culture: new threats to national security. Knowl. Underst. Skill **2**, 104–111 (2005)

21. Barbashin, M.Yu.: Theoretical aspects of culture transformation. South Russ. Rev. **23**, 26–50 (2012). (in Russian)

22. Leach, E.: Culture and Communication: The Logic by Which Symbols are Connected. An Introduction to the use of Structuralist Analysis in Social Anthropology. Cambridge University Press, Cambridge (1981). https://doi.org/10.1017/CBO9780511607684.001

23. Kryukova, N.A.: Media culture and its role in the modern information society. Sci. Bull. **5**(112), 226–228 (2013). (in Russian)

24. Kirillova, N.B.: Media Culture: Theory, History, Practice. Academic Project, Moscow (2008).(in Russian)

25. Digitalization moves the economy. Kazakhstanskaya Pravda **18**, 1 (2018)

26. Nurbergen, A.: On the Digital Wave. Kazakhstanskaya Pravda, 3 (2018). https://www.kazpravda.kz/articles/view/na-trifrovoi-volne. Accessed 8 May 2021

27. What was discussed at the digital forum in Almaty? https://www.inform.kz/kz/o-chem-gov orili-na-cifrovom-forume-v-almaty_a3141750. Accessed 8 Feb 2021
28. Kaminsky, A.: Silicon Steppe. Express K, 1 (2018). https://exk.kz/news/44404/silikonovaia-stiep. Accessed 8 May 2021
29. Naisbitt, D., Naisbitt, J.: Mastering Megatrends: Understanding and Leveraging the Evolving New World. G&D Media, New York (2019)
30. Kachkaeva, A.: Editor's Note, Marginal Notes. Journalism and Convergence: Why and How Traditional Media Are Turning into Multimedia. Focus-Media, Moscow (2010). (in Russian)
31. Ilyin, A.: Subject in the Mass Culture of the Modern Consumer Society (Based on Kitsch Culture): Amfora, Omsk (2010). (in Russian)
32. Volkovsky, N.: One Hundred and Eleven Tales for Journalists. Peter, St. Petersburg (2013). (in Russian)
33. Dictionary of foreign words and expressions. Comp. Zenovich, Olympus, Moscow (2006)
34. Altybaev, A.: Strategic tasks of the Address - 2018: new opportunities. Evening Almaty **8** (2018)
35. Novozhenina, O.: The Internet as a New Reality and a Phenomenon of Modern Civilization. The Influence of the Internet on Consciousness and the Structure of Knowledge. IP RAS, Moscow (2004)
36. Jasanoff, S., Kim, S.-H.: Containing the atom: sociotechnical imaginaries and nuclear regulation in the U.S. and South Korea. Minerva **47**(2), 119–146 (2009)

Art and Technology

"Progress" in Art in Terms of Semiotics of Creativity

Anna I. Demina[⊠] [iD]

Samara National Research University, 34, Moskovskoye Shosse, 443086 Samara, Russia

Abstract. The article gives insight into the universal principles of artistic and technical creativity. General semiotics is our methodology in discovering creative process as the activity that induces the shift of semantic, syntactic or pragmatic rule at the levels of sensory perception, mind or reason, as a result of which the new appears. The research follows the ontological framework of Friedrich Dessauer's concept of the potential cosmos, which is a modern version of Platonism. The concept of creativity as a projective semiosis implies two theses, defended in the article. The first thesis: a work of art is a specific technical object, both incomplete and augmented, incomplete due to the basic intangibility of the artistic language, augmented - in the sense of changed teleological causality. The second theses: the developmental history of literature and the process of changing artistic styles can be described in terms of general semiotics as a shift of a specific semiotic rule.

Keywords: Creativity · Projective semiosis · Invention · Technical object · Incomplete technical object · Augmented technical object · Artistic creativity · Aesthetic object

1 Introduction. Creative Process as Projective Semiotic Activity

A human being becomes a person and a part of culture while learning to operate signs. Every human activity is a sign-related activity. Cognitive activity can be defined as reception and interpretation of signs. It involves the knowledge of a pragmatic rule (distinguishing something as a sign against the background of environment), a semantic rule (correlation of the sign and its meaning) and a syntactic rule (determining the meaning of the sign, that is, its position in the system of other signs). In Kantian tradition (Kant, in his turn, develops the teaching of Nicholas of Cusa), the process of cognition includes three successive stages: perception, mind and reason. At each of these stages, the three specified rules of semiosis are working. However, human activity is not limited to cognition; it has productive and projective types. Moses Kagan identifies five basic types of human activity: cognitive, transformative, value-oriented, communicative, and artistic [1]. The semiotic nature of communicative and value-oriented activities remains beyond the scope of our research.

Using the methodology of general semiotics, the Kantian scheme of cognition and the triact theory by Petr Engelmeyer, we propose to build the theory of creativity as a model of projective semiosis [2–4], namely, to understand any productive activity as the process of creating a new material object, a new object of the mind or a new idea of the reason by transforming or shifting, one (or the sum) of the rules of semiosis in one (or a number) of layers of consciousness. In this case, we can talk about technical activities that create so-called complete technical objects, and about artistic, aesthetic activities that create works of culture that can be called incomplete or augmented technical objects.

Speaking about a complete technical object, we refer to creation of a new device that has not previously existed in reality, solving a certain task, satisfying a certain human need. Such an object is called complete because in the process of its creation it has passed the complete circle of semiosis. Trying to solve a certain problem the inventors organize their perception of reality in a specific way, subordinating the reception to this task - the stage that many psychologists and creative theorists distinguish as the first stage of invention, the purposeful selection of material before the insight. According to the triact theory of Petr Engelmeyer, this is the first act of invention, which is forming the idea, the design of the future technical object [5]. Creativity begins at the point when reception passes into projection. In other words, through intuition (the "reassembly of experience" by human reason), a new idea, a new sign is generated, distinguishable as such due to the work of the pragmatic rule of semiosis. The next step is the second act of creativity - the transition to the level of mind, the construction of a scheme, model, design, determining the place of the sign in the system, building a system of signs. And, finally, the third step that makes the technical object complete is the execution, the embodiment of the scheme in the material environment, the creation of a sensually perceived object, a new meaning as a new object of reality that has not existed before, the fulfillment of the semantic rule of semiosis of sensory perception. These new objects, being embodied and constituting a part of reality, enter as artifacts into the culture, become signs, objects of reception, thus passing the complete semiotic cycle.

2   Methods. Incomplete and Augmented Technical Object

Following Petr Engelmeyer and Alexander Nesterov, we argue that all types of creative processes, including artistic endeavours, are regulated by the laws of technical creativity, because they are also the types of semiotic activity. The specific understanding of technology as an objectifying activity seems to be most relevant, due to its "ability to expediently influence the matter" presenting the "real creativity" [6, pp. 44, 46]. According to Friedrich Dessauer, the technical creative process is characterized by teleology, involves the processing of material (manually or with the help of tools) and is subject to the laws of nature: "…strictly obeying the law of nature, as far as it is known, but beyond the natural datum; always finalistic, purposeful, always first immanent or intramental in the world of thinking and representation of the reflecting spirit and soul, and only then by processing (spiritual and manual) – transferred to the experimentally given world" [7, p. 87]. To create a technical object means to find a pre-established form of solution to the problem in the "fourth kingdom" (the Platonic world of ideas) and to implement it in the material world according to the laws acting there. The first and essential criterion for distinguishing activities is their teleology: what needs are being met and what tasks are being solved. After Kant, Engelmeyer speaks about the four basic human values that determine the spectrum of human activity: truth, goodness, beauty and benefit [5].

Activity aimed at benefit, at solving a certain range of tasks related to the human life, is technical activity; activity aimed at truth is cognitive activity, activity aimed at good is moral and practical activity. And, finally, activity aimed at beauty, at satisfying an aesthetic need, is artistic activity.

There is no doubt that artistic creativity is semiotic activity, where the nature of sign is most evident. Huge amount of research is devoted to the specifics of the artistic sign. The artistic language can be defined in the terminology of Jurij Lotman as a secondary semiotic system ("Since human consciousness is a linguistic consciousness, all types of models built on top of consciousness – including art - can be defined as secondary modelling systems. So, art can be described as a secondary language, and a work of art – as a text in this language" [8, p.22]), or according to Roland Barthes – as a "secondary semiological system" [9]. As noted by Lotman, the signs of art are characterized by an iconic character, "the sign models its content" [8, p. 33]. In other words, the artistic language is characterized by the specifics of semantic and syntactic connections, "the semantization of syntactic elements" [8, p. 34]. The situation when the meaning of an artistic sign is determined by its place in the syntactic system and requires the imagination of the recipient for its "completion" is described by the concepts of "quasi-judgment" by Roman Ingarden, "pseudo-superimposition" by Rudolf Carnap, "empty spaces", "lacunae" of the text by Wolfgang Iser, "autoreferential negation" by Alexander Nesterov [10, pp. 37, 38].

The principal feature of the artistic sign is its intangibility, that is, the absence of the third act as an embodiment in the material environment as an independently existing artificial object that solves a utilitarian task ("A literary text does not enter into a referential connection with the "world", as often happens with phrases from our everyday speech, it "represents" only itself; a literary text is characterized by tautology: it denotes itself" [11]). This is true primarily for fiction, painting, theatre, and music. Stanisław Lem describes this situation through the distinction between a conventional and a fictional figure: "A conventional figure is one whose name (as long as it is "empty") can perform the corresponding functions in the designative plane. The "observer of the apple tree" in my garden – at the moment when I am writing this - is a purely conventional figure, but not a fictional one, because any person who goes to the garden (where there is no one yet) can become such an "observer". On the contrary, a fictional figure (as in a literary work) is such that no meaningful content of its name is possible, either now or ever. There were and won't be any designation of the name "Pan Volodyevsky" from Sienkiewicz's trilogy, "Bohumil" from Dąbrowska's "Nights and Days", or "the Devil" from "Dr. Faustus" by Thomas Mann. The Devil is equal to Pan Volodyevsky as fictional figure, because neither one nor the other can really exist. Fictionality means such a sealing when a name loses any possibility of ever being filled with a designative objectification" [12, p. 311]. The intangibility of art, mostly relating to literature, is denoted by the concept of fiction, and is most clearly manifested in fairy tales, and, in the twentieth century, in various types of fantasy literature, from fantasy to science fiction. A striking example of feasibility problem in science fiction is the "Ware Tetralogy" of Rudy Rucker, where "alla" is an analogue of a magic wand that allows its owner to circumvent the laws of nature and instantly embody any desire. However, even in the fictional world, the magic wand evolves disastrous consequences, serving as an argument in favour of Friedrich

Dessaure's theory of technology: firstly, only those phantasms that conform to the laws of nature can be executed in the form of a technical object; secondly, man-made technical objects have their own power, and the power of technology goes beyond the expected [13]. One of the ways to access the world of pre-set forms of solutions is imagination, fantasy and intuition. In the twentieth century, science fiction itself becomes a form of vision of the possible future variants, some of which can be realized, but this will be more a by-product than the original intention of the authors.

The artistic image works at the level of the mind and reason, without being embodied in a material environment accessible to sensory perception. In this sense, we define a work of art as an incomplete technical object [14], having fundamental impracticability, fictionality, and autoreference. More complex and thus more accurate, in our opinion, is the definition of a work of art as an augmented technical object. The illustrative example of this concept is architecture that combines all the properties of a complete technical object with a well-defined utilitarian function with the properties of an artistic object – an aesthetic function embedded at the level of the design, in its syntax. Sculpture as a form of art also falls under the criteria of a complete technical object, but at the level of design it satisfies the need not utilitarian, inherent in a technical object, but aesthetic. Since the twentieth century, new types of art actively work with ready-made technical objects by re-designating them, removing the utilitarian function and replacing it with an aesthetic one, placing them in an artistic context [15, 16].

Thus, the specificity of an artistic sign (image, symbol) is related to its teleological causation by aesthetic sense. An artistic utterance meaning is always revealed in the act of reception, since it contains "places of uncertainty" that are specified by each recipient. A work of art is an expression of the inexpressible in the act of communication between the author and the recipient. Due to the specifics of the artistic language, its intangibility, which entails freedom from the laws of nature, it is possible to create many alternative worlds that are able to reassemble the recipients' experience when they are getting in contact with each other.

3   Results. Advancement of Technology as Transformation of the Rules of Semiosis

Let us take it as an axiom that every human activity is a semiotic activity, and reality is available to a person only as a semiosis, that is, as a set of rules: pragmatic - recognition of something as a sign against the background of the environment, syntactic - interaction of signs, semantic - correlation of the sign and what it designates. Then, answering the question of how the new is possible, how development in general and human creativity in particular is possible, we come to the conclusion that the new appears as a shift of a particular rule of semiosis at the level of sensory perception, mind or reason. We understand progress as movement and development, that is, the emergence of the new and increasing of complexity. In recent years, quite a large number of authors made attempts to clarify the concept of scientific progress [17–23]. Kulzhanova, Zh.T., Kulzhanova, G.T., Mukhanbetkaliyev, Y.Ye., Kakimzhanova, M.K., and Abdildina, Kh.S, Gurevich, V., Shestakova, I.G. and others write about the current consequences and prospects of technological progress [24–26].

Alexander Nesterov convincingly depicts [27, 28] the technical progress as the development of technical environment through the creation of a sum of artifacts, technical objects, in the sphere of sensory perception (from the first stone tools to the nuclear bomb – all technical objects that meet certain material humans' needs, that is, serving as an extension of their bodies). He calls this phenomenon the first artificial nature. The transition to the second artificial nature begins with the emergence of technical objects that operate in the sphere of human mind, performing intellectual work for it (from a calculator to a neural network). As V. V. Mironov points out, "The necessary operational processing, i.e., the search of huge amounts of information, is no longer available to the human consciousness, the artificial intelligence is doing this" [29, p. 16]. Research in the field of artificial intelligence, attempts to create a strong artificial intelligence have the vector of development towards the third artificial nature – the creation of technical objects working in the field of reason – goal-setting, decision-making, reflection.

Thus, development is a transformation of reception and projection procedures, for which the natural human environment serves as a background that is transformed (there is a shift in the pragmatic rule, a change in the opposition of the background and figure). At the first step, Nature is the background for humanity, where the emergence and formation of culture is possible. When culture becomes the "first artificial nature", it in turn becomes the background, the environment for the development and complication of culture towards the "second artificial nature". The concept of a singularity describes the situation where the second artificial nature becomes the background for the emergence of the third one. At the same time, there is an expansion of the set of rules that a person extracts from reality and applies when creating artificial objects: first within the influence on sensually perceived matter, then in structures of rational thinking, and in the future – in reflection.

At the beginning of the twentieth century, an outstanding Russian philosopher I. I. Lapshin showed that progress of thought is a consistent occurrence of three types of philosophical invention: aphorism, dialogue and system. Aphorism as the first form of reflection arises from folklore, that is, from metaphor. We are dealing with signs that have a fundamentally unobservable entity as their meaning (see, for example, the concepts of water in Thales and fire in Heraclitus). The dialogue generates a dialectical division of thought (the Socratic maieutics), which subsequently forms the idea of the subject. The system as "an ordered set of cognitive or axiological (evaluative) thoughts" [30, p. 164] becomes a condition for the emergence and development of scientific and technical knowledge. Alexander Nesterov sees the environment as a possible fourth philosophical invention [28], which manifests itself in the post-non-classical type of rationality, reflexive theory of management, and general semiotics.

4 Discussion. Transformation of Semiotic Rules as a Way of Developing Art

Let us return to artistic creativity as a specific projective semiotic activity that creates incomplete or augmented technical objects that affect our mind and evoke the aesthetic feeling.

Following the tradition of classical aesthetics, we proceed from the fact that the aesthetic sense, the aesthetic need of a person is universal. One can define aesthetic feeling after Hegel as "the sensuous glow of an idea" ("The beautiful should be defined as a sensuous phenomenon, the sensuous appearance of an idea. In beauty, the sensuous and the objective do not retain any independence, but must renounce the immediacy of their being, since this being is only the existence and objectivity of the concept and is posited as a reality that embodies the concept as being in unity with its objectivity and which therefore embodies the idea itself in this objective existence" [31]). Or, after Kant, it is an experience of "the expediency of nature" [32].

The question arises: if the aesthetic need is universal, then what is the reason for the constant change in artistic styles, the transformation of the artistic language, the emergence of new works of art?

We hypothesize that aesthetic feeling arises first at the level of reception as an ability, as a skill in reference to the perception of nature, as the development of the pragmatic rule of perception of a natural object as the beautiful - an isolated sign that demonstrates harmony, the rules of the universe as a whole. The first works of art arise in the same way as the first technical objects (and are technical objects in a broad sense) by mastering the rules of "natural" nature and their reproduction. This phenomenon is described by the ancient concept of "mimesis" - imitation. Aristotle also notes in Poetics that art imitates not nature as the sum of objects, but rules and principles: "… the task of the poet is not to talk about what happened, but about what could happen, about what is possible in probability or necessity" [33]. Plato also states the imitative nature of art, placing artistic creativity at the lowest rung of the ladder, after craft. If the philosopher is able to know the true essence, the eidos, through pure mathematical knowledge, the ordinary person deals with the shadows of the eidos embodied in the material world, the artisan can, having mastered the rules of matter, reproduce a material object; the artist generates copies of copies, reproducing the shadow of the shadow. Plato illustrates this idea in the "Republic" with an example of three beds: "there are three kinds of beds: one exists in nature itself, and we would recognize it, I think, as the work of God. … The other is the work of a carpenter. … The third is the work of a painter <…> What task does painting set itself each time? Does it seek to reproduce the actual being, or only the appearance? In other words, is painting a reproduction of ghosts or reality? - Ghosts. - So imitative art is far from reality. That's why it seems to me that it can reproduce anything, because it only touches a thing a little, and even then only a ghostly representation of it comes out. It is worth considering whether people were deceived when they met these imitators, whether they noticed, looking at the creations, that such things are far removed from the real being and are easily feasible for someone who does not know the truth, because here the ghosts are created, and not the real thing. If he were truly versed in what he imitates, then, I think, all his efforts would be directed to creation, and not to imitation" [34]. A modern illustration of Plato's thesis can be found in Bioy Casares' novel "The Invention of Morel", in which the hero encounters ghosts – complete copies of people, perfect holograms that are perceived by all the senses. The character of Casares, Morel, invents a machine that can record on film and endlessly reproduce a kind of slice of the sensually perceived world – the entire complex of sensory impressions of any object, including living beings and people. He is driven by a Faustian impulse – to stop time, to make himself and his beloved immortal. However, it turns out that everything that falls under the beam of his machine is soon destroyed and ceases to exist, only a copy, an imitation, a ghost remains.

5 Conclusion

We assume that in art the transition from reception to projection is primary, it happens through intuition as a reassembly of experience and mastering the skill of distinguishing a sign as artistic one, referring to an aesthetic object. The development of art, considered as development of artistic language, involves the increasing complexity and change of its syntax while maintaining the pragmatic skill of distinguishing a sign as an artistic sign. Lotman and Uspensky describe the shift of the pragmatic rule in terms of conventionality in art, which is perceived in the usual state within the framework of a particular cultural tradition as the norm, that is, as a kind of "naturalness". "The very antithesis of "naturalness-conventionality" appears in an era of cultural crises, sharp shifts, when a system is viewed from the outside - through the eyes of another system. That explains the cultural and typological conditionality of the periodically arising desire to turn to art that is not normalized and "strange" from the point of view of the usual norms of convention (for example, children's, archaic, exotic), which is perceived as "natural", and the usual systems of communication connections appear as "abnormal", "unnatural" <...> At the same time, "strange" or alien art, playing a revolutionizing role in the formation of a new artistic norm, can be perceived from the canonical point of view either as more primitive, or as more complicated" [35, pp. 375–376]. In our opinion, any change in the syntax of artistic language is preconditioned precisely by the purpose of training a pragmatic skill for its perception. At the same time, in different periods, either conservative or avant-garde tendencies come to the fore, depending on how far it is necessary to shift the border of perception so that an artistic utterance is being perceived as an artistic sign and generating an aesthetic feeling.

Progress in aesthetics is impossible, aesthetic objects are unchangeable. Following Dessauer, we believe that the development of humanity in all kinds of activities occurs as the development of ways and forms of appeal to the "fourth kingdom" of nature, the potential cosmos containing the full set of the forms of pre-established solutions. Progress in thought can be considered as the development of pragmatic, syntactic and semantic rules for the reception of signs referring to the fourth kingdom. Progress in technology is the application of these rules in order to create new material objects, intellectual objects and reflexive objects (concepts, phantasms) to meet certain needs. From this point, we make a conclusion that progress in art is the application and transformation of the syntactic, semantic, and pragmatic rules of an artistic sign to meet the aesthetic need.

Acknowledgments. The reported study was funded by RFBR, project number 20-011-00462 A.

References

1. Kagan, M.S.: Human Activity. (Experience in Systems Analysis). Politizdat, Moscow (1974). (in Russian)

2. Nesterov, A.Yu., Demina, A.I.: Imagination in the semiotics of creativity. Vestnik Tomskogo gosudarstvennogo universiteta – Tomsk State Uni. J. **460**, 84–89 (2020). (in Russian). https://doi.org/10.17223/15617793/460/10

3. Demina, A.I., Nesterov, A.Yu.: Semiotic approach to analysis of the concept of creativity. In: Prokhorov, S.A. (ed.) Advanced Information Technologies and Scientific Computing (PIT 2020), pp. 429–433. Samara Scientific center of RAS Publishing House, Samara (2020). (in Russian)

4. Nesterov, A.Yu., Demina, A.I.: Tasks of the semiotic theory of creativity. In: Volf, M.N. (ed.) Philosophy, Sociology, Law: Traditions and Perspectives, pp. 38–40. Ofset, Novosibirsk (2020). (in Russian). https://doi.org/10.47850/S.2020.1.10

5. Engelmeyer, P.K.: Creativity Theory. Book House "Librokom", Moscow (2010). (in Russian)

6. Engelmeyer, P.K.: Philosophy of Technology. Lan', St. Petersburg (2013). (in Russian)

7. Dessauer, F.: The Dispute About Technology. Publishing House of the Samara Humanitarian Academy, Samara (2017). (in Russian)

8. Lotman, Yu.M.: The Structure of the Artistic Text. About Art. Iskusstvo-SPb, St. Petersburg (1998). (in Russian)

9. Barthes, R.: Mythologies: The Complete Edition. Farrar, Straus and Giroux, New York (2012)

10. Nesterov, A.Yu.: The problem of defining the concept of "fantastic". Vestnik Tomskogo gosudarstvennogo universiteta – Tomsk State Uni. J. **305**, 35–41 (2007). (in Russian)

11. Todorov, T.: The Fantastic: A Structural Approach to a Literary Genre. Cornell University Press, Cornell (1975)

12. Lem, S.: The Philosophy of Chance. AST, Moscow (2007). (in Russian)

13. Mitcham, C.: Thinking through Technology: The Path between Engineering and Philosophy. University of Chicago Press, Chicago (1994)

14. Nesterov, A.Yu., Demina, A.I.: Artwork as a technical object. Mirgorod **1**(13), 48–74 (2019). (in Russian)

15. Irvin, S., Dodd, J.: In advance of the broken theory: philosophy and contemporary art. J. Aesthetics Art Criticism **75**(4), 375–386 (2017). https://doi.org/10.1111/jaac.12412

16. Lindberg, S.: Liberation of art and technics: artistic responses to Heidegger's call for a dialogue between technics and art. J. Aesthetics Phenomenol. **4**(2), 139–154 (2017). https://doi.org/10.1080/20539320.2017.1396700

17. Bird, A.: What is scientific progress? Noûs **41**(1), 64–89 (2007)

18. Bird, A.: Scientific progress as accumulation of knowledge: a reply to Rowbottom. Stud. Hist. Phil. Sci. **39**(2), 279–281 (2008)

19. Dellsén, F.: Scientific progress: knowledge versus understanding. Stud. Hist. Phil. Sci. **56**, 72–83 (2016)

20. Dellsén, F.: Scientific Progress, Understanding, and Knowledge: Reply to Park. J. Gen. Philos. Sci. **49**(3), 451–459 (2018). https://doi.org/10.1007/s10838-018-9419-y

21. Saatsi, J.: What is theoretical progress of science? Synthese **196**(2), 611–631 (2016). https://doi.org/10.1007/s11229-016-1118-9

22. Park, S.: Does scientific progress consist in increasing knowledge or understanding? J. Gen. Philos. Sci. **48**(4), 569–579 (2017). https://doi.org/10.1007/s10838-017-9363-2

23. Niiniluoto, I.: Optimistic realism about scientific progress. Synthese **194**(9), 3291–3309 (2015). https://doi.org/10.1007/s11229-015-0974-z

24. Kulzhanova, Zh.T., Kulzhanova, G.T., Mukhanbetkaliyev, Y.Ye., Kakimzhanova, M.K., Abdildina, Kh.S.: Impact of technology on modern society – a philosophical analysis of the formation of technogenic environment. Media Watch **11**(III), 537–549 (2020). https://doi.org/10.15655/mw/2020/13082020

25. Gurevich, V.: Technical progress and its consequences the philosophy behind technical progress. In: Protection of Substation Critical Equipment Against Intentional Electromagnetic Threats, pp. 1–24. Wiley, New York (2017). https://doi.org/10.1002/9781119271444

26. Shestakova, I.G.: To the question of the limits of progress: is singularity possible? Vestnik Sankt-Peterburgskogo Universiteta-Filosofiya I Konfliktologiya **34**(3), 391–401 (2018). https://doi.org/10.21638/11701/spbu17.2018.307
27. Nesterov, A.: Technology as semiosis. Technol. Lang. **1**(1), 71–80 (2020). https://doi.org/10.48417/technolang.2020.01.16
28. Nesterov, A.Y.: Clarification of the concept of progress through the semiotics of technology. In: Bylieva, D., Nordmann, A., Shipunova, O., Volkova, V. (eds.) PCSF/CSIS -2020. LNNS, vol. 184, pp. 3–11. Springer, Cham (2021). https://doi.org/10.1007/978-3-030-65857-1_1
29. Mironov, V.V.: Plato and the modern cave of big data. Vestnik of Saint Petersburg University. Phil. Conf. Stud. **35**(1), 4–24 (2019). (in Russian). https://doi.org/10.21638/spbu17.2019.101
30. Lapshin, I.I.: The Philosophy of Invention and Invention in Philosophy. Respublika, Moscow (1999). (in Russian)
31. Hegel, G.W.F.: Aesthetics: Lectures on Fine Art. Oxford University Press, Oxford (1988)
32. Kant. I.: Critique of the Power of Judgment. Cambridge University Press, Cambridge (2000)
33. Aristotle: Poetics (Oxford World's Classics). Oxford University Press, Oxford (2013)
34. Plato: Republic. Penguin Books, New York (2012)
35. Lotman, Yu.M., Uspenskij, B.A.: Conventions in Art. About Art. Iskusstvo-SPb, St. Petersburg (1998). (in Russian)

Transformation of "Alien" Text as a Technology for Generating the New in Russian Drama of the 1990s–2010s

Larisa Gennadyevna Tyutelova[1] (ID), Elena Nikolaevna Sergeeva[1] (ID),
Kseniya Alekseevna Sundukova[1(✉)] (ID), and Daria Dmitrievna Moroseeva[2] (ID)

[1] Samara National Research University, 34, Moskovskoye shosse, Samara 443086, Russia
[2] Osnabrück University, 29, Neuer Grahen, 49074 Osnabrück, Germany
dmoroseeva@uni-osnabrueck.de

Abstract. The essay examines the "alien" text as a technology for generating something new within the concept of dialogic nature of creativity. This technology is regarded as an algorithm that authors use when an "other" appears – another's individual vision and understanding of reality. The study includes the dramatic texts relating to the latest developmental period of an artistic language, defined as the author's period. In particular, we consider the works of Vladimir Sorokin, Lyudmila Ulitskaya, Boris Akunin and Asya Voloshina to be representational for this period. We have described the most frequent techniques of using the "alien" text in postmodern (1990s–2000s) and recent (2010s) drama and found the examples of generating the new with one and the same technique but in absolutely different manner predefined by the author's intention. One method presents the interaction with "other" artistic systems that are fundamentally "alien", in the form of their demolition or disintegration inspired by the author's general sense of life, which allows us to speak about situation of epistemic doubt. Another method involves the "alien" text technology when there is a need to overcome a crisis, to open up a productive dialogue with the "other".

Keywords: Humanitarian technology · "Alien" text · Epistemic doubt · Dialogue · Dramatic practice · Poetics of drama

1   Introduction

For the history of literature, including Russian, an important point is the successive change of three poetic eras: syncretic, traditionalist (or eidetic), individual-author's (the era of artistic modality, according to Broitman [1]). Within these periods, we can talk about dialogical relationship specific for the concept of subject in the artistic work and represented by the scheme "self – the other" (in accordance to M.M. Bakhtin, who calls the "other" that limit, without which the understanding of the boundaries of the "self" is not possible [2]).

If we stick to Bakhtin's concept, then the problem of the "other" (where the "other" is not just unbelonging to me, but capable of seeing and understanding the world differently from the "self") is not characteristic of the syncretism era.

© The Author(s), under exclusive license to Springer Nature Switzerland AG 2022
D. Bylieva and A. Nordmann (Eds.): PCSF 2021, LNNS 345, pp. 376–389, 2022.
https://doi.org/10.1007/978-3-030-89708-6_32

In the traditionalist era, when the boundaries of personality are clearly defined, but (after S.N. Broitman) "non-autonomous involvement of a person with God and the world" [1, p. 256] is fundamentally important, the "alien" text just like a text expressing an individual point of view is also of no interest. It is the question of the author's purposes. The author creating the text turns out to be a translator of some absolute truth belonging to God, whose idea about what is created the artist must reveal. The author's individuality is manifested in the ability to reveal the absolute, but not in its interpretation. At the same time, it is important that a word in the traditionalist era turns out to be identical to what it names. This word is creative, and its energy of creation is not associated with the one who uses this word. Therefore, in relation to the author, any text created in the traditionalist era becomes both "alien" and "the author's". As noted by S.N. Broitman, the text was created with the help of the transpersonal, authoritative, truth-proclaiming word.

But the situation is different with the "alien", who is able to designate the existence of the "other" and his or her vision and understanding, in the era of individual authorship that has revealed the personality in its intrinsic value, singularity and "unique uniqueness" (Bakhtin [3]). The aesthetic orientation towards originality is important for the new poetic era, and therefore, the "alien" arises as a fundamentally different one that does not coincide with my personal individuality. This problem can be solved, if a person is understood through the dual unity of "self - other", where "the other" becomes a condition for understanding the "self".

The word in the individual author's era aims to "eroding" authority. It is inherently dialogical (M.M. Bakhtin), that is, it is focused on someone else's word (the word of the "other") [2].

2 Problems and Hypothesis

In the new poetic era, the area of meaning origin lies in the field of tension between the "self" and the "other," and therefore the formation of meaning is impossible without a dialogue between these two perceptions. Orientation to the "other", in contrast to previous eras, is now becoming an artistic gesture, a type of technology for generating meaning.

Let us stipulate the justification for using the term "technology" in relation to the sphere of spiritual production, which is art. One of the researchers, V.G. Gorokhov, believes that the concept of "technology" is applicable to any kind of human activity, that is, not only to specifically technical activity or to production as everyday consciousness often perceives it. In this sense, technology is a representation of the process of activity, during which its source material is transformed into a result, a product" [4, p. 123] To implement this process, a certain algorithm of action and tools are required.

The algorithm and tools are clearly evidenced by the result of their use. In case of art, that result is the artistic work. Thus, in Russian drama of the late 20th and early 21st centuries, the active interaction of playwrights with other authors through referring to "alien" texts can be regarded as a sign indicating to the use of one technology.

At the same time, a "text, when denoting the creative technology of generating some-thing personal, unique, and therefore new, is regarded as "the minimum unit of speech communication, possessing relative unity (integrity) and relative autonomy (separate-ness)" [5, p. 436]. As a sign of speech, the text is created in artistic practice in order to realize a creative intention, to express the author's sense of life as a whole.

The peculiarity of this experience, which inspires modern works of art, is episte-mological doubt. In literary practice, it is indicated by ironic reflection on all forms of thinking, boundaries and models of cultural consciousness. All values and forms of life are put to the test, go through ironic attitude. One of the technologies for checking values is the technology of using "alien" text. Moreover, not the entire text as complete sign, but any of its parts that exists in the mind of the "other" as a part of the whole is used. Thus, working with elements of an "alien" text is at the same time a work with its integrity, or rather, with the values underlying its integrity.

Drawing on the dramatic practice of the turn of the 21st century, we assume that introduction of an "alien" text as a technology for generating the new - the result of checking the value foundations of life and artistic experience of the past and present - is relevant in moments of personal identity crisis and doubt. This assumption entails the question of quality and essence of the new that is generated through the use of "alien" text. The purpose of this study is to answer the question posed.

3   Research Methods

The research methodology presents a complex combination of various methods and strategies, specific for the material under study, namely, the comparative typological method; systemic-holistic method evaluating the structural links in the text; hermeneutic method for interpretation of these connections relying on their interaction in specific and general contexts, and historical poetics.

The plays of the last decades have been selected as the research material, in which the authors refer to the classical texts of Russian literature. Modern playwrights use the same technology for working with an "alien" text, but similar techniques serve to achieve different artistic goals. For the analysis, we have chosen dramatic texts where the considered connection with the classical tradition is explicated in the title (V. Sorokin "Dostoevsky-trip", 1997; B. Akunin "The Seagull" 2000, A. Voloshina "The Lady with the Dog", 2016, "Gogol's Overcoat", 2016, "Souls of Gogol", 2018) or organizes the plot of the play (L. Ulitskaya "Russian jam", 2003, V. Sorokin "Dysmorphomania", 1990 and "Anniversary", 1993).

4   Results and Discussion

As indicated by Yu. V. Domansky, "the milestone state inherent in calendar boundaries, provokes the search for new formats of existence and, accordingly, the comprehension of art. And of the traditional types of literary creativity (and possibly creativity in general), it is the drama that is the most sensitive of all to such a process of renewal" [6, p. 143]. In modern Russian drama, there is a steady tendency towards genre diversity, polydis-cursiveness and a revision of the traditional categories of "hero", "conflict", etc. [7–11]. All of these phenomena in terms of historical poetics can be considered as attempts to search for the "self" in the process of researching the "other" and entering into contact with it and/or opposition. Very often, such attempts are associated precisely with the use of an "alien" text [12–18]. We will show below different options for using an "alien" text in the plays of chosen authors.

Vladimir Sorokin in his play "Dysmorphomania" (1990) refers to the works of Shakespeare by using the technology of disintegration of Shakespeare's text.

Let us consider a scene of a performance based on Shakespeare's plays, which is given by Sorokin's heroes, patients of a psychiatric clinic. The performance is composed of two Shakespeare's plays ("Hamlet" and "Romeo and Juliet"). In one of the scenes, the heroes meet the ghost of Hamlet's father. More precisely, the ghosts are orderlies of the clinic who pronounce the monologue of the Shadow of Hamlet's father in two voices. The monologue has been translated grammatically from the 1st person to the 3rd and is broken into remarks by the two orderlies (hereinafter, the quotes from plays are given in our translation):

First orderly: He is a spirit, he is your father.
Second orderly: Sentenced to wander at night.
First orderly: And languish in the middle of the fire during the day.
Second orderly: While the sins of his earthly nature.
First orderly: They will not burn out to the ground [19, p. 189].

Thus, one of the brightest and most psychologically tense, "personal" monologues of Shakespeare's tragedy separates from the hero, whose image has been associated with it for many centuries, and any character can utter it. In a similar way, the remark of Mercutio is transmitted to Horatio, Ophelia's replica to Juliet, etc. The characters of the two plays utter other people's lines, thus, we can conclude that the connection between the words and the hero uttering this line has been destroyed. The element that Sorokin manipulates when creating a new text is the actual utterance, cleared of both the context and the subject of speech.

The remarks themselves undergo transformations in two directions.

First, the remarks are reduced when repeated. So, for example, in the final of "Dysmorphomania", the same sequence of lines spoken by Hamlet, Juliet, King, Queen, Nurse, Tybalt and Horatio is repeated five times, but from repetition to repetition the lines are mechanically shortened, this can be demonstrated in the following example:

"Juliet: My prince, how have you been all these days?" - "My prince, how were you?" - "My prince". - "My" [19, p. 195].

Moreover, the lines are distorted. Thus, Hamlet's remark begins with the famous phrase "to be or not to be," as if marking his connection with Shakespeare, and then the hero describes his predicament in a deliberately lowered, vulgar tone: "To be or not to be—these are two questions. What is better for a person, for a normal person - to agree to do everything as it should, or not to agree and not to do it as it should? Do it on the sly, not look, just cry and that's it, or do it loudly, so that everyone can see and say: does he live loudly, does he live widely? Not wanting especially? Just sleep and dream? What kind of dreams? Various? I'm afraid of dreams a little. Sometimes you may dream of something bad …" [19, p. 193].

In the play "Anniversary" the technology of dealing with an "alien" text appears to be less radical. The pretext (in this case it is Chekhov's drama), seemingly, undergoes fewer transformations, but the institution of literature as such is subjected to deconstruction.

The play presents a performance timed to the anniversary of the "Chekhov-Protein Combine", where, according to the writer's idea, the classics are being processed into a kind of "cultural substance". The writer enters into an ironic game with the expression of Ap. Grigoriev "Pushkin is our everything", which has become common place in Russian culture. Sorokin literally realizes Grigoriev's metaphor in relation to Chekhov and the classic Chekhov also becomes "our everything": "The whole decoration, including the smallest details (apples on the table, leaves, etc.) is made from the insides of A.P. Chekhov" [20, p. 131].

Similarly to the "processing" of the physical body of the classical writer, the text of Chekhov's plays is also being processed: the actors on the stage present a performance composed of the replicas of the heroes of different plays by A.P. Chekhov (in the fragment below, there are replicas from the plays "Three Sisters", "Uncle Vanya", "Ivanov" and "The Seagull"):

Irina: Leave for Moscow. Sell the house, end everything here - and to Moscow.

Astrov. No, I don't drink vodka every day. And besides, it's muggy. Nanny, how long have we known each other?

Ivanov: Enough talking nonsense …

Uncle Vanya: Gentlemen, let's drink tea!

Nina: There is little action in our play, just a reading. And in the play, in my opinion, there must certainly be love … [20, p. 131].

For example, from "The Seagull" Sorokin takes Nina's remarks, from "The Cherry Orchard" - the remarks of Firs's servant, omitting the remarks of their interlocutors. At the same time, the author deliberately frees the replica from semantic clutches (Astrov's appeal to the nanny who is not on stage in the quoted fragment). The replica is taken out of the context of the dramatic situation. Sorokin interleaves the remarks, as if making up the fabric of the new text – "Chekhov-protein". The recipient (reader, viewer) is presented with a "Chekhov's play in general": the recipient guesses the lines, realizing them as Chekhov's, but cannot connect them with the context of the situation [more about this: 14]. The final scene of "Anniversary" is a collage of the final remarks of Chekhov's plays (those with which the heroes either leave the stage or end the action of the play).

Existing in the new artistic whole of the "Chekhov-protein text, all the characters in all Chekhov's plays are aesthetically equal, and therefore the choice of characters that gives the impression of being accidental is an aesthetic gesture of the author.

In the center of the play "Anniversary" is the dismemberment of the body: the body of Chekhov - the body of the text - the body of meaning. And more broadly - the body of literature, since in the artistic world there are also "Class-Protein Combines for the Processing of A.S. Pushkins, M.Yu. Lermontovs, I.S. Turgenevs, N.V. Gogols, L.N. Tolstoys, F.M. Dostoevskys and A.P. Chekhovs" [20, p. 127]. Thus, the object of separation and deconstruction can be the work of any author, more broadly – any "alien" text (not only belonging to classic authors, but also taken from a newspaper, as in the play "Dugout"). Moreover, it should be noted that Sorokin likens the creation of a new text to a technological, production process, the material for which is the Russian classics: "We owe our success to the Chekhov Protein Combine named after A.D. Sakharov, who has been supplying us with high-quality products for six years now" [20, p. 129], it is said in the play. High-quality production is the "substance of literature" that Sorokin processes, overcoming the power of literary discourse.

Thus, the technology of working with an "alien" text in V. Sorokin's plays includes such elements as combining a new text from pretext elements, separating a replica from the plot context, as well as from the subject of speech, transforming replicas, creating a "quasi-author's" text.

Another variant of application of the technology of using an "alien" text appears in B. Akunin's play "The Seagull" (2000).

The first act of Akunin's "The Seagull" accurately reproduces the text of the last, fourth act of Chekhov's "The Seagull". However, despite the significant volume of the quoted text, the pretext in its semantic integrity and polysemy is not in demand by Akunin (one can note here the similarity with the technology of working with an "alien" text in Sorokin's plays).

The literal citation of Chekhov's text is at the same time his interpretation: not allowing himself to transform the characters' lines, Akunin works with stage directions, adding new, detailed ones to Chekhov's laconic remarks, and accompanying them almost every line:

Nina (in a trembling voice). My horses are at the gate. Don't see me off, I'll go myself … (Can't stand it, nervous tears.) Gi-give me some water …

Treplev (gives her a drink; he again passes from excitement to distant absent-mindedness, even coldness). Where are you going now?

Nina (knocking her teeth on the glass). To town [21, p. 13].

Let's compare with Chekhov's text:

Nina. My horses are at the gate. Don't see me off, I'll go myself … (Through tears.) Give me some water…

Treplev (gives her a drink). Where are you going now?

Nina. To town [22, p. 58].

In Chekhov's work in this scene the tension that is growing between Treplev and Nina can be interpreted in any way, depending on the will of the director or the perception of the reader. Akunin, however, concretizes the description of the actions and emotions of the heroes with remarks, reducing the freedom of interpretation to a minimum.

The abundance and detail of the remarks bring the text closer to what the reader is used to seeing in a detective story - after all, it is important for the author of a detective story to immerse the reader as much as possible in the specific details. So, the scarf mentioned in the remark, which fell from Nina's shoulder (Chekhov does not have this detail at all), will later be used as evidence against the heroine.

The second act of the play, where the Chekhovian text is no longer present, is a kind of investigative experiment in which eight versions of Treplev's death are worked out - according to the number of main characters. Chekhov's heroes find themselves in the artistic space of a classic detective story, where the subject of aesthetic interest is, first of all, the investigation of a crime. And as a result of this, Chekhov's text within the framework of the new artistic whole of Akunin's play becomes only "preparatory material" for the second action, the investigation. Thus, the "alien" text is included in a new context, the context of the genre, as a result of which its perception changes: Akunin opposes detective motives to Chekhov's motives of actions and words of the heroes, that is, ironically offers readers a different mode of perception of the text.

The ironic play of the modern writer has also manifested into the very gesture of the nomination of the work: "The Seagull" by B. Akunin. Akunin leaves the new text with the name of the pretext, thereby actualizing the conflicting relations between the two texts. However, a genuine dialogue with an "alien" text, as shown above, does not arise in the play. Both the formal (only the last act is relevant) and semantic (all heroes are potential murderers) integrity of the pretext is subjected to ironic doubt.

It can be concluded that in Akunin's "The Seagull" the use of technology for working with an "alien" text is opposed to Chekhov's drama and detective story. A new text is created by including elements of an "alien" text, in which, with the help of some remarks, new semantic accents are placed and the text is embedded in a new genre context.

Although in the plot of L. Ulitskaya's play "Russian jam" (2003) there are many allusions to "The Cherry Orchard", the author actually refers to the text of Chekhov's play only in interludes. Here, the lines from Chekhov's plays are interspersed with the lines of the heroes of Ulitskaya, the ridiculous Moscow intellectuals of the turn of the 21st Century, who, according to the plot of the play, are descendants of Lopakhin and Ranevskaya (the heroes say that Chekhov described their ancestors in his play).

It is important to note an interesting aspect: there is no indication of who exactly utters the line in the interludes, the lines are cut off from the characters, and therefore the words of Chekhov's heroes and their "heirs" (Lyubov Andreevna Ranevskaya and Natalya Ivanovna Dvoryankina, Petya Trofimov and Dudi Lepekhin), are aesthetically and meaningfully equalized.

– We need to call the typewriter repairman! This is a catastrophe! I can not work!
– It is necessary to work! I work like a draft horse! ... <...>
– You live monotonously, you say a lot of unnecessary things!
– The country house is falling apart! Well, really nobody will do anything?
– Should I call a person? Where is Semyon?
– Let Rostislav take care of the house in the end!
– We must stop admiring ourselves! It is necessary to work! You have to work hard!
    [23, p. 127].

Thus, the spiritual worries of Chekhov's heroes find themselves inscribed in the context of everyday problems and concerns of modern people and lose their existential meaning. Pretext elements are present in the new text without transformation, but they are devalued by the context. It is worth pointing out that the interludes contain phrases not only from The Cherry Orchard, but also from other plays of the "Chekhov's Canon": "Three Sisters" and "Uncle Vanya". But this becomes fundamentally unimportant: in this world, "Chekhov in general" turns out to be irrelevant. It is no coincidence that the critic Anna Kislova notes: "Russian jam is a play about Chekhov's death in our minds" [24].

In L. Ulitskaya's play, both the boundaries of Chekhov's images, as well as the boundaries of plots have already been destroyed. Everything is mixed up: in modern times, there is only one characteristic of the world – chaos and nonsense (we are led to this understanding by the final scene of the play, which depicts a catastrophe, the destruction of a house – and of the world).

The clash of different aesthetic positions, the deconstruction of an "alien" text is also characteristic of Asya Voloshina's drama, but the creative tasks of using this technology are completely different here: the chosen pretext is perceived not as an object for destruction and an instrument of postmodern play, but as a fundamentally different statement about the world.

Taking as a basis the classical works of N.V. Gogol ("Dead Souls", "Overcoat"), A.P. Chekhov ("The Lady with the Dog"), L.N. Tolstoy ("The Devil"), Voloshina creates plays "based on" stage production, but this is not just an adaptation of an epic text for the theater. This is the creation of a new author's work, in which the integrity of the depicted world is restored on the basis that the author's idea is acquired through a dialogue with the "other" on the basis of an ironic verification of his position.

In "The Lady with the Dog" (2016), while the plot of Chekhov's story is generally preserved, the principle of duality operates: the heroes of Chekhov's text seem to split into several characters (along with Gurov, Burov and Vurov act and Anna - the Phantom of Anna). What Chekhov considered to be the content of the protagonist's inner world is revealed through the words of his counterparts. At the same time, these characters replace all other characters on the stage, they personify the Yalta and Moscow society. For this, their remarks are made up of various Chekhovian phrases.

In Chekhov: "Her expression, walk, dress, hairstyle told him that she was from a decent society, married, in Yalta for the first time and alone, that she was bored here…" [25, p. 129].

In Voloshina:

BUROV. One. Her expression, walk, dress, hairstyle speak decisively: one.

VUROV. And from a decent society.

BUROV. Well, this, my mommy, is transparent like a cucumber. In Yalta for the first time. Her expression, walk, dress …

GUROV. How does it all speak to you?

BUROV. Don't be mischievous, my mommy. It is old.

VUROV. And she's bored here.

GUROV. Oh, come on! Don't talk to me about boredom! I knew one such woman [26].

Another example is the play "Gogol's Souls" (2018), where "alien" text also influences the formation of the character structure of the play. Bogdan and Selifan, who are called "servants of the proscenium", appear from Gogol's phrase "There is a kind of people known by the name: people are so-so, neither this nor that, neither in the city of Bogdan, nor in the village of Selifan" [27, p.146]. Since in this phrase one of the names coincides with the name of the coachman Chichikov, the second either replaces the name of Gogol's lackey Petrushka, - or means that we have several hypostases of the same image: Selifan - Bogdan, Gurov - Burov - Vurov.

Such a blurring of the boundaries of one image is an articulation of what was already in the classical play of the turn of the 20th Century, in particular - in Chekhov (according to Innokenty Annensky's remark, "three sisters <…> are similar to each other, which seem to be one soul which only took three forms" [28]). Actually, already in the era of Russian Modern, the image of the "other" appears as our guess about him. Therefore, everyone has what should be in everyone. Consequently, on the basis of interaction with the "other", our understanding of this "other" comes to the fore. At one time, this is how "my Pushkin" Tsvetaeva appeared. Now there is "my Chekhov," "my Gogol," "my Tolstoy," etc. Voloshina.

The technology of an "alien" text makes it possible to reveal a number of intertextual connections, which were only indirectly implied in the original work of Chekhov. Thus, the Phantom of Anna appears with the book "Anna Karenina", and Burov and Gurov, repeating the same lines and changing roles, call themselves "Moscow Hamlet" and complain of boredom. Gurov admits: "You are right: I am from Moscow. You always recognize a Muscovite, right? And they are right again: I could have been, perhaps, a professor. I am a philologist by education. And I was once preparing to sing in a private opera. But I quit it and serve in a bank. Also, in a way, a boring story" [26]. These words refer us to Chekhov's feuilleton "in Moscow," to "Ivanov," as well as "Three Sisters," "A Boring Story," which Chekhov would create after "The Lady with the Dog".

As a result, the problem of the open clash of different discourses is solved, as it literally becomes the voices of different characters. The dialogue that took place in Chekhov at a subtext level is explicated and concretized by Voloshina in accordance with the worldview of the author of the play. The playwright offers her own reading of the story of Gurov and Anna. The final remark "They diverge and converge... There is no end to this" [26], combined with the system of repetitions that permeate the series of meetings and partings of the heroes, much more clearly and unambiguously than Chekhov's, emphasizes the impossibility of resolving the current situation and finding the heroes of happiness.

The technology of using an "alien" text as a means of dialogicity includes, first of all, in the circle of "others" with whom the author conducts his conversation and writers. Phrases from their books become a part of replicas or author's directions of Voloshina's play. And this circle can be quite wide. For example, in Gogol's Overcoat (2018) it is stated that not only Gogol's St. Petersburg story, but also A. Bely's Petersburg and Shakespeare's Macbeth will be the pretext [29]. Moreover, the selected Shakespearean material ("bubbles of the earth") in the Russian tradition is closely connected with the poetic cycle of the same name by Alexander Blok. As a result, in "Gogol's Overcoat", due to the "foreign text, Voloshin's dialogue space includes various author's statements, which are actualized in the minds of the reader and interpreter of the "Overcoat" in connection with the concept of "Petersburg text (and this is the difference from "Lady with a Dog", where, as shown above, the clash of different discourses is built on the basis of allusions actually presented in the original Chekhov's text).

The expansion of the circle of authors, with whom the playwright interacts, by Voloshina, can also occur due to the inclusion in their number of those who seek to understand and evaluate the position of the author of the precedent text, hence in "Gogol's Souls" the Entertainer's remarks are a reminder of Pierre from "Invitation to execution" Nabokov:

CONFERENCE. Gogol was the first to see the devil without a mask, - Merezhkovsky writes, - he saw his true face, terrible not for its extraordinary, but ordinary, vulgarity [30].

It is interesting that the theme of the devil also gives rise to the main motive of Voloshina's play, which is ironically developed throughout the entire action, since, probably, it is this theme that she perceives as the main one in Gogol: "Chichikov's main mortal fear is not for himself, but for his future genus; for his family, for his "seed". Disappearing like a ghost is scary. Chichikov's striving for self-multiplication in descendants is precisely the striving of the devil, the most illusory of the ghosts, "towards earthly realism". "Imagine: millions are happy. Millions of mediocre ones, in which, like the sun in the drops of the Pacific Ocean, the single founder of this kingdom is repeated!" [30].

As a result, on the one hand, figures of world culture of different times are included in the circle of those discussing the problems of the world and man. On the other hand, through play practice in its broad sense - acting, speaking in a stranger's voice, addressed to the audience - the author achieves the presentation of his own as a continuation of the stranger, on its basis and through its continuation. This tactic becomes possible due to the fact that the same Gogol helps both the playwright and his audience to understand the role of the "alien" text as a text that generates a special reality. That is why Gogol's remark is heard in "Gogol's Souls":

THIRD SOUL. Scenes and conversations ... In a word, those words that will suddenly pour over some dreamer twenty-year-old boy, when, returning from the theater, he carries in his head a Spanish street, night, a wonderful female image with a guitar and wonderful curls. What is not and what is not dreaming in his head? He is in heaven and has stopped by to visit Schiller - and suddenly fatal words are heard over him like thunder, and he sees that he is back on earth, and even on Hay Square, and even near a tavern, and again went to flaunt before him is life. Life [30].

Voloshina's character who speaks the language of the "other" finally loses his objectivity. He is no longer led by the playwright, the character seems to be creating his own story, leading both the author and the reader like any "other" who tries to see an independently existing personality within him and either see him as someone else:

CONFERENCE. Readers should not be indignant at the author if the persons who have hitherto appeared did not suit his taste; this is Chichikov's fault, here he is a complete master, and wherever he pleases, there we must drag ourselves along [30].

But not only the tasks of expressing one's understanding of the predecessor and oneself are solved when working with an "alien" text. The Voloshina case shows that an "alien" text is also an opportunity for a playwright to express an understanding of modernity. Therefore, Voloshina's heroes live their lives outside the events of Gogol's novel, this is precisely a dramatic, but not traditionally eventful, scene of their interaction. The characters primarily observe and evaluate what they are talking about and intend to do. They exist, as it were, outside the will of Gogol's plot. This is a new plot – Voloshina's plot. Her goal-setting – through the heroes of Gogol, is understood in one way or another by the author, to build an image of her event, to create her own world in which there are precisely her Gogol characters, there are ironic references to Gogol, whose authority allows, through the pictures of the supposedly Gogol novel, characterization of the present:

CONFERENCE. The absurdity thickens. What to do? Gogol [30].

Actually, it is these plays that are also of interest to modern theater. The theater speaks through the author's text about what is happening here and now, and how this is understood by the creators of the play – the playwright and director [31]. That is why, analyzing the Petersburg theater experience, critic Elena Levinskaya notes that "only that which personally excites the creator and he is not afraid to broadcast it" [32].

At the same time, the "alien" text – recognizable to those to whom the statement is directed, helps to establish contact with the audience, to invite them into dialogue. It is the recognition and understanding that are important here, the guarantee of two-way communication as well as irony, which in the case of the author's dramatizations "is designed to bring down the high calmness of the blurred-classical text, to bring it closer to us" [32].

Thus, in Voloshina's interpretation, the texts of Russian classics become a field for a productive dialogue with tradition, in which a certain truth about modernity should be born, the author's vision of ways out of the ideological crisis. This technology can be found in other works in addition to that of Asya Voloshina. Therefore, according to O.V. and T.V. Zhurchevy, filling in the "empty spaces" by the author, "areas of uncertainty of the precedent text that provoke artistic consciousness to fill them in" [33, p. 99] can be seen in Vadim Levanov's trilogy (Death of Firs, Apocalypse from Firs, Slavic Bazaar) and the play "Sakhalin's wife" by E. Gremina. Like Voloshina, "comprehending different texts of Chekhov and individual facts of his biography, traits of his personality, using different receptive strategies, both playwrights, each in their own way, model the logic of the interaction of Russian culture of the 20th century with Chekhov's images" [33, p. 107].

Thus, in a number of modern dramatic texts, the technology of using an "alien" text is used to overcome the crisis, open up the possibility of a productive dialogue with the "other", and create a new statement about modernity.

5   Conclusion

Comparing various phenomena of the dramatic process of recent decades, we have identified a certain similarity in the technology of the authors' work with an "alien" text as a carrier of a different consciousness and perception of the world. Interest in this kind of practice is explained by the laws of the creative process of the era of artistic modality, when the subject is aware of himself in a dialogue with the "other", in this case this "other" is the works of the classics. The elements of this technology are an indication of the source text, the removal of the "alien" word" from the context of the original artistic whole and its inclusion in a new unity.

At the first stage, within the framework of his own text, a modern author-playwright constructs the figure of another, for which he most often turns to the well-known texts of the classics of world and Russian literature (W. Shakespeare, N.V. Gogol, A.P. Chekhov, L.N. Tolstoy, etc.). The designation of the source text can take place in the title of the work. In some cases, the title of the pretext is retained in the title of a derivative text written by a contemporary author – which can be read as a gesture of game assignment. In other cases, only recognizable elements of the title or the name of the author of the pretext are exploited. In a number of cases, recognition of the "alien" occurs due to toponyms, names of characters in a new play, or the use of textbook quotations.

Subsequently, the "alien" text, placed in the context of a new work and torn away from the original integrity, is divided into separate elements. The composition of the heroes changes, remarks can be transmitted from one character of the play to another or borrowed from another source. Causal relationships recorded at the plot level of the original text can be blurred and distorted, sometimes beyond recognition. In another case, on the contrary, there is an explication of meanings that were present in the original text only implicitly, thus an artistic interpretation of the pretext takes place.

Depending on the author's intention, the use of this technology leads to the fixation of the situation of epistemological doubt, or to the search for a way out of it. Modern playwrights use the same technology for working with an "alien" text, but similar techniques serve to achieve different artistic goals.

Research prospects can be (1) analysis of specific manifestations of the technology of working with an "alien" text in individual author's styles; (2) comparison of historical variants of this technology (the specifics of interaction with an "alien" text in romanticism, modernism, postmodernism, etc.). The results obtained in the course of the research can be used to comprehend actual dramatic practice and may interest both researchers and theater critics.

References

1. Broitman, S.N.: Historical Poetics. Russian State University for the Humanities Publ., Moscow (2001)
2. Bahtin, M.M.: The Aesthetics of Verbal Art. Iskusstvo, Moscow (1979)
3. Bahtin, M.M.: Problems of Dostoevsky's Poetics, 4th edn. Sovetskaya Rossiya, Moscow (1979)
4. Gorohov, V.G.: The notion "Technology" in philosophy of technology and the peculiarity of social technologies. Epistemol. Philos. Sci. **28**(2), 110–123 (2011). https://doi.org/10.5840/eps201128235
5. Gindin, S.I.: Text. In: Nikoljukin (ed.) Literary Encyclopedia by Terms and Definitions. Intelwak, Moscow (2001)
6. Domanskii, Yu.V.: Conceptual space of the drama. The today's day. Review of 'Experimental dictionary of modern drama' (Siedlce 2019). RSUH/RGGU Bull. "Literary Theory. Linguist. Cult. Stud." Ser. **2**, 141–148 (2020). https://doi.org/10.28995/2686-7249-2020-2-141-148
7. A chronology of new Russian drama. In: Hanukai, M., Weygandt, S. (eds.) New Russian Drama. An Anthology, pp. xxxiii–xxxiv. Columbia University Press, New York (2019). https://doi.org/10.7312/hanu18510
8. Lyubimtseva-Natalukha, L.: Supertexts in the Russian drama at the turn of the XX–XXI centuries. Izvestia Smolensk State Univ. **1**(49), 73–85 (2020). https://doi.org/10.35785/2072-9464-2020-49-1-73-85
9. Shunikov, V.: Discursive and genre experiments in the contemporary Russian drama. RSUH/RGGU Bull. "Literary Theory. Linguist. Cult. Stud." Ser. **9**(2), 152–160 (2020). https://doi.org/10.28995/2686-7249-2020-9-152-160
10. Maliti, R.: Word manipulation: Ivan Vyrypaev's textual and authorial performativity. Svet Literatury **30**(61), 17–32 (2020). https://doi.org/10.14712/23366729.2020.1.2

11. Yevhenii, V.: Genre conversion in contemporary European drama. Literary Process: Methodol. Names Trends **13** (2019). https://doi.org/10.28925/2412-2475.2019.132

12. Autant-Mathieu, M.C.: Palimpsests of Cexov's drama or the rewriting as a mirror of soviet Union-Russia's mutations. Revue des Etudes Slaves **84**(1–2), 57–69 (2013). https://doi.org/10.4000/res.1069

13. Kowalska, M.: Remake as a form of the dialogue with the classics (Nikolai Gogol's "The Overcoat" as an inspiration in Russian literature in the end of the 20th century and the beginning of the 21st century). Politeja **2**(59), 327–351 (2019). https://doi.org/10.12797/Politeja.16.2019.59.19

14. Maslenkova, N.A., Sergeeva, E.N.: Dialogue with the classics in contemporary Russian culture: Chekhov as a precedent text. In: Europe Reads Chekhov. International Scientific Conference. Reports and Messages, pp. 135–142, University Publishing House "St. St. Cyril and Methodius", Veliko Tarnovo (2012)

15. Meyer-Fraatz, A.: Transfer of Chekhov to another present. Russ. Lit. **69**(1), 97–108 (2011). https://doi.org/10.1016/j.ruslit.2011.02.009

16. Pieczyński, M.: Images of writers classics in modern Russian drama. Vestnik Samara Univ. Hist. Pedagogics Philol. **24**(2), 93–100 (2018). https://doi.org/10.18287/2542-0445-2018-24-2-93-100

17. Shamina, V., Nesmelova, O., Shevchenko, E.: Hamlet revisited: adaptations of Shakespeare in recent Russian drama. Space Cult. India **6**(5), 29–38 (2019). https://doi.org/10.20896/saci.v6i5.422

18. Sikora-Krizhevska, P.: "Someone Else's Word" in contemporary Russian political drama. Slavia Orientalis **67**(4), 691–702 (2018). https://doi.org/10.24425/slo.2018.125407

19. Sorokin, V.: Dysmorphomania. In: Capital. Complete Collection of Plays, pp. 154–198. Zaharov, Moscow (2007)

20. Sorokin, V.: Anniversary. In: Capital. Complete Collection of Plays, pp. 126–144. Zaharov, Moscow (2007)

21. Akunin, B.: The Seagull. Publishing House "Neva", St. Petersburg (2001)

22. Chehov, A.P.: The seagull. In: Complete Works and Letters, vol. 13. The Science, Moscow (1986)

23. Ulitskaya, L.: Cherry jam. In: Cherry Jam and Other, pp. 75–190. Eksmo, Moscow (2008)

24. Kislova, A.: The end of time "Ch". Petersburg Theater Mag. **1**(51) (2008). http://ptj.spb.ru/archive/51/premieres-51/konec-vremenich/

25. Chehov, A.P.: The lady with the dog. In: Complete Works and Letters, vol. 10. The Science, Moscow (1986)

26. Voloshina, A.: The Lady with the Dog. http://mythos.spb.ru/wp-content/uploads/2017/05/Dama_s_sobachkoy-1.doc. Accessed 24 Mar 2021

27. Gogol, N.V.: Dead souls. In: Selected Works in 2 Volumes, vol. 2. Khudozhestvennaja literatura, Moscow (1978)

28. Annenskiy, I.: Mood Drama "Three sisters". http://annenskiy.lit-info.ru/annenskiy/articles/annenskij/kniga-otrazhenij/drama-nastroeniya-tri-sestry.htm. Accessed 24 Mar 2021

29. Voloshina, A.: Gogol's Overcoat. http://mythos.spb.ru/wp-content/uploads/2017/05/Shinel_Gogolya.doc. Accessed 24 Mar 2021

30. Voloshina, A.: Gogol's Souls. http://mythos.spb.ru/wp-content/uploads/2019/05/Dushi_Gogolya.doc. Accessed 24 Mar 2021

31. Tyutelova, L.G.: "Commentator" as an organizer of a communicative event in Asya Voloshina's play "Gogol's Overcoat". Cult. Text **1**(44), 56–66 (2021). https://doi.org/10.37386/2305-4077-2021-1-56-66

32. Levinskaja, E.: Where is it - modernity? Petersburg Theater Mag. **2**(60) (2010). http://ptj.spb. ru/archive/60/vstrechi-v-rossii-60/v-chem-ona-sovremennost/. Accessed 24 Mar 2021
33. Zhurcheva, O.V., Zhurcheva, T.V.: Creative reception of Anton Chekhov's works and personality in the newest Russian drama (Gremina's Sakhalin wife and Llevanov's Chekhov trilogy). Acta Universitatis Lodziensis. Folia Litteraria Rossica **1**(9), 97–108 (2019). https://doi.org/ 10.18778/1427-9681.09.09

Composition and Symmetries - Computational Analysis of Fine-Art Aesthetics

Olga A. Zhuravleva[1] , Andrei V. Komarov[1] , Denis A. Zherdev[1,2] ,
Natalie B. Savkhalova[3] , Anna I. Demina[1] , Eckart Michaelsen[4] ,
Artem V. Nikonorov[1,2] , and Alexander Yu. Nesterov[1(✉)]

[1] Samara National Research University, Moskovskoye Shosse 34, 443086 Samara, Russia
komarov@timilon.ru
[2] IPSI RAS – Branch of the FSRC "Crystallography and Photonics" RAS, Molodogvardeyskaya 151, 443001 Samara, Russia
[3] International Public Organization "Center of Spiritual Culture", Michurina St. 23, 443110 Samara, Russia
[4] Fraunhofer-IOSB, Gutleuthausstr. 1, 76275 Ettlingen, Germany
eckart.michaelsen@iosb.fraunhofer.de

Abstract. This article deals with the problem of quantitative research of the aesthetic content of the fine-art object. The paper states that a fine-art object is a conceptually formed sequence of signs, and its composition is a structural form, that can be measured using mathematical models. The main approach is based on the perception of the formal order as a determinant of the aesthetic category of beauty. The composition of the image is directly related to the formation of aesthetic sensations and values, since it performs the function of controlling the viewer's perception of a work of art. The research is based on the studies of computational aesthetics by G. D. Birkhoff and M. Bense, as well as the studies of the receptive aesthetics of R. Ingarden, W. Iser, H. R. Jauss and Ya. Mukarzhovsky. The computational aesthetics methods, such as CNN-based object detectors, and gestalt-based symmetry analysis, are used to detect symmetry axes in fine-art images. Experimental analysis demonstrates that the applied computational approach is consistent with the philosophical analysis and the expert evaluations of the fine-art images, therefore it allows to obtain more detailed fine-art paintings description.

Keywords: Computational aesthetics · Composition · Symmetry · Neural networks · Object detection · Fine-art paintings

1 Introduction

The problem of the beauty interpretation has always been the interdisciplinary subject of different sciences due to the numerous reasons an art object may have been characterized as referring to the sphere of aesthetics: positive ontological essence, the initial impression and emotions of the viewer, the special effect produced by the fine-art painting on the human brain (the image of the mental camera actualizing via cortex cerebri), the structural organization of the beauty, etc.

As a philosophical and aesthetic paradigm, the beauty emerged in ancient Greece and before the beginning of the twentieth century it was perceived as a central and the only properly aesthetic category related to an idea of embodiment of a positive ontological being in a subject. Before the eighteenth century, in the European philosophy the beauty was perceived as a sensual visibility of an idea, the truth in its structural expression, and was associated with such notions as moderation, harmony, symmetry, order (ordo). At the boundary of the eighteenth-nineteenth centuries, the aesthetics as a theory of beauty evolved in the philosophy of I. Kant and G. W. F. Hegel. Within the context of their concepts the problem of correlation of the beauty with the contents and the subject's form became central and received two resolution tendencies. According to the aesthetics of form, the beauty is put down to the subject by its form and represents a "consistency of the diversity in the whole" [1, p. 96], "the infinite expressed in finite" [2, p. 479]. According to the Hegel's aesthetics of content, the beauty and the truth are equivalent, as the contents and the form should be and the beauty is the materially shaped spiritual substance available for the perception, and therefore the purpose of the art is to "reveal the divine and raise it to the level of consciousness" [3, p. 4], "discover the truth in the spiritual form" [4, p. 61]. In the art, which is intended to organically synthesize the form and the context, the truth is not capable to shape up through the accidental matter: a formal order is the mandatory condition for the embodiment of the positive ideal content.

In pictorial art, the order is implemented in the compositional principle. The composition is the structural form of an artistic work, connecting the image constituents – figures, subjects, spots, colors, light and shadows – in plain of the pictorial representation and formally expressing the content. The composition represents an artistically-imaginative integrity – a complex structure based on arrangement and mutual correlation of the depicted units. Semiotically, an artistic text represents some conceptually formed sequence of signs, where the composition plays the role of syntax and is built as a system of positions – mutual appositionality of functionally and structurally dissimilar dominants. As a result of work of the author's creative mind, the composition is based on two types of sign relations – a syntagmatic one and a paradigmatic one, respectively, an artistic work emerges as a result of artistically-imaginative reinterpretation of the contensively-formal integrity of text constituents.

An artistic work operates as a sign, since it is represented as a medium, refers to some object(s) and implements an interpretational relation at the same time. In such a way, in the pictorial art the line is a basic constituent of an image and serves for forming a contour of anything. The contour, while being filled with color, turns into a cell; a multitude of such cells forms an image system. Each cell of an image is correlated with an object, i.e., it expresses such a thing which expresses itself by something and can be subjected to interpretation by itself. For example, some red shape in the system of a pictorial representation, where the color acts as a signal of some frequency, may be interpreted as an apple with respect to a particular object. Further, this apple, depending on the recipient and the context, may be interpreted as a symbol of wisdom, love or death. Objectively, there is no such an attribute according to which an apple would relate to these phenomena, however it can be suggested that such a form has a number of characteristics of a particular quality (color), which represent signals contributing to the fact of acquiring its value (including a connotative one) by an apple in the interpretation act.

The problem of composition of a pictorial representation and its perception by a viewer is a problem of correlation between the significant and the significatum. The text as an integral structure is based on the capacity of its constituents for valency and setting both implicit connections between themselves and explicit connections between the text and the recipient. According to the point of view of a receptive aesthetics, an aesthetic subject is subjective and is freely created by the mind [5, pp. 133–135] possesses no priory aesthetic properties before being in contact with mind [6], and its perception depends both on a variety of non-aesthetic factors, and signals embedded in the text itself [7, p. 196]. Therefore, within the framework of concepts of an implicit reader and "narrative gaps" by Iser [8], the recipient's expectations create a notional structure of an artistic text being formed via the system of "ambiguity points" or "blank spaces" – notional gaps, notional inconsistencies, non-interpretable symbols or speech figures. Furthermore, a recipient is placed to the aesthetic pole of the artistic work, who is responsible for the text realization, and the aesthetic problem lies in the implementation of the text valency, its turning from an aggregate of propositional functions to a proper artistic expression. Composition is a method of grouping and arranging objects depicted in plane of a pictorial representation relating to both the layer of depicted objects and the layer of reconstructions expressing the visual images [5, p. 330]. The space of a pictorial representation is oriented [9, p. 207], i.e., the depicted objects are arranged in accordance with the contensive requirements and perspective laws with respect to the orientation center. Structural constituents of a pictorial representation bear a valuable "charge", i.e., they are capable of enriching the pictorial text with aesthetic value. While grouping or dispersing, the objects or reconstructed types of subjects (color spots, shapes, lines, etc.) representing particular valuable qualities, constitute an aesthetic value of the entire artistic work, i.e., they form a composition as a valuable image of the pictorial representation.

An object's aesthetic value is a judgement based on and resulting from the aesthetic process. Such a judgement cannot be obtained while avoiding one of the forms of the aesthetic feeling – the aesthetic feeling, the process continuing in time and consisting of a series of deliberate acts, among which the sensual observation of aesthetic object represents a basis only. According to the logic of semiotics, the aesthetic feeling may be understood as a subjective designation process based on the obtained viewing experience, and a composition – as a designation experience control tool possessed by a viewer [10]. Could it be possible to express a judgement about the text and determine its aesthetic value by only relying on the mutual correlation of the constituents? Depending on a particular signs' arrangement and inter-sign communications in the system we can suggest the nature of a particular text and the effect produced by it. In this way, a principle of alliteration used in versification for giving a sound expression to the text is based on repeating combinations of identical or uniform consonants at a particular part of the text. A sequence technique (sequential repetition of a melodic phrase or harmonic configuration at another pitch) in music creates an effect of dynamic change of a composition. Using a particular spectrum of colors in pictorial arts allows the artist inducing a particular psychological sensation with the viewer. Consequently, it is not only the quality but also the relation of constituents of an artistic text that can have an aesthetic effect on the viewer, cause the expected psychical processes – the emotions.

An approach to determining an aesthetic value of the art object based on learning the organization of the means of expression and a psychical effect caused by them, represents an objective analytical method of analysis of the psychology of art – from the form through the functional analysis of components and structure to simulating an aesthetic reaction and setting general laws. One of the founders of the approach L.S. Vygotsky insisted on ambivalence of artistic work since its structure is based on a particular correlation of the material and form. So, in the artistic work, the relations between the material and the form are the dialectic relations between the plot and narrative, where the narrative form overcomes the plot material. The essence of aesthetic influence of art on the viewer consists in the fact that emotions caused by the material and emotions caused by the form may be distinguished in every artistic work. These two types of emotions are heterodromous and contain a cathartic affect "developing in two opposite directions and annihilating in the finite point..." [11, p. 269]. This approach in psychology of art allows to state the existing laws in perceiving the pictorial constituents.

Therefore, a composition may be considered as syntax of an artistic pictorial text and an aesthetic value subject not only to philosophic reflection but also to mathematical calculus.

We claim that the increased interest in computational aesthetics as well as in the digital humanities may be explained via the digital transformation of the mankind, exponential growth of knowledge and the desire to find out a relatively objective tool and instrument regarding the establishment of the "beauty" criteria. Moreover, the twenty-first century has been marked by the development of the scientific anthropocentric paradigm that could not but influence the present research fields, filling them with a larger degree of "subjectiveness" and additional complexity due to the dualistic and even polysemic essence of the present concepts and research trends. In this paper we propose to analyze the composition axes using two different approaches concerning detecting main composition axes and symmetries in the fine-art images. We use linear fit of silent objects detections done by neural network to obtain the main composition axes, also we use gestalt analysis for detection of composition and symmetry axes.

2   Methods of Computational Aesthetics

According to the information theory by Shannon [12] every message is a result of such a selective process, at which the signs form an ordered temporal or spatial sequence. The information as a notion deals not with the content or sense of a message; it only relates to the signs appearing in the message at a particular frequency or probability. Any sequence of symbols, e.g., an artistic text may serve as an example of such a statistical message. An artistic work follows particular mathematical and physical laws regarding its structure since (a) it is an object of reality, (b) it is built in accordance with the laws of geometry, (c) represents an aggregate of signs (or numbers). It means that there exists the possibility of mathematization of aesthetic categories, including them into the scheme categorizing different aesthetic levels. An approach to the artistic work as a sequence of signs was named in twentieth century as computational aesthetics (CA). The computational aesthetics proceeds from the point of view that the aesthetic object is subjected not only to subjective interpretation resulting in a set of descriptive aesthetic notions and categories, but may and should be comprehended via a system of meanings that correlate similarly to numerical value in mathematics and physics: "A decisive fact for the computational aesthetics is that it attempts to consistently replace all traditional notions with controlled and methodically clear mathematical concepts (numbers)" [13].

In 1928 G. Birkhoff suggested that there exist the constituents of order such as symmetry, rhythm, repetition, contrast, etc., which psychologically cause a positive tone of sense as well as the constituents, which cause negative tone, such as ambiguity or excessive repetition [14]. The formula of aesthetic evaluation by Birkhoff sets a relation between the order and complexity in aesthetic perception: the energy of perception is directly proportional to the order and inversely proportional to the complexity, in this case the model of aesthetic perception complies with the rule in aesthetics: the unity in diversity, the "simplicity" and the "symmetry" are the inherent universal form of every aesthetic perception in general. Another formula of aesthetic perception developed by H. Eysenck says that a measure of aesthetic perception is the product of the order and the complexity, in other words, the energy of aesthetic perception is a result of both high level of order and high level of complexity [15, pp. 35–49]. In the texts by Eysenck of 1960–1970-ies aimed at revealing the aesthetic emotions, the author's position comes down to that the beauty is objective, invariant for the entirety of objects, it correlates with some average, pan-human median of aesthetic taste and this taste is genetically determined. The existence of individual differences in aesthetic evaluations are connected by the Eysenck not with demographic factors but such personal traits as intro- and extroversion, the degree of neuroticism and psychoticism. However, the disadvantage of the models suggested by Birkhoff and Eysenck consists in the fact that they could be applied neither for describing the content of works of art nor for evaluating the proper aesthetic feeling. A measure of ordering the Birkhoff's components or the Eysenck's aesthetics of "simple forms" are capable of describing the macroaesthetic and partially microaesthetic essence of the object but are incapable of considering it entirely, i.e., describing or forecasting and effect produced by an object on a recipient, is impossible. In addition, the notions "simplicity", "naturalness", "passiveness" are the categories of a particular cultural and historical period and cannot be considered as comprehensive or objective.

M. Bense, while attempting to determine theoretic provisions of computational aesthetics, criticized the Birkhoff's formula since it describes the perceived image but not the essence of the artistic work and suggests distinguishing between the macro- and microaesthetic levels of aesthetic perception. Speaking about the possibility of quantitatively measuring the art, Bense put forward four theses: on the materiality, on ordering, on communicativeness, on sign's nature. In his theory Bense proceeds from the following:

1. Every object of art consists of a substance, and two types of processes or states exist in the nature – physical or aesthetic; aesthetic processes, as opposed to physical ones, are non-determined, cannot be described by the cause-and-effect relationships and are therefore unpredictable;

2. The order is manifested in the process of creating the structure (composition) of an artistic work when the author selects and distributes its components between each other.
3. The communicativeness of an aesthetic object is connected with the informative relations of transmitter-receiver type between the producer and consumer of that object;
4. An artistic work represents a sign system, it functions inside a sign system, the sign processes proceed within the sphere of an artistic work.

However, the Bense's approach is rather constatative but not interpretative: it hardly establishes the goals of setting the objective content of the notions of "the beauty", "the ugliness", "the sublime", it just describes their structure instead.

Since it is namely the measure of order that is the measure of beauty as the category of aesthetics and can be subjected to numerical measurement, the category of "beauty" may be applied to a formally ordered, symmetrically positioned objects, even if deprived of any contents imported by any external subject. In his article "Defining Computational Aesthetics" F. Hoenig said: "It is of utmost importance that the studies are focused on the aesthetics on the point of view of the form but not contents […] On the contrary, any strictly theoretical result or reflections on the value of art are rather senseless taking into consideration the philosophic problems which the researcher will face" [16]. Therefore, according to this approach, the content of an aesthetic object is sort of excluded from its structure or understood as a secondary component. However, such an approach to artistic work contradicts one of the basic theses of aesthetics which consists in the assumption that the aesthetic is always connected with the expression of objectness, i.e., "the internal life of an object" should be "given externally as well" [17, pp. 391–392]. It is deemed that such a formal approach, firstly does not always allow adequately describing the art object existing as a synthesis of form and content, secondly, impoverishes the aesthetics as a philosophic discipline, and thirdly reduces the aesthetic judgement by turning it into the judgement about the object's external features. As a result, neither the aesthetic feeling, based on the subjective perception, nor the aesthetic value, composed during evaluation of an art object based on the judgement regarding its form and content, may be revealed. In our opinion, computational aesthetics, by being focused on the structure of an art object, its physical parameters, allows extending the perception of aesthetics, which is particularly relevant in the time when the immediate content of a work and the author's intent ("the sense of an artistic work") are not the key factors in evaluating the aesthetic object. In this connection, the complication of mathematical models employed together with the interpretation of results on the point of view of semiotics, philosophy and art studies, as well as involvement of subjective viewer evaluations of aesthetic objects allow finding a compromise decision. In our view, the problem of numeric analysis of an artistic text consists in such an improvement of mathematical models considering the rules of semiotics and subjectivism of human perception.

During the XX–XXI centuries, the development of numerical approaches to evaluation of aesthetic objects has been based on the concepts suggested in works by [14, 18] and [19], and the progress is implemented due to the improving simple models suggested in earlier works so as to enhance their correspondence to the human aesthetic perception.

3 Computational Analysis of Composition and Symmetries

Generally, the artwork can be described via a system of signs, which directly implies the interpretation presence. In painting signs are various forms of depicted objects. The examples include lines, outlines, shapes, specific known objects of reality, as well as biological objects. The line is the basic element of the image and is able to form the outline of the subject matter. The outline turns into a cell when it has been filled with color. Moreover, a set of cells constitutes an image system. Any form in the image can be interpreted with a certain probability degree as some subject or the object. Similarly, such a process can be performed by the subject both independently and using known detection and recognition tools. The task of the CA is to define the interpretation probabilities of this kind for various symbols from the system of image signs. This could be done using like conventional image processing methods applied to fine art pictures [20, 21] or by modern deep learning approach.

The study [22] describes the modern approaches and a review of CA methods. A feature of neural network-based approaches in CA is the estimation of the aesthetic value, perception and memorization of art [23]. Common practice is predicting the integral estimation of the image's aesthetic contents, based on a neural network model, previously trained on a set of assessments assigned to a set of images [22, 24].

Despite the successful use of the features generated by neural network models [25], in order to predict how an image will be perceived by a human being, there is a need for the formation of computable image attributes associated with its semiotics [22]. The purpose of such computable image attributes' formation is a deeper estimation of the image's aesthetic contents. The article [26] describes the basic image attributes in terms of CA. These attributes are divided into three groups: content attributes, composition attributes, and attributes related to lighting and color composition of an image. According to [26], the content attributes include the objects presence of a special type in the image – faces, people, and animals. Composition attributes mean compliance with basic rules of photographic composition, such as the rule of thirds. However, such basic attributes can be extended for the deeper image description.

In this paper we propose to analyze the composition axes using two different approaches concerning detecting main composition axes and symmetries in the fine-art images. The first step is to detect main composition axes and proportions in the image by fitting the silent objects in the image by linear model. We use CNN-based object detectors [27] to detect silent elements in the image, then we fit the line over these detections. This approach allows to detect vertical and horizontal composition axes in different commonly used ratios: ½, rule of thirds, golden section, etc. The second approach is to use gestalt-based symmetry analysis [28] for more precise detection of the symmetry axes in the images.

While in first step we work with high-level image structural elements, like detected faces or persons, gestalt-based approach works with fine-grained image elements – superpixels, computed by SLIC algorithm [29]. According to the gestalt approach, image is split into grid of superpixels, and then these superpixels are hierarchically grouped into gestalts. This grouping is conducted according to six gestalt laws: mirror reflection, good continuation, rotational symmetry, good linear continuation, lattice, or grid grouping. More details on gestalt approach are provided in [28, 30, 31]. In the listed papers, high-level gestalt groups are used for detecting symmetry axes in remote sensing or regular images. In this paper we used gestalt approach for precise detection of symmetry and composition axes of arbitrary direction, in addition to vertical and horizontal axes, detected on the first step.

4   Results

To evaluate the proposed approaches, we conducted the following experiments. We take fine-art paintings from the classical renaissance era as the initial data for the experiments. For the first step of proposed algorithm, we take CNN-based detector in order to detect the silent elements in the image. We take popular YOLO v4 [27] detector model pre-trained on the OpenImages dataset [32] to detect bounding boxes of the people's faces. At the next stage, they were processed in order to fit vertical or horizontal line as the composition axis of the image.

Figure 1 depicts "The Last Supper" of Leonardo da Vinci, with linear fits of the centers of the CNN-detections showing the horizontal and vertical axes of the ½ ratio. Lines obtained by linear fit of face's centers are shown in white and corresponding axes are shown in red. Linear fit of the centers detected faces in Bottcelli's "The Adoration of the Magi" corresponds to ½-ratio vertical axis and golden ratio horizontal axis (Fig. 2). The same ½ and golden ratio detected in "The Spring" painting (Fig. 3).

Gestalt analysis conducted according to [30] allows to detect more sophisticated axes and composition lines in the images. Due to the gestalt approach, we analyze the image structure in more details than using only detected objects. Resulting symmetry lines have arbitrary directions. It allows us to obtain more complex axis. This equates to detect vertical golden ratio axis in "The Birth of Venus" (Fig. 4).

Combining these two approaches allows us to detect primary and secondary composition axes, like it has been shown for "The Adoration of the Magi" in Fig. 5. Gestalt analysis appends diagonal symmetry axis to the vertical and horizontal, detected on the first step.

Experiments show that the proposed approach of the composition analysis works for selected fine-art pictures allows obtaining primary and secondary composition axes. However, further research needed for wider application in the scope of computational aesthetics.

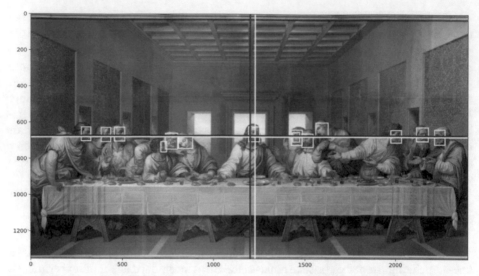

Fig. 1. "The Last Supper" image with linear fit of the detected faces (white) and ½-ratio axes (red).

Fig. 2. "The Adoration of the Magi" image with linear fit of the detected faces (white) and ½-ratio vertical axis and golden ratio horizontal axis (red).

Fig. 3. "The Spring" image with linear fit of the detected faces (white) and ½-ratio vertical axis and golden ratio horizontal axis (red).

Fig. 4. "The Birth of Venus" with axis detected by gestalt analysis close to golden ratio.

Fig. 5. "The Adoration of the Magi" image with horizontal axis obtained by fitting the centers of detected faces and diagonal axes, obtained by gestalt analysis.

5   Conclusion

This study is devoted to the capabilities of computational analysis of aesthetic contents of an artistic text. Our approach is based on the perception of the formal order as a determinant of the aesthetic category of beauty. Since the composition of an art object represents its syntax where particular constituents are appositional signs interconnected by inter-sign relations, it is suitable for study from the perspective of numerical analysis (numerical study). In this connection we state that the composition of an artistic text is an aesthetic value subjected to mathematical measurement through which a quantitative representation of an aesthetic content of an object is possible. The purpose of the study is determining such symmetry lines of an art object, which may be obtained using the image detecting means. Since an art object represents a system of signs subject to both philosophic interpretation and mathematical measurement, for an enhanced efficiency it has been decided to combine these approaches.

To analyse the image symmetry using CNN-based object detectors, and gestalt-based symmetry analysis, we have selected five pictures by artists of the Italian Renaissance. The experiments showed that the offered approach enables determining primary and secondary axes of a composition. Thus, the aesthetic content of an artistic text is potentially subject to interpretation on the point of view of philosophy and art study as well as a neural network analysis. The importance of the study consists in extending the basic calculating attributes of the image, which, in connection with the philosophic analysis and involving of expert evaluations, gives a more adequate representation of artistic text and allows obtaining its more detailed description. We hope that future studies in computational aesthetics related to expanding the attributes of the analysis of images, in synthesis with philosophic and art-studying approach will enable improving the mathematical models of analysis of characteristic nature of art objects and bring them closer to human perception. Further research in algorithmic part will be devoted to complete integration of the CNN-based silent object detection with the gestalt approach, and possible use of attention mechanism to choose appropriate composition structure of the image.

Acknowledgments. The research was funded by RFBR projects 20-011-00462 A, 19-29-01235 MK.

References

1. Kant, I.: The Critique of Judgement. Iskusstvo, Moscow (1994).(in Russian)
2. Schelling, F.: Literary works. In: 2 Volumes, vol. 1. Mysl, Moscow (1987). (in Russian)
3. Hegel, G.W.F.: Lectures on the Philosophy of Art. Felix Meiner Verlag Publishing House, Hamburg (2003)
4. Hegel, G.W.F.: Aesthetics. In: 4 Volumes, vol. 1. Iskusstvo, Moscow (1968). (in Russian)
5. Ingarden, R.: Aesthetic Studies. Foreign Literature Publishing House, Moscow (1962).(in Russian)
6. Mukarzhovsky, Ya.: Studies in Aesthetics and Theory of Art. Iskusstvo, Moscow (1994). (in Russian)
7. Jauss, H.R.: Literary History as a Challenge to Literary Theory. Modern Literary Theory. Antology. Flinta, Nauka, Moscow (2004). (in Russian)
8. Iser, W.: Process of Reading: Phenomenological Approach. Modern Literature Theory. Antology. Flinta, Nauka, Moscow (2004). (in Russian)
9. Husserl, E.: Ideas for a Pure Phenomenology and Phenomenological Philosophy. First Book: General Introduction to Pure Phenomenology. Jahrbuch für Philosophie und phänomenologische Forschung, vol. 1, no. 1 (1913)
10. Nesterov, A., Demina, A.: Artistic work as a technical object. Mirgorod **1**(13), 48–74 (2019). (in Russian)
11. Vygotsky, L.S.: Psychology of Art. Iskusstvo, Moscow (1986).(in Russian)
12. Shannon, C.E.: Works on Theory of Information and Cybernetics. Foreign Languages Publishing House, Moscow (1963).(in Russian)
13. Bense, M.: Einführung in die informationstheoretische Ästhetik. Grundlegung und Anwendung in der Texttheorie. Rowohlt, Reinbek bei Hamburg (1970)
14. Birkhoff, G.D.: Aesthetic Measure. Harvard University Press, Cambridge (1933)
15. McWeeney, G.: The Review on Aesthetic Measurements. Artmetry. The Methods of Exact Sciences and Semiotics. Editorial URSS, Moscow (2020). (in Russian)
16. Hoenig, F.: Defining computational aesthetics. In: Neumann, L., Sbert, M., Gooch, B., Purgathofer, W. (eds.) Computational Aesthetics in Graphics, Visualization and Imaging. CompAesth 2005: Workshop on Computational Aesthetics. The Eurographics Association (2005). https://doi.org/10.2312/COMPAESTH/COMPAESTH05/013-018
17. Losev, A.F.: The history of antique aesthetics. The results of millennial development. In: 2 Volumes, vol. 1. LLC AST Publishing House, Moscow (2000). (in Russian)

18. Fechner, G.T.: Vorschule der Aesthetik, vol. 1. Breitkopf & Härtel, Leipzig (1876).(in German)
19. Cupchik, G.C.: A decade after Berlyne: new directions in experimental aesthetics. Poetics **15**, 345–369 (1986). https://doi.org/10.1016/0304-422X(86)90003-3
20. Bibikov, S.A., Nikonorov, A.V., Fursov, V.A.: Correction of shadow artifacts on colorful digital images. Comput. Opt. **34**(1), 124–131 (2010)
21. Bibikov, S., Zakharov, R., Nikonorov, A., Fursov, V., Yakimov, P.: Detection and color correction of artifacts in digital images. Optoelectron. Instrum. Data Process. **47**(3), 226–232 (2010)
22. Brachmann, A., Redies, C.: Computational and experimental approaches to visual aesthetics. Front. Comput. Neurosci. (2017). https://doi.org/10.3389/fncom.2017.00102
23. Cetinic, E., Lipic, T., Grgic, S.: A deep learning perspective on beauty, sentiment, and remembrance of art. IEEE Access **7**, 73694–73710 (2019). https://doi.org/10.1109/ACCESS.2019.2921101
24. Talebi, H., Milanfar, P.: NIMA: neural image assessment. arXiv:1709.05424 (2018). https://doi.org/10.1109/TIP.2018.2831899
25. Zhang, R., Isola, P., Efros, A.A., Shechtman, E., Wang, O.: The unreasonable effectiveness of deep features as a perceptual metric. In: 2018 IEEE/CVF Conference on Computer Vision and Pattern Recognition (CVPR), pp. 586–595 (2018)
26. Dhar, S., Ordonez, V., Berg, T.L.: High level describable attributes for predicting aesthetics and interestingness. IEEE Comput. Vis. Pattern Recogn. 1657–1664 (2011). https://doi.org/10.1109/CVPR.2011.5995467
27. Bochkovskiy, A., Wang, C.-Y., Liao, H.-Y.M.: YOLOv4: optimal speed and accuracy of object detection. arXiv:2004.10934 (2020)
28. Michaelsen, E., Vujasinovic, S.: Estimating efforts and success of symmetry-seeing machines by use of synthetic data. Symmetry **11**(2), 227 (2019). https://doi.org/10.3390/sym11020227
29. Achanta, R., Shaji, A., Smith, K., Lucchi, A., Fua, P., Susstrunk, S.: SLIC superpixels compared to state-of-the-art superpixel methods. IEEE Trans. Pattern Anal. Mach. Intell. **34**(11), 2274–2281 (2012). https://doi.org/10.1109/TPAMI.2012.120
30. Michaelsen, E., Meidow, J.: Hierarchical Perceptual Grouping for Object Recognition. Springer, Cham (2019). https://doi.org/10.1007/978-3-030-04040-6
31. Michaelsen, E.: On the depth of Gestalt hierarchies in common imagery. In: Del Bimbo, A., et al. (eds.) ICPR 2021. LNCS, vol. 12665, pp. 30–43. Springer, Cham (2021). https://doi.org/10.1007/978-3-030-68821-9_3
32. Kuznetsova, A., et al.: The open images dataset V4: unified image classification, object detection, and visual relationship detection at scale. Int. J. Comput. Vis. **128**, 1956–1981 (2020). https://doi.org/10.1007/s11263-020-01316-z

Methodological Procedures of the Russian Avant-garde and Use in Modern Costume Design

Tatyana A. Petushkova[✉] [ID], Larisa P. Smirnova[ID], and Luibov E. Yakovleva[ID]

Federal State Budgetary Educational Institution of Higher Education, Russian State University named after A.N. Kosygin, (Technology. Design. Art), Sadovnicheskaya st., 33, bldg. 1, 117997 Moscow, Russia

Abstract. The purpose of this study is to generalize and use the methodological developments of the Russian avant-garde in modern design practice as a stimulus for the development of the fashion industry. In their research, the authors relied on the works of Khan-Magomedov, Lavrentiev, Beschastnov, Matyavina and others on the assessment of the role and significance of the Russian avant-garde in the development of the design culture of the 21st century. The creative concepts of avant-garde artists are considered in the article as a combination of scientific, technical and artistic creativity, a synthesis of figurative-associative and logical thinking, science and art. In their work, the authors relied on experimental research on the use of the methodological heritage of the Russian avant-garde of the 20th century in the educational design practice of the Russian State University named after A. N. Kosygin. The authors come to the conclusion that in modern culture, form-making impulses can come both from the functional and constructive structure of products, and "from formal aesthetic searches." The article focuses on the formal and aesthetic search for new technologies for creating fashionable clothes. The legacy of the Russian avant-garde was used as the substantive basis of the creative process, a way of designer's "reflection" and the main motive for the revival of the Russian light industry.

Keywords: Costume design · Russian avant-garde · Style module of Suprematism · Constructivism · Form-building technologies · Creative form creation

1 Introduction

The value of the work increasingly depended not only on the skill and talent of the artist, but also on the original methods of shaping he used [1–5].

In the modern postmodern situation in the Russian fashion industry, much attention is paid to the introduction of new technologies into production, an increase in the number of new fashion brands, the development of digital sales channels, and the simplification of global logistics, which causes "the need to form a sustainable creative sector" and the development of the online industry [6]. In modern science, there has been a tendency to study the value orientation of scientific knowledge in such areas as sociology, pedagogy, philosophy, social psychology and philology, new approaches to solving general theoretical issues have been outlined. In art criticism and the field of artistic creativity of modern postmodernism, the next "revaluation of values", there are similar processes in the development of form-making tendencies of the early twentieth century, which laid the foundations of avant-garde art. The famous researcher of the Russian avant-garde Selim Khan-Magomedov notes: "The further the 20s of our century go into the past, the more obvious it becomes that then something very important was laid for the entire subsequent development of the subject-artistic sphere of creativity" [1].

Today, these trends have become the subject of research by many specialists in order to determine the current situation in the design culture of the 21st century [7–9]. Table 1 shows the main areas of research in the field of philosophy, aesthetics, art criticism.

Table 1. Areas of research of the avant-garde art of the twentieth century [7]

Knowledge area	Direction of research
Philosophy and aesthetics	1. The worldview component of avant-garde artists 2. Evaluation of reflective thought about the era of the avant-garde
Art criticism	1. Own theoretical reflection on the programmatic articles and manifestos of the avant-gardists 2. Research of artistic artifacts as holistic cultural phenomena

In general, the art of the avant-garde of the twentieth century has developed a pluralism of value systems as the basis for future postmodernism, namely: spirituality and cosmism (in abstractionism); balance of the spiritual and sensual principles (in Suprematism); decrease in author's subjectivity (in aleatoric); freedom of creativity (in Dadaism); scientific and artistic reality within the framework of technical progress (in kinetics), the union of art and life (in constructivism), artistic reality based on dissonance (in cubism); mechanistic civilization of the future (in futurism).

The experience of the artistic avant-garde was the forerunner of the aesthetic mastering of the language of computer graphics. So, the artistic direction of conceptualism stimulated the development of the technocratic paradigm of computer art in the 1960s [10–12]. The direction of constructivism contributed to the formation of computer technologies for artists of the 50–70s. XX century; surrealism, in the form of all kinds of collages, photomontages, various photo manipulations, anticipated the special effects of computer graphics to create a fantasy and real world [13–17]. For a practical understanding of the substantive basis of the avant-garde, we further consider the features of creative concepts of the most significant trends in painting, which had access to the objective world.

2   Creative Concepts of the Russian Avant-garde of the 20th Century in Educational Design Practice

"Avant-garde trends in art" of the early twentieth century replaced realism and symbolism "as a paradigmatic leap from Newtonian mechanics to Einstein's theory of relativity and quantum concepts" [8, 9]. Avant-garde artists combined the principles of art and science. The first was ensured by the use of the simplest geometric elements, which varied endlessly in various combinations, forming new compositions and new meanings with a creative attitude inherent in each author. The second borrowed from science the principle of continuous self-renewal and dynamism. As a result, the boundaries of the application of the simplest elements were established, the meanings of "scientific" and "artistic" analytics were clarified, which anticipated the onset of the postmodern situation.

As a result, "… the convergence of scientific, technical and artistic creativity, which was of fundamental importance for the formation of a new attitude to the processes of shaping the subject-spatial environment," took place, and the experiments of the artists "… reached such a level of shaping, where, in fact, priority tasks arose. The value of the work increasingly depended not only on the skill and talent of the artist, but also on the original methods of shaping used by him" [1, p. 9].

To characterize the creative concepts of the avant-garde in its early stages of development, two directions - Suprematism and Constructivism - had a significant impact. The founder of Suprematism, Kazimir Malevich, is an avant-garde artist, "… to the highest degree he concentrated in himself the properties of a style-forming talent. He consistently and extremely thoroughly went through a number of successive artistic trends, assimilating from them, first of all, what determined the main style-forming line of the new time [1].

The style module of Suprematism included "a combination of simple geometric planes with space (or with a white background as its symbol). The combination of these very elements of the style-forming core of Suprematism turned out to be the most common stylistic feature", while the main role in the compositions was played by colour. Any expansion and concretization of the style module took place in each subject area, depending on its specifics [1, p. 51]. The search followed the path of combining various Suprematism planes and volumes in the composition of the product, and the artistic logic was based on the "logic of perception".

A group of constructivists was formed in February 1921, which included Alexei Gan, Alexander Rodchenko, Valentina Stepanova, Vladimir Stenberg, Georgii Stenberg, Konstantin Medunetsky, Karlis Johanson [1, 17, 18].

The stylistic features of constructivism were expressed in the identification of functionally significant volumes and details in architectural and design projects, the use of such archetypal compositional construction schemes as: cruciform intersection of elements at right angles, zigzag and diagonal constructions, the simplest types of symmetry. Photomontage was widely used. "Textile drawing was understood as an autonomous colour-decorative-graphic art system, operating in sharp conflict with the volumetric form of clothing" [1, 17, p. 188]. As a result, new stylistic devices were born that reflected the socio-political needs of the time.

Modern scholars view the artistic avant-garde as a complex of various types of arts in connection with ideological and creative trends that influence the art of the 21st century [19]. The use of the methodological apparatus of historical, art history, philosophical, aesthetic and axiological knowledge made it possible to assert that the avant-garde of the twentieth century is:

- a well-established conceptual term and a single code of modern culture;
- independent language practice with its own grammar;
- combination of scientific, technical and artistic creativity as a synthesis of figurative-associative and logical thinking [13, 20–23].

This provides a methodological basis for using the developments of the avant-garde artists of the twentieth century as a methodological and substantive basis for the creative process in the development and implementation of new technologies in design practice. And here the thought of Khana-Magomedov that: modern stylistic changes in culture can come both from the functional and constructive structure of products, and "from formal aesthetic searches" [1].

In support of this idea, we conducted an experiment on the use of the methodological heritage of the Russian avant-garde of the 20th century in educational design practice. It is understood as:

- the substantial basis of the creative process;
- the way of "self-existence and reflection" of the designer;
- the ability to produce and reproduce the specific sensory imagery of the style;
- a powerful motivator for the revival of modern Russian light industry.

3   Creative Field of Experimental Shaping in Instructional Design

The creative field of experimental shaping in educational design is represented by a formal-aesthetic experiment [24], which is based on the style modules of Suprematism and Constructivism. The experiment is designed for initial educational courses, when the student is still free from the baggage of professional knowledge and boldly trusts his creative intuition.

The method of work on the creation of the costume series is based on the use of the scientific method of analysis and synthesis with the aim of novelty and originality of the artistic language of form creation.

The main stages of work

- selection of a close-minded artist and his painting. Rationale for this choice.
- analytical analysis and search for primary elements, compilation of the alphabet.
- creative interpretation of the source.
- use of elements of technological thinking (assembly, programming of operations, serial production).
- modelling of the context in the information field of a fashionable modern suit [21, 25].

3.1 Style Modules of Suprematism

Figure 1 an example of the "exit" of a Suprematism composition by Klyuna "from the picture plane" into the object space of a fashionable suit. Ivan Vasilyevich Klyun (real name Klyunkov, 1873–1943), painter, graphic artist, sculptor, art critic, a prominent representative of several modern trends, including Suprematism, a special offshoot of Russian abstract art in the first half of the 20th century. A companion and friend, as well as a follower of Kazimir Malevich, who remained in his shadow and was even unjustly considered a "second-tier avant-garde", was one of the most distinctive masters in both Cubo-Futurism and Suprematism.

The basis for choosing a source is the presence of bright colours, clear geometric shapes, the energetic content of the image. At the stage of analysis, the percentage of the colour characteristics of the original composition and a set of geometric planes were determined, the original alphabet was compiled, which determined the creative field of subsequent transformations.

Further, the author's search options are shown, which provided the basis for entering the design of fabric patterns, mono-compositions, prototypes of accessories and jewellery. Special attention should be paid to the search for options that approach the shape of the suit, as a dynamic system, expressed in the poses of the human figure in combination with geometric volumes (Fig. 2). This stage is especially important, as it allows you to completely abstract from the specifics of "styles» and focus on the dynamic expressiveness of the style image.

At the stage of context modelling, creative imagination is aimed at preserving the dynamics of the prototype and its ornamental expressiveness in the silhouette and determining the assortment groups of the costume and accessories, colour planes that increasingly reveal the specificity and diversity of the composition. The stylistic module of Suprematism dictates the combination of simple geometric planes with the specifics of the costume space, when the creative mind is immersed in the atmosphere of discovering new formative possibilities of the method. The "Suprematism Universe" reconciles the sensual and the spiritual in the design of endless series.

In Fig. 3 shows a fragment of a series of sketches using the composition of Kazimir Malevich, his color palette and original geometric elements. On their basis, variants of patterns for fabrics and silhouettes of a fashionable suit in the dynamics of a figure are proposed. In a series of computer sketches, the search for color spots within the silhouettes was carried out, observing the intensity of color combinations and the general dynamism of the form.

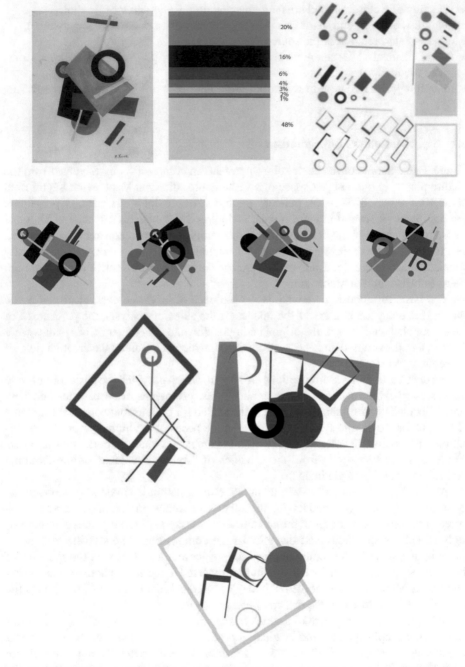

Fig. 1. Determination of the initial provisions of the form-making concept (Student E.A. Mokrinskaya, leaders L.P. Smirnova, T.A. Petushkova)

Fig. 2. Searches for the shape of a suit within the style code of Suprematism (Student E.A. Mokrinskaya, leaders L.P. Smirnova, T.A. Petushkova)

Fig. 3. Form-making searches within the style code of Suprematism (Student A.V. Vasilieva, leaders L.P. Smirnova, T.A. Petushkova)

3.2   Style Modules of Constructivism

In Fig. 4 shows a series of developments, sustained within the framework of constructivism style settings, where the priority was to identify the form, design, functionally significant volumes and details, texture of materials.

Typical technological methods of photomontage with portraits of the authors, computer variations of ornamental compositions within the identified alphabet were used. On this basis, a series of patterns for fabrics and mono compositions for the accessory range were developed. In the silhouettes of the clothes, a characteristic "conflict" of the textile pattern with the volumetric shape was observed. In the experiment, two directions in the presentation of sketches were tested: modern fashionable stylistics and the general, conditional-shaped transmission of the avant-garde style code, in particular, the characteristic "cross-shaped intersection of elements at right angles, zigzag and diagonal constructions" determined the design features of the series. The textured development of colour spots was supposed to convey the peculiarities of the reading of materials: knitwear, drape, leather materials, heels, etc. Collage techniques were widely used in manual and computer execution.

In Fig. 5 shows the options for the development of a style image in the technique of collage and computer graphics.

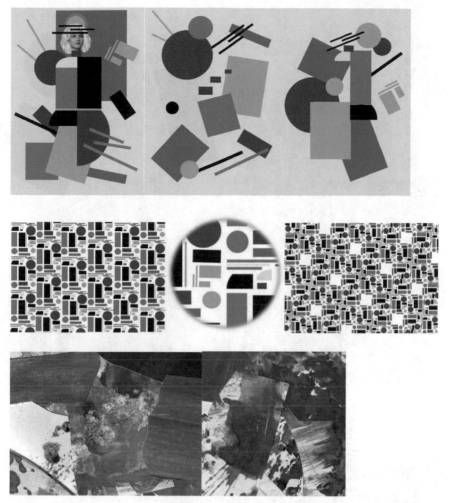

Fig. 4. Determination of the initial provisions of the form-making concept in the style of constructivism (Students E.A. Samokhvalova, K.V. Smetanina, leaders L.P. Smirnova, T.A. Petushkova)

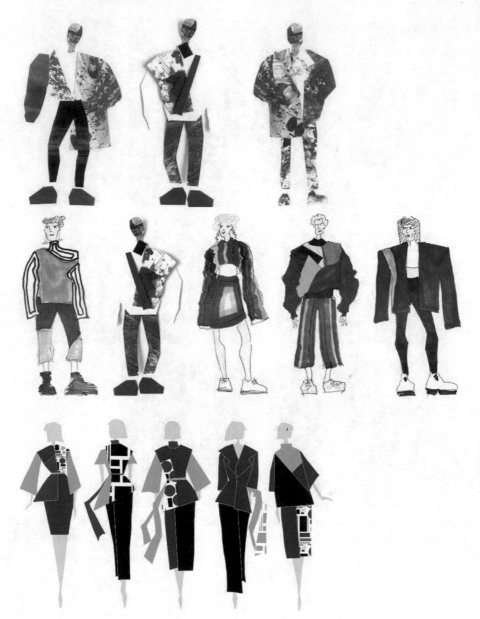

Fig. 5. Designing a suit within style code of constructivism (Students K.V. Smetanina, E.A. Samokhvalova, leaders L.P. Smirnova, T.A. Petushkova)

4   Conclusions

1. The use of the methodological apparatus of the philosophical, aesthetic, historical, art criticism and axiological knowledge made it possible to establish that the art of the avant-garde is an integral cultural phenomenon that largely determined the development of postmodern art in the 21st century.
2. The language of geometric shaping of the Russian avant-garde has made it possible to introduce into the educational practice of costume designers a new algorithm of thinking and shaping activity, based on a holistic understanding of all elements of the costume in a single style key.
3. The conducted formal aesthetic experiment in the early years of study showed that it promotes the development of students' creative abilities, stimulates the development of digital technologies, and contributes to the formation of a sustainable creative sector in the Russian fashion industry capable of responding to the challenges of the time.

References

1. Khan-Magomedov, S.O.: Pioneers of Soviet Design. Galart, Moscow (1995)
2. Glade-Wright, R.E.: New insights effectively shared: originality and new knowledge in Creative Arts postgraduate degrees. Qual. Res. J. **17**, 89–98 (2017). https://doi.org/10.1108/QRJ-04-2016-0023
3. Acar, S., Burnett, C., Cabra, J.F.: Ingredients of creativity: originality and more. Creat. Res. J. **29**, 133–144 (2017). https://doi.org/10.1080/10400419.2017.1302776
4. Serkova, V.A.: The digital reality: artistic choice. IOP Conf. Ser.: Mater. Sci. Eng. **940** 012154 (2020). https://doi.org/10.1088/1757-899X/940/1/012154
5. Ammon, S.: Epilogue: the rise of imagery in the age of modeling. In: Ammon, S., Capdevila-Werning, R. (eds.) The Active Image. PET, vol. 28, pp. 287–312. Springer, Cham (2017). https://doi.org/10.1007/978-3-319-56466-1_12
6. Sedykh, I.A.: Fashion industry. Institute "Development Center" of the National Research University Higher School of Economics (2019). https://dcenter.hse.ru/data/2019/06/03/149 5959454/Industria%20fashion-2019.pdf. Accessed 1 Jan 2020
7. Morgunov, A.P.: Valuable landmarks of the avant-garde art of the twentieth century. Historical, philosophical, political and legal sciences, cultural studies and art history. Quest. Theory Pract. **2–2**(16), 141–146 (2012)
8. Popov, D.A.: Avant-garde as a "scientific" study of art. Izvestia of Saratov University. New Ser.: Philos. Psych. Pedag. **15**(1), 58–61 (2015)
9. Popov, D.A.: Avant-garde art in its relation to science: borrowing and overcoming them. J. Basic Res. **2**(part 16), 3619–3623 (2015)
10. Ammon, S.: Image-based epistemic strategies in modeling: designing architecture after the digital turn. In: Ammon, S., Capdevila-Werning, R. (eds.) The Active Image. PET, vol. 28, pp. 177–205. Springer, Cham (2017). https://doi.org/10.1007/978-3-319-56466-1_8
11. Bylieva, D., Lobatyuk, V., Ershova, N.: Computer technology in art (Venice Biennale 2019). In: Proceedings of Communicative Strategies of the Information Society (CSIS 2019), p. 18. ACM, Saint–Petersburg (2019). https://doi.org/10.1145/3373722.3373785

12. Kazaryan, R., Pogodin, D., Andreeva, P., Galaeva, N., Tregubova, E.: System approach to using information modelling technology in sustainable construction production development. E3S Web Conf. **258**, 09009 (2021). https://doi.org/10.1051/e3sconf/202125809009

13. Margie, N.M.: The art of the avant-garde is the key to the secret. Bulletin of the Peoples' Friendship Uni. Rus. Series: Philos. **4**, 95–99 (2014)

14. Zhuriha, A.M.: Avant-garde in the visual arts. Young Sci. **30**(134), 413–415 (2016)

15. Turlyun, L.N.: Avant-garde at the origins of computer art. Cult. Heritage Siberia **4**(22), 24–35 (2017)

16. Matyavina, A.E.: Artistic avant-garde and the image of culture of the XX century. Abstract of the dissertation for the degree of candidate of philosophical sciences. Lomonosov Moscow State University, Moscow (2004)

17. Lavrent'ev, A.N.: Constructivism laboratory. The constructivist laboratory. Study guide on the history of graphic design: graphic modeling experiments. Grant, Moscow (2000)

18. Beschastnov, N.P., Lavrent'ev, A.N.: Avant-garde fabric. RIP-holding, Moscow (2020)

19. Lodder, C.: Conflicting approaches to creativity? Suprematism and constructivism. In: Celebrating Suprematism, pp. 259–288. Brill, Leiden (2018). https://doi.org/10.1163/978900438 4989_015

20. Matyavina, A.E.: The artistic avant-garde and the image of the 20th century culture. Doctoral dissertation. Lomonosov Moscow State University, Moscow (2004)

21. Petushkova, T.A.: Design of visual-graphic communications of fashion brands. Doctoral dissertation. Russian State University named after A.N. Kosygin (Technology. Design. Art), Moscow (2020)

22. Petushkova, T.A., Smirnova, L.P., Belgorodsky, V.S.: Stylistics of the Russian avant-garde: educational design of industrial collections in costume design. Kostyumologiya **3**, 1–12 (2020)

23. Karpova, Y.: 'A thing of quality defies being produced in quantity': Suprematist porcelain and its afterlife in Leningrad design. In: Celebrating Suprematism, pp. 198–220. Brill, Leiden (2018). https://doi.org/10.1163/9789004384989_012

24. Löytönen, T.: Educational development within higher arts education: an experimental move beyond fixed pedagogies. Int. J. Acad. Dev. **22**, 231–244 (2017). https://doi.org/10.1080/136 0144X.2017.1291428

25. Kurasov, S.V., Lavrent'ev, A.N., Zaeva-Burdonskaya, E.A., Sazikov, A.V.: Stroganovka: 190 Years of Russian Design. Russian Mir, Moscow (2015)

Quantum Fashion as a New Technology in Costume Design

Elena V. Makhinya$^{(\boxtimes)}$ ⓘ and Luibov E. Yakovleva ⓘ

Russian State University named after A.N. Kosygin (Technology. Design. Art),
Sadovnicheskaya Street 33/1, 115035 Moscow, Russia

Abstract. The purpose of this article is an attempt to introduce such new concepts as "quantum fashion", "clothing-cultural artifact" into the modern theory and practice of costume. These concepts are considered against the background of historically replacing fundamental scientific paradigms or "pictures of the world", as a result of which the value attitudes of people change, which cannot but be reflected in fashionable clothes. The theoretical basis of the article is the concept of the scientific picture of the world and its main historical forms introduced by Academician V.S. Stepin, as well as the principle of complementarity, which was simultaneously introduced by N. Bohr and R. Jacobson in the field of humanitarian knowledge. In their work, the authors use a combination of historical and logical research methods. The connection between the change in scientific pictures of the world and the evolution of fashionable processes is traced. The article provides a definition of the concept of quantum fashion and substantiates its applicability in social sciences and humanities. When applied to design, quantum fashion acts as a set of a wide variety of clothing styles, determined not by social factors, but by the creative interaction of a clothing designer with a consumer. Based on the creation of the Astromod brand (by Elena Makhinya), a number of empirical studies have been carried out, which have shown that quantum fashion is based on the synthesis of knowledge about the inner world of a person, his character, motivation and desires. As a practical result, a collection of clothes as an artifact was created, which is an additional source of energy for a person in the modern world.

Keywords: Quantum fashion · Clothing-artifact · Scientific pictures of the world · The principle of complementarity · Creativity · Technology · Interdisciplinary research

1 Quantum

The term "quantum" in modern scientific discourse is used so widely that it is necessary to clearly define the possibility of its application in the field of social and humanitarian knowledge. In particular, to figure out how correct it will be to apply the definition "quantum" to such a social phenomenon as fashion, which is not included in the natural science field.

© The Author(s), under exclusive license to Springer Nature Switzerland AG 2022
D. Bylieva and A. Nordmann (Eds.): PCSF 2021, LNNS 345, pp. 415–428, 2022.
https://doi.org/10.1007/978-3-030-89708-6_35

The use of this term in the humanities may seem especially strange - some trope, euphemism or metaphor [1]. However, the situation changes if we recall that the principle of complementarity, formulated by Niels Bohr back in 1927, lies at the heart of quantum mechanics. This principle, as a principle of the metatheoretical level of knowledge, was brought to life not only by the development of physics of the 20th century, but also by the humanities, whose representatives have always shunned both physics and exact natural science in general [2–4].

"Nevertheless, N. Bohr himself and a number of prominent specialists in the field of humanitarian knowledge, such as, for example, Roman Yakobson or M. Bakhtin, immediately realized the general scientific significance of the principle of complementarity. N. Bohr, in particular, perfectly understood that we are talking about a new methodology, about a new understanding of cognition in general, and that the subject boundaries of physics do not play any significant role here" [5].

The generalized essence of the principle of complementarity is that in any experiment the researcher receives information not about the properties of objects themselves, but about the properties of objects in connection with a specific situation, including, in particular, measuring instruments. As applied to quantum mechanics, the principle of complementarity means that when describing a phenomenon, it is necessary to apply two mutually exclusive ("additional") sets of classical concepts, for example, to consider them in a space-time and energy-impulse field. In this case, it becomes possible to obtain more complete information about this phenomenon.

Generalizing the principle of complementarity, N. Bohr gives it epistemological and ideological significance, trying to apply it to various fields of knowledge - biology, psychology, linguistics, culture in general, defining their "additional pairs" for them. This gave an impetus for the development of the theory of knowledge in a new direction.

"Among the concepts or situations that require an "additional way of description," are indicated, for example, such as reason and instinct, free will and determinism in psychology; concept and sound background in linguistics; mechanism and vitalism in biology; personal freedom and social equality in sociology; justice and mercy in jurisprudence, etc." [6]. As you can see, these pairs inevitably fall into the categories of rational and irrational. It is these two ways of knowing the world, according to A.P. Klimets, lead to the fullness of truth. "The unity of truth is the basic postulate of the human mind, expressed in the principle of complementarity of the rational and irrational aspects in the cognition of nature." [7].

Returning to our main topic, we can state that for successful marketing in the fashion industry (determining fashion trends of the future, developing costume design), it is precisely the quantum approach that is needed: a combination of rational analysis of many categories and circumstances (the emergence of new technologies and their development; social situation with change parameters of public censure or approval; cyclical evolution of fashion processes, etc., etc.) and reliance on the subconscious desires and needs of potential customers (as you know, subconscious motives are much stronger than conscious ones, with which they often enter into cognitive dissonance).

This principle of the fashion industry allows us to consider the term "quantum fashion" not only correct, but also natural. In predicting fashion trends and developing costume design, one more important principle of quantum mechanical analysis is also needed, about which the well-known philosopher of science M.A. Rozov wrote about its existence in the humanities. We are talking about the principle of inseparability, in which "… the perception of the whole does not consist of the perception of individual elements, on the contrary, the perception of elements is determined by the nature of the whole <…> The formulated principle is equally applicable in the field of linguistics, and in semiotics, and in the study of human activity in general" [5]. That is, the principle of inseparability means that it is difficult to define a particular phenomenon (statement, action) outside the general context. Fashion, the application of this principle is more than necessary: outside the context - historical, social, gender, age - a successful marketing policy in general and the development of a costume design is simply impossible. Suffice it to say that the evening dresses of the ladies of the Royal House of Windsor and the ladies who walk on the red carpet at the Oscars not only belong to different categories, but are also not compatible in one space (if many individual elements coincide: fabrics, jewelry, patterns, cost, etc.).

Already by its origin, the costume is an artifact, like a man-made object, the appearance of which in natural nature - without human intervention - is simply impossible. However, in cultural studies, the term "artifact" has acquired an additional definition ("cultural artifact") and additional properties. According to B.I. Kononenko, in cultural studies – "an artifact is a carrier of socio-cultural information, life-meaning meanings, a means of communication." The three ways of being a cultural artifact are material culture, spiritual culture and the area of human relations, "the main modalities of existence include: material (a form of objectification of an artificial object), functional (the amount of modifications during its use); semantic (its meanings, meanings, value in the contexts of sociocultural communication)".

The costume has always existed (and still exists today) in all three mentioned modalities; it marks not only the individual characteristics and the social, status position of its bearer, but in one way or another reflects the general picture of the world, the general principles of understanding the world order that are characteristic of humanity at one or another historical stage. This reflection is not always striking; it can be latent. Moreover, it is not always laid down (or even realized) by the creator and bearer of the costume. Nevertheless, careful analysis allows us to make sure that the costume is in direct connection with the general picture of the world, thus fully complying with the concept of a "cultural artifact".

Thus, artifact suits created on the principles of complementarity (rational and irrational) and inseparability fall under the definition of quantum fashion.

2 Historical Retrospective of the Costume as an Artifact

It is the semantic modality of the costume that shapes the fashion of each historical period, and, accordingly, largely (if not predominantly) determines consumer demand. The key to the successful functioning of an enterprise associated with the creation, production and sale of fashion products can be the correct definition of the dominant of the semantic concept of a modern suit. At the same time, it becomes extremely important that the semantic modality at different times is determined by various factors (ideological, religious, social, technological), which is determined by the picture of the world of a particular historical period [9, 10]. We can see confirmation of this even in a cursory diachronic survey of the existence of a costume.

Let us trace how the suit changes depending on the change in the scientific picture of the world as "an integral image of the subject of scientific research in its main systemic and structural characteristics, formed by means of fundamental concepts, ideas and principles of science at every stage of its historical development." [11].

The picture of the world in Ancient Greece was formed by both theogony and cosmological motives. This made the picture of the world discrete, internally contradictory: a priori, the unlimited power of the gods, their fatal influence on human life was recognized; but at the same time, ancient Greek science - and primarily philosophy - strove for knowledge of the laws of nature, space, man. The sovereignty of natural necessity was the ideological axiom of ancient philosophy. Let us analyze how the cosmological motives of the picture of the world were reflected in the ancient Greek artifact costume.

Let's remember that over the centuries - with all the changes in the ancient Greek costume - one thing remained unchanged: these clothes were never cut and almost never sewn. It consisted of rectangles; and the drapery brought to the level of high skill by the Hellenes gave it individuality and variety. It was the art of drapery and the variety of fabrics that marked the individuality and social status of the wearer of the costume. However, if we remember that the clothes of the ancient Greeks were decorated with ornaments (more often - with a border along the bottom, but sometimes completely), then we will be convinced of the semantic modality of the costume.

The three most common types of ancient Greek ornament that adorned not only decor items, but also costumes - spiral, meander and palmette. Let's leave aside the palmette, which performed rather decorative functions, and turn to the spiral and meander.

The spiral is one of the most important sacred symbols, a sign of creative life force both at the level of the cosmos and at the level of the microcosm. The movement within the spiral, this endless labyrinth, is the idea of eternal and continuous movement - whether the sun, human life; the idea of interconnection and identification of man with the cosmos.

The meander is essentially the same labyrinth, only translated into a different geometric form: from a closed circle to a linear dimension. Straight lines and right angles of the meander labyrinth semantically complement the philosophical meaning of the symbol, introducing into it a value orientation - the striving for the straight path, that is, for virtue. The movement along the meander includes sharp turns, retreats, looping. This is a sign of a difficult and ambiguous path, which is difficult, but necessary, to follow.

In the picture of the world of Ancient Rome, cosmological motives are reduced and practically ousted from public consciousness: they are replaced by ideas about the sacred mission of the Eternal City to carry civilization, the deification of emperors, a pronounced sociality of the cult. And accordingly, this is reflected in the semantics of the costume: sacred ornaments disappear from clothing; togas, tunics and other garments in Ancient Rome are either one-color or decorated with colored stripes; and in some cases the purple toga of the triumphant was adorned with scenes from Roman life embroidered in gold. One of the fashion trends is the desire to shock those around you with the excessive luxury of a suit, fabrics, and accessories. That is, social determination is clearly visible in the semantic modality of the ancient Roman costume.

The picture of the world in the Middle Ages was built on creationism, natural phenomena were interpreted as symbols of a certain relationship of God to man. The Christian anthropological concept is contradictory: a mortal material body and an immaterial immortal soul. This consistent and principled separation of the bodily and the spiritual was never overcome in medieval philosophy. The distinction between the internal and the external person, associated with the emerging Christian personalism in this era, led to the fact that "the internal space in the territorial, social and spiritual relation was privileged in relation to the external space" [12].

With all the national differences in the European medieval costume, general tendencies are evident, reflecting the picture of the world of that time - the subordinate, secondary position of man with his inner world in relation to God. When marking a sharp, contrasting difference in clothing of different social strata (fabrics, decor, abundance of jewelry or their complete absence, etc., etc.), all clothing was built according to a single layering principle. The natural lines of the "sinful body" were carefully hidden; incorporeal, refinement was emphasized, therefore - decency, purity and piety. Accordingly, the modality of the medieval costume is determined religiously and ideologically.

However, by the Late Middle Ages (XIV–XV centuries), the religious-ascetic dominant of the medieval costume began to erode and decrease. This was facilitated by at least two factors, moreover, from fundamentally different spheres: from the spiritual - the flourishing cult of the Beautiful Lady and from the material - the development of industrial production of fabrics and the improvement of the quality of their dyeing.

The neck gradually opens, small decollete appears; the waist is indicated. According to MN Mertsalova, it was in the Late Middle Ages that such a social phenomenon as "fashion" appeared, which in those conditions had "... only a few features: a relatively frequent change in the shape of a costume, a fascination with any novelty in those circles where it was created, excitement imitation" [13].

The semantic modality of the costume of the Late Middle Ages becomes the forerunner of the dominant costume of the next historical period, the Renaissance.

The Renaissance epoch covers a rather serious time period - almost three centuries. Of course, during this time the costume underwent modification and development; moreover: in the costume of different countries of Western Europe (Italy, Spain, France, England, Germany) there are very serious differences. Nevertheless, one cannot fail to see the general tendencies characteristic of this time, and the cardinally distinguishing the Renaissance costume from the medieval or the costumes of subsequent centuries. The theocentric dominant of the Middle Ages was contrasted with humanistic anthropocentrism.

Of course, this is reflected in the costume, where the reviving ideal of a harmonious, natural and beautiful person is clearly visible. Vertical lines, elongated shapes are replaced by horizontal breakdowns of the shape. The shapeless multi-layered robes that hide and deny the bodily origin of a person are replaced by clothes that emphasize (or even exaggerate) the real proportions of the human body: women have lush breasts and hips, a thin waist - visually emphasized by an extended dome-shaped skirt, neckline; in men - a powerful, convex, - often overhead, - chest, tight-fitting pants-stockings - sometimes with also overhead calves, a convex codpiece.

Man, his integral image, was again placed in the center of the picture of the world, the creative nature of man was affirmed. It was in this era that the prerequisites were created for the emergence of science as an independent social institution. At the beginning of the 17th century, during the scientific revolution, a mechanistic picture of the world was created.

The costume and fashion of this period underwent several drastic transformations; the semantic modality traces not only social determinism, but also technological, economic and political.

With the formation and strengthening of absolutism, the main trends in costume reflect a sharp demarcation of the fashion of the aristocrats and the lower estates, class isolation. Initially, this tendency manifested itself gradually: bright colors, which used to be the main tones of ceremonial and everyday high-society costumes, are leaving the costumes of aristocrats; pastel colors come into fashion - pink, blue, pearl, light yellow. And this change is not accidental, but natural: the development of trade relations led to the massive appearance in Europe of inexpensive fabric dyes - indigo, cochineal, - which leads to a reduction in the cost of production. Therefore, in the toilets of the upper strata of society, "… exquisite pastel shades, new, unusual halftones and nuances that would allow them to differ from the representatives of the middle classes, who have recently gained access to bright, rich, persistent tones, previously unattainable because of their high cost." [14].

In the further transformation of the aristocratic costume, the impossibility of engaging in any kind of physical labor was emphasized in every possible way - and, as a result, caricatured. The lines emphasized monumentality, majesty. A large powdered wig, a frock coat strewn with a huge amount of jewelry, and a lush lace frill made the men move slowly and sedately. A frame skirt is fixed in a women's suit - crinolines become more massive. This trend was especially clearly manifested in women's hairstyles - huge structures on a frame made of iron or wooden rods, which was disguised with its own and fake curls, lace, stuffed birds, fresh and porcelain flowers, feathers, models of ships, etc.

Under the influence of the mechanistic picture of the world, according to which the main criterion of beauty is order, eccentric, exaggerated forms leave the costume; there is a return to strict simplified lines.

The mechanical picture of the world remained relevant until the end of the 19th century, until the new discoveries of Albert Einstein in the field of theoretical physics; which marked the transition to the electromagnetic picture of the world. During this time (late 17th - early 19th centuries), with all the changes and transformations of the costume, the main trends persisted: simplification of cut and lines; gradual approach to naturalness, convenience and comfort. And, perhaps, the main thing in all these transformations was the preparation for a real revolution in fashion that took place at the turn of the 19th and 20th centuries.

At the origins of these revolutionary transformations were the Russian woman Nadezhda Lamanova and the Frenchman Paul Poiret. In their creative destinies - in the proximity and subsequent cardinal divergence - the interconnection of the world-view picture of the world and the artist's creative method is very clearly presented; conditionality of the costumeological paradigm by a specific cultural and semantic field.

The functional modality of the female theatrical costume, to a certain extent, became for both designers a kind of bridge thrown over the abyss dividing the costume-artifact of the 19th and 20th centuries to the everyday costume.

The lightweight silhouette of both designers was compensated by luxury: luxury of fabrics, finishes, accessories … It is difficult to say whether this was a deliberate decision to soften the impression of abrupt and unusual design decisions, or - a reflection of aesthetic ideas.

In the post-revolutionary years, N. Lamanova set herself a more difficult task of "feedback": the education of the artistic taste of the masses through the high aesthetics of clothing.

"Only through art can we come to the creation of new and better forms of life. It should penetrate into all areas of everyday life, developing artistic taste and flair among the masses. Clothing is one of the most suitable guides. We rarely get to exhibitions, museums, theaters, and everyday life, like furniture, dishes, clothes, surrounds us, every day with us; the eye gets used to both the good and the bad, therefore, artists should take the initiative in the field of clothing, working on creating from simple materials the simplest, but beautiful forms of clothing suitable for the new way of working life." [15].

Thus, Lamanova gives the costume a function that was not characteristic of him before, the function of an active participant in the formation of the spiritual world of a new person, an educator of his artistic taste [16, 17]. The costume as a cultural artifact is endowed by Lamanova with additional functions: it should not only demonstrate the individual and social-status characteristics of a particular wearer; not only reflect the ideological principles underlying this picture of the world, but also correspond to the nature of its activity; help a person, support him, filling him with additional positive energy.

At the same time, Lamanova in her theoretical developments already at the end of the 1920s. Anticipated the role of the costume-artifact, which is becoming relevant in the modern world, not only passively ascertaining, illustrative, but also active, effective.

This is especially important today, when the world in its modern quantum picture - somewhat unexpectedly for us - has turned out to be fragile; requiring not only active and thoughtless consumption of its natural resources, but also care, protection, careful and diligent attitude.

3   Quantum Picture of the World and Modern Fashion

As we could see, the change in the picture of the world occurred in mankind in leaps and bounds: new fundamental discoveries in the field of natural sciences produced a revolution in man's ideas about the world around him, about its structure.

The origins of the modern quantum picture of the world, formed only thanks to the latest discoveries of the natural sciences, can be found even in ancient Greek natural philosophy, which addressed the topic of nature, its principles and elements. Natural philosophers strove to comprehend the integrity of nature and its origin; to reveal the internal principles of the relationship of various levels of nature as a whole - from inorganic elements to human life. In the depths of natural philosophy, cosmocentrism was formed - the idea of the surrounding world as a huge and ordered whole - the

Cosmos. With all the differences between natural philosophy and the quantum picture of the world, it turns out that they are ontologically close to each other. Both pictures of the world are fully aware of the integrity of nature and man.

It became extremely clear that the only possible path of mankind in the conditions of unpredictable scientific and technological progress is the path of unity with nature, its preservation, respectful and loving attitude.

The quantum picture of the world was based on the theory of relativity of Albert Einstein and quantum mechanics, formed on the study of objects of the microworld - atoms and their constituent elementary particles.

The quantum picture of the world today is not fully formed - and, according to scientists from various fields of scientific knowledge, it will not be fully and consistently formed soon. However, new branches of quantum sciences have already appeared and are actively developing - quantum physics, quantum psychology, quantum biology, etc. and their philosophical foundations.

Most of the discussions - both in physics, and in the philosophy of consciousness, and in psychology - are caused by the topic of the quantum structure of the brain and the quantum nature of consciousness. In 1998, Roger Penrose presented an extraordinary hypothesis to the scientific community: the human brain is a superpowerful quantum computer, and its activity is determined by the laws of quantum physics [18].

Penrose came to these conclusions while working on problems of artificial intelligence. It is significant that Penrose not only supported, but entered into practical cooperation with the neurobiologist and anesthesiologist Stuart Hameroff, so this hypothesis was supported by both biology and medicine. And, although the theory of quantum consciousness is still considered a marginal area of science, its popularity is quite large. Such a young branch of science as quantum psychology is based on this theory. Let's use some postulates of the controversial theory of the quantum theory of consciousness and quantum psychology of Penrose-Hameroff, which help explain some of the modern paradoxes of social processes.

Among the postulates of this theory is the idea of the fundamental nature of consciousness, which acts as a property of the Universe. It consists in an organized self-collapse of wave functions that occurs in the brain at a very high frequency [12]. Hence, we can conclude that not all of a person's experience and preferences are shaped by the environment; social factors can be reduced, in some circumstances, - to a minimum. Each person creates his own model of reality. This postulate is present in both psychology and physics: "… the psychologist knows that each nervous system creates its own model of the world, and today's physics students know that each instrument also creates its own model of the world. In both psychology and physics, we have already outgrown the medieval Aristotelian concepts of "objective reality" and entered the non-Aristotelian world." [13].

Thus, we come to the conclusion that in the formation of a personal subjective picture of the world, value orientations and preferences, the dominant role is played by a person's individuality, his psychological makeup and nervous system. "Perception is not about passive reception of signals, but active interpretation of signals. Or, in a slightly different form: perception is not in passive reactions, but in active, creative trans-actions".

Accordingly, today a person has reached that level of development (and according to quantum theory, a person, as an inseparable part of the Universe, is in constant progressive movement towards improvement), - when he moves from the role of a contemplator to active cooperation with nature; creating a comfortable living environment combined with respect for it [19–22].

The reliance on these postulates radically changes the approach to the study of human needs. And, of course, this approach is of primary interest to us in the study of such a social institution as fashion, in which, as in a mirror, the general trends in the development of modern society are reflected.

A review of modern sources of fashion trends shows the absence of any dominant. If we take any other historical period, then the main trend of mass fashion one way or another could always be determined. It could be global; could be local; nevertheless he was. In the line of a cut (remember the fitted dresses with fluffy skirts of the 1950s or shoulder pads of the 80s); in skirt length (mini 1970s); in fabrics (nylon shirts and Bologna raincoats, which quickly burst into fashion and just as quickly left it), etc. Social determination is read in the existence of a dominant trend - for example, the desire to be modern; demonstrate their belonging to a particular social group. Of course, under the main dominant, distinctly sounding other tendencies arose, and here we can already speak of an ideological determinant. At the same time, the tendencies that existed in opposition showed a struggle. So, for example, in the antagonistic dialogue there were strict suits with ties of party and Komsomol functionaries and - rocker "leather jackets" or jeans sets, demonstrating the adherence of their owner to the Western ideals of freedom; "Decent" crimplen dresses and - deliberately careless hippie clothes; strict "office" style and - uniforms of informal gangs, goths, rappers, skinheads, etc.

Today, a variety of trends that are not combined with each other are in fashion side by side. And they are by no means in dialogue, but exist in parallel, on their own; and often - in one bow (one of the most common combinations is a romantic airy skirt and rough massive boots). In the public spaces of modern cities, we encounter a varied variety of styles and trends. Military and boho; romance and brutality; classic and freaky outrageous; tight knitwear and oversized clothing; glamor and grunge; the brightest "neon" disco style and calm casual; finally, the homeless style, in which expensive branded clothes are not only artificially aged and decorated with torn holes, but also "made up" to look dirty. Like the postmodern thesis about the equality of all discourses in modern culture, all trends are equal. It seems that the era of consumption is coming to an end, under the sign of which passed at least the last quarter of the 20th century and the beginning of the 21st century. If in the semantics of everyday life (and clothing, of course, also) of the consumer society, social determination is extremely strong (status, a clear demonstration of one's belonging to a certain social stratum - predominantly higher than the real one), then today the social determinant is clearly reduced. Preferences and value orientations in the choice of clothing are shifting to another area. And for the successful functioning of a fashion enterprise, it is imperative to define this area - in order to meet the needs of potential buyers. According to our hypothesis, at the present stage, the left-hemispheric model of thinking is replaced by a quantum one, conditioned not so much by social, common for a certain group, reasons, as by individual-psychological, emotional foundations. The left-brain and right-brain models of thinking appear in unity.

4 Experimental Justification of a New Approach to the Development of Costume Design

The study of individual preferences of potential customers was used as the basis for our experimental survey to adjust the development strategy of the Astromoda women's clothing brand. However, it is clear that a strictly individual approach is real with a system of individual tailoring, and is impracticable not only in mass, but also in small-circulation clothing production. In any case, a classification, grouping of individuals according to one or another characteristic is necessary. In this context, it seems optimal to classify our target audience by temperament, that is, the aggregate of individual personality traits that determine the dynamics of her behavior and mental activity.

Four classic types of temperament were used as the basis for targeted model development. However, for the convenience and simplicity of communication with our target audience, we used terminology taken from natural phylophy: the elements of water (melancholic), air (sanguine), earth (phlegmatic), fire (choleric) (Fig. 1). This terminology brings us back to the origins - in aggregate, it was in these four elements that the ancient Greek natural philosophers saw the basis of the existence of the Cosmos and man, as a part of this Universe. The philosophical ideas of the ancient Greeks about the structure of the cosmos found their continuation in Schelling's philosophy of nature and Russian cosmism. According to the greatest philosopher of science of the 20th century Paul Feyerabend, for the development of scientific theory, it is useful to compare scientific concepts with mythological natural philosophy and other forms of extrascientific knowledge.

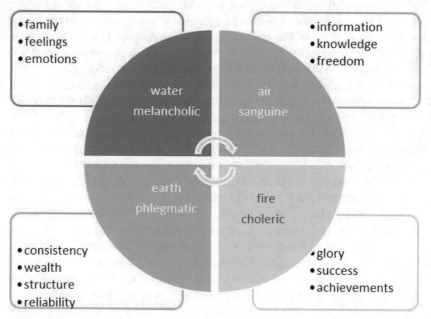

Fig. 1. Structural model of the relationship between character, elements and human motivation

Thus, we assume that for each type the source of energy and harmony will be its own specific set of sensations; accordingly, clothes for each character should fulfill their own, differently interpreted, energetic function.

In accordance with this hypothesis, the following models were developed within the Astromoda brand (Fig. 2).

The element of water. An ensemble of women's clothing, in which the predominant gamut - blue and deep blue - is 50%; complementary colors of the "airy" range - from gray to turquoise - about 30%; and the accent is from the colors of the elements of earth and fire. That is why a bright accessory (lemon hat) is appropriate in the ensemble created for the water sign, which will emphasize the weightless tenderness of the image of this element.

The element of the earth. For this sign, we mainly use green tones and all shades of the natural range - from brown to brick (50%), complementing the overall color with deep shades of blue (30% - water), and using accents of colors of fire or air.

The element of fire. In this case, the primary colors are yellow and red (50%); color companion - turquoise (element of air, 30%); accent accessories - colors of water or earth.

The element of air. The main range of the costume is all shades of the sky, from steel gray to deep turquoise (50%); accompaniment - the color of the element of fire (30%); and only complementary, accent accessories - colors of water and earth.

The refinement of this hypothesis is the results of a survey of our customers and an analysis of the dynamics of sales of experimental models.

Of course, scientific and practical work in this area is carried out not only in terms of color parameters; the preferences of customers and their comparison with the signs of the elements and other components are studied - in particular, prints, accessories, etc. Data collection (questionnaire; analysis of declared preferences and their material confirmation through monitoring sales results in these categories) continues. Conclusions can be made after the collection of a representative data set.

However, we can already formulate the tasks of quantum fashion: determining the internal balance of the elements of a particular person, his life motivations; transfer of an additional source of energy through clothing, which allows you to harmonize the internal structure of the personality. For a practicing designer, the following positions come out on top:

- determination of the type of temperament/element of a person;
- determination of the character of a person;
- selection of appropriate compositional means: colors,
- forms, materials for the development of the project.

As a result of the conducted research, we can assert:

1. That a person has an active creative role in the modern quantum picture of the world.
2. Art and fashion carry out an energetic function, are the basic needs of a person, the subject of the first and urgent need for the realization of his creative abilities. The purpose of quantum fashion: harmonization of the internal structure of a person.

Fig. 2. Samples of experimental models developed in accordance with the principles of quantum fashion Astromoda brand

3. The tasks of the designer are to reveal the internal balance of the elements in the chain: the psychotype of a person, character, value orientations, which determines the project search.

References

1. Waldner, D.: Schrödinger's cat and the dog that didn't bark: why quantum mechanics is (probably) irrelevant to the social sciences. Crit. Rev. **29**, 199–233 (2017). https://doi.org/10.1080/08913811.2017.1323431
2. Faye, J., Jaksland, R.: Barad, Bohr, and quantum mechanics. Synthese (2021). https://doi.org/10.1007/s11229-021-03160-1
3. Wendt, A.: Quantum Mind and Social Science. Cambridge University Press, Cambridge (2015). https://doi.org/10.1017/CBO9781316005163
4. Thomas, P.: Quantum Art & Uncertainty. Intellect, Bristol (2018)
5. Rozov, M.A.: The phenomenon of complementarity in the humanities. In: Lektorskiy, V.A., Ozeyrman, T.I. (eds.) Theory of Knowledge, vol. 4, pp. 208–227. Mysl, Moscow (1995)
6. Konstantinov, F.V. (ed.): Philosophical Encyclopedia. In: 5 Volumes, Soviet encyclopedia, Moscow (1960–1970)
7. Klimets, A.: Science and irrationalism. Quant. Magic **1**(3), 3220–323 (2004). http://quantmag.ppole.ru/quantmag/volumes/VOL132004/p3220.html. Accessed 14 May 2021
8. Kononenko, B.I.: A Large Explanatory Dictionary of Cultural Studies. Veche AST, Moscow (2003)
9. Kang, E.J.: Fashion history in light of Hegel's philosophy of history. In: A Dialectical Journey through Fashion and Philosophy, pp. 155–173. Springer, Singapore (2019). https://doi.org/10.1007/978-981-15-0814-1_10
10. Kang, E.J.: Fashion as a utopian impulse: the inversion of political economy via the consumption of fashion. In: A Dialectical Journey through Fashion and Philosophy, pp. 127–141. Springer, Singapore (2019). https://doi.org/10.1007/978-981-15-0814-1_8
11. Stepin, V.S.: History and philosophy of science. Academic project, Moscow (2011)
12. Hameroff, S.: Conversation with M. Karpov. https://lenta.ru/articles/2016/10/23/hameroff/. Accessed 30 Mar 2021
13. Wilson, R.A.: Quantum Psychology: How Brain Software Programs You and Your World. Hilaritas, Grand Junction (2016)
14. Pasturo, M.: Red: The History of a Color. Princeton University Press, Princeton (2017)
15. Philosophical aspect of the concept of quantum. http://philosophystorm.org/node/3706?fbclid=IwAR0omdFcLjF5bR5a3L2OyZhBQuWFyGepi_Sr8ZRh1nhHSStTZaZ4F6oRVgE. Accessed 20 Mar 2021
16. Bartlett, D.: Nadezhda Lamanova and Russian Pre-1917 modernity: between Haute Couture and Avant-garde Art. Fash. Theory. **21**, 35–77 (2017). https://doi.org/10.1080/1362704X.2016.1144953
17. Spieker, S.: Embedding constructivism. Art J. **79**, 116–118 (2020). https://doi.org/10.1080/00043249.2020.1750854
18. Penrose, R.: The Emperor's New Mind: Concerning Computers Minds and The Laws of Physics. Oxford University Press, Oxford (2002)
19. Zheng, S., Han, B., Wang, D., Ouyang, Z.: Ecological wisdom and inspiration underlying the planning and construction of ancient human settlements: case study of Hongcun UNESCO world heritage site in China. Sustainability **10**, 1345 (2018). https://doi.org/10.3390/su1005 1345

20. Krasnoshchekov, V., Arseniev, D., Rud', V., Switala, F, Chetiy, V.: Improving the quality of pre-master training of foreign students in the field of environment. IOP Conf. Ser. Earth. Environ. Sci. **39**, 012017 (2019). https://doi.org/10.1088/1755-1315/390/1/012017
21. Bylieva, D.S., Lobatyuk, V.V, Rubtsova, A.V: Information and communication technologies as an active principle of social change. IOP Conf. Ser. Earth Environ. Sci. **337**, 012054 (2019). https://doi.org/10.1088/1755-1315/337/1/012054
22. Pavlenko, A.: Technology as a new language of communication between the human being and the world. Technol. Lang. **1**, 91–96 (2020). https://doi.org/10.48417/technolang.2020.01.19

Movement Technology: From Kinetic Art to Digital Art

Maria Fedyukovskaya[1] ⓘ, Alexander Fedyukovsky[2](✉) ⓘ, and Denis Shatalov[2] ⓘ

[1] St. Petersburg University of Management Technologies and Economics, Lermontovsky, 44, 190103 St. Petersburg, Russia
[2] Peter the Great St. Petersburg Polytechnic University (SPbPU), Polytechnicheskaya, 29, 195251 St. Petersburg, Russia

Abstract. Movement has attracted artists for a very long time, however it became possible to apply movement in art only in the 20th century, when rapid development of technologies coincided with destructing the limits of artistic means to express creative thought. The authors consider kinetic art in terms of using various technologies, applied in artworks, such as the wind force, motors and electricity. Movement in Kinetic Art can be treated not only in a broad philosophical understanding as any change, but also in a phenomenological way as a fact sensually perceived, showing static kinetic sculptures paradoxically. Digital technologies have enormously expanded possibilities to create changing objects, but, at the same time, increased expectations from artwork. Demanding some knowledge of certain scientific laws and their technological use for creation, technological artworks are becoming a technology language. Having refused the role of dumb servants, the technologies can appear as a miracle or curse.

Keywords: Kinetic Art · Art · Kineticism · Movement · Technological art

1 Introduction

Ancient relationship of technology and art dates back to the Greek root "τέχνη", meaning "art, skill, craft", which essence, according to Martin Heidegger, consisted in aletheia, i.e., disclosure [1]. Being closely connected with creativity, artifacts' creation, and opposed to the nature, technology, and art dispersed far only in their contemporary understanding. Alfred Nordmann says, that it is possible to consider technology as language [2]. In the widest comprehension, technology can be represented as a relevant language of creating artificial environments [3, 4], in the narrowest one, it can speak to the person [5, 6].

Christopher Coenen and Alexandra Kazakova claim, that the machines, which movement is domesticated or civilized, represent self-expression of things [7]. Kinetic Art, in this regard, represents a specific way of demonstrating the movement created on the basis of science and technology. Initially, the word "kinetics" was not relevant to art. It was usually used only for defining the events relating to movement in physics and chemistry, and only since 1954 the term has been in the field of arts [8].

© The Author(s), under exclusive license to Springer Nature Switzerland AG 2022
D. Bylieva and A. Nordmann (Eds.): PCSF 2021, LNNS 345, pp. 429–437, 2022.
https://doi.org/10.1007/978-3-030-89708-6_36

The desire to subordinate movement to art, in order to provide the latter with the fourth dimension, has had a long history. There remained some evidence of creation of moving statues, figures of birds and animals by ancient Greek masters. Mechanical toys and tools can be considered as proceeding dreams of embodiment of movement in an art form until the early 20th century. In the 20th century, there were some experiments to create dynamic plasticity in Futurism, Dadaism, Bauhaus, Russian Constructivism, etc. E.g., it is possible to remember Duchamp's Bicycle Wheel (1913) and Rotary Glass Plates (1920) or an ambitious Vladimir Tatlin's project of the seven-story building with independently rotating floors of the Monument to the Third International (1920).

However, the official birth of Kinetic Art happened in 1955. "Le Mouvement" exhibition at the Denise Rene Gallery was an important event of art life. Here, performing kinetic and optical art was presented for the first time: Yaacov Agam (born in 1928); Pol Bury (1922–2005); Jean Tinguely (1925–1991); Jesús Rafael Soto (1923–2005). Jean Tinguely's artworks reminded suprematistic pictures, but all their parts moved. Around the gallery a car moved. It consisted of wheels and gears, and its details spinned too. McHale writes, that "If Dadaism, Surrealism, Constructivism, and their later variants have sensitized the contemporary vision to the metamorphosis of cultural values, often through a savage and corrosive irony, they have also provided a usable mythology of the machine and an insight into its creative potentialities" [9]. Burnham, however, saw in Kinetic Art 'pseudo machines', which, he considered failed to reconstruct or mimetically perform life successfully [10]. Chau sees the contemporary Kinetic Art only as a historical, machinic tendency [11].

Kinetic Art is formed in parallel with technical progress, engineering thought movement and science development. And the further this direction develops, the more it amazes with its technology to execute and realize the artist's plan. Therefore, Bariş Yilmaz considers, "… Kinetic art is the perfect art discipline in which 'artist engineer' and 'engineering of art' terms is used appropriately" [12]. Weibel considers Kinetic Art as the forerunner of Digital Art [13], however, it is possible to see, on the contrary, in contemporary moving self-organizing "robots" some logical continuation of the Kinetic Art, moved by modern technologies.

2   Kinetic Artworks in Terms of Technologies Used

Interest in Kinetic Art in the 20th century is dictated by both rapid development of technologies, admiring scientific and technical progress, and the expansion of art's limits to express the creative thought. As Popper notes, in the 20th century, there is a deep synthesis of technology and art [14]. Further we consider the technology solutions lying beyond kinetic artworks of different types.

The term "kinetic" defines belonging to the movement in physics and makes an impression that Kinetic Art surely means movement in space. But it is not so, the movement in Kinetic Art cannot be treated in a broad philosophical understanding only as change, but also paradoxically represent static kinetic sculptures.

2.1 Kinetic Art as Movement

The kinetic sculpture, the first in the art history, is considered to be *Standing Wave* (1920) by Naum Borisovitch Pevzner (better known by his pseudonym Naum Gabo). At its core there is a phenomenon of waves' interference which propagate in opposite directions, while their energy transfer is weakened or absent.

Reuben Margolin also uses waves in his sculptures, but adheres to the following principle: some set of elements is mono-mobile, i.e., they can only rotate or move strictly along the same axis. But when there appear to be plenty of elements, and they work not simultaneously but with a certain delay, such systems can imitate various types of waves, whether they are flat or spatial.

Therefore, for the sculptures working on the wave movement principle, all installation elements are to be strictly synchronized inter se, or else, the wave movements will be greased and indistinct, and at worst, no wave movement will be achieved at all.

One of the most ancient options to set sculptures in motion, which is used today, has been based on using wind force. Wind is capable to cause the movement of an object in any direction, even around its own axis, force it to plan or provide some impetus. This is what various trends to produce artworks are based on. Some of them rotate around their own axes, so-called "wind-driven generators". Having wide surfaces, they are set in motion by sufficient wind pressure. Similar sculptures have a uniform principle of work: they are moved by the pressure which forms wind on rather wide elements. Therefore, in such structures, the surface area calculation and this surface angle to air flow play a key role. E.g., the sculptures by Anthony Howe and Lyman Whitaker, which consist of a set of curvilinear elements set in motion even by a light breeze, are excellent illustrations of this principle.

Wind is capable not only to rotate, but also to move some objects in space. This is the core idea of the plantigrade car proposed by the Russian scientist Paphnutius Chebyshev in 1878. He was the first who embodied this idea and created the car on its basis, calling it plantigrade. The main unit of this machine transforms rotary motion to forward one; Theo Jansen' works are bright representatives of this trend.

Another option to create moving sculptures on the basis of wind is an exact calculation of the center of each element mass. George Rickey' works became a distinctive symbiosis of rotation and movement by means of wind. They are based on the following principle. In each of those works, the joints are most close to the center of gravity of each structural element. It allows all the elements to rotate freely around this joint and the sculpture, consisting of a set of such elements, to create complex trajectories of rotation in several directions. In these works, wind is used as the force which compensates the energy spent for overcoming friction force in the bearings. In Alexander Calder' works, it is this principle based on calculating the center of each element mass is used, which allows the master's works to move just from the air flow created by the viewers passing by.

Anyway, using wind as an engine imposes certain restrictions for sculptures: they are to be light, but, at the same time, strong and to maintain various weather conditions.

The spring mechanism is one of options to start artworks. The main objective is to provide long movement of the object with one spring windup. The movement in this case provides the energy release from the springs. The less energy expense there is in one fluctuation period, the better; for this purpose, the symmetry is used very often. This approach allows to balance the energy, required for one direction movement, and the energy, returning the element to its initial position. In order to minimize these energies, all kinematic couples are placed, whenever possible, to be closest to the masses' center to minimize the force moment. David C. Roy's works are an excellent illustration of such approach.

The similar principle is used also by all prototypes of "perpetual motion machines" which can also be classified as kinetic art-sculptures. Certainly, such artifacts as Newton's *Cradle*, cannot work eternally, and are only calculated so that, for every period of rotation, the energy would be minimally spent, which gives an impression that the mechanism can work eternally. Actually, due to the friction force, terrestrial gravitation, and heat released during the collision, it does not occur.

Therefore, in all sculptures based on the principle of energy release from something, whether it is a spring, cargo system or something else, the main requirement, which all these mechanisms have to fulfill, is to minimally spend energy to make movements. Therefore, while creating the objects light materials are used, and also symmetry bases and the correct arrangement of each element in the design are applied.

Using the engine in Kineticism opened a new era of development. In such works, the main idea was to transform simple rotary motions, most often from one engine, into complex progressive and rotary trajectories of some elements. For this purpose, sculptors use various junctions and ways of movement transfer. (Bob Potts's *Auspicious Messenger*, 2015, Derek Hugger's *Colibri,* and Tamara Kvesitadze's *Ali and Nino*). Fifteen motors in J. Tinguelyeight's *Homage to New York* worked with a completely different effect, creating irregular, sharp movements with sharp gnash and led to slow self-destruction [15].

One of the earliest electrically powered kinetic sculptures, *Light Prop for an Electric Stage*, was created by Laszlo Moholy-Nagy (together with an engineer and a technician) in 1930 [16]. It represented a metallic box with a circular opening at its front side and a second board with a parallel circular opening inside the box. It is interesting that the mechanical motion and lighting effects here played a role of the virtual volume creator. The most known sculpture by Shoffer, CYSP-11 (abbreviated from *Cybernetic and Spatiodynamic*). In 1956 it was presented for the first time at the poetic evening in Sara Bernhardt's theater in Paris, where the choreographer Maurice Bizhar staged a dancing performance with the use of electronic music. Made of steel and aluminum, and equipped with four motors allowing the sculpture to rotate and move, it reacted to intensity and color of external lighting and also to the acoustic environment nature.

2.2 Kinetic Art as Change

Knowledge of science laws allows to create artworks which change in time "by themselves" without any direct external influence. Unlike more effective moving artworks, changes here happen slowly but, therefore, life reveals more naturally in them.

The artist Hans Haake's works became a striking example of applying physical laws in art. In the period of his early works, Hans Haake modelled natural closed systems, using natural materials and science methodology. The artist created ice, water, earth sculptures. His famous works are *Condensation Cube* (1963–1965), *Ice Stick* (1964–1966), *Ice Table* (1967), *High Voltage Discharge Traveling* (1968) based on thermodynamics laws. E.g., *Condensation Cube* is a plastic cube, where little water is located, and towards the artwork a bright electric light is turned. When the temperature in the cube increases, the water starts to evaporate, but the moment the steam becomes saturated, it condenses – again turns into water. Hans Haake would say about this work, "What looks like a box, slowly but constantly, changes and never recurs. These conditions are similar to a living organism which flexibly reacts to the environment" [11].

Another work by Hans Haake, based on physical phenomena, is *Ice Stick*. The artwork is a long copper coil containing a refrigerator and transformer connected to a power source. The coil extracts moisture out of the museum air, converting quickly into ice. Slowly but persistently, the coil is covered with a non-transparent opaque ice cover of its own production, but when the humidity in the room decreases, the amount of ice on the stick lessens.

The artwork *High Voltage Discharge Traveling* is also simulated on the basis of physical phenomena, but, compared to the other early works by Hans Haake, it is based on electrodynamics laws. The crown discharge which is generated by a strong sharply non-uniform pulse electric field. In the area, directly adjoining the electrode with a small curvature radius, there is a partial discharge – the corona. The temperature increase leads to an intensive luminescence including ultraviolet radiation [11].

Robert Rauschenberg's *Mud Muse* (1966) is 1,000 gallons of bentonite clay mixed with water, which, in response to the level of music or sounds loudness, bubbles, and gushes forth with different intensity, stimulated by pulsing air valves beneath its surface.

2.3 Kinetic Art as Statics

Frequently in Kineticism statics is used. When movement is expressed in statics, it represents an art expression of phenomenological comprehension of movement as a sensually perceived fact. The movement in similar sculptures exists only in the viewer's perception.

E.g., Vyacheslav Koleychuk (*Atom*, 1967) uses some strong cables keeping all the system in balance, but since the cables are thin, they create the feeling of tension in this sculpture. Therefore, such structures demand special rigidity and durability of junctions and materials to be applied. Due to this reason, while producing similar sculptures light, and at the same time strong materials are used.

Balancing of stones, the art which is popular in different countries of the world, is one of the most ancient manifestations of static Kineticism. In Llano (Texas) they even regularly hold corresponding world championships. In "Kislovodsk" national park (Russia) Vitaly Sapozhnikov creates stone pyramids to inevitably be scattered in front of the viewer. The structure of variegated stones is rather steady due to very exact locating the mass centers of stones, one above the other, thus there is an ideal balance.

Another direction of statics in Kineticism is creation of illusory movement (*Permeable* by Rafael Soto). In this case, it is necessary to arrange all the elements in a special way that the viewer, in different positions regarding the installation, i.e., interacting with the object, perceived the object movement, though, in this case, there is no kinetic movement per se.

Kinetic column Lumidyne system, created by Frank Malina in 1961, is a back board for supporting fluorescent or incandescent lights, an electric motor for driving the Rotor, and reflecting surfaces, such as mirrors, allowing to create the object which surface constantly changes [17]. His other light kinetic artwork, *Reflectodyne*, consisted of polished rods or mirrors, which were driven by a chain or gear drive, or could be controlled by sound, thanks to a reversible motor, which was fed through a circuit with a microphone, an amplifier and a two-way relay [18].

In Gregory Barsamian's *Juggler* (1997) there is some movement, but it is not perceived what it is. A large cylindrical frame, quickly rotating and invisible in the darkness, creates the illusion of the fact that the people, standing around it, juggle with the objects (as material embodiment of serial photographs or pictures in a flip book).

3   What "Movement" in Artworks Means

In Fig. 1 there is a scheme showing the actual dynamism and complexity of the technologies applied.

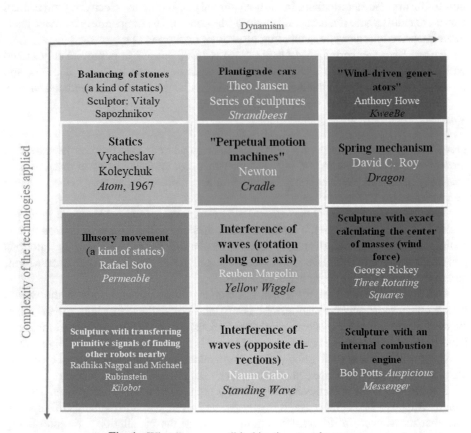

Fig. 1. What "movement" in kinetic artworks means.

Balancing of stones is the least dynamic and technologically simple artwork. Such sculptures are not mobile, the only energy spent is creating the object by the master. Static sculptures demand minimum external influence for little movement. The mechanisms using wind force are known for a long time and do not require complex technology solutions. The wind moving objects can transform the little rotary motion from the wind force to the forward one (e.g., *Plantigrade car* by Paphnutius Chebyshev). But movement provides less dynamism than change on the spot, e.g., Alexander Calder's "mobiles" move and rotate due to an easy whiff from opening a door or from a passer-by. The artworks created according to thermodynamics laws start moving after a long period, but a large amount of energy is required for this purpose. In Hans Haake's *Condensation Cube* the motionless system obtains its mobility due to transforming the lamp heat power into the water phase transition. The figures created by the interference of waves are to be exactly synchronized, which demands complex technological calculations, and each element being mono-moving. Therefore, such sculptures are more energy-intensive but slow-moving. *Standing Wave* by Naum Pevzner, where there is no energy transfer, imitates various types of waves. The sculptures created according to the principle of "perpetual motion machine" demand a considerable external influence, and their movements become more dynamic. Newton's *Cradle* is based on transformation of the kinetic energy to the potential one which the viewer initially gives to the system by withdrawing a marginal ball. The stronger the initial impulse is, the more dynamic image there is. The spring mechanism operates according to the same mechanisms, but in order to deform the spring it is necessary to make more efforts, and also the sculpture movement looks "more alive" since the system attenuation is more obvious than a monotonous *Cradle's*. While all preceding structures have faltering actions, the artworks which structures include electric motors and engines usually smoothly and evenly move, since energy supply is constant (until the device failure or the fuel supply interruption).

4   Modern Options of Movement and Changes in Artworks

Though movement and change become more and more demanded characteristics of artworks [19], the term "Kinetic Art" remained in the past. On the one hand, digital technologies have created unique opportunities to create changing sculptures. On the other hand, using these technologies everywhere increases expectations for artworks. A modern option of the changing artwork can be called *Volumetric Solar Equation* (Rafael Lozano-Hemmer, 2018) which is a huge chandelier of LED lights on battens, which changes reflect the turbulence, flares, and spots visible on the surface of the sun from the latest photos of NASA's Solar Dynamics Observatory (SDO) by means of reaction-diffusion, Navier-Stokes, Perlin noise, and fractal flames [20]. The possibility of reaction to the viewer's actions is a special feature of moving artworks where digital technologies are involved. According to researchers, contact with works make viewers realize the joy of discovery and creativity via mutual interaction [21, 22]. In *Pareidolium* (Rafael Lozano-Hemmer, 2018) the visitor can see his or her face mid-air with clouds of vapor that ascend from the water basin by means of computer-controlled ultrasonic atomizers and the facial recognition system.

The simplest option is to implement the projects providing the generation of changing images or volumes on surfaces. E.g., in Maurizio Bolognini's *Collective Intelligence Machines* (CIMs), commenced in 2000, the image on urban surfaces is generated by means of SMS from mobile phones [23]. Refik Anadol in *Melting Memories* uses EEG data collected on the neural mechanisms of cognitive control of the viewer in order to generate the volume changing digital artworks.

In the meantime, it is impossible to state that nowadays any interest in self-organizing movement is completely lost. The Harvard scientists Radhika Nagpal and Michael Rubinstein's *Kilobot* is an example of using modern technologies. The basic principle used in this project is transferring primitive signals of finding other robots nearby by means of infrared signals. The use of new technologies in similar projects opens infinite opportunities to realize the human thought in artworks. Here the use of various achievements in science, including the latest developments in the field of artificial intelligence and neural networks, is possible.

5   Conclusion

Kinetic Art is a common ground of art, technology, and philosophy. In order to be created, artworks demand some knowledge of certain scientific laws and their technological use, and often force technologies to start talking. Having refused a role of dumb servants, technologies can appear as a miracle or a curse.

The characteristic of classical kinetic artworks is their autonomous existence. They exist per se without direct external influence of the person. Changes or seeming changes in their existence occur "per se", it seems to have been their main, most fascinating feature.

Digital technologies have enormously expanded the possibilities to create changing objects, and, in the meantime, wide circulating of these technological "miracles" increases expectations for the artwork.

References

1. Heidegger, M.: Holzwege. Vittorio Klostermann, Frankfurt a/M (1977)
2. Nordmann, A.: The grammar of things. Technol. Lang. **1**, 85–90 (2020). https://doi.org/10.48417/technolang.2020.01.18
3. Nesterov, A.: Technology as semiosis. Technol. Lang. **1**, 71–80 (2020). https://doi.org/10.48417/technolang.2020.01.16
4. Pavlenko, A.: Technology as a new language of communication between the human being and the world. Technol. Lang. **1**, 91–96 (2020). https://doi.org/10.48417/technolang.2020.01.19
5. Bylieva, D.: The language of human-machine communication. Technol. Lang. **1**, 16–21 (2020). https://doi.org/10.48417/technolang.2020.01.04
6. Coeckelbergh, M.: When machines talk: a brief analysis of some relations between technology and language. Technol. Lang. **1**, 28–33 (2020). https://doi.org/10.48417/technolang.2020.01.05
7. Coenen, C., Kazakova, A.: Utopian grammars of human-machine interaction. Technol. Lang. **2**, 67–80 (2021). https://doi.org/10.48417/technolang.2021.01.06
8. Germaner, S.: Art After 1960. Kabalci Publishing House, Istanbul (1997)

9. McHale, J.: The Future of the Future. George Braziller, New York (1969)
10. Burnham, J.: Beyond Modern Sculpture: the Effects of Science and Technology on the Sculpture of this Century. G. Braziller, New York (1968)
11. Chau, C.: Systems aesthetics: a key polemic in contemporary kinetic art history. In: Chau, C. (eds.) Movement, Time, Technology, and Art. Springer Series on Cultural Computing SSCC, pp. 57–73. Springer, Singapore (2017). https://doi.org/10.1007/978-981-10-4705-3_4
12. Yilmaz, B.: Art engineering and kinetic art. J. Arts Humanit. **3**, 16–21 (2014). https://doi.org/10.18533/journal.v3i12.628
13. Weibel, P.: It is forbidden not to touch: some remarks on the (Forgotten Parts of The) history of interactivity and virtuality. In: Grau, O. (eds.) Media Art History, pp. 21–43. MIT Press, Cambridge (2007)
14. Popper, F.: From Technological to Virtual Art. The MIT Press, Cambridge, MA (2007)
15. Folland, T.: Jean Tinguely, Homage to New York. https://www.khanacademy.org/humanities/art-1010/post-war-european-art/postwar-art-in-switzerland/a/jean-tinguely-homage-to-new-york. Accessed 4 Apr 2021
16. From the Harvard Art Museums' collections Light Prop for an Electric Stage (Light-Space Modulator). https://harvardartmuseums.org/collections/object/299819. Accessed 4 Apr 2021
17. Malina, F.J.: Kinetic painting: the Lumidyne system. Leonardo **40**, 81–90 (2007)
18. Malina, F.J.: Electric light as a medium in the visual fine arts: a memoir. Leonardo **8**, 109–119 (1975)
19. Bylieva, D., Lobatyuk, V., Ershova, N.: Computer technology in art (Venice Biennale 2019). In: Proceedings of Communicative strategies of the information society (CSIS'19), p. 18. ACM, Saint – Petersburg (2019). https://doi.org/10.1145/3373722.3373785
20. Rafael Lozano-Hemmer - Project "Volumetric Solar Equation". https://www.lozano-hemmer.com/volumetric_solar_equation.php. Accessed 1 Apr 2021
21. Chen, G.-D., Lin, C.-W., Fan, H.-W.: The history and evolution of kinetic art. Int. J. Soc. Sci. Humanit. **5**, 922–930 (2015)
22. Ammon, S.: Epilogue: the rise of imagery in the age of modeling. Philos. Eng. Technol. **28**, 287–312 (2017). https://doi.org/10.1007/978-3-319-56466-1_12
23. Bolognini, M.: The SMSMS project: collective intelligence machines in the digital city. Leonardo **37**, 147–149 (2004)

Role of Digital Technologies in the Conceptual Transforming Foundations of Artistic Creativity

Sergei Mezentsev(✉)

National Research Moscow State University of Civil Engineering, 129337 Moscow, Russia

Abstract. Technology in this article refers the transformation of substance, energy and information. Particular attention is paid to digital technologies, which were preceded by analog technologies. If analog technologies were at the heart of socialist realism, then digital technologies became the foundation for postmodernism. In socialist realism, the dominant positions were occupied by communist ideology and propaganda, and technologies in artistic creation performed mainly an auxiliary function. Their main task was to replicate artistic creations. Innovation and creativity in artistic creation were strictly limited by the party and political structures of the Soviet state. In postmodernism, in contrast to socialist realism, external restrictions are insignificant. The state allows the pluralism of opinions and the variety of artistic approaches, and digital technologies make art accessible to everyone. For artistic creation, there are ample opportunities to demonstrate innovation and creativity. The main thing here is the subject of creativity, the artist. This article defends that the thesis that the transition from socialist realism to postmodernism was primarily due to the development of technologies from analog to digital. The decisive role was played by digital technologies, and not by the sociopolitical changes that took place in the Soviet Union and other socialist countries at the turn of the 80s–90s of the twentieth century. The study employs methods drawn from conceptual analysis, hermeneutics, phenomenology, observation, comparison, as well as visual methods.

Keywords: Digital technologies · Artistic creativity · Innovation · Art transformation

1 Introduction

Technology plays a leading role in the information society. Technology in the broadest sense of the word is a set of processes, actions, operations and principles that forms the technosphere, the state of which is determined both by the achieved level of technical development and by socio-cultural factors. If technology in the traditional sense is a technical and organizational projection on production activities, then technology in the modern sense is a technical and organizational projection on the social system and of society's culture on the whole. Technology as a set of rules, techniques, methods and methods, for example, extraction and processing of raw materials are studied by technical sciences, decision-making, management of social processes are studied by social sciences and humanities. Technology in a philosophical sense is a set of rational and methodological methods of systemic management or optimal organization of purposeful processes of transformation of matter, energy and information [1].

Digital technologies are one of the widespread technologies today. Digital technologies are technologies that are used by electronic computers for recording code pulses in a strict sequence and with a certain frequency [2].

Digital technologies have not only replaced analog technology, which use hardware to record a signal described by continuous functions. They changed the conceptual foundations of artistic creation. The digital revolution in the second half of the twentieth century spread not only to society, culture, but also to art, and had a significant impact on its formation and development. Since ancient times, technology and art have interacted with each other, but in different ways. If in the past technology with art predominantly applied in nature, was an intermediary between the artist's creative intention and his material embodiment, nowadays technologies are transforming traditional art, changing its meanings and purpose. The role of the artist is now to create the art of technology [3].

The transformation of the conceptual foundations of artistic creativity was most clearly manifested during the transition of world culture and art from modernism to postmodernism through socialist realism in the Soviet Union and other former socialist countries [4].

Socialist realism is the artistic method of literature and art, which is an aesthetic expression of the Marxist-Leninist dialectical-materialist concept, due to the era of Bolsheviks' struggling for the socialist society's founding and construction. The portrayal of socialist ideals determined both the content and the basic artistic and structural principles of literature and art.

The term "socialist realism" first appeared in print in 1932 and already in 1934 was approved as the basic constructive method of Soviet literature at the First All-Union Congress of Writers. In the Union of Writers' Charter they noted that the Soviet Union writer's task was depicting reality in its revolutionary process of developing, and combining the historical exactness of artistic depiction with the ideological and political education of the working masses in the spirit of socialism [5]. By the principle of ideology, the writers had to show the peaceful life of the people, the search for ways to a new, better, happy life, the heroic deeds of the workers and peasants.

Soon, the principle of nationality was added to this principle, the meaning of which consisted in the comprehensibility and accessibility of art for the perception of ordinary people, in reflecting their lives and interests, in using folk vernaculars, proverbs, sayings, etc. The third principle was the principle of concreteness, according to which, in the depiction of reality, writers and artists had to show the process of historical development, corresponding to the dialectical-materialist understanding of history. As a result, literature and art turned into a powerful tool of ideological influence, and writers were called "engineers of human souls" [6].

Socialist realism was at that time a completely new phenomenon in human society and world culture, it was, in fact, the social norm of creativity. It meant not an only imitation of the great examples of the world heritage of realistic art, but also the use of a creative, innovative, in modern language, pioneering approach. Artists were supposed to connect their works with modern and active participation in the construction of socialist society. The method tasks required from each society member: philosopher, writer, the artist a "true" understanding of the meaning of the transformations taking place in the country, the ability to give a highly qualified assessment of the phenomena of social life in their complex dialectical interaction and development [7].

The Soviet state asserted socialist realism not only to establish control over creative individuals but also to strengthen the propaganda of its policies. During the period of collectivization and industrialization of Soviet power, they needed art to encourage people to perform feats of labor. During the Great Patriotic War, socialist realism was called upon to raise the people to fight the enemy both at the front and in the rear, to inspire and lead to a great victory.

To realize these goals, the Soviet state was very interested in the development of analog technologies, which were used then in radio, television, telegraph, telephone and other media. Technologies did not play an important role in artistic creation, not only because of its low level of development but also for the simple reason that from the mid-30 s to the mid-50 s, it was completely under state control and depended on communist ideology and propaganda.

However, the further development of technologies created more and more opportunities for freedom of creativity within the USSR and acquaintance with the achievements of culture and art of foreign countries. Technological advances inevitably destroyed the Iron Curtain. This was especially evident in the music. For example, the legendary British rock band The Beatles enjoyed fantastic popularity among Soviet music fans [8]. With the advent of radios, tape recorders, and clandestine studios, the demand for Western music increased. And not only to music but also to other works of Western and world art.

In the process of the formation and development of digital technologies, socialist realism had no prospects. An attempt by M.S. Gorbachev's liberation of socialist realism from ideology and propaganda failed. But the rapprochement of the USSR with the technological West, the idea of universal human values, as well as the failure of the August coup in 1991 turned out to be similar to the "Window to Europe" cut through by Peter the Great at the beginning of the eighteenth century.

Socialist realism was replaced by postmodernism, a large role in the emergence of which digital technologies played, – a broad trend in a culture that has replaced modernism in the West and is caused by the so-called "death of super foundations": God, man and author. Postmodernism is characterized by the following features: (1) unlimited, absolute freedom in the choice of methods of self-expression; (2) rethinking traditional images, including them in a new context of artistic depiction; (3) syncretism, the combination into a single whole of heterogeneous elements, even contradicting to each other; (4) dialogical, a look from different sides, creating a polyphonic "symphony"; (5) game form of presentation, invitation of the audience to join the game with meanings; (6) the shocking nature of creativity, demonstrative scandalousness; (7) irony and self-irony [9].

The use of digital technologies over the past decades has significantly changed the process of creating artistic works, which, thanks to the Internet, is now available to everyone.

2 Methods

The study used methods such as analysis, hermeneutics, phenomenology, observation, comparison, as well as visual method.

Thanks to the method of analysis, it was possible to identify and evaluate the philosophical content of the studied scientific texts, literary works and art phenomena. The use of the hermeneutic method made it possible to reveal and understand their meaning and to make an interpretation from the standpoint of a realistic perception of reality. The phenomenological method contributed to the discovery of the original meanings in the process of creation of the construction of ideal objects of the world. With the help of the observation method, facts, knowledge, life experience of different generations and the personal experience of the author of this article were accumulated and comprehended. Through the method of comparison, the similarities and differences were established between socialist realism and postmodernism. The visual method made it possible to represent visually the transformation of the conceptual foundations of artistic creation using the example of works of art.

3 Results

The formation of socialist realism in the 30s and 40s was closely connected with the pathos of the establishment of a new life, with the praise of the Great October Socialist Revolution, of Bolsheviks and Red Army soldiers' glorification, because they showed heroism during the past Civil War and the intervention of imperialist states, the glorification of labor exploits in the process of collectivization and industrialization, a celebration of victory in the Great Patriotic War, post-war restoration of the national economy. Philosophy, literature, the art of those decades created the socialist image of a positive hero – a leader, fighter, defender, builder, worker, mother. The means of artistic expression were novels, poems, verses, songs, music, films, sculpture, architecture. Printed publications (books, brochures, newspapers, posters, photos, portraits), video and audio equipment (television, theater, cinema, radio, gramophones) were mainly used as technological means. These are traditional information technologies aimed only at ways of fixing, replicating and disseminating information.

Fig. 1. "Lenin on the podium".

Fig. 2. "The Motherland Calls!".

Fig. 3. "5-year-old in 4 years – we will do it!".

Vivid examples in the field of literature of that time are the works of M. Gorky's works "Mother", N.A. Ostrovsky's "How the Steel Was Tempered", M.A. Sholokhov's "Virgin Soil Upturned", A.T. Tvardovsky "Vasily Terkin" [10]. In the field of art, it should note the works of such Soviet artists as, for example, A.M. Gerasimov ("Lenin on the podium") (Fig. 1), I.M. Toidze ("The Motherland Calls!") (Fig. 2), V.S. Ivanov ("5-year-old in 4 years – we will do it!") (Fig. 3) [11].

One of the creative approaches in the artistic depiction of reality was the use of prototypes. Thus, Pyotr Zalomov, a worker from Sormovo, and his mother were the prototypes of the main characters in M. Gorky's novel "Mother": Pavel Vlasov and Pelageya Nilovna. According to Gorky, the art of verbal creativity, the creation of characters and types of artistic personages requires imagination, guesswork, "invention" based on the principles of abstraction and concretization [12], highlighting and generalizing in the person of one artistic character, the most characteristic features characteristic of certain people: merchants, nobles, workers, etc. Among the artists must be named I.M. Toidze, who used his wife Tamara as a generalized image of a woman-mother calling every Soviet person to fight against the enemy.

Initially, socialist realism was aimed at the artistic depiction of the best human qualities, manifesting in the revolutionary struggle, in the struggle against internal and external enemies, in labor activity, in public and personal life. This had to serve to the building socialism and communism. The real reality, not embellished, had no interest for adherents and supporters of socialist realism. Not what it is, but how it should be – this is the basic premise of socialist realism. Socialist realism obliged Soviet philosophers, writers, artists and other representatives of creative professions to extol heroic deeds, social, collective and personal achievements of workers in the development of industry, agriculture, transport, education, science, in the study of outer space. Everything that did not serve to the socialist revolution, did not correspond to the communist ideology, was indiscriminately declared by the representatives of socialist realism to be bourgeois, vulgar, unscientific. "Bourgeois philosophy" and "bourgeois science" are very widespread propaganda labels that were glued to all non-Marxist doctrines in Soviet times [13].

Following the policy of the Soviet state, all non-Marxist trends in philosophy, literature, art and the media were also subjected to harsh criticism and prohibitions. In disgrace were not only those philosophers who had previously emigrated or were exiled from the

Fig. 4. "Theater Square and its surroundings".

country as part of the "philosophical steamer" [14], but also those who remained inside the country, in particular, P.A. Florensky and A.F. Losev. Books by M.A. Bulgakov ("The Master and Margarita", "Heart of a Dog"), B.L. Pasternak ("Doctor Zhivago"), A.I. Solzhenitsyn ("The Gulag Archipelago", "In the First Circle", "Ivan Denisovich's one day"), etc. [15]. Even such artists who were hard to suspect in anti-Soviet activities, as F.I. Veseli, author of many cartoons published in Soviet newspapers, were subjected to merciless criticism, persecution and execution (Fig. 4) [16]. It was in these forbidden works, painting, that a genuine reflection of reality took place, but not in socialist realism, which had become detached from reality and ultimately lost its former creative character.

The Soviet state dealt the main blow to modernism, which became the main enemy of socialist realism, it's class enemy, standing in the way of artistic and social progress. Although, in terms of its content, modernism was supposed to be supported by the Soviet state, because this trend in art was characterized by the denial of its predecessors, the destruction of established ideas, traditional ideas, forms, genres, and the desire to search for new ways of perceiving and reflecting reality. Socialist realism shared these positions, but only about the bourgeois class, the capitalist structure of society, but for the builders of socialism, modernism turned out to be unacceptable. Only Soviet modernism was allowed as a style in architecture that emerged during the "Khrushchev thaw" [17]. Soviet modernism includes, in particular, the building of the TASS Information Agency (now ITAR-TASS). The most unusual feature of this building for the Soviet era was a large number of rounded windows. They meant television screens from which Soviet people look at the world around them [18]. This creative idea of architectural art was creative, futuristic and anticipated our information century (Fig. 5).

Fig. 5. The ITAR-TASS building in Tverskoy Boulevard

The further, the more socialist realism became detached from reality. Reality did not keep up with socialist realism, which raised the bar of society and Soviet people' idealization higher and higher. This became especially noticeable in the 70–80s. By the mid-1980s, the gap between socialist realism, on the one hand, and reality, on the other, had reached alarming proportions. Socialist realism, based on the principle of partyness, has become an element of communist ideology, socialist mythology and propaganda. The Soviet Union ended up in a deep crisis, which far-sighted thinkers had long warned about. So, for example, the philosopher N.O. Losskiy argued that in the USSR dialectical materialism turned into a party philosophy of Bolshevism, and it was dogmatized, began to serve not the purposes of the search for truth, but the practical needs of the revolution. Based on this, he made a prophetic conclusion that while Soviet power is reigning, which suppresses the freedom of creativity, research, discussion, Marxism-Leninism is not capable of fruitful development [19]. From this point of view, the collapse of socialist realism, as well as Soviet socialism, seemed inevitable. Gorbachev's policy of perestroika and glasnost, which was aimed, among other things, at the abolition of censorship restrictions and the establishment of the principles of freedom of speech and exchange of information [20], was not crowned with success.

After the failure of the August putsch, Marxism-Leninism was removed from the agenda. Russian philosophy found itself in a situation of uncertainty. The Soviet Union has sunk into oblivion. Many post-Soviet countries, including Russia, moved towards postmodernism, borrowing phenomena from Western social life and culture of the second half of the twentieth century [21]. To a great extent, it was largely facilitated by the information and computer revolution that took place in the 90s.

The information and computer revolution is a qualitative transition to the creation and large-scale implementation in all spheres of society and the activities of every person of computer equipment and information and communication technologies used for the acquisition, storage, production, transformation and transmission of information.

The information and computer revolution is a qualitative transition to the production and mass implementation in all spheres of society and the activities of each individual of computer equipment and information and communication technologies to create, acquire, storing, transforming and transmitting information [22]. Thanks to the information and computer revolution, completely new technologies appeared, including digital ones, which began to have a huge impact on the process of creating, processing and semantic transformation of artistic creations.

The introduction of digital technologies in post-Soviet Russia occurs in two ways: (1) digitalization of products of artistic creativity previously created or currently being created traditionally and (2) using them in the process of creating artefacts.

The political and cultural changes in Russia that occurred as a result of the collapse of the USSR also left their mark on artistic creativity. In the 90s, almost anything could be done. If earlier postmodern works were published by Russian authors abroad or in samizdat, now they began to be published without censorship and restrictions. These are, for example, the works of V.V. Nabokov ("Pale Fire", "Lolita"), A.V. Sokolov ("School for Fools", "Palisandria"), V.V. Erofeeva ("Moscow – Petushki"), E.V. Limonova (Savenko) ("The Loser's Diary", "The Last Days of Superman"), V.O. Pelevin ("Chapaev and Emptiness", "The Caretaker"), V.G. Sorokina ("Blue bacon"), L.S. Petrushevskaya ("Love", "Wild Animal Tales") and others [23]. In the literature, such features as fragmentation, irony, black humor, play, parody, meta-prose, fabulation, blurring the boundaries between elite and mass culture began to be asserted [24].

Postmodernists include such contemporary Russian artists as, for example, L.A. Purygin ("At the Cemetery") (Fig. 6), D.G. Gutov ("Study of a Man in a Turban") (Fig. 7), A.B. Shubin ("Still life with a horse in the window") (Fig. 8) [25].

Fig. 6. "At the Cemetery".   **Fig. 7.** "Study of a   **Fig. 8.** "Still life with a
                                      Man in a Turban".   horse in the window".

Postmodern art styles include collage-based Dadaism; pop art that synthesized popular culture and art; conceptualism, which prioritizes the transmission of ideas; abstraction (post-pictorial and lyrical); assemblage – a collage of volumetric elements; color field; process art, in which the presentation of the work to the viewer is an integral part of the creative process. Modern artists not only provide an opportunity for the viewer to see traditional paintings in three-dimensional measuring, holographic images but also to plunge into the past or move to the alternative reality with the help of VR technologies, to look at familiar things in a new way [26].

Thanks to digital technology, innovative art forms such as net art, bio-art, nano-art, digitograph, WEB-design, and augmented reality (AR) technology have appeared. They are digital technologics that are changing today the conceptual foundations of artistic creativity, forming new, digital arts, such as video art, digital painting, computer graphics, computer animation, electronic music, digital literature [27, 28]. At the same time, digital art continues to interact actively with analog art and to exert influence on it, especially in areas such as graphics, sculpture and painting [29].

Fig. 9. "Bank Bridge in St. Petersburg". Art.   **Fig. 10.** "Masyanya". Art. O. Kuvaev
B. Slobodan.

Digital technologies are used both for the representation of images characteristic of traditional art (Fig. 9) and for the creation of new, virtual images (Fig. 10). In the latter case, one of the striking examples of innovation and creativity is the artistic image of Masyanya, the heroine of flash cartoons created by the artist O. Kuvaev [30].

With the advent of digital technologies, interactivity has become possible, in other words, the opportunity for the viewer to come into contact with the artist and participate in the creation of works, network art. Currently, network art is not limited by anything and anyone. It becomes reality, forming networks that unite people, creating new horizontal connections, which make it possible not to depend on the propaganda machine of the state and the mass media.

4 Discussion

Technologies and artistic creation have always had a mutual influence on each other in those areas where they intersected [31]. But if before artistic creation was primary concerning technological means and conditioned their improvement, now technologies determine the meaning and content of artistic creativity.

In previous eras, technologies were viewed as a simple set of recipes, were predominantly of an applied nature and an intermediary between the artist's idea and its material embodiment. In the twentieth century, the realization of the fact has come that new forms of artistic creativity are coupled with the birth of fundamentally new technologies. It is no coincidence that representatives of socialist realism widely used new technologies not only in replication but also in the process of creating artistic works. That was done in every possible way by the Soviet state for the development of Soviet art and propaganda of the superiority of socialism over capitalism.

However, the development of technologies, their application in artistic creation led to the emergence of new trends in culture and art, destroying the ideological and propaganda foundations of the Soviet state. The emerged dialectical contradiction was eliminated by replacing socialist realism, which remained as a whole within the framework of traditional artistic creativity, with postmodernism.

The most significant result of the introduction of modern technologies and, above all, digital technologies was a revolutionary change in the conceptual foundations of artistic creativity. The main characteristic features of modern artistic creation are the subject's leading role, freedom of creativity, improvement of old technologies and creation of new ones, including by the artist himself.

The use of modern technologies in artistic creation is in constant search of new means of expression, changes the appearance of works of art. Digital technologies become an independent object of artistic creation and create new aesthetic values. They turn into new forms of art [32]. With the help of digital technologies in the field of video art, we can contemplate, for example, democratic or independent cinema.

There are not only advantages in digital technologies, but also disadvantages. Digital technologies form virtual reality, which is fundamentally different from true reality and which develops according to different laws. In this virtual reality, many traditional principles, including aesthetic ones, turn out to be inapplicable [33]. Moreover, digital technologies do not just penetrate the daily life of people but subjugate it to themselves. Besides, not everyone has time to adapt quickly to technological changes.

Digital technologies contribute to technological breakthroughs into the future. The development and continuation of their improvement open the way for the art of artificial intelligence [34]. This means that in the future we are expected a new innovative and creative artistic creation, a new transformation of its foundations.

5 Conclusions

In the process of its development, technologies played a very important role in the transforming of conceptual foundations of artistic creation during the transition from socialist realism to postmodernism. If socialist realism was based on analog technologies, then postmodernism is based on digital technologies. If socialist realism was based on the real world, then in postmodernism it is a virtual one. If socialist realism was under the yoke of ideology and propaganda and creative people were limited by rigid frameworks, then postmodernism creates maximum opportunities for self-expression, self-realization, and creativity. Unlike socialist realism, postmodernism provides a much wider scope for creativity, creation works.

The transformation of conceptual foundations of artistic creativity was expressed in the transition from social norms, social order, determined by the state, to norms established by the creative individual himself, aimed at innovation and creativity in his activities. The further transformation consists of the growing role of digital technologies and the formation of the creative potential of artificial intelligence.

References

1. Mezentsev, S.D., Gustov, Yu.I.: Philosophy. Technics. Science **1**, 96 (2006)
2. Mashevskaya, O.V.: Digital technology as the foundation of digital transformation of modern society. Bull. Polesie State Univ. A Ser. Soc. Sci. Humanit. **1**, 37–44 (2020)
3. Skolota, Z.N.: Contemporary art: forms and technologies. Young Sci. **11**, 852–856 (2013)
4. Kirillova, N.B.: Media Culture: From Modern to Postmodern. Academic Project, Moscow (2006)
5. First All-Union Congress of Soviet Writers: Verbatim Record. Imaginative Literature. Moscow (1934)
6. Faybyshenko, F.: From one engineer of the soul to many: the history of a fabrication. https://www.nlobooks.ru/magazines/novoe_literaturnoe_obozrenie/152/article/20026/. Accessed 21 Apr 2021
7. Sarabyanova, D.V. (ed.): History of Russian and Soviet Art. Higher School, Moscow (1979)
8. Anipchenko, D.: The history of the song "Back in the U.S.S.R". The Beatles. https://song-story.ru/back-in-the-ussr-the-beatles/. Accessed 30 Apr 2021
9. Postmodernism: an abyss of new meanings under the guise of irony and shocking. https://veryimportantlot.com/ru/news/blog/postmodernizm. Accessed 21 Apr 2021
10. Shcherbina, V.: About socialist realism. Quest. Lit. **4**, 3–31 (1957)
11. Manin, V.S.: Russian Painting of the Twentieth Century. Aurora, St. Petersburg (2006)
12. Gorky, M.: About how I learned to write. http://gorkiy-lit.ru/gorkiy/articles/article-351.htm Accessed 28 Apr 2021
13. Classical and modern bourgeois philosophy. Quest. Philos. **12**, 23–38 (1970). **4**, 58–73 (1971)
14. Glavatsky, M.E.: "Philosophical Steamer": 1922. Historiographic Studies. Publishing House of the Ural University, Yekaterinburg (2002)
15. Goryaeva, T.M.: Political Censorship in the USSR. 1917–1991, 2nd edn. Russian Political Encyclopedia (ROSSPEN), Moscow (2009)

16. Veseli Franz Iosifovich. https://yarwiki.ru/article/2654/veseli-franc-iosifovich. Accessed 21 Apr 2021
17. Efimov, D.D., Fakhrutdinova, I.A.: Origins and trends of Soviet modernism. Bull. KGASU **1**(43), 28–40 (2018)
18. The architecture of Soviet modernism: vladimirtan – LiveJournal. https://homyrouz.ru/raz noe-2/sovetskij-modern-arxitektura-sovetskogo-modernizma-vladimirtan-livejournal.html. Accessed 28 Apr 2021
19. Lossky, N.: Dialectical Materialism in the USSR. YMKA-PRESS, Paris (1934)
20. The policy of "glasnost" proclaimed by the general secretary of the CPSU Central Committee Mikhail Gorbachev. https://tass.ru/info/2691509. Accessed 21 Apr 2021
21. Epstein, M.: Postmodernity in Russia: Literature and Theory. R. Elinin's Edition, Moscow (2000)
22. Inozemtsev, V.A., Udovik, V.E.: Information and computer revolution and the formation of the information society. https://gramota.net/materials/3/2011/8-4/18.html. Accessed 28 Apr 2021
23. Galinskaya, I.L.: Postmodernism in Russian literature. Bull. Cult. Stud. **2**, 47–53 (2001)
24. Zakhovaeva, A.G.: The phenomenon of "luxury", "elite" and "mass" in the culture of postmodernism. Bull. Cult. Stud. **3**(78), 21–28 (2016)
25. Nesmeyanova, O.: Postmodernism in art. https://klauzura.ru/2012/02/olga-nesmeyanova-pos tmodernizm-v-iskusstve/. Accessed 21 Apr 2021
26. Krylova, T.: Hi-tech in art: how modern technologies help artists work. https://vc.ru/future/ 80558-hi-tech-v-iskusstve-kak-sovremennye-tehnologii-pomogayut-hudozhnikam-rabotat. Accessed 27 Apr 2021
27. Marcos, A.F., Branco, P., Carvalho, J.A.: The computer medium in digital art's creative process. In: Braman, J., Vincenti, G., Trajkovski, G. (eds.) Handbook of Research on Computational Arts and Creative Informatics, pp. 1–25. IGI Global, New York (2009). https://doi. org/10.4018/9781605663524.ch001
28. Szecheny, E.: Information technologies in art. https://sites.google.com/site/secenelena/inform acionnye-tehnologii-v-iskusstve. Accessed 27 Apr 2021
29. Volozhanina, E.A.: Problems of digital painting. Archit. Des. **1**, 9–13 (2019). https://doi.org/ 10.7256/2585-7789.2019.1.29622
30. Masyanya is a reflective heroine. https://www.kommersant.ru/doc/339826?stamp=637554 141444485842. Accessed 30 Apr 2021
31. Guk, A.A.: Technique and art: explication of interaction in the context of the development of artistic and creative activity. Bull. Kemerovo State Univ. Cult. Arts **9**, 45–53 (2009)
32. Koroleva, L.A.: Integration of the development of art and digital technologies in the space of media culture. Inf. Soc.: Educ. Sci. Cult. Technol. Future **2**, 169–176 (2018)
33. Erokhin, S.V.: Digital technologies in contemporary fine arts. Bull. Volgogr. State Pedagogical Univ. **8**, 145–149 (2008)
34. Krasotkina, E.: Contemporary art: why does it need technology and who will explain it. https://trends.rbc.ru/trends/innovation/6019739e9a794737983f7c04. Accessed 27 Apr 2021

Analysis of Human Behavior as a Condition for Creative Artificial Storytelling

Andrei E. Serikov(✉) 📵

Samara National Research University, 34, Moskovskoye Shosse, 443086 Samara, Russia
serikov.ae@ssau.ru

Abstract. On the one hand, there is currently no convincing example of AI writing successful fiction stories. For computers to write good stories, they need to understand human behavior. But most contemporary systems, designed for automatic extracting storylines from stories and producing new plots, ignore knowledge about human behavior and its circumstances. On the other hand, theoretical understanding of human behavior requires empirical material, and stories analyzed by computers can be such material. If the problem of theoretical explanation of behavior and the problem of computational story understanding and storytelling are considered from a single point of view, then there is a possibility of significant progress in solving the both problems. A method to analyze descriptions of human behavior in fiction stories, that is based on Lewinian formula of behavior, as well as some ideas about possible ways to implement this method of text analysis into computing systems are suggested. If computers can automatically analyze stories not only at the level of linguistic grammar, but also at the level of behavioral grammar, they will reach a deeper understanding of human behavior and the stories people tell. This will allow computers to create meaningful new stories that are truly interesting to people.

Keywords: Computational creativity · Human behavior understanding · Behavioral grammar · Fiction stories analysis

1 Introduction

There are two issues that are discussed independently in the scientific literature. One of them is the problem of computational story understanding and storytelling, the other is the problem of theoretical conceptualization of human behavior. Usual methods for collecting scientific information about human behavior are observations and surveys. But the most attentive observers and the best experts in human behavior probably are fiction story writers. Therefore, for further progress in the theory of behavior, it is necessary to learn how to extract knowledge about behavior from fiction, and in order to be able to analyze large volumes of literature, it is necessary to teach this to computers. In this context, the two problems in question turn out to be interrelated and must be considered from a unified point of view. The fiction stories parsed on the basis of the behavioral analysis will allow teaching computer programs to this analysis, and the automatic analysis of tests will allow using almost infinite fiction corpora as an empirical basis

D. Bylieva and A. Nordmann (Eds.): PCSF 2021, LNNS 345, pp. 449–461, 2022.
https://doi.org/10.1007/978-3-030-89708-6_38

for further research of the patterns of human behavior. At the same time, the results of the analysis can be used by computers to generate more sensible and interesting stories. This is the main idea of this article.

In the next section, the problem of computing creativity will be discussed, the essence of which is that artificial systems still cannot produce good enough stories. Then, in the third section, the topic of discussion will be the importance for computers to understand common sense and use that understanding when writing stories. In the fourth section, an observation will be made that the most advanced contemporary systems designed for automatic extracting storylines from stories and producing new plots, unfortunately, ignore much of the knowledge about human behavior and its circumstances. In the fifth section, the potential of Kurt Lewin's ideas for the further development of the theory of behavior will be discussed. In the sixth section, a method for analyzing fiction stories based on these ideas will be presented. And in the last section, the results of applying this method to a sample story and further research tasks will be discussed.

2 The Lovelace Tests of Artificial Creativity and the Problem of Computational Storytelling

In 2001, Bringsjord et al. proposed a new test for AI, they called *the Lovelace Test*. The general idea was that an artificial agent A built by a human H should be regarded as human-level intelligent agent if it could produce truly creative outputs, and if H could not account for how A did it [1]. Later Riedl argued that such a test was unbeatable because if H had enough resources to build A and sufficient time to analyze its work, H would be able to explain the A's outputs. That is why he proposed an updated version of the test, *the Lovelace Test 2.0*. The main idea was asking "an artificial agent to create a wide range of types of creative artifacts (e.g., paintings, poetry, stories, architectural designs, etc.) that meet requirements given by a human evaluator." [2].

Du Sautoy's book "The Creativity Code" (2019) gives some examples of computers successfully solving creative problems that he managed to collect at the time of writing the book. One of them is a portrait of Edmond Belamy created by an algorithm and sold at Christie's for $432,500 in 2018 [3, pp. 132–133]. Another impressive example of creative programs is a jazz improvising algorithm built by François Pachet and known as the Continuator, because it can continue the music started by a live musician in the same style or, in another version, it can accompany the live musician by guessing the right chord to play. Describing the Continuator's achievements du Sautoy sa is it seems to him "like the moment when the Lovelace Test got passed" [3, p. 205]. There is an even more convincing example of AI success in music creation: "In 2016, an algorithm called AIVA was the first machine to have been given the title of composer by the Société des Auteurs, Compositeurs et Éditeurs de Musique (SACEM), a French professional association in charge of artists' rights." [3, p. 215]. Computational music composition seems to be one of the most developed areas in the field of AI creativity. Moruzzi even argues that we can consider some AI composers intentionally creative [4].

However, there is no convincing example of AI writing successful fiction stories. In this context, it seems important and not at all accidental that Bringsjord et al. explained their idea of the Lovelace Test by the example of originating a story. Creating stories was for them a prototype of intellectual creation. The same can be said about the Lovelace Test 2.0 because most of Riedl's work was devoted to the development of computational storytelling. And the subsequent development of computational creativity showed that the task of providing original sensible narratives proved to be one of the most difficult for AI systems.

But what is the fundamental difficulty in storytelling? What is the main difference between painting a picture, composing music or writing a poem, on the one hand, and generating a story, on the other? The difference is that the story needs a sensible storyline, a plot understandable in terms of human conditions, feelings and behavior.

In lately published articles on computational storytelling authors usually admit that previous work in the field either relied on human annotations of plots or was restricted to narrow domains. That is why they see their contemporary challenge in completely automatic open story generation [5, 6]. When one reads about this it looks like contemporary artificial neural networks could be taught to produce quite good stories, even though they are limited to domain models, genres and sample cases they were trained on. It looks like the general task of teaching computers to tell sensible stories is almost completed and what is left is just to make them possible to tell stories on any topic without re-learning the domain model. But is it really so? Can the best of contemporary computational narrative systems write sensible and really interesting stories?

In 2017, Riedl admitted that "computers still cannot reliably create and tell novel stories, nor understand stories told by humans. When computers do tell stories, via an eBook or computer game, they simply regurgitate something written by a human" [7]. In 2020, Bringsjord discussed this topic with R.J. Marks and they also concluded that, at the time of the discussion, no AI had passed the Lovelace Test [8].

3   The Importance of Common-Sense Knowledge for Generation of Stories

In order for artificial systems to understand human stories and create sensible plots, they must have access to human common sense. Pérez y Pérez admits that most of contemporary automatically produced narratives are rigid and predictable. "An alternative to overcome this shortcoming is the study of how to represent common-sense knowledge (CSK) in computer models of plot generation that allows more flexible and interesting storytellers to be produced." [9, p. 256].

To date, the main known approaches to represent CSK are knowledge-based ontologies and networks-based representations [10, pp. 143–215]. Many of these representations not only contain semantic information about words and things they denote, but as well lexical and syntactic information. It makes possible to use the representations for lexical and grammatical parsing, word sense disambiguation, and other tasks of natural language processing (NLP).

Completing these NLP tasks is often a prerequisite for automatic understanding or creating stories. For example, Harrison et al. introduced an algorithm for open story generation which had been trained on movie plots from Wikipedia, but in order to train it they first had to transform original stories into simplified and generalized event representations by using Stanford CoreNLP toolkit, WordNet and VerbNet [5, p. 195]. WordNet is an online ontology that organizes words into synonym sets, providing information about hyponymy/hypernymy and meronymy/holonymy [11, 12]. VerbNet is an online verb lexicon that classifies verbs on the basis of their syntactic and semantic similarities [13] and probably has strong cross-lingual potential [14]. Stanford CoreNLP toolkit is a NLP system that can tokenize an input text into a sequence of tokens, split a sequence of tokens into sentences, label tokens with their part-of-speech, add likely gender information to names, do syntactic and coreference analysis, etc. [15]. Hand-annotated language corpora such as FrameNet that are used for supervised teaching of automatic semantic role labeling systems [16] are also very important part of CSK.

So our common sense is partly about understanding simple things like "a boy is a youthful male person" or "a schoolboy is a boy attending school", and partly about understanding the familiar structures of the natural languages. However, CSK is not all about them. There is another piece of common sense which is necessary for successful story understanding and storytelling, that is knowledge of behavioral rules, knowledge of how people feel and what they usually do in certain circumstances.

Pérez y Pérez has used the last type of knowledge in his story generator MEXICA. In the system, a story is understood as a sequence of *actions* performed by characters. An action take place in some *preconditions* and changes them into *postconditions*. The user has to supply MEXICA with a list of all possible actions and associated with them preconditions and postconditions written in a special computer language. Preconditions and postconditions codify *social CSK* about rules that depend on social and cultural context, and *logical CSK* about rules that depend on nature of the story world. Social CSK can be represented in terms of *emotional links* between characters such as "character A is in love with character B", and logical CSK is associated with *tensions* such as "the life of a character is at risk" [9].

MEXICA is an example of how CSK about possible actions, their emotional reasons and results can be encoded by hand in order to specify a domain of possible story. The benefit of this approach is that in the resulting stories, people's actions appear meaningful in terms of their dependence on different circumstances and emotional relationships between the characters. The disadvantage is that possible actions as well as their preconditions and postconditions must be explicitly expressed by the users and there is no way for the program to learn some new rules automatically. Another approach is to build a neural network being able to learn new information about actions by analyzing open text sources, and to use it for producing new interesting plots. Currently, such systems exist, but their drawback lies in the rather primitive presentation of stories, which are reduced simply to a sequence of actions.

4   Conceptualizations of Plots, Events and Actions

In 2009, Chambers and Jurafsky introduced an unsupervised system for learning narrative schemas from news articles. A *narrative schema* is a plot-like model of events that merges several *typed narrative event chains* together. "An event is a verb together with its constellation of arguments. An event slot is a tuple of an event and a particular argument slot (grammatical relation), represented as a pair $<v, d>$ where v is a verb and $d \in \{\text{subject, object, prep}\}$." [17, p. 603]. A typed narrative event chain is a sequence of events associated with a role of a certain type. A type is automatically constructed as a set of lexical units related to events in a chain. The model "does not assume the set of roles is known in advance, and it learns the roles at the same time as clustering verbs into frame-like schemas." [17, p. 607]. For example, the system extracted out of a news article three event chains. One chain includes the verbs "arrest" and "charge" associated with the type represented by nouns "police" and "agent", the second chain – the verbs "convict" and "sentence" associated with the nouns "judge" and "jury", and the third chain – the verb "plead" associated with the nouns "criminal" and "suspect". By merging the three chains into one narrative schema the system would produce the resulting plot of a form "The police (agent) arrest and charge a criminal (suspect); the criminal (suspect) pleads guilty (innocent); the judge (jury) convicts and sentences the criminal (suspect)".

The story generating system built by Harrison et al. in 2017, that was mentioned in the previous section, is also based on a conception of stories as *sequences of events*. The system uses a mathematical technique known as Markov Chain Monte Carlo sampling, and after being trained on movie plots from Wikipedia the system can produce a new plot about any suggested topic. "In this work, each event consists of the following elements: <subject, verb, object, modifier> . That is, every sentence in the original story corpus is reduced to a 4-word sequence of this form. Sentences with more than one verb or more than one subject are broken up into multiple events." [5, p. 193]. Accordingly to the model, any new plot produced by the system also consists of such sentences. *Events are the 4-word sentences*. The central element of each event is the verb, and the other three words could be understood as its arguments, which makes the model similar to the abovementioned narrative schemas.

In an open story generator constructed by Yao et al. in 2019, a story is also understood as a sequence of events. But the model of events is even simpler than in the previously described systems. The system uses existing story corpora to automatically extract one keyword from each sentence and constructs *the storyline as the sequence of the keywords*. So *an event is represented by a separate word*. Given any word as a title the system uses learned probabilities in order to produce a new sequence of words, and then uses the plot to fill it with appropriate sentences [6].

So, in contemporary work on automatic learning of plots and open-domain computational storytelling, an event is often understood as an action represented by a verb together with its minimal set of arguments, or by just a word. Accordingly, a story-line is usually defined as a sequence of verbs or other keywords. This type of conceptualization provides very little scope for representing CSK hidden in the context of actions, comparing to rich representations in such systems as MEXICA.

Apparently, a future successful computing storyteller must combine the ability to automatically learn from test corpora and to produce new plots with the ability to base the plots on rich CSK. We shall teach computers to extract plots from analyzed stories not only on the basis of verbs and their arguments, but also on the basis of descriptions of various properties of characters, their states and the circumstances in which they operate. It is necessary to teach computers to perform not only linguistic, but also behavioral analysis of events, while conceptualizing events not only as actions, but also as various circumstances and the possible results of these actions understood on the basis of common sense.

5 Explanation of Human Behavior According to K. Lewin's Formula

If we consider human language competence as a part of human CSK, we must recognize that most people do not explicitly know lexical or grammar rules of their native language and have never heard of semantic roles. They just feel how to speak correctly. For example, a native speaker easily detects and avoids inappropriate collocations but in order to teach computer algorithms how to do this, special sources of CSK are used, such as "texts written by second language learners, incorporating corrections by professional English instructors" or the British National Corpus [18, p. 51]. So, professional language teachers know a lot about appropriate and inappropriate collocations. But for lay speakers this knowledge is implicit. You don't even have to know that the grammar exists in order to speak a native language, you just speak it and that is it.

Likewise, there must be something like cultural behavioral grammar. Experts in AI who write about CSN focus on behavioral rules being very simple and understandable to everybody (like 'people have breakfast in the morning, not in the evening'). But there are many CSN aspects not so obvious for most of people. Linguists know very much about language grammars, and they use the knowledge in order to teach people and computers languages. Anthropologists, psychologists, and other experts in behavior know not very much about behavioral grammars, but they do know a thing or two. And their knowledge should be used in conceptualization of human stories just like the knowledge of linguists.

In behavioral sciences there are several competing explanations of what people usually do and why they do it. On one hand, behaviorists associate behavior with the influence of the environment and conditions of upbringing, and on the other hand, sociobiologists, ethologists and evolutionary psychologists explain it by innate predispositions [19]. Anthropologists usually associate behavior with culture, sociologists – with the social environment. Some theorists combine different types of explanation within a single conception. For example, Runciman describes "three different types of behaviour: evoked behaviour, where the agent is responding directly and instinctively to some feature of the environment; acquired behaviour, where the agent is imitating or has learned from some other agent, whether directly or indirectly; and imposed behaviour, where the agent is performing a social role underwritten by institutional inducements and sanctions." [20, p. 8].

At the end of the last century, the "person–situation debate" took place between personality psychologists who explained human actions by dependence on personality traits, and social psychologists who explained human actions by dependence on situations. In 2009, Funder called the debate a "war" and suggested an agenda for psychology in the "postwar era" – to research systematically how persons, situations and behaviors interact with each other. His proposal was based on Kurt Lewin's idea that behavior can be understood as a function of both persons and situations. "Lewin's formula provides a way to give both persons and situations their due while avoiding fruitless arguments over which one matters more." [21, p. 123].

Lewin himself used a slightly different terminology. In his article "Environmental Forces in Child Behavior and Development" (1933) he introduced the aforementioned formula that later became famous:

$$B = f\ (PE).\tag{1}$$

where B stands for *behavior*, P stands for *a person*, and E stands for *environment*. The equation means that "to understand or predict the psychological behavior (B) one has to determine for every kind of psychological event (actions, emotions, expressions, etc.) the momentary whole situation, that is, the momentary structure and the state of the person (P) and of the psychological environment (E)." [22, p. 79].

One of the main ideas of later Lewin's field theory was that "the person and his environment have to be considered as one constellation of interdependent factors." [23, pp. 239–240]. From this point of view, in order to explain behavior it is necessary not only take into account all external and internal factors that matter, but also consider how these factors influence each other.

In general, such a conceptualization takes into account the most diverse aspects of behavior and various factors that affect it. Which makes this conceptualization very suitable to be used as a theoretical basis for the behavioral analysis of fiction stories.

6   A Method to Analyze Descriptions of Human Behavior in Fiction Stories

All of the models of storytelling mentioned above have been built on understanding stories as sequences of events. By follow this general idea, it is possible to conceptualize events on the basis of the Lewinian theory. According to the conceptualization, *an event is a pattern of behavior implemented in a certain situation, i.e. performed by a person in certain psychophysiological states and certain circumstances*. After the behavior is implemented, the situation changes. Many events and their sequences are typical for a given culture and era, and some events may be typical for people in general. The task of detecting and describing typical events, as well as the connections between them, could be designated as the task of describing the *grammar of behavior*.

An acting person, on one hand, has certain social statuses, such as gender, age, marital status, education, occupation, etc., on the other hand, this is a person who is in a certain physiological and psychological state, for example, he may be sick or healthy, determined to do something or in doubt, experience various emotions, etc. Therefore, the Lewinian person is divided into two components: the status and the psychophysiological state. As a result, an event is characterized by the following minimal set of variables: (1) the status of the character (who acts), (2) her/his psychophysiological state (what s/he thinks, how s/he feels), (3) the circumstances of behavior (in what conditions s/he acts), (4) the pattern of behavior (what s/he says and does).

The main task of the text analysis is to obtain generalized descriptions of typical events. The story is continuously broken down into successive segments, each of which corresponds to one event. A segment contains a non-generalized description of an individual event and is treated as a token, which then receives a zero-level generalized description. This description is made, if possible, in the same words that were used by the author in the original text, with minimal generalizations, such as replacing the character's name with his social status. Similar descriptions of the zero level are generalized by descriptions of the 1st level, similar descriptions of the 1st are generalized by descriptions of the 2nd level, etc.

This behavioral model of an event is not tied to the grammar of the language. In this model, verbs do not necessarily express a person's behavior, they can also express his state and circumstances of action. A single segment can consist of a clause of a complex sentence or of several simple sentences. The main thing is that this segment should describe the above four variables. An event as a whole can be considered as described when each of its variables is described. When there is no explicit description of the corresponding variable in the text, this description is extracted from the context on the basis of CSK. It can be something taken for granted or inferred from an explicit description. It should be understood that the description of an event includes only those aspects of statuses, states, circumstances and actions that are related to each other and form the event as a whole. That is, it is usually possible to extract a fairly detailed description of the characters and circumstances from the context, but only those that explain the actions of the characters in a given situation and the connections between events are important in the description.

7 Sample Story Analysis Results and Discussion

In order to demonstrate how the proposed method works, a short story had to be selected from one of open access resources. The story by A. Kaczmarek "Ollie is Scared" has been chosen from the site "Storyberries" [24]. Then this story has been broken down into 26 fragments corresponding to token events as follows:

1 // "I don't want to go to school today, I feel a bit sick…" Ollie told his Mom at breakfast.

2 // Mom frowned. "Have you got a math test?".

3 // "No, of course not." Ollie pushed his cereal away. "I just feel sick."

4 // Ollie sometimes felt sick before school. Mom wondered why. "If you have a school problem, just tell me about it…"

5 // His mother hugged him real tight.

6 // "When I was a little girl, sometimes the other children called me horrible names."

7 // Ollie looked up. "Why did they do that?".

8 // Some children can be mean, especially if they don't feel so good about themselves."

9 // Mom looked at Ollie. Is that your problem maybe?".

10 // Ollie took a long time to answer. "Yes, it is. How did you guess?" A tear ran down his cheek.

11 // "Because you often feel sick before school, so, moms guess these things…".

12 // "They call me Itsy-Bitsy Ollie, Mini-Me, mean things like that. Now all the other kids call me things too. Just because I'm a bit small. And the first two now often shove me about as well… and it hurts."

13 // Mom was quiet, but then asked, "Did you tell your teacher?".

14 // "No, of course not; that would make them worse, I know it…".

15 // "Bullying is wrong, Ollie. Maybe we can talk to your teacher together; would that be OK?".

16 // Ollie had to think for a long time; he didn't like the idea at all.

"It will make things worse…" he frowned.

17 // But then he thought it was worth a try.

18 // Ollie and his Mom met up with Ollie's teacher after school, with nobody around.

19 // Ollie's teacher was happy that Ollie told her all about the nasty bullying…

"I had an idea that something was going on..." Ollie's teacher told them. "But the important thing is, Ollie, that you tell a grown-up. First your Mom, and then me."

20 // The next day at school, Ollie's teacher told the class: "A lot of schools have No Bullying Projects, and with ALL your help, I'd like to start one in our school. What do you think? Will all you children help me?".

21 // Ollie's teacher started their project in Art Class: Speak Up – Stand Up!

22 // ALL the children agreed that the rules were good:

1. Tell a grown up if you see bullying.

2. Stand Up for others.

23 // They did paintings for the rules, and they made badges.

24 // "It's really quite easy," all the children agreed.

25 // Nobody ever knew that it was because Ollie and his Mom talked to the teacher.

26 // But now their school is a much nicer place for ALL the children: name-calling stopped, and, that's how it should be!

For each of the 26 events, zero level values of all variables have been described. Where in the text the values of variables were implicitly assumed at the level of common sense, they have been spelled out explicitly. As an example, the analysis of the first three events is presented in Table 1, including associated with the events CSK that was inferred from the context, zero-level descriptions of characters' statuses and psychophysiological states, circumstances and performed patterns of behavior, as well as zero-level descriptions of the whole events (Table 1).

It is possible to guess what the next level generalizations might look like for these events. For example, if we came across a similar story about a girl, then the 1st level generalized description for the opening event would be: *when children have to go to school but don't want to do it, they tell their mom that they feel sick.*

The experience of analyzing even such a very simple history shows that it is possible to break the history into events only after identifying the values of all variables, which is not always possible to do. For example, fragment 25 has been singled out as a separate event, although it only deals with the psychological state of the characters and the circumstances, but does not describe any actions that change the situation Since this state and these circumstances are not connected in some necessary way with the actions of the characters in the previous and subsequent fragments, fragment 25 was not combined with one of them into a single event. But in this case, can this fragment be considered as a description of an event? The decision taken was to allow the possibility of an "incomplete" event, i.e. one in which the values of some variables are absent. A possible interpretation is that this fragment describes the missing events that could have taken place if the state of the characters and circumstances were different.

Table 1. The analysis of the first three events

No	CSK	Status	Psychophysiological state	Circumstances	Behavior	Zero-level description of event
1	If someone tells his mom he doesn't want to go to school, he's a schoolboy	A schoolboy	It's a weekday morning and the boy has to go to school	He doesn't want to go to school	He tells his mom that he feels sick	When a schoolboy has to go to school but doesn't want to do it, he tells his mom that he feels sick
2	If a woman frowns, she's upset	The mother of the schoolboy	Her son has told her he doesn't want to go to school	Mom is upset and suspects that her son doesn't want to go to school, not because of illness, but for some other reason	Mom asks if the math test is the reason why the son doesn't want to go to school	Mom is upset because she suspects that her son doesn't want to go to school for some unknown reason, and asks if a math test is the reason
3	If a boy pushes his breakfast away he's upset and doesn't want to eat	The schoolboy	Mom doesn't believe him	He is upset and doesn't want to eat	He says the reason is not a math test, but just because he feels sick	When mom asks her boy if he doesn't want to go to school because of a math test, he repeats that he just feels sick

Some ideas about possible ways of implementing this method of text analysis into computing systems can be suggested. At the initial stage, these should be artificial neural networks, trained on the basis of text corpora analyzed by hand, and then these can be unsupervised learning systems. It will be necessary to clearly define what exactly they need to learn. For example, they must learn to match text fragments with the values of behavioral variables, and, in particular, identify the values based on common sense. Then, based on these values, the systems will be able to create generalized descriptions of zero-level events, identify their similarities and group them into sets, described as typical events. Probably it should be a neural network that uses methods similar to vord2vector or sentence2vector, but instead of vector representation of just words and sentences the neural network will have to learn how to represent text segments associated with events and their variables.

8 Conclusions

If the problem of theoretical explanation of behavior and the problem of computational story understanding and storytelling are considered from a single point of view, then there is a possibility of significant progress in solving both problems. On one hand, based on the automatic analysis and annotation of a large volume of fiction stories, one can describe and compare the behavioral grammars of different cultures as well as identify universal patterns of behavior. On the other hand, if computers can automatically analyze stories not only at the level of linguistic grammar, but also at the level of behavioral grammar, they will be able to understand much deeper the human behavior and the stories people tell. This will allow computers to create meaningful new stories that are truly interesting to people.

References

1. Bringsjord, S., Bello, P., Ferrucci, D.: Creativity, the turing test, and the (Better) Lovelace test. Minds Mach. **11**, 3–27 (2001).https://doi.org/10.1023/A:1011206622741
2. Riedl, M.: The Lovelace 2.0 test of artificial intelligence and creativity. In: Proceedings of the AAAI Workshop: Beyond the Turing Test, Austin, Texas (2015). https://arxiv.org/pdf/1410.6142v3.pdf. Accessed 10 Mar 2021
3. Du Sautoy, M.: The Creativity Code: Art and Innovation in the Age of AI, 2nd edn. Harvard University Press, Cambridge, MA (2020). https://doi.org/10.4159/9780674240407
4. Moruzzi, C.: Creative AI: music composition programs as an extension of the composer's mind. In: Müller, V. (eds.) Philosophy and Theory of Artificial Intelligence 2017. PT-AI 2017. Studies in Applied Philosophy, Epistemology and Rational Ethics, vol. 44, pp. 69–72. Springer, Cham (2018). https://doi.org/10.1007/978-3-319-96448-5_8
5. Harrison, B., Purdy, C., Riedl, M.: Toward automated story generation with Markov Chain Monte Carlo methods and deep neural networks. In: AAAI Publications, Thirteenth Artificial Intelligence and Interactive Digital Entertainment Conference, Papers from the 2017 AIIDE Workshop, pp. 191–197, The AAAI Press, Palo Alto, CA (2017). https://aaai.org/ocs/index.php/AIIDE/AIIDE17/paper/view/15908/15241. Accessed 10 Mar 2021

6. Yao, L., Peng, N., Weischedel, R., Knight, K., Zhao, D., Yan, R.: Plan-and-write: towards better automatic storytelling. In: Proceedings of the AAAI Conference on Artificial Intellegence, vol. 33, no. 01: AAAI-19, IAAI-19, EAAI-20, pp. 7378–7385. AAAI Press, Palo Alto, CA (2019). https://www.aaai.org/ojs/index.php/AAAI/article/download/4726/4604. Accessed 10 Mar 2021

7. Riedl, M.: Computational narrative intelligence: past, present, and future (2017). https://mark-riedl.medium.com/computational-narrative-intelligence-past-present-and-future-99e58cf25ffa. Accessed 10 Mar 2021

8. Mind Matters: Thinking machines? Has the Lovelace test been passed? (2020). https://mindmatters.ai/2020/04/thinking-machines-has-the-lovelace-test-been-passed. Accessed 10 Mar 2021

9. Pérez y Pérez, R.: Representing social common-sense knowledge in MEXICA. In: Veale, T., Cardoso, F. (eds.) Computational Creativity. Computational Synthesis and Creative Systems, pp. 255–274. Springer, Cham (2019). https://doi.org/10.1007/978-3-319-43610-4_12

10. Chowdhary, K.R.: Fundamentals of Artificial Intelligence. Springer, New Delhi (2020). https://doi.org/10.1007/978-81-322-3972-7

11. Miller, G.A., Beckwith, R., Fellbaum, C., Gross, D., Miller, K.J.: Introduction to wordnet: an on-line lexical database. Intern. J. Lexicogr. **3**(4), 235–244 (1990). https://doi.org/10.1093/ijl/3.4.235

12. Miller, G.A.: WordNet: a lexical database for English. Commun. ACM **38**(11), 39–41 (1995). https://doi.org/10.1145/219717.219748

13. Brown, S.W., Bonn, J., Gung, J., Zaenen, A., Pustejovsky, J., Palmer, M.: VerbNet representations: subevent semantics for transfer verbs. In: Proceedings of the First International Workshop on Designing Meaning Representations, pp 154–163. Association for Computational Linguistics, Florence, Italy (2019). https://doi.org/10.18653/v1/W19-3318

14. Majewska, O., et al.: Investigating the cross-lingual translatability of VerbNet-style classification. Lang. Resour. Eval. **52**, 771–799 (2018). https://doi.org/10.1007/s10579-017-9403-x

15. Manning, C.D., Surdeanu, M., Bauer, J., Finkel, J., Bethard, S.J., McClosky, D.: The stanford CoreNLP natural language processing toolkit. In: Proceedings of 52nd Annual Meeting of the Association for Computational Linguistics: System Demonstrations, pp. 55–60. Association for Computational Linguistics, Baltimore, MD (2014). https://doi.org/10.3115/v1/P14-5010

16. Gildea, D., Jurafsky, D.: Automatic labeling of semantic roles. Comput. Ling. **28**(3), 245–288 (2002). https://doi.org/10.1162/089120102760275983

17. Chambers, N., Jurafsky, D.: Unsupervised learning of narrative schemas and their participants. In: Proceedings of the 47th Annual Meeting of the ACL and the 4th IJCNLP of the AFNLP, pp. 602–610. Association for Computational Linguistics, Suntec, Singapore (2009). https://www.aclweb.org/anthology/P09-1068.pdf. Accessed 19 Mar 2021

18. Tandon, N., Varde, A.S., de Melo, G.: Commonsense knowledge in machine intelligence. ACM SIGMOD Rec. **46**(4), 49–52 (2017). https://doi.org/10.1145/3186549.3186562

19. Naour P.: E.O. Wilson and B.F. Skinner. A Dialogue Between Sociobiology and Radical Behaviorism. Springer, New York, NY (2009). https://doi.org/10.1007/978-0-387-89462-1

20. Runciman, W.: The Theory of Cultural and Social Selection. Cambridge University Press, Cambridge, UK (2009). https://doi.org/10.1017/CBO9780511819889

21. Funder, D.C.: Persons, behaviors and situations: an agenda for personality psychology in the postwar era. J. Res. Pers. **43**(2), 120–126 (2009). https://doi.org/10.1016/j.jrp.2008.12.041

22. Lewin, K.: A Dynamic Theory of Personality. Selected Papers. McGraw-Hill, New York and London (1935). https://archive.org/details/dynamictheoryofp032261mbp. Accessed 22 Mar 2021
23. Lewin, K.: Field Theory in Social Science: Selected Theoretical Papers. Harper & Brothers, New York (1951). https://archive.org/details/in.ernet.dli.2015.138989/. Accessed 15 Mar 2021
24. Kaczmarek, A.: Ollie is Scared (2021). https://storyberries.com/bedtime-stories-ollie-is-scared-stories-about-bullying/. Accessed 15 Mar 2021

Artistic Virtual Reality

Daria Bylieva[✉] [iD]

Peter the Great St. Petersburg Polytechnic University (SPbPU), Polytechnicheskaya 29,
Saint Petersburg 195251, Russia
bylieva_ds@spbstu.ru

Abstract. The concept of virtual reality has transformed from a philosophical concept which implies an intangible reality realized in subjective human experience, into a widely used term that refers to a technologically realized simulated experience. The new technology of virtual reality erases the boundaries of subjective and objective, real and fictional, between cinema, theater, video games, art. Being a play as the basic form of human activity, VR immerses a person in an alternative artistic artificial environment. As a result of the analysis of more than two hundred creative projects using virtual reality, several properties have been identified that contribute to the experience of "virtuality": unusual experience, activity, naturalism, interaction, emotional engagement. "Unusual experience" is the characteristic that most distinguishes artistic projects (theater, cinema, art, video games) from other non-creative uses of VR. In virtual reality, the role of the audience is becoming more and more active, which is achieved by technical (360-degree perspective or free movement) and artistic means (transformation of the viewer into a character or creator). The use of film and animation techniques expands the boundaries of what is possible in VR. The ability to convey experience at the level of actuality demonstrates the great potential of VR technology in the construction of reality.

Keywords: Virtual reality · VR · Art · Cinematic virtual reality

1 Introduction

New technologies not only rapidly change a person's life in various aspects [1–5] but also blur the boundaries of phenomena, making us change our attitude to the usual ways of being. Art as a specific form of comprehending the world finds itself in a special position in relation to technology. Changing like other spheres of human existence under the influence of new technologies [6, 7], art at the same time finds itself on the non-technical, non-instrumentalist pole of understanding the world. Gareth Polmeer notes that "self-understanding in a world in which the technical development of knowledge in the rationalist and materialist modes, or scientism, has become distant from the poetic dimension of experience" [8]. At the same time, revealing reality by its own means, art provides a special experience that opens up new perspectives, including technological ones. As Douglas Kellner points out "by its very nature, art pertains to another world and can thus speak truths other than the conventional wisdom" [9]. Art is one of the keys to resolving the question posed by Owen Barfield "How is it that the more able man becomes to manipulate the world to his advantage, the less he can perceive any meaning in it?" [10]. Because, as Herbert Marcuse writes, the world intended in art is neither the given world of everyday reality, nor a world of mere illusion. The reality of art reflects "the actions, thoughts, feelings and dreams of men and women, their potentialities and those of nature", making the illusory authentic, and making the real fictional [11].

Today, the process, begun with photography and cinematography, increases immersive effects so as to produce a powerful illusion of experiencing art authentically. Walter Benjamin in 1935 shows how technological solutions that made it possible to capture the environment on film put the viewer in a new position – "he enters into this work of art the way legend tells of the Chinese painter when he viewed his finished painting" [12]. The world through the camera reveals a different nature - if only because an unconsciously penetrated space is replaced by a space consciously explored by humans [12]. Michael McLuhan writes that the advancement towards more technological communication channels - from writing to movies and TV - should be accompanied by a partake of the character of "that interior artifice" by which people "incarnate the exterior world" [13]. To recreate within themselves the exterior world is "the work of the nous poietikos or of the agent intellect – that is, the poetic or creative process" [13].

Virtual Reality (VR) technology goes one step further in this direction. Since 2016, the mastering of virtual reality technology by artistic professionals has begun. The term "Virtual Reality" is not part of the digital age. Its Latin origin goes back to "potentiality" [14]. However, today the concept of VR is undergoing a technologically supported transformation from a philosophical concept, which most often implies an intangible reality (objectively existing or realized in subjective human experience), into a technical term representing a technologically realized simulated experience. This extension of the subjective "human virtual" to the shared "digital virtual" (in the terminology of Frankel and Krebs [15]) is the quintessence of the digitalization of life. In his 1938 book "The Theater and Its Double", Antonin Artaud described theater as "la realite virtuelle" (in English translation "virtual reality"), that is, a purely fictitious and illusory world, created by true characters, objects, images [16]. This is one of the few past examples where the concept of virtual reality was used for non-individual artistic simulated experience. The new technology of virtual reality erases the boundaries of subjective and objective, real and fictional, possible and impossible. Lifting "the inner and outer world into… spiritual consciousness," which, according to Hegel, is a "universal human need for art" [17] is embodied in virtual reality in an original way.

2   Virtuality in VR

Today, the active development of virtual reality begins. /Developers are looking for a variety of purposes for its application. Specialists working in completely different fields (engineering, education, journalism, medicine, commerce, entertainment, art, cinema, theater, management, science, etc.) find solutions to certain problems in virtual reality. The most common application of technically organized virtual reality can be subdivided into having specific practical goals (for example, in engineering or education) or not having them (which we will here consider as a sign of the artistic, although it is clear that creativity can serve, for example, ideological or political goals). Some VR projects are aimed at triggering action [18, 19].

Applying the classic division of human activities into learning, play, and work, artistic virtual reality falls under the concept of play game. Herbert Spencer writes that activities, called games, are combined with aesthetic activities by the fact that "neither one nor the other helps in any direct way the processes that serve life" [20]. In a broad sense, it is very difficult to define a game, since it includes many dissimilar activities. The departure from everyday life, flowing in a specially allotted space and time is an interesting feature of the game that was highlighted by Huizinga [21]. Herodotus has a passage dedicated to a difficult period in the history of Lydia: one day people immersed themselves in games so much that they forgot about hunger, and the next day they ate, avoiding games [22]. Thus, the game is most naturally and essentially related to VR. And the most important feature of its implementation will be the immersion of the individual in an alternative reality.

Usually, the term telepresence is used to describe immersion in virtual reality for different purposes [23–26]. According to Jonathan Steuer, telepresence is the experience of presence in an environment by means of a communication medium [27]. Now, this term is used for a wide range of digital experiences: visual storytelling on social media [28], websites [29], videochat [30], telepresence robot [31, 32], etc. Despite the diversity of the experience of a digital presence in a place other than the physical body, in most cases, it is a completely realistic environment. Therefore, it seems necessary to use a different term for immersion in a creatively created reality. Robert Hassan uses the term "digitality" to distinguish the "alienating sphere" of digital media from the analogue world [33]. However, it seems to us that such a cardinal division of the worlds is not entirely correct. While it is easy to separate virtual and physical reality, it is not so easy to understand what reality a person belongs to at the moment. (more precisely, to what extent to each of them). As Rachel Falconer notes, the assumed binary relationship between the virtual and the physical "stands on shaky ground and is treated with increasing suspicion because of our technology-driven soporific cross-inhabitation of both lived states" [34]. Considering the existence in an alternative artistic artificial environment, the term "virtuality" seems to be more appropriate. "Virtuality" characterizes the degree of an individual's belonging to virtual reality.

Considering artistic projects with VR, it is possible to assess the potential for revealing a virtual personality in them. As a result of the analysis of more than two hundred artistic projects using virtual reality, several criteria have been identified that are most capable of "virtualizing" a person, involving her or him in virtual reality: naturalism, unusual experience, activity, emotional engagement, interaction (Fig. 1).

The naturalism of VR, realized in theatrical performances, video games, art projects, films, is becoming more and more prominent. One of popular applications of virtual reality during epidemic times (2020–2021) is "social", when a person gets the opportunity to "visit" both the most ordinary and hard-to-reach places [35, 36]. At the same time, VR is not limited by space or time. For example, "*1st Step - from Earth to the Moon*" (Jörg Courtial, Maria Courtial, 2019) offers a journey into space following in the tracks of the Apollo missions. The eye is less and less able to find signs proving the "inauthenticity" of the surrounding, imitating the physical world. The realism of the environment is an

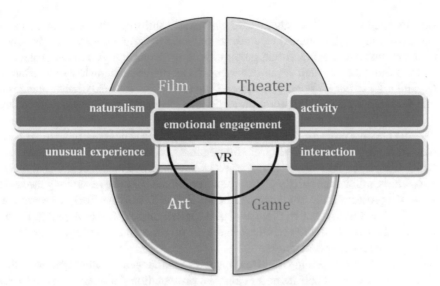

Fig. 1. Artistic virtual reality: features of becoming

important plan for immersion in virtuality, but a naturalistic chair by the aquarium does not have the powerful ability to "move a person" into virtual reality since there is no involvement other than contemplation.

Not limited to a naturalistic environment, virtual reality expands its possibilities with the animation language of creating worlds. In this transition, one can see the rejection of the frightening dictate of the camera, in which "art of imitating… is reduced to the technique of copying" [10]. For example, in the play *Draw Me Close: A Memoir* (Jordan Tannahill, 2019), the original technique using a motion capture system translates the movements of the actress interacting with the viewer into a painted picture. Although some art critics believe that this project should be classified rather not as an interactive theater, but as a media artifact [44]. *The Hangman at Home* (Michelle Kranot and Uri Kranot, 2020), which won the Grand Jury Prize at the Venice Biennale 2020, showcases stories of hand-drawn characters who are suddenly interrupted when they indignantly notice the viewer. Author Michelle Kranot defines his work as something between installation and performance, "no spectacle, but an experience - and a theatrical one" [45]. Drawn virtual reality worlds can take very different forms, documenting changes as seen standing in front of a house in Jakarta in *Replacements (Penggantian)* (Jonathan Hagard, 2020) or the *Hominidae* eco-system visible in x-ray (Brian Andrews, 2020) or mixing light, color, and sound as in *Soundself: a Technodelic* (Robin Arnott, 2020), guided by the participant's voice.

"Unusual experience" is the criterion that most distinguishes artistic projects (theater, cinema, art, video games) from other non-creative uses of VR (even in games). For example, a game of table tennis in virtual reality (*Eleven: Table Tennis*) will have the highest level of interaction with others and high realism, however, the degree of emotional involvement, exciting and immersive in another world, will be average, and the uniqueness of the experience is very low. A team-based virtual reality shooter on a physically large venue will surely involve players, but the unusualness of such an experience exists only at the moment when the experience of virtual reality itself is news to the players. Just like playing virtual guitar or having virtual sex. In artistic projects, on the other hand, "the unusualness of experience" can be the main task, as, for example, in *Quantum Tesseract - A Folding in Space Time* (Kris Pilcher, 2020) that visualizes the fourth dimension through sculptural projections of "HyperCubes" (fourth dimensional shapes).

Within creative projects, the difference between genres is not easy to spot. As a rule, it is easiest to separate VR games from other genres, since they involve created worlds in which the player has a significant amount of variability of actions leading to certain consequences, rather than strictly defined plot development. However, among the newest creative VR projects, there are those within which the viewer can freely move and act. So in the VR part of the Venice Biennale 2020. Of the projects that took part in it, some can be found on video game sites, others in the cinema or animation sections, and others exist only as an independent genre.

Being a new phenomenon, virtual reality mixes familiar genres, which can sometimes be distinguished only by self-name and place of presentation. 360-degree virtual reality means that the viewer is freed from the dictates of the camera and can choose the direction of the gaze. This brings cinema closer to the theater. In addition, it turns out to be a new challenge for filmmakers, returning to the realities of film cinematography, when the final result is visible only at the end of the shooting. Director and producer David Hansen of "*Jesus VR - The Story of Christ*" says "It's akin to shooting a live play - you don't have as many edits (…). The Last Supper is a 14-min scene, and it's one take" [37]. Controlling 360-degree reality can become game mechanics, as in *Down the Rabbit Hole* (Ryan Bednar, 2020), where roots protrude from the walls of the 3D diorama, by grabbing them, the player can rotate the whole world.

It would seem that theater is easily distinguishable from the cinema by the presence of actors. Indeed, theatrical experimentation remains intimately connected with individuals, which emotionally draw a person into the world of the characters. However, whether even in theater the actors should be physically present at the same time with the audience is an open question. For example, in the National Theater of Great Britain production of *All Kinds of Limbo*, the heroine - singer Nabia Brandon - appears in the form of a hologram.

Creative experiments in recent years show that the audience is increasingly transforming from a passive spectator into a subject. It would seem that in terms of the degree of activity there is no match for video games, since these exhibit rich possibilities for manipulating objects and interacting with the outside world. However, from a virtual reality perspective, activity should be closely related to engagement. By itself, the transfer of human actions to the digital world is not a guarantee for the virtualization of consciousness. It requires emotions, empathy for what is happening. And in this regard, a person who falls to his knees from the screams of border guards in the desert, created by Alejandro G. Iñárritu and Emmanuel Lubezki (*Carne y Arena*, 2017), could be more active and engaged than the player who went through the entire zombie apocalypse.

The creators of VR projects claim that they not only immerse the audience in a new reality but also force them to become "someone in it" [38]. For example, in the play

I Killed the Tsar (Patlasov, 2019), the viewer finds himself in the place of Emperor Nicholas II at the time of the execution. The used effect of influencing the viewer of frightening realism in virtual reality brings us back to the history of cinema - the panic from *The Arrival of a Train at La Ciotat Station* (*L'arrivée d'un train en gare de La Ciotat*, Auguste and Louis Lumière 1895) or from the shot point-blank to the camera in *The Great Train Robbery* (Edwin S. Porter, 1903).

It is believed that VR increases audience engagement [39–41], the emotional effect [42], and empathy [43]. For example, *The Enemy* (2018) is a project created by war journalist Karim Ben Khelifa who was dissatisfied with the lack of influence on people by war photographs. The game offers a virtual meeting with men fighting on opposing sides of armed conflicts in El Salvador, Israel-Palestine, and the Democratic Republic of Congo [44]. There are researches confirming the ability of VR to place a person "into the shoes of those whose feelings and experiences are distant to us" [43, 45]. The virtual reality documentary is usually a call for sympathy, e.g. *Daughters of Chibok* (2019) which presents the story about the mother of one of the girls who were abducted in Chibok, Nigeria. However, emotional engagement and empathy will not be the same for a variety of VR experiences. The possibilities of VR as an "Empathy machine" are limited. Empathy will depend on many factors related to technology (naturalism, habituation, the degree of cognitive overloading, etc.), the characteristics of the experience and personality traits.

In addition to being able to direct the gaze in any direction, for greater "virtualization", immersion is assigned a certain role to the viewer. Often this is visually reinforced by the ability to see the character's body as one's "own". Sometimes the viewer even turns out to be the central character, which of course requires a specific plot that does not require activity from such a character. Such a "successfully silent" central character is the jester Yorick, who dies at 90 s (*The Story of One Jester 360*, Alexey Bystritsky, 2017). In the play *"A Cage with Parrots"* (Maksim Didenko, 2017), the viewer turns out to be the main character, not particularly successfully passing the test to go to Mars, and the inactive silent position is associated with a special drug introduced to the hero. An interesting feature of the performance is its beginning on a square near Moscow City, both in fact and virtually. In *Hamlet 360: The Father's Spirit* (2019) director Steven Maler places the viewer in the ghost of Hamlet's father "as an omniscient observer, guide and participant" [46]. Nevertheless, since Father's Ghost still answers his son according to the script, there is here an actor who voices these answers.

At the same time, the theater, keeping its own traditions, sometimes uses virtual reality not as a substitute for theatrical space, but as its extension, when virtual pictures complement the actions of the actors (for example, showing the underwater world (*Frogman*, 2018), or the psyche of a sick person (*Cosmos Within Us*, Tupac Martyr, 2019). Mixed-reality environments can allow the public to interact with their virtual selves, the actors, and even the projection effects, as in *Inner Awareness: The Dream of Du Linang* (Qianhui Feng, 2017).

Not limited to naturalistic images, virtual reality expands its capabilities with the animation language of world creation. Here one can see the rejection of the frightening dictate of the camera, in which 'art of imitating… is reduced to the technique of copying' [10]. For example, in the play *Draw Me Close: A Memoir* (Jordan Tannahill, 2019), the original technique using a unique motion capture system translates the movements of the actress interacting with the viewer into a painted picture. Although some art critics believe that this project should be classified not as an interactive theater, but as a media artifact [47]. In *The Hangman at Home - An Immersive Single User Experience* (Michelle Kranot and Uri Kranot, 2020), which won the Grand Jury Prize at the Venice Biennale 2020, stories of drawn characters are suddenly interrupted when they indignantly notice the viewer. Author Michelle Kranot defines his work as something between installation and performance, "no spectacle, but an experience - and a theatrical one" [48]. Drawn virtual reality worlds can take very different forms, from documenting changes as seen standing in front of a house in Jakarta in *Replacements (Penggantian)* (Jonathan Hagard, 2020) or the *Hominidae* eco-system visible in x-ray (Brian Andrews, 2020) to mixing light, color, and sound as in *Soundself: a Technodelic* (Robin Arnott, 2020), guided by the participant's voice.

3   Between Theater and Videogame

Who is playing with whom in new virtual environments? Becoming more and more real from the point of view of the surrounding world, virtual reality requires an increasingly active subject.

Theater and games have many intersections in the digital age. Video games, as an integral part of modern life, appear in theatrical performances. And the theater itself can come into play, creating a layering of realities. So during the Covid-19 pandemic in 2020, the Bolshoi Drama Theater (St. Petersburg, Russia) constructed a building and staged three Minecraft performances of Russian classics that belong to the school curriculum (Anton Chekhov *The Cherry Orchard*, Alexander Pushkin *Mozart and Salieri* and Denis Fonvizin *The Minor*). This required a search for completely new ways of artistic expression. The actors controlled their characters in real time and spoke the text. But just in this case, the players were assigned the role of spectators with some of them intervening in the production, flying over the stage, against the wishes of the organizers, who were not as experienced in mods as some players [49]. Figure 2 shows a scene from the play *The Minor* (there is a Jeopardy-like game in the background and a herd of pigs in the foreground), and the finale of the performance, when the pig-actors, dressed in costumes, leave the stage.

Some projects seek to combine the active existence of several people in virtual reality. There are art projects that are social experiments that unite several "players", somehow influencing each other. For example, the *Metamorphic virtual environment* (Matthew Niederhauser, Wesley Allsbrook, Elie Zananiri, and John Fitzgerald) is a drawn natural environment where participants, with a decaying visual identity, can interact. In *Zikr: a Sufi Revival* project, four participants can work on each others' environment through a rope of prayer beads and other objects in the dance.

Having become the mainstay of numerous theatrical experiments over the past few years, virtual reality not only provides a director with new opportunities for interacting with the audience but in an interesting way brings theatrical art closer to video games.

The fantasy world of *The Under Presents* (Tender Claws, 2019) combines two traits that are key to video games and theatrical productions. There is a detailed game universe

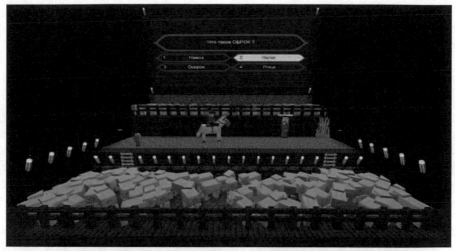

Fig. 2. Scene from the Bolshoi Drama Theater Minecraft staging of Denis Fonvizin's play *The Minor*, also the end of the play (director Edgar Zakaryan, artist-architect Andrey Voronov) [50]

in which players move on a non-linear quest, and a live performance of several real actors, making the gaming experience unique. This experience reveals the differences and interpenetration of theater and videogame. It is clear that the "virtuality" of the player's personality will be diminished when live actors are absent from the fantasy world. It can be assumed that in this case we are no longer dealing with a theater, but with a game that imitates an immersive theater (which is served by the classical structure of the genre and specific techniques). And the question of difference goes beyond the behavior of characters - such a fine line between non-player character, which is increasingly capable of variation in behavior, and a person whose ability to diversity is limited. The player's perceived freedom versus predetermination by a script comes to the fore, and the emotional component of contact with the actor (in particular, taken from the immersive spectacles "one-on-one" experience).

4   Conclusion and Discussion

Artistic VR strives to immerse the audience into the narrative, deepening their interaction with the environment and the characters. VR projects achieve strong virtuality of personality, by artistic and technical means strengthening the immersive personality traits of "Homo Virtualis". Such features as naturalism, unusual experience, activity, emotional engagement, interaction with others affect the degree of movement from physical reality to virtual reality. Each of these characteristics can be represented to a greater or lesser extent, creating a genre resemblance to films, theatrical productions, art objects, or video games. From the beginning artistic VR was designed to surprise and shock, offering to experience the unknown and the impossible. A little later, projects began to express the various experiences of people in a historical, political, psychological, or fantasy frame.

Modern creators do not yet fully understand the possibilities hidden in virtual reality. However, it is already obvious how powerful this technology could be. VR is able to give a person a deep and multifaceted experience of reality. As Anne Sauvagnargues notes, the main thing in art is not that it "gives us a more profound, or more spiritual way of understanding our world, or because it is a mystical door to a better reality", but that "it is the only way we have to construct in a collective way the modes of subjectification trying to find the way of transforming reality" [51]. Transforming reality, expanding the possibilities of "true experience" gives virtual reality the potential of a powerful technology for transforming the future.

References

1. Samorodova, E.A., Belyaeva, I.G., Bakaeva, S.A.: Analysis of communicative methods effectiveness in teaching foreign languages during the coronavirus epidemic: distance format. XLinguae **14**, 131–140 (2021). https://doi.org/10.18355/XL.2021.14.01.11
2. Kazaryan, R.: The concept of anthropotechnical safety of functioning and quality of life. In: Murgul, V., Pukhkal, V. (eds.) International Scientific Conference Energy Management of Municipal Facilities and Sustainable Energy Technologies EMMFT 2019. EMMFT 2019. Advances in Intelligent Systems and Computing, vol. 1258. pp. 759–767. Springer, Cham (2021). https://doi.org/10.1007/978-3-030-57450-5_64
3. Pozdeeva, E., et al.: Assessment of online environment and digital footprint functions in higher education analytics. Educ. Sci. **11**, 256 (2021). https://doi.org/10.3390/educsci11060256
4. Pokrovskaia, N.N., Korableva, O.N., Cappelli, L., Fedorov, D.A.: Digital regulation of intellectual capital for open innovation: industries' expert assessments of tacit knowledge for controlling and networking outcome. Future Internet **13**, 44 (2021). https://doi.org/10.3390/fi13020044
5. Tretyakov, I.L.: Information networks and manipulative technologies in the arsenal of extremists. In: Bylieva, D., Nordmann, A., Shipunova, O., Volkova, V. (eds.) Knowledge in the Information Society. PCSF 2020, CSIS 2020. LNNS, vol. 184. pp. 125–135. Springer, Cham (2021). https://doi.org/10.1007/978-3-030-65857-1_13

6. Serkova, V.: The digital reality: artistic choice. IOP Conf. Ser. Mater. Sci. Eng. **940**, 012154 (2020). https://doi.org/10.1088/1757-899X/940/1/012154

7. Du Sautoy, M.: The Creativity Code: Art and Innovation in the Age of AI. Harvard University Press, Cambridge, MA (2019)

8. Polmeer, G.: Historical questions on being and digital culture. In: Giannini, T., Bowen, J. (eds.) Museums and Digital Culture. Springer Series on Cultural Computing, pp. 49–62. Springer, Cham (2019). https://doi.org/10.1007/978-3-319-97457-6_3

9. Kellner, D.: Herbert Marcuse and the Crisis of Marxism. Macmillan Education, UK, London (1984). https://doi.org/10.1007/978-1-349-17583-3

10. Owen Barfield: The Rediscovery of Meaning, and Other Essay. Barfield Press, Oxford (2013)

11. Marcuse, H.: The Aesthetic Dimension. Beacon Press, Boston (1978)

12. Benjamin, W.: The work of art in the age of mechanical reproduction. In: Illuminations. Schocken Books, New York (1969)

13. McLuhan, M.: The Medium and the Light: Reflections on Religion and Media. Wipf and Stock Publishers, Eugene (2010)

14. Massumi, B.: Envisioning the virtual. In: Grimshaw, M. (ed.) The Oxford Handbook of Virtuality, pp. 55–71. Oxford University Press, New York (2014). https://doi.org/10.1093/oxfordhb/9780199826162.013.010

15. Frankel, R., Krebs, V.J.: Human Virtuality and Digital Life: Philosophical and Psychoanalytic Investigations. Routledge, London (2021)

16. Artaud, A.: The Theater and Its Double. Grove Press, New York (1994)

17. Hegel, G.W.F.: Hegel's Aesthetics: Lectures on Fine Art, vol. 1. Oxford University Press, Oxford (1975)

18. Gillespie, C.: virtual humanity—access, empathy and objectivity in vr film making. Glob. Soc. **34**, 145–162 (2020). https://doi.org/10.1080/13600826.2019.1656173

19. Gruenewald, T., Witteborn, S.: Feeling good: humanitarian virtual reality film, emotional style and global citizenship. Cult. Stud. 1–21 (2020). https://doi.org/10.1080/09502386.2020.1761415

20. Spencer, H.: First Principles. University Press of the Pacific, Stockton (2002)

21. Huizinga, J.: Homo Ludens: A Study of the Play-Element in Culture. Martino Fine Books, Eastford (2014)

22. Herodotus: History of Herodotus: A New English Version. D. Appleton, New York (1861)

23. Han, S.-L., An, M., Han, J.J., Lee, J.: Telepresence, time distortion, and consumer traits of virtual reality shopping. J. Bus. Res. **118**, 311–320 (2020). https://doi.org/10.1016/j.jbusres.2020.06.056

24. Schnack, A., Wright, M.J., Holdershaw, J.L.: Immersive virtual reality technology in a three-dimensional virtual simulated store: investigating telepresence and usability. Food Res. Int. **117**, 40–49 (2019). https://doi.org/10.1016/j.foodres.2018.01.028

25. Kang, S., O'Brien, E., Villarreal, A., Lee, W., Mahood, C.: Immersive journalism and telepresence. Digit. J. **7**, 294–313 (2019). https://doi.org/10.1080/21670811.2018.1504624

26. Yepez, J., Guevara, L., Guerrero, G.: AulaVR: virtual reality, a telepresence technique applied to distance education. In: 2020 15th Iberian Conference on Information Systems and Technologies (CISTI), pp. 1–5. IEEE (2020). https://doi.org/10.23919/CISTI49556.2020.9141049

27. Steuer, J.: Defining virtual reality: dimensions determining telepresence. J. Commun. **42**, 73–93 (1992). https://doi.org/10.1111/j.1460-2466.1992.tb00812.x

28. Lim, H., Childs, M.: Visual storytelling on Instagram: branded photo narrative and the role of telepresence. J. Res. Interact. Mark. **14**, 33–50 (2020). https://doi.org/10.1108/JRIM-09-2018-0115

29. Ongsakul, V., Ali, F., Wu, C., Duan, Y., Cobanoglu, C., Ryu, K.: Hotel website quality, performance, telepresence and behavioral intentions. Tour. Rev. **76**, 681–700 (2021). https://doi.org/10.1108/TR-02-2019-0039

30. Fox Tree, J.E., Whittaker, S., Herring, S.C., Chowdhury, Y., Nguyen, A., Takayama, L.: Psychological distance in mobile telepresence. Int. J. Hum. Comput. Stud. **151**, 102629 (2021). https://doi.org/10.1016/j.ijhcs.2021.102629

31. Cortellessa, G., et al.: ROBIN, a telepresence robot to support older users monitoring and social inclusion: development and evaluation. Telemed. e-Health. **24**, 145–154 (2018). https://doi.org/10.1089/tmj.2016.0258

32. Keller, L., Gawron, O., Rahi, T., Ulsamer, P., Müller, N.H.: Driving success: virtual team building through telepresence robots. In: Zaphiris, P., Ioannou, A. (eds.) Learning and Collaboration Technologies: Games and Virtual Environments for Learning. HCII 2021. LNCS, vol. 12785. pp. 278–291. Springer, Cham (2021). https://doi.org/10.1007/978-3-030-77943-6_18

33. Hassan, R.: Digitality, virtual reality and the 'empathy machine.' Digit. J. **8**, 195–212 (2020). https://doi.org/10.1080/21670811.2018.1517604

34. Bowen, J.P., Giannini, T., Falconer, R., Magruder, M.T., Marconi, E.: Beyond human: arts and identity between reality and virtuality in a post-Covid-19 world. In: Proceedings of EVA London 2021, pp. 7–11. BCS Learning and Development Ltd., London (2021). https://doi.org/10.14236/ewic/EVA2021.2

35. Spiegel, B.: Virtual Reality and the COVID Mental Health Crisis. https://www.scientificamerican.com/article/virtual-reality-and-the-covid-mental-health-crisis/. Accessed 2 May 2021

36. Rubin, B.M.: Virtual reality brings joy to people in assisted-living facilities. https://www.wsj.com/articles/virtual-reality-brings-joy-to-people-in-assisted-living-facilities-11616760002. Accessed 2 May 2021

37. Klett, L.M.: Virtual reality film brings viewers into life of Christ. https://jesusvr.net/2017/03/08/virtual-reality-film-brings-viewers-into-life-of-christ/. Accessed 2 Apr 2021

38. Smagen af Sult (A Taste of Hunger). https://www.labiennale.org/en/cinema/2020/venice-vr-expanded/smagen-af-sult-taste-hunger. Accessed 2 Apr 2021

39. Sagnier, C., Loup-Escande, E., Lourdeaux, D., Thouvenin, I., Valléry, G.: User acceptance of virtual reality: an extended technology acceptance model. Int. J. Hum.–Comput. Interact. **36**, 993–1007 (2020). https://doi.org/10.1080/10447318.2019.1708612

40. Kim, Y., Lee, H.: Falling in love with virtual reality art: a new perspective on 3D immersive virtual reality for future sustaining art consumption. Int. J. Hum.–Comput. Interact. 1–12 (2021). https://doi.org/10.1080/10447318.2021.1944534

41. Leontyeva, V.L., Pokrovskaia, N.N., Ababkova, M.Y.: Intellectual networking in digital education – improving testing for enhanced transfer of knowledge. In: Bylieva, D., Nordmann, A., Shipunova, O., Volkova, V. (eds.) Intellectual Networking in Digital Education – Improving Testing for Enhanced Transfer of Knowledge, pp. 171–191. Springer, Cham (2021). https://doi.org/10.1007/978-3-030-65857-1_16

42. Ding, N., Zhou, W., Fung, A.Y.H.: Emotional effect of cinematic VR compared with traditional 2D film. Telemat. Inform. **35**, 1572–1579 (2018). https://doi.org/10.1016/j.tele.2018.04.003

43. Bujić, M., Salminen, M., Macey, J., Hamari, J.: "Empathy machine": how virtual reality affects human rights attitudes. Internet Res. **30**, 1407–1425 (2020). https://doi.org/10.1108/INTR-07-2019-0306

44. Kabiljo, L.: Virtual reality fostering empathy: meet the enemy. Stud. Art Educ. **60**, 317–320 (2019). https://doi.org/10.1080/00393541.2019.1665401

45. Barbot, B., Kaufman, J.C.: What makes immersive virtual reality the ultimate empathy machine? Discerning the underlying mechanisms of change. Comput. Hum. Behav. **111**, 106431 (2020). https://doi.org/10.1016/j.chb.2020.106431

46. Be the ghost of 'Hamlet' in new virtual reality video by commonwealth Shakespeare Company. https://www.wickedlocal.com/entertainmentlife/20190125/be-ghost-of-hamlet-in-new-virtual-reality-video-by-commonwealth-shakespeare-company. Accessed 21 Apr 2021

47. Thompson, K., Bordwell, D.: Venice 2017: sensory saturday; or, what puts the virtual in VR?. http://www.davidbordwell.net/blog/2017/09/03/venice-2017-sensory-saturday-or-what-puts-the-virtual-in-vr/. Accessed 11 April 2021

48. Team, Z.: The Hangman at home XR by Michelle and Uri Kranot. https://www.zippyframes.com/index.php/production/the-hangman-at-home-by-michelle-and-uri-kranot. Accessed 2 Apr 2021

49. Lodz: A real performance in a virtual environment: the Bolshoi Drama theater opened a branch in minecraft. https://habr.com/ru/company/selectel/blog/509766/. Accessed 2 Apr 2021 (In Russian)

50. Zakaryan, E.: Minecraft performance "The Minor" on the stage of the virtual BDT. https://www.youtube.com/watch?v=xZM1ZR04vo4&feature=youtu.be. Accessed 2 April 2021 (In Russian)

51. Sauvagnargues, A.: Deleuze and Guattari's digital art machines. In: Assis, P. de and Giudici, P. (eds.) The Dark Precursor: Deleuze and Artistic Research, vol. 2. pp. 309–314. Leuven University Press (2017). https://doi.org/10.2307/j.ctt21c4rxx.29

Computer Modeling in Musical Creative Work: An Interdisciplinary Research Example

Sergey Chibirev and Irina Gorbunova(✉)

Herzen State Pedagogical University of Russia, 48 Moika Embankment, St. Petersburg, Russia
gorbunovaib@herzen.spb.ru

Abstract. Currently, when planning research, we can use the most active research approaches, such as experiment and modeling. Taking as an example a model of a musical fragment based on the statistics of notes, this article presents a sample of an interdisciplinary study. The idea of the study is to develop several equations that can quantify some of the verbal characteristics of a musical fragment. To prove the relevance of these equations, we found out the relationship between the parameters of the integral model and the values assigned by experts characterizing the psychological impression from music of a certain musical fragment. We used the group of 70 students as experts who evaluate the music fragments for several verbal values. Then we used several formulas on the model parameters (probability matrix determinants). Taking each pair of expert and calculated values we try to find the correlation. If the correlation is high for some pair of calculated data and expert data, we can use this equation to estimate the verbal characteristic of a musical fragment. As the result of this study we have proven formulas (on the music fragment probability matrix) that can evaluate some verbal characteristics of any music fragment. The development of musical computer technologies (MCTs) and digital musical instruments allows for appropriate experiments, also related to the field of music education. The results of modeling the process of musical creativity using MCTs are of particular importance for the implementation of various types of inclusive music education.

Keywords: Modeling of fuzzy subjects · Music computer technologies · Music modeling

1 Building the Model

1.1 Approaches to Building the Model

European and American scientists attempt to digitally synthesize, analyse, and transform sounds using music computer technologies (MCTs) [1]. The latter implies the automatic creation of the object format from the others of similar format. There are some other systems (Amin et al. [2]; Blenk et al. [3]; Blenk et al. [4]) and some works connected with comprehensive survey and performance evaluation network (Guck et al. [5]; Guck et al. [6]; Guck et al. [7]; Li et al. [8]; Thyagaturu [9]; Thyagaturu et al. [10]; Thyagaturu et al. [11]).

In recent years, Russian scientists have also become more interested in music programming (Gorbunova and Chibirev [12], Gorbunova and Chibirev [13]) and modelling the process of musical creativity (Zaripov [14]; Filatov-Beckman [15], Inatyev and Makin [16]; Gorbunova and Zalivadny [17]). Some authors propose to use a continuous wavelet transform apparatus as a mathematical tool for generating an amplitude-frequency-time representation of a signal. The bases of continuous wavelet transform representing the sound of musical instruments allowed the researcher (Fadeev [18]) to configure the continuous wavelet transform itself.

The results of scientific research in the field of development the integrative model for the semantic space of music are important for the creation and development of the model (Inatyev et al. [19]; Gorbunova et al. [20, 21, 22], as well as developments in the field of specialized software, which is the basis for various types of musical creative activities (see, for example, works: [23, 24], etc.).

The approaches to model development of some fuzzy subject (such as arts) are described in details in article [23], the process of creating models is based on:

- measure
- quantization
- screening
- statistics
- finding loops
- finding logic
- iterative

Measure. This is a complicated question for the most of arts, for example how to recognize and measure the elements of paintings. For music it is simple question, we take for analysis the musical text.

Quantization. Been measured, the data should be categorized and converted to the form, that good for analysis. Input data will be bind to the certain measuring scale. It could be linear or logarithmic or even any custom, the main idea: it should reflect all possible values but without excess. For two close values we must decide: if it is significant for the result - we need to have higher density scale, if no - we can take these values as the same.

Screening. Usually input data is too complicated to find any regularity in it. Screening help to find logic by removing from analysis the dependencies that already found. For example, we already found that every second value is significant, so just removing every first value will reveal the hidden regularity.

Considering music:
Taking as input data the frequencies of musical sounds of some music fragment. We see no logic in this stream of values.

We noticed that not all the frequencies of input sound are used in music but only ones that can be described with formula.

$f = n * 2^{(1/12)}$, where n is natural (**01.** Quantization of frequency).

(This is the frequencies of equal-tempered music scale (12-ET). n [1 .. 12] for this scale we have 12 threats in octave).

Then, we will consider as input data the sequence of natural values (n from the formula above). It is easier to fond regularity in it.

We noticed that not all 12 values are used in one music fragment! For any European melody only 7 of 12 values used. (For Asian and blues scales it is usually 5).

After next screening, we will analyze only these 7 values.

Next it is easy to perform statistical analysis of it. All the redundant information is dropped and will not spoil the statistics.

The similar situation with durations: not all the durations are used in music, but only proportional to natural numbers in powers of two.

$T = To/(2^n)$, where n is natural [1..8] (02. Quantization of duration).

Statistics. Using the statistical methods to analyse the parameters of the input sequence.

Finding Loops. The repeating patterns of data can be found sometimes only after screening.

Using **Error! Reference source not found.** above and analysing the duration as natural values we can found the loop length.

To find the loops we may use several hypotheses. Let us build the regression function to measure the success. Assuming that the length of the loop is some certain value, calculates the result of the function. The best result gives us the right length of the loop.

There are several repeating entities in music: meter, phrase, verse. These periods are typical for song-like structures, but in other music styles there are similar structures that may be formalized in the same way.

Finding Logic. The input sequence may be described by some laws, for example recurrent formula or have some trends depends on some external or internal values. The main idea is to come up with some hypothesis and then prove it or not and calculate its coefficients to make hypothesis act similar as the real sequence. It would be good using the regression analysis for it.

Iterations. After finding the logic, it gives the good result to analyse the data got on the previous stage. Using for analysis the results of previous layer is similar to N-layer Neuron Network. N is the number of iterations in the process of finding the logic.

1.2 Building the Model of Musical Fragment

Applying this approach described above to musical events sequence, recorded as musical text in MIDI format [12], we can create the model based on statistical analysis. The process of building the model of musical fragment is described in following works [13, 25] here we show only the (little simplified) result model, for details please reference to corresponding articles.

For the musical fragment the input sequence consists of Note Events that has 2 attributes: tone and duration.

Model considers the sequences of values of tone and duration separately, but on the stages of Finding Loops (and also Finding Logic) it appears that both sequences has the same structure.

Tone Value Sequence. The tone values sequence is described by transient probability Matrix. To simplify the idea, we can imagine that each row of the Matrix corresponds the current tone, and each value in the row shows the probability of next tone, having the current tone as a row.

There are several matrices of such kind used in the model, for each position in the musical fragment structure [3].

Duration values sequence is described also as probability vector, each value of it shows the probability of certain duration. Of course, all durations meet the requirements of **Error! Reference source not found.** Also there is some additional logic that prevent the breaking the structure of musical periods (found on Finding Loops stage): tacts, musical phrases, verses (for song-like musical forms) etc.

Conclusion. So the data that we have for investigation are:

- transient probability Matricies (for tones).
- probability vector (for durations)
- the musical fragment structure (common for tones and durations)

2   Sample Research Based on Model

2.1   The Purpose of the Research

The idea of the sample research is to develop several formulas that can describe quantitatively some verbal characteristics of musical fragment.

For the research 2 types of data were taken: expert data and calculated data from the model. Then, we calculated the correlation of them. If the correlation be high for some pair of calculated data and expert data – this means that *we can use this formula for evaluating the verbal characteristic of musical fragment.*

First of all, we developed the questionnaire for musical fragment verbal characteristics and prepares 15 musical fragments in various genres. All fragments were stored as MIDI-sequences for easy analysis and playback. All fragments were refined, melody is played with standard sound samples of MIDI-standart instruments with no sound effects and articulation.

2.2   Gathering of Expert Data

About 100 experts from students evaluated each of 15 musical fragments by 8 criteria. The verbal criteria are:

Positive - music evokes the positive emotions (negative is opposite).

Emotional - the strength of emotions, nevertheless the positive or negative (calm is the opposite).

Open - extraversive emotions, music provoces the sociality (intraversive is the opposite).

Creative - the music is variegated and complicated (monotonic and boring is the opposite).

Haste - the music is restless, bustling, stirring (measured and regular is the opposite).

Irrational - illogical, unpredictable (logical, understandable, predictable is the opposite).

Catchy - easy-to-remember, obsessive (unmemorable is the opposite).

Modern - the music seems to be modern, new and fashionable (old-style is the opposite).

The questionnaire can be accessed by the link http://quest.wannatea.com

The quiz results are grouped by fragment and can be accessed by the link http://quest.wannatea.com/cgi-bin/result.pl

2.3   Calculating of Model Data

The tone main probability matrix is taken as input values for formulas. To simplify analysis we use here 12×12 matrix dimension. (we can use other dimensions if melody Scale is calculated before, see [12]) Following formulas proposed for the investigations:

a. goingUp

For all steps where the next tone is **higher** than previous - the value included to the summ. So, to the summ will be included the values above matrix diagonal.

$$F_{up} = \sum_{j>i}^{i,j \,\in\, [0,11]} v_{i,j} \quad / \quad \sum^{i,j \,\in\, [0,11]} v_{i,j} \qquad \text{(03. "goingUp" determinant)}$$

Value is normalized, dividing by summ of all the values in the matrix.

b. goingUp with weight

Same as previous, but the far the value from the diagonal, the more weight it has.

$$F_{up} = \sum_{j>i}^{i,j \,\in\, [0,11]} v_{i,j} * (j-i) \quad / \quad \sum^{i,j \,\in\, [0,11]} v_{i,j} \qquad \text{(04."goingUp"determinantwith weight)}$$

Value is normalized, dividing by summ of all the values in the matrix with the same calculated weight.

c. monotonic

For all steps where the next tone is **equals** to previous - the value included to the summ. So, to the summ will be included the values on the matrix diagonal.

$i,j \in [0,11]$ $i,j \in [0,11]$

$$F_{up} = \sum_{j=i} v_{i,j} \Big/ \sum v_{i,j} \qquad \text{(05. "monotonic" determinant)}$$

Value is normalized, dividing by summ of all the values in the matrix.

c. monotonic with weight

Same as previous, but the far the value from the diagonal, the less weight it has.

$i,j \in [0,11]$ $i,j \in [0,11]$

$$F_{up} = \sum v_{i,j} * (12 - abs(j-i)) \Big/ \sum v_{i,j} \qquad \text{(06."monotonic"determinant with weight)}$$

Value is normalized, dividing by sum of all the values in the matrix with the same calculated weight.

It is expected that these values could correlate with some expert data.

2.4   Finding the Correlation

The result data (expert and calculated) of the research are presented in the Table 1.

Table 1. Expert values and model calculated values.

Fragments:	Fr1	Fr2	Fr3	Fr4	Fr5	Fr6	Fr7	Fr8	Fr9	Fr10	Fr11	Fr12	Fr13	Fr14	Fr15
Expert values															
Catchy	60.3	47.3	64.1	46.3	52.4	58.8	50.5	43.4	41.0	55.9	57.6	51.5	55.2	64.6	57.4
Emotion	78.8	46.8	64.9	47.1	48.8	72.0	56.8	60.7	36.6	43.0	65.8	72.5	84.6	75.9	65.6
Open	64.2	57.8	73.4	51.9	46.8	71.9	54.9	60.0	36.6	59.0	60.8	76.2	67.6	82.4	73.4
Modern	44.4	36.8	37.4	51.3	42.0	43.2	56.1	42.7	41.5	62.2	70.0	58.1	43.7	46.9	53.5
Creative	55.6	35.9	57.4	51.9	49.0	54.2	60.7	57.9	39.0	54.1	51.0	63.2	77.6	75.1	54.7
Positive	55.1	56.3	77.3	52.0	40.3	72.4	43.9	60.2	30.0	44.4	60.5	78.8	57.8	83.0	75.4
Haste	71.3	36.8	46.9	37.3	39.5	52.9	54.9	54.6	26.1	28.0	51.2	62.5	73.7	56.4	50.7
Irrational	45.8	32.5	46.1	49.5	44.2	41.3	53.9	49.0	38.6	46.1	51.0	53.7	64.7	61.9	46.3
Calculated values															
goingUp	38.9	40.3	54.9	38.5	41.4	39.5	52.9	53.3	47.9	41.1	25.9	57.0	42.1	43.6	43.0
goingUpW	47.2	44.1	58.2	50.7	46.1	48.4	61.3	60.1	53.1	44.7	65.2	62.8	43.3	58.5	54.3
Monot	7.0	14.9	8.5	16.6	6.6	6.8	10.1	11.2	13.2	8.1	47.4	9.5	3.9	16.8	14.5
monotW	63.4	67.1	66.0	66.2	76.2	66.9	67.6	71.7	76.9	63.2	82.5	65.5	66.7	71.0	69.1

To calculate the correlation, we take the pairs (one line from Expert values and one from calculated) and check the penalty function.

$i[0,15]$

$$F = \sum abs(E_i - C_i) \Big/ 15 \qquad \text{(07. Penalty function)}$$

E_i - expert values

C_i - model's calculated values

The example of calculations of penalty formula is below in Table 2 (values are per cent).

Table 2. Sample calculation of correlation based on penalty formula.

Catchy	60.3	47.3	64.1	46.3	52.4	58.8	50.5	43.4	41.0	55.9	57.6	51.5	55.2	64.6	57.4
goingUpW	47.2	44.1	58.2	50.7	46.1	48.4	61.3	60.1	53.1	44.7	65.2	62.8	43.3	58.5	54.3
penalty=8.9	13.1	3.2	5.9	4.4	6.3	10.4	10.8	16.7	12.1	11.2	7.6	11.3	11.9	6.1	3.1
Emotion	78.8	46.8	64.9	47.1	48.8	72.0	56.8	60.7	36.6	43.0	65.8	72.5	84.6	75.9	65.6
goingUpW	47.2	44.1	58.2	50.7	46.1	48.4	61.3	60.1	53.1	44.7	65.2	62.8	43.3	58.5	54.3
penalty=12.0	31.6	2.7	6.7	9.9	2.7	23.6	4.5	0.6	16.4	1.7	0.6	9.7	41.3	17.4	11.3
Open	64.2	57.8	73.4	51.9	46.8	71.9	54.9	60.0	36.6	59.0	60.8	76.2	67.6	82.4	73.4
goingUpW	47.2	44.1	58.2	50.7	46.1	48.4	61.3	60.1	53.1	44.7	65.2	62.8	43.3	58.5	54.3
penalty=14.2	17.0	13.7	15.2	1.2	0.7	23.5	6.4	0.1	16.5	14.3	4.4	32.9	24.3	23.9	19.1
Modern	44.4	36.8	37.4	51.3	42.0	43.2	56.1	42.7	41.5	62.2	70.0	58.1	43.7	46.9	53.5
goingUpW	47.2	44.1	58.2	50.7	46.1	48.4	61.3	60.1	53.1	44.7	65.2	62.8	43.3	58.5	54.3
penalty=7.6	2.8	7.3	20.8	0.6	4.1	4.8	5.2	17.4	11.6	17.5	4.8	4.7	0.4	11.6	0.6
Creative	55.6	35.9	57.4	51.9	49.0	54.2	60.7	57.9	39.0	54.1	51.0	63.2	77.6	75.1	54.7
goingUpW	47.2	44.1	58.2	50.7	46.1	48.4	61.3	60.1	53.1	44.7	65.2	62.8	43.3	58.5	54.3
penalty=7.9	8.4	8.2	0.8	1.2	2.9	5.8	0.6	2.2	14.1	9.4	14.2	0.4	34.3	16.6	0.4
Positive	55.1	56.3	77.3	52.0	40.3	72.4	43.9	60.2	30.0	44.4	60.5	78.8	57.8	83.0	75.4
goingUpW	47.2	44.1	58.2	50.7	46.1	48.4	61.3	60.1	53.1	44.7	65.2	62.8	43.3	58.5	54.3
penalty=12.8	7.9	12.2	19.1	1.3	5.8	24.0	17.4	0.1	23.1	0.3	4.7	16.0	14.5	24.5	21.1
Haste	71.3	36.8	46.9	37.3	39.5	52.9	54.9	54.6	26.1	28.0	51.2	62.5	73.7	56.4	50.7
goingUpW	47.2	44.1	58.2	50.7	46.1	48.4	61.3	60.1	53.1	44.7	65.2	62.8	43.3	58.5	54.3
penalty=11.5	24.1	7.3	11.3	13.4	6.6	4.5	6.4	5.5	27.0	16.7	14.0	0.3	30.4	2.1	3.6
Irrational	45.8	32.5	46.1	49.5	44.2	41.3	53.9	49.0	38.6	46.1	51.0	53.7	64.7	61.9	46.3
goingUpW	47.2	44.1	58.2	50.7	46.1	48.4	61.3	60.1	53.1	44.7	65.2	62.8	43.3	58.5	54.3
Penalty=8.3	1.4	11.6	12.1	1.2	1.9	7.1	7.4	11.1	14.5	1.4	14.2	9.1	21.4	3.40	8.00

Catchy	60.3	47.3	64.1	46.3	52.4	58.8	50.5	43.4	41.0	55.9	57.6	51.5	55.2	64.6	57.4
monotW	63.4	67.1	66.0	66.2	76.2	66.9	67.6	71.7	76.9	63.2	82.5	65.5	66.7	71.0	69.1
Penalty=15.5	3.1	19.8	1.9	19.9	23.8	8.1	17.1	28.3	35.9	7.3	24.9	14.0	11.5	6.4	11.7
Emotion	78.8	46.8	64.9	47.1	48.8	72.0	56.8	60.7	36.6	43.0	65.8	72.5	84.6	75.9	65.6
monotW	63.4	67.1	66.0	66.2	76.2	66.9	67.6	71.7	76.9	63.2	82.5	65.5	66.7	71.0	69.1
penalty=14.7	15.40	20.3	1.1	19.1	27.4	5.1	10.8	11.0	40.3	20.2	16.7	7.0	17.9	4.9	3.5
Open	64.2	57.8	73.4	51.9	46.8	71.9	54.9	60.0	36.6	59.0	60.8	76.2	67.6	82.4	73.4
monotW	63.4	67.1	66.0	66.2	76.2	66.9	67.6	71.7	76.9	63.2	82.5	65.5	66.7	71.0	69.1
Penalty=12.3	0.8	9.3	7.4	14.3	29.4	5.0	12.7	11.7	40.3	4.2	21.7	10.7	0.9	11.4	4.3
Modern	44.4	36.8	37.4	51.3	42.0	43.2	56.1	42.7	41.5	62.2	70.0	58.1	43.7	46.9	53.5
monotW	63.4	67.1	66.0	66.2	76.2	66.9	67.6	71.7	76.9	63.2	82.5	65.5	66.7	71.0	69.1
Penalty=20.6	19.0	30.3	28.6	14.9	34.2	23.7	11.5	29.0	35.4	1.0	12.5	7.4	23.0	24.1	15.6
Creative	55.6	35.9	57.4	51.9	49.0	54.2	60.7	57.9	39.0	54.1	51.0	63.2	77.6	75.1	54.7
monotW	63.4	67.1	66.0	66.2	76.2	66.9	67.6	71.7	76.9	63.2	82.5	65.5	66.7	71.0	69.1
Penalty=15.5	7.8	31.2	8.6	14.3	27.2	12.7	6.9	13.8	37.9	9.1	31.5	2.3	10.9	4.1	14.4
Positive	55.1	56.3	77.3	52.0	40.3	72.4	43.9	60.2	30.0	44.4	60.5	78.8	57.8	83.0	75.4
monotW	63.4	67.1	66.0	66.2	76.2	66.9	67.6	71.7	76.9	63.2	82.5	65.5	66.7	71.0	69.1
Penalty=16.6	8.3	10.8	11.3	14.2	35.9	5.5	23.7	11.5	46.9	18.8	22.0	13.3	8.9	12.0	6.3
Haste	71.3	36.8	46.9	37.3	39.5	52.9	54.9	54.6	26.1	28.0	51.2	62.5	73.7	56.4	50.7
monotW	63.4	67.1	66.0	66.2	76.2	66.9	67.6	71.7	76.9	63.2	82.5	65.5	66.7	71.0	69.1
Penalty=21.8	7.9	30.3	19.1	28.9	36.7	14.0	12.7	17.1	50.8	35.2	31.3	3.0	7.0	14.6	18.4
Irrational	45.8	32.5	46.1	49.5	44.2	41.3	53.9	49.0	38.6	46.1	51.0	53.7	64.7	61.9	46.3
monotW	63.4	67.1	66.0	66.2	76.2	66.9	67.6	71.7	76.9	63.2	82.5	65.5	66.7	71.0	69.1
Penalty=21.0	17.6	34.6	19.9	16.7	32.0	25.6	13.7	22.7	38.3	17.1	31.5	11.8	2.0	9.1	22.8

Since result values are normalized in per cent, penaly ∈ [0, 100], we can set some acceptance criteria for correlation. For example, penalty less than 10% consider as acceptable for the formula.

The results are following:

The good correlation of "going Up" determinant is "creative" and "modern". This is expected result for "creative" but not obvious for "modern". As it described above, the "modern" is very complex characteristic, so we need more deep analysis with much more number of factors that is not included to this sample research for the proof of concept of the music model.

There is no acceptable result for "monotonic" discriminant, but the closest verbal characteristic is got for "open".

3   Conclusion

3.1   Results and Summary

This sample research proves that using of modeling in investigation of fuzzy subjects such as arts could be useful and fruitful.

The model, described in articles [13, 25] could be useful instrument for research.

Formulas, developed in this sample research could be used for evaluation of musical fragment, and they are equivalent of expert verbal values.

3.2   Doubts and Limitations

Social-Dependent Experts. For the experts the close group of students were used, so there's high risk that formulas, developed in this research could work acceptable only for this layer of experts and will give different result for any other group of people.

Yes, this is so. But the formulas are developed for this group. If we need to evaluate the music verbal characteristics for another group of users, we can organize another research with other experts from this group. The idea is: experts and the end users of worked out solution should be from the same social group.

Evaluation criteria in the questionnaire could correlate to each other, so, the evaluation made by formula could reflect not exactly one criteria. For example, "open" or "creative" criteria could also be considered as "positive". Also "modern" and "catchy" are very complex values, depended from the number of attributes of musical fragment.

Yes, this is so. But the algorithmic evaluation works even in this complex case, because the correlation is developed and proved during the progress of experiment, it is designed on a base of expert evaluation, so since it worked during expert quiz, it has been "trained" for these formulas, and this should give the same result when operating the model.

The model developed by us is the basis for creating (and is already being used - see, for example, works:) as a creative environment in the system of contemporary inclusive education [26] and inclusive musical education [27]. Thanks to the use of the developed model, it was also possible to study a number of psychological and pedagogical features of the development of information technologies and MCT by people with deep visual impairments [28].

References

1. Gorbunova, I.B.: Music computer technologies in the perspective of digital humanities, arts, and researches. Opcion **35**(Special Edition 24), 360–375 (2019)
2. Amin, R., Reisslein, M., Shah, N.: Hybrid SDN networks: a survey of existing approaches. IEEE Commun. Surv. Tutor. **20**(4), 3259–3306 (2018). https://doi.org/10.1109/COMST.2018.2837161

3. Blenk, A., Basta, A., Johannes, Z., Reisslein, M., Kellerer, W.: Control plane latency with SDN network hypervisors: the cost of virtualization. IEEE Trans. Netw. Serv. Manag. **13**(3), 366–380 (2016). https://doi.org/10.1109/TNSM.2016.2587900
4. Blenk, A., Basta, A., Reisslein, M., Kellerer, W.: Survey on network virtualization hypervisors for software defined networking. IEEE Commun. Surv. Tutor. **18**(1), 655–685 (2016). https://doi.org/10.1109/COMST.2015.2489183
5. Guck, J.W., Reisslein, M., Kellerer, W.: Model-based control plane for fast routing in industrial QoS network. In: Proceedings IEEE Internationmal Symposium on Quality of Service (IWQoS), pp. 65–66. IEEE, Portland, OR (2015). https://doi.org/10.1109/IWQoS.2015.740 4708
6. Guck, J.W., Reisslein, M., Kellerer, W.: Function split between delay-constrained routing and resource allocation for centrally managed QoS in industrial networks. IEEE Trans. Ind. Inform. **12**(6), 2050–2061 (2016). https://doi.org/10.1109/TII.2016.2592481
7. Guck, J., Vanbemten, A., Reisslein, M., Kellerer, W.: Unicast QoS routing algorithms for SDN: A comprehensive survey and performance evaluation. IEEE Commun. Surv. Tutor. **20**(1), 388–415 (2018). https://doi.org/10.1109/COMST.2017.2749760
8. Li, J., Veeraraghavan, M., Reisslein, M., Manley, M., Williams, R.D., Amer, P., Leighton, L.: A Less-Is-More Architecture (LIMA) for a future internet. In: Proceedings of IEEE Global Internet Symposium (Infocom Workshop), pp. 55–60. IEEE, Orlando, FL (2012). https://www.eecis.udel.edu/~amer/PEL/poc/pdf/GlobalInternet2012-JieLi-LIMA.pdf. Accessed 20 May 2021
9. Thyagaturu, A.: Software defined applications in cellular and optical networks (Doctoral Dissertation) Arizona State University, Tempe (2017)
10. Thyagaturu, A., Mercian, A., McGarry, M.P., Reisslein, M., Kellerer, W.: Software Defined Optical Networks (SDONs): a comprehensive survey. IEEE Commun. Surv. Tutor. **18**(4), 2738–2786 (2016). https://doi.org/10.1109/COMST.2016.2586999
11. Thyagaturu, A., Dashti, Y., Reisslein, M.: SDN-based smart gateways (Sm-GWs) for multi-operator small cell network management. IEEE Trans. Netw. Serv. Manag. **13**(4), 740–753 (2016). https://doi.org/10.1109/TNSM.2016.2605924
12. Gorbunova, I.B., Chibirev. S.V.: Modeling the process of musical creativity in musical instrument digital interface format. Opcion **35**(Special Issue 22), 392–409 (2019)
13. Gorbunova, I., Chibirev, S.: Algorithmic modeling of arts and other hard-to-formalize subjects. Int. J. Recent Technol. Eng. **8**(6), 2655–2663 (2020)
14. Zaripov, R.H.: Cybernetics and Music. Nauka Publishing, Moscow (1971). (in Russian)
15. Filatov-Beckman, S.A.: Computer Music Simulation. Publishing House "Urait" Moscow (2015). (in Russian)
16. Idnatyev, M.B., Makin, A.I.: Linguistic-combinatorial modeling of music. In: Gorbunova, I. (ed.) Contemporary Musical Education: Creative, Research, Technologies. CME 2018. Proceedings of the 17th International Research and Practical Conference, pp. 143–149. Publishing House of the Herzen State Pedagogical University of Russia, St. Petersburg (2019). (in Russian)
17. Gorbunova, I.B., Zalivadny, M.S.: Leonhard Euler's theory of music: its present-day significance and influence on certain fields of musical thought. Music Scholarsh. **3**(36), 104–111 (2019).https://doi.org/10.17674/1997-0854.2019.3.104-111
18. Fadeev, A.S.: Identification of musical objects with continuous wavelet transform. Doctoral Dissertation, Tomsk Polytechnic University, Tomsk (2008). (in Russian)
19. Ignatyev, M.B., Zalivadny, M.S., Reshetnikova, N.N.: An integrative model of a semantic space of music. In: An Integrative Model of a Semantic Space of Music: A Collection of Articles, pp. 9–12. Publishing House of the Herzen State Pedagogical University of Russia, St. Petersburg (2016)

20. Gorbunova, I.B., Zalivadny, M.S.: The integrative model for the semantic space of music: perspectives of unifying musicology and musical education. Music Scholarsh. **3**, 55–64 (2018). https://doi.org/10.17674/1997-0854.2018.4.055-064
21. Gorbunova I.B.: The integrative model for the semantic space of music and a contemporary musical educational process: the scientific and creative heritage of Mikhail Borisovich Ignatyev. Laplage em Revista **6**(S), 2–13 (2020)
22. Gorbunova, I.B., Zalivadny, M.S., Tovpich, I.O.: On the application of models of the semantic space of music in the integrative analysis of musical works and music education with music computer technologies. Apuntes Universitarios **10**(4), 13–3 (2020)
23. Robert, F.: Software as a basic music platform. In: 14th Conference. Giuseppe Verdy-Richard Wagner-Moor Emanuel, 22nd February 2014, pp. 119–127. Szeged, Hungary (2014)
24. King, A., Himonides, E., Ruthmann, S.A. (eds.): The Routledge Companion to Music, Technology, and Education. Taylor & Francis Group, New York and London (2017)
25. Gorbunova, I.B., Chibirev, S.V.: Music computer technologies and the problem of modeling the process of musical creativity. In: Regional Informatics "RI-2014", Proceedings of the 14th St. Petersburg International Conference, pp. 293–294. Publishing House of the St. Petersburg Society of Informatics, Computer Technology, Communication and Control Systems, St. Petersburg (2014). (in Russian)
26. Kouroupetroglou, G.: Welcome to ICCHP 2018! (2018). http://www.icchp.org/welcome-chair-18. Accessed 20 Apr 2021
27. Gorbunova, I.B., Zakharov, V.V., Yasinskaya, O.L.: Development of applied software for teaching music to people with deep visual impairments based on music computer technologies. Revista universidad y sociedad **12**(3), 77–82 (2020). https://rus.ucf.edu.cu/index.php/rus/article/view/1559. Accessed 24 Apr 2021
28. Voronov, A.M., Krivodonova, J.E.: Psychological and pedagogical features of the development of information technologies by persons with visual impairments. In: Child in the Modern World: Proceedings of the International Scientific Conference, pp. 251–256. Publishing House of the Herzen State Pedagogical University of Russia, St. Petersburg (2013)

Computer Technologies as Creative Interaction Tools Between Far East and Chinese Musical Cultures

Irina Gorbunova[1](✉) [iD] and Svetlana Mezentseva[2] [iD]

[1] Herzen State Pedagogical University of Russia, 48 Moyka Embankment,
St. Petersburg 191186, Russia
gorbunovaib@herzen.spb.ru
[2] Khabarovsk State Institute of Culture, st. Krasnorechenskaya,
112, Khabarovsk 680045, Russia

Abstract. The article deals with the regional folklore as a field of interaction of academic musical culture in the Far East of Russia and China. There is an analysis of both the aborigines' culture and the Eastern Slavic immigrating culture of the Russian Far East as well as of the indigenous culture outside Russian borders in Asia and the Pacific. Some common features of the Russian Far East and Chinese folklore are found out. There is a description of the concept of "academic musical culture" which comprises a composer's work, successively associated with the foundations of Western European music that were formed during the period of 17th - 19th centuries (including the 20th century composers' work as well as contemporary composers' techniques), musical performance, music and performance infrastructure, educational space and academic musicology. The article highlights fundamental research in the field of studying the traditional musical culture of the Far East and China, research in the field of musical culture of the region along with the use of musical computer technologies in various fields of musical art and education.

Keywords: Far eastern composers · Academic musical culture · Music computer technologies · Creative interaction

1 Introduction

1.1 Research in the Fields of Studying of Traditional Musical Culture of the Russian Far East and Chine

The musical culture of the Far East comprises components of various ethnic cultures. In the Far East it is the aborigines' culture and the Eastern Slavic immigrating culture. Outside Russian borders it is the indigenous culture of Asia and the Pacific (further AP). Historically and, above all, geographically the constant synthesis of these cultures and sometimes assimilation is contingent. Political processes also play a significant role in the formation of the musical picture, the "sound world" of the modern Far East, since music is included in universal deep socio-cultural processes.

Another problem is presented by the formation of Far Eastern academic musical culture with its distinctive features. The concept of "academic musical culture" in this article comprises professional composers' work, successively associated with musical genres, principles of formation, melodic, harmonic, instrumental foundations, developed in Western Europe during the period of 17th - 19th centuries (including the 20th century composers' work alongside with modern composers' techniques) as well as musical performance and music and performance infrastructure. The concept of "academic musical culture" includes academic musicology and the educational process in musical educational institutions that are centers of classical art.

The work highlights the basic research in the fields of studying of traditional musical culture of the Russian Far East and Chine, research in musical culture of the region along with the sphere of musical technologies. The beginning of systematic study of academic musical culture of the Russian Far East is connected to the foundation of the Far Eastern Branch of the Union of Composers of Russia. The work defines the regional composers and the main direction of their creative work. Researchers of the academic musical culture of the region were noted, whose works are a significant addition in understanding the processes of modern domestic musical culture development.

There has been covered the professional composers' work of the academic orientation of China known in Russia alongside with the scientific interests of Russian orientalist researchers and researchers from China.

The need for cultural understanding of the declared problem through academic musical art, traditional musical culture, musical science and musical education is recognized. The special role of music computer technologies in musical culture and education in the Far East of Russia and China is regarded as the most important component for interaction in the field of academic musical culture, the problems of informatization of contemporary musical education are emphasized.

The conclusion is made about the unique experience of composer's creative work in China on the basis of traditional music of the Russian Far East. The anhemiton pentatonic basis of Chinese music is especially regarded as being close to the free organization of Far Eastern ethnic groups' music and as forming the basis of Russian Far Eastern composers' folklore music. In such a scale and tonal proximity, the author sees the ground for the interaction of cultures of the Far Eastern region. This aspect is recognized as important from the point of view of creating an integral multicultural space based on the principles of humanism and tolerance.

1.2 The Purpose of This Study

The purpose of this study is to identify the state of contemporary academic musical culture of the Far East on the example of the Russian Far East of Russia and China in terms of possible development prospects in the field of composers' creative work and academic musical education and to define the role of music computer technologies (further MCT) in modern academic musical culture and educational practices of the Far Eastern region.

1.3 Level of Prior Studies of the Problem

The problem statement of the present study involves a multidimensional cultural understanding of the latest processes taking place in the world regarding the problems of dialogical contacts in musical culture, the interaction of cultures of different peoples through academic musical art, traditional musical culture, musical science and musical education. Within the framework of this issue, the main researchers in three areas: research on the traditional musical culture of the Far East of Russia and China, research on the academic musical culture of the region, research in the field of MCT will be found.

2   Studies of the Academic Musical Culture of the Far Eastern Region

The beginning of a systematic study of the academic musical culture of the Far East was laid in the late 1950s–1960s when the Far Eastern Branch of the Union of Composers of Russia was formed here, and in musical life, a solid place was taken by the forms of philharmonic performance which had been developing before, but from now on it was on an ongoing basis. In the central press there were clearly delineated the topics of touring and "their" musical performance, all-Russian (all-Soviet) and Far Eastern composer's creative work that later became significant. In the research of A. M. Eisenstadt, attention is paid to the stylistic features of folk songs, musical instrumentation [1]. Important for the study and propaganda of musical folklore of the Far East are the works of I. A. Brodsky (Bogdanov), in particular, the publication of a valuable record with the annotation "Music of the Peoples of the Far East of the USSR" [2] and others. Currently, a great contribution to the study of musical folklore of the indigenous peoples of the Far East was made by N.A. Solomonova [3], Yu. I. Sheikin [4], N. A. Mamcheva [5], T. R. Bulgakova [6] and others.

In the works of N. A. Solomonova, on the basis of rich expeditionary material, the traditional musical culture of the peoples of the Far East of Russia was investigated over a long historical period (mid-19th - 20th centuries) [7]. The genre typology of instrumental music of the ritual culture of the Tungus and Manchus was presented by S. V. Mezentseva [8]. S. P. Galitskaya [9], V. N. Yunusova [10], T. I. Naumenko [11], S. B. Lupinos [12], E. M. Alkon [13], A. G. Alyabyeva [14], J. K. Mikhailov [15] et al. devoted their scientific research to the traditional musical culture of the countries in the East.

In articles by B. D. Napreev (about N. Mentser), N. A. Solomonova (about Yu. Vladimirov and E. Kazachkov), P. Volkhin (about song and cantata-oratory creative work) of the 1970s and 80s, the research attention was focused on individual works and names of composers, and on the problems of developing a regional composers' creative work in general. Moreover, the core issues for these authors were the formation and general orientation of the Far Eastern development of composer organization, the issues of genre and style trends [16].

Inevitably, at the next stage of the 1990-2010s in Far Eastern musicology there was a further differentiation of research issues, which was marked by the emergence of a number of fundamental works. Thus, N. A. Solomonova's dissertation "Musical

Culture of the Peoples of the Russian Far East of the 19th–20th centuries." [7] covered a large complex of issues of the functioning of traditional and modern musical folklore and the composer's creative work that arose on its basis. Far Eastern piano education, performance and creative work received a comprehensive analysis in L. A. Matveeva's monograph "Piano Culture of Siberia and the Far East of Russia (late 18th century - 1980s)" [17]. The general processes of genre and style development of the regional music and its special sphere devoted to the interpretation of Far Eastern folklore by composers are presented in T. V. Leskova's publications [16]. A number of private studies complement this picture in terms of specifying some genre and style "branches" of Far Eastern music, in terms of analysing individual works and creative biographies of such composers as Yu. Ya. Vladimirov, N. N. Mentser, A. V. Novikov, S. S. Moskaev, A. T. Goncharenko. Thus, in more than half a century, the Far Eastern music science has developed a regional, local history field of knowledge, which is considered to be a significant addition to the understanding of the development process of modern national musical culture.

The Chinese professional composer's creative work of the academic orientation is not well-known in Russia. Of the latest works, I. A. Zhen's study is worth noting [18]. It is devoted to the professional musical culture of the Far Eastern countries. The scientific works of Russian researchers are more focused on musical performance and musical education, performing schools in China and other AP countries (S. A. Eisenstadt [19], U Gen-Ir [20], et al.). Furthermore, modern national works cover the problems of mentality and Chinese students' social adaptation in the conditions of a Russian university, problems of intercultural communication and others.

The research interests of Chinese investigators cover various perspectives of music education in China and Russia: Linli Fan, Li Xueyan, Li Min. The number of scientific dissertations of authors from China, defended in the Russian dissertation councils, which reveal various aspects of the academic musical culture of China, is growing: Luo Zhihui, Juan Song, Day Yu, Wang Ying, Bian Men, etc. (see abstracts of the respective authors' theses).

3   Influence of the Music Computer Technologies on Contemporary Musical Culture and Education

Currently, the influence of the MCT on contemporary musical culture and education is being actively studied. Especially significant in this sense are the works of V. O. Beluntsov [21], I. M. Krasilnikov [22], I.B. Gorbunova, who has justified the term "music computer technologies" itself and under whose leadership the concept of "Music Computer Technologies in Education" was developed and the education and methods laboratory "Music Computer Technologies" was created at the Herzen State Pedagogical University of Russia in 2002 [23].

The scientific understanding of the researchers includes a variety of aspects related to MCT and the informatization of modern society: issues of the influence of MCT on the professional musicians' and composers' creative work, problems of informatization of contemporary musical education and others. The researcher and the composer, I. M. Krasilnikov, devoted his research and methodical work to developing of e-music concepts in art education. Issues of MCT influence on professional musicians' and composers' creative work by composers themselves are studied in works of G. G. Belov [24], A. Kameris [25], I. B. Gorbunova [26, 27]. The problems of modelling creative processes using music and computer technologies were investigated in the works of M. S. Zalivadny [28], E. V. Kibitkina [29], S. V. Chibirev [30], I. O. Tovpich [31, 32] et al. The computer analysis of sound in music science is devoted, among other things, to A. V. Haruto's research [33], revealing the experience of the Moscow Conservatory.

Thanks to the use of the basic principles of the formation of post-material values in the conditions of the functioning of a high-tech information educational environment, as well as taking into account the principles of attitude to values in the practice of social management and communication [34], it was also possible to study a number of l features of the use of information technologies and MCT as creative interaction tool between Russian Far East and China cultures.

3.1   The Current State of Academic Musical Culture of the Russian Far East

Currently, the academic musical culture of the Far Eastern region of Russia is a valuable conglomerate of the richest ethnic heritage and music of professional composers of the Far East. The works of N. Mentzer, Yu. Vladimirov, V. Rumyantsev, S. Tombak, F. Sadovoy, I. Belitsa, V. Ipatov, P. Mirsky, G. Ugryumov demonstrate the original synthesis of traditional and European intonation cultures, based on specific scale and metrorhythmic, textural laws of Far Eastern folklore, which peculiarly refracted the unique intonation features of traditional folklore. Obviously, Far Eastern composers also created music of non-folklore style and European orientation.

Along with the music of Far Eastern composers, Western European classics always remain popular in concert halls and at creative venues in the region. Recently, works by Chinese composers have also begun to be performed.

In musicology, there was also a significant shift in the study of the musical culture of the East. In particular, V. N. Yunusova says: "In the range of problems that modern researchers of Chinese music are working on, it should be noted: the history of Chinese music, musical and aesthetic issues, the reflection of the picture of the world and the model of the world in the realities of Chinese music and culture in general, study of traditional hieroglyphic notations, traditional orchestras, Chinese musical drama jingju, treatises on music and information about Chinese culture in Japanese sources "[10, p. 784]. In general, the researcher notes the transition from abstract works to the deep development of the problems of Chinese musical culture and the formation of: "theoretical, historical, cultural, ethnomusicological areas of research on Chinese music at the modern stage" [10, p. 786].

3.2   The Academic Musical Culture of China at the Present Stage

The formation and development of the Chinese composer school is associated with the reliance on rich traditional musical art in combination with European compositional technique. The specific features of academic music in China are the reliance on anhemitonic pentatonics as the basis of national melodics, the predominance of images of nature, its descriptiveness, Western European harmonization and principles of composition, as well as American composers (since many modern Chinese composers study in the United States). These features are recognized as the main characteristics of the professional composer's creative work of the academic orientation of China.

Currently, the professional composer work of the academic orientation of China is relatively poorly studied. Nevertheless, the music of academic composers who created the music of major musical forms is quite known - symphonies, concerts, large-scale choral works, operas, as well as chamber vocal and instrumental compositions. The creative work of Chinese composers is studied within the framework of the specialized complex direction "Musical Culture and Musical Performance in the Far East, History of Music" in the scientific laboratory "Problems of Higher Education in the Field of Culture and Art in the Far East" of the Khabarovsk State Institute of Culture (Chen Xiaohui, S.V. Mezentseva).

Composers of China are actively developing new modern techniques of Western European music of the 20th century. Chinese researchers discover such phenomena as "Chinese dodecaphony," aleatorics, polytonality, electronic music using MCT [33]). The Chinese traditional scale system in music of modern Chinese composers (Wang Lisangya, Pang Chimingya, Wang Tszyanchzhuna, Zhu Jiangera, Zhongrong's Luo, etc.), while having unique potential opportunities of synthesis with such technicians as a new modality, polytonality, a seriality, dodecaphony, also needs new methods of the scale and harmony analysis [35].

In terms of musical performance, one cannot fail to note the huge surge in interest in academic music at present. Concert halls are being built everywhere in China, educational institutions have excellent concert venues equipped with the latest technology and musical acoustics. A large role in the promotion of academic musical culture is played by both local and invited performers from other countries. In addition, various symposia, conferences, competitions, including at the international level, are initiated in China, which makes it possible to establish creative contacts and, undoubtedly, enriches the cultural level of the country.

It is important to point out the huge interest of Chinese students in higher education in creative institutes and conservatories of Russia. There has significantly grown the number of foreign scientists from China who are seeking to improve their scientific qualifications by taking a post-graduate course in Russia and defending their scientific research on the academic musical culture of the Far Eastern region. Piano and orchestral string instruments in China are a kind of symbol of European spirituality and culture. Training in playing these instruments is in great demand among foreign students and in creative universities in Russia.

Foreign students are happy to master the music of Far Eastern composers at the Khabarovsk State Institute of Culture, because the pentatonic basis of Chinese music is close to the free organization of music of Far Eastern ethnic groups, which also underlies the works of the folklore orientation of the music of Russian Far Eastern composers. It is thought that it is this proximity (in this case, scale and harmony) of the musical foundations that can and should be the basis for the interaction of various cultures.

3.3   The Role of Music Computer Technology in Musical Academic Culture in the Far East

The musical culture of the Far East in the modern information world, undoubtedly, comes under influence of new digital technologies: composers of Russia and China enrich the technicians with new genres, creative forms, ways, techniques and methods, the pace of newly recorded and broadcast information exchange is accelerated. China's academic music at the present stage actively interacts with modern Western composing techniques and new intonation patterns.

In music and pedagogical education, the role of the MCT is strengthened, which is one of the forms of modern culture existence [36] in general and it is due to the widespread introduction of innovative technologies. But, despite the fact that the educational and pedagogical sphere is an extremely conservative system, the very forms of music education are undergoing a significant change: along with traditional musical specialties and training areas, directed at the development of musical and instrumental performance, vocal art, musicology, composition, conducting, directions related to MCT are growing (for example, the profile "Music and computer technologies" within the areas of "Pedagogical Education" and "Art Education" at the Herzen State Pedagogical University of Russia, many pedagogical universities and academies of music in Russia). A new technology for submitting educational material and new genres of educational literature are developing, for instance, electronic textbooks. Thus, the education system is faced with new tasks related to the development and implementation of modern information competencies, projected on the formation of a new type of musician and their adaptation in the professional and creative sphere.

New research schools are emerging, which, within the framework of a modern digital society, are acquiring new directions of their development related to the use of MCT [37]. The phenomenon of the most electronic-computer musical instrumentation and work is born and rapidly developing [38, 39]. MCTs are a modern instrument of interaction in the field of academic musical culture in the Far East [40–42]. The rapid development of electronic musical instruments, specialized software and hardware designed to teach music, music creation, sound processing, arrangement, create the ground for the emergence of new forms of performing and a composer's creative work, including regional folklore of the Far East (for example, creating arrangements and original works using combinations of European timbres and styles with traditional Chinese instruments and manner of performance). The role of the MCT in preserving and translating the traditional culture of the world's peoples is invaluable [43–45]. In the contemporary musical culture and practice of the Far Eastern region, music computer technologies and - more broadly - digital technologies can and should be culturally understood as a new unique method of broadcasting and interaction between cultures of different countries and peoples.

The development of musical heritage was considered by us in a narrow sense as a direct process of studying musical works, obtaining musical-theoretical and musical-historical knowledge in musical-performing, musical-pedagogical areas of training in creative universities, cultural institutes. However, in a broader sense, the development of musical heritage is seen by us as a cultural phenomenon, expressed in composing, performing music, collecting, preserving and cataloguing folk music, as well as in music education in a broad sense. As a result of these processes, the cultures of the peoples interact (in the perspective of our study, the peoples of the Far East of Russia and China). By musical heritage, we mean the heritage of musical culture accumulated by humanity, an integral part of human culture; in this dissertation research, it is academic musical culture and musical folklore.

Cataloging musical folklore with the use of MCT is a particularly important part of mastering the musical heritage of the past, which will also affect the existence of folklore in the present and future. In particular, compositional, anarchistic activity with the use of MCT is based on the musical origins of traditional music, musical folklore, allowing it to be preserved and giving a second life in the works of compositional folklore. Back in 1975, a seminar on machine aspects of algorithmic analysis of musical and folklore texts held at the Dilijan House of Composers ' Creativity in Yerevan raised the problem of the Armenian Universal Structural and Analytical Catalog of Musical Folklore, the Armenian Universal Analytical Map; the principles of segmentation and algorithmic analysis themselves were discussed: All the participants in the discussion agreed that it was high time to start the experiment "Musicologist—COMPUTER", which, due to its very formulation, will be a strong impetus for the development of musicology and bringing its methods closer to those of the exact sciences. There can be no doubt that the computer will provide a great service to folklorists in solving search and analytical problems. This was the harbinger of a huge serious work, the result of which was a collection of articles "Quantitative methods in folklore and Musicology", published based on the materials of the Yerevan seminar.

The project described in the work "On the project of creating an intellectual system for cataloging and analyzing the music of the peoples of the world" [46], is unique in this sense, it deals with the problem of the need to create a "music bank", a kind of unified catalog of samples of the musical culture of the peoples of the world, which are currently scattered and disunited.

The introduction of information technologies, MCT in performing practice (concert, competition, festival activities, master classes, etc.) allows us to observe the special processes of interaction between the cultures of various peoples of the Far Eastern region.

For example, for more than 30 years, Khabarovsk has hosted the International Festival of Artistic Creativity of Children and Youth "New Names of the Countries of the Asia-Pacific Region", which consolidates musical, performing and artistic achievements. The festival aims to support young talents who show themselves in the field of music and visual arts, the latest information technologies (the competition "Creativity and Technology of the 21st century"). In 2021, 850 people (including more than 120 Khabarovsk residents) from China, North Korea, Japan, the Republic of Korea, Mongolia and for the first time in 2021 – the United States (Oregon) took part in the festival. The Far East of Russia is represented by participants from the Khabarovsk and Primorsky Territories, the Amur and Jewish Autonomous Regions, and the Republic of Sakha (Yakutia). A distinctive feature of the festival in 2021 was an unprecedented increase in the number of participants due to the possibilities of remote communication technologies and participation in the festival in an online format. Preliminary auditions are also held online. "New technologies allow everyone to shine on the festival stage (and, of course, those who are sufficiently prepared and worthy). Contemporary MCTs allow you to make recordings of the participants' performances of the highest quality and open up new opportunities in the field of interaction between the cultures of different countries, in our case, the countries of the Asia-Pacific region, the most fully represented soloists and groups from China. Such a format (full-time and remote) is very relevant today and it is planned to use it in the future. It opens up prospects and expands the geographical framework for cooperation in the field of art culture between different countries of the Far Eastern region. Modern technologies give talented young people a chance to demonstrate their talents, regardless of their geographical location, financial condition, or other circumstances, and get a ticket to the future.

Over the years of the festival, there was an opportunity to observe the accelerating processes of interaction between the cultures of the peoples of the Far Eastern region and the place of digital technologies and MCTs in this process. On the one hand, it is an increasingly active introduction of these technologies into creativity, on the other- the expansion of the possibilities (and quality) of recording and transmitting the results of audiovisual results of creativity. In addition, the interaction of cultures of different peoples of the region gives interesting results in terms of the exchange of repertoire, manner of intonation, playing, stage behavior, energy of musical performance, etc.

The emergence of digital technologies, information and communication methods of broadcasting contributed to the change of the "folklore paradigm". Approaches to understanding modern folklore are also changing. There is a new way to record, store, create and broadcast musical folklore, as well as a new way to interpret musical folklore. The emergence of post-folklor is associated with the invention, first of all, of sound recording, as well as network methods of broadcasting it (and composing it); the network existence caused the formation of a special communication environment: the appearance of these fundamentally different (in comparison with writing) information and communication channels was of crucial importance in the new change of the "folklore paradigm".

The development of the MCT allows not only to create new works of music (music and computer) art, but also gives an impetus to the development of scientific thought and understanding of the latest processes taking place, to the search for methods of music research, which are also of lasting importance for the organization of a system of training a contemporary musician in the conditions of using digital high-tech teaching tools. Thus, an activity-based, personality-oriented approach to teaching, based on constant contact with the teacher through the use of MCT, allows us to overcome a special lightness in relation to the world, a reduced ability of the leading imagination [47] of a young specialist and the previously identified disturbing basis for a critical attitude related to the history of online education in the United States (see, for example: [48]).

4   Conclusion

It is very important to continue to explore the music of Chinese composers in order to identify the identity of the use of folklore materials or the embodiment of Western musical techniques. Now, when we increase the pace of knowledge of Chinese musical culture with great speed, increase the exchange of fruits of scientific research, a unique field arises for interaction and deep understanding of cultures of regions of our countries.

The ways of interaction in the field of musical culture of the Far Eastern region, in addition to the already traditional practices (continuity in the field of musical performance and education), can and should include interaction in the field of composer creativity of our regions. Composers of China can use folklore samples of the natives of the Russian Far East in their work, which are very close in the scale and harmony sense to Chinese folklore. National origins are widely represented in the work of Chinese composers, but in the names of works such origins are not always "declared" by the composer, national "addresses" are not always given. For example, such an "address" is clearly read in Ben Yungju's work "Festive Evenki Quartet" for four clarinets. But such works are isolated (at least those that are known to us). At the moment, such works are most likely not available due to the unknown samples of our folklore by Chinese composers, which, fortunately, is still alive and preserved by the forces of performers, composers and scientists of the Far Eastern region. By providing materials to the Chinese composers, we can create the ground for a unique musical "dialogue of cultures," having gained invaluable experience and development prospects in the field of composer creativity and academic music education, including the implementation of MCT. The search for intonational prototypes and ethnic interaction in the works of composers in China is one of the promising and interesting problems for further study.

Nowadays, it is necessary to understand the importance of MCT, electronic musical creativity in contemporary musical culture and educational practice of the region, understand the importance and prospects of this kind of activity for the preservation and translation of traditional culture, accept the focus of the new generation on interaction with information technologies and understand the dynamism of the latest processes in academic musical culture.

References

1. Eisenstadt, A.M.: Musical folklore of the peoples of the lower Amur region. In: The Musical Folklore of the Peoples of the North and Siberia, pp. 5–53. Publishing House "Sovetsky Composer", Moscow (1966). (in Russian)
2. Brodsky, I.A.: Abstract the Album "Music of the Peoples of the Soviet Far East". Publishing House "Melody", Moscow (1973). (in Russian)
3. Solomonova, N.A.: The Musical Folklore of the Nanai, Ulchi, Nivkhs (Musical-Ethnographic Essay). Doctoral Dissertation. Tchaikovsky Moscow State Conservatory, Moscow (1981). (in Russian)
4. Sheikin, Yu.I.: History of Musical Culture of the Peoples of Siberia: Comparative-historical Research. Publishing House "East. Lit.", Moscow (2002). (in Russian)
5. Mamcheva, N.A.: Musical Instruments in the Traditional Culture of the Nivkhs. Yuzhno-Publishing House "Sakhalin. Obl. Tipografiya", Sakhalinsk (2012). (in Russian)
6. Bulgakova, T.D.: Kamlania nanayskikh shamanov: monograph. Foundation for the Culture of the Peoples of Siberia. VerlagderKulturstiftungSibirien, Furstenberg (2016)
7. Solomonova, N.A.: Musical Culture of the Peoples of the Russian Far East of the 19th–20s Centuries. (Ethnomusicological essays). (Doctoral Dissertation). Tchaikovsky Moscow State Conservatory, Moscow (2000). (in Russian)
8. Mezentseva, S.V.: Genre Typology of Instrumental Music of the Ritual Culture of the Tunguso-Manchus of the Russian Far East. (Doctoral Dissertation). University of the Humanities and Social Sciences, St. Petersburg (2006). (in Russian)

9. Galitskaya, S.P., Palakhova, A.Yu.: Monody: Problems of Theory. Publishing House "Akademiya," Moscow (2013). (in Russian)

10. Yunusova, V.N.: The study of Chinese music in Russia: on the problem of the formation of Russian musical sinology (in Russian). https://china.ivran.ru/f/YUnusova_V.N._Izuchenie_kitajskoj_muzyki_v_Rossii.pdf. Accessed 3 Mar 2021

11. Naumenko, T.I.: National themes in musicological research: history and modernity. In: Musical Art of Eurasia. Traditions and Modernity, pp. 28–34. Publishing House of the Tchaikovsky Moscow State Conservatory, Moscow (2020). (in Russian). https://eurasia.mosconsv.ru/?page_id=154

12. Lupinos, S.B.: Extramusical/intramusical semantics of terms in the musical cultures of China, Korea, Japan: antiquity - the middle ages. In: Gorbunova, I. (ed.) Contemporary Musical Education: Creative, Research, Technologies. CME 2018. Proceedings of the 17th International Research and Practical Conference, pp. 43–47. Publishing House of the Herzen State Pedagogical University of Russia, St. Petersburg (2019). (in Russian)

13. Alkon, E.M.: Musical thinking of the East and West-continuum and discrete. Doctoral Dissertation, Russian Institute of Art History, St. Petersburg (2002). (in Russian)

14. Alyabyeva, A.G.: Traditional Instrumental Music of Indonesia. Lap Lambert Academic Publishing, Saarbrucken (2011)

15. Mikhailov, J.K.: On the problem of the theory of musical and cultural traditions. In: Tsytovich, T. (ed.) Musical Traditions of the Countries of Asia and Africa, pp, 3–20. Publishing House of the Tchaikovsky Moscow State Conservatory, Moscow (1986). (in Russian)

16. Leskova, T.V.: Composer's folklore of the far east of Russia in the genre-style context of regional music. Fundament. Res. 2(10), 2258–2266 (2015). (in Russian)

17. Matveeva, A.: Piano Culture of Siberia and the Russian Far East (Late 18th century - 1980s). Publishing House "Private Collection", Khabarovsk (2009). (in Russian)

18. Zheng, I.Ya.: Professional Musical Culture of the countries of the Far East in the Context of Intercultural Interactions (Doctoral Dissertation). Zabaykalsky State University, Chita (2006). (in Russian)

19. Eisenstadt, S.A.: Piano Schools of the Countries of the Far Eastern Region (China, Korea, Japan). Problems of theory, history, and performing practice (Doctoral Dissertation). Novosibirsk State Conservatory named after M. I. Glinka, Novosibirsk (2015). (in Russian)

20. Wu, G.-I.: Traditional music of the Far East - China, Korea, Japan: historical and theoretical analysis (Doctoral Dissertation). Herzen State Pedagogical University of Russia, St. Petersburg (2012). (in Russian)

21. Beluntsov, V.O.: Computer for the Musician. Publishing House "Peter", St. Petersburg (2001). (in Russian)

22. Krasilnikov, I.M.: Electronic Musical Creativity in the System of Art Education. Publishing House "Phoenix+," Dubna (2007). (in Russian)

23. Gorbunova, I.B.: Music computer technologies as a new educational and creative environment. In: Gorbunova, I. (ed.) Contemporary Musical Education: Creative, Research, Technologies. CME 2002. Proceedings of the International Research and Practical Conference, pp. 43–47. Publishing House of the Herzen State Pedagogical University of Russia, St. Petersburg (2002). (in Russian)

24. Belov, G.G.: Inevitability of Computer Technology in music (meditation composer). In: Gorbunova, I. (ed.) Contemporary Musical Education: Creative, Research, Technologies. CME 2002. Proceedings of the International Research and Practical Conference, pp. 57–61. Publishing House of the Herzen State Pedagogical University of Russia, St. Petersburg (2002). (in Russian)

25. Kameris, A.: Ways of Implementing the Concept of Musical and Computer Education in the Training of the Teacher-musician (Doctoral Dissertation). Herzen State Pedagogical University of Russia, St. Petersburg (2007). (in Russian)

26. Gorbunova, I.B., Kameris, A.: The concept of musical computer education of teachers: problem statement. Int. J. Latest Technol. Techniq. **8**(2S4), 913–918 (2019). https://doi.org/10.35940/ijrte.B1181.0782S419

27. Gorbunova, I.B.: The integrative model for the semantic space of music and a contemporary musical educational process. The scientific and creative heritage of Mikhail Borisovich Ignatyev. Laplage em Revista, **6**(S), 2–13 (2020)

28. Gorbunova, I.B., Zalivadny, M.S.: Integrative model of the semantic space of music: prospects for combining musicology and music education. Music Sch. **4**(33), 55–64 (2018). https://doi.org/10.17674/1997-0854.2018.4.055-064

29. Gorbunova, I.B., Zalivadny, M.S., Kibitkina, E.V.: Musical programming. Publishing House of the Herzen State Pedagogical University of Russia, St. Petersburg (2012). (in Russian)

30. Gorbunova, I.B., Chibirev, S.V.: Modeling of the process of musical creativity in the format of the digital interface of a musical instrument. Option, **35**(Special Issue 22), 392–409 (2019)

31. Gorbunova, I.B., Zalivadny, M.S., Tovpich, I.O.: Mathematical methods of research in musicology: an attempt to analyze the material from the modern historical heritage (Reflections on the Xenakis' book Musiques Formelles). In: Ahmadi, R., Maeda, K., Plaisent, M. (eds.) 15th International Conference on Education, Economics, Humanities and Interdisciplinary Studies (EEHIS-18). ICASET-18, ASBES-18, EEHIS-18 International Conference Proceedings, pp. 134–138. International Association of Humanities, Social Sciences & Management Researchers, Paris (2018). http://uruae.org/siteadmin/upload/9491AE06184022.pdf. Accessed 2 Apr 2021

32. Gorbunova, I.B., Zalivadny, M.S., Tovpich, I.O.: On the application of models of the semantic space of music in the integrative analysis of musical works and music education with music computer technologies. Apuntes Universitarios **10**(4), 13–23 (2020)

33. Kharuto, A.V.: Computer Analysis of Sound in Music Science. Publishing House of the Centre "Moscow Conservatory", Moscow (2020). (in Russian)

34. Shipunova, O.D.: The role of relation to values principle in the social management practices. The existential-communicatory aspect. Middle-East J. Sci. Res. **19**(4), 565–569 (2014). https://doi.org/10.5829/idosi.mejCsr.2014.19.4.123684

35. Peng, Ch.: The Yun-gong-diao fret System and its Implementation in the Works of Chinese Composers of the Twentieth Century. Publishing House "Composer," St. Petersburg (2013). (in Russian)

36. Gorbunova, I.B.: Music computer technologies in the perspective of digital humanities, arts, and researches. Opcion **35** (Special Edition 24), 360–375 (2019)

37. Gorbunova, I.B.: The concept of music computer pedagogical education in Russia. Int. J. Adv. Sci. Technol. **29**(6s), 600–615 (2020)

38. Gorbunova, I.B., Petrova, N.N.: Digital musical instrument as a sociocultural phenomenon. Universidad y Sociedad. **12**(3), 109–115 (2020)

39. Gorbunova, I.B., Petrova, N.N.: Music computer technologies, supply chain strategy and transformation processes in socio-cultural paradigm of performing art: using digital button accordion. Int. J. Supply Chain Manag. **8**(6), 436–445 (2019)

40. Gorbunova, I.B., Mezentseva, S.V.: Russian far East - China: ways of interaction in the field of academic musical culture. Psychol. Edu. **57**(8), 920–933 (2020). https://doi.org/10.17762/pae.v57i8.1193

41. Gorbunova, I.B., Mezentseva, S.V.: The far-eastern ritual "bear holiday" in the space of computer musical technologies. Music Sch. **2**(43), 34–42 (2021). (in Russian). https://doi.org/10.33779/2587-6341.2021.2.034-042

42. Gorbunova, I.B., Mezentseva, S.V.: Music computer technologies in modern culture: classification problems. In: 32nd Pattaya International Conference on Literature, Languages, Humanities and Social Sciences, pp. 1–6. International Association of Humanities, Social Sciences & Management Researchers, Pattaya (2020)

43. Alieva, I.G., Gorbunova, I.B., Mezentseva, S.V.: Musical computer technologies as an instrument of transmission and preservation of musical folklore (by the example of the Russian Far East). Music Sch. **1**, 140–149 (2019). (in Russian). https://doi.org/10.17674/1997-0854. 2019.1.140-149

44. Alieva, I.G., Gorbunova, I.B., Mezentseva, S.V.: Music computer technologies as a worthwhile means of folklore studying, preserving and transmission. Utopia y Praxis Latinoamericana **24**(Extra 6), 118–131 (2019). https://produccioncientificaluz.org/index.php/utopia/art icle/view/30065

45. Gorbunova, I.B., Zalivadny, M.S.: The complex model of the semantic space of music: structure and features. Music Sch. **4**(41), 20–32 (2020). (in Russian). https://doi.org/10.33779/ 2587-6341.2020.4.020-032

46. Alieva, I.G., Gorbunova, I.B.: On the project of creating an intellectual system for the catalogization and analysis of the music of the peoples of the world. Soc. Philos. Hist. Cult. **9**, 105–108 (2016). (in Russian)

47. Romanenko, I.B., Romanenko, Y.M., Voskresenskiy, A.A.: Formation of post material values in conditions of transversality of educational paradigms. Eur. Proc. Soc. Behav. Sci. **35**, 1116–1121 (2018). https://doi.org/10.15405/epsbs.2018.02.131

48. Mureyko, L.V., Shipunova, O.D., Pasholikov, M.A., Romanenko, I.B., Romanenko, Y.M.: The correlation of neurophysiologic and social mechanisms of the subconscious manipulation in media technology. Int. J. Civil Eng. Technol. (IJCIET) **9**(9), 1–9 (2018)

Analysis of Emotional Connotations in Russian Language Texts

Andrey Mileshin[ID], Evgenij Tsopa[ID], Serge Klimenkov[ID], Aleksandr Slapoguzov[ID], and Anastasia Kharitonova[✉][ID]

Faculty of Software Engineering and Computer Systems, ITMO University, Kronverksky Pr. 49, bldg. A, 197101 Saint Petersburg, Russia

`{andrey.mileshin,evgenij.tsopa,serge.klimenkov,`
`aleksandr.slapoguzov,nasty}@cs.ifmo.ru`

Abstract. The article is focused on the problem of identifying the sentiment of texts in Russian. This problem has a significant impact on a wide range of natural language processing tasks such as online feedback processing and review analysis. There are many approaches for solution of this problem, such as using fuzzy logic techniques, keyword techniques, machine learning, and knowledge-based techniques. All these approaches are described and comparatively analysed in the article. As a result of the research, a new model for sentiment extraction from natural language text was proposed and a software module was developed to prove the proposed method. The developed module was integrated into an existing software system based on the semantic network and frame semantics. The effectiveness of the developed system was tested on the previously data set that was evaluated by experts. The test results show the competitiveness of the proposed method in comparison with previous data.

Keywords: Tonality of the text · Sentiment analysis · Natural language text · Text tonality processing · Natural language processing · OCC model

1 Introduction

At the present time, the information space of the modern world is filled with various reviews and opinions. In many ways, it was facilitated by the rapid development of Internet technologies and social activity on the World Wide Web. Today it is already difficult to find a product that does not have reviews on the Internet. People are increasingly openly expressing their opinions and sharing their impressions on social media. Also, microblogging is gaining more and more popularity. With the development of technology, any statement of a public person or a statement of a well-known company becomes available to millions of Internet users around the world. It is not uncommon for such statements to become the cause of various scandals and censures from society. In such conditions, media persons have to be careful with the wording of their thoughts, and companies must strictly monitor the political correctness of their statements. In addition, different brands need to respond quickly to any criticism or praise for their product.

D. Bylieva and A. Nordmann (Eds.): PCSF 2021, LNNS 345, pp. 497–511, 2022.
https://doi.org/10.1007/978-3-030-89708-6_42

Based on this, such a task as determining the tonality and emotional coloring of the text is more urgent than ever in our time.

Sentiment analysis process is used to extract emotion from text. The task of analyzing the sentiment of a text is reduced to the classification of a document according to sentiment estimates. The problem of determining the tonality of the text has been dealt with for a long time. Despite this fact, most of the existing methods of tonality of text in Russian use a scale consisting of only two states, positive and negative. Such a scale cannot reflect the full range of human emotions. This gives rise to frequent errors and leads to multiple difficulties while analyzing the sentiment of the text. Ignoring the semantic relationships between words in the methods of determining the emotional coloring of words also often leads to an incorrect result. E.g. the presence of negation or sarcastic statements in the text can lead not only to errors when using these methods, but even to the opposite of the true result. Among other things, problems can arise while trying to analyze the sentiment of a text that is small or not emotionally charged at all.

Emotional coloring of a text can be perceived in different ways by different people. Its perception is largely influenced by a personal social experience, its environment and level of development. The subject area to which the text belongs, and the time of its writing can also largely affect the result of the analysis of the sentiment of the text. For example, the word "target" has completely different meanings depending on context and subject area the text belongs to. So, in the novel about Robin Hood, "target" will mean a shooting target. In a text about advertising and marketing, "target" will most likely denote the target audience. And in the text concerning the IT sphere, the term "target" will mean a task. An equally striking example is the word "Awful". Initially, this adjective was characterized by something majestic, worthy of reverence and worship, but now it is used to denote something scary, terrible, or completely disgusting. It is an extremely difficult task to give a correct emotional assessment of texts with similar words without taking into account semantic relations and subject area.

This article will consider the existing methods for assessing the sentiment of the text and their shortcomings. In addition, a new method for assessing the emotional coloring of a text and its implementation that does not have these shortcomings will be presented.

2 Targets and Goals

The aim of this work is to develop a model of emotions suitable for analyzing the sentiment of a text in Russian.

The problem of determining the sentiment of a natural language of a text is not new. Various methods and algorithms already exist in this area. Therefore, if a suitable method already exists, it needs to be adapted and, if possible, improved. If a suitable method cannot be found, it must be developed. To achieve this goal, the following set of tasks was defined:

1. Review existing patterns of emotion.
2. Design models for describing emotions.
3. Develop a system for analyzing the sentiment of text.
4. Test the proposed system.

3   Existing Solutions Review

3.1   Keyword Methods

When using the keyword method, the emotional coloring of each word is considered separately. However, not all words have an emotional connotation. Based on a 2003 study by Pennebaker, J. W., Mehl, M. R., & Niederhoffer K, only 4% of words used in a written language have emotional connotations [1]. These methods ignore semantic relationships between words. It greatly complicates the determination of the emotional coloring of the text, which contains stylistic figures of speech. Such stylistic figures of speech as oxymoron contain words that are opposite to each other in meaning. It may be incorrect to determine the emotional coloring of an oxymoron not entirely, but after dividing it into parts. An example of such an oxymoron is the common expression "horribly beautiful". The word "horribly" in lexical dictionaries has a negative connotation, however, in conjunction with "beautiful", the connotation is clearly positive. In addition, the presence of negation in the sentence also negatively affects the accuracy of determining the emotional color. Due to these problems, methods based on the analysis of emotional words in the text have low accuracy. Also, such methods are not suitable for small texts [2].

3.2   Fuzzy Logic Methods

Like the previous method, methods based on the principles of fuzzy logic do not consider semantic connections between words. These methods do not analyze the entire sentence, but only consider the emotional coloring of adjectives and verbs. When using such methods for verbs and adjectives, the type of emotions and their intensity are determined [3]. The use of methods based on fuzzy logic often gives an incorrect result on texts that lack emotional coloring and on overly short texts.

3.3   Machine Learning

The methods for determining the sentiment of the constructed text are divided into two types: machine learning with a teacher and machine learning without a teacher. In the first type, a machine classifier is trained [4]. A large amount of labeled data is required to train a classifier. At the same time, for correct training the data must be tied to the subject area in which the classifier will subsequently be applied [5]. So, a classifier trained on a data set compiled from classical literature will not accurately determine emotions when analyzing reviews from an online store and vice versa. Such approaches are based on the definition of emotional expressions [6, 7]. At the same time, the emotions themselves are often contained not in the vocabulary but in the sense of what is written. Therefore, only lexical analysis is not enough to determine the sentiment of the text.

In unsupervised machine learning, the terms that appear in the text have weight. So, for the most frequent words in the text and at the same time, rarely found in the texts of the entire collection the greatest weight is given [4]. Further, for words with the highest weight the sentiment is determined based on which sentiment of the entire text it is determined. With this approach, a lot of words that can bring emotions into themselves are simply not omitted. Also, semantic relationships between words are not considered. All this can lead to inaccuracies and errors during the analysis of the sentiment of the text.

3.4 Method Based on Graph-Theoretic Models

As in the unsupervised machine learning method in this method, weights are assigned to different words. The more weight, the more the word affects the overall tone of the text. Based on the text, a graph is built the vertices of which are subsequently sorted depending on the weight [8]. Because of the constructed graph a classification of words is carried out in which each word is assigned one of two ratings "positive" or "negative". The emotional coloring of the text itself is defined as the quotient of dividing the sum of positive ratings considering their weight by the sum of negative ratings considering their weight. If the result of this division exceeds one, then the text is considered positive, and the larger the result, the more positive the text. If the result is less than 1, then the text is considered negative, and if the value is near zero, the text is neutral.

The main disadvantage of this method is that it uses only a binary state scale.

3.5 Disadvantages of Existing Methods

Most of the existing solutions for determining the tonality of a text in Russian suggest the use of only two types of emotions: positive and negative. However, these emotions themselves, in turn, can be divided into a larger number of more complex emotions. Without considering these complex emotions, such decisions cannot always correctly determine the meaning of the emotional coloring of the text. Also, most models do not consider the semantic relations in the text that can lead to inaccuracies when analyzing the sentiment of the text.

Considering the above disadvantages of existing approaches, a solution was developed that does not have these problems. This solution will allow for a deeper analysis of the text based on its semantics to determine the emotions present in it.

4 Proposed Solution Design

As mentioned earlier, it is necessary to consider the semantic relations of the words contained in text to determine the emotions contained in it accurately. To account for these relationships a semantic network was used in the work. The Semantic Web is a directed graph. The vertices of such a graph are values that define a specific subject area [9, 10]. Semantic relations in the graph are displayed using edges connecting such values [11]. The use of the semantic network allows one to operate not only with words but also with their meanings during the analysis of the sentiment of the text [12].

The developed solution uses the OCC model developed by Ortony A., Clore G. L. and Collins A [13]. This model was proposed in 1988 and served as the basis for constructing many other models. Today, the OCC model is one of the most popular for the cognitive assessment of emotions. The OCC model considers anticipated events and the emotions they might lead to. However, it does not consider events that a person could not expect.

To account for this kind of emotion, another group of emotions "Unexpected event" was added to the model. This group contains only two emotions: surprise and shock. This paper presents an adapted version of the OCC model designed for processing text in Russian.

As mentioned earlier, in order to accurately determine the emotions contained in the text, it is necessary to take into account the semantic relations of words. To account for these relationships, a semantic network was used in the work. The Semantic Web is a directed graph. The vertices of such a graph are values that define a specific subject area [14, 15]. Semantic relations in the graph are displayed using edges connecting such values [16]. The use of the semantic network allows you to operate not only with words, but also with their meanings [14, 17].

Using the Semantic Web allows you to interact directly with meaning. Also, the semantic network will allow you not to compile in advance a list of tokens that contains words classified as belonging to a certain type of emotion. This is possible due to the fact that the nodes in the semantic network are interconnected by semantic relations (HYPONYM, HYPERNYM, MERONYM, HOLONYM, SYNONYM, ANTONYM). In particular, such relationships in the network as SYNONYM make it possible to define words that are synonyms to the name of emotions and, accordingly, such words can be attributed to this type of emotion. And the ANTONYM connection allows you to identify words that are antonyms, to the name of an emotion, and if there is a denial of a given word in a sentence, then it can also be attributed to the type of this emotion. Finally, the use of the semantic web will allow us to classify various idioms and phraseological units as belonging to someone of the type of emotion.

There are three different classes of emotion in the OCC model. Each class reflects a separate aspect of the environment to which an emotional response occurs from a person. These environmental aspects are summarized below:

- consequences of events of concern to oneself
- actions of other agents
- objects of concern

In addition to classes, the OCC model also contains 6 different groups, which in turn contain 22 different emotions. Groups, emotions, and their descriptions are presented in Table 1. The complete hierarchy of the model is shown in Fig. 1.

When analyzing the sentiment of a text using the OCC model, the valence response to events, agents or objects is determined [18]. These reactions are lined up in a clear structure based on this model. This allows other known natural language processing methods to be applied during text analysis.

Each type of emotion in the OCC model has its own distinctive characteristics. These characteristics are set by a unique combination of parameters. There are sixteen such parameters in total and they are divided into four types: agent-oriented, object-oriented, event-oriented, and variable intensity of emotion.

When setting the value of the variables, the polarity of the system plays an important role. So, from the point of view of a negative entity, negative events can be regarded as positive. To avoid such a situation, it is necessary to clearly establish how the model will view events. The developed system considers events in relation to itself. At the

Table 1. Emotion groups in the OCC model

Groups	Emotions	Descriptions
Simple	Joy	Pleasure from the event you want
	Distress	Displeasure from an event you do not want
Reaction to actions of other objects	Happy-for	Pleasure in an event desired by another
	Pity	Dissatisfaction with an event that is undesirable for another
	Resentment	Displeasure from an event that is desirable for another
	Gloating	Pleasure in an event unwanted by another
Assumption and confirmation	Hope	Pleasure of the anticipated pleasant event
	Fear	Dissatisfaction with an unpleasant alleged event
	Satisfaction	Pleasure of a pleasant, confirmed event
	Fears-confirmed	Displeasure from a confirmed unpleasant event
	Relief	Pleasure from an unpleasant event that has not been confirmed
	Disappointment	Displeasure from an unconfirmed pleasant event
Reaction to an unexpected event	Shock	Emotion "Distress" with the unexpected unwanted event
	Surprise	Emotion "joy" with an unexpected, desired event
Assessment of actions	Pride	Positive assessment of your actions
	Shame	Negative assessment of your actions
	Admiration	Positive assessment of other people's actions
	Reproach	Negative assessment of other people's actions
Complex emotions	Gratification	Joy + Pride
	Remorse	Shame + Distress
	Gratitude	Joy + Admiration

(continued)

Table 1. (*continued*)

Groups	Emotions	Descriptions
	Anger	Reproach + Distress
Assessment of the object	Hate	Unpleasant attitude towards the object
	Love	Pleasant attraction to the object

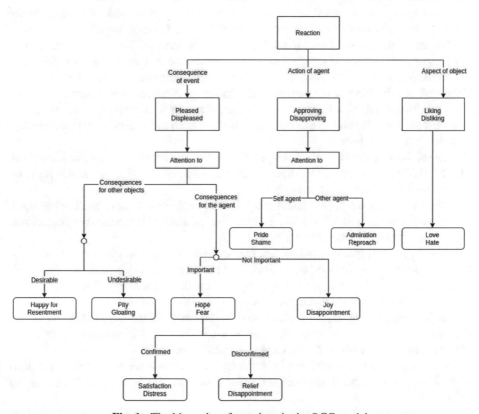

Fig. 1. The hierarchy of emotions in the OCC model

same time, the system considers itself a positive subject and, accordingly, has a negative attitude towards negative events and positively towards positive ones.

Self_Presumption (SP) and **Self_Reaction (SR)** are set depending on the relationship to the event. Can be *desirable/undesirable* and *pleased/displeased*, respectively.

Other_Presumption (OP) is set depending on the naturalness of the agent's action. May be *desirable/undesirable*. It is assumed that positive agents do positive things and negative agents do negative ones. That is, with positive actions of the agent which is considered positive the variable will take the value desirable. When performing negative actions, the variable is set to the value undesirable. The opposite is true for agents who are viewed as negative.

Direction_of_Emotion (DE) defines the belonging of an emotion to the "Reaction to actions of other objects" group. May be *self/other*. If the emotion is the result of one's own actions, the variable takes on the value self. If a reaction occurs to actions of other agents, the value of other is set to the variable. When setting this variable, the system relies on the presence of animate nouns and personal pronouns in the text.

Prospect (pros) is set only if action is expected in the future. Can be *positive/negative*. The semantic orientation of the action is used to set this variable. That is, in case of positive consequences caused by the action, the value of the variable will be set to the value positive. For negative consequences the value will be set to negative.

Status (stat) is set depending on the form of the verb present in the text. Can be *unconfirmed, confirmed, disconfirmed*. The variable takes on the value unconfirmed if the verb is modal or if it has a future tense form. The variable takes on the value confirmed if the verb has the form of the past tense and the event it describes is positive. Also, the value confirmed will be set if the verb is in the past tense form and describes a negative event but the particle "not" is used before it. If there is no such particle, the variable will take the value disconfirmed.

Agent_Fondness (af) defines the positiveness of the agent. May be *liked/not liked*. The value of this variable will be "liked" if the agent associated with the event is positive. Otherwise, the variable is set to not liked.

Object_fondness (of) determines the positivity of the object. May be *liked/not liked*. The value of this variable will be "liked" if the object associated with the event is positive. Otherwise, the variable is set to not liked.

Self_Appraisal (sa) determines whether an event is praiseworthy or blameworthy. May be *praiseworthy/blameworthy*. This variable takes the value praiseworthy if the event has a positive meaning. Blameworthy is set if the event is negative. In this work the variable is set by the semantic orientation of the verb.

Object_Appealing (oa) determines the attractiveness of the object. Can be *attractive/not attractive*. The value of attractive is set if the object is positive but not very common. In all other cases, the variable is not attractive.

Valenced_Reaction (vr) determines whether the proposal contains emotions in principle. Can be *true/false*. If the analyzed offer is neutral, this variable will take on the value false.

Unexpectedness (unexp) defines the suddenness of an event. Can be *true/false*. This variable defines the belonging of the emotion to the "Reaction to an unexpected event" group.

Event_Deservingness (ed) defines the desirability of the event. It can be *high/low*. This variable takes a high value if the agent wants events for himself or for others. Otherwise, the variable is set to low.

Effort_of_Action (eoa) defines the intensity of the action. Can be *obvious/not obvious*. The variable takes the obvious value if the sentence contains an adverb signaling high intensity.

Table 2. Cognitive variables in the OCC model

Emotions	Rules
Joy	vr = true & sr = 'pleased' & sp = 'desirable'
Distress	vr = true & sr = 'displeased' & sp = 'undesirable' & de = 'self'
Happy-for	vr = true & sr = 'pleased' & sp = 'desirable' & af = 'liked' & de = 'other'
Pity	vr = true & sr = 'displeased' & op = 'undesirable' & af = 'liked' & de = 'other'
Resentment	vr = true & sr = 'displeased' & op = 'desirable' & af = 'not liked' & de = 'other'
Gloating	vr = true & sr = 'pleased' & op = 'undesirable' & af = 'not liked' & de = 'other'
Hope	vr = true & sr = 'pleased' & pros = 'positive' & sp = 'desirable' & status = 'unconfirmed' & de = 'self'
Fear	vr = true & sr = 'displeased' & pros = 'negative' & sp = 'undesirable' & status = 'unconfirmed' & de = 'self'
Satisfaction	vr = true & sr = 'pleased' & pros = 'positive' & sp = 'desirable' & status = 'confirmed' & de = 'self'
Fears-confirmed	vr = true & sr = 'displeased' & pros = 'negative' & sp = 'undesirable' & status = 'confirmed' & de = 'self'
Relief	vr = true & sr = 'pleased' & pros = 'negative'' & sp = 'undesirable & status = 'disconfirmed' & de = 'self'
Disappointment	vr = true & sr = 'displeased' & pros = 'positive' & sp = 'desirable' & status = 'disconfirmed '& de = 'self'
Shock	vr = true & sr = 'displeased' & sp = 'undesirable' & de = 'self' & unexp=true
Surprise	vr = true & sr = 'pleased' & sp = 'desirable' & unexp = true
Pride	vr = true & sr = 'pleased' & sa = 'praiseworthy' & sp = 'desirable' & de = 'self'
Shame	vr = true & sr = 'displeased' & sa = 'blameworthy' & sp = 'undesirable' & de = 'self'
Admiration	vr = true & sr = 'pleased' & sa = 'praiseworthy' & op = 'desirable' & de = 'other'
Reproach	vr = true & sr = 'displeased' & sa = 'blameworthy' & op = • 'undesirable' & de = 'other'
Gratification	vr = true & sr = 'pleased' & sp = 'desirable' & sa = 'praiseworthy'
Remorse	vr = true & sr = 'displeased' & sp = 'undesirable' & sa = 'blameworthy'
Gratitude	vr = true & sr = 'pleased' & sp = 'desirable' & sa = 'praiseworthy' & op = 'desirable'
Anger	vr = true & sr = 'displeased' & sp = 'undesirable' & op = 'desirable' & sa = 'blameworthy'
Hate	vr = true & sp = 'undesirable' & sr = 'displeased' & of = 'not liked' & oa = 'not attractive' & event valence= 'negative' & de= 'other'
Love	vr = true & sp = 'desirable' & sr = 'pleased' & of = 'liked' & oa = 'attractive' & event valence = 'positive' & de = 'other'

Expected_Deviation (edev) defines the naturalness of the event for the agent. It can be *high/low*. The variable takes on the value high if the event is typical for the given agent and low if not typical.

Event_Familiarity (ef) defines the naturalness of the event for the object. Can be *common/uncommon*. The variable takes on the value common if the event is typical in relation to the object. If the event is not typical, then the variable takes on the value uncommon.

As stated earlier, each emotion in the OCC model has a unique set of cognitive variables. Therefore, a rule was drawn up for each type of emotion based on the variables described above. These rules are shown in Table 2:

5   Implementation

The work uses a semantic network called Jackalope, developed by Tune-IT. This network is filled with data from the open dictionary Wiktionary [19]. This dictionary is built on a wiki engine and is publicly available on the Internet. The dictionary is freely expandable, which allows any registered user to supplement and improve it. The Wiktionary is not a semantic network, but in the articles that describe tokens, there is information about semantic meanings, which makes it possible to build a semantic network based on it [20, 21]. In addition to the data from the dictionary, nodes denoting emotions and their closest synonyms were additionally manually added to the network. This was done to avoid possible errors and inaccuracies that may be present in the Semantic Web.

Each object in this network has a unique identifier. The identifier is generated in such a way that when the object is created, this identifier does not coincide with the existing one. In the nodes of the graph of the semantic network, there are various semantic concepts [5, 22]. The edges of this graph reflect the semantic relationships between different concepts. Semantic link - describes the relationship between two concepts, in which the first concept can have an attribute (characteristic), which is the second concept. The following types of semantic relations are distinguished: hyponyms, hyperonyms, meronyms, holonyms, synonyms, paronyms and antonyms. Any concept can have one or more specific instances [16, 23]. Each concept has an instance table associated with it. This table consists of 2 columns: key and value. The key is the instance identifier, which is represented by the sin_id structure. In addition, the table stores information about the concept of which this record is an instance. Any concept and its copy can be expressed in many different word forms that represent this concept or its instance in text form. Word forms include various words and their possible forms or expressions that reflect the meaning of a given concept. One and the same word form can express several different concepts or their instances.

In addition to the text itself, many modern text document formats contain meta information. Such information does not contain important information for the analysis of the sentiment of the text. However, its presence can negatively affect the correctness of the result. To solve this problem, a converter with a built-in tokenization module was used in the work. Such a module allows one to split the text into minimal logical units [24]. The text parsed by tokens is more convenient to process in the future [25]. Also, the work used grammatical and morphological parsers. During the analysis of the sentiment of the text in the developed system, the following actions are performed:

1. Tokenization of the parsed text. Getting rid of metadata in the text.
2. Conducting grammatical analysis. Revealing the agent-action-subject link in the text.
3. Morphological analysis:

   a. Arrangement of parts of speech in a sentence to further define the cognitive variables in a sentence.
   b. Analysis of the tense forms of verbs. Required to set the cognitive variable Status.
   c. Categorization of nouns in the text into animate/inanimate categories. Used to set the direction of emotions in the text.
   d. Search for modal verbs in the text. Needed to set the value of variables such as Status, Prospect or Self_Presumption.
   e. Search for adverbs in the text. Used to set the Unexpectedness variable in case of unexpected action and to set the Effort_of_Action variable in case of high action intensity.

4. Determining the valency of a word using the semantic network.
5. Setting the values of the Event_Familiarity and Expected_Deviation variables based on the data on the popularity of relationships between words in the text.

The architecture of the developed solution is shown in Fig. 2. The first stage of the system operation is text tokenization. The file containing the text to be examined is sent via a POST request to the input of the converter. The converter removes metadata from the text and breaks it down into tokens. After processing the text, the converter sends the resulting list of tokens as input to the semantic parser. The second stage of processing begins. The semantic parser refers to the semantic web and receives information about the meaning of words and their emotional coloration. Also, at this stage, the popularity of connections between words is established to determine the naturalness of phrases in a sentence. The third stage is grammatical analysis of the text. During this stage agent-event-object triplets are selected from the previously processed text. The resulting list of triplets is passed to the morphological parser performing the fourth stage of the work. At this stage, morphological analysis is performed. It defines the parts of speech of words, the temporal forms of verbs, the animate/inanimate of nouns, adverbs, and modal verbs are found. After that, the fifth final stage of the system's operation begins. Based on the previously obtained data, the values of the cognitive variables are established. Based on the resulting list of variables, emotions that fit the rules are determined. After processing all the text, the system outputs a list of extracted emotions. This list is returned to the user.

6 Results

To test the developed system, a corpus of emotional colors marked texts was needed. Unfortunately, we did not find a suitable corpus in Russian. Therefore, testing was carried out on a sample of reviews from the Yandex.Market service. The sample size was 100 reviews. Reviews were marked up using the peer review method. When conducting an expert assessment, it is important to notice that the level of coincidence in the decisions

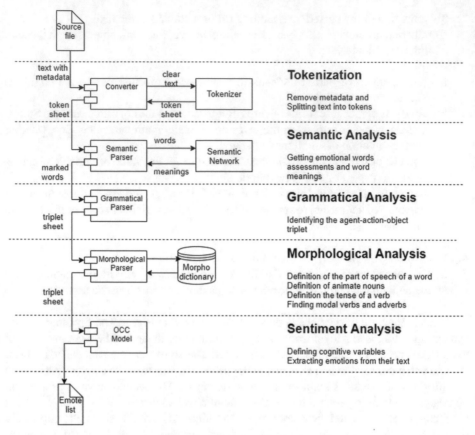

Fig. 2. System architecture

of experts is high. To measure the consistency, the Fleiss kappa metric was used [26]. If the value of the flux kappa is greater than 0.7, it can be considered that the assessment was carried out correctly. In our case, this value was 0.84 which means that when assessing emotions, the experts came to an agreement.

The test results of the developed system were compared with the results of the interpretation of the OCC model for the English language. Testing of the system for the English language was carried out on the text corpora ISEAR, SemEval, Alm's. More information about the device of this system can be found in the work of Orizu Udochukwu "A Rule-Based Approach to Implicit Emotion Detection in Tex" [26]. The accuracy of our system for the Russian language was 71%. It is 13% more than the equivalent system for English [17].

The system developed within the framework of this study allows one to move away from the usual binary scale of emotions when analyzing the sentiment of the text and extract a wide range of emotions from the text. This method will allow for a deeper analysis of the sentiment of the text and improve its quality. The use of the OCC model in the proposed approach makes it possible to use with it other methods of processing text in natural language. And the modular architecture of the system will make it possible to combine this system with other systems for processing text in natural language.

Using the Semantic Web allows the system to set a value for a variety of variables used to analyze the sentiment of a text. This is possible due to the fact that the semantic network allows you to operate not only with words, but also with their meanings. In addition, thanks to the use of the semantic web, it was possible to avoid the need to compile in advance a list of tokens containing words classified as belonging to a certain type of emotion. Also, the use of the semantic network allows you to determine the naturalness of phrases present in the text. Finally, the semantic network allows you to establish the valence of words present in the text, which is a key step in analyzing the sentiment of the text. All this made it possible to increase the efficiency of finding emotions in the text.

The developed solution is based on the OCC model extended and adapted for the Russian language. The presented solution uses a document converter with a built-in tokenizer, as well as morphological, syntactic, and semantic parsers for text analysis. All this made it possible to achieve high accuracy in the analysis of the sentiment of the text in natural language. Errors that occurred during the analysis of the sentiment of the text are due to the imperfection of the grammatical and morphological parser. Further development of the system is represented by improving the grammatical and morphological parsers, as well as improving the work of the tokenizer and expanding the list of document formats supported by the converter.

References

1. Pennebaker, J.W., Mehl, M.R., Niederhoffer, K.G.: Psychological aspects of natural language use: our words: our selves. Annu. Rev. Psychol. **54**, 547–577 (2003). https://doi.org/10.1146/annurev.psych.54.101601.145041
2. Ortony, A.: On making believable emotional agents believable. In: Trappl, R., Petta, P., Payr, S. (eds.) Emotions in Humans and Artifacts, pp. 189–211. The MIT Press, Cambridge (2003)
3. Liu, H., Lieberman, H., Selker, T.: A model of textual affect sensing using real-world knowledge. In: Proceedings of the 8th International Conference on Intelligent User Interfaces - IUI 2003, pp. 125–132. ACM Press, New York (2003). https://doi.org/10.1145/604045.604067
4. Kotelnikov, E., Klekovkina, M.: The automatic sentiment text classification method based on emotional vocabulary. CEUR Work. Proc. **934**, 118–123 (2012)
5. Slapoguzov, A., Malyuga, K., Tsopa, E.: Word sense induction for russian texts using BERT. In: Proceeding of the 28th Conference of Fruct Association, pp. 621–627. Fruct, Helsinki (2021)
6. Kim, S.-M., Hovy, E.: Identifying and analyzing judgment opinions. In: Proceedings of the Main Conference on Human Language Technology Conference of the North American Chapter of the Association of Computational Linguistics, pp. 200–207. Association for Computational Linguistics, Morristown (2006)
7. Strapparava, C., Valitutti, A., Stock, O.: Dances with words. In: Proceedings of the 20th International Joint Conference on Artificial Intelligence, pp. 1719–1725. DBLP, Hyderabad (2007)
8. Ustalov, D.A.: Extraction of terms from Russian-language texts using graph models. In: Graphs Theory and Applications: Comference Proceedings, pp. 62–69. Ural University Publishing House, Ekaterinburg (2012). (in Russian)

9. Pismak, A., Klimenkov, S., Tsopa, E., Yarkeev, A., Nikolaev, V., Gavrilov, A.: Method of semantic refinement for enterprise search. In: Proceedings of the 12th International Joint Conference on Knowledge Discovery, Knowledge Engineering and Knowledge Management, pp. 307–312. SCITEPRESS - Science and Technology Publications, Setubal (2020). https://doi.org/10.5220/0010159703070312

10. Osika, V., Klimenkov, S., Tsopa, E., Pismak, A., Nikolaev, V., Yarkeev, A.: Method of reconstruction of semantic relations using translingual information. In: Proceedings of the 9th International Joint Conference on Knowledge Discovery, Knowledge Engineering and Knowledge Management, pp. 239–245. SCITEPRESS - Science and Technology Publications, Setubal (2017). https://doi.org/10.5220/0006516602390245

11. Klimenkov, S., Tsopa, E., Pismak, A., Yarkeev, A.: Reconstruction of implied semantic relations in Russian Wiktionary. In: Proceedings of the 8th International Joint Conference on Knowledge Discovery, Knowledge Engineering and Knowledge Management, pp. 74–80. SCITEPRESS - Science and and Technology Publications, Setubal (2016). https://doi.org/10.5220/0006038900740080

12. Dubenetskiy, E., Tsopa, E., Klimenkov, S., Pokid, A.: Designing a model of contexts for word-sense disambiguation in a semantic network. In: CEUR Workshop Proceedings, vol. 2590, p. 100. CEUR-WS, St. Petersburg (2020)

13. Colby, B.N., Ortony, A., Clore, G.L., Collins, A.: The cognitive structure of emotions. Contemp. Sociol. **18**, 957 (1989). https://doi.org/10.2307/2074241

14. Pismak, A.E., Pokid, A.V., Kharitonova, A.E., Klimenkov, S.V., Nikolaev, V.V., Gavrilov, A.V.: Using semantic network for storing semi-structured data. Engineer. J. Don **2**. (2020). (in Russian). http://www.ivdon.ru/uploads/article/pdf/IVD_50__2_klimenkov_nikolaev.pdf_d07f3ecbbe.pdf. Accessed 21 Mar 2021

15. Modoni, G.E., National, I., National, I., Terkaj, W., National, I.: A semantic framework for graph-based enterprise search. Appl. Comput. Sci. **10**, 66–74 (2014)

16. Ogarok, A.L.: Method of semantic search and analysis of information. Informatiz. Commun. **1**, 75–80 (2020). https://doi.org/10.34219/2078-8320-2020-11-1-75-80

17. Shaikh, M.A.M., Prendinger, H., Ishizuka, M.: A linguistic interpretation of the OCC emotion model for affect sensing from text. In: Tao, J., Tan, T. (eds.) Affective Information Processing, pp. 45–73. Springer London, London (2009). https://doi.org/10.1007/978-1-84800-306-4_4

18. Pismak, A.E., Klimenkov, S.V., Tsopa, E.A., Slobodkin, A.Y., Nikolaev, V.V.: Merging of semantic networks based on equivalence of topologies. Izv. vysših učebnyh Zaved. Priborostr. **62**, 50–55 (2019). https://doi.org/10.17586/0021-3454-2019-62-1-50-55

19. Pokid, A.V., Klimenkov, S.V., Tsopa, E.A., Zhmylev, S.A., Tkeshelashvili, N.M.: Quick search method for nodes of a semantic network by exact word forms matching. Izv. vysših učebnyh Zaved. Priborostr. 932–939 (2017). https://doi.org/10.17586/0021-3454-2017-60-10-932-939

20. Bi, K., Ai, Q., Croft, W.B.: Iterative relevance feedback for answer passage retrieval with passage-level semantic match. In: Azzopardi, L., Stein, B., Fuhr, N., Mayr, P., Hauff, C., Hiemstra, D. (eds.) ECIR 2019. LNCS, vol. 11437, pp. 558–572. Springer, Cham (2019). https://doi.org/10.1007/978-3-030-15712-8_36

21. Solskinnsbakk, G., Gulla, J.A.: Contextual search navigation using semantic tag signatures. In: Proceedings of the 11th International Conference on Knowledge Management and Knowledge Technologies, vol. 34. ACM Press, New York (2011). https://doi.org/10.1145/2024288.2024329

22. Formica, A., Pourabbas, E., Taglino, F.: Semantic search enhanced with rating scores. Futur. Internet. **12**, 67 (2020). https://doi.org/10.3390/fi12040067

23. Tkeshelashvili, N.: Spreadsheet data extraction using semantic network. In: International Multidisciplinary Scientific GeoConference Surveying Geology and Mining Ecology, Management, pp. 637–644. SGEM, Sofia (2019). https://doi.org/10.5593/sgem2019/2.1/S07.083

24. Tkeshelashvili, N.M., Klimenkov, S.V., Dergachev, A.M.: Table structure recognition method for spreadsheets data software and systems. **29**, 107–112 (2016). https://doi.org/10.15827/0236-235X.116.107-112
25. Carletta, J.: Assessing agreement on classification tasks: the kappa statistic. Squibs Discuss. Assess. Agreem. Classif. Tasks Kappa Stat. **22**(2), 249–254 (1996)
26. Udochukwu, O., He, Y.: A rule-based approach to implicit emotion detection in text. In: Biemann, C., Handschuh, S., Freitas, A., Meziane, F., Métais, E. (eds.) NLDB 2015. LNCS, vol. 9103, pp. 197–203. Springer, Cham (2015). https://doi.org/10.1007/978-3-319-19581-0_17

Creativity and Emotions in the Digital World

Julia Petrova⬤ and Olga Vasichkina⁽⊠⁾ ⬤

Rostov State University of Economics (RINE), B. Sadovaya str., 69, Rostov-on-Don, Russia

Abstract. The relationship between creativity and emotions can be understood from different angles. Though researchers recognized the mutual influence of creativity and emotion long ago, the digital world is leading to new ideas for solving problems through technology. The authors of the article suggest a new model understand the global challenges of the digital world, creativity, and emotions. The improved method of conceptual and empirical research made it possible to put forward a new hypothesis of digital well-being, based on the relationship between new forms of creativity and emotional effect, and to confirm the logical conclusions drawn from the results of the survey. The purpose of the article is to study the problem of the relationship between creativity and emotions in the digital world and how personal characteristics, user activity, and situational factors influence this relationship. The fact that even the shortest time spent on creativity and the emotional state of people is interrelated with well-being increases the relevance of the study of creativity in any form of its manifestation. The importance of the digital world in people's lives increases the relevance of the study of emotions and such forms of creativity as the creation of user-profiles and Internet memes.

Keywords: Creativity · Critical thinking · Digital world · Emotional effect · Creative approach

1 Introduction

At the turn of the century, the Internet changed the way our society functions. We have witnessed a technological revolution and some fundamental changes. Digital technologies, on the one hand, have created a discursive opportunity to rethink the interaction of people, which is caused both by the development of the Internet and by the increase in the number of users and changes in their preferences. On the other hand, using both individual and social levels of creativity, digital technologies have become the basis for the development of new collective creativity and its productive use in new activities. The scale of the problem of the relationship between creativity and emotions is shown in several previous studies. The purpose of the article is the relationship between creativity and emotions of the digital world, which affects the well-being of people and is influenced by personal characteristics, user activity and situational factors. Digital well-being increases the relevance of the study of emotions and creativity considered in the study on the examples of creating the user profiles and distributed Internet memes. Profiles of people on the Internet containing information about the person (nickname,

D. Bylieva and A. Nordmann (Eds.): PCSF 2021, LNNS 345, pp. 512–521, 2022.
https://doi.org/10.1007/978-3-030-89708-6_43

photo, comments, etc.) and Internet memes (image, video, text fragment, etc.) are represented by a new form of creativity, demonstrating the creative thinking of their creators and distributors and the emotional phenomenon created by them. The relevance of the research is to study both a new form of creativity and the emotional effect of the digital world.

2 Methodology

Such terms as "research" and "scientific method" are closely related. Although research is the study of an object or phenomenon to learn something new about it or to study it from a new point of view, and the scientific method is just a way of conducting research, both terms refer to the search for truth, determined by logical considerations. Interactions in the digital world form a rather vast, chaotic space that is constantly expanding by people and technologies. Therefore, methodological understanding, development of methods of analysis and selection of technologies is necessary. The technologies that have created the digital world have given countless opportunities for creativity and testing of new emotions to all Internet users. Technologies have made it possible to apply improved conceptual and empirical methods. The conceptual basis of the method is a synthesis of studies on how the specific relationship between emotions and creativity is explained, how they mutually influence each other, how a profile in the social networks affects emotions, how creative thinking affects the attitude to the spread of Internet memes, the creation of profiles, etc. The empirical basis of the research method is a quantitative assessment of the respondents' behavior about user-profiles and Internet memes as a manifestation of creativity and their digital well-being.

At first sight "creativity" seems an ability that is intrinsic only to artists, designers, writers or marketers, but the truth is that creativity is necessary for all. Creative thinkers can look at things in a new way, out of the box, and find solutions that no one has ever thought of. Creativity is what drives innovation and progress. Of course, the main elements of creativity are professional and technical skills and knowledge, special talent and abilities, but desire and motivation are no less important. Unrealized or partially realized creativity testifies to the absence of one of the elements [1].

A similar conclusion was reached in the study of M. Baas K. De Dreu & B. Nijstad. The authors noted that the enhancement of creativity is influenced by positive moods, which activate and are associated with the motivation of the approach and the focus on promotion [2].

Creativity is often a process that takes place in the social networks through the dynamic behavior of social media users, which includes innovation and entrepreneurship. The increased role of creativity in social media is a response to rapid change and fierce competition. Creativity is indispensable for the development and implementation of innovation [3].

The same social media sites play different roles for people. T. Gazit's study "Tell me who you are and I'll tell you which of the social networks you use: Social media engagement" confirms the statistics and gives a more detailed description of the user. WhatsApp, the most commonly used platform, is mainly used by women and people with an internal locus of control. Facebook is more often used by open people; Instagram is used by women, young people and neurotics. Twitter is more often used by men. In addition, for all social networks, the higher the social and informational usage, the more important social networks become to the users, which largely explains the frequency of participation [4].

Social networks allow not only to get first-hand knowledge of thoughts and feelings but also fully unlock the creative potential of the users. Monitoring of social media reveals the creativity of their users at the social level. While individual creativity is usually associated with the working environment or everyday activities, social creativity in digital space does not only contain a fundamental feature of human intelligence - creativity but also combines it with inventions and programs aimed at studying new activities [5].

Considering how various aspects of mood influence the creative processes, R. Szakacs and Z. Janka, note that creativity is both positively and negatively influenced by the mood. In other words, there are different emotional, motivational and cognitive correlations of the creative process depending on the mood. [6] But the ability to find humorous points of view in negative meanings of memes can "indicate a useful strategy in the context of cognitive reassessment" [7].

A new study conducted by Iraklis Mutidis and Hiuel T. P. Williams, "Good and Bad Events: Combining Network Event Detection with Sentiment Analysis", extends their previously developed event detection and description system by adding a sentiment analysis component and improving ambiguity resolution. The authors demonstrate the advantages of sentiment analysis as a method of event characterization. The methodology proposed by the researchers can determine the opinion of the users involved in each of the events. According to the researchers, "topic detection and tracking, combined with sentiment analysis, can be useful for the research of human analysts in many areas, who can automatically detect emerging events in real-time, understand what is being discussed, and observe the impact on public sentiment...." [8]. This methodology demonstrates one approach to the broad problem of extracting and interpreting the vast information that is spread by memes becoming increasingly available. Identification of such emotions, such as joy, trust, anger, expectations, fear can allow this approach to understanding the mood of the users better, which encourages the creative thinking of the meme-makers and can also influence inter-group relationships in social networks [9].

Creative thinking is a skill that allows you to look at things from a new perspective and different angles. It is an inventive process of thinking that leads to some unexpected conclusions and new ways of doing business. Brainstorming or non-standard thinking can help creative thinking generate ideas. The working definition of creative experience proposed by V. Glăveanu and R. Beghetto includes principled interaction with the unfamiliar and readiness to approach the familiar in an unfamiliar way. In other words, the creative experience of researchers is defined as meetings with a new human being in the world based on meaningful actions and interactions that are marked by the principles of unlimited, non-linear, multiple perspectives and future-oriented view. The authors' definition has some similarities with standard definitions, in that it still recognizes the importance of novelty and efficiency, but goes beyond the existing definitions, arguing that novelty and relevance are not sufficient to characterize the creative experience. The authors affirm a creative experience marked by a set of principles that are intrinsic to creative action [10].

3 Research Results and Discussion

The digital world has created a discursive opportunity to rethink the interaction as a result of the information and communication technologies development. Based on data forecast from the Advertising Research Foundation (ARF) for 2025, the number of social media users will grow to almost 4.41 billion, [11] and the social media usage study represents a study of one of humanity's most popular online activities of the 21st century.

As a result of technological progress, the created wearable devices (tablets, smartphones, smartwatches, etc.) are changing the rules of the game. Their ease of use and popularity is determined by the achievements in the field of touch screen technology of the 21st century [12]. The sustainable direction of technology development allows users to interact anywhere and at any time, promotes the creative thinking of users. The expression of creativity and individuality can be seen from the moment of the creation of an account in a social network. Users of the network create their image, choose their own: nickname, avatar photo, biographical profile text, this is projected, based on what they publish, on the style and content of what they share regularly. Thus, creating the profile represents: "self-building, which modulates the association between creativity and social networking, modulates the relationship between creativity and functional connectivity, and thus classifies people with high and low creativity" [3]. So digital identity defines what a user does on the Internet - language, images, videos, opinions, sharing memes, linguistic style and emotional charging - which is the main way to their success [13].

The set of links that users save online (followers and friends) is a social graph that reveals an incredible amount of information about their creativity and identity. A profile in a sociotechnical environment creates such perceptions that directly affect one's reputation. Virtual characters of people, as well as people in real life, do not always look like the persons they are. In the social context, individuals are always performers, with the correction that it is impossible to look into a person's head and understand what role the individual plays. You can only observe the role he/she plays. The focus on the importance of the role performance, on how the individual expresses his role served as the basis for the development of the concept of the "dramaturgical theory of personality and society", the author of which is E. Goffman [14]. The concept is a reflection on a famous line from a Shakespearean play. "The whole world is a stage, and all the men and women are just players." [14]. The scientist uses theatre as the analogy for social interaction, understanding that people play their roles and participates in theatrical interaction, follow common scenarios of social networks, and use creativity to support their roles. People present a "face" that shows how they want others to see them. They present themselves to others in the way they hope to be perceived. Therefore, first impressions are crucial to how events happen during social interaction. The user of the social network project his/her image after it has been proposed throughout the existence of the created account. The mode of self-presentation through social interaction is difficult to change halfway, and the individual's self-presentation itself is promising, which is either "confirmed by subsequent interactions or discredited." The interlocutor refuses to interact, in an improperly defined situation, or when he cannot obtain the expected emotional component. The exchange of symbolic signs – memes sometimes even lacking actual content, indicates sufficient mutual concern for the other.

In Goffman's analysis, as in symbolic interactionism, the emphasis is on the fact that unpredictable social reality is based on a continuous process of mutual interpretation of transmitted and received signs. It is not predetermined by structures, functions, roles, or history, but often relies on them in the same way that actors use their basic knowledge and experience to create a credible character. Both openness of experience and creativity mediate an association of authenticity and emotion, and the association itself is mediated by openness for changes that are associated with creative thinking. [15]. Events that discredit or question the performer threaten to disrupt the social encounter. If this happens, it leads to a kind of anomie at the micro-level, an abnormality, when all interacting parties are commonly expecting an unpleasant situation to happen. The consequence of communicating with a fake identity on social media is the risk of frustration, just as over-engagement with social media is a rejection of the opportunity to build deeper friendships and express your true "self". Proponents of symbolic interaction Blumer, 1962; Cooley, 1902; Mead, 1934, Blumer, 1962, etc. have long considered the identification of "I" and "Other" a key feature of social interaction. Establishing your own identity for yourself is just as important in interaction as establishing it for others [16]. Modern users of popular platforms are not always the people they pretend to be, so a popular Japanese Twitter user @azusagakuyuki, who represented himself as a young biker with subscribers of more than 16 thousand people, upset his followers with an unexpected disclosure in March 2021. The user of the network turned out to be a fifty-year-old man who processed his photos in the popular FaceApp application. It is reported that the man said that no one wants to see "uncle", and so he turned into a "beautiful woman" [17]."… readiness" to show high creative behavior characteristic of digital society. [18]. In today's society, it is increasingly difficult to believe what we see in online profile content when the word "fake" is becoming increasingly popular in the 21st century.

Nowadays, those who use modern information and communication technologies try to choose the content according to their needs, regardless of whether these needs were the intentions of the producers or not. Communication technologies contribute to this choice, connect and organize people in different contexts. Common patterns that persist over time and become habitual and even routine at micro-level interactions, or are institutionalized at the macro or global levels of interaction, represent structures. Critical and creative thinking is also important for the development of creativity and is a decisive factor for the development of creativity. [19] Systems thinking developed by G. Forrester and members of the Massachusetts Technological University (MIT) Society for Organizational Learning present the idea of systems thinking as a discipline for the vision of integrity. "…a structure that allows you to see relationships, not individual things, to see patterns…" [20].

Recent developments in cyberspace, particularly in social media, are rapidly changing our lives, including the way we communicate and make decisions…. predicting personality from social networks is a powerful tool for personality analysis and information technology [21].

Human interaction shapes the development of our society. The creativity of meme propagators can tell a lot about the emotional component and its evolution over time. This is the reason why sociotechnical environments and social networks in particular have drawn our attention to the study of the role and the development of creative thinking.

The authors asked the students of Rostov State University of Economics to fill in a questionnaire in May 2021, the questions of which embraced meme-engagement and creativity as a social form of digital society among students.

The questionnaire was offered to 95 respondents and consisted of three questions. Two of them were basic to determine whether a person was subscribed to a meme account and which digital platforms were used to exchange memes. The third question was more complicated and had the aim to find out the information about the emotional component of the Internet meme to identify the purpose of sharing memes.

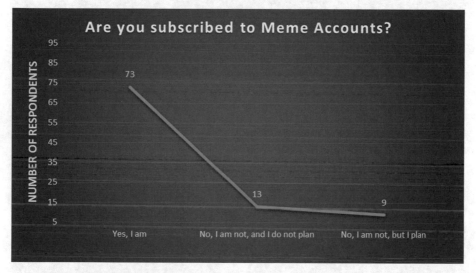

Fig.1. Social questionnaire of the students. Meme Accounts.

Figure 1 reflects the respondents' replies - 95 students, 76 of whom subscribe to the meme accounts, 9 of the total number plan to subscribe, which indicates that 86% of meme users recognize meme as creativity and use them.

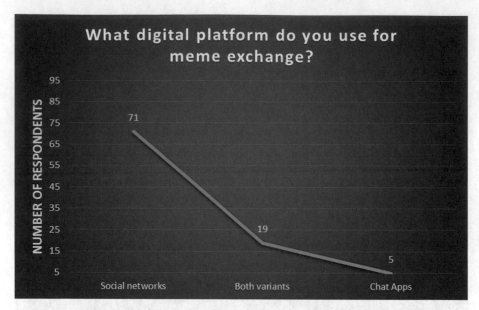

Fig. 2. Social questionnaire of the students. Digital Platforms.

Figure 2 shows that the majority of respondents - 71 students - send memes using social media.

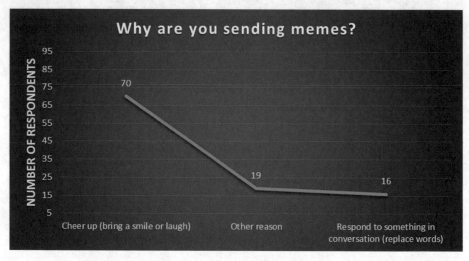

Fig. 3. Social questionnaire of the students. Reasons for sending memes.

The basic indicator that meme, as a creative unit is distributed by respondents is reflected in Fig. 3. At the same time, 70 respondents share memes to raise the mood, which mainly reflects the emotional aspect of memes.

A meme is an idea or ideas that can be reproduced by future generations. It can also be seen as the basic unit of reproduction of a new kind of creativity. Because memes are malleable and immaterial, they are difficult to study. However, the study attempted to draw some general conclusions about what predicts the preservation of memes, at least part of them, and that is their emotional appeal. Our questionnaire also shows that memes, in some cases, already serve as a substitute for words. This replacement of words speaks of the desire for visual communication with the results of creativity in the digital world, which has a growing trend.

Researchers see creativity as a process of taking steps to solve problems or to invent creatively new products that are represented in the study of memes that affect the mood of users in social networks and the entire digital space. In the first stage - which is characterized by a keen interest and curiosity, a creative personality collects information. The second stage consists of problem identification and solution-seeking and involves the processing of large amounts of information; this can occur at both a conscious and unconscious level. The third stage is characterized by divergent thinking, openness and excitement. In the fourth stage, the users of the memes compare the memes they create with those they already have online. In the fifth stage, they distribute the memes, making them available to assess their creativity. In the sixth stage, their work is either recognized or rejected. Shifting their point of view is closely linked to cognitive flexibility, a major aspect of creative ideas. This combination can be explained by the dual process of creative learning. In addition, the viewpoint of the others influences the judgment or appreciation of the creative abilities of the meme creator. This explains the discrepancies between meme creators and sharing [22].

The six-stage model developed by the authors supports a systematic view of the creative process, focusing on the social test that occurs when the work is supported. So the mental processes of a creative person, and the recognition of society, have come together to produce the meme as creativity, which demonstrates how this unpredictable component of human behavior contributes to human development and the promotion of memes on the Internet and social networks. The abstract thinking of creators and distributors of memes allows them to be creative, to develop tools and new ideas [23] that affect the emotional state of all users of the digital space.

The digital world is inherently a dynamic environment that evolves, as connections between users are continuously created and destroyed. This time-changing nature of the digital world influences the dynamic processes taking place in any society and the creativity and emotions of the users.

4   Conclusion

Having analyzed the research of creativity, emotions, statistics and questionnaire conducted to examine the growing influence of meme creativity on the attitudes of social networks users and all members of the digital community, we came to the following conclusions. It must be admitted that we have a complicated emotional life and the age of digitalization is not an exception. A person draws creative ideas from nostalgia, disappointment and sympathy and as a rule, can overpower his emotions. He or she can regulate them using such tools as thinking, behavioral strategies, and strengthen them by implementing his or her creative ideas. One can say that the process of developing, valuing, and translating an idea into the end product has ups and downs and lots of emotions associated with it. Creativity is a myriad of different emotions, both positive and negative, from anxiety in front of an empty computer screen and disappointment from an unsuccessfully created profile, to the delight of insight and pride in your accomplishments, accounts, memes, edited photographs and other creative ideas. Exploring the role of emotions in creativity has taught us a lot. The most important fact is that we have come to the conclusion that creative ideas are best created and shared when people experience positive emotions. The conclusion of a causal relationship between creativity and emotion has already been confirmed by studies which in further research continue to control what is being studied. We do not exclude the possibility that our study may not be continued and will be a simple addition to the already available ones. Nevertheless, we hope that the study of the influence of personal characteristics of the user activity, situational and other factors will be continued. The examples of the user profiles and Internet memes and other pieces of art of the digital world affect emotions and will be considered as some of the components of human well-being.

References

1. Janka, Z.: The impact of mood alterations on creativity. Ideggyogy Sz. **59**(7–8), 236–40 (2006). https://pubmed.ncbi.nlm.nih.gov/17076301/. Accessed 24 Apr 2021
2. Baas, M., De Dreu, C.K., Nijstad, B.A.: A meta-analysis of 25 years of mood-creativity research: hedonic tone, activation, or regulatory focus? Psychol. Bull. **134**(6), 779–806 (2008). https://doi.org/10.1037/a0012815
3. Lee, D.S., Lee, K.C., Seo, Y.W., Choi, D.Y.: An analysis of shared leadership, diversity, and team creativity in an e-learning environment. Comput. Hum. Behav. **42**, 47–56 (2015). https://doi.org/10.1016/j.chb.2013.10.064
4. Gazit, T., Aharony, N., Amichai-Hamburger, Y.: Tell me who you are and I will tell you which SNS you use: SNSs participation. Online Inf. Rev. **44**(1), 139–161 (2019). https://doi.org/10.1108/OIR-03-2019-0076
5. Boden, M.A.: Creativity and artificial intelligence. Artif. Intell. **103**(1–2), 347–356 (1998). https://doi.org/10.1016/S0004-3702(98)00055-1
6. Szakacs, R., Janka, Z.: Mood as a locked gate canalizing multiple creativity: a heuristic single-case study. Neurol. Psychiat. Brain Res. **32**, 48–54 (2019). https://doi.org/10.1016/j.npbr.2019.04.001
7. Perchtold, C.M., et al.: Humorous cognitive reappraisal: more benign humour and less "dark" humour is affiliated with more adaptive cognitive reappraisal strategies. PLoS One **14**(1) (2019). https://doi.org/10.1371/journal.pone.0211618.
8. Moutidis, I., Williams, H.T.P.: Good and bad events: combining network-based event detection with sentiment analysis. Soc. Netw. Anal. Min. **10**(1), 1–12 (2020). https://doi.org/10.1007/s13278-020-00681-4
9. Groyecka-Bernard, A., Karwowski, M., Sorokowski, P.: Creative thinking components as tools for reducing prejudice: evidence from experimental studies on adolescents. Think. Skills Creativit. **39**, 100779 (2021). https://doi.org/10.1016/j.tsc.2020.100779
10. Glăveanu, V.P., Beghetto, R.A.: Creative experience: a non-standard definition of creativity. Creativ. Res. J. **33**(2), 75–80 (2020). https://doi.org/10.1080/10400419.2020.1827606
11. Tankovska, H.: Number of social network users worldwide from 2017 to 2025 (2021). https://www.statista.com/statistics/278414/number-of-worldwide-social-network-users/. Accessed 24 May 2021

12. Nguyen, T.C.: The Most Important Inventions of the 21st Century (2019). https://www.tho ughtco.com/the-most-important-inventions-of-the-21st-century-4159887. Accessed 24 Apr 2021

13. Leea, M.T., Theokaryb, C.: The superstar social media influencer: exploiting linguistic style and emotional contagion over content? J. Bus. Res. **132**, 860–871 (2020). https://doi.org/10. 1016/j.jbusres.2020.11.014

14. Goffman, E.: The Presentation of Self in Everyday Life. Doubleday, Scotland (1959)

15. Xuab, X., Xiac, M., Zhaob, J., Pangb, W.: Be real, open, and creative: How openness to expe-rience and to change mediate the authenticity- creativity association. Think. Skills Creativ. **41**, 100857 (2021). https://doi.org/10.1016/j.tsc.2021.100857

16. McCall, G.J.: The me and the not-me. In: Burke, P.J., Owens, T.J., Serpe, R.T., Thoits, P.A. (eds.) Advances in Identity Theory and Research, pp. 11–25. Springer, Boston (2003). https:// doi.org/10.1007/978-1-4419-9188-1_2

17. Ong, T.: Young female Japanese biker is really 50-year-old man with luscious hair using FaceApp (2021). https://mothership.sg/2021/03/japanese-biker-actually-man/. Accessed 24 May 2021

18. Ogbeibua, S., Pereirab, V., Emelifeonwuc, J., Gaskind, J.: Bolstering creativity willingness through digital task interdependence, disruptive and smart HRM technologies. J. Bus. Res. **124**, 422–436 (2021). https://doi.org/10.1016/j.jbusres.2020.10.060

19. Maksic, S., Smiljana, J.: Scaffolding the development of creativity from the students' perspective. Think. Skills Creativ. **41**, 100835 (2021). https://doi.org/10.1016/j.tsc.2021. 100835

20. Senge, M.P.: The Fifth Discipline: The Art & Practice of the Learning Organisation. Currency Doubleday, New York, London (2006)

21. Mori, K., Haruno, M.: Differential ability of network and natural language information on social media to predict interpersonal and mental health traits. J. Personal. **89**(2), 228–243 (2020). https://doi.org/10.1111/jopy.12578

22. Long, H., Runcob, M.A.: Creativity and perspective. Reference module in neuroscience and biobehavioral psychology. In: Encyclopedia of Creativity, 3rd edn., pp. 246–249 (2020). https://doi.org/10.1016/B978-0-12-809324-5.23672-9

23. Petrova, J., Vasichkina, O.: Computer aspect of interdisciplinary research in philosophy of education and computer science. In: SHS Web Conference, vol. 109, p. 01029 (2021). https:// doi.org/10.1051/shsconf/202110901029

Integrated Use of Data Mining Techniques for Personality Structure Analysis

Elena Slavutskaya[1]($\boxtimes$) (iD), Leonid Slavutskii[2] (iD), Anna Zakharova[2] (iD),
and Evgeni Nikolaev[2] (iD)

[1] Chuvash State Pedagogical University, K. Marx, 38, 428000 Cheboksary, Russia
[2] Chuvash State University, Moscow Prospect, 15, 428015 Cheboksary, Russia

Abstract. Data mining techniques include traditional statistical estimation theory and machine learning methods. They have been intensively developed in recent years, and are increasingly used in a variety of fields. The present research suggests the simultaneous, complex use of different data mining methods. The approach is applied to psycho diagnostic data processing. The traditional methods of factor analysis and cluster analysis are used together, along with artificial neural networks as one of the foundations of artificial intelligence. The possibilities of this approach are demonstrated by the example of adolescents' psycho diagnostic data processing. The study is based on structural Cattelian Personality Theory as the most well-known example using the factor model of personality. Based on the complex processing of psycho diagnostic data, it is shown that each of the data mining methods separately does not provide complete information about the relationship between the psychological characteristics of the respondents. Comparing the results obtained by different processing methods allows us to obtain information about the intra-system relationships of psychological qualities, to assess the hierarchy and significance of personal psychological characteristics. It is shown that the analysis can be carried out with widely used software and on a small sample of respondents of several dozen people. This allows psychologists to analyze the psychological characteristics of homogeneous social groups of respondents whose psycho diagnostic results were obtained under experimental conditions at the same time.

Keywords: Data mining · Psycho diagnostic · Cattelian Personality Theory · Personal psychological characteristics

1 Introduction

The data mining and knowledge discovery in databases (KDD) have been intensively developed in recent years and involve a system analysis of multidimensional random data [1–6], including the so-called "big data" [2, 7]. These methods allow one to solve a wide range of problems in the social sphere, education [8–11], healthcare, optimization of internet traffic, etc. [1, 11–16]. Data mining techniques often have different classifications depending on the tasks to be solved. Most often, these are predictive, classification, approximation methods [14, 15]. These methods include both fairly common statistical approaches, such as regression, factor, cluster, and discriminant analysis [16], and machine learning, such as artificial neural networks (ANN) [17, 18], the "decision tree" [19], etc.

Of the above, traditional correlation and factor analysis are most often used in psychology. In this paper, using concrete examples, we show the possibilities of complex use of traditional methods and modern ANN techniques for psycho diagnostic data processing in the framework of the personality structural analysis based on its factor models. Such an integrated approach does not require a large sample of respondents and allows using widely available software to carry out a psychological analysis of homogeneous social groups.

1.1 Background

The factor models of personality are constantly being improved and transformed. The most well-known and frequently used is the factor model based on the R. Cattell's structural theory of personal traits [20]. In recent decades, the dispositive personality model "Big Five Inventory" has become very widespread [21]. This model offers a description of the persons' behavior and psychological characteristics in the sense of his perception by other people or by himself [22]. The model is intensively adapted for different language and age groups of respondents [23–26]. For the different age groups, it may include a different number of traits, as the Cattells' questionnaires (12-, 14-, and 16-factor questionnaires).

The key problem is that the factors in such models must be independent. And how is this independence possible to ensure? In essence, the independence of factors is most often understood as the absence of correlation. And the absence of correlation means that there is no linear relationship (if the parametric correlation coefficient is considered) or there is no monotonic relationship (for the Spearman correlation coefficient). In fact, the relationships between factors in personality models are present [27, 28], which allows us to move, for example, from the 16-factor Cattell's model to the "Big Five" model. One of the most important tasks is to show that in the factor models of personality there are nonlinear and mediated connections between personality traits and these connections are constantly in dynamic transformation [29].

Personal traits and qualities are the result of multiple factorization, that is, combining a significant number of signs into personal traits that are given a certain interpretation. Each of the traits, therefore, is a synthesis of a certain psychological qualities set, which, in turn, can be individual psychological characteristics or the qualities of another level [30]. Consequently, the traits themselves can be considered as characteristics of different mental levels. Some of them are closer to individual psychological characteristics, some largely describe the mental state, and some refer to psychosocial personality characteristics [12, 19, 31].

1.2 Problem Statement

The purpose of the work is to show the data mining capabilities for the analysis and interpretation of personality traits by the specific examples. Even within the framework of the personality factor model (and the corresponding psycho diagnostic tools), a comprehensive, comparative data analysis using different processing methods allows us to identify fundamentally new features of the psychological characteristics' structure.

2 Data Mining Methods

The study used the psycho diagnostic tools of Cattelian Personality Theory, as the most proven, most characteristic example of a factor model for personality description. Cattell's questionnaires are designed to process the results using factor analysis based on the Pearson correlation coefficients matrix. In addition, hierarchical clustering was carried out, where the same correlation coefficients were used as the link's measure. That is, a dendrogram was constructed based on the values of personal traits from the psycho diagnostic results. Thus, factor analysis and cluster analysis were used as traditional statistical methods based on linear correlations.

The neural algorithm was applied to evaluate nonlinear and latent connections. The neural network used was simple, feedforward, based on the elementary Rosenblatt perceptron [32]. A non-large number of neurons and 1–2 hidden layers in the ANN architecture allow to process data of a relatively small sample of respondents. The criteria for evaluating the quality of the ANN model can be the percentage of link recognition and the errors' level during ANN training and testing [28].

The scheme of the neural network algorithm for evaluating the relationships between the psycho diagnostic data is shown on Fig. 1. The algorithm consists in the sequential replacement of the target function (OUT) at the ANN output with the input features. One of the features is used as the target function, the other personal traits values are fed to the perceptron input. The ANN training quality is evaluated, then the appropriate procedure is performed for the next personal trait, etc. The results of training and quantitative quality control of the neuro-model allow to identify the psychological characteristics that have the most structured and stable links with the rest of the psycho diagnostic data [28]. Psychological characteristics that do not demonstrate such connections may be excluded from consideration in the future.

The possibilities of an integrated approach to the analysis of the personality traits structure were studied on the psycho diagnostic data of adolescents aged 14–15 (questionnaire 14-PF/HSPQ). The sample of respondents was 81 people. The maximum value of the correlation coefficient between the interrelated features reached 0.82, which corresponds to the Pearson correlation coefficient significance level of less than 0.0005. Thus, on the one hand, the sample allows for factor and cluster analysis, on the other hand, it significantly exceeds the number of computational paths (links) in the ANN at $N = 14$ (see Fig. 1).

It will be shown below how the simultaneous processing of the same data by the above methods allows obtain significant additional information for psychological analysis.

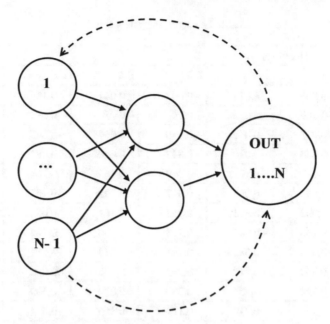

Fig. 1. Algorithm for training and testing a neural network

3 Examples of Data Processing (Comprehensive Analysis of the Adolescents' Personal Traits)

Table 1 shows the factor analysis data of the adolescents' personal traits based on the using the R. B. Cattell's 14-factor questionnaire. According to the Kaiser criterion [33], 6 factors are identified whose eigenvalues $E > 1$. The total contribution of these factors S to the total variance is more than 77%. Such a contribution of six factors from the fourteen initial characteristics can be considered significant and indicates a certain structuring of personal qualities. The factor loadings corresponding to the most significant traits for each factor are highlighted in bold. This grouping of personal traits by factors is characteristic of this age adolescents and allows psychological interpretation for applied purposes.

The frustration Q4 and the degree of emotional stability C are identified as separate factors, which is associated with traditional personal adolescents' problems at this age. The important fact is the allocation of the intellectual indicator B in adolescents as a separate, unrelated factor, that is also characteristic of personal development by the age of 14–15 in connection with the formation of verbal - logical thinking [28].

The F3 factor includes interrelated characteristics D (anxiety-phlegm), J (neurosis-Hamlet's factor). This can be interpreted as a characteristic feature of the studied age, regardless of the type of upbringing, family, or social adaptation. In adolescents of this age, who show great inertia, are reserved, are constant in their interests (D-), prefer group actions, accept social norms and assessments, for them the totality of people is important (J-). This fact confirms the need for a reference group for the normal socialization and development of the adolescents' personality, regardless of the influence and nature of

Table 1. Factor analysis results

Traits	Factor loadings					
	F 1	F 2	F 3	F 4	F 5	F 6
A	**−0,835**	0,014	−0,002	0,041	0,126	0,119
B	−0,089	−0,054	−0,163	−0,056	−0,034	**0,827**
C	0,025	−0,040	−0,050	0,054	**−0,865**	0,083
D	−0,223	−0,020	**0,847**	0,040	−0,052	−0,112
E	0,117	**0,773**	0,249	0,266	0,267	0,187
F	**−0,602**	0,083	0,047	0,422	−0,481	0,055
G	0,003	0,391	0,255	0,561	−0,055	0,514
H	0,130	**−0,845**	0,218	−0,060	0,143	0,037
I	**−0,668**	0,386	−0,281	−0,273	−0,051	0,285
J	0,389	−0,210	**0,739**	−0,253	0,050	−0,070
O	0,112	**0,676**	−0,388	0,087	0,409	−0,218
Q2	0,139	−0,210	0,179	−0,456	−0,524	−0,183
Q3	**0,716**	0,199	−0,128	0,250	0,001	0,131
Q4	−0,197	−0,094	0,137	**−0,870**	0,011	0,136
E	2,331	2,221	1,770	1,754	1,544	1,225
S,%	16,65	15,87	12,65	12,53	11,03	8,75

Note to the table:
E, S are the factors eigenvalues and their contribution to the total variance.
The group of personal communicative characteristics by R. B. Cattell:
A – sociability (gregariousness - isolation); H – courage-timidity in contacts; E – independence - obedience; Q2 – autonomy (degree of group dependence).
The group of intellectual traits: B – verbal intelligence (abstract - concrete thinking).
The group emotional traits: C – degree of emotional stability; D - anxiety-phlegm; F – carefree - concern (levity - care); I – emotional sensitivity (sensitivity - realism); J – (neurosis, «Hamlet's factor»); O – anxiety (the tendency to guilt - self-confidence); Q4 – tension-relaxation (degree of frustration).
The group of behavioral regulatory traits: Q3 – degree of self-control; G –moral normativity (high-low discipline).

the social environment. It can be assumed that the nature of the social environment does not affect and does not change the relationship of these characteristics, in the future, they are rebuilt and change with the age. The reference group is the platform on which the processes of growing up are developed and practiced, which are necessary for entering the period of adolescence with their own tasks of individualization and self-determination.

Factor F2 includes: E (passivity-dominance); O self-confidence - a tendency to guilt (anxiety); H timidity, shyness – courage, adventurism (in contacts). The higher the dominance, the desire for self-affirmation (E+), the more pronounced the fearfulness and subconscious feeling of inferiority (H -), at the same time self-confidence, trustfulness and lack of a sense of threat (O+). The interrelation of these traits is a sign of the internal contradictions of adolescence and describes "aggressive anxiety" as a component of the "disaptation syndrome" [29].

The first most important factor F1 (S% = 16.65) in the group of adolescents demonstrates an interrelated communicative qualities (A) and emotions (F), (I) with the volitional characteristic Q3. Q3+ (self-control of behavior) is inversely related to the personality trait F – (caution-levity). For example, the higher the self-control, the less carefree in communication, frivolity, and more pronounced caution, pessimism, impressionability. This may indicate that, regardless of the type of family and upbringing, those adolescents who have developed a high volitional control of their behavior with a focus on social norms will demonstrate such characteristics as taciturnity, caution, impressionability, gloominess with a tendency to subdepression. And vice versa: with low volitional control and no desire to comply with social requirements, adolescents of both study groups will be active, talkative, lively, easy to switch from one activity to another, they are often the "soul of the company".

Figure 2 shows the dendrogram obtained from the same psycho diagnostic data as the result of cluster analysis. In this case, the dendrogram graphically displays the inverse of the Pearson correlation coefficient.

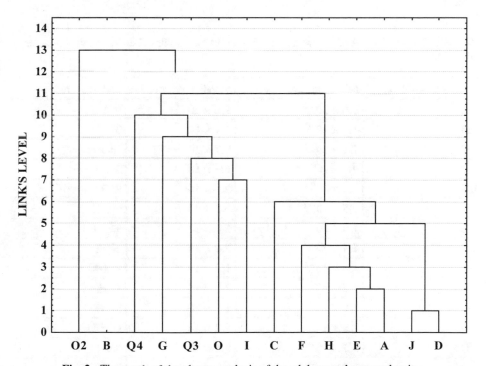

Fig. 2. The result of the cluster analysis of the adolescents' personal traits.

When comparing the data in Fig. 2 with the 14 personality traits factor analysis shown in Table 1, it can be found that, in general, the structure and classification of relationships is preserved. In both cases, the features are grouped according to the level of correlation, the most closely related features are either in one cluster or in the corresponding group of the hierarchical chain of clusters.

To get additional information about the personal features significance, a comparison of the factor and cluster analysis results can be carried out. An example is verbal intelligence (B), which is not closely related to any of the diagnosed traits, it is located at the top of the dendrogram (Fig. 2), but is a separate factor with more than 8% contribution to the overall variance (Table 1). Signs J, D have the high correlation between them and are located in the lower part of the dendrogram, but make up the only third factor contributes to the overall variance. Signs A, E, H and F are also located in the lower part of the dendrogram of Fig. 2, and are among the first two most significant factors in terms of their contribution to the overall variance (Table 1). However, their relationships in the dendrogram do not correspond to the fact that the factors in Table 1 should be independent. These facts once again confirm that the analysis of data based on correlation coefficients does not provide comprehensive information about the intra-system relationships of personality traits. Analysis of latent, nonlinear and mediated connections is required. For these purposes, the ANN was used.

Table 2 shows the results of training the ANN according to the algorithm shown in Fig. 1.

Table 2. The results of ANN training with different personal traits (14-PF) as a target function at its output

Traits at ANN out	Maximum error σ_{max}	Standard error σ_s	Link recognition S
A	$3,6 \times 10^{-1}$	$1,6 \times 10^{-2}$	94,6%
B	$1,5 \times 10^{-1}$	$1,4 \times 10^{-2}$	89,3%
C	$1,1 \times 10^{-1}$	$9,7 \times 10^{-3}$	96,5%
D	$9,0 \times 10^{-2}$	$6,3 \times 10^{-3}$	96,4%
E	$1,4 \times 10^{-1}$	$2,0 \times 10^{-2}$	92,9%
F	$9,5 \times 10^{-2}$	$9,2 \times 10^{-3}$	94,6%
G	$1,0 \times 10^{-1}$	$5,9 \times 10^{-3}$	96,4%
H	$4,4 \times 10^{-2}$	$7,0 \times 10^{-3}$	100%
I	$1,8 \times 10^{-1}$	$1,3 \times 10^{-2}$	91,0%
J	$4,2 \times 10^{-2}$	$7,3 \times 10^{-3}$	100%
O	$1,8 \times 10^{-1}$	$1,8 \times 10^{-2}$	87,5%
Q2	$7,8 \times 10^{-2}$	$1,2 \times 10^{-2}$	96,5%
Q3	$1,1 \times 10^{-1}$	$9,6 \times 10^{-3}$	96,4%
Q4	$3,0 \times 10^{-1}$	$2,2 \times 10^{-2}$	89,3%

The table shows three numerical criteria for the quality of ANN training: maximum, standard error, and the link recognition percentage. The most significant of them are the root-mean-square error σ_s and the percentage of recognition S. To conclude that it is possible to use the ANN model in principle, the indicator σ_s should not exceed units of percent (up to about 5%), and S – not less than 90%. The maximum error in the processing of psycho diagnostic data can be very high: in this case, for the sign A at the output of the ANN, it reaches 36%. Such a high deviation may correspond to non-standard sets of characteristics for individual respondents. These respondents can be selected from the general sample, their test results can be rechecked, and so on. In a statistical sense, the numerical criteria for the quality of ANN training are connected. The correlation coefficients of the 14 values in the columns of Table 2 are: between σ_{max} and σ_s $R_{12} = 0.73$, and between σ_{max} and S $R_{13} = -0.59$. The significance level of these correlation coefficients is less than 0.01 and 0.02, respectively. According to the table, the highest quality of training corresponds to ANN models with target functions in the form of personality traits H, J. In these cases, $S = 100\%$ and learning errors are minimal (see Table 2). This suggests that for this sample of respondents, these traits are most closely related to other personality traits. For signs B, E, I, O, Q4, the quality of ANN training is low, they do not demonstrate the presence of stable links with the aggregate set of other personal traits. This does not mean that there are no such connections. The connections between the data of psychological testing can be indirect (hidden), moreover, the values of some signs can "hinder" the ANN training. Therefore, the impact of each feature on the quality of ANN training should be evaluated selectively. For this purpose, the values of the corresponding signs at the input of the ANN can be excluded (or reset to zero). The results of such a procedure with the personality trait J as the target function at the ANN output are shown in Fig. 3. Here, the ANN training was carried out sequentially on 12 input out of 13 traits. That is, 1 feature out of 14 is used as the target function, and one feature at the input is excluded.

Excluding the values of F, O, Q3, Q4 does not affect the ANN training quality (100% recognition and small errors). If all these features are excluded from the ANN model, the structure of the neural network takes the form shown in Fig. 3. The training quality of this ANN does not "deteriorate" much and turns out to be acceptable: $\sigma_{max} = 0,064$, $\sigma_s = 0,0098$, $S = 92,86\%$.

According to the results of the neural network analysis, the J attribute has the most structured links with the personal traits indicated at the input of the ANN in Fig. 3. At the same time, if the values of J and H are zeroed (excluded) as the personal traits most closely related to the others (see Table 2), then the quality of ANN training in assessing the interconnections of other personal traits "deteriorates". That is, their exclusion violates the general structure of relations in the psychological pattern of adolescents' personality. This fact confirms the presence of indirect cross-functional connections between the personality traits of adolescents. Personality traits J (neurasthenia, Hamlet's factor) and H (courage - timidity in communication) can be considered in this context as the most significant traits that have a influence on the structure of the links of personal traits. This does not contradict the accepted ideas about the psychological characteristics of adolescents at this age.

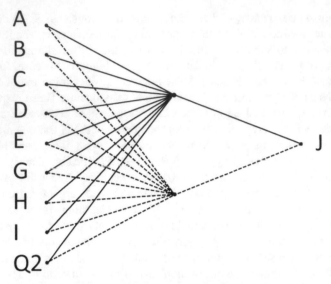

Fig. 3. Structure of ANN after excluding 4 features from its input

4 Discussion

Using the example of the above psycho diagnostic results, we will show some features of the personality structure that are revealed during a comparative analysis using all three Data Mining methods.

In the significant factors (Table 1), there are no features of G, Q2, but both of these features appear in the ANN training, as associated with the Hamlet's factor. Verbal intelligence B is weakly associated with the rest of the personal traits in all three methods, but when using the ANN, it is also found in the traits associated with J.

The trait D (phlegmaticity – excitability) refers closely to the properties of the nervous system as the individual psychological characteristics of the basic level. It has a very high level of connection with J in all three methods, but does not show such a stable connection with the rest of the personality traits that is found for J.

The Hamlet's factor occupies a special place among the personal traits of adolescents. This trait is only found in the 14-factor questionnaire. That is, it just refers to the peculiarities of adolescence. It turns out that in adolescence, J is a kind of connecting links in the Cattell's structure of personal traits. At the same time, the Hamlet's factor is not the most "weighty" factor and is associated, according to the results of correlation data processing (both factor and cluster analysis), only with D. According to neural network analysis, it is interconnected with a significantly larger number of traits. This confirms that the connections between personality traits can be highly nonlinear and mediated (latent). It is very important and characteristic that when comparing the results of neural network and factor analysis, the Hamlet's factor at the ANN training is associated with 5 of the 6 factors in Table 1. There is no such connection only with the factor that separately includes frustration Q4. J is related with the all remaining factors.

The Q4 trait can be attributed to emotional states and, according to all three methods of Data Mining, appears in adolescents as poorly related to other traits: it is allocated as a separate factor in the factor analysis (Table 1), it is located in the upper part of the dendrogram (Fig. 2), and the poor quality of the corresponding ANN training (Table 2).

According to R. Cattell, personal traits are divided into four groups: communicative, intellectual, strong-willed, and emotional. Traits belonging to different groups are related to each other in different factors. Such relationships are transformed, and the number of factors with a statistically significant contribution may change [29]. For psychological research, the study of the cross-functional connections of personal traits often requires a more in-depth analysis on small samples, when respondents of the same social group, the same age, in approximately equal conditions of psycho diagnostics are involved. The approach presented in this paper allows the analysis to be carried out on a relatively small sample of respondents.

5 Conclusions

Thus, the present paper discusses the possibilities and shows the effectiveness of an integrated approach to the psycho diagnostic data processing. On the example of adolescents' psycho diagnostic data processing the effectiveness of simultaneous use of such methods as traditional factor and cluster analysis, together with machine learning methods, in particular, the artificial neural networks, is demonstrated. Each of the methods allows to get some information about the psychological characteristics of the respondents, but this information is slightly different for different data processing methods. The analysis and interpretation of such results and their comparison allow us to obtain much deeper and more adequate information on the psycho diagnostic results, to assess the hierarchy of relationships and the structure of personal traits. It is shown that such information can be obtained on a relatively small sample of respondents of several dozen people. The used data mining methods are present and included in most modern software packages, such as the Russian packages DEDUCTOR and STATISTICA. That is, they are available for practical psychologists.

References

1. Shitikov, V.K., Mastitskiy, S.E.: Classification, Regression and Other Data Mining Algorithms Using R. Heidelberg (2017). https://ranalytics.github.io/data-mining/. Accessed 21 May 2020
2. Mei, J.-P., Lv, H., Yang, L., Li, Y.: Clustering for heterogeneous information networks with extended star-structure. Data Min. Knowl. Disc. **33**(4), 1059–1087 (2019). https://doi.org/10.1007/s10618-019-00626-2
3. Kantardzic, M.: Data Mining: Concepts, Models, Methods, and Algorithms. Wiley, New York (2011)
4. Ababkova, M., Leontyeva, V.: Neurobiological studies within the framework of highly technological teaching. In: SHS Web Conference, vol. 44, p. 00002 (2018). https://doi.org/10.1051/shsconf/20184400002
5. Mureyko, L.V., Shipunova, O.D., Pasholikov, M.A., Romanenko, I.B., Romanenko, Y.M.: The correlation of neurophysiologic and social mechanisms of the subconscious manipulation in media technology. Int. J. Civ. Eng. Technol. **9**, 2020–2028 (2018)

6. Almazova, N., Bylieva, D., Lobatyuk, V., Rubtsova, A.: Human behavior as a source of data in the context of education system. In: SPBPU IDE 2019: Proceedings of Peter the Great St. Petersburg Polytechnic University International Scientific Conference on Innovations in Digital Economy, vol. 37. ACM, Saint – Petersburg (2019). https://doi.org/10.1145/3372177.3373340

7. Slonim, N., Atwal, G.S., Tkachic, G., Bialek, W.: Information-based clustering. Proc. Natl. Acad. Sci. **102**, 18297–18302 (2005)

8. Shipunova, O.D., Berezovskaya, I.P., Mureyko, L.M., Evseeva, L.I., Evseev, V.V.: Personal intellectual potential in the e-culture conditions. Espacios **39**, 15 (2018)

9. Bylieva, D., Almazova, N., Lobatyuk, V., Rubtsova, A.: Virtual pet: trends of development. In: Antipova, T., Rocha, Á. (eds.) DSIC 2019. AISC, vol. 1114, pp. 545–554. Springer, Cham (2020). https://doi.org/10.1007/978-3-030-37737-3_47

10. Bylieva, D., Lobatyuk, V., Safonova, A., Rubtsova, A.: Correlation between the practical aspect of the course and the e-learning progress. Educ. Sci. **9**, 167 (2019). https://doi.org/10.3390/educsci9030167

11. Han, H., Soylu, F., Anchan, D.M.: Connecting levels of analysis in educational neuroscience: a review of multi-level structure of educational neuroscience with concrete examples. Trends Neurosci. Educ. **17**, 100113 (2019). https://doi.org/10.1016/j.tine.2019.100113

12. Rosenberg, S., Van Mechelen, I., De Boeck, P.: A hierarchical class model: theory and method with applications in psychology and psychopathology. In: Arabie, P., Hubert, L., De Soete, G. (eds.) Classification and Clustering, River Edge, pp. 123–155. World Scientific, New York (1996). https://doi.org/10.1142/9789812832153_0004

13. Holena, M., Pulc, P., Kopp, M..: Classification Methods for Internet Applications. Studies in Big Data, vol. 69. Springer, Cham (2020). https://doi.org/10.1007/978-3-030-36962-0

14. Christensen, A.P., Golino, H.: On the equivalency of factor and network loadings. Behav. Res. Methods **53**(4), 1563–1580 (2021). https://doi.org/10.3758/s13428-020-01500-6

15. Witten, I.H., Frank, E., Hall, M.A., Kaufmann, M.: Data Mining: Practical Machine Learning Tools and Techniques, 3rd edn. Elsevier, Amsterdam (2011)

16. Glass, J.V., Stanly, Y.C.: Statistical Methods in Education and Psychology. Prentice-Hall, Upper Saddle River (1970)

17. Grossberg, S.A.: Path toward explainable AI and autonomous adaptive intelligence: deep learning, adaptive resonance, and models of perception, emotion, and action. Front. Neurorobot. **14**, 36 (2020). https://doi.org/10.3389/fnbot.2020.00036

18. Schmidhuber, J.: Deep learning in neural networks: an overview. Neural Netw. **61**, 85–117 (2014). https://doi.org/10.1016/j.neunet.2014.09.003

19. Slavutskaya, E.V., Slavutskii, L.A., Abrukov, V.S., Bichurina, S.U., Sadovaya, V.V.: Vertical system analysis of students' psycho diagnostic data using the 'Decision Tree' method. Sci. Educ. Today **10**(3), 87–107 (2020). https://doi.org/10.15293/2658-6762.2003.05

20. Cattell, R.B.: Advanced in Cattelian Personality Theory. Handbook of Personality. Theory and Research. The Guilford Press, New York (1990)

21. McCrae, R.R., Costa, P.T.: Understanding persons: from Stern's personalistics to five-factor theory. Pers. Individ. Differ. **169**, 109816 (2021). https://doi.org/10.1016/j.paid.2020.109816

22. Mõttus, R., Wood, D., Condon, D.M., Zimmermann, J.: Descriptive, predictive and explanatory personality research: different goals, different approaches, but a shared need to move beyond the big few traits. Eur. J. Pers. **34**(6), 1175–1201 (2020). https://doi.org/10.1002/per.2311

23. Soto, C.J., John, O.P.: Short and extra-short forms of the big five inventory 2: the BFI-2-S and BFI-2-XS. J. Res. Pers. **68**, 69–81 (2017). https://doi.org/10.1016/j.jrp.2017.02.004

24. Shchebetenko, S., Kalugin, A.Y., Mishkevich, A.M., Soto, C.J., John, O.P.: Measurement invariance and sex and age differences of the big five inventory-2: evidence from the Russian version. Assessment **3**, 472–486 (2020). https://doi.org/10.1177/1073191119860901

25. Iimura, S., Taku, K.: Gender differences in relationship between resilience and big five personality traits in Japanese adolescents. Psychol. Rep. **121**(5), 920–931 (2018). https://doi.org/10.1177/0033294117741654

26. Krampen, D., Kupper, K., Rammstedt, B., Rohrmann, S.: The German-language short form of the big five inventory for children and adolescents – other-rating version (BFI-K KJ-F). Eur. J. Psychol. Assess. **2**(37) (2020). https://doi.org/10.1027/1015-5759/a000592

27. Golino, H., Christensen, A.P., Moulder, R.G., Kim, S., Boker, S.M.: Modeling latent topics in social media using Dynamic Exploratory Graph Analysis: the case of the right-wing and left-wing trolls in the 2016 US elections. PsyArXiv (2020). https://doi.org/10.31234/osf.io/tfs7c

28. Slavutskaya, E., Slavutskii, L., Nikolaev, E., Zakharova, A.: Neural network models for the analysis and visualization of latent dependencies: examples of psycho diagnostic data processing. In: Bylieva, D., Nordmann, A., Shipunova, O., Volkova, V. (eds.) PCSF/CSIS -2020. LNNS, vol. 184, pp. 61–70. Springer, Cham (2021). https://doi.org/10.1007/978-3-030-658 57-1_7

29. Kolishev, N.S., Slavutskaya, E.V., Slavutskii, L.A.: The dynamics of personality traits structuring during student's transition to secondary school. Integr. Educ. **23**, 390–403 (2019). https://doi.org/10.15507/1991-9468.096.023.201903.390-403

30. Zwir, I., Del-Val, C., Cloninger, C.R.: Three genetic-environmental networks for human personality. Mol. Psychiatr. (2019). https://doi.org/10.1038/s41380-019-0579-x

31. Kuznetsova, V.B.: Personality traits as mediator of interrelationship between upbringing methods and behavioral problems of children. Psikhologicheskii Zhurnal **38**(1), 31–40 (2017)

32. Rosenblatt, F.: Principles of Neurodymamics. Spartan Books, Washington, D.C. (1962)

33. Kaiser, H.F.: The application of electronic computers to factor analysis. Educ. Psychol. Measur. **20**, 141–151 (1960)

Creativity in Education

Educational Technologies for the Development of Creative Thinking of the Future Engineer

Yulia Grishaeva[1,2(✉)] ⬤, Iosif Spirin[2] ⬤, Svetlana Morgunova[3] ⬤,
and Zhanna Yarullina[3] ⬤

[1] Moscow Pedagogical State University, MPGU, Malaya Pirogovskaya, 1 building 1,
119991 Moscow, Russian Federation
[2] SC Scientific and Research Institute of Motor Transport, Geroyev Panfilovtsev Str., 24,
125480 Moscow, Russian Federation
[3] National Research University "Moscow Power Engineering Institute",
Krasnokazarmennaya Str., 14, 111250 Moscow, Russia

Abstract. The article reveals the content of the concept "creativity", "creative activity", "engineering creativity" and "creative thinking". Currently, there is a problem of mismatch between the content of vocational training and the tasks of the labor market for the engineering industry. Graduates of technical universities experience difficulties associated with the lack of development of creative thinking, research skills that provide the productive potential of their future professional activities. The system of professional training in technical universities is often focused on the formation of a reproductive way of thinking and does not take into account the individual specifics of the cognitive abilities of students, thanks to which it is necessary to develop a special creative style of the future engineer. It is necessary for educators to use both universal general cultural and professional competencies of students in the educational process in order to achieve the goals of developing their creative thinking. The paper considers the relevant current educational technologies and the conditions for their implementation in the practice of training future engineers in the conditions of a technical university.

Keywords: Professional education · Educational technologies · Training of engineers · Creative thinking of a specialist

1 Introduction

The professional competencies of engineering graduates are usually limited to the technical and scientific knowledge used in the relevant industry. The creative development of the professional qualities of these graduates has significant gaps. Educational programs are focused on the theoretical study of materials. Practical skills that provide an opportunity for creativity in the profession are not sufficiently instilled. New programs tend to replicate this lack of technical education [1]. The study of a variety of educational programs has shown that the connection between the subject approach used by teachers of engineering disciplines and educational goals (the ability to creative and innovative

D. Bylieva and A. Nordmann (Eds.): PCSF 2021, LNNS 345, pp. 537–547, 2022.
https://doi.org/10.1007/978-3-030-89708-6_45

independent activity of students) remains problematic. For example, only 2% of electrical engineering programs focus on students' creative solutions to tasks [2]. In this regard, there is a need to justify, systematize and identify the most effective educational technologies for the development of creative thinking of future engineers-graduates of a technical university.

2 Materials and Methods

The research is based on theoretical analysis and comparison of classical and modern approaches to the problem under study. The following methods are used in order to execute the purpose of investigation:

- synthesis and generalization of the results obtained (for theoretical categories and concepts specification);
- abstraction for determination of essential aspects of the research subject;
- historical and logical methods (for disclosure of specific conditions and prerequisites for the ideal object embodiment in practice);
- reflection as a method of metatheoretical knowledge.

3 Theoretical Background, Review and Research Context

3.1 The Phenomenon of Creative Activity for the Engineering Sphere

Creativity is the level of creative talent, the ability to create, which is a relatively stable characteristic of the individual; the ability to make, implement something new: a new solution to a problem, a new method or tool, a new work of art [3].

Let's define the content of the concept of "creative activity". According to S. L. Rubinstein, "activity" is the process by which a person's attitude to the surrounding world, to other people, to the tasks that life puts before him is realized [4]. In turn, "creative activity" is the attitude of the subject of the activity to his work (satisfaction with work, the desire for independence in its performance; positive motivation in the course of its solution) and the process of solving creative problems (independent transfer of previously acquired knowledge, skills, ways of activity in a new situation, vision of the problem, vision of a new function of a known object). Creative activity is an activity in which creativity, as the dominant component, is included in the structure of either its goals or methods. Creative activity is the result and at the same time an important condition for the further development of the individual, the development of his creative potential [5].

The concept of creative activity is inextricably linked with the concept of "creative abilities", which is understood as a synthesis of the properties and characteristics of a person that characterize the degree of their compliance with the requirements of a certain type of educational and creative activity and determine the level of its effectiveness [6].

In the format of our research, the acmeological interpretation of the concept of culture is of interest, namely, culture as the appropriation by the individual, community, organization of the achievements of society, humanity; professional culture, denoting the ability to optimally implement their professional knowledge, skills, and skills in activities, professional self-determination and self-realization [7].

Of interest for our research is the definition of the culture of technical creative activity of the individual, formulated by Zagutin D. S. (2012) in his doctoral dissertation, where this concept is understood as an absolute that combines knowledge/creativity/innovation and creates conditions for the birth of material and spiritual values through the activity of the individual [8].

Consistently considering the concept of "culture of creative activity", we turn to the concept of "creativity". Creativity should be understood as the highest form of activity and independent activity of a person [9]. Creativity is an activity that results in the creation of new, original and more perfect material and spiritual values that have objective or (and) subjective significance [10]. So, "creativity" is an activity that generates something qualitatively new and distinguished by uniqueness, originality and socio-historical uniqueness. Creativity is specific to a person, because it always involves the creator – the subject of creative activity [11].

Thus, creativity is the process of creating a new subject of creative activity. It is necessary to consider separately the concept of "technical creativity", which is understood as a type of student activity, the result of which is a technical object that has signs of utility and subjective (for students) novelty [12].

The concept of "engineering creativity" is synonymous with the concept of "technical creativity". The social significance of engineering creativity is primarily due to the needs of society for innovative technical solutions that ensure the development of scientific and technological progress in its cultural (socio-humanitarian, value) significance for human society [13].

Thus, the culture of creative activity of the future engineer is a socially significant characteristic of the individual, which determines the presence of the potential for creative activity, creativity, and developed creative abilities to achieve the projected results of scientific and technological development in the interests of technosphere security.

3.2 Effective Methodical Conditions for the Formation of Creative Thinking of the Future Engineer

Research results of some authors [14, 15], pay attention to the atmosphere and creative climate during the educational process. The experiments of these researchers were focused on practice and conditions close to real production. The involvement of students in the experimental process for two semesters, as well as the application of SCAMPER's non-standard thinking practices, showed a significant increase in students' creativity in terms of cognition, motivation and personal qualities. Researchers [16, 17] also believe that when performing training tasks for the development of creative abilities, it is important for teachers to focus on the process of creating a new product and on its originality, rather than on the functional characteristics of the final result. The emphasis on encouraging risky, original and unconventional approaches in the design process leads to an increase in the creative component. Other authors [18] emphasize the importance of working on practical skills in the classroom. The method of "inverted learning" is described as effective in developing the creativity of students of technical specialties – the theoretical part is mainly mastered independently; practical tasks are solved in the classroom.

Our experience of working with undergraduate and graduate students speaks of the high efficiency of conducting master classes. Leading scientists and specialists in the relevant fields of knowledge were invited to conduct master classes. A comparative assessment of the effectiveness of traditional training with training in combination with master classes showed an increase of 30–40% in the appearance of creative solutions in the project tasks performed by students.

Also, the increase in practice hours in the curriculum, close cooperation with enterprises and the involvement of students in the production process is discussed in the works of many researchers [19, 20]. According to the results of their research, the "3+1" curriculum leads to a significant increase in innovative engineering solutions for students. In this program, the first three years are devoted to theoretical foundations, and one year is devoted to practical skills in real production.

Cognitive and metacognitive pedagogical approaches are used to improve the creative skills of future technical specialists. Test results [21] show an increase in such indicators as flexibility of thinking, freedom of action, originality and effectiveness after applying these approaches.

Many authors [22–24] substantiate the effectiveness of the application of CSIT theory in training in the development of creative abilities in solving technical problems.

Interactive museums are a type of experimental sites described by some researchers [25] as a way to develop creative thinking in the training of technical specialists. Museums that have numerous technical exhibits on display can successfully serve as cognitive creative platforms for the connotation of technical solutions, conducting field experiments, laboratory work, and practical classes with students. At the same time, such museums can serve as a base for conducting professional development classes for teachers and specialists [26]. The Museo Nazionale Scienza e Tecnologia Leonardo da Vinci (Ital.) in Milan, which has 15.000 exhibits (from models of the inventions of the great engineer, after whom the museum is named, to a submarine), 50.000 videos and photographic materials, and 40.000 books, can be cited as examples of active use of the possibilities of museums. Each section of the museum is equipped with laboratories where students develop creatively, study technology, conduct experiments in the field of genetics, physics, chemistry, robotics and other fields of science and technology [https://italyme.ru/museo-della-scienza-e-della-tecnologia/]. According to a similar concept, which provides for the possibility of conducting scientific discussions, creative experiments and practical classes, the work of the Polytechnic Museum in Moscow, which should be reconstructed in 2021 will also be organized.

In addition, according to the research results [27] there are difficulties of a subjective nature. Teachers' personal beliefs and beliefs about creativity and values directly influence the approach to teaching and developing students' creative abilities. The authors draw attention to the following four facts: the focus of teachers on the created design product, the education and training of teachers, the subjective nature of creativity and their ideas about it, as well as the way of thinking of teachers. The article [28] discusses the systematization of knowledge in this issue and the consistency in pedagogical approaches. The authors write about the importance of appropriate development of the competencies of educators of engineering specialties and the need to introduce the course "Development of technical creativity" for educators.

Some scientific papers [29] talk about the importance of including graphic experience in the engineering learning process. According to the measurement results, this increases the creativity of students. Drawing is a basic element of the professional language of engineers and one of the most important results of their creativity. Also, the graphic means of the engineer to express his thoughts are sketches, photographs, animation and images of industrial design. All these things develop graphic thinking. Here it is appropriate to draw an analogy with artists and sculptors, the result of whose work is the works of painting and three-dimensional works of art. The skills of a future engineer in the use and creation of graphic materials are crucial in the implementation of creative ideas.

It is statistically confirmed that seminars on graphic creativity of engineering students significantly affect the development of their creative potential [30]. An experiment with the introduction of seminars on 3D modeling in CAD was conducted by a group of researchers [31]. The transition from flat graphic design to three-dimensional models in various graphics packages has contributed to a significant increase in the creative approach to solving engineering problems.

Other studies [32] confirm the degree of importance of graphic subjects, such as "engineering graphics" and "graphic design" for the development of creative potential. Graphic representation of objects, freehand drawing, sketches contribute to the development of spatial imagination and visual thinking.

In addition, the style of leisure and extracurricular activities of students affect their creativity in solving engineering problems. So, as a result of research [33, 34] it was found that with artistic experience, students are more creative in solving technical problems and tasks. Such students are distinguished by a developed intuition, understanding of themselves, a desire to challenge their capabilities and a less negative perception of the difficulties encountered. The intersection of science with musical and artistic aesthetics is described in [35]. Their research confirms an increase in the creative abilities of students of technical universities after working with interactive musical and artistic works.

According to many researchers, the development of creativity is hindered by low personal motivation and interest. To solve this problem, the authors [36–38] suggest that the learning process should include play (game-related) activities, group collaboration, and case studies – competition between groups of students encourages personal motivation and creative expression. At the same time, there are opinions [39, 40], that it is the individual approach that develops independence and initiative, which in turn stimulates the innovative abilities of future engineers.

Simulation models of complex production situations stimulate students' creative thinking in solving production management problems, innovative design (for example, for the development of railway infrastructure), and land use rationalization. These and other issues were discussed at the symposium of specialists and educators [41].

The technology of teaching students using a game approach is of interest. A comparison of two methods of simulation training of engineering students in supply chain management: a) individual and b) team showed that it is statistically justified to obtain the best results when these methods are combined. At the beginning of working with a computer model, it is better to use an individual method. Then, as students master the basic methods of solving problems, they need to move on to the team method of solving problems [42]. Computer games provide students with theoretical and practical competencies. The game model of teaching safe design of structures according to the assessment of students and teachers of construction specialties turned out to be useful and attractive. It is confirmed that playing the game led to an increase in subjects' declarative knowledge by 22% and procedural knowledge by 37% [43, 44].

The business game "Passenger Transportation Quality Management" was developed and is used by us for training students of transport specialties. The assessment of the quality of various elements of transport services is performed on a computer or manually based on the task received by students using the Harrington desirability function [45]. After evaluating these elements, students perform the creative part of the game, which consists of choosing a set of measures to improve transportation, ensuring a rational combination of positive changes in quality indicators, the costs of their implementation and the timing of possible project implementation. The game lasts for one academic hour (45 min), runs groups of 2–3 students and ends with an analysis of the results obtained. Comparative pedagogical measurement of the depth of the competencies acquired by students (correctness and validity of answers to questions), compared with lectures, showed an increase in knowledge and skills by about 40–70%. At the same time, it was noted that the students creatively and independently justified the selected activities.

3.3   Educational and Professional Communication in the Development of Creative Thinking of the Future Engineer

Technical support of the 21st century society cannot function without engineering work. When developing objects engineers must possess not only professional technical knowledge, but also take into account many other factors, such as: the purpose of the object, its efficiency, cost, reliability and safety. However, the fact that engineering activity is a creative process remains unchanged. An engineer (from Latin *ingeniare* – "to create") is a creator of new technology, an outside-the-box thinking inventor, designer and technologist. Thus, we can consider the work of an engineer, as creative. John R. Dixon in his work Design Engineering: Inventiveness, Analysis, and Decision Making described the most important qualities required for an engineer. And the first necessary quality that the author highlights is ingenuity. This is a quality that helps to develop new ideas and unconventional, creative thinking [46].

According to Dan, L., Fei, G the modern rapidly developing society «needs citizens not only with rich knowledge, but also with creative ability. The importance of creative thinking cultivation has been widely recognized in various disciplines of education, especially in the field of teaching English as a foreign language (EFL)» [47].

Various digital technologies are used to develop the creative potential and creative thinking of future engineers in the process of teaching a foreign language. Students are actively engaged in practical learning by creative and motivational activities [48]. One of these activities is the creation of an online project, which develops not only the independent creative thinking of students, but also professional and communication skills as it is necessary to present your project to the colleagues.

The technology of teaching using the project method includes the following stages: collecting information, gaining knowledge, discussing solutions, working together to create an online project, presenting the project for public discussion [49].

A study by Kato, F et al., on the creation of an online project and conducted jointly on the basis of institutes in America and Japan, shows that the number of students who participated in the project increased the length of utterance in their target language, which is the main indicator of improving the communication skills of students [50].

The use of mobile learning with gamification and augmented reality also contributes to the development of creative thinking, engagement and interest in learning. The study was carried out by Durao, N. et al. at higher education institutions in Southern Europe, South America and Asia. This study was based on a study of the views of professors of higher education institutions of the continents and found that, according to them, students will be more interested in learning if they use mobile devices [51].

Another study by Jahnke, I., Liebscher, J. found that developing courses using mobile devices contribute to the development of students' creative abilities, which means that learning through mobile devices encourages students' creativity or fosters a creative learning environment [52].

Thus, the use of digital technologies in the learning process develops not only the communication skills of students, but also their creative thinking. Digital technologies, applied methodically competently, increase the cognitive activity of students, which undoubtedly leads to an increase in the effectiveness of training.

4 Conclusions

The development of students' engineering creativity is determined by the combination of various pedagogical approaches combined with innovative pedagogical technologies and the expansion of students' horizons beyond the narrowly professional orientation of educational goals.

The main pedagogical approach to training future engineers is based on the complex use of various "private" approaches studied by numerous researchers. In a real situation, it is impossible to imagine a situation where an engineer constantly approaches the problems he faces from the point of view of one academic subject during the working day. Therefore, an interdisciplinary approach to training is relevant. It is necessary to cross-use by all teachers the students' competencies obtained in the study of various disciplines, both general cultural and professional orientation. The problems of sustainable development of civilization actualize the acquisition of environmental, economic and social competencies by students.

Innovative pedagogical technologies are characterized by a wide use in the educational process of computer models and other IT, including the CSIT platform and modeling in CAD systems. The development of such models requires significant efforts of the pedagogical community. It is effective to use of new platforms for the development of students' creative skills (practical tasks in production conditions, interactive textbooks, and educational opportunities of technical museums, students' participation in science and culture festivals, etc.). In modern conditions, the requirements for mastering engineering graphics by students are increasing. It is well known that engineering graphics are the basis of the professional interface of engineering personnel in communicating with each other and successful production and scientific activities. Computer technologies have become the main means of graphical modeling of objects designed by engineers. These technologies multiply and rediscover the creative potential of the modern engineer. Effective for the development of students' creative engineering skills is the holding of master classes, in which leading scientists and specialists should be involved. Master classes are now widely practiced in the development of creative specialties. A similar method of mentoring is used in graduate school. It is advisable to transfer this experience to the pedagogical practice of teaching students in engineering specialties. It is necessary to recall the "well-forgotten old": the main goal of educational activity in higher education is the comprehensive development of the personality of students, and not to give them only a limited set of narrowly professional competencies (the latter is typical for obtaining working professions). It is proved that the introduction of students to various cultural values significantly develops creative abilities in their future professional activities.

References

1. Martin-Erro, A., Espinosa, M.M., Dominguez, M.: Creativity in the formative curriculum of our industrial engineers. In: 11th International Conference of Education, Research and Innovation, Book Series: ICERI Proceedings, pp. 8926–8933. IATED, Seville (2018). https://doi.org/10.21125/iceri.2018.0645
2. Valentine, A., Belski, I., Hamilton, M., Adams, S.: Creativity in electrical engineering degree programs: where is the content? IEEE Trans. Educ. **62**(4), 288–296 (2019). https://doi.org/10.1109/TE.2019.2912834
3. Shashenkova, E.A.: Research Activities. Perspectiva, Moscow (2010). (in Russian)
4. Leontiev, A.N.: Activity. Consciousness. Personality. Politizdat, Moscow (1977). (in Russian)
5. Ryndak, V.G.: Creation. Brief pedagogical dictionary. Teaching aid. Pedagogical Bulletin, Moscow (2001). (in Russian)
6. Oleshkov, M.Yu., Uvarov, V.M.: Modern Educational Process: Basic Concepts and Terms: Short Terminological Dictionary. Sputnik + Company, Moscow (2006). (in Russian)
7. Derkach, A.A.: Acmeological Dictionary. RAGS, Moscow (2004). (in Russian)
8. Zagutin, D.S.: The culture of a person's technical creative activity: strategic directions, social technologies, a model of formation (Doctoral Dissertation). Krasnodar University of the Ministry of Internal Affairs of Russia, Krasnodar (2012). (in Russian)
9. Batyshev, S.Ya.: Encyclopedia of vocational education. Russian Academy of Education, Moscow (1998). (in Russian)
10. Voronin, A.S.: Glossary of terms on general and social pedagogy. GOU VPO USTU-UPI, Yekaterinburg (2006). (in Russian)
11. Klybin, A.Yu.: A set of methodological support for the educational discipline "Pedagogical technologies". VGIPA, N. Novgorod (2003). (in Russian)
12. Panov, V.G.: Russian Pedagogical Encyclopedia in 2 volumes. Big Ros. Encyl., Moscow (1993). (in Russian)
13. Sazonova, Z.S., Grishaeva, Y.: On the problem of forming the creative thinking of the future engineer. Primo Aspectu **2**, 87–93 (2016). (in Russian)
14. Wu, T.-T., Wu, Y.-T.: Applying project-based learning and SCAMPER teaching strategies in engineering education to explore the influence of creativity on cognition, personal motivation, and personality traits. Thinking Skills Creativity **35**, 100631 (2020). https://doi.org/10.1016/j.tsc.2020.100631

15. Liu, W.-S., Wu, Y.-T., Wu, T.-T.: The study of creativity, creativity style, creativity climate applying creativity learning strategies - an example of engineering education. In: Wu, T.-T., Huang, Y.-M., Shadieva, R., Lin, L., Starčič, A.I. (eds.) ICITL 2018. LNCS, vol. 11003, pp. 490–499. Springer, Cham (2018). https://doi.org/10.1007/978-3-319-99737-7_52
16. Tekmen-Araci, Y., Kuys, B.: The impact of excessive focus on performance during engineering design process on creativity. Int. J. Eng. Educ. 35(6), 1618–1629 (2019)
17. Tekmen-Araci, Y.: Teaching risk-taking to engineering design students' needs risk-taking. Art Des. Commun. High. Educ. 18(1), 67–79 (2019). https://doi.org/10.1386/adch.18.1.67_1
18. Liu, X., Wang, L., Cao, J., Xia, X.: Flipped training-a new teaching model for engineering training. In: Advances in Social Science Education and Humanities Research, vol. 72, pp. 310–313 (2017). https://doi.org/10.2991/icmess-17.2017.74
19. Yao, C.L.: Research on Automobile Service Engineering specialty based on talents training of excellent engineers. In: International Conference on Arts, Management, Education and Innovation, pp. 599–603. Clausius Scientific Press, Taiyuan (2019). https://doi.org/10.23977/icamei.2019.117
20. Hu, Z., Sha, L., Zhang, X., Chen, H.: Research on training mode for excellent engineer plan in Light Chemical Engineering specialty. In: Advances in Social Science, Education and Humanities Research, vol. 182, pp. 218–221 (2018). https://doi.org/10.2991/iceemr-18.2018.47
21. Morin, S., Robert, J.M., Gabora, L.: How to train future engineers to be more creative? An educative experience. Thinking Skills Creativity 28, 150–166 (2018). https://doi.org/10.1016/j.tsc.2018.05.003
22. Zhang, X., Hu, H.: Research on cultivation mode of undergraduates' creativity based on technology literacy - taking the technology education base of Hubei University of Arts and Science as an example. In: Advances in Social Science, Education and Humanities Research, vol. 213, pp. 383–388 (2018). https://doi.org/10.2991/ichssr-18.2018.73
23. Lu, J., Xue, X.: Training mode of innovative talents of civil engineering education based on TRIZ theory in China. Eurasia J. Math. Sci. Tech. Educ. 13(7), 4301–4309 (2017). https://doi.org/10.12973/eurasia.2017.00835a
24. Fiorineschi, L., Frillici, F.S., Rotini, F.: Enhancing functional decomposition and morphology with TRIZ. Lit. Rev. Comput. Ind. 94, 1–15 (2018). https://doi.org/10.1016/j.compind.2017.09.004
25. Qian, P.: The application of Engineering Cognitive Museum in engineering training teaching. In: Advances in Social Science, Education and Humanities Research, vol. 181, pp. 352–355 (2018). https://doi.org/10.2991/icsshe-18.2018.86
26. Zhang, H., Hu, Q., He, L.: Construction and implementation of teaching quality evaluation and performance assessment system on engineering training. In: Advances in Social Science, Education and Humanities Research, vol. 336, pp. 59–61 (2019)
27. Tekmen-Araci, Y., Mann, L.: Instructor approaches to creativity in engineering design education. Proc. Inst. Mech. Eng. Part C-J. Mech. Eng. Sci. 233(2), 395–402 (2019). https://doi.org/10.1177/0954406218758795
28. Byvalkevych, L., Yefremova, O., Hryshchenko, S.: Developing technical creativity in future engineering educators. Revista Romaneasca Pentru Educatie Multidimensionala 12(1), 162–175 (2020). https://doi.org/10.18662/rrem/206
29. Melian, D., Saorin, J., De La Torre-Cantero, J., Lopez-Chao, V.: Analysis of the factorial structure of graphic creativity of engineering students through digital manufacturing techniques. Int. J. Eng. Educ. 4(SI), 1151–1160 (2020)
30. Drange, T., Irons, A., Drange, K.: Creativity in the digital forensics curriculum. In: Proceedings of the 9th International Conference on Compute Supported Education, vol. 2, pp. 103–108 (2017). https://doi.org/10.5220/0006294101030108

31. Carbonell-Carrera, C., Saorin, J., Melian-Diaz, D., de la Torre-Cantero, J.: Enhancing creative thinking in STEM with 3D CAD Modelling. Sustainability 11(21), 6036 (2019). https://doi.org/10.3390/su11216036

32. Martin Erro, A., Espinosa Escudero, M.M., Dominguez Somonte, M.: Spanish engineering graphic expression subjects and its relation to creativity competence. In: 10th International Conference of Education, Research and Innovation. ICERI Proceedings, Seville, Spain, pp. 2292–2296. IATED Academy (2017). https://doi.org/10.21125/iceri.2017.0662

33. Jongho, S., Eune, J., Cho, E., Yeon, J.S., Lee, B.: The characteristics of artistic experience affecting on engineering creativity. J. Creativity Educ. 17(1), 1–25 (2017)

34. Elisondo, R.C., Chiecher, A.C., Paoloni, P.V.R.: Creativity, leisure and academic performance in engineering students. 7, 28–42 (2018)

35. Chang, H.-Y., Lin, H.-C., Wu, T.-T., Huang, Y.-M.: The influence of interactive art of visual music on the creativity of science and engineering students. In: Proceeding of 2019 IEEE Global Engineering Education Conference (EDUCON), Dubai, UAE, pp. 1087–1092. Institute of Electrical and Electronics Engineers (IEEE) (2019). https://doi.org/10.1109/educon.2019.8725233

36. Gomes, M.M., de Sousa Pereira-Guizzo, C.: Intervention for the development of the creativity of engineering students. Revista De Ciencias Humanas Da Universidade De Taubate 12(3), 80–87 (2019). https://doi.org/10.32813/2179-1120.2019.v12.n3.a484

37. Lu, Y.: Research on the cultivation of college students' engineering ability based on innovative training program. In: Advances in Social Science, Education and Humanities Research, vol. 83, pp. 205–209 (2017)

38. Mu, H., Wang, H., Liu, G.: Design and practice of integrated engineering training project based on CDIO mode. In: Advances in Social Science, Education and Humanities Research, vol. 61, pp. 589–594 (2017). https://doi.org/10.2991/isss-17.2017.87

39. Gong, Y., Mei, D.: Research on individualized training model of college students' engineering innovation ability. In: Advances in Social Science, Education and Humanities Research, vol. 61, pp. 86–90 (2017). https://doi.org/10.2991/isss-17.2017.16

40. Noga, H., Garbarz-Glos, B., Pytel, K., et al.: Conditions of technical creativity at various stages of education. In: Scientific Conference on Society, Integration, Education, vol. 4, pp. 109–120 (2017). https://doi.org/10.17770/sie2017vol4.2278

41. Bekebrede, G., Lo, J., Lukosch, H.: Understanding complexity. The use of simulation games for engineering systems. Simul. Gaming 46(5), 447–454 (2015). https://doi.org/10.1177/1046878115618140

42. Tzimerman, A., Herer, Y.T., Shtub, A.: Teaching supply chain management to industrial engineering students: mixed vs. pure approaches in simulation based training. Int. J. Eng. Educ. 31(6), 1688–1700 (2015)

43. Dib, H., Adamo-Villani, N.: Serious sustainability challenge game to promote teaching and learning of building sustainability. J. Comput. Civil Eng. 28(5), SI, A4014007 (2014). https://doi.org/10.1061/(ASCE)CP.1943-5487.0000357

44. Din, Z.U., Gibson, G.E., Jr.: Serious games for learning prevention through design concepts. An experimental study. Saf. Sci. 115, 176–187 (2019). https://doi.org/10.1016/j.ssci.2019.02.005

45. Harrington, E.C.: The desirable function. Ind. Qual. Control 21(10), 494–498 (1965)

46. Dixon, J.R.: Design Engineering: Inventiveness, Analysis, and Decision Making. McGraw-Hill (1966)

47. Dan, L., Fei, G.: Creative thinking cultivation in EFL teaching in Junior high school in English. In: Cox, L.H., Fan, Q., Sun, L., Zhang, J. (eds.) Proceedings of the 8th Northeast Asia International Symposium on Language, Literature and Translation, Dalian, China, pp. 611–615. American Scholars Press (2019)

48. Gonzalez-Vera, P.: New technologies in the ESP class for mechanical engineers. Argenti. J. Appl. Ling. **7**(2), 52–71 (2019)
49. Lu, Q.: A new project-based learning in English writing. Int. J. Emerg. Technol. Learn. **16**(5), 214–227 (2021). https://doi.org/10.3991/ijet.v16i05.21271
50. Kato, F., Spring, R., Mori, C.: Incorporating project-based language learning into distance learning: creating a homepage during computer-mediated learning sessions. Lang. Teach. Res. (2020). https://doi.org/10.1177/1362168820954454
51. Durao, N., Moreira, F., Ferreira, M.J., Pereira, C.S., Annamalai, N.: The state of mobile learning supported by gamification and augmented reality in higher education institutions across three continents. Revista edapeci-educacao a distancia e praticas educativas comunicacionais e interculturais **20**(1), 130–147 (2020). https://doi.org/10.29276/redapeci.2020.20.112211. 130-147
52. Jahnke, I., Liebscher, J.: Three types of integrated course designs for using mobile technologies to support creativity in higher education. Comput. Educ. **146** (2020). https://doi.org/10.1016/j.compedu.2019.103782

Cultivating Creativity of Technically-Minded Students

Elena V. Carter[ID] and Sofia A. Pushmina[(✉)][ID]

Saint Petersburg Mining University, Saint Petersburg 199106, Russia

Abstract. Creativity is coming to the forefront in various spheres of modern life, with technically-relevant educational environment being no exception. Given the fact that it is a rather ambiguous concept, the creative development of engineering students in the course of English for Specific Purposes (ESP) allows fostering professional competence supplemented with efficiency and productivity of a generation Z student. The paper seeks to discuss the integration of creative-imposed methodology in an ESP course. The research provides a pilot three-componential model aimed at cultivating intrinsic motivation of technically-minded students stemming from comfortable and creative educational environment that simulates true-to-life situations in the English language course. To reach the aim, firstly, the relevant career-related material was compiled. Secondly, students were divided in a baseline group and an experimental one with latter undergoing additional creativity-related training. Finally, the authors assessed students' performance and determined the correlation between creativity and productivity of future petroleum specialists. The findings reflecting the benefits of creativity in the ESP course are highlighted.

Keywords: Creativity · ESP · Generation Z · Productivity coefficient · Translation

1 Introduction

Introducing creativity-supporting environment for students of petroleum engineering requires some understanding of the creativity process and its impact on cognition, motivation and performance of future specialists. Modern world requires "a brand-new approach" [1] (p. 275) to cultivating a successful personality. Being a rather ambiguous concept, creativity has led its way to the educational system and now is being delivered not only through art and cultural theories, but also through developing novel and effective ideas, artifacts, or solutions in other domains [2]. Creativity is in great demand in numerous fields such as science, technology, economy, and politics [3]. The debatable area has been flourishing over the recent decade, providing increasing amount of contributions to the concept definition [4, 5], creative cognition [6], technology-enhanced creativity [7], the mechanisms of creative activity [8], giving rise to myths around the term [9].

The definition of creativity is already a highly debatable area, with some scientists advocating it being 'the competency of inventing", while others suggesting the idea of "divergent thinking", and "exercise to imagination" [10, 11]. To immerse into a creative process, a student should have a major motivation, as the key factors are individuals' interest for work, arousing expectations and the pleasure of the result [12].

As the capacity for innovation is considered essential for many industries to survive in a fast-changing world influenced by world lockdowns, imposed cross-country distance education and work, political decisions, universities have to prepare future specialists who are trained to develop "independence and critical thinking" [13] (p. 769) and ready to react to the challenges with creative solutions.

Some scholars "treat creativity as a single, relatively homogeneous phenomenon" [14] (p. 639). However, there is a view that "people are creative when they can solve problems, create products, or raise issues in a domain in a way that is initially novel but is eventually accepted in one or more cultural settings" [15], (p.116). This assumption seems to endorse the idea that creativity can take various forms in different domains. Thus, it may be conceptually and practically essential to learn about peculiarities of creativity among engineers in comparison with the representatives of other professions.

There was an attempt to understand commonalities and dissimilarities regarding artistic versus scientific creativity by focusing on personality features [14, 16–18]. "Certain personality traits consistently covary with creativity, yet there are some domain specificities" [17] (p. 289). The defining social and nonsocial characteristics in art and science were identified by Feist [17]. Social features for the artist included nonconformity, norm doubting, hostility, independence, unfriendliness, aloofness, and lack of warmth. Dominance, arrogance, hostility, and high self-confidence were named as social traits of the scientific creative personality. It is noteworthy "aesthetic taste and a lack of conventionality are common characteristics of creativity in both domains" [16] (p. 214).

According to Weisberg [19], the basis for creativity is problem solving. Though this problem-solving process is similar in artistic and scientific areas, it may take different forms. MacKinnon [20] maintained that the creator in scientific creativity acts as a mediator between designated needs and goals. Particularly important in this regard is Larson, Thomas, and Leviness' [21] observation concerning creativity in engineering domain. They claimed "a distinguishing feature is that the engineer has an eye on function and utility" [21] (p. 2).

While creative thinking by definition individual, the English for Specific Purposes (ESP) instructor cannot predict the outcome of the educational process, however, should aim at language improvement acquiring the skill of quick and creative decision-making in a situation that imitates the real-life working process of a specialist. We should mention here the importance of "mobile-based apps that help implement authentic and up-to-date data," stimulating interest and increasing motivation" [22] (p. 476).

A project-based learning strategy is one of the teaching methods that integrates assignments that intend to develop critical thinking, problem-solving competency, metacognition. Students of the petroleum-engineering department explore and analyze real-life experience task, learn to apply and organize knowledge through data collection, discussion, translation practices, improving cognitive and critical thinking capability while performing a creative task.

People who are skilled in practical, industrial, or mechanical arts can be defined as technically-minded personalities. Features for technically-minded students include, first of all, a high degree of a technical skill: plenty of natural aptitude and a kind of knack for technology are required to bring together the knowledge learned through structured technology training and real-world experience. Among other essential characteristics are the understanding of technical processes, analytical thinking, the ability to thrive under pressure, and sharp problem-solving skills.

The importance of the development of creativity in technically-prone students could not be overestimated [23]. The increasing public demand for engineers who are able to provide creative solutions to career challenges may be explained with the necessity to "provide a proper response to rapidly changing conditions in modern society development" [24] (p. 1). The growing knowledge content in production and the engineer's responsibility for decision-making consequences impose new, heightened requirements to the quality of professional training [25]. In fact, there is a need of awareness-raising instruction on the appropriate and sufficient application of different methods and techniques to foster creativity of students at technical universities [26].

L.N. Kharchenko considers that practical forms (trainings, discussions, role-playing games) are "the most effective forms of education, according to 71.2% of students" [27] (p. 159). However, they should "be focusing on the demands of the market" [28] (p. 745), thus contributing to career proficiency of students. Thus, the interrelation between technical and creative domains is shown in Fig. 1.

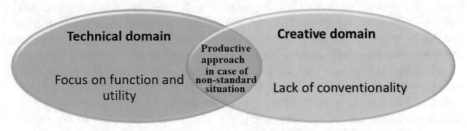

Fig. 1. Interrelation between technical and creative domains.

Therefore, the research lies on the hypothesis that the creative development of engineering students at English classes when they work with career-related assignments would be efficient if:

1. Learning comes from real-life processes and phenomena that might occur in careers.
2. A set of principles to build and implement the procedural guidelines for English instruction is updated. It includes the following: availability, visualized simulation, variability, focus on career and subject-information content.
3. The three-componential model for the development of technically-minded students' creativity is introduced.

Thus, this study deals with the elaboration of techniques for cultivating creativity of technically-minded students when they work with career-related assignments in English classroom.

2   Methods

There are various methodological approaches to creativity that consider a creative personality (traits, characteristics), creative process (cognition), creative products (patents, inventions, ideas, insights) and creative persuasion (historical impact) [29]. Given the fact that students have innate creative potential, an ESP instructor should provide such an environment so as to fulfill it into a creative performance producing a potential creative solution, consequently, product. "Science is the last and highest level of continuous cultural development" [30] (p. 621) that requires creative process implied, thus, the need to introduce creativity to the educational process has come to the front. However, the goal of an ESP instructor is not only to introduce creativity into a learning environment, but also to "harmoniously combine the technological and aesthetic components of professional activity" [31] (p. 111).

The study provides a componential pilot model of creativity implemented on an ESP course of future petroleum engineers.

The model is based on 3 main components to make it work:

1. Intrinsic motivation to complete the tasks and improve the language learning.
2. Creativity-relevant tasks.
3. Educational resources.

67 petroleum-engineering students of 20–21 years old (44 males and 23 females) from Saint Petersburg Mining University participated in the research. There were two groups: an experimental group and a baseline one. The experiment covered the second year of study (4th semester) and included the introduced cycle of six resource-based classes with the expanded set of career-related and creativity-relevant tasks. The students initially had intermediate level according to CEFR.

The researchers assumed field-related knowledge accumulation and integration during the period of two years of study in Russian. When the learners had accumulated theoretical knowledge in petroleum engineering, they were provided resource-based classes in English. At a resource-based class, the students worked individually, in pairs, and in small groups of 5–6 people. The learners were proposed to solve career-related assignments coming from reflection and communication, integration of knowledge in the chosen field with a particular focus on the development of their creativity.

The verification of the research included a test for both experimental and baseline groups. It was aimed at indicating the learners' academic progress and the level of their creativity that triggered better productivity. The main difficulties encountered by the participants while working on career-related and creativity-relevant assignments were defined and analyzed.

3   Results

In the course of a predetermined period, students' performance in ESP was under supervision with an attempt to explore the impact of creativity tasks on productivity and the improvement of language skills.

Based on theoretical and empirical analyses, the necessity to exploit tasks to develop creativity skills in ESP course is substantiated. However, we should adapt the model to a new generation of students.

Generation Z that was a predominant focus of our research combines numerous features that have to be taken into consideration while encouraging them to be creative in learning. Giving a closer look at these students, we may state that they are more technologically savvy, work at their own pace, require feedback from a course instructor and may get easily anxious while performing the task [32]. They have their unique worldview [33]. Being entrepreneur-driven and goal-oriented, they need to be surrounded by a learning environment that will help acquire the skill of creativity so as to succeed in their professional sphere.

ESP technical learning environment in petroleum education is a practice experience that encourages students comprehend and refine professional petroleum competencies in ESP.

One of the ways to define a technical learning environment is to look at it as an interactive network of tasks that may affect students' learning achievements. Generation Z expects personal development, increase for self-esteem, support, peer cohesion, affiliation and involvement. All of these may be achieved through creative learning process. S.A. Rassadina offers in this case "the trending concept of edutainment" [34] (p. 498).

As generation Z prefers practical, real-world learning experiences [35–38], an experimental group was offered a role play that the students found applicable to their role in petroleum industry.

The challenges that the learners faced were lack of skills of professional communication and skills of teamwork, not to mention translation adversities requiring processing, critical analysis ad encoding career-related data. To foster perseverance through creative tasks, the students were told that their work would be accounted to their performance assessment for the period.

To illustrate creativity as a factor that affects productivity not only of an individual, but also of a group as a whole, we developed a formula to evaluate students based on the results of the studied academic period. The formula takes into consideration that the highest score during the final test can be achieved, regardless of acquiring additional skills while completing tasks on creativity (Formula 1). The final test consists of 4 tasks with a maximum score equaling 40 points. The first three assignments for 20 points in total represent career-related tasks on the covered topic, the fourth task is creative, and it is evaluated in the following four aspects: content, vocabulary, grammar and structure with 5 points each. We have introduced a decreasing coefficient precisely in the interests of the study, as it is necessary to demonstrate that the students of the experimental group can get the highest score under all other equal conditions with students of a baseline group.

$$Final\ test\ score = A_1 + A_2 + A_3 + \left(B_{content} + B_{vocabulary} + B_{grammar} + B_{structure}\right) * k_{productivity}, \quad (1)$$

where

A_1, A_2, A_3– assignments to check the knowledge of the career-related topic "Drilling Rigs" with an integrated technical translation task;

$B_{content} + B_{vocabulary} + B_{grammar} + B_{structure}$ – an additional creativity-relevant assignment, where content, vocabulary, grammar and structure are assessed;
$k_{productivity}$ - the coefficient of correlation of creativity tasks imposed in a predetermined period on an experimental group and productivity in a career-related topic; the higher the coefficient value, the lower the productivity.

During the period in question, the students of the experimental group completed 3 tasks acquiring 1 point for each, or participated in a career-related role-play to get 3 points. Creativity-Relevant Assignments in the technical domain are the tasks when students have to find ways to repurpose existing equipment to remedy an issue, or they have to devise tactics to diagnose an ongoing issue. For example, fix a broken bit or a clogged pump or decide what can be done in case of power outrage or uncontrolled oil flow. Based on the points received during the predefined period, we may calculate the productivity coefficient, that is, the correlation of creative tasks and productivity. Thus, we subtracted 3 points from 100%. This coefficient in the final test for the course indicates how productive creative classes are for students. Table 1 shows how the productivity coefficient was achieved.

Table 1. Calculation of the productivity coefficient

Students	Points for creative tasks				$k_{productivity}$
	A_1	A_2	A_3	Final score	$(100 - final)/100$
Average score of students of an experimental group	1	1	1	3	0,97
Average score of students of a baseline group	0	0	0		1

The data given in Table 2 show the comparative assessment of the final test of both groups.

Table 2. Comparative assessment of students' language skills and the correlation between creativity and productivity

Students	Final test score							Total score
	Career-related ESP tasks			Creativity-related ESP task				
	A_1	A_2	A_3	$B_{content}$	$B_{vocabulary}$	$B_{grammar}$	$B_{structure}$	
Average score of students of an experimental group	9,9	4,9	4,9	5	5	5	5	38,4
Average score of students of a baseline group	9,0	4,5	4,5	4,4	3,9	4,2	4,1	34,6

The final test evaluation with the mentioned coefficient of productivity was aimed to prove the importance of creative task in the ESP course. The comparative analysis gave rise to the results shown in Table 2.

The findings show that the proposed methodology and materials achieve their objective and contemplate the correlation between creativity and productivity, which is of great importance for future specialists.

Given the distinctive feature of modern petroleum education, that is "global focus of training" [39] (p. 372), ESP instructors should address competencies [40, 41] and improve soft skills of students [42], what may be acquired through creativity-related assignments. Obviously, this requires more work not only on the part of a student, but an ESP instructor as well. However, we observe its advantages illustrated in Fig. 2.

The accurate study of the completed test showed that the students of both groups managed to complete career-related ESP tasks, involving assignments to fill-in the gaps in an ESP text, matching the words/phrases with their definitions and naming parts of the drilling rig on a provided picture. Although the vast majority of students decided to complete optional creativity-related task, the results varied. That may account for the fact that the students of the baseline group were not ready for divergent and creative thinking. The task involved providing a supplementary text to the video demonstrating the work on a drilling rig. The majority of students confirmed that the assignment was both compelling and challenging to accomplish.

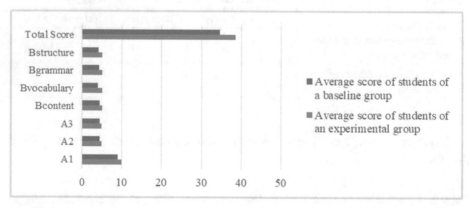

Fig. 2. Comparison of average final test results of both experimental and baseline ESP groups.

4   Conclusions and Recommendations

The study definitely showed the effectiveness of creativity in an ESP course. The research was carried out with the aim to foster creativity among technically-minded students and proved its feasibility. In order to rectify the lack of experience in creative problem-solving (that is idea recognition, idea selection, searching for an effective solution), the authors have chosen two groups to supervise: an experimental group and a baseline one.

The tasks given to the experimental group were in general alignment with a real-life situation in a petroleum sphere.

Assessment has been considered as a key criterion to illustrate the importance of creativity-related tasks. The coefficient of productivity was developed to show the correlation between future success/productivity of a petroleum specialist and creativity trained during educational process.

Another important factor for students to improve their language skills in ESP is to boost motivation given the fact that they belong to a self-reliant, stand-alone generation Z. Thus, the framework of the material provided should be aligned with the practical real-world earning experiences. Being proficient in digital technology, they still require support to master language skills and productivity.

When challenged to perform a creative task, the students demonstrated perseverance, thus proving that case-studies enhance motivation. Both groups were provided with a basic career-related material and showed excellent results in this part of the test. However, only the experimental group could apply what was learned in an ESP course in a simulated real-life role-play that assisted in bolstering professional confidence.

Considering the fact that creativity-related tasks are an effective supplementary method to provide career-immersion for technically-minded students, this area certainly demands further attention.

References

1. Mikheev, A.I.: Formation of a new type individuality under the conditions of generation of integration of interdependent communities. J. Min. Inst. **187**, 275–278 (2010)
2. Runco, M.A.: Creativity: Theories and Themes: Research, Development, and Practice. 2nd edn. Academic Press, New York (2014). https://doi.org/10.1016/C2012-0-06920-7
3. Akyıldız, S.T., Çelik, V.: Thinking outside the box: Turkish EFL teachers' perceptions of creativity. Thinking Skills Creativity **36**, 1–14 (2020). https://doi.org/10.1016/j.tsc.2020.100649
4. Harrington, D.M.: On the usefulness of "value" in the definition of creativity: a commentary. Creat. Res. J. **30**, 118–121 (2018). https://doi.org/10.1080/10400419.2018.1411432
5. Weisberg, R.W.: Response to Harrington on the definition of creativity. Creat. Res. J. **30**, 461–465 (2018). https://doi.org/10.1080/10400419.2018.1537386
6. LeBoutillier, N., Barry, R.: Psychological mindedness, personality and creative cognition. Creat. Res. J. **30**, 1–29 (2018). https://doi.org/10.1080/10400419.2018.1411440
7. Bereczki, E.O., Kárpáti, A.: Technology-enhanced creativity: a multiple case study of digital technology-integration expert teachers' beliefs and practices. Thinking Skills Creativity **39**, 1–27 (2021). https://doi.org/10.1016/j.tsc.2021.100791
8. Zobnina, L.Y.: Psychology of creativity. J. Min. Inst. **187**, 182–186 (2010)
9. Burkus, D.: The Myths of Creativity: The Truth About How Innovative Companies and People Generate Great Ideas. Jossey-Bass, San Francisco (2014)
10. Baer, J., McKool, S.S.: Assessing creativity using the consensual assessment technique. In: Schreiner, Ch.S. (ed.) Handbook of Research on Assessment Technologies, Methods, and Applications in Higher Education, pp. 65–77 (2009). https://doi.org/10.4018/978-1-60566-667-9.ch004
11. Torrance, E.P.: Education and the Creative Potential. Modern School Practices Series. The University of Minnesota Press, Minneapolis (1963)

12. Henle, M.: Some effects of motivational processes on cognition. Psych. Rev. **62**(6), 423–432 (1955). https://doi.org/10.1037/h0042646
13. Pukshansky, B.Ya.: On the role of enlightenment in the modern education. J. Min. Inst. **221**, 766–772 (2016). https://doi.org/10.18454/pmi.2016.5.766
14. Simonton, D.K.: Creativity and genius. In: Pervin, L.A., John, O.P. (eds.) Handbook of Personality: Theory and Research, pp. 655–676. Guilford, New York (1999)
15. Gardner, H.: Intelligence Reframed: Multiple Intelligence for the 21st Century. Basic Books, New York (1999)
16. Charyton, C., Snelbecker, G.E.: General, artistic and scientific creativity attributes of engineering and music students. Creativity Res. J. **19**(2), 213–225 (2007)
17. Feist, G.J.: The influence of personality on artistic and scientific creativity. In: Sternberg, R.J. (ed.) Handbook of Creativity, pp. 273–296. Cambridge University Press, Cambridge (1999)
18. Sternberg, R.J.: Intelligence, wisdom and creativity: three is better than one. Educ. Psychol. **21**, 175–190 (1986)
19. Weisberg, R.W.: Case studies of creative thinking: reproduction versus reconstructing in the real world. In: Smith, S.M., Ward, T.B., Finke, R.A. (eds.) The creative Cognition Approach, pp. 53–72. MIT Press, Cambridge (1995)
20. MacKinnon, D.W.: The nature and nurture of creative talent. Am. Psychol. **17**, 484–495 (1962)
21. Larson, M.C., Thomas, B., Leviness, P.O.: Assessing the creativity of engineers. Des. Eng. Div.: Successes Eng. Des. Educ. Des. Eng. **102**, 1–6 (1999)
22. Pushmina, S.: Teaching EMI and ESP in Instagram. In: Anikina, Z. (ed.) IEEHGIP 2020. LNNS, vol. 131, pp. 475–482. Springer, Cham (2020). https://doi.org/10.1007/978-3-030-47415-7_50
23. Lesher, O.V., Sarapulova, A.V.: The development of communicative creativity of students of a technical university in the process of intercultural communication (on the example of the discipline "Foreign Language"). FSBEI HE "MGTU named after G.I. Nosov", Magnitogorsk (2017)
24. Zubova, E.A.: Developing engineering student creativity in Mathematics classes at technical university. Int. Trans. J. Eng. Manag. Appl. Sci. Technol. **11**, 1–9 (2020). https://doi.org/10.14456/ITJEMAST.2020.261
25. Maklakova, N.V., Khovanskaya, E.S., Senchenkova, E.A.: Role and place of life-long learning in the university education system. Int. Trans. J. Eng. Manag. Appl. Sci. Technol. **10**(16), 1–8 (2019). https://doi.org/10.14456/ITJEMAST.2019.214
26. Alcalde, P., Nagel, J.: Does active learning improve student performance? A randomized experiment in a Chilean university. J. Eurasian Soc. Dialog. **1**(2), 1–11 (2016)
27. Kharchenko, L.N.: Efficiency of innovative and traditional forms of teaching through the eyes of students. J. Min. Inst. **193**, 159–160 (2011)
28. Sazonova, N.N.: Challenges and prospects of management of quality of educational services in Russian technical university. Adv. Soc. Sci. Educ. Hum. Res. **333**, 744–747 (2019). https://doi.org/10.2991/hssnpp-19.2019.142
29. Runco, M.A., Kim, D.: The four Ps of creativity: person, product, process, and press. In: Pritzker, S. (ed.) Encyclopedia of Creativity, 3rd edn., pp. 516–520. Academic Press, New York (2018). https://doi.org/10.1016/B978-0-12-809324-5.06193-9
30. Valiev, N.G., Shorin, A.G.: Pedagogical experiment of the first rector of the Ural State Mining Institute P.P. Von Weymarn as an effort to reform the higher education institution in 1917–1920. J. Min. Inst. **228**, 616–623 (2017). https://doi.org/10.25515/PMI.2017.6.616
31. Grakhov, V.P., Gislyakova, Y., Simakova, U.F.: Formation and development of students' creative potential at a technical university. J. Min. Inst. **213**, 110–115 (2015)
32. DiMattio, M.J.K., Hudacek, S.S.: Educating Generation Z: psychosocial dimensions of the clinical learning environment that predict student satisfaction. Nurse Educ. Pract. **49**, 1–28 (2020). https://doi.org/10.1016/j.nepr.2020.102901

33. Seibert, S.A.: Problem-based learning: a strategy to foster Generation Z's critical thinking and perseverance. Teach. Learn. Nurs. **16**(1), 85–88 (2020). https://doi.org/10.1016/j.teln.2020.09.002

34. Rassadina, S.A.: Culturological elements of edutainment employed in basic cultural formation of non-humanities students. J. Min. Inst. **219**, 498–503 (2016). https://doi.org/10.18454/PMI.2016.3.498

35. Chicca, J., Shellenbarger, T.: Connecting with Generation Z: approaches in nursing education. Teach. Learn. Nurs. **13**(3), 180–184 (2018). https://doi.org/10.1016/j.teln.2018.03.008

36. Hampton, D., Keys, Y.: Generation Z students: will they change our nursing classrooms? J. Nurs. Educ. Pract. **7**(4), 111–115 (2017). https://doi.org/10.5430/jnep.v7n4p111

37. Seemiller, C., Grace, M.: Generation Z: educating and engaging the next generation of students. About Campus **22**(3), 21–26 (2017). https://doi.org/10.1002/abc.21293

38. Schmitt, C.A., Lancaster, R.J.: Readiness to practice in Generation Z nursing students. J. Nurs. Educ. **58**(10), 604–606 (2019). https://doi.org/10.3928/01484834-20190923-09

39. Kazanin, O.I., Drebenstedt, C.: Mining education in the 21st century: global challenges and prospects. J. Min. Inst. **225**, 369–375 (2017). https://doi.org/10.18454/pmi.2017.3.369

40. Barinova, D., Ipatov, O., Odinokaya, M., Zhigadlo, V.: Pedagogical assessment of general professional competencies of technical engineers training. In: Katalinic, B. (ed.) Proceedings of the 30th DAAAM International Symposium, Vienna, Austria, pp. 0508–0512. DAAAM International (2019). https://www.daaam.info/Downloads/Pdfs/proceedings/proceedings_2019/068.pdf

41. Baranova, T., Kobicheva, A., Olkhovik, N., Tokareva, E.: Analysis of the communication competence dynamics in integrated learning. In: Anikina, Z. (ed.) IEEHGIP 2020. LNNS, vol. 131, pp. 425–438. Springer, Cham (2020). https://doi.org/10.1007/978-3-030-47415-7_45

42. Valieva, F.: Soft skills vs professional burnout: the case of technical universities. In: Anikina, Z. (ed.) IEEHGIP 2020. LNNS, vol. 131, pp. 719–726. Springer, Cham (2020). https://doi.org/10.1007/978-3-030-47415-7_76

Creative Projects as a Link Between Theory and Practice

Liudmila V. Daineko(✉) ⓘ, Inna I. Yurasovaⓘ, and Natalia M. Karavaevaⓘ

Ural Federal University named after the first President of Russia B.N. Yeltsin, 19 Mira St., 620002 Yekaterinburg, Russia
l.v.daineko@urfu.ru

Abstract. The Russian national innovation system is based on the creativity of the economy, the value of which is expressed in new ideas, not traditional resources. Innovative ideas change the economy as a whole, transforming the approach to production, trade, and the provision of services, while preserving the local identity of the territory in the era of globalization. Educational institutions that train intellectual and creative personnel for the new economy play a special role in the creative economy. The article describes the experience of organizing interdisciplinary projects according to the instructions of employers by students of different study areas and different levels of training for the development of professional competencies and the development of creative thinking. The conclusion is made about the success of the proposed tool for obtaining practice-oriented knowledge. The results of the study can be used as a justification for the implementation of the principles of interdisciplinary project-based learning in higher education institutions.

Keywords: Project-based learning · Non-standard management solutions · Creative approach · Innovative activity · Creative personnel

1 Introduction

Worldwide the development of the innovative component in all spheres, especially professional, is now given great attention at the highest levels of management. In this regard, the concept of the creative economy, which has been gaining momentum in the world for the past twenty years, is becoming particularly relevant. It should be noted that the UN General Assembly has named 2021 the "International Year of the Creative Economy for Sustainable Development". The resolution states that "creative economy is based, in particular, on knowledge-based economic activities and the relationship between human creativity and ideas, knowledge and technology, as well as on cultural values or artistic creativity, cultural heritage and other forms of individual or collective creative expression" [1]. Creative economy promotes the development of entrepreneurship at a qualitatively different level, expanding the boundaries of the country's economic potential.

An important place in the formation of creative personnel is occupied by universities, which task now is to include creative components in the educational process by providing a support system and creating favorable conditions for the development of non-standard thinking and creative skills among students. The Ural Federal University did not stand aside, implementing in 2017 the project-based learning program, according to which, based on the joint interaction of business representatives, teachers and students, practice-oriented solutions are developed with the possibility of further integration in the professional space of the real sector of economy. It should be noted that the Institute of Economics and Management has extensive experience in organizing project work, however, most student projects are educational in nature. The introduction of real employers' cases into the educational process stimulates the development of the necessary competencies in the labor market, develops creative thinking of students, and teaches them to make non-standard management decisions.

2 Problem Statement

Creative projects are a way to get away from routine processes [2], they help to introduce innovations in operational activities. According to Claxton, Edwards & Scale-Constantinou the development of creative thinking is the foundation of education. These researchers identified the following characteristics of creative people: curiosity, patience, love of experimentation, mindfulness, thoughtfulness, and the ability to change the environment [3]. Runco notes the need to consider creativity as thinking or as creation of a new meaning, every person, regardless of his talent and mental abilities, is able to think creatively [4]. Davies, Jindal-Snape, Collier, Digby, Hay & Howe, on the basis of a 2013 review of educational, policy, and professional literature research on the creative learning environment, concluded the importance of developing creative skills and highlighted the factors for developing these skills: flexible use of space and time, the ability to work outside of school, play practices, respect for others, partnerships, and flexible planning. The review reveals evidence of the influence of the creative environment on student academic performance and the development of teachers' professional skills [5].

It is possible to use different approaches for the development of students' creativity. For example, ah Kim, yun Ryoo & joo Ahn, developed an individualized innovative model through the combination of internal and external university support for design students, that reached the commercialization of the results of design work [6]. Karakas considers the positive side of management education based on the development of integrative and holistic thinking, the formation of a sense of community, the development of creative brainstorming, the acquisition of skills through innovative projects, the expansion of opportunities, as well as the creation of training platforms [7]. Shapiro, Nguyen, Mourra et al., looking at the relationship between creative projects in anatomy and the acquired competencies of medical students, note the opportunities for improving professional skills of students at an early stage of training, as well as the impact of participation in creative projects on the stress level of students [8]. One of the innovative tools that develop students' creative abilities can be a three-dimensional interactive visualization of virtual reality. Researchers have proven the development of creative abilities, communication skills, problem solving, and teamwork [9]. The use of game methods in teaching allows to increase the level of theoretical knowledge of students [10], to facilitate cooperation in mixed teams in the implementation of project activities [11], to develop the ability to make strategic decisions, to improve management skills [12], to make people think and to analyze situations [13]. Students who have mastered various types of business games, such as short games [14], strategy games [15], games for training and selecting staff, games for recruiting employees, games for adapting beginners, games for engagement and motivation [16], etc., try to create [17], to improve and to evaluate the effectiveness of a survival game with an unpredictable outcome [18]. It should be noted that in training with the use of game methods, various options for organizing the educational process are possible [19] - traditional, online and mixed. Project-based learning can act as a tool that, on the one hand, develops students' creative skills, and, on the other, increases the level and quality of mastering professional competencies [20]. Moreover, it is possible to organize project-based learning of students on the complete educational program [21], and in individual disciplines [22]. It should be noted that project-based learning is becoming increasingly popular [23]. Project-based method allows teachers to implement interdisciplinary research, which is a modern trend. Matheson, describing the impact of creative industries on the education of designers, notes that a new holistic approach based on the unity of social, cultural and economic development creates the concept of interdisciplinary design education and is a model of the new economy [24]. Comunian, Gilmore & Jacobi emphasize the importance of interdisciplinary research at the intersection of higher education, creative industries, and the growth of regional economy creativity. Researchers also pay special attention to the role of universities in embedding creative graduates in the region and in the economy in general [25].

3 Methodology

The Department of Economics and Management of Construction and the Real Estate Market of the Institute of Economics and Management of the UrFU for its 50-year history has accumulated enormous experience in creating and evaluating the effectiveness of investment and construction projects. Graduates of the department are specialists, bachelors and masters who maintain close ties with the university, acting later as employers, inviting students to practice and work, as chairmen and members of state examination commissions and as customers of research. In March 2021, two large construction companies in Yekaterinburg "Mayak Corporation" and "MarketMall", headed by graduates of the department, as part of project training requested the development of two projects for the renovation of cultural heritage objects on the portal https://teamproject.urfu.ru/. Projects should have taken into account current trends in the development of the city territory and the needs of residents and guests of the capital of the Urals. The value of these requests lies in the ability of students to gain practical experience in implementing their theoretical knowledge on a real socially significant project. For customers, there is an opportunity to get the opinion of young people about the direction in which the territory of real estate should be developed is valuable.

3.1 Description of Real Estate Objects

The objects proposed for renovation are located in the center of Yekaterinburg (old industrial region) and are objects of cultural and industrial heritage of regional significance

[26]. In general, the objects could have been of interest to small and medium-sized businesses, but this is hindered by the low pedestrian and commercial activity of the historical center of the city and the state of the buildings themselves. In order for small businesses to develop in this location, it is necessary to combine enterprises into clusters to create a point of attraction for citizens and guests of the city.

The architecture of the objects uses classical motifs typical of Yekaterinburg in the second half of the XIX century. But today, there are only walls stretched along the street, several buildings in poor condition, and some internal elements, such as stucco, remained from the original objects. The external and internal appearance of the buildings is unattractive, but renovation can correct this situation.

Table 1. Characteristics of the Nurov's estate cultural heritage object

Features	Parameters
Address	Ekaterinburg, Chapaeva St., 1
Land plot area	1 899 sq.m.
Size of the main building and outbuilding	850 sq.m.
Construction	Stone, brick, bar.
Facades	Smooth plaster; coloring of facades in light color, coloring of decorative elements in white color; joinery of windows; parapet posts; fence with colonnade; decorative decoration of facades
Finishes	Plaster, paint, metal decorative grids, ceiling and wall bars, arcades, stucco ornaments of ceilings and vaults, pilasters, all kinds of niches
Remarks on internal networks	Ventilation, utility system
Parking	Spontaneous parking
Appearance	

To work on the projects, the customers provided the students with information for the design – security obligations for the immovable object of cultural heritage, technical reports on the state, general plans, floor plans, historical references, etc.

The main characteristics of the "Nurov's Estate" are presented in Table 1, the object "Territory of the plant "Tonus" (formerly the complex of the Grebenkov and Kholkin's brewery) in Table 2.

Table 2. Characteristics of the cultural heritage object "Territory of the plant "Tonus" (formerly the complex of Grebenkov and Holkin honey brewery)

Features	Parameters
Address	Ekaterinburg R. Luxemburg str. 62
Land plot area	13 432 sq. m.
The area of the three buildings	About 4000 sq.m.
Construction	Stone, brick, bar.
Facades	Smooth plaster; light color varnishing of facades, window joinery; fence with colonnade; architectural and decorative decoration of facades
Finishes	Plastering, painting, arcades, stucco ornamentation of ceilings and vaults, all kinds of niches
Remarks on internal networks	Ventilation, utility systems
Parking	Spontaneous parking
Appearance	

3.2 Project Participants

In total, 188 students, 6 teachers, and 4 representatives of enterprises participated in the development of renovation projects. Students, depending on their field of study, year and level of study, were divided into groups to differentiate the task, as shown in Table 3.

3.3 Structure of Creative Projects

To complete the task, the students had to develop a project for the renovation of a cultural heritage object, consisting of three sections:

1. Justification for the need for renovation of the selected territory (analysis of the location of the renovation object and the existing use of the object);
2. Development of the renovation concept depending on the requirements for a comfortable urban environment (the concept of the best option usage; positioning of the object on the market; proposals for the use of the object and possible functional zoning of the object's areas; a portrait of possible tenants and / or operators; determination of rental rates according to the current market situation);
3. Financial and economic justification of the renovation project (development of the financial model of the project); conclusions and recommendations on the concept.

3.4 Work Plan on Creative Projects

To get feedback on the projects being created, a phased defense was organized – for students of Group 1 in practical classes, for students of Group 2 with the use of remote technologies in the process of distributed practice, for students of Groups 3 and 4 – weekly by means of remote technologies with the participation of representatives of the customer. During the defense of the projects, a cross-evaluation of each project by group mates was provided. It was organized by filling out a questionnaire and setting the number of points according to the quality criteria of the speech and the presentation material. Students of Group 1, who were in Kazakhstan, and were not present in person in the audience, defended their projects online using the Microsoft team's service. Based on the results of the preliminary defense, the best projects were selected in each academic group. Based on the received comments these projects were finalized for participation in the research battle "Urban Renewal", organized on April 24, 2021 as a part of the International Conference of Students and Young Scientists "Spring Days of Science 2021" of the Institute of Economics and Management of the Ural Federal University (Yekaterinburg). The committee consisted of the head and teachers of the Department of Economics and Management of Construction and Real Estate Market, as well as of the representatives of customer enterprises. The best projects were awarded with diplomas.

Two best student projects of Group 2 took part in the competition of students' works for the best solution of practical cases "Urban Renovation" within the framework of the VIII on-line International Scientific and Practical Conference "Problems of Economics and Construction Management in the conditions of environmentally oriented development" of BrSTU, BSU, TSUAB (Irkutsk, Bratsk, Tomsk) on April 23, 2021 and took the 3rd place in the competition.

Table 3. Student distribution

Indicator	Group № 1	Group № 2	Group № 3	Group № 4
Institute	Institute of Construction and Architecture of Ural Federal University	Institute of Economics and Management (UrFU)		
Direction of preparation	08.03.01 Construction	38.03.02 Management		38.04.02 Management
Level of program	Bachelor's Degree			Master's Degree
Course	2	3	4	**1**
Name of educational program	Construction of buildings, structures and territory development	Industrial management and investment and construction business		Territorial Development and Realty Management
Name of subject	Economics of the construction industry (practical classes)	Dispersed practice	Pre-graduation practice	Project Workshop
Nurov's Estate Project				
Number of students, people	73	13	3	4
Number of projects, pcs	13	3	3	1
"Territory of the plant "Tonus" Project				
Number of students, people	73	15	-	4
Number of projects, pcs	17	3	-	4
Total:				
Number of students, people	188			
Number of projects, pcs	40			

Students of Group 1, inspired by the work on the project, additionally took part in the Regional Competition of the best Youth projects "Take your Height"; in addition, they took the 2nd place in the case study tournament "Business League Cup", held jointly with the "Development Corporation of Sverdlovsk Region". As a result of the participation, students were invited to present the developed renovation project at the 11th International Industrial Exhibition "Innoprom-2021", which will be held on July 5–8, 2021 in Yekaterinburg.

The stages of work on renovation projects and their results are presented in Table 4.

Table 4. Stages and Evaluation of Creative Projects

Indicator	Group № 1	Group № 2	Group № 3	Group № 4
Stage I. Pre-selection of projects				
Number of students/projects	146/30	28/6	3/3	7/4
Frequency of contacts	1x per week			
Responsible person	The teacher	Practice Supervisor	Practice Supervisor	Practice Supervisor, partner (employer)
Forms of interaction used	Offline, online			
Project evaluation methods	Supervisors and classmates put points on a questionnaire and the projects were ranked based on the points			
Number of people, people/projects passed to the II stage	**18/4**	**10/2**	**3/3**	**7/4**
Stage II. Participation in contests				
Planned contests	Research Battle "Renovation of Urban Areas". (Ural Federal University)			
Result	-	-	**1st place**	**2nd and 3rd places**
Unplanned contests	Regional contest of the best junior projects "Take Your Height" Business League Cup Case Tournament	Student case study contest "Renovation of Urban Areas" (Irkutsk)	-	-
Result	**2nd place**	**3rd place**		
Stage III. Presentation of results to employers				
Number of projects, pcs	1	1	3	3

4 Results

According to the results of the evaluation and rating, the customers were presented with four concepts of renovation of the "Nurov's Estate" object (Tables 5, 6) and four concepts of renovation of the "Territory of the plant "Tonus" object (Table 7, 8).

Table 5. Concepts of renovation projects for the Nurov's Estate presented to the customer

Concept	Description of concept
Art cluster	Tenants: workshops and photo studios, exhibition areas, creative co-working, retail space
Restaurant with an immersive museum-theater	Tenants: there is a restaurant on the second floor; there is an immersive museum-theater on the first floor, where theatrical actions are held to reconstruct historical events, as well as mystical stories both for children and adults The audience of the project will be people of different ages who can easily get acquainted with the history of the development of Yekaterinburg through the life story of the owner of the estate, a famous industrialist and merchant who made a significant contribution to the development of the city
Art-hotel	Tenant: A 4-star, 10-room hotel with an authentic 19th-century atmosphere, for guests who want to immerse themselves in the past
Cyberclub	Tenant: computer club (youth community), which provides opportunities to take part in modern activities: esports, streaming, blogs The territory of the facility includes a space for organizing events and filming, platforms for broadcasting game tournaments, rooms for training, events, and webinars

The idea and the developed concept of a "Restaurant with an immersive museum-theater" were of particular interest to the representatives of the customer (Mayak Corporation). It was noted as a completely new format for the development of a cultural heritage object in Yekaterinburg.

Representatives of the customer ("MarketMall") pointed out the shortcomings of the concept of Sports complex (the price hike of the redevelopment project by 50% compared to the new construction and the complexity of the organization of space, due to restrictions on the reconstruction and redevelopment of internal premises). The customers appreciated the project to create Retro-futuristic brewery as a project that supported the history of the place. The students were asked to continue working on it in

Table 6. Financial and economic indicators of renovation projects "Nurov's Estate".

Indicator/concept		Art cluster	Restaurant with an immersive museum-theater	Art-hotel	Cyberclub
Overall indicators	Estimated costs for the whole object, thous.rub	20 713	32 790	39 703	35 800
	Estimated cost per 1 m^2, rub./sq.m	24 394	31 259	37 848	42 117
	Term of reconstruction, years	1,75	1,75	1,5	2,5
	Forecast estimated period, years	24	24	24	24
	The rate of discounting without inflation, %	8,23	8,85	8,85	8,6
	Average rental rate, rub/m^2/month	830	920	720	940
Investment efficiency indicators	NPV (net present value), thousand rubles	2 097	2 380	3 299	1 179
	IRR (internal rate of return), %	11,3	10,0	11,1	9,3
	Profitability index of investment, PI, %	122,3	108,3	116,8	105,3
	Simple payback period	9 years 6 months	10 years	9 years 3 months	11 years
	Discounted payback period	15 years 2 months	19 years 3 months	16 years 4 months	21 years

Table 7. Concepts of renovation projects for the "Territory of the plant "Tonus" presented to the customer (formerly the complex of the Grebenkov and Kholkin's brewery)

Concept	Description of the concept
Food mall	Tenants: restaurants-corners, bars, and stores
Sports complex	Tenants: fitness center operators
Art cluster	Tenants: beauty-coworking, event venues, workshops, cafes, retail space
Retro-futuristic brewery	Opening of the brewery and bars Reconstruction of the external and internal interiors of the brewery with elements in the "Steampunk" style

the next academic year and evaluate the cost-effectiveness of the investment. The Food mall project was indicated by the customers as potentially interesting for implementation.

The customers noted the high level and diversity of the proposed options for the renovation of cultural heritage sites. The ideas of the students will be taken into account in the work on the creation and evaluation of the effectiveness of the integrated development of territories.

5　Discussion

The trend towards the use of project-based methods in teaching is gradually gaining strength, and a greater number of educational institutions are using this approach in their work. The success of using project-based learning is due to the practical orientation of the training, implemented on real-world examples and evaluated by practitioners as experts. The involvement of students in solving the pressing problems of the city helps to increase social responsibility of young people. Using the project-based learning method allows students to study consistently and to delve into the details of the problem under study. In the process of working on a project, there is a consistent movement towards a result that can be applied in practice. An additional motivational factor in the projects was the element of competition, which stimulated students to get not only a higher assessment of the proposed solution, but also to convey their ideas to a wider public, for example, through the desire to participate in tournaments and contests on different levels.

It should be noted that project activities require special attention from a teacher. It is a mistake to think that the teacher is only an outside observer. Even without taking an active part in the direct work on the project, the teacher is responsible for organizing project activities, consulting, and helping to build communication with employers and experts. Students in the process of project activity conduct independent research, based on the available theoretical knowledge, look for possible solutions, and summarize their work. The practical experience gained by students, forms not only professional competencies, but also self-organization skills, the ability to work in a team and to be responsible for the results of their work.

Table 8. Financial and economic indicators of renovation projects the "Territory of the plant "Tonus"

Indicator/concept		Food mall	Sports complex	Art cluster	Retro-futuristic brewery
Overall indicators	Estimated costs for the whole object, thous.rub	350 215	350 215	80 324	No calculations
	Estimated cost per 1 m^2, rub./sq.m	50 000	50 000	20 081	No calculations
	Term of reconstruction, years	2	2	1,25	No calculations
	Forecast estimated period, years	25	25	20	No calculations
	The rate of discounting without inflation, %	10	10	10	No calculations
	Average rental rate, rub/m^2/month	1 274	1 171	725	No calculations
Investment efficiency indicators	NPV (net present value), thousand rubles	67 812	26 228	18 417	No calculations
	IRR (internal rate of return), %	13,5	11,4	13,9	No calculations
	Profitability index of investment, PI, %	124	109	123	No calculations
	Simple payback period	9	10	7,5	No calculations
	Discounted payback period	14	19	12,2	No calculations

No calculations - calculations were not carried out, because for the third-year students, the financial and economic justification of the project was not required.

It should also be noted that for employers, the combination of highly qualified personnel, well-established business processes within the organization and strategic relationships with all stakeholders together produce a synergistic effect [27].

The first experience of organizing research battles "Urban renewal" was considered to be a success. The work of the students on the renovation projects of cultural heritage sites was positively evaluated by the customers and the public.

In the future, it is planned to continue cooperation with employers and include battles in the number of permanent events of the conference "Spring Days of Science" with a number of improvements:

– A deeper differentiation of tasks for the development of interdisciplinary creative projects is planned for students of different study areas and different levels of training, depending on their existing competencies.
– It is planned to establish a strict period for the development of each stage of the project, taking into account the time for its preparation by students of different study fields and different levels of training.
– A more detailed specification of the project structure regarding the scope of the issues under consideration is planned.
– A unified interim defense is planned for all teams involved in project work using remote technologies to improve the objectivity of assessments and exchange experience and ideas.
– It is planned to unite students of different fields of training in teams to work on projects to improve the level of projects (detailed design solutions from students of the Institute of Construction and Architecture, economic justification of the project implementation from students of the Institute of Economics and Management, joint work on the project estimate).

6   Conclusion

The desire to introduce new world trends to the traditional educational process leads to the need to transform the existing educational model and to introduce new system elements into it. One of these elements is practice-oriented learning, which is carried out through various non-standard, socially significant tasks that involve students' creative thinking, and imagination and the ability to create an innovative product. We can say that all participants in the development of urban renovation projects highly appreciated the effectiveness of such methods of work. Students received practical knowledge and the necessary professional competencies, business and university mutually gained beneficial cooperation, the public received the solution of a socially significant problem of the city.

Based on the consideration of the practice of organizing project-based learning for undergraduate and graduate students, the use of the principles of project-based learning as a basis for obtaining practice-oriented education and developing creative thinking is proposed. The results of the research can be used as a justification for the implementation of the principles of interdisciplinary project-based learning for students.

References

1. Resolution adopted by the General Assembly on 19 December 2019. No. 74/198 "International Year of Creative Economy for Sustainable Development, 2021". https://undocs.org/en/A/RES/74/198. Accessed 5 May 2021
2. Obstfeld, D.: Creative projects: a less routine approach toward getting new things done. Organ. Sci. **23**(6), 1571–1592 (2012). https://doi.org/10.1287/orsc.1110.0706
3. Claxton, G., Edwards, L., Scale-Constantinou, V.: Cultivating creative mentalities: a framework for education. Thinking Skills Creativity **1**(1), 57–61 (2006). https://doi.org/10.1016/j.tsc.2005.11.001
4. Runco, M.A.: Education for creative potential. Scand. J. Educ. Res. **47**(3), 317–324 (2003). https://doi.org/10.1080/00313830308598
5. Davies, D., Jindal-Snape, D., Collier, C., Digby, R., Hay, P., Howe, A.: Creative learning environments in education – a systematic literature review. Thinking Skills Creativity **8**, 80–91 (2013). https://doi.org/10.1016/j.tsc.2012.07.004
6. Kim, S., Ryoo, H.Y., Ahn, H.J.: Student customized creative education model based on open innovation. J. Open Innov. Technol. Market Complex. **3**(1), 1–19 (2017). https://doi.org/10.1186/s40852-016-0051-y
7. Karakas, F.: Positive management education: creating creative minds, passionate hearts, and kindred spirits. J. Manag. Educ. **35**(2), 198–226 (2011). https://doi.org/10.1177/1052562910372806
8. Matheson, B.: A culture of creativity: design education and the creative industries. J. Manag. Dev. **2**(1), 55–64 (2006). https://doi.org/10.1108/02621710610637963
9. Shapiro, J., Nguyen, V.P., Mourra, S.: Relationship of creative projects in anatomy to medical student professionalism, test performance and stress: an exploratory study. BMC Med. Educ. **9**, 65–74 (2009). https://doi.org/10.1186/1472-6920-9-65
10. Abulrub, A.H.G., Attridge, A.N., Williams, M.A.: Virtual reality in engineering education: the future of creative learning. In: 2011 IEEE Global Engineering Education Conference (EDUCON), vol. 6, no. 4, pp. 751–757. IEEE, Amman (2011). https://doi.org/10.1109/EDUCON.2011.5773223
11. Bareicheva, M., Larionova, V., Stepanova, N., Davy, Y.: Educational games in training. In: Fotaris, P. (eds.) The 14th European Conference on Game Based Learning, (ECGBL), pp. 47–54. Academic Conferences and Publishing International Limited, Brighton (2020). https://doi.org/10.34190/GBL.20.080
12. Stepanova, N., Larionova, V., Drozdova, A., Brown, K.: Obtaining experience in change management using sprint games. In: Elbaek, L., Majgaard, G., Valente, A., Khalid, S. (eds.) The 13th International Conference on Game Based Learning, pp. 694–704. ECGBL, Odense (2019). https://doi.org/10.34190/GBL.19.021
13. Kubina, E., Stepanova, N., Davy, Y., Kondratyeva, L.: Universal strategy game. In: Fotaris, P. (ed.) The 14th European Conference on Game Based Learning (ECGBL), pp. 323–330. Academic Conferences and Publishing International Limited, Brighton (2020). https://doi.org/10.34190/GBL.20.086
14. Kubina, E., Stepanova, N., Larionova, V., Davy, Y.: First experience in game design for students: case study. In: Elbaek, L., Majgaard, G., Valente, A., Khalid, S. (eds.) The 13th International Conference on Game Based Learning (ECGBL), pp. 414–422. ECGBL, Odense (2019). https://doi.org/10.34190/GBL.19.029
15. Bareicheva, M., Stepanova, N., Bochkov, P., Davy, Y.: Using games to develop personal skills required for strategic decision making. In: Elbaek, L., Majgaard, G., Valente, A., Khalid, S. (eds.) The 13th International Conference on Game Based Learning (ECGBL), pp. 58–66. ECGBL, Odense (2019). https://doi.org/10.34190/GBL.19.028

16. Stepanova, N., Larionova, V., Davy, Y., Brown, K.: Effect of using game-based methods on learning efficiency: teaching management to engineers. In: Ciussi, M. (eds.) The 12th European Conference on Game-Based Learning (ECGBL), pp. 660–668. Academic Conferences and Publishing International Limited, Sophia Antipolis (2018)

17. Stepanova, N., Davy, Y., Bochkov, P., Larionova, V.: Gamification in project management training. In: Ciussi, M. (eds.) The 12th European Conference on Game-Based Learning (ECGBL), pp. 653–659. Academic Conferences and Publishing International Limited, Sophia Antipolis (2018)

18. Stepanova, N., Davy, Y., Bochkov, P., Larionova, V.: Game-based management for students: Ural Federal University Taken as example. In: The 11th European Conference on Games Based Learning (ECGBL), pp. 628–633. Academic Conferences and Publishing International Limited, Graz (2017)

19. Comunian, R., Gilmore, A., Jacobi, S.: Higher education and the creative economy: creative graduates, knowledge transfer and regional impact debates. Geogr. Compass 9(7), 371–383 (2015). https://doi.org/10.1111/gec3.12220

20. Larionova, V., Stepanova, N., Shalina, D.: Management games: organizational behavior and emotional intelligence. In: European Proceedings of Social and Behavioural Sciences EpSBS, vol. 98, pp. 259–270 (2020). https://doi.org/10.15405/epsbs.2020.12.03.27

21. Daineko, L., Davy, Y., Larionova, V., Karavaeva, N., Yurasova, I.: Comparative analysis of the practice of applying the principles of project-based learning in Russia. In: The 14th International Technology, Education and Development Conference (INTED), pp. 6113–6116. International Academy of Technology, Education and Development (2020). https://doi.org/10.21125/inted.2020.1655

22. Daineko, L., Davy, Y., Larionova, V., Yurasova, I.: Experience of using project-based learning in the URFU hypermethod e-learning system. In: The 18th European Conference on e-Learning, pp. 145–150. Academic Conferences and Publishing Limited, Valencia (2019). https://doi.org/10.34190/EEL.19.066

23. Daineko, L.V., Reshetnikova, O.E.: Project method – an effective instrument for developing competencies of future professionals. In: European Proceedings of Social and Behavioural Sciences (EpSBS), vol. 98, pp. 221–230. (2020). https://doi.org/10.15405/epsbs.2020.12.03.23

24. Daineko, L.V., Goncharova, N., Larionova, V.A., Ovchinnikova, V.A.: Fostering professional competencies of students with the new approaches in higher education. In: European Proceedings of Social and Behavioural Sciences (EpSBS), vol. 98, pp. 231–239. (2020). https://doi.org/10.15405/epsbs.2020.12.03.24

25. Bylieva, D.S., Lobatyuk, V.V., Nam, T.A.: Serious Games as innovative tools in HR policy. In: IOP Conference Series: Earth Environmental Science, vol. 337, p. 012048 (2019). https://doi.org/10.1088/1755-1315/337/1/012048

26. Zapariy, V., Zaitseva, E.: Industrial heritage as a component of the Urals' attractive image. In: The 11th International Days of Statistics and Economics, pp. 1873–1882. Melandrium, Prague (2017)

27. Kelchevskaya, N.R., Pelymskaya, I.S., Hani Deghles, S.M., Goncharova, N.V., Chernenko, I.M.: The impact of intellectual capital on the performance and investment attractiveness of Russian companies. In: IOP Conference Series: Earth Environmental Science, vol. 666, no. 6, p. 062076. (2021). https://doi.org/10.1088/1755-1315/666/6/062076

Metacognitive Strategies of Social Intelligence and Creativity Through Digital Communication Tools

Nadezhda N. Pokrovskaia[1,2,3]([✉]) [ID], Vladimir A. Spivak[4] [ID],
Svetlana O. Snisarenko[3] [ID], and Maxim A. Petrov[4] [ID]

[1] Higher School of Media Communications and Public Relations of the Institute of Humanities, Peter the Great St. Petersburg Polytechnic University, St. Petersburg 195251, Russia
nnp@spbstu.ru
[2] Department of Innovation Management, St. Petersburg State Electrotechnical University "LETI", St. Petersburg 197022, Russia
[3] Department of Economics and Management of Social-Economic Systems, St. Petersburg University of Management Technologies and Economics, St. Petersburg 190103, Russia
[4] Department of Sociology and Human Resources Management, St. Petersburg State University of Economics, St. Petersburg 191023, Russia

Abstract. Metacognitive skills help students better understand and master the subjects of study. The efficiency of metacognitive strategies depends on the nature of the task to be accomplished. Exploratory research at 5 Russian universities during the 2020–2021 academic year, under conditions of social isolation, showed preference for a visual expression of emotional support as evidence of social intelligence for routine tasks and exercises, and the predominance of articulation in verbalized form of ideas for creative and collective tasks. Focus-groups, discussions and observations of students' group project defenses allowed researchers to draw conclusions regarding the impact of digital tools used for everyday communication respective to diversity perception and creative growth, digital tools contribute to the broader search for various sources of information, as well as leading to the interest to check, confirm or reject information found on the internet. The metacognitive strategies concern the generation of individual knowledge, but also the construction of social structures and relationships that are crucial for the motivation and efficiency of collective creative activities.

Keywords: Digital communication · Creativity · Collective creative projects · Metacognitive strategies · Communication strategy · Social intelligence

1 Introduction

Intellectual activity plays a dramatic role in the growth of the knowledge-driven economy and is studied so as to develop educational institutions and business organizations. Organizational learning [1, 2] learning organizations [3, 4] and spiral of organizational learning [5–7] are several examples of the ample bulk of conceptual approaches in this field.

Knowledge creation and exchange are widely analyzed as cognitive process [8, 9] that combine a repeating routine with intuitive insights and creativity [10–12]. Metacognition is defined as "thinking about thinking" (or "knowing knowledge" [13]) and is investigated as a tool that helps "to learn to learn" [14, 15].

Metacognitive strategies describe the ways in which people construct their personal procedures for the acquisition, assimilation, implementation and enrichment of knowledge [16, 17]. For a joke, teachers often describe a frequently used students' metacognitive strategy as "let's start and see what we get", which represents chaotic activity with texts, tables, diagrams, numbers, pictures, infographics. It includes the search engine applied as a guide for the learning process and Wikipedia that serves an essential source of information. This joke, in fact, reflects the difficulty that students usually experience to construct the order of actions for learning, to build the clear path for knowledge creation. This metacognitive strategy refers to Napoleon Bonaparte' sentence: "On s'engage et puis on voit" [18, p. 453] ("First engage in and then see what happens"), used also by V.I. Lenin [19, p. 4, 20, p. 381] to describe the unfolding of any complex social change.

Serious analyses discovered the structural approach to the metacognitive strategies that are built in the off- and online environment to organize individual and collective learning [21–24]. First of all, scholars examine the metacognitive strategies in the field of knowledge creation and exchange as a threefold subject:

– as a process of understanding the methods of thinking and learning;
– as capacities and skills that can be developed and will help to better acquire and assimilate knowledge;
– as personal features that determine the potential of a person to create and exchange knowledge, these features are to be detected, diagnosed and taken into account to build, i.e., the person's educational trajectory or career plan.

The pragmatic approach helps to construct teaching techniques using roadmaps, checklists, mind-mapping, etc., to help students to assure to steps of the acquisition of knowledge and practical competences [25–28]. The inventory of metacognitive skills and capacities includes the assessment of abilities (depending on the personality and on the concrete set of circumstances, a situation) and of regulation (intentions and goal setting, actions to be planned and accomplished, monitoring and assessment) that help to increase the effectiveness of the learning process [29, 30].

Metacognitive strategies are widely investigated from the point of view of rational elements of the cognitive process; but several studies have been carried out to better understand the influence of emotional evaluation and psychomotor competence [31]. A systemic model BACEIS relates to the essential internal factors of psychological diagnostics of a personality (Behavior, Affect, Cognition) and external impact of socio-cultural and community patterns (Environment, Interacting, Systems) [15] that includes academic and non-academic milieu of communication and reciprocal exchange of models of reasoning, the ways people use to cope with tasks, to solve problems, and to perceive others and to express themselves.

The impact of the social environment is taken into account especially due to the covid-19 pandemic [32]. The quarantine and isolation made people to re-construct their home space for work and learning activities, and demonstrated the significance of technical support (especial importance of the telecommunication infrastructure for broadband internet access) and of the social connections and interactions that were ruptured in the physical world and were transferred to the online environment [33].

The interactions usually create specific advantages for groups and communities, such as use of accumulated experience. This additional value of collective activity is represented in the concepts of social intelligence [34], collective wisdom [35], group or co-intelligence [36–38]. Economic behavior includes social intelligence into the calculation of rationality and planning purposive behavior [39, 40], economists investigate the realization of social intelligence concept in clustering and marketing. The industrial clusters concentrate in a narrow territory the people with specific competencies who exchange their knowledge and ideas at the spontaneous interactions (in common public spaces, at institutions, parks, etc.) [41]. For the market promotion purposes, the social intelligence has a tacit form of reputation (of a brand) and explicit form of reviews, the consumer behavior dynamics demonstrates the importance of the review for a good or service with decreasing impact of brand [42]. The reviews are analyzed as partly created by sales managers (usually, the positive evaluations of a product, service or company) and customers read, moreover, the negative reviews, that are perceived as real reviews of real individual independent consumers. The sophisticated automated recommending systems of club filtering are developed to overcome the problems of collapse, of boring, mistaken items or mistrust to a system [43].

For the educational context, the particular learning space is built by students for "co-configuration, co-creation and co-design" of knowledge creation [44]. The social intelligence helps students to organize their learning process due to the interaction, mediation, explaining, structuring the material. Before the covid-19 the spontaneous communications took place in university's cafe or hall. After the introduced measures of social isolation all the communications are especially planned and organized. The accidental meetings disappeared, the organization of a meeting through social media, messengers or pedagogical platforms requires conscious efforts to schedule, invite participants, to check the correct functioning of equipment (the famous question "can you hear me?" became phrase for memes). This administrative work to prepare and to manage online communication reveals transaction costs that were previously zero or negligible (due to the natural place for conversations in classrooms or university' halls that were just passed when talking) and were ignored. Social intelligence had been perceived as "free of charge" and was fostered, in a large scale, as an unconscious process [45, 46] with mental mapping [27, 47–49]. The collective wisdom of a group of students had been appeared due to the everyday meetings without peculiar efforts.

Online education tools were massively introduced due to the pandemic, this change produced the specific effects that depend on the nature of knowledge to be acquired. If the routine tasks are digitized, the creative kind of activities requires the specific efforts to communicate, to exchange the new appearing ideas and to select among them the most pertinent to realize as a collective project. The lack of the creativity at the standardized pedagogical environment relates to a clear list of competencies to be taught and learnt [51].

The experience of preparing group projects demonstrates the dynamic evolution of metacognitive skills of students that assure the social intra-group interaction and data exchange, the inter-communities contacts (between students of different background and citizenship) with higher cross-cultural communication (even if they a passing through the additional difficulties due to the necessity to implement more frequently the explicit verbal knowledge).

The difficulties of studying under the remote regime of online learning include the narrow particular problematic field that relates to the collective creative activities. The purpose of this research is to deepen the understanding of the ways and metacognitive strategies that help students to cope with creative tasks that are to be committed through the digital communication, in the specific social isolated context that is imposed by the pandemic and, potentially, will stay the particular condition for several categories of students (with limited capacities of physical assistance at classroom).

This research aims to find the essential strategies that helped the most successful groups to develop the most interesting projects that shift the fixed standard conditions and requirements, in the context of the digital tools used to communicate within the online education.

2   Materials and Methods

The main research question is to identify the enhancement of students' metacognitive strategies during the social isolation. Students were forced to suddenly change the ways they had been learning, to take responsibility for the increased volume of the organizational function required to manage the study process. This new context required improving the knowledge of knowing. The question is if the students develop their metacognitive strategies with digital tools of communication, and if they implement the same skills and methods to organize the social interaction for routine learning activities (such as multiple-choice tests) and for creative autonomous works based on their own choice and responsibility (group projects).

This question includes two essential perspectives: to detect the most successful metacognitive strategies helping for creative tasks and to evaluate the change related to the transfer of the learning process into digital education regime.

The study of the metacognitive strategies that help to develop group creative projects the set of measuring axes is proposed: awareness and regulation, internal – external orientation, emphasis on past – present – future, rationalization of modalities and their articulation and negotiation, verbalization vs visualization. Metacognitive awareness is associated with reflexivity and self-organization [52–54], efficiency of metacognitive strategies in educational activities correlates to the emotional intelligence [55], problem solving instruction outside of regular lessons [56], individual goal-setting and the desire for personal growth [55–59], ontogenetic, sociogenic, liberative, transpersonal, and spiritual models of personal development [60], the roles assumed [61]. For our research, we have limited the metacognitive components set to examine to the use of specific language elements (signs and symbols) replacing complex notions or phenomena (emoticons and emojis for emotions, etc.) and of roles indications (nicknames), as well as to the actions of systematizing and structuring materials, division of labor and planning, and to the commenting about collecting data, working on the presentation and reporting.

The essential elements of the social intelligence include tact and common sense, collaboration and reciprocity, emotional solidarity and mutual support and inspiration, patience and recognition, sense of humor [62–66].

Tact and common sense can be taken into account as an edge on the axes of regulation, external listening and understanding, emphasis on future ("smart streets" and collective wisdom are deeply imbued in the past, but the making decisions are directed to the future). Collaboration and cooperativeness rely on the awareness, external interest, future planning, articulation of modalities and negotiation. The sense of solidarity and certainty in the emotional support is based on the awareness, internal orientation, clear presence in the moment of "now and here", perception of modalities without arguing, visualization. Recognition of other persons and their contribution requires balanced awareness and regulation, clear orientation to external space, rationalization and negotiation and verbalization (textual and/or numeric presentation). Sense of humor helps to cope with non-standard problems and situations and is based on the regulative approach with external orientation, future-driven logics and verbal rational articulation.

Modality as the form of social meaning and interaction represents the interesting aspect of the analysis of the higher school students' communication. The language represents the core essence of the message through the form used: quality of the presentation (e.g., apologizing for the quality of sound or for the blurring reflects), importance of the various forms (text and numeric, visual and audio), the substantial content of communication and its plan (preventing audience about the essence and explaining why should the audience spend their time for which purport), what is the substance the interaction is focused on.

The effort of verbalization is investigated in this research according to qualitative methodology (interviewing and focus groups) for the analysis of the forms of modalities used, reciprocal moral support, mutual assessment of intermediate results and helping each other with difficult tasks, of the forms to share the responsibilities for the improvement of the final version of a project to be defended. During the defence of the projects several questions were asked and discussed within groups.

The everyday consumption of information produced the specific metacognitive skills of looking at metadata firstly and at the content secondly, of checking the data found in diverse sources (from sources of opposite or divergent nature to confirm the truth or to reject false information). This skill should be better developed in cyberspace due to the habits of digital communication.

In this study, the metacognitive strategies representing the efficient implementation of social intelligence are considered through the use of following tools:

- the use of symbols (emoticons, emojis, stickers, pictures etc.), which replace an answer (to accelerate reply) or add a nuance to the written text;
- the use of words oriented to organize a better mutual understanding – to generalize the functions, to structure and systematize the activities and relationship inside small groups working on project and between students' groups in the cohort of the year;
- the use of nicknames, describing the status and role of a person in the group;
- the comments about the plan, the quality of sound or video, the approach to obtain data, etc.;

- the remarks about the changes that concerned the participants' dialogues and behavior within the online space of communication.

The task of this research is to reveal the components of learning behavior that can be identified as signs of the metacognitive strategies chosen by students for their collective creative works.

The hypotheses of the research include the following:

h1 – verbalization witnesses of higher level of metacognition implemented to the creative project, the visual emotional symbols (emojis, emoticons, likes, etc.) are useful for iterative standard tasks, but are less used for the creative projects;

h2 – students using specific nicknames for their colleagues in chats, communities and groups at social media, produce more creativity than their colleagues using "standard" first names or family names – if the dialogue include the names and nicknames that reflect the essential roles, this clear "appointment" helps to translate the person's actions within the network;

h3 – the questions about "how to start", "what is our goal and its different aspects", "which objectives should be achieved with which priority" help to build the clear path to construct the project and to direct the intuitive inventing search;

h4 – the students can describe the evolution of their practices of learning process organization from the point of view of the ways to build their social interactions for the improvement of the common creativity' enhancement.

To test the hypotheses, the researchers elaborated a set of assessment tools that combined "ordinary" control exercises (tests with multiple-choice questions and with short answers) with a collective creative work (autonomous group projects to prepare, present and defend). The co-authors of this paper are teaching similar social-economic disciplines in 5 universities: Peter the Great St. Petersburg Polytechnic University ("Organization of work of APR departments"), Herzen State Pedagogical University of Russia ("Management of communications of enterprise"), St. Petersburg State Electrotechnical University "LETI" ("Management of innovative projects"), St. Petersburg University of Management technologies and economics ("Management of personnel") and St. Petersburg State University of Economics ("Human resources management"). At all these universities, the tasks were addressed to the undergraduate students of the economics and management faculties.

During the academic year 2020–2021 the set of these routine and creative tasks were imposed to students, in total, 37 project groups from 7 students' cohorts were included into research (N = 129, average age is 21.2 years, sd = 0.966, in the total sample 74 females, 57.4%, and 55 males, 42.6%, that represents the typical distribution students by sex at social-economic faculties, except the group in Herzen university, where are only 3 men and 11 women, or 21.4% and 78.6%).

The sample is restricted by number due to the selection of the similar groups of undergraduates that would be comparable by demographic criteria of sex, age and by socio-cultural and linguistic parameter (some cohorts included more than a half of foreign students that would influence the results, these cohorts were not included into the research sample). The sample permits to make conclusions for typical Russian university situation for students of social, economic and management faculties, but the cross-cultural, age and gender differences are not studied because they were not included into the research aims (for these issues, the significant results are obtained in other studies [65–71]).

3 Results

The results were quantified according to the following approaches:

- number of persons who mentioned the use of several tools of interaction (see Table 1);
- number of persons who were observed commenting their actions and the construction of their presentations during reporting – comments were fixed by teachers (see Fig. 1);
- remarks of respondents during focus-groups on the concerns discussed and the statements supported.

The symbols and signs used represent the clear support for the hypothesis that the metacognitive skills necessary for the organization of social interaction and based on the social intelligence, differ and depend on the nature of knowledge assimilated (executing standard tasks or inventing new ideas due to creativity and divergent thinking). The Table 1 contains the data of number of persons who mentioned the use of symbolic forms to express ideas during different students' works:

Table 1. Distribution of the use of expression forms during different kind of work (control exercises, tests with multiple-choice questions, preparing and presentation of projects), number of persons

Tools used to express or react	Control exercise	MCQ test	Preparing project	Project defense	Total persons
Emoji	24	25	7	2	25
Stickers	31	27	75	71	77
Hearts	29	37	46	48	54
Emoticons	30	33	34	36	39
Cute animals	9	3	11	15	18
Flowers	4	2	13	12	14
Facepalm (FP)	32	38	26	16	46
Likes	20	29	10	4	28
Brief answers (OK, LOL…)	68	76	79	84	98
Text	52	54	128	110	128

(*continued*)

Table 1. (*continued*)

Tools used to express or react	Control exercise	MCQ test	Preparing project	Project defense	Total persons
Correct punctuation	42	37	119	34	124
Screenshots	25	23	59	13	64
Pictures/videos	3	8	12	16	20
Soundtracks	2	0	14	11	15

According to the data of the Table 1, the highest results (almost unanimity) are registered for the verbalized tools to communicate about the creative project – 128 of 129 persons (99.2%) attributed the explanation at the stage of preparing the collective project in textual form, with high attention paid to the literacy and, especially, to punctuation (which is often ignored by students at social media or messengers) – 119 respondents (92.3%).

Regarding the standardized knowledge (MC questions or control exercises), the higher degree of use is fixed for emojis and signs (e.g., facepalm as an expression of self-irony, boring or fatigue). This data can be compared with the respondents' choices which were investigated in 2015–2018, when the first attempt to introduce the collective creative project was carried out [72] (see Table 2):

This comparison partly supports the h4 about the dynamics of the tools used for the social interaction. It is a clear evolution from the broad use of the visual instruments towards the verbal, textual, numeric and audio tools of digital communication. It is necessary to add, that the impact of the Clubhouse and of the appearance of audio-messages in the social media (i.e., Vkontakte) and messengers (i.e., WhatsApp) make it impossible to conclude with validity about the enlargement from visual form to the sound, but verbalization is presented even in the use of Facepalm, which had been introduced as a picture into message just several years ago, and now, several students told that they use the abbreviation FP instead of the visual analogue.

According to the discussions in the focus-groups, the widespread opinion is obtained that the nicknames give an additional helpful tool to describe the positioning of a person in the social structure of the small project' group and of the larger cohort community of students. During the online discussion, students mentioned that in life dialogue the role of a person can be shaped with nuances of voice, tones, mimics and gestures, but the digital communication tools do not permit this delicate distinction, and the use of nicknames for designate roles and status accelerate the interaction and increase the efficiency. This comment can be combined with the comparison of the scores of the assessment that allow researchers to conclude that the higher is the level of the originality of a project, the more frequently the participants of the groups used specific names. This data should be checked with a better analysis of its statistical significance, but it is worthy to be mentioned for further detailed research.

The specific change of 2020–2021 was mentioned in three universities from the sample: the students were asked if they had discussed the goals before preparing the

Table 2. Comparison of the use of some symbols and forms of expression, in 2015–2018 and in post-pandemic academic year 2020–2021, % of respondents.

Tools used to express or react	2015–2018	2020–2021
Emoji	54.4	19.4
Stickers	36.8	59.7
Hearts	38.6	41.9
Emoticons	100.0	30.2
Cute animals	56.1	14.0
Flowers	28.1	10.9
Facepalm (FP)	19.3	35.7
Likes	100.0	21.7
Brief answers (OK, LOL…)	-*	76.0
Text	80.7	99.2
Correct punctuation	-*	96.1
Screenshots	-*	49.6
Pictures/videos	43.9	15.5
Soundtracks	-*	11.6

*No data were collected.

project, and the teachers noted the remarkable increase in the post-covid academic year of share of the groups who had discussed the field of the business (choice of sector, industry, activities), but who had targeted the complex and multifaceted purposes such as: "To prepare the best project", "to achieve the high score", "to make a useful work for other disciplines", "to create a project that can be useful for the future job", "to solve a problem at a job place of one of the group members", "to solve a problem in the business owned by one of the team members" (even, four projects were prepared by students who were fostering their own businesses), "to help family business with some ideas", "to accumulate information for the future search of employment", "for fun" (this was the case of group of three men who prepared an imagined virtual business project with extraterrestrial partners), "to better understand the industry" (to penetrate to the real processes in a field of activities, to approach a real professional community, and some similar answers about specific sectors that were of interest for team members and they wanted to deepen in the research and knowledge about this sector), "to find more information about regulation of small business" (of specific fields of activities in the cases of medical services, drugstore or organization of a café with alcohol, all of them have legislative restrictions), "to try if we are able to face this challenge" and "to create closer friendship within a new group of team members" (in the case of the new students integrating in the cohort).

This avalanche of answers about goal-setting was surprising for us, because usually students just reply that the objectives of the project are enumerated by the exercise description and they follow the tasks proposed by teacher "to obtain a mark". The academic year 2020–2021 demonstrated the abrupt change in this narrow pragmatic orientation ("to do what is required"), and the mass occurrence of the diversity of aims and goals with various degree of distance from the educational context as it is. In the both technical universities (Polytech, LETI) and in the university of Management technologies and economics (SPbMTE) the teachers revealed the deep change in the consciousness of doing tasks – students were looking for the sense and meaning of the whole project, and perceived a list of objectives proposed by teacher as a basis to construct their own "building" of a project. This holistic vision was the case of particular teams in 2015–2018 and it became a mass phenomenon after the social isolation and the transfer of communication in the online sphere, which requires articulation and reciprocal motivation, the construction of a space that imitates the real interaction when common interests and values are "caught" by non-verbal signals.

The conscious use of the modalities during the presentation and defence of the group projects is developed in all cohorts, the clear dominance is registered for the substance of the material presented, such as the structured plan of the report (56.5% or 69 persons mentioned the plan, all the groups presented a slide with the plan of presentation), the sources of data (e.g., statistics, analysis of labor market, etc.) (65.1%, 84 respondents) and the steps of the fulfilment of all activities to prepare the collective project (53 respondents, 41.1%) (see Fig. 1):

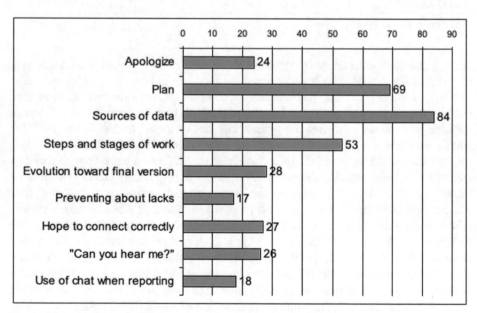

Fig. 1. Metacognitive modality comments fixed during students' reports (number of students who were remarking about reporting and presenting).

Meanwhile at their everyday life students are always accessible through their messengers and social media, but at the moment of reporting and defending the project they were not able to react to the chat, where their colleagues and teacher were asking questions or putting remarks. The omnipresent digital communication is switched off during the group project defence that demonstrates the higher level of concentration on the collective activity.

The similar results were obtained in all 5 universities, except a slight difference of the group at the Herzen Pedagogical University, where the education is oriented to prepare teachers while the other universities intend to prepare professionals for the labor market and business companies. The pedagogical education seemed to be more oriented to the metacognitive skills development, and the students demonstrated a little higher level of the internal awareness and regulation (the discussion was focused on the goals to be achieved, on the describing steps and selected and rejected versions), better understanding of reasons and meanings (what and why they do, how they could use the project for the future profession). But they could co-operate with a weaker degree of collaborative strategy, they experienced difficulties to indicate roles during the division of functions. This remark has no statistical significance, because of the number of respondents in this sample of only 14 persons, but the researchers were surprised with this result because it was partly contradictory to what had been previously expected. The development of metacognitive skills is the core profession of teachers, and we expected the excess of the metacognitive components of the future teachers over the future economists, managers or marketing specialists. The slight increment in the internal metacognition is combined with the preliminary detected deficiency with the external components of the metacognitive strategies and the lack of the social intelligence of the pedagogical university' students.

The social learning enriches the cognitive outcome [73–75] due to the broader and deeper exchange of knowledge. The social intelligence plays a pragmatic role to cope with affective issues to commit the routine tasks, and it creates the fundamental incentives for the intellectual growth in the field of creating inventions, discoveries, innovations, that are based on the creativity [76, 77]. The students prefer the remote learning regime to receive the duplicated knowledge [78] and they are seeking for the assistance of teachers and colleagues for the creative components of learning.

4   Conclusion

"Meta" from Greek "beyond" refers to a higher level of conscience, the generalized, structured and systematized knowledge about physical or virtual reality, within real and symbolic space, in socio-cultural and digital environment. Metacognition and metadata are widely appropriated by specialists in practical craft of organization of cognitive processes and learning, in educational institutions as well as in business organizations. The fundamental philosophical approach to the cognition discovers the threefold structure of symbolic space (semantics, pragmatics and syntax) that refers to the metaphysics of meanings, mathematical formulation of purposes and intentions, and IT constructions of flows and communications in the fuzzy topology [50, 79].

The social intelligence can cultivate or block creativity. The obtained results allow researchers to support the hypotheses about the impact of the social isolation on the change of the core components of the social intelligence that plays a dramatic role for the creative projects. The social interactions help students to cope with the routine standardized tasks, but the collective wisdom and social intelligence are particularly crucial to enhance creativity, to stimulate the production of new ideas, to foster and unfold the process of generating knowledge, of creating new proposals and visions, when the self-confidence and trustful relationship among group members are of high importance.

The qualitative exploratory analysis of the undergraduate students' communicative behavior in online space through social media and messengers during pandemic 2020–2021 demonstrated the decrease of "emojis" and "emoticons" use and the increase of the psychological observance and cognitive remarks. An ironic sentence "now should I tell you that we are Super Pro!" is a rhetorical question that implies several layers of a group's cognitive process, including moral support and the intellectual appraisal. A settlement "tomorrow let's see and wait for the score" intends to relax everyone and to provide recognition that the colleagues have done all their best.

The pandemic demonstrated a clear evolution of strategies of intra-group behavior and inter-groups contact and communication: the use of emojis is changed, the words are more often accompanied with emojis or stickers than replaced by any kind of symbols or pictures, the verbalized form of communication is more frequent and the nicknames are used more often in the reports of projects, the modalities used underline the intentions to apologize, ameliorate and describe the context and circumstances of the choices made.

These changes can be examined in further research that could bring forth question of the main reasons and symptoms of the changes – the growth of digital tools, the transfer to the digital environment or the isolation do play the crucial role for the transformation and differentiation of the metacognitive strategies used by students.

The results of this exploratory research are useful for the distinctive diagnostic of the metacognitive competency that is required for routine and for non-standard tasks: a clear difference is revealed between the articulated expression of thoughts in verbal form for the creative projects and the visual form that corresponds well to the iterative tasks of executing nature.

References

1. Argyris, C.: Double loop learning in organizations. Harvard Bus. Rev. **77502**, 115–124 (1977)
2. Argyris, C., Schön, D.: Organizational Learning. A Theory of Action Perspective. Addison-Wesley, Reading (1978)
3. Senge, P.M.: The Fifth Discipline. The Art and Practice of the Learning Organization. Currency, Doubleday, New York (1990)
4. Senge, P., Kleiner, A., Roberts, C., Ross, R.B., Smith, B.J.: The Fifth Discipline Fieldbook: Strategies and Tools for Building a Learning Organization. Currency, Doubleday, New York (1994)
5. Takeuchi, H., Nonaka, I.: The new new product development game. Harv. Bus. Rev. **64**(1), 137–146 (1986)
6. Nonaka, I.: The knowledge creating company. Harvard Bus. Rev. **69**, 96–104 (1991)
7. Nonaka, I., Takeuchi, H.: The Knowledge Creating Company: How Japanese Companies Create the Dynamics of Innovation. Oxford University Press, New York (1995)
8. Pokrovskaia, N.N.: Global models of regulatory mechanisms and tax incentives in the R&D sphere for the production and transfer of knowledge. In: GSOM 2017 International Conference "GSOM Emerging Markets Conference 2016", pp. 313–315. St. Petersburg University Graduate School of Management, GSOM, St. Petersburg, Russia (2016)

9. Volkov, A., Chulkov, V., Kazaryan, R., Sinenko, S.: Acting adaptation and human parity in the triad "man - Knowledge - Methods." Appl. Mech. Mater. **584–586**, 2681–2684 (2014)
10. Carlile, P., Rebentisch, E.: Into the black box: the knowledge transformation cycle. Manag. Sci. **49**, 1180–1195 (2003)
11. Bradler, C., Neckermann, S., Warnke, A.J.: Incentivizing creativity: a large scale experiment with performance bonuses and gifts. J. Labor Econ. **37**(3), 793–851 (2019)
12. Asanov, I.A., Pokrovskaia, N.N.: Digital regulatory tools for entrepreneurial and creative behavior in the knowledge economy. In: Shaposhnikov, S. (ed.) 2017 International Conference Quality Management, Transport and Information Security, Information Technologies, IT&QM&IS, St. Petersburg, Russia, 23–30 September 2017, pp. 43–46. IEEE, New York (2017). https://doi.org/10.1109/ITMQIS.2017.8085759
13. Siemens, G.: Knowing Knowledge. Lulu.com (2006)
14. Argyris, C.: Teaching smart people how to learn. Harvard Bus. Rev. **69**, 99–109 (1991)
15. Hartman, H., Sternberg, R.J.: A broad BACEIS for improving thinking. Instr. Sci. Int. J. Learn. Cogn. **21**(5), 401–425 (1993)
16. Matlin, M.W.: Memory strategies and metacognition. In: Matlin, M.W. (ed.) Cognition. Wiley, New York (2005)
17. Schraw, G.: Promoting general metacognitive awareness. Instr. Sci. **26**, 113–125 (1998)
18. Baumgarten, N.K.: Importance de la poudre sans fumée à la guerre (fin); traduit du russe et résumé par le capitaine d'artillerie Ollivier. Revue d'artillerie (Nancy) **39**, 441–462 (In Fr.) (1891). https://gallica.bnf.fr/ark:/12148/bpt6k6469352c/f459.item. Accessed 12 Apr 2021
19. Lenin (Ulyanov), V.I.: About our revolution. Pravda 117, 30 May 1923. (in Russian)
20. Lenin (Ulyanov), V.I.: About our revolution. In: Lenin, V.I. (ed.) Collected Works, vol. 45, pp. 378–382 (1970). (in Russian). http://lenin.rusarchives.ru/dokumenty/statya-vi-lenina-o-nashey-revolyucii-po-povodu-zapisok-n-suhanova. Accessed 12 Apr 2021
21. Pokrovskaia, N., Margulyan, Ya., Bulatetskaia, A., Snisarenko, S.: Intellectual analysis for educational path cognitive modeling: digital knowledge for post-modern value creation. Wisdom **14**(1), 69–76 (2020). https://doi.org/10.24234/wisdom.v14i1.305
22. Kasyanik, P.M., Gulk, E.B., Olennikova, M.V., Zakharov, K.P., Kruglikov, V.N.: Educational process at the technical university through the eyes of its participants. In: Auer, M.E., Guralnick, D., Uhomoibhi, J. (eds.) ICL 2016. AISC, vol. 544, pp. 377–388. Springer, Cham (2017). https://doi.org/10.1007/978-3-319-50337-0_36
23. Shipunova, O., Evseeva, L., Pozdeeva, E., Evseev, V., Zhabenko, I.: Social and educational environment modeling in future vision: infosphere tools. E3S Web Conf. **110**, 2011 (2019). https://doi.org/10.1051/e3sconf/201911002011
24. Suleimankadieva, A.E., Petrov, M.A., Popazova, O.A.: Strategic prospects for the development of human capital in the context of singularity and intellectualization of the Russian economy. IOP Conf. Ser.: Mater. Sci. Eng. **940**, 012092 (2020) https://doi.org/10.1088/1757-899X/940/1/012092
25. Ko Koleňáková, R.Š., Kozárová, N.: The impact of mental mapping on pupils' attitudes to learning. CBU Int. Conf. Proc. **7**, 464–471 (2019). https://doi.org/10.12955/cbup.v7.1402
26. Leontieva, V.L., Ababkova, M.Yu.: Image of disciplines and university: a new approach to research. EpSBS **94**, 874–881 (2018). https://doi.org/10.15405/epsbs.2018.12.02.94
27. Tolman, E.C.: Cognitive maps in rats and men. Psych. Rev. **55**, 189–208 (1948)
28. Schraw, G., Dennison, R.S.: Assessing metacognitive awareness. Contem. Educ. Psych. **19**(4), 460–475 (1994). https://doi.org/10.1006/ceps.1994.1033
29. Pokrovskaia, N.N., Petrov, M.A., Molodkova, E.B.: Organizational management factors for universities and business infrastructure communication: Russian-Italian partnership case. In: Shaposhnikov, S. (ed.) Proceedings of XVII Russian Scientific and Practical Conference on Planning and Teaching Engineering Staff for the Industrial and Economic Complex of the

Region (PTES, St. Petersburg, LETI), pp. 205–208 (2018). IEEE, New York. https://doi.org/10.1109/PTES.2018.8604220

30. Oakley, B., Sejnowski, T., McConville, A.: Learning How to Learn: How to Succeed in School Without Spending All Your Time Studying. Tarcher Perigee, New York (2018)

31. Ababkova, M.Y., Cappelli, L., D'Ascenzo, F., Leontyeva, V.L., Pokrovskaia, N.N.: Digital communication tools and knowledge creation processes for enriched intellectual outcome – experience of short-term e-learning courses during pandemic. Future Internet. **13**, 43 (2021). https://doi.org/10.3390/fi13020043

32. Almazova, N., Krylova, E., Rubtsova, A., Odinokaya, M.: Challenges and opportunities for Russian higher education amid COVID-19: teachers' perspective. Educ. Sci. **10**, 368 (2020)

33. Asanov, I., Flores, F., McKenzie, D., Mensmann, M., Schulte, M.: Remote-learning, time-use, and mental health of Ecuadorian high-school students during the COVID-19 quarantine. World Dev. **138**, 105225 (2021). https://doi.org/10.1016/j.worlddev.2020.105225

34. Thorndike, E.L.: Intelligence and its use. Harper's Mag. **140**, 227–235 (1920)

35. D'Ascenzo, F., Golohvastov, D.V., Pokrovskaia, N.N.: Market agents' industrial regulation and cultural inertia in smart community: social engineering or collective wisdom. In: Korableva, O.N., Korablev, V.V. (eds.) Technological Perspective Within the Eurasian Space: New Markets and Points of Economic Growth: Proceedings of the 4th International Scientific Conference, pp. 87–91. Asterion, St. Petersburg (2019)

36. Ober, J.: Democracy's wisdom: an Aristotelian middle way for collective judgment. Am. Pol. Sci. Rev. **107**(1), 104–122 (2013). https://doi.org/10.1017/S0003055412000627

37. Vygotsky, L.S.: Mind in Society: The Development of Higher Psychological Processes. Harvard University Press, Cambridge (1978)

38. Lévy, P., Farley, A., Lollini, M.: Collective intelligence, the future of internet and the IEML. Hum. Stud. Digit. Age **6**(1), 5–31 (2019). https://doi.org/10.5399/uo/hsda.6.1.2

39. Laan, A., Madirolas, G., Polavieja, G.G.: Rescuing collective wisdom when the average group opinion is wrong. Front. Robot. AI **4**, 56 (2017). https://doi.org/10.3389/frobt.2017.00056

40. Goldstein, D.G., McAfee, R.P., Suri, S.: The wisdom of smaller, smarter crowds. In: Proceedings of the 15th ACM Conference on Economics and Computation – EC 2014, pp. 471–488. ACM Press, New York (2014). https://doi.org/10.1145/2600057.2602886

41. D'Ascenzo, F., Pokrovskaia, N.N.: The education in quality and innovation economy as a competitive advantage for high technology clustering in industry and ITC-sector. In: Pokrovskaia, N.N. (ed.) Russian-Italian Strategic Collaboration in Transfer of Knowledge: Quality and Innovation Performance in Knowledge-Driven Economy, pp. 43–49. SSUE, St. Petersburg (2016)

42. Wright, P.: Marketplace metacognition and social intelligence. J. Consum. Res. **28**(4), 677–682 (2002). https://doi.org/10.1086/338210

43. Cerri, S.A., Lemoisson, P.: Serendipitous learning fostered by brain state assessment and collective wisdom. In: Frasson, C., Bamidis, P., Vlamos, P. (eds.) BFAL 2020. LNCS (LNAI), vol. 12462, pp. 125–136. Springer, Cham (2020). https://doi.org/10.1007/978-3-030-60735-7_14

44. Luckin, R., et al.: Using mobile technology to create flexible learning contexts. J. Interact. Media Educ. **22**, 21 (2005). https://doi.org/10.5334/2005-22

45. Marlowe, H.A.: Social intelligence: evidence for multidimensionality and construct independence. J. Educ. Psych. **78**(1), 52–58 (1986). https://doi.org/10.1037/0022-0663.78.1.52

46. Goleman, D.: Social Intelligence: The New Science of Human Relationships. Bantam Books, New York (2006)

47. Blumson, B.: Mental maps. Philos. Phenomenol. Res. **85**(2), 413–434 (2012)

48. Pokrovskaia, N.N.: Tax, financial and social regulatory mechanisms within the knowledge-driven economy. Blockchain algorithms and fog computing for the efficient regulation. In: Shestopalov, M. (ed.) Proceedings of the 2017 20th IEEE International Conference on Soft Computing and Measurements, pp. 709–712. IEEE, New York (2017). https://doi.org/10.1109/SCM.2017.7970698

49. Callon, M., Cohendet, C., Eymard-Duvernay, D., Foray, D.: Réseau et Coordination. Economica, Paris, France (1999). (in France)

50. Brusakova, I.A.: About problems of management of knowledge of the digital enterprise in fuzzy topological space. In: Shestopalov, M. (ed.) Proceedings of the 2017 20th IEEE International Conference on Soft Computing and Measurements, pp. 792–795. IEEE, New York (2017). https://doi.org/10.1109/SCM.2017.7970726

51. Pokrovskaia, N.N., Spivak, V.A., Snisarenko, S.O.: Developing global qualification-competencies ledger on blockchain platform. In: Shaposhnikov, S. (ed.) Proceedings of XVII Russian Scientific and Practical Conference on Planning and Teaching Engineering Staff for the Industrial and Economic Complex of the Region (PTES, St. Petersburg, LETI), pp. 209–212. IEEE, New York (2018). https://doi.org/10.1109/PTES.2018.8604177

52. Wilson, N.S., Bai, H.: The relationships and impact of teachers' metacognitive knowledge and pedagogical understandings of metacognition. Metacogn. Learn. **5**(3), 269–288 (2010). https://doi.org/10.1007/s11409-010-9062-4

53. Chen, J., Wang, M., Kischner, P., Tsai, C.-C.: The role of collaboration, computer use, learning environments, and supporting strategies in CSCL: a meta-analysis. Rev. Educ. Res. **88**(6), 799–843 (2018). https://doi.org/10.3102/0034654318791584

54. Mokos, E., Kafoussi, S.: Elementary students' spontaneous metacognitive functions in different types of mathematical problems. J. Res. Math. Educ. **2**(2), 242–267 (2013). https://doi.org/10.4471/redimat.2013.29

55. Perikova, E.I., Byzova, V.M., Loviagina, A.E.: Metacognitive strategies of decision making in educational activities: efficiency in higher education. Sci. Educ. Today. **9**(4), 19–35 (2019). https://doi.org/10.15293/2658-6762.1904.02

56. Scherer, R., Siddiq, F., Sánchez-Viveros, B.: A meta-analysis of teaching and learning computer programming: effective instructional approaches and conditions. Comput. Hum. Behav. **109**, 106349 (2020). https://doi.org/10.1016/j.chb.2020.106349

57. De Boer, H., Donker, A.S., Kostons, D.D.N.M., van der Werf, G.P.C.: Long-term effects of metacognitive strategy instruction on student academic performance: a meta-analysis. Educ. Res. Rev. **24**, 98–115 (2018). https://doi.org/10.1016/j.edurev.2018.03.002

58. Siddiq, F., Scherer, R.: Revealing the processes of students' interaction with a novel collaborative problem solving task: an in-depth analysis of think-aloud protocols. Comput. Hum. Behav. **76**, 509–525 (2017). https://doi.org/10.1016/j.chb.2017.08.007

59. Bernard, M., Bachu, E.: Enhancing the metacognitive skill of novice programmers through collaborative learning. In: Peña-Ayala, A. (ed.) Metacognition: Fundaments, Applications, and Trends. ISRL, vol. 76, pp. 277–298. Springer, Cham (2015). https://doi.org/10.1007/978-3-319-11062-2_11

60. Ardelt, M., Grunwald, S.: The importance of self-reflection and awareness for human development in hard times. Res. Hum. Dev. **15**(3–4), 1–13 (2018). https://doi.org/10.1080/15427609.2018.1489098

61. Lehrer, R., Lee, M., Jeong, A.: Reflective teaching of logo. J. Learn. Sci. **8**(2), 245–289 (1999). https://doi.org/10.1207/s15327809jls0802_3

62. Guilford, J.P.: The Nature of Human Intelligence. McGraw-Hill, New York (1967)

63. Bandura, A.: Self-efficacy mechanism in human agency. Am. Psychologist. **37**(2), 122–147 (1982)

64. Zimmerman, B.J., Bandura, A., Martinez-Pons, M.: Self-motivation for academic attainment: the role of self-efficacy beliefs and personal goal setting. Am. Educ. Res. J. **29**(3), 663–676 (1992)
65. Ganaie, M.Y., Mudasir, H.: A study of social intelligence & academic achievement of college students of district Srinagar, J&K, India. J. Am. Sci. **11**(3), 23–27 (2015)
66. Tversky, B.: Cognitive maps, cognitive collages, and spatial mental models. In: Frank, A.U., Campari, I. (eds.) COSIT 1993. LNCS, vol. 716, pp. 14–24. Springer, Heidelberg (1993). https://doi.org/10.1007/3-540-57207-4_2
67. Ashraf, N., Banerjee, A., Nourani, V.: Learning to teach by learning to learn. Job market paper (2020). https://economics.mit.edu/files/20802. Accessed 12 Apr 2021
68. Anosova, N., Dashkina, A.: The teacher's role in organizing intercultural communication between Russian and international students. In: Anikina, Z. (ed.) IEEHGIP 2020. LNNS, vol. 131, pp. 465–474. Springer, Cham (2020). https://doi.org/10.1007/978-3-030-47415-7_49
69. Pokrovskaia, N.N.: Leisure and entertainment as a creative space-time manifold in a postmodern world. In: Ozturk, R. (Ed.) Handbook of Research on the Impact of Culture and Society on the Entertainment Industry, pp. 21–38. IGI Global, Hershey (2014). https://doi.org/10.4018/978-1-4666-6190-5.ch002
70. Ababkova, M.Yu., Leontieva, V.L.: Metaphor-based research for studying Russian and Chinese students' perception of the university. In: Shipunova, O., Bylieva, D. (eds.) PCSF 2020 & CSIS 2020. European Proceedings of Social and Behavioural Sciences EpSBS, vol. 98, pp. 89–98. European Publisher, London (2020). https://doi.org/10.15405/epsbs.2020.12.03.9
71. Almazova, N., Rubtsova, A., Eremin, Y., Kats, N., Baeva, I.: Tandem language learning as a tool for international students sociocultural adaptation. In: Anikina, Z. (ed.) IEEHGIP 2020. LNNS, vol. 131, pp. 174–187. Springer, Cham (2020). https://doi.org/10.1007/978-3-030-47415-7_19
72. Ababkova, M.Yu., Fedorov, D.A., Leontieva, V.L., Pokrovskaia, N.N.: Semantics in e-communication for managing innovation resistance within the agile approach. In: Chernyavskaya, V., Kuße, H. (eds.) 18th PCSF - Professional Culture of the Specialist of the Future. The European Proceedings of Social & Behavioural Sciences EpSBS, vol. 51, pp. 1832–1842 (2018). https://doi.org/10.15405/epsbs.2018.12.02.194
73. Lewin, K.: Field Theory in Social Science. Harper and Row, New York (1951)
74. Bandura, A.: Social Learning Through Imitation. University of Nebraska Press, Lincoln (1962)
75. Ababkova, M.Y., Pokrovskaia, N.N., Fedorov, D.A.: Educational services for intellectual capital growth or transmission of culture for transfer of knowledge – consumer satisfaction at St Petersburg Universities. Edu. Sci. **9**, 183 (2019). https://doi.org/10.3390/educsci9030183
76. Baranova, T., Kobicheva, A., Tokareva, E.: The impact of an online intercultural project on students' cultural intelligence development. In: Bylieva, D., Nordmann, A., Shipunova, O., Volkova, V. (eds.) PCSF/CSIS -2020. LNNS, vol. 184, pp. 219–229. Springer, Cham (2021). https://doi.org/10.1007/978-3-030-65857-1_19
77. Cappelli, L., Fedorov, D.A., Korableva, O.N., Pokrovskaia, N.N.: Digital regulation of intellectual capital for open innovation: industries' expert assessments of tacit knowledge for controlling and networking outcome. Future Internet **13**(2), 44 (2021). https://doi.org/10.3390/fi13020044
78. Shearer, R.L., Aldemir, T., Hitchcock, J., Resig, J., Driver, J., Kohler, M.: What students want: a vision of a future online learning experience grounded in distance education theory. Am. J. Dist. Educ. **34**(1), 36–52 (2020). https://doi.org/10.1080/08923647.2019.1706019
79. Troussas, C., Krouska, A., Sgouropoulou, C.: Collaboration and fuzzy-modeled personalization for mobile game-based learning in higher education. Comput. Educ. **144**, 3698 (2020). https://doi.org/10.1016/j.compedu.2019.103698

An Interdisciplinary Approach to Training Teaching Skills for Intercultural Communicative Competence

Anna Kuzmichenko⬭, Arina Mikhailova⬭, Irina Korenetskaya$^{(\boxtimes)}$⬭, and Svetlana Matsevich⬭

Pskov State University, Lenin Square, 2, Pskov 180000, Russia

Abstract. Although intercultural interaction permeates the entire educational system, the introduction of intercultural education at all educational levels is a slow and complex process. Particularly, the intercultural communicative competence (ICC) development among preschool and primary school children requires peculiar teachers' knowledge and skills to form pupils' ICC. The purpose of this article is to analyze the process of developing undergraduates' ICC, and their methodological knowledge and skills in ICC formation in preschool and primary school children. The methodological basis of the work consists in axiological, interdisciplinary, and communicative approaches. The common goal of the research is to develop intercultural skills of future teachers who are ready, on the one hand, to deal with children from interculturally diverse back-grounds, and on the other hand, to prepare children to act in multiculturally diverse society. The study involved 167 students enrolled in the Bachelor programs of Pskov State University. The empirical research included three stages of the experiment. The content of the examination included complexes of interactive methods, with the help of which it was supposed to strengthen the formation of the ICC of bachelors, the ability to work in a team and the development of a set of instrumental competencies together with the development of the methods of teaching ICC to schoolchildren. The practical result of the study is a set of business games facilitating the undergraduates' intercultural communicative competence development as well as their ability to develop the ICC in preschoolers.

Keywords: Intercultural competence · Interdisciplinary approach · Professional development · Professional and pedagogical culture · Teaching skills

1 Introduction

The concept of Russian education and the diversity of socio-economic demands pose bold challenges at different levels of the system to the value meanings of the content, technological approaches and the universality of educational activities. The requirements for the future teachers, who will not only be able to accept these challenges in training practice, but also to transform their needs into their unique individual educational trajectory, become obvious.

© The Author(s), under exclusive license to Springer Nature Switzerland AG 2022
D. Bylieva and A. Nordmann (Eds.): PCSF 2021, LNNS 345, pp. 589–605, 2022.
https://doi.org/10.1007/978-3-030-89708-6_49

The problem is that many future teachers graduate from educational institutions and settle into careers without the requisite competencies to ensure the educational equity that enables all students to attain their personal goals in the global, postmodern world.

In our study we'll highlight one of the most significant competences in global society, i.e. intercultural communicative competence (ICC), which is considered as the basis for the prospective teachers' training. The lack of teachers' ICC may result in dropout and discipline problems [1].

In the framework of our research, we consider not only the approaches and ways of forming ICC among the bachelors themselves, future professionals of pedagogical activity, but also teaching methods by which the future specialist will solve educational problems of developing intercultural communication in preschoolers and children of primary school in pedagogical practice.

To achieve the stated goal in the most effective way we consider it possible to use interdisciplinary approach in the course of undergraduates' educational process.

2   Literature Review

Throughout the XX century, there has been a growing trend in science to synthesize knowledge within the frame of related subjects. It was based on the paradigm of integrity, when the interdisciplinary approach to research and education is formed on the basis of problem-solving and project methods in the disciplinary organization of science and education.

However, the notion of interdisciplinary approach can be variously interpreted. We would like to focus on the definition, suggested by L.P. Repina who views it as the complex of sciences which provides an extensive research space consisting of enlarged fields of knowledge [2].

Nowadays two basic approaches to interdisciplinarity have been formed. The first one suggests the relationship of two or more disciplines with related terminology, research systems, objects of these studies, etc. This combination helps to study the problem of research more widely and thoroughly. Whereas the second approach allows expanding the areas of knowledge that cannot be fully explored by existing scientific disciplines. Thus, at the junction of two disciplines, a new one arises [3].

The first of the mentioned approaches seems to be appropriate for our study as it helps to create a vast field for scientific projects as well as to consider the ICC development from different points of view.

The concept of interculturality is not new for education. It was Comenius in 1600s who suggested the ideas of pedagogical universalism where multiplicity of perspectives was foundational for both knowledge acquisition and for mutual understanding between people of differing backgrounds.

Nowadays the Comenius's principles of peace and international organization of education are conceptualized in terms like intercultural education and intercultural competence [4–8]. In scientific studies we can see such terms as multicultural education [9], global or international education [10], peace education [11] and culturally relevant or responsive education [12].

In the Handbook of Research on Multicultural Education by J.A. Banks the author defines intercultural education as a trend intended to facilitate immigrant students adaptation to a different life in a foreign country, preserving the aspects of their unique ethnic identity, and become the citizens of their new motherland [9].

It is obvious that intercultural education is impossible without some basic competences including intercultural competence.

Now let us dwell upon the notion of competence which can be appropriate for our research.

W. Lonner and S. Hayes state that the concept of competence implies the ability to select one style of behavior as a response to the various challenges of everyday life, dealing with social and work-focused relationships, and searching for solutions to a set of human problems [13]. In case we support this definition, it is obvious that intercultural competence should become a central dimension of teacher training in the context of intercultural education. Intelligent selection of culturally competent teacher behaviors would enable educators to facilitate the teaching of children from diverse cultural backgrounds while providing them with the skills to succeed in an increasingly culturally diverse world. Competent teachers lead students to be competent in the current content of curriculum and performance standards set by the Ministry of Science and Higher education of the Russian Federation and local educational bodies.

The Professional Educator Standards Board mentions the following skills among the necessary ones: the ability to develop in children tolerance and skills of behavior in the changing polycultural environment; the ability to take into account children's cultural differences, to communicate with children accepting their dignity, understanding them; the knowledge of the basics of polycultural education.

The issue of intercultural sensitivity becomes the issue of primary concern and the skill that teachers need to develop. However, there is the question of how to measure teacher intercultural competence.

Searching for criteria and indicators to measure intercultural competence we considered the study of W. Lonner and S. Hayes where there are emotional, contextual and interpersonal intelligence aspects [13]. In the further studies the educators offer adaptation skills for intercultural success such as positive attitude and flexibility, cross-cultural skills (realism, cultural involvement, political astuteness), partnership skills (openness, problem solving).

Studying the pedagogical culture as a set of values allows us to understand how teachers' ideas about the goals, content and methods of the pedagogical process are transferred to students and develop their belief of the teaching approaches. It involves not only creativity and culture but professional development as well. Educators, policymakers, and researchers are looking at teacher's professional development as an important strategy to meet the needs of XXI century children, students and teachers. Professional development is a crucial component of school improvement efforts. It is "a must" or a practical solution to the problem of correction and development of the communicative culture of future teachers which includes ICC as an integral part. The use of a set of core features: duration, collective participation, logic, focus on content, active cognition and a common conceptual framework in professional development of future teachers should be accompanied by the teacher's proficiency and competence. Both being essential parts of the general culture of the teacher, they manifest themselves as a set of values, mastery of techniques and methods of creative pedagogical activity, prerequisites for effectiveness, goal of professional self-improvement and generalized indicators of the teacher's professional competence. They include mastery of the speech culture and methods of effective communication and interaction which in their turn include ICC under study.

Competence biased techniques for the development and correction of the communicative culture of future teachers in the learning process at a pedagogical department of the university may undoubtedly cope with the challenges of the XXI century and lead to further formation and development of ICC of preschool and primary school children.

Intercultural Competence Assessment project [14] framework provides a solid basis for assessment of ICC in professional settings. It is based on the work of M. Byram [15] as well as the key aspects of the integrative multifunctional model of bilingual teacher education, suggested by N. Almazova et al. [16], and a number of target components, elements of constructive management, indicators of effective management model, developed by A. Rubtsova [17]. The great advantage of the presented model is that it is based on several scholars' perspectives, rather than just one scholar's perspective.

Having analyzed the ICC theories, we can identify several components they have in common: the *attitudes* (interest and a strong motivation to get to know representatives with a different ethnic background; ability to value cultural diversity and variety of views; respect people of different cultural affiliation); the *knowledge* and *understanding* (know the peculiarities of the representatives of different cultural affiliation; ability to understand and accept the diversity of cultural groups; be aware and understand certain cultural stereotypes and common prejudices); the *skills* (ability to decentre from one's own perspective and to take other people's perspectives into consideration in addition to one's own; to discover information about other cultural affiliations and perspectives; to interpret other cultural practices, beliefs and values and relate them to one's own; be empathetic, etc.); *actions or activities* (ability to use any opportunity to interact with people who have different cultural affiliation; to interact and communicate appropriately, effectively and respectfully with people of a different cultural background, etc.).

The successful acquisition of these components by students within the ICC may provide a solid basis for global citizenship.

3   Methodology

The first and foremost methodological cornerstone of the study is the axiological approach. So, the human being with individual spiritual values, interpersonal sensitivity and empathy, understanding and acceptance of another human being should be in the center of our pedagogical efforts and should be considered as a philosophical and pedagogical strategy that shows the ways of developing professional skills, and suggests prospects for improving the educational system.

The second methodological basis which was applied primarily at the initial stage is the context-activity approach. It is accompanied with interdisciplinary principles in teaching the discipline, the inclusion of interactive forms and methods of teaching aimed at the development of pedagogical skills in students, early contextual immersion in the professional activities of the future specialists of preschool education with early learning of a foreign language.

The third methodological idea concerns the interdisciplinary approach to develop ICC. We suppose that such disciplines as *"Preschool pedagogy and multicultural education", "EFL", "Theory and Practice of Intercultural Communicative Competence"* with synchronized goals and approaches can provide effective ICC development and it will raise the initial level of ICC in the students, especially since the integrated learning in higher professional education has proved its effectiveness [18]. EFL acquires a special importance in this set of disciplines due to the need to solve the challenges of today's globalized world [19].

The common goal is to develop intercultural skills of future teachers who are ready, on the one hand, to deal with children from interculturally diverse backgrounds, and on the other hand, to prepare children to act in multiculturally diverse society.

The study involved 167 students enrolled in the Bachelor programs of Pskov State University, Russia (Table 1).

Table 1. Distribution of students by departments and training courses

Field of study 44.03.05 Pedagogical education with two profiles of training			
Profiles of training			
"Preschool education and a foreign language"	"Preschool education and Social Pedagogy"	"Primary education and correctional pedagogy"	"Russian and Foreign (English) Language"
1st course 21 students	2nd course 18 students; 3rd course 14 students; 4th course 17 students	2nd course 25 students; 3rd course 27 students	2nd course 10 students; 3rd course 11 students; 4th course 11 students 5th course 13 students
Total number of participants: 167 students			

4 Results and Discussion

The idea of the study was the assumption that interactive teaching methods are a means of forming competencies, both universal, general cultural, and professional, provided by the FSES. A set of interactive methods in the teaching practice of disciplines will contribute to the formation of ICC, development of the ability to work in a team, formation of instrumental competencies while learning about the methods of teaching intercultural communication to children of preschool and primary school.

The logic of the coordination of empirical research was subordinated to the three stages of the experiment. At the first stage the students were examined. The content of the examination included complexes of interactive methods, with the help of which it was supposed to strengthen the formation of the ICC of bachelors, the ability to work in a team and the development of a set of instrumental competencies together with the development of the methods of teaching ICC to school children (Table 2, Fig. 1).

Table 2. Disciplines within which the general idea of empirical research was implemented

Profiles of training		
"Preschool education and a foreign language"; "Preschool education and Social Pedagogy"	"Primary education and correctional pedagogy"	"Russian and Foreign (English) Language"
Disciplines		
"Preschool education and multicultural education"	"Education and socialization of primary school children in a multicultural educational space"	"English as a Foreign Language"; "Theory and practice of Intercultural Communicative Competence"

It was necessary to analyze the complex of the most frequently and effectively used interactive methods in the professional education of teachers of pedagogical departments (Fig. 1).

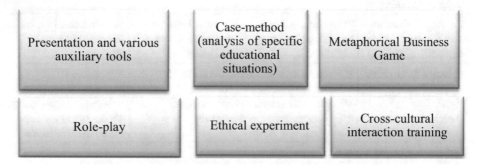

Fig. 1. A set of interactive methods implemented within the disciplines

Then, at the next stages of the empirical study, the following goal was set: to find out the degree of the effect, produced by the methods of interactive teaching the undergraduate students while developing their ICC and competence to work in a team of other undergraduate students of a pedagogical department.

The assumed consequence of the result after the educational co-activity of the teacher and undergraduates was predicted by the pedagogical task that while studying these courses, bachelors will acquire the necessary "knowledge" foundations, the base of pedagogical instrumental skills for further self-education in the field of Pedagogics, and design their original methods for the development of ICC in preschool and primary school children.

The literature analysis helped us to determine that in the formation of competencies in bachelors during the pedagogical interdisciplinary training and the development of pedagogical skills by students, the methodology for the development of ICC in pupils is determined in the target composition of the planned educational process by three bases (Fig. 2).

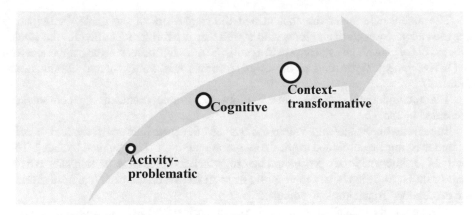

Fig. 2. The target composition of the experimental educational process

The given structure of the experimental educational process, the activity-problematic basis of the target composition will provide a practical basis for the formation of pedagogical skills, tools for the methodology of ICC in the education of preschoolers, and junior schoolchildren.

The need to highlight this basis is important in order to immerse students in the subject knowledge of the phenomenon of Childhood through specially created problem-contexts of the theory and methodology of preschool and primary education, and by means of EFLT as well.

The cognitive basis in the target composition of bachelor's training has determined a series of interactive methods in teaching disciplines, when the teacher can differently determine the range of co-activity with undergraduates, taking into account their educational level at a particular stage of learning, and facilitating the individual educational path of the student.

The context-transforming basis for preparing bachelors for the application of the methodology for the formation of ICC in children of preschool and primary school determines the following points:

- high level of acquisition of teaching skills by the future teachers in the sphere of modern disciplines, and strong motivation to transform the methods of ICC developing in preschool and primary school children, following their individual needs, and creating the successful trajectory of the educational route of pupils in intercultural communication.

Then the process tasks and stages of the educational process were designed. Students had to pick out, "collect", and develop the forms of interactive learning used in the lecture-seminar system of teaching the interdisciplinary subjects mentioned above.

As one of the "cross-cutting – frequently used" interactive methods of the educational process for mastering the content of the disciplines mentioned above, the method of metaphor business game has been chosen.

A business game includes a certain model or pattern and a game itself. Actions of participants are determined by the key aspect of the game framework. Roles mark the prerequisites of the game, they are determined by the script, or the moderator (a teacher).

The model determines the content and the framework of the game. Participants choose and perform particular roles with which they are further identified for the whole process of the game. The game is provided with the rules that reflect all the components of ICC development. These rules are borrowed from the social context of real-life situations (Table 3).

The methodology of organizing the business game included the components presented in Table 3.

Together with the students within the framework of the interactive method of contextual learning we developed the business game-marathon "Pedagogical kitchen". The idea of the "kitchen" is of a procedural nature, which requires from participants to have certain skills from elementary to virtuosic or professional on the principle of mastering the profession "from simple to complex".

"Pedagogical Kitchen" as an interactive metaphorical business game includes a series of professional tests that are the basis for the formation of pedagogical skills and instrumental competencies for team work and the acquisition of methods for the development of ICC in preschoolers and children of primary school.

Methods and techniques of interaction with students that provide a contextual learning process in a marathon game, as well as practical exercises, may be the following:

- working in dyads overcoming the obstacles in the travelling metaphorically called "Direct and difficult questions about the preschool education system in the countries of travel".

Organization: students join in dyads for solving a given practical problems.

Task: to create and present the evidence base: "Pedagogical menu" (the day regime, features of organizing classes with children, etc.) of foreign experience in the application of Waldorf pedagogy, Montessori Pedagogy in the preschool education system.

Task: to develop and present an advertising booklet (as an exquisite pedagogical dish with one educational ingredient) for parents about the advantages of author's pedagogical systems in the education of preschool children.

Task: to develop educational situations for preschool children in Russian and English languages for the methodical student manual using 3 ingredients (excursion, native speaker, interaction), present the originality of the idea.

During the lecture-conversation and discussion-dialogue the teacher discusses with the students the topic: "*To love not to bring up. Or how to accompany the formation of the experience of social behavior of preschoolers?*".

In the dyad students are divided into "*chefs*" and "*sushi chefs*" of pedagogical kitchen.

Organization: the chef determines the theoretical basis of the meeting with the teacher: *(plans the menu and the composition of the pedagogical dishes, applying the principle of different preferences, i.e. theoretical points of view)*;

the sushi chef finds practical examples, experience, and evidence, integrating in a single concept of the presentation of the prepared pedagogical product for the meeting with the teacher (demonstrates *ready-made dishes*, tells about the composition of *pedagogical dishes (methodology)*, *applying the principle of different preferences, i.e. specific practical examples to the topic of discussion*).

Table 3. Components of students' and teachers' activities in the interactive game "Educational kitchen"

The metaphor of the game:
A skilled cook is not someone who has a lot of seasonings in the kitchen, but someone who does not add anything unnecessary to the dish
A real chef is someone who can cook a dozen dishes out of one ingredient
General high culture is one of the indispensable conditions for the formation of a high-class cook

Components of activities	Characteristics of students and teachers activity
Motivational and productive component of activity	It involves a joint choice of the method of teaching with students, the implementation of this "co-activity" with students takes place through a series of extracurricular meetings using the method of "brainstorming". Students are offered a variety of interactive forms of interaction on the development of the content modules of the discipline, but with a brief description of the idea in order to include in the problem-activity circumstances of the choice of mastering the topic
Cognitive-professional component of activity	It determines the process of the entire series of the organization of the learning process (time, methods, responsibility, the participation of all subjects of "co-activity", the development of contextual situations)
The reflexive component of activity is designated by the students as "the territory of truth"	The main idea of the reflexive stage involves a content analysis of all the stages of training in the content modules of the discipline, with a focus on identifying problems in the development of content material by students, determining ways to further search for ways of practice-oriented knowledge, acquiring pedagogical skills, and inter-disciplinary experience in the development of methods for the ICC formation in preschool and primary school children

The actual business game is aimed at a detailed study of typical situations, the main goal of such a game is to find non-standard ways to solve the problems

Task: development of experience of participation in practical trials and search for joint solutions, interaction with the priority of using analytical exercises at seminars that activate students' thinking, aimed at finding new knowledge, ways to solve the problems posed due to the formation of their own conclusions and generalizations.

Using the chart below students analyze the given cases, taking into account the categories and indicators (Table 4):

Table 4. ICC attitude development categories and indicators [20]

Category	Indicator
Tolerance/acceptance	Openness or welcoming reactions toward people, objects or behavior from a different culture; respect of rules of intercultural situation
Interest	Sometimes being disinterested in topics or other newly introduced features
Motivation for contact	Being eager or becoming involved or to be in contact with students from other cultural background
Motivation for language	Willing to learn EFL and show appreciation for language skills
Factual knowledge	Reproducing and recounting facts relating to national or ethnic culture, identity, habits and rules
Language knowledge	Reproducing and recounting words and phrases in EFL
Lack of knowledge	A deficit in factual knowledge on culture-related issues or language knowledge
Meta-linguistic knowledge	Reproducing assumptions or factual knowledge about language, language construction, communication
Communication strategies	The adequate language when interacting with chosen interlocutor, active mime or body language to react and interact with the interlocutor, some fail in interaction; some use negative strategies when the intention is not to enhance the communication but to stop it
Skill of discovery	Asking questions to find out something, are inquisitive
Fear/rejection	Avoiding contact, showing signs of discomfort when exposed to manifestations of cultural differences; refusing to contact with students who have higher level of ICC or speaking on the topics related to other cultures
Hesitation	Avoiding or being cautious and shy speaking on cultural issues
Regret	Expressing sadness and disappointment about certain intercultural situations

Below we present a couple of case tasks, which were offered to students for further analysis (Fig. 3):

The students interpret the cultural, psychological, pedagogical, and methodological aspects of this case, answering the following questions:

– What differences can there be in the ways of communicating in different cultures?
– What are the peculiarities in the way children of different national affiliation read, write, speak?

> *Case task 1:*
> *In some kindergarten groups of the city of Pskov, Russia, we can meet besides Russian children, the kids from, Estonian, Belorussian, Latvian families (due to the border location of Pskov). There are also representatives from Armenia, Georgia as well. How should the kindergarten teacher plan the work in these multicultural preschool groups?*

Fig. 3. Example of case task 1

- What ways can be suggested to cooperate with the family in order to facilitate the adaptation process of a child of a different ethnic background?
- Should adaptation programs for boys and girls be different? Why?
- Point out the aspects of gender differences that need to be taken into account by the teacher of the multicultural children's group.

The students are given some practical-oriented tasks as well:

- to give some options of joint events for multicultural kindergarten groups of different ages.
- to suggest some problem situations or events for preschoolers on the topic "United in Diversity".

Following the plan, students analyzed the following situation (Fig. 4):

> *Six-year-old boys Antanas (from Estonia) and Dima (from Russia) decided to play animal trainers. Antanas says to Dima: "I will be a trainer, and you will be a tiger. I'll train you, and you won't listen. Then I will scold you, beat you with a whip, and I will not feed you."*

Fig. 4. Example of case task 2

The students explained this situation from a psychological, pedagogical, and methodological point of view, taking into consideration the following points:

- What is the definition of the concepts of "emotions" and "feelings"?
- What is the essence of the emotional development of preschoolers?
- What feelings can children have in this game?
- What is the "information space" and how does it affect the content side of the game?
- How should children's games with negative content be treated? Is it permissible for an adult to directly interfere with children's play?
- What methods and techniques can be used to switch the children's attention from this game to a game with positive content.
- What should be done if children abandon other story lines in the game, but insist on their own spontaneous plot?

Later on the students performed two practice-oriented tasks:

- Offer your own versions of the game of animal trainers with positive content and questions of the questionnaire for parents on the topic "What games do your children play?".
- Make recommendations for parents to organize joint games with preschoolers.

Here are other examples of individual meaningful interactive practices with students in mastering the content of the first module "Pre-school pedagogy" in the game-marathon *"Pedagogical Kitchen"* 1.

Educational game tasks: choose the path for the "European Express" to the countries of the express map to the territory of Childhood.

Goal: to continue developing the language component of the ICC in bachelors in the aspect of "professional vocabulary" of preschool pedagogy *(to make a travel route in Russian and English to countries where the main route is connected with the "territory of Childhood")*.

Tasks:

Make a decision about how to answer the questions after the trip:

- Where *(in which countries)* the priority of pre-school Childhood care is the landmark? (collective decision of the team expedition report), (*oral presentation in several languages*).
- Arguments, facts, journalistic investigation: "What is the framework of pre-school education in the countries that you will visit once?" (*oral presentation in several languages*).
- Encrypt an electronic report in two languages to your fellow students about your journey, ask them what they know about the principles, ethics, and content of education in European countries.
- Use not only written, verbal communication systems, but also think about other systems of cross-cultural communication. Determine the rotation of age boundaries in your travel reports.
- Ask yourself, which recipe for each of the countries is authentic and universal for the education of children? Formalize your response as a blogger-user of social networks. Think through and organize a group project of other forms of interaction with the parent community: prepare a two-languages report on the topic.
- Make an independent decision in the design of your practical and scientific observations, use the collective experience of your fellow students in forming an opinion about the trip.
- Make a recipe for the education of preschool children, following the example of education in the countries on the way of your journey.

Game-marathon "Pedagogical kitchen" 2.

Educational game tasks: The menu of children's subculture in classical pedagogy, in the countries of the world: theory and practice.

Task: Complicated context of the game situation: immersion in a professional situation, "student players" do not know the recipe-composition of the dish called "preschool subculture".

Purpose: to teach how to make up the professional map of the teachers' support for the development of children's subculture (in Russian and English).

At the third stage of the study, it was taken into account that interactive teaching methods contribute to the formation of universal, general cultural and professional competencies. Educational activities at the previous stages with students using the above set of methods were aimed at the formation of universal competencies according to the FSES.

The research method is a survey. The research methodology is the assessment called (self-analysis) "My experience of competency", developed by the authors of the article.

Introspection was conducted in the student group: the actual introspection, as well as the assessment of each student, was carried out by all participants.

The final figures of the assessment is a level of the development of the respondent's competencies. The scores are set in the introspection card, and introspection should be conducted anonymously, which significantly increases the reliable correlation of the survey. Then the arithmetic average of each position is calculated.

As a result of the assessment, we can evaluate the self-development of the students, the development of competencies, the improvement of relations in a group, as well as the choice of teaching methods. The evaluation was carried out according to the algorithm:

- firstly, the competencies that are formed in students in the learning process were defined;
- secondly, the map for the self-analysis for assessing competence was worked out with a scale from 1 to 4 points (Table 5);
- thirdly, we informed the students beforehand about the assessment, due to that they managed to develop the right attitude to the procedure;
- fourth, the students are informed about the results of the study, in comparison of self-assessment and external assessment, and the overall results are analyzed.

Introspection was organized for students after the first and last experimental games. The following competencies given in the FSES of Russia were taken into account: "ability to work in a team", "ability to carry out cross-cultural communication", and they were represented by the following skills and abilities, which became the content of the introspection map (Table 5).

The work with the self-analysis maps of the formation of the competence "ability to work in a team" determined the following conclusions. In the initial survey, the average indicator of the formation of the competence "ability to work in a team" was 2.75 points.

At the end of the semester (after 3 months of conducting the interdisciplinary game "Educational kitchen"), according to the results of the final self-analysis, the indicator had a positive dynamics, changes amounted to 0.625 points, the final average result showed 2.125 points (maths average).

The block of self-analysis of the formation of the competence "the ability to implement intercultural communication" showed the following results:

Table 5. Self-analysis map: "My experience of competence" ("ability to work in a team", "ability to implement intercultural communication")

№	Questions of introspection	1	2	3	4
	Source of information-questions on the ability to work in a team				
1	I adapt to new circumstances, perform my work in a usual way				
2	A constructive dialogue with any student of my group is the principle of my way of communication				
3	I suggest my own ideas about the way how to solve the task				
4	I present strong arguments to convince other students				
5	I am able to be either the leader or obeying, depending on the task				
6	If I am mistaken. I can admit it, and accept another view				
7	I am interested in the point of view of the rest of the group				
8	I know how to controlmy emotions, abstracts from personal likes/dislikes				
	Source of information-questions on the ability to carry out cross-cultural communication				
1	I can form a different cultural identity, i.e. I can learn the foreign language, other values, norms, patterns of behavior				
2	I apply the methods of acquiring of the maximum amount of information and the necessary knowledge of another culture				
3	I can achieve success in interaction with representatives of a different cultural group, even with little knowledge of the culture of the partners				
4	I am able to evaluate the situation and interact in it,following the particular norms and values of a foreign culture				

Indicators for evaluation: from 1 to 4 points
1 point-always expressed;
2 points-expressed in most cases;
3 points-rarely expressed;
4 points - never manifests itself

1) The ICC is fully developed: 75% of students believe that participation in interactive forms for studying sections of the discipline really makes it possible to understand the meanings, ways of implementing intercultural communication, contributes to improving the level of foreign language acquisition, students and teacher had positive emotions during the experiment, "ice" in the interrelation between a student and EFL has been broken;

2) The ICC is poorly developed: 10% of students remain skeptical about the experimental work, they had difficulties due to the lack of vocabulary, so that and they had to consult the dictionary most of the time, consequently, they spent time searching for the meaning of new words and that distracted students from free expression of

their thoughts. This group of students are sure that our experiment is not a good way to conduct the classes;

3) The ICC is partly developed: 15% of students believe that it is always possible to engage in self-education, this is necessary in order to solve sudden problems of interactive interaction, they were interested in what they could evaluate, correct each other

The distribution of ICC level development is shown in Fig. 5:

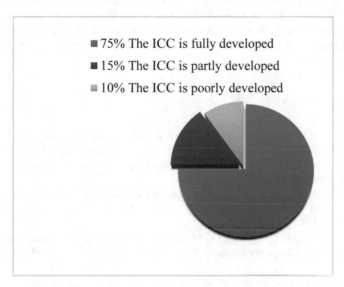

■ 75% The ICC is fully developed

■ 15% The ICC is partly developed

▨ 10% The ICC is poorly developed

Fig. 5. The results of the self analysis of the ICC formation

5 Conclusions

According to the conception of being a global citizen, it seems that the interdisciplinary approach provides an opportunity to consider complex phenomena like ICC from the positions of various subjects.

To apply interdisciplinary approach we have designed the objectives and stages of the educational process on the basis of the following combination of subjects: "Preschool education and multicultural education", "Education and socialization of primary school children in a multicultural educational space", "English as a Foreign Language"; "Theory and practice of Intercultural Communicative Competence".

Our research resulted in "Educational kitchen" (an interactive form of contextual learning) which is determined by the ideas of "co-activity" of bachelor students and teachers in the implementation of complex and ambitious challenges in the preparation of a professional who is able to realize the educational needs of students, taking into account their individual and personal needs in the trajectory of successful development.

The results of the experimental study show a number of trends in the application of methods of interactive teaching of students.

The first tendency: positive effect of the set of interactive methods on the formation of the ICC.

The second trend: the acquisition of skills of self-analysis and evaluation of students can be successfully applied in the educational practice of teaching other disciplines with the construction of feedback "student-teacher-student".

The third trend: the use of interactive teaching methods, business games, in teaching of pedagogical disciplines, allows to cover the issues of professional training of bachelors in pedagogical education, to raise the interest of students in the educational process, which raises their cognitive, emotional and behavior levels, to form the team working skills, to analyze the problem and generate non-standard solutions taking into account the principles of ICC.

References

1. Davis, J.E., Jordan, W.J.: The experience of school context, structure, and experiences on African American males in middle and high schools. J. Negro Educ. **63**, 570–587 (1994)
2. Repina, L.: Interdisciplinary Approaches to the Studies of the Past. Aspekt Press, Moscow (2003).(in Russian)
3. Kreps, T.V.: Interdisciplinary approach in research and teaching: advantages and problems of application. Sci. Bull. South. Inst. Manag. **1**, 115–120 (2019). https://doi.org/10.31775/2305-3100-2019-1-115-120. (in Russian)
4. Fantini, A.E.: Reconceptualizing intercultural communicative competence: a multinational perspective. Res. Comp. Int. Educ. **15**, 52–61 (2020). https://doi.org/10.1177/174549992090 1948
5. Álvarez, L.F.C.: Intercultural communicative competence: in-service EFL teachers building understanding through study groups. Profile Issues Teach. Prof. Dev. **22**, 75–92 (2020). https://doi.org/10.15446/profile.v22n1.76796
6. Byram, M., Golubeva, I., Byram, M., Golubeva, I.: Conceptualising intercultural (communicative) competence and intercultural citizenship. In: The Routledge Handbook of Language and Intercultural Communication, pp. 70–85. Routledge (2020). https://doi.org/10.4324/978 1003036210-6
7. Byram, M.: Teaching and Assessing Intercultural Communicative Competence. Multilingual Matters (2021). https://doi.org/10.21832/9781800410251
8. Valieva, F., Sagimbayeva, J., Kurmanayeva, D., Tazhitova, G.: The socio-linguistic adaptation of migrants: the case of Oralman students' studying in Kazakhstan. Educ. Sci. **9**, 164 (2019). https://doi.org/10.3390/educsci9030164
9. Banks, C.A.: Intercultural and intergroup education, 1929–1959: linking schools and communities. In: Banks, J.A., McGee Banks, C.A. (eds.) Handbook of Research on Multicultural Education, pp. 753–769. Jossey-Bass, San Francisco (2004). https://www.researchgate.net/publication/240702888_Book_Review_Handbook_of_Research_on_Multicultural_Education_2nd_ed. Accessed 17 Apr 2021
10. Merryfield, M.M.: Making Connections Between Multicultural and Global Education: Educators and Teacher Education Programs. American Association of Colleges of Teacher Education, Washington, DC (1996)
11. Stomfay-Stitz, A.M.: Peace Education in America, 1928–1990: Source Book for Education and Research. Scarecrow Press, Metuchen (1993)

12. Gay, G.: Connections between classroom management and culturally responsive teaching. In: Evertson, C.M., Weinstein, C.S. (eds.) The Handbook of Classroom Management: Research, Practice, and Contemporary Issues, pp. 343–370. Lawrence Erlbaum, Mahwah (2006). https://doi.org/10.4324/9780203874783.ch13

13. Lonner, W.J., Hayes, S.A.: Understanding the cognitive and social aspects of intercultural competence. In: Sternberg, R.J., Grigorenko, E.L. (eds.) Culture and Competence: Contexts of Life Success, pp. 89–110. American Psychological Association, Washington, DC (2004). https://doi.org/10.1037/10681-004

14. Intercultural Competence Assessment (INCA). INCA: Assessor manual (2004). https://ec.europa.eu/migrant-integration/librarydoc/the-inca-project-intercultural-competence-assessment. Accessed 17 Apr 2021

15. Byram, M.: Teaching and Assessing Intercultural Communicative Competence. Multilingual Matters, Clevedon (1997)

16. Almazova, N., Eremin, Y., Kats, N., Rubtsova, A.: Integrative multifunctional model of bilingual teacher education. IOP Conf. Ser. Mater. Sci. Eng. **940**, 012134 (2020). https://doi.org/10.1088/1757-899X/940/1/012134

17. Rubtsova, A.V., Almazova, N.I., Bylieva, D.S., Krylova, E.A.: Constructive model of multilingual education management in higher school. IOP Conf. Ser. Mater. Sci. Eng. **940**, 01213 (2020). https://doi.org/10.1088/1757-899X/940/1/012132

18. Baranova, T.A., Kobicheva, A.M., Tokareva, E.Y.: Does CLIL work for Russian higher school students? In: Proceedings of the 2019 7th International Conference on Information and Education Technology, ICIET 2019, pp. 140–145. ACM Press, New York (2019) https://doi.org/10.1145/3323771.3323779

19. Aronin, L.: Multilingualism in the age of technology. Technol. Lang. **1**, 6–11 (2020). https://doi.org/10.48417/technolang.2020.01.02

20. Gerlich, L., Kersten, H., Kersten, K., Massler, U., Wippermann, I.: Interculturalencounters in bilingual preschools. In: Kersten, K., Rodhe, A., Schelletter, C., Steinlen, A. (eds.) Bilingual Preschools. Volume I: Learning and Development, pp. 137–173. WVT, Trier (2010)

The Issue of Adaptive Learning as Educational Innovation

Elizaveta Osipovskaya[1]([✉]) [iD] and Svetlana Dmitrieva[2] [iD]

[1] Peoples Friendship University of Russia (RUDN University), 6 Miklukho-Maklaya Street, Moscow 117198, Russian Federation
[2] Perm National Research Polytechnic University, Komsomolsky Avenue 29, Perm, Russia

Abstract. Covid-19 has catalyzed positive disruptive change in the contemporary education system. It has accelerated the necessity of an innovation implementation process, technology-enhanced learning environment, customized and student-centered learning. It has enabled educators to tailor course content and pathways to students' strengths, skills, interests and learning profiles. Adaptive learning is a contemporary method orchestrating human and digital resources to meet the needs of individual learners. The paper gives a review of the innovation process in education and the implementation of theoretical adaptive learning framework in Russian universities. It identifies innovators and early adopters who can assist two Russian Universities in attempts to implement new educational approaches. The authors also consider if there are any correlations between professors' self-categorization and their practical application of adaptive learning in terms of the pace of learning, type of content presentation, knowledge state and students' interests. Moreover, the researchers investigate and evaluate the Public Relations students' perceived importance of factors affecting their learning: subject choice, subject knowledge acquisition, a type of subject content (text, audio, video), an individual pace in studying the subject, a lecturer's guidance and group size.

Keywords: Innovation in education · Diffusion of innovations · Adaptive learning

1 Introduction

The biggest challenge for any education system is the ability to respond to the demands of the twenty-first-century labor market. Universities are increasingly investing in innovative technologies and call for teachers to implement them immediately, overlooking the importance of organizational culture change and new ways of thinking [1]. Innovative thinking is highly essential when it comes to coping with such environmental turbulence as the COVID-19 pandemic. A recent study showed that Russian universities with a low degree of flexibility and responsiveness and a high level of bureaucracy had a hard time during the coronavirus outbreak [2].

Some studies equate the concept of "innovative thinking" with creative thinking, but that is a misperception since innovation is a manifestation of creativity. The cutting-edge developments are the consequence of creative thinking. The term "innovation" comes from the Latin word "novatio" (the change and renewal). However, not every change means innovation. What makes it different? Firstly, it radically transforms the normal state of affairs (market situation, university structure and process, internal communications with students). Secondly, innovation is not the result of coincidence it is led purposefully. And, thirdly, innovation creates added value for universities in the increased number of applicants, brand awareness or social contribution [3]. There are several types of innovations: 1) technological—realization of potential scientific-and-technological advance into real new products and services; 2) social—renovation in certain areas of society; 3) product—a creation of products with new characteristics that boost their value; 4) organizational—transformations in the enterprise management systems and personnel management strategy [4].

Moreover, this phenomenon means not only the invention of a new product or service that has not existed before but the constant implementation of improvements in certain processes. Thus, innovation can be attributed both to radical and gradual (incremental) changes in the products or strategies.

2 Literature Review

2.1 Implementing Educational Innovations

The innovation implementation process in education encompasses the following steps [5]: 1) defining the goal and conducting a situation analysis at university, city, region, country or international levels; 2) choosing the resources (personnel, logistical or financial) to create a concept and determine the scope and extent of innovation application; 3) creating a basic experimental plan and defining expected outcomes; 4) getting innovation into practice and making it as a tradition

One of the contemporary approaches to the innovation implementation process belongs to the Lifelong Learning Lab. It could be applied either to the business or education industry. The laboratory develops the framework that includes three major areas of focus: 1) the organizational culture of thinking, 2) the key part of the efforts, 3) the scope of innovation application. The culture of thinking describes the way a company operates and its innovation-oriented focus. The key part of the efforts means the starting area of the innovation implementation process: 1) product, 2) model, 3) communication process, 4) technology. The third area is the scope of innovation application that determines the extent of innovation dissemination that covers 1) an entire company, 2) a specific department and a program, 3) a particular subject or a customer (student) journey map.

Thus, we have determined that the innovation implementation process in education has several entry points, however, it is not always successful. Then why do some innovations fail? The diffusion of innovations could be the key answer. The diffusion of innovation theory proposed by Gabriel Tarde and popularized by Everett Rogers includes five stages of the individual technology adoption process: 1) innovators—2.5%, 2) early adopters—13.5%, 3) early majority—34%, 4) late majority—34%, and 5) laggards—16% [6]. Porter and Graham [7] criticize Rogers' ideas because the results of their study are significantly different: innovators—2.9%, early adopters—30%, early majority—49.5%, later majority—16.7% and lagging—1%. We assume that it depends on the innovation adoption stage of each university. Nevertheless, in general, Rogers' classification corresponds to the typical behavior of people when introducing innovative ICT in the education process. This is confirmed by the studies of Baikal et al. [8], Lugma et al. [9], Rusek et al. [10]. We conducted a survey to verify Rogers' theory, its results are presented in the Methodology section.

2.2 The Adaptive Learning Systems

The attitude of educators toward ICT and innovative technology is a crucial factor for introducing new approaches and learning frameworks in any university. In our study, we focus more on adaptive learning systems. Adaptive technologies are called magic bullet that solve all the educational problems [11]. It is also called smart learning that encompasses such features as adaptive support when a system is based on individual needs from learning performance, behaviors, profiles to personal factors, etc.; and adaptive interface, when the system changes the ways of presenting information [12].

Essa claims that next-generation adaptive learning system has seven substantial characteristics: 1) cost-effective to build, maintain, and support; 2) accurate in its assessment of learner knowledge state; 3) efficient in carrying out decisions and recommendations; 4) able to support a vast number of students; 4) flexible in integration; 5) generalizable to domains beyond Science, Technology, Engineering, and Mathematics (STEM) disciplines; 7) learner control, self-regulated learning. The author also presents a formal framework for an adaptive learning system that encompasses five interacting models: 1) domain model specifies what the learner needs to know (knowledge space); 2) learner model represents what the learner currently knows (knowledge state); 3) assessment model is how we estimate a learner's knowledge state; 4) pedagogical model specifies the activities to be performed by the learner to attain the next knowledge state (watching, listening, reading, collaborating, playing, etc.); 5) transition model determines what the learner is ready to learn next [13]. Essa uses an intelligent tutoring system ALEKS (Assessment and Learning in Knowledge Spaces), which is based on KST, in implementing evidence into practice.

Another interesting example of adaptive learning is Lifelong Learning Lab's framework that refers to knowledge space theory. The researchers write the detailed theoretical model based on a practical subject (project management) but do not apply it to the system due to the absence of an appropriate one [14].

In contemporary scientific papers, three types of adaptive learning interventions are discussed: system, framework and model. Most scientific papers use the term "system" [15–21]. It stands for a group of interacting elements that act according to a set of rules in a technology-enhanced learning environment. The most identified kind of system is the intelligent tutoring system. The framework [12, 22, 23] provides a set of guidelines (concepts, practices, values) on how to achieve a better adaptive learning experience. Meanwhile, some papers use the term "model" [24–27], which refers to the pattern of something or description.

According to the literature review, adaptive learning systems are most commonly used in mathematics, physics, computer literacy and biology. It was also found that they are designed as platforms to teach and learn English and German languages.

Further, we examine the well-established adaptive learning systems Alta (Knewton), Canvas, Blackboard Learn, Kahoot! and Schoology by five key features.

Table 1. The most popular adaptive learning systems.

	Alta	Canvas	Blackboard learn	Kahoot!	Schoology
Question variety	+	+	+	+	+
Real-time assessment	+	+	+	+	+
Multimedia content	+	+	+	+	+
Gamification	+	–	–	–	–
Personalized learning paths	+	+	+	–	–

As indicated in Table 1 Alta (Knewton) meets all the educational needs whereas Canvas, Blackboard Learn, Kahoot! and Schoology do not support gamified features. All systems are flexible, generate diverse learning experiences and have a variety of layouts, quiz questions. Moreover, they have their own multimedia content, content authoring tools and provide real-time student assessment. It should be mentioned that despite its adaptability and scalability, not every platform has personalized paths for students.

In the language teaching field, adaptive learning technologies had not attracted the attention of the English language educators until 2013 when Knewton signed contracts with Macmillan and Cambridge University Press, two large English Language Teaching (ELT) publishers [28].

Today, the interest in the use of adaptive technologies in ELT is rapidly growing. However, language learning being a social and interactive discipline poses additional challenges to both researchers who aim to develop innovative products and to EL teachers who should have competencies to apply them to the teaching process.

The researchers identified three areas where adaptive technologies can be best applicable: teaching vocabulary, testing, and developing teaching platforms (Learning Management Systems).

Teaching vocabulary: the application of adapting technology for vocabulary acquisition can introduce variety, alleviate the boredom of drilling new words, and meet the needs of learners with multiple types of intelligence. Learners experience the language in audio and visual formats working at their pace in big groups. Gamification elements turn language learning into a pleasant adventure and enable learners to compete with each other [29]. There is some evidence [30] that such apps lead to a better memorization of vocabulary, though other researchers [31] claim that they facilitate learning of high-frequency words for receptive use while learning for communicative purposes requires meaningful context.

Testing: computer-adaptive tests measure the language proficiency of students and place them in the classes according to their level. The benefits of computer-based tests include greater precision of scoring and reduction of time and cost of testing, while the key challenge remains the security of data [32].

The use of a Learning Management System: ideally such a platform should replace everything that happens in a real-life classroom where a teacher guides students through a given textbook. A completely adaptive course is a future goal for big publishing companies; today online English language courses include only elements of adaptive learning. The major challenge is the complex nature of the English language which cannot be broken down into "knowledge chunks" and studied linearly [33]. This is consistent with Thornbury [34] who supposed that the complexity of language as multiple interrelated subsystems of grammar, lexis, phonology, discourse, pragmatics might be a challenge to develop an adaptive language learning platform. Other challenges include learner to learner interaction, students' control over the learning process, receiving teachers' feedback. Teachers should acquire new skills to cope with adaptive technologies and modify old routines [33].

3   Methodology

In November 2020, we conducted a study to measure the attitude toward ICT innovations among professors of RUDN University and PNRP University and evaluate their preparedness to create personalized learning pathways. The sample included a total of 73 teachers from both universities. The response rate of the online survey amounted to 93% (68 people), 71% women. The average age of educators was 49 years, the youngest respondent was 24, the oldest 66 years old. More than a third of the respondents are associate professors (35%) and assistants (34%), a significantly less percentage of participants are professors (17%) and lecturers (14%).

The questionnaire included sections that were used for other projects (for instance, students' and teachers' perception of learning management systems, video conferencing platforms and obstacles they faced during online education) [35]. Further, we will consider the questions only related to this particular topic.

The participants were asked to choose one of five judgments, which reflect their attitude toward ICT innovations in the educational process. According to Ajzen's Theory of Planned Behavior, this self-categorization can provide reliable information about the actual or future behavior of educators [36].

Statements that were used for the categorization [6]:

1. I actively follow the innovations in the field of ICT, I often test new ICTs before others and adapt them to didactic use even before they become widely available (innovators).
2. I regularly follow innovations in the field of ICT, and I carefully tailor/adapt them to didactic use. I am often among the first ones who use new ICTs, so colleagues often perceive me as a role model and follow my recommendations (early adopters).
3. I prefer to wait and start using new ICTs only when their value becomes tangible or my colleagues recommend them. Normally, I'm not among the first to start using new tools, however, I am not among the last (early majority).
4. I have nothing against new ICTs, but I am cautious and start using them in the pedagogical process when it becomes required (late majority).

5. I am aware that my colleagues appreciate new ICTs, but prefer to use traditional teaching methods. I intend to use these methods also in situations when pressure on teachers to use new ICTs will be applied (laggards).

Further, to comprehend the factors influencing adaptive learning we considered the students' attitude toward this type of learning. The survey aimed to indicate the need for innovations; to reflect students' attitudes to adaptive learning; to indicate students' readiness to changes in the education process.

The survey was anonymous, conducted in November 2020 on Google Forms. A total of 279 Public Relations students in the undergraduate program completed the survey (156 from RUDN University and 123 from PNRPU). They were 2nd, 3rd, 4th-year students at RUDN University and 1st, 2nd, 3rd, 4th-year students at PNRP University. The total response rate amounted to 96% at RUDN University and 99% at PNRPU. At RUDN, the share of 2nd, 3rd, 4th-year participants accounted for 38%, 24%, 38%, respectively; while at PNRPU, the share of 1st, 2nd, 3rd, 4th-year participants amounted to 39%, 33%, 12%, 17%, respectively.

Additionally, the survey about students' perception of factors related to adaptive learning was to confirm the need for personalization/adaptation of the educational process. The survey evaluated the importance of the following factors by a five-point Likert-type scale (5 "essential" to 1 "not at all"): 1) pace of learning, 2) type of content presentation (text, audio, video), 3) knowledge state, 4) learners' guidance, 5) group size, 6) subject choice. The results are presented in Table 3.

4 Results

4.1 Innovation Adoption Distribution Among Teachers and Their Tendency to Use Adaptive Learning

The study on teachers' attitude toward ICT innovations revealed 7% of innovators, 20% of early adopters, 29% of the early majority, 29% of late majority and 15% of laggards (Table 2). The distribution of the sample did not differ much from the results in Roger's survey. Interestingly, the category of innovators included 1% of professors, 2% of associate professors and 4% of assistants. There is a clear correlation between the period of pedagogical activity and the predisposition to new applications of ICT in the educational process. The innovative potential is a characteristic of young teachers (assistants) who master new online services with ease and interest. The majority of categories "early majority" and "late majority" are professors and associate professors who have six or more years of experience, consequently, they are more conservative and tend to use traditional forms of education.

As we wanted to discover the correlation between teachers' self-categorization (Table 1) and practical application of adaptive learning, we asked respondents the question "How often do you use adaptive teaching, namely the pace of learning, the type of content presentation (audio, video, text), the knowledge state (students' qualification, background knowledge, skills, learning experience) and the students' interests (learning objectives and outcomes)?" Educators of RUDN University and PNRP University graded each factor of adaptive learning on a scale from one to five, where 1 is "I never adapt

Table 2. Innovation adoption distribution among teachers of RUDN University and PNRP University

	Innovators	Early adopter	Early majority	Late majority	Laggards
Lecturers	4%	12%	3%	–	–
Assistant Professors	–	4%	4%	–	–
Associate Professors	2%	2%	5%	7%	4%
Professors	1%	2%	17%	22%	11%
Total	**7%**	**20%**	**29%**	**29%**	**15%**

teaching to the students' needs" and 5 – "I always adjust teaching to the learners' preferences". The results are presented in Table 3.

Table 3. The use of factors influencing the adaptive teaching by Innovators and Early adopters among educators of RUDN University and PNRP University.

	All	Innovators	Early adopters
Pace of learning	1.8	4.4	2.6
Type of content presentation	4.1	4.9	3.3
Knowledge state	1.4	4.0	1.9
Students' interests	3.8	4.4	2.1

The survey reveals that the perceived behavioural control does not coincide with the actual behaviour. The answer about teachers taking into account learners' preferences appeared in contradiction to the answer about their self-categorization. In Table 2, 20% of the teachers described themselves as Early Adopters, which means they have a strong predisposition to adapt their teaching methods to students' needs. However, Table 3 shows that on average, Early Adopters apply adaptive teaching only in half of the cases (from 1.9 to 3.3.). As expected, Innovators adopt adaptive learning methods in 90% of cases.

Results in Table 4 show the factors that affect learning ranked by the average level of perceived importance. Students ranked the Pace of Learning as the most important factor. It shows that studying at the university is a challenge for a lot of students, many of them struggle or need more time or revision to cope with the requirements of programmes.

In the second place, students put Lecturer's guidance and Subject Choice. Traditionally, the role of a teacher has been central in the educational process in Russia [37]. Teachers at school and professors at university have been the people who spread knowledge, have authority. Today the educational paradigm is changing from teacher-centred to

Table 4. The RUDN and PNRP students' perceived importance of factors affecting their learning.

Factors	Average answer
Pace of learning	4.9
Type of content presentation	4.1
Knowledge state	3.8
Lecturer's guidance	4.7
Group size	2.3
Subject choice	3.9

student-centred; teachers perform the roles of an educator/a mentor/a facilitator/a supervisor. The survey indicated that the students highly value their lecturers/professors' input and assign them priority; students trust their educators to teach them knowledge and life competencies, i.e. communication skills, cultural awareness, critical thinking, required in their future employment.

As for Subject Choice, until recently, students could not choose the subjects to study at the majority of Russian universities; the educational syllabus was determined by the Ministry of Education. At present, students have become active participants in learning and take responsibility for their decisions.

The Type of Content Presentation was the number three choice. Studying the subject with one-dimensional textbooks fails to give the full experience of its mastering. The educational process should meet the needs of students with multiple intelligence. We, teachers, should be encouraged to focus on the needs of students by diversifying and digitalising the presented material.

The factor Group Size has the least effect on productive knowledge acquisition, in students' opinion.

Therefore, the results represent a vital necessity in designing adaptive learning pathways for students enrolled at different major programmes at different universities. The following frameworks of adaptive learning for Business English and Graphic Design courses have been created. They are the first small step and are going to be developed.

4.2 The Graphic Design Course

The Lifelong Learning Lab's framework was taken as a point for the Graphic Design course model of adaptive learning. The subject involves practical skills training. We selected the topic "Adobe Photoshop. Basic operations" as a smaller topic within the field of computer design (see Fig. 1). As the body of knowledge, we selected a list of 17 subtopics that embraced: basic navigation, document size, layers, a combination of images, levels, vibrance for enhancing the colours, colour change, black and white image, gradient, text, shapes, etc. Further, we compiled a series of questions for every subtopic, such as 'is it possible to be familiar with layers without knowing about the text?'. Then we drew 68 links between subtopics. The next step was to create learning content, references and two assignments for each subtopic. Afterwards, we developed the guidance for transition from one subtopic to another. For instance, if a student fails an assessment, they receive a second assessment and so on.

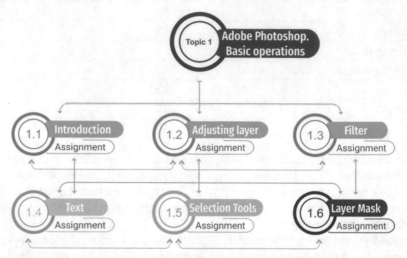

Fig. 1. The visualization of the framework of the Graphic Design course (Topic 1).

4.3 The Business English Course

Adaptive learning has found its implementation in teaching the Business English vocabulary and assessment. The lexis of the course is grouped according to the studied topics: Networking, Presentation, Management, Advertising and Branding, and Social Responsibility. A list of words is compiled for each topic and students have to study and memorise these words. The words are linked to dictionary entries where one can listen to the pronunciation of a word, read its definition in English and translate it into Russian. In the next step, students play with words on Quizlet, do crosswords, listen to audio, watch a video or read a short text to study how the words are used in a context. Each topic has its formats, so students never feel bored by doing similar exercises throughout the year. They can do these activities as long as they feel confident to take an assessment test.

The assessment test indicates the lacunas in students' knowledge of vocabulary and links them to the particular activity for revision and memorisation. If students fail, they go back to online activities and revise words again until they have passed. If they pass the test, they move on to the next topic and the next word list.

While adaptive learning is emerging as a promising technology there is no established system for applying it to every subject. That is why we are currently looking for more reliable and customised platforms for the Graphic Design and Business English courses.

5   Conclusion

Adaptive learning aims to facilitate online instruction that is personalized to the needs of individual learners. This paper investigated the perspectives and limitations of adaptive learning in teaching Graphic Design and Business English.

Today, higher education should be approached from a learner-centred perspective. Constant changes in the business world dictate new rules of the game to universities that have to adapt to a new reality and meet the demands of employers. These two demands of the present time can be satisfied with the introduction of innovations and adapting technologies into the educational process. The arising issue is if the educational system is ready to meet the challenge.

The research found that only 7% of educators are ready to embrace the challenge and describe themselves as "innovators" and 20% as "early adopters" of innovations. However, further research into actual behaviour revealed that only "innovators" use adaptive teaching. It brings us to the conclusion that teachers need training in digital competencies, redesigning their programmes for the online environment, new tools, models, and methods of education.

Students' survey revealed that they connect innovations with their lecturers' guidance, subject choice and pace of learning.

The most crucial choice to make is an adaptive learning platform. Even if such platforms exist, they often fail to completely adapt a course due to the complex nature of a subject (e.g. teaching English).

The current research remains only the first small step into the field of adaptive learning which has a huge potential for the education system.

References

1. Pozdeeva, E.G., Shipunova, O.D., Evseeva, L.I.: Social assessment of innovations and professional responsibility of future engineers. IOP Conf. Ser. Earth Environ. Sci. **337**, 012049 (2019). https://doi.org/10.1088/1755-1315/337/1/012049
2. Stress testing lessons. Universities during the COVID-19 and after it. Ministry of Science and Higher Education (2020). (in Russian). https://www.hse.ru/data/2020/07/06/159528 1277/003_%D0%94%D0%BE%D0%BA%D0%BB%D0%B0%D0%B4.pdf. Accessed 21 Apr 2021
3. Sasaki, M.: Application of diffusion of innovation theory to educational accountability: the case of EFL education in Japan. Lang. Test. Asia **8**(1), 1–16 (2018). https://doi.org/10.1186/s40468-017-0052-1
4. Kiseleva, E.M., Artemova, E.I., Litvinenko, I.L., Kirillova, T.V., Tupchienko, V.A., Bing, W.: Implementation OF innovative management in the actions of the business enterprise. Intern. J. Appl. Busin. Econ. Res. **15**(13), 231–242 (2017)
5. Ma, J., Cai, Y.: Innovations in an institutionalised higher education system: the role of embedded agency. High. Educ. (2021). https://doi.org/10.1007/s10734-021-00679-7
6. Pucer, P., Godnič Vičič, Š., Žvanut, B.: ICT adoption among higher education teachers: a case study of a university in the awareness/exploration stage of blended learning adoption. In: Podovšovnik, E. (ed.) Examining the Roles of Teachers and Students in Mastering New Technologies, pp. 299–314. IGI Global, Hershey (2020). https://doi.org/10.4018/978-1-7998-2104-5.ch016

7. Porter, W.W., Graham, R.C.: Institutional drivers and barriers to faculty adoption of blended learning in higher education. Brit. J. Educ. Technol. **47**(4), 748–762 (2016). https://doi.org/10.1111/bjet.12269

8. Baykal, U., Sokmen, S., Korkmaz, S., Akgun, E.: Determining student satisfaction in a nursing college. Nurse Edu. Today **25**(4), 255–262 (2005). https://doi.org/10.1016/j.nedt.2004.11.009

9. Loogma, K., Kruusvall, J., Ümarik, M.: E-learning as innovation: exploring innovativeness of the VET teachers' community in Estonia. Comput. Edu. **58**(2), 808–817 (2012). https://doi.org/10.1016/j.compedu.2011.10.005

10. Rusek, M., Stárková, D., Chytrý, V., Bílek, M.: Adoption of ICT innovations by secondary school teachers and pre-service teachers within chemistry education. J. Baltic Sci. Edu. **16**(4), 510 523 (2017). http://oaji.net/articles/2017/987-1503904959.pdf. Accessed 21 Apr 2021

11. EdSurge: Decoding Adaptive. Pearson, London (2016). https://www.pearson.com/content/dam/corporate/global/pearson-dot-com/files/innovation/Pearson-Decoding-Adaptive-v5-Web.pdf. Accessed 3 June 2021

12. Gros, B.: The design of smart educational environments. Smart Learn. Environ. **3**(15), 1–11 (2016). https://doi.org/10.1186/s40561-016-0039-x

13. Essa, A.: A possible future for next generation adaptive learning systems. Smart Learn. Environ. **3**(1), 1–24 (2016). https://doi.org/10.1186/s40561-016-0038-y

14. Rymshina, A., Pivovarova, A., Borodin, N., Gizatulina, R.: Adaptive Learning, https://lllab.online/adaptive-learning. Accessed 3 June 2021

15. Almohammadi, K., Hagras, H., Alghazzawi, D., Aldabbagh, G.: A survey of artificial intelligence techniques employed for adaptive educational systems within e-learning platforms. J. Artif. Intell. Soft Comput. Res. **7**(1), 47–64 (2017). https://doi.org/10.1515/jaiscr-2017-0004

16. Berry, P.M., Donneau-Golencer, T., Duong, K., Gervasio, M., Peintner, B., Yorke-Smith, N.: Evaluating intelligent knowledge systems: experiences with a user-adaptive assistant agent. Knowl. Inf. Syst. **52**(2), 379–409 (2016). https://doi.org/10.1007/s10115-016-1011-3

17. Chou, C.Y., Lai, K.R., Chao, P.Y., Tseng, S.F., Liao, T.Y.: A negotiation-based adaptive learning system for regulating help-seeking behaviors. Comput. Edu. **126**, 115–128 (2018). https://doi.org/10.1016/j.compedu.2018.07.010

18. Cui, W., Xue, Z., Thai, K.P.: Performance comparison of an AI-based adaptive learning system. In: China Proceedings of the 2018 Chinese Automation Congress, CAC 2018, Xi'an, pp. 3170–3175. IEEE (2019)

19. Hou, M., Fidopiastis, C.: A generic framework of intelligent adaptive learning systems: from learning effectiveness to training transfer. Theor. Iss. Ergon. Sci. **18**(2), 167–183 (2017). https://doi.org/10.1080/1463922X.2016.1166405

20. Imhof, C., Bergamin, P., McGarrity, S.: Implementation of adaptive learning systems: current state and potential. In: Isaias, P., Sampson, D.G., Ifenthaler, D. (eds.) Online Teaching and Learning in Higher Education. CELDA, pp. 93–115. Springer, Cham (2020). https://doi.org/10.1007/978-3-030-48190-2_6

21. Wang, S., et al.: When adaptive learning is effective learning: comparison of an adaptive learning system to teacher-led instruction. Interact. Learn. Environ. (2020). https://doi.org/10.1080/10494820.2020.1808794

22. Costa, J., Silva, C., Antunes, M., Ribeiro B.: Adaptive learning for dynamic environments: a comparative approach. Eng. Appl. Artif. Intell. **65**(C), 336–345 (2017). https://doi.org/10.1016/j.engappai.2017.08.004

23. Zawacki-Richter, O., Marín, V.I., Bond, M., Gouverneur, F.: Systematic review of research on artificial intelligence applications in higher education – where are the educators? Int. J. Educ. Technol. High. Educ. **16**(1), 1–27 (2019). https://doi.org/10.1186/s41239-019-0171-0

24. Dargue, B., Biddle, E.: Just enough fidelity in student and expert modeling for ITS. In: Schmorrow, D.D., Fidopiastis, C.M. (eds.) AC 2014. LNCS (LNAI), vol. 8534, pp. 202–211. Springer, Cham (2014). https://doi.org/10.1007/978-3-319-07527-3_19

25. Doroudi, S.: Mastery learning heuristics and their hidden models. In: Bittencourt, I.I., Cukurova, M., Muldner, K., Luckin, R., Millán, E. (eds.) AIED 2020. LNCS (LNAI), vol. 12164, pp. 86–91. Springer, Cham (2020). https://doi.org/10.1007/978-3-030-52240-7_16

26. Idris, N., Hashim, S.Z.M., Samsudin, R., Ahmad, N.B.H.: Intelligent learning model based on significant weight of domain knowledge concept for adaptive e-learning. Int. J. Adv. Sci. Eng. Inf. Technol. 7(4–2), 1486–1491 (2017). https://doi.org/10.18517/ijaseit.7.4-2.3408

27. Liang, Q., Hainan, N.C.: Adaptive learning model and implementation based on big data In: 2nd International Conference on Artificial Intelligence and Big Data, ICAIBD, Chengdu, pp. 183–186. IEEE (2019). https://doi.org/10.1109/ICAIBD.2019.8836984

28. Personalization of language learning through adaptive technology. Cambridge University Press (2017). https://www.cambridge.org/us/files/1915/7488/5061/CambridgePapersinELT_AdaptiveLearning_2017_ONLINE.pdf. Accessed 6 Apr 2021

29. Osipovskaya, E., Miakotnikova, S.: Using gamification in teaching public relations students. In: Auer, M.E., Tsiatsos, T. (eds.) ICL 2018. AISC, vol. 916, pp. 685–696. Springer, Cham (2020). https://doi.org/10.1007/978-3-030-11932-4_64

30. Nakata, T.: Computer assisted second language vocabulary learning in a paired-associate paradigm: a critical investigation of flashcard software. Comp. Assist. Lang. Learn. 24(1), 17–38 (2011). https://doi.org/10.1080/09588221.2010.520675

31. Nation, I.S.P.: Learning Vocabulary in Another Language, 2nd edn. Cambridge University Press, Cambridge (2013)

32. Larsen-Freeman, D., Cameron, L.: Complex Systems and Applied Linguistics. Oxford University Press, Oxford (2008)

33. McCarthy, M.: The Cambridge Guide to Blended Learning for Language Teaching. Cambridge University Press, Cambridge (2016)

34. Thornbury, S.: Educational technology: assessing its fitness for purpose. In: McCarthy, M. (ed.) The Cambridge Guide to Blended Learning for Language Teaching, pp. 25–35. Cambridge University Press, Cambridge (2016)

35. Osipovskaya, E., Dmitrieva, S., Grinshkun, V.: Examining technology and teaching gaps in Russian universities amid coronavirus outbreak. In: Auer, M.E., Rüütmann, T. (eds.) ICL 2020. AISC, vol. 1328, pp. 764–774. Springer, Cham (2021). https://doi.org/10.1007/978-3-030-68198-2_72

36. Ajzen, I.: The theory of planned behavior. Org. Behav. Hum. Decis. Process. 50(2), 179–211 (1991). https://doi.org/10.1016/0749-5978(91)90020-T

37. Anosova, N., Dashkina, A.: The teacher's role in organizing intercultural communication between Russian and international students. In: Anikina, Z. (ed.) IEEHGIP 2020. LNNS, vol. 131, pp. 465–474. Springer, Cham (2020). https://doi.org/10.1007/978-3-030-47415-7_49

Technical University Students' Creativity Development in Competence-Based Foreign Language Classes

Yulia V. Borisova⬛, Anna Y. Maevskaya[✉]⬛, and Elvira R. Skornyakova⬛

Saint-Petersburg Mining University, Vasilyevsky Island, 21 Line, 2, St. Petersburg, Russia
{Borisova_YuV,Maevskaya_AYu,Skornyakova_ER}@pers.spmi.ru

Abstract. Technical university students should master general competencies which are as follows: critical thinking, project and teamwork, leadership, communication, cross-cultural interaction and self-development. In modern society graduates have to be highly-competitive, ready to further career development, capable of quick adaptation in constantly changing world. The article demonstrates the application of various creative learning technologies (participation at the Olympiads and forums; use of project technologies; debates; use of game techniques; listening and using various authentic sources) at foreign language classes based on a competence approach and it shows how learning English through creative technologies helps students better master the foreign language. The authors suggest the idea that creatively thinking teachers using the different creative learning activities at the lessons can contribute to formation of creatively thinking students capable to act at different language environment adequately and to become critically thinking specialists at their field, good professional team members and constantly developing personalities with high level of both language and soft skills. The results of the creative techniques usage were illustrated with the help of a pedagogical experiment showing substantiation of using creativity development technologies. The students at the experimental group, in which creative techniques were used, showed greater progress at language and soft skills mastering. The correlation between the development of competencies and the use of some creative techniques was summarized.

Keywords: Competence-based approach · Creativity · Foreign language classes

1 Background

1.1 Competence-Based Approach: Present and Future

Higher professional education in Russia has undergone the transition to the new style of Federal State Educational Standards of Higher Education. This situation provided transformation in assessment process of education results from the concepts of "knowledge", "abilities", and "skills" to the concepts of "competence" and "competency", i.e., the focus is on the competence-based approach.

The analysis of the literature reveals the complexity of interpreting the concepts of "competency" and "competence". Competence-oriented education was formed in the 1970s in the USA. It was developed within the general context of the approach suggested by N. Chomsky in 1965. The term "competence" (from Latin competere – to be capable of anything) introduced by Chomsky [1] referred to the ability necessary to perform a certain, predominantly linguistic activities in the native language. British psychologist Raven [2] proposed the interpretation of this definition as a specific ability to carry out exact actions at a special subject area including understanding arising responsibility.

In the domestic pedagogical theory and practice the concepts "competency" and "competence" are in a sense synonymous, but, nevertheless, they are not the same thing. Competency is more often used to denote an educational result, expressed in a graduate's mastery of certain knowledge and skills, and the ability to cope with the tasks necessary for qualified productive activity. "Competence" can be understood as a person's possession of appropriate competencies, which includes his or her personal attitude towards them and the ability to make exact decisions.

Nowadays there are some research papers dealing with the analysis of competence-based education in the modern pedagogics. Vasetskaya et al. [3] studied the model of competences of the research and teaching university staff, Pokrovskaia et al. [4] carried out diagnostics of professional competencies and motivation of engineers in the knowledge of economics.

But unfortunately, there are practically no research papers dealing with the opportunities of the use of creative ways of teaching foreign languages by creatively thinking teachers at a technical university which can contribute greatly to the development of the necessary competencies of a future engineer.

The theoretical and methodological basis of our research was the system of the theory of personal learning, the general theory of competence-based approach at foreign language classes, modern scientific concepts of creativity learning strategies while forming the model of a specialist.

The main methods used for the research were: analysis of literary sources and publications of recent years on the issues of foreign language creativity teaching in higher education and the issues devoted to the competence-based approach while teaching English to Saint-Petersburg Mining University students and carrying out a pedagogical experiment with the help of statistical methods.

The study was conducted in the Saint-Petersburg Mining University, Russia. It covered two groups of students. The experimental group, i.e. the group in which the creative activities mentioned above were widely and constantly used, involved 112 students' assessment tests and 83 students' exam papers at different stages. The control group, i.e. the students, which were taught mostly with the help of traditional teaching techniques, included the same number of students' assessment tests and exam papers. We focus on assessment tests and exam papers done by students in 2019–2020 and 2020–2021 academic years.

1.2 Competence Development by Means of Foreign Languages and Creativity

According to the new Russian Standards of Higher Education it is necessary to develop universal and general professional competencies. The subject "Foreign language" is taught at every non-linguistic higher education institution, being the obligatory disci-pline. Today "English is necessary to be employed, get promoted and perform profes-sional duties to the highest standards" [5, p. 1]. Rogova, Sveshnikova & Troitskaya [6, p. 375] mentioned: "A traditional approach to ESP teaching in technical universities does not comply with the up-to-date federal educational standards of higher vocational education in Russia which aim at communicative competence acquisition in all lan-guage activities". Numerous researchers [7–9] share this opinion introducing different technologies and methods such as CLIL [10, 11] into educational process.

Teaching foreign languages at a technical university contributes greatly to the devel-opment of a number of general competencies [12]. Goman [13, p. 1] says that "since we plan to teach would-be oil and gas engineers, it is essential to refer to the list of compe-tencies which they are supposed to acquire". Let us consider the main ones based on the new Russian Standards of Higher Education for specialty 18.03.01 "Chemical technolo-gies": systematic and critical thinking, development and realization of projects, team-work and leadership, communication, cross-cultural interaction, self-organization and self-development, life-safety, inclusive competence, economic culture, including finan-cial literacy, and civic stand. In our paper we concentrate on such soft skills development as critical thinking, project and team work, leadership, communication, cross-cultural interaction and self-development [14].

All these competencies can be achieved by students through creative learning at foreign language classes.

By the end of the foreign language course students will have studied essential gram-mar and vocabulary necessary to communicate within the international professional community as well as will have practiced their listening, reading, speaking and writ-ing skills to the extent of profound knowledge [15]. According to Sishchuk, Oblova, & Mikhailova [16, p. 810] "there is a strong need to develop students' competence based on the accepted professional standards in order to adapt or relate expertise needed for future job with the graduates' competence assessment".

To meet the requirements of Standards learning process must involve the modern technological environment. Vinogradova, Kornienko & Borisova [17, p. 401] pointed out that in "the educational standards of the third generation is included in a number of both professional and universal competencies". Murzo, Sveshnikova & Chuvileva [18, p. 143] provide professionally oriented online course "which is focused on information technologies in learning" and is developed for "accessing to learning materials, such as texts, audio, video materials".

In our paper we analyze the foreign language learning role for the development of creativity with the help of competencies mentioned above.

2 Questions and Purposes

2.1 Creative Pedagogy

Creativity at English classes increases students' motivation and fosters their soft skills. In course of teaching the language we understood that creativity of a teacher enhances creativity in students no matter what level of mastering the language they have. Numerous researches prove this idea [19–21]. It should be taken into consideration that teachers may apply creativity methods much more often than students at foreign language classes [22–24]. Specialist who is thinking creatively can be ready to transformation of reality which is going on constantly, working in unusual environment and dealing with arising challenges [25].

Numerous researchers described application of different technologies at English for specific purposes classes, for example, teaching critical thinking classrooms [26], creative methodologies for engineering students [27], new pedagogical framework teaching English for academic purposes using four typical genres in scientific communication, that is, science popularizations, semi-popularizations, technical reports, and research articles [28], using films and TV-series as a multimodal perspective [29]. Pushmina discusses "the integration of social media on the example of Instagram as a teaching tool and ESP/EMI training" [30, p. 475].

Some research papers of the Saint-Petersburg Mining University also indicate that the quality of students' professional education is connected to such aspects as: "academic staff; content of courses; educational environment" [31, p. 264]. Rassadina, Professor of the Mining University, majored in Culturology, studies "edutainment (a portmanteau of 'education' and 'entertainment')" [32, p. 503]. Sishchuk describes "German borrowings in mining, geological and metallurgical terminology" [33, p. 507].

While teaching a foreign language using some forms of creative activities the authors carried out a kind of a pedagogical experiment which confirmed our suggestion that the use of creative techniques leads to development of creatively thinking students and in turn to the improvement of competencies mastering. These results were high enough for technical university students showing their great progress.

There are different ways to develop creativity of students in higher educational institutions. Among them we can mention *Olympiads*, which are aimed at developing students' interest in scientific activities, promotion of scientific and professional knowledge; *organizing debates* at the lessons; the use of *game techniques* and *group chats* make possible to fully reveal all the potential capabilities of a personality such as visual, auditory and kinesthetic systems. Creative work in foreign language classes necessarily involves *listening and using various authentic sources* such as video films, audio materials. Musical materials stimulate creative activity of students.

At this research the authors paid their attention to the use of only some of techniques provoking creativity. Without doubt, the list of the used ones was not fully sufficient; but it was decided to consider those methods of formation creativity, which contribute more to the development of required competences.

All these "methods, materials and tasks are needed to develop independent, unconventional and curious learners, who are willing to take risks, and being flexible and collaborative" [34, p.27]. In modern market and information-oriented society graduates have to be highly-competitive, capable of quick adaptation in constantly changing world [35].

2.2 Creativity-Oriented Methods at Foreign Language Classes

We now describe some technologies developing creativity in detail.

Project work is extremely important while teaching a foreign language. For example, first-year C1 level students of Economics Faculty were supposed to make a project on

"The happiness formula". Students answered the questions how happiness is measured, which places in the world are the happiest/least happy, which factors lead to happiness. Students worked in a team, developed their *systematic and critical thinking*, stuck to time-management.

Also, first-year students of B2 level chose people to go on a space mission. The project called "Mission 2050" developed students' teamwork skills, their ability to choose people who they will get on well in personal and professional spheres.

The result of students' participation at project activities was stimulation to express the opinion of those, who are usually shy, the ability to deal with different sources of information and to organize it in a proper way and to work in teams with their group-mates. For some of them such experience was for the first time in their students' lives and it was *contribution to development of critical thinking competence, leadership and teamwork*.

Concerning *communication skills development* along with critical thinking forma-tion it is worth mentioning the *use of debating technology* at English classes. Numerous researches study this technology as a teaching strategy [36–39]. Throughout the previous and present academic years, the authors of the paper have conducted several debates on controversial topics: "LGBT-community condemns teenagers' coming-outs", "Govern-ment will ban dating websites", "If scientists find out about the Earth destruction in the near future the Government will hide this fact from citizens", "The main mission of a woman is to be a housewife and bring children up".

It was quite exciting for teachers and students themselves to watch and listen to their logically organized speeches, their argumentation was strong, they were able to show advantages and disadvantages which could be found analyzing both sides of situation and the desire to communicate and express their opinions increased dramatically.

It is not a secret that there are lots of opportunities to *use video-films* at foreign language classes, and English lessons at the Mining university are not an exception. For instance, while learning the topic "Childhood: specific moments" our students watched episodes from the movie "Charlie and the Chocolate Factory" with Johnny Depp. There are opportunities to watch films connected with their professional areas. The students of Mining and Geological Prospecting faculties watch "All about Rocks and Minerals", Oil and Gas engineering Faculty students are happy to watch "Fossil Fuels and what not. After viewing the students took part in discussions willingly. They have demonstrated genuine interest and then have written comprehensive film-reviews. Films' watching contributes to development of listening skills which are an *integral part of communicative competence*.

It should also be noted that students have a great variety of opportunities to participate in *All-Russian or International translation competitions, technical forums and different level language Olympiads*. For the past two years students of the paper authors have participated in seven competitions. Two students received the second and the third prize in All-Russian competition conducted by Ogarev Mordovia State University for translating the text "Why the U.S. Dollar Is the Global Currency".

Furthermore, our students have twice participated in competitions of video-presentations organized by Saint-Petersburg State University of Economics. In 2019 students prepared the presentation "Economic security as a factor of globalization", in 2020 students discussed the issue "Black Friday and Cyber Monday Sales".

Every year the foreign languages department of the Saint-Petersburg Mining University conducts an Olympiad consisting of reading, vocabulary and grammar tasks as well as questions about English-speaking countries. Those students, who had been motivated and had taken part in the language competitions and Olympiads of different levels, *contributed to formation of self-organization and self-development*. The opportunity to check themselves and the achievement of successful results helped them to realize that everything was at their hands and they are the persons who can cope with any difficulties and can meet any challenges.

3   Competencies Mastering

The correlation between the development of competencies and the use of some creative techniques can be summarized in a table. Here we can observe the direct connection

Table 1. The use of creative techniques and competence mastering.

Competencies	Used creative techniques	Competencies mastering
Systematic and critical thinking	Critical reading and writing with a number tasks provoking analytical thinking; the use of different projects technologies; organization of debates	The ability to choose information resources in compliance with the given task, to assess this information according to its integrity, completeness and authenticity, to systematize this information
Project and team work, leadership	The use of a range of projects tasks ("The happiness formula", "Mission 2050, etc.); participation in technical conferences (Russian and International Conferences and Forums of Young Researchers of Subsoil Usage), All-Russian or International translation competitions and different level language Olympiads	Comprehension of the main task and its implementation in particular, search of specifications and technical documentation, choice of appropriate tackling the problem, perception of a team's functions and goals. Demonstration of the results of the competence was presenting reports and scientific papers at the lessons and technical conferences (participation in 3 All-Russian and 2 International conferences at the Mining University (4 diplomas); participation in 7 competitions (2 prize-winners)

(continued)

Table 1. (*continued*)

Competencies	Used creative techniques	Competencies mastering
Communication	Organization of debates on controversial topics: "LGBT-community condemns teenagers' coming-outs", "Government will ban dating websites", "If scientists find out about the Earth destruction in the near future the Government will hide this fact from citizens", "The main mission of a woman is to be a housewife and bring children up"	Ability to read and understand (with a dictionary) information in a foreign language on topics of everyday and business communication, to conduct a dialogue in a foreign language of general and business nature. Demonstration of the results of the competence was logically organized speech, strong argumentation, ability to show advantages and disadvantages analyzing both sides of situation and the desire to communicate and express their opinions
Cross-cultural interaction	Movie watching, listening to songs; self-teaching with the help of different website educational channels such as BBC learning, etc	Perception of cross-cultural interaction values, awareness of cultural and social varieties, choice of conflict solving technique, choice of interaction way in a team while tackling professional issues
Self-development	Participation in All-Russian and International translation competitions and different level language Olympiads (competitions of video-presentations "World economy: main trends, risks and challenges" organized by Saint-Petersburg State University of Economics; All-Russian competition conducted by Ogarev Mordovia State University; the Saint-Petersburg Mining University Olympiad, etc.)	Contribution to self-organization (time-management, self-organization path) and self-development; opportunity to check themselves; achievement of successful results; Demonstration of the results of the competence was ability to cope with any difficulties and meet any challenges (4 diplomas, 2 prize-winners)

between the use of certain creativity provoking methods of teaching and learning and their implementation at the educational process at the Mining University (Table 1).

A great variety of creativity development technologies used at foreign language classes at the Saint-Petersburg Mining University have shown that students become much more motivated in learning a foreign language. They demonstrate a genuine interest in the topics they study whether there are general or specific topics connected with their future profession. It also should be noticed that Saint-Petersburg Mining University as all universities of our country organized educational process in online-format from April till in August 2020 due to covid-19 pandemic, ant it returned to the usual organization of education on September 1, 2020. In this connection it is necessary to pay attention that during the period of distance-learning competitions, project work, Olympiads were transformed into distant format of implementation.

The results of our research can be illustrated with the help of some kind of a pedagogical experiment showing substantiation of using creativity development technologies. While teaching creatively a foreign language we have achieved the following results.

Experimental group involved 112 students' assessment tests and 65 students' exam papers at different stages. Control group also included 112 students' assessment tests and 65 students' exam papers. In September 2019, out of 112 first-year students passed self-assessment test with the following results: 13 students (11,6%) – C1 level, 47 students (41,9%) – B1 level, 29 students (25,8%) – A2 level, 23 students – A1 level (20,7%). The results of the tests of the students at the control group were the following: 12 students (10,7%) – C1 level, 49 students (43,7%) – B1 level, 32 students (28,6%) – A2 level, 19 students – A1 level (17%). In September 2020, the same experimental 112 already second-year students improved their language skills in the following way: 17 students (15,2%) – C1 level, 54 students (48,2%) – B1 level, 41 students (36,6%) – A2 level. As for the students of the control group here the results were the following: 14 students (12,5%) – C1 level, 52 students (46,5%) – B1 level, 37 student (33%) – A2 level, 9 students – A1 level (8%).

It should be mentioned that all 23 students of the experimental group of A1 level improved their language skills to A2 level. In the subgroup of one of the authors of the present paper there were no students of B2 level.

As we can see, these results are high enough for technical university students showing greater progress at the experimental group in which creative techniques mentioned above were used at the educational process.

It would be also important to analyze the exam results. In 2020 summer session, 65 s-year students from the experimental group and also 65 students from the control group passed obligatory English exam. At the experimental group there were only two satisfactory marks and academic performance could be estimated as 97%. As for the control group, there the results were not as high as we can observe at the experimental group: there were nine satisfactory marks and academic performance was only 93%.

The authors think that the obtained results at the assessment tests and the examination sessions and the increased number of the participants with a diploma and winners of national and international forums, conferences and language and technical competitions could serve as a reliable indicator of the progress of competence development with the help of creativity developing techniques.

4   Conclusions

The general conclusions based on the given research make it possible to admit that competence-based approach at foreign language classes improves students' language and soft skills dramatically. We can confirm that indicators of general competencies achievement are much more clearly demonstrated while teaching English with the help of creative technologies.

To sum it up, creativity can be enhanced by motivating students to participate in various translation competitions, Olympiads, essay competitions, debates, movie watching, listening to songs, critical reading and writing, project work, self-teaching with the help of different website educational channels such as BBC learning, for example.

Some difficulties may arise while teaching students with different level of English. However, teachers of the Saint-Petersburg Mining University overcome these difficulties adapting tasks motivating each student to participate in the process of learning.

5   Further Studies

We consider our further studies in working on subsequent ideas aimed at creativity development and combining them with traditional methods for formation of non-standard thinking in future mining engineers and specialists in different technical areas.

The educational system is faced with such tasks of its organization, in which each student would be provided with the opportunity not only to master the profession, but also to develop creative thinking. Today we live in net generation and, undoubtedly, there is a need for research on the role of technology in enhancing creativity and thinking skills, from both a teaching and learning perspective.

References

1. Chomsky, N.: Aspects of the Theory of Syntax. MIT Press, Cambridge (1965)
2. Raven, J.: Competence in Modern Society: Its Identification Development and Release. Royal Fireworks Press, New York (1994)
3. Vasetskaya, N.O., Glukhov, V. V, Burdakov, S.F.: The elaboration of the model of competences of the research and teaching university staff. In: 2018 XVII Russian Scientific and Practical Conference on Planning and Teaching Engineering Staff for the Industrial and Economic Complex of the Region (PTES), pp. 98–101 (2018). https://doi.org/10.1109/PTES.2018.860 4215
4. Pokrovskaia, N.N., Petrov, M.A., Gridneva, M.A.: Diagnostics of professional competencies and motivation of the engineer in the knowledge economy. In: 2018 Third International Conference on Human Factors in Complex Technical Systems and Environments (ERGO)s and Environments (ERGO), pp. 28–31. IEEE, St. Petersburg (2018). https://doi.org/10.1109/ERGO.2018.8443851
5. Oblova, I., Gerasimova, I., Sishchuk, J.: Case-study based development of professional communicative competence of agricultural and environmental engineering students. E3S Web of Conf. **175**, 15035 (2020). https://doi.org/10.1051/e3sconf/202017515035

6. Rogova, I.S., Sveshnikova, S.A., Troitskaya, M.A.: Designing an ESP course for metallurgy students. In: 20th Professional Culture of the Specialist of the Future (PCSF 2020) & 12th Communicative Strategies of Information Society (CSIS 2020). The European Proceedings of Social and Behavioural Sciences EpSBS, vol. 98, pp. 372–385 (2020). https://doi.org/10.15405/epsbs.2020.12.03.38

7. Almazova, N.I., Beliaeva L.N., Kamshilova, O.N.: Towards textproductive competences of language worker and novice researcher. In: 18th PCSF 2018 - Professional Culture of the Specialist of the Future. The European Proceedings of Social & Behavioural Sciences EpSBS, vol. LI, pp. 103–109 (2018). https://doi.org/10.15405/epsbs.2018.12.02.11

8. Pokrovskaia, N.N., Ababkova, M.Y., Fedorov, D.A.: educational services for intellectual capital growth or transmission of culture for transfer of knowledge—consumer satisfaction at St Petersburg universities. Educ. Sci. 9(3), 183 (2019). https://doi.org/10.3390/educsci9030183

9. Valieva, F.: Soft skills vs professional burnout: the case of technical universities. In: Anikina, Z. (ed.) IEEHGIP 2020. LNNS, vol. 131, pp. 719–726. Springer, Cham (2020). https://doi.org/10.1007/978-3-030-47415-7_76

10. Baranova, T.A., Kobicheva, A.M., Tokareva, E.Y.: Does CLIL work for Russian higher school students? The Comprehensive analysis of Experience in St-Petersburg Peter the Great Polytechnic University. In: ICIET 2019: Proceedings of the 2019 7th International Conference on Information and Education, pp. 140–145. ACM, New York (2019). https://doi.org/10.1145/3323771.3323779

11. Khalyapina, L.P.: Current trends in teaching foreign languages on the basis of CLIL. Teach. Methodol. High. Educ. 6(20), 56–52 (2017). https://doi.org/10.18720/HUM/ISSN2227-8591.20.5

12. Trostinskaia, I., Popov, D., Fokina, V.: Communicative competence of engineers as a requirement to the future professions. In: 18th PCSF 2018 Professional Culture of the Specialist of the Future. The European Proceedings of Social and Behavioural Sciences EpSBS, vol. LI, pp. 1191–1199 (2018). https://doi.org/10.15405/epsbs.2018.12.02.128

13. Goman, Yu.V.: Development of the dialogue skills in a foreign language in comparing of oil benchmarks. In: International Conference "Complex equipment of quality control laboratories", Journal of Physics: Conference Series, vol. 1384, p. 012013. IOP Publishing, St. Petersburg (2019). https://doi.org/10.1088/1742-6596/1384/1/012013

14. Federal State Educational Standards of Higher Education bachelor's degree for specialty 18.03.01 "Chemical technologies". http://fgosvo.ru/uploadfiles/FGOS%20VO%203++/Bak/180301_B_3_23082020.pdf. Accessed 10 Mar 2021

15. Semushina, E., Valeeva, E., Kraysman, N.: Intensive language learning at technological university for integrating into global engineering society. MJLTM 9(10) (2019). https://doi.org/10.26655/mjltm.2019.10.2

16. Sishchuk, J., Oblova, I., Mikhailova, M.: The comparative analysis of the United Kingdom and the Russian federation occupational standard development. In: Anikina, Z. (ed.) IEEHGIP 2020. LNNS, vol. 131, pp. 804–811. Springer, Cham (2020). https://doi.org/10.1007/978-3-030-47415-7_86

17. Vinogradova, E., Kornienko, N., Borisova, Y.: Cat and Lingvo tutor tools in educational glossary making for non-linguistic students. In: 20th Professional Culture of the Specialist of the Future (PCSF 2020) & 12th Communicative Strategies of Information Society (CSIS 2020). The European Proceedings of Social and Behavioural Sciences EpSBS, vol. 98, pp. 400–409 (2020). https://doi.org/10.15405/epsbs.2020.12.03.40

18. Murzo, Y., Sveshnikova, S., Chuvileva, N.: Method of text content development in creation of professionally oriented online courses for oil and gas specialists. Int. J. Emerg. Technol. Learn. (iJET) 14(17), 143–152 (2019). https://doi.org/10.3991/ijet.v14i17.10747

19. Nazzal, L.J., Kaufman, J.C.: The relationship of the quality of creative problem solving stages to overall creativity in engineering students. Think. Skills Creat. **38**, 100734 (2020). https://doi.org/10.1016/j.tsc.2020.100734

20. Gralewski, J., Karwowski, M.: Are teachers' ratings of students' creativity related to students' divergent thinking? A meta-analysis. Think. Skills Creat. **33**, 100583 (2019). https://doi.org/10.1016/j.tsc.2019.100583

21. Tsai, M.-N., Liao, Y.-F., Chang, Y.-L., Chen, H.-C.: A brainstorming flipped classroom approach for improving students' learning performance, motivation, teacher-student interaction and creativity in a civics education class. Think. Skills Creat. **38**, 100747 (2020). https://doi.org/10.1016/j.tsc.2020.100747

22. Pozdeeva, E.G., Shipunova, O.D., Evsеcva, L.I.: Social assessment of innovations and professional responsibility of future engineers. IOP Conf. Ser.: Earth Environ. Sci. **337**, 012049 (2019). https://doi.org/10.1088/1755-1315/337/1/012049

23. Razinkina, E., Pankova, L., Trostinskaya, I., Pozdeeva, E., Evseeva, L., Tanova, A.: Student satisfaction as an element of education quality monitoring in innovative higher education institution. E3S Web Conf. **33**, 03043 (2018). https://doi.org/10.1051/e3sconf/20183303043

24. Yu-Hsiu, L., Yi-Ling, C., Hsueh-Chih, C., Yu-Lin, C.: Infusing creative pedagogy into an English as a foreign language classroom: learning performance, creativity, and motivation. Think. Skills Creat. **29**, 213–223 (2018). https://doi.org/10.1016/j.tsc.2018.07.007

25. Gabdulchakov, V.F.: Pedagogical skills of the teacher and the degree of creative self-realization of a student in conditions of university educational space. Procedia. Soc. Behav. Sci. **146**, 426–431 (2014). https://doi.org/10.1016/j.sbspro.2014.08.149

26. Dwee, Y.C., Anthony, M.E., Salleh, M.B., Kamarulzaman, R., Kadir, A.Z.: Creating thinking classrooms: perceptions and teaching practices of ESP practitioners. Procedia. Soc. Behav. Sci. **232**(14), 631–639 (2016). https://doi.org/10.1016/j.sbspro.2016.10.087

27. Rus, D.: Creative methodologies in teaching english for engineering students. Procedia Manuf. **46**(2020), 337–343 (2020). https://doi.org/10.1016/j.promfg.2020.03.049

28. Ye, Y.: EAP for undergraduate science and engineering students in an EFL context: what should we teach? Ampersand **7**, 100065 (2020). https://doi.org/10.1016/j.amper.2020.100065

29. Bonsignori, V.: Using films and TV series for ESP teaching: a multimodal perspective. System **77**, 58–69 (2018). https://doi.org/10.1016/j.system.2018.01.005

30. Pushmina, S.: Teaching EMI and ESP in Instagram. In: Anikina, Z. (ed.) IEEHGIP 2020. LNNS, vol. 131, pp. 475–482. Springer, Cham (2020). https://doi.org/10.1007/978-3-030-47415-7_50

31. Goldobina L.A., Orlov P.S.: BIM technology and experience of their introduction into educational process for training bachelor students of major 08.03.01 «Construction. Zapiski Gornogo Institute, **224**, 263–272 (2017). https://doi.org/10.18454/PMI.2017.2.263

32. Rassadina, S.: Cultural foundations of the concept of edutainment as a strategy for the formation of common cultural competence in universities of a non-humanitarian profile. Zapiski Gornogo instituta **219**, 498–503 (2016). https://doi.org/10.18454/pmi.2016.3.498

33. Sishchuk, Y.: Borrowings from German in Russian mining and geological terminology. Zapiski Gornogo instituta **219**, 504–507 (2016). https://doi.org/10.18454/pmi.2016.3.504

34. Li, L.: Thinking skills and creativity in second language education: where are we now? Think. Skills Creat. **22**, 267–272 (2016). https://doi.org/10.1016/j.tsc.2016.11.005

35. Almazova, N., Krylova, E., Rubtsova, A., Odinokaya, M.: Challenges and Opportunities for Russian higher education amid COVID-19: teachers' perspective. Educ. Sci. **10**, 368 (2020). https://doi.org/10.3390/educsci10120368

36. Cariñanos-Ayala, S., Arrue, M., Zarandona, J., Labaka, A.: The use of structured debate as a teaching strategy among undergraduate nursing students: a systematic review. Nurse Educ. Today **98**, 104766 (2021). https://doi.org/10.1016/j.nedt.2021.104766

37. Cole, J.D., Ruble, M.J., Povlak, A., Nettle, P., Sims, K., Choyce, B.: Self-directed, higher-level learning through journal club debates. Health Prof. Educ. **6**(4), 594–604 (2020). https://doi.org/10.1016/j.hpe.2020.05.007
38. Kim, W.-J., Park, H.-J.: The effects of debate-based ethics education on the moral sensitivity and judgment of nursing students: a quasi-experimental study. Nurse Educ. Today **83**, 104200 (2019). https://doi.org/10.1016/j.nedt.2019.08.018
39. McGee, E.U., Pius, M., Mukherjee, K.: Assessment of structured classroom debate to teach an antimicrobial stewardship elective course. Curr. Pharm. Teach. Learn. **12**(2), 220–227 (2020). https://doi.org/10.1016/j.cptl.2019.11.016

Pedagogical Creativity vs Academic Dishonesty in Teaching University Mathematics

Victor Krasnoshchekov$^{(\boxtimes)}$ (iD) and Natalia Semenova (iD)

Peter the Great St. Petersburg Polytechnic University, Polytechnicheskaya, 29, 195251 St. Petersburg, Russia

krasno_vv@spbstu.ru

Abstract. The authors address the problem of academic dishonesty, which is a concern of universities around the world. An analysis of the literature shows that there are two main types of counteraction to academic dishonesty. The first is punishment; the second is the cultivation of value of academic integrity. Each of the types has well-reasoned advantages and disadvantages. The authors propose to counteract cheating manifestations when solving tasks in mathematics by introducing non-traditional tasks based on mathematical creativity. Mathematical creativity is one of the main means of motivating students to study mathematics. The authors assess the possibilities of mathematical creativity to counter academic dishonesty. Based on an analysis of the literature and communication with colleagues, the authors identified two main types of non-traditional tasks aimed at cheating manifestation. The first ones are tasks with graphic content. The second ones are constructive tasks for which students build an object based on its description. Both types of tasks do not take a long time to receive an answer and to make cheating impossible, even with the help of modern information and communication technologies. The authors conducted research among academic groups of 1 year of study in economic areas at Peter the Great St. Petersburg Polytechnic University. The research involved 171 Russian students and 61 Chinese students. The results showed that Russian students adapt to introduction of non-traditional assignments in about 1 semester, Chinese students take a little longer. The results indicate that use of tasks those exclude academic dishonesty does not reduce average results of students' academic success.

Keyword: Academic dishonesty · Mathematical creativity · Mathematics at the University · Russian and Chinese students

1 Introduction

Pedagogical creativity is one of the crosscutting topics of discussion of both professional theorists and practicing teachers [1]. Outstanding Russian scientists S.L. Rubinstein, N.V. Kuzmina, V.A. Kan-Kalik developed theoretical foundations of pedagogical creativity. Among those closer to the modern pedagogical discourse of foreign scientists at the turn of the XX-XXI centuries, we can single out Anna Kraft, who studied various aspects of pedagogical creativity [2].

© The Author(s), under exclusive license to Springer Nature Switzerland AG 2022
D. Bylieva and A. Nordmann (Eds.): PCSF 2021, LNNS 345, pp. 630–645, 2022.
https://doi.org/10.1007/978-3-030-89708-6_52

The authors focus on pedagogical creativity in the field of mathematics. The great Henri Poincaré raised the problem of mathematical creativity in a special report of 1908. Jacques Hadamard developed the theme in his 1945 work on the psychology of mathematical thinking. Dutch mathematician Gontran Ervink [3] and his followers [4, 5] reflected the modern understanding of mathematical creativity and its characteristics is in their works. This article touches upon a narrower topic of pedagogical creativity in teaching mathematics [6]. There are several directions of this issue. For example, there are many works on disclosing the creative potential of students in the process of their mathematical training [7]. The range of discussions here extends from the use of complex intellectual-computer schemes [8] to the approaches of "ethno mathematics" [9], which is a reflection of the archetypal feeling of historical guilt of the descendants of the colonialists and oppressors towards the descendants of the colonized and oppressed societies. Such aspects of pedagogical mathematical creativity, in many respects, are in connection with school mathematics. Therefore, the authors excluded this discourse from this work.

Researchers usually carry out the analysis of university pedagogical creativity of mathematicians in two aspects. In the studies of the first group, they consider the pedagogical creativity of teachers as a means of increasing student motivation [10–14]. They show, that methodical thinking, non-standard thinking [15], the ability to innovate [16], intuition and heuristic approach [17] are the basis for the formation of pedagogical creativity of mathematicians. Much work in the aspect of motivation concern the means of forming a creative approach in teaching mathematics. Researchers consider elective courses [18] as the main forms of organizing the educational process focused on mathematical creativity, which help to increase students' interest in mathematics. Among the methods of conducting classes the researchers call the project method [19], case technologics [20], as well as active and collective teaching methods [21, 22], in particular, the REACT (Relating, Experiencing, Applying, Cooperating, data Transferring) strategy [23] and methods using information technology and special software [24]. Finally, the content of lessons in mathematics, contributing to the formation of a creative approach, bases on the so-called open problems [25, 26], synonyms or similar terms for which are incorrectly posed problems [27] or poorly structured problems [28].

The works of the second group concern the study of the conditions for the formation of a creative approach among mathematicians themselves [29, 30], as well as the analysis of conditions that facilitate or hinder the use of creative methods by teachers in mathematics classes. Among the factors hindering the manifestation of teachers' creative abilities are the lack of trusting relationships between teachers and students, the lack of contact hours in mathematics, as well as the overload of teachers with pedagogical and organizational-methodological works [31]. Some studies suggest methods for studying the creative process [32] and measuring the creative potential [33] of pedagogical mathematicians.

It is important that the authors were unable to find works in which researchers consider the pedagogical creativity of mathematicians as one of the mechanisms for counteracting academic dishonesty. Since academic dishonesty has become widespread in modern education [34–36], it is obvious that there is a search for ways to prevent its manifestations. There are general recommendations for preventing academic dishonesty by introducing creative assignments [37], and varying the wording of open-ended questions in tests [38]. Some researchers have found a negative correlation between creativity and the propensity for academic dishonesty [39]. Although other studies give the opposite result regarding creativity and dishonesty in psychological mean of these terms [40, 41].

Discussions on academic dishonesty are currently focusing on two areas [42, 43] enshrined in university ethics codes [44]. We can call the first direction as punitive. Students those commit academic dishonesty are subject to disciplinary action [45]. The punitive direction has both supporters [46] and constructive critics [47]. We describe the second direction as educational. The university encourages academic integrity and culti-vates positive values [48]. They consider codes of ethics themselves as either a universal remedy [49] or an insufficient remedy [50] for overcoming academic dishonesty.

Usually they consider high student as one of the main psychological barriers to academic dishonesty [51, 52]. Analyzing the literature on mathematical creativity, the authors showed above that the most important goal of the creative influence of teachers is precisely to increase the motivation of students to study a certain discipline. Thus, the problem of using mathematical creativity to prevent academic dishonesty is usually resolved indirectly through increased motivation. In this paper, we concretize and supple-ment the relationship "mathematical creativity - prevention of academic dishonesty" by practical methods that reduce the possibility of manifestations of academic dishonesty.

2 Methods

University math teachers have faced an increasing number of academic dishonesty since the early 2000s. At this time, computer aided systems (CAD) and numerous Internet services appeared for quickly solving mathematical problems. The most famous of these services are WolframAlpha, Slader, Symbolab, Desmos, and Photomath [53]. Every year these systems and services are becoming perfect, they use technologies related to artificial intelligence. For example, they can easily decipher handwritten text. Math educators, of course, are looking for ways to counter this form of academic dishonesty. They offer tasks that students cannot solve correctly with services like WolframAlpha. In addition, they have Ideas exchange on specialized blogs such as the Mathematics Educators Stack Exchange, or Hacker News.

Some discrepancies in the assessment of academic dishonesty by mathematicians arise from different interpretations of the very concept of "cheating". This ambiguity depends on what kind of actions the teacher considers illegal. Some mathematicians do not allow students to use any materials, even their own notes, requiring, for example, knowledge of elementary integrals by memory or the formula for calculating the gradient. Others allow students read manuals and other paper-based materials. We give further consideration from the standpoint of a "liberal" teacher who prohibits students only from working with electronic assistants and any communicators.

Based on the study of the opinions of mathematicians on the Internet, and oral communication with colleagues, and analysis of a few literary sources, the authors iden-tified some types of tasks those university mathematics teachers use to counter student academic dishonesty. We have grouped the assignments by sections of the course in Calculus, referring to the level of an American college or of the first year of study at a

Russian university. We also commented on all assignments, identifying their source, if possible, as well as the presence of a creative approach to their design and the prospects for use against manifestations of academic dishonesty.

The aim of this part of the study is to identify types of assignments that use the creative insights of teachers preventing manifestations of academic dishonesty. In addition. In addition, students should not spend too much time solving the tasks in consideration.

Topic "Introduction to Calculus" ("Limits")

Ex. 1. Find the limit of the expression:

$$\lim_{x \to 2} \frac{x^3 + 2x^2 - 2x - 12}{x^3 - 3x^2 + 5x - 6} \cdot \frac{\arcsin^2(3tg^3(x-2))}{\sqrt[3]{1 + 6(x-2)^6} - 1} \cdot \frac{\sqrt{x^2 + 5} - 3}{4 - \sqrt{6x + 4}} \cdot (9 - 2x^2)^{\frac{3x^2 + x + 1}{x^2 + 4x - 8}}$$

The source is oral talk with G.A. Rastopchina. Teachers used such assignments at the beginning of the process of massive manifestation of academic dishonesty by students, when CAD was the main "assistant" of students. Task includes too many characters typing by hand, and possibly, it is inconvenient for photographing. Using internet platforms usually does not result in the response. The assignment is traditional but creative. As a means of counteracting cheating, the authors rate it at 5 points out of 5 possible. The main disadvantage of the assignment is that it takes a long time to complete by students who do not use cheating. Thus, by punishing dishonest students, the teacher significantly underestimates the grades of honest students. In addition, the teacher spends too much time checking the assignment. The authors recommend such tasks to students in the physics and mathematics areas of training only.

Ex. 2. Find the value of the parameter α at which the value of the limit is equal to − 2:

$$\lim_{x \to 1} \frac{x^2 - \alpha x + \alpha - 1}{x^2 + x - 2}.$$

The source is article [54]. The assignment is unconventional and creative. As a means of counteracting cheating, the authors rate it at 5 points out of 5 possible. The authors recommend such tasks to students in the physics and mathematics areas of training mainly.

Ex. 3. Construct undefined expression [0/0] when $x \to 5$.

The source is article [54]. The assignment is unconventional and creative. As a means of counteracting cheating, the authors rate it at 5 points out of 5 possible. The main advantage of the assignment is that Internet platforms and services do not yet produce mathematical objects from descriptions. Moreover, honest students can solve the problem very quickly. The main disadvantage of the assignment is that it is depends from the context and interpretation of the teacher. It is highly likely that students will follow the patterns shown in the classroom. Thus, being a product of the teachers' creativity, this task may not give the expected creative effect in the case of passive students. The authors recommend such tasks to students in the physics and mathematics areas of training mainly.

Ex. 4. We give the graph of the function (Fig. 1). Find one-sided limits at break points and limits at $\pm\infty$.

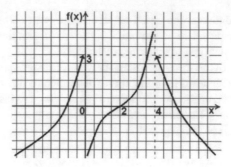

Fig. 1. One-sided limits (Ex. 4)

The task is the authors' own example. The assignment is unconventional and creative. As a means of counteracting cheating, the authors rate it at 5 points out of 5 possible. The main advantage of the assignment is that Internet services are not yet able to cope with graphic information of this kind. Another important benefit of the assignment is that honest students can solve the assignment very quickly. The authors recommend offering the task to students of all fields of study.

Ex. 5. Display the graph of the function having a limit of $-\infty$ at $-\infty$; and limit $-\infty$ at $+\infty$; and one-sided limits $+3$ and $-\infty$ when $x = 0$; and one-sided limits $+\infty$ and $+3$ when $x = 4$.

The task is the authors' own example. The assignment is unconventional and creative. As a means of counteracting cheating, the authors rate it at 5 points out of 5 possible. The advantages and disadvantages of the task are the same as in Ex. 3. The authors recommend such tasks to students in the physics and mathematics areas of training mainly.

Topic "Differentiation Calculus of function of one variable" ("Derivatives")

Ex. 6. Find the derivative of the function:

$$y = 10 sin^{3/5}\left(4\sqrt{3 - \frac{2}{x}} + 5^{\frac{\left(3tg(x/6)+1^{4x^2} - \sqrt[7]{x}+\ln^8\left(1-cos^5\left(1-\frac{1}{x}\right)\right)\right)}{arcsin^6\left(3x-5arctg\sqrt{1-2x}\right)}}\right)^2$$

The assignment is traditional and non-creative. As a means of counteracting cheating, the authors rate it at 4 points out of 5 possible. Recording the task usually requires enumerating options for the correct solution by Internet services. The advantages and other disadvantages are the same as for Ex. 1. The advantages and disadvantages of the task are the same as in Ex. 1. In addition, the answer turns out to be too cumbersome, not convenient for verification. The authors recommend such tasks to students of physical and mathematical areas of training only and for independent work, and not at control events.

Ex. 7. Find the signs of the first and second derivatives on intervals, as well as points with zero values of the first and second derivatives for the function defined by graph (Fig. 2).

The assignment is traditional but creative. As a means of counteracting cheating, the authors rate it at 5 points out of 5 possible. The advantages and disadvantages of the task

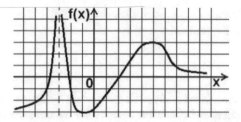

Fig. 2. Convexity and concavity (Ex. 7)

are the same as in Ex. 4. Many colleagues consider this task too simple, not appropriate for a university mathematics course. The authors recommend offering such tasks to students of all areas of training, with the exception of future physics and mathematics.

Ex. 8. Draw a graph of the function increasing in the interval from $-\infty$ to -3, having an infinite discontinuity at x $= -3$; further decreasing in the interval from -3 to -1; further increasing in the interval from -1 to 5 and decreasing in the interval from 5 to $+\infty$. For x $= -1$ the function has a smooth minimum, for x $= 5$ the function has a smooth maximum. On the intervals from $-\infty$ to -3 and from -3 to $+2$ the function is concave, on the interval from $+2$ to $+7$ the function is convex. Function has inflection points at x $= 2$ and at x $= 7$.

The task is the authors' own example. The assignment is unconventional and creative. As a means of counteracting cheating, the authors rate it at 4 points out of 5 possible. The advantages and disadvantages of the task are the same as in Ex. 5. Even students those demonstrate honest behavior usually do poorly on this assignment. The authors recommend such tasks to students in the physics and mathematics areas of training only.

Ex. 9. Find one-sided derivatives of the function at $x = 0$ (Fig. 3).

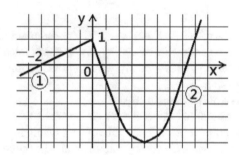

Fig. 3. One-side derivatives (Ex. 9)

We define the function on intervals: when $x < 0 : y = x/2 + 1$; when $x > 0 : y = (x - 2)^2 - 3$.

The task is the authors' own example. The assignment is unconventional and creative. As a means of counteracting cheating, the authors rate it at 5 points out of 5 possible. The advantages and disadvantages of the task are the same as in Ex. 4. To complicate the task, we replace the expression for the function with a description. The authors recommend offering the task to students of all fields of study.

Topic "Integral Calculus of function of one variable" ("Integrals").
Ex. 10. Find indefinite integral:

$$\int \frac{dx}{x^4 + 1}$$

G.K. Ermolaeva constructed many such integrals. The authors cited only the simplest of them. The assignment is traditional but creative. As a means of counteracting cheating, the authors rate it at 1 points out of 5 possible. If the student manages to use the Internet services, then he/she can obtain the solution very quickly. The honest student has to take too long to solve. Possible errors in cumbersome calculations may lead to an incorrect answer. Integration is the subject of the main discussions of mathematicians in forums focused on counteracting academic dishonesty. The authors do not recommend offering such assignments to students.

Ex. 11. Find indefinite integral:

$$\int \left(\frac{x^2 + 1 + e^x}{\sin x} + 2\cos x \right) \cdot \sin x \, dx$$

The source is oral talk with E.S. Edinova. The assignment is unconventional and creative. As a means of counteracting cheating, the authors rate it at 3 points out of 5 possible. Recording the task usually requires enumerating options for the correct solution by Internet services. The authors recommend offering the task to students of all fields of study.

Ex. 12. Find the derivative of the definite integral with variable limits:

$$\int_{1-cos2x}^{3tg(x/6)} \arcsin\left(\frac{3}{\sqrt{x} + 1} \right) dx$$

The source is oral talk with I.A. Komarchev. The assignment is unconventional and creative. As a means of counteracting cheating, the authors rate it at 1 poins out of 5 possible. Internet services do this task easily. The authors recommend offering the task to students of all fields of study.

Ex. 13. Find the area of the domain (Fig. 4).

The assignment is traditional but creative. As a means of counteracting cheating, the authors rate it at 5 points out of 5 possible. The advantages and disadvantages of the task are the same as in Ex. 4. To complicate the task, the authors propose to replace expressions for functions with their verbal descriptions. The authors recommend offering the task to students of all fields of study.

Topic "Differentiation Calculus of function of several variables".
Ex. 14. Find the limit of the expression:

$$\lim_{(x,y)\to(1,-2)} \frac{x^2 - y^2 + 2x + 1}{x + y + 1}$$

The source is article [54]. The assignment is unconventional and creative. As a means of counteracting cheating, the authors rate it at 1 point out of 5 possible. Internet services

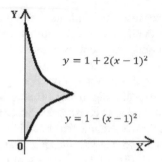

Fig. 4. Area of domain (Ex. 13)

do this task easily. The authors recommend such tasks to students in the physics and mathematics areas of training mainly.

Ex. 15. Construct formulas for chain rule for partial derivatives:

$$z = z(x, y, t); x = x(u, v, w); y = y(v, w); t = t(w);$$

$$z'_u =?; z'_v =?; z'_w =?$$

The task is the authors' own example. The assignment is unconventional and creative. As a means of counteracting cheating, the authors rate it at 5 points out of 5 possible. Internet services are not suitable for constructing formulas. One of the advantages of the assignment is the ability for quick solution for honest students. The authors recommend offering the task to students of all fields of study.

The authors cited examples of creative and non-traditional assignments that enable teachers to prevent student academic dishonesty at test events. The authors have also developed a system of unconventional problems in probability theory in their other works.

3 Results and Discussion

In the Spring Semester of the 2018/19 academic year, changes took place in the organization of training for 1st year students at the Institute of Industrial Management, Economics and Trade of Peter the Great St. Petersburg Polytechnic University. The management of the institute decided to merge two academic streams, with a total number of about 250 people. Before the merger, both streams trained Russian students and foreign students, the majority of whom were citizens of the People's Republic of China (Continental China).

Before the merger, teachers taught mathematics in the two streams in different ways. In the stream, which we will conditionally call No. 1, the teacher, using unconventional tasks from the ones listed above, conducted three tests with a time limit. Thus, students of stream 1, both Russian and foreign, are to a certain extent accustomed to performing non-traditional tasks and minimizing opportunities for academic dishonesty. Unfortunately, the teachers did not save the data for detailed monitoring of the performance of control works in the fall semester, because they did not know anything about the merger of streams. In stream, we provisionally call No. 2, another instructor taught the class using traditional test materials, without implementing creative assignments in order to prevent cases of academic dishonesty.

After the merger of streams, all students in the Spring Semester listened to lectures by the teacher using non-traditional assignments. Teachers conducted three tests based on non-traditional assignments and time constraints. Since the teachers noticed different reaction of the students of the former streams No. 1 and No. 2 to the implemented scheme of control activities, we decided to analyze the quantitative results in order to obtain recommendations for improving the preparation in mathematics using non-traditional tasks.

As the main indicator, we used the percentage of successfully completed works in the total number of works for each of the three control events. We consider this indicator among other things, as an assessment of the level of adaptation of students to the model of control activities using non-traditional creative tasks that reduce the possibility of academic dishonesty. The authors have considered the problem of the criteria for the successful completion of the test in other articles.

78 Russian students (RS1) and 37 Chinese students (CS1) studied in the former stream No. 1. In addition, foreign students from Kazakhstan, Uzbekistan, Morocco, and Vietnam studied in the former stream No. 1, but their groups were insignificant, and therefore they could not deliver representative material for statistical conclusions. 93 Russian students (RS2), and 24 Chinese students (CS2) and a number of citizens of other countries studied in the former stream No. 2. Table 1 contains the number and percentage of successfully completed tests by Russian and Chinese students of the combined lecture stream.

Table 1. Number and percentage of successfully done control works for different types of students

Type/number of students	Control work 1		Control work 2		Control work 3	
	Total number of works	Number/percentage of successful works	Total number of works	Number/percentage of successful works	Total number of works	Number/percentage of successful works
RS1/78	76	64/84,2%	78	66/84,6%	73	64/87,7%
RS2/93	90	65/72,2%	91	69/75,8%	86	69/80,2%
CS1/37	37	21/56,8%	36	23/63,9%	34	24/70,6%
CS2/24	21	0/0,0%	19	5/26,3%	16	8/50,0%

There are several facts. The first fact is the difference between the percentages of successful tests completed by students of streams No. 1 and No. 2 The second one is the difference between the percentages of successful works completed by Russian and Chinese students. The third one is the difference between the percentages of successful papers Chinese students, increasing from the first test to the third one, which is especially noticeable for the CS2 group. These considerations led us to the decision to focus on the separate analysis and explanation of the results specifically for the categories of students presented. To show only the most impressive differences, we omitted in this study a detailed statistical analysis of each of the groups of students for each test, but added the summary (final) data on the success of all three works. Among other things, these aggregated data increase the confidence of the conclusions due to the increased sample sizes. For the summary data and for the results of the third control, we calculate confidence intervals at the significance level of 0.1, which corresponds to a reliability of 90% (Table 2).

Table 2. Number and percentage with confidence intervals of successfully done control works for different types of students

Type/number of students	Control work 3			Practical training final		
	Total number of works	Number/percentage of successful works	Confidence intervals for successful works percentage	Total number of works	Number/percentage of successful works	Confidence intervals for successful works percentage
RS1/78	73	64/87,7%	81,4%-94,0%	227	194/85,5%	81,4%-89,4%
RS2/93	86	69/80,2%	73,1%-87.3%	267	203/76,0%	71,7%-80,3%
CS1/37	34	24/70,6%	57,8%-83,4%	107	68/63,6%	55,9%-71,3%
CS2/24	16	8/50,0%	–	56	13/23,2%	12,9%-32,5%

For Russian students of the RS1 group, the confidence interval of the final success (81.4%–89.4%) with a margin includes the percentage of success in completing the third control, 87.7%. This indicates the successful adaptation of students in this group to the performance of non-traditional tasks that reduce the level of academic dishonesty. Indeed, for the RS1 group, the adaptation process has already passed in the fall semester.

For Russian students of the RS2 group, the confidence interval of the final success (71.7%–80.3%) includes at the limit the percentage of success in completing the third control 80.2%. This indicates an active course of the process of adaptation to the new model for these students of conducting control activities in the Spring Semester. Namely, there is a not too pronounced dynamics of increasing the percentage of success in students' completing control from the first work to the third one.

It is especially important to compare the cross-group success percentages of RS1 and RS2 students. The confidence interval of the final success of RS1 students (81.4%–89.4%) obviously does not include the percentage of the final success of RS2 students, equal to 76.0%. This indicates a statistically confirmed greater adaptation of RS1 students to the introduction of non-traditional tasks those minimize the manifestation of academic dishonesty in relation to RS2 students. At the same time, the percentage of successful completion of the third control by RS2 students 80.2% approaches the confidence interval of the percentage of success in completing the third control by RS1 students (81.4%–94.0%). This speaks not only of the positive dynamics of students of the RS2 group in relation to the acceptance of non-traditional tasks during the Spring Semester, but also of the approach of the degree of their adaptation to the new model of control activities to the indicators of students of the RS1 group.

Unfortunately, the small sample size of Chinese students in the CS2 group does not provide an opportunity for a detailed analysis of the degree of adaptation of these students to the model of conducting control activities that minimizes the manifestation of academic dishonesty. Therefore, it is not possible to construct a confidence interval at a significance level of 0.1 for projecting the success of the third test by CS2 Chinese students. An increase in the level of significance would mean a decrease in the reliability of the study, and we could not allow this one. It is evident only that the percentage of successful work of students in the CS2 group, equal to 50%, is much higher than the upper limit of the confidence interval (12.9%–32.5%) for the percentage of the total success of students in this group. This indicates a significant increase in the degree of adaptation of the Chinese students of the CS2 group to the new method of conducting control activities for them.

On the other hand, the final success percentage of Chinese students in the CS2 group, equal to 23.2%, is significantly less than the lower limit of the confidence interval for the final success of Chinese students in the CS1 group (55.9%–71.3%). This leads to the conclusion that there is a need for a longer period of adaptation of Chinese students to new methods of conducting control activities than for Russian students of the RS2 group put in similar conditions, whose indicators during the spring semester, as shown above, approached the indicators of Russian students in the RS1 group.

Comparison of the confidence intervals for the final success of Chinese students in both groups CS1 and CS2 and Russian students in both groups RS1 and RS2 show that they do not overlap. This indicates a higher degree of adaptation of Russian students to the introduction of non-traditional methods of conducting control activities those reduce the possibility of manifestations of academic dishonesty. It is possible to explain this phenomenon by a number of characteristic features of the Chinese mentality in relation to education [55, 56]. Among these characteristics, the main ones in the aspect of perception of innovation are unconditional obedience to order, and adherence to established patterns, and the priority of everything Chinese over everything foreign [57]. In particular, this means that it is more difficult for Chinese students to readjust when the teacher's requirements for testing are changing. In addition, Chinese students always rely on the opinions and recommendations of compatriots. Usually those are students of past years studied in the same areas of training. If predecessors described certain models of the organization of the educational process, then current students would simply ignore most innovations, perceiving them as incorrect, temporary or accidental information. Chinese students consider information received from compatriots as more consistent with the truth. Teachers can be encouraged to follow the established new rules without making concessions and indulgences to Chinese students. If such an opportunity exists, teachers should explain the meaning and essence of innovations to representatives of the Chinese student community, or, better, to undergraduates and graduate students, whose opinion is important for the students of the first years of study [58].

At the same time, the percentage of successfully completed works of the third control for Chinese students of the CS1 group, equal to 70.6%, is quite close to the lower limit of the confidence interval for the percentage of successfully completed works of the third control for Russian students of the RS2 group (73.1%–87.3%). This indicates a convergence in the degree of adaptation of these groups. It also allows us to estimate the time required for students to adapt to the introduction of innovative techniques. Thus, on average, Russian students need one semester to adapt to innovative methods, while Chinese students need at least two semesters. It is possible to explain this one by the difficulties in learning in a foreign language in the conditions of functioning of pedagogical system that is unusual for them.

We find a certain optimism comparing confidence intervals for the percentage of final success of Russian students in the RS2 group (71.7%–80.3%) and Chinese students in the CS1 group (55.9%–71.3%). These intervals practically overlap, which, in turn, suggests the possibility of a further increase in the degree of adaptation of Chinese students to innovative methods in subsequent semesters of their studies.

4 Conclusion

Academic dishonesty as a negative phenomenon in the global educational environment has become widespread in the last 30 years. This is due, in particular, to the revolutionary improvement of information and communication technologies. Academic responses to academic dishonesty develop in two ways. Representatives of the paradigm of rigid pedagogy reduce the decision to identifying students those commit academic dishonesty, with their subsequent punishment. Supporters of the paradigm of liberal pedagogy suggest strengthening educational work, promoting positive values, including academic integrity. Many researchers have analyzed the merits and demerits of both mechanisms, but have not yet come to an agreement.

The authors proposed to solve the problem of counteracting academic dishonesty by using the potential of creativity. As mathematicians, the authors naturally turned to the potential of mathematical creativity. Many fellow mathematicians are trying to invent increasingly difficult and cumbersome traditional tasks those students cannot solve using Internet platforms focused on solving mathematics tasks quickly. These platforms, created with different intentions, have now become practically patrons of dishonest students. The authors, on the other hand, use non-traditional tasks based on the principles of mathematical creativity. These are tasks of a graphic and constructive nature. Of course, it is best to offer students creative assignments, incorrectly posed tasks and tasks that allow ambiguous solutions. Solving such problems requires a significant amount of time, including contact hours of communication between students and teachers, which is not always possible.

The authors conducted a study of more than 200 first year students those are future economists. Some of them worked with non-traditional assignments from the very beginning of their studies, the rest ones from the Spring Semester. The authors obtained statistically distinguishable data demonstrating the process of adaptation of students to innovations in the methods of teaching mathematics. In particular, they show a slower degree of this adaptation in Chinese students.

At first glance, the authors have partially solved the problem of introducing elements of non-traditional and creative assignments. The authors received positive, statistically valid results. However, questions remained. It is clear that the main source of the rise in academic dishonesty is the massification of higher education. Online platforms for a solution are only the answer to a public need. The massification has dealt an irreparable blow not only to the idea of Humboldt model of University, but also to the entire liberal educational model. Unfortunately, our society has failed to achieve mass motivation to reach the heights of human thought. This turned out to be somewhat analogous to the communist utopia. A likely, albeit unattractive for intelligent people, way out is level education and university stratification.

References

1. Simonton, D.K.: Teaching creativity: current findings, trends, and controversies in the psychology of creativity. Teach. Psychol. **39**(3), 203–208 (2012). https://doi.org/10.1177/0098628312450444
2. Kraft, A.: Creativity, education and society. In: Chappel, K., Cremin, T., Jeffrey, R. (eds.) Writings of Anna Craft. Trentham Books, London (2015). https://www.ucl-ioe-press.com/books/schools-and-schooling/creativity-education-and-society/. Accessed 10 Apr 2021
3. Ervynck, G.: Mathematical creativity. In: Tall, D. (ed.) Advanced Mathematical Thinking, pp. 42–53. Springer, Heidelberg (1991). https://doi.org/10.1007/0-306-47203-1
4. Sriraman, B.: The characteristics of mathematical creativity. ZDM: Int. J. Math. Educ. **41**(1), 13–27 (2008). https://doi.org/10.1007/s11858-008-0114-z
5. Sriraman, B., Yaftian, N., Lee, K.H.: Mathematical creativity and mathematics education. In: Sriraman, B., Lee, K.H. (eds.) The Elements of Creativity and Giftedness in Mathematics. Advances in Creativity and Giftedness, vol. 1, pp. 119–130. Sense Publishers, Rotterdam (2011). https://doi.org/10.1007/978-94-6091-439-3_8
6. Haavold, P., Sriraman, B., Lee, K.-H.: Creativity in mathematics education. In: Lerman, S. (ed.) Encyclopedia of Mathematics Education, pp. 145–154. Springer, Cham (2020). https://doi.org/10.1007/978-3-030-15789-0_33
7. Leikin, R., Pitta-Pantazi, D.: Creativity and mathematics education: the state of the art. ZDM – Int. J. Math. Educ. **45**(2), 159–166 (2013). https://doi.org/10.1007/S11858-012-0459-1
8. Madden, S.R.: Impacting mathematical and technological creativity with dynamic technology scaffolding. In: Freiman, V., Tassell, J.L. (eds.) Creativity and Technology in Mathematics Education. MEDE, vol. 10, pp. 89–124. Springer, Cham (2018). https://doi.org/10.1007/978-3-319-72381-5_4
9. Rigney, L., Garrett, R., Curry, M., MacGill, B.: Culturally responsive pedagogy and mathematics through creative and body-based learning: urban aboriginal schooling. Educ. Urb. Soc. **52**(8), 1159–1180 (2020). https://doi.org/10.1177/0013124519896861
10. Sergeeva, E.V.: How to Interest a Student in Mathematics? Probl. Modern Pedag. Educ. **59**(4), 244–247 (2018). (in Russian). https://elibrary.ru/item.asp?id=35130349. Accessed 15 Apr 2021
11. Bylieva, D., Bekirogullari, Z., Kuznetsov, D., Almazova, N., Lobatyuk, V., Rubtsova, A.: Online group student peer-communication as an element of open education. Futur. Internet. **12**, 143 (2020). https://doi.org/10.3390/fi12090143
12. Shipunova, O.D., Evseeva, L., Pozdeeva, E., Evseev, V.V., Zhabenko, I.: Social and educational environment modeling in future vision: infosphere tools. E3S Web Conf. **110**, 02011 (2019). https://doi.org/10.1051/e3sconf/201911002011
13. Rubtsova, A.V., Almazova, N.I., Bylieva, D.S., Krylova, E.A.: Constructive model of multilingual education management in higher school. IOP Conf. Ser. Mater. Sci. Eng. **940**, 012132 (2020). https://doi.org/10.1088/1757-899X/940/1/012132
14. Grégoire, J.: Understanding creativity in mathematics for improving mathematical education. J. Cogn. Edu. Psych. **15**(1), 24–36 (2016). https://doi.org/10.1891/1945-8959.15.1.24

15. Levenson, E.: Tasks that may occasion mathematical creativity: teachers' choices. J. Math. Teacher Educ. **16**, 269–291 (2013). https://doi.org/10.1007/s10857-012-9229-9
16. Rodrigues, A., Catarino, P.M.C., Aires, A.P., Campos, H.: Conceptions of students about creativity and mathematical creativity: two cases studies in vocational education. Proceedings **2**(21), 1357 (2018). https://doi.org/10.3390/proceedings2211357
17. Hoon, T., Kee, K., Singh, P.: Learning mathematics using heuristic approach. Procedia – Soc. Behav. Sci. **90**, 862–869 (2013). https://doi.org/10.1016/j.sbspro.2013.07.162
18. Leikin, R., Subotnik, R., Pitta-Pantazi, D., Singer, F., Pelczer, I.: Teachers' views on creativity in mathematics education: an international survey. ZDM Math. Educ. **45**(2), 309–324 (2013). https://doi.org/10.1007/s11858-012-0472-4
19. Sergeeva, E.V.: Role of projects in the study of mathematics. In: The Interdisciplinary Approach in Humanities and Social Sciences on Current Issues of Linguistics and Didactics (CILDIAH-2018), SHS Web Conference, vol. 50, p. 01157. EDP Sciences, Les Ulis (2018). https://doi.org/10.1051/shsconf/20185001157
20. Sarvanova, Zh.A., Kochetova, I.V., Dorofeev, S.N., Porvatkin, A.V.: Case technologies in interactive teaching of mathematical disciplines of students of natural-technical profiles. Modern High Tech. **12**(1), 195–199 (2019). http://www.top-technologies.ru/ru/article/view?id=37858. Accessed 15 Apr 2021
21. Molad, O., Levenson, E.S., Levy, S.: Individual and group mathematical creativity among post–high school students. Educ. Stud. Math. **104**(2), 201–220 (2020). https://doi.org/10.1007/s10649-020-09952-5
22. Aljarrah, A.: Exploring collective creativity in elementary mathematics classroom settings (Unpublished doctoral thesis). University of Calgary, Calgary, AB (2018). https://doi.org/10.11575/PRISM/31863
23. Qadri, L., M. Ikhsan, M., Yusrizal, Y.: Mathematical creative thinking ability for students through REACT strategies. Intern. J. Educ. Vocat. Stud. **1**(1), 58 (2019). https://doi.org/10.29103/ijevs.v1i1.1483
24. Freiman, V., Tassell, J.L.: Leveraging mathematics creativity by using technology: questions, issues, solutions, and innovative paths. In: Freiman, V., Tassell, J.L. (eds.) Creativity and Technology in Mathematics Education. MEDE, vol. 10, pp. 3–29. Springer, Cham (2018). https://doi.org/10.1007/978-3-319-72381-5_1
25. Yuniartia, Y., Kusumaha, Y.S., Suryadia, D., Kartasasmitaa, B.G.: The effectiveness of open-ended problems based analytic-synthetic learning on the mathematical creative thinking ability of pre-service elementary school teachers. Intern. Elec. J. Math. Educ. **12**(3), 655–666 (2017). https://www.iejme.com/article/the-effectiveness-of-open-ended-problems-based-analytic-synthetic-learning-on-the-mathematical. Accessed 1 Apr 2021
26. Suastikaa, K.: Mathematics learning model of open problem solving to develop students' creativity. Intern. Elect. J. of Math. Educ. **12**(3), 569–577 (2017). https://www.iejme.com/article/mathematics-learning-model-of-open-problem-solving-to-develop-students-creativity. Accessed 15 Apr 2021
27. Nadjafikhah, M., Yaftian, N., Bakhshalizadehc, Sh.: Mathematical creativity: some definitions and characteristics. Proc. – Soc. Behav. Sci. **31**, 285–291 (2012). https://doi.org/10.1016/j.sbspro.2011.12.056
28. Abdillah, A., Ajeng, G.M., Muhajir, A.R.: Ill-structured mathematical problems to develop creative thinking students. In: Fitriyati, N., Asyhar, A.H., Susanti, E., Juhari, Jauhari, M.N., Hafiyusholeh, M. (eds.) Proceedings of the International Conference on Mathematics and Islam (ICMIs 2018), pp. 28–33. Science and Technology Publications, Setúbal (2018). https://doi.org/10.5220/0008516700280033
29. Scherbakova, S.Yu.: Formation of creative thinking as a condition of professional and personal coming-to-be of future teachers. Herald of Tver State University. Series: Pedagogy and Psychology **4**, 118–127 (2009). https://www.elibrary.ru/item.asp?id=12977541

30. Mastuti, A.G., et al.: Interpretation awareness of creativity mathematics teacher high school. Int. Educ. Stud. **9**(9), 32-41 (2016). https://doi.org/10.5539/ies.v9n9p32
31. Mann, E.L.: Creativity: the essence of mathematics. J. Educ. Gifted **30**(2), 236–260 (2006). https://doi.org/10.4219/jeg-2006-264
32. Schindler, M., Lilienthal, A.J.: Students' creative process in mathematics: insights from eye-tracking-stimulated recall interview on students' work on multiple solution tasks. Int. J. Sci. Math. Educ. **18**(8), 1565–1586 (2019). https://doi.org/10.1007/s10763-019-10033-0
33. Andrade, R.R., Pasia, A.E.: Mathematical creativity of pre-service teachers in solving non-routine problems in State University in Laguna. Univ. J. Educ. Res. **8**(10), 4555–4567 (2020). https://doi.org/10.13189/ujer.2020.081024
34. Marsden, H., Carroll, M., Neill, J.T.: Who cheats at university? A self-report study of dishonest academic behaviours in a sample of australian university students. Aust. J. of Psych. **57**(1), 1–10 (2005). https://doi.org/10.1080/00049530412331283426
35. Maloshonok, N., Shmeleva, E.: Factors influencing academic dishonesty among undergraduate students at Russian universities. J. Acad. Ethics **17**(3), 313–329 (2019). https://doi.org/10.1007/s10805-019-9324-y
36. Bylieva, D.S., Lobatyuk, V.V., Nam, T.A.: Academic dishonesty in e-learning system. In: Soliman, K.S. (eds) Proceedings of the 33rd International Business Information Management Association Conference, IBIMA 2019: Education Excellence and Innovation Management through Vision 2020, pp. 7469–7481. International Business Information Management Association, Granada (2019)
37. Bylieva, D., Lobatyuk, V., Tolpygin, S., Rubtsova, A.: Academic dishonesty prevention in e-learning university system. In: Rocha, Á., Adeli, H., Reis, L.P., Costanzo, S., Orovic, I., Moreira, F. (eds.) WorldCIST 2020. AISC, vol. 1161, pp. 225–234. Springer, Cham (2020). https://doi.org/10.1007/978-3-030-45697-9_22
38. Toro, S.: 8 Ways to Reduce Student Cheating. Edutopia. https://www.edutopia.org/article/8-ways-reduce-student-cheating. Accessed 21 Apr 2021
39. Adyasha, R., Duraipandian, R.: Relationship between creativity and academic integrity of students: an empirical study of management students in India. Manag. Stud. Econ. Syst. **2**(4), 255–262 (2016). https://doi.org/10.12816/0035633
40. Gino, F., Ariely, D.: The dark side of creativity: Original thinkers can be more dishonest. J. Pers. Soc. Psych. **102**(3), 445–459 (2012). https://doi.org/10.1037/a0026406
41. Beaussart, M.L., Andrews, C.J., Kaufman, J.C.: Creative liars: the relationship between creativity and integrity. Think. Skills Creat. **9**, 129–134 (2013). https://doi.org/10.1016/j.tsc.2012.10.003
42. Kara, F., MacAlister, D.: Responding to academic dishonesty in universities: a restorative justice approach. Contemp. Just. Rev. **13**(4), 443–453 (2010). https://doi.org/10.1080/10282580.2010.517981
43. Dremova, O.V.: Russian university policies on students' academic dishonesty: punishment or ethical training. Uni. Manag.: Pract. Anal. **24**(4), 30–45 (2020). https://doi.org/10.15826/umpa.2020.04.033
44. Sutherland-Smith, W.: Crime and punishment: an analysis of university plagiarism policies. Semiotica **2011**(187), 127–139 (2011). https://doi.org/10.1515/semi.2011.067
45. Chirikov, I., Shmeleva, E., Loyalka, P.: The role of faculty in reducing academic dishonesty among engineering students. Stud. High. Educ. **45**(12), 2464–2480 (2019). https://doi.org/10.1080/03075079.2019.1616169
46. Shmeleva, E.: Plagiarism and cheating in Russian universities: the role of the learning environment and personal characteristics of students. Educ. Stud. **1**, 84–109 (2016). https://doi.org/10.17323/1814-9545-2016-1-84-109
47. Higbee, J.L., Thomas, P.V.: Student and faculty perceptions of behaviors that constitute cheating. NASPA J. **40**(1), 39–52 (2002). https://doi.org/10.2202/1949-6605.1187

48. McCabe, D.L., Butterfield, K.D., Treviño, L.K.: Cheating in College: Why Students Do It and What Educators Can Do About It. The Johns Hopkins University Press, Baltimore (2012)

49. Ely, J.J., Henderson, L., Wachsman, Y.: Testing the effectiveness of the University honor code. Acad. Educ. Leadersh. J. **18**(3), 1–10 (2014). https://www.abacademies.org/articles/ael jvol18no32014.pdf

50. Bath, M., Hovde, M.P., George, E., Schulz, K.: Academic integrity and community ties at a small, religious-affiliated liberal arts college. Int. J. Educ. Integr. **10**(2), 31–43 (2014). https://doi.org/10.21913/IJEI.v10i2.1005

51. Smith, K.J., Davy, J.A., Rosenberg, D.L., Haight, G.T.: The role of motivation and attitude on cheating among business students. J. Acad. Bus. Ethics **1**, 12–37 (2009). https://www.aabri.com/manuscripts/08005.pdf. Accessed 3 Apr 2021

52. Krou, M.R., Fong, C.J., Hoff, M.A.: Achievement Motivation and academic dishonesty: a meta-analytic investigation. Educ. Psychol. Rev. **33**(2), 427–458 (2020). https://doi.org/10.1007/s10648-020-09557-7

53. Williams, R.: The use of digital applications and websites in completing math assignments (Thesis, 30.03.2020). Concordia University St. Paul, Portland, USA (2020). https://digitalco mmons.csp.edu/cup_commons_grad_edd/452. Accessed 3 Apr 2021

54. Dorofeev, S.N., Ivanova, T.A., Uteeva, R.A., Shabanov, G.I., Derbedeneva, N.N.: Succession in the training creative activity of the future bachelor of pedagogical education (profile "Mathematics"). Human. Sci. Educ. **9**(4–36), 25–30 (2018). (in Russian)

55. Gu, M.: An analysis of the impact of traditional Chinese culture on Chinese education. Front. Educ. China **1**(2), 169–190 (2006). https://doi.org/10.1007/s11516-006-0001-8

56. Okhorzina, Y.O., Ma, J.: Chinese educational traditions and their impact on the process of learning Russian. Procedia. Soc. Behav. Sci. **215**, 79–83 (2015). https://doi.org/10.1016/j.sbs pro.2015.11.577

57. Wang, T.: Understanding Chinese culture and learning. In: Jeffery, P. (eds) International Education Research Conference, AARE 2006, pp. 1–14. Australian Association for Research in Education, Adelaide (2006). http://pandora.nla.gov.au/pan/24691/20070301-0000/www. aare.edu.au/06pap/wan06122.pdf. Accessed 3 Apr 2021

58. Lin, C.: Culture shock and social support: an investigation of a Chinese student organization on a US campus. J. Intercult. Commun. Res. **35**(2), 117–137 (2006). https://doi.org/10.1080/17475750600909279

Cheating and Plagiarism Among University Students: Ways of Solving the Problem

Natalia E. Anosova(✉) ⓘ and Anna V. Gavrilova ⓘ

Graduate School of Applied Linguistics, Interpreting and Translation, Peter the Great
Saint-Petersburg Polytechnic University, Saint-Petersburg 195251, Russia

Abstract. Today, various electronic devices and gadgets have become an integral
part of university students learning. However, the use of modern technologies can
significantly complicate the work of a teacher, as it becomes difficult to determine
the degree of students' independence and authorship when assessing their tests
and papers. The purpose of this study is to define the strategies that develop
students' responsibility and conscientiousness. The objectives of the study are the
following: to conduct a survey among the teachers to determine the degree of their
concern regarding this problem; to conduct a survey among the students to find out
their attitude towards cheating and plagiarism, and the ways they resort to while
cheating. The survey allowed the authors to identify the reasons for students'
cheating during tests. The study shows that it is the foreign language teachers
who are most concerned about the problem of cheating and plagiarism, while
most students questioned consider cheating normal. The authors conclude that it
is vital to address the problem of cheating and plagiarism by conducting regular
educational sessions to develop their sense of responsibility.

Keywords: Foreign language learning · Testing · Electronic devices · Students'
cheating · Plagiarism · Students' responsibility · Educational sessions

1 Introduction

Digitalization in the higher education system has led to the wide-spread use of various
electronic devices and the Internet. New electronic resources take the learning process to
a new level, making it more mobile, informative, and motivating [1]. At the same time,
life of people is becoming increasingly dependent on the data found in the Internet,
which is especially obvious in higher education. The students' behavioral patterns in the
online environment can be predicted to a certain extent as noted by some researchers
[2]. The pandemic caused by COVID-19 forced universities to switch to distance learning using information and computer technologies. The advantages and disadvantages of
online learning are discussed in a number of works [3, 4]. Almazova et al. state that it is
vitally important to develop technical competence of the academic staff, provide necessary technical facilities for the educational process and encourage students to switch to
online learning when necessary [5]. The study carried out by Rubtsova et al. highlight
the problems that students face when switching to distance learning [6]. The researchers

note that the following factors can hinder the process of learning: the lack of information support for students, lack of clear instructions for using digital educational tools, and problems with students' time management when they take courses online. These problems ought to be addressed urgently as digital technologies have already become an integral part of learning at a university.

However, the use of electronic devices in the learning process does not always lead to a positive result. Such negative phenomena as cheating and plagiarism arise due to the wide range of modern technologies. Hence, we can consider the use of electronic devices in education from an ethical point of view.

Today, ethical concepts and moral categories affect all areas of life, including family, education, and social sphere. Quite often students of Russian universities demonstrate irresponsible behavior, which can also be caused by a lot of negative information that young people receive from mass media. Internet websites focus mainly on entertainment, while the educational potential is abandoned [7]. According to E.S. Malkov, "the wide-spread academic fraud, as well as the predominance of a tolerant attitude towards it among students, entails many negative consequences, both for the entire higher education system and for the individual educational experience of students" [8, p.85]. The importance of considering ethical aspects and awareness of the impact of digital technology on the process of learning is brought up by the researchers of Peter the Great St.Petersburg University Pozdeeva, Shipunova and others [9].

The communicative and social competences of students are highlighted by the researchers [10] as vitally important for achieving the planned learning outcome. The fact that the number of international students is increasing at our universities also contributes to the necessity of developing intercultural competences and creating the conditions of learning where cheating and plagiarizing are unacceptable.

Teachers have always been faced up with the problem of cheating. However, with a wide variety of electronic devices available today, this problem is becoming almost unsolvable. A. Milkus in his article "What we've come to! Dismissal for cheating is now an outstanding event?", gives the example of the two students who were expelled from the Higher School of Economics for examination fraud. The reaction of one of the students caught cheating to the incident was striking. He said: "The university is not a parochial school, students here are given not morality but life lessons, and we strive to obtain the knowledge necessary for real life. The expulsion of an intelligent student for their immoral behavior is ridiculous, as honest people in Russia either do not work in the economic sphere or never achieve anything working there" [11].

E. V. Shmeleva in her research reports that only 12% of students have negative attitude towards cheating and only 2% consider punishment for cheating justifiable [12]. Guest lecturers are amazed at the extent of cheating in Russian universities. N. Martin, a teacher from Scotland, states that it is absolutely impossible for the student to cheat at a British university. Cheating can entail expulsion without the right to appeal and continue studies at another university [13].

Cheating is regarded as a consequence of the students' irresponsible behavior. According to V. P. Pryadein, the reasons behind the refusal to act responsibly may also stem from the under-developed sense of responsibility and infantile personality. Thereby, to deal with this problem, it is necessary to pay attention to the development of these personal qualities [14]. The author concludes that students either do not have a sense of responsibility, or they do not realize that cheating is a fraud.

M.V. Bortsova defines responsibility as "an integral personality trait, manifested in conscious, proactive, independent behavior, which serves as the basis of personality, a mechanism for organizing a person's life" [15, p. 5]. In this interpretation, the key concept is "independent behavior". In the interpretation of K. Muzdybaev, responsibility is one of the indicators of personal maturity, and the process of developing a personality will not be complete if there is no place for a sense of duty and responsibility [16]. Hence, we should develop in students a sense of responsibility, and understanding that cheating is a deception.

To develop responsibility in students, it is essential to understand what feelings should be evoked. According to V. A. Sukhomlinsky, "the upbringing of a teenager—both younger and older minors—is accomplished only by self-education, which forms human dignity. A true master skill in teenager education is to let the child choose the ways of self-education, self-improvement, self-defence by overcoming difficulties and experiencing the joy of surmounting them" [17, p. 49]. Self-education is based on self-love and self-respect. "Without self-respect, there is no personality development" [17, p. 32]. It is obvious that a sense of self-respect needs to be instilled in students, as it conveys the idea that cheating is disrespectful not only towards the teacher, but also to themselves. The ideas of V. A. Sukhomlinsky are supported by Solso R. L. and McLean M. N., who believe that self-control, self-esteem and self-management are essential parts of the internal responsibility of the individual for their behavior and thoughts" [18, p. 295].

Unfortunately, the problem of cheating is not the only ethical problem in university studies. Plagiarism is another serious problem, which manifests itself when students write abstracts, term papers, reports and final qualification works. These problems cannot be solved by technical means alone, since the technology will be forever advancing [19]. It is obvious that there should be ad hoc academic elements designed to prevent academic dishonesty such as the courses with automatic verification of academic results.

Every year students of the Institute of Humanities, majoring in Linguistics are involved in writing a final qualification work, which is subject to verification through the Antiplagiat system. The verification reveals quite a number of cases of plagiarism to be revised. To combat the problem, a special course was designed to familiarize students with the techniques and rules of writing academic papers and essays. The course teaches students how to use quotations and references while writing essays and papers avoiding plagiarism. Upon completion of the course, students are to write an academic paper or essay that is tested for anti-plagiarism.

In addition, the course "Guidelines for writing a dissertation" has been included in the program of the second year of study, and much attention is paid to citing sources in the final qualification work. So students receive versatile knowledge on the issue of anti-plagiarism, however, every year about half of the final qualification works have to be reviewed.

2 Problem Statement

Taking into account that the global task of the university is training highly qualified personnel which could contribute to bringing industrial production to a higher level of efficiency, the problem of cheating and plagiarism in the student environment is very acute. Many years of teaching experience at the university allows us to conclude that the fight against cheating by the usual means of control, such as taking away students' phones and other "gadgets" is not effective. Now, university teachers should change the students' perceptions of cheating and plagiarism, to persuade them that cheating is a reflection of their irresponsibility, of disrespect for themselves and the others, and to encourage them to perceive learning as their main activity at this stage of life.

3 Research Questions

For a comprehensive study of the problem of cheating and plagiarism, a survey of teachers and students was carried out, and the results of final qualification paper verification by the "Antiplagiat" system were analyzed. The questionnaires were supposed to answer the following questions:

- Is cheating a problem? Is cheating a consequence of students' irresponsibility?
- Do students cheat? Do they use the Internet while passing examinations and taking tests? Do they feel ashamed when cheating?

4 Purpose of the Study

The aim of the research is to look into the problem of cheating and plagiarism from the point of view of teachers and students. An attempt has been made to solve the problem of cheating and plagiarism by introducing relevant tasks and exercises into learning process and conducting regular educational sessions with students. The research is aimed to solve the following tasks: to reveal the attitude of teachers to the problem of cheating; to determine the reasons that would encourage students to stop cheating; to suggest ways of solving the problem of cheating and plagiarism by introducing relevant tasks and exercises into the process of learning.

The object of the research is the process of teaching students at a technical university.

The subject of the research is the educational activity of the teacher aimed at developing a sense of responsibility in students.

5 Research Methods

The material for the research was the questionnaires of 38 teachers of Peter the Great St. Petersburg Polytechnic University teaching physics, mathematics and a foreign language. The study contains the results of testing theoretical concepts through empirical applications. 187 students of the 1st and 2nd years of study at the Institute of computer science and technology, Institute of physics, nanotechnology and telecommunications and Institute of applied mathematics and mechanics of Peter the Great St.Petersburg Polytechnic University took part in the survey based on the first questionnaire, and 151 people from these institutes took part in the second questionnaire. The average age of students is from 18 to 24 years old. In addition, the results of Antiplagiat system verification of 120 final qualification works of Master students majoring in Linguistics were analyzed.

6 Findings

In order to tackle the problem of cheating, several steps have been made. First, it was necessary to determine teachers' attitudes to the problem. Teachers from different departments were asked to fill out Form 1.

Their responses to the two most important questions are presented in Table 1.

Table 1. Teachers' responses to the questions of Form 1

	Teachers of Physics and Mathematics	Teachers of Chemistry and Biology	Teachers of Humanities
Cheating is an urgent problem	86%	54%	95%
Cheating is attributed to an undeveloped sense of responsibility	71%	54%	80%

The second objective of the study was to determine students' actual level of cheating and use of electronic devices during tests. To this end, an anonymous survey of students was conducted. 187 students of the 1st and 2nd year of study took part in the survey. Students were asked to fill out Form 2.

The authors paid special attention to several questions related to the problem of cheating and responsibility awareness. The following are the responses to these questions (Table 2):

Table 2. Students' responses to the main questions of Form 2

Question Set	Yes	No	Not sure
I feel I am doing something wrong when I cheat		51%	28%
I use my group mates' works during tests, because I just need to pass the exams	43%		32%
Using gadgets during tests helps a lot to get good marks	32%		28%
University is the place where I can get a graduate diploma, all necessary knowledge and experience can be acquired at work or during traineeship	16%		48%
I am responsible for what I am doing in the learning process	59%		20%

151 people were interviewed to analyze the students' answers about the reasons why they would stop cheating. The results of the survey are presented in Table 3.

Table 3. Statistics on reasons to write tests without resorting to cheating that were chosen by students

Appropriation of intellectual property	Negative reaction from the teacher	Fear of becoming under-qualified
35%	14%	77%

To understand the depth of the plagiarism problem, student's final qualification works were verified by the Antiplagiat system and we got the results presented in Table 4.

Table 4. The results of the Antiplagiat system verification of students' final qualification works

The results of the first verification, average %	The result of the last verification, average %
39,68	80,04

Then, we studied in detail the reasons for the low percentage. The system distinguishes two main parameters: borrowings and citations. The analysis of the results has shown that borrowings prevail in the works (Table 5).

Table 5. Borrowings and citations, %

Borrowings, average %	Citations, average %
28,9	8,83

Another step made to tackle the problem of cheating and plagiarism included the pilot experiment. The experimental Module "Summary translation of articles on professional topics" was introduced into the discipline of technical and business translation for students majoring in Linguistics. The Module included assignments for the abstract translation of articles from English to Russian and from Russian to English, as well as assignments for summarizing articles written in English and Russian. Thus, students were taught to paraphrase and rewrite large amounts of information within a given time period. The assignments implied processing ideas proposed in the scientific articles published recently. Each student was given an individual assignment, which was to rewrite a scientific paper containing approximately two thousand and a half printed symbols within two academic hours without using grammar structures, stylistic devices or fixed phrases from the original text. Scientific papers in the English language were chosen from DOAJ – directory of open access journals. 35 students took part in the experimental study of the Module. The pilot experiment lasted one month. Students were also given similar assignments to do on their own, as part of their homework. During the training, the teacher checked the students' processed texts using the Antiplagiat program installed at Peter the Great St.Petersburg Polytechnic University. At the end of the training, a questionnaire was conducted, which showed a positive assessment of the students of the experimental Module "Abstract translation of papers on professional topics". 89% of the students admitted that the module developed their abilities to analyze and synthesize new information and to express their own ideas more concisely. At the end of the course students submitted a course paper on translation and the results of the Antiplagiat verification showed that the majority of the students who took part in experimental learning were completely "cured" of cheating as 96% of their works demonstrated a high level of originality (about 91%, on the average). The authors plan to continue the experimental training based on the above study with a large number of participants, students majoring in Linguistics and Engineering.

7 Discussion

The study proves that teacher's role is crucial in shaping a student's sense of responsibility. Testing is the most common form of control in teaching foreign languages. Tests are designed by the teachers and are regularly updated, in accordance with the curriculum. However, the keys to the tests can become available in social networks very soon as students can upload the results in the process of taking a test, so that their group mates could use them later on.

It is quite obvious from the above that such means of control as taking phones and other devices away from students are not effective in the fight against cheating. It is necessary to raise students' awareness that cheating is a pattern of irresponsible behavior and disrespect towards themselves and the others, and to motivate them to be conscientious during the process of learning, since it is their main activity at this stage of life. This task also rests with the teacher. Teachers are the main actors in the development of an educational environment that does not accept academic fraud.

The second goal of the study was to assess the degree of cheating and using electronic devices when doing tests. The study shows that students still do not associate irresponsible behavior with cheating during learning process. The data obtained are consistent with the previous studies. A comprehensive study conducted by the teachers of the Financial University is another proof of it: most respondents accept cheating during University studies, and many students confirm that they cheat during exams [20, p. 11].

The study also highlights the ways to reduce the level of cheating and educate students in terms of moral values such as responsibility, self-respect, respect for intellectual property, respect for academicians and other students.

Students ought to understand that the appropriation of intellectual property or copyright abuse is actually immoral. Any teacher can predict the performance of their students at the exam by more than 90%. If a student who does not know the subject well suddenly shows a high score during the test, the teacher can examine the student in the class, in order to confirm the student's mark, or reveal their true state of knowledge.

The results of the study have shown that the fear of leaving the university underqualified is the main reason for most students. The results obtained can be used while preparing for educational sessions and other activities to develop students' sense of responsibility.

The results (Table 5) of the analysis on plagiarism show that a large proportion of the works is borrowed incorrectly. The phenomenon of incorrect borrowing can be explained as follows. First, some students do not take the trouble to process information correctly. Secondly, in the works with a high percentage of plagiarism, students rely too much on authenticity of Internet resources.

These two problems can be approached in different ways. Inappropriate processing of information can be overcome in the process of regular educational sessions during which students should be involved in relevant discussions and related tasks. However, elimination of plagiarism requires more serious attempts. It is necessary to do everything possible to convey to students the idea that such behavior is unethical in the scientific environment, to explain the concept of "intellectual property", to explain that "borrowing" is, in fact, a kind of theft. The "role-playing" technique can facilitate this understanding. Students can be encouraged to put themselves in the shoes of an inventor or researcher who has written an outstanding scientific paper and suddenly found that it was assigned by another author, without referring to their original source.

8 Ways of Solving the Problems of Cheating and Plagiarism

As the research has shown, there are students who lack the responsible attitude toward learning, and the task of developing such attitude should be solved by the university teachers. The joint efforts can give positive results, since, from the point of view of L.I. Bozhovich, responsibility can be developed in the situations requiring its manifestation [21].

Basing on our research, we deem it appropriate to recommend certain measures at all levels of university life.

At the university administrative level, it is necessary to consider the possibility of students signing an agreement on responsibility for the use of someone else's intellectual property (cheating) when doing tests and passing exams. This is already a practice in some colleges in the United States. At St. Olaf's College in Minneapolis (Minnesota) students sign an agreement which obliges them not to cheat in the process of learning.

They are also obliged to inform the college administration about the irrelevant behavior of their peers [22]. If the latter can hardly be introduced in our universities, the agreement stipulating students' responsibility while learning, in our opinion, is quite relevant.

– At the level of institutes, higher schools and departments, it is necessary to support teachers who encourage students' honest presentation of their learning outcomes and those who oppose to cheating and plagiarism. Teachers ought to eliminate the irresponsible attitude of students towards learning, and develop a sense of responsibility among students through educational work.
– It is necessary to bring students down to the idea that gadgets can only help in finding information during research, but they should be put aside in the process of writing a test or passing an exam.

To optimize the course on technical and business translation, an experimental Module "Summary translation of articles on professional topics" was introduced into the discipline of technical and business translation for students majoring in Linguistics. The Module included assignments for the translation of articles from English to Russian and from Russian to English, as well as assignments for summarizing articles written in English and Russian.

In addition to the usual tasks for teaching summary and full translation, students in the classroom were offered creative tasks to develop interpreting competences and to master the skills for gist reading and writing synopsis. Here are examples of such tasks.

1. Students work in mini-groups of 4 people. Each student is given the task to rephrase a mini-text. Then, in turn, he gives the paraphrased text to his group-mate with the same task and so on 4 times. The results obtained after paraphrasing four mini-texts are compared with the original. This task is fulfilled with texts in both the Russian and the English languages.
2. Students work with mini-texts. Their task is to find synonyms to highlighted relevant words, first without a dictionary, and then using a dictionary. This task is performed in the form of a competition. Further, students are invited to make up their own sentences using the selected synonyms.
3. "Chinese whispers". Students work in groups of 4 people. The first student writes a sentence on any topic and whispers it in the ear of their group-mate, who, in turn, passes this message in a paraphrased form to another student. When a variant paraphrased three times reaches the first student, this message is compared with the written original.
4. Paraphrasing by the type of antonymic translation. Students work in mini-groups of 4 people. Each student is given the task to rephrase a mini-text giving it an adverse meaning. Then, in turn, he gives the paraphrased text to his group-mate with the same task and so on 4 times. The results obtained after paraphrasing four mini-texts are compared with the original. This task is fulfilled with texts in both the Russian and the English languages.

In the course of training on the "Module" for a month, a pilot experiment was conducted in which 35 students took part. Students at home completed assignments to abstract and summarize articles in both Russian and English, and the teacher checked their work using the Antiplagiat program. The indicators of originality of the work performed gradually improved. At the end of the training, a questionnaire was conducted, which showed a positive assessment by the students of the experimental Module "Summary translation of articles on professional topics". The authors plan to continue the research with a large number of participants on a diverse thematic of teaching aids.

9 Conclusion

The main conclusion of our study is that more attention should be paid to educational work in the university. The results of the pilot experiment show that the problem of cheating and plagiarism can be approached through the introduction of ad hoc teaching modules aimed at raising students' responsibility and awareness of academic ethics and developing students' analytical competencies in processing large volumes of information.

Unconventional and creative approaches to teaching students how to avoid cheating and plagiarism are likely to encourage their conscientiousness and diligence and develop their self-esteem. The university is the most influential educational institution that can ensure the development of the competences necessary for the highly qualified and responsible specialist of the future.

References

1. Antropova, M.: Mobile technologies in the educational process (on the example of the Chinese WeChat. Cross-Cult. Stud.: Educ. Sci. (CCS & ES) **3**(3), 218–224 (2018). (in Russian)
2. Almazova, N., Bylieva, D., Lobatyuk, V., Rubtsova, A.: Human behavior as a source of data in the context of education system. In: SPBPU IDE'19: Proceedings of Peter the Great St. Petersburg Polytechnic University International Scientific Conference on Innovations in Digital Economy, p. 37. ACM, Saint – Petersburg (2019). https://doi.org/10.1145/3372177.3373340
3. Gonzalez, T., et al.: Influence of COVID-19 confinement on students' performance in higher education. PLOS ONE **15**(10), e0239490 (2020). https://doi.org/10.1371/journal.pone.0239490
4. Kawasaki, H., Yamasaki, S., Masuoka, Y., Iwasa, M., Fukita, S., Matsuyama, R.: Remote teaching due to COVID-19: an exploration of its effectiveness and issues. Int. J. Environ. Res. Public Health **18**(5), 2672 (2021). https://doi.org/10.3390/ijerph18052672
5. Almazova, N., Krylova, E., Rubtsova, A., Odinokaya, M.: Challenges and opportunities for Russian higher education amid COVID-19: Teachers' perspective. Educ. Sci. **10**(12), 368 (2020). https://doi.org/10.3390/educsci10120368
6. Rubtsova, A., Odinokaya, M., Krylova, E., Smolskaia, N.: Problems of mastering and using digital learning technology in the context of a pandemic. In: Bylieva, D., Nordmann, A., Shipunova, O., Volkova, V. (eds.) PCSF/CSIS -2020. LNNS, vol. 184, pp. 324–337. Springer, Cham (2021). https://doi.org/10.1007/978-3-030-65857-1_28
7. Glukhov, D.N.: Problems with responsible attitude of students towards learning. Young Sci. **16**(96), 408–410 (2015)
8. Malkov, E.S.: Cheating is not good, but not in Russia. https://iq.hse.ru/news/177671066.html

9. Pozdeeva, E.G., Shipunova, O.D., Evseeva, L.I.: Social assessment of innovations and professional responsibility of future engineers. IOP Conf. Ser. Earth Environ. Sci. **337**, 012049 (2019). https://doi.org/10.1088/1755-1315/337/1/012049

10. Rubtsova, A.: Socio-linguistic innovations in education: productive implementation of intercultural communication. IOP Conf. Ser. Mater. Sci. Eng. **497**, 012059 (2019). https://doi.org/10.1088/1757-899X/497/1/012059

11. Milkus, A.: We Lived Up! The write-off charge is now an outstanding event? (in Russian). http://www.examen.ru/news-and-articles/news/dozhili!-otchislenie-za-spisyivanie-teper-vyidayushheesya-sobyitie. Accessed 10 Mar 2021

12. Shmeleva, E.D.: Academic fraud in modern universities: an overview of theoretical approaches and empirical research results. J. Econ. Soc. **16**, 55–80 (2015)

13. Martin, N.: Cheat sheets and smartphones: is cheating in exams incurable? http://www.bbc.com/russian/blogs/2015/06/150624_blog_strana_russia_exams_cheating. Accessed 10 Mar 2021

14. Pryadein, V.P.: Responsibility as a systemic quality of personality. UrGPU, Yekaterinburg (2001)

15. Bortsova, M.V.: Development Factors of the Initial Forms of Personal Responsibility: Theoretical Aspect. Publishing Center SGPI, Slavyansk-on-Kuban (2007)

16. Muzdybaev, K.: Psychology of Responsibility. Nauka, Leningrad (1983)

17. Sukhomlinsky, V.A.: The birth of a citizen. Molodaya gvardiya, Moscow (1971)

18. Solso, R.L., McLean, M.N.: Experimental psychology. Prime-EVROZNAK, Saint Petersburg (2006)

19. Bylieva, D., Lobatyuk, V., Tolpygin, S., Rubtsova, A.: Academic dishonesty prevention in e-learning university system. In: Rocha, Á., Adeli, H., Reis, L.P., Costanzo, S., Orovic, I., Moreira, F. (eds.) WorldCIST 2020. AISC, vol. 1161, pp. 225–234. Springer, Cham (2020). https://doi.org/10.1007/978-3-030-45697-9_22

20. Slastenin, V.A.: Pedagogy. School-Press, Moscow (1997)

21. Bozhovich, L.I.: Personality and its formation in childhood. Psychological research/Sciences of the USSR. "Education", Moscow (1968)

22. Sivanich, J., Popova, N.V., Vessart, O.V.: United States of America: Past and Present: A Reader in Regional Studies, 2nd edn. Polytechnic Publishing House, Saint Petersburg (2011)

Students' Preferences in Choosing the Form and Means of Interaction with Professors: Innovation or Tradition?

Dmitrii Popov⬡, Veronika Fokina⬡, Iuliia Obukhova(✉)⬡, and Anastasia Nekrasova⬡

Peter the Great St. Petersburg Polytechnic University,
Polytechnic Street, 29, St. Petersburg, Russia
obuhova_yuo@spbstu.ru

Abstract. This article deals with the preferences of students in choosing forms of communication in the educational process, as well as ways to consult with professors. The relevance of the study is due to attempts to change traditional forms of education at universities on distance learning and to introduce creative corporate educational platforms for communication in the educational space. The purpose of the study is to investigate the opinion of the student audience about the transition to ICT for study and interaction in higher education and to identify the importance of personal contacts and consultations with professors. The study was conducted among students of leading Russian universities through an online sociological survey. Respondents (126 students who study mainly undergraduate and graduate programs) were asked to answer questions related to the most convenient channels of interaction with professors, preferred methods of passing exams and tests, assessment of work on corporate educational platforms, forms of creative study, etc. The integration of distance learning and corporate educational platforms is currently directive in nature, which in many ways leads to a chaotic process, where is underestimated for students the importance of traditional forms of education, including personal consultations, communication in contact student groups, lecturers' comments during the work etc. Students prefer to use Internet technologies, such as e-mail, social networks and messengers, not for educational purposes, but for advisory purposes and perceive involvement of professors into such format as an element of creative approach in educational process in opposite to familiar forms.

Keywords: Higher education · Students' preferences · Creativity in education · Distance education · Communication channels in the field of education · Personal consultations with professors

1 Introduction

One of the current trends in Russian higher education is attempts to integrate traditional and distance educational technologies. This process affects different areas and aspects of

educational process management, such as issues of creativity and readiness for transformations in higher education. In this regard, it is of interest to determine the boundaries of
such integration, which aspects of education should be left traditional, which should be
transferred to remote platforms. The most important for this study seems to determine
the priority channels and formats of interaction with the professor in the educational
process for the student. In this regard, the object of the study is student preferences
when choosing a channel for interaction with a professor for consultation outside the
classroom.

The general hypothesis of the study is based on the fact that the most significant
formats for a student to communicate with a professor are: personal contact communication and personal virtual communication in social networks, in this regard, their poor
representation in distance learning leads to the unavailability of Russian students at this
stage completely or partially switch to distance learning.

The following tools of traditional and distance learning can be distinguished:

1. The traditional format of interaction:

    providing the professor with the author's interpretation and comments as part of
    lectures and seminars;
    personal consultations;
    Passing the exam/test or retake.

2. Remote interaction format:

    the provision of educational materials of scientific sources of a general, universal
    nature (textbook, links to network resources);
    sending works via e-mail from the university account;
    sending works via personal mail;
    e-learning platforms (for example, LMS or Moodle), cloud technologies within
    faculties;
    online consultation;
    passing the exam/test.

In accordance with the goal and hypothesis, the following tasks were set in the study:

1. Identify the channels of interaction with the professor.
2. To identify the most convenient channels for the student to communicate with the
    professor.
3. Determine the format of the most effective communication with the professor.
4. Evaluate the convenience of communication through the e-mail service and social
    networks.
5. Evaluate the convenience of working on corporate educational platforms and identify
    the reasons for the rejection of their use by students and professors.
6. Assess the importance of personal meetings for the student in consultation with the
    professor.

7. To consider the influence of the factor of personal interaction on the student's choice of the form of education (distance or traditional).
8. Examine the weaknesses of distance learning.

This study examines the traditional and distance formats of interaction between professors and students of Russian universities, evaluates the effectiveness of personal and virtual communications, explores university communication platforms and identifies the reason for the rejection of their use by students and professors, explores alternative forms of communication, in particular, communication through personal consultations and social networks, which can foster creativity in educational process (for example, we may remember the case, in which the professor decided to spend lessons in a virtual «room» in popular computer game, here all students may use game's avatars and discussed programming, sitting around virtual bonfire.

Modern universities are understood as a network community, the main participants of which are professors and students, in this community both personal and virtual communication channels are practiced, while the widespread distribution of e-mail, social networks and other Internet technologies, as the study shows, does not mean abandonment personal communication of subjects in the educational space.

2 Literature Review

The problems of choosing forms and channels of interaction in the educational space are of interest to many modern researchers. Moreover, the educational process includes not only the educational process, but also the educational policy of the state and universities, the search for new approaches, technologies and teaching models, means of extracurricular work in the form of meetings and consultations of students with professors, etc.

The first group of authors considers the communication of students within the student community in the online environment as a tool to increase student engagement, studying distance learning models in traditional universities, including corporate educational platforms, in particular, the works of Bryan, Lutte, Lee [1], Gueye, Mballo, Kasse [2], Vargas, Kalman [3], Gkamas, Rapti [4], Dağhan, Akkoyunlu [5], Lahuerta -Otero, Cordero-Gutiérrez [6], Pérez [7] et al.

The second group of authors studies the readiness of professors for teaching and learning through distance learning, considers the problems and difficulties of implementing these processes, including Biryukova, Kolomiets [8], Ouma [9], Egorov, Melanina [10], Borgobello, Madolesi [11], Fernández-Alemán, Sánchez García [12] et al.

The third group of researchers is engaged in the analysis of national educational policies, including, with the aim of identifying factors affecting the practice of distance learning within the framework of a traditional university, as well as identifying communicative competencies in the process of preparing students, including the work of Makoe [13], da Silva, Amaro [14], Trostinskaia, Popov, Fokina [15] and others.

The work of Xiao [16] and Atanasova [17], Shishakly [19] tell us about the relationship between quality of distance education and quality of educational management.

Anggiani, Heryanto study student's attitude to distant forms of education depending on availability of attractive spaces for informal educational duties on campus [18].

The next group of researchers studies the types of communication platforms for professors and students, considers models and forms of convergence of traditional and distance learning, identifies the most preferred from the point of view of students, such as Lemoine, Hackett [20], Chen [21], Ghareb, Ahmed [22], Patton [23]. They study different forms of distant technologies such as smartphones, social networks (for example, Facebook), online classes.

Alaneme, Olayiwola [24] are interested in the problem of creative combining traditional learning and the E-learning methods in higher distance education.

Razinkina, Pankova [25] and others consider the problem of students satisfaction by replacing traditional forms of education on more creative and innovative, but less familiar (and students as not satisfy enough).

The last group of researchers studies the feasibility and quality of distance education, including to assess students' ability to perceive information in different formats, in particular, the work of Crisp [26], Ugolnova [27], Gravani [28], Park [29], Chernik, Lambden [30], Da Silva, Behar [31], Baruah [32], Ababkova, Leontyeva [33] et al.

3   Materials and Methods

The following methods were used to realize the goal of the study and solve the problems posed, as well as verify the hypothesis put forward: empirical research; quantitative and qualitative analysis of the results; diagnostic complex.

The empirical base of the study: On-line survey of students of leading Russian universities of different educational levels (bachelor's degree, specialty, master's program). A total of 172 people took part in the online survey. For analysis, 126 questionnaires were selected. Among the total sample, undergraduate students make up 72%, undergraduates - about 7%, 94.4% are students in humanitarian areas, 69% are women. Among the respondents are students of The North-West Institute of management Russian Presidential Academy of National Economy and Public Administration (NWIM RANEPA) - 63.2%, Peter the Great St.Petersburg Polytechnic University (SPbPU) - 20.8%, Saint-Petersburg State University of Aerospace Instrumentation (SUAI) - 7.2%, etc.).

4   Results

Based on the study, data were obtained on the attitude to distance forms of the educational process. The lack of feedback on the organization of communication between professors and students does not allow a qualitative assessment of the ongoing changes, which caused the scientific and practical interest of the authors of the study. The problem of introducing distance learning forms and the prescriptive nature of changes at this stage in the development of the educational space of the Russian Federation leads to a chaotic process, where there is an underestimation of the importance of traditional personal consultations for students and there are obstacles to building effective corporate educational platforms.

When asked about how to interact with professors, 66.7% of respondents said they actively use the Internet (Fig. 1). In total, 97.7% of respondents use the Internet as a means

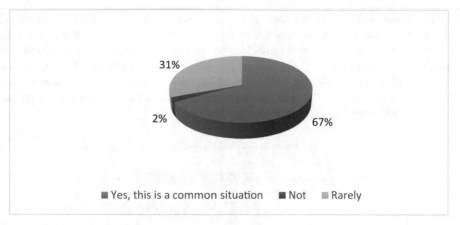

Fig. 1. Do you interact with your University's professors through the Internet?

of communication. In a modern university, the Internet allows establishing an effective interconnection between participants in a communication exchange, but at the same time, it also gives rise to a number of issues requiring discussion. Are both parties ready for active communication? What is the time and place of this communication? Are there any restrictions related to the personal "Internet space" of the subjects of communication? Does a modern university require a formalized way of communication between students and professors through intra-university communities or corporate educational platforms? Answers to these questions will be presented below (Fig. 2).

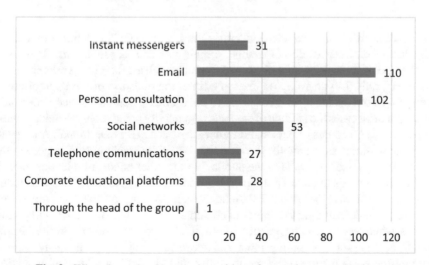

Fig. 2. What channels of interaction with professors do you currently use?

Interviewed students noted that they currently use mainly e-mail (110 answers) and personal consultations (102 answers), use social networks and instant messengers (84 answers). The ability to choose channels of interaction largely solves the problem of lack of personal communication, but does not cancel or replace it. Correspondence by e-mail allows you to provide a mass response to typical calls within the educational process. Effective personal communication and dialogue in social networks allows you to answer specific individual questions, reduces the time spent in communication. Corporate educational platforms (28 answers) have not yet become the basis for communication, and telephone communication (27 answers) largely depends on the individual desire of the professor for this method of communication (Fig. 3).

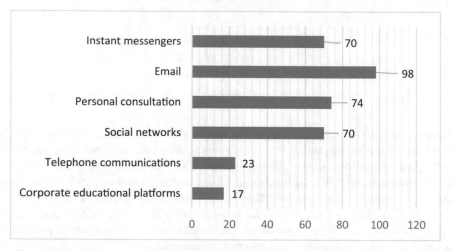

Fig. 3. What channels of communication with professors do you find most convenient?

It is noteworthy that, provided students choose their own communication channel, the leaders of preferences do not change: e-mail, personal consultations. At the same time, the student audience would like to use channels such as: social networks and instant messengers, only 140 answers. At the same time, the question of the appropriateness of admitting both students and professors to each other's personal space is debatable, however, in most cases the attitude of students is generally positive. In the modern world, the openness or closeness of personal Internet space is largely conditional. An essential issue is the willingness or unwillingness on the part of the teaching staff to present their life outside the university walls to the public. On the other hand, sufficient openness in these issues allows us to identify compatible interests. Many respondents from students and professors follow the global HR trend. When deciding on the appointment of a supervisor or in determining the ability of students to participate in project and research activities, when recommending practice, they actively study each other's profiles. Profiles in social networks are becoming more and more informative and in many ways are becoming the basis for the selection of "staff" and "leader" in a kind of educational activity market (Table 1).

Corporate educational platforms are currently in little demand by representatives of the student audience, due to a number of shortcomings in their functioning and problems of distance learning. At the same time, the ubiquitous transition to such interaction

Table 1. How do you feel about the fact that professors can study your profiles on social networks?

OK, everybody does it	66,7%
I'm good, it allows professors to get to know me better	12,5%
I don't like it, my profile is "my" space	20,8%

platforms dictated by legislative decisions (273-FL of the Russian Federation "On Education") requires universities to provide a software base and forms a massive influx of users (Fig. 4).

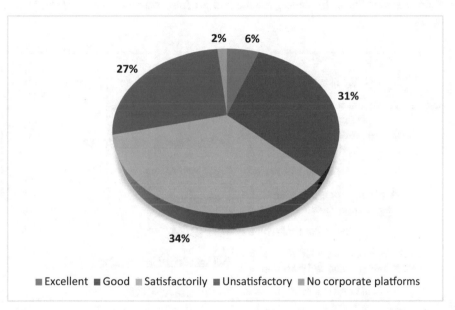

Fig. 4. How do you estimate the convenience of working on corporate educational platforms (student's personal account, Moodle, etc.)?

Corporate platforms, and all educational platforms of leading universities are related to them - this is an analogue of management and marketing tools that are actively used in the private sector and in the public services market. In many ways, the results of the study show an early stage in the formation of relations to similar platforms, which are in the process of development and in many ways encounter negative user experience. Corporate platforms as a whole are successfully used in business and the public sector, providing employees and management with a single point of access to data, tools for managing business processes and tools for collaboration and information exchange. The success of such projects provides only ergonomics for users, quick access to content, stability of the web application, which integrates other corporate applications, knowledge bases, social networks and other software modules. If the user is experiencing difficulties and believes that the platform takes his time and creates problems in the implementation of core activities, then such platforms are doomed to failure. Educational platforms should be convenient for students and professors, including reducing labor costs and contributing to the effectiveness of the educational process. One of the necessary components in improving the work of educational platforms is the provision of communication within and between the teaching and student community, as well as user education programs.

In many respects, the creation of corporate educational platforms by modern universities is determined by the requirements of the time, the tasks of meeting educational standards and the need to increase competitiveness. But the main task is to increase the effectiveness of the distance learning process, primarily for part-time students. The introduction of distance learning can be considered depending on the education format. But, first of all, the authors of the study were interested in the question of perception of distance learning forms by undergraduate and graduate students of full-time education (Fig. 5).

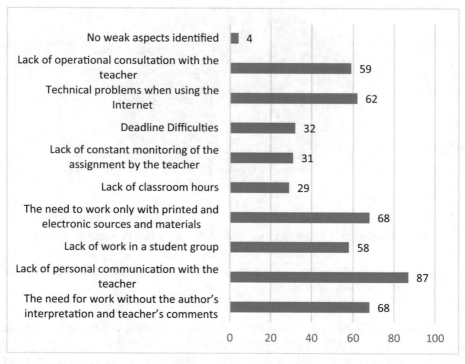

Fig. 5. What are the weaknesses of distance learning that you can highlight?

According to students, the most important factors hindering the effective use of distance learning are:

1. Lack of personal communication with the professor 69.04% (87);
2. The need to work only with printed and electronic sources and materials, as well as the need to work without an author's interpretation and comments of the professor for 53.96% (68)

3. Technical problems when using the Internet 49.20% (62)
4. Lack of operational consultation with a professor 46.82% (59)
5. Lack of work in the student group 46.03% (58) and others.

It is noteworthy that difficulties of a technical nature, despite their importance, are nevertheless inferior to the needs of students in interpersonal and group contact communication. Students also express their unwillingness to refuse important personal meetings and consultations with professors, academic leaders, practice leaders, etc. (Table 2).

Table 2. Are personal meetings with a professor important for your education?

Yes	84,10%
Not	5,60%
Difficult to answer	10,30%

It is important to note that for students, the concepts of "the convenience of educational communication" and "the effectiveness of educational communication" are fundamentally different. Convenience is perceived, first of all, on the basis of time and energy costs, schedule and lifestyle. At the same time, efficiency is largely based on understanding, applicability, practical utility and the possibility of increasing individual competitiveness (Fig. 6).

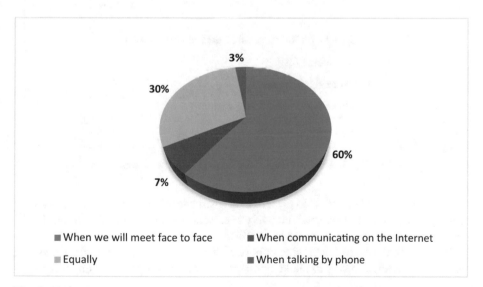

Fig. 6. What channel of communication with the professor, in your opinion, is more effective?

As the study shows, despite the convenience of electronic communication systems, communication in a personal meeting seems to be more effective and efficient for students (60.30%).

Among the advantages of personal consultations are:

– the possibility of two-way communication (dialogue mode);
– personification (personality factor);
– efficiency in solving problems;
– perception of the process of transferring knowledge and experience based on the pedagogical tradition of universities;
– increasing the value of individual communication in the context of digital globalization.

At the same time, students are ready to combine traditional and distance learning and take some elements in distance learning (Fig. 7).

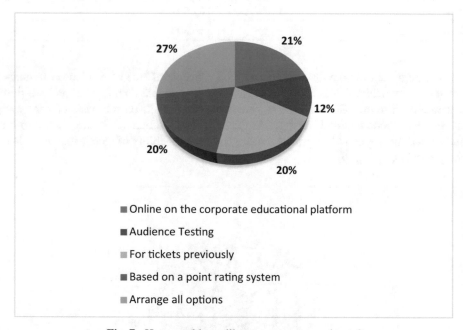

Fig. 7. How would you like to pass exams and tests?

In particular, students have no preferences when choosing the method of exams and tests, the answers were distributed relatively evenly, only testing in the classroom slightly deviated from the average value (only 11.9%).

In this regard, one can note wide opportunities for combined or remote assessment systems. However, as a way to retake online assessment is not effective enough, as:

– without a positive assessment, students who are unprepared for the remote system (non-tracking deadlines, technically non-equipped, unable to work in the question-answer mode with fixing the response time, etc.);
– it is necessary to rediscover the system, which in the face of individual appeals and the imperfection of individual platforms requires considerable time;

– consultations are already possible for those who have already passed, even if tasks
  have been updated, which reduces the ability to assess real competencies.

Assessing students' readiness for a combination of education forms, in general, one
can identify positive trends (Fig. 8).

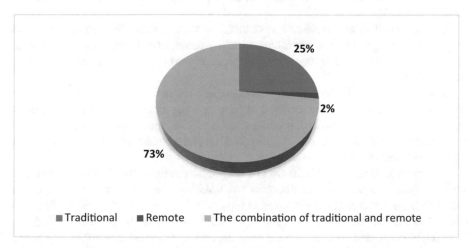

Fig. 8. In your opinion, which format of higher education is more productive?

Most students (73%) accept the opportunity to combine traditional and distance
elements in the course of their studies. However, while traditional education is considered productive by 25.4% of respondents, only 1.6% of respondents choose distance
education.

5   Discussion and Conclusion

The modern educational process is currently difficult to imagine without the use of Internet technology and creative approaches in the usage of mobile and distant technologies.
The World Wide Web has a strong place in the life of modern man, performing a variety of functions, including information, entertainment, educational, communicative, etc.
Technologies (e-mail) and Internet resources (sites, social networks) are actively used
by representatives of different fields of activity for active work.

In the modern world, it is important to understand what forms of communication are
preferred by both students and professors, what channels are used in communication,
how this interaction affects the educational process. The fast pace of informatization and
computerization of education, as the main trend of modern higher education, actualizes
a number of problems in the organization of the educational process, among which the
leading place is occupied by the problem of productive and effective interaction of its
participants.

The study shows us a high degree of student's involvement and interest in the processes of transition to new forms of interaction with professors based on educational

platforms, as well as their positive attitude to the use of more creative approach to the teaching format in comparison with the familiar and traditional.

It is equally important to understand the difference between communication with professors in the new environment (new formal communication in university) and usual format of online forums (Bylieva, Lobatyuk [34]). It is also important to realize, that creative forms of communication present in open education projects today (Bylieva, Bekirogullari [35]).

The key result of this investigation is a fact, that we can mark high degree of students' readiness for integrated forms of the educational process, as well as a clear understanding of the necessity for new and more creative forms of communication between professors and students.

The maintenance of effective personal dialogue between the professor and the student ensures the stability of the educational system and the significance of this process of gaining knowledge. Technological innovations and platforms only provide this process and affect the possibilities of building an effective dialogue, but at this stage they cannot replace personal communication.

At the same time, the introduction of corporate platforms (with their multidimensional technical capabilities for data transfer, building communication and organizing the educational process in the most diverse forms), for example, on the basis of Peter The Great Polytechnic University, made it possible to respond quickly and effectively for the COVID-19 crisis and establish the learning process in the most comfortable way, with creative and sustainable mode. Moreover, the readiness to use creativity for lining up the interaction on the basis of corporate platforms has become one of the requirements of the time and professional training of both teachers and students (Evseeva,, Shipunova [36]).

Different models of education specify a different degree of independence, activity, creativity, manifestation of individuality, student's internal involvement in the interaction and characterize the whole learning process as a whole.

Often, when introducing distance technologies into the educational process, students's readiness to take on a number of tasks and problems, the solution of which was previously left to the professor, is underestimated. The main problem, in our opinion, is the loss of personification in the interaction of the professor and student. If this is applicable when the student makes a conscious choice of part-time or distance learning, then for full-time students, the priorities are contact interaction.

It seems advisable not the directive introduction of distance learning, but a dialogue with the teaching and student communities, receiving feedback and, as a result, taking into account the views of all interested parties.

References

1. Bryan, T.K., Lutte, R., Lee, J., O'Neil, P., Maher, C.S., Hoflund, A.B.: When do online education technologies enhance student engagement? A case of distance education at University of Nebraska at Omaha. J. Public Affairs Educ. 24(2,3), 255–273 (2018). https://doi.org/10.1080/15236803.2018.1429817
2. Gueye, A.D., Mballo, M.H.W., Kasse, O., Gueye, B., Ba, M.L.: Model of integration of distance education in a traditional university: migration of cross-cutting courses to distance

learning. In: Auer, M.E., Tsiatsos, T. (eds.) ICL 2018. AISC, vol. 917, pp. 13–24. Springer, Cham (2019). https://doi.org/10.1007/978-3-030-11935-5_2

3. Vargas, R.M., Kalman, J.: The construction of flexibility in a virtual program at a public Colombian distance education university. Revista de Educacion a Distancia. **58** (2018). https://doi.org/10.6018/red/58/8

4. Gkamas, A., Rapti, M.: Lessons learned from synchronous distance learning in university level at Congo. In: Auer, M.E., Tsiatsos, T. (eds.) IMCL 2017. AISC, vol. 725, pp. 522–530. Springer, Cham (2018). https://doi.org/10.1007/978-3-319-75175-7_51

5. Dağhan, G., Akkoyunlu, B.: An examination through conjoint analysis of the preferences of students concerning online learning environments according to their learning styles. Int. Educ. Stud. **5**(4), 122–138 (2012). https://doi.org/10.5539/ies.v5n4p122

6. Lahuerta-Otero, E., Cordero-Gutiérrez, R., Izquierdo-Álvarez, V.: Like it or die: using social networks to improve collaborative learning in higher education. In: 6th International Conference on Technological Ecosystems for Enhancing Multiculturality. TEEM 2018, pp. 930–935. ACM Digital Library, Salamanca (2018). https://doi.org/10.1145/3284179.3284336

7. Pérez, A.G.: Social networks as tools to enrich learning environments in higher. Revista de Pedagogia. **70**(4), 55–71 (2018). https://doi.org/10.13042/Bordon.2018.60579

8. Biryukova, N.A., Kolomiets, D.L., Kazarenkov, V.I., Sinagatullin, I.M.: Preparing university educators for tutoring adult learners in distance education. In: Issues and Trends in Interdisciplinary Behavior and Social Science. Proceedings of the 6th International Congress on Interdisciplinary Behavior and Social Sciences, ICIBSoS 2017, pp. 275–282. Taylor & Francis Group, Bali (2017)

9. Ouma, R.: Transforming university learner support in open and distance education: staff and students perceived challenges and prospects. Cogent Educ. **6**(1), a1658934 (2019). https://doi.org/10.1080/2331186X.2019.1658934

10. Egorov, G., Melanina, T.V., Roberts, J.J.: Teaching theology from a distance: experiences of the institute of distance learning at St Tikhon's orthodox university in Moscow Russia. HTS Teologiese Stud. **75**(1), a5343 (2019). https://doi.org/10.4102/hts.v75i1.5343

11. Borgobello, A., Madolesi, M., Espinosa, A., Sartori, M.: Use of ICT in pedagogical practices of professors of the Faculty of Psychology of a public university in Argentina. Revista de Psicologia. **37**(1), 279–317 (2019). https://doi.org/10.18800/psico.201901.010

12. Fernández-Alemán, J.L., Sánchez García, A.B., López Montesinos, M.J., Marqués-Sánchez, P., Bayón Darkistade, E., Pérez Rivera, F.J.: Exploring the use of information and communication technologics and social networks among university nursing faculty staff. An opinion survey. Investigación y educación en enfermería. **32**(3), 438–450 (2014). https://doi.org/10.1590/S0120-53072014000300009

13. Makoe, M.: Avoiding to fit a square peg into a round hole: a policy framework for operationalising open distance education in dual-mode universities. Dist. Educ. **39**(2,3), 159–175 (2018). https://doi.org/10.1080/01587919.2018.1457945

14. Da Silva, W.B., Amaro, R., Mattar, J.: Distance education and the Open University of Brazil: History, structure, and challenges. Int. Rev. Res. Open Dist. Learn. **20**(4), 100–115 (2019). https://doi.org/10.19173/irrodl.v20i4.4132

15. Trostinskaia, I., Popov, D., Fokina, V.: Communicative competence of engineers as a requirement to the future professions. In: Chernyavskaya, V., Kuße, H. (eds.) The European Proceedings of Social and Behavioural Sciences, 18th PCSF, vol. 51, pp. 1191–1199 (2018). https://doi.org/10.15405/epsbs.2018.12.02.128

16. Xiao, J.: On the margins or at the center? Distance education in higher education. Dist. Educ. **39**(2,3), 259–274 (2018). https://doi.org/10.1080/01587919.2018.1429213

17. Atanasova, I.A.: University knowledge management tool for the evaluation of the efficiency and quality of learning resources in distance e-learning. Int. J. Knowl. Manag. **15**(4), 38–55 (2019). https://doi.org/10.4018/IJKM.2019100103

18. Anggiani, M., Heryanto, B.A.: Study of informal space on campus by looking at student preferences. In: IOP Conference Series: Materials Science and Engineering of 1st International Conference on Design, Engineering and Computer Sciences, ICDECS 2018, vol. 453, no. 1, p. an012029. IOP Science, Jakarta (2018). https://doi.org/10.1088/1757-899X/453/1/012029
19. Shishakly, R.: Smartphones enhance the management of learning processes in Higher Education: a case study in Ajman University, United Arab Emirates. In: 10th International Conference on E-Education, E-Business, E-Management and E-Learning, IC4E 2019, pp. 63–69. Waseda University Tokyo (2019). https://doi.org/10.1145/3306500.3306513
20. Lemoine, P.A., Hackett, P.T., Richardson, M.D.: Handbook of research on mobile devices and applications in higher education, pp. 373–401 (2016). https://doi.org/10.4018/978-1-5225-0256-2.ch016
21. Chen, M.-M.: Students' perceptions of the educational usage of a Facebook group. J. Teach. Travel Tour. **18**(4,2), 332–348 (2018). https://doi.org/10.1080/15313220.2018.1434448
22. Ghareb, M.I., Ahmed, Z.A., Ameen, A.A.: The role of learning through social network in higher education in KRG. Int. J. Sci. Technol. Res. **7**(5), 20–27 (2018)
23. Patton, B.A.: Synchronous meetings: a way to put personality in an online class. Turk. Online J. Dist. Educ. **9**(4), 18–29 (2008)
24. Alaneme, G.C., Olayiwola, P.O., Reju, C.O.: Combining traditional learning and the E-learning methods in higher distance education: assessing learners' preference. In: 4th International Conference on Distance Learning and Education, ICDLE 2010, San Juan, pp. 187–190 (2010). https://doi.org/10.1109/ICDLE.2010.5606008
25. Razinkina, E., Pankova, L., Trostinskaya, I., Pozdeeva, E., Evseeva, L., Tanova, A.: Student satisfaction as an element of education quality monitoring in innovative higher education institution. E3S Web Conf. **33**, 03043 (2018). https://doi.org/10.1051/e3sconf/20183303043
26. Crisp, B.R.: From distance to online education: two decades of remaining responsive by one university social work programme. Soc. Work. Educ. **37**(6), 718–730 (2018). https://doi.org/10.1080/02615479.2018.1444157
27. Ugolnova, L.: Distance learning in professional higher education: characteristics of students. Voprosy Obrazovaniya **2012**(4), 200–211 (2012)
28. Gravani, M.N.: Use of technology at the open university of Cyprus (OUC) to support adult distance learners: to what extent is being informed by the learner-centred education (LCE) paradigm? In: Technology for Efficient Learner Support Services in Distance Education: Experiences from Developing Countries, pp. 173–188 (2018). https://doi.org/10.1007/978-981-13-2300-3_9
29. Park, H.W.: Examining academic Internet use using a combined method. Qual. Quant. **46**(1), 251–266 (2012). https://doi.org/10.1007/s11135-010-9344-6
30. Chernik, P., Lambden, J., Hay, G., Svrcek, W., Young, B.: Experiences in process control web-based learning. In: ASEE Annual Conference Proceedings 2003, pp. 5123–5128. American Society for Engineering Education, Nashville (2003)
31. Da Silva, K.K.A., Behar, P.A.: Digital competence model of distance Learning students. In: 14th International Conference on Cognition and Exploratory Learning in the Digital Age, CELDA 2017, pp. 109–116. Vilamoura, Algarve (2017)
32. Baruah, T.D.: E-learning as a medium for facilitating learners' support services under open and distance learning: an evaluative study. In: Anjana (ed.) Technology for Efficient Learner Support Services in Distance Education: Experiences from Developing Countries. pp. 93–112. Springer, Singapore (2018). https://doi.org/10.1007/978-981-13-2300-3_5
33. Piskun, O.E., Ababkova, M.Y., Leontyeva, V.L.: Biological feedback method to facilitate academic progress. Teoriya i Praktika Fizicheskoy Kultury **10**, 45–47 (2018)
34. Bylieva, D., Lobatyuk, V., Safonova, A.: Online forums: communication model, categories of online communication regulation and norms of behavior. Humanit. Soc. Sci. Rev. **7**, 332–340 (2019). https://doi.org/10.18510/hssr.2019.7138

35. Bylieva, D., Bekirogullari, Z., Kuznetsov, D., Almazova, N., Lobatyuk, V., Rubtsova, A.: Online group student peer-communication as an element of open education. Futur. Internet **12**, 143 (2020). https://doi.org/10.3390/fi12090143

36. Evseeva, L.I., Shipunova, O.D., Pozdeeva, E.G., Trostinskaya, I.R., Evseev, V.V.: Digital learning as a factor of professional competitive growth. In: Antipova, T., Rocha, Á. (eds.) DSIC 2019. AISC, vol. 1114, pp. 241–251. Springer, Cham (2020). https://doi.org/10.1007/978-3-030-37737-3_22

A Medical University School of Pedagogical Excellence as the Environment for Creativity

Alexandra I. Artyukhina[1]([✉]) [iD], Nina V. Ivanova[2,3] [iD], Olga F. Velikanova[1] [iD],
Svetlana V. Tretyak[1] [iD], Vasiliy V. Velikanov[3] [iD], and Viktor B. Mandrikov[1] [iD]

[1] Volgograd State Medical University, Pavshikh Bortsov sq., 1, Volgograd, Russia
[2] Volgograd State Agrarian University, Universitetsky pr., 26, Volgograd, Russia
[3] Volgograd State Technical University, V.I. Lenina pr., 27, Volgograd, Russia

Abstract. The article deals with the integration of formal and informal education of university instructors within the framework of the School of Pedagogical Excellence (SPE) as an environment for pedagogical creativity. The authors analyze the general cultural competencies in the educational standards of higher education, reflecting the requirements for the creative training of university graduates. A brief overview of researches on educational technologies, creativity and professional development of university instructors, including medical ones, is given. The article presents the approaches to solving the problems of continuous pedagogical training of medical university instructors in terms of the SPE potential. The structure and activities of the SPE are demonstrated. The University instructors are trained in 11 additional professional programs of advanced pedagogical education and retraining. This diversity gives a possibility to satisfy various educational needs of staff members. The Center for Pedagogical Innovations of the University performs the function of psychological and andragogical support for teaching doctors in creativity and distribution of innovative experience. It also accumulates the experience and options for non-formal education. The results of the study can be used to create a system for continuous pedagogical training of university instructors of medical and other professional areas.

Keywords: School of Pedagogical Excellence · Creativity · Formal and informal training

1 Background

The rapid development of production technologies in response to the fast changing economic, financial and informational needs of the market necessitate the improvement of the quality of university graduates training. It means constant improving of teaching and applying new educational technologies in higher education. Special attention is paid to the training of personnel for healthcare system. Such factors as the avalanche growth of scientific knowledge, the emergence of new medical and educational technologies, the transition from routine training to the creative development of intelligence and focusing on personalized training have actualized the necessity for professional development of university instructors, perfection of their expertise and implementation of creative potential.

Since ancient times, philosophers, scientists, physicians, and ordinary people have debated whether medicine is a science or an art. Nowadays, it is generally accepted to consider medicine as an object that has a holistic character, presenting an inextricable relationship between science and art. Medical technologies are the result of scientific research, while their successful application in the treatment of a particular patient is due to the clinical thinking of the doctor. The development and course of the same disease progress differently in different patients. The art of healing and the talent of a doctor are manifested in his creative approach to the choice and use of modern diagnostic and therapeutic technologies for the benefit of each patient.

Thus, a university instructor has a threefold task. The first aspect is to teach modern technologies of examination, diagnosis, treatment, rehabilitation of patients. The second is to introduce students to professional creativity. And the third aspect is to impart the ability to clearly understand and distinguish between the situations where creativity is desirable and the situations in which strict following the algorithm is required. The fact that creativity is not always appropriate is evidenced by a simple everyday example: if you are driving a car on a high-speed highway, you will not like the drivers of car moving nearby be creative in interpreting the traffic rules.

To solve this problem, a teacher of a higher medical school should be a creative person and be able to develop the creative potential of the students. Higher medical education has a number of specific features. First of all, it is flexibility, dynamism, a combination of tradition and innovation. The development of the higher medical school in Russia is conditioned both by constant reforming of the national healthcare system, on the one hand, and by transformations in the pedagogy of higher education, on the other hand.

Attention paid to creative thinking in higher professional education grows in the second decade of the XXI century [1–3]. It can be substantiated by the changes in the system of higher professional education. The teaching community comes to the conclusion that to equip a specialist with technologies, tools and approaches is not enough. What is more important is the desire and readiness to use the obtained knowledge and experience consciously and creatively [4, 5]. In scientific researches the pedagogical creativity is thoroughly studied in connection with the educational process. The creativity of instructors is manifested in development and implementation of active learning strategies in higher education [6, 7]; providing the substantiation for the moderating effect of instructional strategy [8]; in technologies, developing thinking skills and creativity [9, 10]; in the system of digital learning [11–13].

The authors research into instructional mastery, interaction between education, quality of work and assessment of sustainability in higher educational establishments [14–16]. The researchers study the issues of vocational training of teachers, their involvement in professional development through informal online communities and networks. The analysis of effective continuing training is also of particular interest [17–19]. The experience of creativity in routine activities of university instructors attracts much attention of the pedagogical community [20, 21].

The problem field of the research in continuous pedagogical development of medical university instructors in the sphere of professional technologies and creativity includes:

– substantiation of integrating formal and informal pedagogical training of higher medical school instructors using an innovative platform - the School of Pedagogical Excellence training;
– involvement of medical university instruction in distributing innovative pedagogical experience.

In contemporary professional education, the potential of continuing training courses is not used fully enough to integrate formal and informal pedagogical training of higher medical school instructors. The development of competences aimed at successful combining of pedagogical techniques and creativity is of great practical importance. The analysis of the contemporary practices of postgraduate training and continuing vocational training shows that little attention is paid to the development of these competences though the problem is theoretically comprehended. To solve the problem of creating a system of continuous pedagogical training of medical university instructor - a School of Pedagogical Excellence - it is necessary to consider a number of issues.

In the theoretical aspect, it is important to substantiate the structure and potential of the School of Pedagogical Excellence for the integration of formal and non-formal pedagogical training of teaching doctors. The system of continuous pedagogical training for teaching doctors exerts sufficient influence on the development of pedagogical mastery and creativity of medical instructors. Consequently, the structure and potential of the School of Pedagogical Excellence should be studied in terms of its role in the solution of psychological and pedagogical problems of formal and informal vocational training.

In practical terms, it is necessary to consider the School of Pedagogical Excellence as the environment for the development of pedagogical mastery and creativity of teaching doctors.

Analysis of the current views on the problem presented in the literary sources and practical experience allowed us to formulate the purpose of the study - to substantiate and generalize our own experience in creating the concept of the School of Pedagogical Excellence as the environment for the development of pedagogical technologies and creativity and its approbation. The purpose sets a number of tasks:

– to identify approaches to solving the problems of continuous pedagogical training of medical university instructors in terms of the opportunities provided by the School of Pedagogical Excellence;
– to structure the perspectives of the School for the integration of formal and non-formal continuing training with careful consideration of the specifics of medical education.
– to define the tasks of the School of Pedagogical Excellence. Successful solution of the tasks will contribute to the development of the creative potential of medical university instructors;
– To create opportunities and the medium for the application of creative projects, strategies and other methodical products and distribution of innovative pedagogical experience.

The stated tasks of the research were solved by joined efforts of the teaching staff members of the Volgograd State Medical University and Volgograd State Technical University.

2 Methods

The empirical base of the research is the Volgograd State Medical University (VolgSMU). The methods of generalization, comparison, observation, systemic analysis, data grouping and theoretical cognition were utilized by the researchers.

Over the years of study at the university, a medical student has to gain a huge array of theoretical knowledge, acquire abilities, master skills and competencies listed in the federal state educational standard of higher education. The competence-activity approach served the methodological basis of the research. According to it, the result of the university students training is the level of competence development. The federal state educational standards of higher education also appoint general cultural competencies, reflecting the requirements for the creative training of medical students. Among them are the ability to act in non-standard situations, to take social and ethical responsibility for the decisions made and readiness for self-development, self-realization, self-education, use of creative potential.

The former traditional pedagogical training of medical university instructors was often situational, the goal-setting was short-term, pedagogical reflection - insufficient. In the report on the educational project EDUCASE, the experts name the problem of integrating formal, non-formal education and self-development of instructors as one of the most significant trends affecting the introduction of technologies into the sphere of higher education [22]. The researchers attribute the problem to solvable ones. They claim that the teaching community is aware of the problem and knows the ways of its solution. One of them is the conception of continuous pedagogical training of university instructors at the "School of Pedagogical Skills". Awareness of the current state of vocational education encouraged us to design and implement the modules of continuing pedagogical training. The modules are aimed at familiarizing teaching doctors with innovative pedagogical technologies and involvement of medical university instructors in the activities of the School of Pedagogical Excellence. Our tasks were to implement innovative educational techniques, tasks for goal-setting and reflection - productive and creative in nature.

The formal component of the continuous pedagogical training is actualized by the School of Pedagogical Excellence of our university. In accordance with the "Law on Education in the Russian Federation", every instructor of a higher educational establishment should improve the vocational and pedagogical mastery at least once every three years. The VolgSMU introduced a course of pedagogy and educational technologies in 11 professional fields. The graduates are qualified as "Pedagogue of vocational education". The program helps to satisfy multidirectional educational needs of teaching doctors. The teaching staff of a medical university is characterized by heterogeneity in their basic professional education. Only a small part of instructors in general junior courses have a pedagogical education, while the majority of the teachers have medical education or specialize in natural sciences.

Practical classes and trainings for the listeners of the course were carried out separately in groups for the instructors of clinical and non-clinical disciplines. During practical classes, active and interactive learning techniques and approaches are used (problem-based approach, project-based and dialogical learning, educational games, case studies, didactic tasks, technologies for the development of critical/clinical thinking, brainstorming, discussions, etc.).

As it was above stated, all the theoretical and practical research activities were driven by competence–activity approach. In terms of it, the university instructors should actualize and implement their the professional and pedagogical competencies to teach students to follow the algorithm required by treatment protocols, as well as to apply integrated thinking, develop a creative approach, and master professional communication. Approaches to integration of formal and informal pedagogical education of higher medical school instructors should take into account two aspects. The first is the potential of a School of Pedagogical Excellence for the solving of the task. The second aspect is the activity of teaching doctors in the distribution of innovative experience.

The researchers actualized the potential opportunities of the School of Pedagogical Excellence to combine the traditional algorithmic approach with the creative approach in teaching students. The Center for Pedagogical Innovation of the VolgSMU provides the medium for the teaching doctors to test different options for combining these approaches.

While developing competence-oriented tasks, we created conditions for actualizing learning strategies, stimulating cognitive and professional motivation, interest, intellectual and practical skills as well as development of skills of implementing the knowledge, working in a team, formulating individual point of view and applying creativity. Speaking about the competence-oriented assignments, we mean the organization of independent activity of students (both classroom and extracurricular), including a system of specific pedagogical (quasiprofessional) situations and tasks aimed at development of different types of learning activities and the implementation of competence-based experience. So, the specificity of the competence-oriented tasks is that they should be interdisciplinary, on the one hand, and be linked with the professional activities on the other. As an example, in the course of Human Anatomy the students and the instructors of several departments (Department of Human Anatomy, Operative Surgery and Topographic Anatomy) and members of the Surgical Club of the university made a project "Online course of video tutorials on surgical skills and operating techniques". The joint work resulted in obtaining a patent for a simulator mastering manual surgical skills on the brain.

Providing an example of event education as another valuable approach, we used cases of different levels of complexity for training and the finals. With varying time intervals the course attendees were offered to solve case on the studied topics being limited in time. To develop professional readiness we used monothematic, complex and interdisciplinary cases which required actualization of knowledge mastered in the course of undergraduate and postgraduate education. There was an essential requirement to add to the portfolio the information about the ways of resolving each case study and the analysis of the likelihood of different solutions. The participants discussed the problems of the pedagogical process of the department and university, exchanged practical experience in teaching, developed educational projects and other pedagogical products. At the stages of knowledge actualization and final control, creative tasks were widely used. Creativity was demonstrated by the course attendees when developing a project for a lesson in different types of educational technologies or creating electronic educational resources (web quest, mental maps, videos). These activities also helped to improve methodological skills.

Besides, the School instructors provided assistance for teaching doctors in overcoming uncertainty in conducting a pedagogical experiment. In the course of continuing pedagogical training, a list of topics for the final qualification papers included the most pressing pedagogical problems facing the majority of teaching doctors. A number of pedagogical projects were suggested to the course attendees. These projects dealt with the implementation of innovative educational technologies and development of electronic educational resources. Much attention was paid to the reflective and evaluative activities of the trained instructors. So at the stage of pedagogical reflection, the course attendees discussed the strong and weak points of the lesson, the creative findings of their colleagues. The teachers analyzed their own creative achievements and limitations in the portfolio.

3 Results

The integration of formal and non-formal pedagogical education within the School of Pedagogical Excellence was one of the tasks of the continuous pedagogical training of medical university instructors. As it was found in the survey of 200 instructors of the VolgSMU (group of comparison) in 2011, the sphere of scientific interest of teaching doctors was more often within their basic medical specialty, and not within the issues of higher education pedagogy. The instructors of the medical university had a stronger motivation for continuing vocational development than for pedagogical development. The activities provided by the School of Pedagogical Excellence actualized various motivational resources and gave the teaching doctors an opportunity to show their creative skills and pedagogical mastery. One of the strategies utilized was the event education.

Designing a scenario for educational events (excursions, conferences, webinars, master classes) required pedagogical creativity. The technology of event education combines emotional arousal, the need for values consideration, joy of participation in valuable activities, self-confidence, strong motives for the further personal and professional development. All the mentioned had a powerful impact on education and personal development of students and teaching doctors.

Non-formal education is the second option provided by the School of Pedagogical Excellence which was organized within the University's Center for Pedagogical Innovation. The interaction of the Course of Pedagogy and Educational Technologies and the Center for Pedagogical innovation within the School of Pedagogical Excellence and the sociocultural and professional environment is presented (See in Fig. 1.).

The teaching doctors were given the opportunity to attend master classes of the university professors at a convenient time, to participate in extracurricular activities. One of the options was sharing personal pedagogical experience (or tips for solving the most common pedagogical problems) in the University social networks. The survey showed that integration of formal and non-formal education was supported and approved by the teaching doctors enrolled in continuing training programs of the VolgSMU.

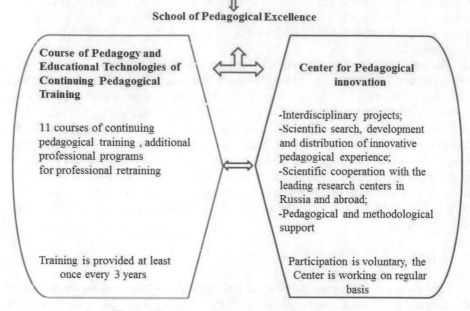

Fig. 1. The interaction of the structural components

Distribution of innovative pedagogical experience occurs in constant interaction with the university professors. Having learned something new at an open university (Massive open online learning) or at a symposium, master classes, or while creating a new teaching technique, the instructor shares the experience with the attendees of the Center for Pedagogical Innovation. They, in turn, bring it to other members of the university teaching staff. Being deeply immersed in pedagogical creativity at the School of Pedagogical Excellence, the instructors more easily involve students in professional creativity.

The outcomes of the School of Pedagogical Excellence activities were analyzed due to the data of surveys performed in 2011 (baseline data, the year of the School foundation, 200 respondents) and in 2020 (650 respondents) using the following criteria: 1) demand for the continuing training course in Pedagogy of higher education (the number of attendees admitted); 2) involvement of teaching doctors in pedagogical investigations and educational projects (in terms of publication activity); 3) attendees' satisfaction with the course content (its practical value and relevance to the current pedagogical challenges). The following Table 1. The results of the School of Pedagogical Excellence activities. gives a summary of the results.

The growth of the attendees' number is more noticeable among the clinical subject instructors as they had scarce pedagogical training at the undergraduate level. The majority of the instructors of theoretical and humanitarian subjects have basic pedagogical education and still they are eager to improve their competence.

Table 1. The results of the School of Pedagogical Excellence activities.

Criterion for comparison	Clinical subjects instructors	Theoretical and humanitarian subjects instructors
Demand for the continuing training course in Pedagogy of higher education (rise, %)	400%	300%
Involvement of teaching doctors in pedagogical investigations and educational projects (rise, %)	15%	25%
Attendees' satisfaction with the course content (rise, %)	23%	18%

To estimate the involvement of teaching doctors in pedagogical investigations and educational projects, we asked the respondents to answer whether the course influenced their interest to pedagogical aspects of their professional activity. We were eager to know if there was an increase in the number of articles dealing with pedagogical problems published by the course graduates.

Attendees' satisfaction with the course content was higher among the clinical subjects' instructors as it was the first experience of pedagogical training for the majority of them. So their pedagogical awareness improved greatly and was highly evaluated.

Such outcomes as improvement of personal pedagogical experience can't be measured, but the course instructors noted that the attendees of the continuing training course act as subjects of their own development, they are able to actualize their creative potential and actively participate in event-driven education. They took part in regional workshops "Pedagogy of Higher Education: Problems and Prospects", "Theoretical Foundations of a Pedagogical Experiment", "Practical Foundations of a Pedagogical Experiment", etc., a number of master-classes and seminars.

A university instructor engaged in pedagogical and scientific research:

- is aware of the theoretical and practical aspects of creativity, which gives the opportunity to teach students to clearly distinguish between the tools of the algorithmic and creative approaches in professional activities;
- involves the students in creativity and serves as an example of enthusiasm and dedication.

In the last two years the creativity of the instructors and students resulted in a number of implemented projects: the School of a young researcher, the Forum of Social and Innovative Development "Projectorium", the Club for Morphologists, the School for Tutors. An important aspect of the distribution of pedagogical experience is the principle of the immediate implementation of the experience into practice. Classes on the psychological, pedagogical, organizational and methodological aspects of the educational process at the School of Pedagogical Excellence are held according to the andragogical model.

The course doesn't interfere with the working schedule, as the classes are given in the evening hours or remotely. It gives the possibility to implement the newly acquired experience right after the training in accordance with the principle of actualizing the learning outcomes.

4 Discussion

The Center for Pedagogical Innovations of the VolgSMU performs the function of psychological and andragogical support for the professional development of the staff members. The Center, as opposed to discretely conducted courses continuing pedagogical training, is a permanently operating University unit which solves the following tasks:

- contributes to the scientific search and studies of the most pressing problems of higher professional education;
- promotes the distribution of pedagogical experience of teaching doctors and other members of the educational community of the University;
- familiarizes the teaching staff of the university with innovative (domestic and foreign) experience in training specialists, promotes the expansion of exchange and cooperation with the leading universities and research centers in the region, Russia and abroad;
- advises the faculty of the University on emerging issues of scientific and methodological advances and the introduction of new approaches, methods, educational technologies and other innovations, including the use of ICT tools;
- develops psychological and professional attitudes to the latest pedagogical technologies in the teaching environment;
- promotes the development, publication and distribution of scientific and methodological manuals on innovative pedagogical technologies.

The tasks solved by the Center for Pedagogical Innovation contribute to the development of the creative potential of medical university instructors. The issues of self-education and pedagogical training of medical university instructors and the need for professional counseling actualize the role of the Center for Pedagogical Innovation. Educational needs of the medical university instructors are met individually and in distant format. Teaching staff members address the specialists of the Center when they need help in introducing new educational technologies, assistance in publications on the subject of higher education pedagogy or in conducting a pedagogical experiment and creating joint projects. University instructors can seek advice on the issues of higher education pedagogy. They can also use the manuals and additional materials for pedagogical self-education posted on the educational portal of the School of Pedagogical Skills.

The continuing training course for medical university instructors provided by the School of Pedagogical Excellence helped to develop situational pedagogical readiness and acquire new experience. The obtained knowledge and experience were implemented in the educational process followed by reflective and evaluative activities.

The School of Pedagogical Skills is a system of continuous pedagogical training of medical university instructors. It integrates formal and non-formal education, develops

creativity and allows accumulating and distributing the experience of pedagogical innovations. The School provides opportunities for teachers to plan their own trajectory of
pedagogical development and leads teachers to personalized learning. Being the subject
of the pedagogical self-development, the teaching doctors determine the strategy and
pace of their learning and development.

5 Conclusion

The search for new principles, promising educational technologies, modern tools in
the activities of a teacher, their analysis, testing, implementation in the pedagogical
practice of higher education are still going on [23, 24]. It should be noted that the
concept of the School of Pedagogical Excellence is more consistent with the idea of
continuous pedagogical development of medical university instructors than traditional,
formal education in the process of continuous training, and allows integrating formal
and non-formal education.

The conceptual ideas of the School of Pedagogical Excellence as the environment for
technology and creativity have been confirmed by practice. Medical university instructors master advanced learning technologies and pedagogical tools of teaching students
in the process of formal and informal training within the framework of the School of
Pedagogical Excellence. It contributes to the development of pedagogical mastery of
university instructors and promotes the transition from differentiated education to personalized teaching of students. The creativity of the instructors is inspired and supported
by the specialists of the Center for Pedagogical Innovation of the University. The joint
efforts result in the development and implementation of innovative technologies, projects
and other pedagogical products.

Since the pedagogical community still lacks a formed system of continuous pedagogical training of university instructors, the presented results of the study can be used
for this purpose.

References

1. Gaspar, D., Mabic, M.: Creativity in higher education. Univ. J. Educ. Res. **3**(9), 598–605
   (2015). https://doi.org/10.13189/ujer.2015.030903
2. Jones, K.: Collaboration, creativity and capital in professional learning contexts. Prof. Dev.
   Educ. **43**(1), 1–5 (2017). https://doi.org/10.1080/19415257.2017.1254371
3. Odinokaya, M., Krepkaia, T., Sheredekina, O., Bernavskaya, M.: The culture of professional
   self-realization as a fundamental factor of students' Internet communication in the modern
   educational environment of higher education. Educ. Sci. **9**, 187 (2019). https://doi.org/10.
   3390/educsci9030187
4. Bylieva, D., Lobatyuk, V., Safonova, A., Rubtsova, A.: Correlation between the practical
   aspect of the course and the e-learning progress. Educ. Sci. **9**, 167 (2019). https://doi.org/10.
   3390/educsci9030167
5. Shipunova, O.D., Berezovskaya, I.P., Mureyko, L.M., Evseeva, L.I., Evseev, V.V.: Personal
   intellectual potential in the e-culture conditions. Espacios **39**, 15 (2018)
6. Harris, A., Jones, M.: Leading professional learning with impact. Sch. Leader Manag. **39**(1),
   1–4 (2019). https://doi.org/10.1080/13632434.2018.1530892

7. Roij, A.B.: The pedagogical legacy of Dorothy Lee and Paulo Freire. In: Active Learning Strategies in Higher Education, pp. 339–359. Emerald Publishing Limited, UK (2018). https://doi.org/10.1108/978-1-78714-487-320181015

8. Hiemstra, D., Van Yperen, N.W., Timmerman, M.E.: Students' effort allocation to their perceived strengths and weaknesses: the moderating effect of instructional strategy. Learn Instr. **60**, 180–190 (2018). https://doi.org/10.1016/j.learninstruc.2018.01.003

9. Artyukhina, A.I., Chumakov, V.I., Knyshova, L.P.: Personalized pedagogical training of highly qualified personnel in residency course. Educ. Bull. "Conscious." **22**(3), 15–19 (2020). https://doi.org/10.26787/nydha-2686-6846-2020-22-3-15-19

10. Elfeky, A.I.M., Masadeh, T.S.Y., Elbyaly, M.Y.H.: Advance organizers in flipped classroom via e-learning management system and the promotion of integrated science process skills. Think Skills Creat. **35**, 100622 (2020). https://doi.org/10.1016/j.tsc.2019.100622

11. Ipatov, O., Barinova, D., Odinokaya, M., Rubtsova, A., Pyatnitsky, A.: The impact of digital transformation process of the Russian university. In: Proceedings of the 31st DAAAM International Symposium, pp. 0271–0275. DAAM, Austria (2020)

12. Muhisn, Z.A.A., Ahmad, M., Omar, M., Muhisn, S.A.: Knowledge internalization in e-learning management system. Telkomnika **18**(3), 1361–1367 (2020). https://doi.org/10.12928/TELKOMNIKA.v18i3.14817

13. Wicaksono, A., Florentinus, T.S., Ahmadi, F.: Development of e-learning in web programming subjects for Moodle based vocational students. IJCET **9**(1), 1–9 (2020). https://doi.org/10.15294/ijcet.v9i1.33095

14. Caeiro, S., Sandoval Hamón, L.A., Martins, R., Bayas Aldaz, C.E.: Sustainability assessment and benchmarking in higher education institutions – a critical reflection. Sustainability **12**(2), 543 (2020). https://doi.org/10.3390/su12020543

15. Mahfud, T., Indartono, S., Saputro, I.N., Utari, I.: The effect of teaching quality on student career choice: the mediating role of student goal orientation. Integr. Educ. **23**(4), 541–555 (2019). https://doi.org/10.15507/1991-9468.097.023.201904.541-555

16. Wicht, A., Müller, N., Haasler, S., Nonnenmacher, A.: The interplay between education, skills, and job quality. Soc. Incl. **7**, 254–269 (2019). https://doi.org/10.17645/si.v7i3.2052

17. Artyukhina, A.I., Velikanov, V.V., Velikanova, O.F., Tretyak, S.V., Chumakov, V.I.: Challenge of digital economy - digital transformation of education. Eur. Proc. Soc. Behav. Sci. **50**, 74–84 (2018). https://doi.org/10.15405/epsbs.2018.12.10

18. Carpenter, J.P., Krutka, D.G.: Engagement through microblogging: educator professional development via Twitter. Prof. Dev. Educ. **41**(4), 707–728 (2015). https://doi.org/10.1080/19415257.2014.939294

19. Macià, M., García, I.: Informal online communities and networks as a source of teacher professional development: a review. Teach. Teach. Educ. **55**, 291–307 (2016). https://doi.org/10.1016/j.tate.2016.01.021

20. Kyndt, E., Gijbels, D., Grosemans, I., Donche, V.: Teachers' everyday professional development. Rev. Educ. Res. **86**(4), 1111–1150 (2016). https://doi.org/10.3102/0034654315627864

21. Rietzschel, E.F.: Freedom, structure, and creativity. In: Reiter-Palmon, R., Kennel, V., Kaufman, J.C. (eds.) Individual Creativity in the Workplace 2018, pp. 203–222. Elsevier Academic Press (2018). https://doi.org/10.1016/B978-0-12-813238-8.00009-7

22. Johnson, L., Adams Becker, S., Estrada, V., Freeman, A.: NMC Horizon Report: 2015 Higher Education Edition. The New Media Consortium, Austin, Texas, The USA (2015)

23. Bylieva, D., Lobatyuk, V., Safonova, A.: Online forums: communication model, categories of online communication regulation and norms of behavior. Humanit. Soc. Sci. Rev. **7**, 332–340 (2019). https://doi.org/10.18510/hssr.2019.7138

24. Majewska, K.: Modern educational tools in the teacher's work. New Educ. Rev. **51**(1), 125–135 (2018). https://doi.org/10.15804/tner.2018.51.1.10

Cross-Disciplinary Code Switching as Means of Encouraging Creativity

Galina A. Dubinina⬛, Larisa P. Konnova⬛, and Irina K. Stepanyan$^{(\boxtimes)}$ ⬛

Financial University Under the Government of the Russian Federation, Leningradsky Ave., 49,
125993 Moscow, Russian Federation
ikstepanyan@fa.ru

Abstract. The prerequisites of the study were the contradiction revealed in the teaching process between the need to form a set of professional competencies that combine different disciplines and their perception by students. The study angle involved economics, mathematics and English - the underlying subjects that are basic for an economist. The purpose of the publication is to develop a particular methodology for encouraging creativity in solving applied problems, developing economists and financiers' professional skills based on interdisciplinary code switching: economics - mathematics – IT - English. Research methods include the study of scientific publications on code-switching theory, pedagogical design and testing created resources. The paper presents a technique for forming a sequence of switching codes, that combine Mathematics, Economics, and English and describe the digital resources created on their basis: a glossary of economic terms and a thesaurus of the R programming language for the discipline "Digital Mathematics". The authors' testing of first-year students and their solution of applied problems to follow shows that the application of interdisciplinary code switching on the basis of specially organized language support contributes to the formation of more stable creative thinking and IT-skills. The novelty of the research lies in the development of the theory of code switching by the authors and the formation of a methodology for its use via interdisciplinary interaction.

Keywords: Creativity · Code-switching · Cross-disciplinary cooperation · Digital educational resources · English medium Instruction

1 Introduction

The rapid introduction of modern computer technologies in all areas of the economy and production significantly changes the set of competencies that a modern specialist should possess. Today, in addition to professional competencies, the most important skills are the confident possession of basic and special IT tools for performing a large number of similar operations, conducting analysis, building models, and a list of such digital skills is formed for almost every specialty, the totality of which forms a whole group of digital economy competencies. These competencies include not only working with information and software products, but also creative, critical thinking, communication skills and the ability to work in a team. Specialists capable of finding efficient, non-standard solutions are in high demand.

The formation of the entire group of digital economy competencies supplemented by the ability to innovate and creative thinking, is becoming one of the crucial tasks in training future specialists in economics and finance.

The authors of this article explore the practice of cross-disciplinary cooperation at the Financial University and focus their research on shaping professional competences through code-switching between different academic disciplines. The Financial University's priority number one is stated to be the training of globally competitive specialists for solving socio-economic problems, who are proficient in modern digital technologies at an advanced, professional level. For many years the University's strategy has focused on early profiling of training and instruction. Consequently, the content of traditional disciplines is updated, new courses are developed, and the educational process itself is changing in this particular direction.

It should be noted that cross-disciplinary products are most effective, be it workshops, practice-oriented projects, courses developed by teachers of various disciplines and delivered in cooperation with business representatives. Researchers infer that the formation of cross-disciplinary teams of educators helps to increase students' sense of membership [1]. This approach allows students to more broadly and comprehensively consider professionally oriented educational cases as well as find new ways to solve them. The formation of skills in applying traditional research methods to various scientific fields is constantly supplemented by developing creative thinking and resourcefulness. Contemporary research [2] shows that creativity in teaching can and should be incorporated into the learning process to help improve educational outcomes. Typically, studies of creativity in education are considered in three main aspects: educational environment, pedagogical practice, and the students' peculiarities [3]. Unfortunately, as it was noted in a systematic review [4], less than 10% of studies on the role in the formation of creative thinking in students of different levels of education contain conclusions and recommendations for teachers.

However, the main element of the educational process for any discipline is still a workshop or a lecture. Therefore, the present-day approach to university instruction must reflect the overall tendency to strengthen the professional component and enhance digitalization. The answer to this challenge is the use of comprehensive case-analyses, where cases are solved using tools from various scientific fields. This approach to teaching is extensively implemented at the Financial University. For instance, based on the concept of early professional orientation, professors from the Department of Mathematics developed a content-contextual model of teaching [5], in which the discipline "Mathematics" is used as the content of the model, computer technologies are used as a means of implementation, and the economic context is used as a means of profilization. Moreover, this stage makes it possible to combine professional context and English Medium Instruction, or EMI [6]. The contextual component enhances the professional orientation, and the content approach allows educators to intensify the process by studying several disciplines at the same time.

The new reality in the academic field is that due to the current high requirements to university graduates the development of digital skills dominates the educational activity at an economic university especially when it concerns the disciplines of the mathematical cycle. Starting from the first year, the propaedeutics of professional digital skills should be provided. The peculiarities of organizing such work are described, for example, in [7]).

An increasingly important role in the training process is played by the study of various IT products [8]. The fact that most of such products are in the English language casts the situation in a different light.

The language training of students of the Financial University offers 3 models of cross-disciplinary educational activity in which English as a foreign language is used as a medium of instruction:

1) Teaching junior students English on the basis of professionally oriented materials and quasi-professional activities, development of oral and written language skills in the educational sphere of communication;
2) Training junior undergraduate students for specialized disciplines with part of the content presented in English;
3) Teaching specialized disciplines and elective subjects fully in English at senior undergraduate Bachelor courses, Master's and postgraduate courses.

Hence the university arranges for the creation of English-language content of the educational process, including English-language web resources, printed materials in English, articles for English-language media, portals and other information platforms.

It should be noted that economic terminology is broadly used in teaching. At the same time, there is a traditional contradiction when, despite the existing knowledge of the English language as a philological discipline, students find it difficult to apply it impromptu, in the actual circumstances. Thus, students often perceive mathematics, economics, and English as parallel, unrelated disciplines. The authors find the task of overcoming this contradiction of paramount importance. This is confirmed by the qualification requirements for graduates, in which the necessary competencies are formulated in the language of professional standards, and not of individual disciplines.

Today, along with the Russian citizens, the Financial University enrolls international students, a sizable part of whom are English-speaking. A cross-disciplinary educational activity combined with EMI is a transnational educational process and takes place in a multilingual environment. Students study their specialty in a language immersion setting, usually in English and the language of the host country. Professional orientation and digitalization of learning complement this list with the language of a particular academic discipline. For instance, at the Financial University under the Government of the Russian Federation the language of mathematics and information technology is lingua franca for most of the academic subjects.

Furthermore, the use of English for IT technologies has a number of specific features, namely, it is saturated with special terms in English. The authors find it necessary to exercise code-switching and to introduce the relevant terminology directly in the process of using programming languages and not within the framework of a foreign language course.

Thus, continuing to develop an integrative approach to learning, the authors believe that from the first year it is necessary to strengthen interdisciplinary interaction: so the study of mathematics should be accompanied by a demonstration of its use in economic tasks, and the process of mastering modern digital technologies can be strengthened by an English-language educational line.

The object of the research is the process of forming professional digital skills in students of an economic university. *The subject* of the research is interdisciplinary code switching in the study of mathematical cycle disciplines by students of the Bachelor's degree program of the Financial University.

The purpose of this research is to develop *the methodology for interdisciplinary code switching: economics – mathematics – IT – English with the aim of shaping economists and financiers' professional skills* intended for developing the creativity and professional digital skills of economists and financiers.

Today, the theory of code switching is actively used in the study of foreign languages, but its potential can be realized to strengthen connections between traditional disciplines. Various interdisciplinary forms of work in the form of projects and workshops are used, as a rule, in senior courses. The technique presented in the article allows already in the first year to form stable cross-disciplinary connections, which contributes to a deeper understanding of the material being studied and the manifestation of creativity and creativity.

To strengthen the mathematical component and the use of digital tools in professional activities, the discipline "Digital Mathematics" has been introduced into the curricula of almost all the areas of training at the Financial University.

The successful experience of teaching this discipline allowed the authors to formulate a ***hypothesis:*** exercising cross-disciplinary code switching on the basis of specially organized language support contributes to the formation of creativity and steady professional skills as early as during the first year at the economic university.

The implementation of this goal is solved by the authors through implementing a number of tasks:

- review of the scope of knowledge on the problem of code-switching, acquaintance with the existing experience of introducing elements of a foreign language in the teaching of other disciplines;
- consideration of the cross-disciplinary cooperation in the application of digital tools;
- theoretical justification of approaches for developing a methodology for cross-disciplinary code-switching;
- creating a database for code-switching chains: economic term - mathematical action - program operator in English – its meaning in Russian;
- development of practice-oriented tasks for the formation of code-switching skills using mathematical calculations and computer implementation;
- analysis of code-switching impact on different disciplines as well as on the effectiveness of the educational process as a whole.

2   Methodical Aspects of Interdisciplinary Code Switching

2.1   Evolution of Approaches to Code-switching

There were periods in professional linguodidactics, when code switching was considered a poor-quality technique in teaching a foreign language, a feature of the teacher's incompetence. Recently, due to the fact that such a phenomenon, despite the efforts to avoid it, is still present in the learning process, educationalists have become more tolerant of bilingualism and even multilingualism [9]). It should be noted that we mean, first of all, English as the language of international communication in all spheres of human life.

In regard to higher education, the following factors are relevant:

- development of content and language integrated learning;
- increasing use of EMI (teaching university subjects partially or completely in English);
- digitalization of all spheres of life, which received a new impetus with the beginning of the covid-19 pandemic and the transition to online education.

The current situation has revealed that the teacher has become one of the most important consumers of digital technologies, and that today we can talk about the emergence of a new culture of digital pedagogical communication. At the same time, adaptation to digital techniques, the language of which is saturated with English-language terms, some of which are not translated into Russian or partially translated, i.e. used in the code switching mode, has become an urgent need.

Code-switching occurs as an injection of a foreign word in the middle of a phrase; between phrases.

Code-switching may occur for the following reasons:

- missing notions in the host language of the discussion;
- more precise expression of the notion in one of the languages;
- aspiration to demonstrate one's affiliation with a specific professional group;
- aspiration to marginalise those who do not speak the second language or are incompetent in the terminology of the field, i. e., to create obstacles to understanding;
- inarticulate terms in one of the languages, etc.

We believe *code-switching* in professional training can be understood as a way to gain linguistic advantages rather than an obstacle for communication. *Code-switching* enables the teacher to build stronger understanding of the analysed phenomena by operating a wider vocabulary pool much as we employ bold fonts in texts or voice modulation to emphasize information in speech. Switching to a different code thus puts stronger emphasis on the teacher's message and helps to drive stronger impact [10].

The transition from traditional instruction toward teaching disciplines in English takes the form of Content and Language Integrated Learning (CLIL) [11]. Progress on the set objectives would be largely assisted by cross-disciplinary cooperation, i.e., focusing on subject-specific topics in teaching English in coordination with the teachers of basic disciplines and studying English-language sources on the core subjects at foreign language workshops [12].

The use of the English language in teaching special disciplines is called English as a Medium of Instruction, or simply English Medium Instruction (EMI), it is a global trend in higher education.

The pedagogical community is tasked with the development of novel educational approaches and solutions engaging digital tools. Notably, digitalization creates a new solution to the individual learning problem, specifically in a foreign language, since it provides better opportunities to chart individual learning trajectories based on the initial competence level and specific needs of the learner.

It, though, poses a challenge for the teacher to preserve the time-proven experience of traditional educational activities in class [13].

Inspired by the analysis of various aspects of cross-disciplinary cooperation and students' expectations, their needs and their satisfaction with the results, we focused on the following most exemplary academic disciplines: business economics, mathematics, IT and English. These disciplines lay the foundation of bachelor training at economic universities from the first year of studies.

The use of digital tools for mathematical calculations and analysis requires the application of special software resources. Most of them operate in English. Thus, continuing to develop an integrative approach to learning, we believe that the process of mastering modern digital technologies can be strengthened by an English-language educational line, since foreign language competence today makes an integral element of professional education. The use of the English language for IT is quite specific and includes multiple special terms.

A major contribution toward the development of the methodology of language support in studies of digital resources can be made by the above mentioned code-switching theory. Research shows [14] that the phenomenon of code-switching is not only about the linguistic aspects but it also opens up a strong cross-disciplinary potential in showing how models from one knowledge domain apply to models in other domains.

2.2 "Digital Mathematics" Workshop as a Ground for Implementing the Integrative Approach

Several years ago, the computer-based practical workshop in "Digital Mathematics" was included in the first year curricula of nearly all specializations at the Financial University with the objective to reinforce the mathematical component and lay the foundation for digital competence in the future financial and economic occupations. Today, in addition to professional competencies, the most important skills are the confident possession of basic and special IT tools for performing a large number of similar operations, conducting analysis and building models. All of them are based on the mathematical apparatus. One of the core objectives of this discipline is to demonstrate the potential of mathematics in describing economic patterns, building models and forecasting outcomes. Accordingly, computer-based practical workshops served as an effective ground to consolidate the development of mathematical knowledge, digital competencies and the propaedeutics of such skills in the future profession in the economic and financial fields.

It should be noted that the new discipline is not meant to replace standalone courses in mathematics and economics but rather supplements them and supports more conscious knowledge acquisition. The course of "Digital Mathematics" presents an open ground to introduce the integrative approach.

The use of digital tools for mathematical calculations and analysis requires the application of special software resources. Most of them operate in English.

The use of the English language for IT is quite specific and includes multiple special terms. We believe that a consistent methodology of language support in teaching and learning a discipline could raise efficiencies in both studying English and computer technology. Besides, solving economic problems in computer-based workshops can improve the acquisition of a language in this domain as a means of international communication.

R is an example of a programming language, representing free software and an international project involving many developers. At the Financial University, its computing capabilities for mathematical analysis and linear algebra applications are taught in the first year workshops on general economic courses with the use of the popular RStudio software environment [15]. In the curriculum, the principles of R are taught in the middle of the first semester after students have mastered the use of graphs and functional analysis in EXCEL. The content of practical workshops is in sync with the content of the course in "Mathematics".

2.3 Creative Approach to Developing the Sequences of Code-switching: Economics – Mathematics – IT – English

In a digital transformation of a society that requires innovation, adaptation and flexibility, creativity must evolve from an inspiration to a way of thinking [16]. The theory of developing methodology for cross-disciplinary code-switching accommodates the following pedagogical approaches:

- *the integrative approach*: the teaching process integrates several disciplines, including mathematics, IT, economics, English through a variety of teaching methods;
- *the activity-based approach:* knowledge acquisition in the course of "Digital Mathematics" occurs fully via practical activities;
- *the contextual approach:* every opportunity is used to incorporate the professional economic context as part of the teaching and learning process;
- *the individual learning concept:* learning aids and teaching arrangements are built around the objective to reinforce students' potential to proceed in line with their individual capabilities, interests and educational needs.

Moreover, the integration of the English-language component as part of the mathematical discipline facilitates the development of the high relevance skills of information coding and transformation.

For economists and specialists in finance, code-switching primarily concerns economic terms and their international equivalents, mathematical apparatus and operators of various programming languages.

The digital mathematical content in EXCEL and the R language presents wide opportunities for sustainable code-switching skill development in the following way: each economic term is taught with its international English-language equivalent, related mathematical formulas and digital toolkit, which also uses operators in English. A sequence of codes is built, and each of them expands and provides deeper visibility into the analyzed concept (see Fig. 1).

An important stage in the basic bachelor training in economic programs is to develop the skills of analysis of various functional dependencies. Thus, the notion of function is a principal concept in the course of mathematical analysis and is widely covered

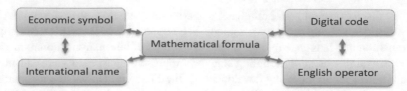

Fig. 1. Sequence of switching codes.

in computer-based practical workshops. The content primarily concerns elementary functions traditionally covered in the course of mathematics.

Applying the economic context to analysed problems means that the main economic functions are introduced early and discussed in a recurring way throughout the course, specifically, the production function, the utility function, the demand function, the supply function, the cost function, the income function, the profit function.

First, let's consider the concept of elasticity to illustrate the sequence of code-switching. Elasticity is a measure of a variable's (function) sensitivity to a change in another variable (argument). Traditionally, the concept of elasticity applies in economics in the analysis of the demand or supply functions, thus, it makes sense to discuss elasticity in mathematics with regard to these functions, too. Next, let's consider a sequence of code-switching in a problem of demand elasticity (see Fig. 2).

The above example demonstrates a complete sequence of code-switching, from the use of economic terms, particularly in English, to mathematical formulas, and to computer implementation. Then, to formulate a solution, an inverse sequence is used, from mathematical results to economic conclusions.

Alongside calculations of elasticity for the demand or supply function at a given point, elasticity is also calculated at the equilibrium point and inferences are drawn whether the function is elastic or not and price levels are determined where elasticity is observed.

Economic theory puts a strong emphasis on the so-called marginal calculations. Mostly, first-year program addresses the functions of marginal revenue $MR(q)$ and marginal costs $MC(q)$, which depend on the amount of output q. The abbreviation used to denote these functions traditionally comes from English. Marginal values show the rate of change in revenue/costs with changes in the amount of sold/manufactured products. The rate of change in the function is calculated by the derivative and the use of digital tools helps to avoid the lengthy process of finding the formula of the derivative based on rules and tables. Similar sequences containing the definition, description and formulas are built for marginal revenue and marginal costs. Similar sequences are proposed for determining the function based on the rate of change.

Economic challenge
Investigate the change in the demand function when the price of a product changes by unit: find the proportionality coefficient between the relative changes in the price of a product and the demand for it.

English / Russian terms:
$E_D(p)$ –elasticity of demand – эластичность спроса
$E_S(p)$ – elasticity of supply – эластичность предложения

Mathematical formula
$$E_D(p) = \frac{p}{D(p)} \cdot D'(p)$$

Fragment of software implementation

```
> p <- 10
> Demand <- 3+143/(5+p)
> Dem  <- expression(3+143/(5+p))
> Dp <- D(Dem, "p")
> Demand_dif <- eval(Dp)
> Elasticity <- function(P) {Demand_dif*p/Demand}
> Elasticity(p)
[1] -0.5070922
```

Digital operators with translation into Russian:
expression – выражение
function – функция
eval – оценить
D – дифференцирование

Math inference
The limit of the ratio of the relative increment of the function to the relative increment of the argument at an infinitesimal increment of the argument at point 10 is approximately -0.5.

Economic conclusion
Under these conditions, a 1% price change will lead to a decrease in demand by about 0.5%, demand is not elastic.

Fig. 2. An example of switching codes when calculating elasticity.

Alongside functional dependencies and marginal calculations, economic analysis also relies on average values. "Digital Mathematics" classes present an opportunity to master mathematical approaches to calculating them so that these techniques would occur automatically in other disciplines as code-switching. Consider an example for calculating the average labor productivity (see Fig. 3).

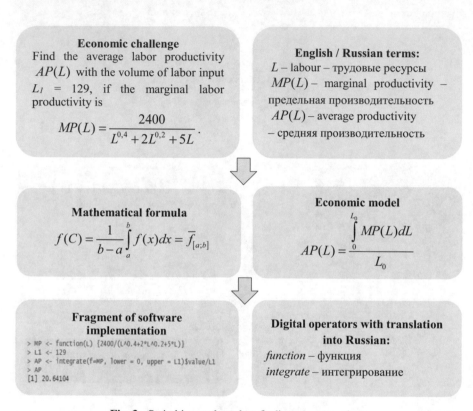

Economic challenge
Find the average labor productivity $AP(L)$ with the volume of labor input L_1 = 129, if the marginal labor productivity is

$$MP(L) = \frac{2400}{L^{0.4} + 2L^{0.2} + 5L}.$$

English / Russian terms:
L – labour – трудовые ресурсы
$MP(L)$ – marginal productivity – предельная производительность
$AP(L)$ – average productivity – средняя производительность

Mathematical formula
$$f(C) = \frac{1}{b-a}\int_a^b f(x)dx = \overline{f}_{[a;b]}$$

Economic model
$$AP(L) = \frac{\int_0^{L_0} MP(L)dL}{L_0}$$

Fragment of software implementation
```
> MP <- function(L) {2400/(L^0.4+2*L^0.2+5*L)}
> L1 <- 129
> AP <- integrate(f=MP, lower = 0, upper = L1)$value/L1
> AP
[1] 20.64104
```

Digital operators with translation into Russian:
function – функция
integrate – интегрирование

Fig. 3. Switching codes when finding average values.

Digital tools employed in the computer-based practical workshops present wide opportunities for the visualization of functional dependencies. Visual data analysis skills make an integral part of an economist's professional digital competencies today. Almost from the very beginning of the course, the analysis of a function includes mapping out a graph, which may further include graphs of the first and second derivatives. Digital tools make it fast and precise and any parameter can be changed. This creates visibility into how changes in a parameter influence the function in general. It is particularly important for economic functions. Such visual analysis is based on code-switching from mathematical dependencies to graphical representations. An example of such code-switching is shown in Fig. 4 illustrating a sequence of code-switching in the "Arachnid market model".

Based on the idea of early profiling of training and instruction and understanding that the principal target for students is professional training, we believe that instruction in basic disciplines, such as mathematics, should necessarily include the economic context.

Economic position of the model

1. The demand for a product in the current year corresponds to the price of this product this year.

2. The supply of a product in the current year is the price of that product in the previous year.

English / Russian terms:
$D(p)$ – demand – спрос
$S(p)$ – supply – предложение
p_0 – equilibrium price – равновесная цена

Mathematical formula

$$p_n = a + k \cdot p_{n-1}, q_n = b + k \cdot q_{n-1}$$

recursive formulas for price and quantity

Fragment of software implementation

	Точки	
	q	p
q0; p0	0,3	9
q1; p0	7	9
q1; p1	7	3,8462
q2; p1	2,3615	3,8462
q2; p2	2,3615	7,4142
q3; p2	5,5728	7,4142
q3; p3	5,5728	4,944
q4; p3	3,3496	4,944
q4; p4	3,3496	6,6541

Visualization

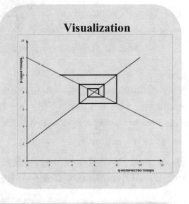

Math inference

The sequence converges $\{(q_0, p_0), (q_1, p_0), (q_1, p_1), (q_2, p_1), (q_2, p_2), ...\}$
towards the market equilibrium point.

Economic conclusion

Market equilibrium can be achieved; the market price tends to the equilibrium price.

Fig. 4. Switching codes using the arachnid market model as an example.

Thus, a program review for the discipline would require mapping out the economic terms and relations that can be illustrated. A recurring emphasis would ensure a proper understanding of the mathematical essence of these concepts and relations. See above for the main economic functions, either single-variable or two-variable functions. Apart from the Russian and English names of these functions and their essence, it is also crucial to build understanding of the relation between a function and its rate of change. In mathematics, this relation is established with the use of the principal operations of mathematical analysis, namely, differentiation and integration (see Fig. 5).

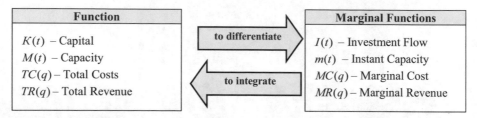

Fig. 5. Relationships between functions and marginal functions.

These sequences of code-switching should be used on a recurring basis to ensure their acquisition by students. This can be facilitated by focusing on applied economic problems and by using a special exercise system.

2.4 Design of Digital Resources for Better Acquisition of Sequences of Cross-Disciplinary Code-switching

The skills of coding and decoding become integral for professionals today in the digital world. They help navigate information flows, structure information and develop mathematical process models and their computer implementation. By applying mathematical modeling to solve an economic or financial problem and selecting R or EXCEL tools for data processing, students switch between the domains of scientific terminology. Meanwhile, the vocabulary of economics and digital mathematics largely rely on English-language terms.

Another important consideration is that English language skills are a requirement for IT specialists given that most programming languages operate with keywords in English and almost all reference sources are written in English. The development of language skills should take into account the specifics of technical English, specifically the vocabulary, grammatical and functional aspects involved. Such training would require special literature and programs.

The foundation of this work to develop skills of cross-disciplinary code-switching comprises resources integrating the principal terms of the studied discipline, digital mathematics in this case, and other related and useful terms from other disciplines. Where it concerns courses of mathematics at an economic university, it primarily applies to economic terms, their international equivalents and operators of a programming language in English.

The use of digital resources in English for problem solving reinforces the importance of studying English. Adequate translations of the digital English terminology into Russian is integral for mapping out students' individual trajectories into their career.

Computer-based practical workshops pose programming challenges for students, on the one hand, and compel them to use English, on the other hand. They also create the need for automatic code-switching: mathematical operation – English operator – digital code.

An electronic thesaurus, or a terminological glossary of the studied discipline, can be indispensable in developing these skills and providing language support in the context of digitalization of learning.

Modern technology enables automated extraction of relevant terms from information massive associated with a specific subject domain. Computing methods are developed for automated extraction of glossaries from information corpora of the taught discipline. Once the domain terminology is determined, one or more definitions are extracted for each of the terms in question.

A convenient tool for creating an electronic glossary is the currently popular learning management system LMS Moodle. Glossaries in it can be created by the teacher and expanded by students independently. Work with glossaries can be part of students' assessment. Apart from terminological systematization, glossaries can be used for tips, commentary and games developed around definitions.

Alongside the main electronic thesaurus of the discipline, a *glossary of economic terms* can also be created and used as an important tool for cementing cross-disciplinary associations. That would be a cross-disciplinary glossary useful in studying mathematics, English and microeconomics. The emphasis in terminology should be on the discipline in question. When the glossary is used in classes of mathematics, students are told they would learn more specific and precise economic definitions in the course of microeconomics. Ideally, they should be already acquainted with economic terms and their international equivalents. Then, the proposed sequences of code-switching would facilitate more thorough and profound understanding of economic processes and their application in the professional context.

3   Code Switching Experience When Studying Digital Mathematics

3.1   R Language Thesaurus for the Discipline "Digital Mathematics"

We and our students created an electronic thesaurus of R as a tool of digital mathematics. The thesaurus includes 13 chapters concerned with nearly all principal topics of the course. Main commands and operators are outlined for each chapter. Links are attached for further reading on the operator and usage examples. A special highlight is command codes that can be run in R-studio. Examples of thesaurus articles are provided in the Fig. 6.

Thesaurus development was arranged in students' group projects. 99 students were divided into 29 mini groups, which represented approximately 60% of the total cohort of first-year students in the financial department.

Each group made a presentation on their work. The main considerations for evaluating the end product were completeness, accessibility and design.

Задание:

Образовать в R три матрицы:

$$S - \text{нулевая матрица, размера } (10 \times 10);$$

$$Q = \begin{pmatrix} 1 & 1 & 1 \\ 1 & 2 & -2 \\ 1 & 3 & 0 \end{pmatrix} \text{ и } P = \begin{pmatrix} -1 & 0 & 1 \\ 2 & 2 & 2 \\ 4 & 5 & 6 \end{pmatrix}.$$

Решение:

Составим следующий текст программы:

```
S <- matrix(0, nrow=10, ncol=10)    # Образовать матрицу из нулей размера 10x10
Q <- cbind(rep(1,3), 1:3, c(1,-2,0))    # Составить матрицу из трех столбцов-векторов
P <- rbind(seq(-1,1,1), rep(2,3), 4:6)    # Составить матрицу из трех строк-векторов
S; Q; P    # Вывести в поле консоли значения объектов S, Q и P
```

Примечание: числовые массивы фактически уже являются матрицами.

```
x <- array(2,dim = c(3,5)); x    # Объявляем одномерный массив x из двоек
s <- as.matrix(x); s    # Объявляем одномерный массив x матрицей s
```

Fig. 6. An example of a chapter from the thesaurus of the "Digital Mathematics".

Eventually, the combined glossary with hyperlinks was conveniently made available for students to ensure easy access whenever they need it.

The use of an electronic glossary and working on it in the first place contribute to individual learning processes. Each student can retrieve useful information depending on the individual competence levels in English, mathematics and IT. Access to the glossary is provided for an unlimited period of time, which is rather convenient if students have difficulty with some of the above mentioned disciplines.

The work on building the glossary was creative. Each student, depending on their abilities and knowledge of English, mathematics and programming, could choose the section they were interested in. In general, it should be noted that modern students are more enthusiastic about solving problems and tasks if they are connected with the digital environment.

Working on a specialized terminological glossary for a discipline and further using it for reference enables better memorization of the operators, their descriptors, syntax and application.

3.2 Terminological Glossary as a Tool to Memorize Code-switching Sequences

To facilitate better memorization of the principal economic concepts and stronger cross-disciplinary associations between economics, English and mathematics, we developed a glossary of economic terms used in the first-year course of mathematics. It covers the main economic functions and their characteristics relating to differentiation and integration. The glossary comprises 25 economic terms in articles arranged in the following way:

Subsequent paragraphs, however, are indented:

- term description;
- definition;
- symbol;
- international equivalent;
- pronunciation in English;
- mathematical formula;
- commentary.

See an example for one of the terms in Fig. 7.

Функция Кобба-Дугласа

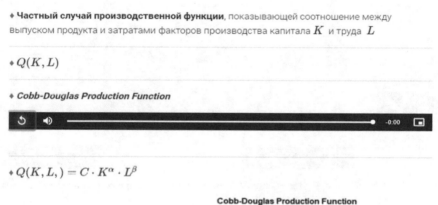

♦ **Частный случай производственной функции**, показывающей соотношение между выпуском продукта и затратами факторов производства капитала K и труда L

♦ $Q(K, L)$

♦ *Cobb-Douglas Production Function*

♦ $Q(K, L,) = C \cdot K^{\alpha} \cdot L^{\beta}$

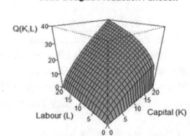

Fig. 7. Example of a glossary article describing the term "Cobb-Douglas function".

The glossary was created in the LMS Moodle system. Students are encouraged to master the glossary in their individual independent work. Individual progress can be tracked in the system. An element is only marked complete when all articles are read up by the student.

To assess the acquisition of code-switching sequences based on the glossary, a test was also developed in the same system.

The test comprises five questions to evaluate the knowledge of definitions of economic terms (filling the gaps), English symbols and pronunciation, and mathematical formulas.

The test offers five tasks on assessing the knowledge of:

- definitions of economic terms (filling the gaps);
- English symbols and their pronunciation;
- mathematical formulas.

See examples of test questions in Fig. 8.

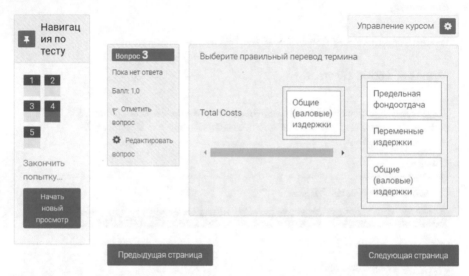

Fig. 8. Question on matching the economic term in English with the economic term in Russia.

The results of the test are laid out in in Fig. 9. The survey was conducted among 192 first-year students of the financial department. Students showed interest in the test and most of them were successful. Such a positive outcome was quite expected as the main objective was to connect the terms covered in different disciplines and reinforce cross-disciplinary associations. Students observed it was the first time that they gained this perspective on the characteristics they knew and they appreciated the clear benefits of this approach.

The somewhat lower result levels for the second task can be explained by two things. Firstly, open questions (when students have to write down rather than select or match answers) are always a challenge and, secondly, symbols of economic terms may differ in various sources.

As we have seen, this practice with an electronic cross-disciplinary glossary improves memorization of important terms. And, more important, references to these terms invoke all the attached associations, specifically international equivalents, symbols and mathematical calculations.

3.3   Using Terminological Sequences in Case Solving

The main objective of a mathematics course at an economic university is to build students' competence in applying the mathematical apparatus for problem-solving in economics.

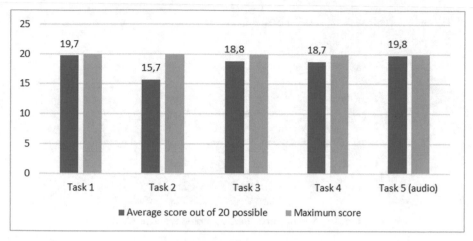

Fig. 9. Test results by the tasks.

Practice-oriented tasks are essential starting from the first year of studies. Given the limited scope of economic knowledge students might have developed by the time, the range of such problems is rather narrow. However, practical problems are set for students and particularly in assessments.

We should note that such problems traditionally pose a challenge. Students find it difficult to switch to microeconomics in mathematics workshops. Thus, we developed and tested training on cross-disciplinary code-switching.

Let's consider next the impact of this training on students' performance in solving one of the economic problems, specifically the analysis of the two-factor production function.

Economic case

For a given production function $Q(K; L) = \frac{L^2 \cdot e^K}{1+e^K}$,, where Q is the level of output, K is the amount of capital assets (capital), L is labor input at $K_0 = 5$, $L_0 = 4$, determine the marginal product of capital and marginal labor productivity, marginal rate of technical substitution, capital and labor elasticity.

Students are asked to find a "manual" and a computer-aided solution to this problem. Each method poses certain difficulties. Students need to:

- recall the mathematical formulas used to calculate the sought values;
- use correct symbols;
- accomplish correct calculations manually or submit a correct software code.

A solution to this problem in the R language is presented in Fig. 10.

Figure 11 outlines the results of solving this problem before the training on code-switching sequences (based on the data for the 2019–2020 academic year, 124 students) and after such training (in the 2020–2021 academic year, 134 students). We considered the task to be 100% completed if the economic values were correctly marked and

```
expression((L^2) * exp(K)/(1 + exp(K)))
> QL <- D(Q, "L"); QL
2 * L * exp(K)/(1 + exp(K))
> QK <- D(Q, "K"); QK
(L^2) * exp(K)/(1 + exp(K)) - (L^2) * exp(K) * exp(K)/(1 + exp(K))^2
> K <- 4
> L <- 5
> MQL <- eval(QL); MQL # Предельная производительность труда при K =4 L = 5
[1] 9.820138
> MQK <- eval(QK); MQK # Предельная фондоотдача при K =4 L = 5
[1] 0.4415677
> MRSLK <- MQL/MQK; MRSLK # Предельная норма замещения труда капиталом
[1] 22.23926
> EQK <- (MQK*K)/eval(Q); EQK # Эластичность выпуска по фондам
[1] 0.07194484
> EQL <- (MQL*L)/eval(Q); EQL # Эластичность выпуска по труду
[1] 2
```

Fig. 10. Solving an economic problem in RStudio.

the program codes for their calculation were correctly drawn up. Completing 75% of the task meant that the student was wrong in calculating the elasticity. Calculating the marginal product of capital and marginal labor productivity was equivalent to 50% task completion. If a student started compiling a code, but the results did not correspond to the correct answers, the task was considered to have been completed by 25%. It is obvious that the results of the students in the 2021 academic year are better than in 2020.

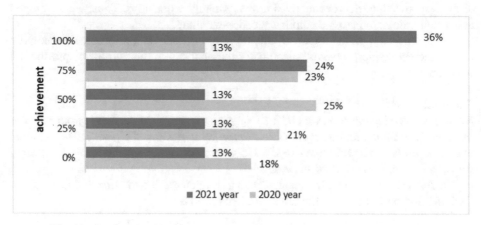

Fig. 11. Students' performance before and after the training on code-switching.

It is certainly not correct to assume based on the above data that there is a strong correlation between students' performance in solving economic problems and the introduction of training on terminological sequences in the learning process. Such calculations would require a wider pool of economic problems and a longer observation period. Moreover, a representative sample must be built and workshop settings must be uniform. Clearly, given the COVID 19 pandemic and intervals of distant learning, these requirements could not be met. However, a comparison of students' performance for the past two years is quite hopeful and substantiates further research concerning the impact of training on cross-disciplinary code-switching on the acquisition of professional skills.

4   Discussion

The aspects of effective organization of cross-disciplinary cooperation are widely discussed by the pedagogical community. A prevalent view is that such cooperation should become the foundation of students' acquisition of professional standards.

An in-depth analysis of this aspect is provided in [17]). The researchers believe that cross-disciplinary cooperation is a requisite condition for the implementation of the competence-based approach and describe the practices of its implementation at a technical university. The paper describes a structural functional model of cross-disciplinary cooperation in the integrated learning information environment built around the MOODLE system. The system enables the implementation of cross-disciplinary products, such as a practical course of programming and a trainer of "computer" English. One advantage of the system is the single underlying database.

We also see the potential of further developing our projects through a wider use of the MOODLE system for cross-disciplinary cooperation.

The paper [18] describes the experience of integration of academic knowledge and cross-disciplinary cooperation in addressing agroecological issues. The author employs the methodology of organic philosophy to combine theoretical and practical subject-specific knowledge. The potential of setting up the learning process in cooperation with production and academic organizations is discussed. It is observed that this work is particularly relevant in educating young researchers (studying in Masters' and PhD programs). We believe this cooperation should begin early on as much as possible. Then it will be possible, even during the first year of studies, to accomplish modest practice-oriented projects.

Practical cases of such cross-disciplinary projects are outlined in [19]. The paper describes the methodology of project work in English on economic problems being the core profile of students' specialization. These activities include three stages: the pre-project stage (preparation of presentations in English, trainings on presentation delivery and discussion); the research stage (analysis of sources in English, project development, delivery of research results); the assessment stage (expert assessment, self-assessment and peer-to-peer assessment). Faculty of economic departments are invited as experts. As a result of such cooperation, certain underlying emerging problems were identified, specifically, the need to develop uniform approaches to assess the results and the need for additional training in the university's core subjects for teachers of the linguistic chairs.

We have the experience of implementing cross-disciplinary projects on economic subjects with the use of the mathematical apparatus. And we support the view that professional cooperation between teachers of various structural departments is rather complicated. It is time-consuming, not fully regulated and requires curricular coordination across various chairs and departments. Meanwhile, the development and use of cross-disciplinary resources places additional demands on the teacher of any discipline. E.g., the development of code-switching sequences in digital mathematics requires some knowledge of economics and English.

An interesting account of cross-disciplinary cooperation practice is presented in [20]. Researchers present a thought-provoking reflection on the practical application of cross-disciplinary cooperation. The paper describes successful implementation of cooperation between teachers of various chairs. Specifically, it concerns a developed course integrating the electronic educational environment, practice-oriented discipline "Introduction to Profession" and English. One of the theoretical pillars of the course is the well-known methodology of teaching English, CLIL, based on teaching professional disciplines by means of a foreign language.

Note that we employed the CLIL methodology when developing the content-contextual model for teaching mathematics. The proposition to teach mathematics using computer technology resources and the economic context is in sync with the idea of [20]. That underscores the promise of the cross-disciplinary approach.

The creation and use of such interdisciplinary products in the educational process makes the educational process more interesting and lively. The creativity and creativity of teachers always serves as a bright example for students, involving them in this process. Certainly, as noted in [4], the teacher's willingness to act as a role model, and collaboration in the classroom contribute to the growth of students' creativity.

5 Conclusions

The conducted research suggests that the principles of code switching used in linguistics can effectively work in the formation of interdisciplinary connections in the educational process, aimed at fostering creativity in the application of digital skills. The use of special exercises integrating terminology from various disciplines makes the learning process more effective and practice-oriented.

To make these examples of code-switching further useful in related disciplines and professional occupation, they should be shaped as specific sequences. E. g., where it concerns mathematics, it is important to emphasize the relation between the economic concept and its mathematical equivalent. Today, it also includes international equivalents of the respective terms and competence in the digital tools for their analysis. In providing the examples of problem solving, we sought to demonstrate exactly these sequences: from an economic problem to a mathematical model to a software implementation, mathematical solution and an economic conclusion.

We developed electronic resources enabling the formulation of cross-disciplinary code-switching sequences for students:

- a glossary of economic terms;
- a thesaurus of the R language for the discipline "Digital Mathematics".

A special test was developed in the MOODLE system to enhance skill acquisition.

The practice of using this system of exercises shows that when they are incorporated in the learning process, students can navigate practice-oriented problems more confidently.

The accomplished work led us to expand the code switching technique with the following points for interdisciplinary education:

1. To make a code-switching sequence complete and operational as a sequence of associations, it should begin and end in the same system of codes.
2. Code-switching between various disciplines would be more effective if students have first acquired stable sequences of such code-switching. Examples of such exercises are presented in this work.
3. Terms used in the course of the discipline in the code-switching system should be previously acquired by the students from the discipline where they belong. E. g., economic functions discussed in the course of Mathematics should be known to students from the course of Economics and their international equivalents, from the course of English.
4. The need to ensure the code-switching system helps to build more in-depth knowledge in students means the demands on the teacher are increasing. Beside their knowledge of the discipline they teach, they need to navigate the disciplines relating to the codes involved.
5. There is no code-switching without cross-disciplinary cooperation between teachers of Economics, Mathematics, Digital Mathematics and English.

The developed training on cross-disciplinary code-switching based on specially arranged language support helps students to shape informal cross-disciplinary associations and expand analytical skills. Going forward, the already built code-switching sequences will occur in studies of economics, English and disciplines of the mathematical cycle, which would facilitate a more creative approach to the analyzed professional problems.

References

1. Stepanyan, I.K., Dubinina, G.A., Nikolaev, D.A., Kapranova, L.D., Pashtova, L.G.: Cross-disciplinary case-analyses of investment optimization in a foreign language applying dynamic programming. Espacios. **38**(62), 19 (2017). http://www.revistaespacios.com/a17v38n62/173 86219.html. Accessed 1 Apr 2020
2. Rinkevich, J.: Creative teaching: why it matters and where to begin. Clear. House **84**(5), 219–223 (2011). https://doi.org/10.1080/00098655.2011.575416
3. Harris, A., De Bruin, L.: Creativity in education. In: Oxford Research Encyclopedia of Education (2018). https://doi.org/10.1093/acrefore/9780190264093.013.383
4. Davies, D., Jindal-Snape, D., Digby, R., Howe, A., Collier, C., Haya, P.: The roles and development needs of teachers to promote creativity: a systematic review of literature. Teach. Teach. Educ. **41**, 34–41 (2014). https://doi.org/10.1016/j.tate.2014.03.003
5. Konnova, L.P., Rylov, A.A., Stepanyan, I.K.: Integrative teaching of mathematics as a means of a forming modern economist. Amazonia Investiga. **9**(26), 486–497 (2020). https://doi.org/10.34069/AI/2020.26.02.56
6. Baranova, T., Khalyapina, L., Vdovina, E., Yakhyaeva, C.: Soft CLIL v.2.0: integrating a mobile app and professional content into the language training. IOP Conf. Ser. Mater. Sci. Eng. **940**, 012140 (2020). https://doi.org/10.1088/1757-899X/940/1/012140
7. Konnova, L.P., Rylov, A.A., Stepanyan, I.K.: Propaedeutics of professional digital skills for first-year students at an economic university. In: Bylieva, D., Nordmann, A., Shipunova, O., Volkova, V. (eds.) PCSF/CSIS -2020. LNNS, vol. 184, pp. 230–244. Springer, Cham (2021). https://doi.org/10.1007/978-3-030-65857-1_20

8. Bylieva, D., Lobatyuk, V., Safonova, A., Rubtsova, A.: Correlation between the practical aspect of the course and the e-learning progress. Educ. Sci. **9**(3), 167 (2019). https://doi.org/10.3390/educsci9030167

9. Ganina, E.V., Dubinina, G.A., Stepanyan, I.K.: Cross-cultural analysis of bilingualism in preparing international students for mastering professionally oriented disciplines. Human. Soc. Sci. J. **9**(4), 142–147 (2019). https://doi.org/10.26794/2226-7867-2019-9-4-142-147

10. Seckin, E.: Code switching: definition, types, and examples (2019). https://owlcation.com/humanities/Code-Switching-Definition-Types-and-Examples-of-Code-Switching. Accessed 7 June 2021

11. Ting, Y.L.T.: CLIL … not only not immersion but also more than the sum of its parts. Eng. Lang. Teach. J. **65**(3), 314–317 (2011). https://doi.org/10.1093/elt/ccr026

12. Goldfarb, V., Krylov, E., Perminova, O., Barmina, N., Vasiliev, L.: Aspects of teaching "advanced gears" for future mechanical engineers within "bachelor of sciences" programs at technical universities. In: Goldfarb, V., Trubachev, E., Barmina, N. (eds.) Advanced Gear Engineering. MMS, vol. 51, pp. 271–287. Springer, Cham (2018). https://doi.org/10.1007/978-3-319-60399-5_13

13. Tareva, E.G., Tarev, B.V.: The assessment of students' professional communicative competence: new challenges and possible solutions. XLinguae. **11**(2), 758–767 (2018). https://doi.org/10.18355/XL.2018.11.02.59

14. Isurin, L., Winford, D., Bot, K.: Multidisciplinary Approaches to Code Switching. John Benjamins Publishing Company, Philadelphia (2009). https://doi.org/10.1075/sibil.41

15. Zadadaev, S.A.: Mathematics in R. Financial University under the Government of the Russian Federation, Moscow (2018)

16. Harris, A.: Research. In: Creativity and Education. Creativity, Education and the Arts, pp. 1–14. Palgrave Macmillan, London (2016). https://doi.org/10.1057/978-1-137-57224-0_1

17. Rzheutskaya, S.Y., Kharina, M. V.: Interdisciplinary cooperation in the integrated information learning environment of technical university. Open Educ. **2**, 21–28 (2017). https://doi.org/10.21686/1818-4243-2017-2-21-28

18. Koptelova, T.I.: Integration of knowledge and interdisciplinary cooperation in solving agro-ecological issues (the case of Nizhny Novgorod State Agricultural Academy). Agricultural **4**, 10–22 (2019). https://doi.org/10.7256/2453-8809.2019.4.30191

19. Kobeleva, E.P.: Teaching foreign languages at universities based on cross-disciplinary integration. Intern. J. Humanit. Natur. Sci. **11–2**(38) (2019). https://doi.org/10.24411/2500-1000-2019-11756

20. Khalyapina, L.P., Popova, N.V., Kuznetsova, O.V.: Interdisciplinary design as a means of implementing content and language integrated learning in higher education. Humanit. Soc. Sci. **8**(3), 145–152 (2017). https://doi.org/10.18721/JHSS.8314

Information Technologies for the Training of Teachers in the Creative Professions

Irina Gorbunova$^{(\boxtimes)}$ (iD) and Anastasia Pankova (iD)

Herzen State Pedagogical University of Russia, 48 Moyka Embankment,
St. Petersburg 191186, Russia
gorbunovaib@herzen.spb.ru

Abstract. The issues considered in the article actualize the problems associated with the characteristic features of teaching and educating students in the context of online educational the process using e-learning systems that are typical for teaching humanities. The authors identify the most appropriate approaches to the formation of educational programs in this area using distance learning technologies. The authors of the article rely on the characteristic educational elements of the Moodle distance learning system, which allows organizing the training of students as future teachers, taking into account the fact that their professional activity is associated with teaching various creative disciplines for children studying in preschool educational institutions and in the lower grades of secondary schools, lyceums, gymnasiums. The article is also devoted to the problems of teaching music informatics, which performs a limiting role and fills up the content components of education in the process of introducing distance learning forms into the pedagogical practice of music teachers.

Keywords: Creative approach · E-learning · Information technology · Music computer technologies

1 Introduction

1.1 The Discipline "Information Technologies in the Artistic and Aesthetic Education of Children"

Module "Information Technologies in the Artistic and Aesthetic Education of Children" is a professional, systematic discipline of the main educational program for the preparation of a bachelor's degree (academic bachelor's degree program) in the direction - Pedagogical education of the training profile "Supplementary Artistic and Aesthetic education of children." At the Herzen State Pedagogical University of Russia, this discipline is studied by students of the Institute of Childhood, combining the preschool department and the primary school department, throughout a year.

The program "Information Technologies in the Artistic and Aesthetic Education of Children" is aimed at the formation of information culture and professional competence of the future specialist, a teacher of supplementary education in the field of art informatics and electronic communication technologies. The module includes the following disciplines: "Fundamentals of Artistic Informatics" and "Music Computer Technologies in the Supplementary Artistic and Aesthetic Education of children." The content of the disciplines provides students with the mastery of information technology as phenomena of culture, art and education, modern electronic communication systems, Internet technologies, information technology in various artistic activities, music and computer technologies in teaching music and computer science. A special emphasis is placed on students' mastering the basis of information technology design in the system of supplementary artistic and aesthetic education.

Training under this module is conducted throughout a year in 7–8 semesters (IV year), when students have already mastered the basic and professional disciplines included in the training program for bachelors specializing in the field of "Supplementary Artistic and Aesthetic Education of Children" within the educational field of "Pedagogical Education."

1.2   The Objective and Expected Accomplishments

The objective and expected accomplishments of the study of the educational program (or vocational-educational module for the preparation of bachelors of pedagogical education) "Information Technologies in the Artistic and Aesthetic Education of Children" are to develop students' competencies necessary to solve problems of development and education of children of preschool and primary school age:

a)   soft skills (SS):

- ready to maintain a level of physical fitness that ensures full-fledged activity (SS-8);
- is able to use first aid techniques, methods of protection in emergency situations (SS-9)

b)   professional (PC):

- is able to use the possibilities of the educational environment, including information, to achieve personal, metadisciplinary and disciplinary results of training and ensure the quality of the educational process by means of the subject taught (PC-4);
- willingness to use systematic theoretical and practical knowledge to set and solve research problems in the field of education (PC-11).

2   Materials and Methods

The overall conclusion on the curricula analysis of "Musical Informatics" of primary, secondary and higher-level institutions [1–5]. In connection with the development of MCT and digital musical instruments [6], the discipline "Musical Informatics" requires a change in content, since new requirements are placed on the music teacher, namely, the ability to conduct educational activities using digital technologies.

The thematic plan should be aimed at developing skills in working with information and communication technologies and motivate the music teacher to independently obtain knowledge in this field using MCT and a musical synthesizer.

3   Research and Discussion

3.1   The Purpose of the Research

Training in the module "Information Technologies in the Artistic and Aesthetic Education of Children" is conducted in the 4th year, when students have already mastered the basic and professional disciplines included in the bachelor's degree program in the specialty "Artistic and Aesthetic Education of children" within the educational direction "Pedagogical Education" (Table 1).

Table 1. Labour input and assessment in the module

Discipline (syllabuses)/semester	Labour input				Teaching load, academic hours			Final assessment/semester
	Course credits in total/Including examinations	Academic hours for theoretical training	Including		Lectures	Workshops	Practicums	
			Teaching load	Independent study				
Fundamentals of Artistic Informatics	4	144	30	114	8	22	–	Pass-fail test
Music and computer technologies in supplementary artistic and aesthetic education of children	4	144	24	120	6	18	–	Pass-fail test
In the module	9/1	324	54	270	10	40	–	Exam

3.2   Logistics for Discipline

Logistics for discipline:

1. Lecture classes:

    a. electronic multimedia presentation kit,
    b. an auditorium equipped with presentation apparatus (multimedia projector, screen, computer/laptop or interactive board)

2. Workshops:

    a. computer class with the Internet access,
    b. presentation apparatus (multimedia projector, screen, computer/laptop or interactive board),
    c. general purpose software packages (text editors, graphic editors, music editors, etc.).

3.3 Final Assessment in the Module of «Information Technology in the Artistic and Aesthetic Education of Children»

Final assessment in the educational module "Information Technologies in the Artistic and Aesthetic Education of Children" is carried out in the form of a cumulative grading system in each discipline: "Fundamentals of Artistic Informatics" and "Music Computer Technologies in the Supplementary Artistic and Aesthetic Education of Children."

At the same time, the assessment of competencies formed in the process of training students under the program "Information Technologies in the Artistic and Aesthetic Education of Children" is carried out based on the criteria for evaluating individual tasks formulated in the technological charts of disciplines (presented in the MOODLE LMS).

Implemented in the form of a pass-fail test, which includes two components:

1) defence of performed individual assignments;
2) presenting and defence of the topic of independent study with online support forms, individually or in a micro-group.

The student presents a network portfolio of practical assignments performed in the course of studying the discipline and the result of independent activities to master the issues and the topics formulated by the teacher; presents the topic and the result of their independent study with an electronic or network form of support (research topic, research problem, description of the content of the problem, ways to solve the problem, annotated catalogue of literary sources on the research problem and the Internet resources), as well as answers the questions on the theoretical and practical material of the course.

3.4 Assessment of Competencies Formed by the Module

Final assessment in the module "Information Technologies in the Artistic and Aesthetic Education of Children" is conducted in the form of an exam and includes:

1. Answer to the question of final assessment (in the discipline "Fundamentals of Artistic Informatics").
2. Answer to the question of final assessment (in the discipline "Music Computer Technologies in Supplementary Artistic and Aesthetic Education of Children").
3. Defence of the previously created information and educational project (Table 2).

In the 2019–2020 academic year, the final assessment in the discipline "Information Technologies in the Artistic and Aesthetic Education of Children" was carried out online. It is especially worth noting that the educational module B.1.21. "Information Technologies in the Artistic and Aesthetic Education of children," consisting of the disciplines "Fundamentals of Artistic Informatics" and "Music Computer Technologies in the Supplementary Artistic and Aesthetic Education of children," is implemented at the Institute of Childhood of the Herzen State Pedagogical University of Russia in the system of correspondence training. The bulk, the vast majority of the students, already work in the specialty as primary school teachers, kindergarten assistants and in other educational

Table 2. Assessment of competencies formed by the module

Competency	Control and measuring materials for assessment of competency formation
(SS – 8)	Within the framework of the cumulative grading system, the competency formation is checked: in the course of practical tasks and independent study, while answering to theoretical questions, based on the design results of the information educational environment
(SS – 9)	Within the framework of the cumulative grading system, the competency formation is checked: in the course of practical tasks and independent study, while answering to theoretical questions, based on the design results of the information educational environment
(PC – 4)	Within the framework of the cumulative grading system, the competency formation is checked: in the course of practical tasks and independent study, while answering to theoretical questions, based on the design results of the information educational environment
(PC – 11)	Within the framework of the cumulative grading system, the competency formation is checked: in the course of practical tasks and independent study, while answering to theoretical questions, based on the design results of the information educational environment

institutions that carry out the educational process related to the education and upbringing of children of preschool and primary school age. Therefore, it was difficult to implement the educational process under the program "Information Technologies in the Artistic and Aesthetic Education of Children" in the context of a sudden coronavirus pandemic and the quarantine with the entire educational process implemented online with the help of the MOODLE LMS: correspondence students master educational programs for about a month and a half, the schedule includes almost daily classes in a number of educational disciplines. Often, these classes were of a daily nature of interaction between the teacher and the students and provided for both the design of each lecture and each workshop, and checking of the assignments completed by the students. The teaching load, which included the preparation and conduct of classes, as well as the checking of completed assignments - for teachers, and, accordingly, the comprehension of new knowledge and the fulfillment of assigned educational tasks, (the performance of practical tasks and tests, independent study and much more) for students, was enormous. It was this circumstance that prompted us, teachers, to very meticulously and qualitatively design each educational element of each educational program and educational module and its constituent disciplines for the most complete presentation of educational materials in the Moodle LMS.

3.5 The Discipline "Musical Informatics"

The discipline "Musical Informatics" in Russia has been gradually introduced into the educational institutions of the country since the late 1990s–early 2000s, but, unfortunately, there has not yet been an orderly system of teaching it, which is influenced by the following factors: lack of technical and teaching facilities in each individual educational institution (availability of an equipped class, which is fitted with specialized music computers with professional software and digital instruments), different level of knowledge in the field of music informatics among teachers of this discipline. The first education and methods laboratory "Music Computer Technologies" in the Russian Federation was established in 2002 in the Herzen State Pedagogical University of Russia The symbiosis of musicians of various specialties, mathematicians and programmers, who work in the EML "Music Computer Technologies" in close cooperation, helped to form the content of the discipline "Musical Informatics."

It can also be noted that there was a need to create a methodology aimed at overcoming formalism in teaching music informatics using the capabilities of modern music computer technologies (MCTs) [7].

3.6 The Exam

The exam in the discipline "Information Technologies in the Artistic and Aesthetic Education of Children" was held in the spring of 2020 in a distance format. For the students, a special forum form was created in order to have the possibility of posting a video of their examination speech (which was prepared in advance and reflected the topics set by the examiner related to the nature of the professional activities of each particular student). In this forum, it was necessary to post the information important for the exam response. To do this, it was essential:

create a new theme;
present a brief annotation in the contents;
prepare an audio tale (it is developed by the students during the course of the program throughout the semester);
attach the answer to the question (the abstract), give the link on the site and the audio tale;
attach the video with the examination speech.

The video should have been recorded using a program that captures a screen image (for example, Bandicam).
The necessary content of the video:

1. Give the full name, indicate the year, the faculty the group, etc.
2. Name the examination question number (on the topic of which the abstract was prepared), talk about the relevance and, briefly, about its content.
3. Show your developed site. Tell about it in free form (for whom, why exactly such an idea arose to create a site, what is special in it, etc.)
4. Show the audio tale. Tell about it in free form.

In general, the video can be of any duration, but, preferably - up to 10 min. This material was proposed to be placed on a remote resource, specifying only the address of the link in the response.

The work on the preparation and conduct of the exam was accompanied by the comments that helped the students in the presentation and placement of the necessary materials. The comments were accompanying, guiding and supportive, for example: "Please be careful about the comments that accompany the forms for posting your exam answers," "First of all, it is necessary to note your presence at the exam. To do this, please go to the section "Registration of those present in online classes" (located below) and mark "I am present," "Next, post the material in TWO forms: Exam answer (for rating); Exam response (to provide material and oral response)," After posting, please stay on the exam page (until it ends) for further interaction with the examination board" and much, much more.

It should be noted that learning in a remote educational process using e-learning systems with using MCTs, characteristic of teaching humanities [7], can be effectively organized if the teacher thinks through and prepares classes, carefully working out various educational elements of the remote education system (for example, LMS Moodle), moreover, the more educational elements of LMS are involved in the educational process, the more effective the student learning process will be. Realizing that such preparation for classes requires a great additional effort on the part of the teacher, we note, however, that only this approach allows organizing student education - future teachers whose professional activities are related to teaching various creative disciplines to children studying in preschool educational institutions and pupils of primary school age, whose education takes place within the framework of general schools, lyceums, gymnasiums, in the conditions of the functioning of the electronic educational process effectively. So we prepared in advance the educational materials necessary to study the disciplines included in the module "Information Technologies in the Artistic and Aesthetic Education of Children." To teach the discipline "Music Computer Technologies in the Supplementary Artistic and Aesthetic Education of Children," lecture educational materials were formed based on our work [8–10], practical tasks and tests, as well as the preparation of mini-studies on a given topic involved the use of textbooks [11–15]. Training in the discipline "Fundamentals of Artistic Informatics" was based on the consideration of a number of fundamental methodological ideas: assessment of m-service acceptance in educational context [16, 17], authentic learning and digital technology in the music classroom [18–20], the integration of information in a digital, multi-modal learning environment [21–23]. Construction of a course on methodological support of the educational process developed by the authors with the participation of the employees of the education and methods laboratory (EML) "Music Computer Technologies" at the Herzen State Pedagogical University of Russia. The approaches developed in EML "Music Computer Technologies" are most important for their implementation in the system of contemporary inclusive musical education [24–27].

In the "Musical Informatics" course of the study, there was identified the thematic plan for training in music informatics containing the following topics:

- Topic 1. The subject of musical informatics
- Topic 2. Music, mathematics, computer science: facets of interaction.
- Topic 3. Architectonics of acoustic and digital musical sound.
- Topic 4. Music synthesizers. from the history of the development of musical synthesizers.
- Topic 5. Sound synthesis technologies.
- Topic 6. musical computer.

- Topic 7. Digital musical synthesizer as a modern software and hardware complex for teaching musical informatics.
- Topic 8. Software for professional activities of the musician.
- Topic 9. Online services to help a teacher-musician. (See: [28]).

When forming the thematic content of the "Musical Informatics" discipline, the following professional activities of music teachers, and set certain tasks were taken into account:

- increase interest in the use of MCT and electronic musical synthesizer (EMS);
- increase the competency (according to professional standards), focus on the formation of its own digital learning environment using MCTs and EMS, focusing on self-education, as the main element of professional development.

4 Conclusion

As the situation has shown, teachers-musicians lack knowledge to organize remote forms of education and form a digital educational environment, although at the moment many resources have appeared on the Internet, which can be additional tools for pedagogical activities. Software applications online, which can be used in their work by teacher-musicians to process musical information (make a voice recording or an instrument using a microphone, process this recording, save and use it if necessary, etc.) remain unclaimed, since teacher-musicians do not assume that these resources are available. If earlier you had to contact special studios to create an arrangement of a musical work for a vocalist or instrumentalist, now every musician who has gained knowledge in the field of musical informatics has learned to work using music computer technologies and digital musical instruments.

A lot of events take place using remote forms - these are online competitions, forums, festivals, teacher training, etc. The situation forces the teacher-musician to know modern digital technologies and techniques for organizing distance learning using various methods of network interaction. When applying a digital educational environment, the content of participants of network interaction is formed, which is aimed at obtaining new knowledge, skills and skills using MCT, EMS.

Thanks to the use of the basic principles of the formation of post-material values in the conditions of transversality of educational paradigms [29], and also taking into account the principles of attitude to values in the practice of social management and communication [30], it was also possible to study a number of psychological and pedagogical features of the development of information technologies and MCT by people with deep visual impairments.

Classical musical education can be effectively supplemented with digital instruments and modern techniques that help realize musical education. Contemporary musicians need knowledge in the field of musical informatics in order to be competitive in modern conditions, to be able to work with musical information.

In conclusion, Built on the basis of a constructivist approach using the methods of context and design training, it was possible to form an effective educational methodology of a creative educational process, when the tutor and the trainee, using remote educational technologies and the advantages of e-learning, could be in continuous interaction, which, as practice has shown, is especially relevant when implementing correspondence forms of training in the course of preparation of specialists for the system of contemporary musical education.

References

1. Ananyev, A.N.: Musical Informatics: the Curriculum of the Discipline. Publishing House of the Scientific and Publishing Center "Moscow Conservatory", Moscow (2020). (in Russian)
2. Korolev, A.A.: Musical and Computer Dictionary. Publishing House "Kompozitor", St. Petersburg (2000). (in Russian)
3. Polozov, S.P.: Training Computer Technologies and Music Education. Publishing House "University Press", Saratov (2002). (in Russian)
4. Rags, Yu.N.: About the course "musical informatics". In: Gorbunova, I.B. (ed.) Contemporary Musical Education – 2003. Proceedings of the 2nd International Research and Practical Conference, pp. 120–125. Publishing House of the Herzen State Pedagogical University of Russia, St. Petersburg (2003). (in Russian)
5. Karuto, A.V.: Musical Informatics. Theoretical Foundations. Publishing House of the Scientific and Publishing Center "Moscow Conservatory", Moscow (2015). (in Russian)
6. Gorbunova, I.B.: Electronic musical instruments: to the problem of formation of performance mastery. In: 16th International Conference on Literature, Languages, Humanities & Social Sciences (LLHSS-18), pp. 23–28. International Association of Humanities, Social Sciences & Management Researchers, Budapest (2018). https://doi.org/10.17758/URUAE4. UH10184023
7. Gorbunova, I.B.: Music computer technologies in the perspective of digital humanities, arts, and researches. Opcion 35(24), 360–375 (2019)
8. Gorbunova, I.B., Chibirev, S.V.: Modeling the process of musical creativity. Opción Año 35(22), 392–409 (2019)
9. Gorbunova, I.B., Pankova, A.A.: Teaching computer science and information technology studies for students of musical and pedagogical specialties. Educacao & Formacao 5(3), 1–17 (2020). https://doi.org/10.25053/redufor.v5i15set/dez.3350
10. Gorbunova, I.B., Plotnikov, K.Y.: Music-related educational project for contemporary general music education of school children. Int. J. Innov. Creat. Change 12(2), 451–468 (2020)
11. Bauer, W.I.: Music Learning Today: Digital Pedagogy for Creating, Performing, and Responding to Music. Oxford Scholarship Online, Oxford (2014). https://doi.org/10.1093/acprof:oso/9780199890590.001.0001
12. Age-Psychological and Psychological-Pedagogical Approaches to Ensuring Children's Information Security. https://rkn.gov.ru/docs/Razdel_9.pdf. Accessed 20 May 2021
13. Liu, M., Su, S., Liu, S., Harron, J., Fickert, C., Sherman, B.: Exploring 3D immersive and interactive technology for designing educational learning experiences. In: Neto, de Souza, R., Gomes, A.S. (eds.) Handbook of Research on 3-D Virtual Environments and Hypermedia for Ubiquitous Learning, pp. 243–259. IGI Global, Hershey (2016). https://doi.org/10.4018/978-1-5225-5469-1.ch051
14. Music Vision International LLC. http://www.softmozart.com. Accessed 20 May 2021
15. Petrovic-Dzerdz, M., Trépanier, A.: Online hunting, gathering and sharing – a return to experiential learning in a digital age. Int. Rev. Res. Open Distrib. Learn. 19(2), (2018). https://doi.org/10.19173/irrodl.v19i2.3732

16. Alasmri, M., Onn, W.C., Hin, H.S.: Social networking framework for learning motivation. J. Southwest Jiaotong Univ. **54**(6), (2019). http://jsju.org/index.php/journal/article/view/450

17. Al-Shaher, M.A.: Assessment of M-service acceptance in educational context. J. Southwest Jiaotong Univ. **54**(5), (2019). https://doi.org/10.35741/issn.0258-2724.54.5.8

18. Calderón-Garrido, D., Cisneros, P., García, I.D., Fernández, D., de las Heras-Fernández, R.: La tecnología digital en la Educación Musical: una Revisión de la Literatura Científica. Revista Electrónica Complutense de Investigación en Educación Musical **16**, 43–55 (2019). https://doi.org/10.5209/reciem.60768

19. Crawford, R.: Authentic learning and digital technology in the music classroom. Monash University, Victoria [Unpublished Ph.D. dissertation] (2007). https://trove.nla.gov.au/work/35149324?q&versionId=43654170. Accessed 15 May 2021

20. Cremata, R., Powell, B.: Digitally mediated keyboard learning: speed of mastery, level of retention and student perspectives. J. Music Technol. Educ. **9**(2), 145–159 (2016). https://doi.org/10.1386/jmte.9.2.145_1

21. Losada, M., Heaphy, E.: The role of positivity and connectivity in the performance of business teams. Am. Behav. Sci. **47**(6), 740–765 (2004)

22. Schüler, A.: The integration of information in a digital, multi-modal learning environment. Learn. Instr. **59**, 76–87 (2019). https://doi.org/10.1016/j.learninstruc.2017.12.005

23. Stensæth, K.: Music therapy and interactive musical media in the future: reflections on the subject-object interaction. Nord. J. Music Ther. **27**(4), 312–327 (2018). https://doi.org/10.1080/08098131.2018.1439085

24. Gorbunova, I., Govorova, A.: Music computer technologies in informatics and music studies at schools for children with deep visual impairments: from the experience. In: Pozdniakov, S., Dagienė, V. (eds.) Informatics in Schools. Fundamentals of Computer Science and Software Engineering. ISSEP 2018. LNCS, vol. 11169, pp. 381–389 (2018). https://doi.org/10.1007/978-3-030-02750-629

25. Hernandez-Ruiz, E.: How is music processed? Tentative answers from cognitive neuroscience. Nord. J. Music Ther. **28**(4), 315–332 (2019). https://doi.org/10.1080/08098131.2019.1587785

26. Jacko, V.A., Choi, J.H., Carballo, A., Charlson, B., Moore, J.E.: A new synthesis of sound and tactile music code instruction in a pilot online braille music curriculum. J. Visual Impair. Blind. **109**(2), 153–157 (2015). https://doi.org/10.1177/0145482x1510900212

27. Gorbunova, I.B., Voronov, A.M.: Music computer technologies in computer science and music studies at schools for children with deep visual impairment. In: 16th International Conference on Literature, Languages, Humanities & Social Research (LLHSS-18). International Conference Proceedings, pp. 15–18. International Association of Humanities, Social Sciences & Management Researchers, Budapest (2018). https://doi.org/10.17758/URUAE4.UH10184021

28. Gorbunova, I.B., Kameris, A., Bazhukova, E.N.: Musical informatics course for musicians with using music computer technologies. Int. J. Recent Technol. Eng. **8**(6), 3040–3045 (2020). https://doi.org/10.35940/ijrte.F7723.038620

29. Romanenko, I.B., Romanenko, Y.M., Voskresenskiy, A.A.: Formation of post material values in conditions of transversality of educational paradigms. Eur. Proc. Soc. Behav. Sci. **35**, 1116–1121 (2018). https://doi.org/10.15405/epsbs.2018.02.131

30. Shipunova, O.D.: The role of relation to values principle in the social management practices. The existential-communicatory aspect. Middle-East J. Sci. Res. **19**(4), 565–569 (2014). https://doi.org/10.5829/idosi.mejCsr.2014.19.4.123684

Challenges of Professional Adaptation: Difference Between Novice University Teachers with and Without Pedagogical Training

Irina Karpovich, Larisa Voronova(✉), Tatiana Ivanova, and Tatiana Krepkaia

Peter the Great St. Petersburg Polytechnic University, St. Petersburg 195251, Russia
karpovich.ia@flspbgpu.ru

Abstract. Novice teachers face challenges in the process of adaptation to professional activity. These problems can be caused by a heavy workload, lack of confidence in the level of professional training, lack of creativity and teaching skills, and the need to cope with a large amount of information that is not directly related to the preparation and conduct of classes. All these difficulties entail an unfavourable balance between work and personal life. The aim of the study is to identify the challenges novice university teachers face in the beginning of their career and compare the frequency of the detected problems within the groups of respondents with and without pedagogical training. 93 novice educators of Peter the Great St. Petersburg Polytechnic University took part in the series of surveys. The researchers applied the quantitative method to collect data using a questionnaire. The study involved quantitative data analysis containing the descriptive statistical analysis and the correlation analysis. It was found out that novice teachers face problems in professional, psychological, and motivational spheres of adaptation. The difference between the values of novice teachers with and without a teaching degree is, in most cases, statistically significant. It makes it possible to conclude that university teachers without pedagogical education need support via additional educational programmes. The research is particularly relevant for improving the professional culture of specialists in the field of education and contributes to the solution of the professional adaptation problem faced by novice educators.

Keywords: Career challenges · Novice university teachers · Professional adaptation · Postgraduate students · Pedagogical education

1 Introduction

A teacher is a person most directly responsible for the learning process and student progress that is why the quality of teaching staff is of great importance for any educational institution. Solving the problems teachers face in the early stages of their careers is an urgent task for education authorities and the pedagogical community [1, 2]. To pay attention to human resource policies and career development is essential to improve novice teachers' transition into the profession [3]. It is an important process of acculturating teachers to their new careers and helping them overcome the hardships of teaching and the accreditation process.

Difficulties novice teachers face at the initial stage of their professional activities can increase the general tension, lead to emotional burnout, and result in intentions to give up teaching. Solving this problem is regarded as an urgent task both for education authorities and the pedagogical community. Recent research shows that 24% of novice teachers leave teaching within the first year, 33% drop out after three years, and between 40% and 50% leave within the first five years [4].

The process of novice teacher self-adjustment is quite problematic. Most novice teachers experience professional, psychological and motivational challenges in adjusting to a new working environment as they have difficulty in arranging and implementing the lesson plan, motivating the students, interacting with colleagues, managing the time and workload, and controlling their emotions [5–7]. These problems may result in work-related stress at the beginning of an instructor's teaching contract and affect not only the quality of teaching but also the teachers' motivation and perception of the programme [8]. Psychological difficulties such as low self-efficacy and burnout are also the consequences of stress experienced by beginning teachers. Results of the study focused on teacher self-efficacy show its positive links with patterns of teacher behaviour, teachers' psychological well-being, including personal accomplishment, job satisfaction and commitment [9, 10]. There is also reverse relationship between self-efficacy and burnout. Findings suggest that teachers who have higher self-efficacy are more likely to have lower burnout at work [11].

Novice university teachers often experience anxiety associated with establishing interpersonal contacts both with colleagues and students. It sometimes manifests itself in the fear of colleagues' opinions on the level of their professional expertise and inability to establish an optimal distance with students, maintain discipline and carry out an objective assessment of students [12, 13]. The need to demonstrate their capabilities, high level of anxiety, lack of confidence in their abilities and weak motivation for self-development and self-improvement, due to a limited amount of time can cause negative emotional states experienced by novice university educators.

Much real insight into the issue of teacher self-adjustment in the beginning of their career lies in comparisons of experienced and novice teachers. There have been several attempts to identify the qualities of expert teachers that separate them from novice teachers. According to the study carried out by D. Westerman (1991) focused on thinking and decision making of expert and novice teachers expert teachers think about learning from the perspective of the student and perform a cognitive analysis of each learning task during planning which they adapt to the needs of students. In contrast, novices use specific lesson objectives to form structured lesson plans that they do not adapt to meet student needs [14].

A more recent study showed similar results. There are three dimensions of differences between expert and non-expert teachers: their ability to integrate aspects of teacher knowledge concerning the teaching act; their response to their contexts of work, and their ability to engage in reflection and conscious deliberation [15]. Another important skill that poses a considerable challenge for beginning teachers is classroom management.

Experience in the classroom influences how teachers interpret problematic classroom events. This includes how they monitor events, how they maintain an ongoing awareness of classroom situations, and how they act in response to events [16]. Findings suggest that novices' interpretations focus on issues of behaviour and discipline. Experts focus on student learning and the teacher's ability to influence learning [17].

State education authorities and teachers around the world concur that inefficient classroom management skills are one of the main problems for teacher retention and effective teaching [18]. Educators who possess creative ability to solve various problems in the classroom are considered to be more successful in teaching and are less likely to leave the profession [19–21]. Researchers found that there is a positive correlation (r = .64) between creative thinking and teacher effectiveness to solve realistic classroom problems [22]. Another research suggests that teacher creativity is useful for generating innovative curriculum and creative activities that will help students learn the content of subjects as expected by state guidelines [23]. Teachers develop their creative problem solving skills during years of practice and this involves learning to combine academic subject knowledge with teaching skills and applying this new knowledge to new conditions that occur every day [24].

The research covering the issue of creativity in teaching states that effective teachers are characterized by using creativity in their teaching practices. Creative teachers have a wide range of methods and strategies to call upon and they are ready to deviate from established procedures and use their own solutions. The researcher posits that novice educators are generally much less prone to creativity than experienced teachers due to the fact that they are knowledgeable about fewer strategies and techniques [25].

Researchers offer different strategies for university teachers training and teacher development programmes to improve the adaptation process. They put much emphasis on the benefits of mentoring of novice teachers, which can increase the rate of novice teacher retention in the teaching profession [26]. New teachers see mentoring as a valuable feature since an expert teacher gives a strong start to a young teacher, sharing not only his classroom and extracurricular experience but also skills and attitudes [27, 28]. Mentors support the teachers in the construction of their professionalism. Doing so requires the right blend of professional, social, and personal support, and depends on a high level of awareness and expertise on the part of the mentor [29].

Beginning teachers, irrespective of their field, have a period of adjustment to go through [30].The research focused on distinctive qualities of language teachers indicates that novice language teachers are less skilled than expert teachers at: thinking about the subject matter from the learner's perspective; having a deep understanding of the subject matter; knowing how to present subject matter in appropriate ways; knowing how to integrate language learning with broader curricular goals [31, 32] and managing behaviour, dealing with unmotivated students and students with learning difficulties [33]. The following features have been identified for science and mathematics novice teachers. Novices are less efficient in planning, encounter problems when trying to be responsive to student needs, stick to their plans regardless of what happens during the actual teaching of the lesson [34, 35].

One can observe this tendency within the groups of beginner specialists who have undergone their professional training and received a degree in Pedagogy [36]. Results of extensive research show that novice teachers of physics; chemistry; physical science; geometry; and biology exhibited steeper growth in effectiveness than novice non–science; technology; engineering; and mathematics teachers did [37]. It means that the pattern of adaptation and the problems novice university educators face can differ depending on whether they have a teaching degree or not and needs further research.

To develop scientific and methodological guidelines on the organisation of pedagogical support, it is necessary to assess the level of readiness for the professional activities of young teachers and postgraduate students without pedagogical education who teach various disciplines at SPbPU.

The focus of our study is the adaptation of novice university teachers with and without pedagogical education. This process has some peculiarities. The problems that novice teachers face at the beginning of their career occur since a lot of university teachers are experts only in their disciplines. Still, when it comes to pedagogy, these teachers are more or less novices because most university teachers have limited knowledge of pedagogical theories and educational sciences [38]. University teachers' current training is primarily scientific and technical and has gaps in the teacher training required today for more effective work in the classroom [39]. Providing novice university educators with pedagogic teaching knowledge is described by researchers as an effective tool for professional development. Teachers reported positive effects from the programmes such as self-confidence, increased reflective skills, and improved knowledge of teaching and learning [40, 41]. Participating in the programmes beginning teachers have the best opportunity to become highly qualified career professionals [42]. As novices gain education and experience in teaching, they observe an increase in quantity and complexity of teaching practices, including content taught and learned, pedagogical processes used and experienced, and basic educational purposes [43].

Evidently, there is a need for well-designed teacher development programmes, however they should be based on extensive research of difficulties novice teachers with and without educational training face. Therefore, the aim of this study is to determine problems novice university teachers without pedagogical education face in professional, psychological, and motivational spheres of adaptation and compare the frequency of the detected problems in these spheres within the groups of novice university teachers with and without pedagogical education.

RQ. Do the problems novice teachers face in professional, psychological, and motivational spheres of adaptation remain the same within the groups of university educators with and without educational training?

2   Methods

The researchers adopted an exploratory, inductive approach in this study aimed at:

1) determining challanges novice university teachers without pedagogical education face at the initial period of their career;
2) comparing the detected problems within the group of novice university teachers with and without pedagogical education.

The researchers applied the quantitative method to collect data using a questionnaire. The questionnaire, we believe, is the most appropriate instrument for the present study, which will ensure insights into the problem, maximise the use of respondents' time, and facilitate the data analysis. The study involved the quantitative data analysis containing the descriptive statistical analysis and the correlation analysis, which provided the researchers with new insights and detailed results. The study consisted of two stages. At the first stage, the researchers identified the problems of adaptation that novice teachers without pedagogical education face. They also evaluated their level of adjustment in the professional, psychological, and motivational spheres. At the second stage, the researchers compared the results obtained in the first stage of the study with the problems that novice teachers with pedagogical education face in professional, psychological, and motivational spheres of adaptation. They acquired these findings from the study carried out in the fall semester of the 2018–2019 academic year [7]. The obtained results made it possible to make recommendations on how to facilitate the process of self-adjustment of the novice educators without pedagogical education.

The study took place at Peter the Great Saint Petersburg Polytechnic University, Russia. The survey involved 53 young educators of engineering departments without pedagogical education and with the work experience in an educational institution less than five years. They teach science disciplines in the field of mathematics, physics, engineering, and technology at the Institute of Computer Science and Technology, the Institute of Metallurgy, Mechanical Engineering and Transport, and the Institute of Applied Mathematics and Mechanics. These young educators are mainly postgraduate students enrolled in PhD programmes at SPbPU and teach classes within the framework of their studies. Of those surveyed, 45 participants (85%) are postgraduate students and 8 participants (15%) are young teachers who already completed their postgraduate education and continued to teach at the university. All respondents volunteered to take part in the survey. The authors informed them about the objective of the study and guaranteed anonymity.

We conducted Stage 1 of the study in February 2020. To address the research question stated above, the researchers used a questionnaire to determine the problems novice educators without pedagogical education face in professional, psychological, and motivational spheres of adaptation. It was the same questionnaire that the authors had applied in the study carried out in 2018–2019 among novice teachers with pedagogical education. The questionnaire consisted of closed-ended multiple-choice questions. Most multiple-choice questions included the option of multiple response and contained the field "Other" so that the respondents could share their views or opinion on the matter in question. An online survey was applied to collect data from the participants using Google Forms, which is an effective and convenient tool for recruiting and processing data. The researchers sent a recruitment email to fifty-three novice and young educators and collected all fifty-three completed sets of responses. The data were coded and entered in SPSS Version 23 (IBM Corp., 2016). Then, researchers carried out the descriptive statistical analysis using frequencies and cross-tabulations to calculate the frequencies in participants' responses, identify the means and standard deviations, and present them in the form of numerical data. The results of Stage 1 allowed the researchers to identify the problems, general patterns, and tendencies associated with the process of novice and young teachers' adaptation.

Stage 2 involved the correlation analysis of the results acquired in Stage 1 and the results obtained from the study carried out in 2018–2019 among 40 novice and young teachers of the Institute of Humanities with working experience of 1–5 years. Of those surveyed, 16 participants (40%) are postgraduate students and 24 participants (60%) are young teachers with teaching experience of 3–5 years. All participants have a pedagogical degree in teaching foreign languages which is their major. Those results revealed the problems the novice teachers with pedagogical education face in professional, psychological, and motivational spheres of adaptation. A series of independent samples t-tests and Fisher's exact test of independence for each questionnaire item were conducted in SPSS to compare the results within the group of novice teachers with and without pedagogical education and identify the relationship between these findings. The t-values with 91 degrees of freedom and the f-test and t-test p-values with the significance level of $\alpha = 0.05$ proved the statistically significant difference between the results of two groups of respondents. Then, the researchers applied the comparative method to analyse the obtained results, which allowed them to make recommendations on how to facilitate the process of self-adjustment of the novice educators without pedagogical education. The results of Stage 2 helped address the research question of the study.

3   Results

The obtained results demonstrate that all novice university teachers who took part in the study face problems of adaptation to professional activity. A comparison of novice university teachers with and without pedagogical education is presented in Table 1.

The results of the t-test and f-test reveal that the difference between the values of novice university teachers with and without pedagogical education is in most cases statistically significant. Although such aspect as "unwillingness to ask for help from more experienced colleagues" is statistically insignificant (f-test p-value $= 0.633$; t-test p-value $= 0.596$; $\alpha = 0.05$). It shows that novice teachers tend to experience apprehension and negative emotions connected to the need for the demonstration of their professional skills. Evidently, this feeling of anxiety is not dependent on novices' educational training.

As we can see from the results presented in Table 1, the majority of novice university educators tend to experience problems entailed by the adaptation process. However, the values differ with regard to whether or not the novice teachers have pedagogical education. Novice university educators with pedagogical education demonstrated better results in professional, psychological, and motivational spheres of adaptation. Possibly, it can be explained by the fact, that the majority of the respondents are postgraduate students enrolled in PhD programmes and teach classes within the framework of their studies. It was not their deliberate choice to become educators and some of them will probably quit teaching as soon as the doctoral study is over.

The most frequent challenges novice without a teaching degree face in professional spheres of adaptation are as follows:

– difficulties with adequate assessment of students (94.3%);

Table 1. Problems that novice and young teachers face in professional, psychological and motivational spheres of adaptation at the initial stage of work.

Problems	Answers (%)			
	Young teachers with pedagogical education (1–5 years)	Young teachers without pedagogical education (1–5 years)	f-test p-value (2-sided) α = 0.05	t-test p-value (2-tailed) df - 91 α = 0.05
Professional spheres of adaptation				
Increased fatigue and decreased performance	41.6%	67.9%	0.02	0.014
Lack of confidence in the level of professional training and creative skills	52.1%	81.1%	0.006	0.006
Problems with students' discipline and managing the classroom activities creatively	34.4%	64.2%	0.007	0.005
Difficulties in working with documentation	52.1%	90.6%	0.000	0.000
Difficulties with adequate assessment of students	75%	94.3%	0.013	0.007
Difficulties in managing the workload	62.5%	83%	0.032	0.025
Psychological spheres of adaptation				
Feeling of anxiety	51%	88.7%	0.000	0.000
Fear of students and difficulties in establishing the optimal distance with them	44.8%	75.5%	0.005	0.002
Fear of communicating with the university administration	37.5%	66%	0.011	0.006

(*continued*)

Table 1. (*continued*)

Problems	Answers (%)			
	Young teachers with pedagogical education (1–5 years)	Young teachers without pedagogical education (1–5 years)	f-test p-value (2-sided) $\alpha = 0.05$	t-test p-value (2-tailed) df - 91 $\alpha = 0.05$
Unwillingness to ask for help from more experienced colleagues	71.9%	77.4%	0.633	0.596
Motivational spheres of adaptation				
Negative change of attitudes towards the chosen profession	18.7%	64.2%	0.000	0.000
Dissatisfaction with the working environment	21.9%	49.1%	0.010	0.009

- difficulties in working with documentation (90.6%);
- difficulties in managing the workload (83%);
- lack of confidence in the level of professional training and creative skills (81.1%);
- increased fatigue and decreased performance (67.9%);
- problems with students' discipline and managing the classroom activities creatively (64.2%).

These challenges are typical for novice university educators. However, novice educators without a teaching degree experience them much more frequently. In some cases, the values are twice as high as those of novice educators with a teaching degree. We can observe a similar trend in the psychological sphere of adaptation. The most common problems experienced by the novice without a teaching degree are:

- the feeling of anxiety (88.7%);
- fear of students and difficulties in establishing the optimal distance with them (75.5%);
- fear of communicating with the university administration (66%).

The last aspect that we took into consideration is the motivational spheres of adaptation. The motivational aspect of adjustment is connected to the positive attitude to work and willingness to carry out professional duties and activities. It includes two items:

- negative change of attitudes towards the chosen profession (64.2%)
- dissatisfaction with the working environment (49.1%).

Evidently, the motivational sphere of adaptation demonstrates better results than professional and psychological ones. Despite the fact, that novice educators experience difficulties at the initial period of adaptation to their profession, half of the respondents are still motivated to continue their work.

4 Discussion

The study detected that at the initial stage of work novice educators face the following problems: bad work-life balance; uncertainty in the level of professional training; the need for a new role distribution and establishment of interpersonal contacts; inability to establish an optimal distance with students and maintain discipline, management of the classroom activities creatively; high level of anxiety and lack of confidence and creative skills; the uncertainty of the career choice motivation, etc. Higher education teachers require confidence and self-efficacy to be more creative and experimental in the classroom. A person's sense of self-efficacy has an effect on their ability to deal with unexpected or challenging situations creatively [44]. It could be supported by the findings of other researchers [7, 44–49], but the rate to which novice educators without a pedagogical degree tend to experience them is much higher. Probably, it can be explained by the fact, that the majority of the respondents did not have any intentions to become teachers. They are postgraduate students enrolled in PhD programmes and teach classes within the framework of their studies.

If novice teachers do not receive appropriate support at the initial period of their career it can lead to the feeling of low self-esteem, personal dissatisfaction, high level of anxiety, increased fatigue and as a result emotional exhaustion and burnout [50, 51]. However, the only item that showed no significant difference between the two groups of respondents is the unwillingness to ask for help from more experienced colleagues. It can be explained by the fact, that novice teachers tend to experience apprehension and negative emotions connected to the need for demonstration of their professional skills. We can conclude that even if novice educators do not ask for help, they still need support and advice. These results support conclusions made by researchers [50, 52, 53].

The results of the study detected the most acute problems faced by novice teachers without pedagogical education. Firstly, it is a negative change of attitudes towards the chosen profession. It supports the research conclusion made by T. S. C. Farrell (2016). According to it, 24% of novice teachers leave teaching within the first year, 33% drop out after three years, and between 40% and 50% leave within the first five years [4]. A. Eteläpelto et al. (2015) examined novice teachers' perceptions of their professional agency during the initial years of work. The study showed that most of the educators who took part in the experiment felt that they would have to renegotiate their professional identities and re-assess their professional ideals and ethical standards [48].

The second problem is the lack of confidence in the level of professional training and creative abilities. This result supports the conclusions made by H. Meyer (2004). Expert teachers have a more complex conception of prior knowledge and make use of their students' prior knowledge in significantly better ways than their novice colleagues [54]. Moreover, novice university teachers often lack competence in providing learners with psychological support [48].

The third challenge novice educators have to face is workload manageability. It is stated as an important problem by E. A. Bettini (2018) - if a novice teacher regards workload as less manageable, he or she is more likely to experience emotional exhaustion and less likely to intend to continue teaching [55].

The next problem is the high level of anxiety. It coincides with conclusions drawn by T. S. C. Farrell (2016) and G. T. Henry et al. (2012) [4, 37]. This state can be explained by the fear of being appraised or criticised by more experienced colleagues. In addition to receiving a lower level of professional support from their superiors, novice teachers are found to lack ways to articulate their own needs to colleagues [56].

One more difficulty is establishing the optimal distance with students. It supports the findings made by other researchers. Classroom self-efficacy and satisfaction with the established relationships influence the relation between the indicators [57]. The inability to establish the optimal distance with students entails one more challenge - adequate assessment of students. The fact, that issue of "fairness" troubles most novice teachers is supported by other researchers. It can be concluded from the data obtained from examining teachers' concerns about fairness. According to the researchers, it refers to such areas as dealing with students and grading [58].

Concluded from the analysis of the detected problems, one of the most effective ways of novice professional adaptation can be the professional and psychological support provided by mentoring and individual assistance performed in the form of on-the-job training, instructional interactions with mentors and experienced colleagues [47, 52, 59]. To overcome the difficulties novice university teachers face it is important to use proper social resources, providing social, professional, and psychological support [55, 60, 61].

It is necessary to carry out work on the adaptation of newly recruited staff, to improve the quality of education. According to some researchers, novice teachers can improve their competence using instructional activities such as a guided public rehearsal [62], or special training programmes. The studies suggest that pedagogical training programmes have the potential to affect participants' interpretations of teaching situations creatively, especially when the participants are not very experienced in teaching [40, 63]. Other researchers point out the importance of including and facilitating creativity in teacher preparation and suggest that teacher education should involve unique experiences to nurture creativity. They state that new teachers should be supported in discovering and developing their creative abilities and becoming more confident in making their own curricular contributions [64–66].

Taking into consideration the fact that most of the participants of the research are the novice university educators involved in doctoral studies it seems reasonable to include the pedagogical disciplines and a course aimed at development of creative skills into the curriculum of their study, which could be supported by the findings of [39, 65]. It is also supported by the positive experience of the Netherlands, where science undergraduates can enroll in a half-year teaching course that leads to the qualification of a teacher for junior secondary education [67].

5   Conclusion

The result of this research demonstrates that novice university teachers face several problems in all the spheres of the adaptation process, i.e. professional, psychological, and motivational spheres of adaptation. These problems remain the same within the groups of university educators with and without educational training, although the participants of the experiment without the degree in teaching tend to face these problems more frequently (the difference between the values of novice university teachers with and without pedagogical education is in most cases statistically significant). These results support conclusions made by L. Postareff et al. (2015) [38]. University teachers are professionals in their field of expertise, but inexperienced in pedagogy. It affects the effectiveness of the educational process because they have only limited knowledge of pedagogical theories and educational sciences [38]. Moreover, the participants of the experiment are mainly the postgraduate students with a teaching experience of less than five years, which causes even more problems in the sphere of professional adaptation to teaching due to the feeling of anxiety, lack of confidence and creative skills, increased fatigue and decreased performance.

We completely agree with the researchers, who conclude, it vital to offer pedagogical training paired with the programme of creative skills development before recruitment to teaching tasks at the university, that is, as part of doctoral studies or at the beginning of a teaching career [63, 68]. This study can be useful in planning the curriculum for postgraduate students. Educational authorities should take into consideration the identified difficulties faced by young teachers without pedagogical education in the process of planning professional development programmes for novice university educators without pedagogical education.

The obtained data refers to the post-graduate students whose teaching experience is a part of doctoral studies lasting less than three years. Further research will focus on the comparison of post-graduate students with other novice university educators (with a teaching experience of up to 5 years) without a pedagogical degree and those who have the pedagogical degree as their minor. Having analysed the obtained data, the authors are planning to design an online training course to support creativity in teaching for the postgraduate students to take during the doctoral studies to facilitate the process of their professional self-adjustment to the teaching process.

References

1. Golikov, A., Kudaka, M., Sergeev, V., Sergeeva, I., Tishin, P., Tumakova, E.: Human capital as a basis for the development of a modern university. In: Mottaeva, A., Melović, B. (eds.) International Scientific Conference Environmental Science for Construction Industry – ESCI 2018, MATEC Web of Conferences 2018, vol. 193, p. 05059. EDP Sciences, Ho Chi Minh City (2018). https://doi.org/10.1051/matecconf/201819305059
2. Odinokaya, M., Krepkaia, T., Karpovich, I., Ivanova, T.: Self-regulation as a basic element of the professional culture of engineers. Educ. Sci. 9(3), 200 (2019). https://doi.org/10.3390/educsci9030200
3. van Velzen, C., van der Klink, M., Swennen, A., Yaffe, E.: The induction and needs of beginning teacher educators. Prof. Dev. Educ. 36(1–2), 61–75 (2010). https://doi.org/10.1080/19415250903454817
4. Farrell, T.S.C.: Surviving the transition shock in the first year of teaching through reflective practice. System 61, 12–19 (2016). https://doi.org/10.1016/j.system.2016.07.005
5. Voronova, L., Karpovich, I., Stroganova, O., Khlystenko, V.: The adapters public institute as a means of first-year students' pedagogical support during the period of adaptation to studying

at a university. In: Anikina, Z. (ed.) IEEHGIP 2020. LNNS, vol. 131, pp. 641–651. Springer, Cham (2020). https://doi.org/10.1007/978-3-030-47415-7_68

6. Stroganova, O., Bozhik, S., Voronova, L., Antoshkova, N.: Investigation into the professional culture of a foreign language teacher in a multicultural classroom from faculty and international students' perspectives. Educ. Sci. **9**(2), 137 (2019). https://doi.org/10.3390/educsci90 20137

7. Karpovich, I. A., Krepkaia, T. N., Voronova, L. S.: Combarros-Fuertes, P.: Novice university educators: professional, psychological and motivational spheres of adaptation. In: IOP Conference Series: Materials Science and Engineering, vol. 940, no. 1, p. 012136 (2020). https://doi.org/10.1088/1757-899X/940/1/012136

8. Rodas Brosam, E.: Innovation: a case study of an English teachers' induction. Chakin: Rev. Ciencias Soc. Hum. **9**, 48–56 (2019). https://doi.org/10.37135/chk.002.09.04

9. Buric, I., Moe, A.: What makes teachers enthusiastic: the interplay of positive affect, self-efficacy and job satisfaction. Teach. Teach. Educ. **89**, 103008 (2020). https://doi.org/10.1016/j.tate.2019.103008

10. Zee, M., Koomen, H.M.Y.: Teacher self-efficacy and its effects on classroom processes, student academic adjustment, and teacher well-being: a synthesis of 40 years of research. Rev. Educ. Res. **86**(4), 981–1015 (2016). https://doi.org/10.3102/0034654315626801

11. Yazdi, M.T., Motallebzadeh, K., Ashraf, H.: The role of teacher's self-efficacy as a predictor of Iranian EFL teacher's burnout. J. Lang. Teach. Res. **5**(5), 1198–1204 (2014). https://doi.org/10.4304/jltr.5.5.1198-1204

12. Karpovich, I.A., Voronova, L.S.: Adaptation of novice university teachers with experience of 1–3 years and 3–5 years. Azimuth Sci. Res. Ped. Psychol. **9**(2), 106–110 (2020). https://doi.org/10.26140/anip-2020-0902-0023

13. Bylieva, D., Lobatyuk, V., Tolpygin, S., Rubtsova, A.: Academic dishonesty prevention in e-learning university system. In: Rocha, Á., Adeli, H., Reis, L.P., Costanzo, S., Orovic, I., Moreira, F. (eds.) WorldCIST 2020. AISC, vol. 1161, pp. 225–234. Springer, Cham (2020). https://doi.org/10.1007/978-3-030-45697-9_22

14. Westerman, D.A.: Expert and novice teacher decision making. J. Teach. Educ. **42**(4), 292–305 (1991). https://doi.org/10.1177/002248719104200407

15. Tsui, A.B.M.: Distinctive qualities of expert teachers. Teach. Teach.: Theory Pract. **15**(4), 421–439 (2009). https://doi.org/10.1080/13540600903057179

16. Wolff, C.E., Jarodzka, H., Boshuizen, H.P.A.: Classroom management scripts: a theoretical model contrasting expert and novice teachers' knowledge and awareness of classroom events. Educ. Psychol. Rev. **33**(1), 131–148 (2020). https://doi.org/10.1007/s10648-020-09542-0

17. Wolff, C.E., Jarodzka, H., Boshuizen, H.P.A.: See and tell: differences between expert and novice teachers' interpretations of problematic classroom management events. Teach. Teach. Educ. **66**, 295–308 (2017). https://doi.org/10.1016/j.tate.2017.04.015

18. Darling-Hammond, L., Bransford, J.: Preparing Teachers for a Changing World: What Teachers Should Learn and be Able to Do. Jossey-Bass, San Francisco (2007)

19. Esquivel, G.B.: Teacher behaviors that foster creativity. Educ. Psychol. Rev. **7**(2), 185–202 (1995). https://doi.org/10.1007/BF02212493

20. Feldhusen, J.F., Hoover, S.M.: A conception of giftedness: intelligence, self concept and motivation. Roeper Rev. **8**(3), 140–143 (1986). https://doi.org/10.1080/02783198609552957

21. Simplicio, J.S.C.: Teaching classroom educators how to be more effective and creative. Education **120**(4), 675–680 (2000)

22. Davidovitch, N., Milgram, R.M.: Creative thinking as a predictor of teacher effectiveness in higher education. Creat. Res. J. **18**(3), 385–390 (2006). https://doi.org/10.1207/s15326934 crj1803_12

23. Chant, R.H., Moes, R., Ross, M.: Curriculum construction and teacher empowerment: supporting invitational education with a creative problem solving model. J. Invitat. Theory Pract. **15**, 55–67 (2009)
24. Kasmaienezhadfard, S., Talebloo, B., Roustae, R., Pourrajab, M: Students' learning through teaching creativity: teachers' perception. J. Educ. Health Commun. Psychol. **4**(1) (2015). http://dx.doi.org/10.12928/jehcp.v4i1.3699
25. Richards, J.C.: Creativity in language teaching. Iran. J. Lang. Teach. Res. **1**(3), 19–43 (2013)
26. Whalen, C., Majocha, E., Nuland, S.: Novice teacher challenges and promoting novice teacher retention in Canada. Euro. J. Teacher Edu. **42**(5), 591–607 (2019). https://doi.org/10.1080/02619768.2019.1652906
27. Petrovska, S., Sivevska, D., Popeska, B., Runcheva, J.: Mentoring in teaching profession. Int. J. Cogn. Res. Sci. Eng. Educ. **6**(2), 47–56 (2018). https://doi.org/10.1080/01626620.2006.10463565
28. Turley, S., Powers, K., Nakai, K.: Beginning teachers' confidence before and after induction. Action Teach. Educ. **28**(1), 27–39 (2006). https://doi.org/10.1080/01626620.2006.10463565
29. de Paor, C.: Lesson observation, professional conversation and teacher induction. Irish Educ. Stud. **38**(1), 121–134 (2019). https://doi.org/10.1080/03323315.2018.1521733
30. Kearney, S., Lee, A.: Understanding beginning teacher induction: a contextualized examination of best practice. Cogent Educ. **1**(1), 967477 (2014). https://doi.org/10.1080/2331186X.2014.967477
31. Borg, S.: Teacher cognition in language teaching: a review of research on what language teachers think, know, believe, and do. Lang. Teach. **36**(2), 81–109 (2003). https://doi.org/10.1017/S0261444803001903
32. Ramming, U.: Calculating with words: perspectives from philosophy of media, philosophy of science linguistics and cultural history. Technol. Lang. **2**(1), 12–25 (2021). https://doi.org/10.48417/technolang.2021.01.02
33. Akcan, S.: Novice non-native English teachers' reflections on their teacher education programmes and their first years of teaching. PROFILE Issues Teach. Prof. Dev. **18**(1), 55 (2016). http://dx.doi.org/10.15446/profile.v18n1.48608
34. Borko, H., Livingston, C.: Cognition improvisation: differences in mathematics instruction by expert and novice teachers. Am. Educ. Res. J. **26**, 473–498 (1989). https://doi.org/10.3102/00028312026004473
35. MacDonald, D.: Novice science teachers learn about interactive lesson revision. J. Sci. Teach. Educ. **3**(3), 85–91 (1992). https://doi.org/10.1007/BF02614746
36. Odinokaya, M., Karpovich, I.: Modern condition of problems of adaptation of Russian teachers at the initial stage of work in the university. Azimuth Sci. Res. Ped. Psychol. **8**(1), 211–214 (2019). https://doi.org/10.26140/anip-2019-0801-0053. (in Russian)
37. Henry, G.T., Fortner, C.K., Bastian, K.C.: The effects of experience and attrition for novice high-school science and mathematics teachers. Science **335**(6072), 1118–1121 (2012). https://doi.org/10.1126/science.1215343
38. Postareff, L., Nevgi, A.: Development paths of university teachers during a pedagogical development course. Educar **51**(1), 37–52 (2015). https://doi.org/10.5565/rev/educar.647
39. Postareff, L., Lindblom-Ylänne, S., Nevgi, A.: The effect of pedagogical training on teaching in higher education. Teach. Teach. Educ. **23**(5), 557–571 (2007). https://doi.org/10.1016/j.tate.2006.11.013
40. Johannes, C., Fendler, J., Seidel, T.: Teachers' perceptions of the learning environment and their knowledge base in a training program for novice university teachers. Int. J. Acad. Develop. **18**(2), 152–165 (2013). https://doi.org/10.1080/1360144X.2012.681785
41. Odinokaya, M., Krepkaia, T., Sheredekina, O., Bernavskaya, M.: The culture of professional self-realization as a fundamental factor of students' internet communication in the modern

educational environment of higher education. Educ. Sci. **9**(3), 187 (2019). https://doi.org/10.3390/educsci9030187

42. Kearney, S.: The challenges of beginning teacher induction: a collective case study. Teach. Educ. **32**(2), 142–158 (2019). https://doi.org/10.1080/10476210.2019.1679109
43. Copeland, W.D., Birmingham, C., DeMeulle, L., D'Emidio-Caston, M., Natal, D.: Making meaning in classrooms: an investigation of cognitive processes in aspiring teachers, experienced teachers, and their peers. Am. Educ. Res. J. **31**(1), 166–196 (1994). https://doi.org/10.3102/00028312031001166
44. Smith, C., Nerantzi, C., Middleton, A.: Promoting creativity in learning and teaching. In: ICED 2014 Conference Proceedings. Stockholm (2014). http://www.iced2014.se/proceedings/1120_Smith.pdf. Accessed 3 Mar 2021
45. Ardisa, P., Wu, Y.T., Surjono, H.D.: Improving novice teachers' instructional practice using technology supported video-based reflection system: the role of novice teachers' beliefs. J. Phys.: Conf. Ser. **1140**, 012029 (2018). https://doi.org/10.1088/1742-6596/1140/1/012029
46. Mathews, H.M., Rodgers, W.J., Youngs, P.: Sense-making for beginning special educators: a systematic mixed studies review. Teach. Teach. **67**, 23–36 (2017). https://doi.org/10.1016/j.tate.2017.05.007
47. Paula, L., Grinfelde, A.: The role of mentoring in professional socialization of novice teachers. Prob. Educ. 21st Cent. **76**(3), 364–379 (2018). http://oaji.net/articles/2017/457-1529089806.pdf
48. Eteläpelto, A., Vähäsantanen, K., Hökkä, P.: How do novice teachers in Finland perceive their professional agency? Teach. Teach. Theory Pract. **21**(6), 660–680 (2015). https://doi.org/10.1080/13540602.2015.1044327
49. Teles, R., Valle, A., Rodríguez, S., Piñeiro, I., Regueiro, B.: Perceived stress and indicators of burnout in teachers at Portuguese higher education institutions (HEI). Int. J. Environ. Res. Public Health. **17**(9), 3248 (2020). https://doi.org/10.3390/ijerph17093248
50. Kim, J., Youngs, P., Frank, K.: Burnout contagion: Is it due to early career teachers' social networks or organizational exposure? Teach. Teach. Educ. **66**, 250–260 (2017). https://doi.org/10.1016/j.tate.2017.04.017
51. Gavish, B., Friedman, I.A.: Novice teachers' experience of teaching: a dynamic aspect of burnout. Soc. Psychol. Educ. **13**, 141–167 (2010). https://doi.org/10.1007/s11218-009-9108-0
52. Löfström, E., Eisenschmidt, E.: Novice teachers' perspectives on mentoring: the case of the Estonian induction year. Teach. Teach. Educ. **25**(5), 681–689 (2009). https://doi.org/10.1016/j.tate.2008.12.005
53. van Ginkel, G., Oolbekkink, H., Meijer, P.C., Verloop, N.: Adapting mentoring to individual differences in novice teacher learning: the mentor's viewpoint. Teach. Teach.: Theory Pract. **22**(2), 198–218 (2016). https://doi.org/10.1080/13540602.2015.1055438
54. Meyer, H.: Novice and expert teachers' conceptions of learners' prior knowledge. Sci. Educ. **88**(6), 970–983 (2004). https://doi.org/10.1002/sce.20006
55. Bettini, E.A., Jones, N.D., Brownell, M.T., Conroy, M.A., Leite, W.L.: Relationships between novice teachers' social resources and workload manageability. J. Spec. Educ. **52**(2), 113–126 (2018). https://doi.org/10.1177/0022466918775432
56. Caspersen, J., Raaen, F.D.: Novice teachers and how they cope. Teach. Teach.: Theory Pract. **20**(2), 189–211 (2014). https://doi.org/10.1080/13540602.2013.848570
57. Canrinus, E.T., Helms-Lorenz, M., Beijaard, D., Buitink, J., Hofman, A.: Self-efficacy, job satisfaction, motivation and commitment: Exploring the relationships between indicators of teachers' professional identity. Eur. J. Psychol. Educ. **27**, 115–132 (2012). https://doi.org/10.1007/s10212-011-0069-2
58. Berry, R.A.W.: Novice teachers' conceptions of fairness in inclusion classrooms. Teach. Teach. Educ. **24**(5), 1149–1159 (2008). https://doi.org/10.1016/j.tate.2007.02.012

59. Almazova, N., Bylieva, D., Lobatyuk, V., Rubtsova, A.: Human behavior as a source of data in the context of education system. In: 2019 International SPBPU Scientific Conference on Innovations in Digital Economy, p. 37. Association for Computing Machinery, New York (2019). https://doi.org/10.1145/3372177.3373340

60. Shin, S.K.: "It cannot be done alone": the socialization of novice English teachers in South Korea. TESOL Q. **46**(3), 542–567 (2012). https://doi.org/10.1002/tesq.41

61. Trostinskaia, I.R.; Safonova, A.S.; Pokrovskaia, N.N.: Professionalization of education within the digital ecnomy and communicative competencies. In: 2017 IEEE 6th Forum Strategic Partnership of Universities and Enterprises of Hi-Tech Branches (Science. Education. Innovations) (SPUE), pp. 29–32. IEEE, St. Petersburg (2017). https://doi.org/10.1109/IVForum. 2017.8245961

62. Kazemi, E., Franke, M., Lampert, M.: Developing pedagogies in teacher education to support novice teachers' ability to enact ambitious instruction. In: Hunter, R., Bicknell, B., Burgess, T. (eds.) 32nd Annual Conference of the Mathematics Education Research Group of Australasia, Crossing divides, vol. 1, pp. 11–30. Merga, Palmerston North (2009)

63. Vilppu, H., Södervik, I., Postareff, L., Murtonen, M.: The effect of short online pedagogical training on university teachers' interpretations of teaching–learning situations. Instr. Sci. **47**(6), 679–709 (2019). https://doi.org/10.1007/s11251-019-09496-z

64. Kimhi, Y., Geronik, L.: Creativity promotion in an excellence program for preservice teacher candidates. J. Teach. Educ. **71**(5), 505–517 (2020). https://doi.org/10.1177/002248711987 3863

65. Barnes, J., Shirley, I.: Strangely familiar: cross-curricular and creative thinking in teacher education. Improv. Sch. **10**(2), 162–179 (2007). https://doi.org/10.1177/1365480207078580

66. Sharma Sen, R., Sharma, N.: Teacher preparation for creative teaching. Contemp. Educ. Dialogue **6**(2), 157–192 (2009). https://doi.org/10.1177/0973184913411185

67. van Rooij, E.C.M., Fokkens-Bruinsma, M., Goedhart, M.: Preparing science undergraduates for a teaching career: sources of their teacher self-efficacy. Teach. Educ. Q. **54**(3), 270–294 (2019). https://doi.org/10.1080/08878730.2019.1606374

68. Vasetskaya, N.O., Glukhov, V., Burdakov, S.F.: The elaboration of the model of competences of the research and teaching university staff. In: Shaposhnikov, S. (ed.) 2018 XVII Russian Scientific and Practical Conference on Planning and Teaching Engineering Staff for the Industrial and Economic Complex of the Region (PTES), pp 98–101. IEEE, St. Petersburg (2018). https://doi.org/10.1109/PTES.2018.8604215

The Experience of Developing Creativity in Future Primary School Teachers

Tatiana Solovyeva$^{(\boxtimes)}$ ⓘ, Natalia Shlat ⓘ, Svetlana Dombek ⓘ,
and Grigoriy Galkovskiy ⓘ

Pskov State University, Lenin Square, 2, 180000 Pskov, Russia

Abstract. Based on the analysis of the world's innovative scientific and methodological approaches, the article presents theoretical and methodological concepts of modeling the development of creativity in students of higher education institutions. The authors have developed and have been testing for several years a didactic model for the development of creativity in future primary school teachers, which includes: an "incubation" stage, during which the students develop the ability to detect contradictions at the unconscious, partially conscious and reflective levels; the stage of stimulating potential creativity, when students learn to recognize minor elements of novelty, methodologically promising, productive differences in didactic material; the stage of designing collective professional methodological products for conducting lessons and extracurricular activities in primary school; the final stage, during which students, relying on their meta-professional instrumental competencies, the knowledge of innovative teaching technologies, use various digital platforms to perform individual creative projects. The article suggests some ways and means of developing creativity in future primary school teachers, as well as describes the psychological and pedagogical conditions for the development of creativity, capable of influencing the need–motivational and emotional spheres of the student's personality: the regular reinforcement of creative manifestations by the teacher, which consists in "centering" the student on the self–achieved result; providing students with freedom of choice in their pedagogical activity when demanding responsibility for it; the evaluation and promotion of creative initiative and high–quality work.

Keywords: Creativity · Didactic model · Development environment · Future teachers'

1 Introduction

At present, creativity plays a key role not only in art, but also in other areas of human activity – in science, production, and even in political life.

D. Bylieva and A. Nordmann (Eds.): PCSF 2021, LNNS 345, pp. 730–746, 2022.
https://doi.org/10.1007/978-3-030-89708-6_59

1.1. If earlier, traditionally, creativity was associated with personal development and the process of individual growth, recently it is increasingly associated with innovation, transformational leadership and effective management, and in a global sense – the phenomena of economic efficiency and social well–being (J. Soffel [1]). According to Z. Fields, and C.A. Bischoff, creativity becomes a force of great value when it is applied to causes that benefit humanity and the world at large [2].

1.2. Modernization of higher education in world pedagogy as a priority trend of rethinking the training of future professional specialists, indicates the search for mechanisms for the formation of supra–professional competencies (universal social and personal qualities, "key competencies", "the 21st–century skills"), to which some researchers add creativity [3–5].

1.3. Since it is the teacher who establishes the conditions for creative (self–) development, encourages and evaluates creativity in students, then the solution to the problem of creativity development in future primary school teachers undoubtedly becomes relevant, since creative activity can only be taught by a creative teacher.

1.4. The relevance of the article is determined by the need for the pedagogical community to acknowledge and make sense of the Russian experience of developing creativity in the personality of future primary school teachers in the process of higher education.

2 Problem Statement

2.1. It is widely believed that modern higher education is in a state of crisis, which is expressed, in particular, by the fact that the educational activities of students are mostly reproductive.

The national monitoring of the educational results of university and college graduates in pedagogical areas shows that future teachers often lack the knowledge of specific strategies for developing creativity and creative thinking skills, since the formalism of the requirements for the achievements of the graduate, the declarative nature of the documents regulating the organization of the bachelor of pedagogy training, often do not correlate with the idea of individual progress of the student.

2.2. The question arises as to how much students attribute certain personal significance to creative activity, which encourages them to act creatively and want others to do the same. The students' ranking of terminal values (goal–values) by the degree of their significance organized by the authors showed that for the majority of future primary school teachers, creative activity is not of particular value: 96.2% of the respondents did not indicate creative activity among the five most important values. Only 3.8% of the students put creativity in the fifth place in the order of importance in their lives.

2.3. It was also necessary to find out to what extent it is important for the students themselves to see the manifestations of creativity in the organization of educational activities at a higher education institution. To answer this question, an absolute evaluation test was conducted using the diagnostic method modified by the authors Ye.M. Bukhvalova, and L.V. Karpushina [6]. The test subjects were asked to rate the importance of the content of several statements on a 7–point scale: categorically unacceptable (1), unacceptable (2), unattractive (3), indifferent (4), attractive (5), important (6), very important (7). A quantitative analysis of the students' answers to the questions is presented in Table 1.

As can be seen from the table, manifestations of creativity in the organization of educational activities at the university are generally important for future primary school teachers. Thus, for 78.8% of students, it is attractive, important and very important for the teacher to encourage the creative activity of students at the lecture; 69.2% – "when non–standard professional tasks are offered in the practical classes"; 84.7% – "when the supervisor allows not only to choose a research problem but also welcomes my creative approach to solving it"; 82.7% – encouraging students' creativity in project activities; 88.5% – "when the teacher allows you to express several points of view, sometimes contradictory, on a particular pedagogical problem", etc. The majority of test participants (84.7%) consider it unacceptable "when the teacher "punishes" for a non–standard answer to a posed question".

Polar opinions were expressed about the need for clear algorithms for solving problems: for 51.9% of the test subjects, it is categorically unacceptable "when a teacher gives a step–by–step algorithm for solving an academic or professional problem and requires strict implementation of it", and for the remaining 48.1% of the survey participants, the presence of such an algorithm is attractive (important, very important).

For 90.4% of students, creative teachers in higher education are attractive, important and very important. Giving reasons for the importance of creative teachers in the university, students point to their "ability to inspire, positively motivate students", "originality of conveying information, which allows better assimilation of knowledge", "focus on fostering creative qualities in students", "optimism", "success in the development of non–standard thinking in students", "interest in their subject", "readiness for a dialogue with students", "desire for new things, for discoveries", etc. 88.5% of students believe that creative graduates of pedagogical universities are attractive (important, very important) for the parents of future first–graders.

However, the "detector" used in the questionnaire, which involves comparing students' answers to questions, also revealed some contradiction: students are attracted to creativity in the organization of the educational process, and on the other, it is important (attractive, very important) for the students that the teacher offers "examples of accomplishing the educational tasks and templates for their presentation" (94.2%). In addition, for 69.2% of students, it is attractive, important, very important, "when the teacher at the lecture provides information in a ready-made form".

2.4. A review of the academic literature has shown that there is no single view among scholars on the definition of the concept of "creativity". In the academic community, creativity is viewed as a multi-faceted phenomenon in the context of studying personality, society and culture. Researchers point to creativity as "creative development", "creative

Table 1. Percentage distribution of the students' answers to the test questions (modified method of Ye.M., Bukhvalova, and L.V. Karpushina [6])

Statement	Number of answers (%)			
	1–3 points (categorically unacceptable, unacceptable, unattractive)	4 points (indifferent)	5–6 points (attractive, important)	7 points (very important)
1. "For me it is… when the teacher at the lecture provides information in a ready-made form"	23,1	7,7	38,5	30,7
2. "For me personally the initiation of creative activity of students by the teacher is…"	13,5	7,7	46,2	32,6
3. "For me now it is… when the supervisor allows not only to choose a research problem but also welcomes my creative approach to solving it"	3,8	11,5	50	34,7
4. "For me, it is… when I am offered non–standard professional tasks in the practical training classes"	19,3	11,5	57,7	11,5
5. "For me it is… when the teacher expects creativity from the students in project activities"	9,6	7,7	61,5	21,2
6. "For me, it is… when the teacher "punishes" for a non–standard answer to a posed question"	84,7	3,8	11,5	0
7. "For me now pedagogical creativity at the internship practice at school is…"	11,5	23,1	46,2	19,2
8. "For me, it is… when the teacher gives a step–by–step algorithm for solving an academic or professional problem and requires strict implementation of it"	51,9	0	42,3	5,8

(*continued*)

Table 1. (*continued*)

Statement	Number of answers (%)			
	1–3 points (categorically unacceptable, unacceptable, unattractive)	4 points (indifferent)	5–6 points (attractive, important)	7 points (very important)
9. "For me, it is… when the teacher allows you to express several points of view, sometimes contradictory, on a particular pedagogical problem"	*0*	*11,5*	*61,5*	*27*
10. "For me, it is… when there is no single correct solution to a pedagogical problem"	*21,2*	*1,9*	*61,5*	*15,4*
11. "For me, the opinion of my fellow students is… if I express an unconventional idea"	*9,6*	*11,5*	*61,5*	*17,4*
12. "For me, it is… to think outside the box"	*17,3*	*15,4*	*46,2*	*21,1*
13. "For me, the creative teachers at the university are …, as they are…"	*0*	*9,6*	*61,5*	*28,9*
14. "For me, it is… when the teacher offers examples of accomplishing the educational tasks and templates for their presentation"	*0*	*5,8*	*53,8*	*40,4*
15. "For me, it is… when the teacher does not add points for the original solution of the problem and does not reduce the score for a trivial answer"	*46,1*	*30,8*	*23,1*	*0*
16. "Creative graduates of pedagogical universities are… for the parents of the future first–graders"	*0*	*11,5*	*38,5*	*50*

abilities", specific (essential) and systemic manifestations of creativity (collision function, divergent thinking, the ability to generate new ideas, the willingness to make quick and reasonable decisions) in the structure of the overall development of the personality (J.R. Sternberg [7]; J.C. Kaufman [8]), the integrity of creative manifestations, that is, the wide range of components of this phenomenon (K.K. Urban [9]), the method of creating knowledge (A. Craft, T. Cremin, and P. Burnard [10], the experience of creative activity (J.C. Kangas [11]).

In contemporary research, it is possible to note the differentiation of the categories "creation" – creativity as a process, the activity-specific in content and result – and "creativity" as a personal component (quality, ability, system of certain skills and abilities). The "humanization" of creativity as an activity is associated with the spread of the views of Renaissance figures (G. Pico Della Mirandola) and the philosophers of subsequent eras (Kant, Nietzsche).

Scholars, describing the versatility of the creativity manifestations, note the external and internal, hidden properties of this phenomenon. For example, the external attributes of creativity are the creative practices, the ability to generate new ideas, think independently, in an original way, quickly navigate in a difficult situation, identify contradictions and resolve them; the internal features of creativity include the qualities of creative thinking, such as inventiveness, the personal need for development, and others.

A.J. Plucker, R. Beghetto, and G. Dow [12] proposed to consider 3 aspects in the essence of creativity: 1. The aspect of the creative process and the product of creativity. 2. The aspect of the creative agent – the personality of the creator. 3. The aspect of the creative environment (microclimate, resources, best practices) and the mentor who organizes the creative process. The most objective approach to evaluating creativity, according to these scholars, is the aspect of the creative process and the product of creativity (the specific solution to a problem or evaluation of a specific product).

2.5. The analysis of literary sources on the development of creativity in the education system allowed us to establish the following. The main characteristic of the creative educational process is the presence of novelty: a new idea, a new method or a new project introduced to the educational system. In creative educational activity, three equally interrelated elements are found – creative teaching (the creative practice of the teacher), training for creativity (organization of the creative process) and creative learning (freedom of creative actions of students) [13].

It is proved that each student can freely develop his or her creativity, and in this case, the key factor in activating creative abilities is the individual himself [14]. According to E.P. Torrance [15], with the help of questions and tasks of an imaginative nature, it is possible to develop fluency, flexibility and originality of creative thinking in students.

Higher education institutions in Spain, China, Lithuania, Malta, and some other countries use a didactic model based on the creative platform methodology developed by Danish researchers C. Byrge, and S. Hansen [16]. According to this model, students' creativity develops: a) in the absence of any judgments (both positive and negative) about their work and the work of other students; b) when immersed in a purposeful creative process in the absence of external stimuli (unnecessary things, devices, and the like); c) when implementing the principle of "parallel thinking", reflecting the idea of concentrating of both an individual member of the student community and of the entire group on solving a single problem; d) when requiring each participant in the creative process to "look" at the situation or problem from different points of view. M. Lund, S. Byrge, and S. Nielsen [17], R. Shaheen [4] point to the possibility of developing creativity by designing special educational content, such as programs using certain active (interactive) methods.

According to scholars, the presence of creativity in students can be proved only under the following three conditions: creative preferences, creative inclinations and abilities, as well as experience of creative activity (M. Csiksentmihalyi [3]). Thus, the research results indicate a complex of systemic components of creativity: the personality (qualities of a creative person), the processes (motivation, activeness, thinking, communication), the product and its characteristics (ideas, scientific, academic and artistic works; novelty, social value, elaboration (how nuanced the idea has been) and the adaptability of the product (its ability to solve the existing problem)), the environment (the socio-cultural context of the creative agent).

2.6. For an empirical study of the development of creativity in future primary school teachers, the work of K.K. Urban is of utmost interest, describing the areas of creativity that are subject to measurement and evaluation: a) cognitive components (divergent thinking and action, the framework of generalized knowledge (the mental outlook) and the knowledge in a specific area, as well as ways of performing mental actions); b) personal qualities (the ability to "dive" into the problem and achieve a goal of solving it, motivational sphere, willingness to accept and produce unconventional solutions).

3   Research Questions

In the course of the study, the following questions had to be answered:

Is the creative activity of value for students – future primary school teachers?

To what extent is it important for students to see manifestations of creativity in organizing educational activities at a higher education institution?

What are the didactic tools for initiating creative cognitive activity of the future primary school teachers in the first years of professional training?

What is the meaning and what is the content of the psychological and pedagogical conditions for developing creativity in students' personalities?

What is the didactic model for the development of creativity in future primary school teachers at a higher education institution?

4   Purpose of the Study

The purpose of the study is to substantiate the strategy and present a didactic model for the development of creativity in future primary school teachers, which would allow us to provide an outline of the conceptual ideas of the real creative–oriented educational process.

The achievement of the goal was accompanied by solving the following tasks.

4.1. Researching into the students' attitude to creative activity as a value and manifestations of creativity in the organization of the educational process at the university.

4.2. Theoretical and empirical substantiation of the effectiveness of the didactic model for the development of students' creativity in the process of higher education.

4.3. Determining the conditions for the development of creativity in the personality of the future primary school teachers in the process of higher education, which we consider as a unity of the objective and the subjective, the internal and the external, the possible and the proper.

5 Research Methods

The conducted research was based on the methods of theoretical analysis of literary sources; a survey of 104 students in the areas of training 44.03.01 Pedagogical education, major in "Primary education" and 44.03.05 Pedagogical education (with two majors of training), major in "Primary education and correctional pedagogy"; a pedagogical experiment and testing of 52 future primary school teachers.

5.1. The empirical part of the study was conducted in 2019–2020 at the Institute of Education and Social Sciences of Pskov State University.

5.2. The study of the problem was carried out in three stages.

The first stage of the study consisted:

– in determining the students' rating of "creative activity" among other universal values;
– in studying the value of "creativity of the educational activities at the university" for students – future primary school teachers – by the method of absolute evaluation;
– in diagnosing the level of creativity in students participating in the pedagogical experiment.

At the second stage of the study, in the process of training future primary school teachers, an experimental didactic model was implemented and the psychological and pedagogical conditions for the development of students' creativity in the educational process were established.

At the third stage, the results of the study were interpreted and summarized.

6 Findings

6.1. To develop creativity in future primary school teachers in the educational process at the university, an acmeological strategy was used, which is essentially a constructive methodological orientation of training, a pedagogical philosophy that meets the principles of "I want", "I perform activities at the maximum level at which I can" and "I get intellectual satisfaction thereby". The strategy is aimed at developing students' ability to see not only the superficial but also the hidden contradictions, inconsistencies in the natural or artificial pedagogical situations, optimally resolve them independently, rising to a new level of their creative abilities disclosure through overcoming professional difficulties.

6.2. To diagnose the level of creativity of students participating in experimental work, we developed a compilation diagnostic methodology: "Six tasks". It was compiled based on the structure and content principles of the well–known tests of creativity (J.P. Guilford

[18], E.P. Torrance [15], F.E. Williams [19]) and included six tasks grouped into verbal and visual batteries (three tasks each).

The verbal tasks at the ascertaining stage of the experiment involved "voicing the idea":

– *You are creating a scientific and methodological journal of psychological and peda-gogical nature. Suggest the name of this journal, as well as the names of the planned 4 rubrics.*
– *In the first art class in the second grade after the summer holidays, you give the task "How did I spend the summer". Reformulate the given title to make it more attractive to second graders.*
– *Makeup one phrase, in which you must use the words: "teacher", "compasses", "sound", "cup", "smile" (the grammatical case of nouns can be changed).*

The visual tasks involved finishing or visualizing the image based on the stimulus material.

– *Complete the drawing of the objects presented in the form of the image associated with the work of a teacher. Name the image.*
– *Examine the object presented in the form. Imagine how a teacher can use it for teaching purposes (not for its intended purpose).*
– *Close your eyes and construct the layout of a training premise in your imagination in as much detail as possible.*

The criteria and indicators of creativity used by us to diagnose the level of its development at the ascertaining and control stages of the experiment in future primary school teachers are presented in Table 2.

The results of the diagnostics of the students participating in the experimental work were ranked as follows (Table 3).

It appeared that before the beginning of the formative stage of the experimental work, at the ascertaining stage, 88.4% of the tested students, future primary school teachers, had an average (42.3%) and below-average (46.1%) levels of creativity. Only 11.7% of the test subjects were able to solve the proposed tasks, showing creativity at a high level (see Fig. 1).

6.3. The didactic model of students' creativity development included stages, psychological and pedagogical conditions, methods and didactic means and was tested in an experimental model. The immersion of the future primary school teachers into the creative educational activities (the first or the "incubation" stage of the proposed model) begins at Pskov State University with the development of students' collisional function of the personality. By the collisional function of the students' personalities, we understand a personal ability realized in the activity of detecting contradictions and "removing" them, which satisfies a person's need for clarity, orderliness, completeness and understanding.

The authors' concept of creativity development in students is based on the psychological postulate that the initiation of the creative cognitive activity of a person is based on the ability to detect inconsistencies. V.E. Klochko's research [20] has shown that the detection of a cognitive contradiction is not a one–time act, but manifests itself

Table 2. Criteria and characteristics of the nominative indicators of creativity

Criteria	Qualitative and quantitative characteristics of the nominative indicators of creativity		
The originality of the idea, i.e. statistical rarity of the answer (E.P. Torrance [15])	Trivial, frequent answers (0 points)	Answers that occur 2–3 times (1 point)	The answer that occurs only once (2 points)
The elaboration of the idea, i.e. the ability to detail the invented ideas (E.P. Torrance [15])	Lack of detail (0 points)	The ideas are repeated, but each has one or two detailed nuances (1 point)	Ideas are not repeated, have more than two detailed nuances (2 points)
The flexibility of thinking, i.e. the speed and ease of finding strategies for solving the problem (J.P. Guilford [18])	New ideas are not produced (0 points)	New solution strategies are developed only with the help of the experimenter's hint (1 point)	New strategies for solving the problem are developed easily, without the help of an experimenter (2 points)
Divergence of thinking, i.e. polyvariance of solving the problem (J.P. Guilford [18], F. E. Williams [19])	The only solution to the problem (0 points)	Two options for solving the problem (1 point)	Three or more options for solving the problem (2 points)

Table 3. Ranking of the creativity in students study results

Rank	Sum of points	Creativity level
I	9 and above	Above average
II	5–8	Medium
III	4 and below	Below average

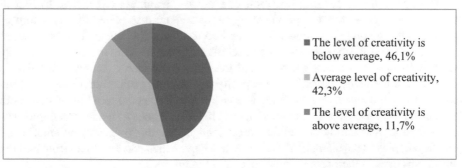

- The level of creativity is below average, 46,1%
- Average level of creativity, 42,3%
- The level of creativity is above average, 11,7%

Fig. 1. The diagram of the students raking according to the levels of creativity development at the ascertaining stage of the experiment (the "Six Tasks" method), %. Source: authors

as a complex process that takes place at all three interacting levels: the unconscious (default settings and attitudes), the partially conscious (emotions), and the reflective (mindfulness).

In the first year, when studying the discipline "Introduction to the Teaching Profession", the teacher, using the specially created didactic materials that create and set the task to identify information and semantic inconsistencies, inviting students to detect them in the behavior, speech, and actions of primary school teachers. It is the goal-setting to detect contradictions in the pedagogical situations suggested for consideration that allows future teachers to avoid reproductive mnemonic activity, generates a desire to "question everything" in the future, teaches to weigh all the arguments "for" and "against" and provides independence of opinion, non–subjection to other people's influences, to "blind copying" of methods and means of pedagogical activity.

In the second year, when studying the discipline "Psychological and Didactic Foundations of the Cognitive Processes Development in Children", the formation of the future primary school teachers' "sensitivity to contradictions, logical errors" and the readiness to detect them continues. At the practical lesson on the topic "Managing the Process of Developing the Logic of Speech in Younger Schoolchildren", after getting acquainted with the concepts of "objective and conceptual consistency" students analyze the written works of younger schoolchildren, identify and substantiate logical errors made by children.

Here are some examples of the children's works offered to students for analysis. "Coniferous trees do not shed their leaves in autumn, except for larch. Hence, larch is a deciduous tree". "Snow and ice are frozen, meltwater" (1st grade). "The number of Laptev walruses in the far North has decreased thanks to the uncontrolled hunting for them." "The still, calm flow of the Svir River, flowing through a hilly plain, fascinates the eye" (2nd grade). "The muscles of the hands are called the obedient muscles because they carry out the commands of the brain. The heart is an unruly muscle, as it works intermittently" (3rd grade). "On the maps of rivers, sailing masters plot the most convenient routes for the movement of ships – fairways. Ship captains try not to enter the fairway area" (4th grade). Students should explain in which cases the laws of logic were violated (the law of identity, the law of non–contradiction), and in which cases the logical rules were violated (the consistency of combining one word with another and avoiding using words not by their meaning).

The opportunities for the formation of students' readiness to respond to logical errors in an original way are provided by the academic discipline "Speech Culture" studied in the second year. Students are invited to creatively "respond" to the speech errors they have found ("in the breast pocket of their trousers"; "in the first grade, every child looks into the teacher's mouth, and then anywhere else"; "animals that live in the reserve are already extinct"; "the capacity of the higher school facilities is not yet enough to incarcerate everyone " (in Russian the word "посадить" stands for both "to incarcerate" and "to accommodate"), etc.) in a playful comic manner. A humorous, caricatured representation of a logical error leads to the emergence of an emotional–gnostic complex in the memory of the future teachers, which prevents them from making similar mistakes in their professional speech in the future.

6.5. The second stage of the proposed didactic model implementation can be called "stimulating the potential creativity" in students. The task of this stage is to develop students' ability to recognize something new, "fresh", to be, as the physicist and neuropsychologist D. Bohm [21] wrote, "attentive, alert and sensitive", even to minor elements of novelty, in other words, to evoke "a certain state of mind".

Thus, in the third year of practical classes in the course of "Theoretical Foundations and Technologies of Primary Education in Natural Science", students are offered several similar methodological situations from a lesson on the same topic but tackled by different teachers using the same pedagogical technique. Students are required to pay special attention to identifying minor, but methodically promising, productive differences in the didactic material used by different teachers.

When studying the academic discipline "Innovative Technologies in Education" during the same year, the practical activity of students becomes more complicated due to an increase in the degree of independence and creativity in conditions of information shortage (T. Solovyeva, and I. Vitkovskaya [22, p. 628]). The student develops a practical lesson based on the principles of the pedagogical technology that is intended for assimilation, and the fellow students, future teachers, isolate and recognize the "new", as if being part of its content.

6.6. The third stage of the didactic model for the development of creativity in future primary school teachers implementation is aimed at creating an environment that allows one to design a collective professional product. At this stage, the "Trello" digital platform is used. Students are divided into groups of four. The teacher creates a brief lesson plan for each group, and each group member on his or her "board" fills it with specific content, approaching the solution of the methodological problem from different standpoints. In the process the materials are preliminarily summarized, complementing each other, and only then, after the analysis and evaluation, the most interesting options for constructing each phase of the lesson are selected (challenging prior knowledge, goal formation, learning new material, organizing independent work, consolidation, generalization and homework).

6.7. At the fourth and final stage of the didactic model implementation, students rely on the previously developed meta-professional instrumental competencies, including the ones from having covered the fundamental disciplines (T. Solovyeva, and I. Vitkovskaya [23]), they practice unconventional (creative) work on their own, individual methodological "products", based on the resource integration of the information and communication technologies and active learning technologies: TRIZ–technologies ("Theory of inventive problem solving"), case technologies, game design, Webquest technologies, art technologies (N. Shlat, B. Borisov, N. Eyliseyeva, and N. Shakirova) [24]. Here we speak about professional projects that are carried out according to a personal plan.

6.8. Any didactic model can function successfully only under the appropriate conditions that create the environment, the setting in which a particular phenomenon or process arises, exists and develops. The implementation of the experimental didactic model for the development of creativity in students – future primary school teachers – in the educational activities of Pskov State University was carried out in compliance with some psychological and pedagogical conditions.

The content characteristic of the psychological and pedagogical conditions for the development of creativity in students is based on the idea that the motivational structure of creativity of the personality, in particular, includes "the desire to create images and ideas that are different from those that exist in a particular situation" motive, the motive for achievement, the indicator of which is performing of creative activities at a high level of difficulty, and the internal locus of control, the indicator of which is the reliance, first and foremost, on one's strengths and abilities. The authors' position coincides with the opinion of the majority of foreign and Russian psychologists who believe that "achievement motivation" can influence the success of any activity (both reproductive and creative), and that the achievement motive arises from the "growing expectations" of a person and can develop at any age due to specially constructed training.

The first psychological and pedagogical condition for the development of creativity in the personality of the future primary school teachers can be described as the regular reinforcement of its manifestations by the teacher in educational activities, which consists in "centring" the student on the independently achieved result. "Centering" the student on the self–achieved the result of the creative activity ensures the strengthening of all three variables of motivation to achieve: the subjective probability of success, the motive for striving for success and the incentive value of success, thereby contributing to the increase in the degree of creativity of the future teacher's personality.

As a second psychological and pedagogical condition for the development of the student's creative personality, it is possible to name "providing students with freedom of choice in pedagogical activity, while demanding responsibility for it". Freedom of creativity as a universal fundamental value, a goal–value or the highest (absolute) value does not exist outside the act of choice. In other words, there are no external reasons for which the student could write his or her creative decision off. The same applies to the student's opinion and the position as a teacher. It all depends solely on his or her own choice. If the student is ready to be responsible for the choice of position made by him or her in the pedagogical activity, thereby the individual "manifests", demonstrates the qualities of a free person.

This condition is based on the ideas of existential philosophy, which defines freedom as a person's "choice" of one of the innumerable possibilities for which he is responsible. From the point of view of existentialism, the freedom of creativity in pedagogical activity is not a fantasy at all, and even more so not a whim of the teacher: even at the stage of choice, not to mention the impact of its results on the consciousness of students, it is connected with other people.

Responsibility in the current study is interpreted not only as a form of self–regulation of a person who can give an account of his single actions and behavior in general, to take the blame for their result but also as a source of his concern for the future. This means that the future primary school teacher at any stage of choice in pedagogical activity should reflect on the consequences of the implementation or rejection of each of the alternatives, and be responsible for them since his or her choice will be directly "reflected" on the children.

The third psychological and pedagogical condition for the development of creativity in the future primary school teachers – "evaluation, promotion of creative initiative and high quality of work" – is based on the thesis of the German–American Gestalt psychologist K. Levin and A. Bandura [25] that the maximum of positive encouragement can positively affect the development of an individual's personality, including such personal characteristics as creativity. The assessment of students' creative activity in the classroom is expressed in the form of a qualitative verbal characteristic, called "quantitatively indeterminate". It is based on the following basic assumptions: the development of clear criteria for evaluating the results of creativity; systematic variation of the types and methods of evaluation, so that there are no occurrences of habituation and fading of reaction to the action of these stimuli; the evaluation is not selective and is not accompanied by offensive remarks or indifference of the teacher to non–standard forms of student activity.

The authors consider the category of "encouragement" to be a method of pedagogical support, designed to perform the following functions: to approve the student's own unconventional opinion; to support and develop his or her desire for fearless discussion of contradictory circumstances of events, allowing the students to creatively assert themselves. The student will always remember the situation in which he or she was praised, without paying much attention to the form of encouragement (even with a gentle smile or an affirmative nod), and will try to stick to this line of behavior. It is necessary to pay attention to the fact that the student on his own does not reflect directly, does not recognize encouragement as the most important factor determining his or her "creative thinking". Only the emotions generated by the encouragement unconsciously direct the student to the mode of activity that caused the encouragement, contributing to the formation of a sense of self–esteem, strengthening the students' desire for the initiative in creative activities. Emotions, being, according to psychologists, a subjective form of the existence of needs (motivation), signal to the student about the need-based significance of creative activity and encourage him or her to continue to act in this way.

6.9. To prove the effectiveness of the proposed didactic model and the conditions for the development of creativity in future primary school teachers, the level of creativity achieved by the participants of the experiment was again diagnosed at the control stage (see Fig. 2).

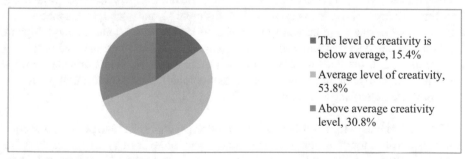

Fig. 2. The diagram of ranking students by levels of creativity development at the control stage of the experiment (the "Six Tasks" method), %.

As can be seen from the comparison of Figs. 1 and 2, the number of students with a level of creativity below average decreased by 30.7%, and the number of participants in the experiment with a level of creativity above average increased by 19.1%.

To assess the statistical reliability of the shift in the values of creativity under the influence of experimental impact, we used the G–criterion of signs. We found that the typical shift is positive; zero reactions − 26; negative shifts − 0. Hence, G–empirical = 0. G–critical (p ≤ 0.01) = 6, G–critical (p ≤ 0.05) = 8. G–empirical < G–critical, therefore, the shift is valid.

7 Conclusion

7.1. The pedagogical philosophy of creativity development in future primary school teachers presented in this article, based on the didactic model and a set of psychological and pedagogical conditions, can be viewed as a personalized route, an individual educational and professional trajectory that allows the student to realize him– or herself. It aims higher education at providing not only students but also teachers with creative freedom, although initially for the majority (96.2%) of the future primary school teachers, creative activity was not a priority terminal value.

7.2. Future primary school teachers are attracted by creative teachers (90.4%) who encourage student creativity in the project (82.7%) and research (84.7%) activities, allow them to express non–standard opinions in the practical training classes (88.5%), but, at the same time, offer samples of educational tasks (94.2%) and provide information in a ready-made form (69.2%) during lecture classes.

7.3. Theoretically thought out and tested for several years at the Pskov State University, the didactic model of creativity development includes the following:

- in the first years of professional training, students are immersed in creative educational activities with the help of methodological tools aimed at developing their ability to detect contradictions at the unconscious, partially conscious and reflective levels;
- further, at the stage of stimulating potential creativity, future teachers are trained to recognize even minor elements of novelty, methodically promising, productive differences in the didactic material used by different primary school teachers;
- then, future teachers are trained to design a collective professional methodological product for conducting lessons and extracurricular activities in primary school;
- at the final stage, the students, relying on their meta-professional instrumental competencies, knowledge of innovative teaching technologies, use various digital platforms to complete individual creative projects.

7.4. At all stages of the described didactic model, psychological and pedagogical conditions established for the development of creativity in future primary school teachers are capable of influencing the need–motivational and emotional spheres of the student's personality: regular reinforcement by the teacher of creative manifestations, which consist in "centring" the student on the self–achieved result; providing students with freedom of choice in pedagogical activities, while demanding responsibility for it; evaluating and promoting creative initiative and high quality of work.

References

1. Soffel, J.: What are the 21 st – century skills every student needs? Future of Jobs report word economic forum (2016). https://www.weforum.org/agenda/2016/03/21st-century-ski lls-future-jobs-students/. Accessed 11 Apr 2021
2. Fields, Z., Bischoff, C.A.: A theoretical model to measure creativity at a university. J. Soc. Sci. **34**(1), 47–59 (2013). https://doi.org/10.1080/09718923.2013.11893117
3. Csiksentmihalyi, M.: A systems perspective on creativity. In: Jane, H. (ed.) Creative Management and Development, 5nd edn., pp. 3–17. SAGE Publications Ltd., Thousand Oaks (2006). http://dx.doi.org/10.4135/9781446213704.n1
4. Shaheen, R.: Creativity and education. Creat. Educ. **1**(3), 166–169 (2010). https://doi.org/10. 4236/ce.2010.13026
5. Starko, A.J.: Creativity in the classroom. In: Schools of Curious Delight, 6th edn. Taylor & Francis Groups, Routledge, London (2017)
6. Bukhvalova, Ye.M., Karpushina, L.V.: Diagnostics of the instrumental values of the personality. Bull. Samara Humanit. Acad. Ser.: Psychol. **2**(4), 46–58 (2008). (in Russian). http://cyberl eninka.ru/article/n/diagnostika-instrumentalnyh-tsennostey-lichnosti. Accessed 03 May 2020
7. Sternberg, J.R.: The nature of creativity. Creat. Res. J. **18**(1), 87–98 (2006). https://doi.org/ 10.1207/s15326934crj1801_10
8. Kaufman, J.C.: Counting the muses: development of the Kaufman domains of creativity scale (K–DOCS). Psychol. Aesthet. Creati. Arts **6**(4), 298–308 (2012). https://doi.org/10.1037/a00 29751
9. Urban, K.K.: From creativity to responsible createlligence. Gift. Educ. Int. **30**(3), 237–247 (2014). https://doi.org/10.1177/0261429413485399
10. Craft, A., Cremin, T., Burnard, P.: Creative Learning 3–11: And How We Document IT. Trentham Books Ltd., London (2008)
11. Kangas, M.: Creative and playful learning: learning through game co–creation and games in a playful learning environment. Think. Skill Creat. **5**(1), 1–15 (2010). https://doi.org/10.1016/ j.tsc.2009.11.001
12. Plucker, A.J., Beghetto, R., Dow, G.: Why isn't creativity more important to educational psychologists? Potentials, pitfalls, and future directions in creativity research. Educ. Psychol. **39**(2), 83–96 (2004). https://doi.org/10.1207/s15326985ep3902_1
13. Robinson, K.: Out of Our Minds: Learning to be Creative. Capstone, Oxford (2011). http:// www.frcdkcmp.com/5365su12/robinsonchpt123.pdf. Accessed 28 Mar 2021
14. Feldman, D.H., Benjamin, A.C.: Creativity and education: an American retrospective. Camb. J. Educ. **36**(3), 319–336 (2006). https://doi.org/10.1080/03057640600865819
15. Torrance, E.P.: The torrance tests of creative thinking norms–technical manual figural (streamlined) forms A and B. Scholastic Testing Service, Bensenville (2008)
16. Byrge, C., Hansen, S.: The creative platform: a new paradigm for teaching creativity. Probl. Educ. 21st Cent. **18**, 33–50 (2009). https://www.scientiasocialis.lt/pec/node/335. Accessed 21 Oct 2020
17. Lund, M., Byrge, C., Nielsen, C.: From creativity to new venture creation: a conceptual model of training for original and useful business modelling. J. Creat. Bus. Innov. **3**(1), 65–88. (2017). https://vbn.aau.dk/ws/files/258628928/Lund_Byrge_Nielsen_2017.pdf. Accessed 21 Oct 2020
18. Guilford, J.P.: Creativity. Am. Psychol. **5**(9), 444–454 (1950). https://doi.org/10.1037/h00 63487
19. Williams, F.E.: Creativity Assessment Packet. CAP. Pro–Ed, Austin (1993)
20. Klochko, V.E.: Initiation of thought activity. Doctoral Dissertation. National Research Tomsk State University, Moscow (1991). (in Russian)

21. Nichol, L.: On Creativity. Routledge, London and New York (2004). https://doi.org/10.4324/9780203194713
22. Solovyeva, T., Vitkovskaya, I.: Philosophy of interpenetration of traditions and innovations in primary school teacher training. In: PCSF & CSIS, European Proceedings of Social and Behavioural Sciences, vol. 98, pp. 624–630. European Publishing Limited, London (2020). http://dx.doi.org/10.15405/epsbs.2020.12.03.63
23. Solovyeva, T., Vitkovskaya, I.: Future teachers' instrumental metaprofessional competencies development as way of their methodical competence establishment. In: Valeeva R (ed.) Arpha Proceedings 1: V International Forum on Teacher Education, Part I: Teacher Education and Training, pp. 665–677. Kazan Federal University, Kazan (2019). https://doi.org/10.3897/ap.1.e0630
24. Shlat, N., Borisov, B., Eyliseyeva N., Shakirova, N.: Conditions of formation of creative activity of students by means of information–communication technologies. In: Society. Integration. Education. Proceedings of the International Scientific Conference 2019, vol. 2, pp. 502–512. Rezekne Academy of Technologies, Rezekne (2019). http://dx.doi.org/10.17770/sie2019vol5.3762
25. Levin, K., Bandura, A.: Gestalt–psychology and socio–cognitive theory of personality. Prajm–EVROZNAK, St. Petersburg (2007). (in Russian)

Forming the Basics of Foreign Language Teachers' Methodological Creativity During Linguistic Training at a University

Nadezhda I. Almazova[1] (iD), Alexey S. Shimichev[2(✉)] (iD), and Olga G. Oberemko[2] (iD)

[1] Peter the Great St. Petersburg Polytechnic University (SPbPU), Polytechnicheskaya, 29, 195251 Saint-Petersburg, Russia
almazova_ni@spbstu.ru
[2] Linguistics University of Nizhny Novgorod, Minin Street, 31a, Nizhny Novgorod, Russia

Abstract. This article considers the issue of indirect methodological training of a foreign language teacher and the formation of the methodological creativity basis in foreign language classes at the university. The authors analyzed the existing research within the framework of the stated problem and found that the process of integrated students' linguistic and methodological training at the university has psychological and pedagogical, socio-psychological, psycholinguistic and creative grounds. The experience of practical work allowed authors to identify some groups of difficulties that students have during their pedagogical practice. They include: difficulties of goal-setting, difficulties in achieving the goals of the lesson, difficulties in reflecting on their own activities. It is argued that their consideration and overcoming are possible due to the reliance on the didactic potential of foreign language classes. From this point of view, the formation of the students' methodological competence takes place, on the one hand, and, the mastery of a foreign language as a means of communication, a means of teaching and a means of organizing pedagogical communication is realized on the other hand. Integrated linguistic and methodological training of future foreign language teachers is organized in two stages: reproductive and reproductive-creative. On the basis of the presented materials, the authors developed a four-staged technology for the formation of students' methodological skills in foreign language classes.

Keywords: Foreign language teaching · Integrated learning · Professional oriented education · Future foreign language teachers training

1 Introduction

The modern system of professional and pedagogical education is characterized by a change in priorities. The traditional way of transmission knowledge to students in a ready-made form gives way to the qualified training of specialists who are able and ready for constructive transformations of the educational space [1], taking into account the needs of both an individual student and the public order. It is a creative-minded teacher who can design and model various educational tracks in foreign languages, as well as predict the results of their own activities, being the main guarantor of achieving the desired results of the educational process.

In such conditions, the primary task is to optimize the methodological training of students in pedagogical areas of training, future foreign language teachers, which is implemented today in two directions: directly through the module of disciplines of the methodological cycle and indirectly in foreign language classes.

Taking into account the indicated objective realities, there is a need to create a comprehensive system of future foreign language teachers' methodological training [2]. This system should be based on cognitive situations, professional cases and problem tasks in the educational and students' cognitive and professional-pedagogical activities. Such approach allows us to give a holistic character to the educational process, provides a personal meaning to the assimilated material and also creates conditions for the development of intellectual creative and professional creative abilities of future teachers, which in turn forms the basis of methodological creativity.

The organization of practice in a foreign language classes, which is based on the principle of professional and pedagogical orientation, can contribute to the transformation of the educational process [3]. Following this principle has an active impact on all the functions of the process of teaching a foreign language, it serves as a means of students indirect methodological training, contributes to the development of students creative thinking. At the same time, conditions are created for the formation of methodological competence by involving students in solving individual methodological tasks in the course of the lesson. It ensures that the future teacher has a professionally oriented command of a foreign language as a means of teaching and pedagogical communication, and also affects the creative orientation of the student's personality. As a result students begin to understand more deeply the essence of his future profession, its values. Sample Heading (Third Level). Only two levels of headings should be numbered. Lower level headings remain unnumbered; they are formatted as run-in headings.

2 Literature Review

The scientometric analysis of the works related to the problem of the presented study led us to the conclusion that in the scientific literature there are many works describing different aspects of professionally oriented foreign languages teaching [4–6]. At the same time, it is emphasized that such work not only does not have a negative impact on the effectiveness of linguistic training, but also has a positive effect on the quality of students' mastery of educational material and contributes to the creation of a cognitive basis for future professional activities. Existing works on the development of professional creativity and professional language education allow us to identify several research areas that determine the vectors of the development of the problem of our research.

Firstly, there are the psychological and pedagogical foundations that describe the implementation of approaches and principles to professionally oriented foreign language teaching of future foreign language teachers in order to form their (teachers') methodological competence. According to Verbitsky [7], indirect methodological training creates a basis for each mastered discipline, provides interdisciplinary connections and the integrative nature of the formation a foreign language teacher professional competence.

The professionalization of students' linguistic training makes it possible to implement a contextual approach by modeling the subject and socio-oriented educational content.

In it turn, research in the field of labor activity psychology proves that the formation of methodological competence of future foreign language teachers corresponds to the generalized scheme of mastering the basics of professional functions. These processes are implemented in foreign language classes as initially set of "correct" algorithms of actions, including extraneous images of the central activity and the development of individual actions. Ilyin [8] argues that this process is achievable with abundant practice and can be provided by indirect methodological training in the course of foreign language practice.

Secondly, the research of the socio-psychological component of the integrated future foreign language teacher training characterize the models of foreign language interaction between the teacher and the students, determine the need to affect the sphere of students' motivation. In this direction, special attention is given to: 1) mastering and working out certain methodological actions that are proper to the teacher's personality and that are realized in educational process under the control of the professor and on a known contingent of students (classmates) [9–11]; 2) appeal to the students' personal motivational sphere as a basis for the development of a motivational and value attitude to the future profession [6]; 3) the formation of students' complete and adequate view about the chosen profession, which serves as a motivating factor for the formation of a positive attitude to learning.

Thirdly, the research of psycholinguistic features of integrated teaching of foreign languages and professional disciplines. These works analyze different aspects, as: 1) classes in a foreign language allow you to create situations of pedagogical communication that should be as close as possible to situations of natural interaction in a professional environment; 2) similar subject content of language material used at school and in junior university courses. It forms the basis for its creative refraction for educational purposes; 3) the professional orientation of classes in a foreign language gives students the opportunity to understand the ways of their own foreign language experience, the stages of forming a multicultural language personality based on the emerging linguodidactic competence; 4) according to Koryakovtseva [12], professionally-oriented training contributes to the formation and development of the student's educational and cognitive competence, primarily its procedural (strategic) component, which involves the choice of strategies and techniques for mastering the studied language and foreign language culture.

Fourthly, the study of the phenomenon of pedagogical mastery and methodological creativity of a foreign language teacher. Within the framework of this direction, Zyazyun, Krivonos, etc. [13] distinguish the main characteristics of the teacher's creative personality, namely: 1) high motivation level to creative self-expression; 2) creative potential treated as a set of knowledge, skills, and creative use of them in the educational process; 3) individual psychological characteristics (strong-willed character traits, self-organization, critical self-esteem, perception of oneself as the creator of material and spiritual values necessary for other people).

Considering the question of the formation and development of foreign language teacher's methodological creativity, it should be noted that the main criteria for the formation of methodological competence is the teacher's professional mastery. According to Galskova it represents the relationship of teacher's professionally significant knowledge and teaching experience, his/her creative and personal qualities [14]. Undoubtedly, in the process of organizing the learning process the teacher saturates the classes with a variety of methods and tools, uses various techniques and ways of working in order to achieve results. But it does not seem possible to call it a creative process, but rather an improvisation. Based on this idea, we consider methodological creativity as the highest degree of foreign language teacher methodological skills and mastery.

3   Materials and Methods

The sources of the research are psychological, pedagogical and methodological works within the framework of the selected problem, the authors' direct pedagogical experience in teaching the disciplines "Methods of teaching foreign languages", "Modern technologies of teaching foreign languages", "Practical course of a foreign language", "The culture of speech communication" for students in the direction of training "Pedagogical education", bachelor's and master's levels.

In order to achieve reliability and reasonableness in the consideration of the problem, they were used systematic, structural, and factor methods. To identify the specific traits of the technology of integrated co-study of a foreign language and disciplines of the methodological cycle, methods of comparative and system-complex analysis of scientific literature on the topic of research, methods of pedagogical analysis and generalization of facts, situations and documents, modeling, and interpretation of the obtained results were employed.

4   Results

As we have found, indirect methodological training of future foreign language teachers carried out in the process of professionally oriented foreign language learning, makes a significant contribution to the formation and development of the students' methodological competence, and hence the elements of methodological mastery and methodological creativity.

However, the practical experience allows us to identify some difficulties that students face during their pedagogical practice and which, in our opinion, determine the organization of the learning process according to the principle of professional pedagogical orientation of the educational process at the university. These difficulties include:

- difficulties of goal setting (analysis of the material presented in the textbooks and forecasting of possible difficulties of schoolchildren; methodological adaptation of the material borrowed from other sources to the learning conditions and schoolchildren needs, etc.);

- difficulties in achieving the goals of the lesson (organization of various forms of pupils' interaction at the foreign language lesson; easy lesson conduction in a foreign language, including the organization of pedagogical communication; creation of situations of intercultural communication at the lesson; identification of difficulties in the implementation of the lesson plan and its immediate adjustment, etc.);
- difficulties in reflecting on students' own activities, expressed in a weak self-analysis of the lesson, which makes impossible the accurate statement of new tasks.

It also should be noted that the lack of students' methodological experience rejects any possibility of methodological creativity [15]. Students-trainees follow step by step the teacher's guide, and also copy the activities of the university foreign language professor, who serves as a model. Such imitation is absolutely unacceptable due to the differences in the psychological and didactic-methodological foundations of teaching schoolchildren and university students.

In this regard, it seems that the didactic potential of foreign language classes is the basis that fully allows for the integrated formation of students' methodological competence, on the one hand, and on the other, it makes possible the mastering a foreign language as a communication means, a tool of teaching and a means of organizing pedagogical communication. At the same time, linguistic and methodological skills allow the teacher to use a foreign language at three levels [16]: communicative (practical foreign language skills), adaptive (analysis of language units and theoretical knowledge to organize learning units, identify possible difficulties and ways to overcome them) and professional modeling (modeling of situations of intercultural communication to organize the presentation of language and speech material to students, as well as methodological algorithms for working with this material).

Integrated linguistic and methodological training of students in pedagogical areas can be implemented in two main stages: reproductive (in junior courses, before mastering the course of methods of foreign language teaching) and reproductive and creative (in parallel with the course of methods of foreign language teaching and other disciplines of the methodological module) [17].

At the first, reproductive stage, it is created in students' minds an indicative basis for performing methodological actions, formed by performing individual action by following the examples, as well as on the basis of schemes, instructions, checklists, algorithms, etc. Mastering this level allows future foreign language teachers to solve methodological problems by reproducing previously learned methods of activity on the basis of a given standard, which does not fully take into account the conditions for solving problems.

In order to ensure the implementation of quasi-professional activities in English classes, we selected texts and developed tasks for them in such a way that the information, firstly, was assimilated in the context of future activities, and secondly, contributed to the formation of the personal meaning of mastering this knowledge. The selected texts are divided into several blocks. It helps students to conceive the content of future professional activity and allow students to get an idea of the integral system of pedagogical activity. "Why do you want to be a teacher?"; "All in one" (it's about the teacher's functions); "Work for heroes? The difficulties of the profession"; "Teacher's speech"; "Teacher's look"; "How to attract attention with non-verbal behaviour"; "The art of praise"; "Classes can be fun"; "How to establish discipline"; "The importance of good communication in class"; "My school internship"; "Learning through failure!"), etc. The algorithm is implemented according to the following scheme: actualization of student's own experience – theoretical conceptualization (how to do it correctly) based on the introduction and assimilation of new information – generalization and acquisition of new experience.

Also, at the reproductive training stage, we often use practice-oriented tasks such as:

- Observe how your professor conducts the lesson and make up an articulatory training;
- Write a dictation on the studied topic being;
- Compose lexical and grammatical exercises similar to those presented in the textbook;
- Remember how your teacher organizes work with grammatical material. What is the algorithm of its operation? etc.

At the second, reproductive and creative stage, the further formation of methodological thinking takes place through the theoretical reinforcement of the existing basis with knowledge and skills formed in the process of mastering the module of methodological disciplines. The achievement of the reproductive and creative level of methodological competence by students means not only the reproduction of the learned methods and teaching tools, but also their combination, transformation, selection of the most optimal and effective techniques for specific learning conditions.

At this stage, when students are studying a foreign language and the course of its teaching methods, we use the following tasks:

- Analyze the organization of work with the studied text, what are these stages?
- Make a complex of exercises to train the grammatical phenomenon of Present Perfect;
- Imagine that you have to organize work with students of the 8th grade on the development of dialog speech skills. What is your algorithm of work and the algorithm of students' work?
- Analyze the content of the textbook section and try to divide it into separate lessons.
- Analyze the content of the textbook section. What goals can I set when working with this section?

The described stages of integrated methodological training of future foreign language teachers correspond to the basic models identified by Wallace [18]: 1) the craft model - in which training is carried out under the guidance of a professor, in the process of observing his/her activities, following his/her advice; 2) the applied science model is the most popular, mastering practical actions and operations is carried out on the basis of existing scientific knowledge. From our point of view there is a third model that combines the two previous. This model includes organized observation of the teacher's activities in combination with empirical knowledge gained during the foreign language learning. As a result, a layer of theoretical knowledge is formed in the students' mind, which is further supported by the students' practical activities that are carried out in close cooperation with the professor and under his/her guidance and control.

In this regard, we are worked out the technology of integrated foreign language teaching and the methodology of teaching. Unfortunately, the scope of this article does not allow us to give a complete description of this progress, so we will limit ourselves to a brief description of its four main stages:

- 1. The stage of motivation and orientation. It aims to create students' cognitive motivation and interest to master a foreign language and carry out professional pedagogical activities. Quasi-professional activity, according to Verbitsky [7], in the process of professionally oriented foreign language teaching involves a combination of students' features of teaching and future independent professional activity. It contributes to the formation of a real system of personal ideas about the methodological activities of a foreign language teacher, which, as scientists emphasize, is of great importance, since the choice of a profession is not always consciously made. In scientific studies, it has been repeatedly stated that as the development of professional activity and under the condition of approaching the educational activity to the professional, there is a change of motives: the cognitive motive transforms into a professional one. As a result, professionally oriented foreign language teaching creates conditions that promote the development of personal qualities necessary for a future foreign language teacher.
- 2. The stage of implementation of the methodological action on the basis of supports and external speech [19], aimed at gaining experience in the implementation of methodological actions. At this stage, students perform reproductive operations using their received basis: memory cards, diagrams, memos, etc. At the same time, external speech is provided with various comments on the teacher's activities, discussion in the process of finding a solution to a methodological problem.
- 3. The stage of performing a didactic action in the mental plan. It means that the previously developed algorithms of activity become internal processes and acquire a personal treats of flow. It should be noted that the individual work becomes especially important at this stage, which is due to the specifics of the teacher's professional activity, the nature of his management of schoolchildren' educational activities.
- 4. The productive and creative stage, in which students try to design and realize their own fragments of foreign language lessons on the basis of individual differentiated tasks. Such organization of students' work allows a) to develop the future teacher's skills of role-based foreign language professional communication (to understand, accept and fulfill a certain functional role; to navigate the roles of communication partners; to regulate their speech behavior in a conflict situation, etc.); b) identify some individual specifics of the skills of professional and pedagogical communication in different students and individualize the learning process; c) develop skills for analyzing situations of pedagogical communication. The analysis and collaborative discussion of the participants' behavior in the role-playing game requires students to update their practical knowledge of pedagogy, psychology, methods of a foreign language teaching and develops their ability to evaluate their own lessons or its fragments. This stage involves the development of students' individual teaching style [20] being an active subject of their own pedagogical activity, able to make professionally competent decisions, having reflexive competence. It is also important to use educational and pedagogical tasks in the classroom that require students to independently explain phenomena and processes, controversial methodological situations, etc.

To sum up, it is possible to present the technology in the form of the following scheme (Fig. 1):

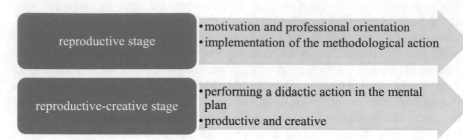

Fig. 1. The technology of integrated foreign language teaching and the teaching methods course.

5 Conclusion

Within the framework of the presented research, it was revealed that it needs a complex, systematic approach to improving the methodological training of future foreign language teachers based on inter-subject interaction, on the one hand, and modern pedagogical teaching technologies, on the other hand, that help to overcome a certain gap between theory and practice. Such approach is an effective way to develop students' professional skills and their methodological creativity. The development of activities in conditions where there is emotional empathy and mental assistance, co-thinking, when the learning process turns into a creative dialog communication, proceeds at the personal level and provides a motive for activity. The classroom activity creates an atmosphere of cooperation in which students acquire personal implication and values.

References

1. Almazova, N., Eremin, Y., Kats, N., Rubtsova, A.: Integrative multifunctional model of bilingual teacher education. IOP Conf. Ser. Mater. Sci. Eng. **940**, 012134 (2020). https://doi.org/10.1088/1757-899X/940/1/012134
2. Almazova, N., Rubtsova, A., Krylova, E., Barinova, D., Eremin, Y., Smolskaia, N.: Blended learning model in the innovative electronic basis of technical engineers training. In: Katalinic, B. (ed.) Proceedings of the 30th DAAAM International Symposium, pp. 0814–0825. DAAAM International, Zadar (2019). https://doi.org/10.2507/30th.daaam.proceedings.113
3. Odinokaya, M., Karpovich, I., Mikhailova, O., Piyatnitsky, A., Klímová, B.: Interactive technology of pedagogical assistance as a means of adaptation of foreign first-year students. IOP Conf. Ser. Mater. Sci. Eng. **940**, 1–13 (2020). https://doi.org/10.1088/1757-899x/940/1/012130
4. Bylieva, D., Lobatyuk, V., Safonova, A.: Online forums: communication model, categories of online communication regulation and norms of behavior. Humanit. Soc. Sci. Rev. **7**, 332–340 (2019). https://doi.org/10.18510/hssr.2019.7138

5. Oberemko, O., Glumova, E., Shimichev, A.: Developing foreign language regional compe-tence of future foreign language teachers: modeling of the process. In: Anikina, Z. (ed.) GGSSH 2019. AISC, vol. 907, pp. 195–209. Springer, Cham (2019). https://doi.org/10.1007/978-3-030-11473-2_22
6. Shimichev, A.S.: Creativity as a component of methodological competence of a foreign language teacher. World Acad.: Cult. Educ. **8**, 81–84 (2020). (in Russian)
7. Verbitsky, A.A., Larionova, O.G.: Personal and competence approaches in education: problems of integration. Logos, Moscow (2009). (in Russian)
8. Ilyin, E.P.: Psychology of creativity and cleverness. PITER, St. Petersburg (2012). (in Russian)
9. Bylieva, D.S., Lobatyuk, V.V., Fedyukovsky, A.A.: Ways of sociotechnical integration of scientists and volunteers in citizen science. IOP Conf. Ser. Mater. Sci. Eng. **940**, 012150 (2020). https://doi.org/10.1088/1757-899X/940/1/012150
10. Maker, C., Sonmi, J., Muammar, O.: Development of creativity: the influence of varying levels of implementation of the DISCOVER curriculum model, a non-traditional pedagogical approach. Learn. Individ. Diff. **18**(4), 402–417 (2008). https://doi.org/10.1016/j.lindif.2008.03.003
11. Shimichev, A.: Development of students' creative abilities in technical university during foreign language training. In: Anikina, Z. (ed.) IEEHGIP 2020. LNNS, vol. 131, pp. 212–221. Springer, Cham (2020). https://doi.org/10.1007/978-3-030-47415-7_22
12. Koryakovtseva, N.F.: Theory of Teaching Foreign Languages: Productive Educational Technologies. Academy, Moscow (2010). (in Russian)
13. Zyazyun, I.V., Krivonos, I.F.: Fundamentals of Teaching Skills. Prosveshchenie, Moscow (1989). (in Russian)
14. Galskova, N.D.: Modern Methods of Foreign Languages Teaching. ARKTI, Moscow (2004). (in Russian)
15. Rogers, C.R.: Towards a theory of creativity. In: Vernon, P.E. (ed.) Creativity, pp. 137–151. Penguin, Harmondsworth (1972)
16. Odinokaya, M., Andreeva, A., Mikhailova, O., Petrov, M., Pyatnitsky, N.: Modern aspects of the implementation of interactive technologies in a multidisciplinary university. In: E3S Web Conference, vol. 164, p. 12011 (2020). https://doi.org/10.1051/e3sconf/20201641201
17. Shipunova, O.D., Berezovskaya, I.P., Smolskaia, N.B.: The role of student's self-actualization in adapting to the e-learning environment. In: TEEM 2019: Proceedings of the Seventh Interna-tional Conference on Technological Ecosystems for Enhancing Multiculturality, pp. 745–750. ACM, León (2019). https://doi.org/10.1145/3362789.3362884
18. Wallace, M.J.: Training Foreign Language Teachers: A Reflective Approach. CUP, Cambridge (1991)
19. Oberemko, O.G.: Axiological bases of future foreign language teacher's communicative competence formation. Prob. Modern Pedag. Edu. **52**(2), 162–169 (2016). (in Russian)
20. Sternberg, R.: The nature of creativity. Creativity Res. J. **1**, 87–98 (2006)

Digital Creative Projects in the Formation of Digital Competence of Teachers of English as a Foreign Language

Nadezhda I. Almazova[iD], Anna V. Rubtsova[iD], Natalia B. Smolskaia[iD], and Antonina A. Andreeva[✉][iD]

Graduate School of Applied Linguistics, Interpreting and Translation, Peter the Great St. Petersburg Polytechnic University, Polytechnicheskaya 29, Saint Petersburg 195251, Russia
lingua@mail.spbstu.ru

Abstract. The article discusses how digital creative projects can be used in the practice of professional training of English as a Foreign Language (EFL) teachers at a higher education institution. As the nature of creativity and creative activity is being examined in the theoretical part of the article, it acquires a definition that any activity, both theoretical and practical, in any professional environment can be creative. Creativity is determined not by the type of activity, but by the presence and degree of manifestation of such qualities as conscious goal-setting, novelty, the significance of the methods of its implementation or result. The need for creative tasks in professional training is determined. Since the object of our study is professional training of future EFL teachers, the structure of professional competence of an EFL teacher is described with a proposal of expanding it with a new digital competence. In the practical part of the article the experimental training conducted at the Graduate school of applied linguistics, interpreting and translation is described. The experimental training consists of studying international regulatory documents on digital competence and completing a digital creative project of an online course on digital literacy development.

Keywords: Creativity · Digital competence · EFL teachers

1   Creativity

Professional pedagogical education is a complex process of developing personal qualities and forming competencies that will help to realize the potential of an individual and achieve his professional goals. Since the profession of an educator requires creativity, creativity plays an important role in the professional training of future English as Foreign Language (EFL) teachers.

Creativity is directly related to human activity. It is this approach to the essence of creativity that reflects its definition as an activity that generates something qualitatively new, distinguished by uniqueness, originality, and socio-historical uniqueness. Creativity is specific for a person, as it always presupposes a creator - a subject of creative activity.

Therefore, creativity should be understood as a special (possessing a special set of qualities) type of activity.

Dobudko et al. define creativity as "the kind of human activity that creates something new, no matter whether it is created by something of the external world or a well-known construction of the mind or feeling that lives and is found only in the person himself" [1].

Khenner proposes to consider creativity as "one of the types of human activity aimed at resolving a contradiction (solving a creative problem), which requires objective (social, material) and subjective personal conditions (knowledge, skills, creativity), a result that has novelty and originality, personal and social significance, as well as progressiveness" [2].

We consider it possible to conclude that creativity is an activity, the product of which is something brand new, original, something that has never happened before. One can generally agree with this definition, but concerning the educational process, the interpretation of the term "creative activity" will be somewhat different, since students often cannot create products that have social novelty and significance.

Thus, Vindaca and Lubkina note in their research that in the psychological aspect there is no difference between the creative activity of a scientist who creates a product of creativity characterized by objective novelty and the creative activity of a student who discovers himself, but not for society [3].

Prestridge et al., analyzing psychological and pedagogical work on the problem of creativity, most fully noted the main signs of educational creative activity [4]. Educational and creative activity - this is one of the types of educational activities aimed at solving educational and creative tasks, carried out mainly in the context of the use of pedagogical means of indirect or prospective management, focused on the maximum use of personal self-government, the result of which has subjective novelty, significance, and progressiveness for personality development and creativity.

Thus, creative activity, organized in the conditions of the educational process of the university, should be focused on solving educational problems. That is, creative educational and cognitive activity is an independent search and creation or construction of some new product (in the individual experience of a student - a new scientific knowledge or method unknown to him, but known, as a rule, in social experience), and therefore, the main criteria of creativity in the cognitive activity of a student are:

- Independence (full or partial);
- Search and enumeration of possible options for moving towards the goal;
- Creation of a new product, previously unknown to him in the process of moving towards the goal.

Creative activity, like any kind of human activity, has its structure. Scientists who have studied the regularity of various types of creativity have come to the general conclusion that the emergence of a problem situation in any sphere of human life, its awareness, the statement of the problem is the beginning of the research, creative search, carried out in the process of thinking, and the solution of the problem is the content of creativity.

Any activity, both theoretical and practical, in any professional environment can be creative. Creativity is determined not by the type of activity, but by the presence and degree of manifestation of such qualities as conscious goal-setting, novelty, the significance of the methods of its implementation or result.

Odinokaya et al. argue that modern higher educational institutions are characterized by the search for forms of psychological and pedagogical conditions that are aimed at revealing individuality, creative growth of personality since today's students lack independence in acquiring knowledge, thinking, and self-expression [5]. Many students are distinguished by a low level of cognitive activity. The situation can be changed by innovations aimed at transforming the traditional educational process, because of which this process acquires research and creative character.

Speaking of the search for psychological and pedagogical conditions that can contribute to increasing the motivation and activity of modern students, one can also refer to the provisions of the sociological theory of generations by Strauss and Howe. According to the theory, today's students born in 1990–2005 belong to generations of Y and Z. Representatives of both generations, according to the sociological theory of generations, are characterized by the need for creative tasks as an opportunity to express themselves. Among the Zs, sociologists note extremely sober ideas about life and about ways to achieve success. Ys dream that the whole world will notice how wonderful they are, on the contrary, the Zs do not have such illusions, they are going to work hard and earn money. Zs are not rebellious teenagers, they are responsible for their lives and are not prone to bad habits. They do not want to destroy the system; they are going to make the world a better place slowly and surely. Ys are the kind of dreamers who strive for beauty and expression through creativity. Zs, in turn, also strive to show their talents, but more assertively - through competition [6]. To meet these needs in the process of professional training of Ys and Zs, it seems important to use: creative tasks; case method; problem-search method; method of projects.

2 Professional Competence

Since the object of our study is the professional training of future EFL teachers, it is necessary to clarify the structure of their professional competence, considering the specifics of their professional activities. In the practice of professional training of EFL teachers over the past two decades, there has been an understanding of their professional competence as a combination of such components as communicative, philological, methodological, psychological-pedagogical, and personal competencies. However, the world has changed, and the education system has changed as well. Teachers of all disciplines, including EFL teachers, today face new challenges in the application of digital technologies in the education process to intensify and optimize it, and these tasks require new specialized knowledge, skills, and abilities that are not included in the listed competencies, but forming a different one - digital competence.

Communicative competence, which is a part of an EFL teacher's professional competence, is invariably relevant with its division into linguistic, speech, and socio-cultural subcompetencies. EFL teachers should strive for free, figurative, and expressive speech, corresponding to a fluent level of proficiency in a foreign language, and should also strive for knowledge of world culture to avoid mistakes of sociolinguistic nature [7, 8].

Psychological-pedagogical competence always remains invariably important for a teacher of any discipline, including an EFL teacher. The ability to take into account the age, psychological and sociological characteristics of students is necessary for every teacher when planning and organizing the learning process [9].

Philological competence, which is also a part of the professional competence of an EFL teacher, in addition to components directly linking it with communicative competence (knowledge of the structure of native and foreign languages, the ability to interpret complex linguistic phenomena in native and foreign languages, etc.), It also has several features that distinguish it (mastery of rhetoric, general knowledge of literature, history of foreign languages and other philological disciplines, the ability to use them in the process of learning and communication) [2].

The methodological competence of an EFL teacher, with all the ambiguity of its assessment by researchers, is invariably associated with the ability to effectively solve the problems of education. The specified competence is not static, but dynamic, and is in the process of constant development. The changing tasks of education and upbringing facing the teacher require him to constantly update and deepen his knowledge, skills, and abilities, as well as revise the individual teaching style [2, 10].

Personal competence is also a part of the professional competence of an EFL teacher and, regardless of the time, includes the willingness to show sufficient exactingness both to students and to oneself; mastery of diction, voice, facial expressions, and gestures, both for establishing interaction with students and as verbal and non-verbal support in teaching foreign languages. The specified competence also presupposes the possession of such personal qualities as tolerance, openness in communication, tact, respect for oneself and others, love for one's profession, activity, and purposefulness [7, 11].

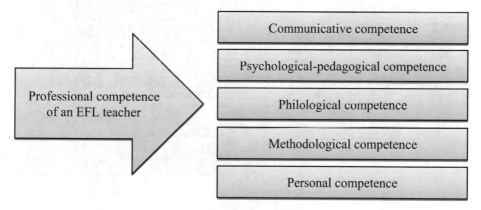

Fig. 1. Professional competence of an EFL teacher structure

As we can see, the listed competencies cover all aspects of the professional activity of an EFL teacher (see Fig. 1), except for the one related to the active and regular use of digital technologies in the educational process on a well-thought-out didactic basis. We agree with Kabanova & Kogan, who indicate the need to expand the structure of professional competence of a modern EFL teacher and supplement it with digital competence [11], as a response to the modern request of society and the state, and as a tribute to the new time [12–14].

We are convinced that the use of creative tasks can make a great contribution to the formation of digital competence since they are mainly focused on solving educational problems and creating a new product. Independent search, conscious goal-setting and novelty characteristics of creative tasks are the qualities necessary for the active and regular use of digital technologies in the educational process. Therefore, we decided to include work on digital creative projects in the process of professional training of future EFL teachers at the Graduate school of applied linguistics, interpreting and translation of Peter the Great St. Petersburg Polytechnic University.

3 Digital Competence

As was mentioned earlier, Khenner indicated communicative, philological, method-ological, psychological-pedagogical, and personal competence as components of the professional competence of an EFL teacher (see Fig. 1) [2]. However, we consider it necessary to expand the list of these components and include digital competence in the structure of the professional competence of an EFL teacher as a response to the request of the modern information society and the requirements of international and domestic regulatory documents (see Fig. 2).

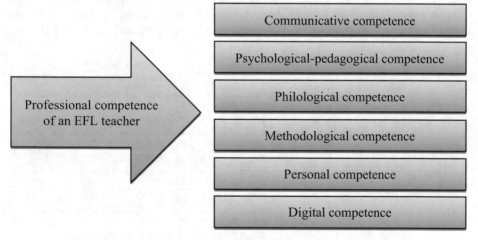

Fig. 2. Expanded structure of an EFL teacher professional competence.

Within the framework of this study, we will define digital competence as a complex of knowledge and skills related to the use of digital technologies in teaching foreign languages, which is a significant part of the professional competence of an EFL teacher [15, 16].

During the professional training of EFL teachers at the Graduate school of applied linguistics, interpreting and translation of Peter the Great St. Petersburg Polytechnic University, we have conducted experimental training of future EFL teachers aimed at their digital competence formation using creative project assignments for the group development of online courses.

The topic of those creative project assignments was not chosen by us by chance, but because of studying international regulatory documentation, namely the document developed by the European Commission - European Framework for the Digital Competence of Educators (DigCompEdu) [17]. The list of requirements presented by the authors of the document discusses the importance of teaching modern educators' digital literacy (see Fig. 3). After exploring the topic of digital literacy with our students, we offer them a group task to develop an online course on digital literacy. Based on the properties of creativity that we described earlier, we assumed that it is the task of a creative nature that will allow our students to form the digital competence necessary for every modern teacher [18–20].

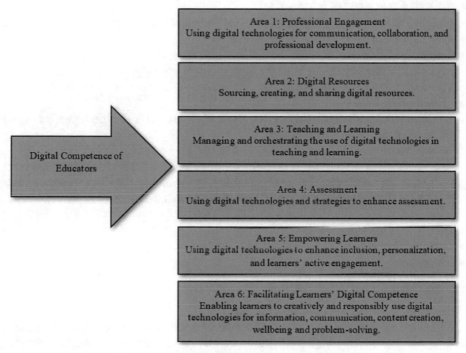

Fig. 3. Digital competence structure according to European Framework for the Digital Competence of Educators (DigCompEdu) [17].

During experimental training, a control group of future EFL teachers studied the structure of digital competence of educators proposed by specialists from the European Commission without completing creative project assignments for the group development of online courses on digital literacy. While the experimental group of students, upon completing the study of the structure of digital competence of educators, was asked to complete a group practical task of a creative nature to develop an online course on digital literacy covering such topics as:

- Topic 1: Digital Hygiene

- Topic 2: Communication Security
- Topic 3: Law in the digital environment
- Topic 4: Psychology of personality in a digital society
- Topic 5: Linguistic personality in the digital environment

While completing this creative practical task to develop an online course on digital literacy, an experimental group of future EFL teachers has performed the following stages of work on each topic:

- Analysis of available information;
- Search for information gaps;
- Collection and study of information;
- Finding the best way to present information to course participants;
- Execution of planned technological operations;
- Current quality control;
- Making (if necessary) changes to the design and technology;
- Preparation of presentation materials;
- Project presentation;
- Exploring the possibilities of using the results of the project (exhibition, sale, inclusion in the bank of projects, publication).

As a result of this practical assignment, our future EFL teachers have presented their group online course "Digital literacy" made via the educational platform Stepik.org. For each topic of the course were prepared the following materials:

- Glossary on the topic;
- Lecture on the topic;
- Presentation on the topic;
- Resources on the topic;
- Online testing on the topic.

The final stage of our experimental training of future EFL teachers was their testing and questioning to determine the formation of their digital competence. The experimental group of students not only have demonstrated a higher level of theoretical knowledge required for digital competence than the control group, but also readiness for the practical application of this competence.

4 Conclusion

We consider it possible to assume that the described experience of using digital creative projects in the process of professional training of future EFL teachers can be useful to educational institutions around the world wishing to prepare modern competitive educators.

The structure of a creative task for a student will contribute to his professional development and digital competence formation if it consists of the following components: the presence of problems and atypical tasks; non-standard approaches to their solution due to the lack of algorithms and the impossibility of using known techniques and actions; uncertainty in the search for a solution; lack of information; development of a solution scheme: plan - program - method - project; determination of tactical actions; using digital technologies.

Independent activity of students is possible when conducting research, performing creative and intellectual tasks, creating projects, participating in role games, preparing essays and reports, writing term papers. Thus, a detailed examination of the content, structure of creative and projective activity allows us to conclude that it is necessary to use them along with other types of educational activities in the educational process of a higher education institution for the creative development of students.

References

1. Dobudko, T., Korostelev, A., Gorbatov, S., Kurochkin, A., Akhmetov, L.: The organization of the university educational process in terms of digitalization of education. Humanit. Soc. Sci. Rev. **7**, 1148–1154 (2019). https://doi.org/10.18510/hssr.2019.74156
2. Khenner, E.: Professional knowledge and professional competencies in higher education. Educ. Sci. J. **20**, 9–31 (2018). https://doi.org/10.17853/1994-5639-2018-2-9-31
3. Vindaca, O., Lubkina, V.: Transformative digital learning in the context of higher education: definition and basic concepts. Rural Environ. Educ. Personal. (REEP) **13**, 177–184 (2020). https://doi.org/10.22616/REEP.2020.021
4. Prestridge, S., Tondeur, J., Ottenbreit-Leftwich, A.: Insights from ICT-expert teachers about the design of educational practice: the learning opportunities of social media. Technol. Pedag. Educ. **28**, 157–172 (2019). https://doi.org/10.1080/1475939X.2019.1578685
5. Odinokaya, M., Andreeva, A., Mikhailova, O., Petrov, M., Pyatnitsky, N.: Modern aspects of the implementation of interactive technologies in a multidisciplinary university. E3S Web Conf. **164**, 12011 (2020). https://doi.org/10.1051/e3sconf/202016412011
6. Howe, N., Strauss, W.: Millennials Rising. Vintage Books, New York (2006)
7. Glukhov, V., Vasetskaya, N.: Improving the teaching quality with a smart-education system. In: 2017 IEEE VI Forum Strategic Partnership of Universities and Enterprises of Hi-Tech Branches (Science. Education. Innovations) (SPUE), pp. 17–21. IEEE, St. Petersburg (2017). https://doi.org/10.1109/ivforum.2017.8245958
8. Rubtsova, A.V., Almazova, N.I., Bylieva, D.S., Krylova, E.A.: Constructive model of multilingual education management in higher school. IOP Conf. Ser. Mater. Sci. Eng. **940**, 012132 (2020). https://doi.org/10.1088/1757-899X/940/1/012132
9. Guillén-Gámez, F.D., Mayorga-Fernández, M.J., Bravo-Agapito, J., Escribano-Ortiz, D.: Analysis of teachers' pedagogical digital competence: identification of factors predicting their acquisition. Technol. Knowl. Learn. **26**(3), 481–498 (2020). https://doi.org/10.1007/s10758-019-09432-7
10. Bylieva, D., Lobatyuk, V., Safonova, A.: Online forums: communication model, categories of online communication regulation and norms of behavior. Hum. Soc. Sci. Rev. **7**, 332–340 (2019). https://doi.org/10.18510/hssr.2019.7138
11. Kabanova, N., Kogan, M.: Needs analysis as a cornerstone in formation of ICT competence in language teachers through specially tailored in-service training course. In: Zaphiris, P., Ioannou, A. (eds.) LCT 2017. LNCS, vol. 10295, pp. 110–123. Springer, Cham (2017). https://doi.org/10.1007/978-3-319-58509-3_11

12. Almazova, N., Bernavskaya, M., Barinova, D., Odinokaya, M., Rubtsova, A.: Interactive learning technology for overcoming academic adaptation barriers. In: Anikina, Z. (ed.) IEE-HGIP 2020. LNNS, vol. 131, pp. 786–794. Springer, Cham (2020). https://doi.org/10.1007/978-3-030-47415-7_84

13. Baranova, T., Kobicheva, A., Tokareva, E.: The impact of an online intercultural project on students' cultural intelligence development. In: Bylieva, D., Nordmann, A., Shipunova, O., Volkova, V. (eds.) PCSF/CSIS -2020. LNNS, vol. 184, pp. 219–229. Springer, Cham (2021). https://doi.org/10.1007/978-3-030-65857-1_19

14. Berezovskaya, I., Shipunova, O., Kedich, S., Popova, N.: Affective and cognitive factors of internet user behaviour. In: Bylieva, D., Nordmann, A., Shipunova, O., Volkova, V. (eds.) PCSF/CSIS -2020. LNNS, vol. 184, pp. 38–49. Springer, Cham (2021). https://doi.org/10.1007/978-3-030-65857-1_5

15. Jeong, K.-O.: Preparing EFL student teachers with new technologies in the Korean context. Comput. Assist. Lang. Learn. 30, 488–509 (2017). https://doi.org/10.1080/09588221.2017.1321554

16. Romero-García, C., Buzón-García, O., de Paz-Lugo, P.: Improving future teachers' digital competence using active methodologies. Sustainability. 12, 7798 (2020). https://doi.org/10.3390/su12187798

17. Punie, Y., Redecker, C.: European Framework for the Digital Competence of Educators (DigCompEdu). Publications Office of the European Union, Luxembourg (2017). https://doi.org/10.2760/159770

18. Bylieva, D., Lobatyuk, V., Fedyukovsky, A.: Ways of sociotechnical integration of scientists and volunteers in citizen science. IOP Conf. Ser. Mat. Sci. Eng. 940, 012150 (2020). https://doi.org/10.1088/1757-899X/940/1/012150

19. Hewagamage, C., Hewagamage, K.: A framework for enhancing ICT competence of universities in Sri Lanka. Int. J. Emerg. Technol. Learn. 10(5), 45 (2015). https://doi.org/10.3991/ijet.v10i5.4802

20. Odinokaya, M., Karpovich, I., Ju Mikhailova, O., Piyatnitsky, A., Klímová, B.: Interactive technology of pedagogical assistance as a means of adaptation of foreign first-year students. IOP Conf. Ser.: Mater. Sci. Eng. 940, 012130 (2020). https://doi.org/10.1088/1757-899X/940/1/012130

The Formation of Analytic Abilities for Coaching Rhythmic Gymnastics

Ali Namazov[1]($\boxtimes$), Elena Medvedeva[2], Aleksandra Suprun[2], and Inna Kivikhariu[2]

[1] Peter the Great St. Petersburg Polytechnic University, ul. Politehnicheskaja 27, St. Petersburg, Russia
kuem@list.ru
[2] Lesgaft National State University of Physical Education, Sport and Health, St. Petersburg, ul. Dekabristov 35, St. Petersburg, Russia

Abstract. Currently gymnast's success depends on the masterly possession of all groups of movements with and without an object. Each competitive composition consists of various technical elements with an apparatus that carries both quantitative and qualitative value. At the same time, in the process of analyzing the content of competitive compositions, it was found that over the past 3 years, the proportion of the evaluated movements by the apparatus has increased almost 2 times. As a result of a consistent change in the requirements of the rules of competitions in rhythmic gymnastics, coaches are forced to devote most of the time of the training process to learning new elements, ligaments, connections with gymnasts and drawing up new compositions of competitive programs. Which requires coaches to constantly search for scientifically grounded creative approaches to the training process and analytical thinking, necessary for the implementation of all types of training of athletes. The professional training of a sports coach is a complex and dynamic pedagogical system, the effectiveness of which depends on many factors that manifest themselves on the basis of the general laws of the pedagogical process and its management. A systematic approach to the formation and assessment of the competence of students allows you to fully reveal the potential of a future coach. At the same time, the basic analytical skills of the future coach serve as the basis for the formation of interdisciplinary knowledge. The article discusses the issues of the effectiveness of the use of analytical activities and the implementation of the thesis in the aspect of the professional competence of future coaches in rhythmic gymnastics.

Keywords: Rhythmic gymnastics · Coach · Professional competence · Analytical skills · Thesis · Testing

1 Introduction

One of the most important tasks of modern Russian education is to ensure the quality of training of young specialists and their competitiveness in the labor market [1–7].

The content of education should meet the specific needs of society and be focused on the formation of competencies that correspond to the functions of the professional

© The Author(s), under exclusive license to Springer Nature Switzerland AG 2022
D. Bylieva and A. Nordmann (Eds.): PCSF 2021, LNNS 345, pp. 765–773, 2022.
https://doi.org/10.1007/978-3-030-89708-6_62

standards [8–13]. Considering that the work of a trainer in rhythmic gymnastics is associated with the manifestation of creativity and a constant search for innovations to solve the promising tasks of training athletes, achieving performance is impossible without the analytical experience and the ability to put forward and prove hypotheses [14–17]. Therefore, the methodology in which the student finds himself in the position of an active participant in the cognition process under the guidance of a teacher is the only constructive one [18–20]. Also, the use of informative tools evaluation results of its natural application.

Rhythmic gymnastics currently demonstrates unique compositions in which complex, original elements and connections are performed. Changes in the main document - the rules of the International Gymnastics Federation - contributed to the progress of the coordination complexity of the competitive composition and the need to simultaneously solve motor tasks at different hierarchical levels of control of building movements. Due to this trend, the requirements for graduates have changed in 2019, which actualized the problem of monitoring the implementation of independent analytical activities, ensuring the progressive competence development of the future coach and professional creativity and the effectiveness of the process it implements [21, 22].

In this regard, there is an urgent need to find ways to create feedback and to quickly assess the preparedness to carry out analytical activities in the process of finishing thesis on modern problems of sports training in rhythmic gymnastics.

The paper is exploring following research questions:

1. To determine the direction of research of students of the Department of Theory and Methods of Gymnastics (Rhythmic Gymnastics) in Lesgaft National State University of Physical Education, Sport and Health in graduation works (Thesis) for the last 5 years (2017–2021).
2. To substantiate the dependence of the quality of the final qualifying work, reflecting the level of students' analytical abilities, on theoretical knowledge determined by test tasks and the compliance of future specialists with the requirements of this sport.

Purpose of the research is to substantiate the possibility of diagnosing analytical capabilities of students as coaches in rhythmic gymnastics, taking into account the requirements of the professional standard "Coach".

2   Methods

To solve the research problems, a set of methods was used:

- theoretical analysis and generalization of special literature, as well as professional standards in the field of physical culture and sports, and education;
- content analysis of the research focus of students of the "rhythmic gymnastics" specialization (2017–2021; n = 108);
- testing;
- methods of mathematical statistics.

The study involved 20 students of 4-year undergraduate studies at Lesgaft University. Diagnostics was carried out according to the results of performing test tasks of five current parts, involving answers to 175 questions. In accordance with each of the five parts, students were asked to answer 20 questions. All obtained data (134 indicators) were subjected to mathematical and statistical processing using the "STATGRAPHICS plus" program (arithmetic mean values, errors of mean values, variation coefficients, rank correlation coefficient).

3 Findings

In order to substantiate the need for purposeful formation of the analytical competence of students as a sports teacher with a higher education, an analysis of the topics of graduation qualification works was made. The analysis was carried out in the specialization of rhythmic gymnastics of the Department of Theory and Methodology of Gymnastics over the past 5 years: 2017 n = 10; 2018 n = 16; 2019 n = 31; 2020 n = 34; 2021 n = 17. It was found that, as a rule, in 50% of cases, students chose the direction of research on their own, based on the experience of their sports and professional activities. The remaining 50% of students, finding it difficult to choose, determined the direction based on the list of topics of graduate qualification works proposed by the department. Most often, this was due to the lack of experience in coaching and sufficient awareness of the problems of sports training in rhythmic gymnastics.

In 99% of cases, the focus of research corresponded to the analytical competence of a trainer for technical and physical training in rhythmic gymnastics.

The following are the topics of theses that the students did not consider in their studies:

- study of the problems of "selection" and "planning" in rhythmic gymnastics;
- the study of problems demonstrating the ability to carry out sports selection and sports orientation in the process of training;
- refereeing of competitions in the chosen sport;
- planning the content of classes taking into account the provisions of the theory of physical culture;
- physiological characteristics of the anatomical, morphological and psychological characteristics of those involved in different sex and age.

The study of these topics at the University is mandatory, as they reflect the future work functions of the coach.

This is due to a number of reasons. Firstly, the fulfillment of the functions of planning, accounting and control, as well as selection is largely regulated by regulatory documents (FSSP in rhythmic gymnastics), and the organization of refereeing and competitive activity - by the rules of the competition. In this regard, students do not have the opportunity to make adjustments to the types of activities associated with the implementation of these functions in order to solve existing problems, and the confirmation of research hypotheses is difficult.

Secondly, 80% of the total number of hours is allocated for mastering the sections "training" and "development" in the chosen sport, which allows us to better understand the existing problems and assess the degree of necessity and possibilities of their solution (Table 1).

In addition, the discipline "professional sports improvement" and "elective disciplines of physical culture and sports" are fully devoted to physical and technical training in sports. It is here, in the content of the limited aspects of training, that students can conduct pilot studies under the supervision of a teacher - supervisor.

However, the analysis of students' choice of the stage of sports training and the age of the subjects for research showed that the training stage ("trainer of the training stage of sports training"; qualification level 6) prevails and, accordingly, the age of gymnasts is 9–12 years old, which is the most favorable for teaching exercise technique and sensitive for the development of most of the motor abilities of athletes. In 2017, the number of topics of graduation qualification works devoted to the problems of sports training of gymnasts of this age was 56%, and in 2021 - 94%.

Table 1. The focus of research on the topics of theses of students according to work functions (n = 108,%).

Naming, age	Naming of the work functions	2017	2018	2019	2020	2021
Preparation of trainees at the stage of initial training (Age 6–8)	Picking groups and sections of the initial preparation phase	0	0	0	0	0
	Planning, accounting and analysis of the results of the training process at the stage of initial training	0	0	0	0	0
	Conducting training activities, sports and outdoor games aimed at developing the personality of those involved, instilling healthy lifestyle skills, fostering physical, moral, ethical and volitional qualities, determining sports specialization	22	13	39	24	6
Preparation involved in the training stage of the sport (sports specialization stage, sports discipline) (Age 8–12)	Selection of trainees in groups and sections of the training stage	0	0	0	0	0
	Planning, accounting and analysis of the results of the training process at the training stage	0	0	0	0	0
	Formation of general and special physical training, technical-tactical training, psychological and theoretical training, and skills of competitive activity in accordance with sports training programs	56	40	45	32	94

(continued)

Table 1. (*continued*)

Naming, age	Naming of the work functions	2017	2018	2019	2020	2021
Preparation of athletes at the stage of improving sportsmanship and the stage of higher sportsmanship in a sport (sports discipline) (Age 13–15)	Selection and assessment of an athlete's prospects for achieving international class results	0	7	0	0	0
	Planning, accounting and analysis of the results of training and competitive activities of athletes at the stage of improving sportsmanship	0	0	0	0	0
	Carrying out training activities in accordance with the individual plans of sports training of athletes	22	40	16	44	0
	Management of the systematic competitive activity of an athlete	0	0	0	0	0

That confirmed the understanding of the students of the significance of this stage for the formation of sportsmanship in rhythmic gymnastics and the vision of problems associated with various aspects of training precisely at the training stage.

Specifying the problems of training in rhythmic gymnastics, the solution of which the theses were aimed at, it was established that from 2017 to 2020, the priority for the study was technical training (from 89–65%). It is this type of training that determines the

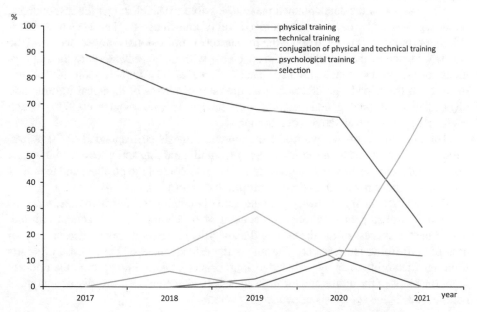

Fig. 1. The choice of research problems by the types of sports training in rhythmic gymnastics by students of the Department of Theory and Methods of Gymnastics (n = 108;%)

technique of motor actions, their complexity, mistakes made in competitive programs, and the performance skills of athletes in general. Therefore, the professionalism of the trainer, first of all, manifests itself in the ability to design learning algorithms, analyze the effectiveness of their implementation, determine the causes of errors and ways to eliminate them (Fig. 1).

It should be noted that in 2020, the topic of the study was the psychological training of gymnasts (11%) and this is due, first of all, to the tendency to increase the technical complexity of rhythmic gymnastics programs and the importance of the psychological readiness of athletes to fulfill them. At the same time, based on the research topics, one of the ways to solve the problem of gymnasts' readiness to master new complex technical actions of the body in combination with the work of the subject was seen in the conjugation of technical and physical types of training (2021 - 65%). This fully reflected the modern tendency in the development of rhythmic gymnastics, indicated the correctness of the trajectory of the development of the analytical competence of the students and the compliance of the competencies being formed with the requirements of the professional standard "Coach".

In order to confirm the directed formation of the students' analytical competence, testing was carried out using a fund of evaluation tools, an expert assessment of the quality of thesis's and a correlation analysis of their relationship. The test tasks, taking into account the professional standard "Coach", corresponded to the generalized function of the coach and made it possible to assess the competencies being formed: the ability to manage the competitive activity of athletes, planning, accounting, analysis of the results of the training process and competitive activity at the stages of sports training, to implement an individual approach in the process of sports training.

The analysis of the data obtained made it possible to establish that the respondents, on average, coped by 67.4% - the average level of knowledge on the issues under consideration, but it can be stated about the formation of professional competencies and the correspondence of the knowledge of a specialist with higher education to the requirements of the professional standard "Coach" (Table 2). The highest score (68%) was recorded in the knowledge of students in the field of means of technical training and physical types of training, and means of folk-characteristic choreography (87%), which is closer to the practical activity of a trainer.

The assessment of the quality of the students' thesis corresponded to 80 points, which, in turn, testified to the ability to analyze, model training activities and plan sports results; to develop means and methods of training, to control the physical and technical readiness of those involved, to draw appropriate conclusions.

Proceeding from this, knowledge in the field of planning training activities, accounting and control in sports training, the content of psychological and tactical training, determine the success of the thesis, and the analytical activity demonstrated in the process of performing the research determines the effectiveness of the process of sports training. The ability to apply knowledge appropriately, taking into account the relevant labor functions of the trainer and the tasks to be solved, which is a derivative of analytical competence, is the core basis of pedagogical creativity.

Table 2. The results of testing the professional competence of students of the specialization "rhythmic gymnastics" in the sections of theoretical training (n = 20; points).

Statistical indicators	CC 1	CC 2	CC 3	CC 4	CC 5
M	2,62	6,27	3,40	5,16	4,1
δ	0,75	1,01	0,65	0,38	0,71
m	0,17	0,23	0,15	0,09	0,16
V (%)	28,70	16,08	19,15	7,40	17,32

Note: CC 4 - methods of teaching folk-characteristic choreography; TQ 5 - psychological and tactical training; CC 1 - planning of training activities and control; CC 2 model characteristics of the effectiveness of training and competitive activities; CC 3 - means of technical and physical training.

4 Conclusion

In the process of research aimed at the formation of students' analytical abilities as a component of the competence of a sports teacher in rhythmic gymnastics, it was determined that:

- in 50% of cases, students chose the direction of research independently, based on the experience of their sports and professional activities;
- in 99% of cases, the focus of research corresponds to the formation of a coach's analytical competence for the implementation of technical and physical training in rhythmic gymnastics;
- for research purposes, students prefer the age of gymnasts 9–12 years old, which is the most favorable for teaching the technique of exercises and sensitive for the development of most of the motor abilities of athletes.

The analysis of the formation of the analytical competence of students made it possible to reveal that:

- the respondents, on average, coped by 67.4%, it can be stated about the formation of professional competencies and the correspondence of the knowledge of a specialist with a higher education;
- the highest score (68%) was recorded in the knowledge of students in the field of means of technical training and physical types of training, and means of folk-characteristic choreography (87%), which is closer to the practical activity of a trainer;

Thus, only in the presence of analytical thinking formed in the process of studying at the University will it allow specialists in the field of rhythmic gymnastics to effectively carry out training and competitive activities, taking into account the development trends of this sport.

References

1. Kostychenko, V.F., Zverev, V.D., Vasiljev, D.A.: Transformation of training of physical culture personnel in the Russian education system. Theory Pract. Phys. Cult. **4**, 50 (2021). (in Russian)
2. Lubysheva, L.I.: Training of sports personnel: a broad professional or a narrow specialist? Theory Pract. Phys. Cult. **9**, 2 (2020). (in Russian)
3. Professional standard Coach. On the approval of professional standard Coach. Ministry of Labour of Russia. http://fgosvo.ru/uploadfiles/profstandart/05.003. Accessed 3 March 2021
4. Barinova, D., Ipatov, O., Odinokaya, M. Zhigadlo, V.: Pedagogical assessment of general professional competencies of technical engineers training. In: Katalinic, B. (ed.) Proceedings of the 30th DAAAM International Symposium, pp. 0508–0512. DAAAM International, Vienna, Austria (2019). https://doi.org/10.2507/30th.daaam.proceedings.068
5. Rules for Rhythmic Gymnastics 2017–2020 (2018). Federation international de gymnastyque, Switzerland. http://vfrg.ru/collegues/documents/pravila/ Accessed 3 March 2021
6. Sullivan, E.E., Ibrahim, Z., Ellner, A.L., Giesen, L.J., Handel, D.A.: Management lessons for high-functioning primary care teams. J. Healthcare Manag. **61**(6), 420–435 (2016)
7. Pokrovskaia, N.N., Petrov, M.A., Gridneva, M.A.: Diagnostics of professional competencies and motivation of the engineer in the knowledge economy. In: 2018 Third International Conference on Human Factors in Complex Technical Systems and Environments (ERGO)s and Environments (ERGO), pp. 28–31. IEEE (2018). https://doi.org/10.1109/ERGO.2018.8443851
8. Jeffress, M.S., Brown, W.J.: Opportunities and benefits for powerchair users through power soccer. Adapt. Phys. Activ. Q. **34**(3), 235–255 (2017). https://doi.org/10.1123/apaq.2016-0022
9. Gruzdeva, M.L., Vaganova, O.I., Kaznacheeva, S.N., Bystrova, N.V., Chanchina, A.V.: Modern educational technologies in professional education. In: Popkova, E.G. (ed.) Growth Poles of the Global Economy: Emergence, Changes and Future Perspectives. LNNS, vol. 73, pp. 1097–1103. Springer, Cham (2020). https://doi.org/10.1007/978-3-030-15160-7_110
10. Loo, L.: Medvedeva, Formation of methodological competencies of female students of the Pedagogical College of Israel, taking into account their cognitive and motor characteristics. Theory Pract. Phys. Cult. **4**, 81 (2021). (in Russian)
11. Medvedeva, E.N., Suprun, A.A., Aizyatullova, G.R., Sakharnova, T.K.: Competence approach to the unified certification system for coaches in rhythmic gymnastics. Uchenye zapiski universiteta imeni P.F.Lesgafta **11**(165), 206–211 (2018)
12. Medvedeva, E.N., Smirnova, N.N., Romasheva, N.V.: Innovative approaches to the analysis of the technical value of equilibria in rhythmic gymnastics. Theory Pract. Phys. Cult. **4**, 6 (2019). (in Russian)
13. Skripleva, E.V., Skoblikova, T.V. Professional training of specialists in physical culture in modern conditions. Uchenye zapiski. Elektronnij nauchnij zurnal Kurskogo Gosudarstvennogo Universiteta, **3**, 72–75 (2008). c
14. Kvashnina, E.V.: Improving the efficiency of the training process of girls involved in rhythmic gymnastics. Pedag. Psychol. Biomed. Probl. Phys. Educ. Sports **12**(3), 53–60 (2017). (in Russian). https://doi.org/10.14526/032017234
15. Mingalisheva, I.A.: Factors for increasing the effectiveness of sports training in fitness aerobics training. Pedag. Psychol. Biomed. Probl. Phys. Educ. Sport **12**(2), 30–38 (2017). (in Russian). https://doi.org/10.14526/01_2017_202
16. Suvorova, O.V., Sorokoumova, S.N., Gutko, A.V., Minaeva, E.V., Ivanova, N.V., Mamonova, E.B.: Experience of emotional and physical violence and psychological boundaries of personality of psychology students. Espacios **38**(56), 34–42 (2017)

17. Trigub, G.Y.: Testing as a method of teaching and knowledge control in a university. Concept **S3**, 66–68 (2017). (in Russian). http://e-koncept.ru/2017/470051.html. Accessed 3 March 2021
18. Fernando, S.Y., Marikar, F.M.: Constructivist teaching/learning theory and participatory teaching methods. J. Curric. Teach. **6**, 110 (2017). https://doi.org/10.5430/jct.v6n1p110
19. Rubtsova, A.V., Almazova, N.I., Bylieva, D.S., Krylova, E.A.: Constructive model of multi-lingual education management in higher school. IOP Conf. Ser. Mater. Sci. Eng. **940**, 012132 (2020). https://doi.org/10.1088/1757-899X/940/1/012132
20. Demidov, V., Mokhorov, D., Mokhorova, A., Semenova, K.: Professional public accreditation of educational programs in the education quality assessment system. E3S Web Conf. **244**, 11042 (2021). https://doi.org/10.1051/e3sconf/202124411042
21. Namazov, A, Kivikhariu, I., Medvedeva, E., Suprun, A.: Intermediate certification of the specialty «rhythmic gymnastics» students. EpSBS **98**, 318–324 (2020). https://doi.org/10.15405/epsbs.2020.12.03.33
22. Santos, F., Gould, D., Strachan, L.: Research on positive youth development-focused coach education programs: future pathways and applications. Int. Sport Coach. J. **6**, 132–138 (2019). https://doi.org/10.1123/iscj.2018-0013

Formation of the Creative Teacher's Personality in the Context of Modern Education

Natalia Kopylova[✉] [iD]

National Research University Moscow Power Engineering Institute, Moscow 111250, Russia

Abstract. In the context of the modernization of contemporary education, pedagogical research is often devoted to an assessment of the competence-based approach. The competence-based approach in education focuses on: 1) achieving a sufficiently high level of knowledge, experience, and awareness for carrying out activities and communicating in various fields, 2) differentiating informational, social, communicative, pedagogical, and other types of competence, 3) serving as a basis for the restructuring the educational process and overcoming a one-sided orientation to the subject's education. Under the conditions of modern creativity, the ability to create should be considered as one of the key competencies of a modern teacher. The creative approach focuses on the formation of creative individuality, the development of a style of creative activity, non-standard solutions of pedagogical problems, and the ability to innovate. In this article, we consider the main definitions of creativity, the pedagogy of creativity, pedagogical creativity and the experience of different techniques of using creativity in the pedagogical process of secondary schools and universities.

Keywords: Creativity · A Student · A Teacher · A Pedagogical Process · A School · A University

1 Introduction

In the context of the modern education modernization, special attention is paid to the competence-based approach to the pedagogical problems' study [1–3]. Competence-based approach in education focuses on achieving a sufficiently high level of knowledge, experience, and awareness for carrying out activities and communicating in various fields; differentiate informational, social, communicative, pedagogical, and other types of competence; can serve as a basis for the restructuring educational process, overcoming one-sided subject education orientation [4].

In the 1920s, the famous Russian psychologist Lev Vygotsky, creating a cultural-historical theory of the human behavior and psychic development, investigating their life activity, mental development in education and upbringing, deduced a certain logic: life develops as a system of creativity aimed at developing individual creative abilities [5, 6].

A well-known teacher, the author of the encyclopedia of educational technologies, G.K. Selevko, represents a person as a set of competencies that represent a person's

D. Bylieva and A. Nordmann (Eds.): PCSF 2021, LNNS 345, pp. 774–782, 2022.
https://doi.org/10.1007/978-3-030-89708-6_63

ability to engage in activities. Key competencies are based on the individual properties and are manifested in certain behaviors that are based on the psychological person's functions, have a wide practical context, and a high degree of universality [7].

2 The Theoretical Conception of a Teachers' Creativity

Under the contemporary conditions of creativity, the ability to create should be considered as one of the key competencies of a modern teacher [8, 9]. The creative approach focuses on the creative individuality formation, the development of a creative activity style, non-standard solution of pedagogical problems, the ability to innovate.

Scientists call creativity in the field of arts, design, creation and sale of new products, scientific knowledge. From the standpoint of philosophy, creativity is the people activity transforming the natural and social world in accordance with their goals and needs on the basis of the objective reality laws.

V.G. Ryndak views creativity as a constructive activity in which human development is unlimited. Creativity is inherent in man, but the level of its realization is determined by value orientations, motives, personality orientation, and its abilities, and the conditions under which it develops. It is creativity that gives a person the opportunity to activate himself. Creativity is the active interaction of the subject with the object, during which the subject changes the world, creates something new, socially significant in accordance with the requirements of the objective laws [10].

According to many scientists, "creativity is a conditional concept, which can be expressed not only in the creation of that which is fundamentally new, not existing before, but also in the discovery of the relatively new (for a given area, a given time, in a given place, for the subject itself)" [7]. With regard to the process of education and upbringing, creativity is considered as a form of human activity aimed at creating values that are qualitatively new for him and have social significance. From these positions, we consider, form and evaluate the experience of creative pedagogical activity of the past and present.

Pedagogical science distinguishes between the pedagogy of creativity. The pedagogy of creativity is the science of the pedagogical system of interrelated types of human activity. The goal of the pedagogy of creativity is the formation of a creative personality, which is characterized by a stable, high-level focus on creativity, a creative style in one or several types of activity.

Pedagogical creativity is considered as a state of pedagogical activity, in which a fundamentally new thing is created in the content, the educational process organization, in solving scientific and practical problems. Studying pedagogical creativity, we became convinced that a modern teacher, a leader, is unthinkable outside of creativity. Creativity is everything that surrounds us, and first of all it is a process and result of creative activity: culture, art, knowledge, labor, "creative thought", beauty. A creative teacher, in the words of N. Roerich, is "the one who opens, makes wise and encourages" [11].

Pedagogical creativity is "the process of development and realization by a teacher in the constantly changing conditions of an educational process, in communication with students, optimal and non-standard pedagogical solutions" [4]. The future teacher's creative self-development is an integrative characteristic that includes system-forming components: self-knowledge, creative self-determination, self-government, self-improvement, and creative self-realization [12].

V.A. Slastenin confirms that "the miracle of the creative personal self-manifestation will happen sooner or later if a teacher masters the creative logic of any pedagogical process. Creativity and innovation have always been in the traditions of Russian teaching. Now, in the conditions of a radical renewal of a society and school, the social nature of pedagogical creativity is changing significantly: it becomes more and more popular. Carrying out the solution of an incalculable set of typical and original pedagogical tasks, a teacher more often acts as a researcher" [13].

In the humanistic system of education and training, creativity has always been an integral part of a pedagogical process. V.A. Slastenin thinks in modern conditions "pedagogical creativity is understood as a process of solving pedagogical problems in changing circumstances. The level of creativity in the teachers' activities reflects the degree to which they use their capabilities to achieve the set goals" [13]. This level identifies the conditions under which the teacher's activity turns into creativity. This is the teachers' pedagogical ideal (the idea of themselves as the creators of a pedagogical process); development of a personal creative concept; the paradigms' adoption of the modern style of scientific and pedagogical thinking, awareness of one's own creative individuality; an emotional and communicative component; creative professional and pedagogical communication; joint activities and interaction of a teacher with students and colleagues, their cooperation and co-creation.

Modernization of education in modern conditions is under the sign of the innovative pedagogy development. In innovation creative initiative is needed more than ever, which is "self-going beyond a given situation, a way of self-affirmation, self-expression, self-realization in a free form that meets the needs of a subject, a desire to express themselves in activities, communication and cognition" [4].

Education Development Program for 2018–2025, prioritized means of modernization and the solution of the following priority tasks are provided: modernization of general and pre-school education as an institution of social development; bringing the content and structure of vocational education and training in line with the needs of the labor market; development of the quality education system and the demand for educational services. Hence, there is an acute problem of modernizing the training of a teacher-creator, a professional teacher, a humanist capable of an innovation, creativity and a skill. The teacher's personality model and reference points for its realization have already been identified. Many aspects of the existing education system are revised from a scientific point of view, which is due to the renewal of state educational standards, the new functions of a teacher, increasing requirements for his/her general cultural and professional competence, as well as the expansion and complication of his/her pedagogical activity, the development of creativity and a skill at the new scientific and technological level continuing education. At the same time, it is necessary to emphasize the difficulties of solving this problem as related to the crisis in society, changes that occur in the educational sphere and pedagogical teams as a result of the complicated socioeconomic living conditions and activities of teachers, an increase in the negative attitude towards the profession, a decrease in labor incentives negative changes in the motivation of their professional activities [14, 15].

From the point of view of modern social order in the field of modernization, namely pedagogical universities and pedagogical departments should play a crucial role in the preparation of a new modern teacher who is able to successfully solve the tasks of advanced education.

When preparing creative teachers' personality in modern conditions, the requirements of the Federal State Educational Standard for Higher Education of a new generation are taken into account, which emphasizes the need for future specialists to develop general cultural and professional competencies based on not only theoretical and empirical knowledge and skills, but also creative qualities, abilities that determine the creative self-development of an individual [16, 17].

Signs of creative teachers' activity are novelty, originality, quality, productivity, efficiency, significance and effectiveness. All of them are realized in the communication of a teacher with students in the course of learning trainees, in the system of interaction with them, the organization of relationships, in cooperation and joint activities [18–21].

To the main areas of teachers and students' cooperation and interaction at a university, we include joint creative activities in a holistic pedagogical process: in teaching, research and extracurricular work [20, 22–24].

The Model of a Creative Teacher

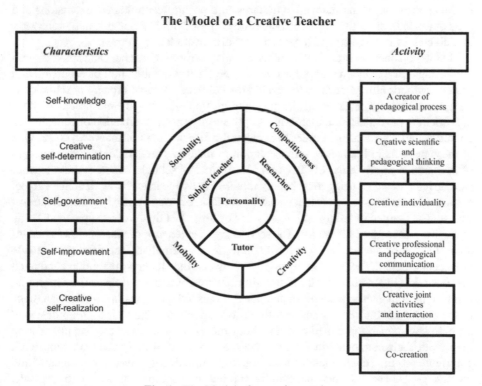

Fig. 1. The Model of a creative teacher.

It was these conceptual ideas that we used as the basis for studying the means of forming the creative personality of a modern teacher and his/her activities in the context of education modernization, and we used as the main ideas in our model of a creative teacher (see Fig. 1). The main aspect in our model is a person (teacher) and his/her creativity. This person should be a subject teacher, a tutor, a researcher with the following qualities: sociability, competitiveness, creativity, and mobility. The main characteristics of a creative teacher should be: self-knowledge, creative self-determination, self-government, self-improvement, and creative self-realization. A creative teacher should be very active. His/her activity should be diversified: a creator of a pedagogical process, creative scientific and pedagogical thinking, creative individuality, creative professional and pedagogical communication, creative joint activities and interaction, and co-creation.

3 Different Examples of a Creative Teacher's Activities

We have examined and generalized the experience of Ryazan secondary schools that function on the basis of the ideas of cooperation and provide joint creative developmental activities for adults and children. The joint creativity of teachers and students is clearly manifested in the educational process, as well as in the field of additional education: in the application of the innovative research and design technologies, excursions and project work, in the activities of the Scientific Schoolchildren's Society, in search work, in club classes, in tourism and local history work, in student government.

Let us consider the organization of the creative activity of teachers and students with the example of excursion and project work. The skillful combination of excursion and project methods allows teachers to develop the value-motivational sphere of schoolchildren, creativity, independence in planning actions, implementing the means and methods of work they have mastered, increasing responsibility for the realization of educational, cognitive and social tasks in order to achieve the effectiveness of their own work. One of these projects in English was an excursion of medical class students to the museum of the famous physiologist I.P. Pavlov in Ryazan. The schoolchildren and the teacher were given the task of using local history material in extracurricular work on the subject to ensure its activation, consistency and the creation of non-standard speech situations. The students united in groups of interests and desires. The first group prepared an essay on the life and work of I.P. Pavlov. The second group acted as journalists and wrote notes about the life of the academician for the newspaper in English. The third group made an excursion plan. The roles were assigned between the students: the guide conducted the excursion in Russian, the translator simultaneously translated the information into English, foreign guests (students of another class) asked questions in a foreign language to the guide, the translator translated the information, etc. The fourth group designed a photo newspaper and edited a video about the visit to the museum, supplementing them with their comments in English. We were convinced that such excursion work contributed to the achievement of teaching, educational, cognitive, developmental and creative goals. The students successfully applied the acquired knowledge in practice, developed the skills of research and group work. The method of projects in extracurricular work on the basis of local history material provides high communication, increases the motivation and cognitive activity of children, influences the formation of their civic position, fosters a sense of pride and respect for the life and activities of famous fellow countrymen, develops the ability of schoolchildren for joint, creative, reflective activity.

As for teachers the project method is used to develop sociability, competitiveness, creativity, and mobility.

Some Ryazan schools work on the basis of creative technologies. The experience of the teaching staff of secondary school № 51 in Ryazan was studied and summarized, which has been working for more than 20 years on the basis of the methodology of collective creative activity of the innovative teacher I.P. Ivanov. All school activities are held in the form of collective creative deeds. The creative nature of the school's work is manifested in the use of innovative humanistic technologies, the creation of an educational and methodological complex (classroom of methodological skills), developing programs for working with gifted children, in the activities of school theaters, in student government, in holding camp gatherings, frank conversations by a candle as a form of a confidential conversation, in organizing the alternation of creative assignments, etc.

Collective creative activity, which plays an important role in the creation and development of a single educational space of the school, is organized by educators and pupils of classes, members of the multi-age children's organization "Unity", duty commanders of units, the units in the following stages:

1. conducting micro-research in classes;
2. creation of a business council;
3. study of innovative experience;
4. conducting "brainstorming";
5. development of projects;
6. analysis-forecast of a social order;
7. discussion and adjustment of a project in classrooms;
8. release of programs-proposals of classes;
9. approval of a project of collective creative work at a council of duty unit commanders;
10. carrying out collective creative work;
11. results' analysis;
12. planning a new business.

Collective creative activity develops in teachers the abilities to be a creator of a pedagogical process, to have creative scientific and pedagogical thinking, creative individuality, to be able to creative professional and pedagogical communication, creative joint activities and interaction, and co-creation with colleagues and students.

A striking example of the creative partnership of teachers and students of the school is the functioning of three school theaters: a drama theater (Melpomene), a miniature theater (Loskut) and an English-language musical theater (TEZA). The activity of the theaters is many-sided: children, together with adults, prepare various performances, sew costumes, and prepare a scenery. They are the multiple winners of city and regional competitions of theatrical skills and the finalists of the All-Russian competition of youth theater groups in 2020.

For example, the musical theater "TEZA" has been operating since 1995. The troupe of the theater (about 30 people) consists of people of different ages from 8 to 20 years old and older. These are pupils and school graduates, teachers, parents, friends, residents of the neighborhood. The theater's list of productions includes performances: "Mary Poppins" by P. Travers, "A Christmas Tale", "Romeo and Juliet" (the performance won the Grand Prix of the Ryazan Festival "Theater Spring – 2019"), "Alice in Wonderland" (the winner of the All-Russian competition of theatrical youth groups). The theater teaches children to speak English fluently, creates positive motivation for learning a foreign language, helps students learn about the culture of the language being studied, express themselves creatively, and get rid of psychological barriers.

Teachers' participation in theatre development with students allows them to cultivate their creative and communicative abilities.

The conceptual ideas of the pedagogy of creativity are applicable to work with university students. So, at Ryazan Universities the principles and technologies of creativity are actively used in the educational process. Creative cooperation of teachers and students is carried out at lectures, seminars, laboratory classes, even during joint trips to various excursions organized by the universities. Many classes in pedagogy are held according to the methodology of the innovative teacher I.P. Ivanov in the form of collective creative activities, at a round table, where the most important issues of modern pedagogy are discussed. Every year, different competitions (for example Mister and Miss of a University) are organized, where students of all faculties, using the technology of collective creative activity, demonstrate their creative abilities in classroom and extracurricular work, in scientific researches, when performing various competitive creative tasks.

So, studying at the universities gives you the opportunity to have self-knowledge, creative self-determination, self-government, self-improvement, and all these qualities will allow you to achieve creative self-realization.

4   Conclusion

We conclude that, in the context of modernizing the holistic pedagogical process, great attention should be paid to the formation of teachers' creative personalities which are capable of adapting to modern conditions and perceiving all innovations in the field of modern innovative education.

In conclusion, it should be said that the main thing in creativity is a person (a teacher). This creative teacher may be a subject teacher, a tutor, a researcher. He/she should be sociable, competitive, creative, and mobile. He/she should try for self-knowledge, creative self-determination, self-government, self-improvement, and creative self-realization. A creative teacher should be a creator of a pedagogical process, have creative scientific and pedagogical thinking, creative individuality, creative professional and pedagogical communication, follow creative joint activities and interaction, and co-creation. Thus a modern teacher should be a very creative person if he/she wants to achieve great results in his/her pedagogical activities.

References

1. Pokrovskaia, N.N., Petrov, M.A., Gridneva, M.A.: Diagnostics of professional competencies and motivation of the engineer in the knowledge economy. In: 2018 Third International Conference on Human Factors in Complex Technical Systems and Environments (ERGO)s and Environments (ERGO), pp. 28–31. IEEE (2018). https://doi.org/10.1109/ERGO.2018. 8443851

2. Baranova, T., Kobicheva, A., Olkhovik, N., Tokareva, E.: Analysis of the communication competence dynamics in integrated learning. In: Anikina, Z. (ed.) IEEHGIP 2020. LNNS, vol. 131, pp. 425–438. Springer, Cham (2020). https://doi.org/10.1007/978-3-030-47415-7_45

3. Demidov, V., Mokhorov, D., Mokhorova, A., Semenova, K.: Professional public accreditation of educational programs in the education quality assessment system. E3S Web Conf. **244**, 11042 (2021). https://doi.org/10.1051/e3sconf/202124411042

4. Zagvyazinsky, V.I., Zakirova, A.F., Strokova, T.A. (eds.): Pedagogical dictionary. Academy, Moscow (2008). (in Russian)

5. Vygotsky, L.S.: Collected works. Vol. 2. Problems of general psychology. Davydov V.V. (ed.). Pedagogika, Moscow (1982). (in Russian)

6. de Champlain, Y., DeBlois, L., Robichaud, X., Freiman, V.: The nature of knowledge and creativity in a technological context in music and mathematics: implications in combining Vygotsky and Piaget's models. In: Freiman, V., Tassell, J.L. (eds.) Creativity and Technology in Mathematics Education. MEDE, vol. 10, pp. 479–505. Springer, Cham (2018). https://doi.org/10.1007/978-3-319-72381-5_19

7. Selevko, G.K.: Encyclopedia of Educational Technologies. Institute of School Technologies, Moscow (2006). (in Russian)

8. Pllana, D.: Creativity in modern education. World J. Educ. **9**, 136 (2019). https://doi.org/10. 5430/wje.v9n2p136

9. Walia, C.: A dynamic definition of creativity. Creat. Res. J. **31**, 237–247 (2019). https://doi. org/10.1080/10400419.2019.1641787

10. Ryndak, V.G.: Creation. A brief pedagogical dictionary. Center OGAU, Orenburg (2001)

11. Roerich, N.K.: Favorites. Pravda, Moscow (1990). (in Russian)

12. Andreev, V.I.: Higher school pedagogy. Innovation and prognostic course: Center for Innovative Technologies, Kazan (2006). (in Russian)

13. Slastenin. Publishing House Magister-Press, Moscow (2000). (in Russian)

14. Grebenkina, L.K., Orekhova, Ye.Yu., Badelina, N.A., Kopylova, M.V.: Interaction of subjects of pedagogical activity in technical university. Int. J. Civ. Eng. Technol. **10**(01), 1241–1252 (2019)

15. Pokrovskaia, N.N., Petrov, M.A., Gridneva, M.A.: Diagnostics of professional competencies and motivation of the engineer in the knowledge economy. In: 2018 Third International Conference on Human Factors in Complex Technical Systems and Environments (ERGO)s and Environments (ERGO), pp. 28–31. IEEE, St. Petersburg (2018). https://doi.org/10.1109/ ERGO.2018.8443851

16. Zaker, A.: Literature and creativity in an ELT context. ASIAN TEFL: J. Lang. Teach. Appl. Linguist. **1**, 175–186 (2016). https://doi.org/10.21462/asiantefl.v1i2.20

17. Abraham, A.: Can a neural system geared to bring about rapid, predictive, and efficient function explain creativity? Creat. Res. J. **19**, 19–24 (2007). https://doi.org/10.1080/104004 10709336874

18. Baer, J., Kaufman, J.C., Gentile, C.A.: Extension of the Consensual Assessment Technique to nonparallel creative products. Creat. Res. J. **16**, 113–117 (2004)

19. Baer, M., Oldham, G.R.: The curvilinear relation between experienced creative time pressure and creativity: moderating effects of openness to experience and support for creativity. J. Appl. Psychol. **91**, 963–970 (2006). https://doi.org/10.1037/0021-9010.91.4.963
20. Basadur, M., Pringle, P., Kirkland, D.: Crossing cultures: training effects on the divergent thinking attitudes of Spanish-speaking South American managers. Creat. Res. J. **14**, 395–408 (2002). https://doi.org/10.1207/S15326934CRJ1434_10
21. Cayirdag, N.: Creativity fostering teaching: impact of creative self-efficacy and teacher efficacy. Educ. Sci. Theory Pract. **17**(6), 1959–1975 (2017). https://doi.org/10.12738/estp.2017.6.0437
22. Becker, G.: The association of creativity and psychopathology: its cultural-historical origins. Creat. Res. J. **13**, 45–53 (2001). https://doi.org/10.1207/S15326934CRJ1301_6
23. Bylieva, D., Bekirogullari, Z., Kuznetsov, D., Almazova, N., Lobatyuk, V., Rubtsova, A.: Online group student peer-communication as an element of open education. Future Internet **12**, 143 (2020). https://doi.org/10.3390/fi12090143
24. Schlabach, J., Hufeisen, B.: Plurilingual school and university curricula. Technol. Lang. **2**(2), 126–141 (2021). https://doi.org/10.48417/technolang.2021.02.12

Indonesian Customs and Excise Training Center During COVID-19 Pandemic: Innovative System of Educational Process Management

Rita Dwi Lindawati[1] [ID], Zhanna N. Maslova[2]([✉]) [ID], and Anna V. Rubtsova[2] [ID]

[1] Customs and Excise Training Center, Bujana Tirta III Street, Rawamangun, East Jakarta, Jakarta 13230, Indonesia
[2] Peter the Great St. Petersburg Polytechnic University (SPbPU), Polytechnicheskaya 29, Saint Petersburg 195251, Russia
`rubtsova_av@sbpstu.ru`

Abstract. The COVID-19 pandemic has affected all aspects of Indonesia that changes various aspects of life. The Customs and Excise and Training Center (CETC) as one of the training organizer under the Financial Education and Training Agency (FETA), Ministry of Finance, also felt the impact. Since March 2020, all classical training activities will be stopped. Every Division in CETC synergized to carry out the strategy for implementing education and training during the pandemic. This research delivered the analysis of the training policy at FETA, the analysis of the strategies and creative decisions which is carried out by CETC in organizing the training program during the pandemic and the analysis of the challenges that CETC faced in organizing training during the pandemic and the ways of solving it. Qualitative Research Methods are used in the research. Minister of Finance Decrees and the other implementing of the decrees were analyzed as a regulatory framework for soft skills development decisions in a pandemic. The result of the study indicates that the training policy at the Customs and Excise Education and Training Center was innovative in organizing the educational process and the creative use of educational technologies, especially, in those training programs where personal contact between lectures and students is needed. Since 2018 FETA has made policies on human resource development and soft skills development by forming a strategy called the Ministry of Finance Corpu (Kemenkeu Corpu). The training policy at the CECT refers to the policy determined by FETA.

Keywords: Education and training practices · E-learning · Creativity · Customs and excise training center · Training strategies · Soft skills · Training policy

1 Introduction

As we know, the COVID-19 pandemic has been infecting the world for many months. Its exposure has crossed various continents with hundreds of countries, including Indonesia. The impact has affected all aspects of the country that changes various sides of life, and not in last place, education and professional training.

D. Bylieva and A. Nordmann (Eds.): PCSF 2021, LNNS 345, pp. 783–794, 2022.
https://doi.org/10.1007/978-3-030-89708-6_64

The Customs and Excise Education and Training Center (CETC) as one of the professional training organizers under the Finance Education and Training Agency of Ministry of Finance also got the significant impact. Various face-to-face trainings, that were supposed to be held, had to be postponed. Even the ongoing training had to be stopped halfway through and hundreds of participants were sent back to their respective units. Of course, that situation required suitable adjustments, either by postponing the schedule or by changing the educational techniques.

The main type of training method carried out during the COVID-19 pandemic is training in the form of distance learning and e-learning. Considering the characteristics of educational process at the Customs and Excise Education and Training Center it should be said that not all of training programmes could be conducted virtually. Thus, for this kind of training sessions the implementation schedule was postponed until 2021. For examples, there are cabin cargo training, safety training, marine patrol training, etc. that cannot be learned virtually.

Training sessions in the form of e-learning had actually been carried out by the Customs and Excise Training Center ever since before the COVID-19 pandemic. It is an implementation of the Ministry of Finance's Strategic Initiative which stated that the educational programmes for human resource development in the form of e-learning were targeted progressively, starting in 2019 as much as 30% of total training sessions per year, 50% in 2020, and 70% in 2021 and beyond. The formula for calculating the percentage of e-learning is as follows: 60% of (number of digital programs / number of total programs) + 40% of (digital participants / total participants).

Actually, e-learning was only allocated for 50% in 2020. However, due to the COVID-19 pandemic in Indonesia since mid-March 2020 all face-to-face trainings have stopped. Since then, all those trainings sessions were transformed into e-learning. Various strategies had been carried out by the Customs and Excise Education and Training Center to find the adaptive educational patterns during the pandemic. All of the team at the Customs and Excise Education and Training Center, who were involved in organizing the training, were trying to formulate strategies to adjust to this pandemic situation. An innovative system for managing the educational process was created, which allowed for the transfer of knowledge and the development of soft skills. All parts of the organization (division of planning and development, division of event organizer, division of evaluation and reporting, division of administration and lecturers) were working altogether in formulating training strategies that could be carried out efficiently during the COVID-19 pandemic.

The purpose of the article is to analyze the educational policy at the Financial Education and Training Agency to work out the strategies which are carried out by CETC in organizing the training program during the pandemic. The article provides an overview of the challenges that CETC faced in organizing the training process during the pandemic and some ways and creative decisions to solve problems are discussed.

2   Theoretical Framework

The Minister of Finance Decree, Number 974/KMK.01/2016 [1], concerning the Implementation of Strategic Initiatives for the Bureaucratic Reform and Institutional Transformation Program of the Ministry of Finance there had been amended several times most recently by the Minister of Finance Decree Number 59/KMK.01/2018. It was related to the Second Amendment to the Minister of Finance Decree, Number 974 /KMK.01/2016 and in some earlier documents [2], on the Implementation of Strategic Initiatives to implement the development of human resource competencies and soft skills through the Ministry of Finance of Corporate University and to improve organizational performance through the development of better human resource competencies. A provision had been made regarding the Ministry of Finance of Corporate University which is contained in the Minister of Finance Decree of the Republic of Indonesia.

The role of the Ministry of Finance Corporate University was as a center of strategic initiatives for implementing human resource competency development which is part of the achievement of the vision and mission of the Ministry of Finance through the realization of the alignment between education, learning, and application of values with targets performance, which is supported by knowledge management.

Another document – The Head of the Financial Education and Training Agency Decree, Number KEP - 82/PP/2020, – contains The Guidelines for the Implementation of Distance Learning in the Financial Education and Training Agency (FETA) [3].

According to The Head of the Financial Education and Training Agency Regulation, Number PER: 2/PP/2019, concerning E-Learning Guidelines within the Ministry of Finance Head of the Financial Education and Training Agency, learning is a mechanism for transferring knowledge, increasing skills and shaping attitudes and behavior for the development of human resources [3]. E-learning should be developed through integrating various methods and resources in the form of competency development other than education and it is carried out through classical and non-classical channels to support achievement of Ministry of Finance performance targets.

E-leaning is competency development that carried out in the form of learning, by optimizing the use of information and communication technology, to achieve learning objectives, performance improvement and public education.

There are several types of e-learning. Synchronous e-leaning is implemented by requiring the support of other parties (facilitated-led) and using two-way communication methods at a certain time (real time), it is tailored to a tight schedule. Facilitation asynchronous c-lcarning is carried out by requiring the support of other parties (facilitated-led) and using two-way communication methods with independent time (time independent) and it is not linked to a tight schedule. Independent asynchronous e-leaning is carried out individually (self-paced) at independent times (time independent) and is not tied to a strict schedule [4].

Distance learning is a learning process that is carried out outside the training venue and emphasizes independent learning which is managed systematically and is not limited by distance and time by using various learning media.

The great difficulty is designing teaching materials for distance learning in the form of modifications to existing classical curricular and e-leaning teaching materials or creating new teaching materials specifically developed for this type of learning. Also, in the development of teaching materials it is necessary to meet the following criteria: they should be easy to access, they can be studied independently. Moreover, they should be simple, well organized and variational.

Last but not least, Regulation of the Head of the Financial Education and Training Agency, Number PER-8/PP/2018, concerns Micro-learning Guidelines in the Ministry of Finance. Micro-leaning is asynchronous e-leaning which is carried out in a focused system in a relatively short time and it emphasizes with the independent learning methods that are supported by easy learning resources [4]. Such as Kemenkeu Learning Center (KLC) – a learning portal that is managed by the Financial Education and Training Agency (FETA).

3   Methods and Materials

This research uses a qualitative method for exploring the existing problems. Qualitative research is used to examine the conditions of natural objects where the researcher is a key instrument [5]. According to S. Sugiyono [6], sample determination in qualitative research is not based on statistical calculations. The selected sample serves to obtain maximum information, not to generalize. Determination of the sample is selected with the snowball sampling technique.

Data resources of this research include primary data and secondary data. Primary data is obtained from interviews. Meanwhile, secondary data is obtained by conducting research on materials accompanying the educational process. The data analysis method used was the data analysis model of Miles and Huberman [7]. According to Miles and Huberman [7], activities in qualitative data analysis are carried out interactively and continue to completion, so that the data is saturated. Activities in data analysis, namely data reduction, display data and conclusion drawing/verification.

On top of that, statutory regulations and statutory implementing regulations were considered as the legal basis for the study.

4   Discussion

The Customs and Excise Education and Training Center is a technical and educational unit under the Financial Education and Training Agency of the Ministry of Finance. The following is the organizational structure (see Fig. 1).

The educational policy at the Customs and Excise Education and Training Center refers to the policy determined by the Financial Education and Training Agency. Since 2018 the Financial Education and Training Agency has made policies on human resource development by forming a strategy called the Ministry of Finance Corpu (Kemenkeu Corpu). Kemenkeu Corpu is Corporate University of the Ministry of Finance that aims to achieve the Ministry of Finance vision and mission through creating link-and-match between learning, knowledge management and internal values in order to meet the overall performance and targets of Ministry of Finance. The following model is the road map of the Ministry of Finance Corpu (see Fig. 2).

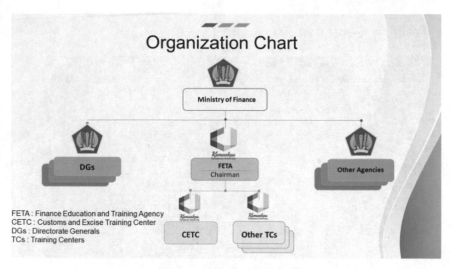

Fig. 1. Organization chart

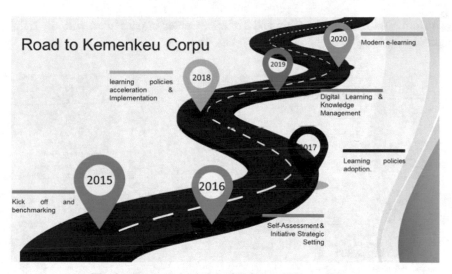

Fig. 2. The road map of the Ministry of Finance Corpu.

It is clear that digital learning and knowledge management are strategic initiatives in the context of the Ministry of Finance Corpu, which has been starting to be implemented in educational policy since 2019. Then, in 2020 the implementation of training policies is gradually changed in the form of modern e-learning. Here are the differences between a training center and a corporate university (see Fig. 3).

It should be said that before the COVID-19 pandemic the e-learning training system had been implemented at the Financial Education and Training Agency in general and

Training Center vs Corporate University

Points of Distinctions	Training Center	Corporate University
Main Focus	Operational/Technical Needs	Business-alligned Learning
Service	Reactives, based on what needed	Proactive
Learning Process	Fragmented	Intergrated
Methods	Mainly classical.	Blended learning, structured learning, learning from others, tacit knowledge, online learning.
Roles	Training, Vocational, Technical	Part of Human Capital Management, Strategic Learning, Learning Journey

Fig. 3. Training center VS corporate university.

at Customs and Excise Education and Training Center in particular, with the training target of 50% full e-learning in 2020 (see Fig. 4).

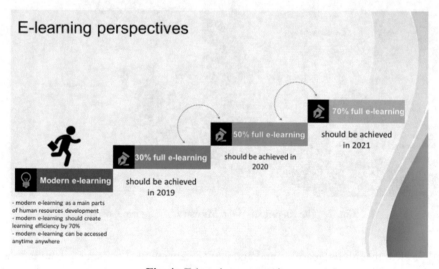

Fig. 4. E-learning perspective.

The Customs and Excise Education and Training Center is the echelon II unit under the Financial Education and Training Agency of Ministry of Finance. The Customs and Excise Education and Training Center has 4 (four) divisions and 1 (one) functional position of lecturer who work together in carrying out their duties as a trainer.

The Customs and Excise Education and Training Center is supported by 83 members with educational backgrounds consisting of 41% Undergraduate, 30% Master, 17% Diploma and 12% Senior High School. This organization has 14 lecturers who are in the functional positions called *"Widyaiswara"*. All lecturers must have at least a postgraduate level of education (see Fig. 5). The following figure shows the number of lecturers at the Customs and Excise Education and Training Center.

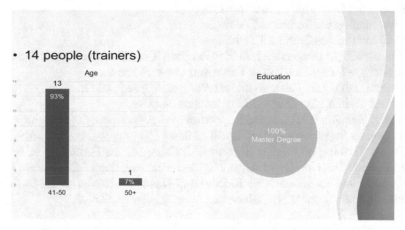

Fig. 5. Trainers at customs and excise education and training center.

The learning needs were analyzed with the ADDIE-model (analysis, design, development, implementation, evaluation). This model consists of several steps, namely: analysis, creating learning designs, developing learning design, implementing, and evaluating. In principle, the process for translating policies into education and training starts from the preparatory stage, which are to gather all information of strategic, policy and other issues that would be reviewed at the training commission meeting which consists of all Directorates-General under the Ministry of Finance. Furthermore, the meeting would collect data as the results to be verified. This verification process produced information of what competencies required by the Directorates-General and the training needs. After that, it was finalized and ratified.

In addition to the need to restructure the teaching process, the educational organization faced the problem of preparing teachers for the new working conditions. And one of the innovative solutions was the customized training sessions of teachers where the educational cases were used.

Strategies carried out by each division at the Customs and Excise Education and Training Center during the COVID-19 Pandemic are described below.

4.1 Division of Planning and Development

This division is in charge of formulating the education and training planning, training development, curriculum preparation and lecturers' professional development. The Division identified the types of training with the Skill Group Owner whether a course can be learned virtually or not. Skill Group Owner is an official/staff appointed as a study group leader based on expertise and competency mastery in carrying out the duties and functions in that study group. He has the duty to assist its group and to analyze the learning needs.

Types of training that are specific to learn virtually are as follows:

- Strategic intelligence workshop.
- Transnational Organized Crimes Workshop.
- Analyst Intelligence Technical Training.
- Tactical Intelligence Technical Training.
- Technical Skill Training on the Use of Hi-co Scan X-ray Container Inspection System.
- Technical Training on the Use of Cabin and Cargo Scanners.
- Technical Training for Customs and Excise Patrol Speedboat Crew.
- Technical Training for Weapons Maintenance and Use.
- Technical Training for Patrol and Inspection of Sea Transportation Facilities.
- Workshop on Supervision of Goods and Means of Transport at the Border.
- Technical Training on Canine: Technical Training for the Formation of Narcotics and Currency Tracking Dogs; Advanced Technical Training of Directorate General of Customs and Excise (DGCE) Bloodhound Handler – Border Dogs; Basic Level Technical Training for DGCE Bloodshot Dog Handler Formation.

These training programmes are difficult to be held virtually because participants need direct practices in the field, not only through simulation videos, in order to obtain the expected competencies. Training materials in the form of practical instructions were created for these courses and a plan for consolidating practical skills was developed. Another creative solution was the use of gamification elements when adapting training courses to the distance form. The combination of emotional appeal and audiovisual, computational, informational and other capabilities of computer technology has great didactic potential, which can and should be implemented in educational practice. The advantage of educational games is that they allow students to learn by doing and without fear of making mistakes. It reacts upon the effectiveness of the formation of practical skills, which are especially important for the above listed practical courses.

Complex conditions have required a creative restructuring of learning management at all levels. A special management system has been developed and responsibilities have been distributed among departments so that the quality of training is maintained. So, the types of distance learning that were organized by the Customs and Excise Education and Training Center are the follows: organizing open access is a distance learning in the form of e-learning [8] and micro-learning, which does not require an assignment letter from the group leader to attend the training; designing webinars for Custom Collage and Ministry of Finance's Corpus Talk [9, 10]; converting the classical (face-to-face) training curriculum into the distance learning curriculum with Skill Group Owner.

4.2 Division of Event Management

This division has a function to organize education and training – opening, holding and closing the education and training events. Implemented strategies were the follows:

creating WhatsApp group consisting of training participants, lecturers and person in charge so that communication was well established; supporting intensive coordination during the implementation of the training; facilitating Zoom application to support the implementation of education and training; managing Community of Practice and Forum Group Discussion.

This department was also responsible for organizing customized training courses for teachers. This innovative approach to staff training was applied in order to improve teachers' digital competences and to improve their skills in working with the corporate training portal in a short period of time. Thus, the overall scheme of innovation was implemented, and the staff was able to quickly and deeply adapt to it. After this training teachers were not only able to deliver training in a remote format, but also they could to participate in the implementation of educational games and the development of a corporate training portal more effectively.

4.3   Division of Evaluation and Reporting

This division is responsible for the evaluation of the training during its implementation, the post-education-and-training evaluation and to follow-up evaluation. Implemented strategies were the follows: training evaluation was carried out by using Ministry of Finance Learning Center (KLC) website, Google-form, e-mail or Zoom for practice exam that require face-to-face meeting (such as information service exams) [11, 12]. KLC – learning center is an online learning media that covers various materials on State Financial Management and it can be accessed by all Ministry of Finance employees and public. KLC's main function was to support the education and training process held within the Ministry of Finance. During the pandemic a number of innovative tools were embedded in the corporate training portal, such as educational games, and the foundation for the use of virtual reality technology in the work of the training center was laid.

4.4   The Trainers

The functional positions of trainers are to educate, teach and coach the training participants. Implemented strategies were the follows: developing a distance learning curriculum with the team; teaching with using the distance learning teaching method; operating the Zoom, Google classroom, WhatsApp application and so on; developing teaching materials that meet the distance learning criteria, namely teaching materials for distance learning [13, 14], which are new and specifically developed for distance learning, studying international experience [15, 16]. The involvement of teachers in the stages of the organization and control of the educational process has become part of the innovative reorganization of the management system.

4.5   Division of Administration

This division supports the implementation of all activities at the Customs and Excise Education and Training Center. Implemented strategies were the follows: providing teaching rooms well-equipped with distance learning facilities, such as computers, head-sets, and microphones; ensuring the smooth access of the internet network for distance learning in coordination with the Ministry of Finance's Information and Technology Center; creating smart classroom; providing internet quota for teachers and participants with a reimbursed system. The administration was also charged with developing reporting forms on remote work.

5 Conclusion

The development of an innovative learning management system with a division of func tions between departments allowed maintaining the quality of training and ensuring the development of soft skills during the pandemic. Every Division in the Customs and Excise Training Center synergized to carry out the strategy for implementing education and training during the pandemic. There are numerous challenges that must be addressed. The management system innovative structural adjustments allowed for the implementation of customized training for trainers and the staff took an active part in the development of the fundamentally new educational content.

The smart solutions allowed us to overcome the obstacles associated with the restructuring of learning: in order to change work patterns and culture all employees could attend seminars; the stuff was instructed in managing the implementation of training; lecturers did sharing-sessions between employees related to the use of applications and devices in the distance learning (Zoom, KLC, etc.); practices in setting and using multimedia devices as well as organizing crews and committees during live broadcast were shown. The needs of infrastructure for the implementation of remote training sessions (cameras, lighting, audio, computer equipment, etc.) were completed and coordinated with Financial Information System and Technology Center (Pusintek) as the manager of the network of every working units throughout the Ministry of Finance in providing adequate network support in distance learning and e-learning. Also, Zoom cloud virtual meeting account was subscribed [17–20].

Regarding the changes of learning culture from classical (face-to-face) to independent learning (e-learning) or distance learning, Customs and Excise Education and Training Center intensively coordinated with users to provide full support in the implementation of both distance learning and e-learning, so that training participants could be given time to spare from their routine activities to participate effectively.

Creative approach to the organization of the learning process has significantly expanded the scope of educational games and it became the groundwork for the widespread use of virtual reality technologies in the practice of vocational training.

References

1. McRae, D., Robet, R.: Don't ask, don't tell: academics and electoral politics in Indonesia. Contemp. Polit. 1(26), 38–59 (2020). https://doi.org/10.1080/13569775.2019.1627736
2. Welch, A., Syafi'i, S.: Indonesia: islamic higher education – periphery within periphery? In: Symaco, L. (ed.) Education in South-East Asia, pp. 95–114. Bloomsbury Academic, London (2013). https://doi.org/10.5040/9781472544469.ch-005

3. Chapman, B., Dearden, L., Doan, D.: Global higher education financing: the income-contingent loans revolution. In: Callender, C., Locke, W., Marginson, S. (eds.) Changing Higher Education for a Changing World, pp. 87–100. Bloomsbury Academic, London (2020). https://doi.org/10.5040/9781350108448.0014

4. Green, F., Henseke, G.: Graduate employment and underemployment: trends and prospects in high participation systems of higher education. In: Callender, C., Locke, W., Marginson, S. (eds.) Changing Higher Education for a Changing World, pp. 173–187. Bloomsbury Academic, London (2020). https://doi.org/10.5040/9781350108448.0022

5. Flewitt, R., Ang, L.: Experimental research and randomized control trials (RCTs). In: Research Methods for Early Childhood Education, pp. 179–201. Bloomsbury Academic, London (2020). https://doi.org/10.5040/9781350015449.0019

6. Sugiyono, S.: Metode Penelitian Kuantitatif, Kualitatif, dan R&D. Alfabeta, Bandung (2018)

7. Miles, M.B., Huberman, A.M.: Qualitative Data Analysis: A Sourcebook of New Methods. SAGE publications Inc., California (1984)

8. Rubtsova, A., Odinokaya, M., Krylova, E., Smolskaia, N.: Problems of mastering and using digital learning technology in the context of a pandemic. In: Bylieva, D., Nordmann, A., Shipunova, O., Volkova, V. (eds.) PCSF/CSIS -2020. LNNS, vol. 184, pp. 324–337. Springer, Cham (2021). https://doi.org/10.1007/978-3-030-65857-1_28

9. Bylieva, D., Lobatyuk, V., Kuznetsov, D., Anosova, N.: How human communication influences virtual personal assistants. In: Bylieva, D., Nordmann, A., Shipunova, O., Volkova, V. (eds.) PCSF/CSIS -2020. LNNS, vol. 184, pp. 98–111. Springer, Cham (2021). https://doi.org/10.1007/978-3-030-65857-1_11

10. Bylieva, D., Bekirogullari, Z., Lobatyuk, V., Nam, T.: How virtual personal assistants influence children's communication. In: Bylieva, D., Nordmann, A., Shipunova, O., Volkova, V. (eds.) PCSF/CSIS -2020. LNNS, vol. 184, pp. 112–124. Springer, Cham (2021). https://doi.org/10.1007/978-3-030-65857-1_12

11. Almazova, N., Barinova, D., Ipatov, O.: Forming of information culture with tools of electronic didactic materials In: Annals of DAAAM and Proceedings of the International DAAAM Symposium, vol. 29, no. 1, pp. 0587–0593. DAAAM International, Vienna (2018). https://doi.org/10.2507/29th.daaam.proceedings.085

12. Castaño-Muñoz, J., Duart, J.M., Sancho-Vinuesa, T.: The Internet in face-to-face higher education: can interactive learning improve academic achievement? Br. J. Edu. Technol. 45(1), 149–159 (2014). https://doi.org/10.1111/bjet.12007

13. Almazova, N., Krylova, E., Rubtsova, A., Odinokaya, M.: Challenges and opportunities for Russian higher education amid covid-19: Teachers' perspective. Educ. Sci. 10(12), 368, 1–11 (2020). https://doi.org/10.13140/2FRG.2.2.30144.76803

14. Shahroom, A.A., Hussin, N.: Industrial revolution 4.0 and education. Int. J. Acad. Res. Bus. Soc. Sci. 8(9), 314–319 (2018). https://doi.org/10.6007/IJARBSS/v8-i9/4593

15. Hussin, A.A.: Education 4.0 made simple: ideas for teaching. Int. J. Educ. Lit. Stud. 6(3), 92–98 (2018). https://dx.doi.org/10.7575/aiac.ijels.v.6n.3p.92

16. Bylieva, D., Lobatyuk, V., Safonova, A.: Online forums: communication model, categories of online communication regulation and norms of behavior. Hum. Soc. Sci. Rev. 1(7), 332–340 (2019). https://doi.org/10.18510/hssr.2019.7138

17. Karpovich, I., Odinokaya, M.: Interactive technology of pedagogical assistance as a means of adaptation of foreign first-year students. In: IOP Conference Series Materials Science and Engineering, vol. 940, p. 012130 (2020). https://doi.org/10.1088/1757-899X/940/1/012130

18. Odinokaya, M., Andreeva, A., Mikhailova, O., Petrov, M., Pyatnitsky, N.: Modern aspects of the implementation of interactive technologies in a multidisciplinary university. In: E3S Web of Conferences, vol. 164, p. 12011 (2020). https://doi.org/10.1051/e3sconf/202016412011

19. Fatima, S.S., Idress, R., Jabeed, K., Sabzwari, S., Khan, S.: Online assessment in undergraduate medical education: challenges and solutions from a LMIC university. Pakistan J. Med. Sci. **37** (2021). https://doi.org/10.12669/pjms.37.4.3948
20. O'Brien, D.J.: A guide for incorporating e-teaching of physics in a post-COVID world. Am. J. Phys. **89**, 403–412 (2021). https://doi.org/10.1119/10.0002437

Best Practice Models

Technology-Based Methods for Creative Teaching and Learning of Foreign Languages

Ekaterina A. Samorodova[1]($\boxtimes$) ⓘ, Irina G. Belyaeva[1] ⓘ, Jana Birova[2] ⓘ,
and Aleksander Kobylarek[3] ⓘ

[1] MGIMO University, Prospekt Vernadskogo, 76, 11945 Moscow, Russia
{e.samorodova,i.beliaeva}@inno.mgimo.ru
[2] University of Ss. Cyril and Methodius in Trnava, Nám. J. Herdu 2, 91701 Trnava, Slovakia
jana.birova@ucm.sk
[3] University of Wrocław, J. Wł. Dawida 1, 50-527 Wrocław, Poland

Abstract. Integrative trends in the world, the expansion of intercultural interaction, and rapid social development contribute to the mastery of one or several foreign languages, not only to communicate with the representatives of other cultures but also to improve professional skills, as modern society requires specialists to solve global problems. The purpose of this study is to analyze the existing and most commonly used methods of teaching a foreign language and methods of learning a foreign language in terms of their effectiveness in the opinion of teachers and students. The combination of creativity and new technologies in teaching and learning a foreign language reach the mentioned aimss. The basis for the current research consists in scientific work by professionals from different countries, covering the issues of teaching foreign languages, as well as the results of a teachers' questionnaire and bachelors degree students. The empirical methods of the present research comprise studying and analyzing the works of some Russian and European scientists and teachers; interviewing teachers and students in the form of an anonymous questionnaire. Theoretical methods include analysis, synthesis, comparison, generalization, deduction, and induction. From among the great variety of methods the ones most commonly used are those related to ICT. Also, the majority of teachers recognize training video creation as most effective for the formation of students' creative abilities. According to the students' opinion, that type of activity has also the greatest learning effect, and stimulates creative abilities, rendering the educative process more interesting and efficient.

Keywords: Communicative competencies · Creative abilities · Innovative teaching methods · Training video creation

1 Introduction

One of the main activities of the future professional foreign specialist educated in higher educational institutions is to establish contacts with foreign partners and to lead effective cooperation on the international level. This task can be fulfilled thanks to the specialists' well-developed communicative competences, which determines the need for their communication training within the framework of foreign language disciplines.

Contemporary higher education requires the quick achievement of maximum results. Thus, foreign language teachers have to look constantly for methods to intensify the linguistic, educational process, to improve its quality, and at least to achieve the goal of preparing a competitive specialist for the world labor market. Foreign language teachers have a large arsenal of methods to accomplish their pedagogical tasks. Foreign language education at different universities and faculties in the world has its own traditions in the teaching process. But one way or another, the goal is the same - to find such methods and ways of teaching a foreign language that, in the context of modern requirements, will help to form strong knowledge and help a student - a beginner specialist - to actively use it in a certain communicative environment. In order to accomplish our research we have set the following research tasks:

1) To identify methods of teaching a foreign language which are used by teachers in bachelor's degree programs;
2) To rank the applied methods of teaching a foreign language in the bachelor 's degree programs according to the frequency of their use by teachers;
3) To rank the ICT methods of teaching a foreign language in the bachelor's degree program according to their effectiveness from the teacher's point of view;
4) To rank the ICT methods of learning a foreign language in the bachelor's degree program according to their effectiveness from the students' point of view.

The results of this study will help foreign language teachers at humanitarian universities to define the most suitable methods of teaching foreign languages for everyday teaching activities. Moreover, they will be able to further modernize, improve and develop those methods which are already successfully used by the colleagues from institutes and universities having participated in research.

2 Materials and Methods

The expansion of intercultural and multilateral interaction contributes to the mastery of one or several foreign languages, not only to communicate with the representatives of other cultures but also to improve professional skills, as contemporary society requires specialists to solve global problems. This fact determines the increasing role of language education which includes the formation of a new personality, the development of international thinking and tolerance towards the representatives of other cultures, and the teaching of a supranational vision of social interactions. This is most clearly observed at the level of higher education, which functions as a public institution preparing specialists for effective activity in any sphere of public life in international modern society.

At the present time, the education system is being modernized. The main goal of teaching foreign languages in higher educational institutions is the formation of both a communicative competence and an independent personality capable of linguistic self-education and self-improvement with a willingness to participate in intercultural dialogue.

D. Hymes, american linguist, sociolinguist, anthropologist was the first researcher who proposed the concept of communicative competence [1]. Gradually, over time this term has changed its meaning - from the knowledge of linguistic forms and social rules to the capability of generating error-free spontaneous speech in a foreign language [2 p. 47–48]. Nowadays, communicative competence is being understood in two senses. On the one hand it refers only to verbal communication, on the other hand to all aspects of communication.

With regard to higher education, the skill of a grammatically correct and effective use of linguistic tools in order to extract, process, and produce information gains practical importance along with the need for professional communication.

The graduate of a bachelor's degree program with a foreign language competence should be able to:

- conduct a conversation in a foreign language, participate in discussions, give public speeches on socio-political, professional, and sociocultural topics;
- use speech etiquette correctly; perceive and process various socio-political, professional, and sociocultural information in a foreign language, obtained from printed, audiovisual, and auditive sources, following the set goal;
- perform an oral interpretation of a written text within the diplomatic sphere of communication (translation and interpretation);
- perform written translation of written texts from a foreign language to Russian vice versa within the framework of the diplomatic sphere of communication (written translation);
- annotate and abstract printed and audio materials within the framework of legal and diplomatic spheres of professional communication.

The successful achievement of these goals requires a careful selection of the learning tools and analyzing the content of the proposed educational information.

Thus, in order to carry out the successful implementation of business or international communication in a foreign language, students need to have a well-developed multicomponent communicative competence. For its formation, first of all, it is necessary to take into account the content and formal features of the linguistic and speech material, which will serve as a basis for building a linguodidactic model of teaching foreign students.

The learning process is aimed at developing students' creative thinking and reasoning, which will help them enrich their knowledge throughout their life, improve their skills and acquire new abilities. For continuous self-development, students should have such personal qualities as initiative, reflexivity, anticipation, and correction of the result, an adequate evaluation and critical analysis of the results, the ability to solve problems. Researchers have derived the following formula for learning a foreign language: cognition - communication - creativity [3].

The development of the necessary abilities and skills is supported by various interactive forms of conducting classes, innovative methods and technologies stimulating the students' learning and cognitive activities: project [4–7], case studies [8], ICT [9–12], game [13–15], tandem [16], podcasts [17], cooperative and collaborative learning [18], discussions [18], dilemma [19], theater staging [20], SCRUM [21], round table [22], peer review [23], brainstorming [13], video blogging [24], interviews, briefings [24, 25], JigsawReading [26], flipped classroom [27], CLIL [28–30].

This study intends to discover which methods of teaching a foreign language are used, which methods the teachers consider the most popular and effective, how great is the didactic potential of the most frequently used methods according to the students, and what difficulties students encounter when learning by these methods. In order to answer these questions, we identified the methods of teaching a foreign language in some bachelor's degree programs. The programs are based on the real practical activities of foreign language teachers of the mentioned faculties. A questionnaire was realized by the specialists, teachers, doctors from some European and Russian Universities. They were invited to point out the methods which they use to teach a foreign language.

The anonymous questionnaire included 150 respondents. This survey has revealed the modern, widely used methods of teaching a foreign language to potential specialists in the humanities. The questionnaire implied selecting methods according to the respondent's personal opinion, based on their personal experience in learning a foreign language in some bachelor's degree programs. The questionnaire suggested a list of modern methods of teaching a foreign language which we selected on the basis of an analysis of the scientific literature (Table 1) and asked the respondent to choose the methods which were used. The number of selected methods was optional.

The respondents also had the opportunity to add other modern methods of teaching a foreign language. Table 1 highlights the most popular methods of teaching a foreign language.

Table 1. Modern methods of teaching a foreign language

1	ICT-involving
1.1	ICT – project
1.2	Interactive video
1.3	Training video creation
1.4	Podcasts
2	Problem-oriented methods
2.1	Discussion
2.2	Dilemma
2.3	Case studies
2.4	Round table
3	Project method
4	Gaming
4.1	Role-play
4.2	Business game
4.3	Imitation game (within the framework of one class)
4.4	Training company

(*continued*)

Table 1. (*continued*)

5	Functional
6	Co-operative learning
7	Tandem
8	Methods of mnemonics
9	Methods of neurolinguistic programming

The first methods grouping represents methods of teaching a foreign language with the involvement of ICT. In the modern ICT world, new technologies play a special role in learning processes. Despite all difficulties it has brought, the current pandemic has also made it possible to master new forms of teaching foreign languages through ICT methods.

The second grouping consists of teaching methods which are problem-based such that the main element of teaching is to find solutions. The most common methods in this group are the case studies method and the round table. These methods are especially popular in teaching a professional language, for example, for international lawyers.

The fourth group of methods of teaching a foreign language is made up of gaming methods often used for larger audiences.

The rest of the methods indicated in the table each represent technology-based methods for a special group and are aimed at both the formation of general linguistic competencies and professional ones. Interactive forms of training create a substantive and meaningful context for future professional activity, teach students to work in a team, solving a common problem.

For the designation of interactive methods we can use the didactic term "active teaching methods", which are characterized by a high degree of student involvement in the learning process and contribute to their cognitive and creative activities in fulfilling the tasks. The alternative term is "interactive teaching methods" - i.e., methods based on students' interaction with each other.

During interactive training, participants act together to find a solution to the problem, simulate situations related to practical activities, evaluate the actions of others and their own behavior, and receive feedback from the teacher and their fellow students. It must be emphasized that "active teaching methods" and "interactive teaching methods" are distinguished by the fact that the latter reflects the students' intensive and effective educational activity as a result of the teacher's mentoring activity.

3   Results and Discussion

The questionnaire served to identify the methods which are used to teach a foreign language in some European and Russian universities, it also allowed us to identify the most commonly used methods in teaching practice (Table 2).

Table 2. The methods of teaching a foreign language in bachelor's' degree program according to the frequency of use, in %

150 Respondents	
Methods	Frequency of use, in %
ICT - project	10%
Interactive video	8%
Training video creation	9%
Podcasts	6%
Discussion	8%
Dilemma	3%
Case studies	9%
Round table	8%
Project	9%
Role-play	8%
Business game	7%
Imitation game (within the framework of one class)	3%
Training company	3%
Functional	1%
Cooperative learning	3%
Tandem	1%
Methods of mnemonics	2%
Methods of neurolinguistic programming	1%

According to the results of the first survey, we see the following result. In their work, university professors use, to one degree or another, all the methods presented.

The methods less used by teachers are functional, tandem, methods of mnemonics, methods of neurolinguistic programming.

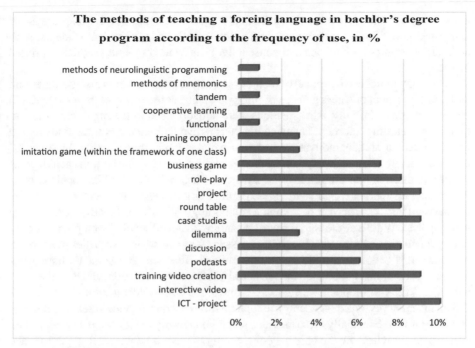

Fig. 1. The methods of teaching a foreign language in bachelor's degree programs according to the frequency of use, in %

This choice is most likely due to the fact that these methods are best suited for teaching a foreign language to another audience. Or the choice is related to the personal preferences of the teacher. Also, one of the reasons for the not very widespread use of some methods could be that they are insufficiently studied and explored at the university.

As far as for **tandem method** having 1% of use, it is important to underline the following. English-language bachelor's degree programs at some of the surveyed universities permit students from all over the world with different learning goals to be present in the same group. For example, some aim at learning the French language, others aim at learning the Russian language, or German. We will call such type of groups a mixed type. The presence of mixed groups has set a new goal for teachers: finding a method of tandem or cooperative learning that would allow students to successfully fulfill their tasks, reaching their goal by acquiring good knowledge of both languages. The heterogeneous composition of the group and the presence of native speakers of the studied languages is not an obstacle but the key to the successful mastery of these foreign languages. The learning process in this bilingual group is managed by a teacher who is a native speaker of the Russian language and has sufficient French language proficiency in teaching it as a foreign language. The presence of native speakers improves the students' phonetic skills. This increases the level of questions for discussions, role-play, and business games. Cultural, social and political aspects of the studied foreign language are studied more deeply. Besides, the students of mixed groups get additional daily listening practice from their classmates. Understanding foreign language speech by ear is a necessary condition for mastering a foreign language at a high level. The presence of native speakers allows practicing the professional skills of international students, such as interpretation (simultaneous and consecutive), interviewing, and producing printed media.

The study indicates the spread of methods used by teachers such as discussion, round table, role-play and business game, a training company, some forms of case studies, and a project method. Initially, these methods were actively used in the mono-environment of students learning a foreign language via their native language. Such methods have a positive effect on student motivation, as their use makes it possible to realize the students' needs in socialization, self-expression through being engaged in the joint activities of the group, through opportunities for creativity, and acquisition of professional skills.

Among all methods, ICT is the most popular among tested teachers, as shown by the survey (Fig. 1). The use of computer technology allows for creating a layer of a learning system both at the initial stage and at an advanced level, developing not only purely linguistic skills but also the skills necessary for professional activities in a foreign language. ICT involves a wide range of activities. Teachers and students have great access to educational audio and video materials, relevant information, and exercises that automate skills via computer systems when control can be carried out automatically. As is shown in the survey, teachers widely use such methods as training video creation, podcast and ICT-project. Especially these methods gained popularity during the pandemic, when online education became the only way of teaching in the whole world. In order to make the learning process productive, effective and interesting, teachers from all over the world have worked on the creation and development and implementation of new technologies and methods. Creation of training videos seems to be one of the most interesting and effective materials for presenting and explaining new material.

The second research result of the present study consists in the identification of particular types of ICT as types of training activities (Fig. 2). The respondents received a questionnaire in which they were asked to evaluate the effectiveness of different ICT-involving activities in a foreign language in the bachelor's degree program, such as podcasts creation, ICT-project, training video creation and interactive video. The evaluation was carried out on a scale from 5 points to 0 points, according to the following criteria:

5 – very high effectiveness;
4 – high effectiveness;
3 – medium effectiveness;
2 – low effectiveness;
1 – very low effectiveness;
0 – non-effective method.

The same number of points can be given to several methods. The methods which are not used are not evaluated. Tables 3 demonstrates the most effective ICT methods and their frequency of use.

Table 3. The effectiveness of ICT-involving activities in teaching a foreign language, according to the teachers, in %

150 Respondents	
Methods	Frequency of use, in %
ICT - project	25%
Interactive video	26%
Training video creation	28%
Podcasts	21%

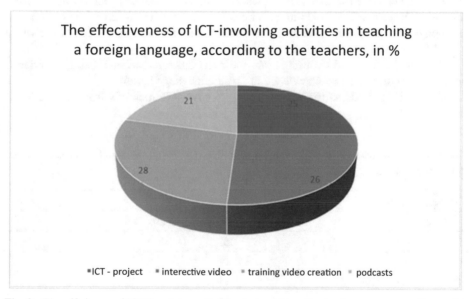

The effectiveness of ICT-involving activities in teaching a foreign language, according to the teachers, in %

- ICT - project
- interective video
- training video creation
- podcasts

Fig. 2. The efficiency of ICT-involving activities in teaching a foreign language, according to the teachers, in %

This choice of methods by teachers is quite understandable. The ICT test method is one of the most widespread methods that teachers from many universities in the world introduce into their work. Explaining new material through Power Point presentations or other platforms has become commonplace. Audio and video lessons based on the ICT project method were a necessary condition for maintaining the quality of education and its continuation (in many respects) in the difficult situation of the pandemic in the world. ICT presentations and projects can be considered classics of modern education.

Interactive videos have also been shown to be effective in developing speaking and writing skills. University teachers enjoy using interactive videos in their work and most often, such tasks are offered to students as homework.

However, the question of this study concerns the effectiveness of ICT methods in working with students. Here, efficiency can be understood as the degree of assimilation of new knowledge through special selected methods and the subsequent implementation of this knowledge in the professional sphere.

Most teachers have shown that making training videos is one of the most effective and interesting ways to work. Creativity, as an integral part of this method, is a powerful catalyst for optimizing the educational process. Again the situation with the pandemic comes to mind, when the world is mired in monochromatic online classes, and the use of the method of creating educational/training videos has enriched the educational process in the best possible way. The undoubted advantages of creating educational videos are: clarity, simplicity of form for perception, and the creative approach adopted - from a selection of material to video editing. It cannot be denied that the information presented in the video is perceived faster than anything else.

The answer to the fourth research question is based on a survey of 200 students from different European and Russian universities. The survey was carried out in the form of an anonymous questionnaire as well. Students from different universities of the world were asked to answer research questions:

What ICT methods do you think are the most efficient and interesting in order to acquire and assimilate new knowledge in foreign language lessons.

Table 4 (Fig. 3) demonstrates the most effective ICT methods from the students' point of view.

Table 4. The effectiveness of ICT-involving activities in learning a foreign language, according to the students, in %

200 respondents	
Methods	Frequency of use, in %
ICT - project	20
Interactive video	26
Training video creation	30
Podcasts	24

According to the survey, training video creation has the highest educative potential from the students' point of view. The creative approach underlying this method, as mentioned above, plays an important motivating role in learning a foreign language. To create an educational video, it is necessary to carefully work out the grammatical and lexical material, in other words, demonstration material, which forms the basis of this video. It is necessary to take different nuances into account when creating such a film. This requires serious independent work if the creation of such a video is given as homework. It can also be a joint project of a whole group of students, which also

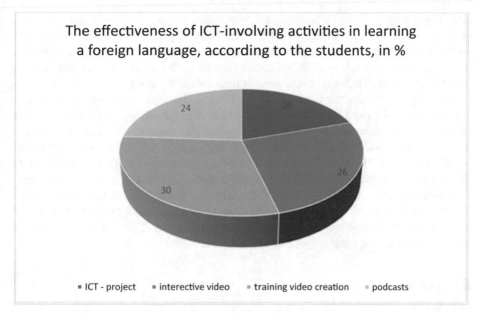

The effectiveness of ICT-involving activities in learning
a foreign language, according to the students, in %

24   20

26

30

ICT - project   interective video   training video creation   podcasts

Fig. 3. The effectiveness of ICT-involving activities in learning a foreign language, according to the students, in %

plays a huge role in the development of a future specialist, namely his ability to work in a team. Interactive video methods are also very effective from the students' points of view. Interactive video materials can be based on authentic audiovisual materials, including original feature films and news, which make it possible to implement the basic principles of teaching a foreign language (communicative approach, visibility, accessibility, consideration of the students' age characteristics, stimulation, and development of thinking, a combination of different educational forms and methods depending on the objectives and content of the training, as well as the connection between theory and practice).

Thus, students are ready to learn with the use of video materials, understanding their educational value.

The survey has shown that creating and using video materials in a foreign language class is common and appropriate for achieving the objectives in question.

4   Conclusion

Integrative trends in the world, the expansion of intercultural interaction, and rapid social development contribute to the mastery of one or several foreign languages, not only to communicate with the representatives of other cultures but also to improve professional skills, as modern society requires specialists to solve global problems. With regard to higher education, the skills of a grammatically correct and effective use of linguistic tools in order to extract, process, and produce information gains practical importance along with the need for professional communication. Learning a foreign language is accompanied by adapting to a non-native language environment. Thus, one of the essential methodical tasks is to think about the application of those methods that will demand the acquisition and further development of knowledge in the field of foreign language. A creative approach to the use of new technologies in teaching and learning foreign languages contributes to the achievement of the set tasks.

Among the training tools aimed at stimulating the students' learning activity in the classroom, the most widespread methods are those involving ICT on the basis of educational and authentic materials. The popularity of these methods among teachers of a foreign language is explained by a high degree of visualization, interaction, and actualization of the learning material presented by the teacher. ICTs help to reveal the cultural potential and to work with non-adapted speech, they contribute to the formation of an independent personality capable of linguistic self-education, they meet such basic didactic criteria as visibility, accessibility, stimulation, and development of thinking. It is supported by the intensification of the learning process and the use of new teaching methods that provide a high degree of students' independence.

References

1. Hymes, D.H.: On communicative competence. In: Pride, J.B., Holme, J. (eds.) Sociolinguistics. Selected Readings, pp. 269–293. Penguin, Harmondsworth (1972)
2. Cuq, J.P.: Dictionnaire de didactique du français langue étrangère et secondaire. Clé International. S.E.J.E.R., Paris (2003). (In Fr.)
3. Deikina, A.D.: Russian language in the modern educational model of bilingualism: cognition - communication - creativity. Dyn. Linguist. Cult. Process. Mod. Russia 5, 1167–1169 (2016). (In Russian)
4. Alipichev, A., Takanova, O.: Independent research activity of MSc and Ph.D. students: case-study of the development of academic skills in FFL classes. XLinguae 13, 237–252 (2020). https://doi.org/10.18355/XL.2020.13.01.18
5. Drovosekov, S., Sakhieva, R.: Peculiarities of using projects in learning English as a foreign language. XLinguae 11, 91–101 (2018). https://doi.org/10.18355/XL.2018.11.01.09
6. Rozhina, V., Baklashova, T.: Teaching English language to young school-age children while making projects, playing games and using robotics. XLinguae 11, 102–113 (2018). https://doi.org/10.18355/XL.2018.11.01.10
7. Bitokova, S., Kardanova, A., Shardanova, M., Efendieva, R., Dzaseszeva, L.: Learning English language and culture through idioms: a case study at Kabardino-Balkarian State University. XLinguae. 11, 28–38 (2018). https://doi.org/10.18355/XL.2018.11.03.03
8. Fesenko, O.P., Fedyaeva, E.V., Bestsennaya, V.V.: Cases in methods of teaching Russian as a foreign language. Lang. Cult. 9, 104–117 (2017). https://doi.org/10.17223/24109266/9/12
9. Karsenti, T., Kozarenko, O., Skakunova, V.: Integration into the world of ICT during the French lessons. XLinguae 13, 292–313 (2020). https://doi.org/10.18355/XL.2020.13.01.22
10. Gracheva, O., Matukhin, P., Komissarova, N., Saltykova, O., Kopylova, P.: Some principles of the computer RFL workbook development: the interactive vocabulary and grammar elements. XLinguae 12, 103–115 (2019). https://doi.org/10.18355/XL.2019.12.01.08
11. Bylieva, D., Bekirogullari, Z., Kuznetsov, D., Almazova, N., Lobatyuk, V., Rubtsova, A.: Online group student peer-communication as an element of open education. Future Internet 12, 143 (2020). https://doi.org/10.3390/fi12090143

12. Bylieva, D., Lobatyuk, V., Tolpygin, S., Rubtsova, A.: Academic dishonesty prevention in e-learning university system. In: Rocha, Á., Adeli, H., Reis, L.P., Costanzo, S., Orovic, I., Moreira, F. (eds.) WorldCIST 2020. AISC, vol. 1161, pp. 225–234. Springer, Cham (2020). https://doi.org/10.1007/978-3-030-45697-9_22
13. Nechayuk, I.A.: Active methods as optimization model for teaching English for special purposes. Interact. sci. **2**(12), 96–99 (2017). https://doi.org/10.21661/r-117837
14. Kalyuzhnaya, T.V., Skorobogatova, E.V., Vlasova, A.V.: Business role-play as a means of competence approach implementation in teaching the English language at higher education institutions. Bullet. Kemerovo State Univ. **3**(2), 52–54 (2015). https://doi.org/10.21603/2078-8975-2015-2-52-54. (In Russian)
15. Prikoszovits, M.: Auf dem Prüfstand – Wie berufsbezogen und praktikabel sind moderne handlungsorientierte DaF-Unterrichtsaktivitäten? Magazin **25**, 34–45 (2017). (in Ger.). http://dx.doi.org/10.12795/magazin.2017.i04
16. Almazova, N., Rubtsova, A., Eremin, Y., Kats, N., Baeva, I.: Tandem language learning as a tool for international students sociocultural adaptation. In: Anikina, Z. (ed.) IEEHGIP 2020. LNNS, vol. 131, pp. 174–187. Springer, Cham (2020). https://doi.org/10.1007/978-3-030-47415-7_19
17. Hasan, Md.M., Tan, B.H.: Podcast applications in language learning: a review of recent studies. Eng. Lang. Teach. **6**(2), 128–135 (2013). https://doi.org/10.5539/elt.v6n2p128
18. Kogan, M., Zakharov, V., Popova, N., Almazova, N.: The impact of corpus linguistics on language teaching in Russia's educational context: systematic literature review. In: Zaphiris, P., Ioannou, A. (eds.) HCII 2020. LNCS, vol. 12205, pp. 339–355. Springer, Cham (2020). https://doi.org/10.1007/978-3-030-50513-4_26
19. Kholod, N.I.: Application of moral dilemmas method for students communicative competence development in classes of foreign language at higher education institution. Bullet. Tomsk State Pedag. Univ. **7**(196), 92–95 (2018). https://doi.org/10.23951/1609-624X-2018-7-92-95. (in Russian)
20. Kungurova, I.M., Voronin, E.V., Dolzhenko S.G.: Art technologies in the formation of innovative pedagogical activity among students (on the example of teaching the discipline "technologies and methods of teaching foreign languages"). Science **6**(25) (2014). (in Russian). http://dx.doi.org/10.15862/31PVN614
21. Jurado-Navas, A., Munoz-Luna, R.: Scrum methodology in higher education: innovation in teaching, learning and assessment. Int. J. High. Educ. **6**(6), 1–18 (2017). https://doi.org/10.5430/ijhe.v6n6p1
22. Rodomanchenko, A.: Roundtable discussion in language teaching: assessing subject knowledge and language skills. J. Lang. Educ. **3**(4), 44–51 (2017). https://doi.org/10.17323/2411-7390-2017-3-4-44-51
23. Sysoyev, P.V., Merzliakov, KA.: Methods of teaching international relations students writing skills using peer review. Lang. Cult. **2**(34), 195–206 (2016). (in Russian)
24. Thamarana, S.: A comparative study of various english language teaching methods, approaches. Eng. Stud. Intern. Res. J. **3**(2), 204–207 (2015). https://doi.org/10.13140/RG.2.1.4026.5046
25. Bidenko, L., Shcherbak, H.: Implementing audio-lingual method to teaching Ukrainian as a foreign language at the initial stage. Adv. Educ. **3**, 23–27 (2017). https://doi.org/10.20535/2410-8286.82711
26. Yuhananik, Y.: Using jigsaw model to improve reading comprehension of the ninth graders of SMPN 1 Karangploso. IJOTL-TL: Indones. J. Lang. Teach. Linguist. **3**(1), 51–64 (2018). https://doi.org/10.30957/ijoltl.v3i1.50
27. Abdullina, L., Ageeva, A., Gabdreeva, N.: Using the "flipped classroom" model in the teaching of the theoretical disciplines (French language) at the university. XLinguae **12**, 161–169 (2019). https://doi.org/10.18355/XL.2019.12.01XL.12

28. Rošteková, M., Palova, M.: Curricular internationalization in higher education in Slovakia through the integration of content and language – experience with teaching in French. XLinguae **13**, 204–224 (2020). https://doi.org/10.18355/XL.2020.13.01.16
29. Baranova, T.A., Kobicheva, A.M., Tokareva, E.Y.: Does CLIL work for Russian higher school students? In: Proceedings of the 2019 7th International Conference on Information and Education Technology - ICIET 2019, pp. 140–145. ACM Press, New York (2019). https://doi.org/10.1145/3323771.3323779
30. Krylov, E.: Engineering education – convergence of technology and language. Technol. Lang. **1**(1), 49–50 (2020). https://doi.org/10.48417/technolang.2020.01.11

Methods of Achieving Successful Professional Communication from the Perspective of Intercultural Communication

Yuri V. Eremin[1] ⓘ, Olga G. Oberemko[2] ⓘ, and Elena A. Maliutina[2(✉)] ⓘ

[1] Peter the Great St. Petersburg Polytechnic University Saint Petersburg, Polytechnicheskaya 29, Saint Petersburg 195251, Russia
[2] Nizhny Novgorod Dobrolyubov State Linguistic University, Minina 31a, Nizhny Novgorod 603155, Russia
exert91@ya.ru

Abstract. Professional communication is often understood as just an ability to read scientific texts and to have terminology knowledge. However, scientists of different nationalities have a different mentality. In order to interact successfully in a professional sphere, it is necessary not only to have the above-mentioned knowledge and skills, but also to be able to communicate with another culture representatives. Understanding the business partner's mentality, as well as presenting your own identity in an accessible way is the key to fruitful cooperation. This is possible only if a Russian partner is creative and able to go beyond traditional ideas, patterns, relationships and create new meaningful ways and interaction techniques. In other words, this can be called creative thinking. The purpose of the study is to determine the levels of students' cultural self-identity, offer the author's approach, principles and assignments for designing a second language course. The leading research method was an experimental training conducted in both the intervention and the control group. The questionnaire method, methods of mathematical statistics and graphical results representation were used as well. Nizhny Novgorod State Linguistics University was the study experimental basis. The experimental training results showed an increase in the level of students' cultural self-identity. The correlation between a sufficient level of cultural self-identity and professional success was proved. Thus, the intercultural dialogue implementation is a factor in the success of second-language professional communication.

Keywords: Teaching foreign languages · Cultural identity · Professional success and creativity

1 Introduction

High competition aggravates national and cultural differentiation. Linguistic and cultural barriers as well as misunderstandings lead to interethnic confrontation. Linguistic education is of particular importance in solving this problem. This is due to the fact that the second language learning allows people to understand the 'world picture', expand their own borders, and arouse interest in cultural differences.

D. Bylieva and A. Nordmann (Eds.): PCSF 2021, LNNS 345, pp. 811–827, 2022.
https://doi.org/10.1007/978-3-030-89708-6_66

According to some researchers, the reason for the failure in communication is the disregard of cultural differences or cultural ignorance. Effective cross-cultural communication in a second language involves not only vocabulary, grammar and phonetics knowledge. Despite the knowledge of lexical and grammatical models, people are not always able to understand each other correctly. Adequate use of language units requires their origin and speech usage knowledge as well as history information, political and cultural life of the target language country without taking into account creative thinking.

Professional communication is often considered an ability to read scientific texts, terminology knowledge, understand scientific reports by ear, and communicate within one's profession. Many researchers offer teaching exact sciences such as mathematics, physics, chemistry, etc. in English even starting from high school. Of course, the language of science is universal. Achieving fluency in foreign languages in the field of professional communication is the main goal of teaching a foreign language in a non-linguistic university. However, scientists of different nationalities have a different mentality. In order to interact successfully in the professional sphere, it is necessary not only to have terminology knowledge, but also to be able to communicate with another culture representatives. There is no doubt that trust and sympathy are essential components of long-term cooperation. Often, the ability to communicate, soft skills are the foundation of business relations. Understanding the partners' mentality, as well as both the ability to present your own identity in an accessible way and manifestation of a creative approach to planning an intercultural dialogue is the key to fruitful cooperation and victory over competitors. Being creative means *being flexible* and ready to accept the others' ideas. By accepting the views of others, students can learn to generate their own ideas. *Be ready for planning.* Due to the large volume of outgoing information, it is difficult to immediately organize the thoughts that arise, so it is important to have the skill of planning the presentation of thoughts. *Be persistent.* When faced with a difficult task, do not put it off for later, but develop your own algorithm by straining your mind. *Be ready to correct your mistakes.* A person with creativity does not ignore and does not justify his\her mistakes, but turns them into experience for further training. To be aware, that is, to have the ability to observe yourself and analyze the course of your own methods, which you resorted to complete the task. Look for compromise solutions. When planning and building the presentation of one's own thoughts, it is necessary to take into account the perception of it by other people, otherwise, such decisions may remain at the level of statements.

The cultures interaction is carried out at various levels: interpersonal, interstate, professional and business. It is not by chance that one's own culture and only one's native language proficiency often makes the illusion of the existence of only one and correct 'picture of the world', excluding others, or considering them obviously wrong. Personal meetings, direct communication, delegations' exchanges, understanding the traditions, customs, and people's everyday life have a huge impact on the positive outcome of transactions, and the positive resolution of various professional issues at any level of interaction. An illiterate, in terms of cultural awareness, dialogue, on the contrary, is more likely to lead to failure, breakdown of agreements, etc. In cross-cultural communication, especially in the professional sphere, it is necessary, in our opinion, to take into account the ethno-cultural and national characteristics of interlocutors, the peculiarities of their mentality, as well as emotional and spiritual structure and way of thinking.

The problems of intercultural communication are widely discussed in the literature. Various aspects of the cultural dialogue have been studied by scientists in different contexts. There is a large number of articles and books related to the problems of interaction between different nations representatives. Scientists have developed the theory of cultural patterns of interaction [1], the theory of cultural dimensions [2], the theory of the linguistic and cultural literacy [3], the theory monoactive, polyactive and reactive cultures [4], linguistic theory of speech [5].

From our point of view, not only cultural awareness of the partner country is the key to success in achieving professional goals. Along with taking into account the peculiarities of the business partner national character and knowledge of cross-cultural business etiquette, it is necessary to be able to talk about your own national traditions, values and views in a foreign language. Business communication, along with the need to master the language in its specialized aspect, also involves a student's ability to present the uniqueness of their own country and culture in an accessible, understandable and commercially attractive form. Thus, in our opinion, for successful implementation of business interaction in the context of foreign-language intercultural communication, a sufficient level of cultural self-identity is necessary (Fig. 1).

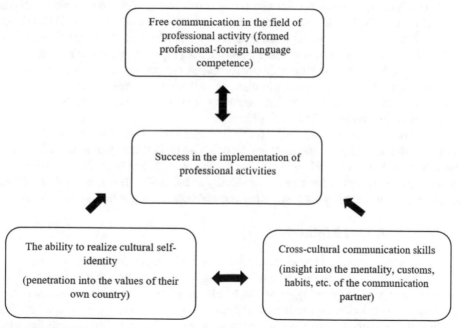

Fig. 1. The scheme of achieving professional success in the context of business cross-cultural communication.

In this article we trace the relationship between a developed cultural self-identity and a deep understanding of the value orientations and motivations of partners belonging to another culture with achieving success in the professional sphere. So, it is not possible to ignore a foreign interlocutor's mentality and their way of thinking. Otherwise, it would not be possible to motivate them or arouse interest in their own culture and effectively present their values to representatives of other cultures.

The ideas of our work are based on theoretical studies in the field of cross-cultural communication (A. Thomas, E. Hall, G. Hofstede, E. Hirsch, R. Lewis), the fundamental ideas of personal identity theories (R. Berne, J. Marsh, E. Fromm, E. Erikson, V. A. Yadov), linguistic and cross-cultural approaches to teaching foreign languages (R. P. Milrud, O. G. Oberemko, P. V. Sysoev, E. M. Vereshchagin, V. G. Kostomarov).

The identification mechanism was revealed in the concept of Z. Freud for the first time. The term "identity" came to linguodidactics from psychology, namely from the works of E. Erickson, published in the 50–60 years of the XX century [6, 7]. As a comparative analysis result of parenting styles in different cultures, E. Erickson concluded that each stage of development has certain society expectations, which an individual can either justify or not. Then he\she is either included in society or rejected by it.

In the humanities, this term is understood in two aspects: personal and social identity.

Personal identity as a psychological category is based on personal qualities and individual characteristics. It is understood as a subjective sense of one's individual identity and integrity, a human's ability to preserve the unity of their "I" and their "self" throughout their lives [7].

Social identity is described by social categories and is based on a group analysis which the individuals belong to and which they identify themselves with. This concept is considered to be the most important socialization mechanism. It is manifested in the individuals' acceptance of a social role when entering a group, and in their awareness of group membership, the formation of social attitudes, etc.; as a process and result of self-identification with another person, group, image or symbol on the basis of an established emotional connection, as well as their inclusion in their inner world and acceptance as their own norms, values and patterns [7].

Both personal and social identity form an inseparable whole and a personality. Cultural self-identity is a component of social identity and is defined by us as an individuals' awareness of themselves belonging to a particular culture, self-identification with their cultural norms, values and mentality, the ability to talk about their country in a foreign language, readiness for intercultural dialogue and overcoming stereotypes.

2   Materials and Methods

The main methodological ideas contributed to the development of the author's ethnically-oriented approach. This approach involves taking into account the ethno-cultural, ethno-linguistic and ethno-psychological students' characteristics and partner countries' representatives. It is based on the ethnos understanding as a stable set of people that has historically developed in a certain territory, having common relatively stable language features, culture and psyche, as well as the consciousness of their unity and difference from other similar entities (self-consciousness), fixed in the self-name [9–13]. In the course of professional interaction with a representative of a multinational state, it is necessary to understand that the communicant may be a representative of a certain ethnic group and have a different mentality from the national one.

The main methods of conducting experimental training were:

- methods of information collection (observation, questioning of teachers and students, the method of competent judges);
- methods of information processing (calculation of results using methods of mathematical statistics).

Communication is not just a verbal process. The success of communication is largely determined by a number of non-verbal factors: the presence of deep background knowledge, the ability to present their own culture, knowledge of the etiquette rules, understanding of the implicit meanings behind the facial expressions, gestures of the communication partner, etc. It is a mistake to believe that only knowledge of the basic language units of a foreign language allows people to communicate effectively with a representative of another culture. This is due to the fact that the language barrier is often not the most significant obstacle to an adequate understanding of each other.

These approaches allowed us to formulate the author's principles, which formed the basis for developing a course of teaching foreign languages to students of non-linguistic universities with the aim of mastering professional communication, as well as adults in foreign language courses for special purposes.

The principle of achieving professional success.

This principle involves the formation of students' awareness of the role of cultural identity in commercial activities, the ability to present their country to a foreign partner. Many companies develop business strategies and enter the global market. The role of cross-cultural interaction is huge. This is due to the fact that the consequences of poor communication can include lack of cooperation, poor product reputation, etc. [15]. Students need to understand that the winner is the one who can present him/herself in the most advantageous light and who is able to attract and interest a potential business partner in the current conditions of fierce competition. It is important to have not only the brand and reputation of the company, but also to create an attractive image of your culture. It is equally important to be able to win over the interlocutor, to show your traditions and customs, to dispel stereotypes, to remove misunderstandings, to overcome intolerance. Note that along with your own self-identity, it is important to clearly understand the representative of which ethnic group you work with, take into account the mentality of the interlocutor.

The teachers must include specific language material.

The principle of parallel inclusion of a foreign language and students' own culture information units.

The features of language and culture become most noticeable in contrast, i.e., they are the most obvious in comparing cultures. It is also important to study students' own culture. Cultural mistakes, unlike language mistakes, are often not ignored [16–20]. An unprepared person, faced with a foreign culture, can experience a "culture shock". This is a phenomenon that characterizes the conflict that has arisen between a new culture and a system of values, worldview, way of thinking of a person and his\her ideas about the rules of social order.

In the process of intercultural communication, difficulties arise that cannot be overcome only by mastering the models that serve to create a speech utterance. It is necessary to possess traditional ideas, images, metaphors, and symbols for a given society. Forming active and receptive lexical foreign language skills and abilities, it is necessary to remember that there can be a whole layer of phenomena reality behind a foreign word. It is called the cultural component of meaning [16]. It is necessary to check not just the word translation into the native language, but to find out what is behind this concept, what is its place in the world of foreign culture. Similarly, in the process of teaching a foreign language, it is necessary to include units of information from one's own culture simultaneously. Foreigners will communicate with us not in order to learn from us about their world, but in order to get information about ours [19].

The principle of accentuation of native culture affective concepts (visual and verbal).

The ability to present your own country to a foreign partner is the key to successful completion of transactions, reaching agreements, and having a positive reputation. Students often experience difficulties in transmitting information related to their native culture in a foreign language. In the process of cross-cultural communication, including in the professional sphere of communication, students are not able to show their cultural self-identity [20].

In our study native culture affective concepts represent the key phenomena of one's own culture that cause the greatest emotional response in communication. Such concepts can include absolutely everything that affects students' emotional sphere: from personal attachments to 'the power places' of their country, region, city.

So, for example, the following lexical units can act as affective concepts of Russian culture: dacha (a country house in Russia), the Ural Mountains (divide Russia into Europe and Asia), 9 time zones in Russia, etc. Moreover, private photos can be affective as well: a personal dog or cat, flowers that grow in their own dacha, beautiful views from the window of their own house, funny photos from professional cases, etc.

The principle of using PR technologies in teaching a foreign language.

Public Relations (PR) is a planned positive effort aimed at creating a favorable communication environment. This principle involves teaching a foreign language through drawing attention to one's own culture, traditions, and customs. As in any PR activity, we are based on the desire to "infect" students with ideas of self-identification, to teach them to convince and to turn over competitors. The influence of PR technologies is enormous: almost everything we know about ancient Egypt, Assyria, and Persia has become widely-known thanks to materials aimed at glorifying the next ruler. A huge part of the materials was created to support and recognize the merits of kings, rulers, and other leaders [20].

This includes jokes, proverbs, sayings that reflect the mentality of people from the target culture. Students can be offered tasks for commenting on the suggested phraseological units. They may compare them with their own culture, and identifying differences. It is also proposed to discuss how to use the acquired knowledge about the culture of the country of the language being studied in the professional sphere. These are the examples of the proverb for discussions.

- The road to hell is paved with good intentions.
- God Created people, and Colt made Them Equal.
- Money does not grow on trees.
- A friend in need is a friend indeed.

- If it isn't broken, don't fix it.
- If you want something done right, do it yourself.

The principle of distinguishing basic categories that reflect the native culture.

Each culture has its own basic categories, which often do not coincide with the culture of the partner country. Language and thought, language and history, language and culture are inseparable. To create a positive image of your country, region, show the main and little-known sights and cultural facts it is crucial to take into account the mentality of foreign partners in order to do business with representatives of a different ethno-cultural group. This principle is based on providing students with tools in the form of language tools that describe their native culture and are understandable for foreign partners.

Thus, the main categories to be assimilated are:

- Geography and climate.
- Key historical events.
- Customs and traditions.
- Food.
- Features of urban and rural life.
- The device of the house, everyday life.
- The conditions in which the professional activity will take place.

Sample tasks for the development of cultural identity and intercultural communication skills.

According to the research of M. Schneider, intercultural learning consists of the following components:

1) Language component (includes realities, background words, non-equivalent vocabulary, and common forms of speech).
2) The pragmatic component (the customs of everyday life).
3) Historical component (updated if the countries have a common history, certain related events in the past, both positive and negative points of intersection).
4) Aesthetic component (views on the appearance of people, buildings, institutions, the structure of everyday life from an aesthetic point of view).
5) The ethical component (concepts of morality and morality).
6) The imagological component (images and stereotypes of the surrounding world).
7) Reflexive component (the student has a new view of both his personality, his culture, and the one being studied, there is an analysis of the realities and events of everyday life, social problems and trends. This component can be attributed to the criterion of the effectiveness of the learning process) [21–23].

According to M. Schneider, ideally organized intercultural education involves the use of materials that would demonstrate in a contrasting form the peculiarities of the mentality and culture of two or more peoples and could serve as an occasion for discussion about the conflicts and contradictions associated with the belonging of participants in the intercultural dialogue [21].

The communicative situations should anticipate the events in which the students will be involved in the real process of communication. Therefore, educational speech situations should put students in conditions similar to natural ones. However, as practice shows, many tasks that at first glance seem quite communicative, do not actually reflect the system of values and relationships in the culture of native speakers and are not based on the models of relevant situations inherent in this language [23, 24].

The set of exercises is aimed at developing the cultural identity of students by means of a foreign language and has a number of features. In the process of adult education, a special place belongs to problem tasks of varying degrees of complexity, since they stimulate the development of communicative and cognitive, general cultural and educational skills of adult students in the process of joint learning of the language and culture of their own and another country. The second characteristic feature of the set of exercises is its focus on taking into account (and overcoming) national stereotypes by comparing the cultural values of one's own and the studied cultures.

We offer sample assignments designed in accordance with the approaches and learning principles described above:

1. Sample tasks aimed at understanding the role of self-identity in achieving professional success:

    Discuss the following questions: what contributes to the success of business negotiations? How important is it to use a foreign language while still being a representative of your country? Is there a relationship between an attractive representation of one's own culture, country, or region for effective cooperation and communication with foreign colleagues? Comment on your ideas.

2. Sample task aimed at familiarizing yourself with the information units of a foreign language and your own culture (on the example of Russia, Great Britain and the United States).

    Role-playing game:

    Problem statement: Students are divided into three groups.

    Group A. Imagine that a group of foreign tourists from the United States and Britain comes to your city. Your task is to tell them about the typical structure of the Russian house, the rules of behavior at a party, and the etiquette traditions.

    Group B. Imagine that you are tourists from the United States who want to visit Russia and learn more about the structure of the Russian home, traditions, etiquette, and related customs. In turn, find and prepare information about the design of typical homes in the United States.

    Group C. Imagine that you are tourists from Britain who want to visit Russia and learn more about the structure of the Russian home, traditions, etiquette, and related customs. In turn, find and prepare information about the design of typical homes in Britain.

    The course of the role-playing game. The participants of group A tell groups B and C about the structure of the Russian home, traditions, customs, and hospitality. Students in groups B and C identify differences between houses in Russia, Britain and the United States.

    Discussion.

    Answer the questions:

1) What features of Russian houses can cause the greatest surprise?
2) What facts can be shocking for a foreigner?
3) What is common in domestic traditions in Russia, Britain and the United States?
4) What is the attitude to the home, hearth in different countries?
5) Buy your own housing or rent. Discuss it. What choice are the representatives of Britain and the United States likely to make?

3. Sample task aimed at accentuating affective concepts (visual and verbal) of the native culture (on the example of Russia):
   Look at the following words. What emotions do they evoke in you?
   Moscow, St. Petersburg, Russian Orthodox Church, New Year, holiday chores.
4. Sample task aimed at using PR technologies:
   Imagine a small group of your foreign business partners is coming to your city. Work in pairs and decide where you would take them for dinner. Choose a menu and a location to give them a "taste" of your city/region/country.
5. A sample task aimed at identifying basic categories that reflect the native culture for foreign partners to familiarize themselves with (for example, Russia and the United States).

Read the text "Discover Swallow's Nest Castle", answer the questions to the text, then find the most interesting facts about the castle and its history. Imagine that you want to recommend this attraction to an American who wants to visit Russia. Write him a letter and tell him why he should visit the "Swallow's Nest".

3   Results

3.1   Initial Characteristics

Note that identity and self-identity are non-identical concepts. Identity is the identification of an object. While self-identity is precisely the identification of oneself with someone or something. In other words, identity, in our opinion, is how others perceive an individual, and self-identity is how an individual perceives him- or herself. Identity and self-identity in different people may or may not coincide. Russian immigrants to the United States, for example, can be identified as "Russians", while they do not consider themselves as such. So, their self-identity is different. Perhaps they are so assimilated, in their opinion, that they consider themselves Americans or, for example, 'people of the world'. That is, in this example, the Russian identity is lost.

Based on the concept of J. Marcia [8], we determine the cultural self-identity levels of development. They are manifested in the skills of cross-cultural communication, and, consequently, contribute to the achievement of professional success.

The starting level is approximately equal to the stage of 'diffuse identity' according to J. Marcia classification. At this stage, students demonstrate a superficial knowledge of their native culture. A person cannot talk about his\her country, its history, heritage, traditions in a foreign language. There is no sense of empathy, activity, or independence. They do not realize the need to present their culture in business contacts with foreign partners. People tend to ignore or downplay the importance of their own culture in the global space, especially in the process of business cross-cultural communication. There is also a denial of cultural differences.

The associative-analytical level of cultural self-identity development is correlated with the stage of "early identification" according to J. Marcia. At this stage, mental reflection on the topic of one's own culture begins. It is characterized by the knowledge of reference samples of their culture. The cultural comparison is carried out under the teacher guidance but not independently. Occasionally, there are manifestations of a sense of empathy, activity, and respect for other cultures.

The verbal-analytical level is correlated with the "moratorium" stage according to J. Marcia. There is a need for constant one's own culture knowledge enrichment, understanding the correlation between professional success and the ability to present oneself to a foreign partner. There is a tendency to creative problem solving, the desire to expand the knowledge of both the native culture and the target one. In the process of cross-cultural communication, the student demonstrates partial knowledge of basic categories of their own culture, partial possession of affective concepts, takes into account the mentality of the interlocutor. At this level, there is an increased sensitivity to the problems of interaction with foreign partners, non-standard situations. There is a need to use unconventional approaches, to interest the partner in unusual ideas that can position him\her, bring them to a new level of relations. At the verbal-analytical level, thinking is perceived as something controlled and purposeful using cognitive techniques, assuming an evaluative component. Reproduction of information according to a template and according to a sample, the use of algorithms in the presentation.

The verbal-creative level corresponds to the stage of 'achieved identity' according to J. Marcia. Students have a deep knowledge of their own culture. They know the information units, the affective concepts of their native culture. They are able to isolate the basic categories without effort. The person shows empathy, tolerance to other cultures and takes into account the interlocutors' mentality. Students are able to achieve professional success due to their ability to express themselves competently and to defend their own point of view in a foreign language. They are able to develop strategies and speech behavior tactics, model intercultural communication situations, search for information in a foreign language related to professional activities without teachers' help. The unity of cognitive, emotional and behavioral principles has been achieved. The personality moves from the search for cultural self-identification to practical professional self-realization.

Let us compare the verbal-analytical and the verbal-creative levels. At the verbal-analytical level there is a great attachment to the context, memorized verbal manifestations of communicative intentions. Moreover, the teacher's instructions play an important role, which is due to the insufficient level of foreign language proficiency. At the verbal-creative level we can observe the ability to achieve a professional goal, to find a way out of a seemingly hopeless situation, using the situation, objects, circumstances, verbal manifestations in an unusual way without relying on educational materials is noted. The evaluative component prevails, students evaluate the logic of their arguments, whether the courses of his thoughts are diverse, confidence in the facts on the basis of which he makes decisions, that is, a critical analysis of information that begins at the third level and fully unfolds at the fourth. Re-organization of the acquired knowledge, the ability to organize the initial information from certain sources in a certain logical order according to the purpose of the activity. Correlation of new information with previously received and learned experience and reflection, i.e., analysis of your work, your own activities, its successes and failures.

To test the effectiveness of these methods of work, we conducted a series of training experiments, implemented in three blocks. The base for the pilot test was the Nizhny Novgorod State Linguistic University named after N. A. Dobrolyubov. The experiment was conducted among adults who study English for special purposes in foreign language courses.

The experiment lasted two years.

The first block. September 2019 – June 2020. The program was "English for tourists" at N. A. Dobrolyubova State Linguistic University of Nizhny Novgorod. In this group, 62 students aged 20–55 years were trained. The level of proficiency was defined as B1 according to the Common European Framework of Reference (CEFR).

The second block. September 2019 - June 2020. The program was "English for Business and Commerce" at N. A. Dobrolyubova State Linguistic University of Nizhny Novgorod. In this group, 64 students aged 19–50 years were trained. The level of training is similar to the first block, namely B1 according to the Common European Framework of Reference (CEFR).

The third block. October 2019 – May 2020. The program was called "English for Medical professionals" at N. A. Dobrolyubova State Linguistic University of Nizhny Novgorod. In this group, 72 students aged 18–56 years were trained. The level of training is similar to the first and second blocks, namely B1 according to the Common European Framework of Reference (CEFR).

This section may be divided by subheadings. It should provide a concise and precise description of the experimental results, their interpretation, as well as the experimental conclusions that can be drawn.

3.2 Course of the Experiment

The experimental study was conducted in the intervention and the control groups of students. Both groups studied a foreign language for special purposes of professional communication. The duration of training, the number of students, the initial level of development of self-identity and communicative competence in a foreign language were identical.

The purpose of the experimental test was to form and further develop the oral inter-cultural speaking skills and the self-identity of each student to achieve professional success. These oral skills are in both monological and dialogical forms.

The structure of the experiment included three stages: the preparatory stage, the experimental training stage, and the final stage. The preparatory stage involved two periods of preparation and experimental training organization: theoretical and practical. Theoretical training included: analysis of scientific literature in the field of intercultural communication, theory of personal identity, approaches to teaching foreign languages; study of pedagogical experience of teaching adults who study a foreign language for special purposes; summarizing our own experience of working with foreign partners, observations during the negotiation process.

We conducted a diagnostic cross-section that demonstrates the initial level of development of cultural self-identity in students in the intervention group (IG) and the control group (CG). Thus, the subjects were asked to perform oral tasks that allow students to determine the level of oral intercultural communication skills. As well as knowledge of Russian culture in terms of the ability to engage in intercultural interaction and the self-identity development.

Determination of the cultural self-identity level of development was carried out on the basis of the following criteria:

- quality of knowledge about your own culture;
- knowledge of lexical and grammatical means that allow you to present your own culture in a foreign language in an attractive way;
- the presence of deep linguistic and cultural knowledge;
- knowledge of speech and ethical methods of intercultural communication in the professional sphere;
- the scope of the statement;
- speech rate;
- consistency of the statement;
- full disclosure of the subject of the message.

The presented criteria for evaluating the oral statements of the subjects clearly demonstrate that the average result in the range of 4.5–5 points indicates the verbal-creative level of self-identity development; the result in the range of 4–4.4 points - the verbal-analytical level; the result in the range of 3–3.9 points-the associative-analytical level; the result of 2.9 points and below - the starting level.

The total number of subjects – 198 people (Table 1).

Table 1. Baseline levels of cultural self-identity development (diagnostic cross-section).

	The starting level	The associative-analytical level	The verbal-analytical level	The verbal-creative level
Number of students in the IG	66 (67%)	20 (20%)	13 (13%)	0 (0%)
Number of students in the CG	59 (60%)	27 (27%)	13 (13%)	0 (0%)

We calculated the success rate for the entire set of tasks for the development of cultural self-identity in IG and CG according to the formula of V. P. Bespalko:

$$k = a/n$$

where a is the number of subjects who showed a positive result (average score of 3 points or higher), n is the total number of subjects in the group (99 students).

The obtained data: $k (IG) = 33/99 = 0.33$; $k (CG) = 40/99 = 0.4$.

Since the minimum permissible coefficient is $k = 0.7$, the results of the calculation show a low level of self-identity development, which requires further work.

The experimental training phase consisted of six series of experiments, including the following topics:

- Geography and climate.
- Key historical events.
- Customs and traditions.
- Food.
- Features of urban and rural life.
- The device of the house, everyday life.
- Topics of professional communication (tourism, business and commerce, medicine).

There is also data of the intermediate cut (Table 2).

Table 2. Intermediate levels of cultural identity development.

	The starting level	The associative-analytical level	The verbal-analytical level	The verbal-creative level
Number of students in the IG	58 (59%)	22 (22%)	10 (10%)	9 (9%)
Number of students in the CG	57 (58%)	35 (35%)	7 (7%)	0 (0%)

In the intervention group, training was carried out based on the principles and approaches described above and on the basis of the proposed set of exercises. In the control group, the training was conducted on regular educational materials, without the use of author's developments.

Here are the results of the final cut (Table 3).

Table 3. Final levels of cultural identity development (final cross-section).

	The starting level	The associative-analytical level	The verbal-analytical level	The verbal-creative level
Number of students in the IG	**13 (13%)**	**20 (20%)**	**39 (40%)**	**27 (27%)**
Number of students in the CG	**46 (47%)**	33 (33%)	13 (13%)	7 (7%)

3.3　Results Analysis

A comparative analysis of the data of the final and diagnostic cross-sections shows an intensive positive dynamic of the development of cultural self-identity. In the absence of students with the verbal-creative level of cultural self-identity development at the pre-experimental stage, the results of the final cross-section show an increase in this indicator by 27% in IG and by 7% in CG. The number of students with the verbal-analytical level in the IG increased by 27%. It did not change in the CG. The associative-analytical level did not change in the IG and increased in the CG by 1%. The number of students with the starting level of cultural identity decreased: in IG by 54% and in CG by 13%.

Success rate based on the results of the final cut:

$$k(IG) = 86/99 = 0.86; \; k(CG) = 53/99 = 0.53$$

The obtained result indicates that the work in the IG on the cultural self-identity development in achieving professional goals was carried out successfully and does not require further teachers' actions. However, in CG, the success rate slightly increased (from 0.4 to 0.53) and remained be3low the acceptable level.

Students focus on professional communication. It is a mistake to think that only a commercial dialogue will bring success. People need to be informed that in order to succeed in business, it is first necessary to establish human relationships with a foreign business partner. It is crucial to understand the partners' mentality and to take into account their norms of conducting business negotiations. This leads to a successful commercial dialogue. The results of the first section of the experiment show that there is no understanding of the need to take into account the mentality. After working on this area, we see the emergence of this understanding among students, but not all. However, students do not sufficiently understand the difference in mentality and identity. After the third block of training, people realized the need to develop cultural identity and penetration into the consciousness of the other.

Thus, a comparative analysis of the results of the diagnostic and final cross-sections based on mathematical processing of statistical data confirms the effectiveness of the proposed methodology for the cultural self-identity development.

We also conducted a survey of students in the IG and CG for achieving success in the professional field, which we attributed to: successful conclusion of a transaction, reaching agreements, receiving recommendations, extending contracts, etc.

The results of the survey showed that the majority of students in the IG achieved success in cross-cultural professional interaction. The data is shown in the table (Table 4).

Table 4. Results of the survey on achieving the goals of intercultural professional communication for the period June 2020-December 2020.

Have you achieved success in cross-cultural professional interaction?	Students of the intervention group	Students of the control group
Definitely yes	32 (33%)	15 (15%)
Rather yes	47 (47%)	16 (17%)
Rather no	9 (9%)	45 (45%)
Definitely no	7 (7%)	20 (20%)
Abstain	4 (4%)	3 (3%)

4 Discussion

Thus, the implementation of cross-cultural dialogue can be a factor in the success of foreign-language professional communication. Cross-cultural communication is a combination of 'the world picture', a representative of an ethno-cultural group psychological characteristics and the accepted form of communication. It is also the way of thinking and the way of people's life. In order to communicate successfully and effectively with representatives of other nationalities, it is necessary: 1) to know the peculiarities of the foreign interlocutor's mentality, lifestyle, and social structure; 2) to achieve the verbal-creative, verbal-analytical or the associative-analytical level of one's own cultural identity. This will help to convey one's point of view. Moreover it will make national identity attractive in the eyes of a foreign partner with taking into account the mentality of the other; 3) to make the educational process in line with an ethnically-oriented approach based on the principles of: achieving professional success; parallel inclusion of information units of a foreign language and one's own culture; accentuation of affective concepts (visual and verbal) of the native culture; the use of PR technologies in teaching foreign languages; the identification of basic categories that reflect the native culture, for familiarization of foreign partners. It is necessary to develop the ability to argue students' opinion. To do this, there is a need to provide both exciting and unusual tasks that contribute to the development of creativity.

References

1. Hall, E.: Beyond Culture. Anchor Books, Garden City (1976)

2. Hofstede, G.: Culture's Consequences: Comparing Values, Behaviors, Institutions and Organizations Across Nations. Sage, Beverly Hills (1980)
3. Hirsch, E., Kett, J.: The New Dictionary of Cultural Literacy: What Every American Needs to Know. Houghton Mifflin Harcourt, Boston (2002)
4. Lewis, R.D.: Cross Cultural Communication: A Visual Approach. Transcreen Publications (1999)
5. Vereshchagin, E., Kostomarov, V.: Linguistic and Regional Studies Theory of Words. Russkiy yazyk Publ, Moscow (1980)
6. Danilova, E., Yadov, V.: Social identifications in post-soviet Russia: empirical evidence and theoretical explanation. Int. Rev. Sociol. **7**(2), 319 335 (1997). https://doi.org/10.1080/039 06701.1997.9971240
7. Erikson, E.: Identity, Youth and Crisis. Norton, Jenkins (1968)
8. Marcia, J.: Identity in adolescence. In: Handbook of Adolescent Psychology. Wiley, New York (1980)
9. Ter-Minasova, S.: Teacher, student, textbook in today's Russia. Bull. Moscow Univ. **4**, 14–23 (2016)
10. Rubtsova, A.: Socio-linguistic innovations in education: productive implementation of intercultural communication. In: IOP Conference Series: Materials Science and Engineering, vol. 497, p. 012059 (2019). https://doi.org/10.1088/1757-899X/497/1/012059
11. Bylieva, D., Lobatyuk, V., Safonova, A.: Online forums: communication model, categories of online communication regulation and norms of behavior. Human. Soc. Sci. Rev. **7**(1), 332–340 (2019). https://doi.org/10.18510/hssr.2019.7138
12. Odinokaya, M., Karpovich, I., Mikhailova, O., Piyatnitsky, A., Klímová B.: Interactive technology of pedagogical assistance as a means of adaptation of foreign first-year students. In: IOP Conference Series: Materials Science and Engineering, vol. 940, p. 1–13 (2020). https://doi.org/10.1088/1757-899x/940/1/012130
13. Presbitero, A., Attar, H.: Intercultural communication effectiveness, cultural intelligence and knowledge sharing: extending anxiety-uncertainty management theory. Int. J. Intercul. Relat. **67**, 35–43 (2018). https://doi.org/10.1016/j.ijintrel.2018.08.004
14. Merkin, S.: Saving Face in Business: Managing Cross-Cultural Communication. Palgrave Macmillan, London (2015)
15. Almazova, N., Rubtsova, A., Eremin, Y., Kats, N., Baeva, I.: Tandem language learning as a tool for international students sociocultural adaptation. In: Anikina, Z. (ed.) IEEHGIP 2020. LNNS, vol. 131, pp. 174–187. Springer, Cham (2020). https://doi.org/10.1007/978-3-030-47415-7_19
16. Maliutina, E., Oberemko, O.: The problem of developing cultural self-identity in adults learning a foreign language. Tomsk State Univ. J. **460**, 202–212 (2020). https://doi.org/10.17223/15617793/460/24
17. Seregina, T., Zubanova, S., Druzhinin, V., Shagivaleeva, G.: The role of language in intercultural communication. Space Cult. **7**(3), 243–253 (2019). https://doi.org/10.20896/saci.v7i3.524
18. Bylieva, D.S., Lobatyuk, V.V., Fedyukovsky, A.A.: Ways of sociotechnical integration of scientists and volunteers in citizen science. In: IOP Conference Series: Materials Science and Engineering, vol. 940, p. 012150 (2020). https://doi.org/10.1088/1757-899X/940/1/012150
19. AlTaher, B., Bassmah, B.: The necessity of teaching intercultural communication in higher education. J. Appl. Res. High. Edu. **12**(3), 506–516 (2020). https://doi.org/10.1108/JARHE-04-2019-0082
20. Yusof, N., Kaur, A., Cheah Lynn-Sze, J.: Post graduate students insights into understanding intercultural communication in global workplaces. Innov. Educ. Teach. Int. **56**(1), 77–87 (2019). https://doi.org/10.1080/14703297.2017.1417148

21. Schneider, M.: Interkulturelles Lernen im Russischunterricht. Handbuch Fremdsprachenunterricht **41**(50), 3, 162–165 (1997). (in Ger.)
22. Othman, A., Norbaiduri, R.: Intercultural communication experiences among students and teachers: implication to in-service teacher professional development. J. Multicult. Educ. **14**, 223–238 (2020). https://doi.org/10.1108/JME-04-2020-0024
23. Odinokaya, M., Andreeva, A., Mikhailova., O., Petrov, M., Pyatnitsky N.: Modern aspects of the implementation of interactive technologies in a multidisciplinary university. In: E3S Web Conference, vol. 164, p. 12011 (2020). https://doi.org/10.1051/e3sconf/202016412011
24. Shipunova, O., Berezovskaya, I., Smolskaia, N.: The role of student's self-actualization in adapting to the e-learning environment. In: 7th International Conference on Technological Ecosystems for Enhancing Multiculturality, pp. 745–750. ACM, New York (2019). https://doi.org/10.1145/3362789.3362884

Study of the Efficiency of a Multilingual Educational Model

Tatiana Baranova⑩, Aleksandra Kobicheva⁽✉⁾ ⑩, and Elena Tokareva⑩

Peter the Great Saint-Petersburg Polytechnic University, St. Petersburg 195251, Russia

Abstract. As new informational conditions contribute to the discovery of new ways to improve the quality of the educational process, a multilingual learning model was elaborated. A creative approach to the educational process stimulates the involvement and active participation of students, which makes the learning process less tiring. The purpose of the paper is to assess the influence of this model on students' foreign language proficiency, formation and development of multilingualism and to examine the attitude of students towards the efficiency of such a model. For our research we used qualitative and quantitative data of students' records of Spanish testing, multilingualism development testing and interviews on students' attitude towards multilingual model and model efficiency from students' perspective ($N = 85$). Results on students' Spanish language proficiency showed positive impact of learning model as scores in all categories (listening, reading, writing and speaking) were substantially improved. Students' testing results on multilingualism development were quite high (most students gave more than 50% right answers) that also confirms the positive effect of such model. According to the interview results, students support the idea of developing multilingualism, but some students believe that this should not be a mandatory university program. Also, students positively assessed the developed educational model.

Keywords: Multilingualism · Multilingual development · Language teaching · Second language acquisition

1 Introduction

The modernized education of the XXI century is based on humanism, openness towards the outside world. Multilingual education, focused on the harmonization of different cultural spaces, is a necessary component of a modern vision of the world. There is a gradual introduction of the student to a full-fledged existence in the conditions of a modern multilingual and heterogeneous society (including the world of modern communications, social networks, e-mail, etc.). The student's self-development and activity of cognition throughout life is encouraged and supported, and learning is understood as a chain of continuous questions to which there are many conflicting answers. At the same time, the language used in the learning process has the function of a communication and cognition tool [1]. The signs of tolerance and cooperation are nurtured, concepts of the continuous field of culture, history and language of mankind are perceived and assimilated, covering different eras, languages, creeds and civilizations [2, 3]. The ability to discern the general develops, recognizing the right of everyone to be different, without emphasizing the other as unacceptable. The ability to be guided in judgments by one's own acquired experience develops, using data from various and often conflicting sources, including in different languages.

A creative approach to the educational process stimulates the involvement and active participation of students, which makes the learning process less tiring. Learning multiple foreign languages is more effective if you make the process more creative. Thus, replenishing one's linguistic stock, learning one language on the basis of several others cannot be standardized, since each student has his own linguistic background. It is the creative aspect that will make it possible to implement a personality-oriented approach, which is important for the introduction of multilingualism into the education system.

Research on multilingualism as a learning and teaching resource has steadily expanded over the past few years, in tandem with initiatives taken by regional blocs and individual countries to promote the learning of multiple languages, especially in secondary education [4]. Such initiatives reflect the increasingly diverse nature of many cities and countries around the world [5], where hundreds of nationalities live together and communicate in different languages. Given the growing importance of developing multilingual citizens in an increasingly multipolar world, countries with multilingual education initiatives need to develop effective curricula that will lead to successful learning outcomes for students [6].

At Peter the Great Polytechnic University an initiative group of professors and researchers developed a multilingual educational model that was introduced in curricula of 2nd and 3rd year undergraduate students. The aim of the current study is to assess the influence of this model on students' foreign language proficiency, formation and development of multilingualism and to examine the attitude of students towards such a model.

1.1 Literature Review

Government advocacy for multilingualism, that is, the use of more than one language by individuals, societies, institutions and groups in everyday life [7, 8], is based in part on the belief that it improves cognitive abilities [9], mathematical knowledge [10], creativity [11] and cultural knowledge [12]. A more multilingual society can also lead to greater understanding, better relationships and openness between countries, both individually and at the national level. When it comes to teaching and learning languages, multilingualism and its potential benefits can be approached from a variety of theoretical frameworks. Jessner [13, 14], applying dynamical systems theory to language learning by multilingual people, writes that such people develop multilingual systems consisting of language-specific and non-language skills and abilities related to language learning, management and maintenance, which not available for monolingual. She argues that these systems have a positive effect on creative thinking, communication sensitivity, flexibility, translation skills, and the interactive and pragmatic competence of multilingual people.

The increased interest and growth in research of multilingualism that has been identified over the past decade has led to the development of this field into a new discipline [8]. As a result of the general trend, some forms of multilingualism are the norm for the world's population [15]. As multilingualism and linguistic diversity spread and became visible in several parts of the world, educational policies were designed to meet the emerging needs for language learning and linguistic diversity, which also took place in the context of the European Union [16].

The desire for multilingualism has been recorded in European educational systems due to the recognized importance for all citizens of being able to speak two European languages in addition to their mother tongue, an action defined as the "2 + 1 formula" [17]. Even though English was good established in many European countries as the first foreign language and remains more or less indisputable due to its influential role of the lingua franca throughout Europe and around the world [18], learning lingua franca alone is no longer considered adequate; the development of multilingualism is considered a basic prerequisite for free movement and intercultural communication in Europe [19].

The development of communication skills in two foreign languages is considered a basic competence for people and has been integrated into the education systems of most European countries [20, 21]. In Russia, a recent initiative by the Ministry of Education (MoE) gives students the opportunity to learn two foreign languages by the time they leave school [22]. And while many multilingual initiatives have traditionally focused on learning English as a foreign language (EFL) or a second language (ESL), this also seems to be changing.

To conceptualize multilingualism, one can use the theory of multi-competence [23, 24], which encompasses a person's or community's knowledge of more than one language, taking into account the total amount, as well as the relationship between them and how they can affect not only the use of the language but also to the mind in general. This multi-competence can provide more cognitive flexibility and metalinguistic awareness that multilinguals will use to explore more deeply how languages function [25]. A study by Kramsch and Whiteside [26] argues that multilingual people exhibit behaviors that exceed simple communicative competence in interaction, which can benefit students not only in terms of language learning, but also in other areas of life.

According to Kramsch and Whiteside, a multilingual person has "a particularly keen ability to play with different language codes and with different spatial and temporal resonances of those codes" [26, p. 664]. They call it "symbolic competence", defining it as "the ability not only to bring closer or appropriate someone else's language, but also to form the very context in which the language is studied and used" [26, p. 664]. Symbolic competence fits well with the adaptation by Singleton and Aronin [27] of Gibson's [28] affordance theory to describe multilingualism. In terms of languages, affordance represents what languages offer, in a material and non-material way, to a person in terms of cognitive and evaluative abilities, emotions and knowledge [29–32]. Taking advantage of these opportunities entails an increased awareness of multilingual people of the many ways in which they can interact with their environment. The more languages a person knows, the more opportunities they have, although empirical research to support this is still difficult to find. In short, all of these frameworks suggest that multi-language students have a potentially rich and complex set of tools that they can use to improve their learning in a variety of ways.

2 Methodology

Our research involved 2nd and 3rd year undergraduate students (N = 85) studying under the program «International Business». This educational program is international, therefore the composition of student groups is multinational. All disciplines are taught in English. In the second year, the study was carried out in 3 groups (15, 12 and 16 students, respectively). In the third year, 3 groups were also selected (12, 14 and 16 students). Groups were selected in which it was possible to divide students, taking into account their native languages, into small subgroups. In addition, there were no Russian and English-speaking students in these groups, which placed all students on an equal footing when studying Spanish in English. For our experiment on the development of multilingualism, we have developed an educational model. The developed model was applied in the lessons of the "Spanish language" discipline.

The developed educational model assumes the study of the Spanish language using the native languages of the students, as well as the English language. English is the common language for the whole group and the teacher. Each group consists of 12–16 students of different nationalities. We deliberately did not include 1st year students in the experiment, since the experiment requires students who have learned the basics of Spanish grammar. In the 2nd and 3rd year, students devote more time to vocabulary topics. The native languages of the students are Arabic, Hindi, Chinese and French.

The experiment lasted one semester (September 2020 - January 2021). Before starting the experiment, we tested students for English proficiency in order to determine the readiness of students to learn according to the proposed model (we considered the B2 level to be sufficient). All students on the preliminary test in English showed good knowledge of the language (B2 and higher). In addition, we conducted pre- and post-test Spanish language tests to analyze the effectiveness of the educational model for learning Spanish. Initial testing in Spanish showed less positive results. So, in the second year, 23% of students did not reach A2. In the third year there were only 3% of such students. Most of the students in the second year were confident in the A2 level, and in the third year - in the initial B1.

After all the initial testing, the students were asked to split into groups according to their native languages in order to work in monolingual teams on a given topic. During the semester, each group studies 4 vocabulary topics. For the second course, the topics were: National Holidays in Spain, Tourism in Spain, Cuisine of Spain. For the third year, the topics were more difficult - Medicine in Spain, Ecology in Spain, Education in Spain and the media in Spain. For each team, a section was created on the Moodle virtual platform, which contained text in their native language on a given topic (Fig. 1). Each team, as a preparation for the classroom, independently studied the proposed text, compiled a thematic vocabulary with the translation of vocabulary into English and Spanish. The texts for each subgroup differ in content, but all are similar in topics.

During the classroom session, each group presented the content of the read text in English and presented a prepared vocabulary with translations of terminology into Spanish. As a result of teamwork, the group had a ready-made English-Spanish dictionary on a given topic, as well as studied basic information in native languages and presented in English. In this way, we develop reading, speaking and listening skills among students. In addition, students learn basic switching between their mother tongue, their first

Fig. 1. Subgroup self-study text in French on Ecology.

foreign language (English) and partly Spanish (vocabulary compilation). This stage is preparatory to the activation of multilingualism in teaching.

Having assimilated the information received, the students had to work on a case related to the topic under study. Students are divided into new subgroups in such a way that students of each nationality are in each subgroup. The case contains a description of the problem situation and requires students to search for a solution to the problem. In the case, information is presented in the form of a text describing a problem situation and a video fragment in which the problem is discussed by several stakeholders, supplementing the material with emotional color and subjective assessments. When working on a case,

Fig. 2. Case study in Spanish on Ecology

students are required to be creative and rely on the cultural characteristics of Spain (all situations occur in Spain). The case is presented in Spanish in the form of text (Fig. 2), but a video clip in English is attached to it (Fig. 3).

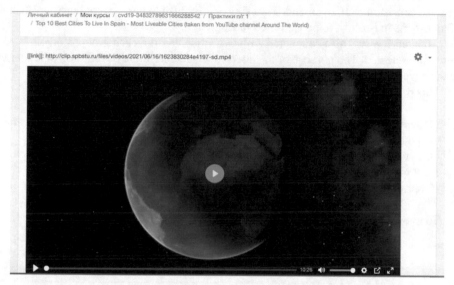

Fig. 3. Ecology video clip in English

An audio recording in Spanish may also be provided for completeness (Fig. 4).

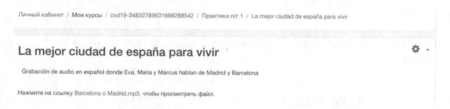

Fig. 4. Spanish audio recording on Ecology

Students in their teams need to accumulate the information received in both languages, discuss, come up with a solution and make a presentation with their proposal in Spanish.

At this stage, students develop listening, reading and speaking skills, as well as writing skills (when drawing up a presentation). The main aspect of this stage is the full-fledged activation of multilingualism, since to solve the problem, students need to use all the information received earlier, namely:

– information from the first texts that each student received in his native language;
– information from the video presented in English;

– information from the text with the case presented in Spanish.

This algorithm of work is repeated when studying each topic. At the end of the experiment, the students were offered a test to determine the development of multilingualism. The test consisted of 10 open-ended questions in English to be answered in Spanish and 10 questions in Spanish to be answered in English.

An interview was also conducted with two students from each group, students were randomly selected. The purpose of the interview is to determine the attitude of students to the development of multilingualism in the educational environment.

This paper is based on the following research questions:

1. Is there a significant difference in the level of Spanish proficiency before and after the course?
2. Does the proposed teaching methodology contribute to the formation and development of multilingualism?
3. If a proposed learning model is efficient from the students' perspective?

To obtain the data we used both quantitative and qualitative data (Table 1).

Table 1. Data collection.

Results	Sort of data collection	Type of data
Development of multilingualism	Scores on testing	Quantitative
Spanish proficiency	Scores on testing	Quantitative
Students' attitude to multilingualism	Interview	Qualitative

For the analysis descriptive statistics and pair-samples Students's t-test were conducted.

3   Results and Discussion

3.1   Spanish Proficiency

Testing on Spanish proficiency consisted of 4 parts: Listening, Reading, Writing, and Speaking. In Part 1, students watched the video on one of the course topic, and then gave short answers to the questions asked. In the Reading section, students were offered a text (case study) with a volume of approximately 500 words; after reading it was necessary to complete 3 tasks: complete text with correct topic sentences for each paragraph, choose whether the statement is True/False, and answer the questions about the main information from the text. Writing skills were evaluated on the base of answers from Listening and Reading sections. For the speaking test, students conducted small individual presentations and verbally answered few questions.

Testing results of 2nd year study students are presented in Table 2.

Table 2. Results of 2nd year study students (N = 41).

Category	Test	Results (average mean)	SD	T-value
Listening	Pre-test	14,4	2,14	5,3***
	Post-test	17,71	2,19	
Reading	Pre-test	14,53	1,89	5,4***
	Post-test	18,21	1,83	
Writing	Pre-test	17,27	1,93	2,3*
	Post-test	18,57	1,97	
Speaking	Pre-test	15,46	2,01	4,1**
	Post-test	17,12	1,97	

Note: * $p < 0.05$; ** $p < 0.01$; *** $p < 0.001$.

Table 3. Results of 3rd year study students (N = 44).

Category	Test	Results (average mean)	SD	T-value
Listening	Pre-test	16,4	2,05	2,2*
	Post-test	17,89	2,02	
Reading	Pre-test	16,33	2,09	3,3**
	Post-test	18,59	1,99	
Writing	Pre-test	17,27	1,97	3,7**
	Post-test	19,95	2,13	
Speaking	Pre-test	17,55	2,01	5,1***
	Post-test	20,12	1,97	

Note: * $p < 0.05$; ** $p < 0.01$; *** $p < 0.001$.

Testing results of 3rd year study students are presented in Table 3.

In general, the overall quality of students' Spanish knowledge (we measured listening, reading, writing, and speaking skills) was improved. The comparison of results of the two tests (before and after the course) of all participants in the experiment shows that the improvements were significant in all categories. Comparing the results of 2nd year students and 3rd year students we can note that on the 2nd year of study they developed mostly listening and reading skills while on the 3rd year of study they improved significantly writing and speaking skills. To conclude we affirm the efficiency of multilingual educational model for Spanish learning purposes.

3.2   Development of Multilingualism

Each open-ended question on multilingualism development was assessed on a 2-point scale, so the maximum result that students could achieve was 40 points. Testing results are presented in Fig. 5.

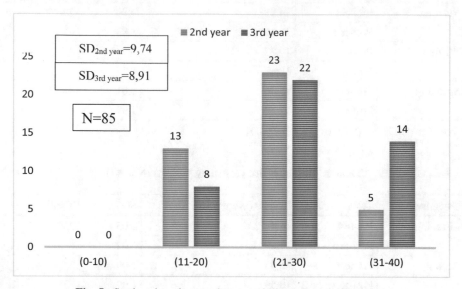

Fig. 5. Students' testing results on multilingualism development.

According to the results, most students gave more than 50% right answers (68% of 2nd year students and 82% of 3rd year students). As students of 3rd year study wrote testing better than students of 2nd year study that also confirms the efficiency of multilingual model that was introduced in curricula.

3.3   Students' Attitude to Multilingualism

To determine the attitude of students towards the development of multilingualism, an interview was conducted. For the interview, 2 students from each group were randomly selected (a total of 3 groups in the 2nd year and 3 groups in the 3rd year). The interview was conducted after the final tests were written, but before the final grades were given, in order to avoid the impact of the assessment on the perception of multilingualism and the educational model. The interviews were in a semi-structured format. Students were asked to answer 5 questions:

1. How do you feel about multilingualism in general?
2. Do you consider it necessary to develop multilingualism in the process of studying at a university?
3. Do you think that learning from the new model has allowed you to develop multilingualism?

4. Describe your impressions of the last semester and learning on the new model?
5. Which learning model is more acceptable for you - traditional or new?

The answers of the 2nd year and 3rd year students differed significantly. All respondents have a positive attitude towards multilingualism in general and believe that the development of multilingualism is necessary in the modern world. However, 3rd year students expressed doubts about the importance of the development of multilingualism in the educational system of the university. In their opinion, not all students studying at a university need knowledge of several foreign languages. Thus, each student decides for himself how many and what languages he needs to learn and does it outside the university. On the contrary, 2nd year students spoke out in support of multilingualism in university education. The respondents noted that the opportunity to study several foreign languages and acquire the skill of switching between languages allows students to be competitive in the labor market after graduation.

All students stressed that the new educational model had a significant impact on their ability to switch between all the languages they speak. However, 2nd year students noted that they felt more confident only towards the end of the course. The reason for this difficulty is the lack of knowledge of the Spanish language and the lack of vocabulary. It was much easier for 2nd year students to communicate in English: "English words came to mind first of all, and this did not allow switching to Spanish". The students also noted that underdeveloped Spanish speaking skills were also a problem for switching between languages. On the contrary, 3rd year students note: "the effect turned out to be unusual, the border between the languages seemed to be erased". Students noted that sometimes they stopped noticing which language the conversation was in and kept up the conversation in a given language.

All students indicated that the educational process during the last semester was more interesting than before. New assignments, group work, as well as project work on cases contributed to greater involvement and motivation of students. In addition, the students noted that the language learning process became invisible, as the students simply worked on projects and solved the assigned tasks, automatically learning the language. Most of the students (8 out of 12 people) said that they would like to continue learning the language in a new way and further. Thus, we can conclude that the students liked the educational model developed by us more than the traditional format.

4 Conclusion

According to the gained results on students' attainments we can affirm that educational process based on multilingual model is effective and learners' records are positive. Having made a t-value analysis of Spanish proficiency it was identified that results in all categories significantly improved after the course based on this model. The results of testing on multilingualism development showed that most students gave more than 50% right answers (68% of 2nd year students and 82% of 3rd year students).

Based on the results of the interviews with students, it can be concluded that the majority of students positively assessed the developed educational model, noted its effectiveness for the development of multilingualism. However, the 2nd year students did not have enough knowledge of the Spanish language to take advantage of all the advantages of this model and the process of developing multilingualism for these students was delayed and proved to be insufficiently effective. Despite this, 2nd year students actively support the idea of multilingualism and express their readiness to develop multilingualism in the process of studying at the university. On the contrary, 3rd year students expressed doubts about the need to develop multilingualism within the framework of university education, since not everyone has a need to study several foreign languages. However, during the experiment, the students noted that the skill of switching from one language to another has developed significantly.

The theoretical work of other scholars [8, 13, 23, 29] in this field has been a useful resource for planning and designing, and we expect that our study will provide something of value for future researchers, too. Of course, there are some limitations in our study, as it does not take into account a novelty effect—students did not have an experience of learning in a multilingual environment. Additionally, the sample size was relatively small because it was the first time, we implemented such an educational model, and the duration of the course was only one semester.

In our further research we are going to evaluate students' satisfaction of multilingual environments in a developed learning model and upgrade it due to responses.

References

1. Volodarskaya, E.B., Grishina, A.S., Pechinskaya, L.I.: Virtual learning environment in lexical skills development for active vocabulary expansion in non-language students who learn English. In: 12th International Conference on Developments in eSystems Engineering (DeSE), pp. 388–392. IEEE, Kazan (2019). https://doi.org/10.1109/DeSE.2019.00077

2. Tatiana, B., Kobicheva, A., Tokareva, E.: Web-based environment in the integrated learning model for CLIL-learners: examination of students' and teacher's satisfaction. In: Antipova, T., Rocha, Á. (eds.) DSIC 2019. AISC, vol. 1114, pp. 263–274. Springer, Cham (2020). https://doi.org/10.1007/978-3-030-37737-3_24

3. Baranova, T., Kobicheva, A., Olkhovik, N., Tokareva, E.: Analysis of the communication competence dynamics in integrated learning. In: Anikina, Z. (ed.) IEEHGIP 2020. LNNS, vol. 131, pp. 425–438. Springer, Cham (2020). https://doi.org/10.1007/978-3-030-47415-7_45

4. Byram, M.: Language education in and for a multilingual Europe. For. Lang. Educ. Multiling. Classrooms 7, 33–56 (2018). https://doi.org/10.1075/hsld.7.02byr

5. Demidov, V., Mokhorov, D., Mokhorova, A., Semenova, K.: Professional public accreditation of educational programs in the education quality assessment system. In: E3S Web Conference, vol. 244, p. 11042 (2021). https://doi.org/10.1051/e3sconf/202124411042

6. Baranova, T., Kobicheva, A., Tokareva, E., Vorontsova, E.: Application of translinguism in teaching foreign languages to students (specialty "Ecology"). In: E3S Web Conference, vol. 244, p. 11034 (2021). https://doi.org/10.1051/e3sconf/202124411034

7. Clyne, M.: Multilingualism. In: Coulmas, F. (ed.) The Handbook of Sociolinguistics, pp. 301–314. Blackwell Publishing, Hoboken (2017). https://doi.org/10.1002/9781405166256.ch18

8. Anosova, N., Dashkina, A.: The teacher's role in organizing intercultural communication between Russian and international students. In: Anikina, Z. (ed.) IEEHGIP 2020. LNNS, vol. 131, pp. 465–474. Springer, Cham (2020). https://doi.org/10.1007/978-3-030-47415-7_49

9. Hirosh, Z., Degani, T.: Direct and indirect effects of multilingualism on novel language learning: an integrative review. Psychon. Bull. Rev. **25**(3), 892–916 (2017). https://doi.org/10.3758/s13423-017-1315-7

10. Dahm, R., De Angelis, G.: The role of mother tongue literacy in language learning and mathematical learning: is there a multilingual benefit for both? Int. J. Multiling. **15**(2), 194–213 (2018). https://doi.org/10.1080/14790718.2017.1359275

11. Fürst, G., Grin, F.: Multilingualism and creativity: a multivariate approach. J. Multiling. Multicult. Dev. **39**(4), 341–355 (2018). https://doi.org/10.1080/01434632.2017.1389948

12. Ellis, E.M.: The Plurilingual TESOL Teacher: the Hidden Languaged Lives of TESOL Teachers and Why They Matter. De Gruyter Mouton, Boston (2016)

13. Jessner, U.: Linguistic Awareness in Multilinguals: English as a Third Language. Edinburgh University Press, Edinburgh (2006)

14. Jessner, U.: A DST model of multilingualism and the role of metalinguistic awareness. Mod. Lang. J. **92**, 270–283 (2008). https://doi.org/10.1111/j.1540-4781.2008.00718.x

15. Bhatia, T.K., Ritchie, W.C.: Handbook of Bilingualism. Blackwell, Oxford (2004)

16. Wilton, A.: Multilingualism and foreign language learning. In: Knapp, K., Seidlhofer, B., Widdowson, H. (eds.) Handbook of Foreign Language Communication and Learning. De Gruyter Mouton, Göttingen (2009). https://doi.org/10.1515/9783110214246.1.45

17. Council of Europe: White paper on education and training. Teaching and learning towards the learning society. European Commission, Brussels (1995)

18. Graddol, D.: English next. The British Council (2006). https://www.britishcouncil.org/learning-research. Accessed 7 Apr 2021

19. Commission of Europe: Action Plan 2004–2006: the promoting of language learning and linguistic diversity. European Commission. Council of Europe, Strasbourg (2003)

20. Eurydice: Foreign Language Teaching in schools in Europe. European Commission, Brussels (2001)

21. Eurydice: Key data on teaching languages at school in EuropeThe European Unit, Brussels (2005)

22. Letter from the Ministry of Education and Science Regarding the Teaching of a Second Foreign Language. The Ministry of Education and Science of the Russian Federation, Moscow (2018). https://zknrf.ru/acts/Pismo-Minobrnauki-Rossii-ot-17.05.2018-N-08-1214/. Accessed 7 Apr 2021

23. Cook, V.: Multi-competence: black hole or wormhole for second language acquisition research. In: Han, Z., Park, E.S. (eds.) Understanding Second Language Process, pp. 16–26. Multilingual Matters, Ontario (2008)

24. Cook, V.: Multi-competence. In: Chapelle, C. (ed.) The Encyclopedia of Applied Linguistics, pp. 3768–3774. Wiley-Blackwell, Chichester (2012)

25. Calafato, R.: Evaluating teacher multilingualism across contexts and multiple languages: validation and insights. Heliyon **6**(8), e04471 (2020). https://doi.org/10.1016/j.heliyon.2020.e04471

26. Kramsch, C., Whiteside, A.: Language ecology in multilingual settings. Towards a theory of symbolic competence. Appl. Linguist. **29**(4), 645–671 (2008). https://doi.org/10.1093/applin/amn022

27. Singleton, D., Aronin, L.: Multiple language learning in the light of the theory of affordances. Innov. Lang. Learn. Teach. **1**(1), 83–96 (2007). https://doi.org/10.2167/illt44.0

28. Gibson, J.: The theory of affordances. In: Shaw, R., Bransford, J. (eds.) Perceiving Acting, and Knowing, pp. 67–82. Lawrence Erlbaum Associates, Hillsdale (1977)

29. Aronin, L.: Multilingualism in the age of technology. Technol. Lang. **1**(1), 6–11 (2020). https://doi.org/10.48417/technolang.2020.01.02

30. Popova, N.V., Almazova, N.I., Evtushenko, T.G., Zinovieva, O.V.: Experience of intra-university cooperation in the process of creating professionally-oriented foreign language textbooks. Vyss. Obraz. Ross. = High. Educ. Russ. **29**, 32–42 (2020). https://doi.org/10.31992/0869-3617-2020-29-7-32-42

31. Almazova, N., Barinova, D., Ipatov, O.: Forming of information culture with tools of electronic didactic materials. In: Katalinic, B. (ed.) Annals of DAAAM and Proceedings of the International DAAAM Symposium, vol. 29, no. 1, pp. 0587–0593. Danube Adria Association for Automation and Manufacturing, DAAAM, Zadar (2018). https://doi.org/10.2507/29th.daaam.proceedings.085

32. Altenbach, H.: Scientific language – a comparative analysis of English, German and Russian. Technol. Lang. **1**(1), 1–5 (2020). https://doi.org/10.48417/technolang.2020.01.01

Video Sketches as the Means to Improve Students' Creativity in Studying Foreign Languages

Daria Burakova$^{(\boxtimes)}$ (iD), Oksana Sheredekina (iD), Maya Bernavskaya (iD), and Olga Mikhailova (iD)

Peter the Great St. Petersburg Polytechnic University,
Polytechnicheskaya 29, St. Petersburg 195251, Russia
`burakova_da@spbstu.ru`

Abstract. The paper covers the application of video sketches as the way to provide out-of-class monitoring that could stimulate students' creativity and motivation to study foreign languages at technical universities. The study introduces the data analysis based on the statistics survey conducted in academic groups of 1-course students of technical specialties of Peter the Great St. Petersburg Polytechnic University during the autumn term of 2020–2021 academic year. Two groups of students were randomly chosen as experimental and control groups. The students of the control group were to present their monologues in a traditional format, whereas students of an experimental group had an opportunity to prepare video sketches instead. The authors emphasize the benefits of recording video sketches compared to face-to-face format of delivering a monologue as well as the possible problems with the ways to overcome them. On the basis of the research and the results of the experiment the authors conclude that the application of video sketches can improve students' speaking abilities, stimulate their creativity and overall make the process of teaching foreign languages at the university more efficient.

Keywords: Foreign language teaching · Information and communication technologies · Out-of-class monitoring · Overcoming psychological barriers · Foreign language anxiety · Monologue · Video sketch

1 Introduction

Teaching foreign languages at the university nowadays cannot be viewed without the use of information and communication technologies which are becoming extremely relevant. While the system of higher education has evolved, the application of computer and Internet technologies appears to be particularly valuable, as it helps teachers improve the learning process, increase students' motivation, which makes the use of classroom hours for foreign language studying more efficient.

As the studying process suggests, the transition to a blended learning system is fulfilled using different Internet technologies, represented by various extracurricular forms of control and all kinds of online courses. At the same time, in accordance with a learner centered approach in teaching a foreign language a teacher has to facilitate teacher-student interaction.

This article introduces one of the forms of task accomplishment within a learner centered approach using information and communication technologies (a video sketch) and emphasizes the relevance of its application in foreign language teaching in a technical university.

The problem of reducing the number of classroom foreign language hours makes the system of out-of-class monitoring of students' assignments extremely needed. The 'Basic Course of English' at Peter the Great St. Petersburg Polytechnic University introduces such type of control as presenting a monologue. This task is supposed to be perceived as the type of control as the content and the structure of the monologue must follow definite requirements. The monologue should contain certain steps, must be based on the materials studied during a special module and has to contain definite vocabulary presented in the module. Thus, the accomplishment of the task will show not only general speaking abilities of the students but also monitors and controls their level of program acquisition. Recording video sketches of the monologues is used as a special way of the task accomplishment, which represents the mentioned type of out-of-class monitoring. Involving such a type of information and communication technology as a video monologue is considered as an opportunity to activate the process of teaching a foreign language in a technical university.

In addition, the problem of psychological barriers or foreign language anxiety arises. Psychological barriers are recognized as some of the biggest obstacles for every foreign language learner, and it is the teacher who should help students overcome them. The authors assume that implementing such creative tasks as recording video sketches could help in achieving this goal. Moreover, the emphasis of the present experiment was laid on the creative constituent of the video sketches, assuming that stimulating students' creativity would facilitate their overcoming foreign language to even greater extend.

Thus, the main points for our consideration in this paper are to define the video sketch as the way to overcome possible psychological barriers and implement the ideas of out-of-class monitoring and blended learning in teaching foreign languages. The basic interest of the present research is highlighted by the experiment organized in St. Petersburg Polytechnic University of Peter the Great during the autumn term of 2020–2021 academic year in order to compare face-to-face presentation of a monologue with a video format of recorded monologues (video sketches), to monitor how introducing creative thinking section to the monologues and stimulating creativity would affect students' success compared to the previous experiments and to verify the effectiveness of the latter based on the students' progress results and their reviews and comments.

2 Literature Review

2.1 Monologue

The curriculum enables to use such a form of control as a monologue assessment on the topic studied during a particular module. In this case, a monologue can be considered as speech prepared in advance on a certain topic, compiled on the basis of several special texts that have a ready-made form. A teacher evaluates the monologue not only by its lexical resource and grammar range and accuracy, as well as accurate word and sound pronunciation, but also from the point of view of the proper structure of the speech. A student has to analyze the studied materials (both in the form of reading and listening), emphasise the main points and make a logically structured monologue with an introduction and conclusion.

The application of video recordings of the monologues has positive features both for students and teachers. First of all, this method provides indisputable advantages as the form of control: it makes a student be more concentrated on the structure and the content of a monologue [1]. This way, students can lay their account only on themselves and, therefore, they are becoming more responsible for the results of their work and, thus, they are trying to make it more qualitative. The process of forming skills is accompanied by students' cognitive activity [2]. In general, we can notice that students' motivation for learning foreign languages is getting higher [3].

Obviously, students are becoming not only careful with the contents and the structure of their monologues, but also trying to be more accurate with the lexical, grammatical and pronunciation components of their speech, thus making the whole presentation more qualitative. Furthermore, there arises the possibility for self-monitoring, self-checking and self-assessment [4–6], as the students are able to make the necessary number of new recordings if they notice any mistakes in their initial video sketch. As the authors consider, with the possibility to realise your strengths and weaknesses, the ability to evaluate yourself while watching your own video monologues, stimulates students for self-learning and helps them get higher level of motivation for further self-development. Teachers should initiate student's self-control, which is highly significant in the process of teaching foreign languages.

Some psychological difficulties may arise when a student delivers his monologue directly to the teacher, as some students are so afraid of failure as well as of making mistakes, that they experience severe stress and for this reason deliver their speech much worse compared to the situation if they have done it in a more relaxed atmosphere. The ability to re-record a video sketch helps students to withstand such difficulties and overcome possible psychological barriers [7, 8].

Video format of presenting monologues also presents some advantages for the teachers as well. Firstly, the application of video sketches in teaching foreign languages helps with the realisation of a learner-centred approach in teaching [9]. According to this approach the studying process should be organized using special methods, techniques and forms, taking into account the individual characteristics of the students. In addition, comparing to the 'face-to-face' delivery of the monologue, teachers have the possibility to re-watch certain extracts in case they didn't understand any point, therefore they are capable of paying greater attention to this format of task accomplishment [10]. If the teacher considers it possible he or she could leave some feedback on a particular video sketch and send it to the student individually. Whereas the student, getting such feedback, can trace his own mistakes and try not to make them further. Moreover, provided that the student records a series of video monologues on various topics covered during the term, the teacher can clearly trace the student's progress, thereby motivating him for further development and success. Finally, it is really stimulating for the teachers to see how their students are trying to make their monologues better, how they are dedicated to further improvement, how creative they could be with such kind of task accomplishment.

Despite the mentioned advantages of using a video monologue as a form of control, there could be some difficulties that arise when using video sketches as an alternative to the monologue delivery face-to-face [11]. First of all, from the psychological point of view, some students feel more comfortable when answering a monologue directly to the teacher, expecting some support from him. Then, these are technical constraints, as some students occasionally face problems to record their monologues, therefore, this format of task accomplishment should be positioned as supplementary one. The teacher should always provide the choice how exactly students could present their monologues - in the traditional face-to-face format or in the form of a video sketch. Still, it is advisable to stimulate recording video sketches with some extra scores, as it makes the students more diligent and hard-working.

We should emphasize that a video format in delivering a monologue should not be considered as the mandatory way to present a speech, it should be used as a supplementary tool in the educational process. It is vital for students to keep a lively interaction between the teacher and the students in the foreign language classes, as well as to develop the ability to speak in front of a large audience. This method should not replace the live communication in the class but it could provide the studying process with new opportunities. Furthermore, we can assume that video format of delivering a monologue can be perceived as the implementation of blended learning approach.

2.2 Blended Learning

Nowadays blended learning is recognized as one of the most rapidly-growing major tendencies and is described as "the single-greatest unrecognized trend in higher education today" [12; p. 2]. A lot of researchers agree that blended learning has already become "common place educational strategy in higher education" [13; p. 2333] and would become 'inevitable step for all universities' [14; p. 104].

Researchers have tried to define this trend that is gaining importance in educational process. Charles R. Graham proposed one of the first scientific definitions of 'blended learning': 'Blended learning is the combination of traditional face-to-face and technology-mediated instruction' [14; p. 4]. Garrison also defined blended learning as 'the thoughtful integration of classroom face-to-face learning experiences with online learning experiences' [15; p. 96]. Most researchers note the crucial role of ICT (Information and Communication Technologies) in the studying process.

The authors of the present research use the term 'blended learning' as the idea of a mixture of traditional in-class lessons and online format of fulfilling home assignments as out-of-class monitoring. The article refers blended learning as digitally-performed type of control, while mostly studying process is carried in a traditional classroom format. Whereas some researchers estimate that via blended learning approach '30–80% of learning/teaching activities are conducted through web-based ICT' [13; p. 2336], according to the statistics obtained in St. Petersburg Polytechnic University in 2019 more that 50% of undergraduate and graduate students were engaged in e-learning [16].

A range of scientific works dedicated to the students' attitude to blended learning approach has been published and most researchers find that students have greater level of satisfaction with blended approach compared with both face-to-face and online formats [17]. Moreover, they found the exact correlation between this satisfaction and their success and grades for the course. Thus, we can suggest that students' attitude and their positive perceptions are highly important.

Various experiments prove that implementation of blended learning has a positive effect 'reducing dropout rates and in improving exam marks' [18; p. 818]. Therefore, in their previous research the authors of the present study had an opportunity to conduct an experiment during the 2019–2020 academic year, which ended with the exam on the Basic course of English as a foreign language and compared the results afterwards.

There should be mentioned one more important factor which has a great impact on students' success is teacher's positive feedback that they can receive both face-to-face and online. Especially this feedback has marked value for less proficient students 'who are hesitant, less motivated, and face difficulty in associating learning with technological applications' [19; p. 1]. 'Feedback is considered one of the most powerful factors in promoting achievement in a variety of contexts' [20; p. 120].

Furthermore, the use of ICT could improve self-regulation 'as a necessary pedagogical condition for the successful professional self-realization of students' [21; p. 150]. Even though the concept of teacher's role is inevitably changing due to online learning, it is still vital in the studying process, and teachers should find the right balance between face-to-face and online modes [22]. Therefore, in our research we consider that overwhelming positive feedback for students' video monologues is truly useful for their further self-development and self-realization.

The authors connect the importance of the present study with researches dedicated to the use of blended learning in teaching English as a foreign language especially in non-linguistic universities [23, 24]. The researchers propose such basic problem as the 'lack of academic hours allocated for learning foreign languages in technical higher schools' [23; p. 335].

However, literature review has revealed the lack of studies on blended learning approach connected with the use of ICT as the form of control or out-of-class monitoring such as video sketches or monologues. Thus, this issue needs to be studied separately.

2.3 Creative Constituent of Video Sketches as the Way to Overcome Psychological Barriers and Reduce Foreign Language Anxiety

While studying a foreign language, every student inevitably starts making mistakes, which could prevent them from developing certain skills they need to improve the knowledge of the language. The basic problem of grown up students is that they tend to hide their incompetence and lack of self-confidence behind the 'walls' that could protect their own pride, ambitions and self-respect. These "walls" are usually defined as psychological barriers and they are considered as some of the biggest obstacles for students, which a teacher should help them overcome.

The psychological barrier is defined as everything that hinders, restrains, and eventually reduces the effectiveness of studying and personal development [25; p. 151]. Psychological barriers as a whole are generally referred to as 'foreign language anxiety' in various resources meaning 'a complex phenomenon that has been found to be a predictor of foreign language achievement' [26; p. 217]. This notion appeared several decades ago starting with 1980-s and it is still getting special attention in the scientific resources of the latest 5 years dedicated to teaching foreign languages. Horwitz elaborated a special Foreign Language Classroom Anxiety Scale (or FLCAS) [27], according to which the students could be divided into three groups: Low Anxiety, Average Anxiety and High Anxiety depending on their foreign language aptitude, proficiency and the amount of stress they experience while practicing different foreign language activities. Researchers frequently use this scale to evaluate the success of teachers in reducing FLA in the students learning foreign languages.

Researchers distinguish different types of FLA such as 'fear of negative evaluation, communication apprehension, and the negative attitudes toward English language class' [28; p. 163] and many others but in the case of the present study the most relevant one is communicative anxiety. This type of anxiety is mostly associated with 'oral performance' and 'oral class activities' [29]. Mostly researchers admit that in this case the level of anxiety depends on various factors (e.g. individual differences [30]), with 'self-confidence in speaking English, gender and proficiency playing an important role in classroom performance' [31; p. 21]. As speaking brings anxiety, and its level rises in stressful situations [32] such as speaking in front of the class, we consider that recording video sketches with monologues could help students avoid such stressful situations and reduce the level of anxiety.

One of the most important parts of the monologue structure we consider the 'Creative Thinking' section, where a student is supposed to introduce his/her own extra ideas on the topic that haven't been mentioned before. Students are motivated to express their creativity in many ways starting with the contents of the monologue and finishing with any other creative ideas.

Many researchers emphasize the importance of creative learning context [33] and prove that there is close relationship between creativity and language anxiety [34] as creativity along with independence provide comfortable educational environment [35] and can potentially reduce the level of language anxiety. This provided comfortable educational environment is perceived as the key factor of a successful academic program [35] and developing creative abilities is essential for students' further self-development and self-education [36]. Thus, these activities are especially important for university education [37].

The authors of the present research presume that recording video sketches with the creative constituent is fundamental in promoting creative skills. Morais and Almeida propose that unfortunately some "students perceive barriers to their creative expression' and point out 'the importance of finding place to creativity in their daily academic lives' [37]. The basic hypothesis of our research is that positive emotions that students can receive by performing such tasks and getting valuable feedback for them could motivate students to participate in further creative development activities [33].

The authors consider that the use of video sketches in foreign language teaching could stimulate foreign language enjoyment and reduce the level of foreign language anxiety.

3 Methodology

The research puts forward a hypothesis that recording video sketches and especially its creative constituent could facilitate the level of students' English language mastery and their speaking abilities in particular. Furthermore, the present article continues the investigation of the issues raised in our previous research based on the experiment carried out in 2019–2020 academic year [38]. The authors decided to emphasise the creative constituent of such form of task accomplishment as a video sketch. For this reason we added 'Creative Thinking' section into the structure of the monologue providing the students with extra scores for including this section and, consequently, overall stimulating creativity of the students. The idea was to see how creativity could influence on the students' motivation to study foreign languages. Thus, the authors raise such research questions as: whether video sketches recording could reduce the level of students' foreign language anxiety; if it could encourage overall students' creativity in approaching foreign language learning; if the provided teacher's feedback and support could break possible psychological barriers in the process of learning a foreign language and whether this type of teacher-to-student cooperation could give rise to intrinsic motivation for studying foreign languages.

3.1 Data (The Present Study)

To prove the effectiveness of introducing video sketches as a form of monitoring the students' success, control and experimental groups were selected during the autumn semester (September–December 2020). As in the previous experiment the participants (with the total number of 250 students) were represented by 1-course students of technical specialties (mechanical and civil engineering departments), of the same age (18–19 year-olds), from different regions of the Russian Federation, mostly boys (73%), because of their specialties and, moreover, had different levels of the English language proficiency.

3.2 Materials

Due to the program, during the term 4 modules of the textbook have to be studied and a monologue must be prepared after each unit; thus, by the end of the term students must prepare 4 monologues, respectively.

Each monologue as the form of control was evaluated within a score system according to the following criteria:

1. Content (maximum 2 points). A monologue is assessed whether its contents matches the theme of the studied module, how the speaker uses the materials of the textbook and whether he or she expresses his or her opinion on these issues.
2. Structure (maximum 2 points). A teacher estimates whether the speaker includes the introduction and conclusion to the monologue, whether the speech is logically presented (coherence) and whether the speaker uses various means of cohesion (for example, linking words).

3. Lexical resource or Vocabulary (maximum 2 points). The teacher estimates how well the speaker uses new lexical units. According to the requirements agreed in advance, it was necessary to use at least 75% of the basic vocabulary and a certain amount of additional vocabulary. The correct use of lexical units and the norms of their pronunciation are also taken into account.
4. Grammatical range and accuracy (maximum 2 points). The student is expected to use various constructions in speech, thus, expanding his grammatical range, and he should use them correctly.
5. Pronunciation (maximum 2 points). The accurate pronunciation of lexical units is estimated, so as the intonation pattern, phrasal division of the statement and fluency.

Moreover, an extra score was additionally provided to those students who would include the so called 'Creative Thinking' section into their video sketch. In this section the students were supposed to introduce their own extra ideas on the topic that hasn't/haven't been mentioned before. Overall, other possible unusual ways to present the material and deliver a speech were encouraged, thus motivating students to be creative in expressing their ideas.

Here is an example of monologue evaluating criteria (Table 1):

Table 1. The assessment criteria of a monologue.

Aspects	Score		
	2 scores	1 score	0 scores
Contents	The content corresponds with the topic of the monologue, all the proposed subtopics are included into the speech, the speaker presents his/her opinion about the topic. The speech should last for 3,5–4 min and contain at least 25–30 sentences	The content mostly corresponds with the topic of the monologue, but one or two points are omitted or there is a lack of personal opinion about the topic. The speech lasts for 2,5–3 min and contains at least 18–24 sentences	The content doesn't correspond with the proposed plan of the monologue and the speaker doesn't introduce his/her opinion about the topic. The speech lasts for less than 2 min and contains less than 17 sentences
Structure (coherence and cohesion)	The speech is logically structured, all the subtopics are connected with each other, and the speaker uses not less than 10 linking words	The order of the presented subtopics is controversial, there is a lack of linking words in the speech (at least there are 5–6 basic ones)	The structure of the speech is un-clear; there is a lack of linking words in the speech (less than 5)

(*continued*)

Table 1. (*continued*)

Aspects	Score		
	2 scores	1 score	0 scores
Vocabulary or lexical resource	The speaker used over 75% of the Basic Vocabulary (more than 26 units) on the topic of the monologue. The words are counted if they are used in the context and pronounced correctly	The speaker used 50–75% of the Basic Vocabulary (17–25 units) on the topic of the monologue	The speaker used less than 50% of the Basic Vocabulary (less than 16 units) on the topic of the monologue
Grammatical range and accuracy	The speaker used 4–5 different grammatical constructions (such as the Passive Voice, Conditional Sentences, the Reported Speech, the Infinitive Constructions, etc.) and made not more than 4 grammar mistakes that don't hinder the understanding of the main idea	The speaker used at least 3 different grammatical constructions (such as the Passive Voice, Conditional Sentences, the Reported Speech, the Infinitive Constructions, etc.) and made 5–6 grammar mistakes that mostly don't hinder the under-standing of the main idea	The speaker used only 0–2 different grammatical constructions and made more than 6 grammar mistakes which hinder the understanding of the main idea
Fluency and pronunciation	The speech is fluent; the speaker pronounces the words correctly with the proper intonation and pauses	The speech is fairly fluent but the speaker makes quite a lot of long pauses and makes several mistakes on pronunciation	The speech is not fluent with many pauses, the speaker makes a lot of mistakes on pronunciation
*Creative thinking	+1 extra score for including Creative Thinking section into the monologue		

Each monologue was then evaluated 10 points maximum (or 11 scores if Creative Section was included), and by the end of the term students got marks according to the scores of all four monologues altogether. Otherwise, students had the possibility to answer a monologue at the credit-lesson, where they were supposed to answer only one of the four monologues at random basis. It is important to mention that the greatest part of the students decided to present all the monologues during the term.

A teacher provided the students with a detailed plan of a monologue when the module was going to end to make sure that students would know what topics they had to cover in their speech. This plan was proposed in a form a table as the one presented in the Table 2 for the Module on the topic 'Work':

Table 2. Appendix on the topic 'Work'.

You are going to give a talk about WORK	Vocabulary*	Linking words and phrases*
REMEMBER! Your speech will be graded according to the following criteria: relevance, coherence, fluency, grammar & vocabulary (see *Parameters and Criteria of the monologue*)	*Fill in the columns with* • *words, collocations and idiom,* • *linking words* *on the topic 'Work'*	
Step 1. Introduction 1. Make up a hook sentence that will attract the listener's attention to your speech (a quote, proverb, tongue-twister, etc.) 2. Lead your speech steadily to the 2nd step 3. Introduction consists of 4–6 sentences		
Step 2. Jobs 1. Speak about the most important job(s). Prove your point of view 2. ... the most important things to you in your future job. Comment on your ideas		
Step 3. Homeworking 1. Speak about the reasons for homeworking, the advantages and disadvantages of working from home 2. Would you like to work from home? Why? What would be your ideal pattern of working hours?		
Step 4. Work Placements 1. What are work placements and internships? 2. Speak about the benefits young people can get from work placements and internships		
Step 5. Creative Thinking Introduce your own extra idea(s) on work that hasn't/haven't been mentioned before. Substantiate your choice		
Step 6. Conclusion 1. Repeat the main idea of the introduction in other words 2. Summarize the ideas of steps 2, 3, 4, 5		

The students were also provided with 'Basic Vocabulary' document for every monologue and the list of linking words which could be used in any monologue. According to

the proposed criteria students were supposed to use more than 75% of the basic vocabulary (the use of advanced vocabulary and collocations was highly recommended as well)
and not less than 10 different linking words to get the maximum score.

3.3 Participants

In the control group (130 students), the students were proposed to present a monologue
within the traditional approach - to answer a monologue directly to the teacher. The
experimental group consisted of 120 students, and they were given the task to make
video recordings of their monologues. The criteria for evaluating the monologues were
similar in both groups. In the experimental group, students had the choice whether to
make a video, to deliver a speech face-to-face to the teacher or as a credit at the last
lesson. As a result, out of the total number of students, 99 students decided to take the
opportunity of sending video monologues, 12 students decided to answer it to the teacher
and other 9 left it for the test lesson.

4 Results and Discussion

At the end of the semester, the authors carried out a questionnaire in the experimental
group in order to determine how the students evaluate the effectiveness and creativity of
video monologues in the process of learning a foreign language (Fig. 1).

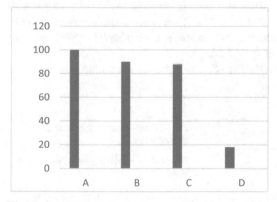

A – 100% - provoke foreign language acquisition

B – 90% - modern method of
interaction

C – 88% - increase creativity

D – 18% - prefer traditional
communication

Fig. 1. Evaluation of effectiveness of video monologues in the process of learning a foreign
language.

All 120 questionnaires (A - 100%) indicate that the video monologue recording helps
students to increase the level of foreign language acquisition in general and to improve
their speaking skills in particular.

108 students (B - 90%) noted that video monologues should be applied within the
framework of a modern university.105 students (C - 88%) stated that they preferred
to make a video of the monologue, instead of answering it directly to the teacher and
described this task as a creative one, and only 15 of them (D - 18%) answered that they

would prefer traditional communication with the teacher. As for the reasons, they indicated: that they feel extremely embarrassed and uncomfortable while making a record; as well as the lack of free time to record a successful attempt.

The questionnaire for teachers contained a request to explain their opinion on applying video sketches in practice (Fig. 2).

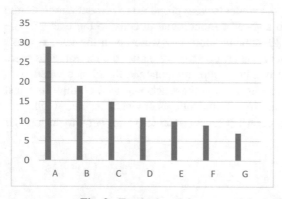

A –29% convenience of evaluating

B – 19% -save time for the classroom activities

C – 15% - promote students' communicative skills

D – 11% - stimulate self-control

E – 10% - encourage the development of creative thinking

F – 9% - increase student's motivation

G - help students overcome the language barrier

Fig. 2. Teacher's opinion on applying video sketches in practice.

The most frequent answers to this question were: - it is more convenient to evaluate a monologue, (A - 29%); - due to the fact that it is possible to save time for the classroom activities (B - 19%); - to develop of students' communicative skills (C - 15%); - to stimulate the improvement of self-control (D - 11%); - to encourage the development of creative thinking (E - 10%); - to increase the student's motivation(F - 9%); - to help students overcome the language barrier (G - 7%).

Students were also proposed to note the all aspects of the video monologue recording that were important for them (Fig. 3).

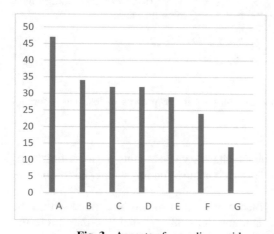

A – 47 % - possibility to make different attempts

B – 34% - convenience of recording time

C – 32% - self-control possibility

D – 32% - improving of communicative skills

E – 29% - opportunity to develop creativity

F – 24% - low level of anxiety

G – 14% - English language competence self-assessment

Fig. 3. Aspects of recording a video monologue in students' opinion.

Based on the results of survey, the following answers were received: the possibility to make a different record of a failure attempt (A - 47%); the possibility to record a video at any convenient time and at any place (B - 34%); the possibility of self-control, analysis of their own mistakes (C - 32%); the development of communicative skills (D - 32%); the ability to develop creativity while recording video monologue (E - 29%); less anxiety compared to the traditional form of control (F - 24%); the opportunity to assess your level of a foreign language (G - 14%).

After the experiment was over, the authors got the results of the exam, which both groups had to pass at the end of the academic year in summer 2020. The curriculum suggests that the exam for the Basic Course of English should consist of two parts: a monologue on one of the topics that students had been provided with in advance and retelling or summarizing of an article. According to the circumstances only a monologue was included in the exam. As the exam was assessed within the 5-band criteria, the students could get the following scores for the monologue: 2 – as 'unsatisfactory', 3 – for 'satisfactory', 4 – for 'fairly good' and 5 – for 'excellent'.

According to the results obtained by the authors all students passed the exams successfully with the following scores (Table 3):

Table 3. The results of the exam for general course of English

Band	The control group	The experimental group
'2' – 'Unsatisfactory	0 (none) out of 130 (0%)	0 (none) out of 120 (0%)
'3' – 'Satisfactory'	28 out of 130 (22%)	11 out of 120 (9%)
'4' – 'Good'	39 out of 130 (30%)	31 out of 120 (26%)
'5' – 'Excellent'	63 out of 130 (48%)	78 out of 120 (65%)

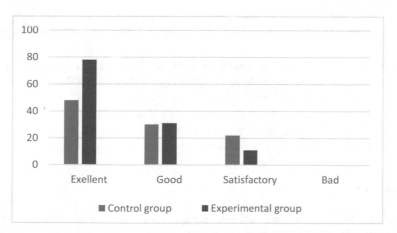

Fig. 4. The results of the exam for basic course of English in the control/experimental group

As we can see from the Table 1 and Fig. 4, the students from the experimental group passed the exam significantly better than the students from the control group:

the percentage of students who got 'excellent' mark in the experimental group is 17% higher than those of the control group, and the distinction between the percentages of the students who got 'satisfactory' mark reaches 13%. Overall, it can be seen that the percentage of the students who got a mark higher than 'satisfactory' is 91% in the experimental group compared to 78% in the control group. These data show that applying video monologues as the means of out-of-class monitoring could help students to get prepared for the exam better and there is a clear connection between general satisfaction with the course and the students' progress. The authors believe that this type of control will stimulate the speaking abilities of the students and is fundamental in promoting creative skills.

5   Conclusions

By conducting the experiment and having analyzed the exam results, the authors came to the following conclusions:

- recording video sketches as the form of out-of-class monitoring is definitely effective;
- recording video sketches can help improve students' speaking abilities;
- creative constituent of the video sketches could influence on the gaining of English language proficiency;
- video sketches recording could minimize the level of students' foreign language anxiety compared to the face-to-face monologue delivery;
- in case of getting sufficient feedback and support from the teacher it is easier for the students to break possible psychological barriers in the process of learning a foreign language;
- such creative tasks could foster overall students' creativity in approaching foreign language learning;
- this type of task accomplishment stimulates better teacher-to-student cooperation, which in turn gives rise to higher level of motivation to study foreign languages.

Suggesting such unusual challenging types of activity, stimulating students' creativity would definitely improve the students' speaking abilities and facilitate the process of teaching foreign languages in non-linguistic universities in general.

Thus, the authors can make a conclusion that this method definitely helps to get students interested, increase their motivation, and engage them in the creative process, and thus make their studying a foreign language more efficient and successful.

References

1. Frobenius, M.: Beginning a monologue: the opening sequence of video blogs. J. Pragmat. **43**(3), 814–827 (2011). https://doi.org/10.1016/j.pragma.2010.09.018
2. Holubnycha, L.O.: Intensification of students' cognitive activity while forming skills of English monologue speech (experience exchange). Bull. Cherkasy Bohdan Khmelnytsky Natl. Univ. Ser. "Pedagog. Sci." **9**, 53–60 (2016). (in Russian). http://ped-ejournal.cdu.edu.ua/article/view/2083. Accessed 3 Mar 2021

3. Ivanova, R.A., Ivanov, A.V.: Second language learners' project activity by developing speaking skills. Educ. Res. **47**(9(2)), 986–996 (2016). https://revistainclusiones.org/index.php/inclu/article/view/1445 Accessed 6 Mar 2021

4. Rich, P., Hannafin, M.: Scaffolded video self-analysis: discrepancies between preservice teachers' perceived and actual instructional decisions. J. Comput. High. Educ. **21**, 128–145 (2009). https://doi.org/10.1007/s12528-009-9018-3

5. McDonald, S.: Building a conversation: preservice teachers' use of video as data for making evidence-based arguments about practice. Mag. Mangers Change Educ. **50**(1), 28–31 (2010). https://eric.ed.gov/?id=EJ890765. Accessed 5 Mar 2021

6. Odinokaya, M., Krepkaia, T., Karpovich, I., Ivanova, T.: Self-regulation as a basic element of the professional culture of engineers. Educ. Sci. **9**(3), 200 (2019). https://doi.org/10.3390/educsci9030200

7. Ahmad, C.N.C., Shaharim, S.A., Abdullah, M.F.N.L.: Teacher-student interactions, learning commitment, learning environment and their relationship with student learning comfort. J. Turk. Sci. Educ. **14**(1), 57–72 (2017). https://tused.org/index.php/tused/article/download/137/93. Accessed 5 Mar 2021

8. Hall, J.K., Walsh, M.: Teacher-student interaction and language learning. Ann. Rev. Appl. Linguist. **22**, 186–203 (2002). https://doi.org/10.1017/S0267190502000107

9. Roter, D.L., et al.: Use of an innovative video feedback technique to enhance communication skills training. Med. educ. **38**(2), 145–157 (2004). https://doi.org/10.1111/j.1365-2923.2004.01754.x

10. Odinokaya, M.A., Barinova. D.O., Sheredekina, O.A., Kashulina, E.V., Kaewunruen, S.: The use of e-learning technologies in the Russian university in the training of engineers of the XXI century. In: IOP Conference Series: Materials Science and Engineering, vol. 940, p. 012131 (2020). https://doi.org/10.1088/1757-899X/940/1/012131

11. Kemp, A.T., et al.: Technology and teaching: a conversation among faculty regarding the pros and cons of technology. Qual. Rep. **19**(3), 1–23 (2014). https://nsuworks.nova.edu/tqr/vol19/iss3/2. Accessed 30 Mar 2021

12. Young, J.R.: "Hybrid" teaching seeks to end the divide between traditional and online instruction. Chronic. High. Educ. **48**(28), A33–A34 (2002). http://eric.ed.gov/ERICWebPortal/recordDetail?accno=EJ645445. Accessed 30 Mar 2021

13. Gikandi, J.W., Morrow, D.A., Davis, N.E.: Online formative assessment in higher education: a review of the literature. Comput. Educ. **57**(4), 2333–2351 (2011). https://doi.org/10.1016/j.compedu.2011.06.004

14. Garrison, D.R., Kanuka, H.: Blended learning: uncovering its transformative potential in higher education. Internet High. Educ. **7**(2), 95–105 (2004). https://doi.org/10.1016/j.iheduc.2004.02.001

15. Graham, C.R., Woodfield, W., Harrison, J.B.: A framework for institutional adoption and implementation of blended learning in higher education. Internet High. Educ. **18**(July), 4–14 (2013). https://doi.org/10.1016/j.iheduc.2012.09.003

16. Bylieva, D., Lobatyuk, V., Safonova, A., Rubtsova, A.: Correlation between the practical aspect of the course and the E-learning progress. Educ. Sci. **9**(3), 167 (2019). https://doi.org/10.3390/educsci9030167

17. Owston, R., York, D., Murtha, S.: Student perceptions and achievement in a university blended learning strategic initiative. Internet High. Educ. **18**(July), 38–46 (2013). https://doi.org/10.1016/j.iheduc.2012.12.003

18. López-Pérez, M.V., Pérez-López, M.C., Rodríguez-Ariza, L.: Blended learning in higher education: students' perceptions and their relation to outcomes. Comput. Educ. **56**(3), 818–826 (2011). https://doi.org/10.1016/j.compedu.2010.10.023

19. Yusoff, S., Yusoff, R., Md Noh, N.H.: Blended learning approach for less proficient students. SAGE Open **7**(3), 1–8 (2017). https://doi.org/10.1177/2158244017723051

20. Prilop, C.N., Weber, K.E., Kleinknecht, M.: Effects of digital video-based feedback environments on pre-service teachers' feedback competence. Comput. Hum. Behav. **102**, 120–131 (2020). https://doi.org/10.1016/j.chb.2019.08.011

21. Odinokaya, M., Krepkaia, T., Sheredekina, O., Bernavskaya, M.: The culture of professional self-realization as a fundamental factor of students' internet communication in the modern educational environment of higher education. Educ. Sci. **15**(3), 150–160 (2019). https://doi.org/10.3390/educsci9030187

22. Huang, Q.: Comparing teacher's roles of F2f learning and online learning in a blended English course. Comput. Assist. Lang. Learn. **32**(3), 190–209 (2019). https://doi.org/10.1080/09588221.2018.1540434

23. Sumtsova, O.V., Aikina, T.Yu., Zubkova, O.M., Voronkova, M.A.: Practical implementation of blended learning in terms of English as a foreign language at technical universities. Ponte. **72**(10), 335–344 (2016). https://doi.org/10.21506/j.ponte.2016.10.29

24. Almazova, N., Rubtsova, A., Krylova, E., Barinova, D., Eremin, Y., Smolskaia, N.: Blended learning model in the innovative electronic basis of technical engineers training. In: Annals of DAAAM and Proceedings of the International DAAAM Symposium, vol. 30, no. 1, 814–825 (2019). https://doi.org/10.2507/30th.daaam.proceedings.113

25. Ivanova, P., Burakova, D., Tokareva, E.: Effective teaching techniques for engineering students to mitigate the second language acquisition. In: Anikina, Z. (ed.) IEEHGIP 2020. LNNS, vol. 131, pp. 149–158. Springer, Cham (2020). https://doi.org/10.1007/978-3-030-47415-7_16

26. Onwuegbuzie, A.J., Bailey, P., Daley, C.E.: Factors associated with foreign language anxiety. Appl. Psycholing. **20**(2), 217–239 (1999). https://doi.org/10.1017/S0142716499002039

27. Horwitz, E.K., Horwitz, M.B., Cope, J.: Foreign Language classroom anxiety. The Mod. Lang. J. **70**(2), 125–132 (1986). https://doi.org/10.1111/j.1540-4781.1986.tb05256.x

28. Alrabai, F.: The influence of teachers' anxiety-reducing strategies on learners' foreign language anxiety. Innov. Lang. Learn. Teach. **9**(2), 163–190 (2015). https://doi.org/10.1080/17501229.2014.890203

29. Léger, D.D.S., Storch, N.: Learners' perceptions and attitudes: implications for willingness to communicate in an L2 classroom. System **37**(2), 269–285 (2009). https://doi.org/10.1016/j.system.2009.01.001

30. Piniel, K., Albert, Á.: Advanced learners' foreign language-related emotions across the four skills. Stud. Second Lang. Learn. Teach. **8**(1), 127–147 (2018). https://doi.org/10.14746/ssllt.2018.8.1.6

31. Matsuda, S., Gobel, P.: Anxiety and predictors of performance in the foreign language classroom. System **32**(1), 21–36 (2004). https://doi.org/10.1016/j.system.2003.08.002

32. Kralova, Z., Skorvagova, E., Tirpakova, A., Markechova, D.: Reducing student teachers' foreign language pronunciation anxiety through psycho-social training. System **65**, 49–60 (2017). https://doi.org/10.1016/j.system.2017.01.001

33. Savage, B.M., Lujan, H.L., Thipparthi, R.R., DiCarlo, S.E.: Humor, laughter, learning, and health! A brief review. Adv. Physiol. Educ. **41**(3), 341–347 (2017). https://doi.org/10.1152/advan.00030.2017

34. Homayouni, A.R., Abdollahi, M.H., Hashemi, S., Farzad, V., Dortaj, F.: Correlation of between creative thinking with language anxiety and learning English in Turkmen bilingual students. Soc. Sci. (Pakistan) **11**(4), 419–421 (2016). https://doi.org/10.3923/sscience.2016.419.421

35. Andreev, V.V., Vasilieva, L.N., Gorbunov, V.I., Evdokimova, O.K., Timofeeva, N.N.: Creating a psychologically comfortable educational environment as a factor of successful academic program acquisition by technical university students. Univ. J. Educ. Res. **8**(10), 4707–4715 (2020). https://doi.org/10.13189/ujer.2020.081040

36. Khompodoeva, M.V., Nikulina, L.P., Shukaeva, A.V.: Students' independent cognitive activity and its formation at universities. In: Anikina, Z. (ed.) IEEHGIP 2020. LNNS, vol. 131, pp. 672–684. Springer, Cham (2020). https://doi.org/10.1007/978-3-030-47415-7_71
37. Morais, M.F., Almeida, L.: "I would be more creative if…": are there perceived barriers to college students' creative expression according to gender? Estud. Psicol. (Campinas) **36**, e180011 (2019). http://dx.doi.org/10.1590/1982-0275201936e180011
38. Burakova, D., Sheredekina, O., Bernavskaya, M., Timokhina, E.: Video sketches as a means of introducing blended learning approach in teaching foreign languages at technical universities. In: Conference: 14th International Scientific Conference "Rural Environment. Education. Personality. (REEP)", vol. 14, pp. 50–58 (2021) https://doi.org/10.22616/REEP.2021.14.005

Digital Pedagogical Cues for the Development of Creativity in High School

Elena I. Chirkova[1] , Elena M. Zorina[1(✉)] , and Liliia V. Rezinkina[2]

[1] Saint Petersburg State University of Architecture and Civil Engineering, Vtoraya Krasnoarmeiskaya, 4, 190005 Saint Petersburg, Russia
zorinaem@bk.ru

[2] Saint Petersburg State University of Industrial Technologies and Design, Bolshaya Morskaya Str., 18, 191186 Saint Petersburg, Russia

Abstract. The article presents an analysis of university students abilities in developing creativity using digital pedagogical cues. Modern students are those of the "digital generation" who require other forms of interaction throughout the entire learning process, they need the widespread use of information technologies, as well. Interest in learning is increasing due to the use of creative assignments for the students of technical specialities. The article shows in detail the use of digital storytelling as a mixed digital pedagogical cue for the development of creativity in the format of project activities and teamwork. All these skills are basic for a 21st-century specialist, therefore they need constant shaping and improvement. Since students are already motivated learners, the final result of their project and the criteria-based assessment of fellow students serve as an additional incentive for the development of creativity. The pandemic and compulsory e-learning conditions provide additional opportunities for students' creative development using ICT resources and tools.

Keywords: Creativity · E-learning · Digital pedagogical cue · Digital native · Skill

1 Introduction

The modern educational paradigm is being modernized. It becomes necessary for future specialists to develop both professional and personal skills (hard and soft ones). Klaus Schwab [1] believes that the 4th technological revolution has begun, changing society from industrial to informational one, the era of knowledge is coming (era as a knowledge society), that is, information is useful in case it is transformed into knowledge. In his turn, Mitchell Resnick [2] suggests that the ability to adapt to rapidly changing conditions of life in general and education, in particular, indicates the need to develop a creative society. Both researchers agree on one thing - creativity, as a basic skill, is needed by any in-demand specialist of the 21st century and it should be developed throughout life, following the paradigm of lifelong learning.

1.1 Research Objectives

American scientists believe that creativeness "involves the development of a novel product, idea, or problem-solution that is of value to the individual and/or the larger social group" [3]. This definition shows that creativity is not a talent, but a skill that can be developed. Mitchell Resnick speaks about it when considering misconception about creativity [2]:

- Creativity is about artistic expression;
- Only a small segment of the population is creative;
- Creativity comes in a flash of insight;
- You can't teach creativity.

Ken Robinson [4] argues that creativity is the most important quality in education and the most serious feature of the 21st-century specialist. However, David Nunan [5] believes that creative, extraordinary solution of theoretical problems and the creation of associations does not help the development of creativity, which is needed for practical implementation in the professional sphere. It is difficult to agree with the last statement because the modern educational process requires students and teachers with a divergent way of thinking, which is not given at birth but is used to shape creativity.

According to Mitchell Resnick, creativity is not a manifestation of innate talent, but a skill that can be taught. We agree with this statement and use digital pedagogical cues during classes with students to develop creativity and divergent thinking.

Creativity can be developed both individually and, in a team, that is, when working on a group project. For example, Taiwanese scientists believe that "team creativity may be defined as stemming from individual team members' creativity or as a culmination of complex interactions among the group as a whole" [6]. Woodman et al. suggested that team creativity consists of the creative process, product and situation, as well as the creative person, and how each component interacts with the others [7, 8]. We agree that in a modern university, it is easier to make creative projects in a team, revealing creativity in each of its members.

The development of creativity, according to Davies et al., is influenced by some features of the organization of the educational space:

- "the physical (flexible use of space, flexibility and free movement, availability of diverse tools, materials, and technology);
- the pedagogical (original and exciting learning activities, authentic and realistic tasks, playful approaches, ensuring time for ideas, allowing students ownership of their learning);
- the psychosocial (trust, mutual respect, collaboration);
- the external (collaboration with outside agencies)" [9].

Especially for students of the "digital generation", it is necessary to lay the groundwork that stimulates the creative process. This skill is part of the pedagogical design that the modern teacher should master.

1.2 Literature Review

Teaching creativity of digital generation students is closely associated with the digital pedagogical cues applying in the educational process. Undoubtedly, to motivate a modern student, it is necessary to use various types of assignments, both project and educational, scientific and creative. This allows you to move from reproductive exercises to heuristic ones that require creative divergent thinking.

For example, e-learning or blended learning combined with brainstorming can be visualized on a virtual whiteboard. The creation of associations on the academic topic, on the one hand, allows the teacher to understand how much background knowledge students have, and on the other hand, to increase learning motivation.

It should be noted that the use of ready-made digital pedagogical cues can serve as an impetus for students to create their educational tool. For example, comics are often used to learn English grammar. Beginning with reproductive reading or phrases, you can move on to creative exercises where students fill in thought bubbles or speech bubbles based on visuals alone. With the help of special sites (for example, Canva), students can create their educational comics with specified characters to practice grammar, vocabulary and communication tasks.

Another example of the use of digital pedagogical cues is digital storytelling - a narration on a specific or random topic, made in a team or individually with the help of computer technology in the educational process. According to the International Society for Educational Technologies (ISTE – www.iste.org), creativity is important for students, because they must "demonstrate creative thinking, construct knowledge, and develop innovate products and processes using technology" [10].

One of the modifications of digital storytelling (DST) is the making of an educational film [11]. "Cinemapedagogics" has long been successfully used in classes in various subjects, but only the independent creation of an electronic educational resource allows students to reveal their full creative potential. A foreign language educational film develops such soft skills as verbalization, communication, collaboration and, of course, creativity. Imagination and creativity that students display when writing scripts and making a film increases their educational motivation and a sense of success, not only as an individual but also as a professional. A special topic helps to reveal the success of mastering a speciality.

The cooperation of teamwork members in the learning process is also changing. The teacher acts as a facilitator or tutor, only guiding and correcting the work of students. In addition, the requirements for the level of proficiency in ICT of the teacher himself, as a carrier of e-didactics, are changing [12, 13]. According to NETS, to prepare teachers to use technology advanced learning in the digital age you should adhere to a standard:

- "Facilitate and inspire student learning and creativity;
- Design and develop digital-age learning experiences and assessments;
- Model digital-age work and learning;
- Promote and digital citizenship and responsibility;
- Engage in professional growth and leadership" [14].

The ICT teacher's competency standards have been prepared by UNESCO and its partners to assist policy-makers and methodologists in the educational space identify the skills teachers need to harness technological development in the education service. It "emphasizes that it is not enough for a teacher to have ICT competencies... teachers should be able to help students learn in the spirit of collaboration, problem-solving, creativity through the use of ICTs" [15].

Making a film is a part of creative technology when students use various devices for filming and showing videos - this is working with music, studying stage skills, knowledge of a foreign language when the film or scene itself is a pedagogical cue for memorizing educational material.

Work on creating a video allows students to develop a need for an independent, free, creatively active approach and understanding knowledge and readiness to introduce innovative approaches under the guidance of a teacher. The use of creative technologies shows that the educational process may not be monotonous, but represent a kind of game (edutainment - education and entertainment), be cognitive and exciting. Creative technologies serve as the basis for the growth of a student's intellectual potential, his/her professional skill in mastering a speciality.

Moreover, making a film is the acquisition of communication skills, which are lacking in the modern world (primarily due to the digitalization of all aspects of life).

The combination of digital technologies and performing arts not only helps navigate in their profession (the creation of such films is very useful in obtaining the skills of joint work in design, construction, etc.) but also form a creative personality capable of finding a solution in the most difficult or non-standard situations, to understand the trends of the future.

Any film is a combination of speech and movement [16]. This combination creates a basis for creativity when making films, which can contain various scenes filled with words and movements, connecting the main thoughts of the characters and complementing them, interpreting them with the help of gestures, facial expressions and expressive body movements.

This arsenal of tools, their combinatorics form a creative potential for making images. The creative content of films (multimedia) also consists in creating various humorous situations, when not only knowledge of a foreign language is required, but knowledge of the nuances of linguistic phenomena that should be studied, memorized and used in a script.

Creating digital storytelling also requires knowledge of non-verbal means of communication (NMC) that are behind any human behavior. NMCs of different cultures will help a student to adapt when the need arises to study at foreign universities, take part in international Olympiads, work on joint projects, and be a participant in any international program.

It should be noted that creativity is an integral part of the pedagogical activity, which ensures its success and effectiveness. You can often hear that the teacher is always on stage (podium), he/she is an actor who uses creativity to make and bring into existence new things in the world around him/her, to interest students, push them to solve simple problems using non-trivial methods.

2   Methods

2.1   Theoretical Basis

The digitalization of the educational process and training of "digital generation" students require changes not only in the presentation of educational material but also in its processing. Each person has his/her preferences in receiving information - the leading channel of perception. Usually, visuals (perception through the eyes), audios (perception through the ears) and kinesthetic learners (perception through tactile touch) are distinguished. However, with the general use of gadgets, a new channel for transmitting information has appeared - digital, and people who prefer to interact with electronic devices are called digital people.

Initially, it was believed that, like a computer, such students love logic and are inclined to execute algorithms. However, according to our research, it was revealed that digital people do not have a leading channel for perceiving information. They do not necessarily have algorithmic thinking, but they are always quick to search for, process and transmit digital information.

The development of creativity in digital people occurs with the help of digital pedagogical cues and, as a basic soft skill, is included in the meta-subject competence.

2.2   Empirical Research Method

Many Russian universities use diagnostics of the dominant perceptual range (according to S. Efremtsev) [17] to determine the psychological type of the student. At the St. Petersburg State University of Architecture and Civil Engineering, 181 students of 1–3 courses of various faculties (architectural, construction, automobile and road-building, IT departments, etc.) were interviewed. Among them, representatives of all 4 types were identified, but almost half of them turned out to be digital, which is undoubtedly typical for students of the "digital generation" (Fig. 1). Previously, it was believed that digital people make only 1%, but widespread digitalization has led to an increase in the number of people with the absence of a pronounced channel of information perception, who freely interact with digital devices.

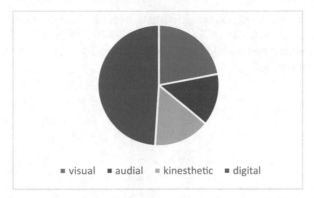

Fig. 1. Distribution of students by channels of information perception.

The preliminary level of creativity was determined by the method of Jerome Bruner [18] (Fig. 2). It assumes that the level of creativity is divided into three intervals: low level (from 0 to 5 points), medium level (from 6 to 9 points), high level (from 10 to 15 points).

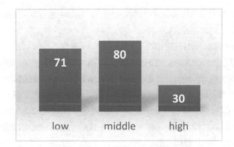

Fig. 2. Preliminary level of creativity by J. Bruner.

In addition, creativity formation level was defined as a soft skill. For this, the author's approach was used to students, when each statement is given points (from 0 to 5, depending on the correctness of the statement: never, rarely, from time to time, often, almost always):

- I can generate ideas adequate to the task at hand;
- I can find original and unexpected solutions;
- I am looking for several solutions to one problem;
- I can experiment and introduce something new;
- I have a creative hobby.

The maximum number of points is 25. Creativity is divided into 5 levels: low (<35%), reduced (35–49%), middle (50–69%), elevated (70–89%), high (90–100%). Most of the students had a reduced and middle level, which practically coincided with the results of the previous test (Fig. 3).

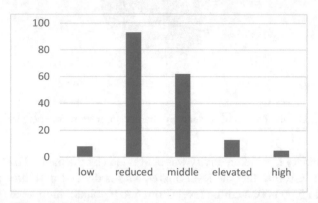

Fig. 3. The preliminary level of creativity of students according to E. Zorina.

Then the students were divided into small groups to create their project - an educational film on their speciality.

So, in one of the groups, when filming a video on the speciality "Theory of working processes and modeling processes in internal combustion engines", a film was created showing how an internal combustion engine works with a description of the movement of a piston, valves, etc. in English. The work of the engine was portrayed by the students themselves. Their movements were accompanied by humorous comments in English.

Another group made a film illustrating the production of concrete and the laying of the strip foundations for the building.

In a group of architects, the theme was a tour of places in St. Petersburg with a certain style of architecture.

At the same time, the teacher only helped with the selection of information and settled conflicts within the groups.

At the final stage, a presentation was made with mutual appreciation of students by students.

3   Results

After all, projects were created and presented for discussion, the students were again measured for their creativity level according to the two tests above (Fig. 4 & Fig. 5).

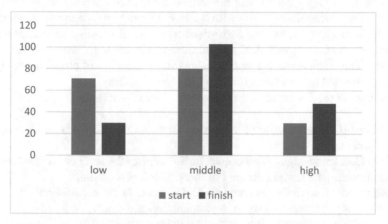

Fig. 4. Comparative histogram of creativity level according to J. Bruner.

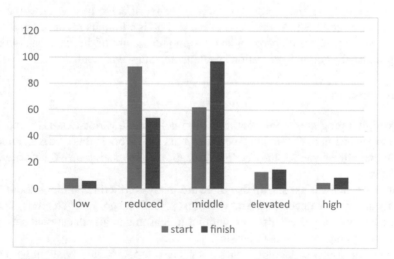

Fig. 5. Comparative histogram of creativity level according to E. Zorina.

There was an increase in creativity, regardless of whether they were students of a technical or humanitarian speciality. The most striking progress was made by those students who were actively leading in their groups, trying to create a new product.

4   Discussion

Based on the experiment, we came to conclude that the application of various digital pedagogical cues - brainstorming and other associationograms, infographics and digital storytelling, Six Thinking Hats project analysis - allows us to develop creativity and creative thinking, and also brings creativity to the educational process, because there are two different types of pedagogical creativity – to teach creatively and to teach for the sake of creativity [19, 20].

The use of pedagogical cues in the university learning process allows you to get changes in the educational strategy with the help of a creative component, which leads to a motivation increase, as well as to the formation of creative special projects by, for example, students of technical specialities, which helps to love more their future profession. The enthusiasm that students show in the creation of electronic educational resources is carried over to further training in their speciality.

However, on the one hand, several problems were identified:

- insufficient ICT competence of teachers, which did not allow the full use of digital pedagogical cues;
- teachers lack knowledge of techniques for measuring the level of creativity and tests for determining the leading channel of information perception;
- limited technical equipment in modern higher educational establishments;
- students fear to express and try to implement unusual solutions;
- the unwillingness of students to work on a team project.

On the other hand, these problems did not seem critical to us and did not affect the results of the experiment. At the Department of Foreign Languages at our university, a master class for teachers on pedagogical design and the use of digital pedagogical cues and especially the "digital generation" students teaching was held.

5 Conclusion

Modern digital programs, computer animation help create various kinds of images that would have been impossible to imagine a few years ago. Work in this area is an interesting material that allows you to study certain educational topics, organize the creative work of students.

Literature analysis has shown that there are many conflicting opinions about what is creativity and how to develop it. And yet, most scientists agree that creativity is a basic skill of a 21st-century specialist, and without it, learning at all educational stages from kindergarten to high school is impossible.

The use of digital pedagogical cues allows develop + creativity along with critical and divergent thinking, as well as the ICT competence of both students and teachers. Such mixed pedagogical cue as digital storytelling helps increase educational motivation, promote love for the profession and a foreign language, and also improves teamwork skills at one project. The actualization of these ideas will be useful to the student in the process of studying at the higher educational establishments, and later in professional activities as well.

We are confident that creativity can be developed in students of any speciality, and a foreign language serves as one more additional cue for creative learning.

The conditions of the pandemic and compulsory e-learning provided additional opportunities for the creative development of students using ICT resources and tools.

References

1. Schwab, K.: The Fourth Industrial Revolution. Crown Business, New York (2017)
2. Resnick, M.: Lifelong Kindergarten: Cultivating Creativity through Projects, Passion, Peers, and Play. MIT Press, Massachusetts (2017)
3. Hennessey, B.A., Amabile, T.M.: Creativity. Ann. Rev. Psychol. **5**, 69–98 (2010). https://doi.org/10.1146/annurev.psych.093008.100416
4. Robinson, K.: Out of Our Minds: Learning to be Creative. Wiley-Capstone, New York (2001)
5. Nunan, D.: English for 21st-century skills: dilemmas, challenges and solutions. In: Mavridi, S., Xerri, D. (eds.) English for 21th-Century Skills, pp. 2–9. Express publishing, Newbury (2020)
6. Liu, H.-Y., et al.: Predictors of individually perceived levels of team creativity for teams of nursing students in Taiwan: a cross-sectional study. J. Prof. Nurs. **37**(2), 272–280 (2021). https://doi.org/10.1016/j.profnurs.2021.01.010
7. Woodman, R.W., Sawyer, J.E., Griffin, R.W.: Toward a theory of organizational creativity. Acad. Manage. Rev. **18**(2), 293–321 (1993)
8. Pillay, N., Park, G., Kim, Y.K., Lee, S.: Thanks for your ideas: gratitude and team creativity. Organ. Behav. Hum. Decis. Process. **156**, 69–81 (2020). https://doi.org/10.1016/j.obhdp.2019.11.005
9. Davies, D., Jindal-Snape, D., Collier, C., Digby, R., Hay, P., Howe, A.: Creative learning environments in education: a systematic literature review. Think. Skills Creat. **8**, 80–91 (2013). https://doi.org/10.1016/j.tsc.2012.07.004
10. Ohler, J.B.: Digital Storytelling in the Classroom. New Media Pathways to Literacy, Learning, and Creativity. Corwin, Thousand Oaks (2013)
11. Wu, J., Chen, D.-T.V.: A systematic review of educational digital storytelling. Comput. Educ. **147**, 103786 (2020). https://doi.org/10.1016/j.compedu.2019.103786
12. Choshanov, M.A.: E-didactics: a new look at learning theory in the digital age. J. Educ. Techno. Soc. **16**(3), 684–696 (2013). (in Russian)
13. Saubern, R., Urbach, D., Koehler, M., Phillips, M.: Describing increasing proficiency in teachers' knowledge of the effective use of digital technology. Comput. Educ. **147**, 103784 (2020). https://doi.org/10.1016/j.compedu.2019.103784
14. Information Society for Technology in Education. The National Educational Technology Standards for Teachers. ISTE. http://www.iste.org/standards/nets-for-teachers. Accessed 10 Feb 2021
15. UNESCO ICT Competency Framework for Teachers. http://www.unesco.org/new/en/unesco/themes/icts/teacher-education/unesco-ict-competency-framework-for-teachers/. Accessed 10 Feb 2021
16. Chirkova, E.I.: The emotional component in teaching language disciplines. In: Davydova, N.V., Vasilyeva, N.Yu. (eds.) Russian Language in the Multiethnic Educational Space of a Military University. Materials of the Interuniversity Scientific-Practical Conference, pp. 248–257, VAS, St. Petersburg (2017). (in Russian)
17. Fetiskin, N.P., Kozlov, V.V., Manujlov, G.M.: Socio-psychological Diagnostics of Personality Development and Small Groups. Publishing house of the Institute of Psychotherapy, Moscow (2002).(in Russian)
18. Bruner, J.S.: Beyond the Information Given: Studies in the Psychology of Knowing. Routledge, London (2010)
19. Farrugia, C.M.: Creativity is intelligence having fun: the random word tools as a practical guide to stimulating serious creativity. In: Mavridi, S., Xerri, D. (eds.) English for 21th Century Skills, pp. 35–42. Express Publishing, Newbury (2020)
20. Maksić, S., Jošić, S.: Scaffolding the development of creativity from the students' perspective. Think. Skills Creat. **41**, 100835 (2021). https://doi.org/10.1016/j.tsc.2021.100835

Mind Mapping Method in Foreign Language Education: Transformative Effects of a Productive Approach

Zhanna N. Maslova[(✉)] [iD], Anna V. Rubtsova[iD], and Elena A. Krylova[iD]

Peter the Great St. Petersburg Polytechnic University (SPbPU), Polytechnicheskaya 29,
Saint Petersburg 195251, Russia
rubtsova_av@sbpstu.ru

Abstract. The study establishes the connection between the need to form soft skills in students and critical thinking skills. The objective of the study was to examine how productive methods of developing soft skills and critical thinking affect the learning and structuring of abstract concepts. The study argue that the formation of soft skills and the development of critical thinking can be effective in foreign language teaching for structuring information and developing creativity. Also, the article is devoted to the research of effectiveness of productive teaching methods originally designed to develop critical thinking (mind maps) in teaching foreign language at university. To find out how the mind mapping method affects the perception, structuring of scientific information and creativity a special study was conducted. In addition, the method of statistical analysis was used in the study. Based on the results of the changes in the level of formation of ideas about scientific concepts are analyzed. Critical thinking in the study appears in a dual role: it is thinking skills, which are supposed to develop, and the scientific concept, which was used to create the research material. Analysis showed a different, more personalized, structuring of scientific information when using the productive method of learning (mind mapping). The results prove that using of productive teaching methods in the foreign language teaching significantly affects the level of adoption and structuring of scientific information, as well as the level of creative thinking.

Keywords: Productive method · Critical thinking · Creativity · Mind mapping · Soft skills · Foreign languages education · Initiating dispositions

1 Introduction

In modern education forming of soft skills or soft competences is crucially substantial. This process is associated with the peculiarities of adoption and structuring of scientific information, as well as creativity. It should be said that soft skills are important for a modern specialist in any field, and modern higher education should be aimed at their formation. Evidently, this process can be successful though using innovative methods of teaching. In general innovative teaching of language means creativity and novelty of the teacher which changes the style and method of teaching. Productive methods of teaching are methods of obtaining new knowledge and skills as a result of creative activity. The condition for the functioning of productive methods is the presence of problem tasks. Problem tasks - it is always a search for a new way of solving. Thus, productive methods of learning are based on a problematic interpretation of educational material. Productive methods of learning are used in the framework of the productive approach that includes a number of components, which in general build the methodological system of modern professionally oriented foreign language education.

The problem of using productive teaching methods is widely discussed. For instance, methods, investigated in the articles, include personal multilinguistic portfolio (PMP) technology [1], experiments, personal projects, and laboratories that are guided by the teacher [2].

Less attention is paid to methods originally designed for the development of critical thinking in higher education. However, these methods can be effective within the framework of productive approach. This study elaborates in detail on the mind mapping method and the question of how the usage of this productive teaching method affects the adoption and structuring of scientific information has not been studied yet.

Mind maps were developed by Buzan T. [3] as a way of helping students make notes that used only key words and images, but mind map can be used by teachers to explain concepts in an innovative way. "The non-linear format of mind maps allows the text reader to view the entirety of his notes at a glance, then easily places new information in the appropriate branch or make connections between ideas" [4] (p. 37). Also, mind mapping has been found to be an effective method for teaching adult learners [3]. According to Holland, Holland & Davis [5] mind mapping made students enthusiastic because it increased the students' sense of competence.

Moreover, this method is perfectly compatible with the use of technology. For example, Al-Jarf R. [6] and others [7] employed software mind mapping for improving students' writing skills and it is reported that the mind mapping tool encouraged creative thinking and the students became faster at generating and organizing ideas for their writing.

However, it should be emphasized that the level of assimilation of scientific information affects soft skills and creativity and the development of soft skills is ineffective with insufficient owing of critical thinking skills, because soft skills (also known as non-cognitive skills) are patterns of thought, feelings and behaviors [8] that are socially determined and can be developed throughout the lifetime to produce value. The issue of forming soft competences is especially vital in teaching a foreign language, as students need to master practical skills of communication and cultural behavior. The prospect is seen in the application of innovative teaching methods related to the development of critical thinking. Soft skills or competences can be formed by developing critical thinking, effective formation of scientific concepts and applying productive teaching methods into the learning process.

Critical thinking we regard as a narrower concept than soft skills and a widely accepted educational goal. Critical thinking is thinking about things in certain ways so as to arrive at the best possible solution in the circumstances that the thinker is aware of. Creative thinking can be seen as part of the critical thinking which lets people consider things from a fresh perspective and different angles. It's an inventive thought process which results in new conclusions and ideas, new ways of doing things, but at its core is the ability to make decisions and think with an open mind – critically.

The question is not only in implementation methods of developing critical thinking in language education but in the following: can critical thinking skills be developed with innovative methods in the teaching of language and whether it can have a positive effect on language education.

The purpose of the study is to apply productive teaching method (mind mapping) originally designed for the development of critical thinking in language education and find out how this method affects the quality of adoption of scientific information. The object of the study is changes in the level of students' understanding of the scientific concept critical thinking after using a productive teaching method mind mapping.

2 Materials and Methods

As mentioned above, the goal of the study is to experimentally prove the hypothesis that the use of productive teaching methods leads to a qualitatively different understanding and perception of complex scientific concepts. Productive teaching methods that were designed to develop critical thinking skills, such as mind mapping, can be effective for teaching foreign languages and they improve both language skills and rational thinking skills.

Critical thinking is one of the central concepts of research. It is counted from 14 to 17 definition of critical thinking [9, 10]. In the article critical thinking is understood as "careful goal-directed thinking" [11] and it has at least three features: it is done for the purpose of making up one's mind about what to believe or do; the person engaging in the thinking is trying to fulfill standards of adequacy and accuracy appropriate to the thinking; the thinking fulfills the relevant standards to some threshold level.

In educational contexts, a definition of critical thinking is a "programmatic definition" [12] (p. 19). It expresses a practical program for achieving an educational goal. For this purpose it is useful articulation of a critical thinking process with criteria and standards for the kinds of thinking that the process may involve. The real educational goal is recognition, adoption and implementation by students of those criteria and standards. Thus, that adoption and implementation in turn consists in acquiring the knowledge, abilities and dispositions of a critical thinker.

It was noticed that combination of separate instruction in critical thinking with subject-matter instruction in which students are encouraged to think critically was more effective than either by itself [13].

In this regard, the concept of critical thinking in the study is used in two ways: as a scientific concept that is learned and structured by students, and as an indicator of the effectiveness of productive teaching method (mind maps).

In the research mind mapping, as a productive teaching method originally designed for the development of critical thinking, was applied in language education and assimilation of scientific information.

It should be said that innovative methods have the potential to improve education as well as they have the potential to empower people. The purpose of education is not just making a students' literate but making them thinking rationally. Foreign (the English language) language learning is a great opportunity for application of pedagogical innovations.

As we observe the activities students are studying language through we will see many similarities between teaching foreign languages and developing critical thinking. Students who studies foreign languages use critical thinking through listening to, reading, viewing, creating and presenting texts, interacting with others. Students critically analyze their opinions when they discuss the aesthetic or social value of different texts. In discussions students develop critical thinking as they share personal responses and express preferences for specific texts, state and justify their points of view and respond to the views of others. Students also use and develop their creative thinking capability when they consider the innovations made by authors or create ideas for their own texts based on real or imagined events. In creating their own texts students also explore the impact of other texts and language, feelings and opinions. In general, students explore the creative possibilities of the English language to represent their ideas.

Methods of developing critical thinking are aimed at the formation of its different sides: how to organize random thoughts, to produce creative solutions, how to argue with own logic and find weaknesses in that of others and so on. Priority focus is to concentrate on how to discover the relationships between the ideas because it includes a complex ability to define ideas, compare view points, find similarities and gauge differences.

One of the productive methods that was first created to develop critical thinking skills, and then was successfully borrowed by didactics, is mind maps or mind mapping.

The key notion behind mind mapping as an productive method of teaching is that students are able to use the full range of visual and sensory tools at their disposal. Pictures, forms, colors play a part in learning armory and can help to recollect information for long time. The key is to build up mind maps that make the most of these things building on students' own creativity, thinking and cross linking between different ideas.

The present study targeted respondents belonging to different cultures. The selected group (n = 17) was consisted from Russian (n = 9) and Chinese (n = 8) students of Peter the Great St. Petersburg Polytechnic University. Research work was conducted during the winter semester of the 2020/2021 academic year.

The first stage of the research involved discussion and representation of the scientific concept 'critical thinking' in the form of an essay in English, which students study as a foreign language. This stage also included a contextual analysis of the essays and keyword analysis.

The second stage consisted of applying the mind map method to explain the concept of critical thinking and analysis of the mind maps.

The results of the study are summarized in statistical tables.

3 Implications

For the analysis eight initiating dispositions [14] (p. 25) for critical thinking were taken as the basis. Students were asked to formulate the concept 'critical thinking' and its features – what is recognized and understood by them – in the form of an essay as a homework assignment when studying a topic related to the problems of human adaptation.

For the key characteristics of the concept of 'critical thinking' the initiating dispositions of critical thinking were taken [14]. He ideas of initiating dispositions were reflected in the essays through keywords and key phrases, which were the units of analysis:

Attentiveness: a critical thinker needs to be attentive to one's surroundings [14] (p. 25), [15].

The analysis showed that it is commonly recognized that critical thinking should be a part of everyday life: «it can be applied to every area of life from schooling to the workplace and even in the relationships with family and friends»; «creating needs to find problems which exist in real environment, only you have questioning eyes, you can find what you are unsatisfied with, what needs to change», «critical thinking skills teach a variety of skills that can be applied to any situation in life», «we need to acquire huge information from outside world, including books, TV media, internet and so on».

The bearer of critical thinking must be attentive not only to the world around him, but also to the professional field in which he is engaged: «the ability to think clearly and rationally is important whatever we choose to do. If you work in education, research, finance, management or the legal profession, then critical thinking is obviously important».

Attention as a quality of critical thinking is well understood and distinguished regardless of cultural background.

Habit of Inquiry: Inquiry is effortful, and a thinker needs an internal push to engage in it. [14] (p. 25), [16] (p. 294).

It should be said here that in the essays of the Russian and Chinese students there was few references that mastery of critical thinking requires effort and deliberate teaching. It is the unconscious domain of the concept. For instance: «you need a good education for critical thinking», «it takes time».

Self-confidence: willingness to think critically requires confidence in one's ability to inquire [14] (p. 25), [15].
Critical thinking as an intellectual ability is partially realized by students. For this representation they use the words such as «ability» and «skills». For example: «critical thinking is the ability to analyze the way you think and present evidence for your ideas, rather than simply accepting your personal reasoning as sufficient proof», «the new economy places increasing demands on flexible intellectual skills, and the ability to analyze information and integrate diverse sources of knowledge in solving problems», «it means the ability to notice and predict opportunities, problems, and solutions. It is closely connected with the skill that involves the gathering, understanding, and interpreting of data and other information – analysis». One's ability to inquire is concretized in such qualities as «gathering», understanding», «interpreting», «analyzing».

Courage: Fear of thinking for oneself can stop one from doing it [17] (p. 16).
The need for fearlessness in the essay is represented «from the opposite» with words indicating courage and independence: «critical thinking creates independence. It means that while making any decision, one relies on their own attitude towards things, instead of listening to other people's opinions», «it provides you with an ability to stand out

of the crowd and to follow your own beliefs and values even when your social circle is totally against it. You become more independent in your views and open to any kind of opinions because every situation has many different points and sides to look from».

Students prefer not to identify fear as a feeling that impedes critical thinking. The position of Chinese students on this disposition is less represented in the essays.

Open-Mindedness: A dogmatic attitude will impede thinking critically [2, 14–16].

Open-mindedness is a complex concept. It is the willingness to search actively for evidence against one's favored beliefs and this willingness can be expressed in a variety of specific actions such as overcoming communication problems, a critical attitude to one's judgments [18].

In the texts analyzed, this disposition is presented broadly: «It entails effective communication and problem solving abilities and a commitment to overcome our native egocentrism and sociocentrism», «it broadens our minds and develops skills».

Analysis of the essays reveals students' desire to present themselves as progressive thinkers; students actively criticize dogmatic thinking: «believe that critical thinking is a purposeful and self-adjusting judgment, manifested in interpretation, analysis, evaluation, and reasoning, as well as explanations of the evidence, concepts, methods, and principles on which the judgment is based, as well as consideration of background factors», «critical thinking is the ability to analyze the way you think and present evidence for your ideas, rather than simply accepting your personal reasoning as sufficient proof».

Other examples: «thinking clearly and systematically can improve the way we express our ideas. In learning how to analyze the logical structure of texts, critical thinking also improves comprehension abilities», «critical thinking related research focuses on how to systematically construct clear ideas and the characteristics of unclear ideas», «if one manages to follow the main principles of critical thinking, they will be able to estimate their current situation and make choices without regrets or hesitations».

In the texts of Chinese students the problem of open-mindedness is embedded in a broader social and scientific context.

They discuss the need to develop children's thinking skills: «you know, children have the most curiosity. As time goes by, they become indifferent to everything around them. Their eyes are full of mess; even they are unwilling to open them. We have killed their curiosity successfully». Some Chinese students draw parallels with contemporary national studies: «this is actually the "speculative" thinking advocated by Zhengxin Zhengju. The so-called "speculation" means to frequently question, reflect frequently, and constantly think and analyze whether the viewpoints you hold are based on scientific cognition».

Willingness to Suspend Judgment: willingness to think critically requires a willingness to suspend judgment while alternatives are explored [14, 19].

The ability to judge consciously is represented as part of critical thinking in most of the essays: «we gather information, consider the choices we can make, explore the possibilities, and form a general idea: what we intend to do and why this is the right choice. In other words, it is to make purposeful and reflective judgments about what to believe or do, and this is the core of critical thinking», «In life, critical thinking can make people calmer and look at problems more sensibly, and will not judge things one-sidedly because of one or two details», «critical thinking skills are particularly important to form independent judgments».

The possibility of alternative judgments is to find logical connections: «critical thinking is a person's ability to think without any predilections and understand the logical connections between ideas. It is regarded as an effective way of thinking because it gives us opportunities to make rational decisions based on facts», «this allows a person to weigh-in on various facts and perspectives while identifying errors in logic and reasoning, which then helps in the problem solving process».

Students are aware of skepticism as part of the ability to think critically: «independent and rational thinking is the core of critical thinking. We need to be skeptical about everything and argue against these views to form our own views on this». Skepticism as a quality is noted by Russian students.

Trust in Reason: critical thinking requires trust in reason, in intuition, imagination, and emotion [20].

This disposition is reflected most prominently in the texts of the essays: «critical thinking can bring a clearer perception that also makes a person capable of examining their own values, convictions, and opinions», «we can't build critical thinking without inference, which means drawing conclusions based on relevant data, information, and personal knowledge and experience».

In describing this disposition students are more likely to refer to their own life experiences Chinese: «We should give them more chances to discuss something, to debate with classmates, to present their point of view and express preferences. We should reduce the amount of exams as soon as possible. Give them free space to fly in the sky of imagination… I don't expect to see them (children) become the slaves of the exam», «If you do not have critical thinking skills, then you will easily be "paced" by the overwhelming new media articles. Reading such an article, you will unknowingly produce "satisfaction", feel that the article "poke your heart", and believe the author's point of view in the article. As everyone knows, in this process, your mind has been manipulated».

Russian students prefer to discuss the problem in the third person: «It is said that well-developed critical thinking helps people analyze their actions, in other words, engage in self-reflection. Critical thinkers are better decision-makers and can maintain new beliefs and views whenever it's practical or rational for them».

Seeking the truth: a critical thinker should care about the truth otherwise he is content to stick with the initial bias on an issue [15, 16, 21].

Truthfulness is understood as the quality of information: «it means you test information by asking whether it is true or not», «when we receive any type of information, we tend to believe it as a matter of fact. However, a person who has developed critical thinking, is prone to put into question any statement or suggestion that seems to be fake or incorrect».

Truth is conceptualized as a solution to a problem: «skill is problem solving which expands the process of gathering and analyzing information and communicating to identify troubleshoot solutions», «creating needs to find problems which exist in real environment only you have questioning eyes, can you find what you are unsatisfied with, what needs to change. Problem is always the source of innovation».

After the distribution of keywords and word combinations, they were counted and each disposition was assigned a value score and statistical formulas were then applied. The value of each disposition ranged from 1 (weak) to 5 (strong) for the essays in the evaluation procedure, the score depended on the total number of words and expressions, as well as the whole context.

After writing the essays, students were asked to present the concept of 'critical thinking' in visual form (mind maps) rather than in textual form. Students could draw mind maps or use special programs. The logic of concept representation in this case is quite different, as textual information is structured with the help of visual blocks and images (Fig. 1).

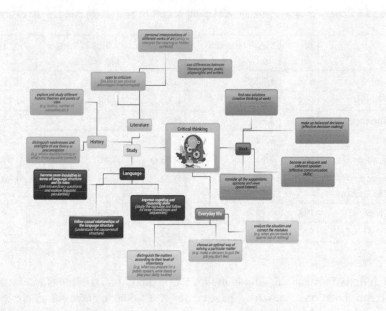

Fig. 1. An example of a mind map (1).

The material was analyzed using Cronbach's Alpha, means and standard deviation. The Cronbach's Alpha (0.81) is showing the reliability of the results. Pearson correlation coefficient is calculated as 0.86. This value can be considered appropriate for the purposes of the current study.

4 Results

The study showed that the use of certain teaching methods affects the perception and structuring of scientific information. The presentation of the scientific concept 'critical thinking' in textual and textual-visual forms allowed us to identify differences in students' perceptions.

According to students' essays, curiosity is opposed to bias: «critical thinking also promotes curiosity, which leads to us having a wide range of interests. It makes us creative. They say that those who have made masterpieces in literature, music, and art were ordinary people, but they had a tendency to think critically». The structuring of the notion of critical thinking in essays regarding initiating dispositions has differences in the works of Russian and Chinese students (Table 1). Among the most important dispositions are Self-confidence, Open-mindedness, Willingness to suspend judgment (Russian students) and Open-mindedness, Self-confidence, Willingness to suspend judgment, respectively. For the Russian students Attentiveness is more important, the Chinese students pay more attention to and Trust in reason Attentiveness.

Table 1. Representation of meaningful components of critical thinking in the essays.

Initiating dispositions	N	Mean (Russian students)	St. Dev.	Mean (Chinese students)	St. Dev.
Attentiveness	9/8	4,32	0,82	4,11	0,79
Habit of inquiry	9/8	1,78	0,64	2,63	0,71
Self-confidence	9/8	4,87	0,81	4,72	0,80
Courage	9/8	3,81	0,56	3,67	0,52
Open-mindedness	9/8	4,61	0,60	4,80	0,64
Willingness to suspend judgment	9/8	4,53	0,72	4,43	0,68
Trust in reason	9/8	4,12	0,62	4,23	0,59
Seeking the truth	9/8	3,66	0,51	3,91	0,58

Description of critical thinking in the form of a mental map demonstrates a slightly different representation of the term. The students were asked to make computerized mind maps or to generate their mind maps using a paper and colored pens. The mind maps presented are of varying completeness and quality. Attention to the world environment is concretized in the areas of "work", "everyday life", "study". "Study" is divided into specific areas (history, language, literature) with critical thinking skills that can be obtained within these subject areas. There are fewer keywords in mental maps. Students discuss less the benefits of critical thinking, but they pay more attention how it can be applied to everyday life. Analysis of the material has shown that the most important thing becomes the ability to make effective decisions. Self-generated computerized mind mapping has affected students' achievement in their comprehension.

Mind maps represent a more personal level of perception of the concept 'critical thinking', because students conceptualize life problems in the form of specific questions. For example, "Could you buy a house downtown?", "What if you under excessive work stress" (Fig. 2) and so on.

Mind maps help to structure scientific information based on personal needs. In this way, they help make abstract concepts practical-oriented (Table 2). As shown in the table Habit of inquiry was not represented in the mind maps, students were concentrated

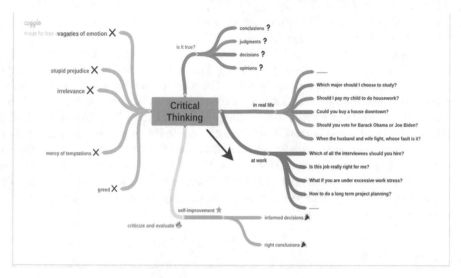

Fig. 2. An example of a mind map (2).

on Self-confidence and Open-mindedness, although, this disposition was reflected in the essays. In the group of Russian students in the third place is Attentiveness instead of Willingness to suspend judgment. Attentiveness is less important for the Chinese students but Courage is more important for them. Self-confidence has become the most important quality for Chinese students, which is represented in the mind maps.

Table 2. Representation of meaningful components of critical thinking in the essays.

Initiating dispositions	N	Mean (Russian students)	St. Dev.	Mean (Chinese students)	St. Dev.
Attentiveness	9/8	4,74	0,80	4,21	0,76
Habit of inquiry	9/8	0		0	–
Self-confidence	9/8	4,85	0,74	4,81	0,78
Courage	9/8	3,57	0,51	3,98	0,59
Open-mindedness	9/8	4,71	0,63	4,79	0,63
Willingness to suspend judgment	9/8	4,23	0,74	4,32	0,69
Trust in reason	9/8	4,15	0,61	4,29	0,54
Seeking the truth	9/8	4,10	0,53	4,10	0,49

Analysis has shown that the productive method of mind maps is effective in structuring scientific information and it makes the information personally oriented, as well as it helps to establish a hierarchy of logical relationships. Mind map method is not a substitute for verbal methods of presenting information, which help to develop abstract thinking. They need to be combined.

Mind mapping helped the students explore the concept 'critical thinking' and its key associations in a logical and organized manner. The perception of this scientific concept has become more structured and personalized. The usage of mind maps fostered students' creativity because they focused on the creative process and decision-making. Using mind maps is a productive method of teaching to facilitate student learning which was originally used to develop critical thinking. Students can illustrate a vision, exhibit their contextual knowledge and creativity, and make associations about a central theme during this activity. This method stimulate learners to 'think hard' about a topic, consolidate a learner's understanding of the material, enable learners to look at a topic from different perspectives, clarify a goal or plan for their own investigations, inspire them to want to find out the answer.

5 Conclusions

Achieving learning objectives through productive teaching method (mind mapping) showed a different level of structuring of scientific information, more effective development of critical thinking. Using of this productive teaching method also allows students to grasp the theoretical aspects in the study of a foreign language better and look at language as a scientific discipline.

Despite the general development of critical thinking in students' adoption of scientific information cultural and national peculiarities persisted [22–24]. When thinking critically about a scientific concept in essay form students correctly identify many of its features but miss some essential characteristics. Russian students tend to discuss the concept in the scientific field: «The principles of principle thinking can be the following: collect complete information, understand and define all terms, question the conclusions and the source of facts, look for hidden assumptions and biases». Chinese students have a strong sense of contrary reasoning.

It should be noted that cultural differences persist in the perception of scientific information when using different teaching methods. But when using active teaching methods, students have the opportunity to form an active research position, be creative and form critical thinking skills, as well as it is the base for soft skills development.

This study is promising for further exploration of effective ways of structuring scientific information in the learning process. This is especially important nowadays, when humanity exist in a dense flow of information, where the selection and structuring of information is crucially important.

References

1. Rubtsova, A., Almazova, Z., Eremin, Yu.: Innovative productive method of teaching foreign languages to international students. In: The European Proceedings of Social & Behavioural Sciences EPSBS 18th PCSF 2018 - Professional Culture of the Specialist of the Future, pp. 1–12. The Future Academy (2019). http://dx.doi.org/10.15405/epsbs.2018.12.02.1

2. Candrasekaran, S.: Productive methods of teaching middle school science. Int. J. Humanit. Soc. Sci. Invent. **3**, 15–25 (2014). http://www.ijhssi.org/papers/v3(7)/Version-2/C03720 15025.pdf. Accessed 10 Mar 2021
3. Buzan, T.: The ultimate book of mind maps. Harper Collins Publishers, Australia and Canada (2005). https://archive.org/stream/pdfy-MEuyCwZKGT3fqH56/The+Ultimate+ Book+of+Mind+Maps+-+Tony+Buzan_djvu.txt. Accessed 10 Mar 2021
4. Rizqiya, R.: The use of mind mapping in in teaching reading comprehension. ELTIN J. **1**(I), 32–43 (2013). https://doi.org/10.33394/jollt.v8i2.2483
5. Davies, M.: Concept mapping, mind mapping and argument mapping: what are the differences, and do they matter? High. Educ. **62**(3), 279–301 (2010). https://doi.org/10.1007/s10 734-010-9387-6
6. Holland, B., Holland, L., Davies, J.: An investigation into the concept of mind mapping and the use of mind mapping software to support and improve student academic performance, Learning and Teaching Projects, 2003/2004, pp. 89–94. https://www.semanticscholar.org/paper/An-investigation-into-the-concept-of-Mind-Map ping-%2F-Holland-Holland/b49eb59c843fb94e5941399d469eebf571ec893d. Accessed on 18 Mar 2021
7. Al-Jarf, R.: Enhancing freshman students' writing skills with mind mapping software. In: The 5th International Scientific Conference e-Learning and Software for Education, pp. 375–382. National Defence University, Bucharest (2009)
8. Bylieva, D., Lobatyuk, V., Safonova, A.: Online forums: communication model, categories of online communication regulation and norms of behavior. Humanit. Soc. Sci. Rev. **1**(7), 332–340 (2019). https://doi.org/10.18510/hssr.2019.7138
9. Ennis, R.H.: An appraisal of the watson-glaser critical thinking appraisal. J. Educ. Res. **52**(4), 155–158 (1958). https://doi.org/10.1080/00220671.1958.10882558
10. Rawls, J.A.: Theory of Justice. Harvard University Press, Cambridge (1971)
11. Hitchcock, D.: Critical thinking. In: Stanford Encyclopedia of Philosophy (2018). https:// plato.stanford.edu/entries/critical-thinking/#DefiCritThin
12. Scheffler, I.: The Language of Education. Charles C. Thomas, Springfield (1960)
13. Abrami, P.C., Bernard, R.M., Borokhovski, E., Waddington, D.I., Wade, A.C., Person, T.: Strategies for teaching students to think critically: a meta-analysis. Educ. Res. Rev. **85**(2), 275 314 (2015). https://doi.org/10.3102/0034654314551063
14. Facione, P.A.: Critical thinking: a statement of expert consensus for purposes of educational assessment and instruction. Research Findings and Recommendations Prepared for the Committee on Pre-College Philosophy of the American Philosophical Association. ERIC Document ED315423 (1990)
15. Facione, P.A., Facione, N.C., Giancarlo, C.: California Critical Thinking Disposition Inventory. The California Academic Press, Millbrae (2001)
16. Bailin, S., Case, R., Coombs, J., Daniels, L.: Common misconceptions of critical thinking. J. Curric. Stud. **31**(3), 269–283 (1999). https://doi.org/10.1080/002202799183124
17. Richard, P., Elder L.: The Miniature Guide to Critical Thinking: Concepts and Tools. Foundation for Critical Thinking, 4th edn. Foundation for Critical Thinking Press, USA (2006)
18. Almazova, N., Bylieva, D., Lobatyuk, V., Rubtsova, A.: Human behavior as the source of data in the education system In: ACM International Conference Proceeding Series, ID 2019: Proceedings of Peter the Great St. Petersburg Polytechnic University International Scientific Conference on Innovations in Digital Economy 2019, pp. 337–340. Association for Computing Machinery, Saint Petersburg (2019). https://doi.org/10.15405/epsbs.2020.12.03.4
19. Halpern, D.F.: Teaching critical thinking for transfer across domains: disposition, skills, structure training, and metacognitive monitoring. Am. Psychol. **53**(4), 449–455 (1998). https://doi. org/10.1037/0003-066X.53.4.449

20. Thayer-Bacon, B.J.: Is modern critical thinking theory sexist? inquiry: crit. Thinking Across Disciplines **10**(1), 3–7 (1992). https://doi.org/10.5840/inquiryctnews199210123
21. Shipunova, O., Berezovskaya, I., Smolskaia, N.: The role of student's self-actualization in adapting to the e-learning environment. In: Garcнa-Pecalvo, F.J. (ed.) TEEM 2019. Proceedings of the Seventh International Conference on Technological Ecosystems for Enhancing Multiculturality, pp. 745–750. ACM, New York (2019). https://doi.org/10.1145/3362789.336 2884
22. Bylieva, D.S., Lobatyuk, V.V., Fedyukovsky, A.A.: Ways of sociotechnical integration of scientists and volunteers in citizen science. In: IOP Conference Series: Materials Science and Engineering, vol. 940, p. 012150 (2020). https://doi.org/10.1088/1757-899X/940/1/012150
23. Odinokaya, M., Andreeva, A., Mikhailova, O., Petrov, M., Pyatnitsky, N.: Modern aspects of the implementation of interactive technologies in a multidisciplinary university. In: E3S Web of Conferences, vol. 164, p. 12011 (2020). https://doi.org/10.1051/e3sconf/202016412011
24. Odinokaya, M.A., Karpovich, I.A., Mikhailova, O.Ju., Piyatnitsky, A.N., Klímová B.: Interactive technology of pedagogical assistance as a means of adaptation of foreign first-year students. In: IOP Conference Series: Materials Science and Engineering, vol. 940, p. 012130 (2020). https://doi.org/10.1088/1757-899X/940/1/012130

The Development of Creative Thinking in Engineering Students Through Web-related Language Learning

Ekaterina V. Vinogradova(✉) ⓘD, Yulia V. Borisovaⓘ, and Natalya V. Kornienkoⓘ

Saint-Petersburg Mining University, Vasilyevsky Island, 21 Line, 2, St. Petersburg, Russia
{vinogradova_ev2,borisova_yuv,kornienko_nv}@pers.spmi.ru

Abstract. This article is devoted to the use of special techniques for the development of creative thinking with engineering students using foreign language acquisition as a tool. The process was exacerbated with the COVID pandemic in 2020 through distance language learning. The authors described the use of "linguistic surfing" method as one of the tasks enhancing non-standard thinking while learning the professional vocabulary in the foreign language in a creative way. The results were analyzed basing on the testing interview and opinion survey comparing two groups of students: those who used the suggested educational technique at the process of online education and the students, who studied online without the above mentioned web-related "linguistic surfing" method. The analysis of the received results of the interview showed the efficiency of using of this technique especially for professional vocabulary mastering at the digital environment educational process. The opinion survey demonstrated that the percentage of the students with positive attitude to online transition was higher among those who studied online with the help of the "linguistic surfing" learning technique. The research proved the suggested technique to be effective in enhancing the creative thinking among technical students through web-related task aimed at vocabulary acquisition within the conditions of forced distance format of education.

Keywords: Creativity thinking development · "Linguistic surfing" · Distance language learning

1 Background Research

1.1 The Latest Transformations in the Educational Area

Nowadays all the environment which surrounds us is transforming, we can observe the new examples of hardware and software in all areas of life, technology and science, introduction of new platforms, applications and what not. It is impossible to cope with all these tasks and to survive in this crazy technological world without good mental health and creative thinking.

We have had an opportunity to assure that creativity is an integral characteristic of every specialist in whatever area of professional sphere it is – even technical engineering.

D. Bylieva and A. Nordmann (Eds.): PCSF 2021, LNNS 345, pp. 881–891, 2022.
https://doi.org/10.1007/978-3-030-89708-6_71

The researchers examine the role of teaching creativity using modern technologies [1] and ascribe to the creative and progressive teaching such methods as scientific "dialogue in spoken and written language" [2; p. 509], various ICT tools, which "help to formulate requirements to professionally oriented language learning curricula" [3, p. 150], "the experiment of using CAT and Lingvo Tutor options by non-linguistic technical students for making glossaries" [4, p. 1],"case study analysis based on real cases from current companies" [5, p. 1], even "integration of social media" [6, p. 1] and so on.

The ability to generate new ideas, use a creative approach in solving problems becomes an integral skill in the process of professional implementation. There are no patterns, no formulas to make a discovery. Our task is to help students develop creative thinking and apply un-orthodox solutions in their studies and everyday life. In this regard, we decided to consider a foreign language as a tool, means and method of developing creative thinking.

Traditional teaching methods, where the main goal was to acquire knowledge, are no longer able to provide a qualitatively new educational product [7]. The requirements of the modern world to the educational process invariably entail the transformations of the educational context itself.

The year of 2020 has presented the unprecedented example of the incredible circumstances and unbelievable experience because of the pandemic situation which has made its own adjustments to the educational process in the country. The Universities in Russia were forced to face the seemingly insurmountable problem of learning with the absence of direct personal contact "teacher-student" by turning to distant learning. This led to transforming curriculum into an online format, which is a challenge for all the educational process participants [8].

1.2 Language Learning as an Incentive to Prosper and Develop

The world faces a lot of new challenges due to the globalization process. Not only has international communication and interaction become an inevitable part of science and technology as necessity, but it also determines the quality of the scientific achievements and their comprehension.

It is important that students (future engineers) speak the English language fluently as it is an international and global language nowadays and it promotes their professional development and provides career prospects, giving opportunities to "obtain the up-to-date information from the foreign literature and exchange data in their professional activities" [9, p. 270] because it is "essential to refer to the list of competencies which they are supposed to acquire" [10, p. 1].

Languages develop your thinking ability, thus it contributes to the creative thought pattern and its enhancement. Language is one of the most important means of organizing your brain and your mind. Future engineers are persons who inevitably experience the necessity of creation of new ideas, techniques and methods in their professional careers reacting to new tasks and challenges. The question of training future engineers capable to understand the basics of their specialty in English was discussed by a number of researchers [11], some researchers investigated the problem of "specific purposes texts translation training" [12, p. 1]. There were papers devoted to the use of modern IC technologies in the educational process of teaching foreign languages and to the use of smart educational systems [13, 14]. Also we can find articles devoted to digital economy and development of communicative competencies [15], formation of ICT competencies in teachers [16].

1.3   Saint Petersburg Mining University's Positive Practice of Transition to ICT Technologies in Language Learning

It is a matter of general observation that teaching staff at the technical departments of the Mining University devote a great deal of attention to teaching students how to use special technologies and applications, for instance, "building information modeling technology (BIM)" [17, p. 264]. S.A. Rassadina has inspected the concept of "edutainment" which involved a web quest during cultural studies course which "increases motivation for independent work in studying non-major subjects" [18, p. 502]. The department of the foreign languages strives to keep contributing to these educational trends and investigating the use of ICT in the course of foreign languages teaching in various aspects [3–6].

Every year, students undergo entry test in the foreign language that comprises 120 questions aimed at vocabulary and grammar knowledge. Moreover, they have mid-term tests and final assessment examination, which allows us to keep track of the education process results. At the end of the course, one of the test assignments is to read and translate professional texts from English into Russian. Traditionally, this task poses a great challenge to a large portion of students whose language is not higher than intermediate level.

The year 2020 was aggravated with partly distant educational process which posed even more difficulty to the students. Some period of studying and testing was totally transferred into online format. Nevertheless, the students successfully fulfilled the final task during the exams that took place in June and December 2020. The results were just as high as they used to be before the pandemic. The diagram below (Fig. 1) represents the results of the final assessment exam among students of the oil and gas faculty of Saint-Petersburg Mining University in 2019 when the students underwent a traditional practical course and in the pandemic year of 2020 when online training prior to the exam amounted to 1/3 of the English language course. The format and the difficulty level of the exams were similar in those years.

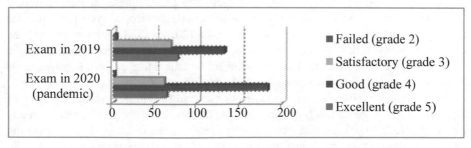

Fig. 1. Comparative statistical analysis of exam results.

Total number of 2019 students of the oil and gas faculty who took English exam amounted to 282 persons and 2020 students – to 309 persons. The grading system in Russia entails four grades as specified in the legend of the diagram (Fig. 1). The diagram shows that there are some distribution fluctuations in the grades: The number of "excellent" grades decreased in 2020, while the number of "good" grades increased.

However, we consider the average results of the two years representative:

Table 1. Average exam grades in 2019 and 2020 (pandemic).

Year	Calculations	Average grade
2019	$(77 \times 5 + 132 \times 4 + 69 \times 3 + 4 \times 2) \div 282 =$	4,0
2020	$(64 \times 5 + 182 \times 4 + 61 \times 3 + 2 \times 2) \div 309 =$	3,99

As can be seen from the calculations above (Table 1), the average grade of the students of the Oil and Gas Faculty in 2019 and 2020 remained the same in spite of the challenging conditions that the pandemic posed. In our opinion, such a result can serve as reliable evidence of the fact that sudden transition to ICT technologies at language learning at the Mining University didn't ruin the educational process of language teaching and learning at the department of foreign languages, and one of the suggested techniques that helped to keep the necessary level of language mastery – a method of "linguistic surfing" will be described in details in this article.

1.4 Challenges Arising from Transition to ICT Technologies

In view of recent events and current pandemic situation, all the above mentioned issues have become even more relevant and have been significantly reconsidered. In these unusual conditions, the teachers have to expand their methodological skills, transform the teaching approach to non-traditional, come up with various activities, which can enhance student learning skills in online classes [19]. Let us consider some of the challenges of current time that are pending for proper mental development of today's students so that in future they could become prominent scientists and engineers that enhance the technological progress.

First of all, "The relevance of using up-to-date distance learning techniques … during COVID pandemic" forced us to reconsider and adapt the developed system of full-time class education to the distant learning within no time [20, p. 145]. How have we managed to implement that task? This question has bothered many researchers across the globe [21–23], but we would like to draw attention to the adaptation process in terms of its difficulty to retain the creative component of education.

Secondly, the problem of academic dishonesty and further inadequate assessing of students' work is urgent nowadays, and some papers describing this problem have already appeared [24, 25]. During the mass e-learning there were reported much more cheating cases. This momentary indulgence may lead to bad consequences in the long-run. So we must think through the tasks and methods of teaching that eliminate any opportunity for cheating.

Third, the problem facing the teachers is how to hold students' attention and motivation to study in a new digital environment, which has never been used in such a way at the educational process before. It is obvious that lack of personal contact and absence of possibility to control the behavior of the students in a customary way could influence the acquisition of foreign languages not in a good way.

Nevertheless, it is a pending task to solve these and further arising challenges through constant modernization and upgrading of educational techniques.

2 Suggested Techniques

In order to solve the discussed problems, it was necessary to elaborate a new system of material presentation, so that the students do not lose any motivation and continue knowledge acquisition. Within a short period of time, we had to adapt to the distant learning and change the proportion of ICT involved in the educational process from 10–30% to 100%. There is some tactics and procedures that had to be imposed.

In Saint-Petersburg Mining University we have an opportunity to work with several ICT tools – the university web portal for communication, sharing information, and sustaining awareness. Moreover, the university provided its students and teaching staff with an access to the videoconferencing software Cisco WebEx – a web tool that has been successfully used for educational and academic purposes over the globe [26–28]. Integration of ICT in educational process has become an inevitable part of the university's work and generates a variety of tasks, gives incentive to both students' and lecturers' creativity [29–31]. Researches show that "digital literacy" enhances professional competence and independence [32]; and, living "in post-information era" [32], humanities in general and foreign languages in particular are taught integrating innovation technologies at the educational process [34–36].

Thus, we have upgraded the abovementioned "web quest" [18] method to use it during the English language practical classes, referred to as a method of "linguistic surfing". We imposed this method in relation to vocabulary acquisition process. This method was used during the online educational process at the Mining University. The research was conducted from April till December 2020 and 26 students of the Oil and Gas faculty studied online using the method of "linguistic surfing". We compared the achieved results with the results of students who didn't apply the abovementioned method during their online vocabulary studies (also 26 students of Oil and gas faculty of the same level of English).

As a rule, the students of the Mining University while studying professional vocabulary are supposed to learn an extensive list of professional terminology in the foreign language and its translation in Russian. Quite often they face difficulty understanding the notion or the item that the term (from the vocabulary list) means. So, in the conditions of Internet-based education process, we gave to the group of the students a task to surf the net and find not only the dictionary definition of the unknown term but also further details, explanations, illustrations and interesting facts about the word. We distributed the words from the vocabulary list among the students; and having conducted their linguistic web surfing independently, the students presented their results to the group through the online class via videoconference. They were encouraged to use any creative ideas they have to present their terms. They used various ICT tools, such as PowerPoint presentations, mind maps, flashcards and demonstrated their findings with the help of Cisco WebEx function of sharing screen. Here are some examples of our students' linguistic web surfing results (Fig. 2):

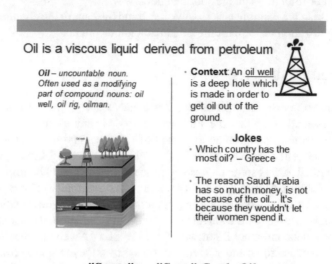

Oil is a viscous liquid derived from petroleum

Oil – uncountable noun. Often used as a modifying part of compound nouns: oil well, oil rig, oilman.

• **Context**: An oil well is a deep hole which is made in order to get oil out of the ground.

Jokes
• Which country has the most oil? – Greece

• The reason Saudi Arabia has so much money, is not because of the oil... It's because they wouldn't let their women spend it.

"Sweet" vs. "Sour" Crude Oil

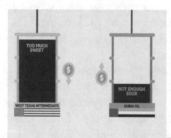

Sweet crude refers to crude oil that is extracted that is found to contain very low amounts of sulfur. Crude oil with low sulfur content is classified as "sweet." Crude oil with a higher sulfur content is classified as "sour". Sulfur content is considered an undesirable characteristic for both processing and end-product quality. Therefore, sweet crude is typically more desirable and valuable than sour crude. West Texas Intermediate (WTI) crude oil is a good example of sweet crude oil, while oil from Canada and the U.S. Gulf Coast tends to be sour.

Fig. 2. Students' presentation as a result of linguistic web surfing

3   Results

3.1   Results of the Testing Interview

In order to assess the results of the integration of the suggested educational technique, the outcome of the integrated task was assessed subsequent to the results of the oral interviewing of the students on their specialization topic in the English language. All the students undergo such an interview prior to their final exam in order to check whether they have acquired the foreign terminology and are able to speak on their professional topic in English. The interview is conducted in the form of discussion between the teacher and the student when the student has to prove their awareness of the topic through well-reasoned answers to the teacher's questions. The example of the questions for the oil and gas faculty: "What is the difference between the onshore and offshore oil and gas production?"; "Who is a roughneck?"; "What are the upstream, midstream and downstream operations?", etc.

During the assessment, the following aspects are taken into account (every aspect is assessed from 1 to 5 points with a maximum of 15): 1) correct usage of professional vocabulary, 2) correct usage of English grammar, 3) coherence and fluency.

In this research we analyzed the results of the two upper-intermediate groups of the oil and gas faculty. One group (26 students) was given the described vocabulary task, while the other group (also 26 students) were not.

Further we conducted an opinion survey among the students whose English language course was expanded with the described creativity task and those ones who studied online without use of the described web-related learning technique. The questionnaire inquired the following two questions for both groups (the experimental and control ones):

1. What is your attitude to the transformation of your studies to the digital educational environment? a. Positive. b. Negative. c. Neutral. 2. Which classes do you prefer? a. In situ classes at the university premises. b. Distant classes online. c. Cannot choose.

There also was the third question asked only at the group, whose educational process was added with the tasks of the "linguistic surfing" method:

3. Was the suggested web-related vocabulary task creative? Give your appraisal of the task. For every question the students could give explanatory comment for clearer representation.

As it was mentioned the main criterion for assessment of successfulness of implementation of the use of "linguistic surfing" technique was the result of our traditional oral interview in English which is conducted among the students of the Mining University as an admission to the final exam task. This interview was assessed in taking into consideration adapted IELTS speaking band descriptors. One of the points to be assessed during the interview is lexical resource. The maximum number of points for vocabulary usage aspect was 5. Among those students who were given vocabulary tasks connected with "linguistic surfing" the results of lexical resource assessment were as follows: 5 out of 5 points (used the vocabulary readily, flexibly, with a precise meaning) – 69% (18 students); 4 out of 5 points (used the vocabulary effectively enough but with some pauses and insignificant inappropriacies) – 23% (6 students); 3 out of 5 points (used only basic professional vocabulary with long pauses) – 8% (2 students). As the results of lexical resource assessment in the group of the students who were not given our tasks they were the following: 5 out of 5 – 53% (14 students); 4 out of 5 – 34% (9 students); 3 out of 5 – 11% (3 students). Speaking about assessment of other aspects during oral interview (fluency and coherence, grammatical range and accuracy, pronunciation) the results in the experimental and control group were equitable and the differences were not as obvious as in case connected with vocabulary usage.

Analyzing the received results, we can see the use of the suggested web-related "linguistic surfing" technique provoking creative thinking proved to be quite successful at the digital environment educational process and helped master the mining University students their professional vocabulary.

3.2 Results of the Opinion Survey

As to our student's attitude to the transformation of the educational process, the opinion survey showed diverse results. Answering the first two questions out of 26 of the respondents taking part in the experimental vocabulary studying, 38,5% (10 students) answered that they preferred class studies, 40,5% (11 students) could not choose which format is better and could see positive aspects in both, 21% (5 students) were totally pleased with the distant format and considered it advantageous. Answering the question about creativeness of the online tasks that had been fulfilled, students showed conformity: 90% (24 students out of 26) appraised the integrated tasks (including "linguistic surfing") as "creatively different" and "peculiar in a positive way", which means that there is positive attitude even from those who adhered to class studies and were against the distance format.

As for the results of the opinion survey among those 26 students, who studied online without use of the described web-related learning technique, the results were the following: 46% (12 students) had negative attitude to the transformation of their studies to the digital educational environment and they answered that they preferred university lessons, 42% (11 students) were neutral to the transition to the online educational environment and 12% (3 students) had totally positive attitude to the distant format of education.

As we can see, the numbers of the students, having negative and neutral opinions about transformation to the digital environment in both experimental and control groups are not so much different, but the percentage of the student with positive attitude to online transition is 9% higher among the students whose process of English language learning was expanded with the described web-related learning technique.

4 Conclusions

The suggested task was aimed to promote development of creative thinking in foreign language classes for future engineers in the proposed conditions and was aimed at development of professional vocabulary at the Oil and Gas faculty students of the Mining University. We used the method of "linguistic surfing" which helped raise motivation in students. Their involvement in fulfilling web-related vocabulary tasks has increased. Moreover, the achieved results proved to be better in comparison with only traditional vocabulary tasks during the period of online education. We should also note that the purpose of mastery in professional terminology in both languages – native and foreign – has been achieved. However, distant learning is not a preferable format of education according to the opinion poll.

Still creativity can be enhanced even if the conditions have changed spontaneously. Moreover, it is even better for non-standard thinking to be developed in extreme and unusual environment.

Since the assessment results have proved to be as high as the results of the previous years, it is reasonable to conclude that it is possible to fulfill the educational purpose even in unexpectedly changeable circumstances of forced online learning.

References

1. Bereczkia, E.O., Kárpáti, A.: Technology-enhanced creativity: a multiple case study of digital technology-integration expert teachers' beliefs and practices. Think. Skills Creat. **39**, 1–27 (2021). https://doi.org/10.1016/j.tsc.2021.100791
2. Shchukina, D.A.: Teoriya i practika nauchnogo dialoga v sovremennom tekhnicheskom vuze. (Theory and practice of modern scientific dialogue in technical higher educational institution). Zapiski Gornogo instituta **219**, 508–512 (2016). https://doi.org/10.18454/PMI.2016.3.508.(in Russian)
3. Murzo, Y., Sveshnikova, S., Chuvileva, N.: Method of text content development in creation of professionally oriented online courses for oil and gas specialists. Int. J. Emerg. Technol. Learn. (iJET) **14**(17), 143–152 (2019). https://doi.org/10.3991/ijet.v14i17.10747
4. Vinogradova, E., Kornienko, N., Borisova, Y.: Cat and Lingvo tutor tools in educational glossary making for non-linguistic students. In: 20th Professional Culture of the Specialist of the Future (PCSF 2020) & 12th Communicative Strategies of Information Society (CSIS 2020). The European Proceedings of Social and Behavioural Sciences EpSBS, vol. 98, pp. 400–409. European Publisher, London (2020). https://doi.org/10.15405/epsbs.2020.12.03.40
5. Oblova, I.S., Sishchuk, J.M.: Case-study based development of professional communicative competence of agricultural and environmental engineering students. E3S Web of Conf. **15035**, 175 (2020). https://doi.org/10.1051/e3sconf/202017515035
6. Pushmina, S.: Teaching EMI and ESP in Instagram. In: Anikina, Z. (ed.) IEEHGIP 2020. LNNS, vol. 131, pp. 475–482. Springer, Cham (2020). https://doi.org/10.1007/978-3-030-47415-7_50
7. Dina, A.-T., Ciornei, S.-I.: The advantages and disadvantages of computer assisted language learning and teaching for foreign languages. Procidea Soc. Behav. Sci. **76**, 248–252 (2013). https://doi.org/10.1016/j.sbspro.2013.04.107
8. Almazova, N., Krylova, E., Rubtsova, A., Odinokaya, M.: Challenges and opportunities for Russian higher education amid COVID 19 teachers' perspective. Educ. Sci. **10**(12), 368 (2020). https://doi.org/10.3390/educsci10120368
9. Lebedeva, I.S.: Some peculiarities of English for specific purposes course development. J. Min. Inst. **187**, 270–271 (2010). http://pmi.spmi.ru/index.php/pmi/article/view/6672. (in Russian)
10. Goman, I.V.: Development of the dialogue skills in a foreign language in comparing of oil benchmarks. In: International Conference "Complex Equipment of Quality Control Laboratories", Journal of Physics: Conference Series, vol. 1384, p. 012013. IOP Publishing, St. Petersburg (2019). https://doi.org/10.1088/1742-6596/1384/1/012013
11. Kogan, M.S., Gavrilova, A.V., Nesterov, S.A.: Training engineering students for understanding special subjects in English: the role of the online component in the experimental ESP course. In: IV International Conference on Information Technologies in Engineering Education (Inforino), pp. 1–6. IEEE, Moscow (2018). https://doi.org/10.1088/10.1109/INFORINO.2018.8581837
12. Borisova, Y., Maevskaya, A., Skornyakova, E.: Specific purposes texts' translation training for mining and civil engineering specialties students. In: 20th Professional Culture of the Specialist of the Future (PCSF 2020) & 12th Communicative Strategies of Information Society (CSIS 2020). The European Proceedings of Social and Behavioural Sciences EpSBS, vol. 50, pp. 130–138. European Publisher, London (2020). https://doi.org/10.15405/epsbs.2020.12.03.13
13. Kogan, M., Zakharov, V., Popova, N., Almazova, N.: The impact of corpus linguistics on language teaching in Russia's educational context: systematic literature review. In: Zaphiris, P., Ioannou, A. (eds.) HCII 2020. LNCS, vol. 12205, pp. 339–355. Springer, Cham (2020). https://doi.org/10.1007/978-3-030-50513-4_26

14. Rubtsova, A.V., Almazova, N.I., Bylieva, D.S., Krylova, E.A.: Constructive model of multilingual education management in higher school. IOP Conf. Ser. Mater. Sci. Eng. **940**, 012132 (2020). https://doi.org/10.1088/1757-899X/940/1/012132

15. Pokrovskaia, N.N., Ababkova, M.Y., Fedorov, D.A.: Educational services for intellectual capital growth or transmission of culture for transfer of knowledge—consumer satisfaction at St. Petersburg Universities. Educ. Sci. **9**, 183 (2019). https://doi.org/10.3390/educsci9030183

16. Valieva, F., Fomina, S., Nilova, I.: Distance learning during the corona-lockdown: some psychological and pedagogical aspects. In: Bylieva, D., Nordmann, A., Shipunova, O., Volkova, V. (eds.) PCSF/CSIS -2020. LNNS, vol. 184, pp. 289–300. Springer, Cham (2021). https://doi.org/10.1007/978-3-030-65857-1_25

17. Goldobina, L.A., Orlov, P.S.: BIM technology and experience of their introduction into educational process for training bachelor students of major 08.03.01 "construction". Zapiski Gornogo instituta **224**, 263–272 (2017). https://doi.org/10.18454/PMI.2017.2.263. (in Russian)

18. Rassadina, S.: Cultural foundations of the concept of "edutainment" as a strategy for the formation of common cultural competence in universities of a non-humanitarian profile. Zapiski Gornogo instituta **219**, 498 (2016). https://doi.org/10.18454/pmi.2016.3.498. (in Russian)

19. Bao, W.: COVID −19 and online teaching in higher education: a case study of Peking University. Hum. Behav. Emerg. Technol. **2**(2), 113–115 (2020). https://doi.org/10.1002/hbe2.191

20. Murzo, Yu., Chuvileva, N.: Use of information technologies in developing foreign language competence for professional interaction of undergraduate and postgraduate students specializing in mineral resources. iJET, **16**(03), 144–153 (2021). https://doi.org/10.3991/inet.v16i03.17875

21. Maican, M.-A., Cocoradă, E.: Online foreign language learning in higher education and its correlates during the COVID-19 pandemic. Sustainability **13**(2), 781 (2021). https://doi.org/10.3390/su13020781

22. Teräs, M., Suoranta, J., Teräs, H., Curcher, M.: Post-Covid-19 education and education technology 'solutionism': a seller's market. Postdigit. Sci. Educ. **2**(3), 863–878 (2020). https://doi.org/10.1007/s42438-020-00164-x

23. Gonzalez, T., et al.: Influence of COVID-19 confinement on students' performance in higher education. PLoS One **15**(10), e0239490 (2020). https://doi.org/10.1371/journal.pone.0239490

24. Pylkin, A., Serkova, V., Petrov, M., Pylkina, M.: Information hygiene as prevention of destructive impacts of digital environment. In: Bylieva, D., Nordmann, A., Shipunova, O., Volkova, V. (eds.) PCSF/CSIS -2020. LNNS, vol. 184, pp. 30–37. Springer, Cham (2021). https://doi.org/10.1007/978-3-030-65857-1_4

25. Bylieva, D.S., Lobatyuk, V.V., Nam, T.A.: Academic dishonesty in e-learning system. In: Soliman, K.S. (ed.) Proceedings of the 33rd International Business Information Management Association Conference, IBIMA 2019: Education Excellence and Innovation Management Through Vision 2020, pp. 7469–7481 (2019)

26. Dames, L.S., Royal, C., Sawyer-Kurian, K.M.: Active student engagement through the use of WebEx, MindTap, and a residency component to teach a masters online group counseling course. In: Keengwe, J., Bull, P.H. (eds.) Handbook of Research on Transformative Digital Content and Learning Technologies, pp. 245–268. IGI Global (2017). https://doi.org/10.4018/978-1-5225-2000-9.ch014

27. Akiyama, Ts., Masuda, H., Yamaoka, H.: Preparation for remote activities in the university using Cisco WebEx education offer. In: ACM SIGUCCS Annual Conference (SIGUCCS 2021), pp. 46–49. ACM, New York (2021). https://doi.org/10.1145/3419944.3441171

28. Poluekhtova, I.A., Vikhrova, O.Yu., Vartanova, E.L.: Effectiveness of online education for the professional training of journalists: students' distance learning during the COVID-19 pandemic. Psychol. Russ. State Art **13**(4), 26–37 (2020). https://doi.org/10.11621/pir.2020.0402

29. Bylieva, D., Lobatyuk, V., Safonova, A., Rubtsova, A.: Correlation between the practical aspect of the course and the E-learning progress. Educ. Sci. **9**(3), 167 (2019). https://doi.org/10.3390/educsci9030167

30. Shipunova, O.D., Berezovskaya, I.P., Mureyko, L.M., Evseeva, L.I., Evseev, V.V.: Personal intellectual potential in the e-culture conditions. Espacios **39**(40), 15 (2018). http://www.revistaespacios.com/a18v39n40/18394015.html Accessed 10 Mar 2021

31. Aladyshkin, I.V., Kulik, S.V., Odinokaya, M.A., Safonova, A.S., Kalmykova, S.V.: Development of electronic information and educational environment of the university 4.0 and prospects of integration of engineering education and humanities. In: Anikina, Z. (ed.) IEEHGIP 2020. LNNS, vol. 131, pp. 659–671. Springer, Cham (2020). https://doi.org/10.1007/978-3-030-47415-7_70

32. Evseeva, L.I., Shipunova, O.D., Pozdeeva, E.G., Trostinskaya, I.R., Evseev, V.V.: Digital learning as a factor of professional competitive growth. In: Antipova, T., Rocha, Á. (eds.) DSIC 2019. AISC, vol. 1114, pp. 241–251. Springer, Cham (2020). https://doi.org/10.1007/978-3-030-37737-3_22

33. Cortada, J.W.: Life in a post-information age era? In: Living with Computers, pp. 59–76. Copernicus (2020). https://doi.org/10.1007/978-3-030-34362-0_6

34. Frolova, V., Chernykh, V., Bykovskaya, G.: Innovative Technologies. Eur. Proc. Soc. Behav. Sci. EpSBS **51**, 110–118 (2018). https://doi.org/10.15405/epsbs.2018.12.02.12

35. Muñoz-Luna, R., Taillefer, L.: Integrating Information and Communication Technologies in English for Specific Purposes. Springer, Cham (2018). https://doi.org/10.1007/978-3-319-68926-5

36. Semushina, E., Valeeva, E., Kraysman, N.: Intensive language learning at technological university for integrating into global engineering society. MJLTM **9**(10), 1–9 (2019). https://doi.org/10.26655/mjltm.2019.10.2

Formation of Translation Competence in the Process of Engineering Education

Anna V. Rubtsova[iD], Maria Odinokaya[iD], Darina Barinova[(✉)] [iD], and Nadezhda I. Almazova[iD]

Peter the Great St. Petersburg Polytechnic University, St. Petersburg 195251, Russian Federation

Abstract. The article presents the development of an innovative methodology of e-learning of foreign languages in the process of engineering education based on the use of professionally-oriented scientific articles and papers. The authors analyze the term 'translation competence' and determine the component structure of this competence. The following components of translation competence were identified: linguistic, communicative, semantic, interpretive, textual, intercultural. In order to form the highlighted dimensions of the translation competence, a number of key factors that affect the success of engineering education have been determined. During the research the system of pre-translation and translation exercises for engineering education was developed, based on the materials of electronic professionally-oriented informational texts for engineering students studying English for a future profession. Within the framework of this research, the methodology of formation of translation competence in foreign language professional training of engineering students was considered. The results of experimental training confirm the effectiveness of the innovative methodology developed, aimed at the formation of translation competence of engineering students mastering a foreign language for professional needs.

Keywords: Innovative methodology · Translation competence · E-learning · Foreign language · Professionally-oriented articles · Engineering education

1 Introduction

The modern information society dictates new standards for teachers and students, stimulating more successful mastering of competencies, the development of personal qualities, the ability to freely realize their educational needs and focus on cognitive interests, as well as to determine professional needs and their future implementation [1–4].

As a result of the obtaining the Master's degree, graduates of engineering programs at universities should develop general and professional competencies established by the program. Within the framework of this study, the method of formation of translation competence in foreign language professional training of engineering students was considered. The relevance of the problem stated in the study is proved by the need to increase the level of students' translation competence with the help of modern technologies and the lack of research in this area when teaching students of engineering specialties [5].

D. Bylieva and A. Nordmann (Eds.): PCSF 2021, LNNS 345, pp. 892–907, 2022.
https://doi.org/10.1007/978-3-030-89708-6_72

Due to the epidemiological situation in the country, the education had to face in 2020, it was difficult for teachers both to adapt to the current situation, and also to maintain the educational process at the proper level. To solve this kind of problem, it is necessary to develop innovative universal e-learning technologies, with the help of which the teaching of a foreign language for the development of translation competence will be effective.

The process of forming translation competence is impossible without the interaction of the teacher and students, as well as students with each other and with the computer. In the new environment, developed modern teaching methods should take into account the use of new information technologies [6, 7].

Translation skills and abilities are created through the use of specially selected teaching e-learning materials. Such materials include translation exercises and educational texts. Exercise is the main way to develop the skills you need. In the course of this work, translation skills are developed and the basis for improving translation skills is created. No matter how good modern textbooks and educational-methodical complexes in general are, the teacher cannot do without additional material if he really intends to educate a student in a student who can easily feel himself in the modern world, a person who can adapt in new society in the e-learning process. Numerous electronic periodicals in a foreign language will help the teacher to cope with this task [8, 9].

2 Background and Means to Resolve the Problem

Professionally-oriented information in newspaper materials in English is a significant part of the electronic supplementary educational material in engineering education. Systematic work with scientific articles and newspapers expands the capabilities of a foreign language teacher in achieving complex general educational and educational goals, especially when teaching engineering students [10].

Acquaintance with modern media in English-speaking countries contributes to the formation of socio-cultural competence, broadening the horizons of students, deepens knowledge about the country of the target language, its customs and traditions, teaches them to compare and evaluate facts and events that occur in the world. Reading newspaper and scientific articles in English in the given specialty in electronic form, students learn to receive new information from the original text, they develop the skill of intuitive perception of a foreign language text with a direct understanding of the content [11–14]. Reading the professionally-oriented articles also contributes to the consolidation and expansion of the vocabulary of students and the mastery of new links for them, more complex grammatical phrases of the English language [15, 16].

When developing a set of exercises based on the material of electronic professionally-oriented informational texts for engineering students studying English for a future profession, we have determined a number of factors that affects the success of the study, namely:

- goal (target setting), speech task - conditional or real;
- speech actions of the student;
- linguistic form and content;

- a certain place in the series of related exercises (according to the principle of increasing difficulty, taking into account the sequence of the formation of speech skills and abilities;
- a certain time allotted for the exercise;
- product (result) of the exercise;
- material (verbal and non-verbal: text, pictures, diagrams, maps);
- way of doing the exercise (orally, in writing);
- organizational forms of implementation (individually, in pairs, in a group).

All the factors mentioned will for the basement of the system of exercises for forming the translation competence in engineering education.

3   The Solution to the Problem

Translation competence is a special set of abilities, knowledge and skills necessary for a successful professional translation activity. Translation competence is considered an integral part of general professional competence and requires constant development and improvement. The task of forming the translation competence of engineering students is especially urgent nowadays [17–19].

In order to form translation competence on the basis of material from English-language periodicals and to develop a set of tasks for students of engineering fields of training in teaching a foreign language, it was necessary to highlight the basic knowledge and skills necessary in the process of working with electronic texts and materials. The following components of translation competence were identified: linguistic, communicative, semantic, interpretive, textual, intercultural [20].

Consider the composition of each component:

- build an utterance in accordance with linguistic norms (linguistic component);
- correlate linguistic means with the tasks and conditions of communication (communicative component);
- extract, memorize and generate the semantic content of the message (semantic component);
- identify the contextual meanings of linguistic means and their transformation (interpretive component);
- distinguish between the type, style and genre of the text, as well as construct and reproduce the text according to the given typology (textual component);
- decode and adequately interpret the speech and non-speech behavior of representatives of different cultures, taking into account the socio-cultural context of a specific communicative situation (intercultural component).

In order to form the highlighted components of translation competence, the following pre-translational and translational exercises were selected for the goal.

1. Pre-translational:

- answers to questions to the text, checking the depth of understanding and the presence of the necessary background knowledge;
- discussion of the concepts underlying the content of the text and related terms and concepts;
- various exercises to improve the knowledge of the target language (drawing up synonyms and differentiating the meanings of synonyms, stylistic assessment of the proposed options, paraphrasing statements).

2. Translational:

- linguistic, developing the ability to solve translation problems associated with the peculiarities of the semantics of units and structures of the source language and the target language;
- operational, practicing the ability to use various methods and techniques of translation;

Communicative, creating the ability to successfully perform the necessary actions at different stages of the translation process [21, 22].

Each of the highlighted exercises contributes to the formation of the highlighted component of translation competence (Table 1).

Table 1. Exercises aimed at developing the components of translation competence

Exercises	Components of translation competence
Pre-translational: answers to questions to the text, checking the depth of understanding and the presence of the necessary background knowledge	Linguistic, interpretive, textual
Pre-translational: discussion of the concepts behind the content of the text and related terms and concepts	Linguistic, semantic, textual
Pre-translational: improving the knowledge of the target language (drawing up synonymous series, differentiating the meanings of synonyms, stylistic assessment of the proposed options, paraphrasing statements)	Semantic, intercultural
Translational: language (solving translation problems related to the peculiarities of the semantics of units and structures of the source and translation languages)	Linguistic, semantic, interpretive
Translational: operational (practicing the use of various methods and techniques of translation)	Linguistic, interpretive
Translational: communicative (performing the necessary actions at different stages of the translation process)	Communicative, intercultural

Within the framework of the study, a method was created for the formation of the selected components of translation competence in the process of working with electronic English-language professional information texts in the given specialty [6, 23, 24]. It consists of five stages:

1. reading and translating the headlines of articles;
2. reading and translating abstracts and keywords;
3. a summary of the content of the newspaper article in Russian or foreign language (depending on the level of English);
4. a brief overview of a number of articles;
5. a brief overview of the entire issue of the newspaper as a whole.

The experiment included three stages: initial, formative, and final. During the experiment, three tests were carried out to control the level of formation of translation competence.

4 Results and Discussions

The experimental base of the research is: Peter the Great St.Petersburg Polytechnic University. The participants of the experiment were groups of first-year graduate Master's students of the Institute of Mechanical Engineering of Materials and Transport, studying the course "Foreign language in professional communication". The average age of students was 22–35 years old. Number of people in both groups - 42. The experimental and control groups consisted of 21 people each. The students of the control and experimental groups were diagnosed with the level of development of translation competence. The training of the experimental group was based on materials from English-language specialized informational texts which they had to find using website http://www.world-newspapers.com/uk, and in the control group, the teaching and learning complex was used (educational-methodical complex: Advanced Masterclass) [4]. The tasks for the tests of the experimental group at all stages were based on electronic English-language newspaper informational materials of the relevant scientific spheres.

To prepare for the project, it was necessary to conduct a theoretical and methodological substantiation of the chosen methodology or technology, having studied the previous experience of teachers and methodologists, and then formulate the goal, objectives and hypothesis of the experiment, determine the basis of the experiment and the composition of the subjects, conduct initial testing to assess the input level of students' translation competence who will participate in the experiment.

The tests at the initial and final stages, carried out to assess the level of students' translation competence within the framework of the experiment, have an identical structure: they include 15 open-type questions aimed at checking the level of formation of all selected components of the translation competence. Evaluation was carried out on a scale from 2 to 5, where 2 is "unsatisfactory" and 5 is "excellent".

The level of formation of students' translation competence was determined on the basis of the analysis of translation competency among the same students at the end of the experimental testing.

The preliminary assessment of the level of students' translation competence was carried out on the basis of diagnostics of the formation of its main components:

- build an utterance in accordance with linguistic norms (linguistic component) (C 1);
- correlate linguistic means with the tasks and conditions of communication (communicative component) (C 2);
- extract, memorize and generate the semantic content of the message (semantic component) (C 3);
- to identify the contextual meanings of linguistic means and their transformation (interpretive component) (C 4);
- to distinguish the type, style and genre of the text, as well as to construct and reproduce the text according to the given typology (text component) (C 5);
- decode and adequately interpret the speech and non-speech behavior of representatives of different cultures, taking into account the socio-cultural context of a specific communicative situation (intercultural component) (C 6).

The tasks for the tests of the experimental group at all stages were based on English-language information materials of the professionally oriented sphere of engineering students from the electronic resource http://www.world-newspapers.com/uk. At the initial stage of the experiment, the tests contained 15 open-ended questions.

In the experiment, the scale of measurements by D. A. Novikov was used, which served as a theoretical basis for the analysis of each of the stages of the pedagogical experiment [25]. For example, in the study of the comparative effectiveness of two teaching methods, A and B (control and experimental), assessments are made on two scales - five-point and ten-point. Grades on a ten-point scale can be converted into grades on a five-point scale: grades "10" and "9" will be assigned to "5", "8" and "7" - to "4" and so on. The numerator indicates the number of students who received the corresponding grade in the group trained according to method A, in the denominator - according to method B. Next, the average score is calculated using the formula.

Based on the data obtained, it can be concluded that the students of the experimental group have a relatively weak degree of translation competence. According to the results, the greatest difficulties were caused by such tasks as:

- find equivalents of newspaper clichés;
- find out in what sense the underlined words are used in the text (literally or figuratively);
- read and translate the signature under the photograph (drawing, caricature).

The experimental group had the least difficulties in such tasks as:

- from these narrative sentences to compose headings containing information in a concise form;
- title the article;
- read the title and make suggestions about the content of the article.

Speaking about the results of the control group, we found that the greatest difficulties arose in:

- express the thought given in the first paragraph in a concise form;
- read the text and find key sentences that convey the main idea.

The least difficulties arose in the examples of such tasks:

- find in the proposed text expressions and clichés characteristic of the business style.

The level of formation of the components of translation competence among students of the experimental group at the initial stage was the following: linguistic component – 90%, communicative component – 35%, semantic component – 82%, interpretive component – 40%, text component – 47%, intercultural component – 51%. The level of formation of the components of translation competence among students of the control group at the initial stage was the following: linguistic component – 79%, communicative component – 81%, semantic component – 84%, interpretive component – 94%, text component – 97%, intercultural component – 100%.

Thus, the experimental group faced difficulties in improving the interpretive, textual and communicative translation components. The control group had problems in improving the linguistic, communicative and semantic translation components.

The figures below (Fig. 1 and Fig. 2) show the levels of formation of the components of translation competence (at the initial stage).

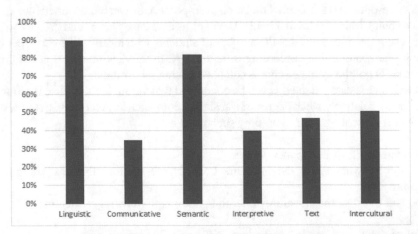

Fig. 1. The level of formation of the components of translation competence among students of the experimental group at the initial stage.

Starting to work with the electronic newspaper, an introductory conversation was held with students about English-language newspaper informational texts [26]. They should also be familiarized with the general structure of newspapers, with the placement of materials published in them, list the headings available in electronic newspapers

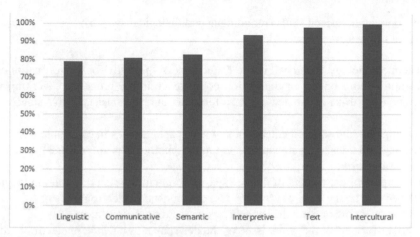

Fig. 2. The level of formation of translation competence among students of the control group at the initial stage.

with which they will work. They should be taught to browse newspapers, while paying attention to the headings of the articles and the blocks of text typed in highlighted type, to get a general idea of the issue.

At the initial stage of working with an electronic newspaper in the website http:// www.world-newspapers.com/uk, the teacher determines the articles that are suitable for students to read and discuss according to their scientific field. Such electronic articles should be:

- not too large in volume (approximately 20 lines of a newspaper column);
- comprehensible in terms of lexical composition and subject matter;
- interesting in content;
- relevant on the subject, rich in socio-political or scientific information;
- having educational value.

At the next stage of working with an electronic newspaper, students can be offered to select texts of interest to them on a scientific topic in an electronic resource [5, 27]. Working with a newspaper is carried out at the level of a prepared speech. The teacher's task is to:

a) ensure the speech interaction of students;
b) bring communication to the level of unprepared speech.

Work on each topic is structured as follows:

- introduction of social and political vocabulary on the topic;
- primary activation of language material through exercises;
- work with articles according to the scheme: from initial acquaintance - to reading and speaking;

- development of speech skills (monologue and dialogical speech);
- control of the learned material.

The methodology developed in the framework of the study for the formation of all the identified components of translation competence in the process of working with electronic English-language newspaper informational texts consists of five stages:

1. reading and translating the headlines of newspaper articles in electronic form;
2. reading and translating signatures, text under photographs and political cartoons in electronic form;
3. a summary of the content of the newspaper article in Russian or a foreign language (depending on the level of preparedness of students) in electronic form;
4. a brief overview of a number of articles in electronic form;
5. a brief overview of the entire issue of the newspaper as a whole in electronic form.

In the first stage of working with an electronic newspaper, students should rely on:

- knowledge of the most important political and social events;
- language guess;
- understanding the lexical and grammatical structure of the phrase.

To work with the newspaper at this stage, the following types of exercises were used:

- Find cases where the source of information is presented in the headings. How is this source indicated?
- The heading includes information in a concise form. Expand the heading.
- Compose headings containing information in a concise form from the narrative sentences given.
- Entitle the article.
- Read the heading and say what the proposed article might refer to.

At the second stage, when working on the translation of signatures under photographic materials of English-language newspaper informational texts, students rely on a linguistic guess, the possibilities of which are expanded due to the greater visibility of the material in electronic form. Selections of newspaper illustrations with visual material, which the teacher selects in electronic form. The goals of working with newspaper illustrations in teaching English can be varied. Illustrations and texts to them can be widely used when working on lexical units, when studying grammatical phenomena, for teaching reading, in working on the development of skills and abilities in oral speech.

So, when working on the translation of the captions under the illustrations, the teacher should first read the caption himself, and then ask one of the students to translate it. Then, after stylistic processing, the final version of the translation is read. Thanks to the regularly carried out work, students develop listening skills and memorizing words and expressions typical of the newspaper information style.

To work with an electronic newspaper at this stage, you can use the following types of exercises:

- Find the most common newspaper expressions and cliches in the captions under the illustrations.
- Find equivalents of the phrases and the newspaper cliches (find equivalents of phrases and newspaper clichés);
- Find words which meaning can be easily guessed from the context.
- Find out in what meaning the underlined words are used in the text (in direct or figurative).
- Find keywords and key phrases.
- Read and translate the caption under the photo (drawing, caricature).

At the third stage of work, when reading electronic information material, the teacher asks students to identify familiar names, abbreviations, typical traditional newspaper phrases or clichés from the web resource chosen [28, 29]. The electronic article is read and translated under the strict guidance of a teacher who helps to eliminate lexical and grammatical difficulties. Students can be trained through questions as well as through translation.

So, for example, when reading an article on a medical topic, it is advisable to study words and phrases associated with it, as well as invite students to make an electronic glossary on the topic, for example:

- vaccination,
- clinical trials,
- immunization campaign,
- jab,
- injection,
- handling,
- efficacy, etc.

After the electronic newspaper article has been read, the teacher may invite students to summarize its content in English. You can also specify introductory and connecting phrases:

1. The headline of the article is…
2. It is written by…
3. The article reviews the latest…
4. The article opens with a description of events in…
5. Then the author gives a detailed of events in…
6. At the end of the article the author draws the conclusion…
7. From my point of view, the most interesting items (facts) in article are the following…

To work electronically with newspaper informational texts at this stage, you can use the following exercises:

- Read the electronic article and answer the questions.
- Express in one sentence the thought contained in the first paragraph of the article.
- Define the terms given in the article.

- Read the article and highlight the sentence that helps to understand the heading/figurative meaning.
- Read the article and find key sentences that convey the main idea.
- Think of a different heading for the article, more fully conveying what is being said; argue your point of view.

At the fourth stage, the main requirement is to develop the ability to compose a brief overview of electronic newspaper articles.

The main methodological principles of work here are:

- careful selection of material, taking into account its relevance, informational significance, suitability for discussion and retelling;
- taking into account the individual characteristics of students;
- a variety of types and techniques of working with newspaper materials.
- To work with a newspaper at this stage, it is proposed to use the following types of exercises:
- View the article and explain what is important and relevant in it.
- Shorten the article to three sentences expressing the main idea; write down your version.
- Browse through several articles and determine which parts contain basic information.
- Prepare a short message on one of the topics of a socio-historical or socio-cultural nature based on press materials.

At the fifth stage, assignments for students' work with English-language newspaper informational texts acquire a creative character.

1. Report. (Students present with a pre-prepared review of the newspaper issue.)
2. Seminar. (Students prepare creative abstracts of articles on international events, sports commentary, medical or political issues.)
3. "Press conference". This creative type of work requires particularly careful preparation and a good knowledge of newspaper material. The teacher distributes among the students the roles of "correspondents" and "international specialists". "Correspondents" should ask questions of "specialists", that is, in fact, interview them in a foreign language. There may be some discrepancy in opinions between the "specialists", as a result of which during the "press conference" active discussions may arise, students actively participate in the creative work.)

Thus, the presented set of exercises is aimed both at communicating the necessary professional knowledge using electronic sources, and at developing translation skills and abilities in general, and at developing creativity. At the same time, skills constitute the ultimate goal of teaching translation, as they ensure the practical professional activity of a translator. Some of the above exercises are multifunctional in nature and develop several skills and abilities at once. The proposed typology of exercises, in our opinion, ensures the formation of all components of translation competence and can be used to work with students of the "Linguistics" training area in the process of working with electronic English-language newspaper information texts.

At the final stage of the experiment, the following tasks were identified:

- conduct control testing to assess the level of translation competence after the introduction of materials in English informational texts into the learning process;
- discuss the work done (what experience was gained), evaluate the personal contribution of each student and the general work of the group;
- summarize and analyze the research results.

The final test we conducted contains 15 questions, they are also open-ended.

According to the test results, the greatest difficulties for the students of the experimental group were caused by such tasks as:

- shorten the article to three sentences expressing the main idea; write down your version;
- come up with a different title for the article, which more fully conveys what is being said; argue your point of view.

The least difficulties arose when completing the following sample assignments:

- give a definition of the terms given in the article in English;
- find equivalents of phrases and newspaper clichés (they are provided to students in Russian).

Speaking about the results of the control group, it was found that the greatest difficulties arose in the following examples of tasks:

- come up with a different name for the text, more fully conveying what is being said; argue your point of view;
- shorten the text to three sentences expressing the main idea; write down your version.

The least difficulties arose in the examples of the following tasks:

- read the text and highlight the sentence that helps to understand
- name
- figurative meaning.

Thus, the experimental group faced difficulties in improving the semantic and intercultural translation components. The control group had problems in improving the semantic and textual components.

The figures below (Fig. 3 and Fig. 4) show the levels of formation of the components of translation competence (at the final stage). The level of formation of the components of translation competence among students of the experimental group at the final stage was the following: linguistic component – 72%, communicative component – 80%, semantic component – 60%, interpretive component – 78%, text component – 68%, intercultural component – 85%. The level of formation of the components of translation competence among students of the control group at the initial stage was the following: linguistic component – 100%, communicative component – 98%, semantic component – 88%, interpretive component – 94%, text component – 93%, intercultural component – 95%.

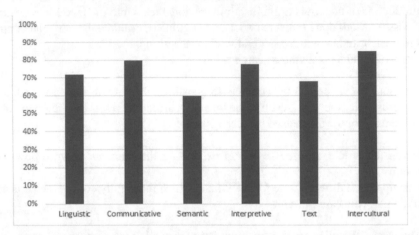

Fig. 3. The level of formation of the components of translation competence among students of the experimental group at the final stage.

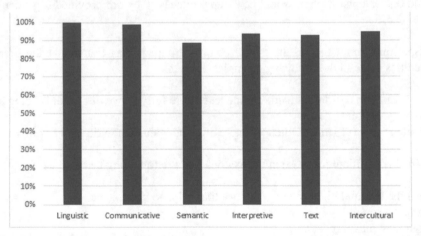

Fig. 4. The level of formation of the components of translation competence among students of the control group at the final stage.

After receiving the test results of the experimental and control groups at the initial and final stages of the experiment, the average scores for the groups were calculated, as well as the average percentage values. These values were obtained by multiplying the group's average score by one hundred percent and dividing the result by 5 (the maximum score that can be obtained for the test). The average score of test results at the initial stage in the experimental group was 3.8 (76%), and in the control group - 4.6 (92%). The average score of test results at the final stage in the experimental group was 4.3 (86%), and in the control group - 4.8 (96%). Figure 5 shows a comparison of the test results of the control and experimental groups at the initial and final stages in percent.

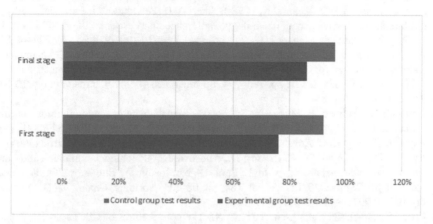

Fig. 5. A comparison of the test results of the control and experimental groups at the initial and final stages.

5 Conclusion

The results of the final stage of the experiment showed significant changes in the level of formation of translation competence in general. A comparative analysis of the dynamics of qualitative changes in the development of the translation competence of engineering students before and after the experiment also showed a different level of training in the control and experimental groups. In the experimental group, at the initial stage, indicators corresponding to the elementary level of development of translation competence prevailed, while in the control group the results had a high level of development.

After the final testing and analysis of the results obtained, we made the following conclusions:

- the level of development of translation competence in the experimental group increased from 76% (initial stage) to 86% (final stage);
- the level of development of translation competence in the control group increased from 92% (initial stage) to 96% (final stage).

The results of experimental training confirm the effectiveness of the methodology developed, aimed at the formation of translation competence of engineering students mastering a foreign language for professional communication. A comparative analysis of the results obtained at the initial and final stages indicates a positive dynamic in the formation and development of translation skills and abilities of the tested students.

References

1. Akhmedyanova, G.: Concepts of oeuvre and algorithmization in modern engineering education. In: Solovev, D.B., Savaley, V.V., Bekker, A.T., Petukhov, V.I. (eds.) Proceeding of the International Science and Technology Conference "FarEastCon 2019." SIST, vol. 172, pp. 911–917. Springer, Singapore (2020). https://doi.org/10.1007/978-981-15-2244-4_87

2. Almazova, N., Bernavskaya, M., Barinova, D., Odinokaya, M., Rubtsova, A.: Interactive learning technology for overcoming academic adaptation barriers. In: Anikina, Z. (ed.) IEE-HGIP 2020. LNNS, vol. 131, pp. 786–794. Springer, Cham (2020). https://doi.org/10.1007/978-3-030-47415-7_84

3. Anderson, L., Hibbert, P., Mason, K., Rivers, C.: Management education in turbulent times. J. Manag. Educ. **42**(4), 423–440 (2018). https://doi.org/10.1177/1052562918779421

4. Aspinall, T., Capel, A.: Advanced Masterclass. Oxford University Press, Oxfrod (2004)

5. Bylieva, D., Lobatyuk, V., Kuznetsov, D., Anosova, N.: How human communication influences virtual personal assistants. In: Bylieva, D., Nordmann, A., Shipunova, O., Volkova, V. (eds.) PCSF/CSIS -2020. LNNS, vol. 184, pp. 98–111. Springer, Cham (2021). https://doi.org/10.1007/978-3-030-65857-1_11

6. Cress, U., Stahl, G., Ludvigsen, S., Law, N.: The core features of CSCL: Social situation, collaborative knowledge processes and their design. Int. J. Comput.-Support. Collab. Learn. **10**(2), 109–116 (2015). https://doi.org/10.1007/s11412-015-9214-2

7. Dascalu, M., Trausan-Matu, S., McNamara, D.S., Dessus, P.: ReaderBench: automated evaluation of collaboration based on cohesion and dialogism. Int. J. Comput.-Support. Collab. Learn. **10**(4), 395–423 (2015). https://doi.org/10.1007/s11412-015-9226-y

8. Davis, F.D.: Perceived usefulness, perceived ease of use, and user acceptance of information technology. MIS Q. **13**(2), 319–340 (1989). https://doi.org/10.2307/249008

9. Dragon, T.: Support of teacher's work in the field of development of computational thinking through e-learning resources. In: Proceedings of the 2019 3rd International Conference on Education and Multimedia Technology, pp. 131–135. Association for Computing Machinery, New York (2019). https://doi.org/10.1145/3345120.3352738

10. Gile, D.: Basic Concepts and Models for Interpreter and Translator Training. John Benjamins Publishing Company, Amsterdam/Philadelphia (2015)

11. Gruzdeva, M.L., Vaganova, O.I., Kaznacheeva, S.N., Bystrova, N.V., Chanchina, A.V.: Modern educational technologies in professional education. In: Popkova, E.G. (ed.) Growth Poles of the Global Economy: Emergence, Changes and Future Perspectives. LNNS, vol. 73, pp. 1097–1103. Springer, Cham (2020). https://doi.org/10.1007/978-3-030-15160-7_110

12. Hamzah, N., Arriffin, A., Hamid, H.: Web-based learning environment based on students' needs. IOP Conf. Ser. Mater. Sci. Eng. **226**(1), 012196 (2017). https://doi.org/10.1088/1757-899X/226/1/012196

13. Hug, B., Krajcik, J.S., Marx, R.W.: Using innovative learning technologies to promote learning and engagement in an urban science classroom. Urban Educ. **40**(4), 446–472 (2005). https://doi.org/10.1177/0042085905276409

14. Hymes, D.: On communicative competence. In: Pride, J.B., Holmes, J. (eds.) Sociolinguistics, pp. 269–293. Penguin, Harmondsworth (1972)

15. Ipatov, O., Barinova, D., Odinokaya, M., Rubtsova, A., Pyatnitsky, A.: The impact of digital transformation process of the Russian University. In: Katalinic, B. (ed.) Proceedings of the 31st DAAAM International Symposium, pp. 0271–0275. DAAAM International, Vienna, Austria (2020). https://doi.org/10.2507/31st.daaam.proceedings.037

16. Köpke, B., Nespoulous, J.-L.: Working memory performance in expert and novice interpreters. Interpreting **8**(1), 1–23 (2006)

17. Korneev, D.G., Gasparian, M.S., Kiseleva, I.A., Mikryukov, A.A.: Ontological engineering of educational programs. Revista Inclusiones: Revista de Humanidades y Ciencias Sociales **7**, 312–324 (2020)
18. Lu, J., Lajoie, S.P., Wiseman, J.: Scaffolding problem-based learning with CSCL tools. Comput. Support. Learn. **5**, 283–298 (2010). https://doi.org/10.1007/s11412-010-9092-6
19. Ludvigsen, S., Law, N., Rose, C.P., Stahl, G.: Frameworks for mass collaboration, adaptable scripts, complex systems theory, and collaborative writing. Int. J. Comput.-Support. Collab. Learn. **12**(2), 127–131 (2017). https://doi.org/10.1007/s11412-017-9257-7
20. Munday, J., Thompson, N., McGirr, M.: Supporting and scaffolding early childhood teachers in positive approaches to teaching and learning with technology. In: MacDonald, A., Danaia, L., Murphy, S. (eds.) STEM Education Across the Learning Continuum, pp. 113–135. Springer, Singapore (2020). https://doi.org/10.1007/978-981-15-2821-7_7
21. Oxford, R.L.: Language Learning Strategies: What Every Teacher Should Know. Newbury House/Harper & Row, New York (1990)
22. Rao, R.V.: Teaching-learning-based optimization algorithm. In: Rao, R.V. (ed.) Teaching-Learning-Based Optimization Algorithm and Its Engineering Application, pp. 9–39. Springer, Cham (2016). https://doi.org/10.1007/978-3-319-22732-0_2
23. Shulus, A.A., Zarudneva, A., Yatsechko, S., Fetisova, O.: The algorithm of modern Russia's transition to the digital economy. In: Popkova, E.G., Sergi, B.S. (eds.) ISC 2019. LNNS, vol. 87, pp. 57–63. Springer, Cham (2020). https://doi.org/10.1007/978-3-030-29586-8_7
24. Sinha, S., Rogat, T.K., Adams-Wiggins, K.R., Hmelo-Silver, C.E.: Collaborative group engagement in a computer-supported inquiry learning environment. Int. J. Comput.-Support. Collab. Learn. **10**(3), 273–307 (2015). https://doi.org/10.1007/s11412-015-9218-y
25. Novikov, D.A.: Statistical Methods in Pedagogical Research (Typical Cases). MZ-Press, Moscow (2004). (in Russian)
26. Baranova, T., Khalyapina, L., Kobicheva, A., Tokareva, E.: Evaluation of students' engagement in integrated learning model in a blended environment. Educ. Sci. **9**, 138 (2019). https://doi.org/10.3390/educsci9020138
27. Valieva, F.: Soft skills vs professional burnout: the case of technical universities. In: Anikina, Z. (ed.) IEEHGIP 2020. LNNS, vol. 131, pp. 719–726. Springer, Cham (2020). https://doi.org/10.1007/978-3-030-47415-7_76
28. Aronin, L.: Multilingualism in the age of technology. Technol. Lang. **1**, 6–11 (2020). https://doi.org/10.48417/technolang.2020.01.02
29. Shipunova, O.D., Berezovskaya, I.P., Mureyko, L.M., Evseeva, L.I., Evseev, V.V.: Personal intellectual potential in the e-culture conditions. Espacios **39**, 15 (2018)

Evaluating the Capacity of Foreign Language Speaking Tasks to Stimulate Creativity

Elena V. Borzova and Maria A. Shemanaeva(✉)

Petrozavodsk State University, 33, Lenin av., 185000 Petrozavodsk, Russia
anat.bor@onego.ru

Abstract. It has always been highlighted by educators that creativity is essential for efficient living. Numerous studies conclude it can be purposefully developed in every academic course, including foreign language education. This study explores the possibility to foster student creativity in foreign language speaking. We analyzed the creative potential of foreign language speaking tasks offered by three university coursebooks for intermediate level students. We regard this concept as the task capacity to engage learners in generating and discussing their own ideas, emotions, and attitudes, providing their *personal investment* which suggests expressing opinions, comparing them, summarizing and drawing conclusions. Thus, while generating their own content, students activate and transform all their resources going beyond the learning context. Therefore, the issue we investigated is whether the tasks that different coursebooks contain may potentially encourage student creativity in foreign language speaking. We believe that such tasks should have a few constraints related either to the content or language or conditions that allow teachers to gradually lead students from guided creativity practice to their free foreign language self-expression. In addition, we questioned university students to find out their opinions about the creative potential of foreign language speaking tasks. The results of the whole research reveal that in the coursebooks we analyzed there is still too much focus on language and a limited variety of foreign language speaking tasks encouraging student creativity. This may hinder the progress of the learners' ability to freely express themselves and eventually decrease the level of their foreign language communicative competence.

Keywords: Creativity · Foreign language speaking · Task creative potential · University course books

1 Introduction

Creativity is always mentioned among the most important 21st century soft skills that every person needs for a meaningful and successful life. "Creativity enters into virtually every aspect of life. It's not a special 'faculty' but an aspect of human intelligence in general: in other words, it's grounded in everyday abilities such as conceptual thinking, perception, memory, and reflective self-criticism" [1]. What follows is that this ability reveals itself in a huge variety of contexts and actions, ranging from simple everyday activities up to the highest level of human inventions and masterpieces of science and arts.

© The Author(s), under exclusive license to Springer Nature Switzerland AG 2022
D. Bylieva and A. Nordmann (Eds.): PCSF 2021, LNNS 345, pp. 908–925, 2022.
https://doi.org/10.1007/978-3-030-89708-6_73

Fostering learner creativity is generally considered a significant mission of education [2]. Each academic course is expected to contribute to mastering varied aspects of student creativity. It is generally acknowledged that classroom creativity promotes students' deep learning, their appreciation for how knowledge is created, and insights into themselves as learners [3]. One common ground that unites creativity development regardless of the course is practicing student divergent thinking and imagination through the specific content and tools that this course can provide [4–6]. Researchers also underline that it is important to work on the development of different thinking modes which are building upon each other, fostering openness and expanding a range of student possibilities [7]. Barbot, Besançon, Lubart (2015) explore student creative potential which is defined as an ability to produce original ideas that have value in their context. The emphasis in their study is placed on the ways of measuring it through examining student accomplishments or abilities and traits underlying creativity, or through contextualized tasks that simulate real-world creative work [7].

Our focus is directed towards foreign language speaking which, in case it reflects student-generated content, can be regarded as a creative process. Speaking in a foreign language classroom possesses certain properties that makes it noticeably creative. It is highly imaginable communication because all those involved can perfectly speak their native language, expressing more sophisticated content at that. Learners need to scan all their resources, which are often rather scarce, especially at the beginning level, to come up with meaningful utterances. It demands the student ability to compensate for the lack of certain language units in their memory and find some ways of expressing themselves. Students have to choose the content that is consistent both with the task and the situation and make decisions. All these abilities are gradually developed in the course of learning in specially designed tasks.

The purpose of the present article is to explore the creative potential of foreign language speaking tasks that are offered by university foreign language coursebooks. The problem questions that we pose for our study are what task factors can either restrict or promote student creative self-expression in foreign language speaking and whether recently published coursebooks for English language university education contain speaking tasks that can potentially foster student creativity in this particular field.

2   Literature Review

The key concepts of the present article are creativity, foreign language speaking, student engagement, and creative potential of foreign language speaking tasks.

Actually, in every paper on education the words "creativity" or "creative" are abundantly used by its authors in different contexts. Generally speaking, the definition of creativity implies originality and effectiveness of some active process and its outcomes [8]. Creativity concerns generating new, previously unknown ideas and behaviours in novel situations or treating familiar situations in new ways [4, 9]. However, in educational literature, the attribute "previously unknown" implies that they are new to the learning student who is going beyond what could be expected [10, p. 55] while acquiring new experiences in various areas to act effectively. As learning relies on the previous experiences, it should be underlined that student creative process in producing something new for himself and herself includes not only generating new ideas and behaviours, but also treating familiar situations in new ways, going beyond stereotypical associations, combining familiar ideas in unfamiliar ways [1], exploring familiar facts and opinions, combining, transforming and adapting them to new contexts, making connections, developing new interpretations or looking for new arguments or solutions, as well as having original and imaginative thoughts and ideas about something; thus working with a broad semantic field [1, 2, 11, 12]. The outcomes of this process are characterized by originality and effectiveness, new interpretations and solutions, a variety of new meanings and relationships in things and connections, which are relevant to goals [4]. Therefore, creativity can be expressed by anyone at different levels of proficiency in a great variety of contexts [10, p. 56]. It is based on activating all aspects of the learner whole personality: cognitive (divergent thinking, analytical thinking, mental flexibility, associative thinking, metaphorical thinking) and on "creative-cognitive high order skills" in general [13, 14]; personal and affective (willingness to take risks, openness to new ideas and experiences, tolerance to ambiguity, intuitive thinking), as well as subject specific (relevant knowledge and skills), recognizing that a problem can be viewed from multiple perspectives [4, 12, 15, 16], which results in producing as many creative, original, and varied responses as possible [17].

From the perspective of the present study, it is necessary to stress that all these acts, as well as their outcomes, though they are performed inside our minds, can be verbally, or by means of other signs or images, expressed, shared, and discussed by the agent. It is here where the concept of speaking comes to the fore. Speaking is described as an oral speech act and its outcome (utterance) which is aimed either at delivering information, or expressing feelings and attitudes, or maintaining relationships, or influencing others (discourse). Speaking is always carried out with the help of a language and addressed to a listener. It is described as "a high complex mental activity" [18], involving thinking and memory [19]. Speaking can accompany both creative acts as well as presentations of their outcomes, further discussions, and final problem solutions.

The authors of the latest version of CEFR do not single out speaking among the main foreign language teaching and learning aims, but those which they suggest (production, interaction, mediation, reception) also implicitly include speaking [20]. They consider learners' progression up the scale of foreign language acquisition as moving towards higher levels from basic information exchange or presentation towards more sophisticated collaborative discussion or sustained monologue production; progressing from reactive to proactive participation, and from simple to complex content, adjusting his/her language appropriately to make his /her communication more effective. It is obvious that, to achieve these goals, learners must be involved in a creative speaking process trying to articulate a personal idea, preference, feeling, or attitude [21], generating their own products of varied degree of complexity, taking their current language resources and molding them to new situations [22].

The main tool that teachers use to develop foreign language speaking skills are tasks [23]. For decades, researchers have emphasized the importance of communicative tasks that promote "skill getting" and "skill using [24, 25]. Communicative speaking tasks, such as discussions, information transformation, simulations, or projects, involve students in creative language use to stimulate genuine communication through conveying and negotiating personal meaning [25].

Among a great variety of task factors, researchers of foreign language speaking creativity enhancement highlight establishing a special learning environment in which "creativity flourishes" [26, p. 35, 27] and which is enjoyable and low-stress [28]; developing learners' creative thinking skills through involving them in the most significant macro-process such as inquiring, doing, imagining, and reflecting [10, p. 57] with a concern with interaction and attention to output [29]; using open communicative activities [30] and a lot of other task conditions. Researchers are unanimous stating that a high level of student engagement in completing tasks sparks their creativity most significantly [31–35].

It is generally acknowledged that engagement is hard to define and measure, but it is clearly seen and felt in the classroom [35]. It cannot be tied to a single factor [32], manifesting itself in a variety of metrics: cognitive, behavioral, social, and emotional [36–38].

Among the factors that drive student engagement, researchers mention the design and form of the information being shared, "utilizing content that is relevant and relatable for students" [35, p. 17]; the classroom culture; the instructional practices implemented by a teacher, "encouraging productive play and creative expression" [35, p. 22]; the structure and activities of a lesson, allowing students to create, encouraging student discussion, enabling student choice, integrating interactive and hands-on learning experiences, and requiring students to carry the cognitive load [35, p. 24]. Recent studies also indicate positive effects of allowing learners some degree of personal investment in task content to achieve a higher level of student engagement [38, 39]. Lambert et al. assume that the most significant factor for student engagement in foreign language speaking is learner-generated content which is defined as content that learners choose and tailor to a specific classroom context [34, p. 393]. Linh Phung's study revealed that the factors determining student task preference are topic familiarity, personal relevance, opportunity to create ideas, and opportunity to address a genuine communicative need by speaking with interlocutors of different L1 backgrounds [39]. Promoting engagement, teachers facilitate opportunities for students to create which "fuels deeper engagement and more investment in the learning process" [35, p. 25]. It is obvious that this is a two-way process: engagement sparks creativity which, in its turn, enhances engagement.

In the context of learning, the degree of student creativity in foreign language speaking tasks depends on the constraints that the task imposes on them. These constraints determine the degree of student autonomy in choosing both the content and language for self-expression [24, 40] as well as the character of their actions. The task complexity depends on both cognitive demands that the student has to fulfill [41] and on the mode of student interaction [42]. All these characteristics, as we discussed above, are related to learner creativity. Therefore, designing creative foreign language speaking tasks, we can vary different task parameters, trying to allow students to exercises different levels of creativity.

Meanwhile, literature on creativity in foreign language education has repeatedly reported teachers' difficulties in integrating creativity into the traditional classroom routines Researchers point out that "most of the activities present tightly controlled or guided situations with no purpose other than to practice specific language forms" [18, p. 23]. The study of five English Language textbooks showed that they do not provide enough communicative activities, interaction and negotiation of meaning [43]. Therefore, it is obvious that designing and using creative foreign language speaking tasks poses a problem issue both for course book writers as well as teachers.

3   Rational for the Study

Keeping in mind the multi-faceted nature of creativity which can manifest itself at different levels of complexity and in an endless variety of activities, the basic assumption for the goals of our research is that in a learning context student creativity can reveal itself even in simple foreign language speaking acts which suggest.

- generating leaner's own authentic content;
- approaching content issues from different perspectives and due to it producing a variety of responses;
- finding appropriate linguistic and non-linguistic means to get the message through to listeners.

In this study we set the purpose of investigating the creative potential of foreign language speaking tasks in university paper and online coursebooks. Based on the literature review, we define the creative potential of foreign language speaking tasks as their capacity to engage learners in productive oral speech acts based on learner generated content presented from different angles and addressed to interested listeners encouraging them to discuss the issue further. We have selected these parameters as the most significant ones for creative foreign language speaking task design because, on the one hand, they allow students to personally, cognitively, emotionally, and interactively engage in speaking, expressing their individual creative abilities in creative performance while, on the other hand, they foster openness and ambiguity for various ideas which students share. Analyzing university coursebooks, we will consider a number of constraints for student creative self-expression.

Let's compare two frequently used speaking assignments:

N1. Give a tip to foreigners coming to our region on what sights they should see here. Explain your choice.
N2. Work out a list of tips to foreigners coming to our region to make their trip here unforgettable.

Task N1 strictly limits the student speaking creativity in terms of the content, language, the volume and structure of the output. Nevertheless, it encourages learners to make their own reasonable choices and present their ideas as they find fit.

Task N2 does not contain constraints giving learners freedom of self-expression, practicing their divergent thinking skills and encouraging their creative approach to the oral output.

It is evident that foreign language speaking tasks can vary in their creative potential. Some of them may tightly guide student utterances through constraints imposed in the assignment. Such constraints are inevitable for designing the foreign language learning process to make it feasible, successful, and manageable.

Through gradually increasing multiple potentials for creativity (ranging from low to high potential) depending on its fit between one's resources and the various creative tasks demands [4], teachers lead students to going beyond their comfortable boundaries to meet the challenge of the task. The learner output in creative speaking tasks may be of different character, depending on whether it is part of strictly guided practice or of free practice. In such tasks, acts of foreign language speaking can be combined with drawing writing, designing computer graphics, graphic organizers and PPP, selecting as well as taking photos or making videos, presenting sketches or drama, reciting, doing project and research work. Therefore, students have an abundance of rich opportunities for creative self-expression both through speaking and other activities as well.

4 Methodology

The research was aimed to find out.

(a) types of speaking tasks present in coursebooks and an online course;
(b) the creative potential of these tasks (criteria-based assessment and students estimates-based assessment;
(c) the proportion of speaking tasks with the highest creative potential in the coursebooks and the online course analyzed in the study.

4.1 Research Design and Methods

During the research, the data were obtained by means of quantitative and qualitative methods: (a) through the analysis of the coursebooks in terms of using speaking tasks in foreign language university education in particular; (b) through classroom observations and analysis of the results of classwork; (c) through conducting a questionnaire among the student participants in the project (46 respondents) which was designed to find out the students' opinion about the creative potential of different speaking tasks; (d) through gathering data related to the university context of foreign language teaching and learning.

The overview of the data gathered in the initial phase allowed us to articulate research questions for the study and to outline a few solutions to consider. The final phase consisted of the analysis of the results obtained.

4.2 Research Materials

During the first stage, we analyzed one unit of each coursebook chosen for the study and one online course unit to single out the speaking task types offered by the authors. This comparative study analyzed two EFL coursebooks and one online course currently being used to teach English in Russia. The current situation with COVID-19 put an additional strain of education and educators all over the world [44, 45] and many of them have turned to resources and courses available online. The impact of e-learning is controversial [46] that is why we have chosen an online course to assess the creative potential of tasks it offers. The names of the textbooks will not be revealed and they will be labelled coursebook 1 (CB1), coursebook 2 (CB2) and online course (OC) so as to keep them anonymous and to avoid influencing the reader with any preconceptions that might be created about the coursebooks evaluated in this study. To consider a wider range of the speaking tasks used, we chose courses designed for different purposes: general English/ESP (EAP)/ESP (BE). CB1and the OC are aimed at integrated skills development within each unit (reading/writing/speaking/listening), while CB2 is structured into 4 modules, each focusing on a certain skill (reading/writing/speaking/listening). Analyzing the latter, we looked at the module aimed at speaking skills development. The speaking module is divided into two units, the first one deals with socializing, while the second one focuses on academic presentations. In contrast to the first unit (socializing skills), the unit devoted to academic presentations is based on the use of an individual learning path with a special emphasis on student learning needs. This focus is implemented through developing learners' skills in expressing personalized content in the foreign language. To look at the creative potential of the use of an individual learning path, [47] we consider these two units of the speaking module separately. To distinguish the two units of CB2 we labelled them CB2 (A) (socializing unit) and CB2 (B) (academic presentation unit). The results of the analysis are presented in Table 1.

Table 1. Coursebooks an online course characteristics.

	Coursebook 1	Coursebook 2		Online course
Content	General English	English for specific purposes (Academic English)		English for specific purposes (Business English)
Level	Intermediate	Intermediate		Mixed
Topic of the unit	Speaking about yourself	Socializing	Academic Presentation (Individual learning path)	Entering the job market
Focus	Integrated skills development	Speaking		Integrated skills development
Number of tasks that require student speaking within the unit	29 out of 43	27 out of 47	39 out of 61	1out of 30
Percentage of tasks aimed at developing speaking skills	67%	57%	64%	3%

The analysis of these courses has revealed several characteristics. CB1, being a blended course which aims at the integrated foreign language skills development through different practical tasks and exercises, is more speaking-oriented in comparison with

CB2. If we look at the percentage of purely speaking tasks within a unit/module, we can see that 67% of the tasks in CB1 encourage student speaking while the units of the speaking module in CB2 (A) and CB2 (B) contain only 57% and 65% speaking tasks. Although the difference accounts only for 2–10%, it matters as the CB2 module is primarily aimed at speaking skills development.

The first conclusion which can be drawn from the analysis is that, in case the course is aimed at developing integrated skills, multifunctional tasks prevail. The emphasis on multifunctional multifaceted tasks saves time and allows to achieve integrated results.

The second conclusion which can be drawn is that the online course considered does not offer a sufficient proportion of tasks aimed at speaking skills development as only one task out 30 is aimed at practicing them. It is not enough to estimate its creative potential as we consider that creativity should be drawn on step-by-step approach. The learner is unlikely to become creative without persistent practicing and developing creativity. Perceptive tasks provide the learner with an opportunity to enrich their knowledge base as well as receptive skills while production (oral and written) implies student active involvement and creativity.

5   Criteria and Results of the Study

5.1   Coursebook Analysis

To assess the creative potential of the speaking tasks in the selected courses, we chose a few criteria, each based on a certain constraint for student creative self-expression.

- Speaking is tightly guided. The content and the language are imposed and often predictable.
- Language aids (vocabulary and/or grammar) are provided.
- A plan (or a detailed algorithm) is provided.
- Time/number of student utterances is limited.
- Speakers go beyond the given text/visual, but the content is based on the ideas provided by the text/visual.
- Students are free to express any content without constraints.

In all these tasks students generate their own content which makes their speaking creative. The difference among them is in the amount of creative efforts that speakers are expected to make in producing their utterances.

The analysis of the speaking tasks used in the courses allowed us to group them according to the criteria above (Table 2):

During the next stage, we assessed the creative potential of the speaking tasks used in the coursebooks.

According to Lambert et al., [34] the creative potential of foreign language speaking tasks depends on three major criteria:

1. A *personal investment* in the content which can be:

   - weak if the speaker is limited by a narrow topic;

Table 2. Speaking Tasks Constraints

Speaking Task Constraints	Examples
Speaking is tightly guided. The content and the language are imposed and often predictable	Practice the questions and answers in pairs (the questions and answers are given) Practice the conversations. Offer other possible responses
Language aids (vocabulary and/or grammar) are provided The content is learner-generated but there are some target language units to use	Tell the class if these adjectives describe you Introduce yourself. Use the phrases below Introduce your partner. Use the phrases below
A plan (algorithm) is provided The content is structured and sequenced according to a plan/algorithm while the language is partially specified either in the example or detailed instructions	Give similar information about you/your partner Ask and answer the same questions about your relative (use the ideas in the box to help) Role-play according to the instruction Prepare a role-play (a step-by-step instruction is given) Give a short talk (a detailed step-by-step instruction (algorithm) is provided) Comment on the quote (speaking is based on the questions which specify the ideas and prompt student responses) Provide feedback on the presentation of your partner (questions are given)
Time/number of utterances is limited	Give a short 3 min talk (a detailed step-by-step instruction (algorithm) is provided)
Speakers go beyond the given text/visual, but the content is based on the ideas provided by the text/visual	*Based on pictures*: Look at the photo. Why are the people laughing?/Look at the pictures and discuss them *Based on texts*: What do you think… (developing the text story)/How can the story develop?/What are advantages/disadvantages?/How does it correlate the quote at the beginning of the unit/Why did it happen? Compare your answers with your partner's answers. Are they similar or different? Answer the questions (based on the text) Discuss the statements. Are they true for you? Comment on the quote. Agree/disagree

(*continued*)

Table 2. (*continued*)

Speaking Task Constraints	Examples
Freedom of expression (speaking about personal experiences): students express any content without constraints	Talking about you. Personal experience questions (Example: Who are you closest to? Why?)
	Work with a partner, have a conversation about a good/bad day
	Discuss an example of (based on learner's experience)
	Compare and discuss your ideas. (based on learner's experience)

- medium if the speaker is not limited by the narrow content but is encouraged to use specific language (vocabulary/grammar);
- strong if the speaker is free to choose the topic and the language.

2. The topic is *relevant* and the speaker feels comfortable to discuss it.

A teacher-generated topic is likely to reduce the creative potential of the task as it can be either not relevant to the speakers' experience or needs or the speakers can feel not ready to discuss it. That is why a learner-generated topic is a much better option because speakers accept responsibility for the choice of the topic, thus empowering themselves and increasing their agency level.

3. The topic is *appropriate* for the immediate context.

Speakers' needs are an important factor while generating the topic for speaking tasks. Every topic should fit into the context and meet the needs of the speakers in some way. A learner-generated topic puts more emphasis on the learners' needs and consequently has a higher creative potential.

In our study, we gave a number of points to each criterion to assess the creative potential of the tasks using a quantitative approach. Thus, student personal investment depending on its strength (weak, medium or strong) is assigned 1/2/3 point respectively. According to the relevance of the topic, the task can be classified as teacher-generated (0 points) or learner-generated (1 point). The appropriateness of the topic is given 0 point or 1 point if the topic is teacher- or learner-generated respectively. The results of the speaking tasks classification by their creative potential can be seen in the following table (Table 3).

5.2 Student Assessment of the Creative Potential of Speaking Tasks

During the research, we also used the questionnaire to find out students' opinions about the creative potential of these types of foreign language speaking tasks. The questionnaire contained several types of speaking tasks and the respondents were asked to rank them

Table 3. Creative Potential of Foreign Language Speaking Tasks

	Within the given content provided by the text or visual	Language aids (vocabulary/grammar) are provided	A plan (algorithm) is provided	Time/number of utterances is limited	The content is based on the ideas provided by the text / visual (going beyond it)	Freedom of expression (no constraints)
Learner *personal investment* in the content	Learner-generated/ <u>weak</u>/ 1	Learner-generated /<u>medium</u> 2	Learner-generated /<u>strong</u> 3	Learner-generated /medium 2	Learner-generated /medium 2	Learner-generated /strong 3
The topic is *relevant, the speaker feels comfortable*	Teacher-generated 0	Learner-generated 1	Learner-generated 1	Learner-generated/ Teacher-generated 1/0	Learner-generated/ Teacher-generated 1/0	Learner-generated 1
The topic is *appropriate*	Teacher-generated 0	Learner-generated 1	Learner-generated 1	Learner-generated/ Teacher-generated 1/0	Learner-generated/ Teacher-generated 1/0	Learner-generated 1
Total	1	4	5	2/4*	2/4*	5

from 1 (not creative) to 5 (very creative). 46 students answered the questionnaire; the results can be seen in Table 4 (Fig. 1).

Table 4. Students' assessment of the task creativity potential results.

Creative potential/task type	1	2	3	4	5	Mean average
Within the given content	26%	26%	35%	9%	4%	2,39
Language aids (vocabulary and/or grammar) are provided	9%	35%	17%	30%	9%	2,96
A plan (algorithm) is provided	0	13%	26%	30%	30%	3,78
Time/number of utterances is limited	0	13%	35%	48%	4%	3,43
Speakers go beyond the given text/visual, but the content is based on the ideas provided by the text/visual	0	4%	4%	44%	48%	4,21
Freedom of expression: students express any content without constraints	0	0	0	13%	87%	4,86

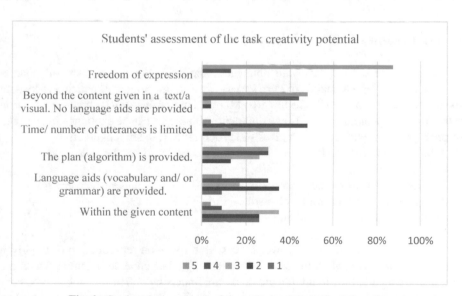

Fig. 1. Students' assessment of the task creativity potential results.

5.3 Cumulative Results

Further analyses of the coursebooks and the online course showed the frequency of these types of speaking tasks and their creative potential (Table 5).

Table 5. Cumulative results

	Creative potential expressed in points (1–5)	Students assessment mean average	Coursebook 1	Coursebook 2		Online course
				A	B	
Within the given content	1	2.39	18(62%)	18(67%)	7 (19%)	1 (100%)
Language aids (vocabulary/grammar) are provided	4	2.96	5 (17%)	7 (25%)	13(33%)	
The plan (algorithm) is provided	5	3.78	0	1(4%)	4(10%)	
Time/number of utterances is limited	2/4	3.43	0	1(4%)	3 (7%)	
Beyond the given content (a text/a visual)	2/4	4.21	2 (7%)	0	-	
Freedom of expression	5	4.86	4 (14%)	0	12(31%)	

6 Discussion

Several trends arise from the results obtained in the study. We can see that both course-books CB1 and CB2(A)mainly (62%/67%) use the foreign language speaking tasks with the lowest creative potential. The first trend shows a major focus on language acquisition rather than on speaking creativity development. Although it can be explained by the need to practice the target language, it gives little room, if any, to student creativity.

The most common tasks in the courses that we analyzed are the following:

- Practice the questions and answers in pairs (the questions and answers are given)
- Role-play the conversations (role cards are given)
- Answer the questions (based on the text)

It is obvious that question-answer tasks with a focus on reproduction of the given content prevail. Meanwhile the tasks with much a higher creative potential which are less restraining and thus inhibiting creativity are underused (17%/25%). It can be con-cluded that the authors' primary emphasis is placed on language accuracy rather than on integrative results such as meaningful self-expression. Plan-or algorithm-based tasks are not present or underrepresented in the coursebooks (0/4%), although the creative potential of these tasks is very high as they provide only subtle guidance related to the logic of the utterance. They still require a student to be independent in the choice of the ideas and language while working on his/her own. The tasks which imply freedom of expression ranking the highest point for their creative potential account for 14%/0%

in CB1 and CB2 respectively. Both CB1 and CB2 make use of 4 out of 6 types of the speaking tasks that we analyzed.

The analysis of the unit based on the use of an individual learning path CB2(B) revealed a different trend. The tasks with the lowest creative potential account only for 19% while similar tasks but with more freedom in terms of the content amount to 33%. Plan- and algorithm-based tasks represent 10% of the speaking tasks which means more focus on creativity and independence. 31% of the speaking tasks in this unit are associated with freedom of expression and this result is twice as high as the results if the units in CB1 and CB2(A).

The final tasks that conclude the unit in CB1 are still based on the previously read text. This task type can inhibit creativity in case the students do not feel ready to discuss the topic offered. The second final task in the unit goes under the heading 'Talking about you' and includes personal questions.

In CB2 the final tasks in both sections (with and without an individual learning path) are of a plan- or algorithm-based task type. They provide guidance specifying a few ideas to be included into the student utterance to maximize its scope but the topic is not specified and is learner-generated.

It is clear that the speaking tasks that are offered at the final stage of the units still restrict student creativity imposing certain constraints on their activities.

Comparing the university students' opinions about the task creativity potential, it can be concluded that although they tend to assess the tasks which imply speaking within the given content from 1 to 3 points, their mean average still accounts for 2.39 which is the lowest. Thus, students' assessment is broadly congruent with the criteria-based assessment. Language aids are seen by the majority of the respondents as inhibiting with their mean average at 2.96, 35% of the students assigned them only 2 points. Surprisingly, 13% and 26% of the students assign 2 and 3 points to the tasks with a plan/an algorithm provided. It is likely they believe the plan (algorithm) to restrain creativity, while 30% of the students attribute them 4 and 5 points, grading them as tasks with a high creative potential. However, the biggest discrepancy is seen in the assessment of plan(algorithm)-based tasks as according to our criteria we regard them to be very creative while 39% of the students do not agree with it. A possible explanation for this trend may be the fact that students do not use this kind of tasks frequently and they do not have enough experience using them in learning. CB1 and CB2 do not offer such tasks. Only the unit based on the use of an individual learning path taps into this task type. 35% and 48% of the students ranked the tasks which limit the time of speaking or the number of utterances 3 and 4 points respectively and they mean average stands for 3.48 which implies quite a high creative potential. The tasks aimed at speaking beyond the given content and the tasks offering freedom of expression are considered by the respondents as the task types with the highest creative potential as their mean average is 4.21 and 4.86 respectively.

As we can see from Table 1, the speaking tasks which provide freedom of expression get the highest creative potential score (5 out of 5). These tasks are believed by many teachers to be highly creative. However, the tasks which provide an algorithm of producing/structuring the utterance but do not specify the exact content, thus giving learners the opportunity to demonstrate their own creative approach, get the same maximum. Even if an algorithm or a plan is offered, more focus is placed on the content and, if the topic is learner-generated, it stimulates the learner to be more open and willing to discuss it without being inhibited by the language barrier. Tasks which encourage speaking beyond the given content (a text or a visual aid) are likely to boost creativity in case the topic is seen by learners as relevant to their experience and appropriate to their needs. Otherwise the impact on the learners' creativity is not high as it is inhibited by irrelevance and inappropriateness of the topic.

The coursebooks analyzed in the study offer a variety of speaking tasks which are focused on both language acquisition practice and partial self-expression. However, the results of the study demonstrate they cannot be relied on entirely to foster learners' creativity for the following reasons:

- Excessive use of tasks aimed at mechanical/reproductive rather than meaningful student practice.
- Insufficient use of plan- or algorithm-based tasks with a high degree of creativity in producing a logically structured product.
- Predominance of teacher/coursebook-generated content with rare cases of learner-generated content.

The online course we analyzed turned out to be unable to boost student speaking creativity primarily due to the insufficient number of speaking tasks.

We can conclude that the foreign language speaking tasks with the lowest creative potential prevail in the courses under study which makes it necessary for university teachers to supply additional tasks to boost student creativity in foreign language speaking.

Creativity in speaking should be nurtured by a step-by-step approach, all types of the tasks should be used with more focus on the tasks which offer more room for freedom of expression. The use of an individual learning path and more emphasis on multifunctional multifaceted tasks is beneficial in terms of fostering learner's agency, creativity and achieving integral outcomes.

The limitations of the study are related to the focus of the research on considering one particular aspect of the foreign language speaking tasks that several coursebooks contain.

7 Conclusion

The analysis of the coursebooks widely used in university foreign language education in Russia shows that the speaking tasks offered by their designers are not sufficient to foster students' creativity while the online course turned out to lack such tasks at all. Due to the generally shared assumption and research evidence that creative tasks appeal to the majority of learners and guarantee a higher level of foreign language acquisition, university teachers need to incorporate creative speaking tasks into their classroom. As creativity is nurtured by a step-by-step approach starting with guided practice when tasks contain certain constraints and gradually moving on to student free expression of real learner-generated content, a system of such tasks with varying constraints is required to boost student foreign language speaking creativity. It is also important to underline that the full realization of the creative potential of such tasks to a great extent depends on the relationships and atmosphere in the classroom.

References

1. Boden, M.A.: The Creative Mind. Myths and mechanisms, 2nd edn. Routledge, Taylor & France Group, London and New York (2004)
2. Richards, J.C.: Creativity in language teaching. Iran. J. Lang. Teach. Res. **1**(3), 19–43 (2013)
3. Doyle, C.L.: Speaking of creativity: frameworks, models, and meanings. In: Mullen, C.A. (ed.) Creativity Under Duress in Education? CTAE, vol. 3, pp. 41–62. Springer, Cham (2019). https://doi.org/10.1007/978-3-319-90272-2_3
4. Lubart, T., Zenasni, F.: Creative potential and its measurement. Int. J. Talent Dev. Creat. **1**(2), 41–50 (2013)
5. Lubart, T., Barbot, B., Besançon, M.: Creative potential: assessment issues and the EPoC battery. Stud. Psychol. **40**(3), 540–562 (2019). https://doi.org/10.1080/02109395.2019.165 6462
6. Brito, S.M.: Teaching and detecting the creative potential — experience and perspectives. In: Sánchez-García, J.C. (ed.) Entrepreneurship Education and Training, pp. 171–199. IntechOpen, London, Rijeka (2015). https://doi.org/10.5772/58993
7. Barbot, B., Besançon, M., Lubart, T.: Creative potential in educational settings: its nature, measure, and nurture. Int. J. Prim. Elem. Early Years Educ. **43**(4), 371–381 (2015). https://doi.org/10.1080/03004279.2015.1020643
8. Runco, M.A., Jaeger, G.J.: The standard definition of creativity. Creat. Res. J. **24**(1), 92–96 (2012). https://doi.org/10.1080/10400419.2012.650092
9. Sternberg, R.J.: Implicit theories of intelligence, creativity, and wisdom. J. Pers. Soc. Psychol. **49**(3), 607–627 (1985). https://doi.org/10.1037/0022-3514.49.3.607
10. Vincent-Lancrini, S., et al.: Fostering Students' Creativity and Critical Thinking: What it Means in School, Educational Research and Innovation. OECD Publishing, Paris (2019). https://doi.org/10.1787/62212c37-en
11. Ritchhart, R.: Intellectual Character: What It Is, Why It Matters, and How to Get It. Jossey-Bass, San Francisco (2002)
12. Bulatova, A., Zhuravleva, N., Melnikova, S.: Creativity as a way to new literacy realization. KnE Soc. Sci. **4**(13), 29–38. https://doi.org/10.18502/kss.v4i13.7694
13. Runco, M.A., Selcuk, A.: Divergent thinking as an indicator of creative potential. Creat. Res. J. **24**, 66–75 (2012). https://doi.org/10.1080/10400419.2012.652929
14. Suma, Z., Miki, O.: Using simulation to develop divergent and reflective thinking in teacher education. Sustainability **12**(7), 2879 (2020). https://doi.org/10.3390/su12072879
15. Kousoulas, F.: The interplay of creative behavior, divergent thinking, and knowledge base in students' creative expression during learning activity. Creat. Res. J. **22**, 387–396 (2010). https://doi.org/10.1080/10400419.2010.523404
16. Barak, M., Levenberg, A.A.: Model of flexible thinking in contemporary education. Think. Skills Creat. **22**, 74–85 (2016). https://doi.org/10.1016/j.tsc.2016.09.003
17. Gallavan, N.P., Kottler, E.: Advancing social studies learning for the 21 century with divergent thinking. Soc. Stud. **103**, 165–170 (2012). https://doi.org/10.1080/00377996.2011.605641
18. Abd El Fattah Torky, S.: The Effectiveness of a Task-Based Instruction program in Developing the English Language Speaking Skills of Secondary Stage Students. Ain Shams University Women's College, Cairo (2006)
19. Jones, R.H.: Introduction: discourse and creativity. In: Rodney, J. (ed.) Discourse and Creativity, pp. 1–14. Pearson, Harlow (2012)

20. Common European Framework of Reference for Languages: Learning, Teaching, Assessment Companion Volume with New Descriptors; Council of Europe (2018). http://www.coe.int/lang-cefr. Accessed 20 Apr 2020
21. Nunan, D.: Designing Tasks for the Communicative Classroom. Cambridge University Press, Cambridge (1989)
22. Larsen-Freeman, D.: On language learner agency: a complex dynamic systems theory perspective. Mod. Lang. J. **103**(S1), 61–79 (2019). https://doi,org/10.1111/modl.12536
23. Ellis, R.: Task-Based Language Learning and Teaching. Oxford University Press, Oxford (2003)
24. Brown, H.D.: Teaching by Principles: An Interactive Approach to Language Pedagogy, 2nd edn. Addison Wesley Longman, White Plains (2001)
25. Nunan, D.: Task-based language teaching in the Asia context: defining 'task.' Asian EFL J. **8**(3), 12–18 (2006)
26. Read, C.: Seven pillars of creativity in primary ELT Seven pillars of creativity in primary ELT. In: Maley, A., Peachey, N. (eds.) Creativity in the English Language Classroom, pp. 26–37. British Council, London (2015)
27. Liao, Y.-H., Chen, Y.-L., Chen, H.-C., Chang, Y.-L.: Infusing creative pedagogy into an English as a foreign language classroom: learning performance, creativity, and motivation. Think. Skills Creat. **29**, 213–223 (2018). https://doi.org/10.1016/j.tsc.2018.07.007
28. Murthaugh, S.M.: Creativity based Instruction for L2 Speaking. Apheit J. **6**(1), 68–82 (2017). http://www.journals.apheit.org/jounal/Inter-vol6-1/p68-82-SamaraMarie-Murtaugh.pdf
29. Robinson, P., Gilabert, R.: Task-based learning: cognitive underpinnings. In: Chapelle, C. (ed.) The Concise Encyclopedia of Applied Linguistics. Blackwell, Oxford (2019)
30. Becker, C., Roos, J.: (2016) An approach to creative speaking activities in the young learners' classroom. Educ. Inq. **7**(1), 27613 (2016). https://doi.org/10.3402/edui.v7.27613
31. Fredricks, J.A., Blumenfeld, P.C., Paris, A.H.: School engagement: potential of the concept, state of the evidence. Rev. Educ. Res. **74**(1), 59–109 (2004). https://doi.org/10.3102/003465 43074001059
32. Curtis, H., Werth, L.: Fostering student success and engagement in a K-12 online school. J. Online Learn. Res. **1**(2), 163–190 (2015)
33. Philip, J., Duchesne, S.: Exploring Engagement in Tasks in the Language Classroom. Ann. Rev. Appl. Linguist. **36**, 50–72 (2016). https://doi.org/10.1017/S0267190515000094
34. Lambert, C., Zhang, G.: Engagement in the use of English and Chinese as foreign languages: the role of learner-generated content in instructional task design. Mod. Lang. J. **103**(2), 391–421 (2019). https://doi.org/10.1111/modl.12560
35. Aguilar, M., Sheldon, K., Ahrens, R., Janowicz, P.: State of Engagement. Report 2020. GoGuardian, Los Angeles (2020)
36. Egbert, J.: A study of flow theory in the foreign language classroom. Mod. Lang. J. **87**(4), 499–518 (2003). https://doi.org/10.1111/1540-4781.00204
37. Bygate, M., Samuda, V.: Creating pressure in task pedagogy: the joint roles of field, purpose and engagement within the interaction approach. In: Mackey, A, Charlene, P. (eds.) Multiple Perspectives on Interaction: Second Language Research in Honor of Susan M. Gass, pp. 90–116. Routledge, New York & London (2009)
38. Aubrey, S.: Measuring flow in the EFL classroom: learners' perceptions of inter- and intra-cultural task-based interactions. TESOL Q. **51**(3), 661–692 (2017). https://doi.org/10.1002/tesq.387
39. Phung, L.: Task preference, affective response, and engagement in L2 use in a US university context. Lang. Teach. Res. **21**(6), 751–766 (2017). https://doi.org/10.1177/136216881668 3561

40. Criado Sánchez, R.: The "Communicative Processes-based model of Activity Sequencing" (CPM): a Cognitively and Pedagogically Sound Alternative to the "Representation-Practice-Production Model of Activity sequencing" (P-P-P) in ELT. Odisea **10**, 33–56 (2009). https://doi.org/10.25115/odisea.v0i10

41. Robinson, P.: Task complexity, theory of mind, and intentional reasoning: effects on L2 speech production, interaction, uptake and perceptions of task difficulty. Int. Rev. Appl. Linguist. **45**(3), 193–213 (2007)

42. Michel, M.C.: Effects of task complexity and interaction on L2-performance. In: Robinson, P. (ed.) Second Language Task Complexity: Researching the Cognition Hypothesis in Language Learning and Performance, pp. 141–174. John Benjamins Publishing Company, Amsterdam/Philadelphia (2011). https://doi.org/10.1075/tblt.2.12ch6

43. Gómez-Rodríguez, L.F.: English textbooks for teaching and learning English as a foreign language: do they really help to develop communicative competence? Educ. Educ. **13**(3) (2010). https://doi.org/10.5294/EDU.2010.13.3.1

44. Almazova, N., Krylova, E., Rubtsova, A., Odinokaya, M.: Challenges and opportunities for Russian higher education amid COVID-19: teachers' perspective. Educ. Sci. **10**, 368 (2020). https://doi.org/10.3390/educsci10120368

45. Almazova, N., Barinova, D., Ipatov, O.: Forming of information culture with tools of electronic didactic materials. In: Katalinic, B. (ed.) Annals of DAAAM and Proceedings of the International DAAAM Symposium, vol. 29, issue 1, pp. 0587–0593. Danube Adria Association for Automation and Manufacturing, DAAAM, Zadar, Croatia (2018). https://doi.org/10.2507/29th.daaam.proceedings.08

46. Bylieva, D., Lobatyuk, V., Safonova, A., Rubtsova, A.: Correlation between the practical aspect of the course and the E-learning progress. Educ. Sci. **9**, 167 (2019). https://doi.org/10.3390/educsci9030167

47. Shemanaeva, M.: Individual learning path as synergy of synchronous and asynchronous learning. Lang. Cult. **39** (2017). https://doi.org/10.17223/19996195/39/20

Using a Creative Approach to Teach Russian as a Foreign Language to International Students Majoring in "Music Education"

Inna V. Sheglova⬥, Anna V. Rubtsova⬥, and Elena A. Krylova⁽✉⁾⬥

Peter the Great St. Petersburg Polytechnic University, St. Petersburg 19525, Russian Federation

Abstract. The study proves that the optimization of the educational process relies on the key characteristics of the creative approach taking into consideration the creative nature of the tasks assigned and the future professional activity of students. A set of exercises aimed at the formation and development of professional knowledge among international students majoring in "Music Education" is presented in the paper. The educational materials are presented step by step: from Elementary level to Certification level I (A1-B1). The tasks are getting more and more complicated with students' acquiring new vocabulary and grammar. The nature of the tasks also undergoes changes: from receptive to productive. The students were offered to keep a thematic glossary "The Russian language and my profession" in two formats: in electronic and handwritten versions. The electronic format allows one to display photographs, save links to sources and, what seems to be important, work with a Russian keyboard. Handwritten format, in turn, involves improving writing skills. Based on the analysis of the students' academic results, the influence of the activity approach on the productive skills and professionally oriented knowledge of international students is studied. The proposed approach allows to solve the urgent pedagogical problem of the formation and development of a student as an active subject of the educational process, capable of self-organization, self-improvement, and self-realization in the future profession.

Keywords: Russian as a foreign language · Creative approach · Professionally oriented teaching · Productive skills · International students

1 Introduction

The creative approach presupposes such an organization of the educational process when the main role is assigned to the active, maximally independent cognitive activity of students organized with the help of creative tasks. We believe that this approach facilitates the learning process, encourages the students to be more active and independent, and motivates them for cognitive activity that results in positive learning outcomes and the formation of the required competencies. In addition, it fits into the methodological doctrine of modern pedagogy that is oriented on the formation of a student as an active subject of the educational environment [1]. Many types of creative tasks are used in Russian as a foreign language educational process. They include essays, presentations, reports, educational games, reading fiction, etc. When completing creative tasks, the attention of students switches from the plane of expression to the plane of content. It is very important that a foreign language from the object of study turns into a way of transmitting information; it helps to express ideas, feelings, intentions and goals. What is more, an important factor contributing to the effectiveness of creative tasks is the emotional involvement of the participants. Thus, creative tasks create unique conditions for meaningful communication and mastering a professionally oriented foreign language.

2 Literature Review

The methodological basis of the study goes in line with the following scientific approaches: co-creation of learning [2], critical thinking [3], competitive learning [4], mobile learning [5], situational interdisciplinarity [6].

Many scholars currently recognize professionally oriented language teaching as high-priority in education. I.V. Aleshanova and N.A. Frolov give the following definition of "professionally oriented approach". It is "a system of didactic means of organizing the educational process in a foreign language, including changing the goal, content, process and form and orienting a foreign language course towards the profession acquired by students and possible areas of its real use in professional activity" [7]. According to A.P. Belyaeva, "professional bias serves as a starting point for the construction of any pedagogical systems, integration and differentiation in higher education" [8].

Professional orientation helps students to understand the necessity of foreign language knowledge for their future profession; it motivates them to acquire foreign language skills to master their profession successfully [9].

Thus, a professionally oriented approach is viewed as a methodological system that determines the effectiveness of teaching a foreign language and assumes an active role of the student. The need for active, motivated and independent students determines the relevance and importance of the professionally oriented creative approach. Different terms can be found in scientific papers: communicative-activity approach [10], active learning [11], problem solving and creative activity [12], a systemic activity based approach [13].

The key characteristics of the professionally oriented creative approach include the following [14]:

- Knowledge is not self-sufficient, it is just a means of performing actions and teaching them. Knowledge acquisition is the first step in explaining and preparing subsequent practical actions;
- Creative educational activities offered to the students simulate future professional activities;
- It is not enough to memorize educational content; one can acquire knowledge only applying it. The best way to apply knowledge is to use creative tasks;
- A teacher should not transfer the knowledge; a teacher's role is to design, organize and manage educational activities;
- Education objectives should be in line with the nature of students' future professional activities.

- In order to develop the students' creative abilities, to activate their need for creativity, a teacher needs to create such educational conditions in which "creation" appears as the need to solve the problem.
- Creative activities are such tasks that require work of both logical and creative types of thinking, and are based on analysis, synthesis, and other cognitive strategies.

It is important that students, acquiring new knowledge and methods of obtaining it during the period of study, understand that only knowledge obtained as a result of their own cognitive activity can be internalized [15, 16]. Thus, self-learning and self-development are of vital importance.

From the whole variety of definitions of the concept of "independent study", we would like to highlight the following [17]:

- Purposeful, internally motivated and self-corrected activity of the individual;
- Educational materials are aimed at applying experience and knowledge to solve new problems and are creative by nature.

Summarizing the above-mentioned we would like to emphasize that "a good teacher differs from an ineffective teacher in that he creates stimulating conditions for the transition of learning to self-education, education - into self-education, development - into creative self-development of the individual" [18].

Thus, we would like to conclude that modeling a student in the context of his future professional activity is an integral part of the planning and implementation of the educational process.

3   Methodology

The creative approach led to the actualization of the problem-solving method. In this regard, new knowledge is represented as a problem that requires the students to solve it independently and creatively. Identifying the educational content, we also relied on the system-structural, structural-functional approaches and emotional-value analysis. The systemic-structural and structural-functional approaches imply the consideration of the content of education as a system of interrelated components. At the same time, "each unit within the system must have the status of a functional one, that is, contribute to the achievement of the goals of this system, helping to optimize it: to make it more stable, to debug the mechanisms of interaction with other systems, to regulate the connections of individual parts within the system" [19, 20]. The emotional-value analysis, in turn, implies the creation of conditions for the creative development of students.

Testing and implementation of the developed learning materials for intensification Russian as a foreign language educational process included several stages of experimental research.

At the first stage of the study, an analysis of the pedagogical context was carried out – we assessed the general level of Russian language proficiency of students using a set of linguistic tests and oral placement tests.

Developing the pretest, we used the testing materials from the State Educational Standard in Russian as a foreign language, the first certification level. We chose these materials because they allow to test almost all the components of a foreign language communicative competence. These tests are developed by Russian native speakers, have been checked for validity and, without a doubt, are reliable. The pretest consisted of the following blocks: reading, writing, listening, grammar and vocabulary, speaking. The highest score for the pretest was 125 points. The following levels of the students' Russian as a foreign language proficiency levels were identified: intermediate level - 125–100 points; basic level - 99–80 points; elementary level - less than 79 points.

At the second stage, a set of the developed learning materials based on the creativeapproach was introduced into Russian as a foreign language educational process.

The final stage of the study dealt with the obtained results analysis and reflection.

4 The Solution to the Problem

Let us illustrate the stages of the work of a teacher of Russian as a foreign language in the context of the creative approach in a group of students majoring in "Music Education". A system of creative exercises was implemented during the experimental training, including the following types:

- vocabulary exercises;
- grammar exercises;
- guided problem solving tasks;
- communicative speaking tasks;
- a set of exercises to develop conscious reading and creative writing.

The educational materials were presented step by step: from elementary level to certification level I (A1-B1). The tasks were getting more and more complicated with students' acquiring new vocabulary and grammar. The nature of the tasks also underwent changes - from receptive to productive.

The tasks for Elementary and Basic level can be characterized as quasi-creative, which can be explained by the lack of the students' vocabulary and grammar knowledge. These tasks had several objectives including the necessary vocabulary and grammar structures introduction and facilitation of the students' engagement into the educational process. To achieve these aims we used the Internet resources and information communication technologies along with traditional paper based exercises.

Examples of the educational materials for Elementary and Basic levels are presented below.

- *Use the example above to make small texts about Russian composers. Learn the texts by heart.*

  – Tchaikovsky.
  – Who is this?
  – This is a composer.

- *Russian composers. Write their full names, learn by heart.*

Peter Ilyich Tchaikovsky.
Peter is a name.
Ilyich is a patronymic.
Tchaikovsky is a surname.

- *Get acquainted with the works of Russian composers/listen to their works, describe them (choose a number of adjectives)/Use the example above to make small texts.*

Pyotr Ilyich Tchaikovsky is a famous Russian composer. The Seasons is his work. "January" - bewitching, intriguing, cold, sparkling music/adjectives are taken from students' works/(Fig. 1).

Fig. 1. An example of a creative vocabulary task

- *Open the link* https://wordwall.net/resource/13142455/untitled44. *Match the words and their definitions* (Fig. 2).

We would like to mention that such interactive tasks (www.wordwall.net, www.learningapps.org, etc.) were used in different situations during the lesson and as a part of the students' hometask: to drill vocabulary and grammar, to introduce new lexical units, to check the understanding of the text for reading/listening, etc.

With the necessary vocabulary and grammar structures acquisition and the enhancement of the students' language proficiency level, the educational materials got more complicated and creative by nature. One of the students' favourite tasks was to create interactive images about famous Russian composers and to present their projects in class. The images were created with the help of the Internet resource https://app.genial.ly/ (Fig. 3). Such tasks gave an opportunity to practice vocabulary and grammar skills and what is more important developed the students' cognitive and creative skills. The reports presented by the students were a subject for peer review and correction enhancing the engagement of all the students into the educational process.

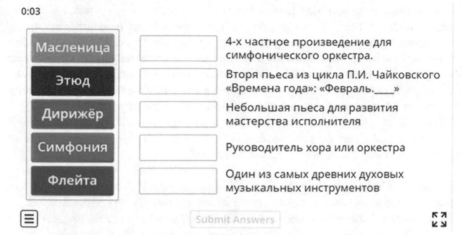

Fig. 2. Vocabulary exercise, basic level

Fig. 3. The example of an interactive image

We would like to give one more example of a creative writing task used at the Intermediate level. The student's creative written speech, being motivated, purposeful and structured is a productive process relying on creative thinking and imagination. Creative written assignments are such tasks that are aimed not only at the development of receptive (vocabulary, grammar) and productive (writing, speaking) skills, but also at the implementation of a personality oriented approach, which, in turn, involves the development of students' independence, motivation and creativity [21, 22].

Study the statements about P.I. Tchaikovsky and his work. Write two essays (20–25 sentences): "I share N's position and believe..." and "I do not share position N, because..."/Choose 2 statements and analyze them/.

- *Our love for Tchaikovsky, for his beautiful music is passed from century to century, from generation to generation, and this is its immortality (Dmitry Shostakovich).*
- *The name of Tchaikovsky became an era in the world development of music, a valuable Russian contribution to it, one of the manifestations of the inexhaustible artistic talent of the Russians with their most humane culture (Boris Asafiev).*
- *Tchaikovsky's First Piano Concerto, like the first pancake, came out lumpy (Nikolai Soloviev).*
- *"Eugene Onegin" is a stupid opera hastily written by Tchaikovsky... everything in it is insulting for a Pushkin masterpiece (Vladimir Nabokov).*
- *The Fourth Symphony is the last word of art, there is no way beyond that, this is the limit of genius, this is the crown of triumph, this is the point of deity, you can give your soul for it, lose your mind, and nothing will be regretted... (Nadezhda Von Meck).*
- *Wonderful music by Tchaikovsky, in which you hear everything, absolutely everything! Peace, and alertness, and thrill of excitement, and grief, and despair, and love, and hope, and ecstasy... (Maya Plisetskaya).*

These examples are a small part of the developed system of creative tasks but they can clearly show that the educational process became more engaging giving the students an opportunity to practice Russian as a foreign language in meaningful context. Besides the above-mentioned tasks, the students were offered to keep a thematic glossary "The Russian language and my profession" in two formats: in electronic and handwritten versions. The electronic format allows you to display photographs, save links to sources and, what seems to be important, work with a Russian keyboard. Handwritten format, in turn, involves improving writing skills.

5   Research Results and Discussion

The implementation of the above-described set of tasks based on creative and professionally oriented approaches made it possible to intensify and activate the educational process, which was proven by the analysis of the results of the experimental work.

Two students' groups majoring in "Music Education" took part in the experimental education: PO-1 (experimental group) and PO-2 (control group), 12 people in each. We conducted linguistic tests and oral placement tests that revealed the students' level of Russian, including reading, listening, writing and speaking skills. Diagnostic tests revealed identical results of the formation of skills in writing, reading, speaking, listening (Table 1).

Table 1. The pretest results

Russian as a foreign language proficiency level					
	Number of students	Elementary level (>70 points)	Basic level (99–80 points)	Intermediate level (125–100 points)	Average group grade (1–5) for the course
Experimental group (PO-1)	12	4	6	2	3
Control group (PO-2)	12	3	6	3	3

However, further on, each group had its own educational trajectory. For the experimental group (PO-1), the educational process was organized on the basis of the described creative approach and the developed set of tasks, the control group experienced traditional lessons organization with no systematic professionally-oriented and problem-solving tasks (in the second term they studied a new discipline "Professionally oriented Russian as a foreign language").

Having finished the experimental education, we conducted the posttest at the end of the term, including the same aspects for assessment (Table 2).

Table 2. The posttest results

Russian as a foreign language proficiency level					
	Number of students	Elementary level (>70 points)	Basic level (99–80 points)	Intermediate level (125–100 points)	Average group grade (1–5) for the course
Experimental group (PO-1)	12	-	3	9	4,8
Control group (PO-2)	12	-	6	6	3,5

The analysis of the factual material (the students' academic results in the discipline "Russian as a foreign language: general proficiency") also found that students of PO-1 group demonstrate better results concerning the formation of foreign language productive skills than PO-2 group. Figure 4 shows the group's average grade, which was gradually improved in experimental (PO-1) group as the result of efficient educational content and approaches used in the classroom.

In the second term, the students of control and experimental groups studied a new discipline "Professionally-oriented Russian as a foreign language course." In terms of

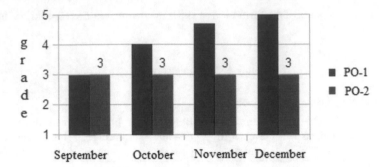

Fig. 4. Students' academic results in the discipline "Russian as a foreign language: general proficiency"

this study it is worth mentioning that the experimental group students showed better results than the students of the control group (Fig. 5). It can be explained by the following: intensive professionally oriented training during the first term in experimental group made it easy for the students to acquire a new discipline.

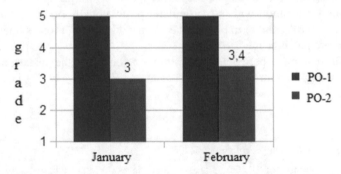

Fig. 5. Students' academic results in the discipline "Professionally-oriented Russian as a foreign language course"

Thus, the study of the influence of the creative approach on the foreign language productive skills and professional knowledge revealed a direct relationship between the organization of classes on the basis of creative approach and the indicators of the formation of relevant knowledge and skills. The positive results of mastering the discipline "Professionally-oriented Russian as a foreign language course" in experimental group are determined by the appropriate training - professionally oriented vocabulary, thematic search work, problem solving tasks, etc.

6   Conclusion

This study aimed to check the impact of the creative approach and the proposed set of tasks on developing international students' productive skills and professionally oriented knowledge. Organizing Russian as a foreign language lessons we relied on using problem-solving tasks, creating communicative situations for professionally oriented vocabulary extensive use thus increasing the levels of international students' involvement levels. International students were motivated to do the proposed tasks as they were in line with their professional interests. Post-experimental results (the formation of the required competencies at the required level), the students' active position in relation to their future professional activities demonstrate the effectiveness of the approach used to organize the educational process in Russian as a foreign language.

References

1. Garanina, R.: Technology of pedagogical assistance to the formation of the student as a subject of the educational space. Eur. J. Contemp. Educ. **9**(4), 751–772 (2020). https://doi. org/10.13187/ejced.2020.4.751
2. Kaminskienė, L., Žydžiūnaitė, V., Jurgilė, V., Ponomarenko, T.: Co-creation of learning: a concept analysis. Eur. J. Contemp. Educ. **9**(2), 337–349 (2020). https://doi.org/10.13187/ ejced.2020.2.337
3. Bećirović, S., Hodžić, F., Brdarević-Čeljo, A.: The problems of contemporary education: critical thinking development in the Milieu of high school education. Eur. J. Contemp. Educ. **8**(3), 469–482 (2019). https://doi.org/10.13187/ejced.2019.3.469
4. Coronado, S., Sandoval-Bravo, S., Celso-Arellano, P.L., Torres-Mata, A.: Competitive learning using a three-parameter logistic model. Eur. J. Contemp. Educ. **7**(3), 448–457 (2018). https://doi.org/10.13187/ejced.2018.3.448
5. Belle, L.J.: An evaluation of a key innovation: mobile learning. Acad. J. Interdiscip. Stud. **8**(2), 39–45 (2018). https://doi.org/10.2478/ajis-2019-0014
6. Reckinger, R., Wille, C.: Situative interdisciplinarity: empirical reflections on ten years of cross-disciplinary research. Acad. J. Interdiscip. Stud. **7**(3), 9–34 (2018). https://doi.org/10. 2478/ajis-2018-0055
7. Aleshanova, I.V., Frolova, N.A.: Professional-oriented approach in teaching foreign languages in a technical university. Mod. Probl. Sci. Educ. **6**, 68–74 (2012)
8. Belyaeva, A.P.: Integrative-modular pedagogical system of vocational education. SPb-Radom: IPTO RAO, Saint-Petersburg (1997)
9. Pyanzina, I.V.: Model of management of educational activity of students in the process of teaching a foreign language. Bull. Tomsk State Univ. **392**, 183–190 (2015). https://doi.org/ 10.17223/15617793/392/31
10. Deryabina, S., Dyakova, T.: Formation of foreign communicative competence in the context of virtual education. In: Joint Conferences: PCSF 2020 and CSIS 2020, European Proceedings of Social and Behavioural Sciences, vol. 98, pp. 186–193. European Publisher, London (2020). https://doi.org/10.15405/epsbs.2020.12.03.19
11. Dashkina, A.I.: Improving foreign language proficiency and cultural competence by involvement in developing assignments. In: 18th Professional Culture of the Specialist of the Future, PCSF 2018, European Proceedings of Social and Behavioural Sciences, vol. 51, pp. 861–873. Future Academy, London (2018). https://doi.org/10.15405/epsbs.2018.12.02.93
12. Gulk, E.B., Tabolina, A.V.: Gaming technologies as a mean of development of motivation of students. In: 18th Professional Culture of the Specialist of the Future, PCSF 2018, European Proceedings of Social and Behavioural Sciences, vol. 51, pp. 1672–1678. Future Academy (2018). https://doi.org/10.15405/epsbs.2018.12.02.179
13. Von Brevern, H., Synytsya, K.: Systemic-structural theory of activity: a model for holistic learning technology systems. Educ. Technol. Soc. **9**(3), 100–111 (2006). https://doi.org/10. 1109/ICALT.2005.250

14. Atanov, G.: Where to start the implementation of the activity approach in teaching. Educ. Technol. Soc. **2**(7), 179–184 (2004)
15. Almazova, N., Eremin, Y., Kats, N., Rubtsova, A.: Integrative multifunctional model of bilingual teacher education. In: IOP Conference Series: Materials Science and Engineering, International Scientific Conference "Digital Transformation on Manufacturing, Infrastructure and Service", vol. 940, p. 012134. IOP Publishing Ltd., St. Petersburg (2019). https://doi.org/10.1088/1757-899X/940/1/012134
16. Bylieva, D., Bekirogullari, Z., Lobatyuk, V., Nam, T.: How virtual personal assistants influence children's communication. In: Bylieva, D., Nordmann, A., Shipunova, O., Volkova, V. (eds.) PCSF/CSIS -2020. LNNS, vol. 184, pp. 112–124. Springer, Cham (2021). https://doi.org/10.1007/978-3-030-65857-1_12
17. Almazova, N., Rubtsova, A., Krylova, E., Almazova-Ilyina, A.: Blended learning as the basis for software design. In: Katalinic, B. (ed.) Annals of DAAAM and Proceedings of the International DAAAM Symposium, vol. 30, pp. 1726–9679. DAAAM International, Vienna, Austria (2019). https://doi.org/10.2507/30th.daaam.proceedings.112
18. Kurzyakova, A.A.: The value of self-realization in the process of becoming a student at a university. Bull. Tomsk State Univ. **396**, 211–218 (2015). https://doi.org/10.17223/15617793/396/37
19. Bylieva, D., Lobatyuk, V., Kuznetsov, D., Anosova, N.: How human communication influences virtual personal assistants. In: Bylieva, D., Nordmann, A., Shipunova, O., Volkova, V. (eds.) PCSF/CSIS -2020. LNNS, vol. 184, pp. 98–111. Springer, Cham (2021). https://doi.org/10.1007/978-3-030-65857-1_11
20. Kuklina, S.S., Tatarinova, M.N.: System-structural analysis of the component composition of the content of foreign language education. Bull. Tomsk State Univ. **460**, 190–201 (2020). https://doi.org/10.17223/15617793/460/23
21. Novikova, I.A., Berisha, N.S., Novikov, A.L., Shlyakhta, D.A.: Creativity and personality traits as foreign language acquisition predictors in university linguistics students. Behav. Sci. **10**, 35 (2020). https://doi.org/10.3390/bs10010035
22. Du, K., Wang, Y., Ma, X., Luo, Z., Wang, L., Shi, B.: Achievement goals and creativity: the mediating role of creative self-efficacy. Educ. Psychol. **40**(10), 1249–1269 (2020). https://doi.org/10.1080/01443410.2020.1806210

The Use of Literary Works for Stimulating Students' Creativity

Natalia A. Katalkina[1] (iD), Nadezhda V. Bogdanova[1](✉) (iD),
and Galina I. Pankrateva[2,3] (iD)

[1] Peter the Great St. Petersburg Polytechnic University, Saint Petersburg, Politekhnicheskaya ul., 29, St. Petersburg 195251, Russia
[2] Emperor Alexander I St. Petersburg State Transport University, Moskovsky pr., 9, St. Petersburg 190031, Russia
[3] Herzen State Pedagogical University of Russia, Naberezhnaya r. Moyki, 48, St. Petersburg 191186, Russia

Abstract. One of the main tasks of the modern higher education system is to develop students' ability to find ways for solving different tasks creatively and independently. Creativity is considered in the article as a training ability inherent in any individual, however, requiring an appropriate social environment for its development, including the original material for further creative processing, and the audience for perceiving and evaluating the product of creativity. The purpose of the study is to explore the possible ways of stimulating students' creativity based on the case study project method. Conducting projects include several stages: reflection and creative processing of foreign language literary works, as well as subjective evaluation of the projects' effectiveness by students. The authors suggest the project work for developing creative products through cooperative learning. The case study concludes the implementation of such projects is consistent with the opportunities and interests of students of higher education and allows activating and improving a whole complex of communicative skills for production and comprehension activities, as well as the so-called soft skills. Creative projects based on the use of literary works can be applied at the early stage of foreign language learning in order to develop creative abilities reflecting students' individual interests and skills.

Keywords: Foreign language learning · Creativity · Cooperative learning · Literature works · Case study · Project method

1 Introduction

The development of the modern information society in recent decades has led to a change in the requirements for specialists in almost all professional areas. At present, employers appreciate not only and not so many employees who have the knowledge, but, first of all, those who have the ability to solve problems. Accordingly, one of the main tasks of higher education is to develop the ability of students to find ways to solve their tasks creatively and independently [1]. It is possible to develop this ability through purposeful

D. Bylieva and A. Nordmann (Eds.): PCSF 2021, LNNS 345, pp. 937–947, 2022.
https://doi.org/10.1007/978-3-030-89708-6_75

activity on creative tasks and projects during studying process. The authors focus on the case study method, which combines project method, case role-playing game method, game design method, discussion method and others.

Many researchers note the positive impact of a creativity-based approach to learning on effect of any foreign language course [2–4] and emphasize the role of inspiration and pedagogical diversity in language learning [5]. The researchers also discussed the relation of creativity to academic achievement both in general education and in language learning [6]. At the same time, creativity is viewed not only as "a natural human tendency", but also as an ability that "can be shaped, visualized, emphasized, trained, and used to achieve specific (earlier-planned) actions" [2, p. 155]. Thus, creativity is a dynamic feature, according to which an individual at each particular moment takes a certain position on the creativity scale from Big 'C', assuming unique "world changing ideas, artistic creations and dynamic inventions of an elite", to mini 'c', which "relates to the more prosaic discoveries and explorations in the world of music, art, industry, technology and all forms of discourse" [7].

An incentive to move up on the creativity scale in addition to personal characteristics, can be environmental impact. It is not accidental that development of creative ideas is considered as part of a broader, socially determined process, composed of individuals, knowledge domains and a field of informed experts [8]. According to this model, the creative personality can be considered as an element of the chain, transforming and processing the experience of previous generations and passing it to subsequent links of the chain. Accordingly, creativity in general and linguistic creativity in particular arise in society, which is confirmed by studies that establish that cooperative learning was the best predictor of creativity and proved to have a significantly positive effect on foreign language learners' motivation, reduction of language anxiety and increase of the confidence level in using the language [9–15, 23].

Interestingly, the components of creative activity are identical to three basic elements of education: the introduction (input), the self-made explanation of the known material segment (intake) and creating a specific material or non-material product on this basis (output) [2]. It can also be assumed that the quality of a product arising from both learning and creativity in most cases depends on the quality of the initial knowledge, which serves as a starting point for creativity and the characteristics of the creative environment perceiving the product.

As part of this study, conducted at a German language lesson, an attempt was made to take advantage of the close relationship between creativity and learning in order to develop them simultaneously. As a qualitative input, authentic German-language literary works were chosen on the basis of which students were invited to create a new creative product.

The purpose of the study is to explore the possible ways of conducting projects based on the reflection and creative processing of foreign language literary works at the early stage of a foreign language learning, as well as subjective evaluation by students of higher educational institutions of such projects' effectiveness in a foreign language classroom.

2 Methodology

2.1 Participants

The research was conducted as part of German language classes at Peter the Great St. Petersburg Polytechnic University among 25 undergraduate bachelor students, including 15 female (60%) and 10 male (40%) ones, aged 19–20. The participants study international relations, specializing in the Germany region, studying German as a second foreign language, with the proficiency level of the language - A2.

2.2 Procedures

As part of the study, a preliminary survey was conducted, the purpose of which was to find out the interest of students in the development of creative abilities during the educational process, as well as their preferences regarding specific types of creativity. Case technologies are one of the mechanisms that let teachers involve communicative and creative abilities of students studying a foreign language. Following the results of the survey, a German language lesson was planned and conducted using the project method, based on a creative interpretation of the stories by a modern German writer Leonhard Thoma, previously read by students. The purpose of the project was to develop creative using of the German language as well as creative thinking.

This research was conducted to find out answers to the following questions:

Q1: Do creative projects based on foreign language literary works meet the interests and needs of students?
Q2: What skills do students learn through creative projects?

Thus, the following cases were set:

Case 1: What traditional and creative forms of presentation can show that the text has been read and new words and expressions have been mastered?
Case 2: Offer your own solution to the reflection of the read text.

In our research the following stages of the training process were provided:

1. *Immersion in cooperative learning:*
    The main task of this stage was the formation of motivation for cooperative learning, the manifestation of initiatives of the project participants. The text of the case was sent for self-study and preparation of answers to the questions. At the beginning of the class, students were tested for their knowledge of the content of the case material "Das Idealpaar" by Leonhard Thoma, and their interest in the discussion.
2. *Organization of cooperative learning:*
    Working groups were formed in accordance with the choice of creative performance: theater play, illustrations, musical, poem. In each small group there was a refinement, the development of a single position, which was drawn up for the presentation. In each group, a "speaker" or "actors" were selected.

3. *Analysis and reflection of cooperative learning:*
    The main task of this stage was to show educational results of working with the case. The performance had to contain the analysis of the situation; the content of the solution, the presentation techniques, as well as the efficiency of using technical means were evaluated. In addition, at this stage, the effectiveness of the training process was analyzed, problems of organizing cooperative learning were identified and tasks for further work were set.

After the completion of the project, students filled in questionnaires, the analysis of which made it possible to assess students' motivation for creative project activities, problems arising during the course, as well as the subjective evaluation of students the sub-skills and skills developed during the implementation of creative projects. The questionnaires for the preliminary and final survey included both closed and open-ended questions.

3   Results and Findings

The results of the preliminary survey showed that 91% of students believe that the creative potential of an individual should be developed in the classroom. At the same time, only 1% of respondents believe that the educational process allows to reveal its creative potential to the full extent, whereas 16.7% of respondents suggest that their creative abilities are not developed at all while studying at the university (Fig. 1).

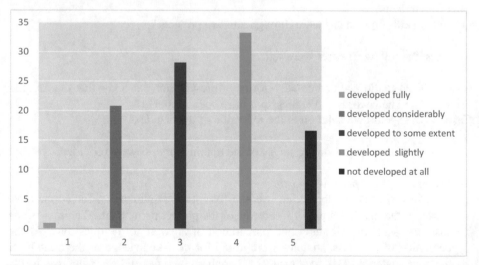

Fig. 1. Descriptive statistics of students' evaluation of their creative abilities' development during the educational process at the university.

In addition, the survey of students revealed the types of creativity that arouse the greatest interest among students. It was found that the most popular among students are cinema and theater (79.2%), music (70.8%) and literature (70.8%) (Fig. 2).

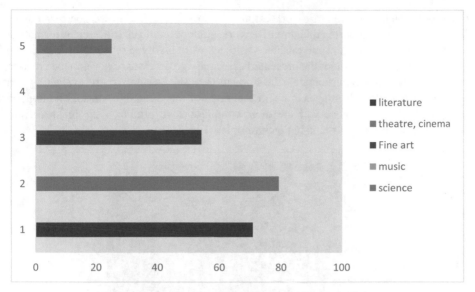

Fig. 2. Students' preferences regarding types of creativity.

Thus, we came to the conclusion that there is a need to reveal and improve the creative potential of students in the educational process, whereas films, theatre plays, musicals and literary works should be preferably used as the input. Assuming that the experiment was conducted with students who speak German at the A2 level, we chose small-format authentic literary works of a modern German-speaking author as the input, since the format of the completed story allows us to include authentic examples of the language in the wider cultural and social context and thus make them more meaningful and easier to understand and to learn [5, 16], and the written form provides a nonlinear perception of the work, with the possibility of repetitions and returns to incomprehensible or most important segments of the text.

After reading the book "Das Idealpaar" by Leonhard Thoma, students were invited to divide into groups according to their individual preferences, choose the story they liked most of all and develop some creative product from its plot. Further, the rules for working on the project were agreed. In particular, students worked in teams discussing everything in German, distributed certain roles among team members, for example, a moderator, a controller, a proofreader, a designer, etc. In case of problems or questions during work on the project, students had the opportunity to contact the teacher at any time. Students could choose the form of the final creative product from the proposed options, such as a short video film, theatre play, song, poem, comic book, television show or choose their own form. The students were given 60 min for discussion and production of the creative product. As a result of the work, 3 theatre plays, 1 musical film, 1 poem and 1 comic book were created and presented.

All works were a free interpretation of the literature work read, including the creation of an audio-visual series and the adaptation of the language material. It is important to note that the most common technique was dramatization, applied in 4 works. For creating new products based on literature works, students chose music, fine art and poetry. Another interesting fact is that during their creative work, students could use different information technologies, although, despite this, technologies were involved only in one work.

Since the primary task of the presented case study project was to develop the creative potential of the project participants, the students themselves were experts evaluating and commenting on the advantages and shortcomings of the created products. The highest rating among the audience was the most comprehensive product - the musical film, combining drama and music using technologies (Fig. 3).

Fig. 3. A scene from a musical film, produced by the students.

Another example of the students' creative works is the following poem:

Das Gedicht «Mozart, sonntags, gratis»

Er war allein bei diesem erstaunlichen Konzert,
Er war allein in einer riesigen Halle,
Er liebt Musik und Mozart und was noch
2 Karten und er ist alleine auf der Straße.
Er wollte dich zu einem Date einladen,
Er rief Lorena an, aber sie sagte
Vielen Dank für deinen Anruf
Schön, von dir zu hören.
Noch drei Tage. Also gut, keine Experimente mehr,
Sonja. Workaholic und gute Kollegin
Sie würde gerne in den Musik-Palast gehen
Aber sie hat einen Plan und er ist ihr wichtig,
Sie wird das ganze Wochenende beschäftigt sein.
Michael, Paula würde gerne gehen
Aber das tolle Wetter hat sie daran gehindert.
Eine einzige Katastrophe
Viele Touristen und sehr unpraktisch,
Er sieht Lorena und versteht nicht
Warum schwieg sie davon?

Er öffnete die Augen und sah die Musiker
Und genoss die Atmosphäre,
Er versprach sich, das nächste Mal,
Morgen kauft er nur eine Karte.

In this poem, the students give the summary of the read story. At the same time, the authors of the poem represent the skill to highlight the main idea of the read text and render the plot of the authentic literature work with comprehensible linguistic means. As an example of the creative interpretation of the literature work' plot with a drawing tool we can present a comic drawn by students. (Fig. 4). The comic demonstrates the narrative thread and presents the key phrases. The ratio of the information compression is higher than in the above-mentioned musical film and poem.

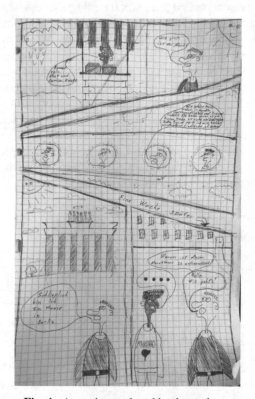

Fig. 4. A comic, produced by the students.

The results of the work on the case study project were analyzed based on the students' answers to the questionnaires. The survey showed that 87.5% of project participants would like to work on designing creative products in the future during studying process. When evaluating the work on the project, 50% of respondents rated the highest score – 5; 31.3% chose 4 and 18.8% - 3.

The first open-ended question offered to choose specific aspects the students liked in the case study project: 48% of respondents indicated the possibility of interaction in the group, 28% –creative process, 12% - possibility of self-expression, 8% - opportunity to discuss, develop and implement their own ideas and 4% - opportunity to play a role, the ability to speak German in everyday situations, as well as the ability to guess which of the stories read is going to be presented by another group.

The second open-ended question specified aspects the students did not like in the project: 80% of respondents did not see any shortcomings, expressing full satisfaction with the project work, 12% noted the format of the work was not suitable for everyone, since not everyone knew, for example, how to paint or get involved in a role, 4% complained about receiving insufficient new information and another 4% about lack of time and details.

The third question discussed the difficulties that arose during the project: 60% of respondents indicated no difficulties, 20% identified the need to speak German throughout the work on the project challenging, some participants also named such problems as difficulty in finding original ideas, difficulties in structuring the final product, lack of time and difficulty in improvising, playing a role, as well as organizing work in a group. When asked to indicate what kind of assistance they needed from the teacher, 82% of students replied they did not need any; 8% of respondents would prefer more detailed explanation about the task being performed and help in choosing a literary work that serves as the starting point of work or the form of presenting the product. 4% of respondents needed help translating some words, and another 4% needed assistance in formulating their ideas in German.

Evaluating creative products presented during the case study project on a five-score scale, 31.3% of respondents expressed complete satisfaction with their work, or scored5, 43.8% of participants rated their works as 4, and 25% as 3.

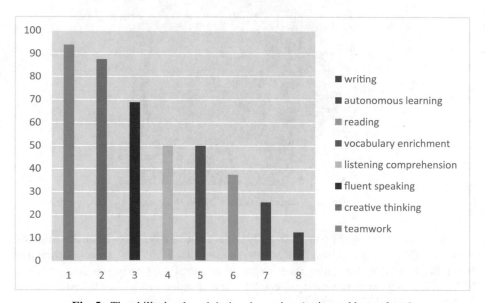

Fig. 5. The skills developed during the project (estimated by students).

According to students, work on the project allowed them to develop such abilities as team working (93.8%), creative thinking (87.5%) and the ability to speak German fluently (68.8%) (Fig. 5).

4 Discussions and Conclusion

Thus, during the study, it was found that in the framework of the educational process in higher education, insufficient attention is paid to the formation of the creative abilities of students. Further, the possibilities of solving this problem by conducting creative projects based on literary works in a foreign language classroom were identified and tested during the experiment. The use of the case study project method provided students with an opportunity to achieve goals in cooperative learning, develop their communication skills and general cultural competencies, and also responsibility for joint activities [17]. The analysis of the experiment's results concluded that work on creative case study projects based on comprehension and creative processing of foreign language literary works can be carried out even at the early stage of a foreign language learning. The study proves that the implementation of such projects corresponds to the university students' abilities, since all project participants coped with the task assigned to them to create and present a creative product, while the vast majority of them, according to their own estimates, did not face any difficulties in the process of work. It can also be argued that work on creative projects based on literary works meets the interests and needs of students, since most of the project participants gave a positive assessment of its results and expressed their readiness to take part in such projects in the future.

Work on creative case study projects at the same time allows to activate and improve a whole range of the skills to face various communicative activities of production and comprehension, both orally and in writing, that is one of the most important objectives in foreign language teaching [18]. The fact that the project is based on the works of a modern German-speaking author allows us to ensure a high-quality and authentic output as a result of the work. Moreover, project activities allow you to develop not only foreign language communication abilities of students, but also general competencies, the so-called soft skills, such as the ability to work in a team, creative thinking and the ability to self-study. This kind of activity increases creativity in a broad sense as a property to recreate, refashion, and recontextualise both the language structures and the content [19]. It "helps create a positive learning environment and involve students in new academic experiences" [20, p. 174], to interpret "universal cultural concepts both from the viewpoint of their similarities in different cultures (for the development of polycultural competence) and differences (for the development of ethnocultural competence)» [21, p. 145], as well as to develop "intellectual potential of an individual, analytical skills, critical thinking, the development of self-analysis skills and awareness of one's own capabilities, creative ability, initiative with a sense of responsibility for one's actions, and interpersonal skills" [13, p. 137]. To get a comprehensive assessment of students' emotional engagement, everybody from considering groups were offered to complete an online motivation questionnaire anonymously before and after the course [22, p.5].

The present study was limited to learners, studying German language majoring in international relations, Germany region. The purpose of further research in this field could be studying how creative projects based on authentic literary works reflect the interests and opportunities of students of other specialties. Moreover, involving more students in the experiment, as well as students with different levels of a foreign language proficiency could provide more detailed and specific information.

References

1. Sagitova, R.R., Fahrutdinov, B.I.: The model of formation of self-directed language learning of university students in a unified higher education area. Humanit. Soc. Sci. Rev. **7**(6), 85–88 (2019). https://doi.org/10.18510/hssr.2019.7618
2. Polok, K., Mala, E., Muglova, D.: On the salience of becoming a creative foreign language teacher. XLinguae **13**(4), 152–162 (2020). https://doi.org/10.1108/10748121111107672
3. Starbuck, D.: Teaching Creatively. Getting it Right. Learning Performance Training, London (2006)
4. Rubtsova, A.V., Almazova, N.I., Bylieva, D.S., Krylova, E.A.: Constructive model of multilingual education management in higher school. IOP Conf. Ser. Mater. Sci. Eng. **940**, 012132 (2020). https://doi.org/10.1088/1757-899X/940/1/012132
5. Rezaabadi, O.T.: A diverse inspirational language-teaching approach. Int. J. Appl. Linguist. Eng. Lit. **5**(7) (2016). https://doi.org/10.7575/aiac.ijalel.v.5n.7p.19
6. Pishghadam, R., Khodadady, E., Zabihi, R.: Learner creativity in foreign language achievement. Eur. J. Educ. Stud. **3**(3), 465–472 (2011)
7. Kobayashi, D.: Creativity through drama in language learning. Mask Gavel **2**(1), 13–19 (2013)
8. Peppler, K.A., Solomou, M.: Building creativity: collaborative learning and creativity in social media environments. On Horizon **19**(1), 13–23 (2011). https://doi.org/10.1108/107 48121111107672
9. Nosratinia, M.: Creativity and language learning strategies: toward a more successful language learning. Int. J. Sci. Eng. Res. **5**, 1156–1170 (2014)
10. Marashi, H., Khatami, H.: Using cooperative learning to boost creativity and motivation in language learning. J. Lang Transl. **7**(1), 43–58 (2017)
11. Jeong, H., Li, P., Suzuki, W., Sugiurab, M., Kawashima, R.: Neural mechanisms of language learning from social contexts. Brain Lang. **212**, 104874 (2021). https://doi.org/10.1016/j.bandl.2020.104874
12. Zhang, L.L.: English flipped classroom teaching model based on cooperative learning. Edu. Sci.: Theory Pract. **18**(6), 3652–366 (2018). https://doi.org/10.12738/estp.2018.6.278
13. Mohamed, M.N.A., Ngadiran, N. Md., Samad, N.A., Powzi, N.F.A.: E-collaboration among students of two regions: impacts on English language learning through peer learning. Int. J. Learn. Teach. Educ. Res. **18**(9), 201–215(2019). https://doi.org/10.26803/ijlter.18.9.11
14. Namaziandost, E., Shatalebi, V., Nasri, M.: The impact of cooperative learning on developing speaking ability and motivation toward learning English. J. Lang. Educ. **5**(3), 83–101 (2019). https://doi.org/10.17323/jle.2019.9809
15. Bylieva, D., Lobatyuk, V., Safonova, A., Rubtsova, A.: Correlation between the practical aspect of the course and the E-Learning progress. Educ. Sci. **9**(3), 167 (2019). https://doi.org/10.3390/educsci9030167
16. Zorina, E.M., Chernovets, E.G., Pashkin, S.B.: Teaching foreign-language authentic reading by means of educational technosphere. Eur. Proc. Soc. Behav. Sci. **73**, 105–113 (2019). https://doi.org/10.15405/epsbs.2019.12.13
17. Daineko, L.V., Reshetnikova, O.E.: Project method – an effective instrument for developing competencies of future professionals. Eur. Proc. Soc. Behav. Sci. **98**, 221–230 (2020). https://doi.org/10.15405/epsbs.2020.12.03.23

18. Sarasa-Cabezuelo, A.: Use of scratch for the teaching of second languages. iJET **14**(21), 80–95 (2019). https://doi.org/10.3991/ijet.v14i21.11217
19. Smith, C.A.: Student creativity and language performance. In: Sonda N., Krause A. (eds.) JALT2012 Conference Proceedings, pp. 285–297. Japan Association for Language Teaching, Tokyo (2013). https://jalt-publications.org/files/pdf-article/jalt2012-030.pdf. Accessed 11 March 2021
20. Almazova, N., Rubtsova, A., Eremin, Y., Kats, N., Baeva, I.: Tandem language learning as a tool for international students sociocultural adaptation. In: Anikina, Z. (ed.) IEEHGIP 2020. LNNS, vol. 131, pp. 174–187. Springer, Cham (2020). https://doi.org/10.1007/978-3-030-47415-7_19
21. Almazova, N., Baranova, T., Khalyapina, L.: Development of students' polycultural and ethnocultural competences in the system of language education as a demand of globalizing world. In: Anikina, Z. (ed.) GGSSH 2019. AISC, vol. 907, pp. 145–156. Springer, Cham (2019). https://doi.org/10.1007/978-3-030-11473-2_17
22. Baranova, T., Khalyapina, L., Kobicheva, A., Tokareva, E.: Evaluation of students' engagement in integrated learning model in a blended environment. Educ. Sci. **9**(2), 138 (2019). https://doi.org/10.3390/educsci9020138
23. Bogdanova, N.V., Katalkina, N.A., Pankrateva, G.I., Afanaseva, E.A.: Tandem language learning: research experience in Russian universities context. In: Anikina, Z. (ed.) IEEHGIP 2020. LNNS, vol. 131, pp. 267–277. Springer, Cham (2020). https://doi.org/10.1007/978-3-030-47415-7_28

Corpus Linguistic Technology as a Tool to Improve Creative Thinking in the Interpretation of English Language Idioms

Ekaterina Osipova$^{(\boxtimes)}$ (ID) and Ekaterina Bagrova (ID)

Peter the Great St. Petersburg Polytechnic University, Polytechnicheskaya 29, 195251 St. Petersburg, Russia
{osipova_es,bagrova_eyu}@spbstu.ru

Abstract. The article validates the possibility and efficiency of improving creative thinking ability while teaching the interpretation of idioms based on corpus linguistic technology to EFL (English as a Foreign Language) students. Creative thinking ability is defined as the ability to generate, develop, and test hypotheses in the process of the interpretation of idioms based on corpus linguistic technology. The goal of the current research is to identify specific features of the students' interpretation activity while working with corpus linguistic technology and, on this basis, substantiate the possibilities of developing creative thinking ability by means of corpus linguistic technology while teaching the interpretation of idioms. The article provides examples of corpus-based exercises aimed at developing students' creative thinking ability while teaching the interpretation of idioms. The methodology has been tested on the EFL students of Bachelor study program in Public Relations implemented by Peter the Great St. Petersburg Polytechnic University in the years 2020–2021. The authors describe a case study to reveal how corpus linguistic technology can be used in foreign language teaching and detail the procedure of the learners' work with the basic corpus tools while interpreting idioms. The result of the experimental training indicates that the average level of creative thinking ability increased, confirming the hypothesis of the research on the effectiveness of the corpus linguistic technology as a tool to improve EFL students' creative thinking ability in the process of teaching the interpretation of idioms.

Keywords: Creative thinking · Corpus linguistic technology · Idiom interpretation · EFL students

1 Introduction

The increasing scale of change and a new line of development establish a new requirement for education, marked by a high level of creativity [1, 2]. To meet the abovementioned request, Russian education system is actively renovating, "substantially transforming the educational environment as a creative, open, accepting and stimulating environment" [3, p. 892].

D. Bylieva and A. Nordmann (Eds.): PCSF 2021, LNNS 345, pp. 948–962, 2022.
https://doi.org/10.1007/978-3-030-89708-6_76

The urgency of the problem of the development of students' creative thinking is due to the social function of education, that is "reliance on talent, creativity and initiative of a person as the most important resource of economic and social development" [3, p. 892]. "The most important requirement for the results of education is a request for the mass of creative competencies" [4, p. 11], "a high level of independence in educational activities" [5, p. 14].

The current educational aim is to "help students shape new thought patterns in order to develop both their critical and creative potential required for the holistic development of a personality" [6]. These conditions are particularly noteworthy for the higher education system, since "selfreliance, creativity, mobility, responsibility, ability for both personal and professional creative self-fulfillment become fundamental for any professional and indicate one's competitiveness in the job market" [7, p. 737].

Corpus linguistic technology can be implemented in the process of language teaching to meet the demand as it has the potential to improve students' creative thinking ability. The purpose of the article is substantiation of the possibilities of developing creative thinking by means of corpus linguistic technology while teaching the interpretation of idioms.

2   Literature Review

The priority goal of modern education is not only the transfer of knowledge, skills and abilities from a lecturer to a student, but the full-fledged formation and development of the student's abilities to independently outline the educational problem, formulate an algorithm for solving it, control the process and evaluate the result. Analysis of studies shows that since the beginning of the last century, numerous attempts have been made to understand the phenomenon of creativity, its components, stages of the creative process, and types of thinking. Given the ambiguity of the phenomenon of creativity, the range of points of view on creativity is quite wide.

Many researchers believe that the ability to create exists in all people, the problem lies in how developed these abilities are.

From a historical perspective, foreign studies of the phenomenon of creativity can be presented as follows.

Starting with Wallas [8] who established the chronological flow of the creative process by proposing the following stages: preparation, incubation, illumination, and verification. Creativity was seen as a concept in a large thought process, and not as a significant independent phenomenon.

Later an American psychologist Torrance [9] defined creativity as a process of sensitivity to problems, gaps in knowledge, identifying difficulties, seeking solutions, proposing hypotheses and testing them repeatedly.

A person who collected and analyzed more than forty definitions of creativity in order to develop the first model of creativity was Rhodes [10]. The model covers four independent variables, including personality, process, product, press. This model has made a significant contribution to the overall fundamental basis of creativity research.

Research by Sperry [11] was based on medical evidence that there is right and left brain thinking, argued that right brain thinking presupposes creativity, while left hemisphere thinking presupposes logic. Another study by the scientist showed that when the two hemispheres are physically separated, it leads to a decrease in creative behavior and achievement.

Amabile [12] proposed innovative ideas, creating a three-component model that combines interrelated concepts such as "domain-relevant skills", "creativity-relevant processes", "task motivation". Domain-relevant skills are associated with the desire to solve a problem or create something new. Creativity-relevant processes involve knowledge of technical, procedural and cognitive aspects, while task motivation encompasses creative processes such as inspiration, imagination, flexibility, and the combination of them into a new idea. The combination of these variables leads to creativity. Later Amabile added to her model one more component – a surrounding environment that can both stimulate and hinder the development of creativity.

Robinson [13] argues that creativity is a quality inherent in every person, and not just a limited circle of talented individuals. Each person, in his opinion, has a colossal creative potential due to the fact that he is already a human being. The creative process does not mean creation alone; the emotional state also plays an important role. The leading idea of Robinson is that in the changing conditions of a changing world, it is necessary to transform the education system, which will be able to develop the creative potential of its subjects.

Different researches analyse creativity from different sides and propose different theories.

According to the behaviorist theory of creativity (Skinner) [14], a person is not the initiator of a creative action, he is rather a focal point at which the forces of heredity and environment converge to achieve a general effect. Skinner argues that it is necessary to analyze the genetic and environmental factors that influence human behavior, and then create an environment for the manifestation of creative behavior.

Humanistic theory by Adler [15], Maslow [16] sees creativity as a mechanism for the development and self-expression of an individual. Maslow, a vivid supporter of this direction, argued that creativity is a universal, natural, inherent personality trait, contributing to self-actualization.

The associative theory of creativity by Mednik [17] is based on the idea that associations are the basis of creative thinking. Creative thinking is formed as a result of new combinations of associations between ideas, moreover, the more distant the ideas between which associations arise, the more creative thinking is considered - provided that these associations meet the requirements of the task.

The focus of cognitive theory (Guildford [18], Gordon [19], Koestler [20], Osborne [21], Wallace [8]) is thinking skills and the thought process.

Psychoanalytic theory (Freud [22], C. Jung [23]) asserts that the true and most powerful source of creativity are unconscious mental processes. In accordance with this theory, all discoveries, new ideas and thoughts are in the realm of the unconscious, thus, creativity is a quality that is given from above and does not lend itself to the influence of either consciousness or the will of a person.

The theories discussed above convincingly prove the versatility and ambiguity of the phenomenon of creativity.

Each of the theories considers a certain layer of the problem, starting from the value priorities of researchers, and, despite the fact that each of them contains a significant layer of discoveries, observations and substantiations, they cannot be called comprehensive. The importance of the formation of a creative personality and the development of creative thinking of university students is noted by many scientists [24–27]. For the development of creative thinking of students, the following requirements can be formulated: to teach them to think critically, to see and solve problems and contradictions, to analyze large amounts of information, to adapt to changing conditions, to show a high level of creativity in solving assigned tasks. Students need to show their creative activity, developing their individual abilities. It is important for teachers and students to understand the interaction between technology and language that reveal itself in many ways [28].

In foreign language teaching creativity is developed by means of different information technologies such as getting creative through photographs and video projects, the makerspace movement, encouraging ideas and growth through social media [29].

In our research corpus linguistic technology is considered a tool to develop EFL students' creative thinking ability. The application of technology of the electronic materials integration can guarantee solid professional knowledge, skills and abilities to the students. Such personal qualities as the desire for creativity, responsibility for the task performed, the ability to organize own educational and cognitive activities and personal initiative are successfully formed in the process of performing tasks using the certain technology [30].

3   Materials and Methods

In the process of the corpus-based teaching the interpretation of idioms to EFL students open-access corpora are used. That is a group of linguistic corpora, created by Mark Davies, Professor of Brigham Young University (BYU) which comprises the British National Corpus (BNC), the Corpus of Contemporary American English (COCA), the Time Magazine Corpus (TMC), the Corpus of Historical American English (COHA). "Corpus linguistics has had a direct impact on teaching foreign languages since the advent of large online corpora. In 1991, Professor Tim Jones from the University of Birmingham introduced the concept of Data Driven Learning (DDL) and was an ardent supporter of the use of corpora by students to test their hypotheses about word compatibility, grammatical constructions, usage contexts, etc., explaining that "research is too important to be left to the researchers"" [31].

A set of exercises aimed at developing students' creative thinking abilities has been devised. They are ranked according to the complexity of the low, medium, and high levels of problem solving. While working with this set of exercises, it is essential to exclude access to any dictionaries or reference books to encourage students' creative thinking. Here are three exercises for hypothesizing the meaning of an idiom in a narrow or broad context: the first refers to a low level of problem solving, the second to a medium level, and the third to a high level (the exercises are given in abridgement).

Exercise 1. Without using a dictionary, choose from the suggested list of words the one that is the most suitable for the context, fill in the gaps and translate the sentence. To clarify the meaning of an idiom, use the electronic corpora of the English language that you know.

straw vote lame duck whispering campaign rip bear wheel man

Congressman Jones was a _____ and did not vote on many issues that were important to his constituents.

Exercise 2. Without using a dictionary, suggest Russian equivalents to the idiom "get rid of", based on the context. Determine the type of professional discourse of the presented fragments and translate them.

A. Before use, the aluminium plates were heated in an exhausted quartz tube in an electric furnace nearly to its melting point *to get rid of* hydrogen and other gases. The films under examination were kept in a dessicator and heated in the electric furnace just before use and transferred at once to the testing vessel.

B. "There are chemicals in some of the parts that come with the iPhone that are well known in California to cause birth defects", Greenpeace said. The center's experience in "hundreds of different cases" is that companies prefer *to get rid of* offending chemicals rather than taint images of brands with health warnings, according to Greenpeace.

C. We believe that the use of these weapons, particularly nuclear weapons, is a crime against humanity, and thus the international community should react collectively by redoubling its efforts *to get rid of* the threats posed by such weapons once and for all. In this context, we believe that the destruction of chemical weapons is and remains the fundamental bedrock of the Chemical Weapons Convention.

Exercise 3. Guess the meanings of the idioms in italics. Use the electronic corpus of the English language and create the concordance strings to find more examples of the use of the words and phrases in italics. Without using a dictionary, translate the following text fragments. Check the correctness of your understanding of abbreviations, using any of the linguistic corpora.

Nick Leeson opened a secret trading account numbered 88888 to facilitate his furtive trading. He lost money from the beginning. Increasing his bets only made him lose more money. By the end of 1992, the 88888 account was **under water** by about GBP 2 MM. A year later, this **had mushroomed** to GBP 23 MM. By the end of 1994, Leeson's 88888 account had lost a total of GBP 208 MM.

To effectively achieve the research objectives and test the research hypothesis, that corpus linguistic technology can be used as a tool to improve critical thinking ability while teaching the interpretation of idioms to EFL students, the experimental training has been conducted. Totally 120 students in four groups are studied. The experimental training lasts for 15 weeks, two hours per week (total 30 h).

It is necessary to mention that during the training students are not allowed to use any Russian-English or English-Russian dictionaries or on-line translators while dealing with the interpretation of idioms and they are asked to write down the algorithm of their decision-making, stages of generating, developing, and testing hypotheses. However, students of the experimental group can use any linguistic corpora and corpus linguistic tools such as LIST, CHART, KWIC, COMPARE.

The LIST tool allows students to find the most frequent collocations and make lists of the contextual usage of the desired idiom. While translating the sentence *Институт по образованию в области водных ресурсов* ***действует под эгидой ЮНЕСКО***, the difficulty is caused by the idiom – *под эгидой.*

Students used the following algorithm: (a) accumulate cognitive resources and brain-storm some ideas based on an intuitive guess, the most common ideas are *under the aegis of...*, *under the auspices of smth.*, (b) access the BNC corpus and use the LIST tool (see Fig. 1),

			FREQ	
1		AEGIS	140	

			FREQ	
1		AUSPICES	351	

Fig. 1. The results of the query in BNC. Tool LIST.

(c) use corpus data to evaluate the hypothesis: as it's seen from Fig. 1 *auspices* is more frequently used, (d) verbalize *auspices* as the translation solution.

The next task to be solved is to think of the appropriate verb which matches the *auspices*. Students generate their hypotheses, then use the LIST tool to check them. Corpus data were analyzed and evaluated to choose the correct verb. Students arose a wide range of hypothetic solutions based on an intuitive guess. 75% of students managed to confirm their hypothesis, *operate under the auspices,* using the LIST tool of BNC corpora (see Fig. 2).

1	CONDUCTED UNDER THE AUSPICES	18	
2	ESTABLISHED UNDER THE AUSPICES	9	
3	PRODUCED UNDER THE AUSPICES	9	
4	PERFORMED UNDER THE AUSPICES	8	
5	WAS UNDER THE AUSPICES	7	
6	DEVELOPED UNDER THE AUSPICES	7	
7	OPERATED UNDER THE AUSPICES	7	
8	HELD UNDER THE AUSPICES	6	
9	OPERATES UNDER THE AUSPICES	6	

Fig. 2. The results of the query in BNC. Tool LIST.

Here is the translation of the sentence offered by most of the students after corpus analysis. *The Institute for Water Education is operating under the auspices of UNESCO.* It's worth mentioning that the name of the institute was easily found in COCA in the academic journal BioScience, 2011, "…the United Nations Educational, Scientific and Cultural Organization's *Institute for Water Education…*".

The CHART tool allows students to chart the frequency of use of a lexical unit by genre and year. Let us illustrate the work with the CHART tool by using the example of translating a sentence in a spoken discourse. *Можете ли назвать* **ориентировочную** *стоимость вашей квартиры?*Students were asked to use an idiom while translating the sentence. Most students were well aware of the idiom, *a ball-park figure,* but they were not sure if this idiom could be used in spoken discourse, some students also suggested that the idiom might be out-of-date. To test their hypotheses, students used the CHART tool of the COCA corpus (see Fig. 3).

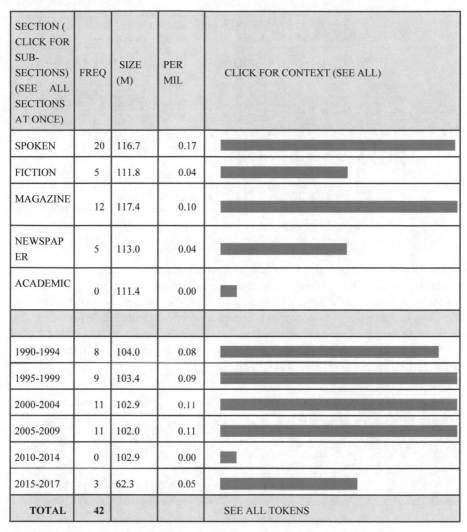

SECTION (CLICK FOR SUB-SECTIONS) (SEE ALL SECTIONS AT ONCE)	FREQ	SIZE (M)	PER MIL	CLICK FOR CONTEXT (SEE ALL)
SPOKEN	20	116.7	0.17	
FICTION	5	111.8	0.04	
MAGAZINE	12	117.4	0.10	
NEWSPAP ER	5	113.0	0.04	
ACADEMIC	0	111.4	0.00	
1990-1994	8	104.0	0.08	
1995-1999	9	103.4	0.09	
2000-2004	11	102.9	0.11	
2005-2009	11	102.0	0.11	
2010-2014	0	102.9	0.00	
2015-2017	3	62.3	0.05	
TOTAL	42			SEE ALL TOKENS

Fig. 3. The result of the query "a ballpark figure" in COCA. Tool CHART.

The corpus analysis proves that the chosen idiom has not fallen out of use and functions in the spoken discourse. 20% of students were satisfied with the results at this stage and started translating the sentence *Can you name/call/suggest a ballpark figure?* and it led to some mistakes in collocation. 80% of students continued their research and tested their hypotheses about the verbs which go with the idiom. To solve this problem, the students turned to the KWIC tool, which makes it possible to generate concordance strings, i.e., word compatibility strings ordered relative to the right and left contexts. The analysis of the corpus data below (see Fig. 4) allowed students to focus on the option *give me a ballpark figure.*

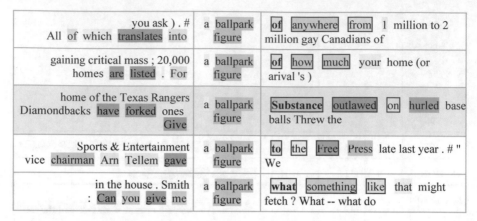

you ask) . # All of which translates into	a ballpark figure	of anywhere from 1 million to 2 million gay Canadians of
gaining critical mass ; 20,000 homes are listed . For	a ballpark figure	of how much your home (or arival 's)
home of the Texas Rangers Diamondbacks have forked ones Give	a ballpark figure	Substance outlawed on hurled base balls Threw the
Sports & Entertainment vice chairman Arn Tellem gave	a ballpark figure	to the Free Press late last year . # " We
in the house . Smith : Can you give me	a ballpark figure	what something like that might fetch ? What -- what do

Fig. 4. The result of the query in COCA. Tool KWIC.

The COMPARE tool is one more helpful tool at the stage of comparing, evaluating, and testing the current hypotheses. It is used for comparing the frequency of the use of two words in relation to the third. This option is essential in cases where a dictionary entry offers two or more matches or two hypotheses have been generated. While translating *звучать фальшиво, неискренне* students were hesitating about the options *it rings hollow/it rings false*. According to the corpus-based results (see Fig. 5) of the COMPARE tool, the most frequent option is *it rings falls*.

WORD 1 (W1): **HOLLOW** (0.34)					WORD 2 (W2): **FALSE** (2.96)				
WORD	W1	W2	W1/W2	SCORE	WORD	W2	W1	W2/W1	SCORE
RING	80	13	6.2	18.2	RANG	12	27	0.4	0.2
RINGS	42	13	3.2	9.6	RINGS	13	42	0.3	0.1
RANG	27	12	2.3	6.7	RING	13	80	0.2	0.1

Fig. 5. The result of the query in COCA. Tool KWIC.

For the solution of the research tasks a series of methods was applied: (a) methods of a theoretical analysis (analysis, synthesis, generalization, modelling, comparative); (b) empirical methods (educational and control experiments); (c) a tabular, graphical and corpus-based presentation of information; (d) mathematical statistics.

4   Results

The interpretation of idioms demands quite a high level of creativity from students. Creativity forms the mental space in which the struggle of hypotheses develops. While teaching the interpretation of idioms to EFL students, we can't help developing creative thinking ability, which is understood as the ability to develop and test new hypotheses.

The interpretation of idioms is a set of cognitive and creative operations for the decision of tasks of varying complexity. Creative thinking is directly related to problem-based learning. The main stages of problem solving comprise (a) an analysis of the problem, (b) the adoption of a plan or a hypothesis, (c) the test of a hypothesis, (d) the comparison with the source conditions, (e) a return to baseline analysis or selection of the final solution. The most important stage in solving the problem is the process of hypotheses formation and hypotheses development that act as one of the main functional formations in the structure of mental activity, being, on the one hand, the result of research activity, and on the other - a means of changing it, directing, regulating, and evaluating subsequent actions.

Low level of students' creative thinking ability leads to frequent mistakes and problems in the interpretation of idioms at the stage of formation and development of the translation hypothesis. This stage includes several steps (a) accumulating cognitive resources, (b) arising a hypothetic solution based on an intuitive guess, (c) an analytical interpretation of the hypothesis, (d) evaluation the hypothesis from the point of view of logical representations, (e) verbalizing the translation solution.

To solve the abovementioned problem, we apply corpus linguistic technology as a tool to improve EFL students' creative thinking ability while teaching the interpretation of idioms. In this research, we regard creative thinking ability as the ability to generate, develop, and test hypotheses in the process of the interpretation of idioms based on corpus linguistic technology.

While working with the interpretation of idioms based on corpus linguistic technology, students were asked to write down the steps they followed to come up with the desired results. Taken together students' notes, we design a practical algorithm for the interpretation of idioms based on corpus linguistic technology: (a) put forward a translation hypothesis, (b) access to the linguistic corpus (one or more), (c) choose the tool LIST to test the hypotheses in terms of the frequency of an idiom in different types of discourse (quantitative parameter) and idiom's compliance with the given context (qualitative parameter), (d) choose the tool CHART to test the hypotheses in terms of the idiom's frequency by genre and year, (e) choose the tool KWIC to generate concordance strings to test the hypotheses in terms of compatibility of an idiom, (f) choose the tool COMPARE if the hypotheses about the usage of two words in relation to the third should be tested, (g) interpret and evaluate corpus data, (d) decide on the interpretation of idioms.

To assess the changes in students' creative thinking ability, we devised the following criteria:

(a) fluency (the number of ideas that occur per unit of time),
(b) originality (the number of "rare" ideas that differ from the generally accepted, typical responses),
(c) flexibly (the number of ideas which are quickly switched from one idea to another),
(d) accuracy of idiom interpretation (the number of correct ideas of idioms interpretation).

Students were asked to interpret the list of 25 idioms in context. 5 min per one idiom were given to come up with the equivalent. 1 min for generating hypotheses and 1 min for developing hypotheses and 3 min for testing and choosing the idiom's equivalent. The test was conducted twice: before the corpus-based training (pre-training stage) and after (post-training stage). The only difference is that at the post-training stage, students could use only corpus linguistic technology while dealing with the interpretation of idioms at the stages of developing and generating hypotheses.

At the stage of generating hypotheses, students wrote the number of ideas that occurred per unit of time to assess fluency. At the stage of developing hypotheses, students wrote a number of ideas which were quickly switched from one idea to another to assess flexibility. The number of "rare" ideas (originality) and correct ideas (accuracy) were assessed by a teacher.

The average rate of fluency, originality, flexibility, and accuracy was calculated by formula:

$$K = \frac{N1}{N2} \qquad x = \frac{(K1 + K2 \ldots + K120)}{n}$$

K – the average rate of the abovementioned criteria of one student,

N1 – the number of ideas,

N2 – the total number of idioms, offered to students for interpretation in the test,

X – the average rate of the abovementioned criteria of the experimental group,

n – the number of students in a group.

Having calculated the changes of the average rate of fluency, originality, flexibility, and accuracy at the pre-training and post-training stages, we can judge about some improvements of the students' critical thinking ability. As seen from the bar chart below, positive changes occurred with the proposed criteria (see Fig. 6).

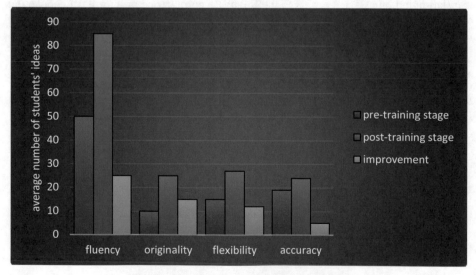

Fig. 6. The level of creative thinking ability

5 Discussion

Corpus technology has already proved its validity for plenty of linguistic studies, which might be separated into several groups: (a) quantitative studies of English lexical and grammatical features [32, 33], (b) studies on corpus methodology and design [34, 35], (c) corpus-based studies on linguistic changes (lexical, syntactical and stylistic) [36, 37], (d) corpus-based discourse analysis [38–43].

Some papers focus on special tools of corpus linguistics, for instance, concordance or KWIC (key words in context), and how to use them to solve linguistic issues. For example, Seretan and Wehrli [44] present "an enhanced type of concordancer that integrates syntactic information on sentence structure as well as statistical information on word co-occurrence".

The advantages of corpus technology for language learning and second language learning are widely presented in the literature [45–49].

The researchers suggest to use corpus technology for the following purposes: a) as an empirical component of lectures, student assignments, and projects; b) to determine the meaning of words and identify differences in usage between synonymous lexical items; c) to study lexical collocations; d) to focus on linguistic evidence that either supports or contradicts the prescriptive grammar rules [50–54].

Although the use of corpus linguistic technology in foreign language teaching is widely acknowledged by specialists in language education, but there is very little discussion on what advantages corpus-based teaching can bring into the improvement of students' creative thinking ability.

6 Conclusion

Corpus linguistic technology can be used as an effective tool to improve EFL students' creative thinking ability while teaching the interpretation of idioms, providing:

1) teaching the interpretation of idioms is based on problem solving activities, which are ranked according to the complexity;
2) students are taught how to generate and develop their hypotheses and test them without using dictionaries, but corpus linguistic technology;
3) the search for a translation solution includes the following stages: (a) accumulating cognitive resources, (b) arising a hypothetic solution based on an intuitive guess, (c) an analytical interpretation of the hypothesis, (d) evaluation the hypothesis from the point of view of logical representations, (e) verbalizing the translation solution;
4) a practical algorithm for the interpretation of idioms based on corpus linguistic technology is used: (a) put forward a translation hypothesis, (b) access to the linguistic corpus (one or more), (c) choose the tool LIST to test the hypotheses in terms of the frequency of an idiom in different types of discourse (quantitative parameter) and idiom's compliance with the given context (qualitative parameter), (d) choose the tool CHART to test the hypotheses in terms of the idiom's frequency by genre and year, (e) choose the tool KWIC to generate concordance strings to test the hypotheses in terms of compatibility of an idiom, (f) choose the tool COMPARE if the

hypotheses about the usage of two words in relation to the third should be tested, (g) interpret and evaluate corpus data, (d) decide on the interpretation of idioms;

5) students master the abilities of working with basic corpus linguistic tools (LIST, CHART, KWIC, COMPARE) to such an extent that their creative thinking operations lead to the correct interpretation.

References

1. Bowen, W.G.: Higher Education in the Digital Age. Publishing House of Higher School of Economics, Moscow (2018). (in Russian)
2. Crowley, E.F., Malmguist, J., Ostlund, S., Broder, D.R., Edstrom, K.: Rethinking Engineering Education. The CDIO Approach. Publishing House of Higher School of Economics, Moscow (2015)
3. Barysheva, T.A., Gogoleva, V.V., Zyabkina, T.F., Maksimova, E.V.: Development of student's creativity by means of reflective technologies in educational information environment. In: Anikina, Z. (ed.) IEEHGIP 2020. LNNS, vol. 131, pp. 891–903. Springer, Cham (2020). https://doi.org/10.1007/978-3-030-47415-7_96
4. Kuzminov, Y., Frumin, I.: Russian education – 2020: model of education for economics based on knowledge. In: Proceedings of the IXth International Scientific Conference "Modernization of Economics and Globalization", pp. 32–64. SU HSE, Moscow (2008). (in Russian)
5. Rudskoy, A.I., Borovkov, A.I., Romanov, P.I., Kiseleva, K.N.: Engineering Education: World Experience in Cultivation of Intellectual Elite. Publishing House of Peter the Great St. Petersburg Polytechnic University, St. Petersburg (2017).(in Russian)
6. Lubart, T., Zenasni, F.: A new look at creative giftedness. Gifted Talent. Int. **25**, 53–57 (2010). https://doi.org/10.1080/15332276.2010.11673549
7. Sigal, N.G., Linyuchkina, E.G., Plotnikova, N.F., Zabolotskaya, A.R., Bagmanova, N.I.: Academic environment for the development of creative fulfillment of innovative students. In: Anikina, Z. (ed.) IEEHGIP 2020. LNNS, vol. 131, pp. 737–744. Springer, Cham (2020). https://doi.org/10.1007/978-3-030-47415-7_78
8. Wallas, G.: The Art of Thought. Jonathan Cape, London (1938)
9. Torrance, E.P.: Creative Learning and Teaching. Dodd, Mead Co, New York (1970)
10. Rhodes, M.: An analysis of creativity. Phi Delta Kappan **42**(7), 305–310 (1961)
11. Sperry, R.W.: Brain mechanisms in behaviour. In: Hutchings, E., Jr. (Ed.) Frontiers in Science, p.48. Basic Books, New York (1958)
12. Amabile, T.M.: The Social Psychology of Creativity. Springer, New York (1983)
13. Robinson, K.: Out of our Minds. Learning to be Creative. Capstone Publishing Ltd, Mankato (2011)
14. Skinner, B.F.: The Technology of Teaching. Appleton-Century-Crafts, New York (1968)
15. Adler, A.: The individual psychology of Alfred Adler. a systematic presentation of selections from his writings. New York, Harper & Row (1956)
16. Maslow, A.: Motivation and Personality. Longman, Harlow (1987)
17. Mednick, S.A.: The associative basis of the creative process. Psychol. Rev. **69**(2), 220–232 (1962)
18. Wreen, M.: Creativity. Philosophia **43**(3), 891–913 (2015). https://doi.org/10.1007/s11406-015-9607-5
19. Gordon, W.J.: Synestetics: The Development of Creative Capacity. Harper, New York (1961)
20. Koestler, A.: The Act of Creation. Hutchinson, London (1964)

21. Osborn, A.F.: Applied Imagination: Principles and Procedures of Creative Thinking. Scribner, New York (1948)
22. Freud, S.: New Introductory Lectures on Psycho-Analysis, The Standard Edition of the Complete Psychological Works of Sigmund Freud. W. W. Norton & Company, New York (1933)
23. Jung, C.G.: Psychology of the Unconscious. Moffat, Yard, and Co., New York (1916)
24. Sica, L.S., Ragozini, G., Di Palma, T., Aleni Sestito, L.: Creativity as identity skill? Late adolescents' management of identity, complexity and risk-taking. J. Creat. Behav. **53**(4), 457–547 (2019). https://doi.org/10.1002/jocb.221
25. Barbot, B., Heuser, B.: Creativity and identity formation in adolescence: a developmental perspective. In: The Creative Self: Effect of Beliefs, Self-Efficacy, Mindset, and Identity, pp. 87–98. Academic Press, New York (2017). https://doi.org/10.1016/B978-0-12-809790-8.00005-4
26. Wang, H.-C.: Fostering learner creativity in the English L2 classroom: application of the creative problem-solving model. Think. Skills Creat. **31**, 58–69 (2019). https://doi.org/10.1016/j.tsc.2018.11.005
27. Li, L.: Thinking skills and creativity in second language education: Where are we now? Think. Skills Creat. **22**, 267–272 (2016). https://doi.org/10.1016/j.tsc.2016.11.005
28. Krylov, E.: Engineering education – convergence of technology and language. Techn. Lang. **1**(1), 49–50 (2020). https://doi.org/10.48417/technolang.2020.01.11
29. Dulksnienė, L., Mačianskienė, N.: Integration of creativity-developing activities in foreign language learning: students' attitude. Sustain. Multiling. **16**(1), 65–90 (2020). https://doi.org/10.2478/sm-2020-0004
30. Almazova, N., Barinova, D., Ipatov, O.: Forming of information culture with tools of electronic didactic materials. In: Proceedings of the 29th DAAAM International Symposium, pp. 0587–0593. DAAAM International, Vienna (2018). https://doi.org/10.2507/29th.daaam.proceedings.085
31. Kogan, M., Zakharov, V., Popova, N., Almazova, N.: The impact of corpus linguistics on language teaching in Russia's educational context: systematic literature review. In: Zaphiris, P., Ioannou, A. (eds.) HCII 2020. LNCS, vol. 12205, pp. 339–355. Springer, Cham (2020). https://doi.org/10.1007/978-3-030-50513-4_26
32. Schmidtke, D., Kuperman, V.: Mass counts in World Englishes: a corpus linguistic study of noun countability in non-native varieties of English. Corpus Linguist. Ling. Theory **13**, 135–164 (2017). https://doi.org/10.1515/cllt-2015-0047
33. López-Couso, M.J.: Continuing the dialogue between corpus linguistics and grammaticalization theory: three case studies corpus linguist. Ling. Theory **12**, 7–29 (2016). https://doi.org/10.1515/cllt-2015-0069
34. Iruskieta, M., Diaz De Ilarraza, A., Lersundi, M.: Establishing criteria for RST-based discourse segmentation and annotation for texts in Basque Corpus Linguist. Ling. Theory **11**, 303–34 (2015). https://doi.org/10.1515/cllt-2013-0008
35. Weisser, M.: DART – the dialogue annotation and research tool corpus linguist. Ling. Theory **12**, 355–388 (2016)
36. Koplenig, A.: Using the parameters of the Zipf-Mandelbrot law to measure diachronic lexical, syntactical and stylistic changes – a large-scale corpus analysis Corpus Linguist. Linguist. Theory **14**, 1–34 (2018). https://doi.org/10.1515/cllt-2014-0049
37. Perek, F.: Recent change in the productivity and schematicity of the way-construction: a distributional semantic analysis Corpus Linguist. Ling. Theory **14**, 65–97 (2018). https://doi.org/10.1515/cllt-2016-0014
38. Baker, P.: Acceptable bias? Using corpus linguistics methods with critical discourse analysis Crit. Discourse Stud. **9**, 247–256 (2012)

39. Brindle, A.: A corpus analysis of discursive constructions of the sunflower student movement in the English-language Taiwanese press. Discourse Soc. **27**, 3–19 (2016). https://doi.org/10.1177/0957926515605957
40. Beliaeva, L., Chernyavskaya, V.: Evidence-based linguistics: methods in cognitive paradigm Vopr. Kognitivnoy Lingvistiki **3**, 77–84 (2016)
41. Chernyavskaya, V.: Corpus-assisted discourse analysis of Russian university 3.0 identity. Tomsk State Univ. J. Philol. **58**, 97 –114 (2019). https://doi.org/10.17223/19986645/58/7
42. Kim, K.H.: Examining US news media discourses about North Korea: a corpus-based critical discourse analysis. Discourse Soc. **25**, 221–244 (2014). https://doi.org/10.1177/0957926513516043
43. Samaie, M., Malmir, B.: US news media portrayal of Islam and Muslims: a corpus-assisted critical discourse analysis. Educ. Philos. Theory **1**, 1–16 (2017). https://doi.org/10.1080/00131857.2017.1281789
44. Seretan, V.,Wehrli, E.: Syntactic concordancing and multi-word expression detection Int. J. Data Min. Model. Manag. **5**, 158–81 (2013). https://doi.org/10.1504/IJDMMM.2013.053694
45. Ackerley, K.: Effects of corpus-based instruction on phraseology in learner English Lang. Learn. Technol. **21**, 195–216 (2017)
46. Kogan, M., Gavrilova, A., Nesterov, S.: Training engineering students for understanding special subjects in English: the role of the online component in the experimental ESP course. In: 2018 IV International Conference on Information Technologies in Engineering Education, pp.1–6. Inforino, National Research University MPEI, Moscow (2018)
47. Kogan, M.: Ways of integrating approaches of corpus linguistics into the translators training programme. In: VII International Conference Translation. Language. Culture, pp. 215–219. Pushkin Leningrad State University, Saint Petersburg (2016)
48. Pavlovskaya, I., Gorina, O.: Corpus-based cognitive oriented methods of lexical analysis in foreign language studies. Cherepovets State Univ. Bull. **1**, 132–138 (2017)
49. Rodriguez-Fuentes, R.: Review of corpus linguistics and linguistically annotated corpora Lang. Learn. Tech. **19**, 56–60 (2015)
50. Derybina, I.: Control verbs studying based on corpus linguistics as prdagogical problem. Bull. Tambov State Univ. **10**, 154–158 (2012)
51. Kokoreva, A.: Parallel corpus in foreign language teaching. Bull. Tambov State Univ. **2**, 57–62 (2013)
52. Kuzminykh, I., Khoroshilova, S.: Investigating the impact of corpus-based classroom activities in English phonetics classes on students' academic progress. Bull. Novosibirsk State Pedag. Univ. **7**, 40–51 (2017)
53. Sosnina, E.: Research of pupils' linguistic errors on the base of learner corpus Izvestiya of the Samara Russian academy of sciences. Sci. Center. Soc. Hum. Medicobiol. Sci. **19**, 39–44 (2017)
54. Sysoyev, P., Evstigneev, M.: Foreign language teachers' competency and competence in using information and communication technologies Procedia -Soc. Behav. Sci. **154**, 82–86 (2016)

The Influence of Digital Transformations on Learners' and Educators' Creativity

Alexandra Dashkina⬤, Alexander Dmitrijev⬤, Liudmila Khalyapina⬤, and Aleksandra Kobicheva⁽✉⁾⬤

Peter the Great Saint Petersburg State Polytechnic University, St Petersburg, Russia

Abstract. The usage of information technology has transformed education systems with significant impacts on academic life and the reorganization of the educational environment. Therefore, it is essential to understand the benefits arising from the adoption of new effective teaching practices connected to the digital world. Among all benefits, we would like to emphasize the influence of digital transformation on learners' and educators' creativity. In our paper, we consider the concepts of "creative thinking" and "creative thinking skills" and analyze the possibilities of student-teacher creative collaboration. Also, we discuss an experiment designed whether corpus technologies influence students' creativity development. The experiment included students ($N = 50$) who learned English idioms in the framework of the corpus approach. The results showed the positive impact of corpus technologies on creative-thinking skills and their efficiency for English learning purposes. In conclusion we recommend such technology as an effective digital tool in the educational process.

Keywords: Creative-thinking skills · Corpus technologies · Digital transformation · Creative collaboration

1 Introduction

Digital transformations in higher education have drastically changed the processes at all levels: planning, governance and enabling (management and business processes, performance assessment, student administration, staffing, procurement, budgeting and funding); learning and teaching (developing educational content, tests, examinations, planning lectures and classes, checking home assignments, teachers' and students' academic mobility), and research (participation in scientific conferences, scientific projects, publishing monographs, scholarly literature, articles in science journals) [1].

This article generally focuses on the influence of digital transformations on teachers' and educators' creativity, in particular, on developing educational materials, doing creative assignments and conducting scientific research.

Digital transformations in education have been occurring at a frenetic rate recently. They have been spurred by distance learning, which involves applying cutting-edge educational technologies. The main advantage of digital transformation is considerable time-saving, which allows both learners and educators to unlock their creative potential.

D. Bylieva and A. Nordmann (Eds.): PCSF 2021, LNNS 345, pp. 963–984, 2022.
https://doi.org/10.1007/978-3-030-89708-6_77

Time is the most valuable resource; therefore, it should not be wasted on routine processes and operations that can easily be automated via using special software. Since modern universities have expanded their capacity as research centers, teachers and students should step up their efforts to develop meaningful intellectual products instead of wasting time on repetitive tasks, which can easily be tackled by software.

In the context of the modern digitally-oriented education most scholars would concur that at present we can elicit three fundamental ways in which digital technologies can be applied in teaching: *blended learning*, which involves the integration of the latest technologies into the traditional learning process to improve the efficiency of teaching various disciplines; *distance learning,* which is based on the use of digital technologies which open the door to the educational process without a teacher being directly involved; *MOOCs*, a new educational phenomenon based on the concept of lifelong learning and mainly used for self-education and professional development.

The term "digital technologies" was first applied in scientific research not so long ago. In the English-language pedagogical works, it is used as a generic term that embraces a variety of information and communication technologies that have recently appeared (cloud, mobile, smart technologies, etc.) and have already become traditional [2].

In the 1990s, W. Warschauer outlined three main stages of using computer technologies in teaching foreign languages – behaviorist, communicative and integrative ones. Later on, this direction was called Computer Assisted Language Learning (CALL), which means learning foreign languages on the basis of computer technologies.

The *behaviorist stage* in the development of computer-assisted learning (1950s – 70s) was based on behavioral theory of teaching, which was popular at the time. Computer exercises trained students' skills (grammar, writing) through repetition. The main principle of developing computer programs was providing "drill and practice". The computer only partially performed functions of a teacher, it was perceived as a device providing only educational material to students, and the student acted as an object of learning.

The advent of personal computers opened up a whole host of new possibilities. This was the beginning of a new *communicative stage* (late 1970s–80s) based on the communication theory popular in teaching in the 1980s. The main principles of the communicative approach to the use of computer technologies were: an emphasis on the use of linguistic forms in speech; implicit teaching of grammar; emphasis on students' creation of their own sentences and texts rather than on the use of ready-made ones; the absence of a traditional assessment system (correct/incorrect), the possibility of several answers; maximum use of the target language in the teaching process; two-way interaction between the student and the computer.

In the 1990s, there was a rapid leap in the development of computer technology associated with the emergence of the Internet, the invention of multimedia and hypertext technologies, and the further improvement of communication technologies. A need for a different approach to the study of foreign languages was felt. The *integrative stage* is characterized by introducing new approaches to teaching foreign languages, which implies the use of the language in a real context, practicing four types of speech activity reading, writing, listening comprehension and speaking), as well as the harmonious integration of ICT (information-communicative technologies) into the learning process [3, 4].

Since the early 2000s a completely new stage in the use of digital technologies in education referred to as a *socially interactive stage* has been underway. This milestone is characterized by users' active social interaction due to the rapid development of social services Web 2.0 (the term is credited to the American scholar Tim O'Reily) [5] and their mobile applications, content aggregation, the rapid development of user content, online collaboration systems, instant access to educational material, etc. At this stage, digital technologies have become an integral part of both the learning process as well as individuals' everyday lives. This concept is extensively studied by E. D. Patarakin, E. N. Yastrebtseva, J. West, S. Boss, J. Brown, T. Burroughs, S. Downs and many others.

Some kernel ideas of this concept underlie the theory of connectivism developed by J. Siemens, who provided a new perspective on ICT-based didactic opportunities which are consonant with modern approaches to teaching with the support of Web 2.0 technologies [6]. Siemens's contribution was later advocated and summarized by S. Hargadon, who referred to this new milestone in digital technologies as *social learning* [7].

We cannot but mention such a field as mobile learning of foreign languages (MALL), which was formed early XXI century and which is closely attached to the platform of various digital technologies. The term "mobile learning" ("mLearning"), which appeared in the English-language pedagogical literature about 15 years ago, has recently been increasingly deployed in Russia on a large scale. Mobile technologies contribute to the modification of the three main components of the pedagogical process – access to teaching tools, forms of implementation of educational interaction and methods of presenting educational material and assignments [8].

Another rapidly developing and increasingly popular approach to implementation of digital technologies is Content and Language Integrated Learning (CLIL) [9, 10]. CLIL implies "subject-language integrated learning" and draws on the dual goal of students to acquire knowledge of a professional discipline and at the same time to develop their foreign language competence [11]. These goals support and facilitate each other. In each case a student draws on diverse digital supporters as linguistic corpora, thesauruses, electronic textbooks, etc. in the course of tackling a problem.

Over the past decade, the number of publications on CLIL methods and their practical application in universities around the world has increased significantly. A large number of publications describe the European and Russian experience of implementing CLIL in the educational process, especially in engineering education. Many studies aim at assessing the effectiveness of CLIL approach in terms of increasing the students' language competence [12–14], some authors are seeking opportunities and ways to stimulate creative thinking through CLIL [15].

The relationship between creativity and technology is well known to educators and noteworthy because both technology and creativity in education are complex areas [16]. In the past few years, there has been an increase in studies on creativity supported by digital technologies [17]. It is important for educators to explore the relationship between technology and creativity in order to discover how creativity can be brought into teaching and learning [18].

2   Creative Thinking Skills

In order to understand how to unlock our students' creative potential it is necessary to analyze such concepts as "creative thinking" and "creative thinking skills" with the help of which it would be clear what real digital technologies can be used in creativity development.

The idea of creative potential development has been known in Psychology and Education for the last decades, but it becomes more and more popular again in the present-day situation connected with new stages of social and economic (free market) development which demands a higher level of personal responsibility and personal active position in any sphere of economics.

The change in the nature of the economy and society in the context of the transition to the informational stage of development of modern civilization has led to the transformation of the labor market and new requirements for the labor force, which, in addition to professional competencies, include a whole range of supra-professional skills called "soft skills" [19, 20].

Soft skills are considered as a system of communicative and personal competencies, which includes any non-professional skills that increase the efficiency of a specialist's labor activity.

This allows us to draw a conclusion about the existing need for the formation of the following group of soft skills among university students as a necessary component of their professional competence:

1) social and communication skills (communication skills, interpersonal skills, group work, leadership, social intelligence, responsibility, communication ethics);
2) cognitive skills (critical thinking, problem solving skills, innovative (innovative) thinking, intellectual workload management, self-study skills, information skills, time management);
3) personality attributes and components of emotional intelligence (emotional intelligence, honesty, optimism, flexibility, creativity, motivation, empathy).

Until now, this task has not been designated as fundamentally significant in the higher education system of the Russian Federation. The issues of methodological organization and purposeful use of the possibilities of academic disciplines of the humanitarian cycle for its formation were not considered in detail.

Understanding the importance of this problem and the significance of its solution for students studying at Peter the Great St. Petersburg Polytechnic University and their further successful implementation in professional activities, the Directorate of Basic Educational Programs developed and proposed an innovative approach to the formation of supra-professional skills among students of all institutes and all directions. preparation.

The idea of this approach is as follows: for second and third year students, already in the next 2021–2022 academic year, it will be given the opportunity to choose on an alternative basis, in accordance with their own interests and their own preferences, those academic disciplines that, on the one hand, will be focused on precisely for the formation of soft skills, and, on the other hand, they will be as interesting and useful to them as possible. In order to implement this approach, a completely new module is introduced into the curriculum - "Self-development module/Soft skills".

In our study we have decided to pay attention to one of this soft-skills – creative thinking.

As it is known that creative thinking is a way of looking at problems or situations from a fresh perspective that suggests unorthodox solutions (which may look unsettling at first). Creative thinking can be stimulated both by unstructured process such as brainstorming, and by a structured process such as lateral thinking (www.businessdictionary. com). In other words, creativity is defined as an ability to make something new and creative thinking as an ability to think differently: to look at the problem from a different angle and to find a new solution for its decision.

The list of scientific works in the sphere of creative thinking development is very big. We can enumerate such very important investigations as: "Teaching critical thinking through engagement with multiplicity" [21], "Mindfulness and creativity: Implications for thinking and learning" [22], "Theoretical foundations of design thinking – A constructivism learning approach to design thinking" [23], etc. For example, Teun J. Dekker is sure that a multidisciplinary curriculum, student-centered pedagogy and a diverse academic community are demonstrating the best results in the process of creative thinking skills development.

All of these studies generate a number of theoretical and practical ideas and suggestions for how to teach creative thinking better, what conceptions to choose and what technologies to try. A notable and substantive role in moulding students' creative thinking is assigned to student-teacher collaboration.

3   Student-Teacher Creative Collaboration

Since distance learning was integrated into the educational process, it has allowed teachers and students to save time because they do not have to commute to university, which means that they can concentrate on creative endeavors. Creativity is an inherent component of higher education in the context of rapidly developing technologies since university graduates have to work towards innovative solutions tailored for non-typical situations. The key components of creativity are the following: a creative personality (a student or a teacher); a creative process, creative skills and a creative atmosphere [24]. Distance learning opens up ample opportunities for creative personalities (teachers and students) to apply their creative skills because they get access to digital tools that save their time as well as to valuable information that they can draw on in their research.

Teachers can dedicate more time to scientific research and developing educational materials, whereas students have a possibility to improve their skills by doing more written assignments. Creating pieces of writing stimulates analytical and critical thinking, improves powers of concentration and the ability to devise a coherent and logical structure, streamlines students' knowledge and can be regarded as a springboard to scientific research.

Due to large-scale digital transformations students and teachers spend more time on their own. Although seclusion can be stressful for people with certain personality traits, especially for extraverts, it is conducive to introspection. For instance, in offline classes some students speak a foreign language fluently, but make lexical and grammatical mistakes, and even when the teacher draws their attention to them, these errors do not disappear because the teacher's remarks are easily forgotten. However, when students make the same mistakes in a piece of writing, for instance, in an essay; the teacher underlines them, and the learners work on these errors, they are less likely to make them in the future. In distance learning error correction can be regarded as a creative endeavor, especially if the teacher partially delegates part of this work to students, and if routine and repetitive tasks are done by software. For example, when students learn to write an essay, they can use online platforms to practice some frequent phrases typical of written discourse and even create pieces of writing on the basis of the readily-available essay patterns. When it comes to creative writing, the teacher can have students check each other's essays and highlight the parts in them that seem incorrect. Then the teacher looks through these essays and uses the review mode to insert the relevant comments. Organizing the writing process in this way gives teachers more time for developing educational materials and doing scientific research. In addition, students gain peer assessment skills and learn to pay attention to mistakes made by other learners. Thus, when some essay-writing steps are automated and peer assessment is applied, it can result in saving considerable amounts of time.

Self-discipline and time management are critical for creativity. "Self-direction and regulation are essential for creative learning environments since planning and managing activities help learners develop a solution and complete a creative project" [23]. Creative assignments, such as projects, various kinds of research work, articles, etc. have to be completed by particular deadlines, which cannot be achieved without such components of self-discipline, as time management, endurance and perseverance. Distance learning on digital platforms has taught everyone involved to value their time. Teachers have to write numerous reports on learners' academic performance, transform offline lectures into digital ones, look for existing educational content for their students on the internet and develop new assignments. On the other hand, students working on digital platforms have to meet tight deadlines; time limits have become less fuzzy whether they do a test or work on a project. When huge chunks of time are not wasted, this time can be redeployed more efficiently, for example, it can be spent on doing research and writing scholarly articles in order to disseminate its results.

In 2020 the students of the Graduate School of Engineering Pedagogy, Psychology and Applied Linguistics submitted 109 scholarly articles (10 of which were indexed in Scopus) in collaboration with their teachers, which was 24 articles more than in 2019. Apparently this dramatic surge in the number of scientific publications, which can also be regarded as a direct consequence of digital transformations. Students are more willing to do research and write academic papers in co-authorship with their teachers. This kind of collaboration turns modern universities into research centers involving students in scientific studies, which have become an inherent part of the educational process.

Digital transformations facilitate the process of collaborative interdisciplinary research, making it less time-consuming and more streamlined. Students and teachers from different departments can communicate on hubs for teamwork, such as MS Teams, in the course of doing scientific research. For example, research often involves distributing a questionnaire or doing tests, the results of which are expected to prove or, on the contrary, debunk a certain hypothesis. Now researchers do not need to spend time printing out tests or questionnaires, giving them to respondents and waiting for them to hand in the written results. Moreover, digital transformation spares the researchers the need to check the questionnaires and tests written in paper format and calculate the results because now it is done by special software. This time can be spent more productively on thorough analysis of the results and making conclusions on their basis.

Another field of student-teacher creative collaboration facilitated by digital transformations is developing educational materials. For instance, some groups of students majoring in linguistics from St. Petersburg Peter the Great Polytechnic University have recently been given a creative assignment which involves writing sentences illustrating certain grammar rules. Students need to apply their creative skills in the course of doing this assignment because the sentences they develop should be related to the topics and cover the vocabulary from their basic course book. Moreover, only catchy and memorable sentences will be later included in a course book, the authors of which will be the teachers and the students working on the project. Apart from grammar, linguistic majors have recently been developing different kinds of educational materials such as lexical exercises (crosswords and multiple choice sentences); reading and listening comprehension assignments, etc. Regular participation of students in developing assignments encourages their creativity and is also conducive to better learning outcomes since learners memorize, analyze and internalize information more efficiently if they act as creators of learning materials rather than passive recipients. Learning materials developed as the result of student-teacher collaboration can be arranged in a digital course book accessible for students in the university electronic library.

The process of developing course books has become more creative since innovative digital tools were integrated into the tertiary education. Conventional textbooks in printed form are being replaced by course books in digital format. It is not just the question of format; developing digital books requires more creativity than conventional ones. On the other hand, an educator developing educational materials has access to many tools that were inconceivable before the arrival of computer technologies. It has become possible to embed powerful interactive simulations into materials thus going beyond the capacity of a conventional textbook to an entirely new experience. Students using digital textbooks can learn in the course of interacting with the educational materials. Teachers can consider digital course books as a platform on which they can share their creative content, which can be customized depending on its users. Interactivity boosts learners' interest and cognitive motivation.

Motivation is vital for creative thinking; therefore, assignments that involve finding solutions to problems facilitate learners' intrinsic motivation [25]. Both teachers and learners feel that their creativity is encouraged by higher levels of motivation, and, instead of wasting time chatting in recesses, they have started to communicate in a more profound and fruitful way, which often results in developing innovative educational content.

They have been using platforms like Zoom and MS Teams on an unprecedented scale for work-related and studies-related communication, which is also conducive to creativity.

Creativity should be encouraged by teachers in many ways: they can use digital platforms to offer students open assignments, ask them to prepare presentations and teach them to think dynamically taking various scenarios into consideration [26]. In the conditions of distance learning on digital platforms students are offered creative assignments, such as preparing presentations, developing tasks for other learners, analyzing a work of literature, a video, an artifact, etc. Such assignments are done in small groups of students who have access to MS Teams or other hubs for teamwork. Thus, students' communication is built around the learning-related topic, which rules out meaningless small talk and encourages their creativity.

4   The Use of Corpora in Developing Students' Creative Thinking

4.1   The Accord of DDL-Approach and Corpus Technologies

In order to clearly exemplify some principles of forming students' creative thinking and how it is developed within the framework of student-teacher creative collaboration, we decided to illustrate it in the context of corpus linguistics (hereinafter CL). Moreover, we consider corpus technologies in the research because they have been increasingly popular, particularly in the pandemic environment and due to the further spread of distance education.

To date, corpora have become an integral part of linguistics, being one of its cornerstones. After the emergence of corpora, the entire linguistic science has become disparate, since all its achievements now need to be interpreted in the context of the "corpus era", and all modern linguistics, according to V.A. Plungyan, should become the linguistics of corpora [27]. Thus, modern CL can be regarded not only as a discipline but also as a special science culturomics which can be defined as "the study of the culture of mankind, the direction of its development in time by means of quantitative analysis of words and phrases in very large volumes of digitized texts" [28, 29].

Since the arrival of corpus technologies, CL has been enriched with its own corpus instrumentation for pedagogically-oriented analysis. This toolkit allows a researcher to analyze lexical chunks, gapaxes, find and distinguish collocations and colligations, lexical bundles, or clusters, master the hedging strategy typical of modern scientific discourse both in the natural sciences [30] and in the humanities [31]. In other words, corpus technologies contribute to the development and strengthening of interdisciplinary ties, which is especially important today in the era of rapidly developing digitalization as one of the indicators of the 4th industrial revolution.

The feasibility of creating and using corpora is determined by the following prerequisites:

1. Since the end of the twentieth century, dictionaries and grammar textbooks have been created on the basis of corpora.
2. CL is used in theoretical linguistics to collect the data necessary for the analysis of certain phenomena in their natural context.
3. Based on case studies, machine learning is deployed in various areas of applied linguistics and information technology.

4. Corpora can be applicable for many other language-related tasks. The created and prepared corpus can be used repeatedly by different researchers and for different purposes, which is one of the criteria for the sustainability of electronic educational resources.

In summary, corpus linguistics enables a researcher:

– to use the corpus as a source and tool for multidimensional lexicographic works for various historical and modern dictionaries;
– to obtain data on the frequency of word forms, lexemes, grammatical categories, to trace various changes in frequencies and contexts in diverse periods of time, to obtain data on the joint occurrence of lexical units, etc.;
– to build a concordance for any word;
– to study the dynamics of the processes of altering the lexical composition of the language, to analyze lexical and grammatical characteristics in different genres and from different authors.

Corpus technologies are actively implemented in teaching bachelor students who major in linguistics starting from the 3rd year, as well as in the course of master's program "Computer Language Teaching". Mastering CL methods contributes to the development of both linguistic/foreign language competence and ICT competence, which is of paramount relevance for our graduates, since it is assumed that on graduating they are supposed to be able to teach foreign languages for special purposes (LSP) to students majoring in engineering, economic and humanitarian areas with the extensive use of ICT. In our opinion, ICTs have not been sufficiently introduced into the professional activity of a linguist-teacher. Despite its value and efficiency, CL approaches to study foreign language are not cross-functionally accepted ones in teaching/learning foreign languages [32, 33]. That is why we consider that it is more vital to focus on the practice of using corpus technologies in the course of teaching students to think creatively.

Due to the fact that pursuing the methodology for compiling corpus-based exercises is directly related to choosing the most effective teaching method, the DDL (data-driven learning) educational format proposed back in 1991 by Tim Jones is considered to be one of the possible solutions. He suggested that students should act as researchers who, with the aid of computer technology and corpora, had to conduct their own linguistic mini-studies [34] Ultimately, the main task for the student is "learn how to learn" [35], which can be achieved, among other things, with the aid of corpus-based assignments, which "allow the student to try on the role of an experimenter who conducts his own unique research and does not put together other people's ideas" [36].

Taking into consideration the analysis carried out above, it is high time we discoursed upon *corpus-based approach*, or *corpus methodology* as a combination of methods of linguistic research based on a corpus of texts, focused on applied language learning, its functioning in real textual environments, which is relevant not only for teaching LSP, but also for proper organization of work in the DDL format for future specialists in the field of computer language teaching [36, 37]. "The corpus method helps to systematize the data obtained by the student as a result of creative search, as mentioned by G. Palmer, who divided the educational vocabulary into two large groups: "strictly selected material, which he called the microcosm, and a spontaneous one" [38].

DDL seems to be a promising method of managing the educational process. This method is characterized by the ability to work with the so-called "raw" language data, which are selected directly from the corpus and which are investigated by students under their own steam. The *observe – hypothesize – experiment* model is based on the idea of independent efficient mastering of the material by students who undoubtedly boost their creative thinking skills in pursuit of a sound solution to a problem. Such training provides students with the opportunity to draw their own conclusions about the meaning of words, phrases and grammatical rules based on the analysis of authentic language material.

The efficiency of the DDL-method of teaching foreign language has been proven several times by domestic researchers [39–43]. Corpus learning based on Web-didactics is one of the reasons for abandoning the traditional explanatory approach in favor of cognitive-communicative practices, that is, the transition to "learning through research", as well as from the learner-as-researcher paradigm to the new learner-as-traveler paradigm. The combination of DDL approach and CLIL technology (Content Language Integrated Learning) also seems promising. Such a digital strategy, in which students independently carries out their mini-research using corpus technologies, certainly contributes to the development of students' personal creativity and creative thinking.

4.2 Modular Plan

We developed a modular plan consisting of stages of sequential mastering of skills in dealing with corpora – from knowledge of the functional and technical capabilities of the corpus search engine and the underlying terminology to the ability to independently develop a set of exercises based on such a corpus. The plan was devised based on the material of the National Russian Corpus, although it can be applied when working with such corpora as BNC, COCA, which are also widely applied by students in complex research work (Table 1).

Table 1. The plan of mastering RNC.

Level of mastering RNC	Aim	Examples of tasks
Elementary	Familiarization with the RNC device, mastering the 'corpus' terminology	Search tasks starting with the phrase 'Find in the main corpus…'
		1) the adjective *самодовольный (complacent)* in the main body, read the passport of the lemma and the full context with this word form;
		2) combinations of the word *бодрый (cheerful)* with male animals; sort alphabetically

(*continued*)

Table 1. (*continued*)

Level of mastering RNC	Aim	Examples of tasks
Intermediate	Search for units by language level – from morphemic to syntactic; mastering the structural and semantic organization of the text	1) set subcorpus X and find in it…;
		2) set the grammatical sign X and check;
		3) select category X among the semantic attributes and analyze
Advanced	Consistent analysis of linguistic realities, the creation of their own subcorpora, the application of a heuristic approach	1) Using a poetic corpus, prove that in the XVIII century the word *приличен* could have stress on the last syllable: *приличЁн*. Give an example (examples)
		2) When did the first examples of the verb *реагировать* appear in the corpus? Give the first two examples. What do they mean? When did the first examples of the verb *реагировать* with any prefix appear? Give the first two examples

The methods employed in the plan are related to the idea that the ability to rationally use the corpus is of particular relevance for performing corpus-based linguistic research. This skill is called *corpus literacy*, which implies the following: "1) knowledge of basic concepts, methods and technologies of corpus linguistics; 2) study and generalization of main methods and stages of creating a corpus; 3) typology of corpora; 4) certification of the text; 5) characteristics of various types of external and internal text markup; 6) The Internet as a corpus, etc. An important practical skill is the ability to correctly set the search parameters in the National Russian Corpus with reference to other corpus resources of the Russian language in the open space of the Internet" [36, p. 160].

All the tasks at the three stages are aimed at teaching students the indispensible skills to work with the structural and semantic organization of the text, which implies not just an isolated analysis of specific units, but the ability to independently formulate relevant and up-to-date conclusions based on in-text deep connections, as well as identified collocations. To this end, a certain motion vector is set, which in a certain way fosters the consistency and sequence of the search. For example, through subcorpus search, students can select examples based on the material within the area of their expertise; this makes it possible to demonstrate language changes even over quite a foreseeable period of time – a fact that is not always obvious and comprehensive to students.

Students are offered such tasks, on the basis of which they learn to evaluate each linguistic phenomenon not discretely, not as a statement of fact, but as an integral part of the language dynamics, the evolution of the language system, and learn to explain the outcomes on the basis of this knowledge. The student should also be able to assess the prospects of using this material in pedagogical activities, especially those related to teaching Russian as a foreign language. Such tasks reveal interdisciplinary connections much more efficiently, on the basis of which pedagogical and ICT competencies are more clearly formed, which together contribute to the improvement of the student's digital creativity.

4.3   Corpus-Based Mastering Idioms for Developing Students' Creative Thinking

Didactic potential of applying corpus approach to teaching can be demonstrated on the example of studying and teaching English phraseology. We believe that one of the prime examples where students can disclose their creative potential is mastering phraseology based on various corpora. In each language regardless of its origin, idioms reflect and convey a holistic picture of the world and are considered to be the best proof of the cultural reality in a particular language community. Mastering phraseology is carried out at the linguistic level and narrowed down to the cultural level, where students get familiar with the national and cultural picture of the world.

We have already obtained some results within this type of exercises, but we did not invite students for carrying out such exercises for our foremost goal was primarily to test our hypothesis whether the use of corpora contributes to developing idiomatic competence. For the verification, totally 535 idioms were selected from the General English textbook *Language Leader Upper-Intermediate, Manual for Translation Practice*, and *Illustrated Russian and English Proverbs and Sayings Manual*. Then we began to search each idiom in COCA. We have shown that the COCA is relevant for this kind of research: of 535 idioms, there were retrieved 262 units in their canonical form and 71 – in modified (distorted) forms. After that we compared different contexts and revealed the following typical modifications:

1) expanding the composition of an idiom by introducing an additional lexical component;
2) reduction in composition of phraseological units;
3) a change in grammatical form;
4) lexical variation of the components of phraseological units.

These results could be considered as the first step to a more complete study of the variation of these phraseological units, to further systematization of knowledge about the relationship between the fixedness of lexemes and the degree of their transformation [44, 45].

We decided to implement our results in the process of teaching our bachelor students. As a material for the experiment, British National Corpus was enlisted. There were selected some idiomatic constructions studied by 4-year linguistic students in preparation for the final exams in the textbook "CPE use of English. Examination practice". Such idioms as *full of beans, out of the blue, break even, take the bull by the horns, lay bare something, behind bars, a brainwave, in cold blood, beating about/around the bush, ring a bell, chip off the old block, have butterflies in one's stomach, on the spur of the moment,* etc. were under consideration.

There were totally selected 50 students and 50 idioms for conducting the survey. Before asking students to search idioms in the corpus, we encouraged them to guess the meaning of an idiom outside the context. For each idiom there were suggested three possible definitions, and one of them was correct. They read the idiom and its definitions in English and were supposed to pick and choose an equivalent in Russian relying on their own knowledge of phraseological units.

Then we got them to conduct their own corpus research. On doing this task, the students selected 5–7 text fragments among search results given that most clearly conveyed the lexical meaning of the idiom illustrating possible collocations and connotative rate. Based on the context, students made assumptions about meanings of phraseological structures and selected the most adequate equivalent in Russian. These assumptions appeared to be more accurate and more toward to the correct meaning of the idiom since they could see the idiom functioning in different, even contrasting, contexts.

After that we had students draw on several dictionaries on English phraseology so as they could assure themselves that even such sources can not facilitate profound search and analysis of obtaining data, because dictionaries do not supply a researcher with so wide range of diverse contexts unlike any corpus (see Fig. 1, 2, 3, 4 and Table 2).

Out of the blue

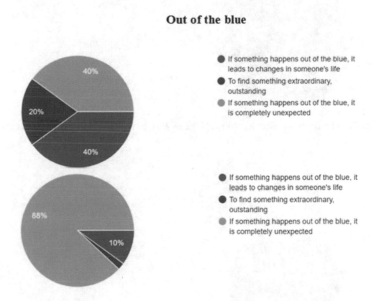

Fig. 1. The results on the idiom *out of the blue* before and after corpus search.

The difficulties that students experience in learning English idioms can be overcome if corpora are involved in the educational process. It is important that students should comprehend how native speakers of the target language use idioms in everyday speech. But the key role is played not so much by the fact that the student knows how to use the corpus in the search for an idiom, but the very process of creative search for a suitable definition in English. This greatly increases the chances of obtaining comprehensive data on idioms and developing their own research skills. The search for the Russian equivalent for the English idiom is carried out at the next stage, when the student, relying on the contexts provided by the corpus, was able to adequately assess the functionality of the idiom and its approximate meaning. After that, as an assignment, students can be

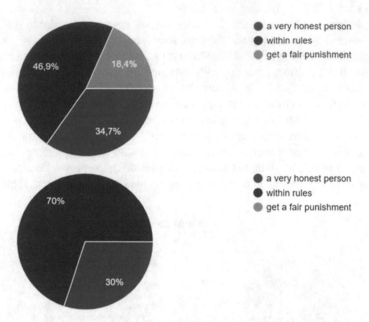

Fig. 2. The results on the idiom *fair and square* before and after corpus search.

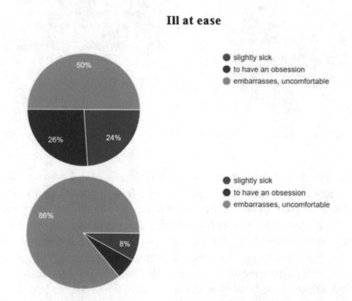

Fig. 3. The results on the idiom *ill at ease* before and after corpus search.

Turn over a new leaf

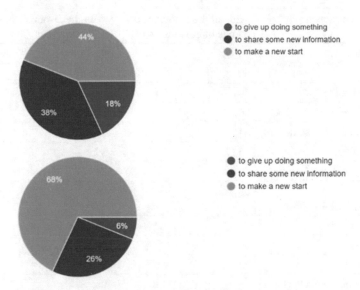

Fig. 4. The results on the idiom *turn a new leaf* before and after corpus search.

Table 2. The results of mastering idiomatic constructions without resources and using corpus linguistics.

Idiomatic contruction	Without the resources of corpus linguistics (%)	With the resources of corpus linguistics (%)
Full of beans	64	92
Out of the blue	40	88
Break even	68	96
Take the bull by the horns	64	100
Lay bare something	54	94
Blue-eyed boy	50	98
Bolt from the blue	64	94
Blessing in disguise	66	94

(*continued*)

Table 2. (*continued*)

Idiomatic contruction	Without the resources of corpus linguistics (%)	With the resources of corpus linguistics (%)
Llost cause	70	96
Off colour	60	100
Fly off the handle	70	98
Hand in glove with somebody	56	88
Make head or tail of something	64	96
The ins and outs	55	89
Keeping up with the Joneses	50	84
Before you can say Jack Robinson	68	94
Behind bars	88	96
A brainwave	86	98
In cold blood	52	90
Beating about/around the bush	66	96
Ring a bell	60	80
Have butterflies in one's stomach	52	90
On the spur of the moment	64	96
Wet blanket	56	94
A red-letter day	79	96
Down in the dumps	56	88
Let sleeping dogs lie	70	91
On the dole	66	98
Fair and square	47	70
Head over heels	62	98
To keep a straight face	64	80
Ill at ease	50	86
Lose heart	75,5	92
Shed light upon	68	96
The life and soul of sth	82	96
Turn over a new leaf	44	68
Every nook and cranny	86	96
Hit the nail on the head	58	96
Null and void	68	92

(*continued*)

Table 2. (*continued*)

Idiomatic contruction	Without the resources of corpus linguistics (%)	With the resources of corpus linguistics (%)
Once in a blue moon	88	96
Pop the question	76	96
In a rut	70	94
A long shot	63	86
Hit the sack	62	87
Out-and-out	74	92
Part and parcel of	75,5	92
To smell a rat	50	94
Red tape	61	92
Dog eat dog	68	96
Donkey work	54	96

given sentences taken from the corpus and containing an idiom, and possible Russian equivalents, which must be selected from the proposed list (Table 3).

Table 3. English idioms and their Russian equivalents.

1) action speak loader than words	a) быть в центре внимания
2) get a word in edgeways	b) сорока на хвосте принесла
3) get straight to the point	c) судят не по словам, а по делам
4) have a quick word with	d) расхохотаться
5) hear it on the grapevine	e) душа компании
6) be on the same wavelength	f) сразу перейти к делу
7) think before you speak	g) перекинуться парой слов
8) burst into laughter	h) быть на одной волне
9) hold centre stage	i) вставить слово
10) run out of things to say	j) исчерпать темы для разговора
11) the life of the party	k) молвишь-не воротишь

By referring to the context, students get the opportunity to analyze various cases of the use of idioms, see stylistic devices in sentences, and better understand the meaning of idiomatic structures, they can trace the syntactic connection with other members of the sentence. In the process of working with these educational materials, students develop their contextual guess and creative thinking, which proves to be a key competence when students apply speech skills in the course of communication. The formation of idiomatic competence is an essential factor in the learning a foreign language. Students who pay due attention to its development demonstrate better proficiency in the language being studied.

4.4 Linguistic Corpus as a Support for Designing Exercises by Students

We teach our students to use the results obtained in compiling their own DDL-exercises. Speaking about working with corpuses as a source of compiling DDL exercises, one cannot but mention the role and place of concordance programs, which help to carry out the analysis of the assembled corpus. Undergraduates get familiar with the functionality of the AntConc concordancer, which belongs to the 3rd generation programs. These programs have such characteristics as the availability of various functions, built-in statistical methods, scalability (i.e., they can be configured to work with larger corpora), the ability to process corpora in different languages.

As a part of the dispersed SRW (scientific research work), students were supposed to assemble their own corpus on the selected topic of a specific textbook in General English or English for Specific Purposes. As a part of the project assignment, it was proposed to analyze the corpus using AntConc, compile a list of keywords and phrases based on them, and develop educational materials. As keywords for searching texts from open sources, it was proposed to use words from the manual, to which students thus developed additional materials. The volume of the corpus was supposed to be in the range of 10–30 thousand words. Students were given various topics for this assignment.

Some of the students built up a corpus on the *Biathlon* topic and included idioms previously found in COCA in this corpus. For this research work, the AntConc program was used to create a corpus on the *Biathlon* topic and then identify the frequency words and expressions on the basis of which the set of exercises was created.

The Biathlon mini-corpus was created for two target audiences. The first target audience was translators who translate biathlon news and documents. The second target audience was interpreters working with teams (interpreter attaché), competition personnel, and news services. The corpus accounted for 57,761 words. The most frequent words and phrases were identified using AntConc. Based on this mini-corpus, a lesson plan and a set of exercises were drawn up. They were designed for students who were willing to familiarize themselves with highly specialized terminology on the Biathlon topic. The duration of the class was 90 min, and the required level of English proficiency was C1.

The defense of the project was a demonstration of the developed exercises for classroom and independent work in accordance with the drawn-up lesson plan on the basis of the chosen topic of the textbook. The authors of the most interesting projects subsequently included them as a part of their final qualifying works.

5   Conclusion

Digital transformation has become a priority for higher education institutions in this second decade of the 21st century, and this is a natural and necessary process for organizations that claim to be leaders of change and be highly competitive in their domain. Information technology usage has transformed education systems with significant impacts to academic life and to the reorganization of the educational environment.

In this paper we considered the impact of digital transformations in educational processes on students' and teachers' creative thinking development. According to many researches [15–17], multidisciplinary curriculum, student-centered pedagogy and a diverse academic community are demonstrating the best results in the process of creative thinking skills development. In turn, we decided to analyze how corpus-based approach as a combination of methods of linguistic research based on a corpus of texts, focused on applied language learning can facilitate the creative thinking skills formation. We believe that one of the prime examples where students can disclose their creative potential is mastering phraseology based on various corpora. The experiment included 4-year linguistic undergraduate students ($N = 50$) who studied 50 idiomatic constructions in preparation for the final exams in the textbook "CPE use of English. Examination practice". The results showed that students much better mastering idiomatic constructions using corpus linguistics than without resources of corpus linguistics that confirms the efficiency of chosen methodology for English learning purposes. Furthermore, conducting their own corpus research students expanded their knowledge and creative thinking skills that helped them to make more clear assumptions about meanings of phraseological structures and selected the most adequate equivalent in Russian.

The theoretical work of other scholars [33–38] in this field has been a useful resource for planning and designing, and we expect that our study will provide something of value for future researchers, too. Of course, there are some limitations in our study, as the sample size was relatively small because it was the first time, we implemented such an experiment, and the duration of the course was only one semester.

In our further research we are going to evaluate students' satisfaction with corpus-based learning.

References

1. Šereš, L., Tumbas, P., Pavlićević, V.: Digital transformation of higher education: competing on analytics. In: Sandkuhl, K., Lehmann, H. (eds.) The Role of Enterprise Architectures and Portals, pp. 9491–9497. Inted, Valencia (2018). https://doi.org/10.21125/inted.2018.2348
2. Dudeney, G., Hockly, N., Pegrum, M.: Digital Literacies: Research and Resources in Language Teaching. Routledge, New York (2014)
3. Warschauer M.: The Internet for English Teaching: Guidelines for Teachers. The TESL Reporter (1997). http://www.aitech.ac.jp/~iteslj/Articles/Warschauer-Internet.html. Accessed 10 April 2021
4. Titova, S.V.: Digital Technologies in Language Teaching: Theory and Practice. Editus, Moscow (2017)
5. O'Reilly, T.: What is web 2.0? Design patterns and business models for the next generation of software (2005). http://www.oreillynet.com/pub/a/oreilly/tim/news/2005/09/30/what-is-web-20.html. Accessed 15 March 2021

6. Siemens, G.: Learning and knowing in networks: Changing roles for educators and designers. Paper 105: University of Georgia IT Forum (2005). http://it.coe.uga.edu/itforum/Paper105/Siemens.pdf. Accessed 17 March 2021
7. Hargadon, S.: Web 2.0. Is the Future of Education. (2008). http://www.techlearning.com/blog/2008/03/web_20_is_the_future_of_educat_1.php. Accessed 22 March 2021
8. Titova, S.V., Avramenko, A.P.: Mobile Technologies in Teaching Foreign Languages. Publishing house of Moscow University, Moscow (2014)
9. Coyle, D., Hood, P., Marsh, D.: CLIL: Content and Language Integrated Learning. Cambridge University Press, Cambridge (2010)
10. Coyle, D.: Meaning-making, language learning and language using: an integrated approach. Inclusive pedagogy across the curriculum. In: Deppeler, J., Loreman, T., Smith, R., Florian, L. (eds.) International Perspectives on Inclusive Education, vol. 7, pp. 235–258. Emerald Group Publishing Limited, Bingley (2015). https://doi.org/10.1108/S1479-363620150000007021
11. Serova, T., Krylov, E.: Integrative teaching foreign language for engineering graduates in context of specialty. Lang. Cult. **2**(6), 50–57 (2015)
12. Dalton-Puffer, C., Smit, U. (eds.): Empirical Perspectives on CLIL Classroom Discourse. Peter Lang, Frankfurt, Vienna (2007). www.univie.ac.at/Anglistik/Dalton/SEW07/Dalton-Puffer.pdf. Accessed 22 March 2021
13. Lasagabaster, D., Sierra, J.: Immersion and CLIL in English: more differences than similarities. ELT J. **64**(4), 367–375 (2010). https://doi.org/10.1093/elt/ccp082
14. Baranova, T., Khalyapina, L., Vdovina, E., Yakhyaeva, K.: Soft CLIL v.2.0.: integrating a mobile app and professional content into the language training. IOP Conf. Ser.: Mater. Sci. Eng. **940**(1), 012140 (2020). https://doi.org/10.1088/1757-899X/940/1/012140
15. Ting, T.: CLIL and neuroscience: how are they related? In: Ruiz de Zarobe, Y., Sierra, M., Gallardo del Puerto, F. (eds.) Content and Foreign Language Integrated Learning, pp. 75–101. Peter Lang, Bern (2011)
16. Mishra, P.: Rethinking technology & creativity in the 21st century: crayons are the future. TechTrends **56**(5), 13–16 (2012). https://doi.org/10.1007/s11528-012-0594-0
17. Mishra, P., Henriksen, D., Mehta, R.: Creativity, digitality, and teacher professional development: unifying theory, research, and practice. In: Niess, M., Gillow-Wiles, H. (eds.) Handbook of Research on Teacher Education in the Digital Age, pp. 691–722. Information Science Referenc, Hershey, PA (2015). https://doi.org/10.4018/978-1-4666-8403-4.ch026
18. Yalcinalp, S., Avci Yucel, U.: Place of creativity in educational technology: a systematic review of the literature. In: 8th annual International Conference of Education, Research and Innovation (ICERI2015), pp. 4716–4717. Inted, Seville (2015)
19. Demidov, V., Mokhorov, D., Mokhorova, A., Semenova, K.: Professional public accreditation of educational programs in the education quality assessment system. E3S Web Conf. **244**, 11042 (2021). https://doi.org/10.1051/e3sconf/202124411042
20. Baranova, T.A., Trostinskaya, I.R., Kobicheva, A.M., Tokareva, E.Y.: Improving the students' big data era skills through the online international project X-culture. In: Proceedings of the 2020 The 3rd International Conference on Big Data and Education (ICBDE 2020), pp. 15–20. Association for Computing Machinery, New York, NY, USA (2020). https://doi.org/10.1145/3396452.3396454
21. Dekker, T.J.: Teaching critical thinking through engagement with multiplicity. Think. Skills Creat. **37**, 100701 (2020). https://doi.org/10.1016/j.tsc.2020.100701
22. Henriksen, D., Richardson, C., Shack, K.: Mindfulness and creativity: implications for thinking and learning. Think. Skills Creat. **37**, 100689 (2020). https://doi.org/10.1016/j.tsc.2020.100689
23. Mishra, P., Fahnoe, C., Henriksen, D.: The Deep-Play research group: creativity, self-directed learning, and the architecture of technology rich environments. TechTrends **57**(1), 10–13 (2013). https://doi.org/10.1007/s11528-012-0623-z

24. Kintoryak, Y.N., Ostapenko, I.N.: Creativity as a component of the university's intellectual capital (in the conditions of distance learning). Econ. Pap. **8**, 237–241 (2012)

25. Pande, M., Bharathi, S.V.: Theoretical foundations of design thinking – a constructivism learning approach to design thinking. Think. Skills Creat. **37**, 100637 (2020). https://doi.org/10.1016/j.tsc.2020.100637

26. Fasco, D.: Education and creativity. Creat. Res. J. **13**(3,4), 317–327 (2001). https://doi.org/10.1207/S15326934CRJ1334_09

27. Kaplan, D.E.: Creativity in education: teach. Creat. Devel. Psy. **10**, 140–147 (2019). https://doi.org/10.4236/psych.2019.102012

28. Plungjan, V.A.: Corpus as a tool and as ideology: on some lessons of modern corpus linguistics. Rus. Lang. Sci. Coverag. **2**(16), 7–20 (2008)

29. Michel, J.-B., et al.: Quantitative analysis of culture using millions of digitized books. Science **331**(6014), 176–182 (2011). https://doi.org/10.1126/science.1199644

30. Masevich, A.Ts., Zakharov, V.P.: Corpus linguistics methods in historical and culturological studies. In: Computational linguistics and computational ontologies, pp. 24–43. ITMO University, Saint Petersburg (2016). (In Russian)

31. Hyland, K.: Writing without conviction? Hedging in science research articles. Appl. Linguist. **17**(4), 433–454 (1996). https://doi.org/10.1093/applin/17.4.433

32. Safronenkova, E.L.: Hedging vs tolerance in presenting the scientific result in research articles (based on English research articles of the humanities field). St. Petersburg State Polytech. Univ. J. Humanit. Soc. Sci. **10**(3) 51–57 (2019) https://doi.org/10.18721/JHSS.10305

33. Cobb, T., Boulton, A.: Classroom applications of corpus analysis. In: Biber, D., Reppen, R. (eds.) Cambridge Handbook of English Corpus Linguistics, pp. 478–497. Cambridge University Press, Cambridge (2015). https://doi.org/10.1017/CBO9781139764377.027

34. Tatiana, B., Kobicheva, A., Tokareva, E.: Web-based environment in the integrated learning model for CLIL-learners: examination of students' and teacher's satisfaction. In: Antipova, T., Rocha, Á. (eds.) DSIC 2019. AISC, vol. 1114, pp. 263–274. Springer, Cham (2020). https://doi.org/10.1007/978-3-030-37737-3_24

35. Johns, T.: Should you be persuaded – two samples of data-driven learning materials. In: Johns, T., King, P. (eds.) Classroom Concordancing, pp. 1–16. Birmingham University, Birmingham (1991)

36. Leech, G.: Teaching and language corpora: a convergence. In: Wichmann, A., Fligelstone, S., McEnery, A. M., Knowles, G. (eds.) Teaching and Language Corpora, pp. 1–23. Longman, London (1997)

37. Dikareva, S.C.: Corpus technology in the dialogue teacher – student researcher. In: Proceedings of the international conference Corpus Linguistics–2011, pp. 157–162. St. Petersburg State university. Faculty of Philology, Saint Petersburg (2011). (in Russian)

38. Dobrushina, N.R.: Corpus methodology of teaching the Russian language. In: RNC. 2006–2008. New results and prospects, pp. 338–351. Nestor-History, Saint Petersburg (2009). (in Russian)

39. Gorina, O.G.: On the issue of corpus selection of key lexical units. Stephanos **1**(21), 111–117 (2017). (in Russian)

40. Agafonova, L.I.: Some issues of using corpus technologies as a factor in improving the quality of teaching foreign languages. Bull. Rus. State Pedagog. Univ. Named After A.I. Herzen **87**, 80–88 (2009). (in Russian)

41. Sadovnikova, O.J.: Direct and indirect use of corpora in foreign language teaching. Sci. Pedag. J. East. Sib. Master Dixit **2**, 152–161 (2013). (in Russian)

42. Dmitriev, A.V.: Specificity of the course computer technologies in linguistic studies for masters of the department of linguistics and intercultural communication. In: Innovative ideas and approaches to integrated teaching of foreign languages and professional disciplines in the

system of higher education. Proceedings of the international school-conference, pp. 128–130. Peter the Great Saint Petersburg State Polytechnic University, Saint Petersburg (2017). (in Russian)

43. Dmitriev, A.V., Kogan, M.S.: Features of the program computer language teaching of training for masters of linguistics. In: International Scientific Conference X anniversary St. Petersburg sociological readings 'Fourth Industrial Revolution: realities and modern challenges', pp. 230–237. Publishing house of Peter the Great St Petersburg Polytechnic University, Saint Petersburg (2018). (in Russian)

44. Dmitrijev, A., Kogan, M.: The role of corpus linguistics in the training of specialists in the field of computer language teaching. In: Anikina, Z. (ed.) IEEHGIP 2020. LNNS, vol. 131, pp. 511–520. Springer, Cham (2020). https://doi.org/10.1007/978-3-030-47415-7_54

45. Kogan, M., Komarova, I.: Exploring English phraseology with corpus linguistics methods. In: Computational linguistics and computational ontologies. Issue 4 (Proceedings of the XXIII International Joint Scientific Conference "Internet and Modern Society", pp. 40-49. ITMO University, St Petersburg (2019)

46. Dmitriev, A., Kogan, M., Komarova, I.: The report Didactic potential of applying corpus approach to English phraseology studies and teaching. In: Teaching and Language Corpora Conference (TaLC2020), Perpignan, France (2020)

Infographics as a Creative Design Method for Foreign Language Teaching

Olga Trubitsina[1] (iD), Tatiana Volovatova[1](✉) (iD), and Yuri V. Eremin[2] (iD)

[1] Herzen State Pedagogical University of Russia, Moika Embankment 48,
191186 St. Petersburg, Russia
truwat@bk.ru, otn09052011@gmail.com

[2] Peter the Great St. Petersburg Polytechnic University, Polytechnicheskaya, 29,
195251 St. Petersburg, Russia

Abstract. This paper considers the conceptual design scopes of infographics. The particular emphasis is on the integration of infographics into the training process of foreign language teachers-to-be. A creative potential of infographics for developing a communicative competence is analyzed. During the research study aimed at identifying awareness of bachelor students of Herzen state pedagogical university about the impact and application of infographics to enhance foreign language teaching. The article focuses on the competence-based approach, and the components of foreign language communicative competence are delineated. It has been revealed that using of infographics in the context of the competence approach makes the process of learning foreign languages practice-oriented and promotes the formation of the necessary skills for foreign language communication. The giving example of infographics provides with formation of a receptive type of speech activity. The data collected via quantitative methodology with a set of questionnaire as the main instrument which has undergone the content validity. The results of survey demonstrate students' opinions about creation a rich environment to foreign language learning through infographics. The results obtained allow making a conclusion about methodological validity and technological relevance of infographics and it can be considered a means of enhancement the pedagogical excellence of foreign language teachers.

Keywords: Infographics · Foreign language teaching · Digital environment

1 Introduction

The technical modernization of Russian education in the 21st century had an impact on the foreign language teachers' occupation and entailed the need to create the ability of controlling the information-intensive flow, transforming it effectively, to implement and to apply information. The training of foreign language teachers-to-be for the design of educational and methodological materials in a digital environment is subordinated to the strategic objective of vocational education. The aim is to ensure self-development and search for ways of enculturation of foreign language teachers at the information - intensive scope of future professional activity.

© The Author(s), under exclusive license to Springer Nature Switzerland AG 2022
D. Bylieva and A. Nordmann (Eds.): PCSF 2021, LNNS 345, pp. 985–999, 2022.
https://doi.org/10.1007/978-3-030-89708-6_78

The onrush of digital technologies that create and develop a digital environment [1], an online behavior model [2]; the emergence of a new generation of students with special socio-psychological characteristics arises a necessity to develop a digital educational process [3, 4], rethink the conceptual features of higher vocational education and training, especially linguistic [5].

The foreign language professional educator should carry out pedagogical activities in a digital environment; apply information and communication technologies effectively to solve the problems of organization and individualization of the learning activities of students, professional direction problems [6–9]. It should be noted that in the world experience of determining indicators of professional competence, the European Center for Foreign Languages has introduced since 2006 the European Portfolio for student teachers of Languages (here in after referred to as EPOSTL). In the provisions for the analysis of professional competence, the statements "*independent learning and the use of a virtual learning environment*" for the professional development of a teacher and motivating students to work in this environment are high-lighted [10]. Due to the fact that in EPOSTL the "*virtual learning environment*" is revealed through the component description of various electronic resources, it allows one to identify this with a digital environment, which consists of information and communication environment. The formation and development of the ordained readiness and abilities of foreign language teacher-to-be requires conscious systematic work to familiarize the student using such an environment.

A foreign language digital environment is distinguished by a special subject orientation, which is manifested in its contents. This is due to the peculiarities of a foreign language as a learning item. The mastering of foreign language does not give the student direct knowledge of reality, but appears as a means of expressing thoughts about objective reality, taking into account the socio-cultural background [11–13]. We follow the views of Velmahos G. C., Toutouzas K. G., Sillin L. F., Chan L., Clark R. E., Theodorou D., and Tikhonova A. L., about the environment uninterruptible updating with a constant set of resources, systematized in the discursive, situational, linguistic, subspace of national background knowledge. A digital support for the educational minimum in the training of foreign language teachers-to-be is provided with a constant increment of the environment due to the expansion of knowledge, contextual speech skills, the available variety of educational materials [14, 15].

Despite the openness and diversity of educational and methodological materials in a digital environment, foreign language teacher-to-be should be able to hold its own to create materials of this nature, which is regulated not only by current European frameworks, but also it is confirmed by data from foreign studies. According to Jimbo H., Hisamura K., Yoffe L., Eliane H. Augusto-Navarro, as the basis for the professional activity of English teacher as a foreign language, the following skills are distinguished: selection and adaptation of educational materials, distinct and creative design of educational materials, educational and methodological complexes based on personal and age features of students [16, 17].

The acquired data demonstrate the contradiction between the urgent need of using infographics as a presentation and activation tool of educational material, and the insufficiency of its application for the implementation of tasks in teacher's training.

2 Results

The submitted research is aimed at theoretical justification, getting developments into actual practical use of infographics in foreign language teaching.

Some researchers from different countries have already taken a scientific interest to the educational and creative opportunities of infographics. For example, the investigators from Malaysia indicate a high suitability of infographics to make information concise and coherent in teaching of graphical design and digital media through infographics [18], a scientist from Australia notes the development of target competencies in the study of the public health course involving infographics [19]. As for the researchers from Cyprus they have observed in the anatomy course that the subjects and visuals are used by infographics are easier to remain in the minds and infographics was handled as an effective class and visual communication material [20, 21]. It is interesting to note that R. A. S. Rueda (Mexico) indicates the facilitation in the assimilation and utilization of knowledge through infographics [22]. And one more study presents the insights for development powerful and creative infographics, it provides evidence for the effectiveness of the infographics assignment and strategies for teaching it [23].

Infographics can be used as a tool for creating educational and methodological materials in a digital environment. We shall consider the activating infographics influence and determine the potential of using infographics in the methodology of foreign language teaching. It should be noted that infographics has multifaceted characteristics, capabilities for analysis and is determined by specialists in various fields of science, areas related to information support of human activity [24, 25]. In modern psycho-linguistic studies infographics is examined as a polycode text, consisting of two inhomogeneous parts: verbal and non-verbal (Chernigovskaya T. V., Dobrego A., Nikolaeva E. I., Shelepin E. Yu., Zashchirinskaya O. V.).

One of the types of communicative design with a special ability to transfer information by visual and verbal-graphic means, infographics is considered by Bertlin J., Cleveland W. S., Laptev V. V., Orlov P. A., Cram R., Tufte E.R., Homes N., and Smiciklas M. The authors emphasize the recent discovery of such infographics property as operational information relaying. Converted into infographics elements of learning material and course content are introduced concisely in an accessible form. According to Bin Dahmash A., Ashwag Al-Hamid A., & Alrajhi M., Lingard H., Blismas N., Harley J., Stranieri A., Zhang R.P. and Pirzadeh, P., Damyanov I., Tsankov N. understanding, extraction, transformation of information encrypted in a polycode, inhomogeneous text leads to the formulation of whole sentences in external speech in foreign language teaching [26–28]. This effect occurs because perception entails active cognitive activity, combined with differentiated types of activity from motor components to logical, inductive, creative thinking in students [29, 30].

An intellectual and educational creativity means that students can experiment with infographics generation and use real communication situations, deliberate on the relationship between language practice and theory. They perceive and interpret information from text and pictograms. It is deliberated to have the ability to independently design infographics based on the keywords and conceptual blocks of a text. Moreover students link conceptions to comprise new substantial ones through deepening thinking to explicate the meanings in the infographics. And it describes a creativity process established on brainstorming by producing new relationships among things to step up with new ideas. Bloom's Taxonomy suggests that designing, creativity and producing information are higher-order thinking skills [31–33]. Teachers can compose higher-order learning assignments by incorporating infographics in the classroom. Researchers from Italy, the USA came to the similar conclusions, the more a student makes switches between text and images in the process of studying the learning material, the better the educational material is ingested [34].

It is important to note that the creation of educational and methodological materials in a digital environment reveals the content of education through the analysis of goals and training needs. A series of versatile procedures should be considered as obligatory when a teacher works in a digital environment. Infographics design model consists of three main components: content generation, visual design generation and digital design [35]. According to the suggested infographics design process, students should first generate content, and then prepare the draft, move on to the visual and digital design phase. The exceptionality of the process of creating educational and methodological materials in a digital environment through infographics has the following structure in our study (Fig. 1):

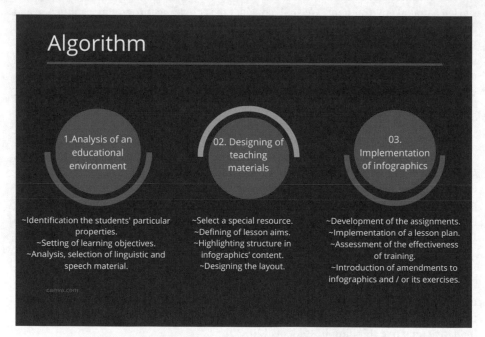

Fig. 1. Infographics' layout algorithm in a digital environment

Stage I. Analysis of an educational environment: 1. Identification the students' particular properties. 2. Setting of learning objectives. 3. Analysis, selection of linguistic and speech material.

Stage II. Designing of teaching materials: 1. Select a special resource with suitable infographics templates. 2. Determination of the purpose of infographics application, the stage of training 3. Highlighting structure in infographics' content. 4. Designing the layout (explication is possible "free hand") and creation of infographics.

Stage III. Implementation of infographics: 1.Development of the assignment's blueprint of the using infographics 2.Implementation of a lesson plan. 3. The effectiveness evaluation of the educational process. 4. Introduction of amendments to infographics and/or its exercises (if it's necessary).

The proposed infographics design process is consisted of three main stages: analysis of an educational environment, designing of teaching materials, implementation of infographics. Teachers-to-be should know well the technology of formation of foreign language communicative competence, theoretical foundations of foreign language teaching in order to effectively visualize information, and they need knowledge of digital tools to create infographics.

The design and further use of infographics is based on a two-way mediated communication between student and teacher, who designs the student's appeal to additional blocks of information. In particular, this property is realized in the organization of infographics' blocks that can serve as a support for the development of skills in the receptive form of speech activity, in the written form of speech communication.

A block system can represent a method for storing and manipulating information, in which it is stored at the form of interconnected infographics' blocks. It is assumed that, a teacher can organize his own scheme of block connections, but based on clear associations. Teachers need to focus on a topic, engage with content critically and support their idea by selecting key illustrative points to create pictorial representations. The use of web-based applications to design infographics facilitates to develop professional digital skills. Creating infographics contributes to locus on how to use digital skills. They learn how to organize, interpret and eliminate data during this process. In an effort to achieve the right combination of text information and illustrations, digital tools allow them to create quickly and enable them to easily edit.

A grouping of blocks can be represented as a node of slots associated with a topmost "core". The upper slot of node is designed as a minimized, represented space-saving schematic diagram of the main topic; the affiliated slots can be text, image, audio and video resources, which represents the "second layer of immersion" and reveal in more detail any fragment of the re-encoded text. Links between blocks, node elements are provided by visual elements.

Node can be vertical, hierarchical, and horizontal, between slots. The possibility of compact presentation of large amounts of information and deployment as the learner "immerses" is an advantage of using block nodes in organizing text for reading. In turns, this contributes to the optimization of the organization of independent creative work of students, the construction of an individual learning path and, provided that the requirements for the minimum level of mastering educational material are clearly formulated, the determination of the individual level of "immersion" for each student.

Navigation through visual elements, hyperlinks allows students to reveal more deeply some fragments of the text and to present others more rolled. It depends on the specific methodical attitude. The representation of educational materials through infographics allows integrating qualitative components (diagram, map, image, mental map, miniatures: pictograms/ideograms, icons, pointers, text information) and quantitative components (graphs, charts, histograms) of a digital environment and is instrumental in the intensification of communication's writing form teaching.

The use of infographics by foreign language teachers-to-be also has significant methodological capacity, providing students with the necessary and modern technological tool that they can use in future occupational activity, adapting it to specific learning environment. Developing educational and methodological materials, the teacher-to-be starts with determining the goals, the tool for achieving which can be infographics and which predict the qualitative and quantitative range of possible pedagogical results.

The study conducted by López Cupita L. A., & Puerta Franco L. M., on the opinions of university students about the usage of infographics as a visual tool were investigated. Additionally, according to the results obtained, it was determined that using infographics as a tool combining text and graphs to help users to get and transmit information was effective on development reading skills [36].

Methodically developed and structured infographics can be embedded in the modules of the chapters of the textbook, included in the thematic subsection. The use of infographics in foreign language classes allows forming, developing and improving the components of communicative competence (**linguistic, speech, sociocultural, compensatory, educational and cognitive competence**). The development of speech competence will be facilitated by the inclusion of infographics in the educational process. Infographics is an effective tool for working on oral and writing speech, while instead of the traditional summarization, annotating content, it becomes possible to detail the embedded in infographics information [37].

As an example, consider the possibility of using infographics in teaching foreign texts reading.

Created by a teacher-to-be the infographics "The land and the people of GB" (see Fig. 2) provides with formation of a receptive type of speech activity. Assignments for students can be formulated as follows. Here are some specific ideas for the tasks:

- Explore the infographics and ask 3 questions.
- Give careful consideration at the illustrations and read the text boxes close to them. Think over and tell about what can unite them.
- Read the text and fill in the missing semantic blocks.
- Find in the text and put in the infographics each country-specific relevant data: the capital, the patron saint, nationality, a big industrial city.
- Resume the sentences. Make findings of the utility of the reading information. What information turned out to be the most interesting?

At the pre-text stage, students explore the infographics and ask 2–3 questions to the text. Moreover students give careful consideration at the illustrations and read the text boxes close to them. They make a decision and say what unites the illustrations and the text boxes.

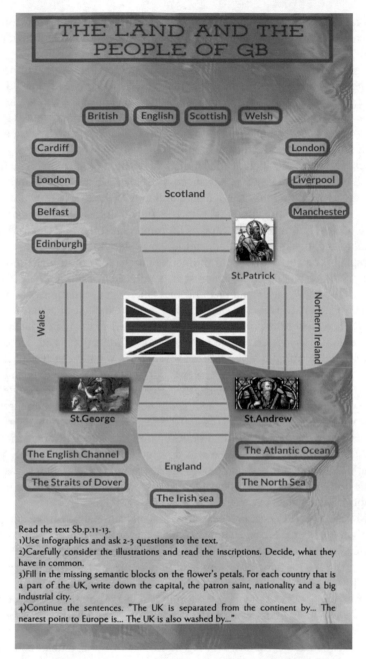

THE LAND AND THE PEOPLE OF GB

British   English   Scottish   Welsh

Cardiff   London

London   Liverpool

Scotland   Manchester

Belfast

Edinburgh

St.Patrick

Wales   Northern Ireland

St.George   St.Andrew

The English Channel   The Atlantic Ocean

The Straits of Dover   England   The North Sea

The Irish sea

Read the text Sb.p.11-13.
1)Use infographics and ask 2-3 questions to the text.
2)Carefully consider the illustrations and read the inscriptions. Decide, what they have in common.
3)Fill in the missing semantic blocks on the flower's petals. For each country that is a part of the UK, write down the capital, the patron saint, nationality and a big industrial city.
4)Continue the sentences. "The UK is separated from the continent by... The nearest point to Europe is... The UK is also washed by..."

Fig. 2. Infographics "The land and the people of GB"

At the text stage, students will read the text and fill in the missing semantic blocks. For each country, they need to find in the text and write down in the infographics: the capital, the patron saint, nationality, a big industrial city.

At the post-text stage, in order to comprehend the text, students will be asked to continue the sentences, express opinions about the usefulness of the reading information.

Infographics can also be used to understand the content of the text while listening. Drawing on visuals and factual information in numbers, infographics can help students become more confident in oral speech. Infographics contributes to the systematization and highlighting the most significant learning elements. It was undoubtedly true, that such aspects as the visualization of educational and methodological material (the availability of illustrations, video and audio accompaniment, hyperlinks in infographics); feedback (the use of infographics as a test material, with the power of which the effectiveness of material assimilation is monitored); the ability to make changes with the emergence of new information, all it plays an important role in perception achieving, understanding and memorization of educational material [38].

Infographics contains elements of problematicity as a form of visual information. According to the results of investigation that is published in the journal "COMPUTERS & EDUCATION", the resolution of problematicity in visual information is carried out on the basis of analysis, synthesis, generalization, reduction or expansion of information [39]. The higher the problematicity in infographics, the higher intensity of the student's mental activity is created. Thus, the creation of educational and methodological information through infographics contributes to the assimilation of the material more intensive, orientates the student to the search for systemic connections and patterns. Each infographics block can be thematic, therefore, it will provide an excellent opportunity to review the learnt vocabulary and study new lexical set, developing linguistic competence. Students will be able to establish independently the truth of presented in infographics statements, find confirmation/refutation in sources of other formats. The educational and cognitive competence will develop this way. After reading presented in infographics information and correlating it, for example, with photographs, students will be able to learn about the realities of the country of the target language, developing socio-cultural competence.

In our research experience, compensatory and educational-cognitive competencies were not highlighted in independent infographics. According to A.N. Shamov, compensatory competence is correlated with linguistic, its aim developing the ability to find synonyms, guess the meaning of a word, and predict the next word/sentence, structure statements [40].

The formation of other components of foreign language communicative competence affects the development of compensatory competence. According to the main theoretical provisions of the competence-based approach, the formation of key competencies, including educational and cognitive, is the main result of educational activities. Special educational and cognitive skills correlate with all components of communicative competence.

There are interesting results of the research carried out by employees of the Cognitive Research Laboratory of St. Petersburg State University using machinery for recording eye movements. Based on the keyword methodology Murzina L.N. and Stern A.S., the authors of the experiment found that texts are rated as "simpler" in infographics format. Also, for texts in the format of infographics, a greater number of correct answers to questions about the content of the text were received from the probationary. The data indicate the advantages in comparison of the usual text format in terms of the simplicity of educational materials perception and the reliability of its assimilation [41].

In order to identify awareness about impact and application of infographics to enhance foreign language teaching, we conducted a survey of 249 students of Institute of Foreign Languages of the Herzen State Pedagogical University of Russia (from 1 to 4 courses). The questionnaire was attended by 188 students of the professional direction "Pedagogical education" (specialization "Education in foreign languages") and 61 students of the professional direction "Linguistics" (specialization "Theory and methods of teaching foreign languages and cultures").

Students 1 through 4 courses were asked to evaluate the effectiveness of using infographics in foreign language teaching. As can we see from Fig. 3, to the question "Does using of infographics help to create a rich environment to foreign language learning?" 78% of freshmen and 37.5% of sophomores answered in the affirmative, 5% and 10% of respondents in the first and the second categories disagreed. 17% of 1st course students found it difficult to answer, 52.5% of 2nd course students found it difficult to answer, too. The scarcity of practice creating infographics in a digital environment led to answers difficulty of 2nd course students.

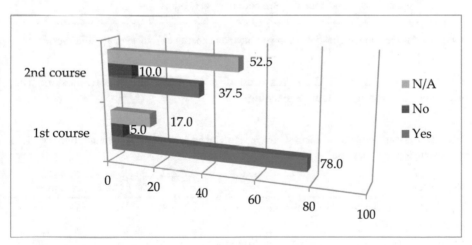

Fig. 3. Answers to the question: "Does using of infographics help to create a rich environment to foreign language learning?

Figure 4 demonstrates that most of the students didn't design online (73–95%). Some the third year students (20%), sophomores (7.5%), freshmen (2%) stated that acquainted Piktochart, also respondents pointed out Canva (7.5%), Vengage (3%), Creately (2–2.5%), Easel.ly (2.5–7%), Visme (2%) and Coreldraw (1%).

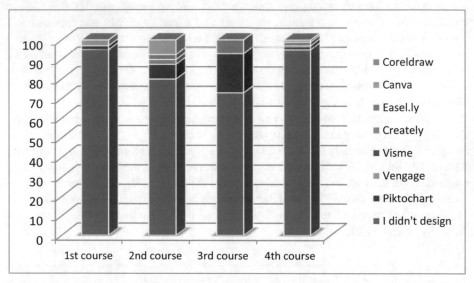

Fig. 4. Answers to the question: "Do you have an experience to design at online services?"

Students of 3rd and 4nd courses needed to categorize the effectiveness of using infographics in foreign language teaching.

Table 1. Answers to the question, 3rd course: Do you agree with the statement "Infographics is effective in teaching because…"?

Responses	Yes,%	No,%	No answer,%
Infographics presents teaching information in an easy-to-read format	87	0	13
Infographics conveys quickly large information scope	80	0	20
Infographics presents capaciously the correlations between several data sets	67	0	33
Infographics passes on teaching information aesthetically attractive	73	7	20
Infographics visualizes information for the further analysis	90	0	10

Students, who are studying on an internal basis, evaluated the properties of infographics, which are presented in Table 1 and 2. The characteristics were formulated taking into account the possible availability of integrating in infographics verbal and non-verbal components that allow presenting complex information firmly and expressively.

Table 2. Answers to the question, 4th course: Do you agree with the statement "Infographics is effective in teaching because…"?

Responses	Yes,%	No,%	No answer,%
Infographics presents teaching information in an easy-to-read format	88	0	12
Infographics conveys quickly large information scope	69	6	25
Infographics presents capaciously the correlations between several data sets	45	7	48
Infographics passes on teaching information aesthetically attractive	79	3	18
Infographics visualizes information for the further analysis	84	3	13

87% and 88% of 3rd and 4th course students, respectively, agreed with the statement that *"infographics presents educational information in an easy-to-read format"*. 13% of third-year students and 12% of fourth-year students found it difficult to answer. The statement *"infographics conveys quickly large information scope"* was positively assessed by 80% of third-year students and 69% of fourth-year students. The results suggest that there is a low total volume of those who disagreed with this statement: 6% of 4th year students. One third and a quarter at each course found it difficult to answer, respectively. The subject of research attention was also the assessment of the *ability of infographics to presents capaciously the correlations between several data sets*. In the total volume of results a prevailing share of affirmative answers is also noted in 3rd course. The results were almost equally divided among fourth-year students. It was possible to reveal that more than half of the respondents in each category confirmed that *"infographics conveys information aesthetically attractive"* and *" infographics visualizes information for the further analysis."*

3 Conclusion

The data presented by Russian [42, 43] and foreign researchers and analysts [44–46], as well as the intermediate results of our research, allow us to assert that designed on the basis of foreign texts infographics makes it possible to intensify the process of foreign language teaching. As well as, it facilitates reducing studying time; infographics properly reflects the important country-specific aspects and increases the efficiency of assimilation of culture-through-language information. In addition by means of infographics it can be recreated real situations of native speakers' communication, and carried out an analysis based on a comparison of cultural realities and characteristics of native and foreign-language societies, that serves as a meaningful basis for students' speech.

The practical significance of the research consists in the design of unique infographics for English language teaching that is partially presented in this publication.

Based on scientific research it was consideration and analysis characteristics of the integration of infographics into the educational process for the formation of foreign language communicative competence. Moreover, the openness of infographics' generation in a digital environment, the ability to interact and function together with other educational and methodological materials for the purpose of foreign language teaching makes it possible to apply it in the professional activities of foreign language teachers.

Studies that will be conducted in the future can consist of the integration infographics to educational and methodological complexes for English teaching, also in the practical use of the theoretical provisions and conclusions of this research by foreign language teachers.

References

1. Bylieva, D., Lobatyuk, V., Kuznetsov, D., Anosova, N.: How human communication influences virtual personal assistants. In: Bylieva, D., Nordmann, A., Shipunova, O., Volkova, V. (eds.) PCSF/CSIS -2020. LNNS, vol. 184, pp. 98–111. Springer, Cham (2021). https://doi.org/10.1007/978-3-030-65857-1_11
2. Bylieva, D., Lobatyuk, V., Safonova, A.: Online forums: communication model, categories of online communication regulation and norms of behavior. Hum. Soc. Sci. Rev. 7(1), 332–340 (2019). https://doi.org/10.18510/hssr.2019.7138
3. Bylieva, D., Bekirogullari, Z., Lobatyuk, V., Nam, T.: How virtual personal assistants influence children's communication. In: Bylieva, D., Nordmann, A., Shipunova, O., Volkova, V. (eds.) PCSF/CSIS -2020. LNNS, vol. 184, pp. 112–124. Springer, Cham (2021). https://doi.org/10.1007/978-3-030-65857-1_12
4. Odinokaya, M., Andreeva, A., Mikhailova, O, Petrov, M., Pyatnitsky, N.: Modern aspects of the implementation of interactive technologies in a multidisciplinary university. E3S Web Conf. 164, 12011 (2020). https://doi.org/10.1051/e3sconf/202016412011
5. Odinokaya, M.A., Karpovich, I.A., Mikhailova, O.J., Piyatnitsky, A.N., Klímová, B.: Interactive technology of pedagogical assistance as a means of adaptation of foreign first-year students. IOP Conf. Ser.: Mater. Sci. Eng. 940, 012130 (2020). https://doi.org/10.1088/1757-899X/940/1/012130
6. Claro, M., Salinas, A., Valenzuela, S.: Teaching in a digital environment (tide): defining and measuring teachers' capacity to develop students' digital information and communication skills. Comput. Educ. 121, 162–174 (2018). https://doi.org/10.1016/j.compedu.2018.03.001
7. Almazova, N., Bernavskaya, M., Barinova, D., Odinokaya, M., Rubtsova, A.: Interactive learning technology for overcoming academic adaptation barriers. In: Anikina, Z. (ed.) IEEHGIP 2020. LNNS, vol. 131, pp. 786–794. Springer, Cham (2020). https://doi.org/10.1007/978-3-030-47415-7_84
8. Almazova, N., Eremin, Y., Kats, N., Rubtsova, A.: Integrative multifunctional model of bilingual teacher education. IOP Conf. Ser.: Mater. Sci. Eng. 940(1), 012134 (2020). https://doi.org/10.1088/1757-899X/940/1/012134
9. Almazova, N., Rubtsova, A., Krylova, E., Almazova-Ilyina, A.: Blended learning as the basis for software design. In: Annals of DAAAM and Proceedings of the International DAAAM

Symposium, vol. 30, no. 1, pp. 806–813 (2019).
https://doi.org/10.2507/30th.daaam.proceedings.112
10. European portfolio for student teachers of languages.
https://www.ecml.at/epostl. Accessed 10 March
11. Kabardov, M.K.: Communicative and Cognitive Components of Linguistic Abilities (Doctorate Dissertation). Psychological Institute of the Russian Academy of Education, Moscow (2001).(in Russian)
12. Marishchuk, L.V.: Ability to master foreign languages and the didactic technology of their development. (Doctorate dissertation). St. Petersburg. State university, St. Petersburg (1999). (in Russian)
13. Zimnyaya, I.A.: Psychological Aspects of Teaching Foreign Language Speaking. Prosveshchenie, Moscow (1978).(in Russian)
14. Velmahos, G.C., Toutouzas, K.G., Sillin, L.F., Chan, L., Clark, R.E., Theodorou, D., et al.: Cognitive task analysis for teaching technical skills in an inanimate surgical skills laboratory. Am. J. Surg. **187**(1), 114–119 (2004).
https://doi.org/10.1016/j.amjsurg.2002.12.005
15. Tikhonova, A.L.: Foreign language teachers-to-be imparting to instructional design of digital resources for language teaching. Bull. Chelyabinsk State Pedag. Univ. Pedag. Psychol. **11**, 177–185 (2010). (in Russian)
16. Jimbo, H., Hisamura, K., Yoffe, L.: Developing English teacher competencies: an integrated study of pre-service training, professional development, teacher evaluation, and certification systems. The English edition of the grant-in-aid for scientific research report. JACET SIG, Tokyo (2010)
17. Augusto-Navarro, E.H.: The design of teaching materials as a tool in EFL teacher education: experiences of a Brazilian teacher education program. Ilha Do Desterro **1**, 121–137 (2015).
https://doi.org/10.5007/2175-8026.2015v68n1p121
18. Noh, M.A.M., et al.: The use of infographics as a tool for facilitating learning. In: Hassan, O.H., Abidin, S.Z., Legino, R., Anwar, R., Kamaruzaman, M.F. (eds.) International Colloquium of Art and Design Education Research (i-CADER 2014), pp. 559–567. Springer, Singapore (2015).
https://doi.org/10.1007/978-981-287-332-3_57
19. Darcy, R.: Infographics, assessment and digital literacy: innovating learning and teaching through developing ethically responsible digital competencies in public health. In: Chew, Y.W., Chan, K.M., Alphonso, A. (eds.) Personalised Learning. Diverse Goals. One Heart, pp. 112–120. ASCILITE, Singapore (2019)
20. Ozdamli, F., Ozdal, H.: Developing an instructional design for the design of infographics and the evaluation of infographics usage in teaching based on teacher and student opinions. Eurasia J. Math. Sci. Technol. Educ. **14**(4), 1197–1219 (2018).
https://doi.org/10.29333/ejmste/81868
21. Ozdamli, F., Kocakoyun, S., Sahin, T., Akdag, S.: Statistical reasoning of impact of infographics on education. Procedia Comput. Sci. **102**, 370–377 (2016).
https://doi.org/10.1016/j.procs.2016.09.414
22. Rueda, R.A.S.: Use of infographics in virtual environments for personal learning process on Boolean Algebra. Vivat Acad. **130**, 64–74 (2015).
https://doi.org/10.15178/va.2015.130.64-74
23. Gallicano, T.D., Ekachai, D., Freberg, K.: The infographics assignment: a qualitative study of students' and professionals' perspectives. Public Relat. J. **8**(4) (2014).
http://www.prsa.org/Intelligence/PRJournal/Vol8/No4/. Accessed 25 May 2020
24. Educational infographics improve sun protection knowledge in patients with nonmelanoma skin cancer. J. Am. Acad. Dermatol. **76** (2017).
https://doi.org/10.1016/j.jaad.2017.04.429

25. Siricharoen, W.V., Siricharoen, N.: How infographic should be evaluated. In: The 7th International Conference on Information Technology. ICIT 2015, pp. 558–564. Alzaytoonah University of Jordan, Jordan (2015).
    https://doi.org/10.15849/icit.2015.0100
26. Bin Dahmash, A., Ashwag Al-Hamid, A., Alrajhi, M.: Using infographics in the teaching of linguistics. Arab World Engl. J. 8(4), 430–443 (2017).
    https://dx.doi.org/10.24093/awej/vol8no4.29
27. Lingard, H., Blismas, N., Harley, J., Stranieri, A., Zhang, R.P., Pirzadeh, P.: Making the invisible visible: stimulating work health and safety-relevant thinking through the use of infographics in construction design. Eng. Constr. Archit. Manag. 25(1), 39–61 (2018).
    https://doi.org/10.1108/ECAM-07-2016-0174
28. Damyanov, I., Tsankov, N.: The role of infographics for the development of skills for cognitive modeling in education. Int. J. Emerg. Technol. Learn. 13(1), 82–92 (2018).
    https://doi.org/10.3991/ijet.v13i01.7541
29. Aldalalah, O.M.A.: The effectiveness of infographic via interactive smart board on enhancing creative thinking: a cognitive load perspective. Int. J. Instr. 14(1), 345–364 (2021).
    https://doi.org/10.29333/iji.2021.14120a
30. Levunlieva, M.: From perception through understanding to creative imagination. Int. J. Sci. Appl. Pap. 10(1), 100–104 (2015)
31. Jones, N.P., Sage, M., Hitchcock, L.: Infographics as an assignment to build digital skills in the social work classroom. J. Technol. Hum. Serv. 37(2–3), 203–225 (2019).
    https://doi.org/10.1080/15228835.2018.1552904
32. Noh, M.A.M., Fauzi, M.S.H.M., Jing, H.F., Ilias, M.F.: Infographics: teaching and learning tool. Malays. Online J. Educ. 1(1), 58–63 (2017)
33. Alikina, E.V., Falko, K.I., Rapakova, T.B., Erickson, S.: Developing infographic competence as the integration model of engineering and linguistic education. In: Anikina, Z. (ed.) IEEHGIP 2020. LNNS, vol. 131, pp. 692–698. Springer, Cham (2020).
    https://doi.org/10.1007/978-3-030-47415-7_73
34. Mason, L., Tornatora, M.C., Pluchino, P.: Integrative processing of verbal and graphical information during re-reading predicts learning from illustrated text: an eye-movement study. Read. Writ. 28(6), 851–872 (2015).
    https://doi.org/10.1007/s11145-015-9552-5
35. Nuhoğlu Kibar, P., Akkoyunlu, B.: Modeling of infographic generation process as a learning strategy at the secondary school level based on the educational design research method. Educ. Sci. 43(196), 97–123 (2018).
    https://doi.org/10.15390/EB.2018.7592
36. López Cupita, L.A., Puerta Franco, L.M.: The use of infographics to enhance reading comprehension skills among learners. Colomb. Appl. Linguist. J. 21(2), 230–242 (2019).
    https://doi.org/10.14483/22487085.12963
37. Becker, B.E., Huselid, M.A., Beatty, R.W.: The Differentiated workforce: Transforming Talent into Strategic Impact. Harvard Business Press, Boston, Boston Mass (2009)
38. Pšenáková, I., Szabó, T.: Interactivity in learning materials for the teaching. In: 16th International Conference on Emerging E-Learning Technologies and Applications (ICETA), pp. 445–450. Starý Smokovec, The High Tatras, Slovakia (2018).
    https://doi.org/10.1109/ICETA.2018.8572208
39. Barzilai, S., Mor-Hagani, S., Zohar, A.R., Shlomi-Elooz, T., Ben-Yishai, R.: Making sources visible: promoting multiple document literacy with digital epistemic scaffolds. Comput. Educ. 157, 162–174 (2020).
    https://doi.org/10.1016/j.compedu.2020.103980
40. Shamov, A.N.: Methodology of teaching foreign languages: theoretical course. NGLU im. N.A. Dobrolyubova: N. Novgorod (2012). (in Russian)

41. Riekhakainen, E.I., Petrova, T.E., Zemskova, T.A., Kuznetsova, A.S., Shatalov, M.A.: Infographics and verbal text: peculiarities of perception. In: Eighth International Conference on Cognitive Science, pp. 887–889. Institute of Psychology RAS, Svetlogorsk (2018). (in Russian)
42. Agaltsova, D.V., Ilyuschenko, N.S.: Applying marketing mix model WebQuest in professional English teaching. Perspect. Sci. Educ. **49**(1), 440–449 (2021).
https://doi.org/10.32744/pse.2021.1.30
43. Almazova, N.I., Rubtsova, A.V., Evtushenko, T.G., Smolskaia, N.B., Radchenko, Y.: Infographics as a tool of improving the quality of foreign languages teaching in a multidisciplinary university. Modern Pedag. Educ. **1**, 67–72 (2021). (in Russian)
44. Montebello, M.: Digital Pedagogies and the Transformation of Language Education. Hershey: IGI Global, Pennsylvania (2021).
https://doi.org/10.4018/978-1-7998-6745-6
45. Bhasin, T., Butcher, C.: Teaching effective policy memo writing and infographics in a policy programme. Eur. Polit. Sci. (2021).
https://doi.org/10.1057/s41304-021-00330-0
46. Avgerinou, M. D., Pelonis, P.: Handbook of Research on K-12 Blended and Virtual Learning Through the i^{2}Flex Classroom Model. IGI Global (2021).
https://doi.org/10.4018/978-1-7998-7760-8

Author Index

Printed in the United States
by Baker & Taylor Publisher Services